MECHANISMS AND MECHANICAL DEVICES SOURCEBOOK

MECHANISMS AND MECHANICAL DEVICES SOURCEBOOK

Fourth Edition

NEIL SCLATER
NICHOLAS P. CHIRONIS

McGraw-Hill
New York • Chicago • San Francisco • Lisbon • London • Madrid
Mexico City • Milan • New Delhi • San Juan • Seoul
Singapore • Sydney • Toronto

The McGraw·Hill Companies

1 2 3 4 5 6 7 8 9 0 BKM/BKM 0 1 2 1 0 9 8 7 6

ISBN-13: 978-0-07-146761-2
ISBN-10: 0-07-146761-0

The sponsoring editor for this book was Larry S. Hager and the production supervisor was Pamela A. Pelton. It was set in Times by International Typesetting and Composition. The art director for the cover was Anthony Landi.

Printed and bound by BookMart Press.

This book is printed on acid-free paper.

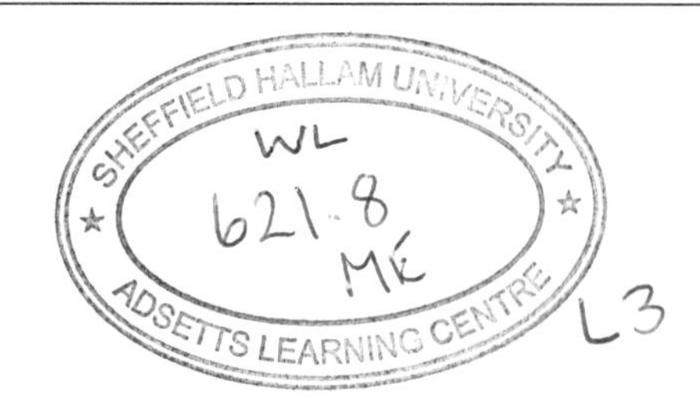

ABOUT THE EDITORS

Neil Sclater began his career as an engineer in the military/aerospace industry and at a Boston engineering consulting firm before changing his career path to writing and editing on electronics and electromechanical subjects. He was a staff editor for *Electronic Design* magazine and McGraw-Hill's *Product Engineering* magazine.

He then started his own consulting business in technical communications. For the next 30 years Mr. Sclater served a diversified list of clients by writing and editing marketing studies, technical articles, brochures, and new product releases. These clients were in the fields of optoelectronics, industrial instrumentation, power supplies, motors, and batteries. During this period he also contributed many bylined columns and technical articles to engineering publications on the subjects of electronic components and systems, motion control, servomotors, and instrumentation.

Mr. Sclater holds degrees from Brown and Northeastern Universities, and he has completed graduate courses in industrial management. He is the author or coauthor of 12 books including 11 engineering reference books published by McGraw-Hill's Professional Book Group on subjects as varied as semiconductors, electronics dictionaries, electronics technology, electrical power, and mechanics. After the death of Nicholas Chironis he became the author/editor of the three subsequent editions of *Mechanisms and Mechanical Devices Sourcebook.*

The late **Nicholas P. Chironis** was the author/editor of the first edition of *Mechanisms and Mechanical Devices Sourcebook.* He recognized the importance of saving a diversified selection of technical illustrations and text on practical mechanisms from out-of-print books and old magazines as an archival resource for engineers and designers; he then put the selection together with later mechanical design articles in a single volume. Mr. Chironis was a mechanical engineer, an instructor in machine design, and an engineering consultant before becoming the Mechanical Design Editor for *Product Engineering* magazine. He held the degrees of bachelor and master of mechanical engineering from Polytechnic University, Brooklyn, N.Y.

CONTENTS

PREFACE

This is the fourth edition of *Mechanisms and Mechanical Devices Sourcebook*, a well-illustrated reference book containing a wide range of information on both classical and modern mechanisms and mechanical devices. This edition contains three new chapters: one on basic mechanisms; the second on mobile robots; and the third on new directions in mechanical engineering. The chapter on basic mechanisms provides an overview of the physical principles of mechanics; the chapter on mobile robots examines existing scientific and military mobile robots and the scientific and engineering research in advanced robotics; the chapter on new directions in mechanical engineering reviews the present status and future prospects for microtechnology, highlighting progress in and acceptance of microelectromechanical systems (MEMS). Also included in the chapter are articles on nanotechnology, focused on the role mechanical engineers are playing in this burgeoning science. The field of nanotechnology now involves several branches of engineering as well as the physical, chemical, biological, and medical sciences. A previous section on rapid prototyping has been updated and upgraded as a separate chapter.

This edition contains a large core of archival drawings and text describing and illustrating proven mechanisms and mechanical devices carried over from previous editions. This core has been reorganized to make topics of interest to readers easier to find. Some previously published pages were deleted because their content was deemed to be of little value in future designs, and some figures have been redrawn to make them easier to understand. An extensive and comprehensive index has been provided to make this core a valuable reference resource for engineers, designers, inventors, students, hobbyists, and all enthusiasts for things mechanical. The 11 chapters in this core illustrate practical design solutions that can be recycled into new products.

The first edition of this book, published in 1991, did not mention the influence of electronics and computer science on mechanical engineering and mechanical design. However, since that time a sea change has occurred in the practice of mechanical engineering; today it is difficult to find any contemporary mechanical system or appliance that does not in some way include electronic components or circuits that improve its performance, simplify its operation, or provide for additional safety features. Those components might be as simple as solid-state rectifiers or light-emitting diodes (LEDs) or as complex as microprocessor-based modules that permit the product or system to operate autonomously.

The chapter on basic mechanisms provides the reader with a useful introduction to much of the content of this book; it will also serve as a refresher tutorial for those who have studied mechanical principles in the past and want to get up to speed on the fundamentals again. Topics covered include the inclined plane, screw jack, levers, linkages, gears, cams, and clutches. A previous tutorial chapter on motion control systems that contained illustrations and text describing control schemes and key components has been retained, and a former chapter on industrial robots has been revised and updated with new illustrations and specifications for some of the latest industrial robots. The new chapter on mobile robots extends the book's coverage of robotics and points out their growing economic and technical importance in scientific exploration and research as well as military missions and emergency services.

The new chapter on rapid prototyping discusses the emerging leaders in the field and reports on the trends: increasing popularity of 3-D plastic, paper, and wax models for engineering and design evaluation, and the extrapolation of existing technologies into the fabrication of functional metal and ceramic products. Replacement metal parts for older out-of-production machines are now being made rapidly and cost-effectively by eliminating the high cost and time delay involved in remaking the metal or ceramic dies or casting molds used in mass-production manufacturing.

The earlier articles on MEMS have been revised by reporting on the new developments and significant gains in the complexity of those devices; some MEMS are now being produced in large commercial volumes in established markets. The choices in material alternatives to silicon are discussed, and new microphotographs show more sophisticated multilayer devices.

The impact of electronic controls and communications circuits on mechanical engineering is nowhere more evident than on the latest motor vehicles. Microprocessors and electronics abound: they now control the engines and transmissions in all kinds of motor vehicles, and they have improved vehicle performance and fuel efficiency. Vehicular safety has also been improved by electronically deployed air-bags, antilock braking (ABS), stability or skid control (ESC), traction control (TC), and tire-pressure monitoring. Communication systems summon aid for drivers involved in accidents or breakdowns, and onboard navigation systems now provide map displays of streets to guide drivers.

With the exception of illustrations generously contributed by corporations, and government laboratories (see Acknowledgments), all of the figures in the tutorial Chapters 1 to 4 and 18 and 19 were drawn by this author on a Dell personal computer with software included in the Microsoft Windows XP package. Also, the five illustrations on the front cover of this book were derived from selected figures in those chapters.

Much of the archival core in this edition was first collected from a variety of published sources by Douglas C. Greenwood, then an editor of *Product Engineering* magazine; it first appeared in three volumes published by McGraw-Hill between 1959 and 1964. Nicholas Chironis edited and reorganized much of this content and supplemented it with contemporary technical articles to form the first edition of this book. In subsequent editions this core has been reorganized and new material has been added. References to manufacturers or publications that no longer exist have since been deleted because they are no longer valid sources for further information. The terms *devices* and *mechanisms* used to describe objects in the core pages have been used interchangeably and only some of them have been changed. However, the comprehensive index accounts for these differences in designation. The names of the inventors of these mechanisms and devices have been retained so that readers can research the status of any patents once held by them.

—Neil Sclater

ACKNOWLEDGMENTS

This author gratefully acknowledges the permission granted by the publisher of *NASA Tech Briefs* (Associated Business Publications, New York) for reprinting two of its recent articles. They were selected because of their potential applications beyond NASA's immediate objectives in space science and requirements for specialized equipment. The names of the scientists/inventors and the NASA facilities where the work was performed have been included. For more information on those subjects, readers can write directly to the NASA centers and request technical support packages (TSPs), or they can contact the scientists directly through the *NASA Tech Briefs* Web site, www.nasatech.com.

I also wish to thank the following companies and organizations for granting me permission to use selected copyrighted illustrations, and providing other valuable technical information by various means, all useful in the preparation of this edition:

- FANUC Robotics North America, Inc., Rochester Hills, Mich.
- Sandia National Laboratories, Sandia Corporation, Albuquerque, N.Mex.
- SolidWorks Corporation, Concord, Mass.

MECHANISMS AND MECHANICAL DEVICES SOURCEBOOK

CHAPTER 1
BASICS OF MECHANISMS

INTRODUCTION

Complex machines from internal combustion engines to helicopters and machine tools contain many mechanisms. However, it might not be as obvious that mechanisms can be found in consumer goods from toys and cameras to computer drives and printers. In fact, many common hand tools such as scissors, screwdrivers, wrenches, jacks, and hammers are actually true mechanisms. Moreover, the hands and feet, arms, legs, and jaws of humans qualify as functioning mechanisms as do the paws and legs, flippers, wings, and tails of animals.

There is a difference between a *machine* and a *mechanism*: All machines transform energy to do work, but only some mechanisms are capable of performing work. The term *machinery* means an assembly that includes both machines and mechanisms. Figure 1a illustrates a cross section of a machine—an internal combustion engine. The assembly of the piston, connecting rod, and crankshaft is a mechanism, termed a *slider-crank mechanism*. The basic schematic drawing of that mechanism, Fig. 1b, called a *skeleton outline*, shows only its fundamental structure without the technical details explaining how it is constructed.

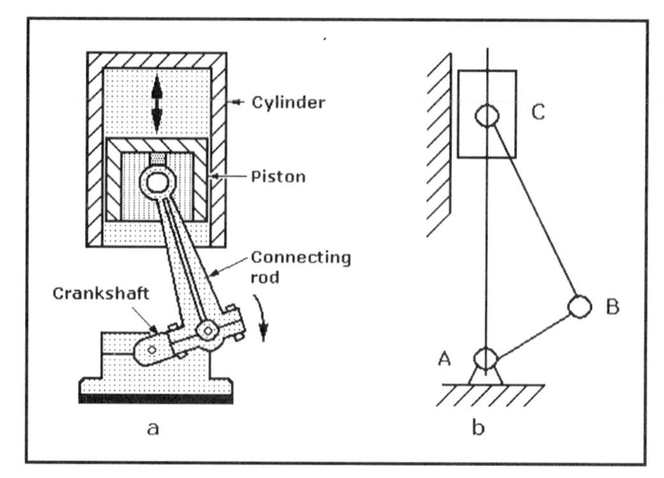

Fig. 1 Cross section of a cylinder of an internal combustion engine showing piston reciprocation (a), and the skeleton outline of the linkage mechanism that moves the piston (b).

PHYSICAL PRINCIPLES

Efficiency of Machines

Simple machines are evaluated on the basis of efficiency and mechanical advantage. While it is possible to obtain a larger force from a machine than the force exerted upon it, this refers only to force and not energy; according to the law of conservation of energy, *more work cannot be obtained from a machine than the energy supplied to it*. Because work = force × distance, for a machine to exert a larger force than its initiating force or operator, that larger force must be exerted through a correspondingly shorter distance. As a result of friction in all moving machinery, the energy produced by a machine is less than that applied to it. Consequently, by interpreting the law of conservation of energy, it follows that:

$$\text{Input energy} = \text{output energy} + \text{wasted energy}$$

This statement is true over any period of time, so it applies to any unit of time; because power is work or energy per unit of time, the following statement is also true:

$$\text{Input power} = \text{output power} + \text{wasted power}$$

The *efficiency of a machine is the ratio of its output to its input*, if both input and output are expressed in the same units of energy or power. This ratio is always less than unity, and it is usually expressed in percent by multiplying the ratio by 100.

$$\text{Percent efficiency} = \frac{\text{output energy}}{\text{input energy}} \times 100$$

or

$$\text{Percent efficiency} = \frac{\text{output power}}{\text{input power}} \times 100$$

A machine has high efficiency if most of the power supplied to it is passed on to its load and only a fraction of the power is wasted. The efficiency can be as high as 98 percent for a large electrical generator, but it is likely to be less than 50 percent for a screw jack. For example, if the input power supplied to a 20-hp motor with an efficiency of 70 percent is to be calculated, the foregoing equation is transposed.

$$\text{Input power} = \frac{\text{output power}}{\text{percent efficiency}} \times 100$$

$$= \frac{20\text{ hp}}{70} \times 100 = 28.6\text{ hp}$$

Mechanical Advantage

The *mechanical advantage* of a mechanism or system is the ratio of the load or weight W, typically in pounds or kilograms, divided by the effort or force F exerted by the initiating entity or operator, also in pounds or kilograms. If friction has been considered or is known from actual testing, the mechanical advantage, MA, of a machine is:

$$\text{MA} = \frac{\text{load}}{\text{effort}} = \frac{W}{F}$$

However, if it is assumed that the machine operates without friction, the ratio of W divided by F is called the *theoretical mechanical advantage*, TA.

$$\text{TA} = \frac{\text{load}}{\text{effort}} = \frac{W}{F}$$

Velocity Ratio

Machines and mechanisms are used to translate a small amount of movement or distance into a larger amount of movement or distance. This property is known as the *velocity ratio*: it is defined as the ratio of the distance moved by the effort per second divided by the distance moved by the load per second for a machine or mechanism. It is widely used in determining the mechanical advantage of gears or pulleys.

$$\text{VR} = \frac{\text{distance moved by effort/second}}{\text{distance moved by load/second}}$$

INCLINED PLANE

The *inclined* plane, shown in Fig. 2, has an incline length l (AB) = 8 ft and a height h (BC) = 3 ft. The inclined plane permits a smaller force to raise a given weight than if it were lifted directly from the ground. For example, if a weight W of 1000 lb is to be raised vertically through a height BC of 3 ft without using an inclined plane, a force F of 1000 lb must be exerted over that height. However, with an inclined plane, the weight is moved over the longer distance of 8 ft, but a force F of only $^3/_8$ of 1000 or 375 lb would be required because the weight is moved through a longer distance. To determine the mechanical advantage of the inclined plane, the following formula is used:

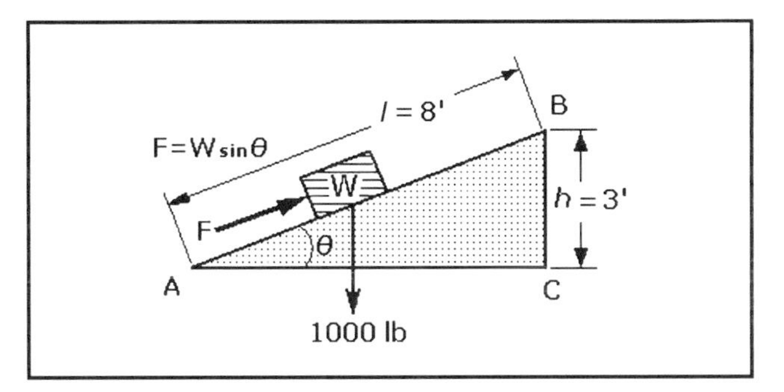

Fig. 2 Diagram for calculating mechanical advantage of an inclined plane.

$$F = W \sin\theta \qquad \sin\theta = \frac{\text{height } h}{\text{length } l}$$

where height h = 3 ft, length l = 8 ft, $\sin\theta$ = 0.375, and weight W = 1000 lb.

$$F = 1000 \times 0.375$$

$$F = 375 \text{ lb}$$

$$\text{Mechanical advantage MA} = \frac{\text{load}}{\text{effort}} = \frac{W}{F} = \frac{1000}{375} = 2.7$$

PULLEY SYSTEMS

A single pulley simply changes the direction of a force so its mechanical advantage is unity. However, considerable mechanical advantage can be gained by using a combination of pulleys. In the typical pulley system, shown in Fig. 3a, each block contains two pulleys or sheaves within a frame or shell. The upper block is fixed and the lower block is attached to the load and moves with it. A cable fastened at the end of the upper block passes around four pulleys before being returned to the operator or other power source.

Figure 3b shows the pulleys separated for clarity. To raise the load through a height h, each of the sections of the cable A, B, C, and D must be moved to a distance equal to h. The operator or other power source must exert a force F through a distance $s = 4h$ so that the velocity ratio of s to h is 4. Therefore, the theoretical mechanical advantage of the system shown is 4, corresponding to the four cables supporting the load W. The theoretical mechanical advantage TA for any pulley system similar to that shown equals the number of parallel cables that support the load.

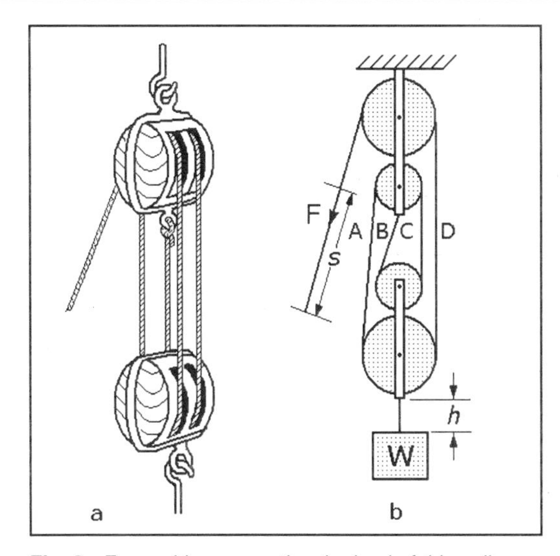

Fig. 3 Four cables supporting the load of this pulley combination give it a mechanical advantage of 4.

SCREW-TYPE JACK

Mechanisms are often required to move a large load with a small effort. For example, a car jack allows an ordinary human to lift a car which may weigh as much as 6000 lb, while the person only exerts a force equivalent to 20 or 30 lb.

The *screw jack*, shown in Fig. 4, is a practical application of the inclined plane because a screw is considered to be an inclined plane wrapped around cylinder. A force F must be exerted at the end of a length of horizontal bar l to turn the screw to raise the load (weight W) of 1000 lb. The 5-ft bar must be moved through a complete turn or a circle of length $s = 2\pi l$ to advance the load a distance h of 1.0 in. or 0.08 ft equal to the pitch p of the screw. The pitch of the screw is the distance advanced in a complete turn. Neglecting friction:

$$F = \frac{W \times h}{s}$$

where $s = 2\pi l = 2 \times 3.14 \times 5$, $h = p = 0.08$, and $W = 1000$ lb

$$F = \frac{1000 \times 0.08}{2 \times 3.14 \times 5} = \frac{80}{31.4} = 2.5 \text{ lb}$$

$$\text{Mechanical advantage MA} = \frac{\text{load}}{\text{effort}} = \frac{2\pi l}{p} = \frac{31.4}{0.08} = 393$$

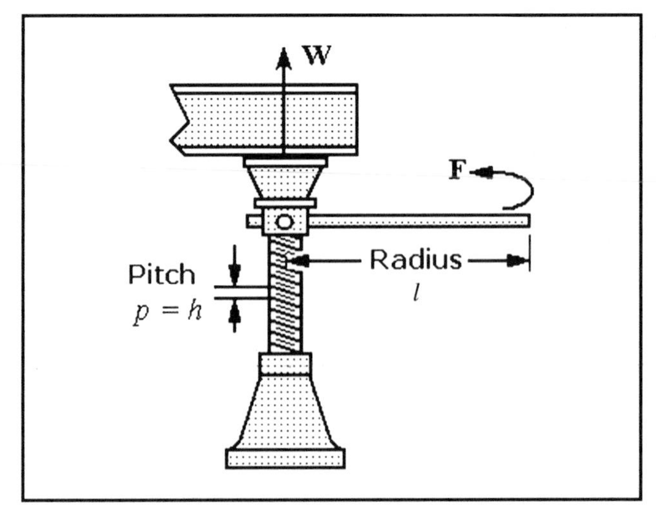

Fig. 4 Diagram for calculating the mechanical advantage of a screw jack.

LEVERS AND MECHANISMS

Levers

Levers are the simplest of mechanisms; there is evidence that Stone Age humans used levers to extend their reach or power; they made them from logs or branches to move heavy loads such as rocks. It has also been reported that primates and certain birds use twigs or sticks to extend their reach and act as tools to assist them in obtaining food.

A lever is a rigid beam that can rotate about a fixed point along its length called the *fulcrum*. Physical effort applied to one end of the beam will move a load at the other end. The act of moving the fulcrum of a long beam nearer to the load permits a large load to be lifted with minimal effort. This is another way to obtain *mechanical advantage*.

The three *classes of lever* are illustrated in Fig. 5. Each is capable of providing a different level of mechanical advantage. These levers are called *Class 1*, *Class 2*, and *Class 3*. The differences in the classes are determined by:

- Position along the length of the lever where the effort is applied
- Position along the length of the lever where the load is applied
- Position along the length of the lever where the fulcrum or pivot point is located

Class 1 lever, the most common, has its fulcrum located at or about the middle with effort exerted at one end and load positioned at the opposite end, both on the same side of the lever. Examples of Class 1 levers are playground seesaw, crowbar, scissors, claw hammer, and balancing scales.

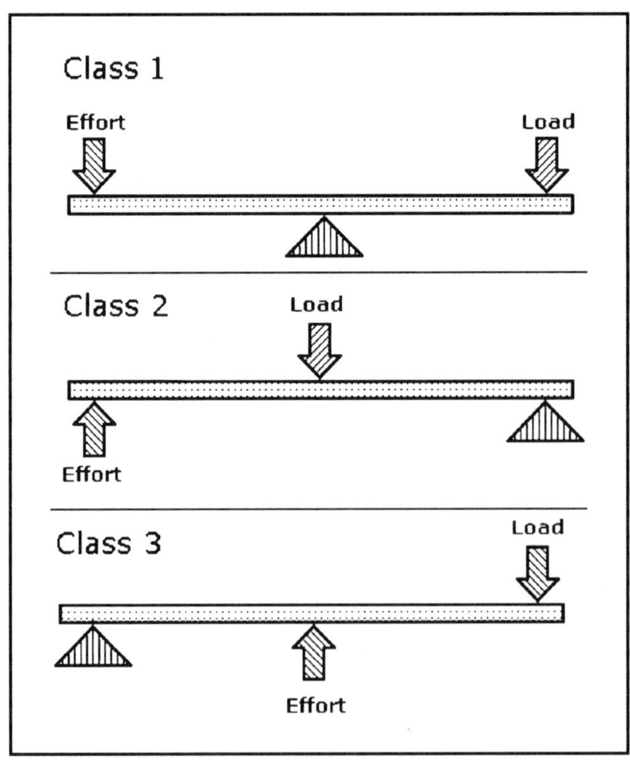

Fig. 5 Three levers classified by the locations of their fulcrums, loads, and efforts.

Class 2 lever has its fulcrum at one end; effort is exerted at the opposite end, and the opposing load is positioned at or near the middle. Examples of Class 2 levers are wheelbarrow, simple bottle openers, nutcracker, and foot pump for inflating air mattresses and inflatable boats.

Class 3 lever also has its fulcrum on one end; load is exerted at the opposite end, and the opposing effort is exerted on or about the middle. Examples of Class 3 levers are shovel and fishing rod where the hand is the fulcrum, tweezers, and human and animal arms and legs.

The application of a Class 1 lever is shown in Fig. 6. The lever is a bar of length AB with its fulcrum at *X*, dividing the length of the bar into parts: l_1 and l_2. To raise a load *W* through a height of *h*, a force *F* must be exerted downward through a distance *s*. The triangles AXC and BXD are similar and pro[illegible] therefore, ignoring friction:

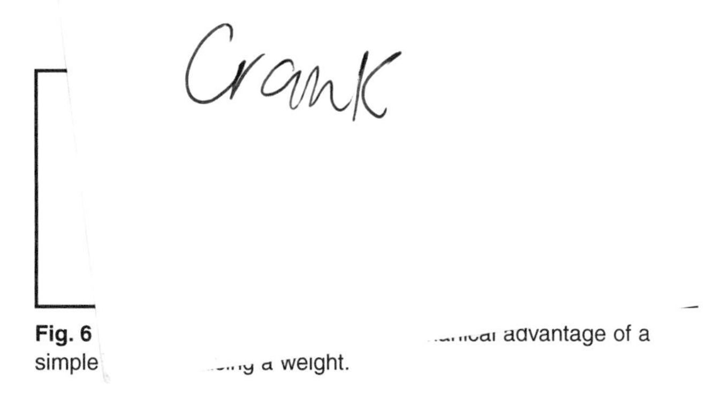

Fig. 6 [illegible] advantage of a simple [illegible] weight.

Winches, Windlasses, and Capstans

Winches, windlasses, and capstans are machines that convert rotary motion into linear motion, usually with some mechanical advantage. These machines are essentially Class 1 levers: effort is applied to a lever or crank, the fulcrum is the center of the drum, and the load is applied to the rope, chain, or cable.

Manually operated windlasses and capstans, mechanically the same, were originally used on sailing ships to raise and lower anchors. Operated by one or more levers by one or more sailors, both had barrels or drums on which rope or chain was wound. In the past, windlasses were distinguished from capstans; windlasses had horizontal drums and capstans had vertical drums. The modern term *winch* is now the generic name for any manual or power-operated drum for hauling a load with cable, chain, or rope. The manually operated winch, shown in Fig. 7, is widely used today on sailboats for raising and trimming sails, and sometimes for weighing anchors.

Ignoring friction, the mechanical advantage of all of these machines is approximately the *length of the crank* divided by the *diameter of the drum.* In the winch example shown, when the left end of the line is held under tension and the handle or crank is turned clockwise, a force is applied to the line entering on the right; it is attached to the load to perform such useful work as raising or tensioning sails.

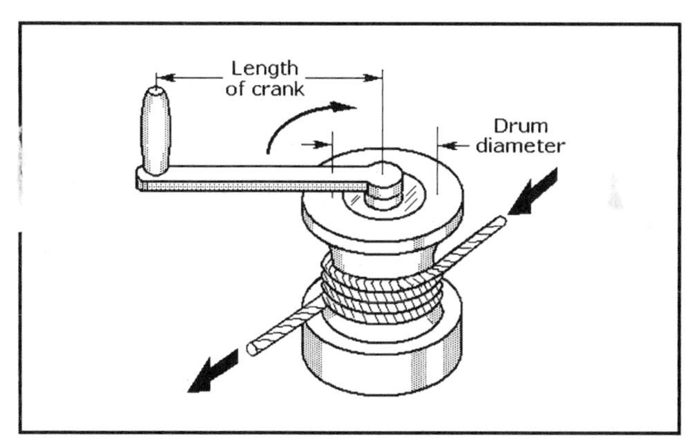

Fig. 7 Diagram for calculating the mechanical advantage of a manually operated winch for raising anchors or sails.

LINKAGES

A *linkage* is a mechanism formed by connecting two or more levers together. Linkages can be designed to change the direction of a force or make two or more objects move at the same time. Many different fasteners are used to connect linkages together yet allow them to move freely such as pins, end-threaded bolts with nuts, and loosely fitted rivets. There are two general classes of linkages: *simple planar linkages* and more complex *specialized linkages*; both are capable of performing tasks such as describing straight lines or curves and executing motions at differing speeds. The names of the linkage mechanisms given here are widely but not universally accepted in all textbooks and references.

Linkages can be classified according to their primary functions:

- *Function generation*: the relative motion between the links connected to the frame
- *Path generation*: the path of a tracer point
- *Motion generation*: the motion of the coupler link

Simple Planar Linkages

Four different simple planar linkages shown in Fig. 8 are identified by function:

- *Reverse-motion linkage*, Fig. 8a, can make objects or force move in opposite directions; this can be done by using the input link as a lever. If the fixed pivot is equidistant from the moving pivots, output link movement will equal input link movement, but it will act in the opposite direction. However, if the fixed pivot is not centered, output link movement will not equal input link movement. By selecting the position of the fixed pivot, the linkage can be designed to produce specific mechanical advantages. This linkage can also be rotated through 360°.
- *Push-pull linkage*, Fig. 8b, can make the objects or force move in the same direction; the output link moves in the same direction as the input link. Technically classed as a four-bar linkage, it can be rotated through 360° without changing its function.

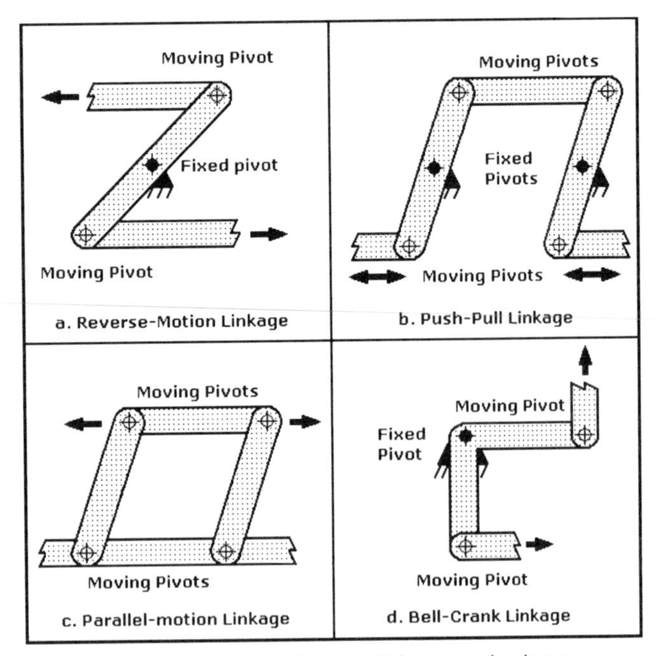

Fig. 8 Functions of four basic planar linkage mechanisms.

- *Parallel-motion linkage*, Fig. 8c, can make objects or forces move in the same direction, but at a set distance apart. The moving and fixed pivots on the opposing links in the parallelogram must be equidistant for this linkage to work correctly. Technically classed as a four-bar linkage, this linkage can also be rotated through 360° without changing its function. Pantographs that obtain power for electric trains from overhead cables are based on parallel-motion linkage. Drawing pantographs that permit original drawings to be manually copied without tracing or photocopying are also adaptations of this linkage; in its simplest form it can also keep tool trays in a horizontal position when the toolbox covers are opened.
- *Bell-crank linkage*, Fig. 8d, can change the direction of objects or force by 90°. This linkage rang doorbells before electric clappers were invented. More recently this mechanism has been adapted for bicycle brakes. This was done by pinning two bell cranks bent 90° in opposite directions together to form tongs. By squeezing the two handlebar levers linked to the input ends of each crank, the output ends will move together. Rubber blocks on the output ends of each crank press against the wheel rim, stopping the bicycle. If the pins which form a fixed pivot are at the midpoints of the cranks, link movement will be equal. However, if those distances vary, mechanical advantage can be gained.

Specialized Linkages

In addition to changing the motions of objects or forces, more complex linkages have been designed to perform many specialized functions: These include drawing or tracing straight lines; moving objects or tools faster in a retraction stroke than in an extension stroke; and converting rotating motion into linear motion and vice versa.

The simplest specialized linkages are four-bar linkages. These linkages have been versatile enough to be applied in many different applications. Four-bar linkages actually have only three moving links but they have one fixed link and four pin joints or pivots. A useful mechanism must have at least four links but closed-loop assemblies of three links are useful elements in structures. Because any linkage with at least one fixed link is a mechanism, both the parallel-motion and push-pull linkages mentioned earlier are technically machines.

Four-bar linkages share common properties: three rigid moving links with two of them hinged to fixed bases which form a *frame*. Link mechanisms are capable of producing rotating, oscillating, or reciprocating motion by the rotation of a crank. Linkages can be used to convert:

- Continuous rotation into another form of continuous rotation, with a constant or variable angular velocity ratio
- Continuous rotation into oscillation or continuous oscillation into rotation, with a constant or variable velocity ratio
- One form of oscillation into another form of oscillation, or one form of reciprocation into another form of reciprocation, with a constant or variable velocity ratio

There are four different ways in which four-bar linkages can perform inversions or complete revolutions about fixed pivot points. One pivoting link is considered to be the *input* or *driver* member and the other is considered to be the *output* or *driven member.* The remaining moving link is commonly called a *connecting link*. The fixed link, hinged by pins or pivots at each end, is called the *foundation link.*

Three inversions or linkage rotations of a four-bar chain are shown in Figs. 9, 10, and 11. They are made up of links AB, BC, CD, and AD. The forms of the three inversions are defined by the position of the shortest link with respect to the link selected as the foundation link. The ability of the driver or driven links to make complete rotations about their pivots determines their functions.

Drag-link mechanism, Fig. 9, demonstrates the first inversion. The shortest link AD between the two fixed pivots is the foundation link, and both driver link AB and driven link CD can make full revolutions.

Crank-rocker mechanism, Fig. 10, demonstrates the second inversion. The shortest link AB is adjacent to AD, the foundation link. Link AB can make a full 360° revolution while the opposite link CD can only oscillate and describe an arc.

Double-rocker mechanism, Fig. 11, demonstrates the third inversion. Link AD is the foundation link, and it is opposite the shortest link BC. Although link BC can make a full 360° revolution, both pivoting links AB and CD can only oscillate and describe arcs.

The fourth inversion is another *crank-rocker mechanism* that behaves in a manner similar to the mechanism shown in Fig. 10,

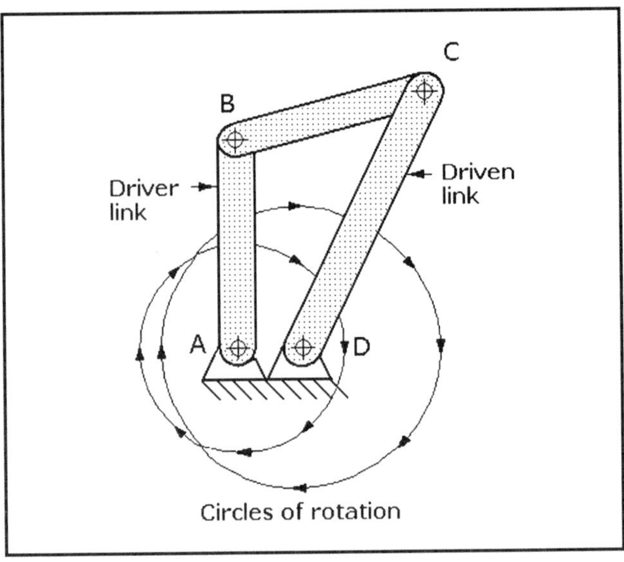

Fig. 9 Four-bar drag-link mechanism: Both the driver link AB and driven link CD can rotate through 360°. Link AD is the foundation link.

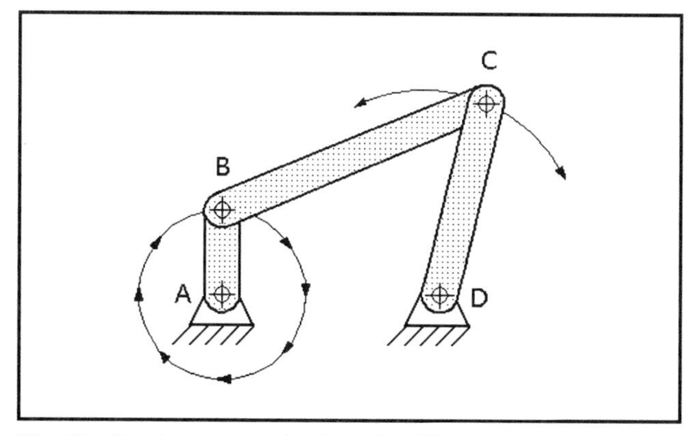

Fig. 10 Crank-rocker mechanism: Link AB can make a 360° revolution while link CD oscillates with C describing an arc. Link AD is the foundation link.

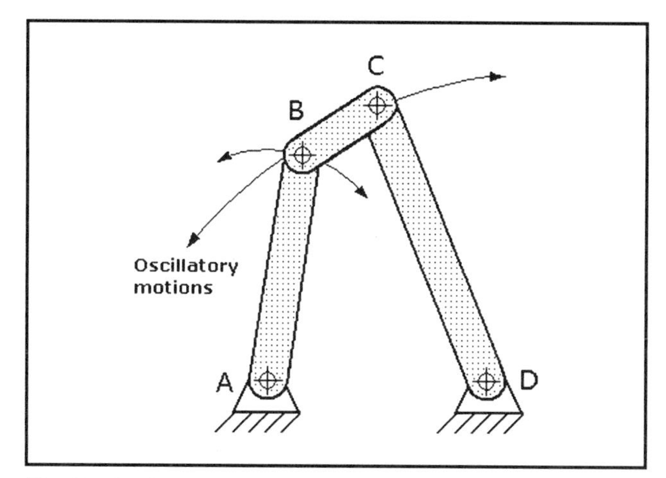

Fig. 11 Double-rocker mechanism: Short link BC can make a 360° revolution, but pivoting links AB and CD can only oscillate, describing arcs.

but the longest link, CD, is the foundation link. Because of this similarity between these two mechanisms, the fourth inversion is not illustrated here. A drag-link mechanism can produce either a nonuniform output from a uniform input rotation rate or a uniform output from a nonuniform input rotation rate.

Straight-Line Generators

Figures 12 to 15 illustrate examples of classical linkages capable of describing straight lines, a function useful in many different kinds of machines, particularly machine tools. The dimensions of the rigid links are important for the proper functioning of these mechanisms.

Watt's straight-line generator, illustrated in Fig. 12, can describe a short vertical straight line. Equal length links AB and CD are hinged at A and D, respectively. The midpoint E of connecting link BC traces a figure eight pattern over the full mechanism excursion, but a straight line is traced in part of the excursion because point E diverges to the left at the top of the stroke and to the right at the bottom of the stroke. This linkage was used by Scottish instrument maker, James Watt, in a steam-driven beam pump in about 1769, and it was a prominent mechanism in early steam-powered machines.

Scott Russell straight-line generator, shown in Fig. 13, can also describe a straight line. Link AB is hinged at point A and pinned to link CD at point B. Link CD is hinged to a roller at point C which restricts it to horizontal oscillating movement.

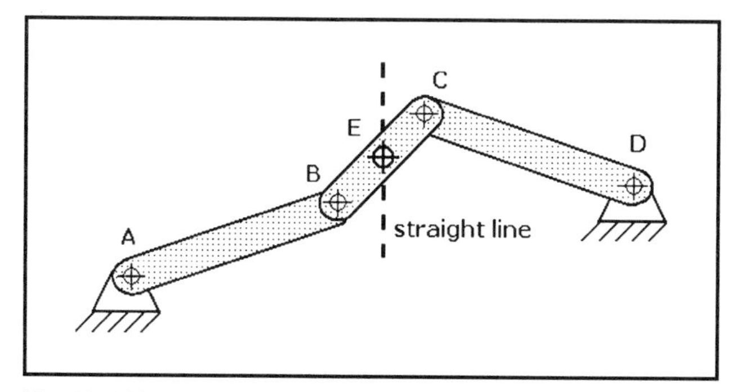

Fig. 12 Watt's straight-line generator: The center point E of link BC describes a straight line when driven by either links AB or CD.

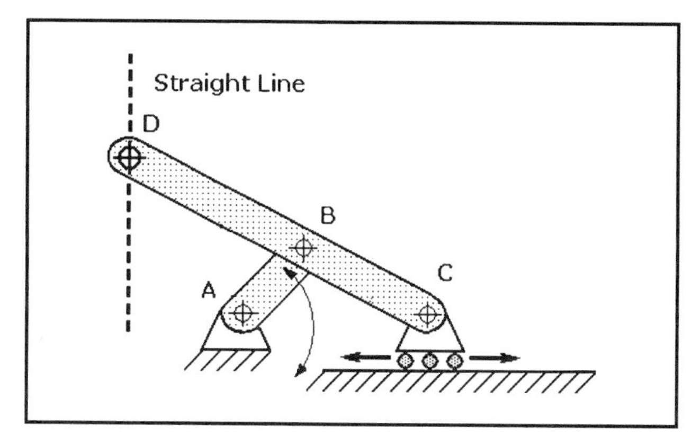

Fig. 13 Scott russell straight-line generator: Point D of link DC describes a straight line as driver link AB oscillates, causing the slider at C to reciprocate left and right.

This configuration confines point D to a motion that traces a vertical straight line. Both points A and C lie in the same horizontal plane. This linkage works if the length of link AB is about 40 percent of the length of CD, and the distance between points D and B is about 60 percent of the length of CD.

Peaucellier's straight-line linkage, drawn as Fig. 14, can describe more precise straight lines over its range than either the Watt's or Scott Russell linkages. To make this linkage work correctly, the length of link BC must equal the distance between points A and B set by the spacing of the fixed pivots; in this figure, link BC is 15 units long while the lengths of links CD, DF, FE, and EC are equal at 20 units. As links AD and AE are moved,

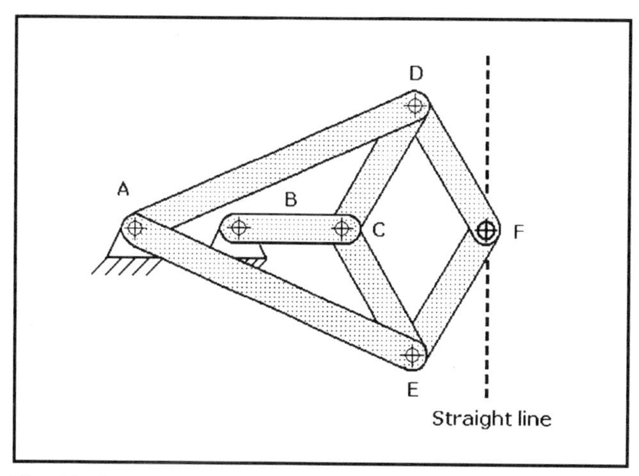

Fig. 14 Peaucellier's straight-line generator: Point F describes a straight line when either link AD or AE acts as the driver.

point F can describe arcs of any radius. However, the linkage can be restricted to tracing straight lines (infinite radiuses) by selecting link lengths for AD and AE. In this figure they are 45 units long. This linkage was invented in 1873 by the French engineer, Captain Charles-Nicolas Peaucellier.

Tchebicheff's straight-line generator, shown in Fig. 15, can also describe a horizontal line. Link CB with E as its midpoint traces a straight horizontal line for most of its transit as links AB and DC are moved to the left and right of center. To describe this straight line, the length of the foundation link AD must be twice the length of link CB. To make this mechanism work as a straight-line generator, CB is 10 units long, AD is 20 units long, and both AB and DC are 25 units long. With these dimensions, link CB will assume a vertical position when it is at the right and left extremes of its travel excursion. This linkage was invented by nineteenth-century Russian mathematician, Pafnuty Tchebicheff or Chebyshev.

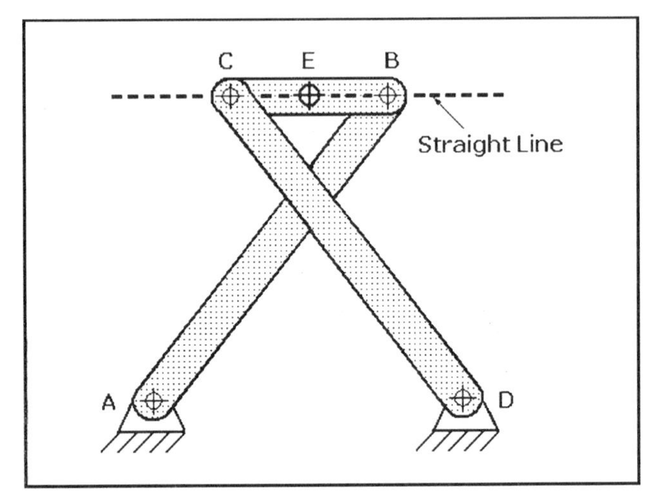

Fig. 15 Tchebicheff's straight-line generator: Point E of link CB describes a straight line when driven by either link AB or DC. Link CB moves into a vertical position at both extremes of its travel.

Rotary/Linear Linkages

Slider-crank mechanism (or a simple crank), shown as Fig. 16, converts rotary to linear motion and vice versa, depending on its application. Link AB is free to rotate 360° around the hinge while link BC oscillates back and forth because point C is hinged to a roller which restricts it to linear motion. Either the slider or the rotating link AB can be the driver.

This mechanism is more familiar as the piston, connecting rod, and crankshaft of an internal combustion engine, as illustrated in Fig. 1. The piston is the slider at C, the connecting rod is link BC, and the crankshaft is link AB. In a four-stroke engine, the piston is pulled down the cylinder by the crankshaft, admitting the air-fuel mixture; in the compression stroke the piston is driven back up the cylinder by the crankshaft to compress the air-fuel mixture. However, the roles change in the combustion stroke when the piston drives the crankshaft. Finally, in the exhaust stroke the roles change again as the crankshaft drives the piston back to expel the exhaust fumes.

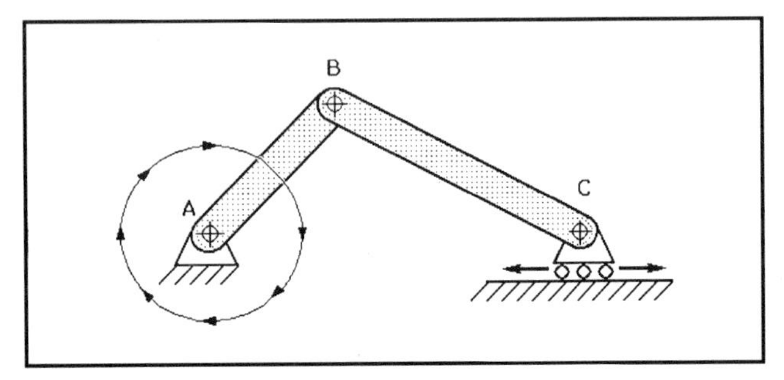

Fig. 16 Slider-crank mechanism: This simple crank converts the 360° rotation of driver link AB into linear motion of link BC, causing the slider at C to reciprocate.

Scotch-yoke mechanism, pictured in Fig. 17, functions in a manner similar to that of the simple crank mechanism except that its linear output motion is sinusoidal. As wheel A, the driver, rotates, the pin or roller bearing at its periphery exerts torque within the closed yoke B; this causes the attached sliding bar to reciprocate, tracing a sinusoidal waveform. Part a shows the sliding bar when the roller is at 270°, and part b shows the sliding bar when the roller is at 0°.

Rotary-to-linear mechanism, drawn in Fig. 18, converts a uniform rotary motion into an intermittent reciprocating motion. The three teeth of the input rotor contact the steps in the frame or yoke, exerting torque 3 times per revolution, moving the yoke with attached bar. Full linear travel of the yoke is accomplished in 30° of rotor rotation followed by a 30° delay before returning the yoke. The reciprocating cycle is completed 3 times per revolution of the input. The output is that of a step function.

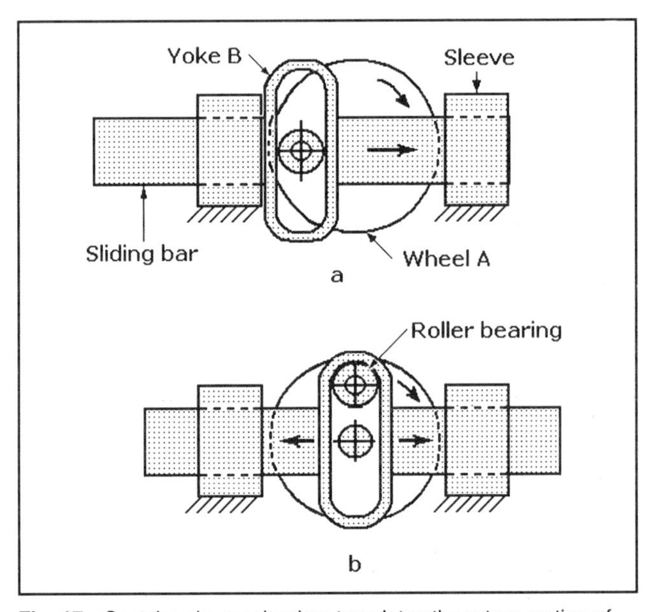

Fig. 17 Scotch-yoke mechanism translates the rotary motion of the wheel with a peripheral roller into reciprocating motion of the yoke with supporting bars as the roller exerts torque within the yoke. The yoke is shown in its left (270°) position in (a) and in its center (0°) position in (b).

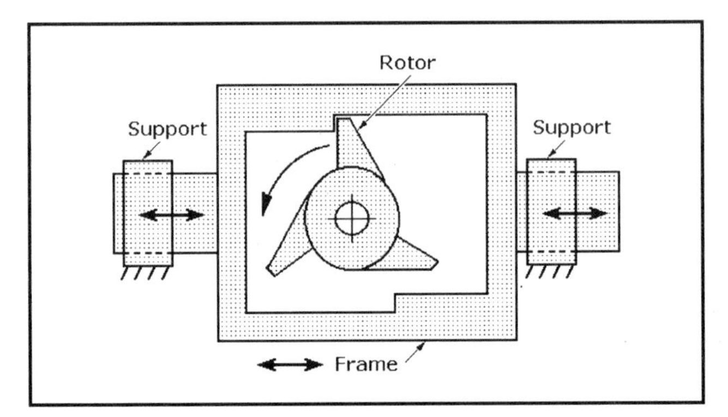

Fig. 18 Rotary-to-linear mechanism converts the uniform rotation of the 3-tooth rotor into a reciprocating motion of the frame and supporting bars. The reciprocating cycle is completed 3 times per rotor revolution.

Geneva wheel mechanism, illustrated in Fig. 19, is an example of intermittent gearing that converts continuous rotary motion into intermittent rotary motion. Geneva wheel C makes a quarter turn for every turn of lever AB attached to driving wheel A. When pin B on lever AB turns clockwise, it enters one of the four slots of geneva wheel C; the pin moves downward in the slot, applying enough torque to the geneva wheel to turn it counterclockwise $^1/_4$ revolution before it leaves the slot. As wheel A continues to rotate clockwise, it engages the next three slots in a sequence to complete one geneva wheel rotation. If one of the slots is obstructed, the pin can only move through part of the revolution, in either direction, before it strikes the closed slot, stopping the rotation of the geneva wheel. This mechanism has been used in mechanical windup watches, clocks, and music boxes to prevent overwinding.

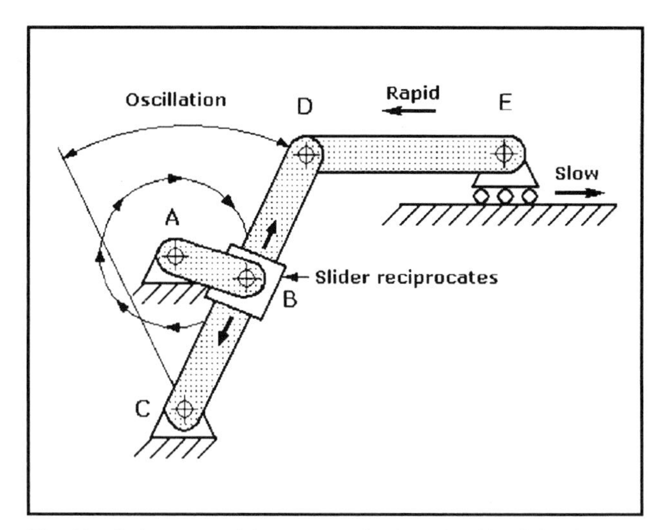

Fig. 20 Swing-arm quick-return mechanism: As drive link AB rotates 360° around A, it causes the slider at B to reciprocate up and down along link CD, causing CD to oscillate though an arc. This motion drives link DE in a reciprocating motion that moves the rolling slider at E slowly to the right before returning it rapidly to the left.

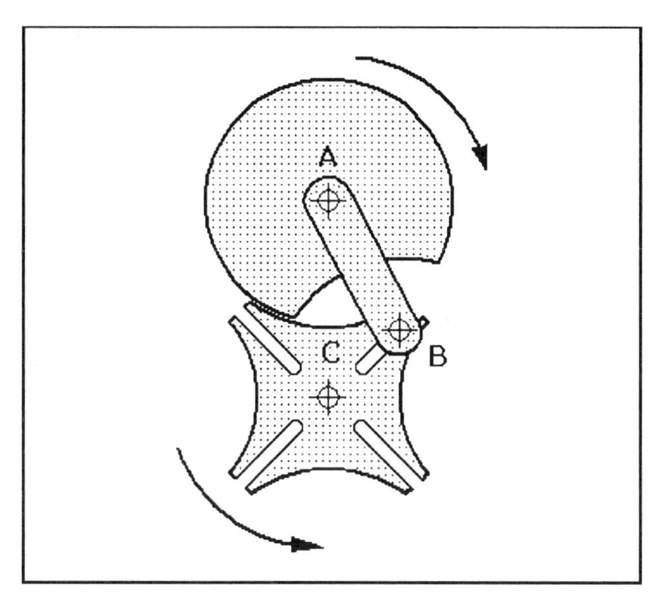

Fig. 19 Geneva wheel escapement mechanism: Pin B at the end of lever AB (attached to wheel A) engages a slot in geneva wheel C as wheel A rotates clockwise. Pin B moves down the slot, providing torque to drive the geneva wheel counterclockwise $^1/_4$ revolution before it exits the first slot; it then engages the next three slots to drive the geneva wheel through one complete counterclockwise revolution.

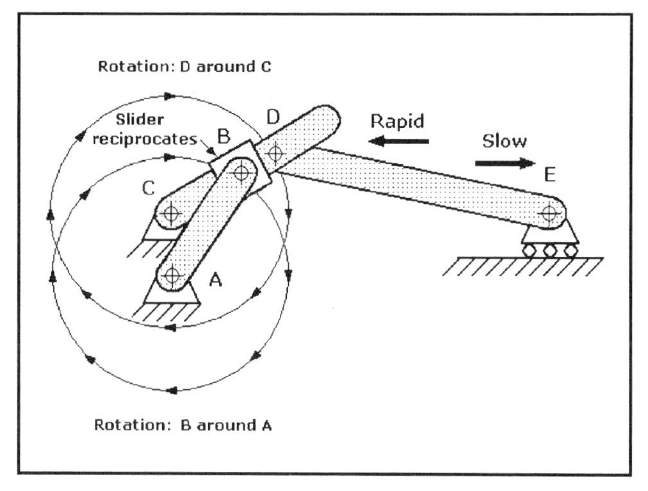

Fig. 21 Whitworth's quick-return mechanism: As drive link AB rotates 360° around A, it causes the slider at B to reciprocate back and forth along link CD, which, in turn causes CD to rotate 360° around C. This, motion causes link DE to reciprocate, first moving rolling slider at E slowly to the right before returning it rapidly to the left.

Swing-arm quick-return mechanism, drawn as Fig. 20, converts rotary motion into nonuniform reciprocating motion. As drive link AB rotates 360° around pin A, it causes the slider at B to reciprocate up and down along link CD. This, in turn, causes CD to oscillate left and right, describing an arc. Link DE, pinned to D with a rolling slider pinned at E, moves slowly to the right before being returned rapidly to the left.

Whitworth quick-return mechanism, shown as Fig. 21, converts rotary motion to nonuniform reciprocating motion. Drive link AB rotates 360° about pin A causing the slider at B to reciprocate back and forth along link CD; this, in turn, causes link CD to rotate 360° around point C. Link DE is pinned to link CD at D and a rolling slider at E. The slider at E is moved slowly to the right before being returned rapidly to the left. This mechanism, invented in the nineteenth century by English engineer, Joseph Whitworth, has been adapted for shapers, machine tools with moving arms that cut metal from stationary workpieces. A hardened cutting tool attached at the end of the arm (equivalent to point E) advances slowly on the cutting stroke but retracts rapidly on the backstroke. This response saves time and improves productivity in shaping metal.

Simple ratchet mechanism, drawn as Fig. 22, can only be turned in a counterclockwise direction. The ratchet wheel has many wedge-shaped teeth that can be moved incrementally to turn an oscillating drive lever. As driving lever AB first moves clockwise to initiate counterclockwise movement of the wheel, it drags pawl C pinned at B over one or more teeth while pawl D prevents the wheel from turning clockwise. Then, as lever AB reverses to drive the ratchet wheel counterclockwise, pawl D is released, allowing the wheel to turn it in that direction. The amount of backward incremental motion of lever AB is directly proportional to pitch of the teeth: smaller teeth will reduce the degree of rotation while larger teeth will increase them. The contact surfaces of the teeth on the wheel are typically inclined, as shown, so they will not be disengaged if the mechanism is subjected to vibration or shock under load. Some ratchet mechanisms include a spring to hold pawl D against the teeth to assure no clockwise wheel rotation as lever AB is reset.

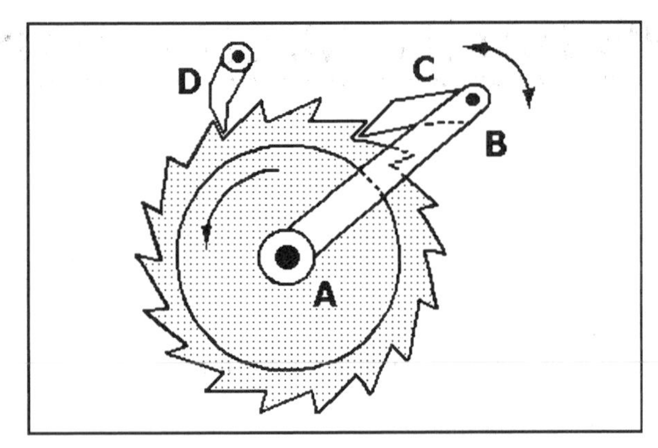

Fig. 22 This ratchet wheel can be turned only in a counterclockwise direction. As driving lever AB moves clockwise, it drags pawl C, pinned at B over one or more teeth while pawl D prevents the wheel from turning clockwise. Then as lever AB reverses to drive the ratchet wheel counterclockwise, pawl D is released allowing the wheel to turn it in that direction.

GEARS AND GEARING

A *gear* is a wheel with evenly sized and spaced teeth machined or formed around its perimeter. Gears are used in rotating machinery not only to transmit motion from one point to another, but also for the mechanical advantage they offer. Two or more gears transmitting motion from one shaft to another is called a *gear train,* and *gearing* is a system of wheels or cylinders with meshing teeth. Gearing is chiefly used to transmit rotating motion but can also be adapted to translate reciprocating motion into rotating motion and vice versa.

Gears are versatile mechanical components capable of performing many different kinds of power transmission or motion control. Examples of these are

- Changing rotational speed
- Changing rotational direction
- Changing the angular orientation of rotational motion
- Multiplication or division of torque or magnitude of rotation
- Converting rotational to linear motion, and its reverse
- Offsetting or changing the location of rotating motion

The teeth of a gear can be considered as levers when they mesh with the teeth of an adjoining gear. However, gears can be rotated continuously instead of rocking back and forth through short distances as is typical of levers. A gear is defined by the number of its teeth and its diameter. The gear that is connected to the source of power is called the *driver*, and the one that receives power from the driver is the *driven gear*. It always rotates in a direction opposing that of the driving gear; if both gears have the same number of teeth, they will rotate at the same speed. However, if the number of teeth differs, the gear with the smaller *r* number of teeth will rotate faster. The size and shape of all gear teeth that are to mesh properly for working contact must be equal.

Figure 23 shows two gears, one with 15 teeth connected at the end of shaft A, and the other with 30 teeth connected at the end of shaft B. The 15 teeth of smaller driving gear A will mesh with 15 teeth of the larger gear B, but while gear A makes one revolution gear B will make only $\frac{1}{2}$ revolution.

The number of teeth on a gear determines its diameter. When two gears with different diameters and numbers of teeth are meshed together, the number of teeth on each gear determines gear ratio, velocity ratio, distance ratio, and mechanical advantage. In Fig. 23, gear A with 15 teeth is the driving gear and gear B with 30 teeth is the driven gear. The gear ratio GR is determined as:

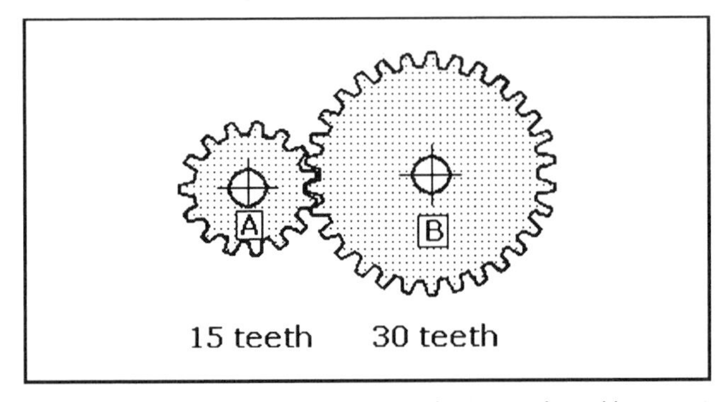

Fig. 23 Gear B has twice as many teeth as gear A, and it turns at half the speed of gear A because gear speed is inversely proportional to the number of teeth on each gear wheel.

$$\begin{aligned} GR &= \frac{\text{number of teeth on driven gear B}}{\text{number of teeth on driving gear A}} \\ &= \frac{30}{15} = \frac{2}{1} \text{ (also written as 2:1)} \end{aligned}$$

The number of teeth in both gears determines the rotary distance traveled by each gear and their angular speed or velocity ratio. The angular speeds of gears are inversely proportional to the numbers of their teeth. Because the smaller driving gear A in Fig. 23 will revolve twice as fast as the larger driven gear B, velocity ratio VR is:

$$VR = \frac{\text{velocity of driving gear A}}{\text{velocity of driven gear B}} = \frac{2}{1} \text{ (also written as 2:1)}$$

In this example load is represented by driven gear B with 30 teeth and the effort is represented by driving gear A with 15 teeth. The distance moved by the load is twice that of the effort. Using the general formula for mechanical advantage MA:

$$MA = \frac{\text{load}}{\text{effort}} = \frac{30}{15} = 2$$

Simple Gear Trains

A gear train made up of multiple gears can have several drivers and several driven gears. If the train contains an odd number of gears, the output gear will rotate in the same direction as the input gear, but if the train contains an even number of gears, the output gear will rotate opposite that of the input gear. The number of teeth on the intermediate gears does not affect the overall velocity ratio, which is governed purely by the number of teeth on the first and last gear.

In simple gear trains, high or low gear ratios can only be obtained by combining large and small gears. In the simplest basic gearing involving two gears, the driven shaft and gear revolves in a direction opposite that of the driving shaft and gear. If it is desired that the two gears and shafts rotate in the same direction, a third *idler gear* must be inserted between the driving gear and the driven gear. The idler revolves in a direction opposite that of the driving gear.

A simple gear train containing an idler is shown in Fig. 24. Driven idler gear B with 20 teeth will revolve 4 times as fast counterclockwise as driving gear A with 80 teeth turning clockwise. However, gear C, also with 80 teeth, will only revolve one turn clockwise for every four revolutions of idler gear B, making the velocities of both gears A and C equal except that gear C turns in the same direction as gear A. In general, the velocity ratio of the first and last gears in a train of simple gears is not changed by the number of gears inserted between them.

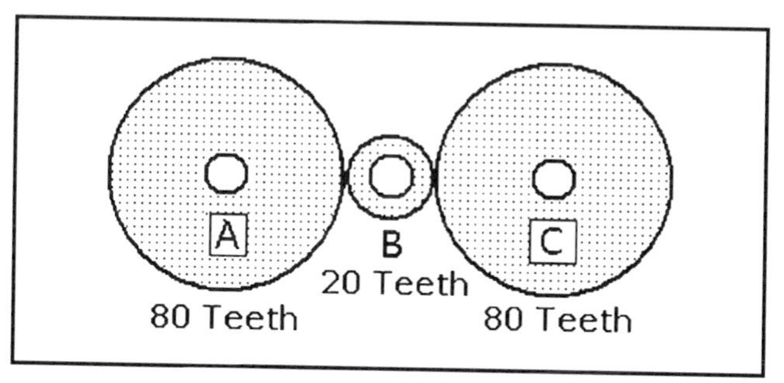

Fig. 24 Gear train: When gear A turns once clockwise, gear B turns four times counter clockwise, and gear wheel C turns once clockwise. Gear B reverses the direction of gear C so that both gears A and C turn in the same direction with no change in the speed of gear C.

Compound Gear Trains

More complex compound gear trains can achieve high and low gear ratios in a restricted space by coupling large and small gears on the same axle. In this way gear ratios of adjacent gears can be multiplied through the gear train. Figure 25 shows a set of compound gears with the two gears B and D mounted on the middle shaft. Both rotate at the same speed because they are fastened together. If gear A (80 teeth) rotates at 100 rpm clockwise, gear B (20 teeth) turns at 400 rpm counterclockwise because of its velocity ratio of 1 to 4. Because gear D (60 teeth) also turns at 400 rpm and its velocity ratio is 1 to 3 with respect to gear C

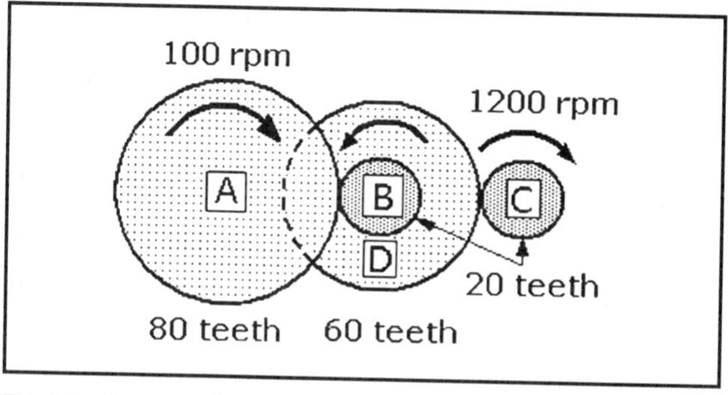

Fig. 25 Compound gears: Two gears B and D are mounted on a central shaft and they turn at the same speed. If gear A rotates at 100 rpm clockwise, gears B and D turn counterclockwise at 400 rpm, and gear C, driven by gear D, turns clockwise at 1200 rpm.

(20 teeth), gear C will turn at 1200 rpm clockwise. The velocity ratio of a compound gear train can be calculated by multiplying the velocity ratios for all pairs of meshing gears. For example, if the driving gear has 45 teeth and the driven gear has 15 teeth, the velocity ratio is ${}^{15}/_{45} = {}^{1}/_{3}$.

Gear Classification

All gears can be classified as either external gears or internal or annual gears:

- *External gears* have teeth on the outside surface of the disk or wheel.
- *Internal* or *annual gears* have teeth on the inside surface of a ring or cylinder.

Spur gears are cylindrical external gears with teeth that are cut straight across the edge of the disk or wheel parallel to the axis of rotation. The spur gears shown in Fig. 26a are the simplest gears. They normally translate rotating motion between two parallel shafts. An *internal* or *annual* gear, as shown in Fig. 26b, is a variation of the spur gear except that its teeth are cut on the inside of a ring or flanged wheel rather than on the outside. Internal gears usually drive or are driven by a pinion. The disadvantage of a simple spur gear is its tendency to produce thrust that can misalign other meshing gears along their respective shafts, thus reducing the face widths of the meshing gears and reducing their mating surfaces.

Rack gears, as the one shown in Fig. 26c, have teeth that lie in the same plane rather than being distributed around a wheel. This gear configuration provides straight-line rather than rotary motion. A rack gear functions like a gear with an infinite radius.

Pinions are small gears with a relatively small number of teeth which can be mated with rack gears.

Rack and pinion gears, shown in Fig. 26c, convert rotary motion to linear motion; when mated together they can transform the rotation of a pinion into reciprocating motion, or vice versa. In some systems, the pinion rotates in a fixed position and engages the rack which is free to move; the combination is found in the steering mechanisms of vehicles. Alternatively, the rack is fixed while the pinion rotates as it moves up and down the rack: Funicular railways are based on this drive mechanism; the driving pinion on the rail car engages the rack positioned between the two rails and propels the car up the incline.

Bevel gears, as shown in Fig. 26d, have straight teeth cut into conical circumferences which mate on axes that intersect, typically at right angles between the input and output shafts. This class of gears includes the most common straight and spiral bevel gears as well as miter and hypoid gears.

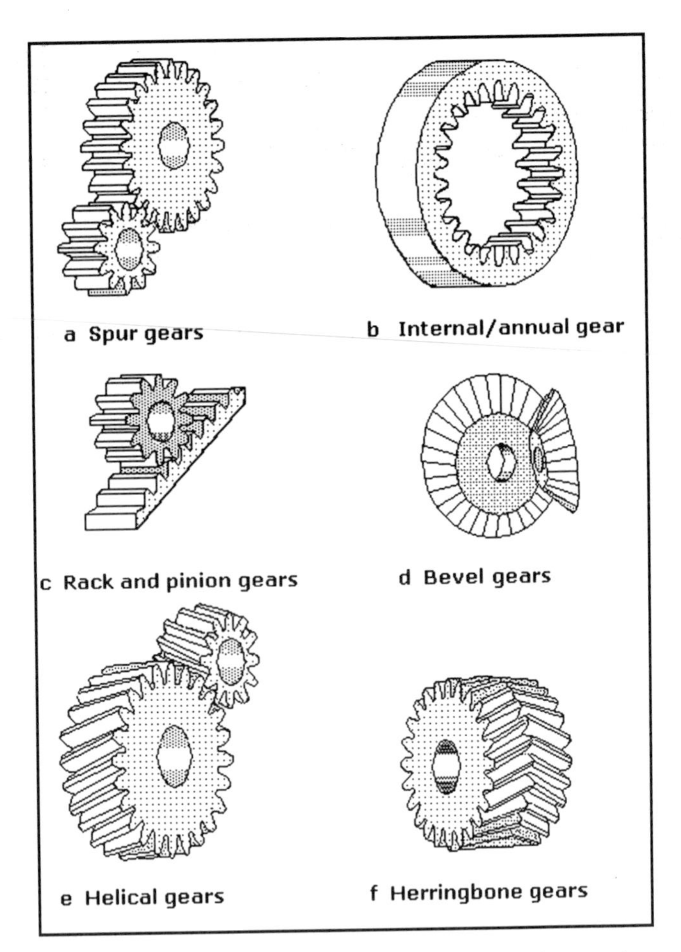

Fig. 26 Gear types: Eight common types of gears and gear pairs are shown here.

Straight bevel gears are the simplest bevel gears. Their straight teeth produce instantaneous line contact when they mate. These gears provide moderate torque transmission, but they are not as smooth running or quiet as spiral bevel gears because the straight teeth engage with full-line contact. They permit medium load capacity.

Spiral bevel gears have curved oblique teeth. The spiral angle of curvature with respect to the gear axis permits substantial tooth overlap. Consequently, the teeth engage gradually and at least two teeth are in contact at the same time. These gears have lower tooth loading than straight bevel gears and they can turn up to 8 times faster. They permit high load capacity.

Miter gears are mating bevel gears with equal numbers of teeth used between rotating input and output shafts with axes that are 90° apart.

Hypoid gears are helical bevel gears used when the axes of the two shafts are perpendicular but do not intersect. They are commonly used to connect driveshafts to rear axles of automobiles, and are often incorrectly called *spiral gearing*.

Helical gears are external cylindrical gears with their teeth cut at an angle rather than parallel to the axis. A simple helical gear, as shown in Fig. 26e, has teeth that are offset by an angle with respect to the axis of the shaft so that they spiral around the shaft in a helical manner. Their offset teeth make them capable of smoother and quieter action than spur gears, and they are capable of driving heavy loads because the teeth mesh at an acute angle rather than at 90°. When helical gear axes are parallel they are called parallel helical gears, and when they are at right angles they are called helical gears. Herringbone and worm gears are based on helical gear geometry.

Herringbone or double helical gears, as shown in Fig. 26f, are helical gears with V-shaped right-hand and left-hand helix angles side by side across the face of the gear. This geometry neutralizes axial thrust from helical teeth.

Worm gears, also called *screw gears*, are other variations of helical gearing. A worm gear has a long, thin cylindrical form with one or more continuous helical teeth that mesh with a helical gear. The teeth of the worm gear slide across the teeth of the driven gear rather than exerting a direct rolling pressure as do the teeth of helical gears. Worm gears are widely used to transmit rotation, at significantly lower speeds, from one shaft to another at a 90° angle.

Face gears have straight tooth surfaces, but their axes lie in planes perpendicular to shaft axes. They are designed to mate with instantaneous point contact. These gears are used in right-angle drives, but they have low load capacities.

Practical Gear Configurations

Isometric drawing Fig. 27 shows a *special planetary gear configuration.* The external driver spur gear (lower right) drives the outer ring spur gear (center) which, in turn, drives three internal planet spur gears; they transfer torque to the driven gear (lower left). Simultaneously, the central planet spur gear produces a summing motion in the pinion gear (upper right) which engages a rack with a roller follower contacting a radial disk cam (middle right).

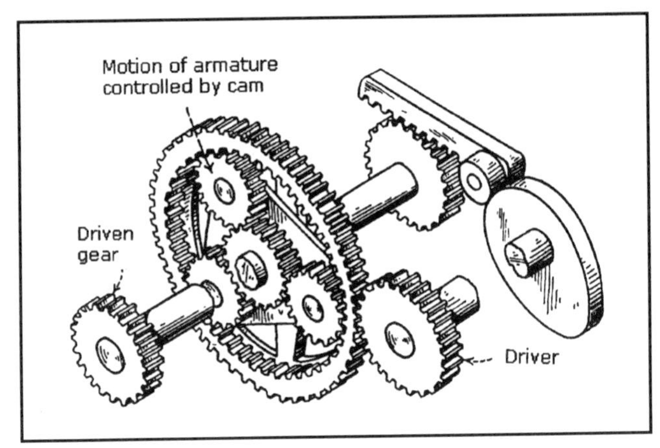

Fig. 27 A special planetary-gear mechanism: The principal of relative motion of mating gears illustrated here can be applied to spur gears in a planetary system. The motion of the central planet gear produces the motion of a summing gear.

Isometric drawing Fig. 28 shows a *unidirectional drive.* The output shaft B rotates in the same direction at all times, regardless of the rotation of the input shaft A. The angular velocity of output shaft B is directly proportional to the angular velocity of input shaft A. The spur gear C on shaft A has a face width that is twice as wide as the faces on spur gears F and D, which are mounted on output shaft B. Spur gear C meshes with idler E and with spur gear D. Idler E meshes with the spur gears C and F. Output shaft B carries two free-wheel disks, G and H, which are oriented unidirectionally.

When input shaft A rotates clockwise (bold arrow), spur gear D rotates counterclockwise and it idles around free-wheel disk H. Simultaneously, idler E, which is also rotating counterclockwise, causes spur gear F to turn clockwise and engage the rollers on free-wheel disk G. Thus, shaft B is made to rotate clockwise. On the other hand, if the input shaft A turns counterclockwise

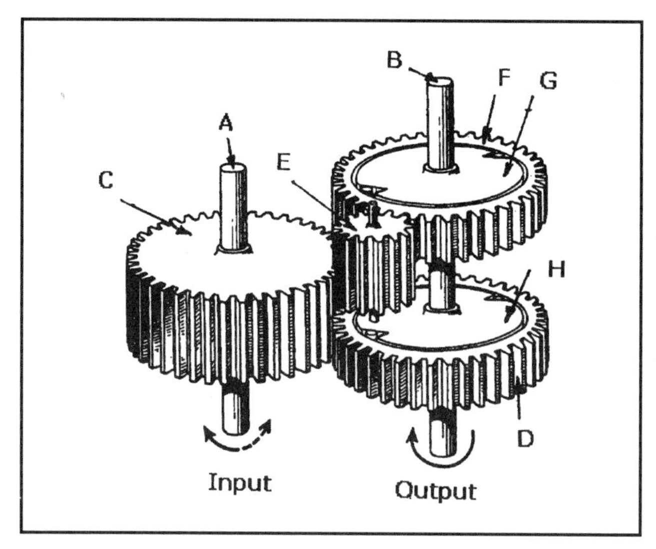

Fig. 28 The output shaft of this unidirectional drive always rotates in the same direction regardless of the direction of rotation of the input shaft.

(dotted arrow), spur gear F will idle while spur gear D engages free-wheel disk H, which drives shaft B so that it continues to rotate clockwise.

Gear Tooth Geometry

The geometry of gear teeth, as shown in Fig. 29, is determined by pitch, depth, and pressure angle.

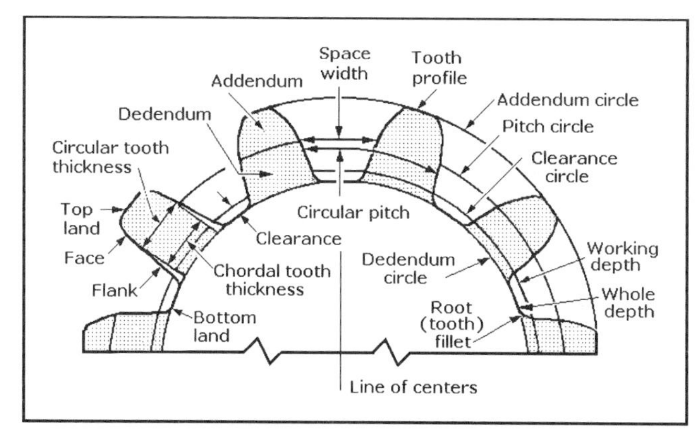

Fig. 29 Gear-tooth geometry.

Gear Terminology

addendum: The radial distance between the *top land* and the *pitch circle.* This distance is measured in inches or millimeters.

addendum circle: The circle defining the outer diameter of the gear.

circular pitch: The distance along the pitch circle from a point on one tooth to a corresponding point on an adjacent tooth. It is also the sum of the *tooth thickness* and the *space width.* This distance is measured in inches or millimeters.

clearance: The radial distance between the *bottom land* and the *clearance circle.* This distance is measured in inches or millimeters.

contact ratio: The ratio of the number of teeth in contact to the number of teeth not in contact.

dedendum: The radial distance between the *pitch circle* and the *dedendum circle.* This distance is measured in inches or millimeters.

dedendum circle: The theoretical circle through the *bottom lands* of a gear.

depth: A number standardized in terms of pitch. Full-depth teeth have a *working depth* of $2/P$. If the teeth have equal *addenda* (as in standard interchangeable gears), the addendum is $1/P$. Full-depth gear teeth have a larger contact ratio than stub teeth, and their working depth is about 20 percent more than stub gear teeth. Gears with a small number of teeth might require *undercutting* to prevent one interfering with another during engagement.

diametral pitch (P): The ratio of the number of teeth to the *pitch diameter.* A measure of the coarseness of a gear, it is the index of tooth size when U.S. units are used, expressed as teeth per inch.

pitch: A standard pitch is typically a whole number when measured as a *diametral pitch (P). Coarse pitch gears* have teeth larger than a diametral pitch of 20 (typically 0.5 to 19.99). *Fine-pitch gears* usually have teeth of diametral pitch greater than 20. The usual maximum fineness is 120 diametral pitch, but involute-tooth gears can be made with diametral pitches as fine as 200, and cycloidal tooth gears can be made with diametral pitches to 350.

pitch circle: A theoretical circle upon which all calculations are based.

pitch diameter: The diameter of the *pitch circle*, the imaginary circle that rolls without slipping with the pitch circle of the mating gear, measured in inches or millimeters.

pressure angle: The angle between the *tooth profile* and a line perpendicular to the *pitch circle*, usually at the point where the pitch circle and the tooth profile intersect. Standard angles are 20° and 25°. It affects the force that tends to separate mating gears. A high pressure angle decreases the *contact ratio*, but it permits the teeth to have higher capacity and it allows gears to have fewer teeth without *undercutting*.

Gear Dynamics Terminology

backlash: The amount by which the width of a tooth space exceeds the thickness of the engaging tooth measured on the pitch circle. It is the shortest distance between the noncontacting surfaces of adjacent teeth.

gear efficiency: The ratio of output power to input power taking into consideration power losses in the gears and bearings and from windage and the churning of the gear lubricant.

gear power: A gear's load and speed capacity. It is determined by gear dimensions and type. Helical and helical-type gears have capacities to approximately 30,000 hp, spiral bevel gears to about 5000 hp, and worm gears to about 750 hp.

gear ratio: The number of teeth in the larger gear of a pair divided by the number of teeth in the *pinion* gear (the smaller gear of a pair). It is also the ratio of the speed of the pinion to the speed of the gear. In reduction gears, the ratio of input speed to output speed.

gear speed: A value determined by a specific pitchline velocity. It can be increased by improving the accuracy of the gear teeth and the balance of all rotating parts.

undercutting: The recessing in the bases of gear tooth flanks to improve clearance.

PULLEYS AND BELTS

Pulleys and belts transfer rotating motion from one shaft to another. Essentially, pulleys are gears without teeth that depend on the frictional forces of connecting belts, chains, ropes, or cables to transfer torque. If both pulleys have the same diameter, they will rotate at the same speed. However, if one pulley is larger than the other, mechanical advantage and velocity ratio are gained. As with gears, the velocities of pulleys are inversely proportional to their diameters. A large drive pulley driving a smaller driven pulley by means of a belt or chain is shown in Fig. 30. The smaller pulley rotates faster than the larger pulley in the same direction as shown in Fig. 30a. If the belt is crossed, as shown in Fig. 30b, the smaller pulley also rotates faster than the larger pulley, but its rotation is in the opposite direction.

A familiar example of belt and pulley drive can be seen in automotive cooling fan drives. A smooth pulley connected to the engine crankshaft transfers torque to a second smooth pulley coupled to the cooling fan with a reinforced rubber endless belt. Before reliable direct-drive industrial electric motors were developed, a wide variety of industrial machines equipped with smooth pulleys of various diameters were driven by endless leather belts from an overhead driveshaft. Speed changes were achieved by switching the belt to pulleys of different diameters on the same machine. The machines included lathes and milling machines, circular saws in sawmills, looms in textile plants, and grinding wheels in grain mills. The source of power could have been a water wheel, windmill, or a steam engine.

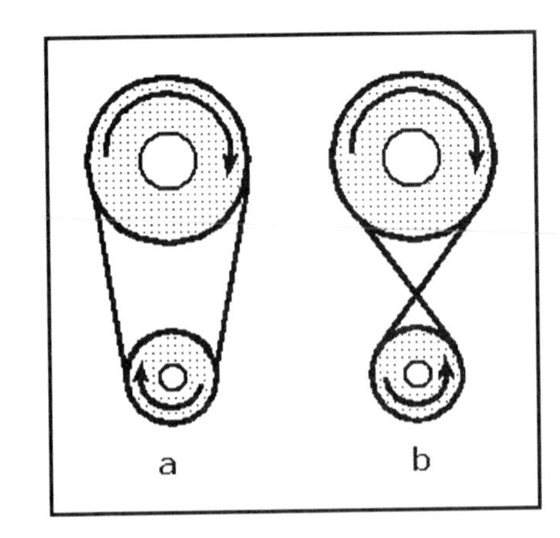

Fig. 30 Belts on pulleys: With a continuous belt both pulleys rotate in the same direction (a), but with a crossed belt both pulleys rotate in opposite directions (b).

SPROCKETS AND CHAINS

Sprockets and chains offer another method for transferring rotating motion from one shaft to another where the friction of a drive belt would be insufficient to transfer power. The speed relationships between sprockets of different diameters coupled by chains are the same as those between pulleys of different diameters coupled by belts, as shown in Fig. 30. Therefore, if the chains are crossed, the sprockets will rotate in different directions. Bicycles have sprocket and chain drives. The teeth on the sprockets mesh with the links on the chains. Powered winches on large ships act as sprockets because they have teeth that mate with the links of heavy chain for raising anchors. Another example can be seen in tracked equipment including bulldozers, cranes, and military tanks. The flexible treads have teeth that mate with teeth on driving sprockets that propel these machines.

CAM MECHANISMS

A *cam* is a mechanical component capable of transmitting motion to a follower by direct contact. In a cam mechanism, the cam is the driver and the driven member is called the *follower*. The follower can remain stationary, translate, oscillate, or rotate. The general form of a plane cam mechanism is illustrated in the kinematic diagram Fig. 31. It consists of two shaped members A and B with smooth, round, or elongated contact surfaces connected to a third body C. Either body A or body B can be the driver, while the other body is the follower. These shaped bodies can be replaced by an equivalent mechanism. Points 1 and 2 are pin-jointed at the centers of curvature of the contacting surfaces. If any change is made in the relative positions of bodies A and B, points 1 and 2 are shifted, and the links of the equivalent mechanisms have different lengths.

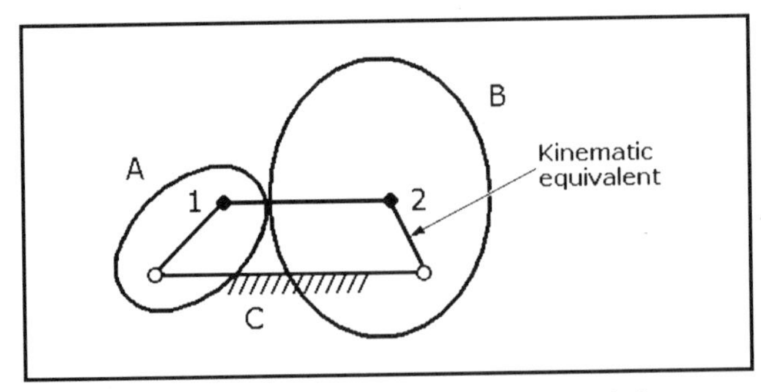

Fig. 31 Basic cam mechanism and its kinematic equivalent. Points 1 and 2 are centers of curvature of the contact point.

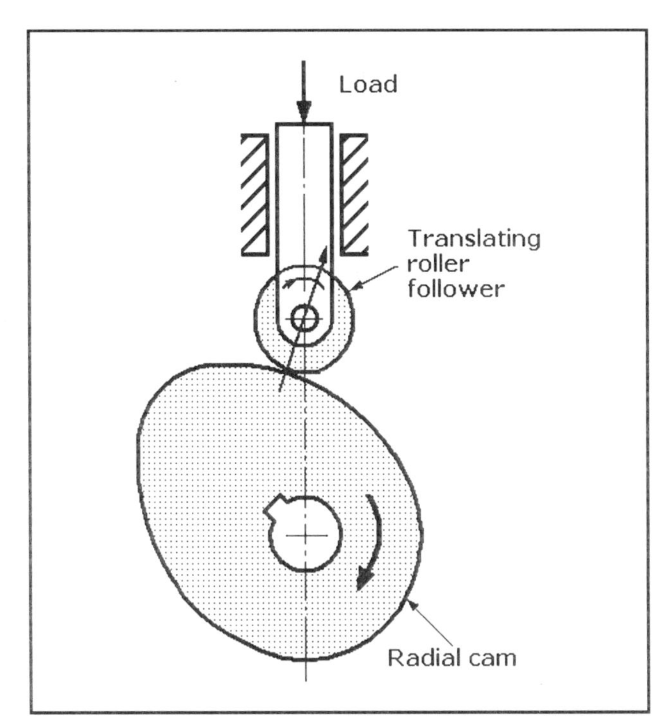

Fig. 32 Radial open cam with a translating roller follower. The roller is kept in contact with the cam by the mass of the load.

A widely used open radial-cam mechanism is shown in Fig. 32. The roller follower is the most common follower used in these mechanisms because it can transfer power efficiently between the cam and follower by reducing friction and minimizing wear between them. The arrangement shown here is called a *gravity constraint cam*; it is simple and effective and can be used with rotating disk or end cams if the weight of the follower system is enough to keep it in constant contact with the cam profile. However, in most practical cam mechanisms, the cam and follower are constrained at all operating speeds by preloaded compression springs. Cams can be designed by three methods:

- Shaping the cam body to some known curve, such as a spiral, parabola, or circular arc
- Designing the cam mathematically to determine follower motion and then plotting the tabulated data to form the cam
- Drawing the cam profile freehand using various drafting curves

The third method is acceptable only if the cam motion is intended for low speeds that will permit the use of a smooth, "bumpless" curve. In situations where higher loads, mass, speed, or elasticity of the members are encountered, a detailed study must be made of both the dynamic aspects of the cam curve and the accuracy of cam fabrication.

Many different kinds of machines include cams, particularly those that operate automatically such as printing presses, textile looms, gear-cutters, and screw machines. Cams open and close the valves in internal combustion engines, index cutting tools on machine tools, and operate switches and relays in electrical control equipment. Cams can be made in an infinite variety of shapes from metal or hard plastic. Some of the most important cams will be considered here. The possible applications of mechanical cams are still unlimited despite the introduction of *electronic cams* that mimic mechanical cam functions with appropriate computer software.

Classification of Cam Mechanisms

Cam mechanisms can be classified by their input/output motions, the configuration and arrangement of the follower, and the shape of the cam. Cams can also be classified by the kinds of motions made by the follower and the characteristics of the cam profile. The possible kinds of input/output motions of cam mechanisms with the most common disk cams are shown in Figs. 33a to e; they are examples of rotating disk cams with translating followers. By contrast, Fig. 33f shows a follower arm with a roller that swings or oscillates in a circular arc with respect to the follower hinge as the cam rotates. The follower configurations in Figs. 33a to d are named according to their characteristics: a *knife-edge;* b, e, and f *roller;* c *flat-faced;* and d *spherical-faced.* The face of the flat follower can also be oblique with respect to the cam. The follower is an element that moves either up and down or side to side as it follows the contour of the cam.

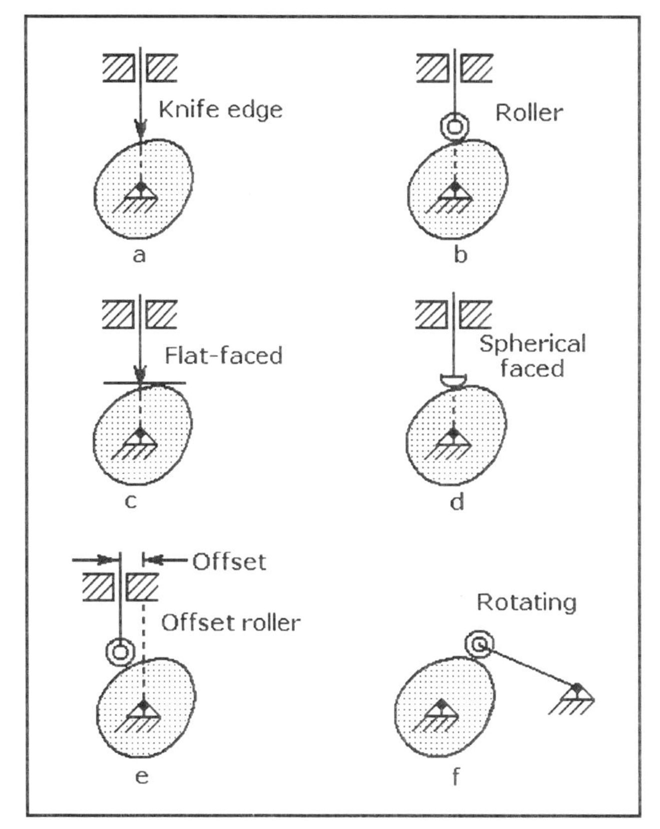

Fig. 33 Cam configurations: Six different configurations of radial open cams and their followers.

There are two basic types of follower: *in-line* and *offset*. The centerline of the in-line follower passes through the centerline of the camshaft. Figures 33a to d show five followers that move in a plane perpendicular to the axis of rotation of the camshaft. By contrast, the centerline of the offset follower, as illustrated in Fig. 33e, does not pass through the centerline of the camshaft. The amount of offset is the horizontal distance between the two centerlines. Follower offset reduces the side thrust introduced by the roller follower. Figure 33f illustrates a translating or swing-arm rotating follower that must be constrained to maintain contact with the cam profile.

The most common rotating disk or plate cams can be made in a variety of shapes including offset round, egg-shaped, oval, and cardioid or heart-shaped. Most cams are mounted on a rotating shaft. The cam and follower must be constrained at all operating speeds to keep them in close contact throughout its cycle if a cam mechanism is to function correctly. Followers are typically spring-loaded to maintain constant contact with the shaped surface of the cam, but gravity constraint is still an option.

If it is anticipated that a cam mechanism will be subjected to severe shock and vibration, a *grooved disk cam,* as shown in Fig. 34, can be used. The cam contour is milled into the face of a disk so that the roller of the cam follower will be confined and continuously constrained within the side walls of the groove throughout the cam cycle. The groove confines the follower roller during the entire cam rotation. Alternatively, the groove can be milled on the outer circumference of a cylinder or barrel to form a *cylindrical* or *barrel cam,* as shown in Fig. 35. The follower of this cam can translate or oscillate. A similar groove can also be milled around the conical exterior surface of a *grooved conical cam.*

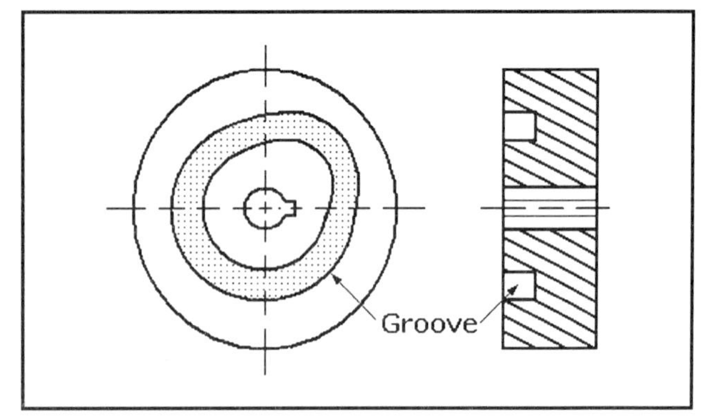

Fig. 34 Grooved cam made by milling a contoured cam groove into a metal or plastic disk. A roller follower is held within the grooved contour by its depth, eliminating the need for spring-loading.

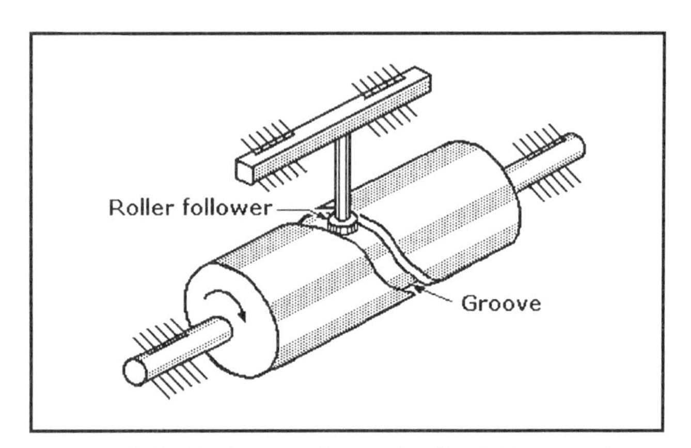

Fig. 35 Cylindrical or barrel cam: A roller follower tracks the groove precisely because of the deep contoured groove milled around the circumference of the rotating cylinder.

By contrast, the barrel-shaped *end cam,* shown in Fig. 36, has a contour milled on one end. This cam is usually rotated, and its follower can also either translate or oscillate, but the follower system must be carefully controlled to exercise the required constraint because the follower roller is not confined by a groove. Another distinct form of cam is the *translating cam,* as shown in Fig. 37. It is typically mounted on a bed or carrier that moves back and forth in a linear reciprocal motion under a stationary vertical translating follower, usually with a roller. However, the cam can also be mounted so that it remains stationary while a follower system moves in a linear reciprocal motion over the limited range of the cam.

The unusual dual-rotary cam configuration shown in Fig. 38 is a *constant-diameter cam*; it consists of two identical disk cams

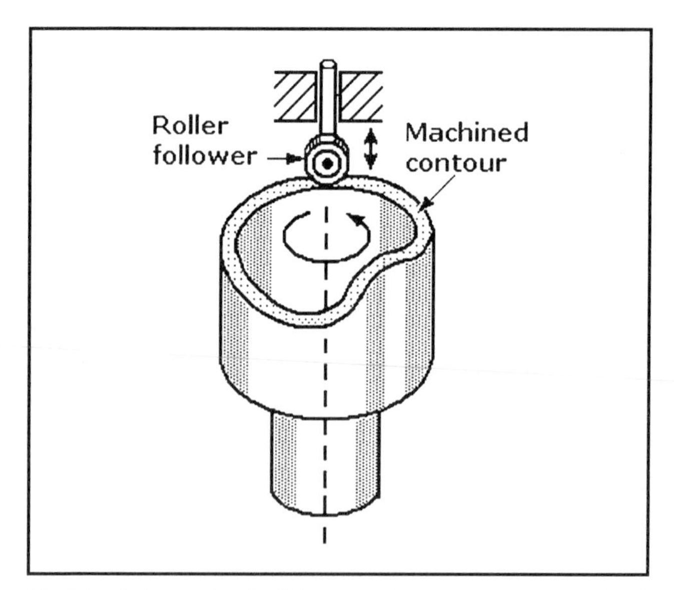

Fig. 36 End cam: A roller follower tracks a cam contour machined at the end of this rotating cylindrical cam.

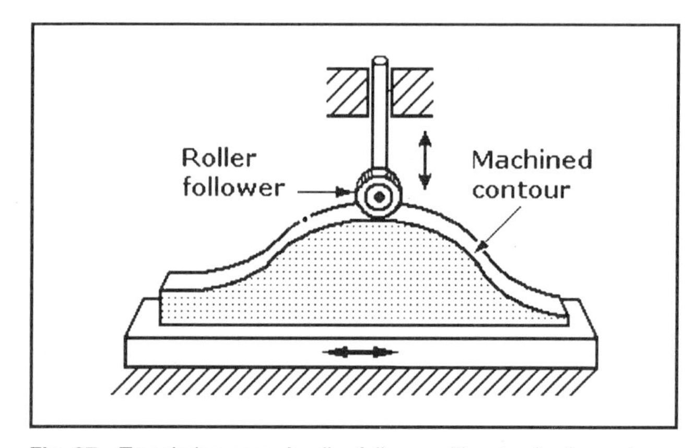

Fig. 37 Translating cam: A roller follower either tracks the reciprocating motion of the cam profile or is driven back and forth over a stationary cam profile.

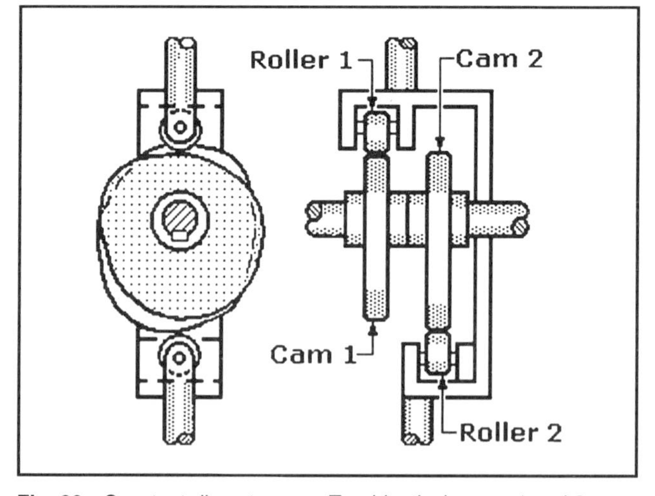

Fig. 38 Constant-diameter cam: Two identical cams, 1 and 2, are separated on the same shaft and offset at an angle that provides a virtual constant diameter. Cam 1 with roller follower 1 is the functioning cam, and cam 2 with roller follower 2 constrains cam 1 to smooth its motion.

with followers mounted a fixed distance apart on a common shaft, but the cams are offset so that if superimposed their contours form a virtual circle of constant diameter. Cam 1 is the functional cam while cam 2 acts as a constraint, effectively canceling out the irregular motion that occurs with a single rotary cam and follower.

The motions of the followers of all of these cam mechanisms can be altered to obtain a different sequence by changing the contour of the cam profile. The timing of the sequence of disk and cylinder cams can be changed by altering the rotational speed of their camshafts. The timing of the sequence of the translation cam can be changed by altering the rate of reciprocal motion of the bed on which it is mounted on its follower system. The rotation of the follower roller does not influence the motion of any of the cam mechanisms.

Cam Terminology

Figure 39 illustrates the nomenclature for a radial open disk cam with a *roller follower* on a plate cam.

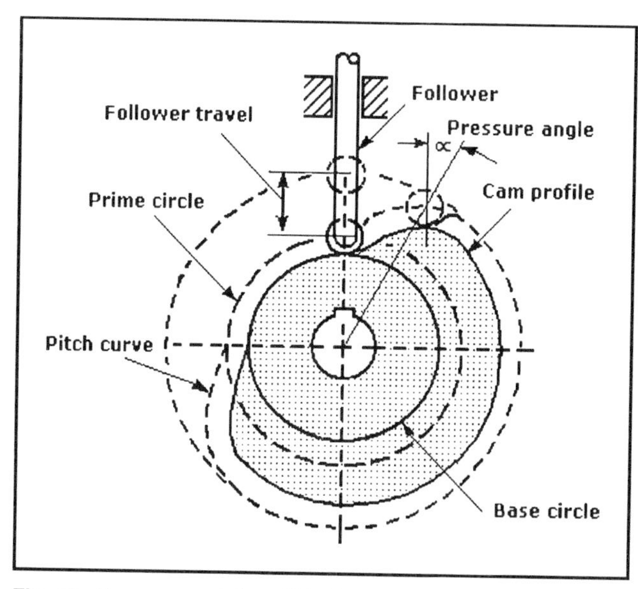

Fig. 39 Cam nomenclature: This diagram identifies the industry-accepted technical terms for cam features.

base circle: The circle with the shortest radius from the cam center to any part of the cam profile.

cam profile: The outer surface of a disk cam as it was machined.

follower travel: For a *roller follower* of a disk cam it is the vertical distance of follower travel measured at the center point of the roller as it travels from the *base circle* to the *cam profile.*

motion events: When a cam rotates through one cycle, the follower goes through rises, dwells, and returns. A *rise* is the motion of the follower away from the cam center; a *dwell* occurs when the follower is resting; and a *return* is the motion of the follower toward the cam center.

pitch curve: For a *roller follower* of a disk cam it is the path generated by the center point of the roller as the follower is rotated around a stationary plate cam.

pressure angle: For a *roller follower* of a disk cam it is the angle at any point between the normal to the pitch curve and the instantaneous direction of follower motion. This angle is important in cam design because it indicates the steepness of the cam profile.

prime circle (reference circle): For a *roller follower* of a disk cam it is the circle with the shortest radius from the cam center to the pitch curve.

stroke or throw: The longest distance or widest angle through which the follower moves or rotates.

working curve: The working surface of a cam that contacts the follower. For a *roller follower* of a plate cam it is the path traced by the center of the roller around the cam profile.

CLUTCH MECHANISMS

A *clutch* is defined as a coupling that connects and disconnects the driving and driven parts of a machine; an example is an engine and a transmission. Clutches typically contain a driving shaft and a driven shaft, and they are classed as either externally or internally controlled. *Externally controlled clutches* can be controlled either by friction surfaces or components that engage or mesh positively. *Internally controlled clutches* are controlled by internal mechanisms or devices; they are further classified as *overload, overriding,* and *centrifugal.* There are many different schemes for a driving shaft to engage a driven shaft.

Externally Controlled Friction Clutches

Friction-Plate Clutch. This clutch, shown in Fig. 40, has a control arm, which when actuated, advances a sliding plate on the driving shaft to engage a mating rotating friction plate on the same shaft; this motion engages associated gearing that drives the driven shaft. When reversed, the control arm disengages the sliding plate. The friction surface can be on either plate, but is typically only on one.

Cone Clutch. A clutch operating on the same principle as the friction-plate clutch except that the control arm advances a cone on the driving shaft to engage a mating rotating friction cone on the same shaft; this motion also engages any associated gearing that drives the driven shaft. The friction surface can be on either cone but is typically only on the sliding cone.

Expanding Shoe Clutch. This clutch is similar to the friction-plate clutch except that the control arm engages linkage that forces several friction shoes radially outward so they engage the inner surface of a drum on or geared to the driven shaft.

Externally Controlled Positive Clutches

Jaw Clutch. This clutch is similar to the plate clutch except that the control arm advances a sliding jaw on the driving shaft to make positive engagement with a mating jaw on the driven shaft.

Other examples of externally controlled positive clutches are the *planetary transmission clutch* consisting essentially of a sun gear keyed to a driveshaft, two planet gears, and an outer driven

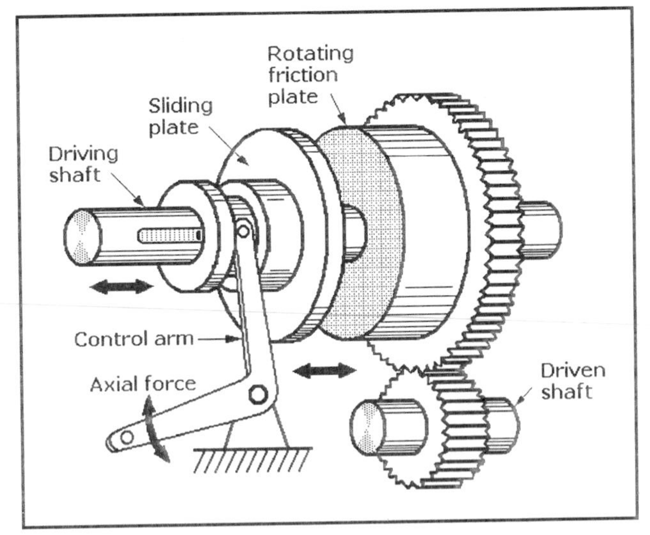

Fig. 40 Friction plate clutch: When the left sliding plate on the driving shaft is clamped by the control arm against the right friction plate idling on the driving shaft, friction transfers the power of the driving shaft to the friction plate. Gear teeth on the friction plate mesh with a gear mounted on the driven shaft to complete the transfer of power to the driven mechanism. Clutch torque depends on the axial force exerted by the control arm.

ring gear. The *pawl and ratchet clutch* consists essentially of a pawl-controlled driving ratchet keyed to a driven gear.

Internally Controlled Clutches

Internally controlled clutches can be controlled by springs, torque, or centrifugal force. The *spring and ball radial-detent clutch,* for example, disengages when torque becomes excessive, allowing the driving gear to continue rotating while the driveshaft stops rotating. The *wrapped-spring clutch* consists of two separate rotating hubs joined by a coil spring. When driven in the right direction, the spring tightens around the hubs increasing the friction grip. However, if driven in the opposite direction the spring relaxes, allowing the clutch to slip.

The *expanding-shoe centrifugal clutch* is similar to the externally controlled *expanding shoe clutch* except that the friction shoes are pulled in by springs until the driving shaft attains a preset speed. At that speed centrifugal force drives the shoes radially outward so that they contact the drum. As the driveshaft rotates faster, pressure between the shoes and drum increases, thus increasing clutch torque.

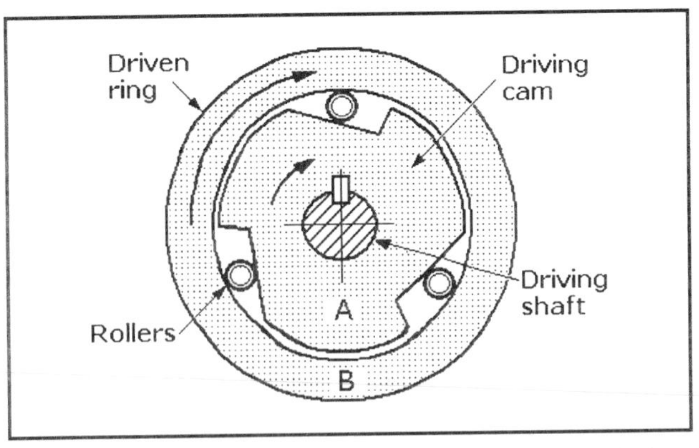

Fig. 41 Overrunning clutch: As driving cam A revolves clockwise, the rollers in the wedge-shaped gaps between cam A and outer ring B are forced by friction into those wedges and are held there; this locks ring B to cam A and drives it clockwise. However, if ring B is turned counterclockwise, or is made to revolve clockwise faster than cam A, the rollers are freed by friction, the clutch slips, and no torque is transmitted.

The *overrunning or overriding clutch,* as shown in Fig. 41, is a specialized form of a cam mechanism, also called a *cam and roller clutch.* The inner driving cam A has wedge-shaped notches on its outer rim that hold rollers between the outer surface of A and the inner cylindrical surfaces of outer driven ring B. When driving cam A is turning clockwise, frictional forces wedge the rollers tightly into the notches to lock outer driven ring B in position so it also turns in a clockwise direction. However, if driven ring B is reversed or runs faster clockwise than driving cam A (when it is either moving or immobile) the rollers are set free, the clutch will slip and no torque is transmitted. Some versions of this clutch include springs between the cam faces and the rollers to ensure faster clutching action if driven ring B attempts to drive driving cam A by overcoming residual friction. A version of this clutch is the basic free-wheel mechanism that drives the rear axle of a bicycle.

Some low-cost, light-duty overrunning clutches for one-direction-only torque transmission intersperse cardioid-shaped pellets called *sprags* with cylindrical rollers. This design permits cylindrical internal drivers to replace cammed drivers. The sprags bind in the concentric space between the inner driver and the outer driven ring if the ring attempts to drive the driver. The torque rating of the clutch depends on the number of sprags installed. For acceptable performance a minimum of three sprags, equally spaced around the circumference of the races, is usually necessary.

GLOSSARY OF COMMON MECHANICAL TERMS

acceleration: The time rate of change of velocity of a body. It is always produced by force acting on a body. Acceleration is measured as feet per second per second (ft/s^2) or meters per second per second (m/s^2).

component forces: The individual forces that are the equivalent of the resultant.

concurrent forces: Forces whose lines of action or directions pass through a common point or meet at a common point.

crank: A side link that revolves relative to the frame.

crank-rocker mechanism: A four-bar linkage characterized by the ability of the shorter side link to revolve through 380° while the opposing link rocks or oscillates.

couple: Two equal and opposite parallel forces that act at diametrically opposite points on a body to cause it to rotate around a point or an axis through its center.

displacement: Distance measured from a fixed reference point in a specified direction; it is a vector quantity; units are measured in inches, feet, miles, centimeters, meters, and kilometers.

double-crank mechanism: A four-bar linkage characterized by the ability of both of its side links to oscillate while the shortest link (opposite the foundation link) can revolve through 360°.

dynamics: The study of the forces that act on bodies not in equilibrium, both balanced and unbalanced; it accounts for the masses and accelerations of the parts as well as the external forces acting on the mechanisms: It is a combination of *kinetics* and *kinematics*.

efficiency of machines: The ratio of a machine's output divided by its input is typically expressed as a percent. There are energy or power losses in all moving machinery caused primarily by friction. This causes inefficiency, so a machine's output is always less than its input; both output and input must be expressed in the same units of power or energy. This ratio, always a fraction, is multiplied by 100 to obtain a percent. It can also be determined by dividing the machine's mechanical advantage by its velocity ratio and multiplying that ratio by 100 to get a percent.

energy: A physical quantity present in three-dimensional space in which forces can act on a body or particle to bring about physical change; it is the capacity for doing work. Energy can take many forms, including mechanical, electrical, electromagnetic, chemical, thermal, solar, and nuclear. Energy and work are related and measured in the same units: foot-pounds, ergs, or joules; it cannot be destroyed, but it can be wasted.

- *Kinetic energy* is the kind of energy a body has when it is in motion. Examples are a rolling soccer ball, a speeding automobile, or a flying airplane.
- *Potential energy* is the kind of energy that a body has because of its position or state. Examples are a concrete block poised at the edge of a building, a shipping container suspended above ground by a crane, or a roadside bomb.

equilibrium: In mechanics, a condition of balance or static equilibrium between opposing forces. An example is when there are equal forces at both ends of a seesaw resting on a *fulcrum*.

force: Strength or energy acting on a body to push or pull it; it is required to produce acceleration. Except for gravitation, one body cannot exert a force on another body unless the two are in contact. The Earth exerts a force of attraction on bodies, whether they are in contact or not. Force is measured in poundals (lb-ft/s^2) or newtons (kg-m/s^2).

fulcrum: A pivot point or edge about which objects are free to rotate.

kinematic chain: A combination of links and pairs without a fixed link.

kinematics: The study of the motions of bodies without considering how the variables of force and mass influence the motion. It is described as the geometry of motion.

kinetics: The study of the effects of external forces including gravity upon the motions of physical bodies.

linear motion: Motion in a straight line. An example is when a car is driving on a straight road.

lever: A simple machine that uses opposing torque around a fulcrum to perform work.

link: A rigid body with pins or fasteners at its ends to connect it to other rigid bodies so it can transmit a force or motion. All machines contain at least one link, either in a fixed position relative to the Earth or capable of moving the machine and the link during the motion; this link is the *frame* or *fixed link* of the machine.

linkages: Mechanical assemblies consisting of two or more levers connected to produce a desired motion. They can also be mechanisms consisting of rigid bodies and lower *pairs*.

machine: An assembly of mechanisms or parts or mechanisms capable of transmitting force, motion, and energy from a power source; the objective of a machine is to overcome some form of resistance to accomplish a desired result. There are two functions of machines: (1) the transmission of relative motion and (2) the transmission of force; both require that the machine be strong and rigid. While both machines and mechanisms are combinations of rigid bodies capable of definite relative motions, machines transform energy, but mechanisms do not. A *simple machine* is an elementary mechanism. Examples are the lever, wheel and axle, pulley, inclined plane, wedge, and screw.

machinery: A term generally meaning various combinations of machines and mechanisms.

mass: The quantity of matter in a body indicating its inertia. Mass also initiates gravitational attraction. It is measured in ounces, pounds, tons, grams, and kilograms.

mechanical advantage: The ratio of the load (or force W) divided by the effort (or force F) exerted by an operator. If friction is considered in determining mechanical advantage, or it has been determined by the actual testing, the ratio W/F is the mechanical advantage MA. However, if the machine is assumed to operate without friction, the ratio W/F is the theoretical mechanical advantage TA. Mechanical advantage and velocity ratio are related.

mechanics: A branch of physics concerned with the motions of objects and their response to forces. Descriptions of mechanics begin with definitions of such quantities as acceleration, displacement, force, mass, time, and velocity.

mechanism: In mechanics, it refers to two or more rigid or resistant bodies connected together by joints so they exhibit definite relative motions with respect to one another. Mechanisms are divided into two classes:

- *Planar*: Two-dimensional mechanisms whose relative motions are in one plane or parallel planes.
- *Spatial*: Three-dimensional mechanisms whose relative motions are not all in the same or parallel planes.

moment of force or torque: The product of the force acting to produce a turning effect and the perpendicular distance of its line of action from the point or axis of rotation. The perpendicular distance is called the *moment* arm or the *lever arm torque*. It is measured in pound-inches (lb-in.), pound-feet (lb-ft), or newton-meters (N-m).

moment of inertia: A physical quantity giving a measure of the *rotational inertia* of a body about a specified axis of rotation; it depends on the mass, size, and shape of the body.

nonconcurrent forces: Forces whose lines of action do not meet at a common point.

noncoplanar forces: Forces that do not act in the same plane.

oscillating motion: Repetitive forward and backward circular motion such as that of a clock pendulum.

pair: A joint between the surfaces of two rigid bodies that keeps them in contact and relatively movable. It might be as simple as a pin, bolt, or hinge between two links or as complex as a universal joint between two links. There are two kinds of pairs in mechanisms classified by the type of contact between the two bodies of the pair: *lower pairs* and *higher pairs*.

- Lower pairs are *surface-contact pairs* classed either as *revolute* or *prismatic*. Examples: a hinged door is a revolute pair and a sash window is a prismatic pair.
- Higher pairs include *point, line, or curve pairs*. Examples: paired rollers, cams and followers, and meshing gear teeth.

power: The time rate of doing work. It is measured in foot-pounds per second (ft-lb/s), foot-pounds per minute (ft-lb/min), horsepower, watts, kilowatts, newton-meters/s, ergs/s, and joules/s.

reciprocating motion: Repetitive back and forth linear motion as that of a piston in an internal combustion engine.

rotary motion: Circular motion as in the turning of a bicycle wheel.

resultant: In a system of forces, it is the single force equivalent of the entire system. When the resultant of a system of forces is zero, the system is in equilibrium.

skeleton outline: A simplified geometrical line drawing showing the fundamentals of a simple machine devoid of the actual details of its construction. It gives all of the geometrical information needed for determining the relative motions of the main links. The relative motions of these links might be complete circles, semicircles, or arcs, or even straight lines.

statics: The study of bodies in equilibrium, either at rest or in uniform motion.

torque: An alternative name for *moment of force*.

velocity: The time rate of change with respect to distance. It is measured in feet per second (ft/s), feet per minute (ft/min), meters per second (m/s), or meters per minute (m/min).

velocity ratio: A ratio of the distance movement of the effort divided by the distance of movement of the load per second for a machine. This ratio has no units.

weight: The force on a body due to the gravitational attraction of the Earth; weight W = mass $n \times$ acceleration g due to the Earth's gravity; mass of a body is constant but g, and therefore W vary slightly over the Earth's surface.

work: The product of force and distance: the distance an object moves in the direction of force. Work is not done if the force exerted on a body fails to move that body. Work, like energy, is measured in units of ergs, joules, or foot-pounds.

CHAPTER 2

MOTION CONTROL SYSTEMS

MOTION CONTROL SYSTEMS OVERVIEW

Introduction

A modern motion control system typically consists of a motion controller, a motor drive or amplifier, an electric motor, and feedback sensors. The system might also contain other components such as one or more belt-, ballscrew-, or leadscrew-driven linear guides or axis stages. A motion controller today can be a stand-alone programmable controller, a personal computer containing a motion control card, or a programmable logic controller (PLC).

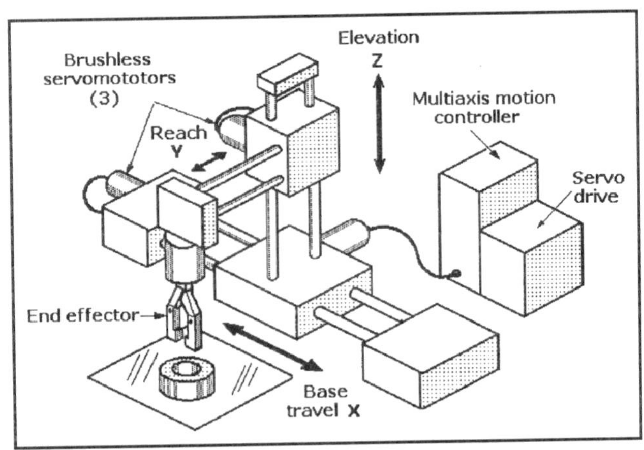

Fig. 1 This multiaxis X-Y-Z motion platform is an example of a motion control system.

All of the components of a motion control system must work together seamlessly to perform their assigned functions. Their selection must be based on both engineering and economic considerations. Figure 1 illustrates a typical multiaxis X-Y-Z motion platform that includes the three linear axes required to move a load, tool, or end effector precisely through three degrees of freedom. With additional mechanical or electromechanical components on each axis, rotation about the three axes can provide up to six degrees of freedom, as shown in Fig. 2.

Motion control systems today can be found in such diverse applications as materials handling equipment, machine tool centers, chemical and pharmaceutical process lines, inspection stations, robots, and injection molding machines.

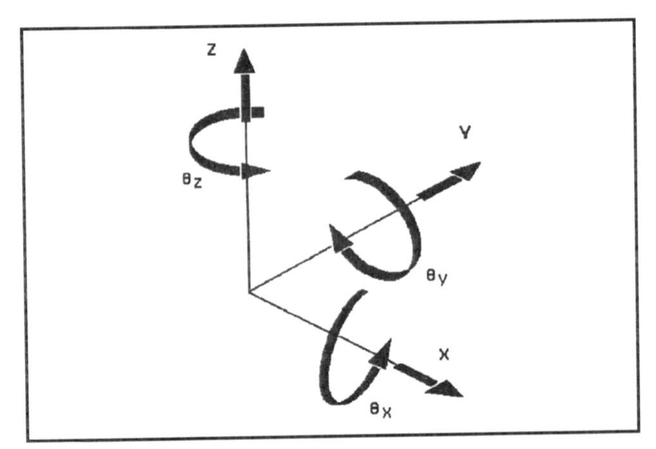

Fig. 2 The right-handed coordinate system showing six degrees of freedom.

Merits of Electric Systems

Most motion control systems today are powered by electric motors rather than hydraulic or pneumatic motors or actuators because of the many benefits they offer:

- More precise load or tool positioning, resulting in fewer product or process defects and lower material costs
- Quicker changeovers for higher flexibility and easier product customizing
- Increased throughput for higher efficiency and capacity
- Simpler system design for easier installation, programming, and training
- Lower downtime and maintenance costs
- Cleaner, quieter operation without oil or air leakage

Electric-powered motion control systems do not require pumps or air compressors, and they do not have hoses or piping that can leak hydraulic fluids or air. This discussion of motion control is limited to electric-powered systems.

Motion Control Classification

Motion control systems can be classified as *open-loop* or *closed-loop*. An open-loop system does not require that measurements of any output variables be made to produce error-correcting signals; by contrast, a closed-loop system requires one or more feedback sensors that measure and respond to errors in output variables.

Closed-Loop System

A *closed-loop motion control system*, as shown in block diagram Fig. 3, has one or more feedback loops that continuously compare the system's response with input commands or settings to correct errors in motor and/or load speed, load position, or motor torque. Feedback sensors provide the electronic signals for correcting deviations from the desired input commands. Closed-loop systems are also called servosystems.

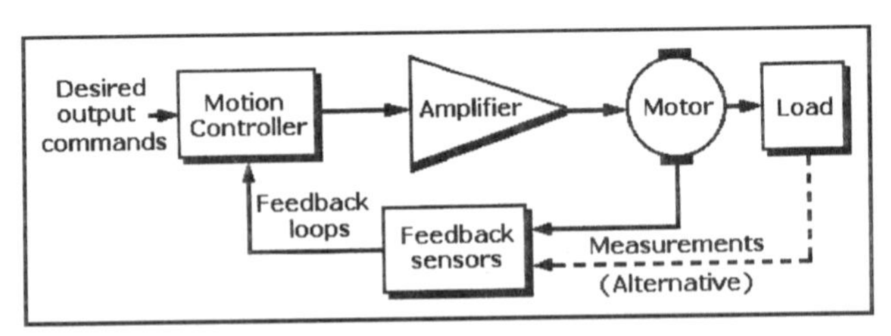

Fig. 3 Block diagram of a basic closed-loop control system.

Each motor in a servosystem requires its own feedback sensors, typically encoders, resolvers, or tachometers, that close loops around the motor and load. Variations in velocity, position, and torque are typically caused by variations in load conditions, but changes in ambient temperature and humidity can also affect load conditions.

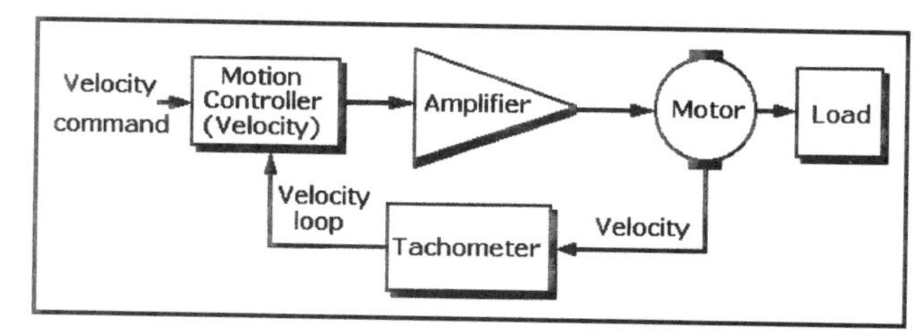

Fig. 4 Block diagram of a velocity-control system.

A *velocity control loop,* as shown in block diagram Fig. 4, typically contains a tachometer that is able to detect changes in motor speed. This sensor produces error signals that are proportional to the positive or negative deviations of motor speed from its preset value. These signals are sent to the motion controller so that it can compute a corrective signal for the amplifier to keep motor speed within those preset limits despite load changes.

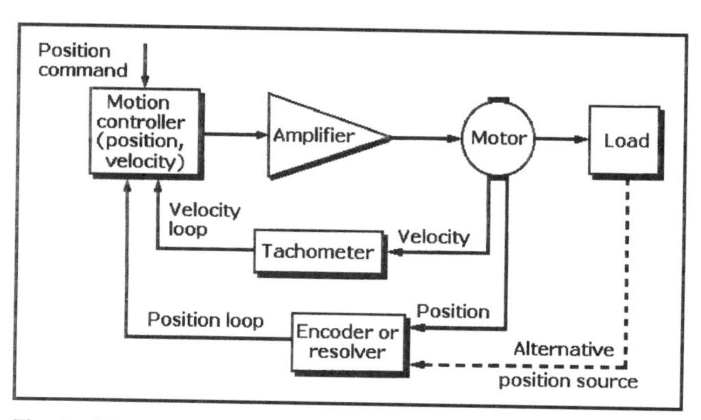

Fig. 5 Block diagram of a position-control system.

A *position-control loop,* as shown in block diagram Fig. 5, typically contains either an encoder or resolver capable of direct or indirect measurements of load position. These sensors generate error signals that are sent to the motion controller, which produces a corrective signal for amplifier. The output of the amplifier causes the motor to speed up or slow down to correct the position of the load. Most position control closed-loop systems also include a velocity-control loop.

The *ballscrew slide mechanism,* shown in Fig. 6, is an example of a mechanical system that carries a load whose position must be controlled in a closed-loop servosystem because it is not equipped with position sensors. Three examples of feedback sensors mounted on the ballscrew mechanism that can provide position feedback are shown in Fig. 7: (a) is a rotary optical encoder mounted on the motor housing with its shaft coupled to the motor shaft; (b) is an optical linear encoder with its graduated scale mounted on the base of the mechanism; and (c) is the less commonly used but more accurate and expensive laser interferometer.

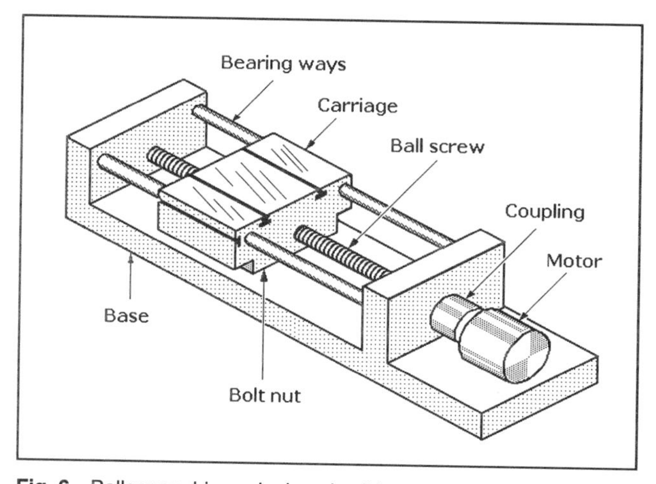

Fig. 6 Ballscrew-driven single-axis slide mechanism without position feedback sensors.

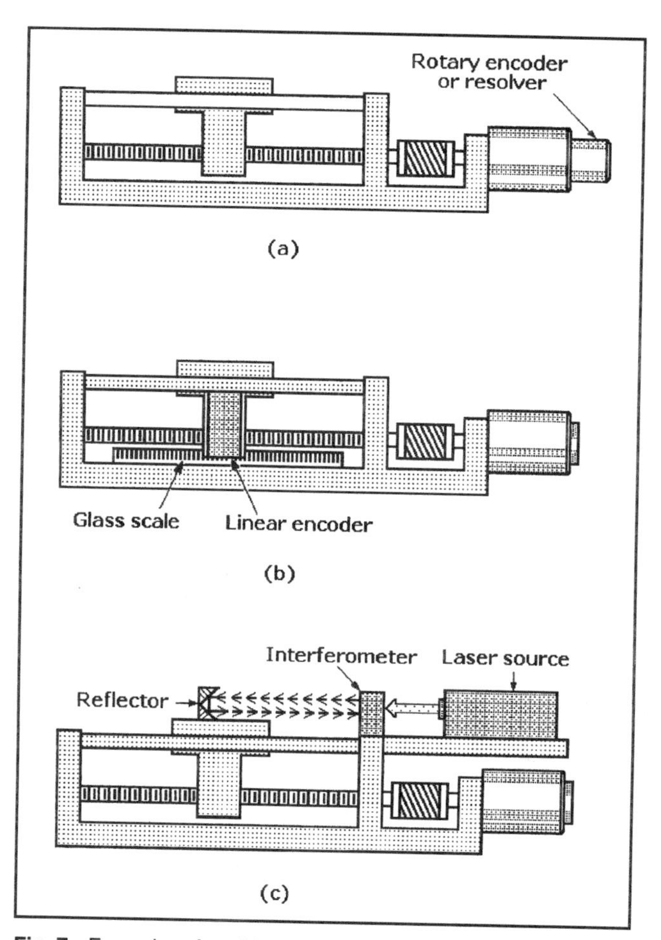

Fig. 7 Examples of position feedback sensors installed on a ballscrew-driven slide mechanism: (a) rotary encoder, (b) linear encoder, and (c) laser interferometer.

A *torque-control loop* contains electronic circuitry that measures the input current applied to the motor and compares it with a value proportional to the torque required to perform the desired task. An error signal from the circuit is sent to the motion controller, which computes a corrective signal for the motor amplifier to keep motor current, and hence torque, constant. Torque-control loops are widely used in machine tools where the load can change due to variations in the density of the material being machined or the sharpness of the cutting tools.

Trapezoidal Velocity Profile

If a motion control system is to achieve smooth, high-speed motion without overstressing the servomotor, the motion controller must command the motor amplifier to ramp up motor velocity gradually until it reaches the desired speed and then ramp it down gradually until it stops after the task is complete. This keeps motor acceleration and deceleration within limits.

The trapezoidal profile, shown in Fig. 8, is widely used because it accelerates motor velocity along a positive linear "upramp" until the desired constant velocity is reached. When the motor is shut down from the constant velocity setting, the profile decelerates velocity along a negative "down ramp" until

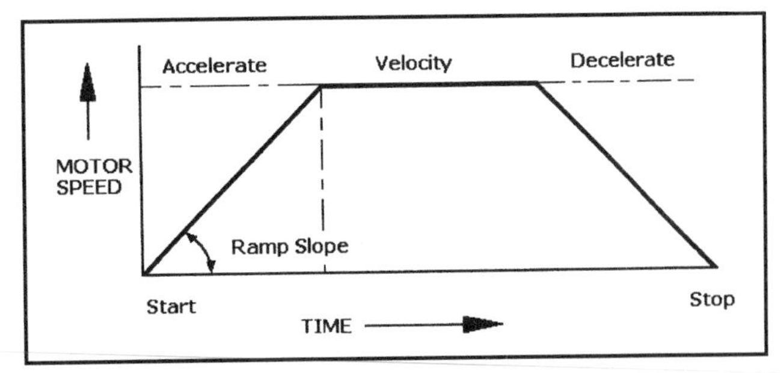

Fig. 8 Servomotors are accelerated to constant velocity and decelerated along a trapezoidal profile to assure efficient operation.

the motor stops. Amplifier current and output voltage reach maximum values during acceleration, then step down to lower values during constant velocity and switch to negative values during deceleration.

Closed-Loop Control Techniques

The simplest form of feedback is *proportional control*, but there are also *derivative* and *integral control* techniques, which compensate for certain steady-state errors that cannot be eliminated from proportional control. All three of these techniques can be combined to form *proportional-integral-derivative (PID) control.*

- In *proportional control* the signal that drives the motor or actuator is directly proportional to the linear difference between the input command for the desired output and the measured actual output.
- In *integral control* the signal driving the motor equals the *time integral* of the difference between the input command and the measured actual output.
- In *derivative control* the signal that drives the motor is proportional to the *time derivative* of the difference between the input command and the measured actual output.
- In *proportional-integral-derivative (PID) control* the signal that drives the motor equals the weighted sum of the difference, the time integral of the difference, and the time derivative of the difference between the input command and the measured actual output.

Open-Loop Motion Control Systems

A typical *open-loop motion control system* includes a stepper motor with a programmable indexer or pulse generator and motor driver, as shown in Fig. 9. This system does not need feedback sensors because load position and velocity are controlled by the predetermined number and direction of input digital pulses sent to the motor driver from the controller. Because load position is not continuously sampled by a feedback sensor (as in a closed-loop servosystem), load positioning accuracy is lower and position errors (commonly called step errors) accumulate over time. For these reasons open-loop systems are most often specified in applications where the load remains constant, load motion is simple, and low positioning speed is acceptable.

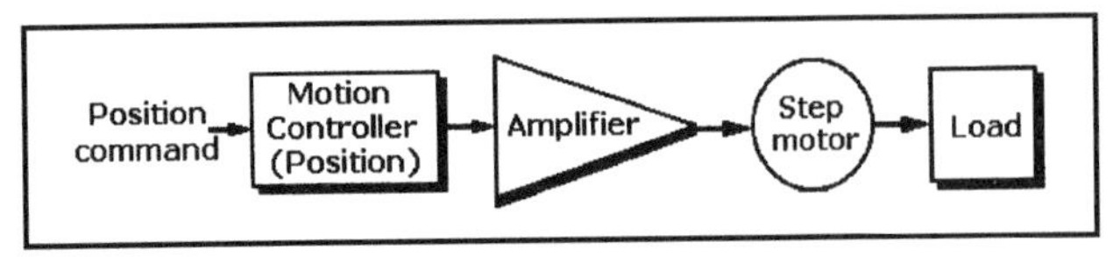

Fig. 9 Block diagram of an open-loop motion control system.

Kinds of Controlled Motion

There are five different kinds of motion control: *point-to-point, sequencing, speed, torque,* and *incremental.*

- In *point-to-point motion control* the load is moved between a sequence of numerically defined positions where it is stopped before it is moved to the next position. This is done at a constant speed, with both velocity and distance monitored by the motion controller. Point-to-point positioning can be performed in single-axis or multiaxis systems with servomotors in closed loops or stepping motors in open loops. X-Y tables and milling machines position their loads by multiaxis point-to-point control.
- *Sequencing control* is the control of such functions as opening and closing valves in a preset sequence or starting and stopping a conveyor belt at specified stations in a specific order.
- *Speed control* is the control of the velocity of the motor or actuator in a system.
- *Torque control* is the control of motor or actuator current so that torque remains constant despite load changes.
- *Incremental motion control* is the simultaneous control of two or more variables such as load location, motor speed, or torque.

Motion Interpolation

When a load under control must follow a specific path to get from its starting point to its stopping point, the movements of the axes must be coordinated or interpolated. There are three kinds of interpolation: *linear, circular, and contouring.*

Linear interpolation is the ability of a motion control system having two or more axes to move the load from one point to another in a straight line. The motion controller must determine the speed of each axis so that it can coordinate their movements. True linear interpolation requires that the motion controller modify axis acceleration, but some controllers approximate true linear interpolation with programmed acceleration profiles. The path can lie in one plane or be three dimensional.

Circular interpolation is the ability of a motion control system having two or more axes to move the load around a circular trajectory. It requires that the motion controller modify load acceleration while it is in transit. Again the circle can lie in one plane or be three dimensional.

Contouring is the path followed by the load, tool, or end-effector under the coordinated control of two or more axes. It requires that the motion controller change the speeds on different axes so that their trajectories pass through a set of predefined points. Load speed is determined along the trajectory, and it can be constant except during starting and stopping.

Computer-Aided Emulation

Several important types of programmed computer-aided motion control can emulate mechanical motion and eliminate the need for actual gears or cams. *Electronic gearing* is the control by software of one or more axes to impart motion to a load, tool, or end effector that simulates the speed changes that can be performed by actual gears. *Electronic camming* is the control by software of one or more axes to impart a motion to a load, tool, or end effector that simulates the motion changes that are typically performed by actual cams.

Mechanical Components

The mechanical components in a motion control system can be more influential in the design of the system than the electronic circuitry used to control it. Product flow and throughput, human operator requirements, and maintenance issues help to determine

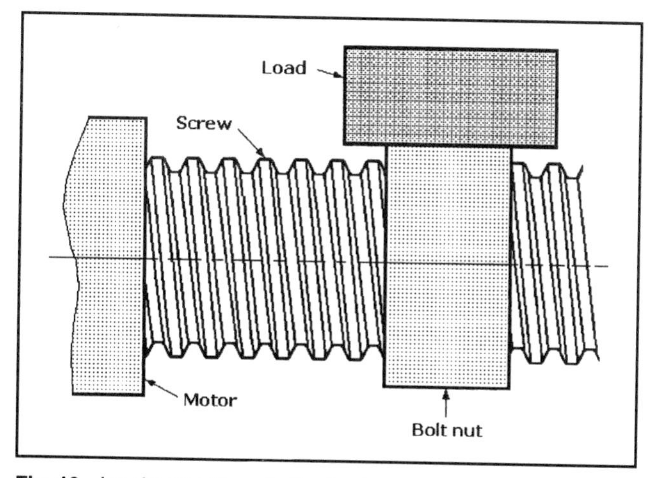

Fig. 10 Leadscrew drive: As the leadscrew rotates, the load is translated in the axial direction of the screw.

the mechanics, which in turn influence the motion controller and software requirements.

Mechanical actuators convert a motor's rotary motion into linear motion. Mechanical methods for accomplishing this include the use of leadscrews, shown in Fig. 10, ballscrews, shown in Fig. 11, worm-drive gearing, shown in Fig. 12, and belt, cable, or chain drives. Method selection is based on the relative costs of the alternatives and consideration for the possible effects of backlash. All actuators have finite levels of torsional and axial stiffness that can affect the system's frequency response characteristics.

Linear guides or stages constrain a translating load to a single degree of freedom. The linear stage supports the mass of the load

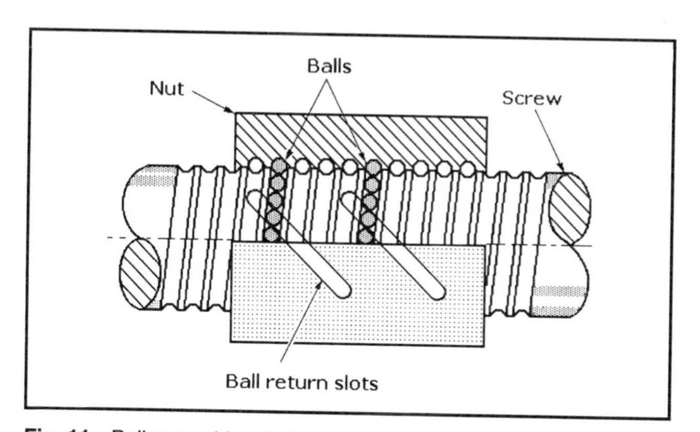

Fig. 11 Ballscrew drive: Ballscrews use recirculating balls to reduce friction and gain higher efficiency than conventional leadscrews.

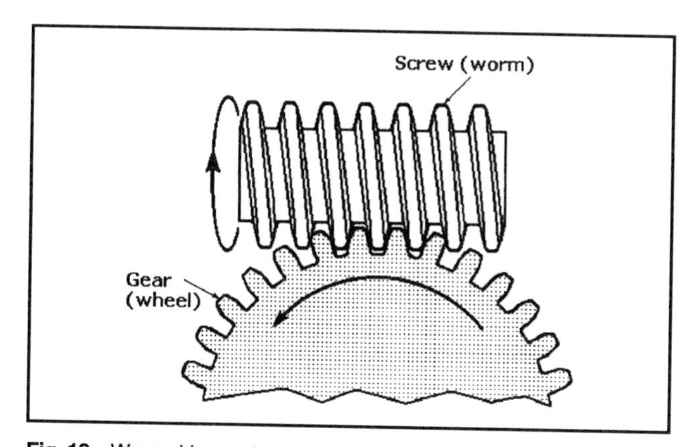

Fig. 12 Worm-drive systems can provide high speed and high torque.

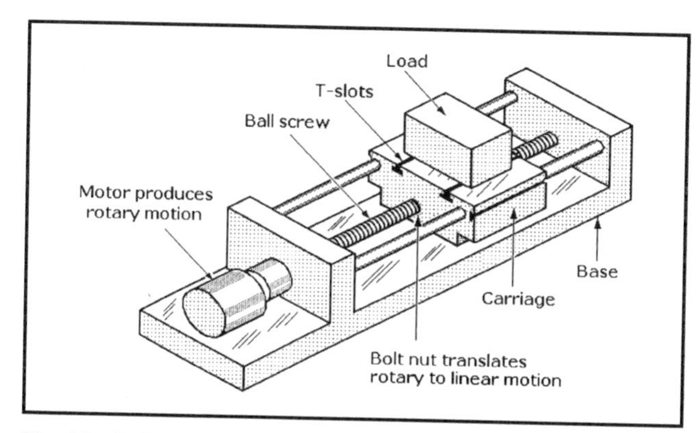

Fig. 13 Ballscrew-driven single-axis slide mechanism translates rotary motion into linear motion.

to be actuated and assures smooth, straight-line motion while minimizing friction. A common example of a linear stage is a ballscrew-driven single-axis stage, illustrated in Fig. 13. The motor turns the ballscrew, and its rotary motion is translated into the linear motion that moves the carriage and load by the stage's bolt nut. The bearing ways act as linear guides. As shown in Fig. 7, these stages can be equipped with sensors such as a rotary or linear encoder or a laser interferometer for feedback.

A ballscrew-driven single-axis stage with a rotary encoder coupled to the motor shaft provides an indirect measurement. This method ignores the tolerance, wear, and compliance in the mechanical components between the carriage and the position encoder that can cause deviations between the desired and true positions. Consequently, this feedback method limits position accuracy to ballscrew accuracy, typically ±5 to 10 μm per 300 mm.

Other kinds of single-axis stages include those containing antifriction rolling elements such as recirculating and nonrecirculating balls or rollers, sliding (friction contact) units, air-bearing units, hydrostatic units, and magnetic levitation (Maglev) units.

A single-axis air-bearing guide or stage is shown in Fig. 14. Some models being offered are 3.9 ft (1.2 m) long and include a carriage for mounting loads. When driven by a linear servomotor the loads can reach velocities of 9.8 ft/s (3 m/s). As shown in Fig. 7, these stages can be equipped with feedback devices such as cost-effective linear encoders or ultrahigh-resolution laser interferometers. The resolution of this type of stage with a noncontact linear encoder can be as fine as 20 nm and accuracy can be ±1 μm. However, these values can be increased to 0.3 nm resolution and submicron accuracy if a laser interferometer is installed.

The pitch, roll, and yaw of air-bearing stages can affect their resolution and accuracy. Some manufacturers claim ±1 arc-s per

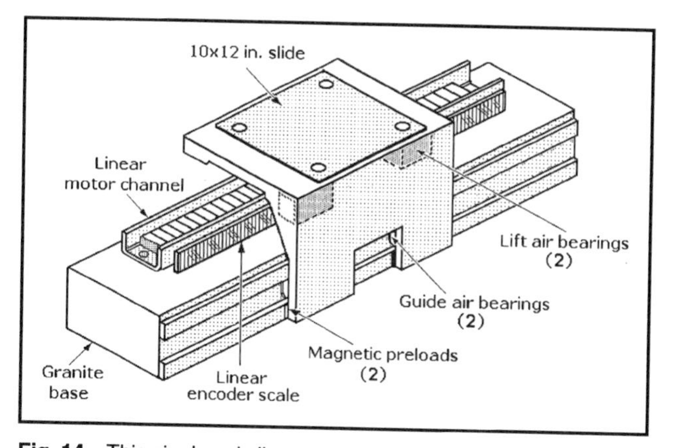

Fig. 14 This single-axis linear guide for load positioning is supported by air bearings as it moves along a granite base.

100 mm as the limits for each of these characteristics. Large air-bearing surfaces provide excellent stiffness and permit large load-carrying capability.

The important attributes of all these stages are their dynamic and static friction, rigidity, stiffness, straightness, flatness, smoothness, and load capacity. Also considered is the amount of work needed to prepare the host machine's mounting surface for their installation.

The structure on which the motion control system is mounted directly affects the system's performance. A properly designed base or host machine will be highly damped and act as a compliant barrier to isolate the motion system from its environment and minimize the impact of external disturbances. The structure must be stiff enough and sufficiently damped to avoid resonance problems. A high static mass to reciprocating mass ratio can also prevent the motion control system from exciting its host structure to harmful resonance.

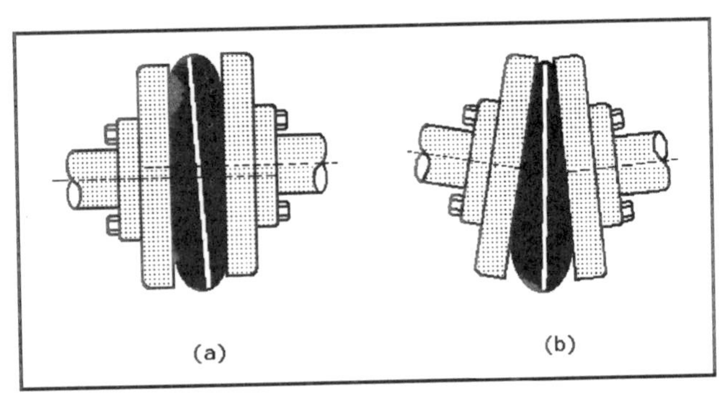

Fig. 15 Flexible shaft couplings adjust for and accommodate parallel misalignment (a) and angular misalignment between rotating shafts (b).

Any components that move will affect a system's response by changing the amount of inertia, damping, friction, stiffness, or resonance. For example, a flexible shaft coupling, as shown in Fig. 15, will compensate for minor parallel (a) and angular (b) misalignment between rotating shafts. Flexible couplings are available in other configurations such as bellows and helixes, as shown in Fig. 16. The bellows configuration (a) is acceptable for light-duty applications where misalignments can be as great as 9° angular or $\frac{1}{4}$ in. parallel. By contrast, helical couplings (b) prevent backlash at constant velocity with some misalignment, and they can also be run at high speed.

Other moving mechanical components include cable carriers that retain moving cables, end stops that restrict travel, shock absorbers to dissipate energy during a collision, and way covers to keep out dust and dirt.

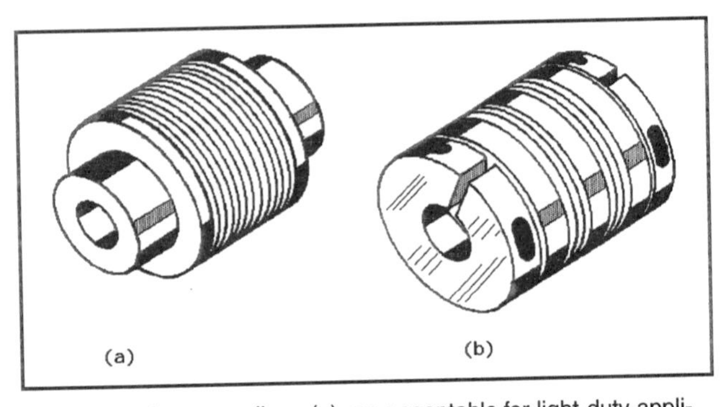

Fig. 16 Bellows couplings (a) are acceptable for light-duty applications. Misalignments can be 9° angular or $\frac{1}{4}$ in. parallel. Helical couplings (b) prevent backlash and can operate at constant velocity with misalignment and be run at high speed.

Electronic System Components

The motion controller is the "brain" of the motion control system and performs all of the required computations for motion path planning, servo-loop closure, and sequence execution. It is essentially a computer dedicated to motion control that has been programmed by the end user for the performance of assigned tasks. The motion controller produces a low-power motor command signal in either a digital or analog format for the motor driver or amplifier.

Significant technical developments have led to the increased acceptance of programmable motion controllers over the past 5 to 10 years: These include the rapid decrease in the cost of microprocessors as well as dramatic increases in their computing power. Added to that are the decreasing cost of more advanced semiconductor and disk memories. During the past 5 to 10 years, the capability of these systems to improve product quality, increase throughput, and provide just-in-time delivery has improved significantly.

The motion controller is the most critical component in the system because of its dependence on software. By contrast, the selection of most motors, drivers, feedback sensors, and associated mechanisms is less critical because they can usually be changed during the design phase or even later in the field with less impact on the characteristics of the intended system. However, making field changes can be costly in terms of lost productivity.

The decision to install any of the three kinds of motion controllers should be based on their ability to control both the number and types of motors required for the application as well as the availability of the software that will provide the optimum performance for the specific application. Also to be considered are the system's multitasking capabilities, the number of input/output (I/O) ports required, and the need for such features as linear and circular interpolation and electronic gearing and camming.

In general, a motion controller receives a set of operator instructions from a host or operator interface and it responds with corresponding command signals for the motor driver or drivers that control the motor or motors driving the load.

Motor Selection

The most popular motors for motion control systems are stepping or stepper motors and permanent-magnet (PM) DC brush-type and brushless DC servomotors. Stepper motors are selected for systems because they can run open-loop without feedback sensors. These motors are indexed or partially rotated by digital pulses that turn their rotors a fixed fraction or a revolution where they will be clamped securely by their inherent holding torque. Stepper motors are cost-effective and reliable choices for many applications that do not require the rapid acceleration, high speed, and position accuracy of a servomotor.

However, a feedback loop can improve the positioning accuracy of a stepper motor without incurring the higher costs of a complete servosystem. Some stepper motor motion controllers can accommodate a closed loop.

Brush and brushless PM DC servomotors are usually selected for applications that require more precise positioning. Both of these motors can reach higher speeds and offer smoother low-speed operation with finer position resolution than stepper motors, but both require one or more feedback sensors in closed loops, adding to system cost and complexity.

Brush-type permanent-magnet (PM) DC servomotors have wound armatures or rotors that rotate within the magnetic field produced by a PM stator. As the rotor turns, current is applied sequentially to the appropriate armature windings by a mechanical commutator consisting of two or more brushes sliding on a ring of insulated copper segments. These motors are quite mature, and modern versions can provide very high performance for very low cost.

There are variations of the brush-type DC servomotor with its iron-core rotor that permit more rapid acceleration and deceleration because of their low-inertia, lightweight cup- or disk-type armatures. The disk-type armature of the pancake-frame motor, for example, has its mass concentrated close to the motor's faceplate permitting a short, flat cylindrical housing. This configuration makes the motor suitable for faceplate mounting in restricted space, a feature particularly useful in industrial robots or other applications where space does not permit the installation of brackets for mounting a motor with a longer length dimension.

The brush-type DC motor with a cup-type armature also offers lower weight and inertia than conventional DC servomotors. However, the tradeoff in the use of these motors is the restriction on their duty cycles because the epoxy-encapsulated armatures are unable to dissipate heat buildup as easily as iron-core armatures and are therefore subject to damage or destruction if overheated.

However, any servomotor with brush commutation can be unsuitable for some applications due to the electromagnetic interference (EMI) caused by brush arcing or the possibility that the arcing can ignite nearby flammable fluids, airborne dust, or vapor, posing a fire or explosion hazard. The EMI generated can adversely affect nearby electronic circuitry. In addition, motor brushes wear down and leave a gritty residue that can contaminate nearby sensitive instruments or precisely ground surfaces. Thus, brush-type motors must be cleaned constantly to prevent the spread of the residue from the motor. Also, brushes must be replaced periodically, causing unproductive downtime.

Brushless DC PM motors overcome these problems and offer the benefits of electronic rather than mechanical commutation. Built as inside-out DC motors, typical brushless motors have PM rotors and wound stator coils. Commutation is performed by internal noncontact Hall-effect devices (HEDs) positioned within the stator windings. The HEDs are wired to power transistor switching circuitry, which is mounted externally in separate modules for some motors but is mounted internally on circuit cards in other motors. Alternatively, commutation can be performed by a commutating encoder or by commutation software resident in the motion controller or motor drive.

Brushless DC motors exhibit low rotor inertia and lower winding thermal resistance than brush-type motors because their high-efficiency magnets permit the use of shorter rotors with smaller diameters. Moreover, because they are not burdened with sliding brush-type mechanical contacts, they can run at higher speeds (50,000 rpm or greater), provide higher continuous torque, and accelerate faster than brush-type motors. Nevertheless, brushless motors still cost more than comparably rated brush-type motors (although that price gap continues to narrow) and their installation adds to overall motion control system cost and complexity. Table 1 summarizes some of the outstanding characteristics of stepper, PM brush, and PM brushless DC motors.

Table 1. Stepping and Permanent-Magnet DC Servomotors Compared.

	Stepping	PM Brush	PM Brushless
Cost	Low	Medium	High
Smoothness	Low to	Good to excellent	Good to excellent
Speed range	0–1500 rmp (typical)	0–6000 rpm	0–10,000 rpm
Torque	High- (falls off with speed)	Medium	High
Required feedback	None	Position or velocity	Commutation and position or velocity
Maintenance	None	Yes	None
Cleanliness	Excellent	Brush dust	Excellent

The linear motor, another drive alternative, can move the load directly, eliminating the need for intermediate motion translation mechanism. These motors can accelerate rapidly and position loads accurately at high speed because they have no moving parts in contact with each other. Essentially rotary motors that have been sliced open and unrolled, they have many of the characteristics of conventional motors. They can replace conventional rotary motors driving leadscrew-, ballscrew-, or belt-driven single-axis stages, but they cannot be coupled to gears that could change their drive characteristics. If increased performance is required from a linear motor, the existing motor must be replaced with a larger one.

Linear motors must operate in closed feedback loops, and they typically require more costly feedback sensors than rotary motors. In addition, space must be allowed for the free movement of the motor's power cable as it tracks back and forth along a linear path. Moreover, their applications are also limited because of their inability to dissipate heat as readily as rotary motors with metal frames and cooling fins, and the exposed magnetic fields of some models can attract loose ferrous objects, creating a safety hazard.

Motor Drivers (Amplifiers)

Motor drivers or amplifiers must be capable of driving their associated motors—stepper, brush, brushless, or linear. A drive circuit for a stepper motor can be fairly simple because it needs only several power transistors to sequentially energize the motor phases according to the number of digital step pulses received from the motion controller. However, more advanced stepping motor drivers can control phase current to permit "microstepping," a technique that allows the motor to position the load more precisely.

Servodrive amplifiers for brush and brushless motors typically receive analog voltages of ±10-VDC signals from the motion controller. These signals correspond to current or voltage commands. When amplified, the signals control both the direction and magnitude of the current in the motor windings. Two types of amplifiers are generally used in closed-loop servosystems: *linear* and *pulse-width modulated* (PWM).

Pulse-width modulated amplifiers predominate because they are more efficient than linear amplifiers and can provide up to 100 W. The transistors in PWM amplifiers (as in PWM power supplies) are optimized for switchmode operation, and they are capable of switching amplifier output voltage at frequencies up to 20 kHz. When the power transistors are switched on (on state), they saturate, but when they are off, no current is drawn. This operating mode reduces transistor power dissipation and boosts amplifier efficiency. Because of their higher operating frequencies, the magnetic components in PWM amplifiers can be smaller and lighter than those in linear amplifiers. Thus, the entire drive module can be packaged in a smaller, lighter case.

By contrast, the power transistors in linear amplifiers are continuously in the on state although output power requirements can be varied. This operating mode wastes power, resulting in lower amplifier efficiency while subjecting the power transistors to thermal stress. However, linear amplifiers permit smoother motor operation, a requirement for some sensitive motion control systems. In addition linear amplifiers are better at driving low-inductance motors. Moreover, these amplifiers generate less EMI than PWM amplifiers, so they do not require the same degree of filtering. By contrast, linear amplifiers typically have lower maximum power ratings than PWM amplifiers.

Feedback Sensors

Position feedback is the most common requirement in closed-loop motion control systems, and the most popular sensor for providing this information is the rotary optical encoder. The axial shafts of these encoders are mechanically coupled to the driveshafts of the motor. They generate either sine waves or pulses that can be counted by the motion controller to determine the motor or load position and direction of travel at any time to permit precise positioning. Analog encoders produce sine waves that must be conditioned by external circuitry for counting, but digital encoders include circuitry for translating sine waves into pulses.

Absolute rotary optical encoders produce binary words for the motion controller that provide precise position information. If they are stopped accidentally due to power failure, these encoders preserve the binary word because the last position of the encoder code wheel acts as a memory.

Linear optical encoders, by contrast, produce pulses that are proportional to the actual linear distance of load movement. They work on the same principles as the rotary encoders, but the graduations are engraved on a stationary glass or metal scale while the read head moves along the scale.

Tachometers are generators that provide analog signals that are directly proportional to motor shaft speed. They are mechanically coupled to the motor shaft and can be located within the motor frame. After tachometer output is converted to a digital format by the motion controller, a feedback signal is generated for the driver to keep motor speed within preset limits.

Other common feedback sensors include resolvers, linear variable differential transformers (LVDTs), Inductosyns, and potentiometers. Less common are the more accurate laser interferometers. Feedback sensor selection is based on an evaluation of the sensor's accuracy, repeatability, ruggedness, temperature limits, size, weight, mounting requirements, and cost, with the relative importance of each determined by the application.

Installation and Operation of the System

The design and implementation of a cost-effective motion-control system require a high degree of expertise on the part of the person or persons responsible for system integration. It is rare that a diverse group of components can be removed from their boxes, installed, and interconnected to form an instantly effective system. Each servosystem (and many stepper systems) must be tuned (stabilized) to the load and environmental conditions. However, installation and development time can be minimized if the customer's requirements are accurately defined, optimum components are selected, and the tuning and debugging tools are applied correctly. Moreover, operators must be properly trained in formal classes or, at the very least, must have a clear understanding of the information in the manufacturers' technical manuals gained by careful reading.

GLOSSARY OF MOTION CONTROL TERMS

Abbe error: A linear error caused by a combination of an underlying angular error along the line of motion and a dimensional offset between the position of the object being measured and the accuracy-determining element such as a leadscrew or encoder.

acceleration: The change in velocity per unit time.

accuracy: (1) absolute accuracy: The motion control system output compared with the commanded input. It is actually a measurement of inaccuracy and it is typically measured in millimeters. (2) motion accuracy: The maximum expected difference between the actual and the intended position of an object or load for a given input. Its value depends on the method used for measuring the actual position. (3) on-axis accuracy: The uncertainty of load position after all linear errors are eliminated. These include such factors as inaccuracy of leadscrew pitch, the angular deviation effect at the measuring point, and thermal expansion of materials.

backlash: The maximum magnitude of an input that produces no measurable output when the direction of motion is reversed. It can result from insufficient preloading or poor meshing of gear teeth in a gear-coupled drivetrain.

error: (1) The difference between the actual result of an input command and the ideal or theoretical result. (2) following error: The instantaneous difference between the actual position as reported by the position feedback loop and the ideal position, as commanded by the controller. (3) steady-state error: The difference between the actual and commanded position after all corrections have been applied by the controller.

hysteresis: The difference in the absolute position of the load for a commanded input when motion is from opposite directions.

inertia: The measure of a load's resistance to changes in velocity or speed. It is a function of the load's mass and shape. The torque required to accelerate or decelerate the load is proportional to inertia.

overshoot: The amount of overcorrection in an underdamped control system.

play: The uncontrolled movement due to the looseness of mechanical parts. It is typically caused by wear, overloading the system, or improper system operation.

precision: See *repeatability.*

repeatability: The ability of a motion control system to return repeatedly to the commanded position. It is influenced by the presence of *backlash* and *hysteresis*. Consequently, *bidirectional repeatability*, a more precise specification, is the ability of the system to achieve the commanded position repeatedly regardless of the direction from which the intended position is approached. It is synonymous with *precision.* However, accuracy and precision are not the same.

resolution: The smallest position increment that the motion control system can detect. It is typically considered to be display or encoder resolution because it is not necessarily the smallest motion the system is capable of delivering reliably.

runout: The deviation between ideal linear (straight-line) motion and the actual measured motion.

sensitivity: The minimum input capable of producing output motion. It is also the ratio of the output motion to the input drive. This term should not be used in place of resolution.

settling time: The time elapsed between the entry of a command to a system and the instant the system first reaches the commanded position and maintains that position within the specified error value.

velocity: The change in distance per unit time. Velocity is a *vector* and speed is a *scalar*, but the terms can be used interchangeably.

MECHANICAL COMPONENTS FORM SPECIALIZED MOTION-CONTROL SYSTEMS

Many different kinds of mechanical components are listed in manufacturers' catalogs for speeding the design and assembly of motion control systems. These drawings illustrate what, where, and how one manufacturer's components were used to build specialized systems.

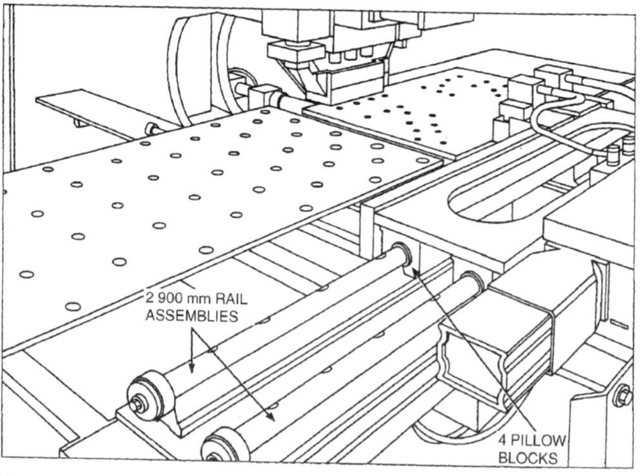

Fig. 1 Punch Press: Catalog pillow blocks and rail assemblies were installed in this system for reducing the deflection of a punch press plate loader to minimize scrap and improve its cycle speed.

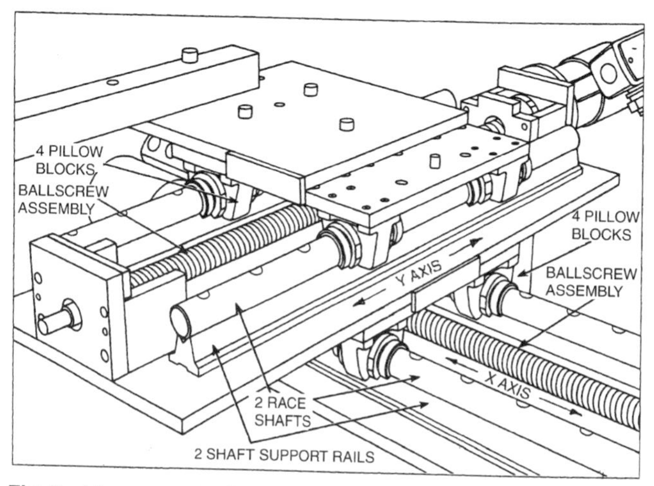

Fig. 2 Microcomputer-Controlled X-Y Table: Catalog pillow blocks, rail guides, and ballscrew assemblies were installed in this rigid system that positions workpieces accurately for precise milling and drilling on a vertical milling machine.

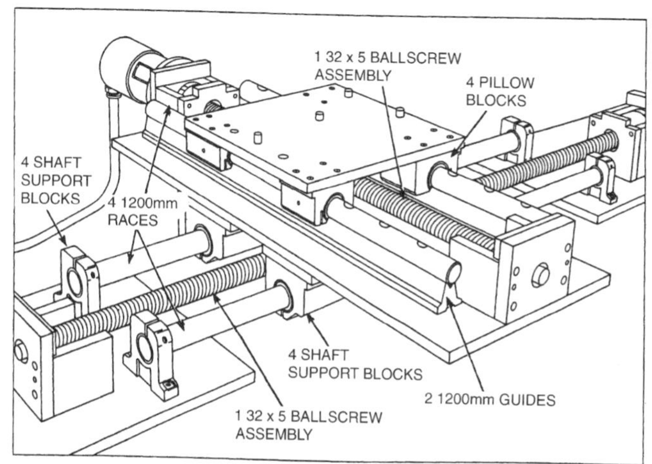

Fig. 3 Pick and Place X-Y System: Catalog support and pillow blocks, ballscrew assemblies, races, and guides were in the assembly of this X-Y system that transfers workpieces between two separate machining stations.

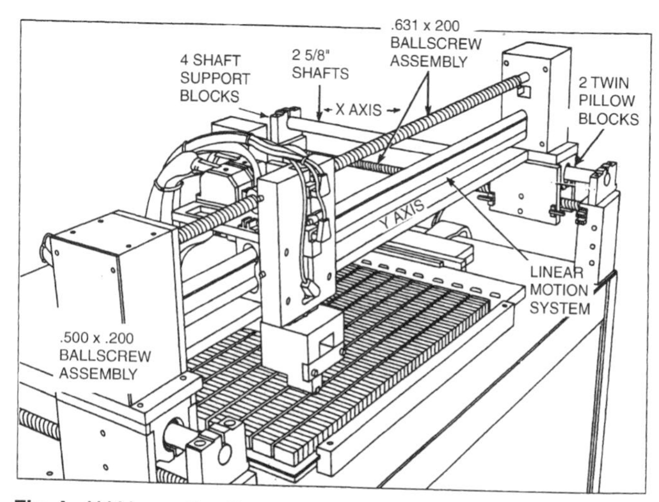

Fig. 4 X-Y Inspection System: Catalog pillow and shaft-support blocks, ballscrew assemblies, and a preassembled motion system were used to build this system, which accurately positions an inspection probe over small electronic components.

SERVOMOTORS, STEPPER MOTORS, AND ACTUATORS FOR MOTION CONTROL

Many different kinds of electric motors have been adapted for use in motion control systems because of their linear characteristics. These include both conventional rotary and linear alternating current (AC) and direct current (DC) motors. These motors can be further classified into those that must be operated in closed-loop servosystems and those that can be operated open-loop.

The most popular servomotors are permanent magnet (PM) rotary DC servomotors that have been adapted from conventional PM DC motors. These servomotors are typically classified as brush-type and brushless. The brush-type PM DC servomotors include those with wound rotors and those with lighter weight, lower inertia cup- and disk coil-type armatures. Brushless servomotors have PM rotors and wound stators.

Some motion control systems are driven by two-part linear servomotors that move along tracks or ways. They are popular in applications where errors introduced by mechanical coupling between the rotary motors and the load can introduce unwanted errors in positioning. Linear motors require closed loops for their operation, and provision must be made to accommodate the back-and-forth movement of the attached data and power cable.

Stepper or stepping motors are generally used in less demanding motion control systems, where positioning the load by stepper motors is not critical for the application. Increased position accuracy can be obtained by enclosing the motors in control loops.

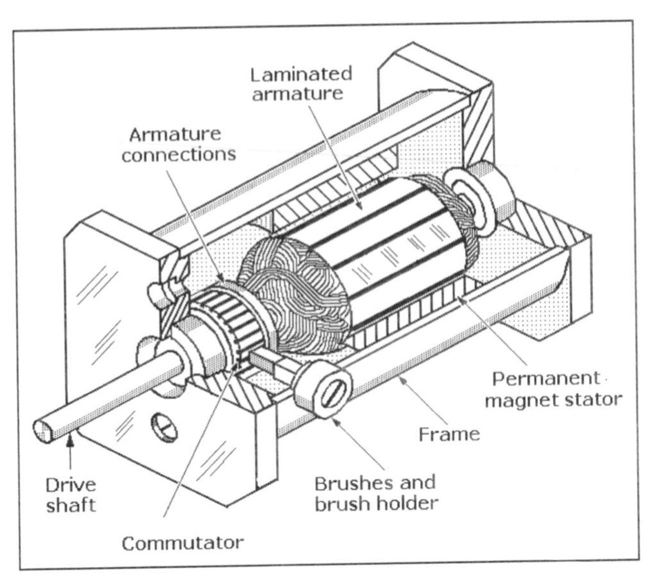

Fig. 1 Cutaway view of a fractional horsepower permanent-magnet DC servomotor.

Permanent-Magnet DC Servomotors

Permanent-magnet (PM) field DC rotary motors have proven to be reliable drives for motion control applications where high efficiency, high starting torque, and linear speed–torque curves are desirable characteristics. While they share many of the characteristics of conventional rotary series, shunt, and compound-wound brush-type DC motors, PM DC servomotors increased in popularity with the introduction of stronger ceramic and rare-earth magnets made from such materials as neodymium–iron–boron and the fact that these motors can be driven easily by microprocessor-based controllers.

The replacement of a wound field with permanent magnets eliminates both the need for separate field excitation and the electrical losses that occur in those field windings. Because there are both brush-type and brushless DC servomotors, the term *DC motor* implies that it is brush-type or requires mechanical commutation unless it is modified by the term *brushless*. Permanent-magnet DC brush-type servomotors can also have armatures formed as laminated coils in disk or cup shapes. They are lightweight, low-inertia armatures that permit the motors to accelerate faster than the heavier conventional wound armatures.

The increased field strength of the ceramic and rare-earth magnets permitted the construction of DC motors that are both smaller and lighter than earlier generation comparably rated DC motors with alnico (aluminum–nickel–cobalt or AlNiCo) magnets. Moreover, integrated circuitry and microprocessors have increased the reliability and cost-effectiveness of digital motion controllers and motor drivers or amplifiers while permitting them to be packaged in smaller and lighter cases, thus reducing the size and weight of complete, integrated motion-control systems.

Brush-Type PM DC Servomotors

The design feature that distinguishes the brush-type PM DC servomotor, as shown in Fig. 1, from other brush-type DC motors is the use of a permanent-magnet field to replace the wound field. As previously stated, this eliminates both the need for separate field excitation and the electrical losses that typically occur in field windings.

Permanent-magnet DC motors, like all other mechanically commutated DC motors, are energized through brushes and a multisegment commutator. While all DC motors operate on the same principles, only PM DC motors have the linear speed–torque curves shown in Fig. 2, making them ideal for closed-loop and variable-speed servomotor applications. These linear characteristics conveniently describe the full range of motor performance.

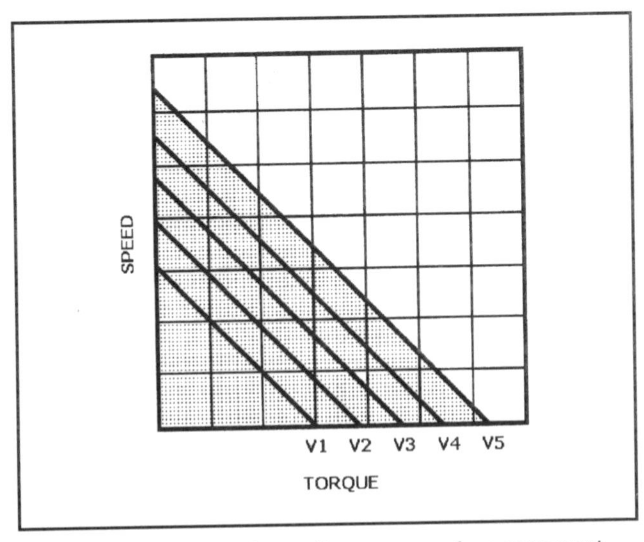

Fig. 2 A typical family of speed/torque curves for a permanent-magnet DC servomotor at different voltage inputs, with voltage increasing from left to right (V1 to V5).

It can be seen that both speed and torque increase linearly with applied voltage, indicated in the diagram as increasing from V1 to V5.

The stators of brush-type PM DC motors are magnetic pole pairs. When the motor is powered, the opposite polarities of the energized windings and the stator magnets attract, and the rotor rotates to align itself with the stator. Just as the rotor reaches alignment, the brushes move across the commutator segments and energize the next winding. This sequence continues as long as power is applied, keeping the rotor in continuous motion. The commutator is staggered from the rotor poles, and the number of its segments is directly proportional to the number of windings. If the connections of a PM DC motor are reversed, the motor will change direction, but it might not operate as efficiently in the reversed direction.

Disk-Type PM DC Motors

The disk-type motor shown in the exploded view in Fig. 3 has a disk-shaped armature with stamped and laminated windings. This nonferrous laminated disk is made as a copper stamping bonded between epoxy–glass insulated layers and fastened to an axial shaft. The stator field can either be a ring of many individual ceramic magnet cylinders, as shown, or a ring-type ceramic magnet attached to the dish-shaped end bell, which completes the magnetic circuit. The spring-loaded brushes ride directly on stamped commutator bars.

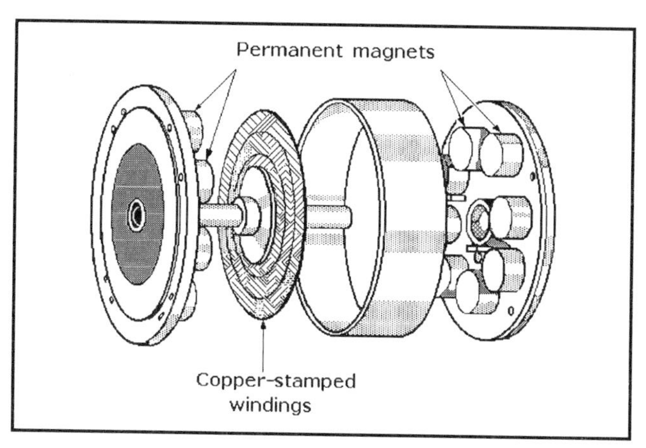

Fig. 3 Exploded view of a permanent-magnet DC servomotor with a disk-type armature.

These motors are also called *pancake motors* because they are housed in cases with thin, flat form factors whose diameters exceed their lengths, suggesting pancakes. Earlier generations of these motors were called *printed-circuit motors* because the armature disks were made by a printed-circuit fabrication process that has been superseded. The flat motor case concentrates the motor's center of mass close to the mounting plate, permitting it to be easily surface mounted. This eliminates the awkward motor overhang and the need for supporting braces if a conventional motor frame is to be surface mounted. Their disk-type motor form factor has made these motors popular as axis drivers for industrial robots where space is limited.

The principal disadvantage of the disk-type motor is the relatively fragile construction of its armature and its inability to dissipate heat as rapidly as iron-core wound rotors. Consequently, these motors are usually limited to applications where the motor can be run under controlled conditions and a shorter duty cycle allows enough time for armature heat buildup to be dissipated.

Cup- or Shell-Type PM DC Motors

Cup- or shell-type PM DC motors offer low inertia and low inductance as well as high acceleration characteristics, making them useful in many servo applications. They have hollow cylindrical armatures made as aluminum or copper coils bonded by polymer resin and fiberglass to form a rigid "ironless cup," which is fastened to an axial shaft. A cutaway view of this class of servomotor is illustrated in Fig. 4.

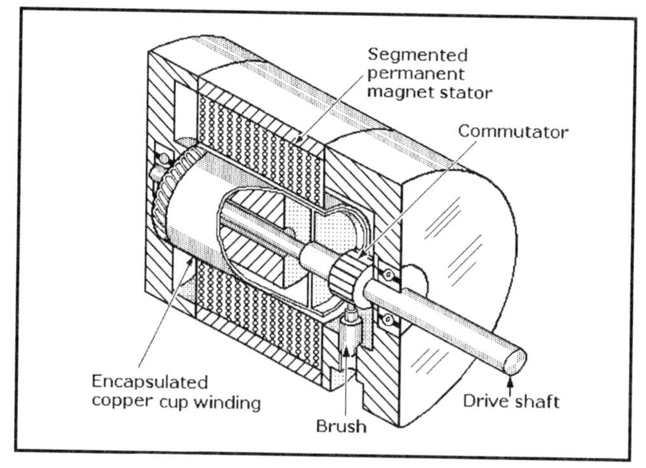

Fig. 4 Cutaway view of a permanent-magnet DC servomotor with a cup-type armature.

Because the armature has no iron core, it, like the disk motor, has extremely low inertia and a very high torque-to-inertia ratio. This permits the motor to accelerate rapidly for the quick response required in many motion-control applications. The armature rotates in an air gap within very high magnetic flux density. The magnetic field from the stationary magnets is completed through the cup-type armature and a stationary ferrous cylindrical core connected to the motor frame. The shaft rotates within the core, which extends into the rotating cup. Spring brushes commutate these motors.

Another version of a cup-type PM DC motor is shown in the exploded view in Fig. 5. The cup-type armature is rigidly fastened to the shaft by a disk at the right end of the winding, and the magnetic field is also returned through a ferrous metal housing. The brush assembly of this motor is built into its end cap or flange, shown at the far right.

The principal disadvantage of this motor is also the inability of its bonded armature to dissipate internal heat buildup rapidly because of its low thermal conductivity. Without proper cooling and sensitive control circuitry, the armature could be heated to destructive temperatures in seconds.

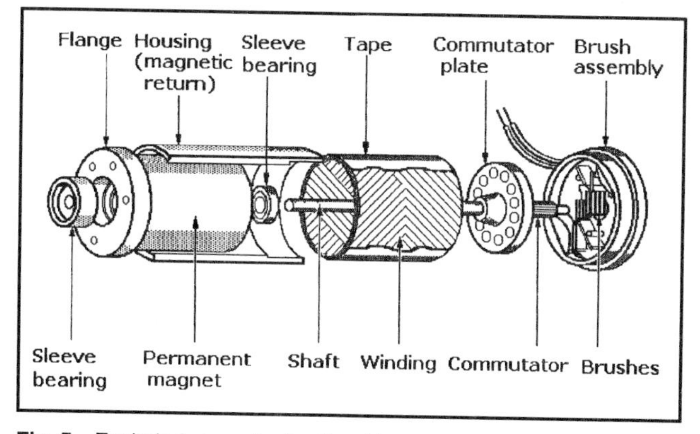

Fig. 5 Exploded view of a fractional horsepower brush-type DC servomotor.

Brushless PM DC Motors

Brushless DC motors exhibit the same linear speed–torque characteristics as the brush-type PM DC motors, but they are electronically commutated. The construction of these motors, as shown in Fig. 6, differs from that of a typical brush-type DC motor in that they are "inside-out." In other words, they have permanent magnet rotors instead of stators, and the stators rather than the rotors are wound. Although this geometry is required for brushless DC motors, some manufacturers have adapted this design for brush-type DC motors.

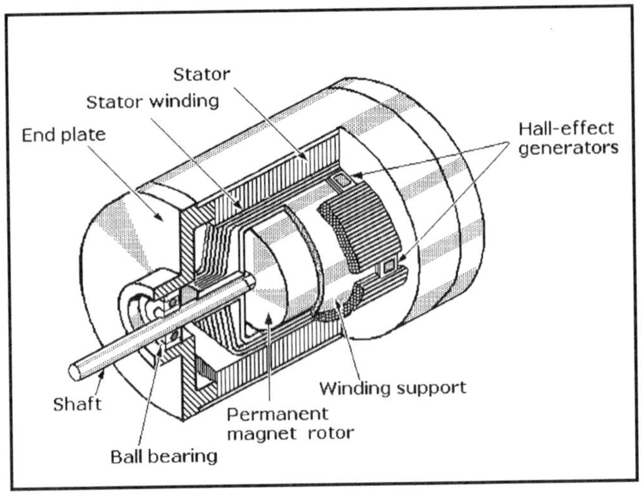

Fig. 6 Cutaway view of a brushless DC motor.

The mechanical brush and bar commutator of the brushless DC motor is replaced by electronic sensors, typically Hall-effect devices (HEDs). They are located within the stator windings and wired to solid-state transistor switching circuitry located either on circuit cards mounted within the motor housings or in external packages. Generally, only fractional horsepower brushless motors have switching circuitry within their housings.

The cylindrical magnet rotors of brushless DC motors are magnetized laterally to form opposing north and south poles across the rotor's diameter. These rotors are typically made from neodymium–iron–boron or samarium–cobalt rare-earth magnetic materials, which offer higher flux densities than alnico magnets. These materials permit motors offering higher performance to be packaged in the same frame sizes as earlier motor designs or those with the same ratings to be packaged in smaller frames than the earlier designs. Moreover, rare-earth or ceramic magnet rotors can be made with smaller diameters than those earlier models with alnico magnets, thus reducing their inertia.

A simplified diagram of a DC brushless motor control with one HED for the electronic commutator is shown in Fig. 7. The HED is a Hall-effect sensor integrated with an amplifier in a silicon chip. This IC is capable of sensing the polarity of the rotor's magnetic field and then sending appropriate signals to power transistors T1 and T2 to cause the motor's rotor to rotate continuously. This is accomplished as follows:

(1) With the rotor motionless, the HED detects the rotor's north magnetic pole, causing it to generate a signal that turns on transistor T2. This causes current to flow, energizing winding W2 to form a south-seeking electromagnetic rotor pole. This pole then attracts the rotor's north pole to drive the rotor in a counterclockwise (CCW) direction.

(2) The inertia of the rotor causes it to rotate past its neutral position so that the HED can then sense the rotor's south magnetic pole. It then switches on transistor T1, causing current to flow in winding W1, thus forming a north-seeking stator pole that

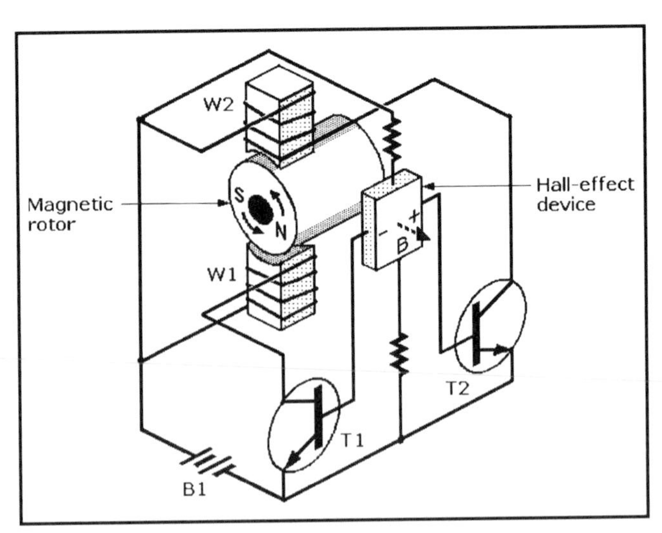

Fig. 7 Simplified diagram of Hall-effect device (HED) commutation of a brushless DC motor.

attracts the rotor's south pole, causing it to continue to rotate in the CCW direction.

The transistors conduct in the proper sequence to ensure that the excitation in the stator windings W2 and W1 always leads the PM rotor field to produce the torque necessary to keep the rotor in constant rotation. The windings are energized in a pattern that rotates around the stator.

There are usually two or three HEDs in practical brushless motors that are spaced apart by 90 or 120° around the motor's rotor. They send the signals to the motion controller that actually triggers the power transistors, which drive the armature windings at a specified motor current and voltage level.

The brushless motor in the exploded view Fig. 8 illustrates a design for a miniature brushless DC motor that includes Hall-effect commutation. The stator is formed as an ironless sleeve of copper coils bonded together in polymer resin and fiberglass to form a rigid structure similar to cup-type rotors. However, it is fastened inside the steel laminations within the motor housing.

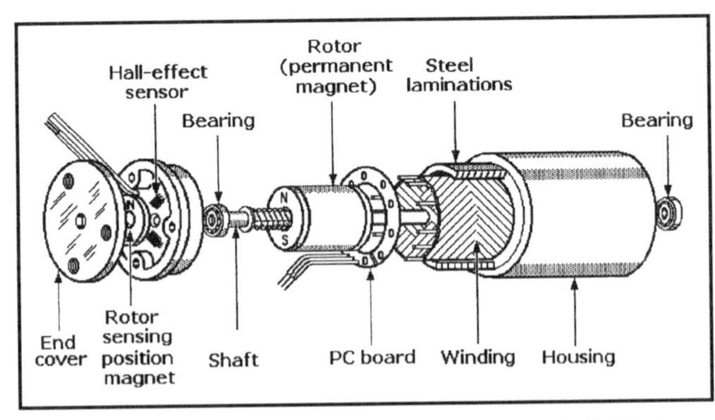

Fig. 8 Exploded view of a brushless DC motor with Hall-effect device (HED) commutation.

This method of construction permits a range of values for starting current and specific speed (rpm/V) depending on wire gauge and the number of turns. Various terminal resistances can be obtained, permitting the user to select the optimum motor for a specific application. The Hall-effect sensors and a small magnet disk that is magnetized widthwise are mounted on a disk-shaped partition within the motor housing.

Position Sensing in Brushless Motors

Both magnetic sensors and resolvers can sense rotor position in brushless motors. The diagram in Fig. 9 shows how three magnetic sensors can sense rotor position in a three-phase electronically commutated brushless DC motor. In this example, the magnetic sensors are located inside the end bell of the motor. This inexpensive version is adequate for simple controls.

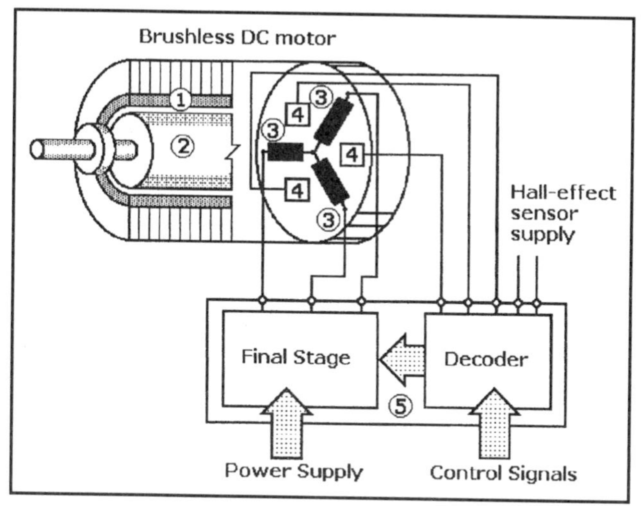

Fig. 9 A magnetic sensor as a rotor position indicator: stationary brushless motor winding (1), permanent-magnet motor rotor (2), three-phase electronically commutated field (3), three magnetic sensors (4), and the electronic circuit board (5).

In the alternate design shown in Fig. 10, a resolver on the end cap of the motor is used to sense rotor position when greater positioning accuracy is required. The high-resolution signals from the resolver can be used to generate sinusoidal motor currents within the motor controller. The currents through the three motor windings are position independent and respectively 120° phase shifted.

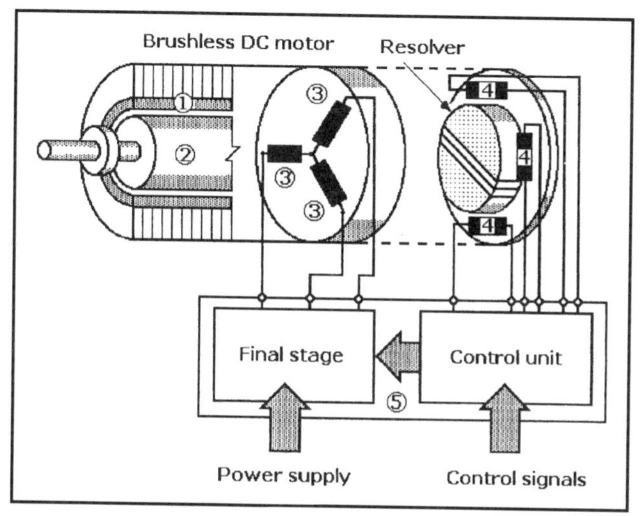

Fig. 10 A resolver as a rotor position indicator: stationary motor winding (1), permanent-magnet motor rotor (2), three-phase electronically commutated field (3), three magnetic sensors (4), and the electronic circuit board (5).

Brushless Motor Advantages

Brushless DC motors have at least four distinct advantages over brush-type DC motors that are attributable to the replacement of mechanical commutation by electronic commutation.

- There is no need to replace brushes or remove the gritty residue caused by brush wear from the motor.
- Without brushes to cause electrical arcing, brushless motors do not present fire or explosion hazards in an environment where flammable or explosive vapors, dust, or liquids are present.
- Electromagnetic interference (EMI) is minimized by replacing mechanical commutation, the source of unwanted radio frequencies, with electronic commutation.
- Brushless motors can run faster and more efficiently with electronic commutation. Speeds of up to 50,000 rpm can be achieved versus the upper limit of about 5000 rpm for brush-type DC motors.

Brushless DC Motor Disadvantages

There are at least four disadvantages of brushless DC servomotors.

- Brushless PM DC servomotors cannot be reversed by simply reversing the polarity of the power source. The order in which the current is fed to the field coil must be reversed.
- Brushless DC servomotors cost more than comparably rated brush-type DC servomotors.
- Additional system wiring is required to power the electronic commutation circuitry.
- The motion controller and driver electronics needed to operate a brushless DC servomotor are more complex and expensive than those required for a conventional DC servomotor.

Consequently, the selection of a brushless motor is generally justified on the basis of specific application requirements or its hazardous operating environment.

Characteristics of Brushless Rotary Servomotors

It is difficult to generalize about the characteristics of DC rotary servomotors because of the wide range of products available commercially. However, they typically offer continuous torque ratings of 0.62 lb-ft (0.84 N-m) to 5.0 lb-ft (6.8 N-m), peak torque ratings of 1.9 lb-ft (2.6 N-m) to 14 lb-ft (19 N-m), and continuous power ratings of 0.73 hp (0.54 kW) to 2.76 hp (2.06 kW). Maximum speeds can vary from 1400 to 7500 rpm, and the weight of these motors can be from 5.0 lb (2.3 kg) to 23 lb (10 kg). Feedback typically can be either by resolver or encoder.

Linear Servomotors

A linear motor is essentially a rotary motor that has been opened out into a flat plane, but it operates on the same principles. A permanent-magnet DC linear motor is similar to a permanent-magnet rotary motor, and an AC induction squirrel cage motor is similar to an induction linear motor. The same electromagnetic force that produces torque in a rotary motor also produces torque in a linear motor. Linear motors use the same controls and programmable position controllers as rotary motors.

Before the invention of linear motors, the only way to produce linear motion was to use pneumatic or hydraulic cylinders, or to translate rotary motion to linear motion with ballscrews or belts and pulleys.

A linear motor consists of two mechanical assemblies: *coil* and *magnet*, as shown in Fig. 11. Current flowing in a winding in a magnetic flux field produces a force. The copper windings

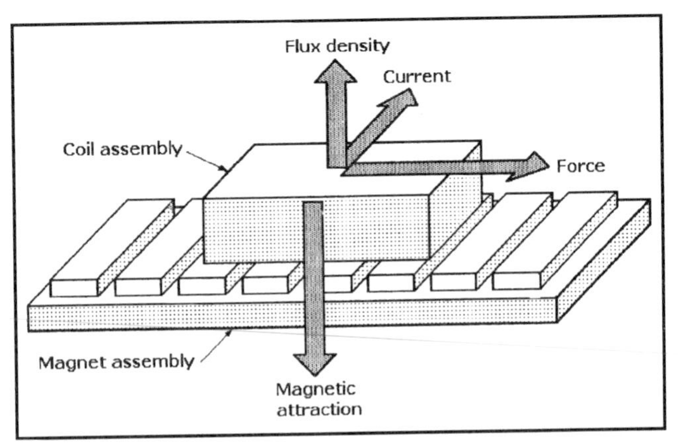

Fig. 11 Operating principles of a linear servomotor.

conduct current (I), and the assembly generates magnetic flux density (B). When the current and flux density interact, a force (F) is generated in the direction shown in Fig. 11, where $F = I \times B$.

Even a small motor will run efficiently, and large forces can be created if a large number of turns are wound in the coil and the magnets are powerful rare-earth magnets. The windings are phased 120 electrical degrees apart, and they must be continually switched or commutated to sustain motion.

Only brushless linear motors for closed-loop servomotor applications are discussed here. Two types of these motors are available commercially—*steel-core* (also called *iron-core*) and *epoxy-core* (also called *ironless*). Each of these linear servomotors has characteristics and features that are optimal in different applications.

The coils of steel-core motors are wound on silicon steel to maximize the generated force available with a single-sided magnet assembly or way. Figure 12 shows a steel-core brushless linear motor. The steel in these motors focuses the magnetic flux to produce very high force density. The magnet assembly consists of rare-earth bar magnets mounted on the upper surface of a steel baseplate arranged to have alternating polarities (i.e., N, S, N, S).

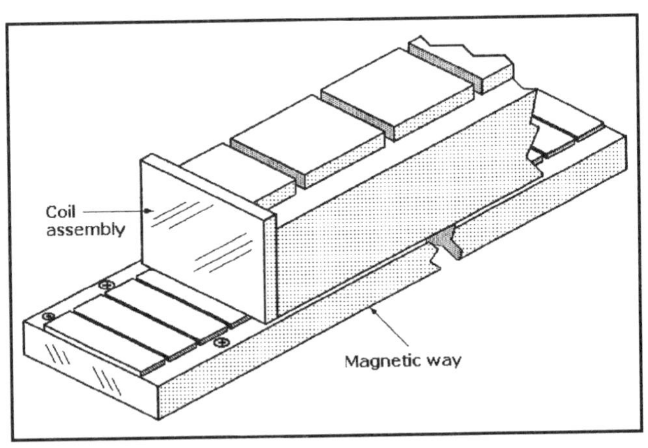

Fig. 12 A linear iron-core linear servomotor consists of a magnetic way and a mating coil assembly.

The steel in the cores is attracted to the permanent magnets in a direction that is perpendicular (normal) to the operating motor force. The magnetic flux density within the air gap of linear motors is typically several thousand gauss. A constant magnetic force is present whether or not the motor is energized. The normal force of the magnetic attraction can be up to 10 times the continuous force rating of the motor. This flux rapidly diminishes to a few gauss as the measuring point is moved a few centimeters away from the magnets.

Cogging is a form of magnetic "detenting" that occurs in both linear and rotary motors when the motor coil's steel laminations cross the alternating poles of the motor's magnets. Because it can occur in steel-core motors, manufacturers include features that minimize cogging. The high thrust forces attainable with steel-core linear motors permit them to accelerate and move heavy masses while maintaining stiffness during machining or process operations.

The features of epoxy-core or ironless-core motors differ from those of the steel-core motors. For example, their coil assemblies are wound and encapsulated within epoxy to form a thin plate that is inserted in the air gap between the two permanent-magnet strips fastened inside the magnet assembly, as shown in Fig. 13. Because the coil assemblies do not contain steel cores, epoxy-core motors are lighter than steel-core motors and less subject to cogging.

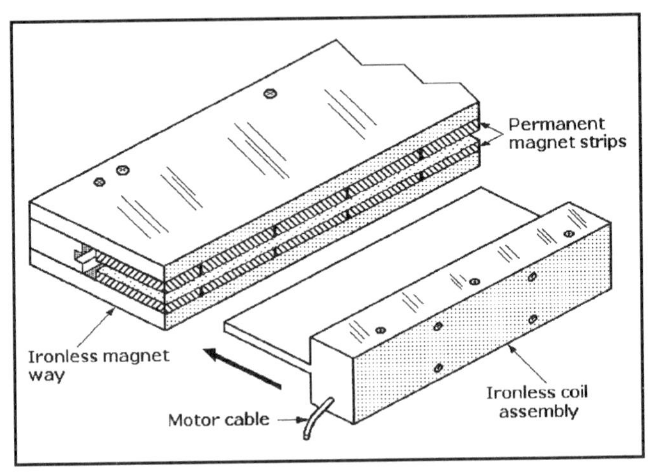

Fig. 13 A linear ironless servomotor consists of an ironless magnetic way and an ironless coil assembly.

The strip magnets are separated to form the air gap into which the coil assembly is inserted. This design maximizes the generated thrust force and also provides a flux return path for the magnetic circuit. Consequently, very little magnetic flux exists outside the motor, thus minimizing residual magnetic attraction.

Epoxy-core motors provide exceptionally smooth motion, making them suitable for applications requiring very low bearing friction and high acceleration of light loads. They also permit constant velocity to be maintained, even at very low speeds.

Linear servomotors can achieve accuracies of 0.1 µm. Normal accelerations are 2 to 3 *g*, but some motors can reach 15 *g*. Velocities are limited by the encoder data rate and the amplifier voltage. Normal peak velocities are from 0.04 in./s (1 mm/s) to about 6.6 ft/s (2 m/s), but the velocity of some models can exceed 26 ft/s (8 m/s).

Ironless linear motors can have continuous force ratings from about 5 to 55 lbf (22 to 245 N) and peak force ratings from about 25 to 180 lbf (110 to 800 N). By contrast, iron-core linear motors are available with continuous force ratings of about 30 to 1100 lbf (130 to 4900 N) and peak force ratings of about 60 to 1800 lbf (270 to 8000 N).

Commutation

The linear motor windings that are phased 120° apart must be continually switched or commutated to sustain motion. There are two ways to commutate linear motors: *sinusoidal* and *Hall-effect device (HED)*, or *trapezoidal*. The highest motor efficiency is achieved with sinusoidal commutation, while HED commutation is about 10 to 15% less efficient.

In sinusoidal commutation, the linear encoder that provides position feedback in the servosystem is also used to commutate the motor. A process called "phase finding" is required when the motor is turned on, and the motor phases are then incrementally advanced with each encoder pulse. This produces extremely smooth motion. In HED commutation a circuit board containing Hall-effect ICs is embedded in the coil assembly. The HED sensors detect the polarity change in the magnet track and switch the motor phases every 60°.

Sinusoidal commutation is more efficient than HED commutation because the coil windings in motors designed for this commutation method are configured to provide a sinusoidally shaped back EMF waveform. As a result, the motors produce a constant force output when the driving voltage on each phase matches the characteristic back EMF waveform.

Installation of Linear Motors

In a typical linear motor application the coil assembly is attached to the moving member of the host machine and the magnet assembly is mounted on the nonmoving base or frame. These motors can be mounted vertically, but if they are they typically require a counterbalance system to prevent the load from dropping if power temporarily fails or is routinely shut off. The counterbalance system, typically formed from pulleys and weights, springs, or air cylinders, supports the load against the force of gravity.

If power is lost, servo control is interrupted. Stages in motion tend to stay in motion while those at rest tend to stay at rest. The stopping time and distance depend on the stage's initial velocity and system friction. The motor's back EMF can provide dynamic braking, and friction brakes can be used to attenuate motion rapidly. However, positive stops and travel limits can be built into the motion stage to prevent damage in situations where power or feedback might be lost or the controller or servo driver fail.

Linear servomotors are supplied to the customer in kit form for mounting on the host machine. The host machine structure must include bearings capable of supporting the mass of the motor parts while maintaining the specified air gap between the assemblies and also resisting the normal force of any residual magnetic attraction.

Linear servomotors must be used in closed loop positioning systems because they do not include built-in means for position sensing. Feedback is typically supplied by such sensors as linear encoders, laser interferometers, LVDTs, or linear Inductosyns.

Advantages of Linear vs. Rotary Servomotors

The advantages of linear servomotors over rotary servomotors include:

- *High stiffness:* The linear motor is connected directly to the moving load, so there is no backlash and practically no compliance between the motor and the load. The load moves instantly in response to motor motion.
- *Mechanical simplicity:* The coil assembly is the only moving part of the motor, and its magnet assembly is rigidly mounted to a stationary structure on the host machine. Some linear motor manufacturers offer modular magnetic assemblies in various modular lengths. This permits the user to form a track of any desired length by stacking the modules end to end, allowing virtually unlimited travel. The force produced by the motor is applied directly to the load without any couplings, bearings, or other conversion mechanisms. The only alignments required are for the air gaps, which typically are from 0.039 in. (1 mm) to 0.020 in. (0.5 mm).
- *High accelerations and velocities:* Because there is no physical contact between the coil and magnet assemblies, high accelerations and velocities are possible. Large motors are capable of accelerations of 3 to 5 *g*, but smaller motors are capable of more than 10 *g*.
- *High velocities:* Velocities are limited by feedback encoder data rate and amplifier bus voltage. Normal peak velocities are up to 6.6 ft/s (2 m/s), although some models can reach 26 ft/s (8 m/s). This compares with typical linear speeds of ballscrew transmissions, which are commonly limited to 20 to 30 in./s (0.5 to 0.7 m/s) because of resonances and wear.
- *High accuracy and repeatability:* Linear motors with position feedback encoders can achieve positioning accuracies of ±1 encoder cycle or submicrometer dimensions, limited only by encoder feedback resolution.
- *No backlash or wear:* With no contact between moving parts, linear motors do not wear out. This minimizes maintenance and makes them suitable for applications where long life and long-term peak performance are required.
- *System size reduction:* With the coil assembly attached to the load, no additional space is required. By contrast, rotary motors typically require ballscrews, rack-and-pinion gearing, or timing belt drives.
- *Clean room compatibility:* Linear motors can be used in clean rooms because they do not need lubrication and do not produce carbon brush grit.

Coil Assembly Heat Dissipation

Heat control is more critical in linear motors than in rotary motors because they do not have the metal frames or cases that can act as large heat-dissipating surfaces. Some rotary motors also have radiating fins on their frames that serve as heat sinks to augment the heat dissipation capability of the frames. Linear motors must rely on a combination of high motor efficiency and good thermal conduction from the windings to a heat-conductive, electrically isolated mass. For example, an aluminum attachment bar placed in close contact with the windings can aid in heat dissipation. Moreover, the carriage plate to which the coil assembly is attached must have effective heat-sinking capability.

Stepper Motors

A *stepper* or *stepping motor* is an AC motor whose shaft is indexed through part of a revolution or *step angle* for each DC pulse sent to it. Trains of pulses provide input current to the motor in increments that can "step" the motor through 360°, and the actual angular rotation of the shaft is directly related to the number of pulses introduced. The position of the load can be determined with reasonable accuracy by counting the pulses entered.

The stepper motors suitable for most open-loop motion control applications have wound stator fields (electromagnetic coils) and iron or permanent magnet (PM) rotors. Unlike PM DC servomotors with mechanical brush-type commutators, stepper motors depend on external controllers to provide the switching pulses for commutation. Stepper motor operation is based on the same electromagnetic principles of attraction and repulsion as other motors, but their commutation provides only the torque required to turn their rotors.

Pulses from the external motor controller determine the amplitude and direction of current flow in the stator's field windings, and they can turn the motor's rotor either clockwise or counterclockwise, stop and start it quickly, and hold it securely at desired positions. Rotational shaft speed depends on the frequency of the pulses. Because controllers can step most motors at audio frequencies, their rotors can turn rapidly.

Between the application of pulses when the rotor is at rest, its armature will not drift from its stationary position because of the stepper motor's inherent holding ability or *detent torque*. These motors generate very little heat while at rest, making them suitable for many different instrument drive-motor applications in which power is limited.

The three basic kinds of stepper motors are *permanent magnet, variable reluctance,* and *hybrid.* The same controller circuit can drive both hybrid and PM stepping motors.

Permanent-Magnet (PM) Stepper Motors

Permanent-magnet stepper motors have smooth armatures and include a permanent magnet core that is magnetized widthwise or perpendicular to its rotation axis. These motors usually have two independent windings, with or without center taps. The most common step angles for PM motors are 45 and 90°, but motors with step angles as fine as 1.8° per step as well as 7.5, 15, and 30° per step are generally available. Armature rotation occurs when the stator poles are alternately energized and deenergized to create torque. A 90° stepper has four poles and a 45° stepper has eight poles, and these poles must be energized in sequence. Permanent-magnet steppers step at relatively low rates, but they can produce high torques and they offer very good damping characteristics.

Variable Reluctance Stepper Motors

Variable reluctance (VR) stepper motors have multitooth armatures with each tooth effectively an individual magnet. At rest these magnets align themselves in a natural detent position to provide larger holding torque than can be obtained with a comparably rated PM stepper. Typical VR motor step angles are 15 and 30° per step. The 30° angle is obtained with a 4-tooth rotor and a 6-pole stator, and the 15° angle is achieved with an 8-tooth rotor and a 12-pole stator. These motors typically have three windings with a common return, but they are also available with four or five windings. To obtain continuous rotation, power must be applied to the windings in a coordinated sequence of alternately deenergizing and energizing the poles.

If just one winding of either a PM or VR stepper motor is energized, the rotor (under no load) will snap to a fixed angle and hold that angle until external torque exceeds the holding torque of the motor. At that point, the rotor will turn, but it will still try to hold its new position at each successive equilibrium point.

Hybrid Stepper Motors

The hybrid stepper motor combines the best features of VR and PM stepper motors. A cutaway view of a typical industrial-grade hybrid stepper motor with a multitoothed armature is shown in Fig. 14. The armature is built in two sections, with the teeth in the second section offset from those in the first section. These motors also have multitoothed stator poles that are not visible in the figure. Hybrid stepper motors can achieve high stepping rates, and they offer high detent torque and excellent dynamic and static torque.

Hybrid steppers typically have two windings on each stator pole so that each pole can become either magnetic north or south, depending on current flow. A cross-sectional view of a hybrid stepper motor illustrating the multitoothed poles with dual windings per pole and the multitoothed rotor is illustrated in Fig. 15. The shaft is represented by the central circle in the diagram.

The most popular hybrid steppers have 3- and 5-phase wiring, and step angles of 1.8 and 3.6° per step. These motors can provide more torque from a given frame size than other stepper types because either all or all but one of the motor windings are energized at every point in the drive cycle. Some 5-phase motors have high resolutions of 0.72° per step (500 steps per revolution).

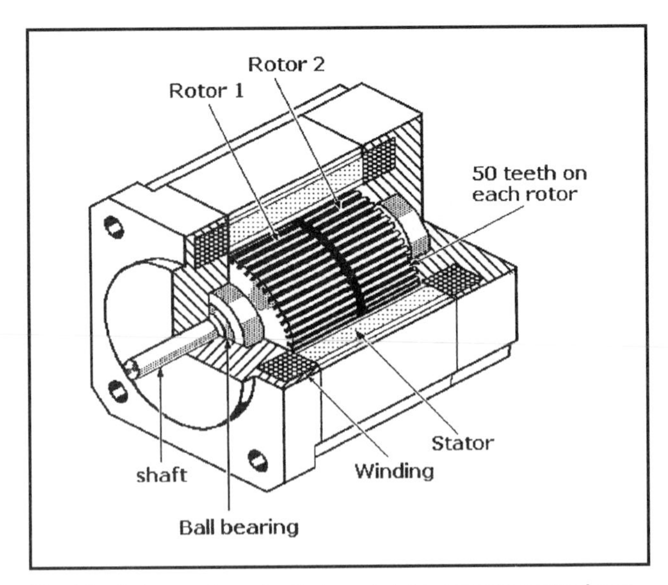

Fig. 14 Cutaway view of a 5-phase hybrid stepping motor. A permanent magnet is within the rotor assembly, and the rotor segments are offset from each other by 3.5°.

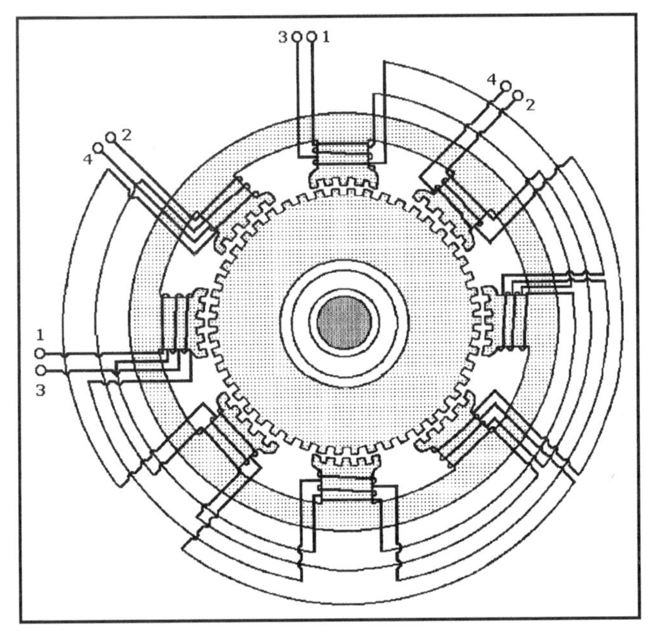

Fig. 15 Cross-section of a hybrid stepping motor showing the segments of the magnetic-core rotor and stator poles with its wiring diagram.

With a compatible controller, most PM and hybrid motors can be run in half-steps, and some controllers are designed to provide smaller fractional steps, or *microsteps*. Hybrid stepper motors capable of a wide range of torque values are available commercially. This range is achieved by scaling length and diameter dimensions. Hybrid stepper motors are available in NEMA size 17 to 42 frames, and output power can be as high as 1000 W peak.

Stepper Motor Applications

Many different technical and economic factors must be considered in selecting a hybrid stepper motor. For example, the ability of the stepper motor to repeat the positioning of its multitoothed

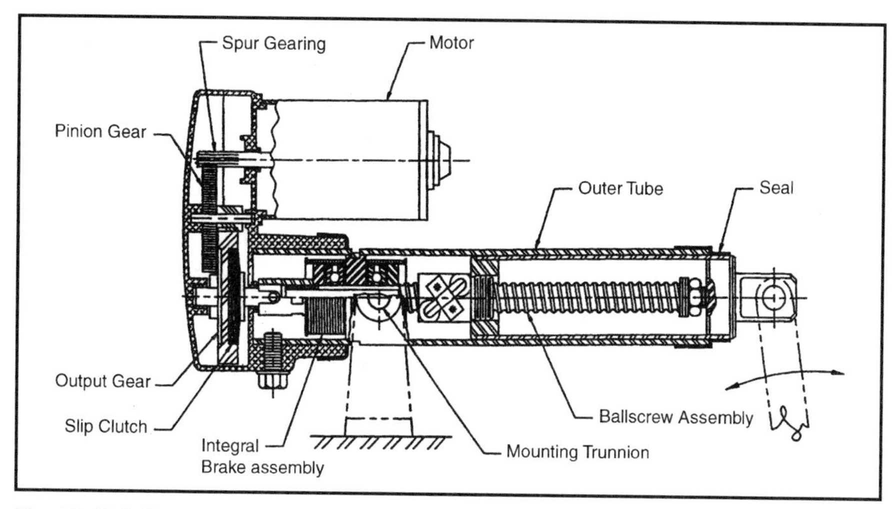

Fig. 16 This linear actuator can be powered by either an AC or DC motor. It contains ballscrew, reduction gear, clutch, and brake assemblies.

rotor depends on its geometry. A disadvantage of the hybrid stepper motor operating open-loop is that, if overtorqued, its position "memory" is lost and the system must be reinitialized. Stepper motors can perform precise positioning in simple open-loop control systems if they operate at low acceleration rates with static loads. However, if higher acceleration values are required for driving variable loads, the stepper motor must be operated in a closed loop with a position sensor.

DC and AC Motor Linear Actuators

Actuators for motion control systems are available in many different forms, including both linear and rotary versions. One popular configuration is that of a Thomson Saginaw PPA, shown in section view in Fig. 16. It consists of an AC or DC motor mounted parallel to either a ballscrew or Acme screw assembly through a reduction gear assembly with a slip clutch and integral brake assembly. Linear actuators of this type can perform a wide range of commercial, industrial, and institutional applications.

One version designed for mobile applications can be powered by a 12-, 24-, or 36-VDC permanent-magnet motor. These motors are capable of performing such tasks as positioning antenna reflectors, opening and closing security gates, handling materials, and raising and lowering scissors-type lift tables, machine hoods, and light-duty jib crane arms.

Other linear actuators are designed for use in fixed locations where either 120- or 220-VAC line power is available. They can have either AC or DC motors. Those with 120-VAC motors can be equipped with optional electric brakes that virtually eliminate coasting, thus permitting point-to-point travel along the stroke.

Where variable speed is desired and 120-VAC power is available, a linear actuator with a 90-VDC motor can be equipped with a solid-state rectifier/speed controller. Closed-loop feedback provides speed regulation down to one-tenth of the maximum travel rate. This feedback system can maintain its selected travel rate despite load changes.

Thomson Saginaw also offers its linear actuators with either Hall-effect or potentiometer sensors for applications where it is necessary or desirable to control actuator positioning. With Hall-effect sensing, six pulses are generated with each turn of the output shaft during which the stroke travels approximately $\frac{1}{32}$ in. (0.033 in. or 0.84 mm). These pulses can be counted by a separate control unit and added or subtracted from the stored pulse count in the unit's memory. The actuator can be stopped at any 0.033-in. increment of travel along the stroke selected by programming. A limit switch can be used together with this sensor.

If a 10-turn, 10,000-ohm potentiometer is used as a sensor, it can be driven by the output shaft through a spur gear. The gear ratio is established to change the resistance from 0 to 10,000 ohms over the length of the actuator stroke. A separate control unit measures the resistance (or voltage) across the potentiometer, which varies continuously and linearly with stroke travel. The actuator can be stopped at any position along its stroke.

Stepper-Motor Based Linear Actuators

Linear actuators are available with axial integral threaded shafts and bolt nuts that convert rotary motion to linear motion. Powered by fractional horsepower permanent-magnet stepper motors, these linear actuators are capable of positioning light loads. Digital pulses fed to the actuator cause the threaded shaft to rotate, advancing or retracting it so that a load coupled to the shaft can be moved backward or forward. The bidirectional digital linear actuator shown in Fig. 17 can provide linear resolution as fine as 0.001 in. per pulse. Travel per step is determined by the pitch of the leadscrew and step angle of the motor. The maximum linear force for the model shown is 75 oz.

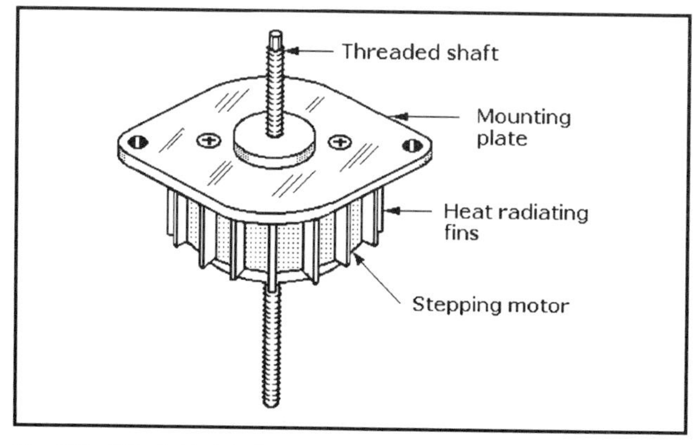

Fig. 17 This light-duty linear actuator based on a permanent-magnet stepping motor has a shaft that advances or retracts.

SERVOSYSTEM FEEDBACK SENSORS

A servosystem feedback sensor in a motion control system transforms a physical variable into an electrical signal for use by the motion controller. Common feedback sensors are encoders, resolvers, and linear variable differential transformers (LVDTs) for motion and position feedback, and tachometers for velocity feedback. Less common but also in use as feedback devices are potentiometers, linear velocity transducers (LVTs), angular displacement transducers (ADTs), laser interferometers, and potentiometers. Generally speaking, the closer the feedback sensor is to the variable being controlled, the more accurate it will be in assisting the system to correct velocity and position errors.

For example, direct measurement of the linear position of the carriage carrying the load or tool on a single-axis linear guide will provide more accurate feedback than an indirect measurement determined from the angular position of the guide's leadscrew and knowledge of the drivetrain geometry between the sensor and the carriage. Thus, direct position measurement avoids drivetrain errors caused by backlash, hysteresis, and leadscrew wear that can adversely affect indirect measurement.

Rotary Encoders

Rotary encoders, also called *rotary shaft encoders* or *rotary shaft-angle* encoders, are electromechanical transducers that convert shaft rotation into output pulses, which can be counted to measure shaft revolutions or shaft angle. They provide rate and positioning information in servo feedback loops. A rotary encoder can sense a number of discrete positions per revolution. The number is called *points per revolution* and is analogous to the *steps per revolution* of a stepper motor. The speed of an encoder is in units of counts per second. Rotary encoders can measure the motor-shaft or leadscrew angle to report position indirectly, but they can also measure the response of rotating machines directly.

The most popular rotary encoders are *incremental optical shaft-angle encoders* and the *absolute optical shaft-angle encoders*. There are also *direct contact* or *brush-type* and *magnetic rotary encoders*, but they are not as widely used in motion control systems.

Commercial rotary encoders are available as standard or catalog units, or they can be custom made for unusual applications or survival in extreme environments. Standard rotary encoders are packaged in cylindrical cases with diameters from 1.5 to 3.5 in. Resolutions range from 50 cycles per shaft revolution to 2,304,000 counts per revolution. A variation of the conventional configuration, the *hollow-shaft encoder,* eliminates problems associated with the installation and shaft runout of conventional models. Models with hollow shafts are available for mounting on shafts with diameters of 0.04 to 1.6 in. (1 to 40 mm).

Incremental Encoders

The basic parts of an incremental optical shaft-angle encoder are shown in Fig. 1. A glass or plastic code disk mounted on the encoder shaft rotates between an internal light source, typically a light-emitting diode (LED), on one side and a mask and matching photodetector assembly on the other side. The incremental code disk contains a pattern of equally spaced opaque and transparent segments or spokes that radiate out from its center as shown. The electronic signals that are generated by the encoder's electronics board are fed into a motion controller that calculates position and velocity information for feedback purposes. An exploded view of an industrial-grade incremental encoder is shown in Fig. 2.

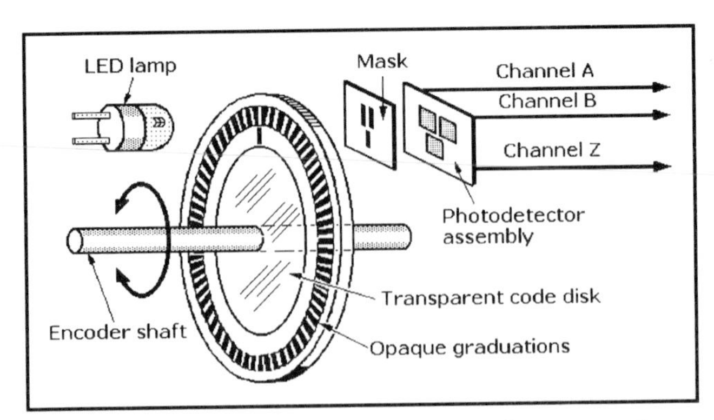

Fig. 1 Basic elements of an incremental optical rotary encoder.

Glass code disks containing finer graduations capable of 11 to more than 16-bit resolution are used in high-resolution encoders, and plastic (Mylar) disks capable of 8- to 10-bit resolution are used in the more rugged encoders that are subject to shock and vibration.

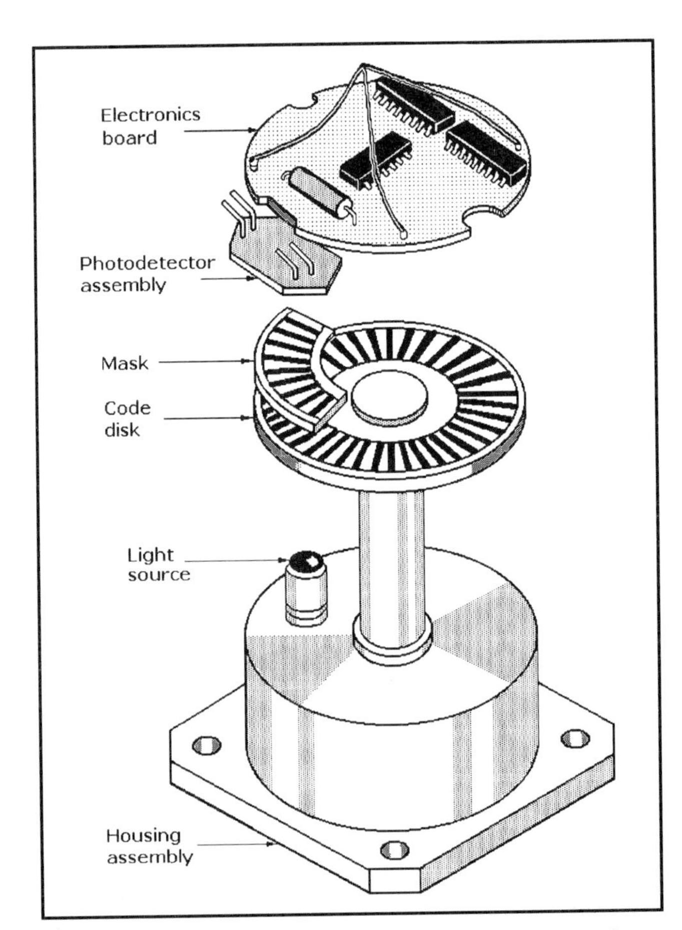

Fig. 2 Exploded view of an incremental optical rotary encoder showing the stationary mask between the code wheel and the photodetector assembly.

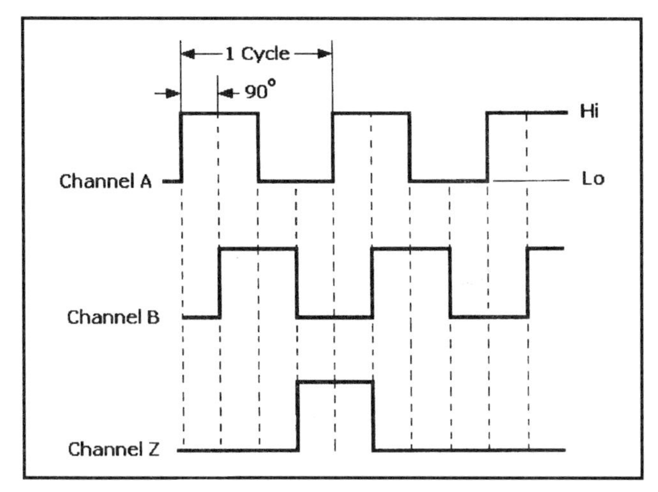

Fig. 3 Channels A and B provide bidirectional position sensing. If channel A leads channel B, the direction is clockwise; if channel B leads channel A, the direction is counterclockwise. Channel Z provides a zero reference for determining the number of disk rotations.

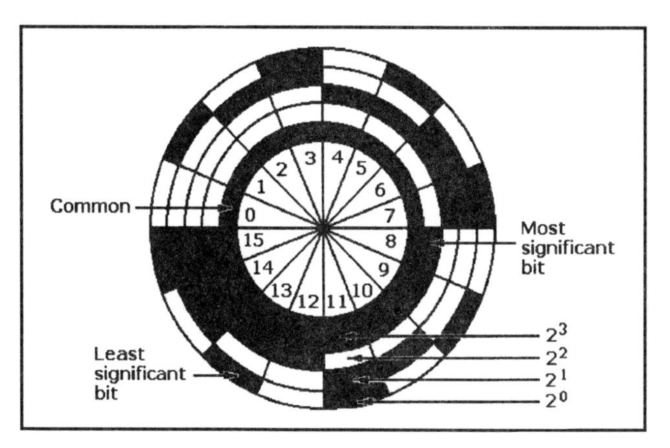

Fig. 4 Binary-code disk for an absolute optical rotary encoder. Opaque sectors represent a binary value of 1, and the transparent sectors represent binary 0. This four-bit binary-code disk can count from 1 to 15.

The quadrature encoder is the most common type of incremental encoder. Light from the LED passing through the rotating code disk and mask is "chopped" before it strikes the photodetector assembly. The output signals from the assembly are converted into two channels of square pulses (A and B) as shown in Fig. 3. The number of square pulses in each channel is equal to the number of code disk segments that pass the photodetectors as the disk rotates, but the waveforms are 90° out of phase. If, for example, the pulses in channel A lead those in channel B, the disk is rotating in a clockwise direction, but if the pulses in channel A lag those in channel B lead, the disk is rotating counterclockwise. By monitoring both the number of pulses and the relative phases of signals A and B, both position and direction of rotation can be determined.

Many incremental quadrature encoders also include a third output Z channel to obtain a zero reference or index signal that occurs once per revolution. This channel can be gated to the A and B quadrature channels and used to trigger certain events accurately within the system. The signal can also be used to align the encoder shaft to a mechanical reference.

Absolute Encoders

An *absolute shaft-angle optical encoder* contains multiple light sources and photodetectors, and a code disk with up to 20 tracks of segmented patterns arranged as annular rings, as shown in Fig. 4. The code disk provides a binary output that uniquely defines each shaft angle, thus providing an absolute measurement. This type of encoder is organized in essentially the same way as the incremental encoder shown in Fig. 2, but the code disk rotates between linear arrays of LEDs and photodetectors arranged radially, and a LED opposes a photodetector for each track or annular ring.

The arc lengths of the opaque and transparent sectors decrease with respect to the radial distance from the shaft. These disks, also made of glass or plastic, produce either the natural binary or Gray code. Shaft position accuracy is proportional to the number of annular rings or tracks on the disk. When the code disk rotates, light passing through each track or annular ring generates a continuous stream of signals from the detector array. The electronics board converts that output into a binary word. The value of the output code word is read radially from the most significant bit (MSB) on the inner ring of the disk to the least significant bit (LSB) on the outer ring of the disk.

The principal reason for selecting an absolute encoder over an incremental encoder is that its code disk retains the last angular position of the encoder shaft whenever it stops moving, whether the system is shut down deliberately or as a result of power failure. This means that the last readout is preserved, an important feature for many applications.

Linear Encoders

Linear encoders can make direct accurate measurements of unidirectional and reciprocating motions of mechanisms with high resolution and repeatability. Figure 5 illustrates the basic parts of an optical linear encoder. A movable scanning unit contains the light source, lens, graduated glass scanning reticule, and an array of photocells. The scale, typically made as a strip of glass with opaque graduations, is bonded to a supporting structure on the host machine.

A beam of light from the light source passes through the lens, four windows of the scanning reticule, and the glass scale to the array of photocells. When the scanning unit moves, the scale modulates the light beam so that the photocells generate sinusoidal signals.

The four windows in the scanning reticule are each 90° apart in phase. The encoder combines the phase-shifted signal to produce

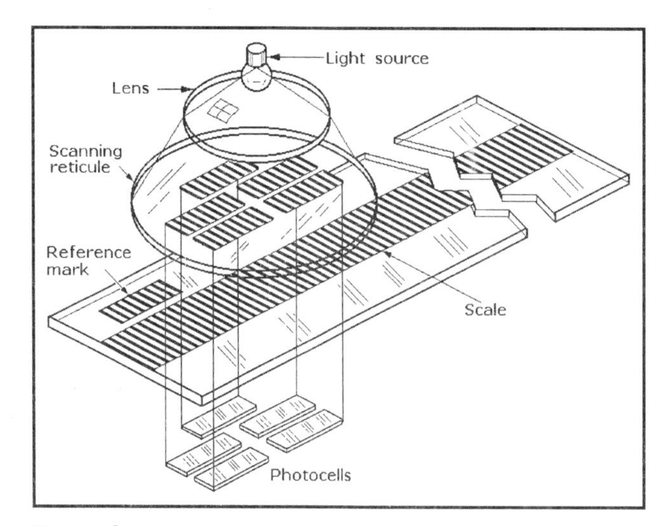

Fig. 5 Optical linear encoders direct light through a moving glass scale with accurately etched graduations to photocells on the opposite side for conversion to a distance value.

two symmetrical sinusoidal outputs that are phase shifted by 90°. A fifth pattern on the scanning reticule has a random graduation that, when aligned with an identical reference mark on the scale, generates a reference signal.

A fine-scale pitch provides high resolution. The spacing between the scanning reticule and the fixed scale must be narrow and constant to eliminate undesirable diffraction effects of the scale grating. The complete scanning unit is mounted on a carriage that moves on ball bearings along the glass scale. The scanning unit is connected to the host machine slide by a coupling that compensates for any alignment errors between the scale and the machine guideways.

External electronic circuitry interpolates the sinusoidal signals from the encoder head to subdivide the line spacing on the scale so that it can measure even smaller motion increments. The practical maximum length of linear encoder scales is about 10 ft (3 m), but commercial catalog models are typically limited to about 6 ft (2 m). If longer distances are to be measured, the encoder scale is made of steel tape with reflective graduations that are sensed by an appropriate photoelectric scanning unit.

Linear encoders can make direct measurements that overcome the inaccuracies inherent in mechanical stages due to backlash, hysteresis, and leadscrew error. However, the scale's susceptibility to damage from metallic chips, grit oil, and other contaminants, together with its relatively large space requirements, limits its applications for these encoders.

Commercial linear encoders are available as standard catalog models, or they can be custom made for specific applications or extreme environmental conditions. There are both fully enclosed and open linear encoders with travel distances from 2 in. to 6 ft (50 mm to 1.8 m). Some commercial models are available with resolutions down to 0.07 μm, and others can operate at speeds of up to 16.7 ft/s (5 m/s).

Magnetic Encoders

Magnetic encoders can be made by placing a transversely polarized permanent magnet in close proximity to a Hall-effect device sensor. Figure 6 shows a magnet mounted on a motor shaft in close proximity to a two-channel HED array which detects changes in magnetic flux density as the magnet rotates. The output signals from the sensors are transmitted to the motion controller. The encoder output, either a square wave or a quasi sine wave (depending on the type of magnetic sensing device) can be used to count revolutions per minute (rpm) or determine motor shaft accurately. The phase shift between channels A and B permits them to be compared by the motion controller to determine the direction of motor shaft rotation.

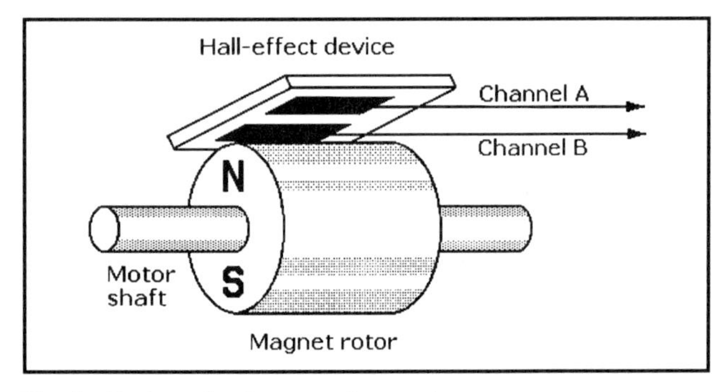

Fig. 6 Basic parts of a magnetic encoder.

Resolvers

A resolver is essentially a rotary transformer that can provide position feedback in a servosystem as an alternative to an encoder. Resolvers resemble small AC motors, as shown in Fig. 7, and

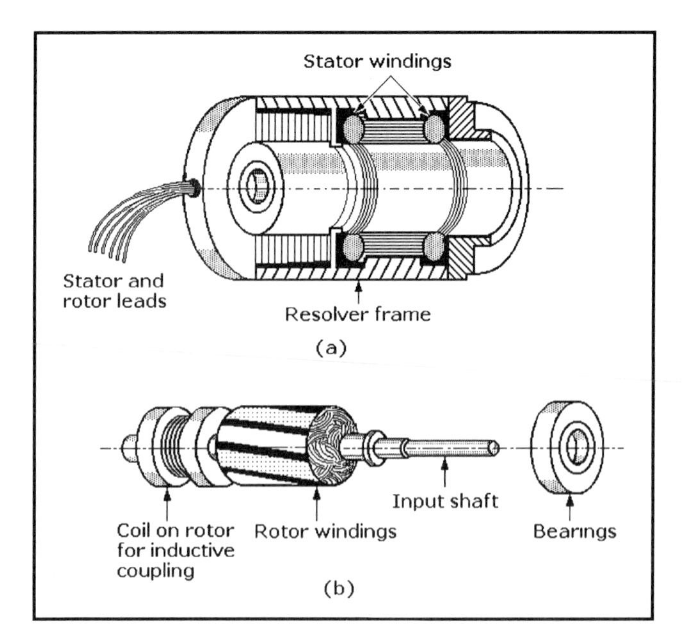

Fig. 7 Exploded view of a brushless resolver frame (a), and rotor and bearings (b). The coil on the rotor couples speed data inductively to the frame for processing.

generate an electrical signal for each revolution of their shaft. Resolvers that sense position in closed-loop motion control applications have one winding on the rotor and a pair of windings on the stator, oriented at 90°. The stator is made by winding copper wire in a stack of iron laminations fastened to the housing, and the rotor is made by winding copper wire in a stack of laminations mounted on the resolver's shaft.

Figure 8 is an electrical schematic for a brushless resolver showing the single rotor winding and the two stator windings 90° apart. In a servosystem, the resolver's rotor is mechanically coupled to the drive motor and load. When a rotor winding is excited by an AC reference signal, it produces an AC voltage output that varies in amplitude according to the sine and cosine of shaft position. If the phase shift between the applied signal to the rotor and the induced signal appearing on the stator coil is measured, that angle is an analog of rotor position. The absolute position of the load being driven can be determined by the ratio of the sine output amplitude to the cosine output amplitude as the resolver shaft turns through one revolution. (A single-speed resolver produces one sine and one cosine wave as the output for each revolution.)

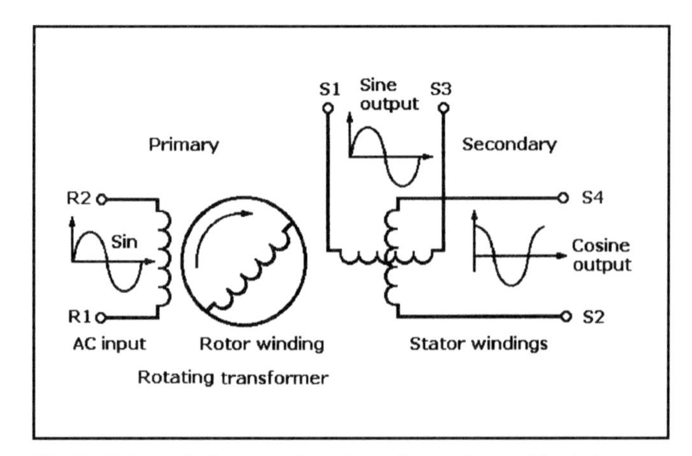

Fig. 8 Schematic for a resolver shows how rotor position is transformed into sine and cosine outputs that measure rotor position.

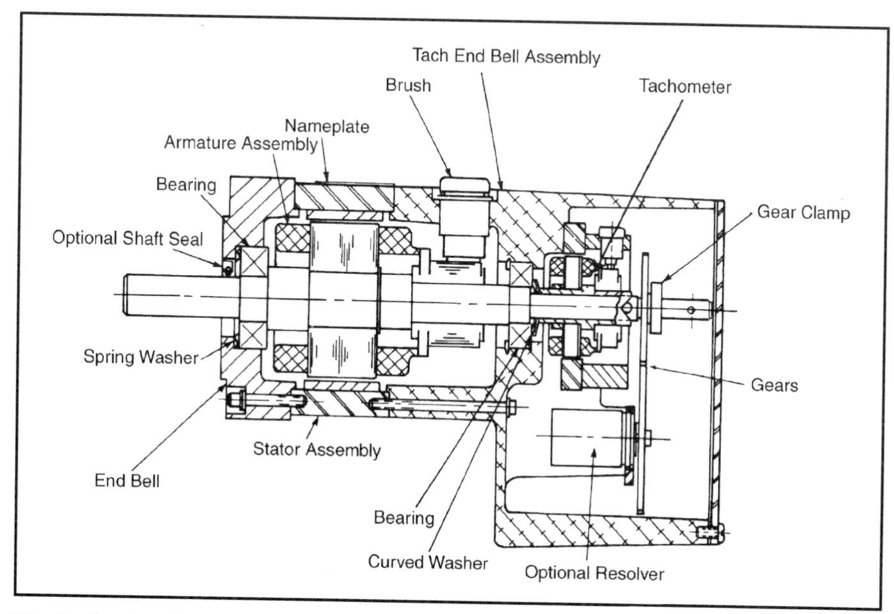

Fig. 9 Section view of a resolver and tachometer in the same frame as the servomotor.

Connections to the rotor of some resolvers can be made by brushes and slip rings, but resolvers for motion control applications are typically brushless. A rotating transformer on the rotor couples the signal to the rotor inductively. Because brushless resolvers have no slip rings or brushes, they are more rugged than encoders and have operating lives that are up to ten times those of brush-type resolvers. Bearing failure is the most likely cause of resolver failure. The absence of brushes in these resolvers makes them insensitive to vibration and contaminants. Typical brushless resolvers have diameters from 0.8 to 3.7 in. Rotor shafts are typically threaded and splined.

Most brushless resolvers can operate over a 2- to 40-volt range, and their winding are excited by an AC reference voltage at frequencies from 400 to 10,000 Hz. The magnitude of the voltage induced in any stator winding is proportional to the cosine of the angle, q, between the rotor coil axis and the stator coil axis. The voltage induced across any pair of stator terminals will be the vector sum of the voltages across the two connected coils. Accuracies of ±1 arc-minute can be achieved.

In feedback loop applications, the stator's sinusoidal output signals are transmitted to a resolver-to-digital converter (RDC), a specialized analog-to-digital converter (ADC) that converts the signals to a digital representation of the actual angle required as an input to the motion controller.

Tachometers

A tachometer is a DC generator that can provide velocity feedback for a servosystem. The tachometer's output voltage is directly proportional to the rotational speed of the armature shaft that drives it. In a typical servosystem application, it is mechanically coupled to the DC motor and feeds its output voltage back to the controller and amplifier to control drive motor and load speed. A cross-sectional drawing of a tachometer built into the same housing as the DC motor and a resolver is shown in Fig. 9. Encoders or resolvers are part of separate loops that provide position feedback.

As the tachometer's armature coils rotate through the stator's magnetic field, lines of force are cut so that an electromotive force is induced in each of its coils. This emf is directly proportional to the rate at which the magnetic lines of force are cut as well as being directly proportional to the velocity of the motor's drive shaft. The direction of the emf is determined by Fleming's generator rule.

The AC generated by the armature coil is converted to DC by the tachometer's commutator, and its value is directly proportional to shaft rotation speed while its polarity depends on the direction of shaft rotation.

There are two basic types of DC tachometer: *shunt wound* and *permanent magnet* (PM), but PM tachometers are more widely used in servosystems today. There are also moving-coil tachometers which, like motors, have no iron in their armatures. The armature windings are wound from fine copper wire and bonded with glass fibers and polyester resins into a rigid cup, which is bonded to its coaxial shaft. Because this armature contains no iron, it has lower inertia than conventional copper and iron armatures, and it exhibits low inductance. As a result, the moving-coil tachometer is more responsive to speed changes and provides a DC output with very low ripple amplitudes.

Tachometers are available as stand-alone machines. They can be rigidly mounted to the servomotor housings, and their shafts can be mechanically coupled to the servomotor's shafts. If the DC servomotor is either a brushless or moving-coil motor, the stand-alone tachometer will typically be brushless and, although they are housed separately, a common armature shaft will be shared.

A brush-type DC motor with feedback furnished by a brush-type tachometer is shown in Fig. 10. Both tachometer and motor rotor coils are mounted on a common shaft. This arrangement provides a high resonance frequency. Moreover, the need for separate tachometer bearings is eliminated.

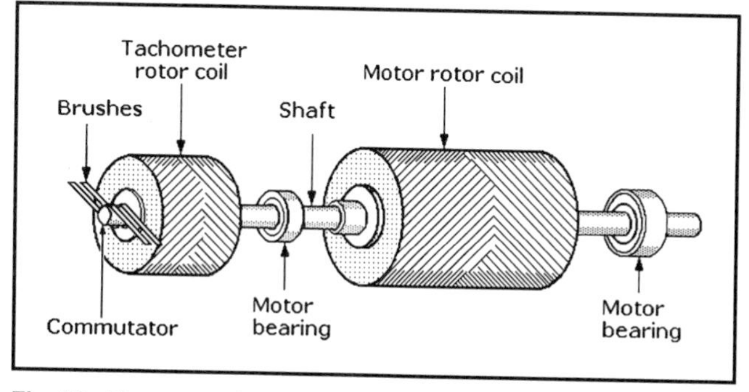

Fig. 10 The rotors of the DC motor and tachometer share a common shaft.

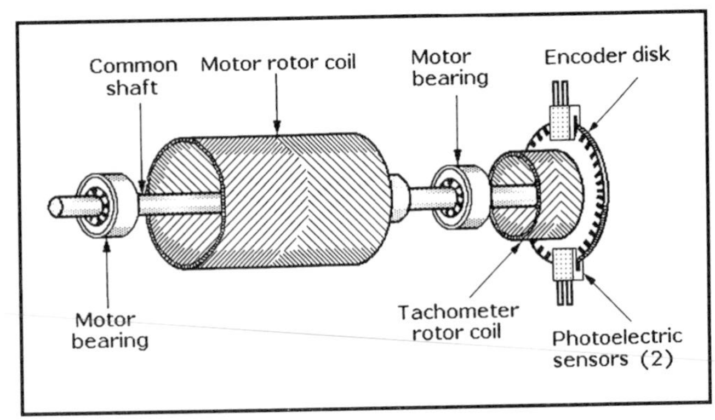

Fig. 11 This coil-type DC motor obtains velocity feedback from a tachometer whose rotor coil is mounted on a common shaft and position feedback from a two-channel photoelectric encoder whose code disk is also mounted on the same shaft.

In applications where precise positioning is required in addition to speed regulation, an incremental encoder can be added on the same shaft, as shown in Fig. 11.

Linear Variable Differential Transformers (LVDTs)

A linear variable differential transformer (LVDT) is a sensing transformer consisting of a primary winding, two adjacent secondary windings, and a ferromagnetic core that can be moved axially within the windings, as shown in the cutaway view in Fig. 12. LVDTs are capable of measuring position, acceleration, force, or pressure, depending on how they are installed. In motion control systems, LVDTs provide position feedback by measuring the variation in mutual inductance between their primary and secondary windings caused by the linear movement of the ferromagnetic core.

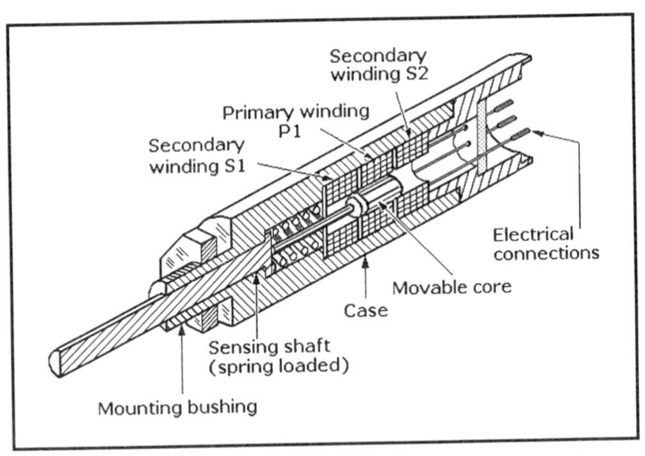

Fig. 12 Cutaway view of a linear variable displacement transformer (LVDT).

The core is attached to a spring-loaded sensing shaft. When depressed, the shaft moves the core axially within the windings, coupling the excitation voltage in the primary (middle) winding P1 to the two adjacent secondary windings S1 and S2.

Figure 13 is a schematic diagram of a LVDT. When the core is centered between S1 and S2, the voltages induced in S1 and S2 have equal amplitudes and are 180° out of phase. With a series-opposed connection, as shown, the net voltage across the secondaries is zero because both voltages cancel. This is called the *null position* of the core.

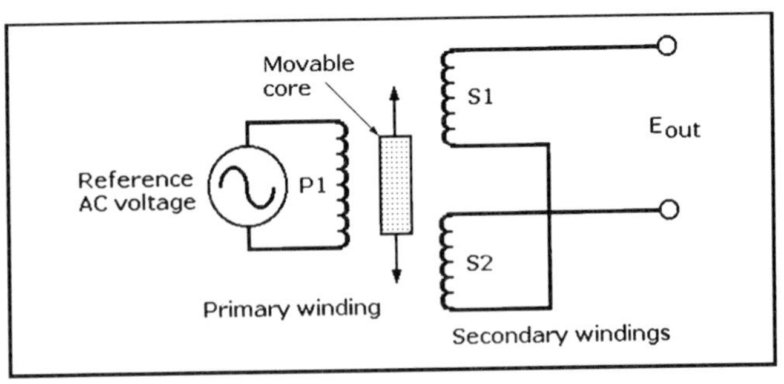

Fig. 13 Schematic for a linear variable differential transformer (LVDT) showing how the movable core interacts with the primary and secondary windings.

However, if the core is moved to the left, secondary winding S1 is more strongly coupled to primary winding P1 than secondary winding S2, and an output sine wave in phase with the primary voltage is induced. Similarly, if the core is moved to the right and winding S2 is more strongly coupled to primary winding P1, an output sine wave that is 180° out-of-phase with the primary voltage is induced. The amplitudes of the output sine waves of the LVDT vary symmetrically with core displacement, either to the left or right of the null position.

Linear variable differential transformers require signal conditioning circuitry that includes a stable sine wave oscillator to excite the primary winding P1, a demodulator to convert secondary AC voltage signals to DC, a low-pass filter, and an amplifier to buffer the DC output signal. The amplitude of the resulting DC voltage output is proportional to the magnitude of core displacement, either to the left or right of the null position. The phase of the DC voltage indicates the position of the core relative to the null (left or right). A LVDT containing an integral oscillator/demodulator is a DC-to-DC LVDT, also known as a DCDT.

Linear variable differential transformers can make linear displacement (position) measurements as precise as 0.005 in. (0.127 mm). Output voltage linearity is an important LVDT characteristic, and it can be plotted as a straight line within a specified range. Linearity is the characteristic that largely determines the LVDT's absolute accuracy.

Linear Velocity Transducers (LVTs)

A linear velocity transducer (LVT) consists of a magnet positioned axially within two wire coils. When the magnet is moved through the coils, it induces a voltage within the coils in accordance with the Faraday and Lenz laws. The output voltage from the coils is directly proportional to the magnet's field strength and axial velocity over its working range.

When the magnet is functioning as a transducer, both of its ends are within the two adjacent coils, and when it is moved axially, its north pole will induce a voltage in one coil and its south pole will induce a voltage in the other coil. The two coils can be connected in series or parallel, depending on the application. In both configurations, the DC output voltage from the coils is proportional to magnet velocity. (A single coil would only produce zero voltage because the voltage generated by the north pole would be canceled by the voltage generated by the south pole.)

The characteristics of the LVT depend on how the two coils are connected. If they are connected in series opposition, the output is added and maximum sensitivity is obtained. Also, noise generated in one coil will be canceled by the noise generated in the other coil. However, if the coils are connected in parallel, both sensitivity and source impedance are reduced. Reduced sensitivity improves high-frequency response for measuring high velocities, and the lower output impedance improves the LVT's compatibility with its signal-conditioning electronics.

Angular Displacement Transducers (ATDs)

An angular displacement transducer (ATD) is an air-core variable differential capacitor that can sense angular displacement. As shown in exploded view in Fig. 14, it has a movable metal rotor sandwiched between a single stator plate and segmented stator plates. When a high-frequency AC signal from an oscillator is placed across the plates, it is modulated by the change in capacitance value due to the position of the rotor with respect to the segmented stator plates. The angular displacement of the rotor can then be determined accurately from the demodulated AC signal.

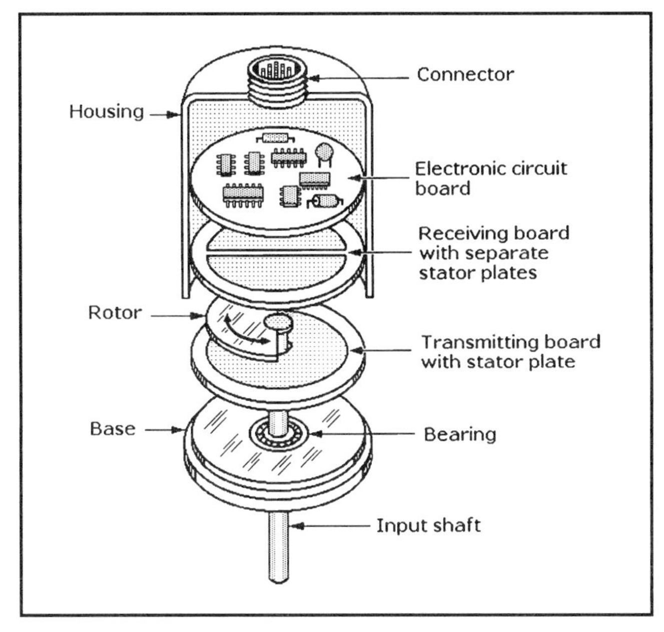

Fig. 14 Exploded view of an angular displacement transducer (ADT) based on a differential variable capacitor.

The base is the mounting platform for the transducer assembly. It contains the axial ball bearing that supports the shaft to which the rotor is fastened. The base also supports the transmitting board, which contains a metal surface that forms the lower plate of the differential capacitor. The semicircular metal rotor mounted on the shaft is the variable plate or rotor of the capacitor. Positioned above the rotor is the receiving board containing two separate semicircular metal sectors on its lower surface. The board acts as the receiver for the AC signal that has been modulated by the capacitance difference between the plates caused by rotor rotation.

An electronics circuit board mounted on top of the assembly contains the oscillator, demodulator, and filtering circuitry. The ADT is powered by DC, and its output is a DC signal that is proportional to angular displacement. The cup-shaped housing encloses the entire assembly, and the base forms a secure cap.

DC voltage is applied to the input terminals of the ADT to power the oscillator, which generates a 400- to 500-kHz voltage that is applied across the transmitting and receiving stator plates. The receiving plates are at virtual ground, and the rotor is at true ground. The capacitance value between the transmitting and receiving plates remains constant, but the capacitance between the separate receiving plates varies with rotor position.

A null point is obtained when the rotor is positioned under equal areas of the receiving stator plates. In that position, the capacitance between the transmitting stator plate and the receiving stator plates will be equal, and there will be no output voltage. However, as the rotor moves clockwise or counterclockwise, the capacitance between the transmitting plate and one of the receiving plates will be greater than it is between the other receiving plate. As a result, after demodulation, the differential output DC voltage will be proportional to the angular distance the rotor moved from the null point.

Inductosyns

The Inductosyn is a proprietary AC sensor that generates position feedback signals that are similar to those from a resolver. There are rotary and linear Inductosyns. Much smaller than a resolver, a rotary Inductosyn is an assembly of a scale and slider on insulating substrates in a loop. When the scale is energized with AC, the voltage couples into the two slider windings and induces voltages proportional to the sine and cosine of the slider spacing within a cyclic pitch.

An Inductosyn-to-digital (I/D) converter, similar to a resolver-to-digital (R/D) converter, is needed to convert these signals into a digital format. A typical rotary Inductosyn with 360 cyclic pitches per rotation can resolve a total of 1,474,560 sectors for each resolution. This corresponds to an angular rotation of less than 0.9 arc-s. This angular information in a digital format is sent to the motion controller.

Laser Interferometers

Laser interferometers provide the most accurate position feedback for servosystems. They offer very high resolution (to 1.24 nm), noncontact measurement, a high update rate, and intrinsic accuracies of up to 0.02 ppm. They can be used in servosystems either as passive position readouts or as active feedback sensors in a position servo loop. The laser beam path can be precisely aligned to coincide with the load or a specific point being measured, eliminating or greatly reducing Abbe error.

A single-axis system based on the Michaelson interferometer is illustrated in Fig. 15. It consists of a helium–neon laser, a polarizing beam splitter with a stationary retroreflector, a moving retroreflector that can be mounted on the object whose position is to be measured, and a photodetector, typically a photodiode.

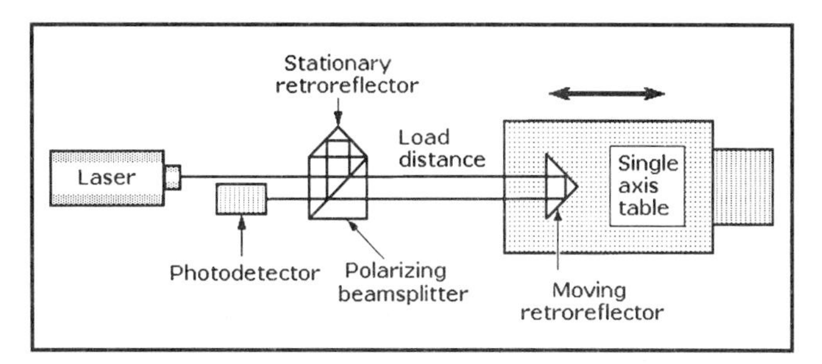

Fig. 15 Diagram of a laser interferometer for position feedback that combines high resolution with noncontact sensing, high update rates, and accuracies of 0.02 ppm.

Light from the laser is directed toward the polarizing beam splitter, which contains a partially reflecting mirror. Part of the laser beam goes straight through the polarizing beam splitter, and part of the laser beam is reflected. The part that goes straight through the beam splitter reaches the moving reflectometer, which reflects it back to the beam splitter, that passes it on to the photodetector. The part of the beam that is reflected by the beam splitter reaches the stationary retroreflector, a fixed distance away. The retroreflector reflects it back to the beam splitter before it is also reflected into the photodetector.

As a result, the two reflected laser beams strike the photodetector, which converts the combination of the two light beams into an electrical signal. Because of the way laser light beams interact, the output of the detector depends on a *difference* in the distances traveled by the two laser beams. Because both light beams travel the same distance from the laser to the beam splitter and from the beam splitter to the photodetector, these distances are not involved in position measurement. The laser interferometer measurement depends only on the difference in distance between the round trip laser beam travel from the beam splitter to the moving retroreflector and the fixed round trip distance of laser beam travel from the beam splitter to the stationary retroreflector.

If these two distances are exactly the same, the two light beams will recombine in phase at the photodetector, which will produce a high electrical output. This event can be viewed on a video display as a bright *light fringe*. However, if the difference between the distances is as short as one-quarter of the laser's wavelength, the light beams will combine out-of-phase, interfering with each other so that there will be no electrical output from the photodetector and no video output on the display, a condition called a *dark fringe*.

As the moving retroreflector mounted on the load moves farther away from the beam splitter, the laser beam path length will increase and a pattern of light and dark fringes will repeat uniformly. This will result in electrical signals that can be counted and converted to a distance measurement to provide an accurate position of the load. The spacing between the light and dark fringes and the resulting electrical pulse rate is determined by the wavelength of the light from the laser. For example, the wavelength of the light beam emitted by a helium–neon (He–Ne) laser, widely used in laser interferometers, is 0.63 μm, or about 0.000025 in.

Thus, the accuracy of load position measurement depends primarily on the known stabilized wavelength of the laser beam. However, that accuracy can be degraded by changes in humidity and temperature as well as airborne contaminants such as smoke or dust in the air between the beam splitter and the moving retroreflector.

Precision Multiturn Potentiometers

The rotary precision multiturn potentiometer shown in the cutaway in Fig. 16 is a simple, low-cost feedback instrument.

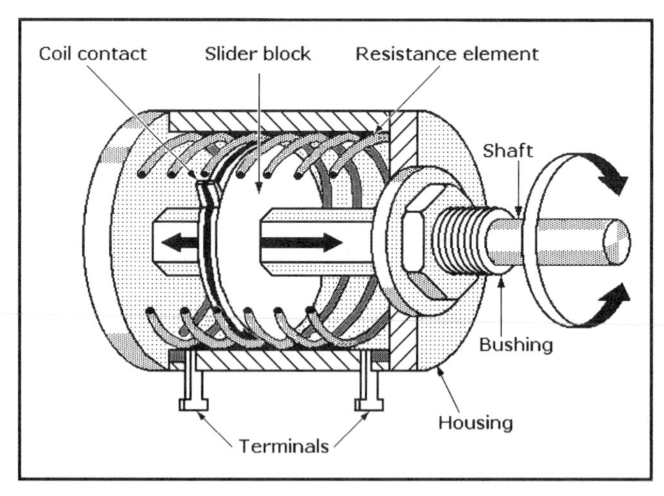

Fig. 16 A precision potentiometer is a low-cost, reliable feedback sensor for servosystems.

Originally developed for use in analog computers, precision potentiometers can provide absolute position data in analog form as a resistance value or voltage. Precise and resettable voltages correspond to each setting of the rotary control shaft. If a potentiometer is used in a servosystem, the analog data will usually be converted to digital data by an integrated circuit analog-to-digital converter (ADC). Accuracies of 0.05% can be obtained from an instrument-quality precision multiturn potentiometer, and resolutions can exceed 0.005° if the output signal is converted with a 16-bit ADC.

Precision multiturn potentiometers have wirewound or hybrid resistive elements. Hybrid elements are wirewound elements coated with resistive plastic to improve their resolution. To obtain an output from a potentiometer, a conductive wiper must be in contact with the resistive element. During its service life, wear on the resistive element caused by the wiper can degrade the precision of the precision potentiometer.

Solenoids: An Economical Choice for Linear or Rotary Motion

A solenoid is an electromechanical device that converts electrical energy into linear or rotary mechanical motion. All solenoids include a coil for conducting current and generating a magnetic field, an iron or steel shell or case to complete the magnetic circuit, and a plunger or armature for translating motion. Solenoids can be actuated by either direct current (DC) or rectified alternating current (AC).

Solenoids are built with conductive paths that transmit maximum magnetic flux density with minimum electrical energy input. The mechanical action performed by the solenoid depends on the design of the plunger in a linear solenoid or the armature in a rotary solenoid. Linear solenoid plungers are either spring-loaded or use external methods to restrain axial movement caused by the magnetic flux when the coil is energized and restore it to its initial position when the current is switched off.

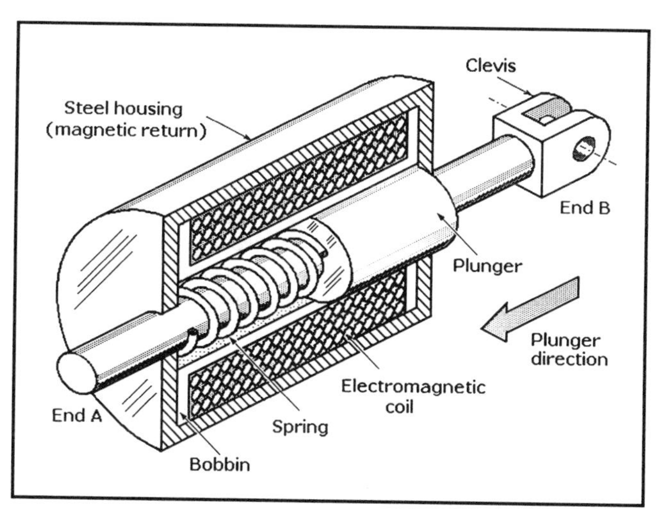

Fig. 1 The pull-in and push-out functions of a solenoid are shown. End A of the plunger pushes out when the solenoid is energized while the clevis-end B pulls in.

Cutaway drawing in Fig. 1 illustrates how pull-in and push-out actions are performed by a linear solenoid. When the coil is energized, the plunger pulls in against the spring, and this motion can be translated into either a "pull-in" or a "push-out" response. All solenoids are basically pull-in-type actuators, but the location of the plunger extension with respect to the coil and spring determines its function. For example, the plunger extension on the left end (end A) provides "push-out" motion against the load, while a plunger extension on the right end terminated by a clevis (end B) provides "pull-in" motion. Commercial solenoids perform only one of these functions. Figure 2 is a cross-sectional view of a typical pull-in commercial linear solenoid.

Rotary solenoids operate on the same principle as linear solenoids except that the axial movement of the armature is converted into rotary movement by various mechanical devices. One of these is the use of internal lands or ball bearings and slots or races that convert a pull-in stroke to rotary or twisting motion.

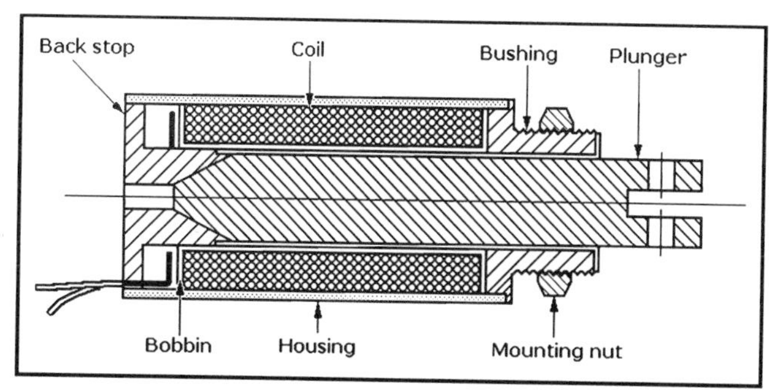

Fig. 2 Cross-section view of a commercial linear pull-type solenoid with a clevis. The conical end of the plunger increases its efficiency. The solenoid is mounted with its threaded bushing and nut.

Motion control and process automation systems use many different kinds of solenoids to provide motions ranging from simply turning an event on or off to the performance of extremely complex sequencing. When there are requirements for linear or rotary motion, solenoids should be considered because of their relatively small size and low cost when compared with alternatives such as motors or actuators. Solenoids are easy to install and use, and they are both versatile and reliable.

Technical Considerations

Important factors to consider when selecting solenoids are their rated torque/force, duty cycles, estimated working lives, performance curves, ambient temperature range, and temperature rise. The solenoid must have a magnetic return path capable of transmitting the maximum amount of magnetic flux density with minimum energy input. Magnetic flux lines are transmitted to the plunger or armature through the bobbin and air gap back through the iron or steel shell. A ferrous metal path is more efficient than air, but the air gap is needed to permit plunger or armature movement. The force or torque of a solenoid is inversely proportional to the square of the distance between pole faces. By optimizing the ferrous path area, the shape of the plunger or armature, and the magnetic circuit material, the output torque/force can be increased.

The torque/force characteristic is an important solenoid specification. In most applications the force can be a minimum at the start of the plunger or armature stroke but must increase at a rapid rate to reach the maximum value before the plunger or armature reaches the backstop.

The magnetizing force of the solenoid is proportional to the number of copper wire turns in its coil, the magnitude of the current, and the permeance of the magnetic circuit. The pull force required by the load must not be greater than the force developed by the solenoid during any portion of its required stroke, or the plunger or armature will not pull in completely. As a result, the load will not be moved the required distance.

Heat buildup in a solenoid is a function of power and the length of time the power is applied. The permissible temperature rise limits the magnitude of the input power. If constant voltage is applied, heat buildup can degrade the efficiency of the coil by effectively reducing its number of ampere turns. This, in turn, reduces flux density and torque/force output. If the temperature of the coil is permitted to rise above the temperature rating of its insulation, performance will suffer and the solenoid could fail prematurely. Ambient temperature in excess of the specified limits will limit the solenoid cooling expected by convection and conduction.

Heat can be dissipated by cooling the solenoid with forced air from a fan or blower, mounting the solenoid on a heat sink, or circulating a liquid coolant through a heat sink. Alternatively, a larger solenoid than the one actually needed could be used.

The heating of the solenoid is affected by the duty cycle, which is specified from 10% to 100%, and is directly proportional to solenoid *on* time. The highest starting and ending torque are obtained with the lowest duty cycle and *on* time. Duty cycle is defined as the ratio of *on* time to the sum of *on* time and *off* time. For example, if a solenoid is energized for 30 s and then turned off for 90 s, its duty cycle is $^{30}/_{120} = {}^{1}/_{4}$, or 25%.

The amount of work performed by a solenoid is directly related to its size. A large solenoid can develop more force at a given stroke than a small one with the same coil current because it has more turns of wire in its coil.

Open-Frame Solenoids

Open-frame solenoids are the simplest and least expensive models. They have open steel frames, exposed coils, and movable plungers centered in their coils. Their simple design permits them to be made inexpensively in high-volume production runs so that they can be sold at low cost. The two forms of open-frame solenoid are the *C-frame solenoid* and the *box-frame solenoid.* They are usually specified for applications where very long life and precise positioning are not critical requirements.

C-Frame Solenoids

C-frame solenoids are low-cost commercial solenoids intended for light-duty applications. The frames are typically laminated steel formed in the shape of the letter C to complete the magnetic circuit through the core, but they leave the coil windings without a complete protective cover. The plungers are typically made as laminated steel bars. However, the coils are usually potted to resist airborne and liquid contaminants. These solenoids can be found in appliances, printers, coin dispensers, security door locks, cameras, and vending machines. They can be powered with either AC or DC current. Nevertheless, C-frame solenoids can have operational lives of millions of cycles, and some standard catalog models are capable of strokes up to 0.5 in. (13 mm).

Box-Frame Solenoids

Box-frame solenoids have steel frames that enclose their coils on two sides, improving their mechanical strength. The coils are wound on phenolic bobbins, and the plungers are typically made from solid bar stock. The frames of some box-type solenoids are made from stacks of thin insulated sheets of steel to control eddy currents as well as keep stray circulating currents confined in solenoids powered by AC. Box-frame solenoids are specified for higher-end applications such as tape decks, industrial controls, tape recorders, and business machines because they offer mechanical and electrical performance that is superior to those of C-frame solenoids. Standard catalog commercial box-frame solenoids can be powered by AC or DC current, and can have strokes that exceed 0.5 in. (13 mm).

Tubular Solenoids

The coils of *tubular solenoids* have coils that are completely enclosed in cylindrical metal cases that provide improved magnetic circuit return and better protection against accidental damage or liquid spillage. These DC solenoids offer the highest volumetric efficiency of any commercial solenoids, and they are specified for industrial and military/aerospace equipment where the space permitted for their installation is restricted. These solenoids are specified for printers, computer disk-and tape drives, and military weapons systems; both pull-in and push-out styles are available. Some commercial tubular linear solenoids in this class have strokes up to 1.5 in. (38 mm), and some can provide 30 lbf (14 kgf) from a unit less than 2.25 in. (57 mm) long. Linear solenoids find applications in vending machines, photocopy machines, door locks, pumps, coin-changing mechanisms, and film processors.

Rotary Solenoids

Rotary solenoid operation is based on the same electromagnetic principles as linear solenoids except that their input electrical energy is converted to rotary or twisting rather than linear motion. Rotary actuators should be considered if controlled speed is a requirement in a rotary stroke application. One style of rotary solenoid is shown in the exploded view in Fig. 3. It includes an armature-plate assembly that rotates when it is pulled into the housing by magnetic flux from the coil. Axial stroke is the linear distance that the armature travels to the center of the coil as the solenoid is energized. The three ball bearings travel to the lower ends of the races in which they are positioned.

The operation of this rotary solenoid is shown in Fig. 4. The rotary solenoid armature is supported by three ball bearings that

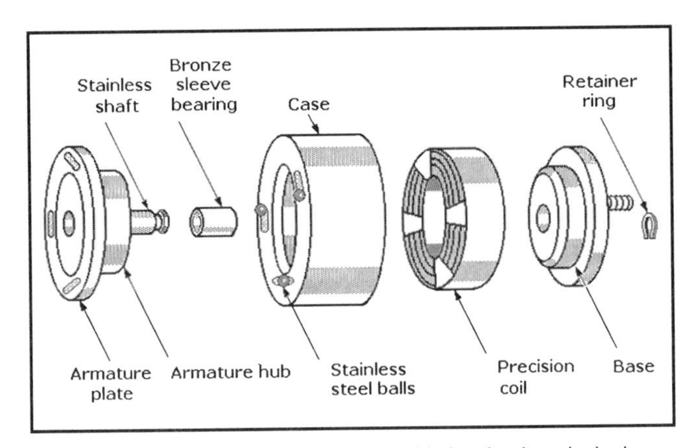

Fig. 3 Exploded view of a rotary solenoid showing its principal components.

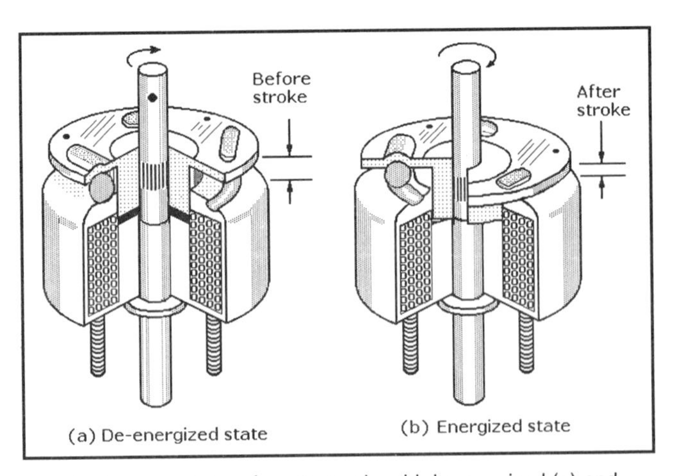

Fig. 4 Cutaway views of a rotary solenoid de-energized (a) and energized (b). When energized, the solenoid armature pulls in, causing the three ball bearings to roll into the deeper ends of the lateral slots on the faceplate, translating linear to rotary motion.

travel around and down the three inclined ball races. The de-energized state is shown in (a). When power is applied, a linear electromagnetic force pulls in the armature and twists the armature plate, as shown in (b). Rotation continues until the balls have traveled to the deep ends of the races, completing the conversion of linear to rotary motion.

This type of rotary solenoid has a steel case that surrounds and protects the coil, and the coil is wound so that the maximum amount of copper wire is located in the allowed space. The steel housing provides the high permeability path and low residual flux needed for the efficient conversion of electrical energy to mechanical motion.

Rotary solenoids can provide well over 100 lb-in. (115 kgf-cm) of torque from a unit less than 2.25 in. (57 mm) long. Rotary solenoids are found in counters, circuit breakers, electronic component pick-and-place machines, ATM machines, machine tools, ticket-dispensing machines, and photocopiers.

Rotary Actuators

The rotary actuator shown in Fig. 5 operates on the principle of attraction and repulsion of opposite and like magnetic poles as a motor. In this case the electromagnetic flux from the actuator's solenoid interacts with the permanent magnetic field of a neodymium–iron disk magnet attached to the armature but free to rotate.

The patented Ultimag rotary actuator from the Ledex product group of TRW, Vandalia, Ohio, was developed to meet the need for a bidirectional actuator with a limited working stroke of less than 360° but capable of offering higher speed and torque than a rotary solenoid. This fast, short-stroke actuator is finding applications in industrial, office automation, and medical equipment as well as automotive applications.

The PM armature has twice as many poles (magnetized sectors) as the stator. When the actuator is not energized, as shown in (a), the armature poles each share half of a stator pole, causing the shaft to seek and hold mid-stroke.

When power is applied to the stator coil, as shown in (b), its associated poles are polarized north above the PM disk and south beneath it. The resulting flux interaction attracts half of the armature's PM poles while repelling the other half. This causes the shaft to rotate in the direction shown.

When the stator voltage is reversed, its poles are reversed so that the north pole is above the PM disk and south pole is below it. Consequently, the opposite poles of the actuator armature are attracted and repelled, causing the armature to reverse its direction of rotation.

According to the manufacturer, Ultimag rotary actuators are rated for speeds over 100 Hz and peak torques over 100 oz-in. Typical actuators offer a 45° stroke, but the design permits a maximum stroke of 160°. These actuators can be operated in an *on/off* mode or proportionally, and they can be operated either open- or closed-loop. Gears, belts, and pulleys can amplify the stroke, but this results in reducing actuator torque.

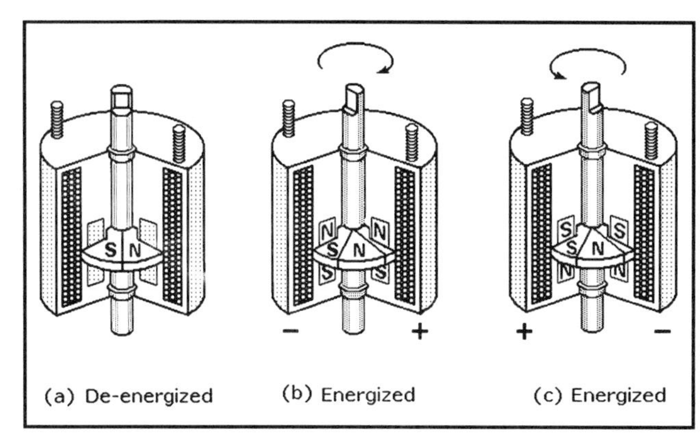

Fig. 5 This bidirectional rotary actuator has a permanent magnet disk mounted on its armature that interacts with the solenoid poles. When the solenoid is deenergized (a), the armature seeks and holds a neutral position, but when the solenoid is energized, the armature rotates in the direction shown. If the input voltage is reversed, armature rotation is reversed (c).

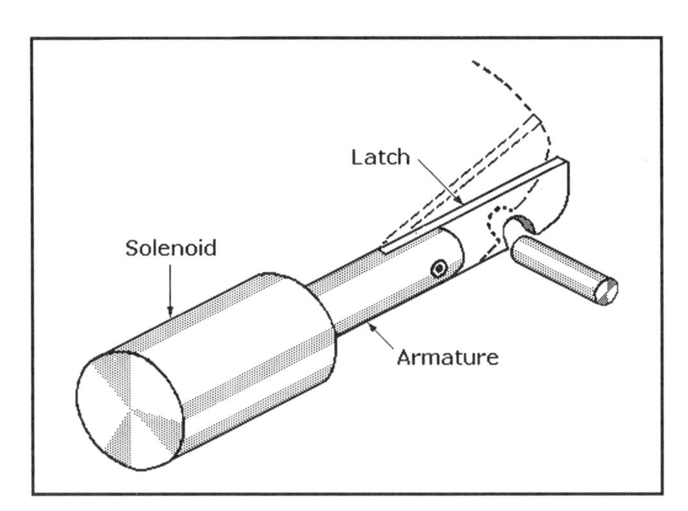

Latching: Linear solenoid push-out or pull-in motion can be used in a wide variety of latching applications such as locking vault doors, safe deposit boxes, secure files, computers, and machine tools, depending on how the movable latch is designed.

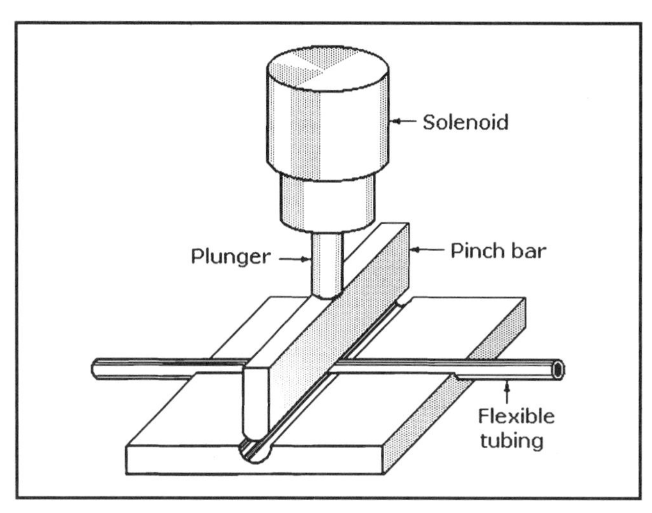

Pinchoff of Flexible Tubing: This push-out linear solenoid with an attached blade can control or pinch off liquid flowing in flexible tubing when energized by a remote operator. This arrangement can eliminate valves or other devices that could leak or admit contaminants. It can be used in medical, chemical, and scientific laboratories where fluid flow must be accurately regulated.

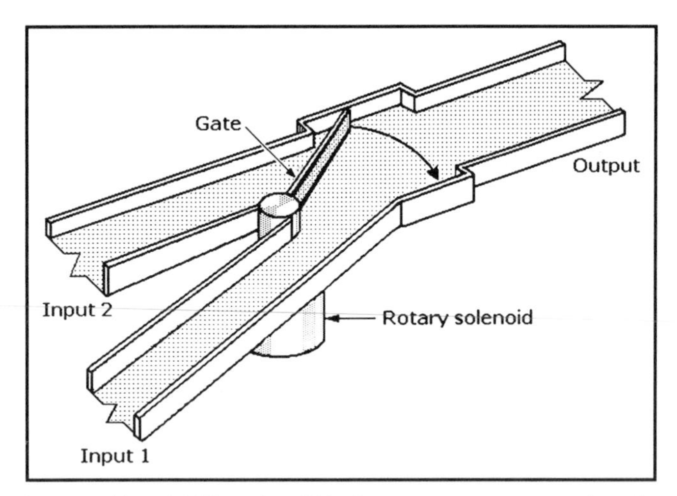

Parts or Material Diversion: This diverter arrangement consists of a rotary solenoid with a gate attached to its armature. The gate can swing to either of two alternate positions under push button or automatic control to regulate the flow of parts or materials moving on belts or by gravity feed.

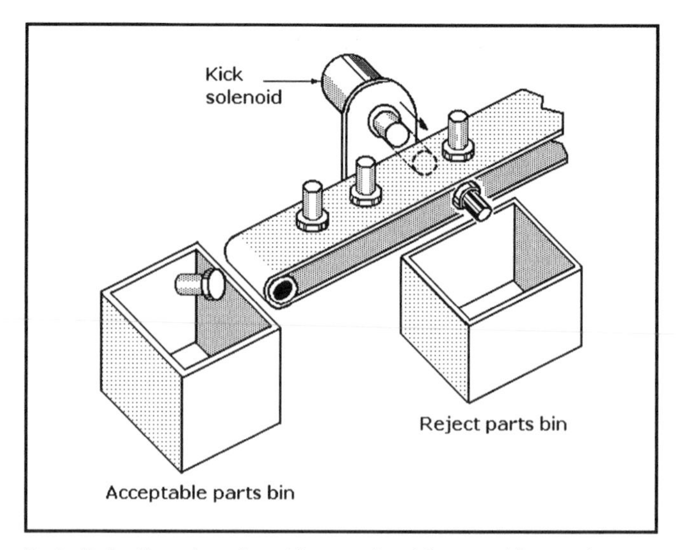

Parts Rejection: A push-out linear solenoid can rapidly expel or reject parts that are moving past it into a bin when triggered. An electronic video or proximity sensing system is required to energize the solenoid at the right time.

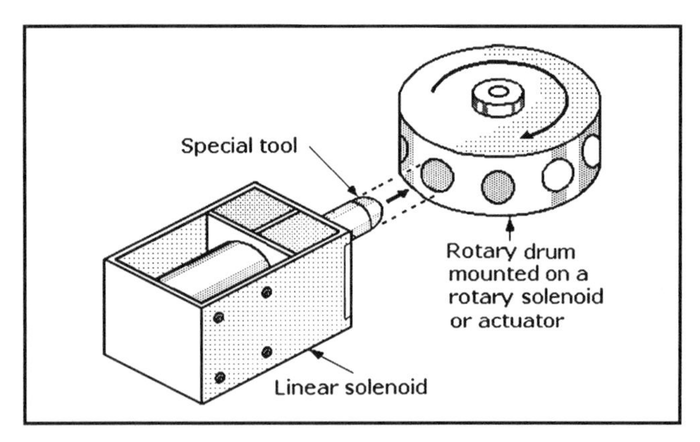

Rotary Positioning: A linear push-out solenoid is paired with a multistation drum containing objects that are indexed by a linear solenoid or actuator. This arrangement would permit the automatic assembly of parts to those objects or the application of adhesives to them as the drum is indexed.

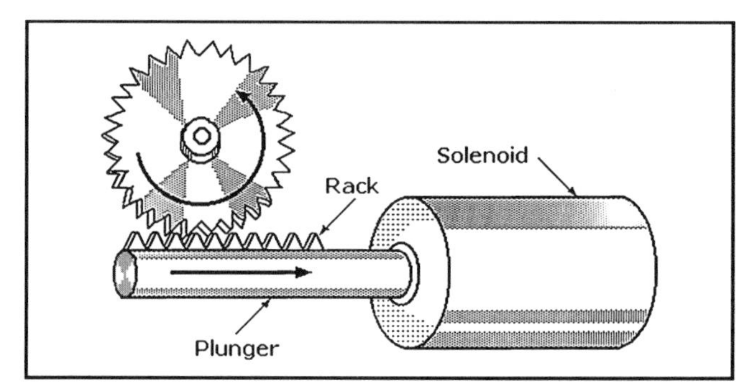

Ratcheting Mechanism: A pull-in solenoid with a rack mounted on its plunger becomes a ratcheting mechanism capable of turning a gear for the precise positioning of objects under operator or automated control.

CHAPTER 3
INDUSTRIAL ROBOTS

INTRODUCTION TO ROBOTS

Any meaningful discussion of robots should include a precise definition of the term because many remotely controlled machines and those with robotic features are incorrectly called robots. The definition used in this chapter and Chapter 4 is: *An electronically reprogrammable, multitasking machine capable of carrying out a range of motions or tasks, typically, but not exclusively by autonomous means.* It applies to all stationary and mobile robots, but excludes *robot-like* automated machines or appliances that are not conveniently reprogrammable. A manipulator that is remotely directed over a wired or wireless links is not a robot unless it can perform some of its functions autonomously under the control of reprogrammable software. This excludes radio-controlled "bots," manually operated manipulators for handling radioactive or toxic materials in laboratories or for collecting biological, geological, or archaeological specimens in the deep ocean from manned undersea platforms.

The primary controller of a modern robot is a central processing unit (CPU), which can be either a microprocessor or microcontroller; it will have its main memory on the same chip or in a separate package and one or more sensors capable of providing feedback during the performance of an assigned task. The CPU will be programmed to assure the safety of nearby persons and the prevention of property destruction by the robot. Miniaturization, reduction in cost, and the improved capability of embedded computer-based electronics have opened new applications for robots and stimulated worldwide growth in all sectors of the robot market. Advances in wireless technology, image processing, speech recognition, motion sensing, and the development of more efficient, compact, lighter, and longer-lived motors and batteries have contributed to that growth.

Robots were first developed in the early years of the twentieth century when autonomous machines were programmed mechanically by clockwork or pegged drums. Others were powered by windup springs or falling weights on the end of a chain. In those days cuckoo clocks, windup music boxes, and mechanical toys qualified as robots. After the introduction of compact electric motors, robots could be powered by batteries or from line power although they were still programmed with rotary-cam or pegged-drum limit switches. Player pianos and textile machines were also considered to be robots because they were controlled by punched tape or cards that could be replaced. While all of these devices were programmable, they were not conveniently or rapidly reprogrammable, the hallmark of today's robots.

All modern robots are considered to be *mechatronic* machines which means that they combine mechanical, electromechanical, and electronic components; they are able to perform functions synergistically, not possible with mechanical or electromechanical programming. As a result, embedded computer circuitry and related sensors have gained equality in importance with the robot's mechanical components.

Most industrial robots are both stationary and autonomous, but some can be moved to different positions to perform assigned tasks such as welding long seams or painting long panels. These robots either have moving parts that can be programmed or they are controlled by a human operator via a cable. However, a robot that is able to move over a limited vertical, horizontal, or angular distance is just a *movable* robot.

By contrast, true mobile robots, regardless of their form or attributes, are capable of moving in three dimensions. Some are autonomous, but most are semiautonomous; that is, they are typically controlled by a human operator in the performance of their missions. Existing mobile robots are capable of moving in many different environments: on land, in the air, on or under the sea, and in space, on the moon or a planet's surface. Control of these mobile robots can be achieved by signals sent by radio or over wires or fiberoptic cable.

A remote operator can direct a mobile robot to start, stop, or maneuver in any of three dimensions to perform a mission, but for meaningful control the operator needs feedback from onboard sensors such as video cameras, radar, or lidar (laser-based light detection and ranging); this information permits the operator to determine the distance and direction the robot has traveled and the presence of any obstacles that would terminate the mission. The feedback signals also permit the mission to be performed more efficiently in a minimum amount of time. Embedded computer systems of some mobile robots can be reprogrammed remotely, if necessary, through their two-way command and control links. Remotely controlled mobile robots of all kinds are more accurately called *telerobots*.

Animatronic and android robotics represent a specialized niche of the robotics industry. These animal and humanlike robots are true robots because they are computer programmed and are, for the most part, autonomic. However, they cannot be evaluated like industrial robots in terms of cost-effectiveness or productivity or like mobile robots for their ability to perform dangerous or life-threatening tasks. Animatronic and android robots are usually custom made by small privately-owned shops or university laboratories. Robotic animals and those with humanlike qualities (also called humanoids) can be found in amusement parks where they function as announcers, actors, or guides; they can also be found in museums where humanoids built to resemble historic figures deliver prerecorded speeches about their lives or significant experiences.

Sophisticated humanoids can move their heads, mouths, eyes, arms, and hands while delivering their presentations. They are made in all shapes and sizes, and they can be mobile but are typically stationary. These robots typically have metal armatures or skeletons with movable levers for arms and forms for mounting movable eyes, mouths, and hands. Flexible "skin" is formed from pigmented silicone molded to form facial features and hands. Wigs and appropriate clothing complete the illusion. Limbs and facial features are animated by electric motors or pneumatic actuators under computer control. Some humanoids have built-in sensors and power supplies, while others depend on external power sources and are activated by external sensors such as video cameras, microphones, or pressure-sensitive switches.

Some androids are being used for serious scientific research in cognitive science, behavioral science, neuroscience, and study of human interactions.

While the importance of embedded computer-based electronics and associated sensors to the successful completion of a mission cannot be underestimated, these subjects are not treated in detail in this book because of its focus on the mechanisms and mechanical devices of robots. Table 1 classifies most robots that now exist or are in development.

Economic Importance of Robots

Market studies carried out in 2006 in Japan and the United States agree that the global market for robotics is now approximately $12 billion. This amount is now about evenly divided between industrial robots and all types of mobile robots combined.

Table 1 Classification of Robots

Industrial: immobile, single-task robots that perform specific tasks (e.g., welding, painting, materials handling, and assembly) but have little interaction with humans or the environment

Service: semi or fully autonomous mobile institutional or government robots that assist humans, service equipment, and perform other autonomous functions

- Medical and healthcare: assisting surgeons in routine procedures and physical therapists in patient rehabilitation
- Industrial inspection and/or cleaning: clean/inspect pipes and ducts and in restricted, hazardous, or radioactive spaces
- Material handling: sorting, storing, retrieving; internal package delivery
- Police and emergency/public services: bomb disposal, surveillance, armed hostage negotiation, firefighting, and victim search in collapsed or burning buildings
- Scientific exploration: explore distant planets, deep ocean, polar regions, caves, proximity to active volcanoes; study tornados and other storm phenomena
- Military applications: unmanned ground vehicles (UGV), unmanned vehicles (UUV), unmanned aerial vehicles (UAV); surveillance, explosive disposal, weapons platforms
- Entertain and educate: androids (humanoids) entertain in amusement parks, give educational presentations in museums, interact with humans

Personal: semi or fully autonomous mobile consumer robots that assist humans, entertain, or perform other routine functions

- Home maintenance: vacuum cleaning, lawn mowing, intrusion detection
- Assistance for disabled and elderly in performing routine tasks
- Hobbyist/education and entertainment: lifelike toys

R&D: robot development in academic, government, and industrial laboratories focusing on new applications while reducing cost, power consumption, and weight; examples—self-replicating and swarming robots

However, this figure does not account for government, corporate, and academic spending on R&D in robotics. It appears that the market for military, law enforcement, and emergency service robots is driving this growth spurt, and it is believed that spending on unmanned robotic platforms by the military services is the major reason why the mobile robot market is expected to surge ahead of the industrial robot market in the coming years.

INDUSTRIAL ROBOTS

Industrial robots are defined by the characteristics of their control systems; manipulator or arm geometry; modes of operation; and their end effectors or the tools mounted on a robot's wrist. Industrial robots can be classified by their programming modes which correlate with their performance capabilities: *limited* versus *unlimited sequence control*. These terms refer to the paths that can be taken by the end effector as it is stepped through its programmed motions. Four classes are recognized: limited sequence control and three forms of unlimited sequence control: *point-to-point, continuous-path*, and *controlled-path*.

Another distinction between industrial robots is in the way they are controlled: either *servoed or nonservoed*. A servoed robot includes a closed-loop which provides feedback and enables it to have one of the three forms of unlimited sequence control. This is achieved if the closed loop contains a velocity sensor, a position sensor, or both. By contrast, a nonservoed robot has open-loop control, meaning that it has no feedback and is therefore a limited sequence robot.

Industrial robots can be powered by electric motors or hydraulic or pneumatic actuators. Electric motors are now the most popular drives for industrial robots because they are the least complicated and most efficient power sources. Hydraulic drives have been installed on industrial robots, but this technology has lost favor, particularly for robots that must work within a controlled and populated environment. Hydraulic drives are noisy and subject to oil leakage, which presents a fire hazard in an enclosed space. Moreover, hydraulic drives are more maintenance prone than electric drives.

Nevertheless, hydraulic-drive robots can handle loads of 500 lb or more, and they can be used safely outdoors or in uncontrolled spaces. They are also used in situations where volatile gases or substances are present; these hazards rule out electric motors because a fire or explosion could be caused by electric arcing within the electric motor. Some limited-duty benchtop robots that are powered by pneumatic actuators, but they are typically simple two- or three-axis robots. On the other hand, pneumatic power is now widely used to operate end effectors such as "hands" or grippers mounted on the wrists of electric-drive robots. An example is a wrist assembly that includes two rotary pneumatic actuators capable of moving a gripper around two axes, roll and yaw.

The term *degrees-of-freedom* (DOF) as applied to a robot indicates the number of its axes, an important indicator of a robot's capability. Limited sequence robots typically have only

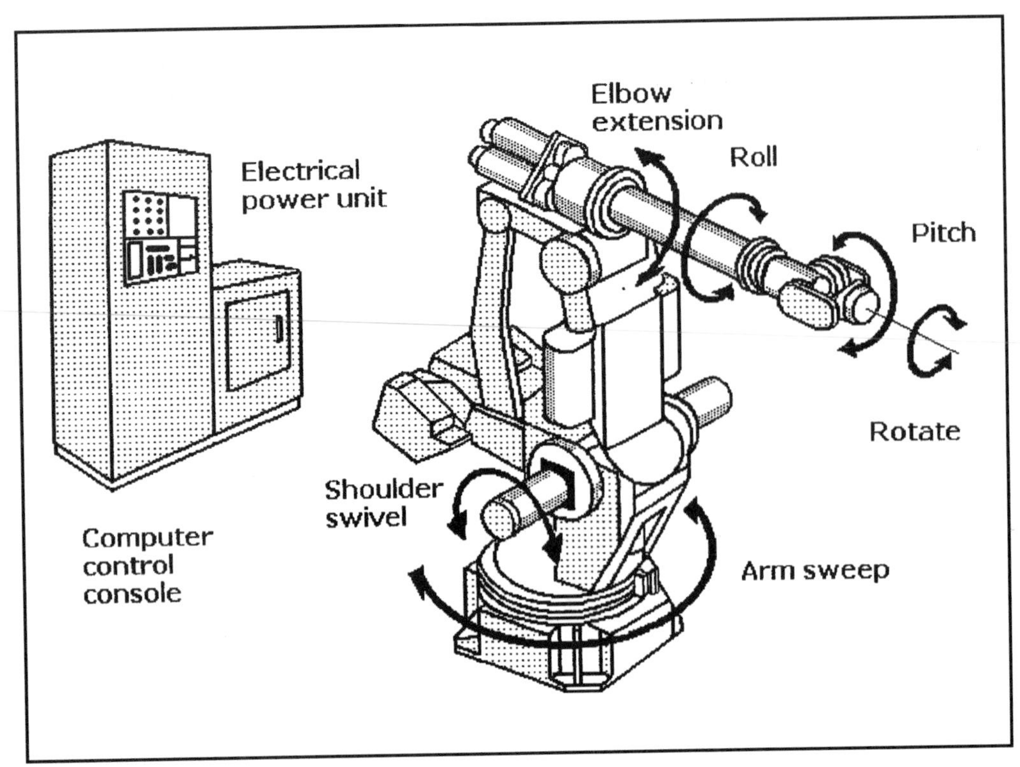

Fig. 1 Components of a floor-standing, six-degree-of-freedom industrial robot.

two or three axes while unlimited sequence robots typically have five or six axes because they are intended to perform more complex tasks. However, the basic robot manipulator arm might have only three axes: arm sweep (base rotation), shoulder swivel (reach), and elbow extension (elevation), but a wrist can provide as many as three additional axes—pitch, roll, and yaw.

The heavy-duty, floor-standing robot shown in Fig. 1 has six principal axes, each driven by an electric motor. The console contains a digital computer that has been programmed with an operating system and applications software so that it can perform the robot's assigned tasks. The operator or programmer can control the movements of the robot arm or manipulator with push buttons on the control console so that it can be run manually through its complete program sequence. During programming, adjustments can be made in the program to prevent any part of the robot from colliding with nearby objects.

Some industrial robots are equipped with training pendants—handheld control boxes that are connected to the computer control console by cable. The pendant typically contains a push-button panel and a color graphic liquid-crystal display (LCD). It permits an operator or programmer to "teach" the robot by leading the wrist and end effector manually through the complete assigned task. The movements of each of the axes in the path sequence are stored in memory so that the robot will play back the routines precisely when commanded to do so.

Some floor-standing industrial robots are built so that they can be mounted upside down, vertically, or at an angle to gain better access to their intended work areas. The inverted robots are typically suspended from structural frames. Those frames might have rails on them to permit the robot to travel over limited distances while engaged in work such as welding long seams or painting long objects. Similarly, the robots might be positioned in a fixed position on a wall or they could be mounted on a vertical rail if vertical movement is required. They could also be mounted on rails set at an angle with respect to the floor for angular excursions.

Industrial Robot Advantages

An industrial robot can be programmed to perform a wider range of tasks than a dedicated automatic machine such as a numerically controlled machining center, even if that machine can accept a wide selection of different tools for doing different jobs. While industrial robots are considered to be multipurpose machines, most manufacturers design robots with certain characteristics that favor specific applications such as welding, painting, loading products or parts in cartons, or performing assembly work.

Manufacturers' literature and specifications sheets list these specialties, but their designations do not mean that the robots are limited to those functions; they will function satisfactorily in the performance of other applications where similar characteristics are required. Those characteristics that make a robot well suited for specific applications are determined by their specifications; these include such factors as size, weight, payload, reach, repeatability, range, speed, and the cost of operation and maintenance.

The decision to purchase a robot should be based on the evaluation of a long checklist of requirements for justifying the purchase. First and foremost of these decisions is the customer's conclusion that a lower cost dedicated machine cannot do the work more cost-effectively than a robot. Other factors to be evaluated, both technical and economic, are

- Ability of the owner to integrate the robot with existing manufacturing facilities
- Cost of training or retraining operators and programmers for the robot
- Cost of writing new applications software to direct the process to be automated
- Estimation of the overhead cost and time lost during downtime while a human operator changes tools between jobs or performs routine robot maintenance

The full benefits of an industrial robot cannot be realized unless it is properly integrated with the other conventional machines, conveyers, and materials handling equipment that form a coordinated work cell. Early robot purchasers learned a costly lesson when they found that isolated robots could not pay for themselves because they were not integrated into the normal workflow of the factory, so they were abandoned. Carefully engineered work cells assure that there is a coordinated and timely flow of work.

Industrial robots have been most cost effective in situations where they perform arduous, repetitious tasks, especially in hostile environments where human operators are exposed to life-threatening environmental conditions. These locations include environments where

- Temperatures or humidity are excessive
- Noxious or toxic fumes can damage the lungs
- Welding arcs can damage unprotected eyes
- Molten metal spray or open flame can burn unprotected skin
- High-voltage sources present a constant electrocution hazard

Nevertheless, robots have frequently proven themselves in work situations where none of these factors were present because they were able to demonstrate more consistent and higher quality workmanship than could be performed by skilled and experienced workers. Examples are found in welding, painting, and repetitive assembly work, even in conditioned indoor environments such as automotive assembly lines and appliance factories.

Industrial robots are now found at work in a wide range of industries from machine tool, automotive, aircraft, and shipbuilding to consumer appliance manufacturing. In addition, many robotic machines that are not easily recognizable perform such nimble tasks as pick-and-place assembly of electronic components on circuit boards. In addition, robots capable of moving rapidly along the length and height of extensive shelving in automated warehouses are storing and retrieving various objects and packages under remote computer control.

Industrial Robot Characteristics

The important specifications to consider in a robot purchase decision are payload, reach, repeatability, interference radius, motion range and speed, payload capacity, and weight. Reach is measured in inches or millimeters, and motion range is determined by the robot's three-dimensional (3-D) semispherical work envelope. This is the locus of points that can be reached by the robot's workpoint when all of their axes are in their extreme positions. Motion speed is measured for each axis in degrees per second. The robot must be able to reach all the parts or tools needed to perform its task, so the working range typically determines the size and weight of the robot required.

Robot axis motion speed is typically in the range from 100°/s to 300°/s. High rates of acceleration and deceleration are favored. Payloads are most important if the robot is to do a significant amount of lifting. These are measured in pounds and kilograms. Some production industrial robots are able to handle maximum loads up to 880 lb or 400 kg, but most requirements are far lower—generally less than 50 lb. A large floor-standing robot can weigh as much as 2 tons.

Stiffness is another important robot specification. This term means that the robot arm must be rigid enough in all of its possible positions to perform its assigned tasks without flexing or shifting under load. If the robot has sufficient stiffness it can perform repetitive tasks uniformly without deviating from its programmed dimensional tolerances. This characteristic is specified as *repeatability*, which correlates with stiffness and is measured in inches or millimeters of deviation.

Industrial Robot Geometry

There are four principal stationary robot geometries: (1) articulated, revolute, or jointed arm; (2) polar-coordinate or gun turret; (3) Cartesian; and (4) cylindrical. A low-shoulder articulated robot is shown in Fig. 2 and a high-shoulder articulated robot is shown in Fig. 3.The articulated robot geometry is the most commonly used configuration today for floor-standing industrial

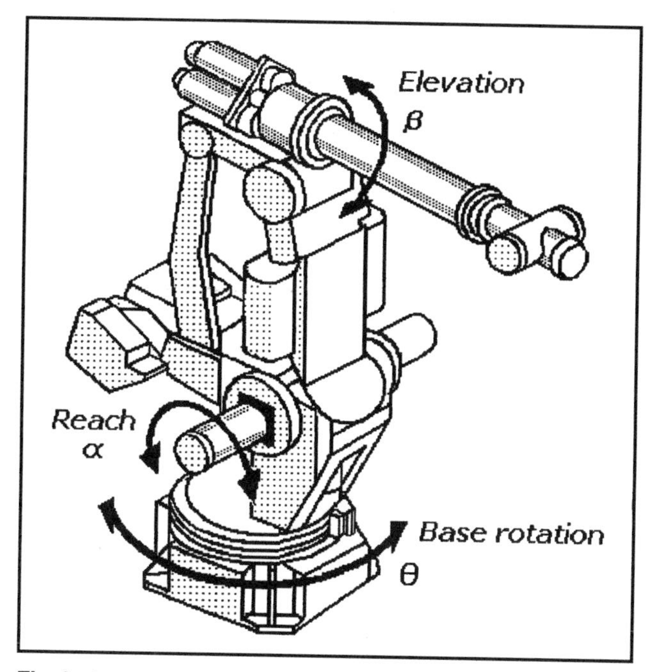

Fig. 2 Low-shoulder, articulated, revolute, or jointed geometry robot.

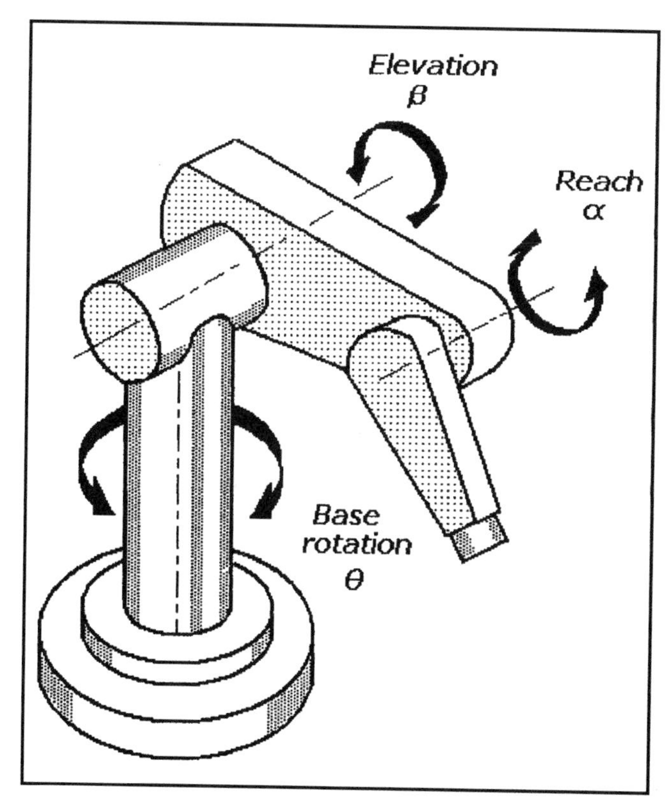

Fig. 3 High-shoulder, articulated, revolute, or jointed geometry robot.

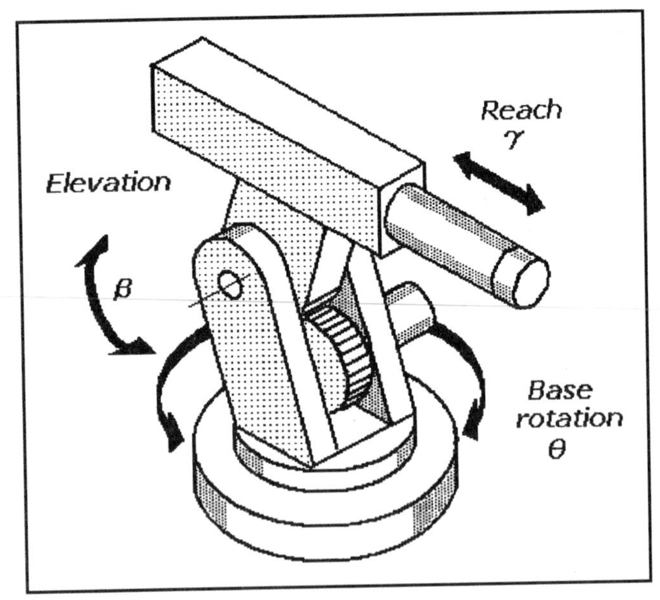

Fig. 4 Polar coordinate or gun-turret geometry robot.

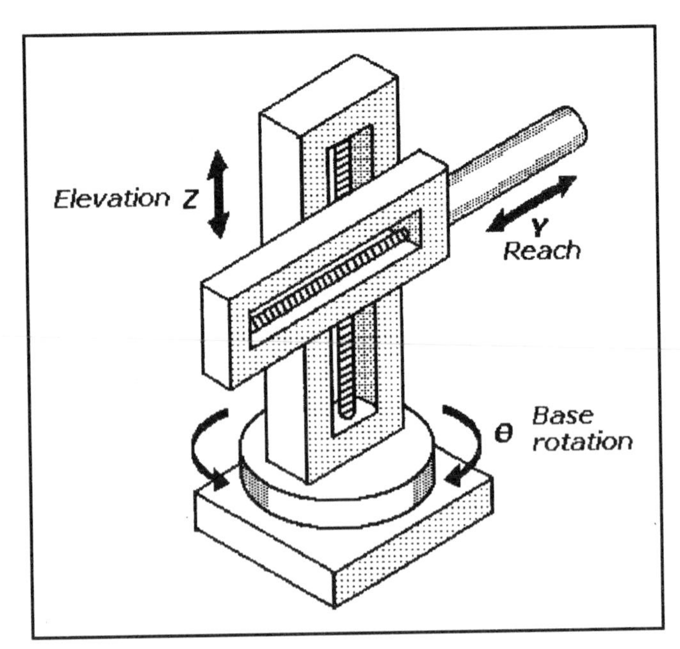

Fig. 6 Cylindrical-coordinate geometry robot.

robots, but there are many variations. Thepolar-coordinate geometry robot is illustrated in Fig. 4; the Cartesian-coordinate geometry robot is illustrated in Fig. 5, and the cylindrical-coordinate geometry robot is illustrated in Fig. 6. Among the variations of these basic designs is the vertically jointed geometry robot shown in Fig. 7.

A robot's wrist at the end of the robot's arm serves as a mounting plate for end effectors or tools. There are two common designs for robot wrists: two-degree-of-freedom (2DOF) and

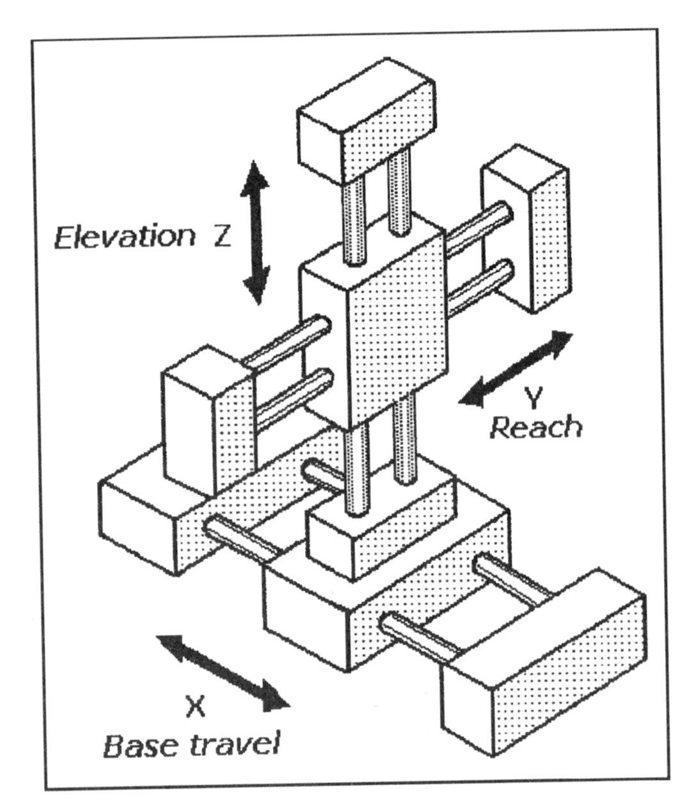

Fig. 5 Cartesian coordinate-geometry robot.

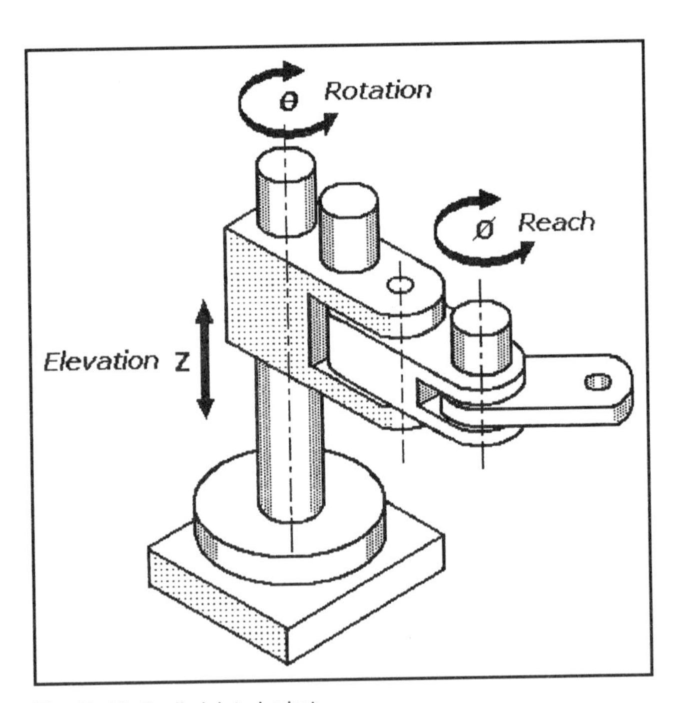

Fig. 7 Vertically-jointed robot.

three-degree-of-freedom (3DOF). An example of a 2DOF wrist is shown in Fig. 8; it is permits roll around the arm axis and pitch around an axis at right angles to the arm axis. Another version of a 2DOF wrist, illustrated in Fig. 9, has the capability of a second independent roll around the arm axis in addition to the pitch around an axis at right angles to the arm axis. A 3DOF wrist is shown in Fig. 10; in addition to roll and pitch it offers yaw around a third axis perpendicular to both the pitch and roll axes. More degrees-of-freedom can be added by installing end effectors or tools that can move around axes independent of the wrist.

Many different kinds of end effectors are available for robots, but among the most common are pincer- or claw-like two-fingered grippers or hands that can pick up, move, and release objects. Some

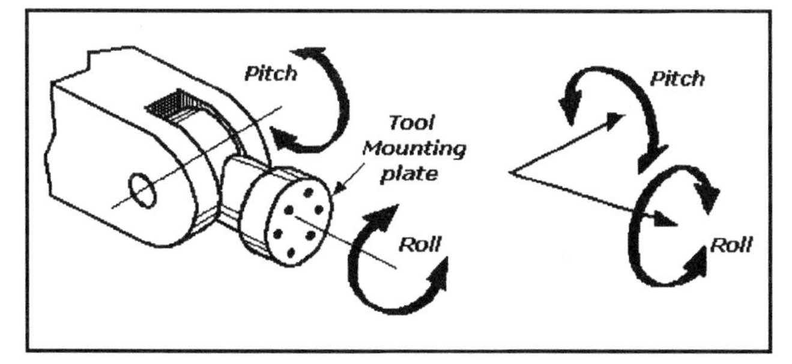

Fig. 8 Two-degree-of-freedom robot wrist can move a tool or end effector attached to its mounting plate around both pitch and roll axes.

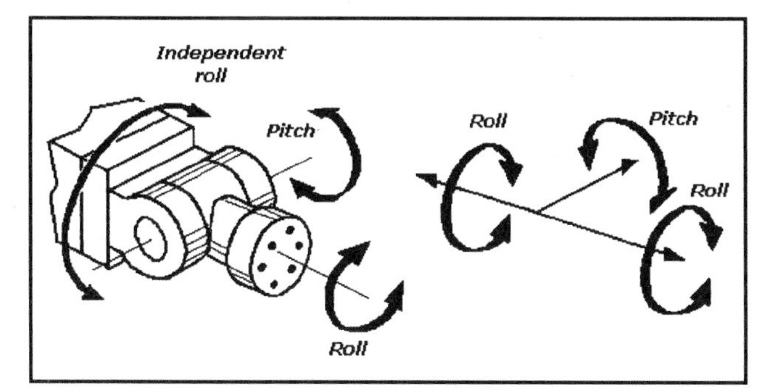

Fig. 9 Two-degree-of-freedom robot wrist can move a tool or end effector attached to its mounting plate around pitch and two roll axes.

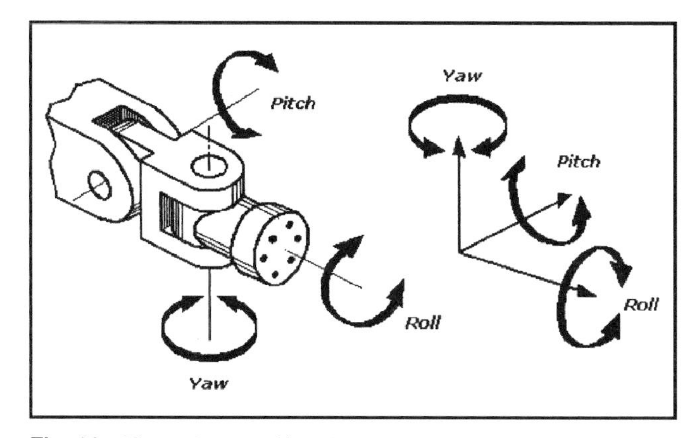

Fig. 10 Three-degree-of-freedom robot wrist can move a tool or end effector attached to its mounting plate around three axes: pitch, roll, and yaw.

of these grippers are general purpose, but others have finger gripping surfaces that have been machined to contours that fit precisely around or inside specific objects oriented in preestablished positions. Fingers fashioned to grasp the outside of objects such as cylinders move inward to grasp and lift the object; fingers fashioned to grasp the inside of objects such as pipes or cylinders move inside the object and expand outward to grasp inside surfaces and lift the object.

The end effectors shown in Figs. 11 and 12 require independent actuators to power them. These are typically electric motors or pneumatic cylinders with pistons that are mounted between the end effector and the robot's wrist. However, the gripper shown in Fig. 13 includes an actuator that could either be a pneumatic or hydraulic piston for opening or closing the gripper fingers.

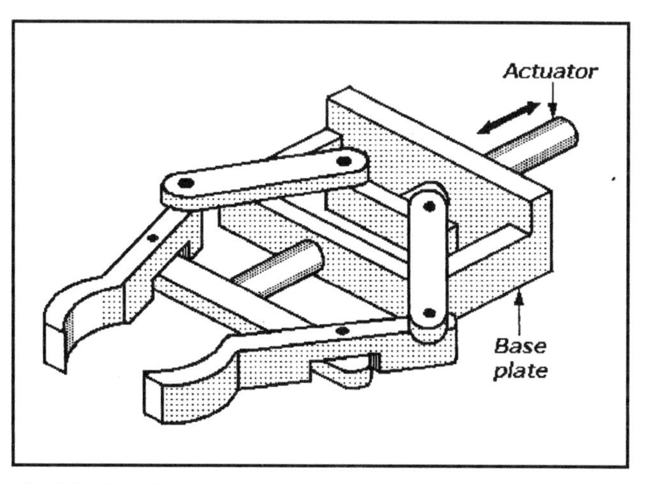

Fig. 11 Robotic gripper operated by a reciprocating mechanism. Links open and close the "fingers" permitting them to grasp and release objects. A separate power source (not shown) is required.

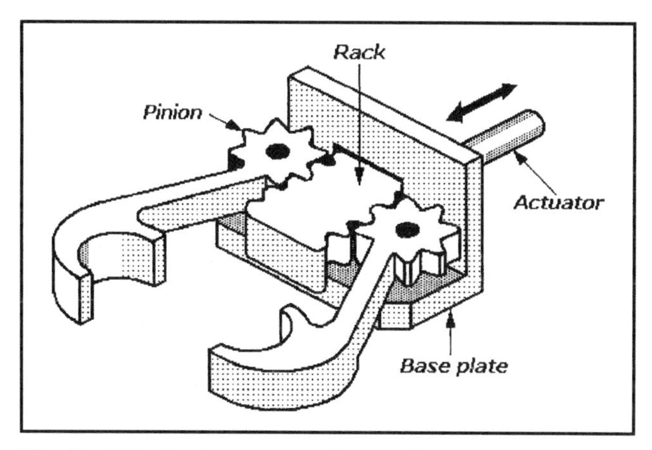

Fig. 12 Robotic gripper operated by a rack and pinion mechanism. Rack and pinions open and close the "fingers" permitting them to grasp and release objects. A separate power source (not shown) is required to operate this gripper.

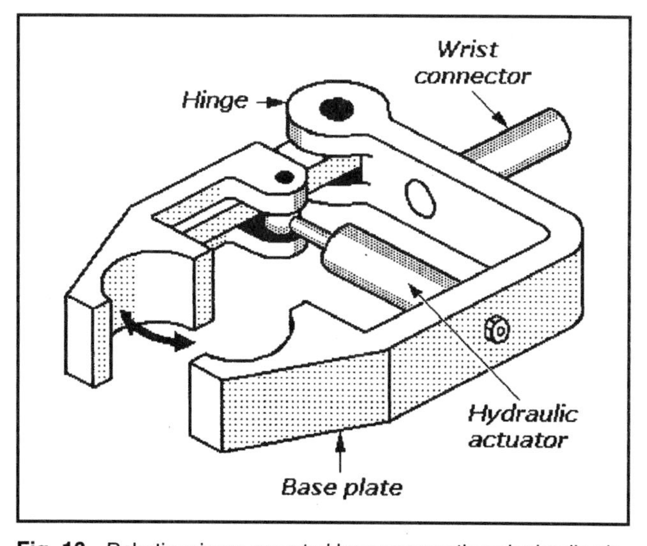

Fig. 13 Robotic gripper operated by a pneumatic or hydraulic piston. Piston opens and closes the "fingers" permitting them to grasp and release objects.

More sophisticated and versatile multifingered robotic grippers are now available, but they must be controlled either by software within the host robot or by an independent controller such as a laptop computer. For example, the gripper can have three fingers and an opposing thumb which can curl around objects of varying sizes, shapes, and orientations when actuated by electric motors to establish a firm grasp on the objects. While these grippers are more expensive, they eliminate the need for custom machining fingers to fit objects, and they can pick up randomly positioned objects.

Industrial Robot Examples

The data sheets for three robots from FANUC Robotics America, Inc., Rochester Hills, Michigan, have been reprinted on the following pages to illustrate the range of size, weight, performance capabilities, and options for industrial robots now in production. The key specifications given here include the number of axes, payload at the wrist, reach, repeatability, and weight.

LR Mate Series Robots

The LR Mate series, as shown in Fig. 14, is a family of three electric servomotor-driven minirobots capable of performing a wide variety of tasks in a broad range of working environments. The LR Mate 100*i*B, 200*i*B and 200*i*B/3L benchtop robots have many construction and performance features in common and are similar in size and weight.

The 100*i*B has five axes, a reach of 23 in. (58.4 cm), and a weight of 84 lb (38 kg), while all versions of the 200*i*B have six axes, and most have reaches of 28 in. (71 cm) and weight of 99 lb (45 kg). All have ±0.04 mm repeatability and wrist payloads up to 11 lb (5 kg). These robots have slim wrists and small footprints permitting them to work in restricted spaces, and all can flip over backward to increase the size of their work envelopes.

All LR Mate robots can be mounted upright or inverted, but the 200*i*Bs can also be mounted on angled or vertical surfaces. Among the many tasks that can be performed by robots in this series are machine tending, material handling and removal, assembly, picking and packing, and testing and sampling. Reliability is enhanced by sealed bearings, brushless AC motors, and gear and/or harmonic drives, eliminating belts, chains, and pulleys.

The 200*i*B can be adapted for clean room operation, industrial washing, and food handling, and the 200*i*B/3L has a longer reach. Options include an Internet-ready teaching pendant with a graphic color LCD display, machine vision for guidance and inspection, and mechanical brakes on all joints.

M-420*i*A Robots

There are two robots in the M-420*i*A/M-421*i*A series. The M-420*i*A, shown in Fig. 15, has four axes while the M-421*i*A has two axes. Both are driven by electric servomotors and are similar in construction, but waist and wrist rotation have been deleted in the 421 for higher speed in loading empty packing cases from the top. The 421 is 220 lb lighter than the 420 which weighs 924 lb (620 kg), but its wrist payload is higher: 110 lb (50 kg) versus 88 lb (40 kg). Both robots have long 6 ft (1.83 m) reaches, and ±0.5 mm repeatability; this makes them well suited for high-speed packaging and palletizing, machine loading, and mechanical assembly.

The M-420*i*A/M-421*i*A series robots are controlled by FANUC intelligent control systems which give them complete control of high-speed packaging lines. A vision system gives the robots the ability to pack randomly-oriented products. Other equipment includes Internet-ready teaching pendants with large LCD color displays to simplify programming and interfacing with the customer's work cells. Linear motion can reach 165 in./s (4.19 m/s) at full speed. Pneumatic and electrical connections at the wrist simplify end-of-arm tool design and integration. The options include a high-speed wrist that boosts speed from 350°/s to 720°/s, but decreases maximum payload by 22 lb (10 kg).

M-900*i*A Series Robots

There are three robots in the M-900*i*A family: M-900*i*A/350, M-900*i*A/260L, and M-900*i*A/600. They are floor-standing , six-axis, heavy-duty robots with maximum payloads at the wrist of 770 lb (350 kg), 260 lb (572 kg), and 1320 lb (600 kg), respectively, and reaches of between 8.75 and 10.2 ft (2.65 and 3.11 m). The 600 model, shown in Fig. 16, has a reach of 9.3 ft (2.84 m), repeatability of ±0.4 mm, and it weighs 3 tons (2722 kg). The robots in this series can perform materials handling and removal, loading and unloading of machines, heavy-duty spot welding, and the handling of large sheets of glass, metal, or wood.

These high-speed robots provide point-to-point positioning and smooth controlled motion. M-900*i*A robots have high-inertia wrists with large allowable movements that make them suitable for heavy-duty work. Their slim outer arm and wrist assemblies minimize interference with system peripherals and permit them to work in confined spaces. Small robot footprints and reduced computer control console size conserve valuable floor space. Many process attachment points are provided on the robot's wrists to make tool integration easier. All axes have brakes, and the 350 and 260L robots can also be inverted and mounted on angled or vertical surfaces.

M-900*i*A robots support standard and distributed input/output (I/O) networks and have standard Ethernet ports. Process-specific software packages are available for a range of applications. Web-based software permits remote connectivity, diagnosis, and production monitoring, and machine vision software is available for robot guidance and inspection. Options include enhanced protection against harsh environments and standard baseplates for quick robot installation. Auxiliary axis packages are available for integrating peripheral servo-controlled devices. In addition, an optional teaching pendant with a LCD color touch screen permits easier programming and interfacing the robot to a user's custom work cell.

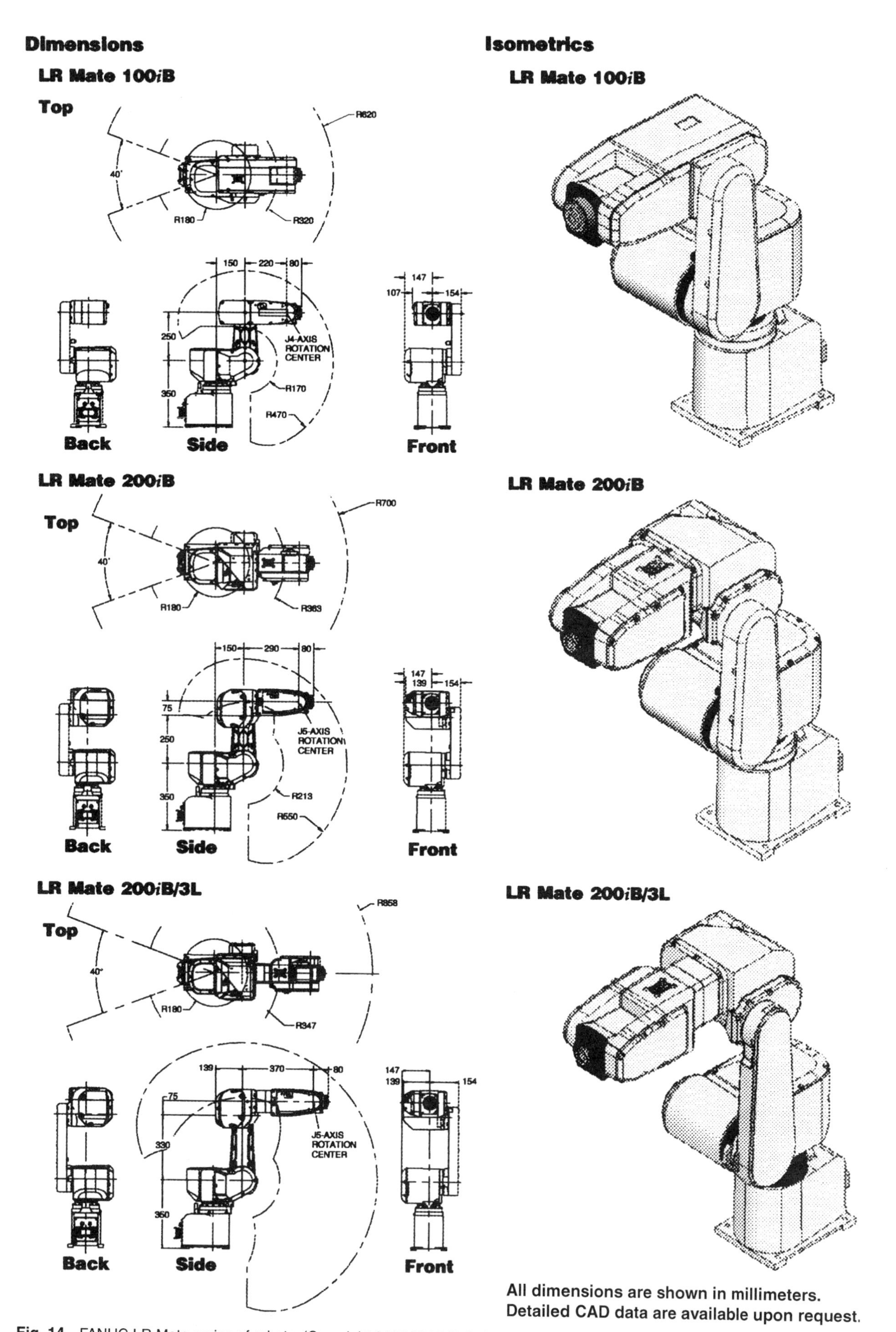

Fig. 14 FANUC LR Mate series of robots. (Copyright 2006 FANUC Robotics America, Inc. Reprinted with permission)

M-420*i*A Dimensions

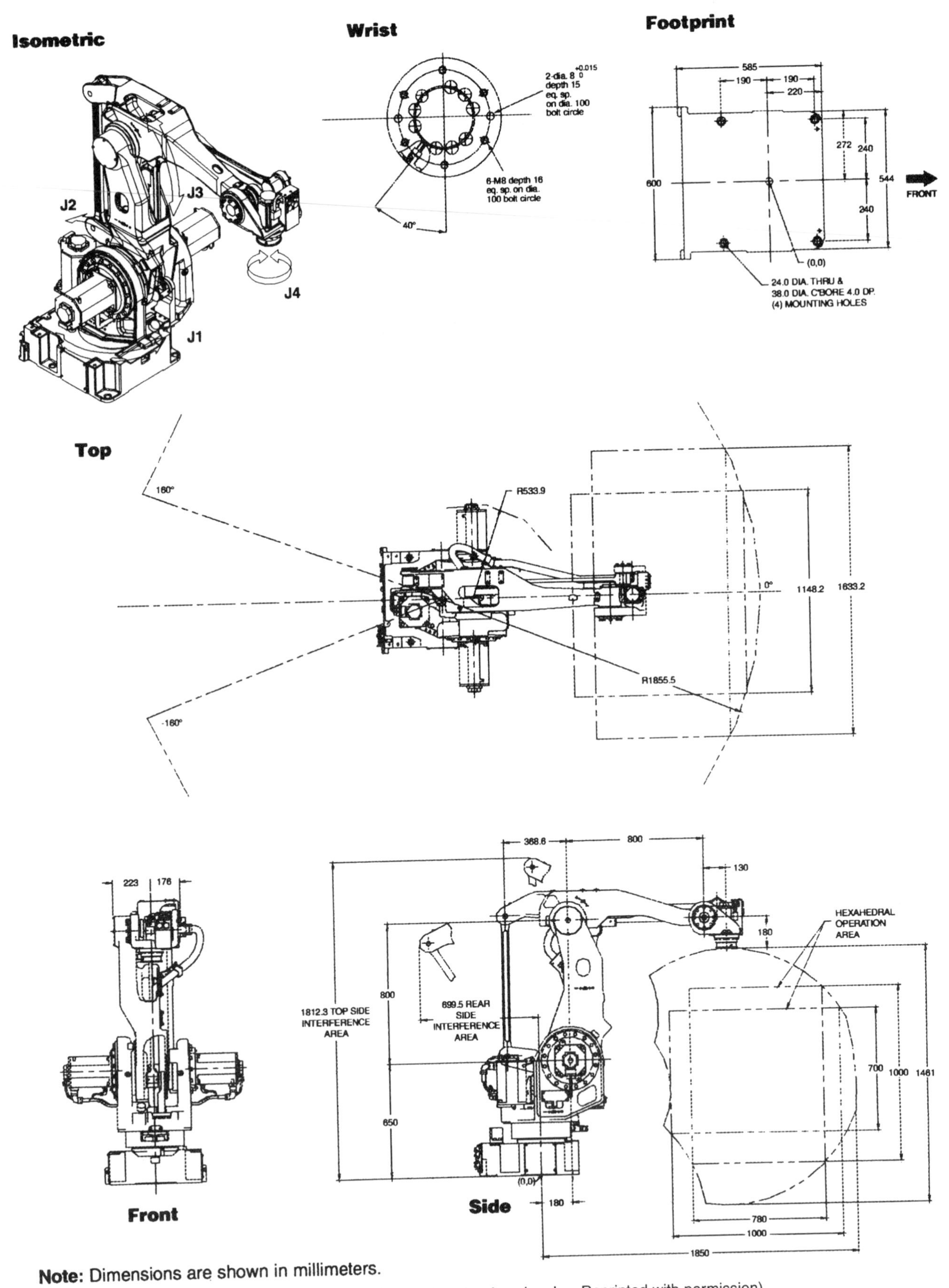

Note: Dimensions are shown in millimeters.

Fig. 15 FANUC M-420*i*A robot. (Copyright 2006 FANUC Robotics America, Inc. Reprinted with permission)

M-900*i*A/600 and 700kg Option Dimensions

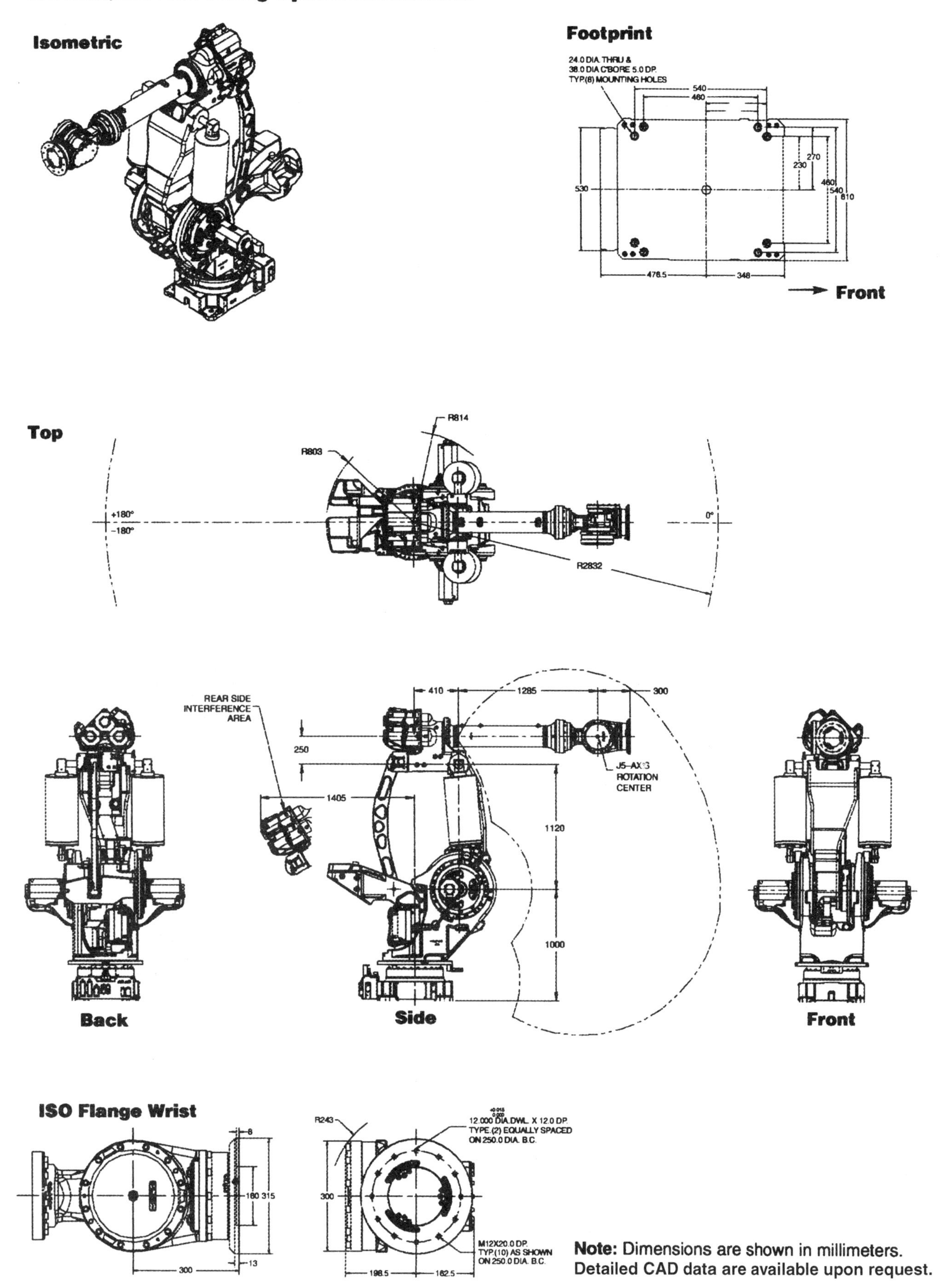

Fig. 16 FANUC M-900*i*A /350 robot. (Copyright 2006 FANUC Robotics America, Inc. Reprinted with permission)

MECHANISM FOR PLANAR MANIPULATION WITH SIMPLIFIED KINEMATICS

Simple combinations of actuator motions yield purely radial or purely tangential end-effector motions.
Goddard Space Flight Center, Greenbelt, Maryland

The figure schematically illustrates three manipulator mechanisms for positioning an end effector (a robot hand or other object) in a plane (which would ordinarily be horizontal). One of these is a newer, improved mechanism that includes two coaxial, base-mounted rotary actuators incorporated into a linkage that is classified as "P4R" in the discipline of kinematics of mechanisms because it includes one prismatic (P) joint and four revolute (R) joints. The improved mechanism combines the advantages of coaxial base mounting (as opposed to noncoaxial and/or nonbase mounting) of actuators, plus the advantages of closed-loop (as opposed to open-loop) linkages in such a way as to afford a simplification (in comparison with other linkages) of inverse kinematics. Simplification of the kinematics reduces the computational burden incurred in controlling the manipulator.

In the general case of a two-degree-of-freedom manipulator with two rotary actuators, the inverse kinematic problem is to find the rotary-actuator angles needed to place the end effector at a specified location, velocity, and acceleration in the plane of motion. In the case of a typical older manipulator mechanism of this type, the solution of the inverse kinematic problem involves much computation because what one seeks is the coordinated positions, velocities, and accelerations of the two manipulators, and these coordinates are kinematically related to each other and to the required motion in a complex way.

In the improved mechanism, the task of coordination is greatly simplified by simplification of the inverse kinematics; the motion of the end effector is easily resolved into a component that is radial and a component that is tangential to a circle that runs through the end effector and is concentric with the rotary actuators.

If rotary actuator 2 is held stationary, while rotary actuator 1 is turned, then link D slides radially in the prismatic joint, causing the end effector to move radially. If both rotary actuators are turned together, then there is no radial motion; instead, the entire linkage simply rotates as a rigid body about the actuator axis, so that the end effector moves tangentially. Thus, the task of coordination is reduced to a simple decision to (a) rotate actuator 1 only to obtain radial motion, (b) rotate both actuators together to obtain tangential motion, or (c) rotate the actuators differentially according to a straightforward kinematic relationship to obtain a combination of radial and axial motion.

This work was done by Farhad Tahmasebi of **Goddard Space Flight Center.**

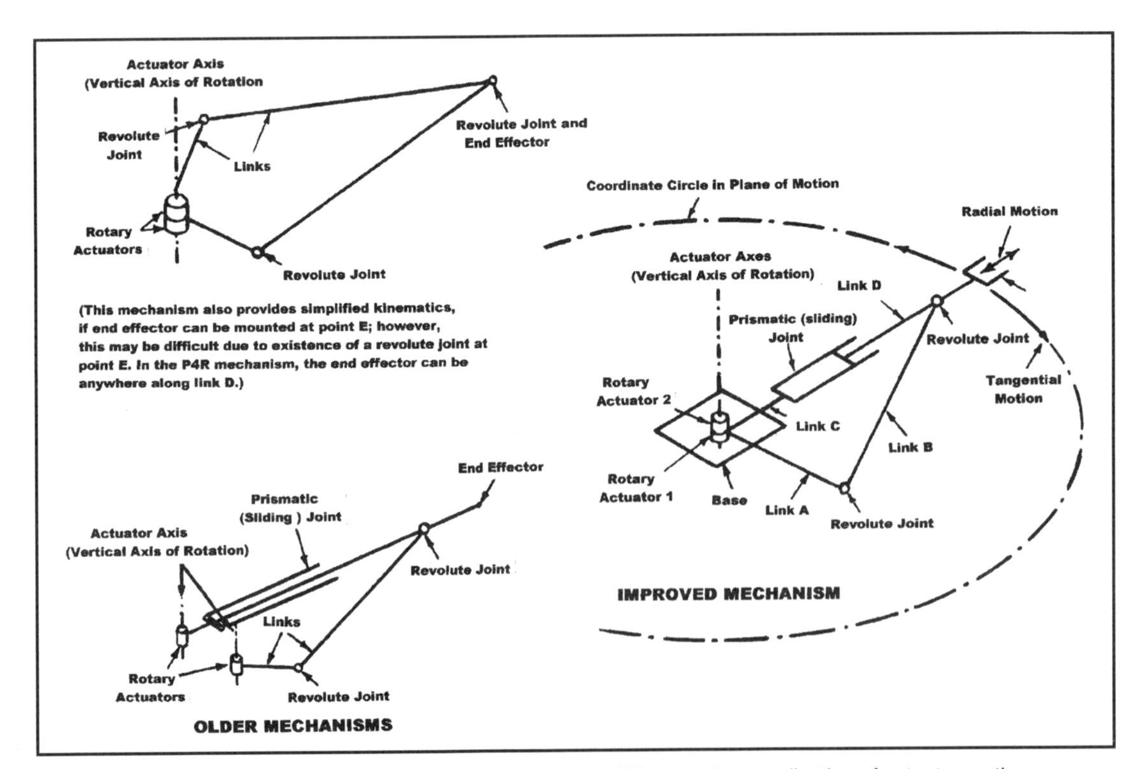

The Improved Mechanism affords a simplification of kinematics: Whereas the coordination of actuator motions necessary to obtain specified end effector motions in the older mechanisms is a complex task, it is a relatively simple task in the improved mechanism.

TOOL-CHANGING MECHANISM FOR ROBOT

A tool is handed off securely between an end effector and a holster.

Goddard Space flight Center, Greenbelt, Maryland

Figure 1 is a partially exploded view of a tool-changing mechanism for robotic applications. The mechanism effects secure handoff of the tool between the end effector of the robot and a yoke in which the tool is stowed when not in use. The mechanism can be operated in any orientation in normal or low gravitation. Unlike some other robotic tool-changing mechanisms, this one imposes fewer constraints on the design of the robot and on the tool because it is relatively compact. Moreover, it does not require the large insertion forces and the large actuators that would be needed to produce them. Also, it can be stored in zero g and can survive launch loads.

A tool interface assembly is affixed to each tool and contains part of the tool-changing mechanism. The tool is stowed by (1) approximately aligning the tips of the yoke arms with flared openings of the holster guides on the tool interface assembly, (2) sliding the assembly onto the yoke arms, which automatically enforces fine alignment because of the geometric relationship between the mating surfaces of the yoke-arm wheels and the holster guide, (3) locking the assembly on the holster by pushing wing segments of a captured nut (this is described more fully later) into chamfered notches in the yoke arms, and (4) releasing the end effector from the tool interface assembly.

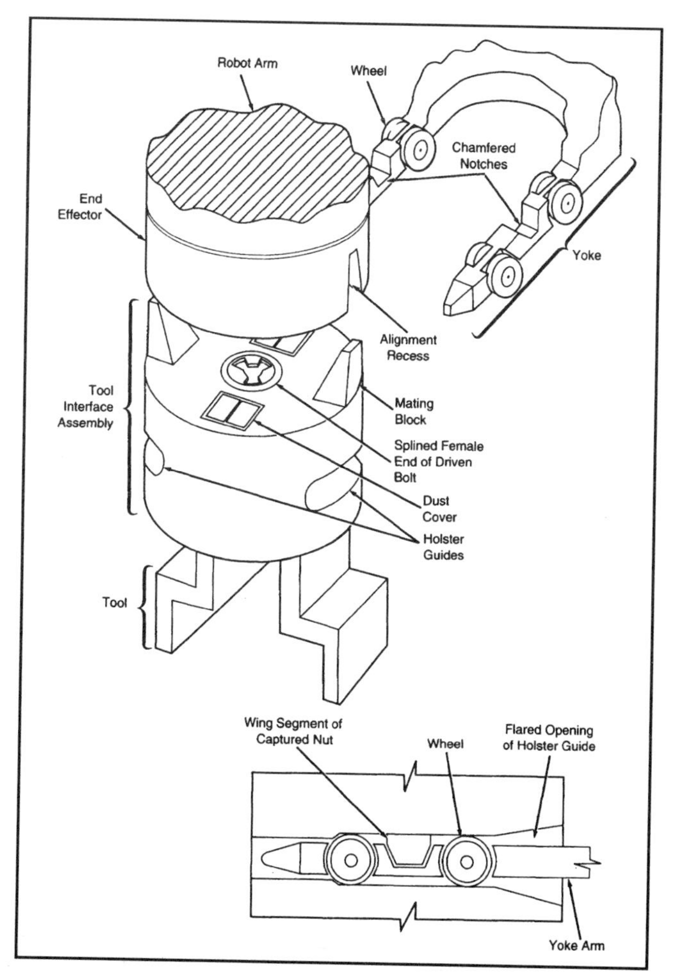

Fig. 1 This tool-changing mechanism operates with relatively small contact forces and is relatively compact.

The end effector includes a male splined shaft (not shown in Fig. 1) that is spring-loaded to protrude downward. A motor rotates the male splined shaft via a splined drive shaft that mates with a splined bore in the shank of the male splined shaft. The sequence of movements in which the end effector takes the tool from the holster begins with the movement of the end effector into a position in which its alignment recesses can engage the mating blocks on the tool interface assembly. The end effector is then pushed downward into contact with the tool interface assembly. Meanwhile, the male splined shaft is rotated until the spring force can push it through the opening in the splined female end of a driven bolt, and an alignment cone at the end of the splined male shaft bottoms in a conical hole in the female end of the driven bolt (see Fig. 2)

Assuming that the thread on the driven bolt is right handed, the male splined shaft is rotated clockwise until a vertical spline on this shaft engages a tab in the driven bolt. At that location the shaft and bolt rotate together. As the rotation continues, the driven bolt moves downward in a captive nut until the mating splined surfaces on the male splined shaft and driven bolt make contact. This prevents further downward movement of the driven bolt.

As the rotation continues, the captive nut moves upward. The wing segments mentioned previously are then pulled up, out of the chamfered slots on the yoke arms, so that the tool interface plate can then be slid freely off of the yoke. Simultaneously, two other wing segments of the captured nut (not shown) push up sets of electrical connectors, through the dust covers, to mate with electrical connectors in the end effector. Once this motion is completed, the tool is fully engaged with the end effector and can be slid off the yoke. To release the tool from the end effector and lock it on the yoke (steps 3 and 4 in the second paragraph), this sequence of motions is simply reversed.

This work was done by John M. Vranish of **Goddard Space Flight Center**.

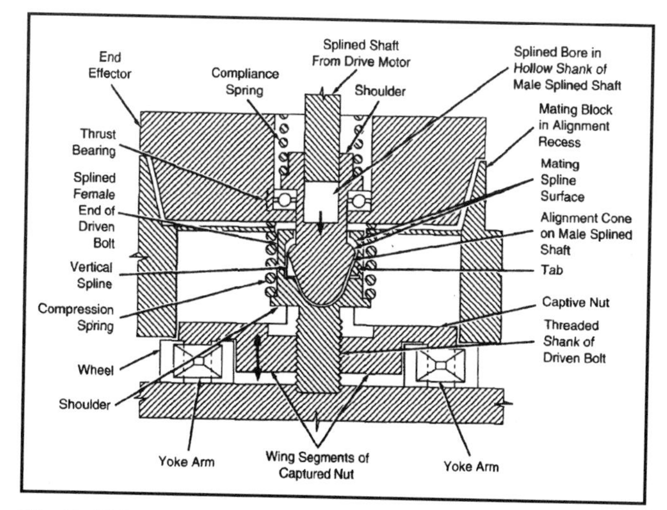

Fig. 2 This end effector and tool interface assembly is shown in its initial mating configuration, immediately before the beginning of the sequence of motions that release the tool from the yoke and secure it to the end effector.

PIEZOELECTRIC MOTOR IN ROBOT FINGER JOINT

A direct drive unit replaces a remote electromagnetic motor.
Marshall Space Flight Center, Alabama

A robotic finger contains an integral piezoelectric motor. In comparison with a robotic finger actuated by remote motors via tendonlike cables, this robotic finger is simpler and can therefore be assembled, disassembled, and repaired more easily. It is also more reliable and contains more internal space that can be allocated for additional sensors and control circuitry.

The finger (see figure) includes two piezoelectric clamps and a piezoelectric-rotator subassembly. Each clamp is composed of a piezoelectric actuator, a concave shoe, and a thin bushing with an axial slit. A finger-joint shaft fits in the bushing. When the actuator in a clamp is de-energized, the shaft is free to rotate in the bushing. When the same actuator is energized, it expands and pushes the shoe against the bushing. This action clamps the shaft. (The slit in the bushing allows it to flex so that more actuator force acts on the shaft and is not wasted in deforming the bushing.)

The piezoelectric-rotor subassembly includes a pair of piezoelectric actuators and a component simply called the rotator, which is attached to the bushing in clamp 2. The upper rotator actuator, when energized, pushes the rotator a fraction of a degree clockwise. Similarly, when the lower rotator is energized, it pushes the rotator a fraction of a degree counterclockwise. The finger-joint shaft extends through the rotator. The two clamps are also mounted on the same shaft, on opposite sides of the rotator. The rotator actuators are energized alternately to impart a small back-and-forth motion to the rotator. At the same time, the clamp actuators are energized alternately in such a sequence that the small oscillations of therotator accumulate into a net motion of the shaft (and the finger segment attached to it), clockwise or counterclockwise, depending on whether the shaft is clamped during clockwise or counterclockwise movement of the rotator.

The piezoelectric motor, including lead wires, rotator-actuator supports, and actuator retainers, ads a mass of less than 10 grams to the joint. The power density of the piezoelectric motor is much grater than that of the electromagnetic motor that would be needed to effect similar motion. The piezoelectric motor operates at low speed and high torque—characteristics that are especially suitable for robots.

This work was done by Allen R Grahn of Bonneville Scientific, Inc., for **Marshall Space Flight Center.**

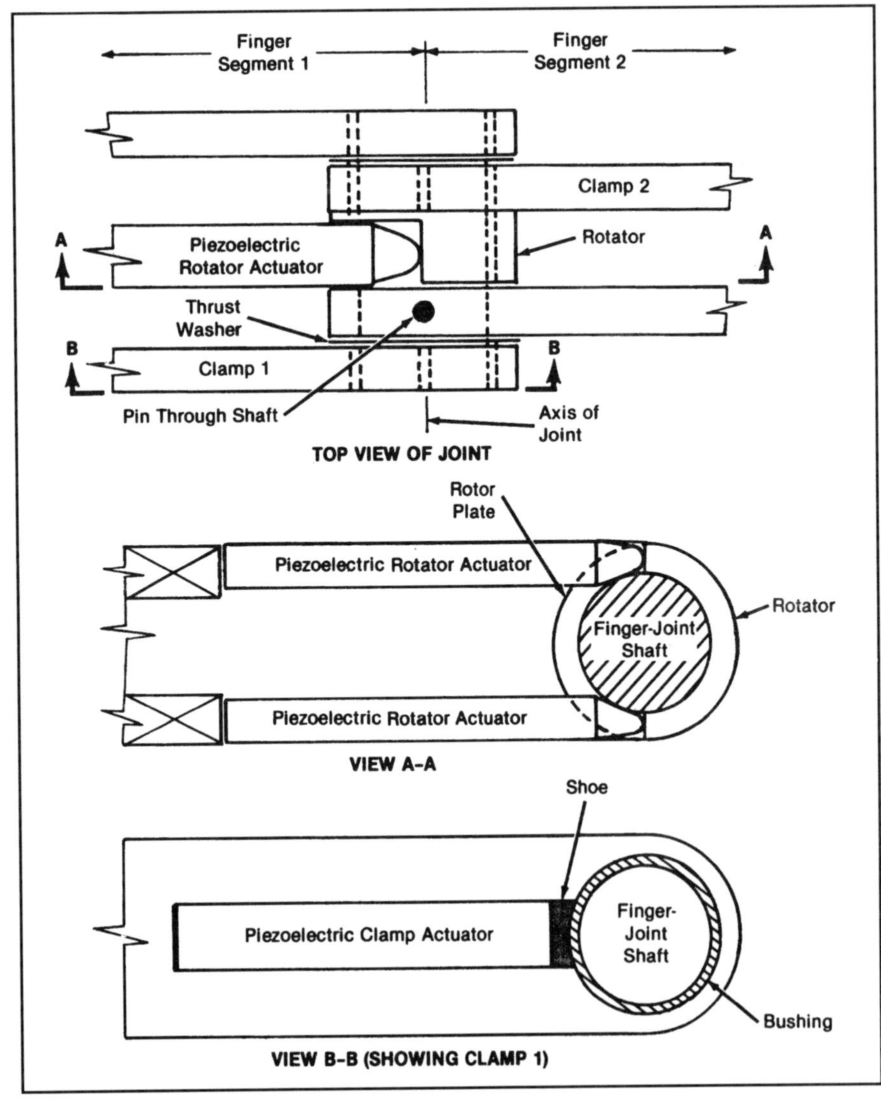

Each piezoelectric clamp grasps a shaft when energized. The piezoelectric rotor turns the shaft in small increments as it is alternately clamped and unclamped.

SELF-RECONFIGURABLE, TWO-ARM MANIPULATOR WITH BRACING

Structure can be altered dynamically to suit changing tasks.
NASA's Jet Propulsion Laboratory, Pasadena, California

A proposed two-arm robotic manipulator would be capable of changing its mechanical structure to fit a given task. Heretofore, the structures of reconfigurable robots have been changed by replacement and/or reassembly of modular links. In the proposed manipulator, there would be no reassembly or replacement in the conventional sense: instead, the arms would be commanded during operation to assume any of a number of alternative configurations.

The configurations (see figure) are generally classified as follows: (1) serial structure, in which the base of arm 1 is stationary, the tip of arm 1 holds the base of arm 2, and the tip of arm 2 holds the manipulated object; (2) parallel structure, in which the bases of both arms are stationary and the tips of both arms make contact with the manipulated object at two different points; and (3) the bracing structure, in which the basis of both arms are stationary and the tip of arm 2 grasps some intermediate point along the length of arm 1. the serial and parallel structures can be regarded as special cases of the bracing structure. Optionally, each configuration could involve locking one or more joints of either or both arms, and the bracing contact between the two arms could be at a fixed position of arm 1 or else allowed to slide along a link of arm 1.

The performances of the various con-figurations can be quantified in terms of quantities called "dual-arm manipulabilities," and "dual-arm resistivities." Dual-arm manipulabilities are defined on the basis of kinematic and dynamic constraints; dual-arm resistivities are defined on the basis of static-force constraints. These quantities serve as measures of how well such dextrous-bracing actions as relocation of the bracing point, sliding contact, and locking of joints affect the ability of the dual-arm manipulator to generate motions and to apply static forces.

Theoretical study and computer simulation have shown that dextrous bracing yields performance characteristics that vary continuously and widely as the bracing point is moved along the braced arm. In general, performance characteristics lie between those of the serial and parallel structures. Thus, one can select configurations dynamically, according to their performance characteristics, to suit the changing requirements of changing tasks.

This work was done by Sukhan Lee and Sungbok Kim of Caltech for **NASA's Jet Propulsion Laboratory.**

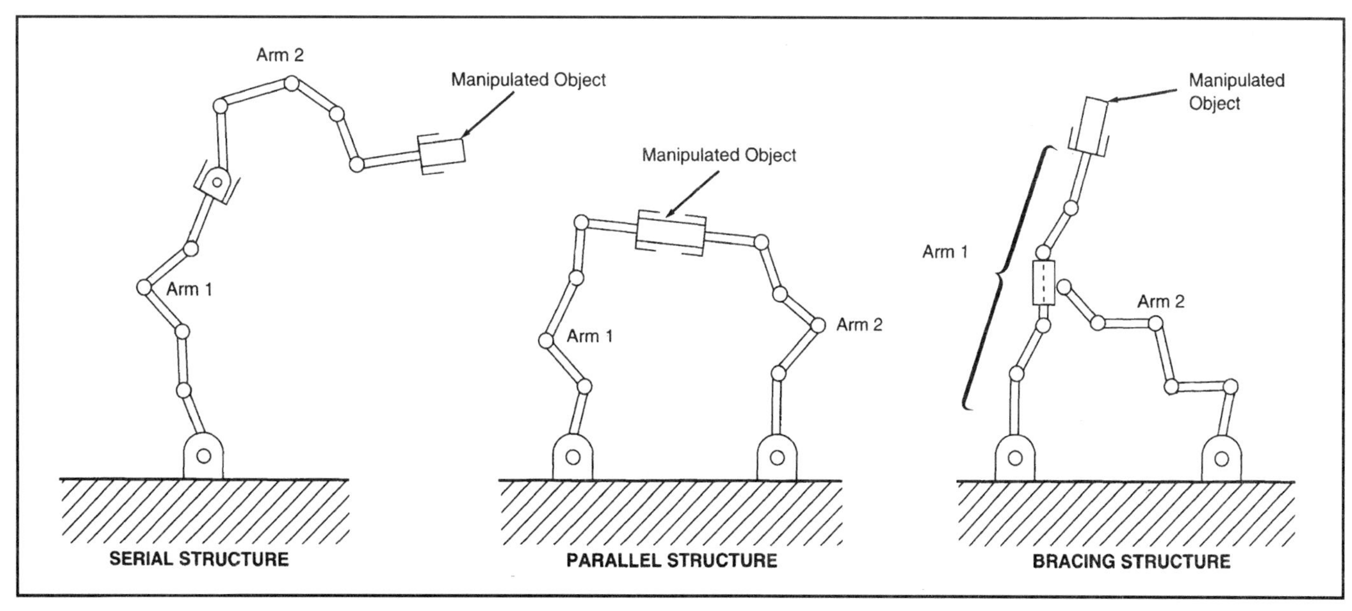

Alternative structures of cooperating manipulator arms can be selected to suit changing tasks.

IMPROVED ROLLER AND GEAR DRIVES FOR ROBOTS AND VEHICLES

One type eliminates stick/slip, another eliminates reaction torques.

Lewis Research Center, Cleveland, Ohio

Two types of gear drives have been devised to improve the performances of robotic mechanisms. One type features a dual-input/single-output differential-drive configuration intended to eliminate stick/slip motions; the other type features a single-input/dual-angular-momentum-balanced-output configuration intended to eliminate reaction torques.

Stick/slip motion can degrade the performance of a robot because a robotic control system cannot instantaneously correct for a sudden change between static and dynamic friction. Reaction torque arises in a structure that supports a mechanism coupled to a conventional gear drive, and can adversely affect the structure, the mechanism, or other equipment connected to the structure or mechanism.

In a drive of the differential type, the two input shafts can be turned at different speeds and, if necessary, in opposite directions, to make the output shaft turn in the forward or reverse direction at a desired speed. This is done without stopping rotation of either input shaft, so that stick/slip does not occur. In a drive of the angular-momentum-balanced type, turning the single input shaft causes the two output shafts to rotate at equal speeds in opposite directions.

The figure schematically illustrates one of two drives of the differential type and one drive of the angular-momentumbalanced type that have been built and tested. Each of the differential drives is rated at input speeds up to 295 radians per second (2,800 r/min), output torque up to 450 N·m (4,000 lb-in.), and power up to 5.6 kW (7.5 hp). The maximum ratings of the angular-momentum-balanced drive are input speed of 302 radians per second (2,880 r/min), dual output torques of 434 N·m (3,840 lb in.) each, and power of 10.9 kW (14.6 hp).

Each differential drive features either (as explained in the next two sentences) a dual roller-gear or a roller arrangement with a sun gear, four first-row planet gears, four second-row planet gears, and a ring gear. One of the differential drives contains a planetary roller-gear system with a reduction ratio (measured with one input driving the output while the other input shaft remains stationary) of 29.23:1. The other differential drive (the one shown in the figure) contains a planetary roller system with a reduction ratio of 24:1. The angular-momentum-balanced drive features a planetary roller system with five first- and second-row planet gears and a reduction ratio (the input to each of the two outputs) of 24:1. The three drives were subjected to a broad spectrum of tests to measure linearity, cogging, friction, and efficiency. All three drives operated as expected kinematically, exhibiting efficiencies as high as 95 percent.

Drives of the angular-momentum-balanced type could provide a reaction-free actuation when applied with proper combinations of torques and inertias coupled to output shafts. Drives of the differential type could provide improvements over present robotic transmissions for applications in which there are requirements for extremely smooth and accurate torque and position control, without inaccuracies that accompany stick/slip. Drives of the differential type could also offer viable alternatives to variable-ratio transmissions in applications in which output shafts are required to be driven both forward and in reverse, with an intervening stop. A differential transmission with two input drive motors could be augmented by a control system to optimize input speeds for any requested output speed; such a transmission could be useful in an electric car.

This work was done by William J. Anderson and William Shipitalo of Nastec, Inc., and Wyatt Newman of Case Western Reserve University for **Lewis Research Center.**

Stationary Ring Rollers
Rotating Ring Rollers
2nd Row Planet Rollers
Input Shafts
Sun Roller
Sun Roller
Output Shaft
Cluster Carrier
First-Row Planet Rollers
DIFFERENTIAL DRIVE

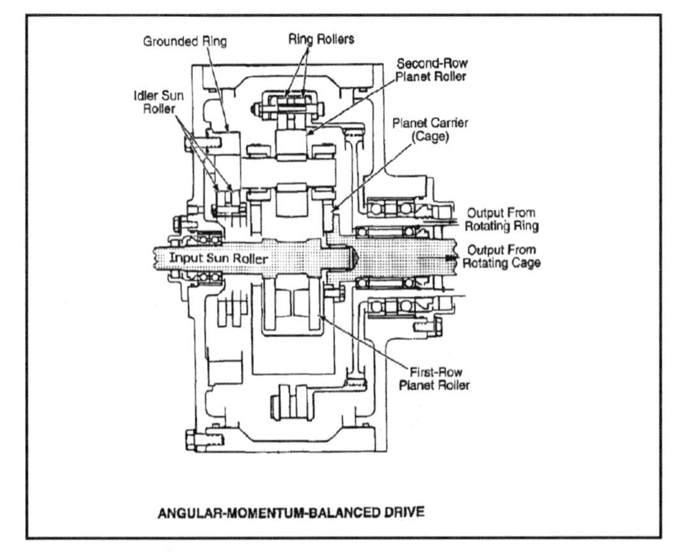

These Improved Gear Drives offer advantages for control of traction and rotary actuation in robots. In addition, drives of the differential type could be used in variable-speed transmissions in automobiles.

GLOSSARY OF ROBOTIC TERMS

actuator: Any transducer that converts electrical, hydraulic, or pneumatic energy into power to perform motions or tasks. Examples are electric motor, air motor, and solenoid.

adaptive control: A method for optimizing performance by continuously and automatically adjusting control variables in response to measured process variables. A robot adaptive control requires two extra features beyond its standard controls: (1) at least one sensor capable of measuring changes in the robot's working conditions and (2) the robot's central processors must be programmed to process sensor information and send signals to correct errors in the robot's operation.

air motor: A device that converts pneumatic pressure and flow into rotary or reciprocating motion.

android: A robot that mimics human appearance and behavior. Other equivalent terms are humanoid or anthropomorphic.

arm: An interconnected set of mechanical levers and powered joints that simulate a human arm and act as a manipulator; it can move an end of arm *wrist* with an attached end effector or tool to any spatial position within its work envelope.

axis: A linear direction of travel in any of three dimensions: axes in Cartesian coordinates are labeled *X,Y*, and *Z* to orient axis directions with respect to the Earth's surface: *X* refers to a directional plane or line s parallel to the Earth; *Y* refers to a directional plane or line perpendicular to *X* and is also parallel to the Earth; and *Z* refers to a directional plane or line vertical to *X* and *Y* and perpendicular to the Earth.

cable drive: A drive that transmits mechanical power by means of flexible cables and pulleys from a motor or actuator to a remote robot joint such as a wrist or ankle. It is also known as a *tendon drive*.

closed loop: A control scheme that compares the output value with the desired input value and sends an error signal when they differ, causing corrective action which restores equality between the values.

collision protection device (CPD): A device attached to a robot wrist that can detect a potential collision between the robot and a foreign object or that a contact has actually been made with one and sends a signal to the robot central processor causing it to stop or divert the motion of the arm before damage can be done. It is also known as a *collision sensor or a crash protection device*.

computer-vision system: An electronic system containing a video camera and computer with a vision program that allows a robot to acquire, interpret, and process visual information. The camera is set to view a restricted field into which parts are moved. The vision system can recognize specific parts in various orientations and locations within the camera's field of view and direct the robot to perform specific operations on that part. The system can be programmed to separate individual parts from a mixed group of parts, grasp a part regardless of its orientation for packaging or assembly, measure or inspect parts, and reject faulty or incomplete parts. A two-dimensional (2-D) vision system can process 2-D images to obtain part identity, position, orientation, and quantity of parts. A 3-D vision has the properties of a 2-D system but also has depth perception which can be used to avoid assembly errors, detect out-of-place objects, distinguish between similar parts, and correct positioning discrepancies.

continuous-path programming: A robot motion control method that maintains absolute control over the entire motion path of the tool or end effector. Programming can be done by manual teaching or moving the robot wrist sequentially through its work cycle. The robot wrist moves to closely spaced positions according to the program. The end effector performs the assigned tasks while the axes of motion are moving. All axes of motion move simultaneously, each at a different velocity, to trace a smooth continuous 3-D path or trajectory. It is recommended in applications where the tool path is critical, such as painting, adhesive placement, or arc welding.

controlled-path programming: A robot motion control method in which all axes move along a straight path between points at a set velocity. Each axis is coordinated so that it accelerates to the specified path velocity and decelerates smoothly and proportionally to provide a predictable, controlled path.

An operator can use a teach pendant to program only the end points of the desired path. This method is used in such applications as parts assembly, welding, materials handling, and machine tending.

degrees-of-freedom (DOF): A value defined by the number of rotational axes through which motion can be obtained by the robot with or without an end effector or tool attached to its wrist. DOF indicates the number of independent ways in which the end effector can be moved.

end effector: Any tool, sensor, or device attached to the robot wrist for performing a task. Examples include grippers or hands, welding torches, paint spray guns, or measuring devices. End effectors are typically powered by pneumatic actuators or electric motors independent of the host robot, and they can add one or more independent degrees of freedom to the robot.

gripper: A mechanical grasping tool or hand attached to a robot's wrist that can pick up and place objects with various shapes and orientations to perform such tasks as assembly, packing, or loading and unloading of parts or materials. Most common grippers have two opposing fingers that are machined to fit specific objects with defined orientations. They can be driven by hydraulic or pneumatic actuators or electric motors powered by supplies that are independent of the host robot. Some versatile grippers with three of more fingers are dexterous enough to grasp objects, regardless of their shape or orientation, when their fingers are directed to close around the object. They are usually computer-controlled by dedicated software either in the host robot's central computer or in an independent notebook computer.

kinematic chain: The combination of rotary and/or translational joints or axes of motion.

kinematic model: A mathematical model used to define the position, velocity, and acceleration of each moving member of a robot without considering its mass and force.

manual teaching: A method for programming a robot by leading the end effector manually through the entire sequence of motions required to perform a task so that all axial positions can be recorded by the operator with a control panel or teach pendant. The position coordinates are stored in the robot's computer memory so that they can be played back automatically to perform the recorded task.

manufacturing cell: A concentrated group of manufacturing equipment typically including one or more industrial robots, a computer-vision system, and ancillary equipment such as parts conveyors, indexing tables, inspection station, end-effector changers, and storage racks dedicated to performing a specific function at one location. All equipment is coordinated and synchronized by a computer to carry out continuous processing. It is also called an *assembly cell* or an *assembly center*.

mobile robot: A robot with its own self-contained means of propulsion: wheels, tracks, propeller, or other mechanism for crawling, climbing, swimming or flying. It contains a central processor and appropriate sensors. It is likely to carry or contain onboard equipment including, tools, manipulators, and sensors for navigation and the performance of tasks. These robots participate in but are not limited to Earth or planetary exploration, surveillance, and the disposal of bombs or other hazardous material. Mobile robots can be partly or completely autonomous, but most are directed by a remote operator via two-way wireless, wired, or fiberoptic links which transmit information to and from the robot. Sensors such as video cameras and distance-measuring instruments provide navigational guidance for the remote operator. The links can also be used to reprogram the robot's computer while it is moving. Most mobile robots are more accurately called *telerobots*; they now include unmanned ground vehicles, water and underwater vehicles, and aircraft or drones.

movable robots: Industrial robots mounted on wheels or rollers that can be driven by a power source but and restricted to travel on rails or tracks. They can move in horizontal, vertical, or angled directions while performing assigned tasks such as painting or welding under manual or autonomous control.

open loop: A control technique in which the robot's tasks are performed without error correction; accuracy typically depends on components including a position motion controller and a stepping motor.

payload: The load that can be lifted by a robot, measured in pounds (lb) or kilograms (kg).

payload capacity: The maximum *payload* that can be handled safely by a robot.

point-to-point programming: A robot motion control method in which a series of numerically defined stop points or positions are programmed along a path of motion. The robot moves to a position where it stops, performs an operation and moves on to the next position. It continues these steps in a sequence performing all operations until the task is completed. This control method is suited for pick-and-place materials handling and other applications where the movements between points need not be controlled.

reach: The maximum distance that can be reached by the *tool point*, a theoretical point beyond the robot's wrist when all axes are extended to their limits. It is measured in inches (in.) or millimeters (mm).

repeatability: The limits of variation or deviation of the robot's tool center point position obtained after repeated cycles under fixed conditions. It is measured as plus or minus millimeters (± mm).

resolution: The smallest incremental motion that can be made by a robot; a measure of robot accuracy.

rotary joint: A mechanism consisting of fixed and rotary components which, when attached to the robot wrist, permit the wrist to rotate through 360° without interrupting the supply of compressed air, water, or electricity required for operating various end effectors; utilities are supplied through a slip ring.

rotational inertia: The property of a rotating body that resists changes in angular velocity around its axis of rotation; it is measured as a mass unit multiplied by the square of the unit length—*moment of inertia*, (lb-ft^2 and kg-m^2).

teach pendant: A handheld control box connected by cable to the computer control cabinet that permits the operator to enter programming data defining stop points while "teaching" the robot.

telerobot: A mobile robot with partial autonomous control of its sensors or functions whose movements must be directed by a remote operator via a wireless, wired, or fiberoptic link.

tool changer: (1) A mechanism with two components (master and tool) for exchanging robotic tooling while it is performing its tasks. The master, attached to the robot wrist, contains half of the coupling; the other half is attached to the end effector or tool to be exchanged. When the two parts of the coupling are mated, they can be locked together automatically by pneumatic or hydraulic pressure. Examples of end effectors that can be exchanged are grippers, welding tools, or deburring machines. (2) A specialized robotic gripper that can pick up, clamp, and release a large variety of end effectors with common arbors or shanks as required. (3) A rotary turret mounted on the robot's wrist containing a selection of tools that can be indexed into a working position as required.

tool center point (TCP): A position between the robot's wrist and that part of the tool or *end effector* that defines the location where tool activity is concentrated; examples are the nozzle of a paint spray gun or the end of an arc-welding electrode.

workspace: The locus of all points or envelope that can be reached by the wrist using all of the robot's available axis motions.

wrist: A set of joints, typically rotational, at the end of the robot arm with a mounting plate for attaching end effectors or tools. Wrists can have two- or three-degrees-of-freedom (DOF): Two DOF wrists can move the tool or end effector around both pitch and roll axes, or pitch and two independent roll axes. A 3DOF wrist can move a tool or end effector around the pitch, roll, and yaw axes.

CHAPTER 4

MOBILE SCIENTIFIC, MILITARY, AND RESEARCH ROBOTS

INTRODUCTION TO MOBILE ROBOTS

For most people, the term mobile robot conjures up the image of a small wheeled or tracked platform darting about while performing some critical task: its robotic arm could be picking up and carrying away a suspicious package—possibly a terrorist bomb or some dangerous radioactive material; on the other hand, it might be boldly entering a building using its video camera to seek out hiding criminals or victims trapped in rubble left by an earthquake or tornado. Clearly, these robots are going into harm's way to do dangerous work, thereby preventing their handlers from risking possible injury or death. However, these are only a few of the many roles undertaken by the latest generations of mobile robots.

Mobile robots take on many forms: most are wheeled or tracked, but some move on synchronized mechanical legs; others swim on the surface of the water or dive deep into the ocean. And there are many different kinds of flying robots or drones. Three different extraterrestrial mobile robots have explored the surface of Mars and sent back valuable scientific information about its terrain and the composition of its rocks, and a terrestrial scientific robot has explored the perilous vents of active volcanoes. Undersea robots have discovered long-lost sunken ships and viewed exotic marine life and volcanic vents thousands of feet down on the sea floor; others have engaged in undersea archeological expeditions and recovered unusual artifacts and sunken treasure well beyond the depths that human divers can safely go. Flying robots in the form of fixed-wing aircraft or helicopters carry out low-altitude aerial surveillance without endangering live pilots; some of these are armed with weapons for attacking targets of opportunity.

Mobile robots are generally small and light, permitting them to be powered by batteries. While some are autonomous, most are semiautonomous; that is, they are directed from a remote location by human operators via two-way wireless, wired, or fiberoptic links. Nevertheless, conventional motor vehicles have been outfitted with multiple sensors, computers, and self-contained motion control systems permitting them to travel autonomously over many miles of rough desert. Some aerial drones intended for long-range surveillance rival conventional manned aircraft in size. Regardless of their size or shape, all robots share many common attributes with static industrial robots: they are reprogrammable and capable of performing a wide range of tasks determined by their onboard instruments, sensors, controls, and software programs.

Directions for the movement of semiautonomous mobile robots are sent to the robot and important guidance information is fed back from onboard sensors and instruments; this information permits the operator to make remote course corrections to avoid obstacles that could stall the robot and defeat the mission. The onboard sensors typically include video cameras, accelerometers, gyroscopes, global positioning system (GPS) receivers, and distance-measuring equipment based on the transmission of ultrasound, microwave frequencies, or laser light energy. Consequently most mobile robots are more accurately called *telerobots*. A true mobile robot should be distinguished from a *movable robot*, defined as a static robot mounted on wheels or rollers and capable of moving only over a relatively short length of track from which it is mounted or suspended.

The economic importance and market size of mobile robots has only recently begun to catch up with that of industrial robots despite the fact that R&D for mobile robots and their components has been carried on for many years in various academic, industrial and government engineering labs. The demand for practical mobile robots has been limited primarily by cost factors because, until recently, most have been custom built as one-off models or in low-volume quantities. The purchase of a mobile robot, unlike an industrial robot, cannot be justified on the basis of productivity or cost-effectiveness; rather it must be based on the ability of the robot to perform unusual tasks, typically dangerous, without the risk of injury or death that would be incurred by a human attempting the same task. Needless to say, some missions such as extraterrestrial exploration cannot now be performed by humans regardless of how well they are protected by special equipment.

Some consumer and commercial mobile robots have been successful in the past, but sales were only a small fraction of the worldwide industrial robot market. The mobile robot industry began to take off within the past few years when technology advanced to the point where the necessary command and control circuitry became available. These included embedded microcomputer-based motion control modules and practical computer software as well as wireless communications protocols developed for computer networking. Other factors were the availability of smaller, lighter, and more efficient electric servomotors with neodymium-iron rare earth magnets and longer-life, more compact batteries that increase a robot's operating time between charges. Most terrestrial mobile robots are now equipped with infrared (IR) or radio frequency (RF) range finders that warn the remote operator of obstructions in the robot's path that should be avoided. A simplified block diagram of a generic mobile telerobot is shown in Fig. 1.

Mobile robots are now able to perform reliably in terrestrial and extraterrestrial scientific exploration, emergency police and public safety activities, but more importantly, in the performance of essential military applications. The industry has been spurred by the recent purchases of different types of mobile robots by the U.S. Department of Defense. There was a need for mobile robots that could disarm munitions, detect mines, and explore caves or tunnels to root out entrenched enemies. There was also a requirement for remote reconnaissance platforms that could surreptitiously gain tactical information and detect the presence of enemy incursions under all weather conditions.

Mobile robots suitable for deployment with emergency public services and military operations are built to conform to demanding standards: they must withstand many environmental extremes without failure; these include shock, vibration, temperature, blowing sand, and salt spray without failing. Some must also function reliably after exposure to intense nuclear radiation and corrosive chemicals. By contrast, despite their heavy-duty work assignments, most industrial robots function within controlled indoor environments.

More than a dozen academic, industrial, and government laboratories in the United States alone are now engaged in R&D of mobile robots: the academic laboratories include Carnegie Mellon University (CMU); California Institute of Technology (Caltech); University of Southern California; University of Pennsylvania; Massachusetts Institute of Technology (MIT); Cornell University; and the University of Georgia. Notable among the industrial laboratories is Xerox PARC (Palo Alto Research Center). In addition, the robot manufacturers themselves have been making significant improvements to their products.

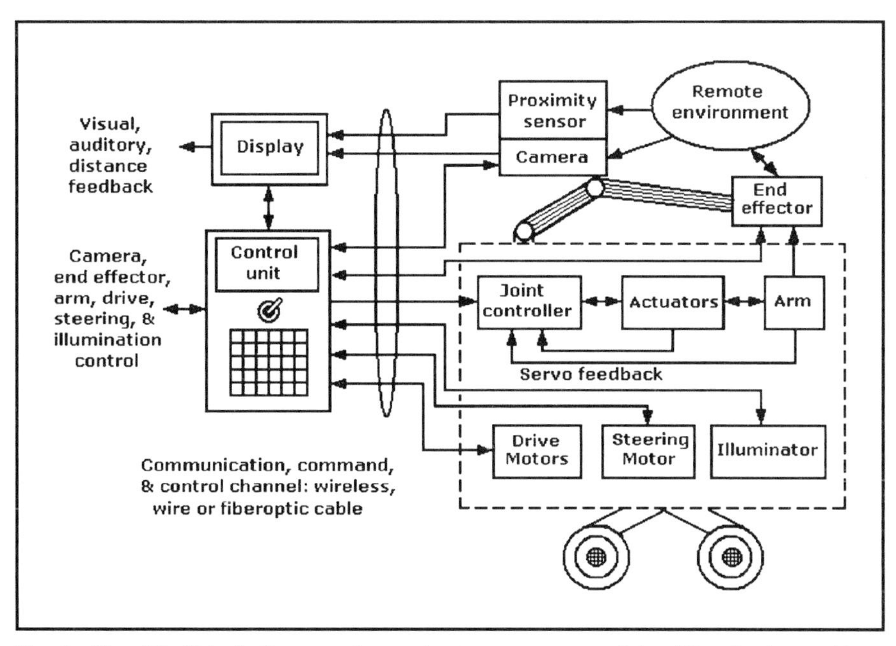

Fig. 1 Simplified block diagram of a semi-autonomous terrestrial mobile telerobot and its communication and control links to a human operator.

SCIENTIFIC MOBILE ROBOTS

Three NASA Mars exploration rovers, *Sojourner, Spirit*, and *Opportunity*, all wheeled, mobile semiautonomous robots, have been successful in their missions to Mars. All have returned vast quantities of scientific information that has advanced our knowledge of the Red Planet. In addition, they have demonstrated the highest levels of reliability, far beyond predictions, because all were able to function for more than a year in the extreme environment of that distant planet.

Spirit landed on Mars on January 3, 2004 and *Opportunity* landed 21 days later on January 24, 2004. Both rovers were still functioning, having passed the two-Earth-year and ten-month mark in October 2006, far exceeding their expected lifetime of 90 days. They have investigated hills, craters, and sandy plains in search of water or evidence of it, the key to past or present life on Mars.

The success of these robots vindicated their design, engineering, and construction decisions; this has paved the way for even more sophisticated robotic rovers for future unmanned missions to Mars and other planets. *Spirit* and *Opportunity* share many of the features of the earlier *Sojourner*, but they have improved auto-navigational driving capability. All three robots have demonstrated the synergy gained by integrating communications and computer-based motion control with a versatile mobile platform. NASA plans to land two more robotic explorers on Mars in this decade. The *Phoenix Mars Scout*, to be launched in 2007, will search for organic chemicals, and the *Mars Science Laboratory* is expected to follow in 2009.

The basic mechanical features common to the three rovers now on Mars are illustrated in Fig. 2. This drawing is based on information obtained from the Personal Rover Project of The Robotics Institute of Carnegie Mellon University (CMU). The

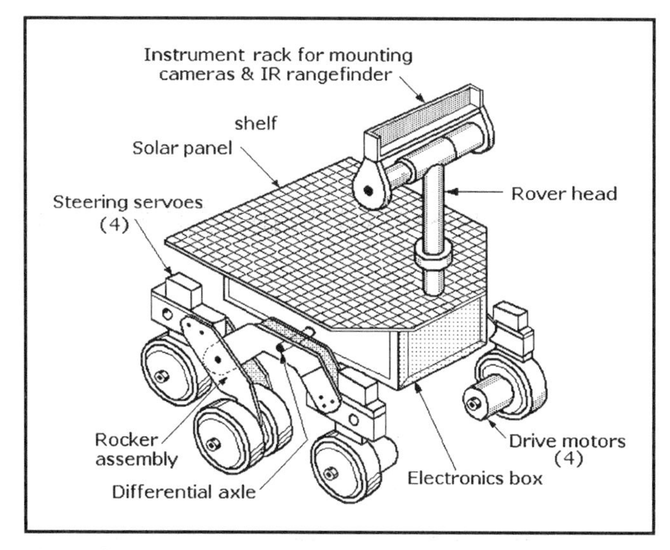

Fig. 2 Scientific planetary rover for training personnel in remote control of a mobile robot on a distant planet when all commands and feedback signals are time delayed because of the vast distances between it and Earth. It is a simplified version of actual Mars rovers.

Personal Exploration Rover 2004 is a simplified version of the Mars exploration rovers intended for educational and training missions here on Earth. This robot is propelled by four steerable wheels, each driven by an electric motor aided by two bogies in a rocker assembly. This arrangement permits the robot to move backward and forward, turn left or right, and climb over rocks or out of shallow pits while avoiding insurmountable obstructions.

The four wheels (two of them bogies) on each side of the robot are attached to rocker assemblies mounted on a differential axle; this allows each wheel assembly to pivot. In addition, the rear wheels and bogies, mounted on a separate arm, are able to pivot around a separate pin, assuring maximum flexibility. Steering servos permit all four wheels to be turned under computer control.

The electronic box forms the main body of this robot; it contains all the electronics which include computers, communications transceivers, instrumentation, and various power supplies. The electronics box is covered with a solar panel which provides additional power for operating the robot and recharging batteries when sufficient solar energy is available.

A retractable mast supports the rover head; this provides a mount for a video camera and instruments. A panoramic servomotor rotates the mast through $\pm 180°$ in azimuth and a tilt servomotor turns the head (45°/− 30° in elevation. For Earthbound experiments and training, a panoramic camera and IR rangefinder are mounted on the rover head, and an ultraviolet (UV) light is mounted on the front of the electronics box. The communications whip and parabolic dish antennas are not shown on the drawing.

Both of NASA's Martian rovers, *Spirit* and *Opportunity*, are far more sophisticated mechanically and electronically than the CMU training robot. They were designed to fold up into small, relatively flat packages to survive the long space voyage to Mars. Once there, both survived their landings and subsequent bouncing motions without impaired functions before being commanded to unfold and activate themselves; this was reminiscent of a butterfly emerging from a cocoon. The wheels and suspension system unfolded and the instrument mast and all onboard antennas were raised in a predetermined order. This feature is unprecedented for autonomous robots here on Earth. The leading specifications for these mobile Mars rovers are given in Table 1.

Table 1 Leading Specifications for Martian Rovers *Opportunity* and *Spirit*

Chassis length	5.2 ft (1.6 m)
Chassis width	7.5 ft (2.3 m)
Chassis height	4.9 ft (1.5 m)
Chassis weight	384 lb (174 kg)
Communication mode	Radio frequency
Key onboard equipment	Panoramic camera, microscopic imager, 3 different IR spectrometers to identify rock composition

The mobility systems on the Mars rovers are located at the backs of each platform. Each wheel has a diameter of about 10 in. (26 cm) and includes a spiral flexure pattern that connects the external part of the wheel with a spoke; this feature absorbs shock, preventing it from being transferred to critical parts of the rover. Each wheel also has cleats which provide traction for traversing soft sand and climbing over rocks.

The rovers have full "toolboxes" equipped with all the apparatus needed to search for the signs of ancient water and climate. The panoramic stereo cameras mounted on their retractable masts and IR spectrometers are able to survey the scenery around the rovers and look for the most interesting rocks and soils. High-gain, low-gain, and ultrahigh frequency (UHF) antennas send and receive communications at various frequencies from Earth. More information on these and other onboard scientific instruments and how they function is available on the NASA's Jet Propulsion Laboratory (Pasadena, CA) Web site www.jpl.nasa.gov/ and as archival documentation. This coverage is beyond the scope of this article.

MILITARY MOBILE ROBOTS

Two mobile robots built to conform to military specifications, *Dragon Runner* and *PackBot,* are described here because they have reached the production stage, and have been tested in the field under combat conditions. However, both of these robots are still subject to modifications and improvements of their chassis and onboard equipment suites to fine tune them for improved performance, increased reliability, and cost reduction.

The U.S. Marine Corps is the sponsor of *Dragon Runner,* a small four-wheeled mobile robot that features four-wheel drive and front-wheel steering (see Fig. 3). Weighing only 15 lb (6.8 kg), it is light enough to be carried by a single marine in a backpack. Configured as a flat box with a wheel on each corner, it has an invertible suspension system that permits it to function normally with either side up. This characteristic permits it to be tossed over walls or other obstacles, thrown out of windows, tossed off of moving vehicles, or heaved downstairs without concern for how it lands. It is tough enough to withstand a fall of 30 ft (10 m), land on its wheels, and move away at speeds up to 20 mph (32 km/h). However, when directed to do so, it can move at a slow, deliberate pace. The leading specifications for *Dragon Runner* are given in Table 2.

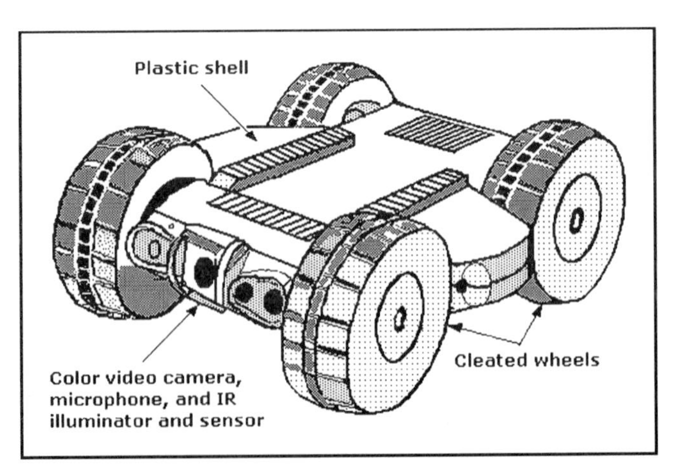

Fig. 3 Dragon Runner, a remote-controlled, throwable robot developed for the U.S. Marines for surveillance in urban areas. It will survive a 20 ft drop and still operate with either side facing up.

Table 2 Leading Specifications for *Dragon Runner*

Chassis length	15.5 in. (40 cm)
Chassis width	15.5 in. (40 cm)
Chassis height	5.0 in. (13 cm)
Chassis weight	15 lb (6.8 kg)
System weight	<30 lb (<13.6 kg)
Speed	20 mph (32 km/h)
Communication mode	Radio frequency
Onboard equipment (typical)	Color video camera, microphone, IR illuminator, IR sensor

Dragon Runner was designed by a staff member of the Robotics Institute of Carnegie Mellon University, Pittsburgh, Pennsylvania. The robot enables marines to gather intelligence without being in harm's way because it reduces the time that they must expose themselves to danger. Its onboard cameras can look for dangerous situations around corners or in locations exposed to heavy gunfire.

Dragon Runner is directed by a handheld control unit with a 4-in. video display that is bright enough to be read in daylight; the control unit permits its marine operator to send it out as an advance scout. The robot is equipped with an IR sensor for obstacle avoidance; a microphone and obstacle sensor return vibrations to the operator when they are triggered. A military frequency radio link connects the robot to the control unit.

The complete system will fit into a lightweight equipment pack easily carried by one person, and the robot can be deployed from its pack in less than a minute. Admittedly, the deployment of this robot must be carefully planned to make sure that its presence does not alert an enemy who could destroy it before it is able to send back vital intelligence to its marine operator.

The *PackBot EOD,* shown in Fig. 4, is a heavier, more versatile mobile robot than *Dragon Runner* that moves on tracks rather than wheels. The EOD version was designed for explosive ordnance disposal and HAZMAT (hazardous materials) removal and cleanup. It can also be organized for other functions such as search, surveillance and reconnaissance, as well as hostage rescue by changing its onboard selection of instruments and sensors.

The EOD version is equipped with a three-link flexible manipulator arm that serves as a support for video cameras and illumination sources as well as a gripper that can rotate through 360° to perform a wide range of manipulation tasks. It has a reel for paying out fiberoptic communication cable under computer control if that operation mode is selected. A fully equipped *PackBot EOD* can be carried and operated by a single person. A separate remote control unit, basically a militarized laptop computer, has a 15-in. screen. The leading specifications for *PackBot* are given in Table 3.

The basic *PackBot* chassis has one pair of main tracks and a second ancillary pair of tracks called *flippers.* The flippers can be raised to allow the robot to climb stairs, maneuver over rocky terrain, and navigate narrow, twisting corridors. All of the endless-belt tracks are made from a polymer plastic; these permit the robot to traverse securely over surfaces as varied as ice, fields strewn with rubble or covered with mud or snow, and sandy beaches and deserts.

The chassis has eight separate payload bays, each with interchangeable payload modules for equipment; this can range from video/audio systems and chemical/biological sensors to mine detectors, ground-penetrating radar (GPR), and extra batteries. Other onboard instrumentation can include a temperature sensor, magnetometer, accelerometer, inclinometer, and a compass. This versatility permits rapid changes to be made in *PackBot's* configuration so that it is prepared to meet the needs of a variety of missions.

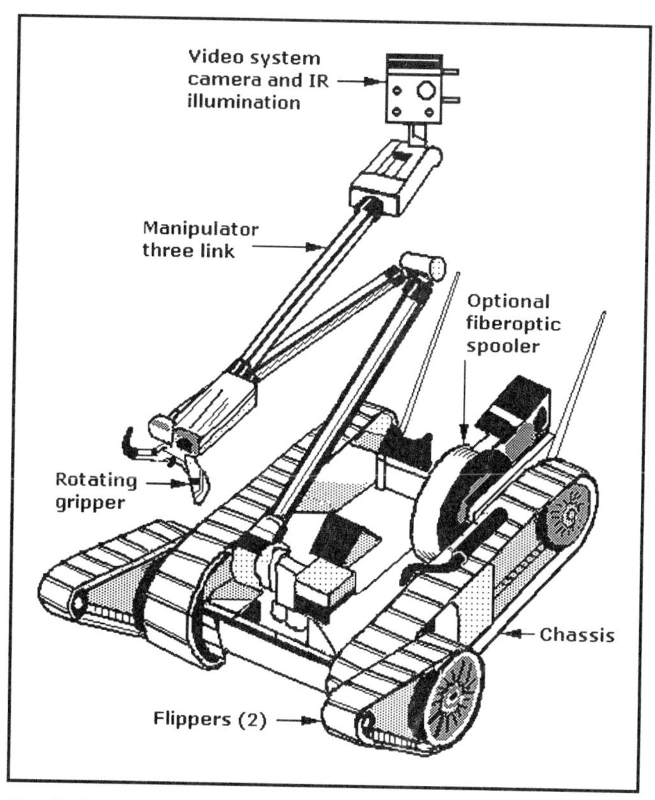

Fig. 4 PackBot, a tracked military mobile robot developed for explosive ordnance disposal (EOD), hazardous materials handling, search and rescue, surveillance, and other functions. Light enough to be hand carried, it has a second set of movable tracks called flippers to enhance its climbing ability.

Table 3 Leading Specifications for *PackBot*

Chassis height	16 in. (41 cm)
Chassis width	20 in. (51 cm) w/flippers
Chassis length	35 in. (88 cm) w/flippers extended
Chassis weight	53 lb (24 kg) fully loaded
Control unit weight	45 lb (20 kg)
Payload capacity	35 lb (16 kg)
Speed	~7 to 12 ft/s (2 to 3.6 m/s) ~5 to 8 mph
Communication mode	Radio frequency or optional fiberoptic cable
Onboard equipment (basic)	Manipulator system w/gripper, microphone, 4 color video cameras, and IR illuminator

The *PackBot* has proven itself with American ground troops in Afghanistan and Iraq by helping to clear caves and bunkers of enemy combatants, search buildings for hidden combatants or munitions, and cross fields strewn with antipersonnel mines. Although designed for military missions, *PackBot* can perform civil emergency service roles: examples include participation in the rescue of hostages, pinpointing the location of hidden snipers, search for trapped persons either under the debris of a collapsed building or in smoke-filled rooms, and investigate and dispose of packages containing explosive or radioactive materials.

Self-Reconfigurable Robots

New classes of versatile self-reconfiguring robots show promise for the performance of a wider range of tasks than conventional purpose-built robots. Now being developed in various laboratories, they are expected to be more adaptable in performing different tasks in more demanding environments than present generations of static and mobile robots. By contrast, most industrial robots being manufactured today are capable of performing only the specific tasks for which they were designed in controlled environments.

The objective of most R&D in self-reconfiguring robots now taking place in industrial, academic, and government laboratories is to demonstrate their feasibility and propose potential applications. It is expected that these versatile robots will be able to assemble themselves into many different practical configurations from self-contained modular building blocks, permitting them to change both their overall size and methods for self-locomotion.

Self-reconfiguring robot systems include modules containing the mechanisms and electrical circuitry that permit them to dynamically and automatically reconfigure themselves into more complex forms. R&D efforts are now focused on designing the most effective modules and their intermodule connection schemes as well as designing the most cost-effective methods for configuring the modules into systems. Existing criteria include making the modules as small, simple, and reliable as possible without restricting their ability to function independently of one another. This calls for minimizing the number of included components, linkages, and functions.

Regardless of design configuration, the modules will permit the robots to metamorphose into different shapes so that they can move efficiently over differing terrain while avoiding or surmounting any obstructions in their paths. For example, the robots might organize themselves to mimic the typical locomotion of such animals as snakes, crabs, or tigers. Most proposed unit-configurable robot systems are actuated by rotating one module relative to the other modules or expanding and/or contracting the module. Connection modules are either magnetic or mechanical and they can all be identical or a mixture of different units.

In addition to changing their means of self-propulsion, the robots will be expected to reconfigure their sensors and end effectors or tools for the performance of different assigned tasks. Some of the more promising modules developed to date consist of two building blocks called "atoms" joined by a right angle connector to form units called "molecules." Each modular section is capable of rotating in two degrees-of-freedom around each other as well as one degree-of-freedom around their mutual connector. The connectors contain electromagnets or grippers to assure structural rigidity after the robot is configured. These self-reconfiguring robots might be assigned to inspect and even make repairs on the outside surfaces of a spacecraft, inspect objects for damage in hazardous or radioactive areas, search for victims in earthquake or tornado rubble, or pick up and move objects within confined spaces.

Other important engineering considerations are communication between the modules, module actuator power requirements, and methods for supplying electric power to the complete robot. Intermodule communication is required for effective module interaction and the control of all of the robot's components. One approach is to use the connection mechanism as both the communications and power link between modules. Actuator power is the amount of force the module's actuators must exert for the modules to organize themselves; it must be at least strong enough to move each module's weight against gravity.

Higher fault tolerance in self-reconfiguring robot systems than is achieved in more conventional robots is expected to be gained by including spare modules in each system so that failed modules can automatically be replaced. One demanding objective for robots in this class is the possibility of making them capable of self-assembly when on site in such remote locations as planets or on the deep ocean floor so that they can adapt to specific tasks not foreseen in mission planning.

Although no robots of this type are yet available as commercial products, research results so far have shown that smart modules, even those with relatively limited capabilities, can autonomously organize or reorganize themselves into different structures; this ability permits them to perform many different assignments over a wide range of environmental conditions. The work of developing self-reconfigurable robots is reported to be taking place at Information Sciences Institute (ISI), University of Southern California (USC), University of Pennsylvania, Johns Hopkins University, Massachusetts Institute of Technology (MIT), Xerox PARC, and the National Institute of Advanced Industrial Science and Technology (AIST) of Japan.

M-TRAN II Robots

An existing example of a self-reconfigurable modular robot is the M-TRAN II (for modular transformer) developed by the Japanese Distributed Systems Design Research Group of the National Institute of Advanced Industrial Science and Technology (AIST). The basic module, as shown in Fig. 5, consists of three components: two semi-cylindrical blocks and a connecting link; one block is passive and the other is active. Each block can be rotated from $-90°$ to $+90°$ independently by a geared motor in the link, and each block has three connecting surfaces with four retractable permanent magnets (shown as four circular spots). The modules can interconnect by magnetic force because the polarities of the magnets in each block differ. The modules measure 2.4 × 2.4 × 4.7 in. (60 × 60 (120 mm) and weigh 14 oz (400 gm).

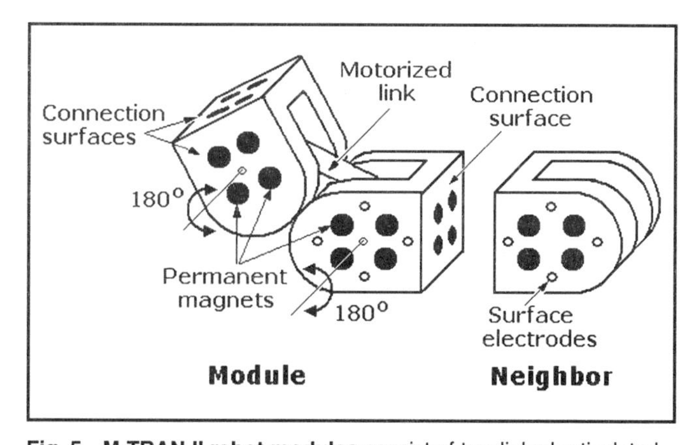

Fig. 5 M-TRAN II robot modules consist of two linked articulated parts containing a motor, communications and control circuitry, and magnets. The resulting robots can self-assemble into optimum locomotion configurations to adapt to a wide range of practical applications.

Each of the block's connecting surfaces can be attached to another connecting surface in every orthogonal relationship. The connection is established when the four magnet disks mounted on a card within the block are advanced to the surface by heated nonlinear springs; when a magnet card is retracted the modules will be released. The blocks in each module move around the blocks in the other modules, while the joints in each module allow them to flex and link together in chainlike arrays. The structure can be reconfigured by changing the positions of the modules and repeating such simple steps as uncoupling the connection, rotating the module, and reconnecting the modules.

The assembled robotic system typically consists of about a dozen modules; it is battery-powered and controlled by RF signals. Each module contains a microcontroller, neuron chip, acceleration sensor and power supply cards, a lithium-ion battery, and a motorized link that rotates the active and passive blocks. Power, signal, and ground electrodes are located on each module's connection surface.

Locomotion patterns for the robot are automatically generated by the host computer and downloaded to the modules. These permit the modules to rearrange themselves into countless different shapes and create dramatically different patterns of movement. M-TRAN II can configure itself to walk steadily on four legs (quadruped motion) as shown in Fig. 6, or shift into a long string of modules that can slither like a snake or crawl like a caterpillar, as shown in Fig. 7. It can also configure itself as a wheel so it can roll or as a spider so it can crawl, using its projecting modules as legs.

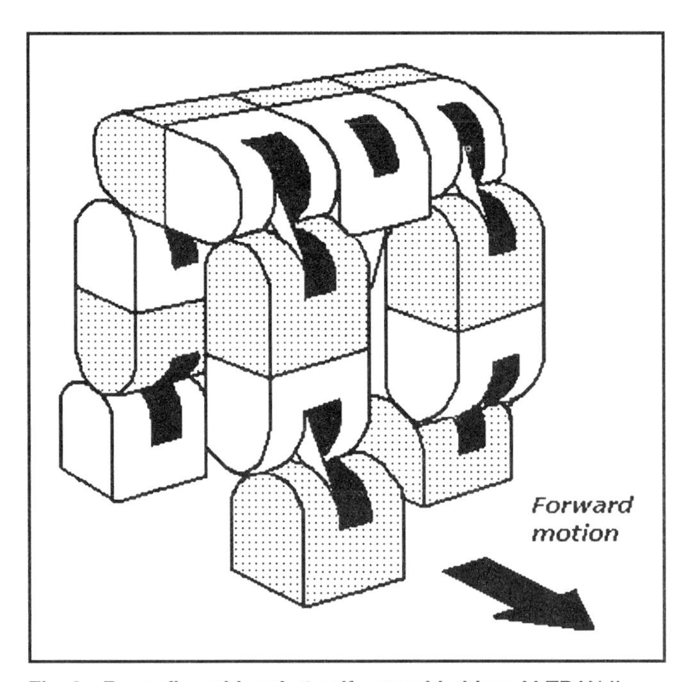

Fig. 6 Reconfigurable robot self-assembled from M-TRAN II modules to demonstrate quadruped locomotion.

Software within the M-TRAN II computer permits the robot to initiate new ways to move on its own as well as be commanded to move in patterns dictated by programmers. The M-TRAN II configuration has an advantage that other robots lack: its ability to complete an assigned mission despite the loss or malfunction of one or more modules. That loss might prevent the robot from continuing to move in the optimum pattern, but it could evolve a new locomotion strategy with the remaining modules to finish the mission, although it might not be as elegant as the initial motion pattern.

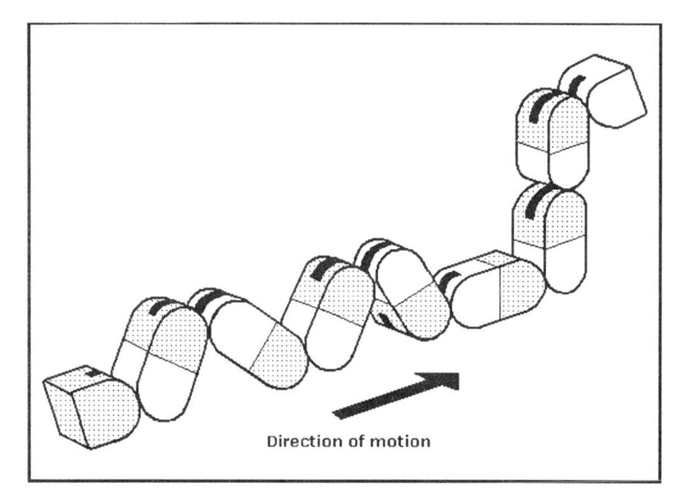

Fig. 7 Reconfigurable robot self-assembled from M-TRAN II modules to demonstrate serpentine locomotion making it capable of entering confined spaces or wall climbing.

PolyBot Robots

Another concept for reconfigurable robots was developed by researchers in the Modular Robotics Laboratory at the Xerox PARC. A stable of versatile robots called *PolyBots* has demonstrated that different groupings of small, identical, interchangeable modules can locomote like a snake, a spider, a lizard, or a wheel, to mention a few. The objective of this research is to show that robots can adapt to new environments and tasks by altering both their behavior and geometry.

A *PolyBot*, as shown in Fig. 8, is made up of a train of many modules; each is essentially autonomous robot because it contains a microcontroller, motor, sensors, and the ability to attach itself to other modules. In some configurations, power is supplied from an external supply and conducted from module to module. These modules attach themselves together to form chains which can function like an arm, leg, or finger, depending on its mission. PARC researchers have demonstrated that *PolyBots* can perform a wide variety of tasks from moving boxes to riding a tricycle.

Fig. 8 PolyBot reconfigurable robot, designed at Xerox PARC, is made up of autonomous modules able to self-assemble into a variety of locomotion configurations.

The first generation G1 prototype introduced the basic motorized-hinge design with all mating surfaces about 5 cm (2 in.) on a side shared in all succeeding generations. These modules were built from radio-controlled servos obtained from hobby shops; power and command signals were supplied externally. The modules had to be screwed together manually, so they were unable to self-reconfigure. These G1 modules were the first examples of robot locomotion by a simple reconfiguration in 1997. This design was the basis for NASA's *Snakebot.*

Succeeding generations of G1 modules were more robust and scalable than the first G1 version because they included more robust servos with more reliable output stages and a higher-voltage power bus. Improved generation II (G2) modules included onboard microcontrollers as well as improvements in the module's docking ability because of the installation of shape-memory alloy-actuated latches. Docking of chains of modules is also helped by IR emitters and detectors. The black cylinder sticking out of the module is the motor with a gearbox. A G2 module's clamping strength is about five times that of the original G1 modules.

Humanoid Robots

Scientists in laboratories around the world have been building prototype android or humanoid robots for years: they typically have computer "brains," twin camera "eyes," speaker "mouths" or voice boxes, and various powered joints that form movable hands, arms, and legs. Some are bipedal and can move about, but other more sophisticated anthropomorphic versions with human-like features have external power supplies and sensors. These human-friendly robots are intended to interact naturally with people and are being used to study human behavior.

A small telerobot built by the Computer Science and Artificial Intelligence Laboratory at MIT has combined concepts from biomechanics, control theory, and machine learning to make a significant advance in robotic technology. The robot, named *Toddler*, is simpler than many other anthropomorphic robots, and it does not attempt to mimic a human in appearance. However, project director Dr. Russ Tedrake reports that *Toddler* does mimic "the natural dynamics of the body."

The simple mechanical design of *Toddler* is shown in Fig. 9. Wireless Ethernet circuit boards mounted on top of the robot give it the appearance of having a head. Its arms and legs are passively hinged to shoulder-high "hips" formed by a wide circular metal band. The arms are mounting panels for the lithium-polymer battery packs that power the robot, and their weights act to counterbalance the opposing legs as they move; each leg can swing freely without power. The ankles contain electric motors that apply power to crank mechanisms which move the feet for each step. The feet are tapered wooden wedges, shaped to allow better foot clearance; they allow *Toddler* to rock from side to side as it walks. Each foot is coated with latex paint for improved traction. The leading specifications for *Toddler* are given in Table 4.

An onboard Intel Pentium microprocessor "learns" to control *Toddler's* locomotion by using information from the sensors that monitor the robot's orientation; the processor then sends corrective signals to the electric motors at each ankle to adjust *Toddler's* gait as it walks to keep it from falling. Wireless Ethernet signals allow a remote operator to start and stop the robot.

Toddler was designed to learn to walk, and this approach will, according to Dr. Tedrake, provide insights into human learning, rehabilitation, and prosthetics. He hopes that this research will lead to the development of robotic companions and helpers. *Toddler's* relatively simple mechanical design permits it to walk down a ramp with its onboard computer turned off. Because the objective of the research was the creation of a robot with passive walking behavior, the ankle motors with their turning cranks are intended only to keep *Toddler* moving on a level surface. The next generation robot with the same design objectives will have jointed knees permitting it to walk over rougher surfaces.

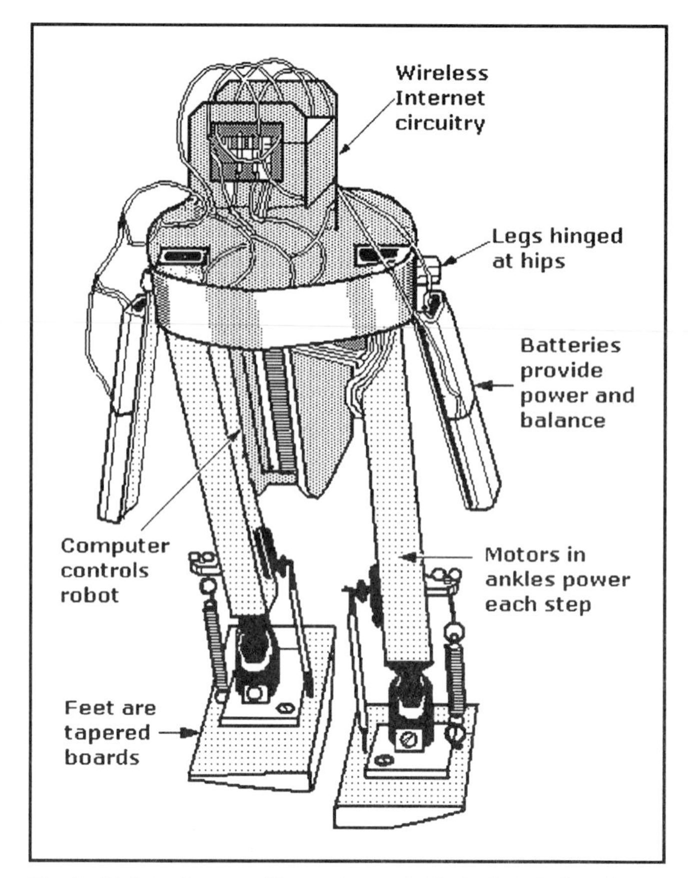

Fig. 9 Toddler humanoid remote-controlled robot designed at MIT's Computer Science and Artificial Intelligence Laboratory walks using the natural dynamics of the body.

Table 4 Leading Specifications for *Toddler*

Weight	≈6 lb (≈2.7 kg)
Height	18 in. (46 cm)
Stride	3 in. (7.6 cm)
Speed	<1 mph (<1.6 km/h)
Motors	Two in each ankle
Onboard equipment	700-MHz processor, 802.11b card, 256 MB RAM, 2 GB flash memory

Swarming Robots

Research is now being carried out on mobile robots called *swarming robots*. Generally small and box-shaped with wheels, these mobile robots are programmed to cluster, disperse, follow one another, and move in orbiting motions. In this way they emulate bees that cooperate to find food and build and maintain hives or dolphins that cooperate, when necessary, to round up fish and herd them into concentrated masses so that many of them can feed easily. While the precise methodology used by these animals to cooperate is not well understood, it is believed to involve inherited traits unique to the species and it is triggered by auditory or other sensory cues.

In experiments conducted so far, swarming robots equipped with sensors and radio equipment have demonstrated that they are capable of acting in concert and mimicking the behavior of swarming bees or dolphins by detecting environmental stimuli and contacting others in the group so that they can collectively accomplish a task that has been preprogrammed into each individual robot.

In a program sponsored by the Defense Advanced Research Projects Agency (DARPA), the robots communicated by IR energy signals that originate from a chosen leader with an antenna on its head. The signals are then distributed to the rest of the robots. In this way, the robots "talk" to each other. When they are talking, the lights begin flashing. Each light has an accompanying sound which tells the robots how far they are away from each other. The objective of this R&D is to demonstrate that robots can cooperate with each other to maintain the proper separation distances and avoid collisions with other objects.

A Hopping Mobile Robot

Researchers at the California Institute of Technology (Caltech), in Pasadena, have developed a novel locomotion concept that permits a mobile robot to roll on wheels and hop. According to the Caltech scientists, a hopping robot has two advantages over conventional wheeled mobile robots: it can hop over ground rapidly to avoid obstacles in its path and then roll slowly and precisely on its wheels to perform its mission. As a result, it acts like a kangaroo able to hop rapidly to escape danger but move slowly to graze on vegetation. Built for NASA's Jet Propulsion Laboratory (JPL), also in Pasadena, the prototype robot demonstrated that it could do scientific exploration on rough terrain on Earth or planets as well as carry out military duties or law enforcement assignments.

The robot both hops and rolls on its wheels, but it can traverse rough terrain and jump over obstacles at least twice its height to position itself rapidly and accurately at a specified location. The robot contains a nonlinear spring mechanism that can be adjusted to deliver the force necessary to hop over any obstacles up to its maximum jumping height. It has two front wheels that permit independent driving and steering control and a third passive rear wheel that keeps it in an upright position.

The Caltech robot can be programmed to hop in a desired direction over a range of takeoff angles with respect to the ground. The hopping distance can be adjusted by selecting an appropriate takeoff angle and adjusting the spring loading. The robot reaches the specified location by making a series of hops and short driving spurts while using its front steering wheels for precise positioning.

When in an upright position, the robot can remain motionless, move on its wheels, or prepare for a hop. After landing from a hop, an autonomous mechanism permits the robot to right itself by using a combination of side-panel actuation and the shifting of its center of mass. The side panel also protects vulnerable onboard equipment from damage by hard obstacles it might encounter upon landing.

An onboard digital video camera permits the remote operator to acquire images for navigation and orientation. Onboard electronics and software can process acquired data and compute paths for hopping and roaming. The robot can be directed either by wired or wireless communication links to the host computer and operator. It weighs about 3.3 lb (1.5 kg), occupies a minimum volume of about 11.8 in.^{3} (193 cm^3), and can jump about 3 ft (1 m) high and 6 ft (2 m) horizontally.

A Single-Wheel Stabilized Robot

Ballbot is a tall, thin, agile mobile robot with a high center of gravity that is propelled by a single spherical-ball wheel. It is dynamically stabilized in an upright position by an onboard closed-loop inertial guidance system. Developed and constructed by scientists at the Robotics Institute of Carnegie Mellon University, it is an alternative to tall multiwheel robots that are dynamically unstable because of their high centers of gravity.

The design objectives for Ballbot were to build a robot tall enough for interaction with people that is structured to permit the easy exchange of onboard components to meet present and future research requirements. Figure 10 shows Ballbot's tower framework built by fastening a stack of round metal decks to three vertical aluminum channels spaced 120° apart; the ball drive system is mounted below this skeletal stack. This robot can move in any direction within confined spaces that would bar most wheeled robots. The researchers recognized the inherent instability of tall wheeled robots that can tip over if they encounter steeply inclined surfaces.

Ballbot's decks are 16 in. (0.4 m) in diameter and it is 4.9 ft (1.5 m) high. Its battery, control computer, battery charger, and inertial measuring unit (IMU) are placed on separate decks and aligned with the robot's axis. A 9 in. (216 mm) diameter drive ball in the drive system is supported by three ball bearings spaced 120° apart. This ball is propelled by servomotor-driven rollers positioned at the ball's equator, and optical encoders measure the robot's travel. Ballbot weighs 100 lb (45 kg). The IMU includes three gyroscopes and accelerometers mounted at right angles to each other to sense motion in the pitch, roll, and yaw directions. The gyros send data to the computer for modulating the motion of the drive rollers to keep the robot upright as it travels; these signals are corrected to provide a constant gravity-seeking reference. Ballbot is self-contained and can operate autonomously, but it can also be controlled by a radio link. When the power to the robot is shut down, legs extend from the three channels to hold the robot in a vertical position.

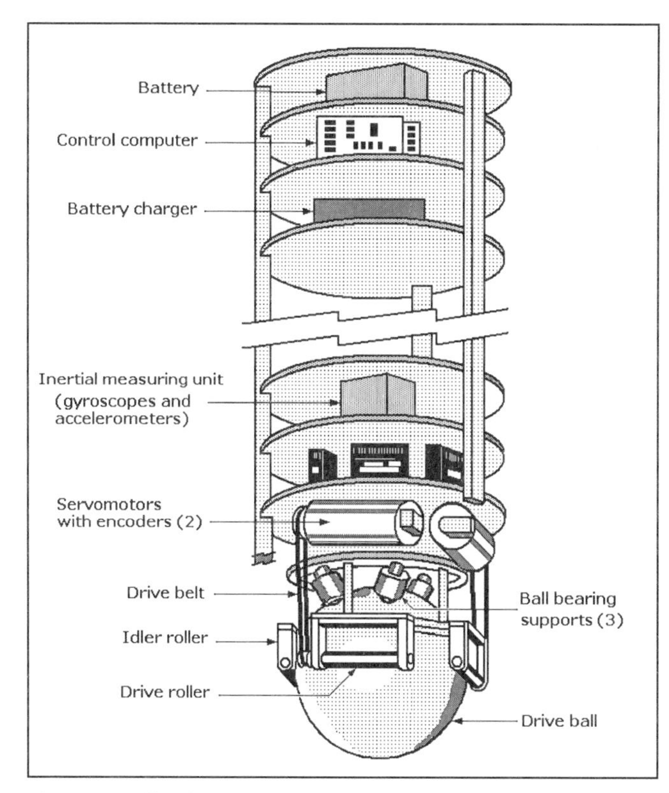

Fig. 10 Ballbot is a tall, thin robot (shown truncated) with a single spherical wheel so it can move easily in narrow spaces. It is kept upright as it travels by a closed-loop control system which includes an inertial guidance unit for providing a true vertical reference.

SECOND-GENERATION SIX-LIMBED EXPERIMENTAL ROBOT

This robot is designed to be more agile and dexterous than its predecessor.

NASA's Jet Propulsion Laboratory, Pasadena, California

The LEMUR II, shown in the figure, is the second generation of the Limbed Excursion Mechanical Utility Robot (LEMUR). It includes improvements and new features that extend its capabilities beyond those of its predecessor now called LEMUR I. LEMUR I is configured in a bilateral layout as a rectangular body with three legs on each side. It was designed to demonstrate robotic capabilities for performing assembly, maintenance, and inspection on earth with a view toward transferring these skills for work on spacecraft in deep space. It is able to walk autonomously along a truss structure toward a site for the performance of mechanical assembly at a specified location and to perform certain operations.

LEMUR I is equipped with stereoscopic video cameras and image-data-processing circuitry for navigation and the performance of mechanical operations. It is also equipped with a wireless modem, permitting it to be controlled from a remote location. When it arrived at the specified work site, it performed simple mechanical operations with one or both of its front limbs. It could also transmit video images to a host computer.

Each of the six limbs of the LEMUR I operated independently. The four rear limbs have three degrees of freedom (DOFs), while the front two limbs have four DOFs. The front two limbs were designed to hold, and operate or be integrated with tools. The LEMUR I includes an onboard computer containing a collection of digital control circuits. These include digital input/output circuits, analog-to-digital converters for input, and digital-to-analog (D/A) converters for output. Feedback from optical encoders in the limb actuators provided closed-loop microcomputer control of the positions and velocities of the actuators. LEMUR II includes the following improvements over LEMUR I:

- The drive trains for the joints of the LEMUR II are more sophisticated, providing greater torque and accuracy.
- The six limbs are arranged symmetrically about a hexagonal body platform instead of in straight lines along the sides. This arrangement permits omnidirectional movement meaning that the robot can walk off in any direction without having to turn its body.
- The number of degrees of freedom of each of the rear four limbs has been increased by one. Now, every limb has four degrees of freedom: three at the hip (or shoulder, depending on one's perspective) and one at the knee (or elbow, depending on one's perspective).
- Now every limb (instead of only the two front limbs) can perform operations. For this purpose, each limb is tipped with an improved quick-release mechanism for swapping of end-effector tools.
- New end-effector tools have been developed. These include an instrumented rotary driver that accepts all tool bits that have 0.125-in. (3.175 mm)-diameter shanks, a charge-coupled-device (CCD) video camera, a super bright light-emitting diode (LED) for illuminating the work area of the robot, and a generic collet tool that can be quickly and inexpensively modified to accept any cylindrical object up to 0.5 in. (12.7 mm) in diameter.
- The stereoscopic cameras are mounted on a carriage that moves along a circular track, thereby providing for omnidirectional machine vision.
- The control software has been augmented with software that implements innovations reported in two prior *NASA Tech Briefs* articles.

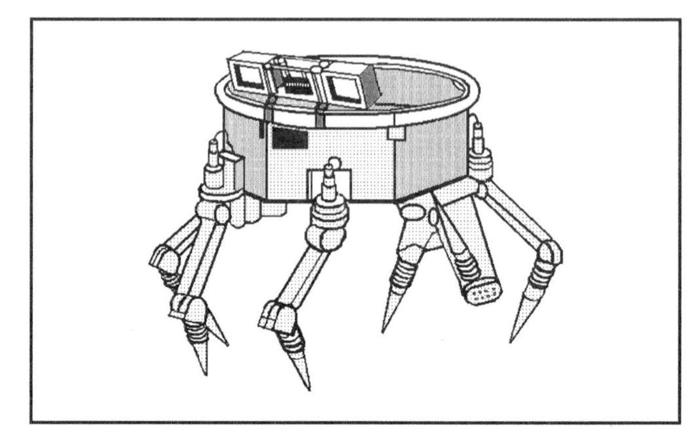

The Lemur II can move its stereoscopic cameras along a circular track to view objects at any angle. Its symmetrical arrangement of six limbs permits motions in any direction. (LEMUR II NASA Robot) NPO 35140, Jan 28, 2006.

This work was done by Brett Kennedy, Avi Okon, Hrand Aghazarian, Matthew Robinson, Michael Garrett, and Lee Magnone of Caltech for NASA's Jet Propulsion Laboratory.

Editor's note The JPL-built LEMURS are to be test beds for a proposed NASA Autonomous Walking Inspection and Maintenance Robot (AWIMR) and are expected to be influential in their design. The AWIMR is a NASA concept calling for a six-limbed, battery-powered robot capable of climbing around on the exteriors of spacecraft looking for meteorite damage, structural faults, or other problems that could pose a threat to astronauts inside. The prime contractor for AWIMR, Northrop Grumman, hopes to have the AWIMR space qualified and ready for deployment on the Crew Exploration Vehicle in 2010.

The AWIMR would also be equipped with tools permitting it to make repairs such as detecting and fixing leaks. It is expected to have a 12 in. (30 cm) central body similar in size to the LEMUR II, but would weigh nearly twice as much [44 vs. 26 lb (20 vs. 12 kg)]. It would house a computer, cameras, and other sensors for conducting inspections, tools for making repairs, and special grippers on its limbs for holding onto spacecraft surfaces. Astronauts will control the robot using wireless links, such as WiFi or Bluetooth, and a joystick interface.

ALL-TERRAIN VEHICLE WITH SELF-RIGHTING AND POSE CONTROL

Wheels driven by gearmotors are mounted on pivoting struts.
NASA's Jet Propulsion Laboratory, Pasadena, California

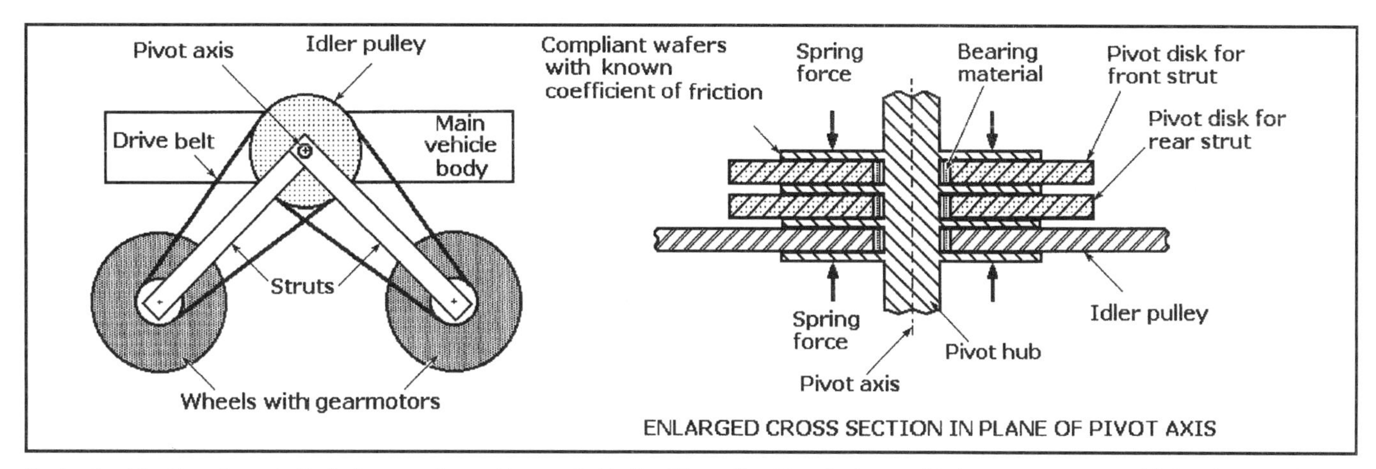

Each wheel Is driven by a dedicated gearmotor and is coupled to the idler pulley. The pivot assembly imposes a constant frictional torque T, so that it is possible to (a) turn both wheels in unison while both struts remain locked, (b) pivot one strut, or (c) pivot both struts in opposite directions by energizing the gearmotors to apply various combinations of torques T/2 or T.

A small prototype robotic all-terrain vehicle features a unique drive and suspension system that affords capabilities for self righting, pose control, and enhanced maneuverability for passing over obstacles. The vehicle is designed for exploration of planets and asteroids, and could just as well be used on Earth to carry scientific instruments to remote, hostile, or otherwise inaccessible locations on the ground. The drive and suspension system enable the vehicle to perform such diverse maneuvers as flipping itself over, traveling normal side up or upside down, orienting the main vehicle body in a specified direction in all three dimensions, or setting the main vehicle body down onto the ground, to name a few. Another maneuver enables the vehicle to overcome a common weakness of traditional all-terrain vehicles—a limitation on traction and drive force that makes it difficult or impossible to push wheels over some obstacles: This vehicle can simply lift a wheel onto the top of an obstacle.

The basic mode of operation of the vehicle can be characterized as four-wheel drive with skid steering. Each wheel is driven individually by a dedicated gearmotor. Each wheel and its gear-motor are mounted at the free end of a strut that pivots about a lateral axis through the center of gravity of the vehicle (see figure). Through pulleys or other mechanism attached to their wheels, both gearmotors on each side of the vehicle drive a single idler disk or pulley that turns about the pivot axis.

The design of the pivot assembly is crucial to the unique capabilities of this system. The idler pulley and the pivot disks of the struts are made of suitably chosen materials and spring-loaded together along the pivot axis in such a way as to resist turning with a static frictional torque T; in other words, it is necessary to apply a torque of T to rotate the idler pulley or either strut with respect to each other or the vehicle body.

During ordinary backward or forward motion along the ground, both wheels are turned in unison by their gearmotors, and the belt couplings make the idler pulley turn along with the wheels. In this operational mode, each gearmotor contributes a torque T/2 so that together, both gearmotors provide torque T to overcome the locking friction on the idler pulley. Each strut remains locked at its preset angle because the torque T/2 supplied by its motor is not sufficient to overcome its locking friction T.

If it is desired to change the angle between one strut and the main vehicle body, then the gearmotor on that strut only is energized. In general, a gearmotor acts as a brake when not energized. Since the gearmotor on the other strut is not energized and since it is coupled to the idler pulley, a torque greater than T would be needed to turn the idler pulley. However, as soon as the gearmotor on the strut that one desires to turn is energized, it develops enough torque (T) to begin pivoting the strut with respect to the vehicle body.

It is also possible to pivot both struts simultaneously in opposite directions to change the angle between them. To accomplish this, one energizes the gearmotors to apply equal and opposite torques of magnitude T: The net torque on the idler pulley balances out to zero, so that the idler pulley and body remain locked, while the applied torques are just sufficient to turn the struts against locking friction. If it is desired to pivot the struts through unequal angles, then the gearmotor speeds are adjusted accordingly.

The prototype vehicle has performed successfully in tests. Current and future work is focused on designing a simple hub mechanism, which is not sensitive to dust or other contamination, and on active control techniques to allow autonomous planetary rovers to take advantage of the flexibility of the mechanism.

This work was done by Brian H. Wilcox and Annette K. Nasif of Caltech for **NASA's Jet Propulsion Laboratory**.

Resources

The following magazine articles and Web sites provided information about robots featured in this chapter:

Magazine Articles

Popular Mechanics (March 2006) All Eyes on Mars, pp. 16–17
Technology Review (August 2005) Machine in Motion, pp. 64–69
Technology Review (May 2002) Reconfigurable Robots, pp. 54–59

Web Sites

Personal Robot
www.cs.cmu.edu/~personalrover/PER

Mars Exploration Rover
marsrovers.nasa.gov/home/

Dragon Runner Robot
www.military.com/soldiertech/0,14632,Soldiertech_Dragon-Robot,,00.html

PackBot EOD Robot
www.army-technology.com/contractors/mines/i_robot/i_robot1.html

M-TRAN II Self Reconfigurable Modular Robot
http://unit.aist.go.jp/is/dsysd/mtran/English/

Modular Robots at Palo Alto Research Center (PARC)
www2.parc.com/spl/projects/modrobots/

Toddler Walking Robot
web.mit.edu/newsoffice/2005/robotoddler.html

Ballbot Single-Wheeled Mobile Robot
http://www.ri.cmu.edu/pub_files/pub4/lauwers_tom_2006_1/lauwers_tom_2006_1.pdf

CHAPTER 5

LINKAGES: DRIVES AND MECHANISMS

FOUR-BAR LINKAGES AND TYPICAL INDUSTRIAL APPLICATIONS

All mechanisms can be broken down into equivalent four-bar linkages. They can be considered to be the basic mechanism and are useful in many mechanical operations.

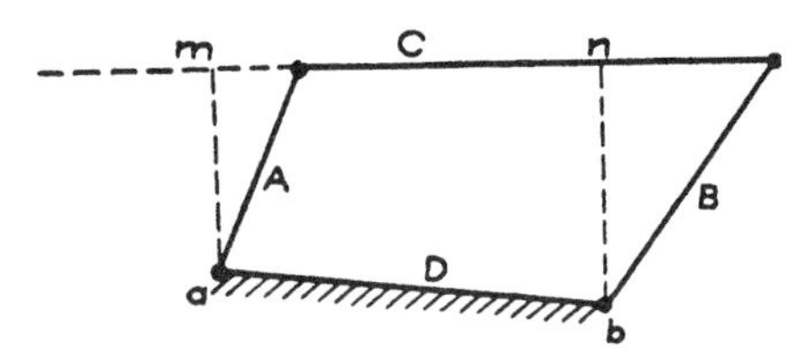

FOUR-BAR LINKAGES—Two cranks, a connecting rod and a line between the fixed centers of the cranks make up the basic four-bar linkage. Cranks can rotate if *A* is smaller than *B* or *C* or D. Link motion can be predicted.

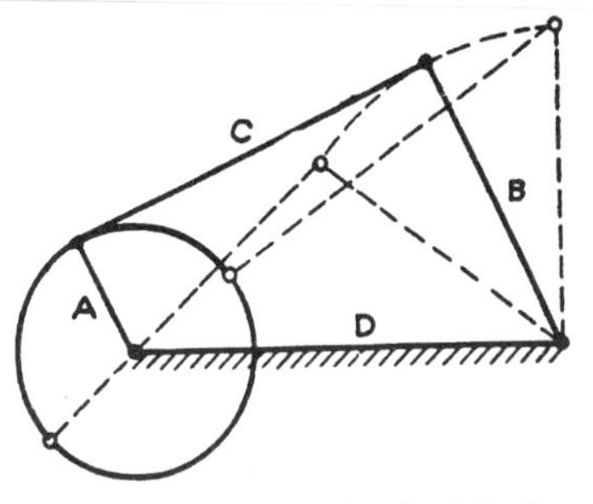

CRANK AND ROCKER—the following relations must hold for its operation: $A + B + C > D$; $A + D + B > C$; $A + C - B < D$, and $C - A + B > D$.

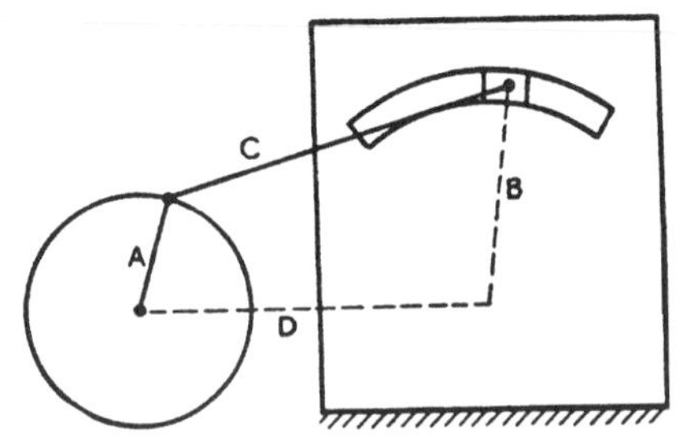

FOUR-BAR LINK WITH SLIDING MEMBER—One crank is replaced by a circular slot with an effective crank distance of *B*.

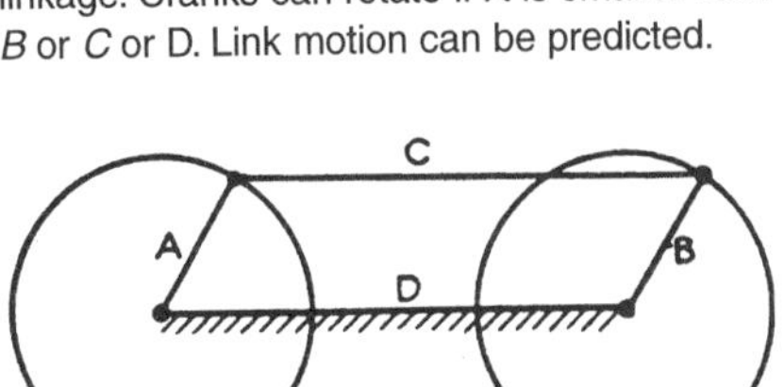

PARALLEL CRANK FOUR-BAR—Both cranks of the parallel crank four-bar linkage always turn at the same angular speed, but they have two positions where the crank can-not be effective.

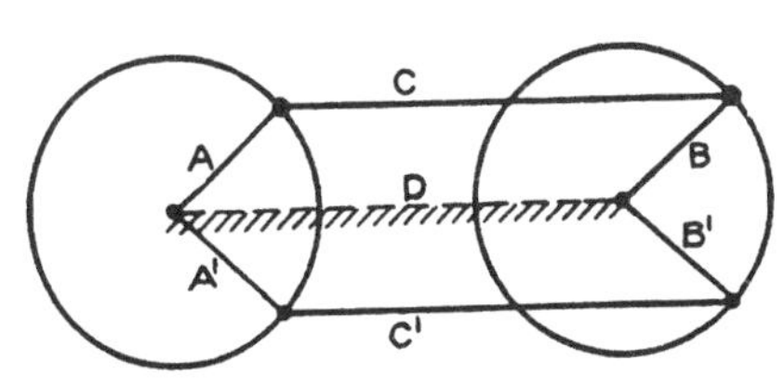

DOUBLE PARALLEL CRANK—This mechanism avoids a dead center position by having two sets of cranks at 90° advancement. The connecting rods are always parallel.

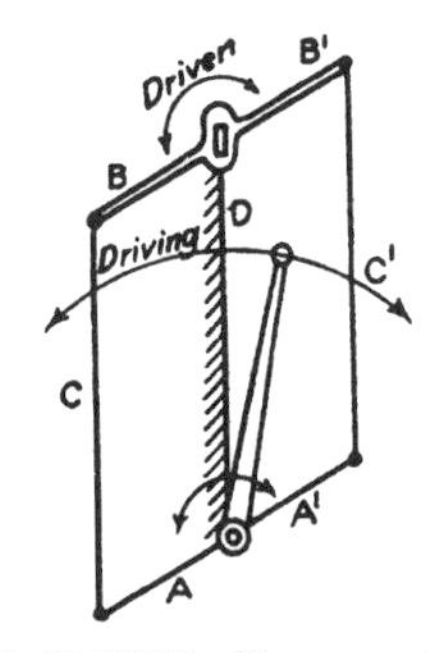

PARALLEL CRANK—Steam control linkage assures equal valve openings.

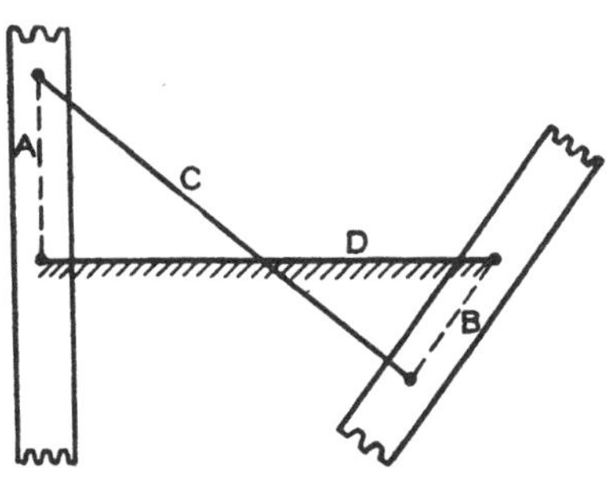

NON-PARALLEL EQUAL CRANK—The centrodes are formed as gears for passing dead center and they can replace ellipticals.

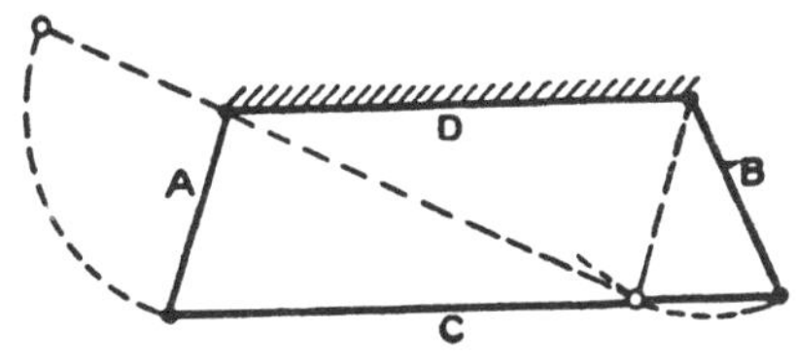

SLOW MOTION LINK—As crank *A* is rotated upward it imparts motion to crank *B*. When *A* reaches its dead center position, the angular velocity of crank *B* decreases to zero.

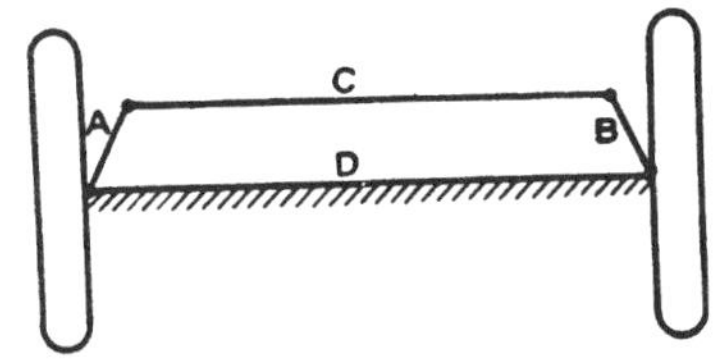

TRAPAZOIDAL LINKAGE—This linkage is not used for complete rotation but can be used for special control. The inside moves through a larger angle than the outside with normals intersecting on the extension of a rear axle in a car.

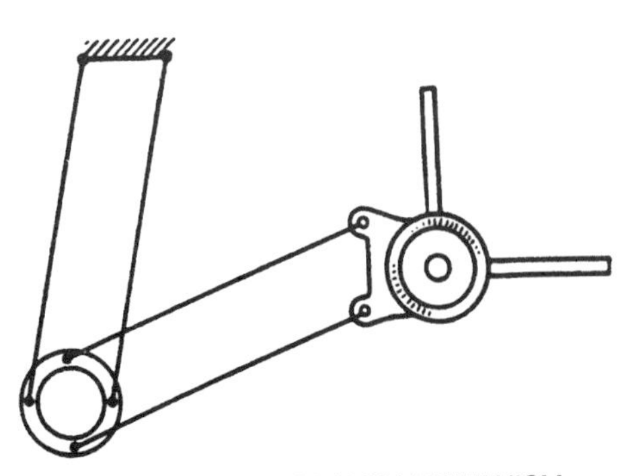

DOUBLE PARALLEL CRANK MECHANISM—This mechanism forms the basis for the universal drafting machine.

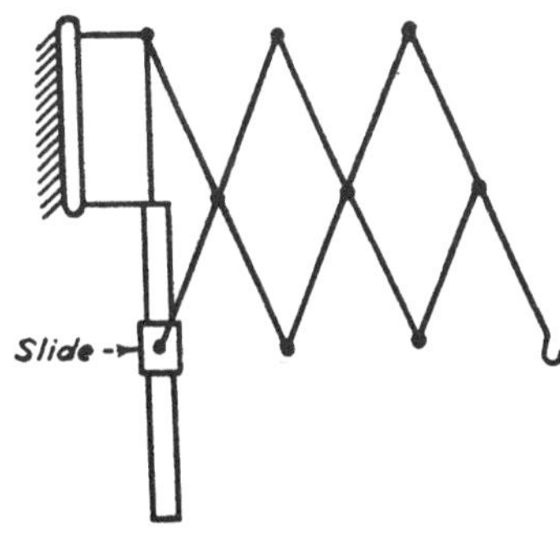

ISOSCELES DRAG LINKS—This "lazy-tong" device is made of several isosceles links; it is used as a movable lamp support.

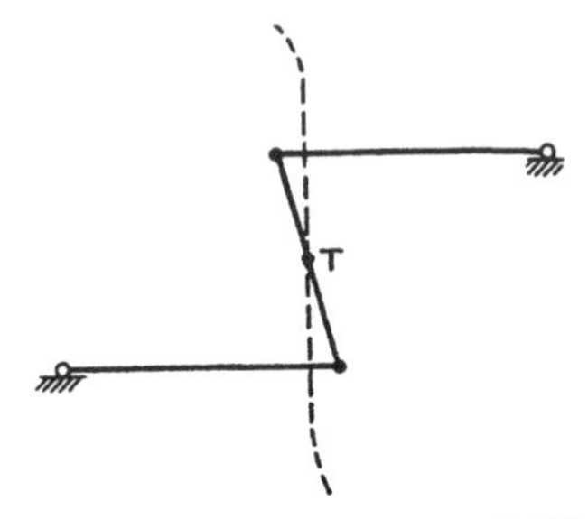

WATT'S STRAIGHT-LINE MECHANISM—Point T describes a line perpendicular to the parallel position of the cranks.

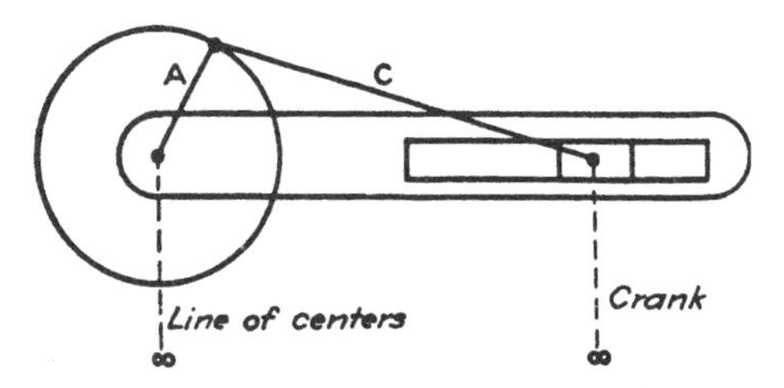

STRAIGHT SLIDING LINK—This is the form in which a slide is usually used to replace a link. The line of centers and the crank *B* are both of infinite length.

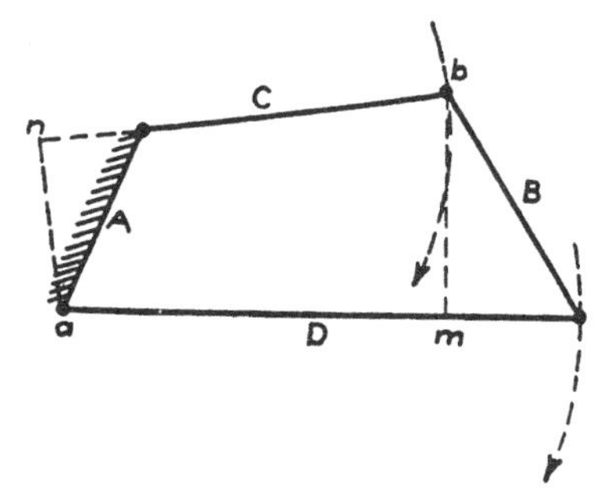

DRAG LINK—This linkage is used as the drive for slotter machines. For complete rotation: $B > A + D - C$ and $B < D + C - A$.

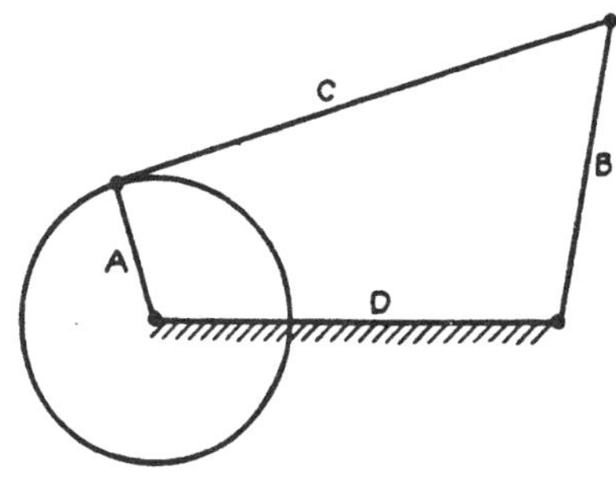

ROTATING CRANK MECHANISM—This linkage is frequently used to change a rotary motion to a swinging movement.

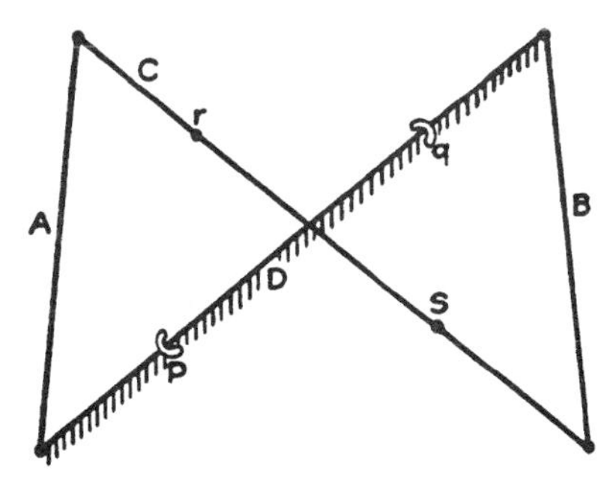

NON-PARALLEL EQUAL CRANK—If crank *A* has a uniform angular speed, *B* will vary.

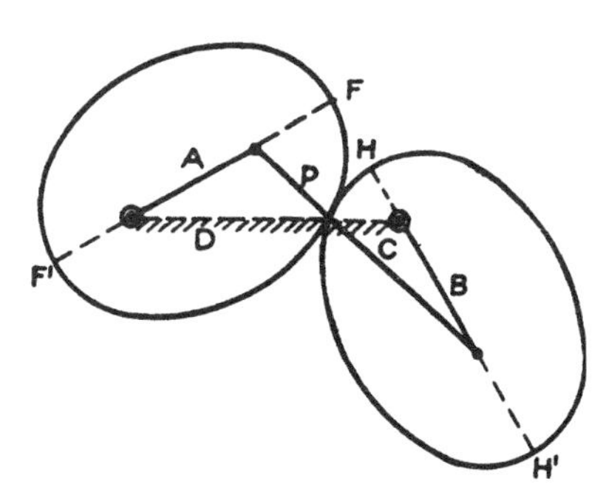

ELLIPTICAL GEARS—They produce the same motion as non-parallel equal cranks.

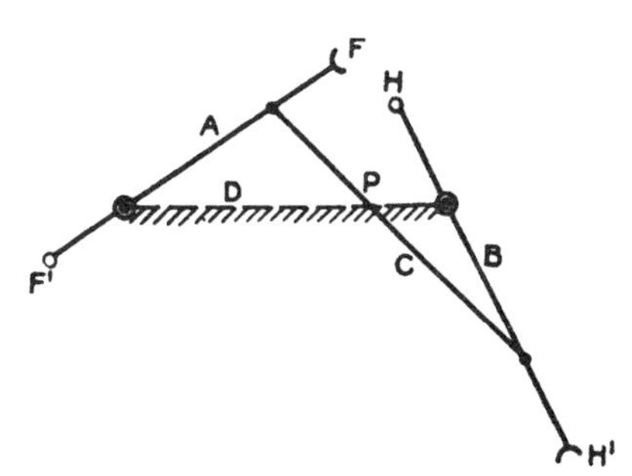

NON-PARALLEL EQUAL CRANK—It is the same as the first example given but with crossover points on its link ends.

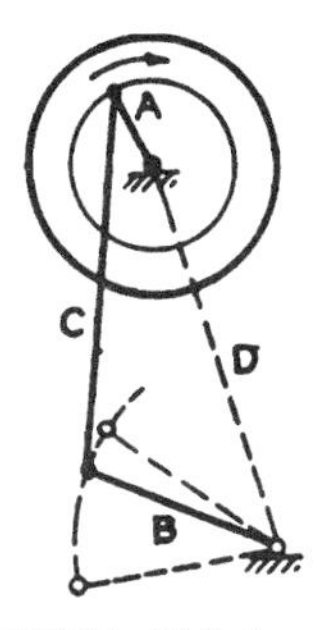

TREADLE DRIVE—This four-bar linkage is used in driving grinding wheels and sewing machines.

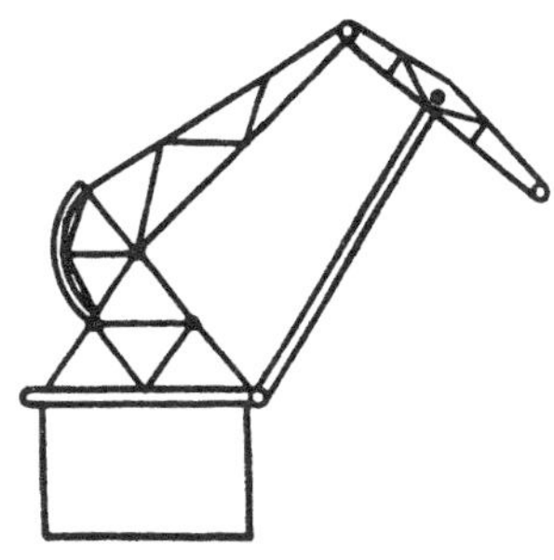

DOUBLE LEVER MECHANISM—This slewing crane can move a load in a horizontal direction by using the D-shaped portion of the top curve.

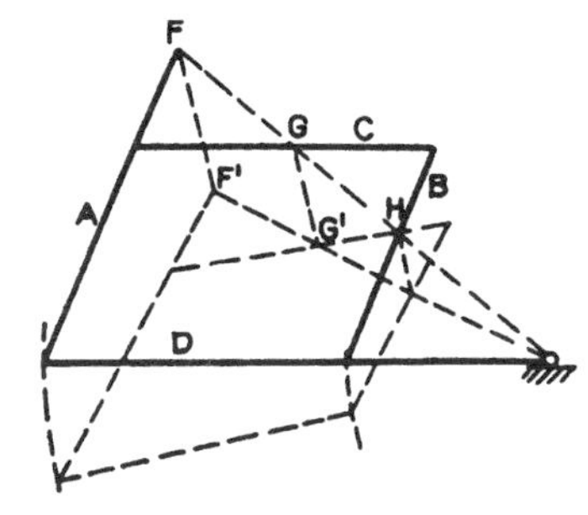

PANTOGRAPH—The pantograph is a parallelogram in which lines through *F*, *G* and *H* must always intersect at a common point.

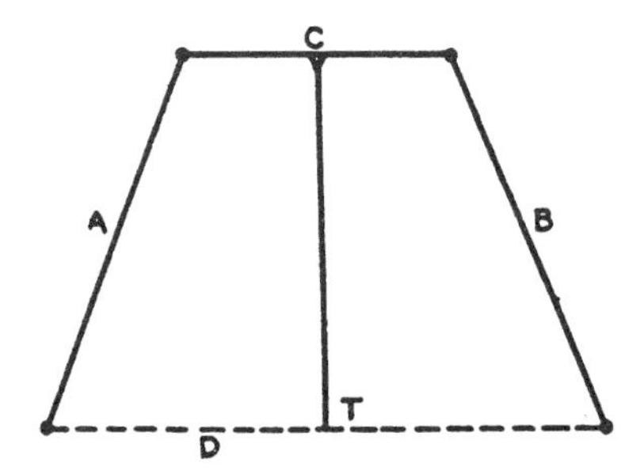

ROBERT'S STRAIGHT-LINE MECHANISM—The lengths of cranks *A* and *B* should not be less than 0.6 *D*; *C* is one half of *D*.

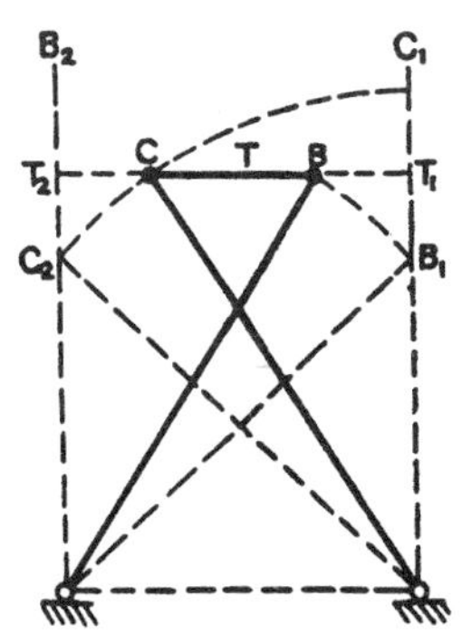

TCHEBICHEFF'S—Links are made in proportion: $AB = CD = 20$, $AD = 16$, $BC = 8$.

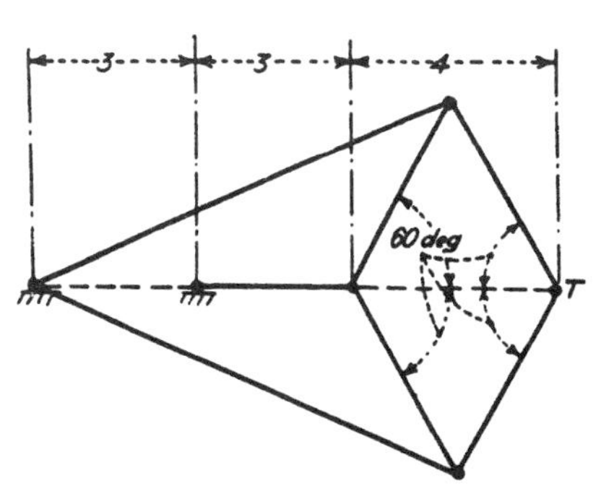

PEAUCELLIER'S CELL—When proportioned as shown, the tracing point *T* forms a straight line perpendicular to the axis.

SEVEN LINKAGES FOR TRANSPORT MECHANISMS

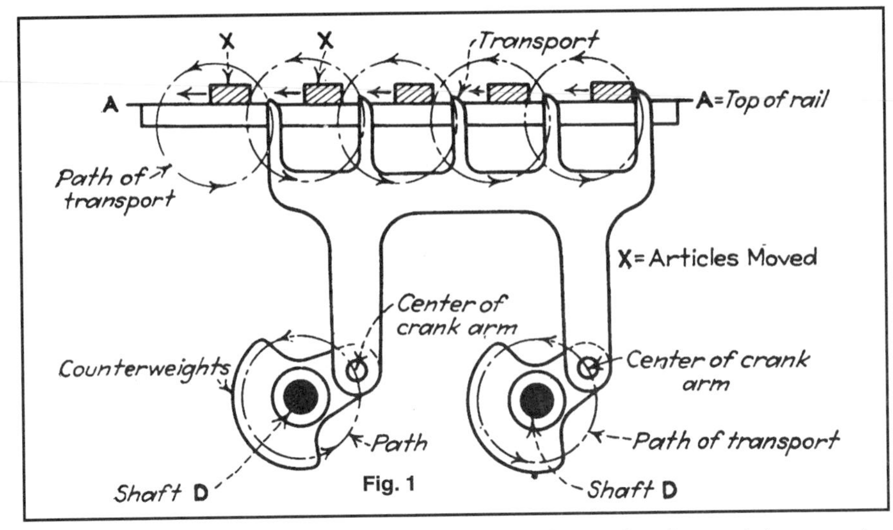

Fig. 1 In this design a rotary action is used. The shafts *D* rotate in unison and also support the main moving member. The shafts are carried in the frame of the machine and can be connected by either a link, a chain and sprocket, or by an intermediate idler gear between two equal gears keyed on the shafts. The rail *A-A* is fixed rigidly on the machine. A pressure or friction plate can hold the material against the top of the rail and prevent any movement during the period of rest.

Transport mechanisms generally move material. The motion, although unidirectional, gives an intermittent advancement to the material being conveyed. The essential characteristic of such a motion is that all points in the main moving members follow similar and equal paths. This is necessary so that the members can be subdivided into sections with projecting parts. The purpose of the projections is to push the articles during the forward motion of the material being transported. The transport returns by a different path from the one it follow in its advancement, and the material is left undisturbed until the next cycle begins. During this period of rest, while the transport is returning to its starting position, various operations can be performed sequentially. The selection of the particular transport mechanism best suited to any situation depends, to some degree, on the arrangement that can be obtained for driving the materials and the path desired. A slight amount of overtravel is always required so that the projection on the transport can clear the material when it is going into position for the advancing stroke.

The designs illustrated here have been selected from many sources and are typical of the simplest solutions of such problems. The paths, as indicated in these illustrations, can be varied by changes in the cams, levers, and associated parts. Nevertheless, the customary cut-and-try method might still lead to the best solution.

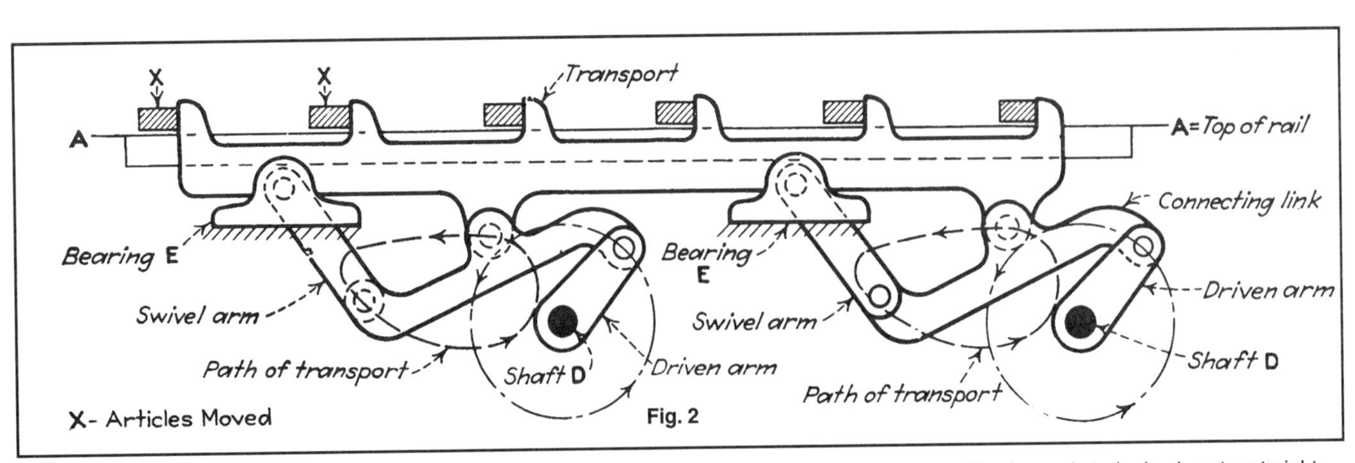

Fig. 2 Here is a simple form of linkage that imparts a somewhat "egg-shaped" motion to the transport. The forward stroke is almost a straight line. The transport is carried on the connecting links. As in the design of Fig. 1, the shafts *D* are driven in unison and are supported in the frame of the machine. Bearings *E* are also supported by the frame of the machine and the rail *A-A* is fixed.

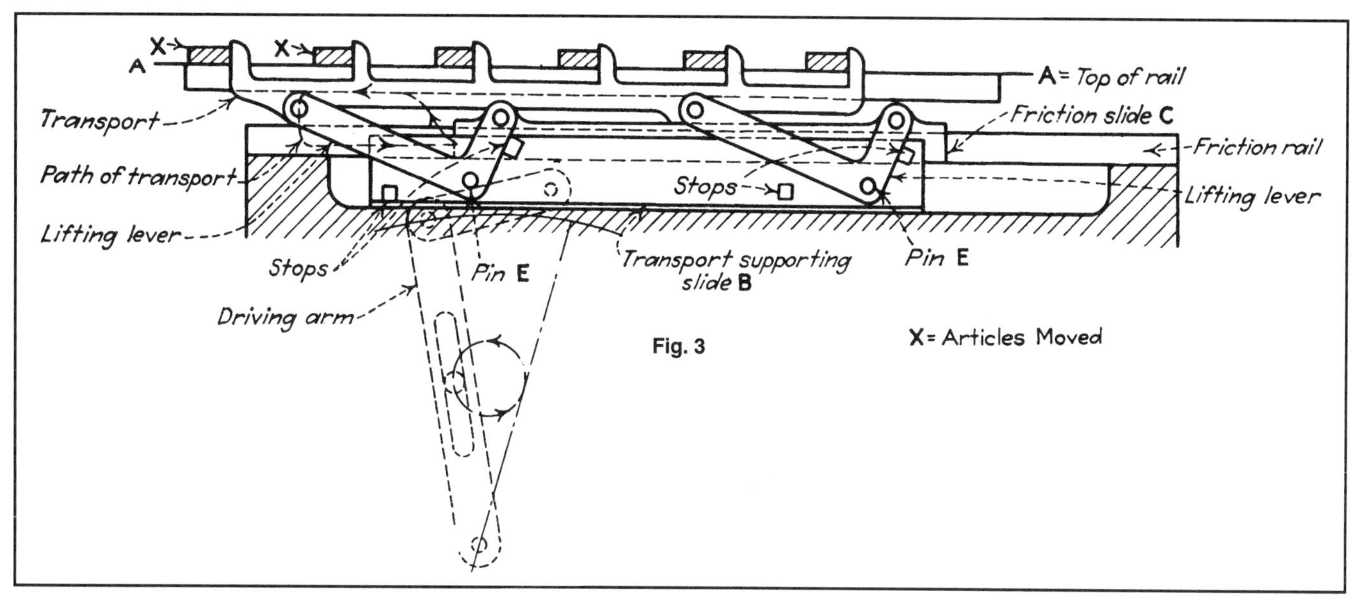

Fig. 3

Fig. 3 In another type of action, the forward and return strokes are accomplished by a suitable mechanism, while the raising and lowering is imparted by a friction slide. Thus it can be seen that as the transport supporting slide *B* starts to move to the left, the friction slide *C,* which rests on the friction rail, tends to remain at rest. As a result, the lifting lever starts to turn in a clockwise direction. This motion raises the transport which remains in its raised position against stops until the return stroke starts. At that time the reverse action begins. An adjustment should be provided to compensate for the friction between the slide and its rail. It can readily be seen that this motion imparts a long straight path to the transport.

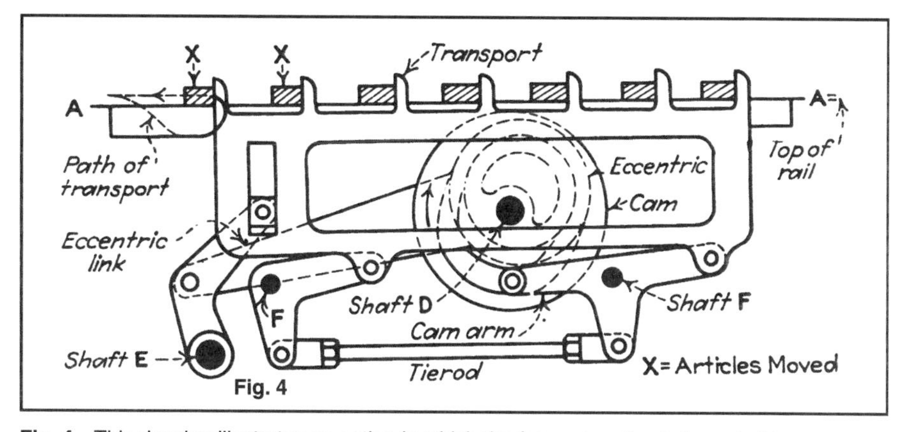

Fig. 4

Fig. 4 This drawing illustrates an action in which the forward motion is imparted by an eccentric while the raising and lowering of the transport is accomplished by a cam. The shafts, *F, E,* and *D* are positioned by the frame of the machine. Special bell cranks support the transport and are interconnected by a tierod.

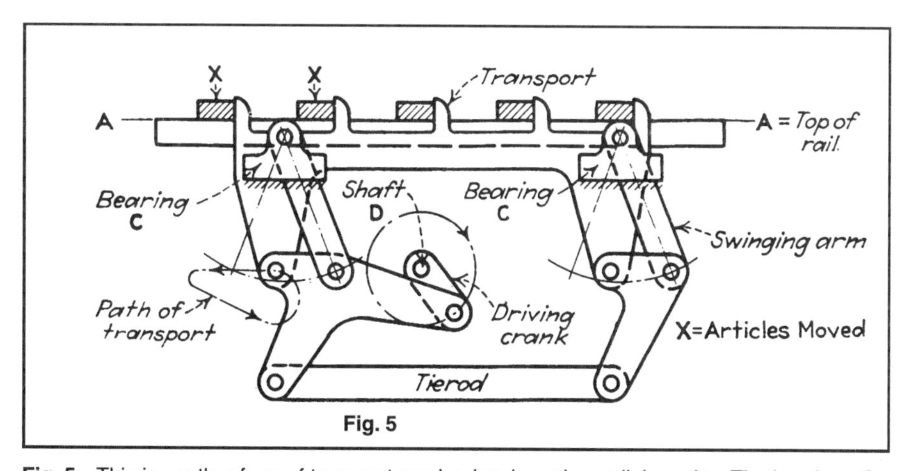

Fig. 5

Fig. 5 This is another form of transport mechanism based on a link motion. The bearings *C* are supported by the frame as is the driving shaft *D.*

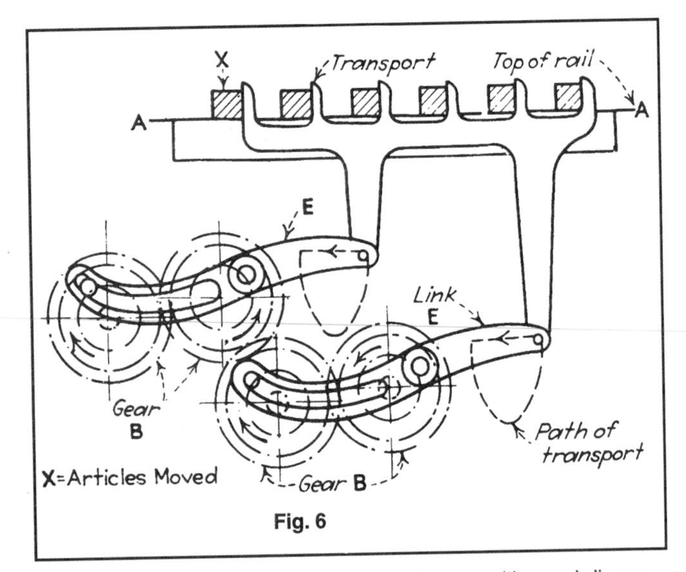

Fig. 6

Fig. 6 An arrangement of interconnected gears with equal diameters that will impart a transport motion to a mechanism. The gear and link mechanism imparts both the forward motion and the raising and lowering motions. The gear shafts are supported in the frame of the machine.

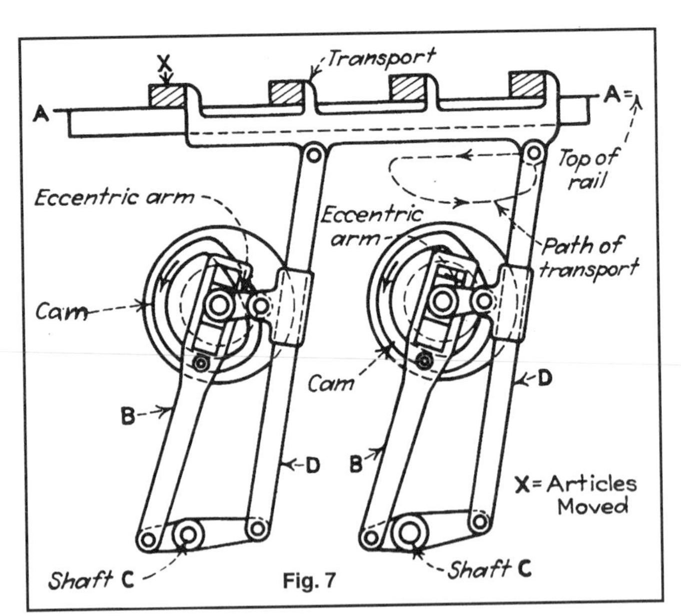

Fig. 7

Fig. 7 In this transport mechanism, the forward an return strokes are accomplished by the eccentric arms, while the vertical motion is performed by the cams.

FIVE LINKAGES FOR STRAIGHT-LINE MOTION

These linkages convert rotary to straight-line motion without the need for guides.

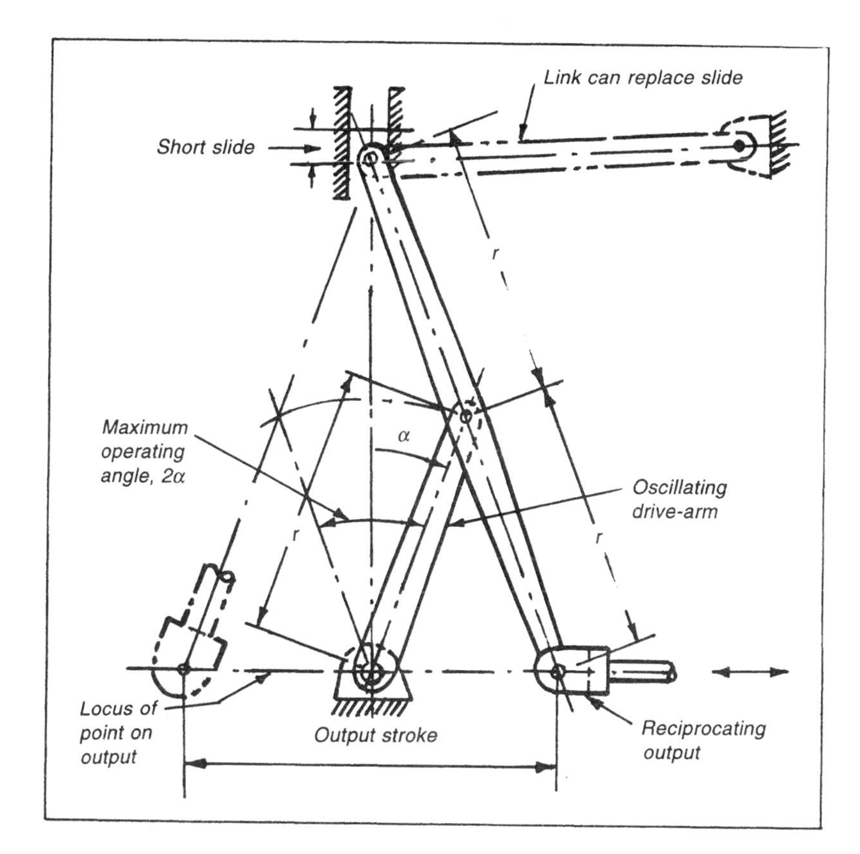

Fig. 1 An Evans' linkage has an oscillating drive-arm that should have a maximum operating angle of about 40°. For a relatively short guideway, the reciprocating output stroke is large. Output motion is on a true straight line in true harmonic motion. If an exact straight-line motion is not required, a link can replace the slide. The longer this link, the closer the output motion approaches that of a true straight line. If the link-length equals the output stroke, deviation from straight-line motion is only 0.03% of the output stroke.

Fig. 2 A simplified Watt's linkage generates an approximate straight-line motion. If the two arms are of equal length, the tracing point describes a symmetrical figure 8 with an almost straight line throughout the stroke length. The straightest and longest stroke occurs when the connecting-link length is about two-thirds of the stroke, and arm length is 1.5 times the stroke length. Offset should equal half the connecting-link length. If the arms are unequal, one branch of the figure-8 curve is straighter than the other. It is straightest when a/b equals (arm 2)/(arm 1).

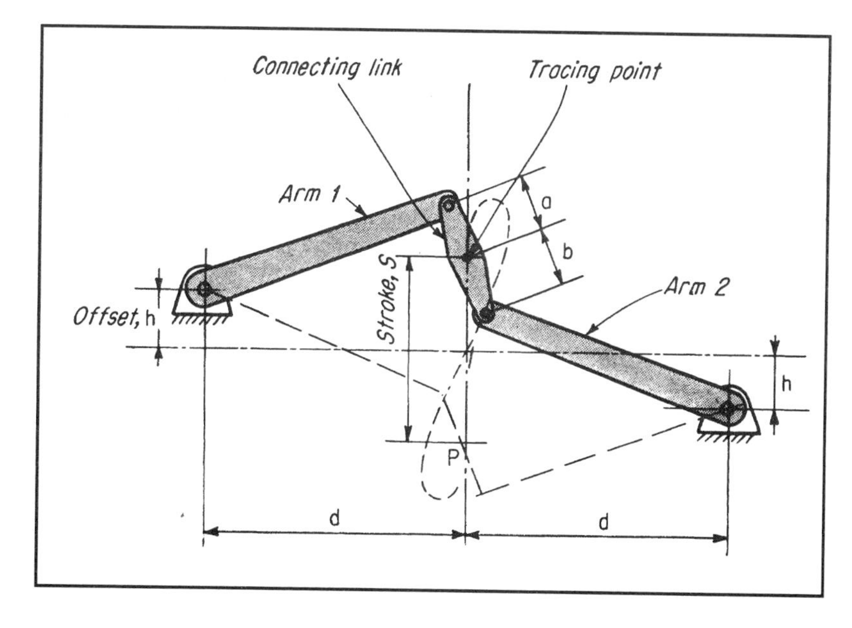

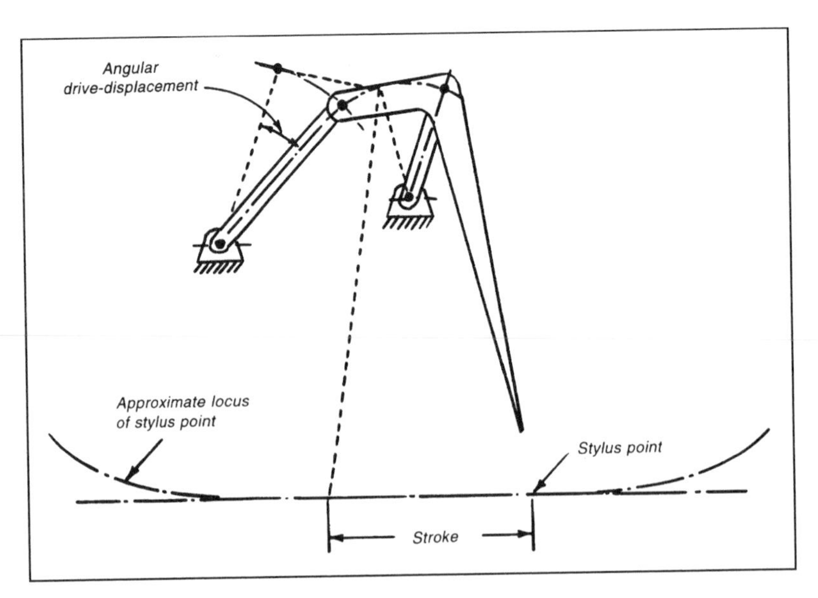

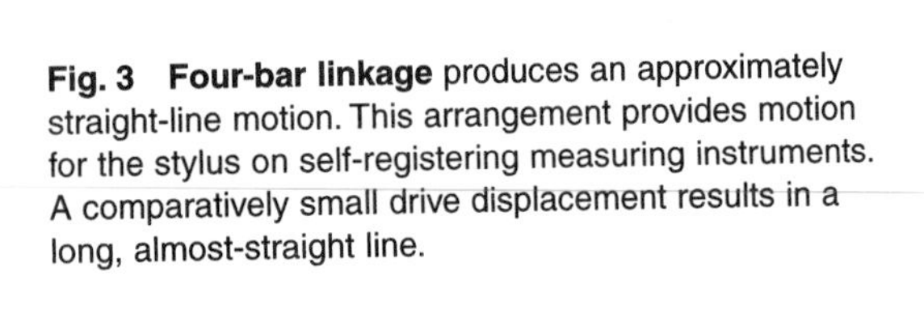

Fig. 3 Four-bar linkage produces an approximately straight-line motion. This arrangement provides motion for the stylus on self-registering measuring instruments. A comparatively small drive displacement results in a long, almost-straight line.

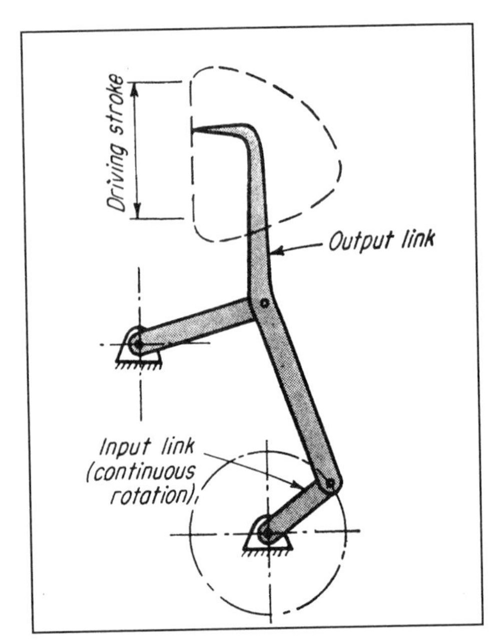

Fig. 4 A D-drive is the result when linkage arms are arranged as shown here. The output-link point describes a path that resembles the letter *D*, so there is a straight part of its cycle. This motion is ideal for quick engagement and disengagement before and after a straight driving stroke.

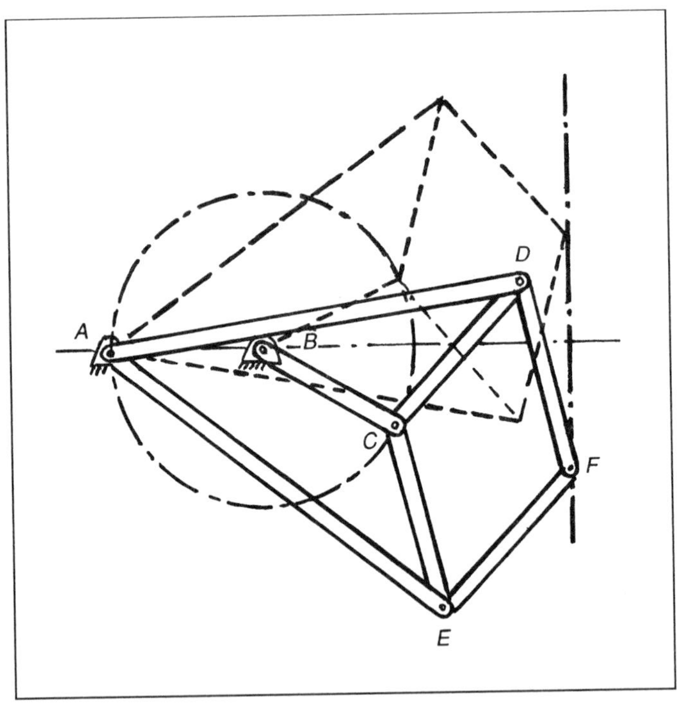

Fig. 5 The "Peaucellier cell" was the first solution to the classical problem of generating a straight line with a linkage. Within the physical limits of the motion, $AC \times AF$ remains constant. The curves described by *C* and *F* are, therefore, inverse; if *C* describes a circle that goes through *A*, then *F* will describe a circle of infinite radius—a straight line, perpendicular to *AB*. The only requirements are that: $AB = BC$; $AD = AE$; and *CD*, *DF*, *FE*, *EC* be equal. The linkage can be used to generate circular arcs of large radius by locating *A* outside the circular path of *C*.

SIX EXPANDING AND CONTRACTING LINKAGES

Parallel bars, telescoping slides, and other devices that can spark answers to many design problems.

Figs. 1 and 2 Expanding grilles are often put to work as a safety feature. A single parallelogram (fig. 1) requires slotted bars; a double parallelogram (fig. 2) requires none—but the middle grille-bar must be held parallel by some other method.

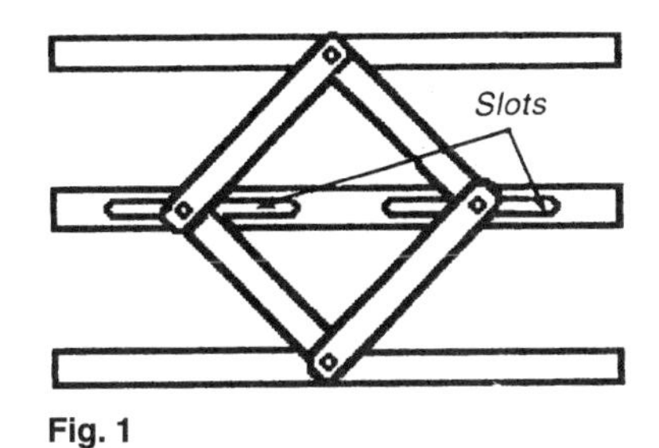

Fig. 1

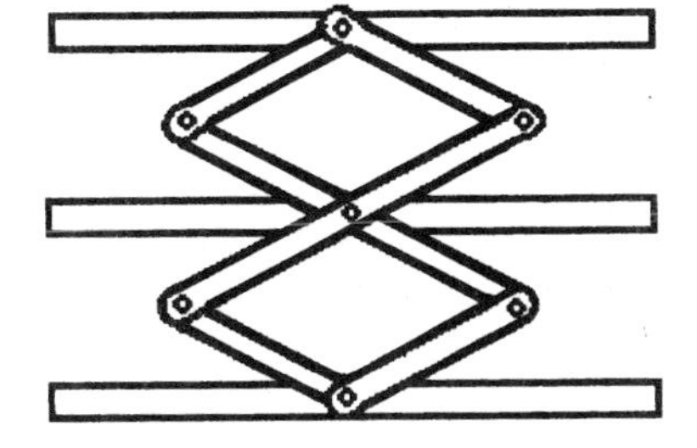

Fig. 2

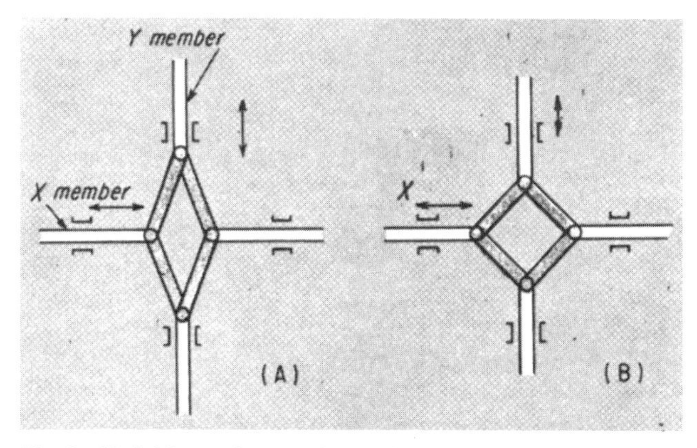

Fig. 3 Variable motion can be produced with this arrangement. In (A) position, the *Y* member is moving faster than the *X* member. In (B), speeds of both members are instantaneously equal. If the motion is continued in the same direction, the speed of *X* will become greater.

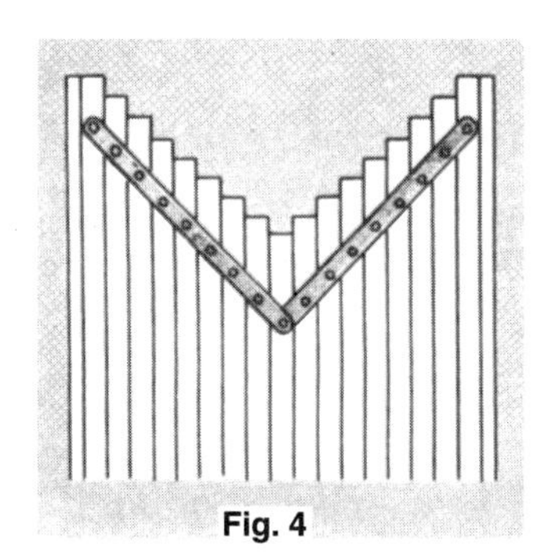

Fig. 4

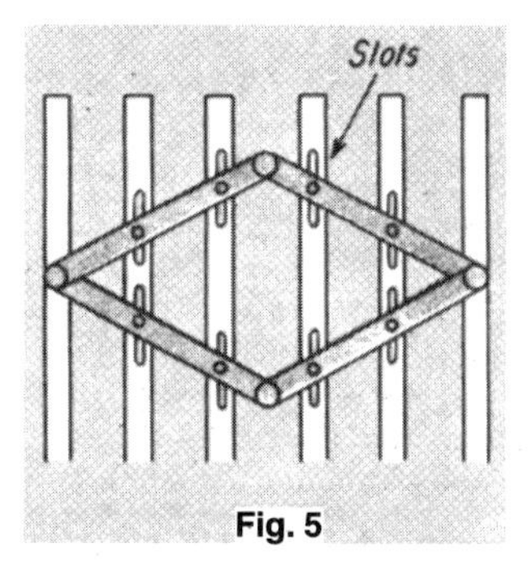

Fig. 5

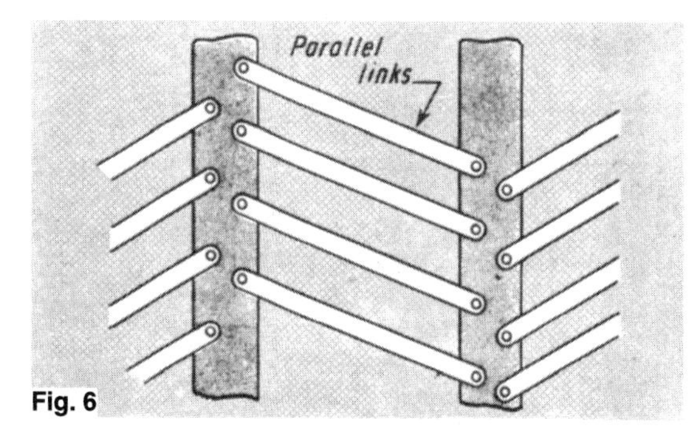

Fig. 6

Figs. 4, 5, and 6 Multibar barriers such as shutters and gates (fig. 4) can take various forms. Slots (fig. 5) allow for vertical adjustment. The space between bars can be made adjustable (fig. 6) by connecting the vertical bars with parallel links.

FOUR LINKAGES FOR DIFFERENT MOTIONS

Fig. 1 No linkages or guides are included in this modified hypocyclic drive which is relatively small in relation to the length of its stroke. The sun gear of pitch diameter *D* is stationary. The drive shaft, which turns the T-shaped arm, is concentric with this gear. The idler and planet gears, with pitch diameters of *D*/2, rotate freely on pivots in the arm extensions. The pitch diameter of the idler has no geometrical significance, although this gear does have an important mechanical function. It reverses the rotation of the planet gear, thus producing true hypocyclic motion with ordinary spur gears only. Such an arrangement occupies only about half as much space as does an equivalent mechanism containing an internal gear. The center distance *R* is the sum of *D*/2, *D*/4, and an arbitrary distance *d*, determined by specific applications. Points *A* and *B* on the driven link, which is fixed to the planet, describe straight-line paths through a stroke of 4*R*. All points between *A* and *B* trace ellipses, while the line *AB* envelopes an astroid.

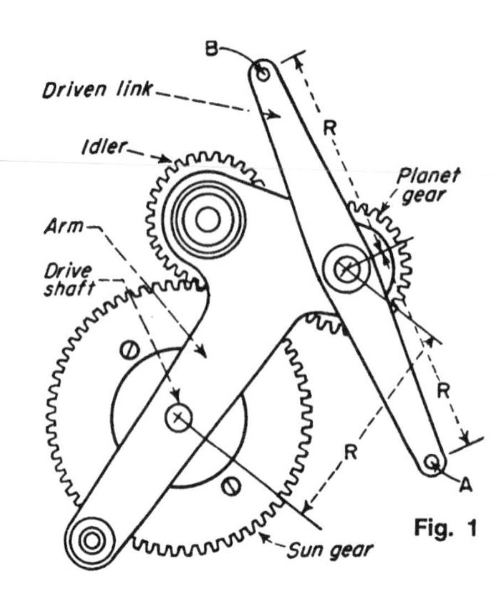

Fig. 1

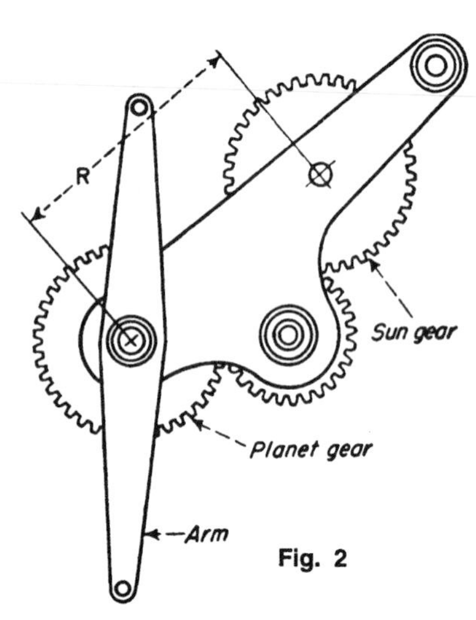

Fig. 2

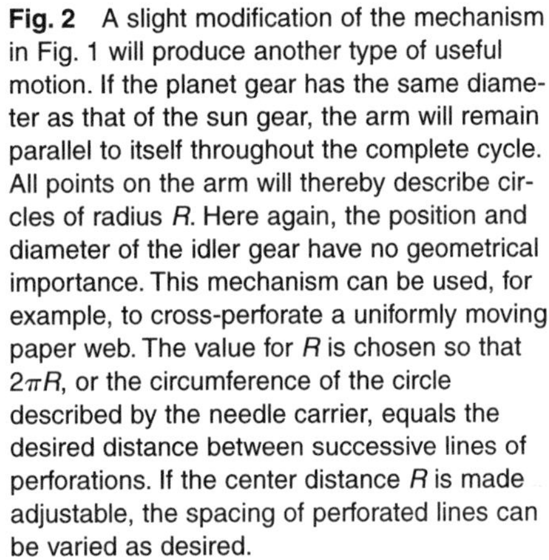
Fig. 2 A slight modification of the mechanism in Fig. 1 will produce another type of useful motion. If the planet gear has the same diameter as that of the sun gear, the arm will remain parallel to itself throughout the complete cycle. All points on the arm will thereby describe circles of radius *R*. Here again, the position and diameter of the idler gear have no geometrical importance. This mechanism can be used, for example, to cross-perforate a uniformly moving paper web. The value for *R* is chosen so that $2\pi R$, or the circumference of the circle described by the needle carrier, equals the desired distance between successive lines of perforations. If the center distance *R* is made adjustable, the spacing of perforated lines can be varied as desired.

Fig. 3 To describe a "D" curve, begin at the straight part of path G, and replace the oval arc of C with a circular arc that will set the length of link DC.

Fig. 3

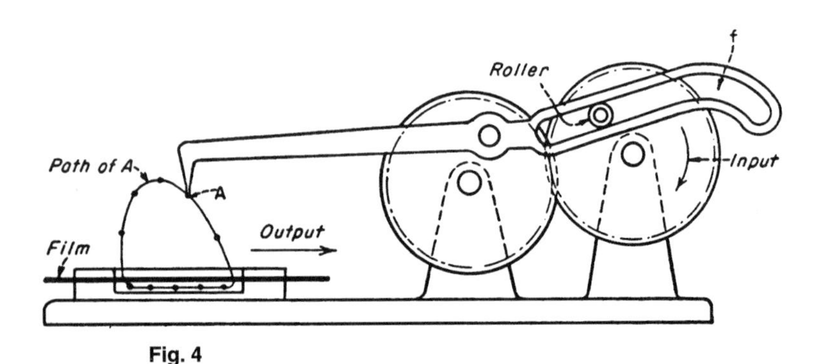

Fig. 4

Fig. 4 This mechanism can act as a film-strip hook that will describe a nearly straight line. It will engage and disengage the film perforation in a direction approximately normal to the film. Slight changes in the shape of the guiding slot *f* permit the shape of the output curve and the velocity diagram to be varied.

NINE LINKAGES FOR ACCELERATING AND DECELERATING LINEAR MOTIONS

When ordinary rotary cams cannot be conveniently applied, the mechanisms presented here, or adaptations of them, offer a variety of interesting possibilities for obtaining either acceleration or deceleration, or both.

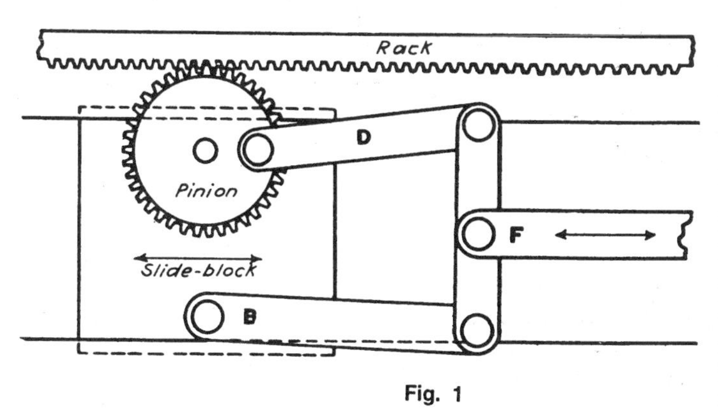

Fig. 1

Fig. 1 A slide block with a pinion and shaft and a pin for link B reciprocates at a constant rate. The pinion has a crankpin for mounting link D, and it also engages a stationary rack. The pinion can make one complete revolution at each forward stroke of the slide block and another as the slide block returns in the opposite direction. However, if the slide block is not moved through its normal travel range, the pinion turns only a fraction of a revolution. The mechanism can be made variable by making the connection link for F adjustable along the length of the element that connects links B and D. Alternatively, the crankpin for link D can be made adjustable along the radius of the pinion, or both the connection link and the crankpin can be made adjustable.

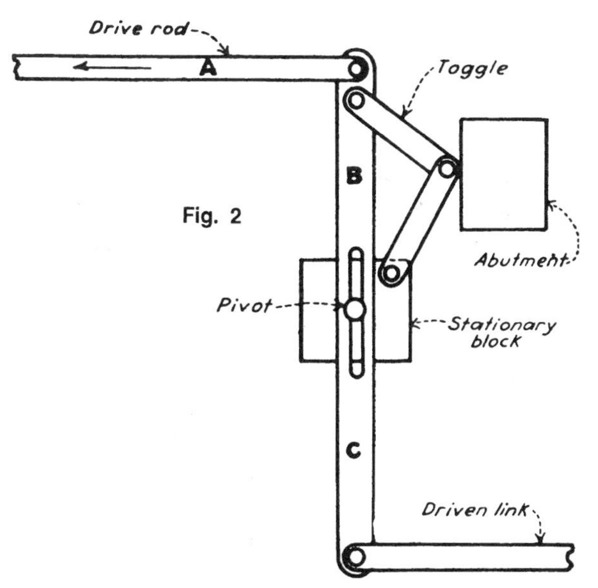

Fig. 2

Fig. 2 A drive rod, reciprocating at a constant rate, rocks link BC about a pivot on a stationary block. A toggle between arm B and the stationary block contacts an abutment. Motion of the drive rod through the toggle causes deceleration of driven link B. As the drive rod moves toward the right, the toggle is actuated by encountering the abutment. The slotted link BC slides on its pivot while turning. This lengthens arm B and shortens arm C of link BC. The result is deceleration of the driven link. The toggle is returned by a spring (not shown) on the return stroke, and its effect is to accelerate the driven link on its return stroke.

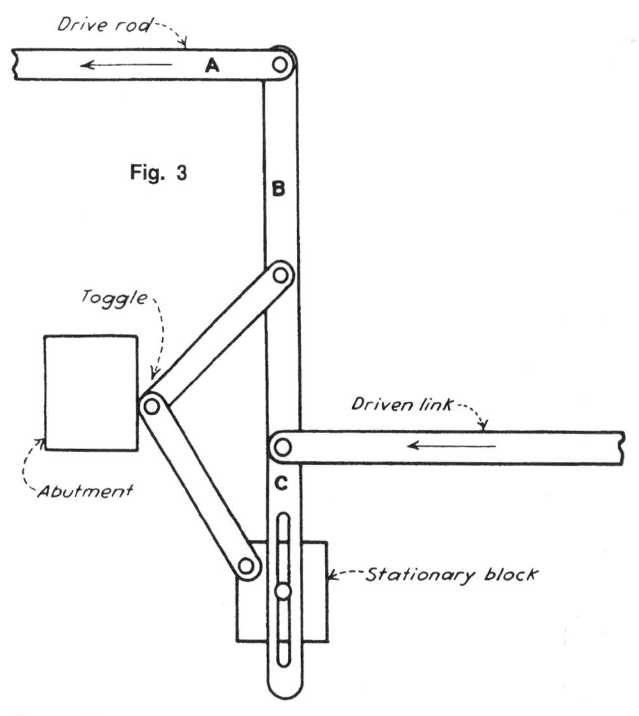

Fig. 3

Fig. 3 The same direction of travel for both the drive rod and the drive link is provided by the variation of the Fig. 2 mechanism. Here, acceleration is in the direction of the arrows, and deceleration occurs on the return stroke. The effect of acceleration decreases as the toggle flattens.

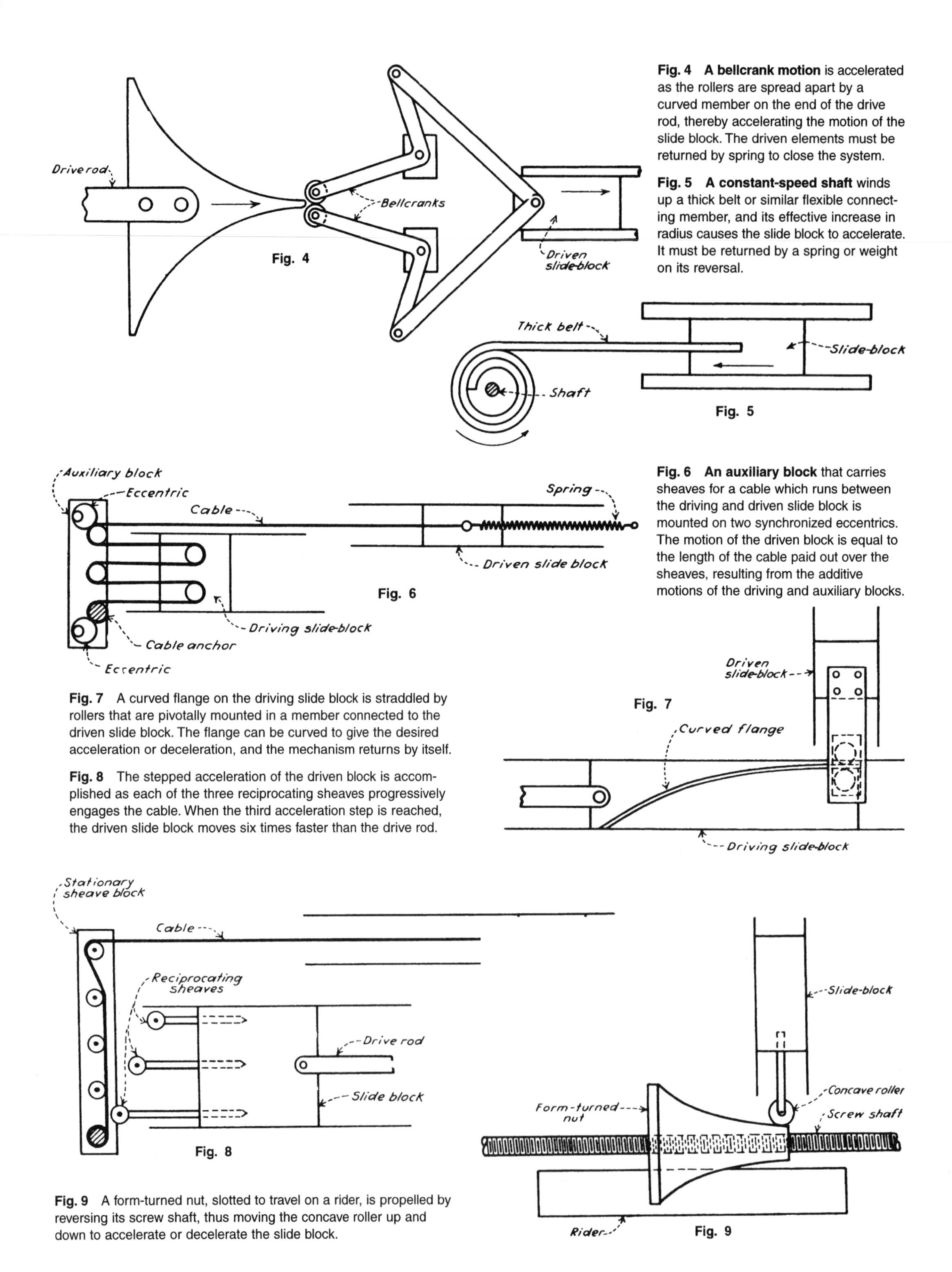

Fig. 4 A bellcrank motion is accelerated as the rollers are spread apart by a curved member on the end of the drive rod, thereby accelerating the motion of the slide block. The driven elements must be returned by spring to close the system.

Fig. 5 A constant-speed shaft winds up a thick belt or similar flexible connecting member, and its effective increase in radius causes the slide block to accelerate. It must be returned by a spring or weight on its reversal.

Fig. 6 An auxiliary block that carries sheaves for a cable which runs between the driving and driven slide block is mounted on two synchronized eccentrics. The motion of the driven block is equal to the length of the cable paid out over the sheaves, resulting from the additive motions of the driving and auxiliary blocks.

Fig. 7 A curved flange on the driving slide block is straddled by rollers that are pivotally mounted in a member connected to the driven slide block. The flange can be curved to give the desired acceleration or deceleration, and the mechanism returns by itself.

Fig. 8 The stepped acceleration of the driven block is accomplished as each of the three reciprocating sheaves progressively engages the cable. When the third acceleration step is reached, the driven slide block moves six times faster than the drive rod.

Fig. 9 A form-turned nut, slotted to travel on a rider, is propelled by reversing its screw shaft, thus moving the concave roller up and down to accelerate or decelerate the slide block.

TWELVE LINKAGES FOR MULTIPLYING SHORT MOTIONS

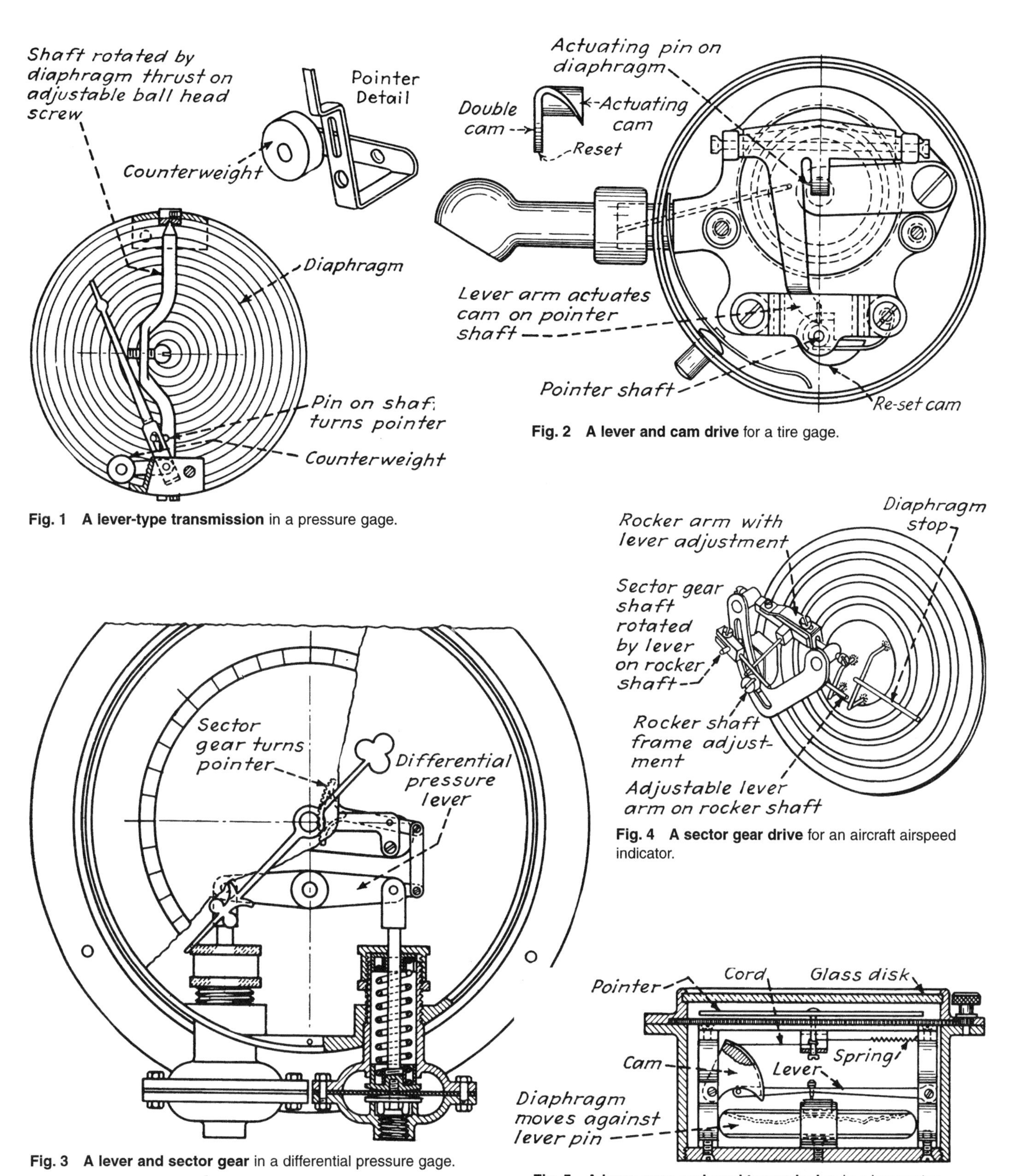

Fig. 1 A lever-type transmission in a pressure gage.

Fig. 2 A lever and cam drive for a tire gage.

Fig. 3 A lever and sector gear in a differential pressure gage.

Fig. 4 A sector gear drive for an aircraft airspeed indicator.

Fig. 5 A lever, cam, and cord transmission in a barometer.

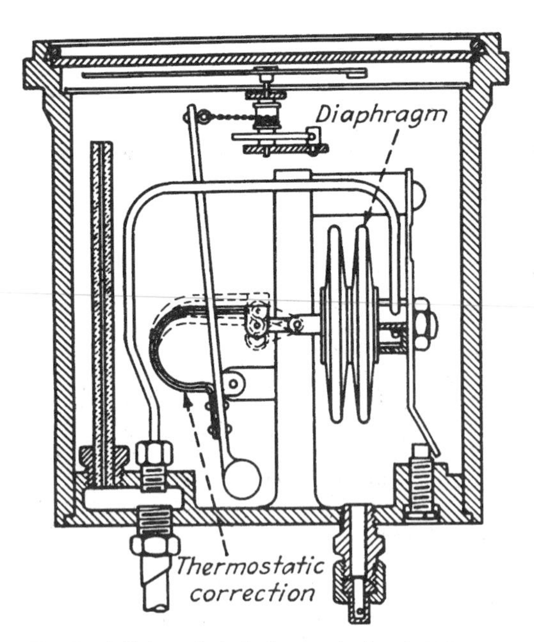

Fig. 6 A link and chain transmission for an aircraft rate of climb instrument.

Fig. 7 A lever system in an automobile gasoline tank.

Fig. 8 Interfering magnetic fields for fluid pressure measurement.

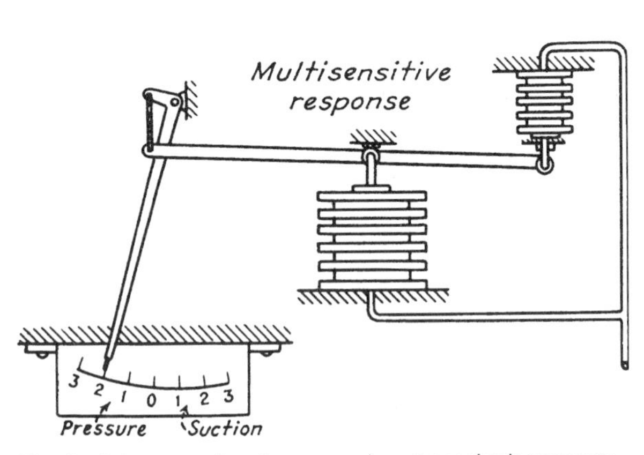

Fig. 9 A lever system for measuring atmospheric pressure variations.

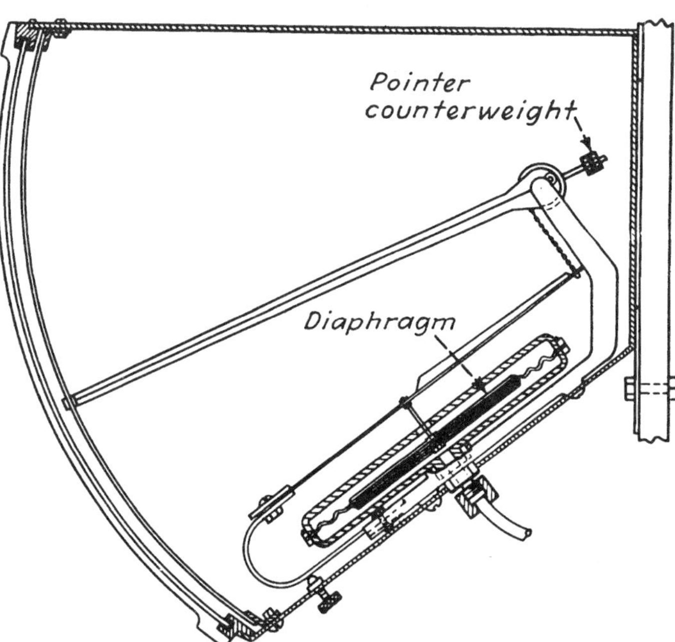

Fig. 10 A lever and chain transmission for a draft gage.

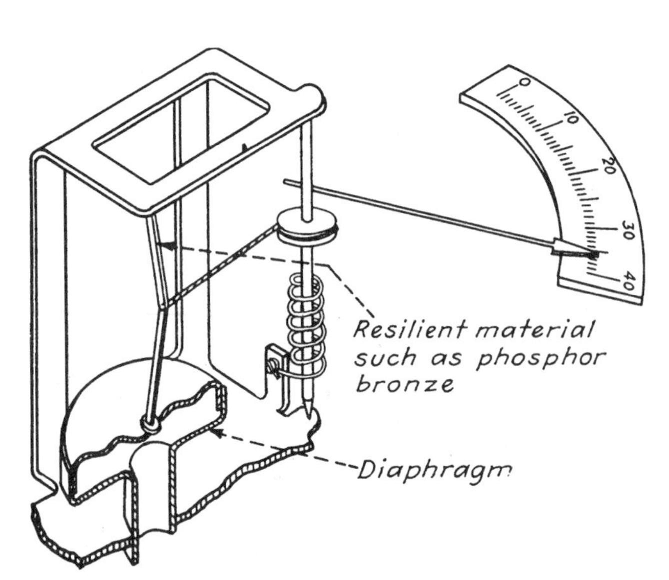

Fig. 11 A toggle and cord drive for a fluid pressure measuring instrument.

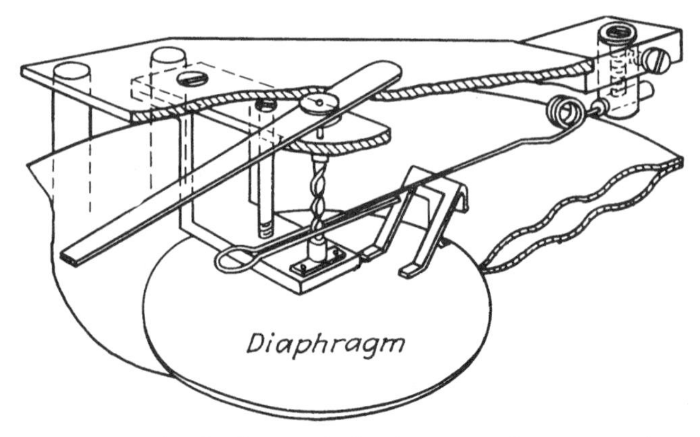

Fig. 12 A spiral feed transmission for a general purpose analog instrument.

FOUR PARALLEL-LINK MECHANISMS

Eight-bar linkage

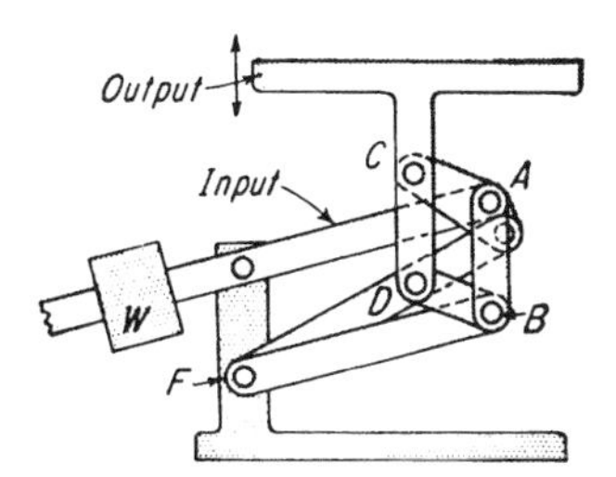

Link AB in this arrangement will always be parallel to EF, and link CD will always be parallel to AB. Hence CD will always be parallel to EF. Also, the linkages are so proportioned that point C moves in an approximately straight line. The final result is that the output plate will remain horizontal while moving almost straight up and down. The weight permitted this device to function as a disappearing platform in a theater stage.

Double-handed screw mechanism

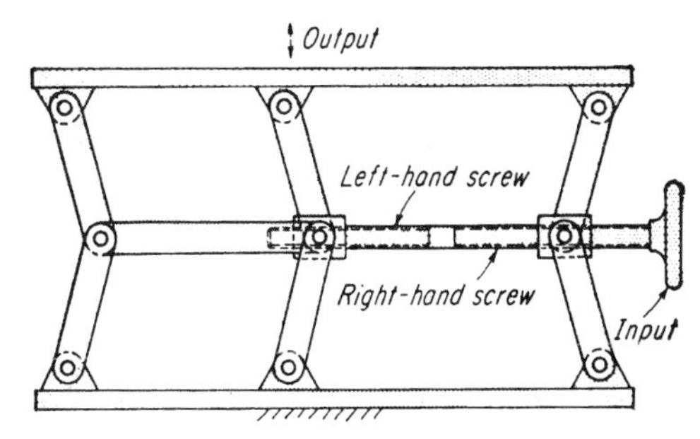

Turning the adjusting screw spreads or contracts the linkage pairs to raise or lower the table. Six parallel links are shown, but the mechanism can be build with four, eight, or more links.

Tensioning mechanism

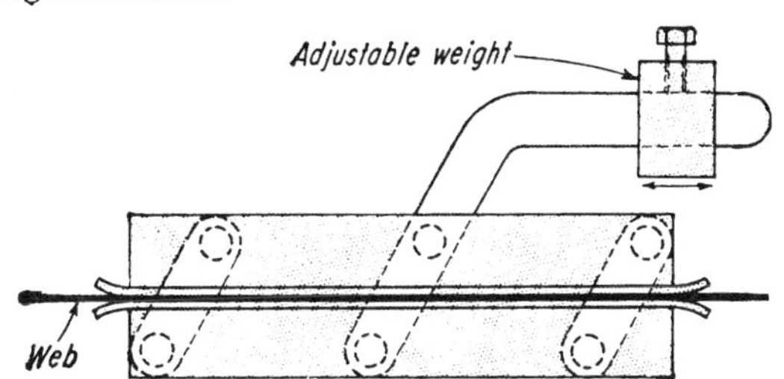

A simple parallel-link mechanism that produces tension in webs, wires, tapes, and strip steels. Adjusting the weight varies the drag on the material.

Triple-pivot mechanism

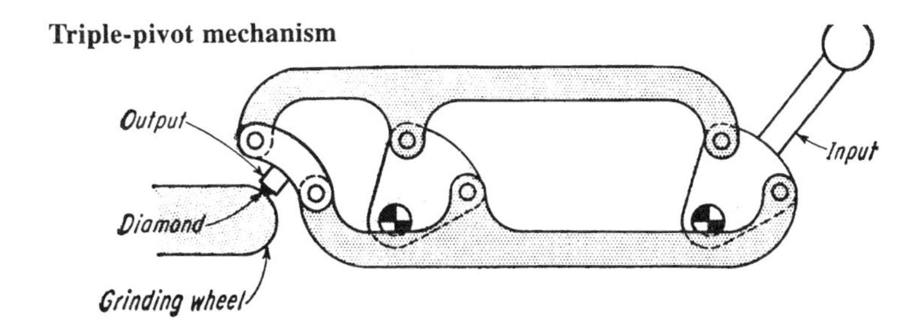

Two triangular plates pivot around fixed points on a machine frame. The output point describes a circular-arc curve. It can round out the cutting surfaces of grinding wheels.

SEVEN STROKE MULTIPLIER LINKAGES

Reciprocating-table drive

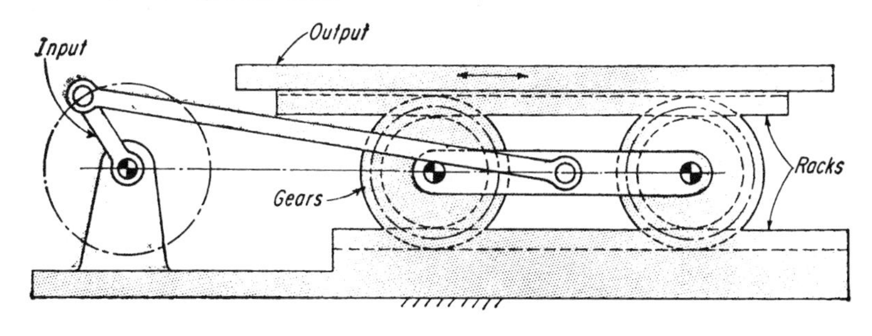

Two gears rolling on a stationary bottom rack drive the movable top rack, which is attached to a printing table. When the input crank rotates, the table will move out to a distance of four times the crank length.

Parallel-link feeder

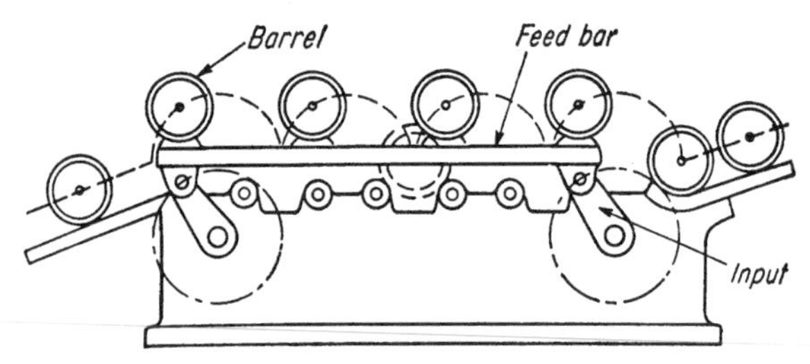

One of the cranks is the input, and the other follows to keep the feeding bar horizontal. The feeder can move barrels from station to station.

Parallelogram linkage

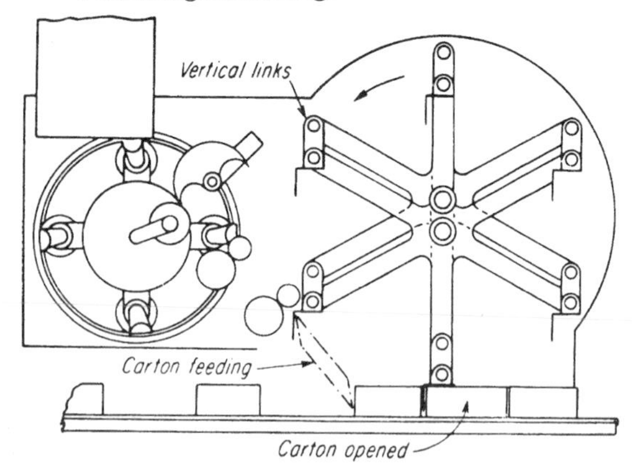

All seven short links are kept in a vertical position while rotating. The center link is the driver. This particular machine feeds and opens cartons, but the mechanism will work in many other applications.

Parallel-link driller

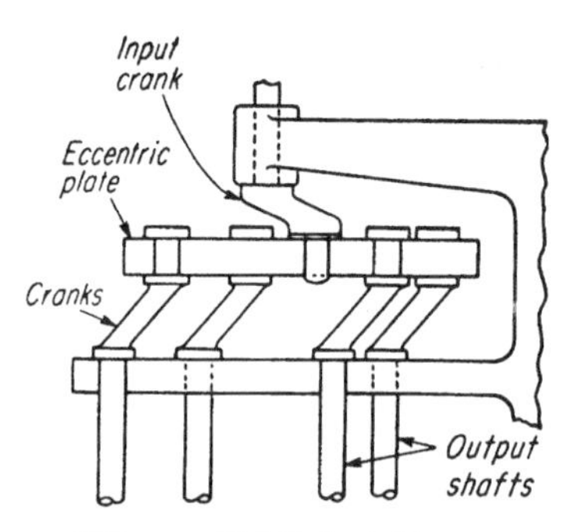

This parallel-link driller powers a group of shafts. The input crank drives the eccentric plate. This, in turn, rotates the output cranks that have the same length at the same speed. Gears would occupy more room between the shafts.

Parallel-plate driver

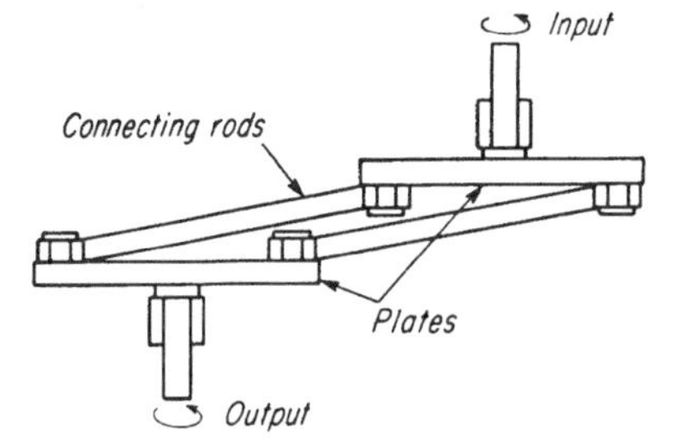

The input and output shafts of this parallel-plate driver rotate with the same angular relationship. The positions of the shafts, however, can vary to suit other requirements without affecting the input-output relationship between the shafts.

Curve-scribing mechanism

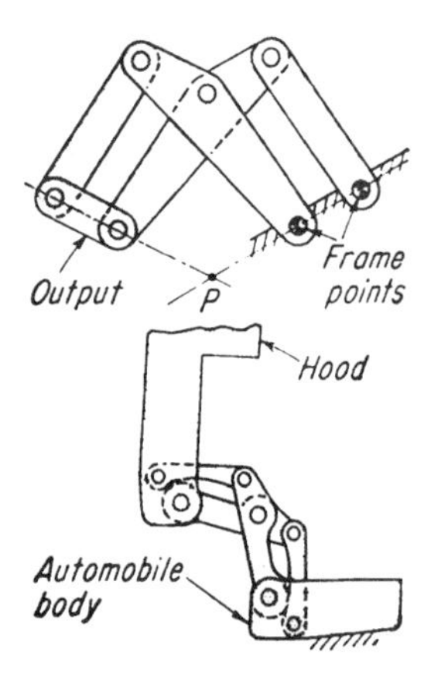

The output link rotates so that it appears to revolve around a point moving in space (*P*). This avoids the need for hinges at distant or inaccessible spots. The mechanism is suitable for hinging the hoods of automobiles.

Parallel-link coupling

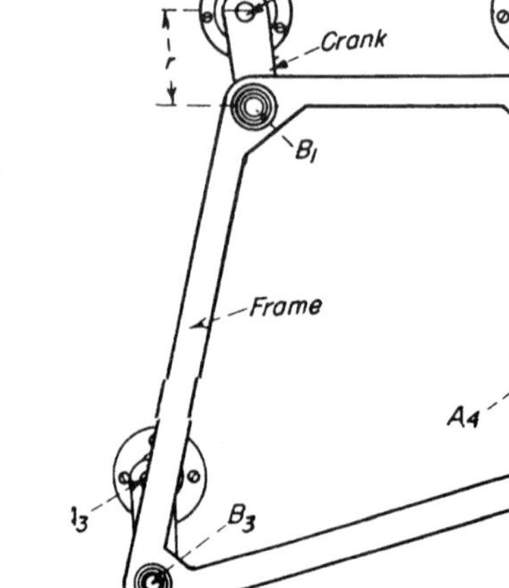

The absence of backlash makes this parallel-link coupling a precision, low-cost replacement for gear or chain drives that can also rotate parallel shafts. Any number of shafts greater than two can be driven from any one of the shafts, provided two conditions are fulfilled: (1) All cranks must have the same length *r;* and (2) the two polygons formed by the shafts *A* and frame pivot centers *B* must be identical. The main disadvantage of this mechanism is its dynamic unbalance, which limits the speed of rotation. To lessen the effect of the vibrations produced, the frame should be made as light as is consistent with strength requirements for the intended application.

NINE FORCE AND STROKE MULTIPLIER LINKAGES

Wide-angle oscillator

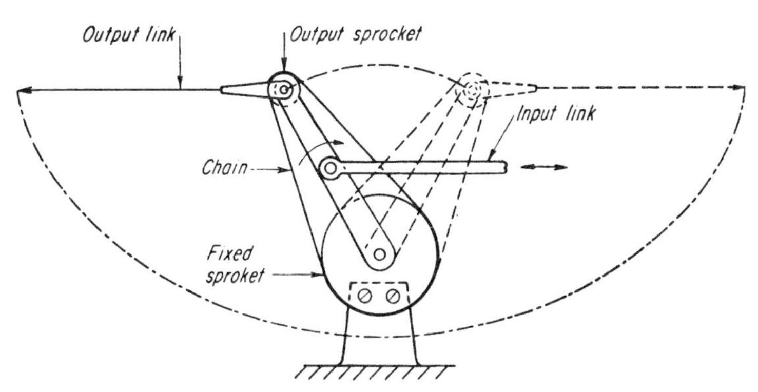

The motion of the input linkage in the diagram is converted into a wide-angle oscillation by the two sprockets and chain. An oscillation of 60° is converted into 180° oscillation.

Gear-sector drive

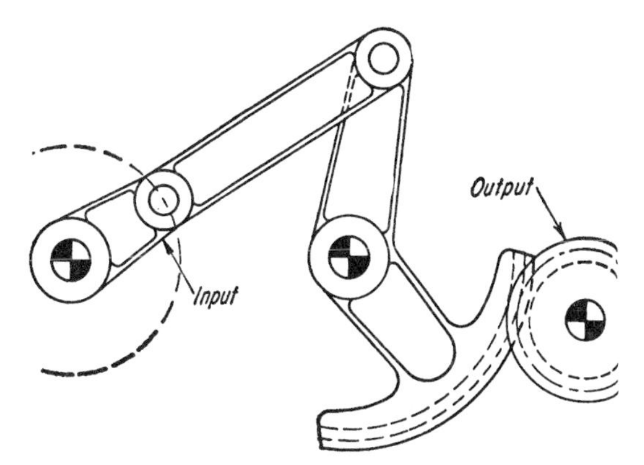

This is actually a four-bar linkage combined with a set of gears. A four-bar linkage usually obtains so more than about 120° of maximum oscillation. The gear segments multiply the oscillation in inverse proportion to the radii of the gears. For the proportions shown, the oscillation is boosted two and one-half times.

Angle-doubling drive

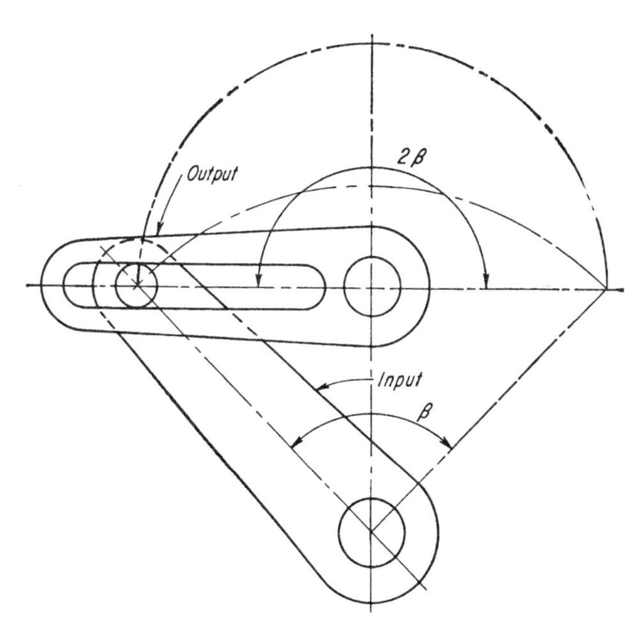

This angle-doubling drive will enlarge the oscillating motion β of one machine member into an output oscillation of 2β. If gears are employed, the direction of rotation cannot be the same unless an idler gear is installed. In that case, the centers of the input and output shafts cannot be too close. Rotating the input link clockwise causes the output to follow in a clockwise direction. For any set of link proportions, the distance between the shafts determines the gain in angle multiplication.

Pulley drive

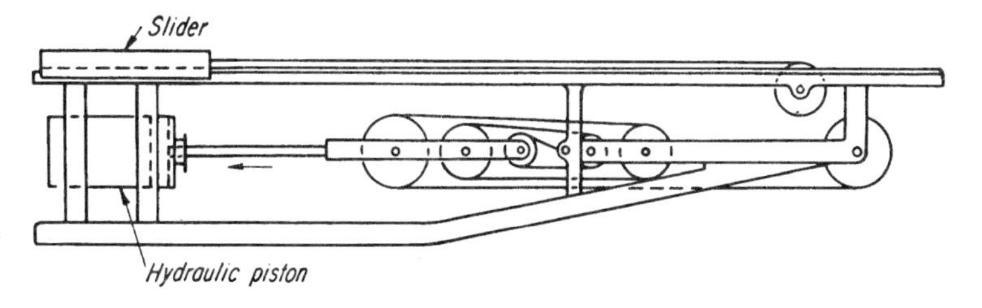

This pulley drive multiplies the stroke of a hydraulic piston, causing the slider to move rapidly to the right for catapulting objects.

Typewriter drive

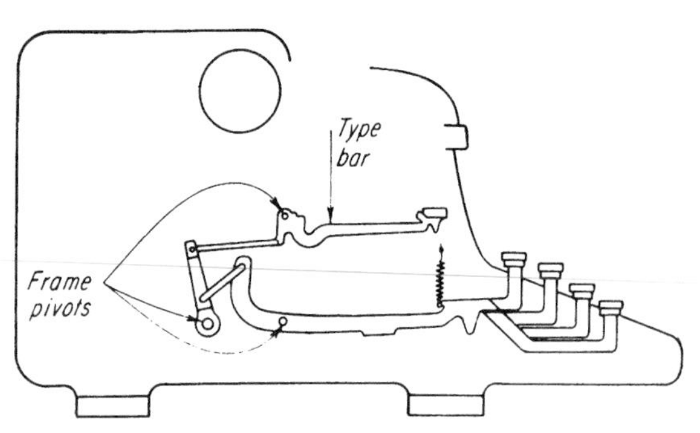

This drive multiplies the finger force of a typewriter, producing a strong hammer action at the roller from a light touch. There are three pivot points attached to the frame. The links are arranged so that the type bar can move in free flight after a key has been struck. The mechanism illustrated is actually two four-bar linkages in series. Some typewriters have as many as four four-bar linkages in a series.

Double-toggle puncher

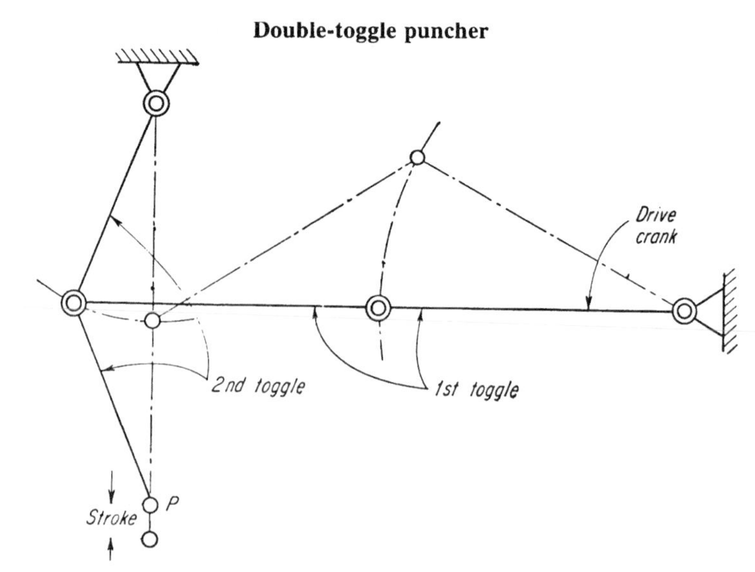

The first toggle of this puncher keeps point *P* in the raised position although its weight can exert a strong downward force (as in a heavy punch weight). When the drive crank rotates clockwise (e.g., driven by a reciprocating mechanism), the second toggle begins to straighten so as to create a strong punching force.

Gear-rack drive

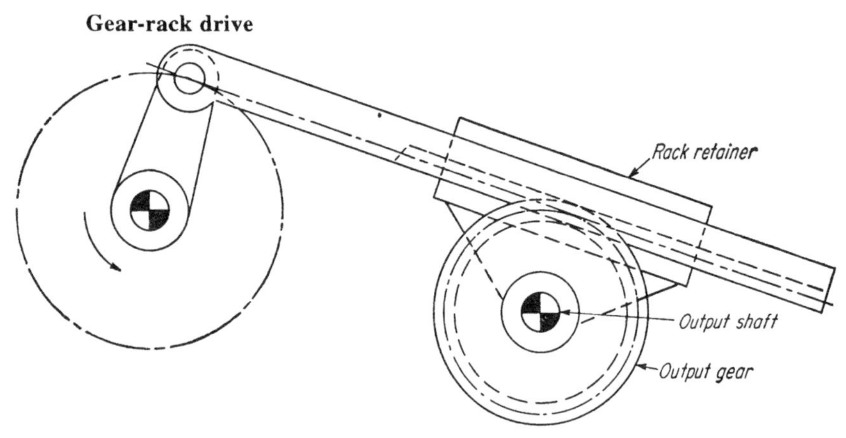

This drive mechanism converts the motion of an input crank into a much larger rotation of the output (from 30° to 360°). The crank drives the slider and gear rack, which in turn rotates the output gear.

Chain drive

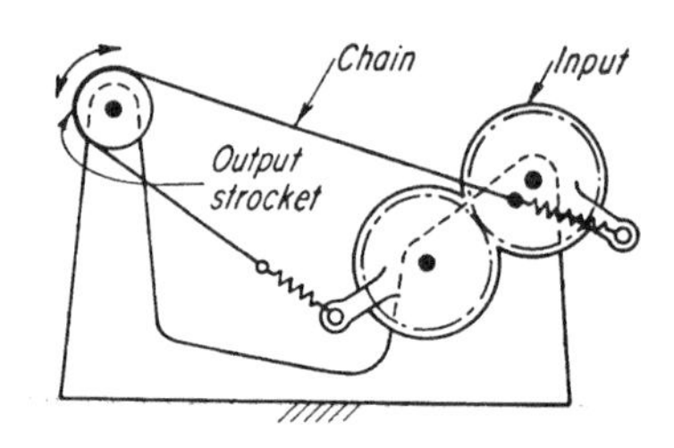

Springs and chains are attached to geared cranks of this drive to operate a sprocket output. Depending on the gear ratio, the output will produce a desired oscillation, e.g., two revolutions of output in each direction for each 360° of input.

Linkage-train drive

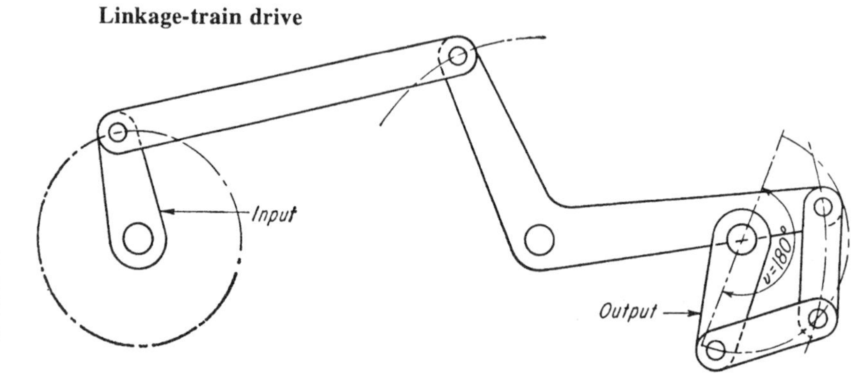

Arranging linkages in series on this drive can increase its angle of oscillation. In the version illustrated, the oscillating motion of the L-shaped rocker is the input for the second linkage. The final oscillation is 180°.

EIGHTEEN VARIATIONS OF DIFFERENTIAL LINKAGE

Figure 1 shows the modifications of the differential linkage shown in Fig. 2(A). These are based on the variations in the triple-jointed intermediate link 6. The links are designated as follows: Frame links: links 2, 3 and 4; two-jointed intermediate links: links 5 and 7; three jointed intermediate links: link 6.

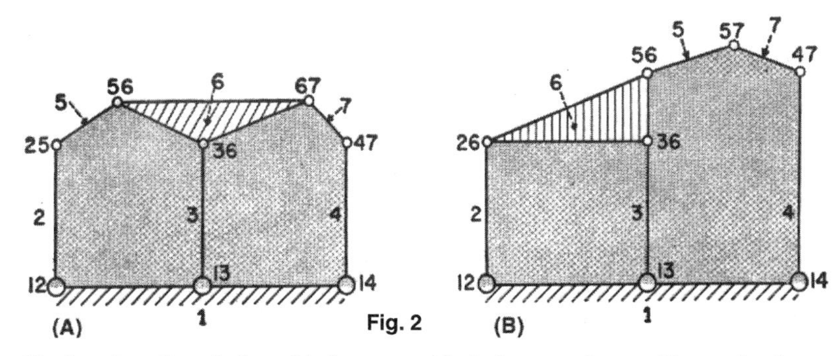

The input motions to be added are a and b; their sum s is equal to c_1a 1 c_2b, where c_1 and c_2 are scale factors. The links are numbered in the same way as those in Fig. 2(A).

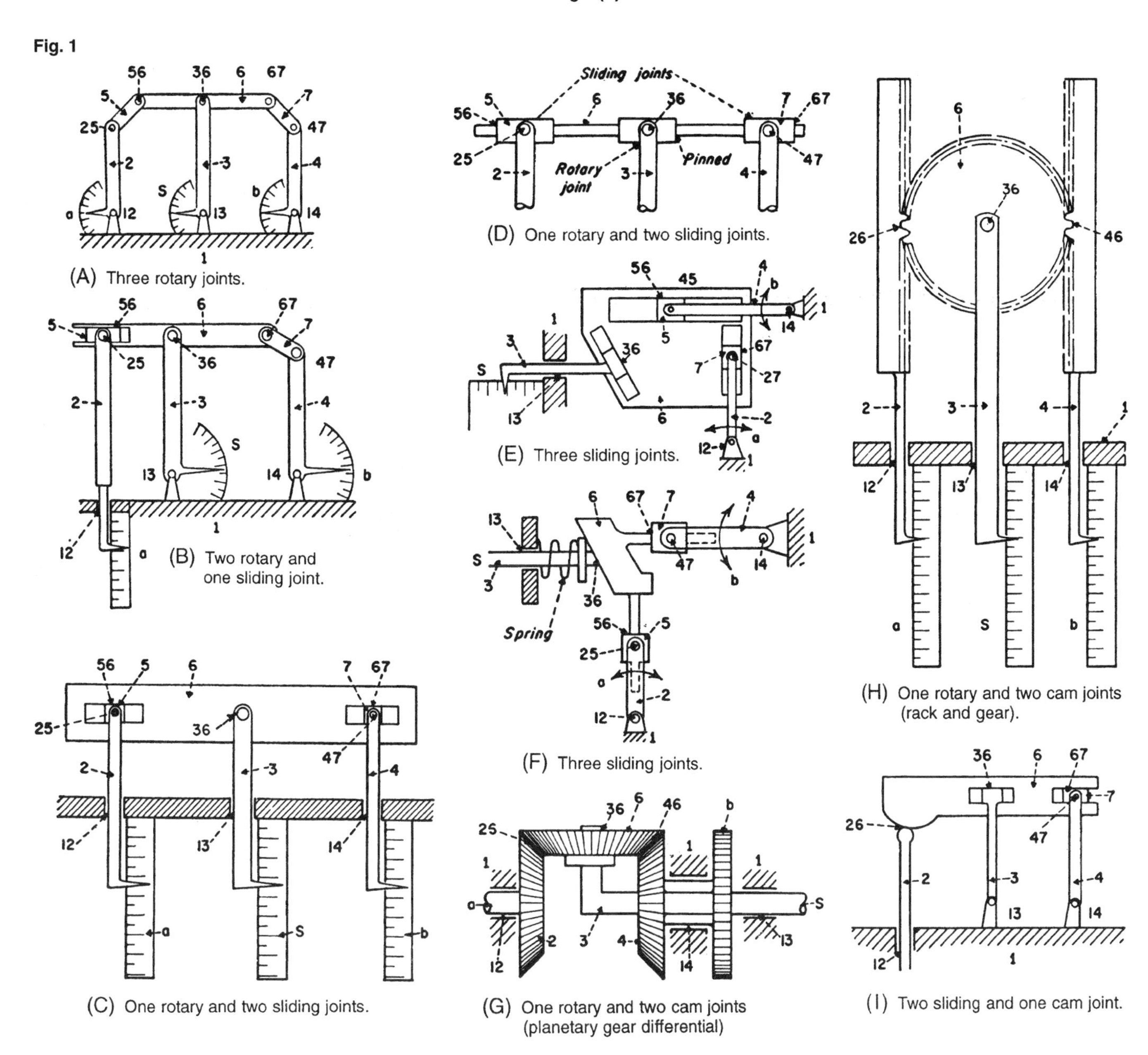

(A) Three rotary joints.

(B) Two rotary and one sliding joint.

(C) One rotary and two sliding joints.

(D) One rotary and two sliding joints.

(E) Three sliding joints.

(F) Three sliding joints.

(G) One rotary and two cam joints (planetary gear differential)

(H) One rotary and two cam joints (rack and gear).

(I) Two sliding and one cam joint.

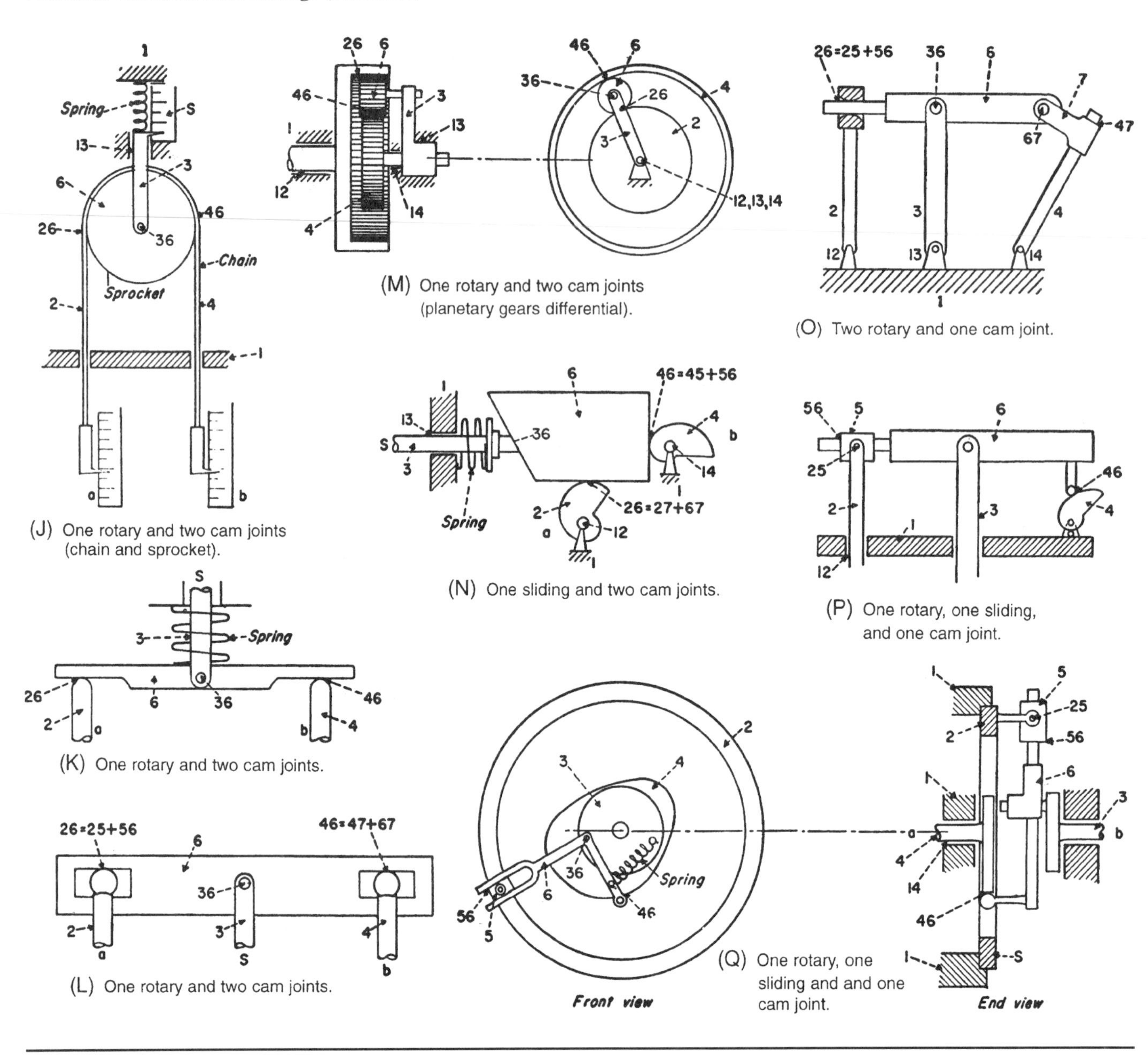

(J) One rotary and two cam joints (chain and sprocket).

(K) One rotary and two cam joints.

(L) One rotary and two cam joints.

(M) One rotary and two cam joints (planetary gears differential).

(N) One sliding and two cam joints.

(O) Two rotary and one cam joint.

(P) One rotary, one sliding, and one cam joint.

(Q) One rotary, one sliding and and one cam joint.

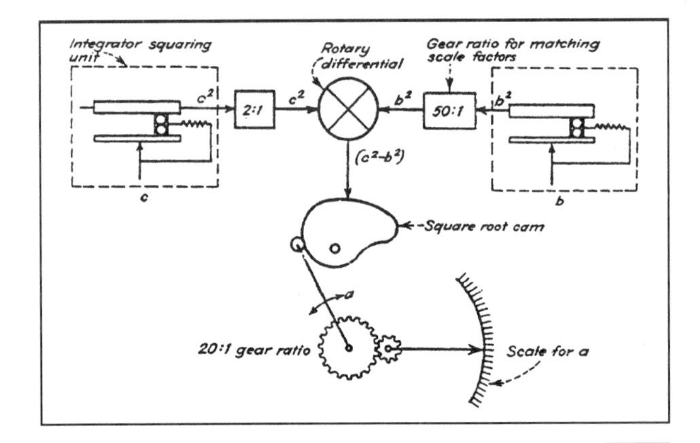

The intergrator method of mechanizing the equation $a = \sqrt{c^2 - b^2}$ is shown in the schematic form. It requires an excessive number of parts.

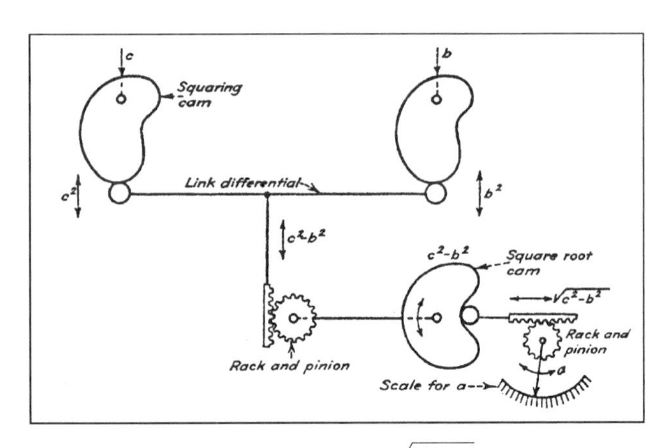

The cam method of mechanizing $a = \sqrt{c^2 - b^2}$ uses function generators for squaring and a link differential for subtraction. Note the reduction in parts from the integrator method.

FOUR-BAR SPACE MECHANISMS

There are potentially hundreds of them, but only a few have been discovered so far. Here are the best of one class—the four-bar space mechanisms.

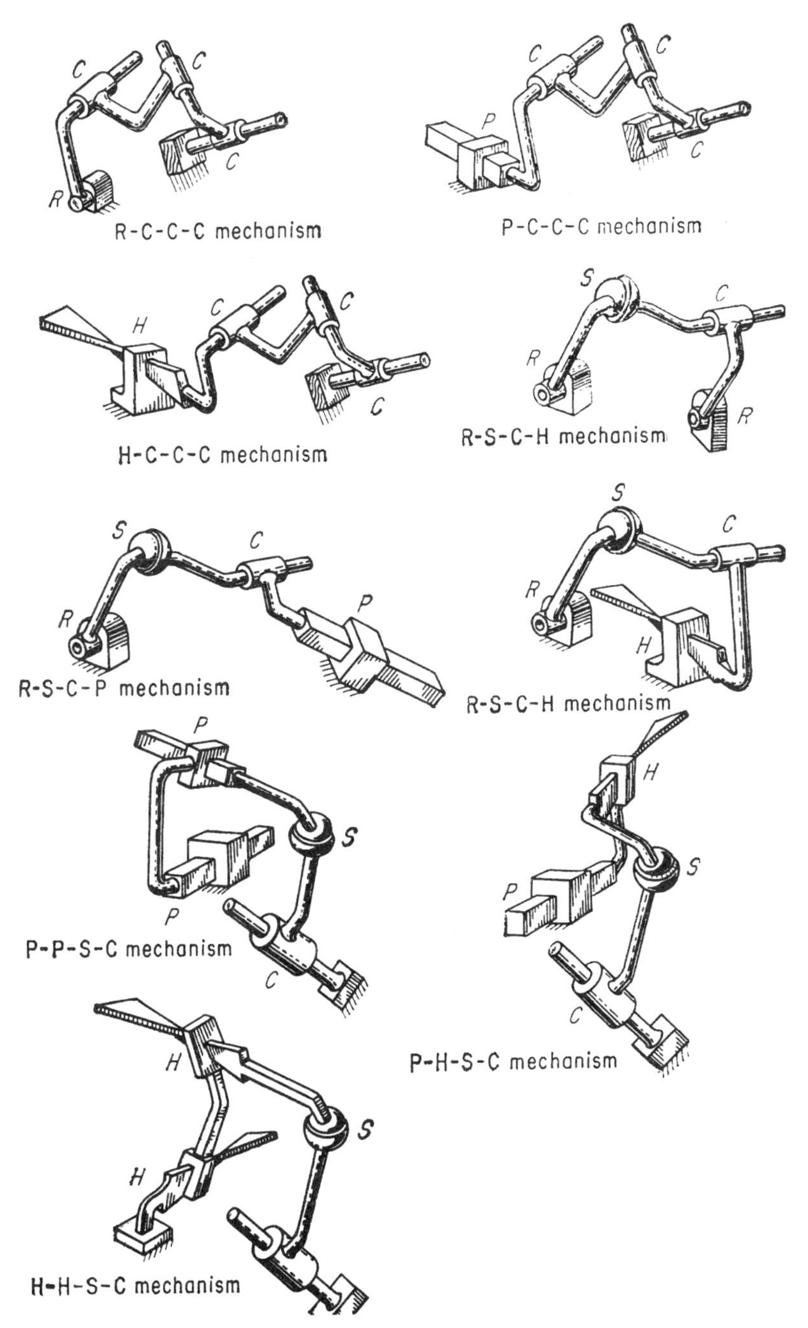

Fig. 1 The nine chosen mechanisms.

A virtually unexplored area of mechanism research is the vast domain of three-dimensional linkage, frequently called space mechanism. Only a comparatively few kinds have been investigated or described, and little has been done to classify those that are known. As a result, many engineers do not know much about them, and applications of space mechanisms have not been as widespread as they could be.

Because a space mechanism can exist with a wide variety of connecting joints or "pair" combinations, it can be identified by the type and sequence of its joints. A listing of all of the physically realizable kinematic pairs has been established, based on the number of degrees-of-freedom of a joint. These pairs are all the known ways of connecting two bodies together for every possible freedom of relative motion between them.

The Practical Nine

The next step was to find the combination of pairs and links that would produce practical mechanisms. Based on the "Kutzbach criterion" (the only known mobility criterion—it determines the degree of freedom of a mechanism due to the constraints imposed by the pairs), 417 different kinds of space mechanisms have been identified. Detailed examination showed many of these to be mechanically complex and of limited adaptability. But the four-link mechanisms had particular appeal because of their mechanical simplicity. A total of 138 different kinds of four-bar mechanisms have been found. Of these, nine have particular merit (Fig. 1).

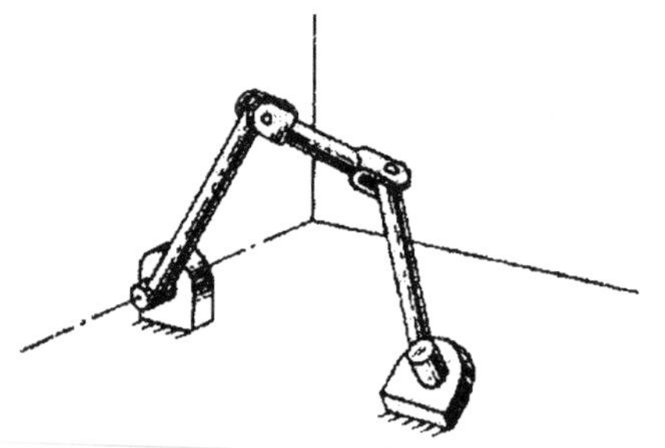

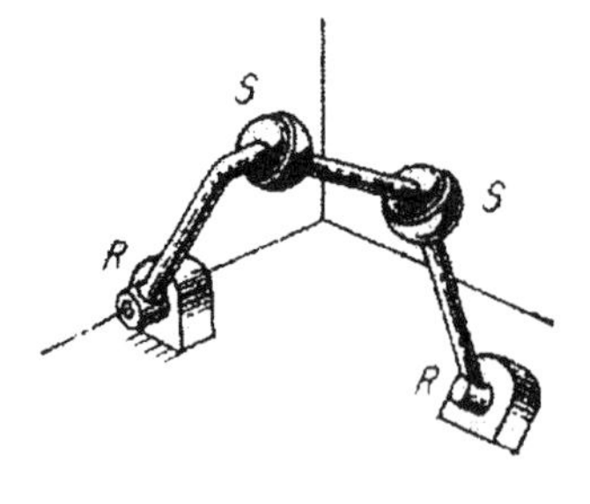

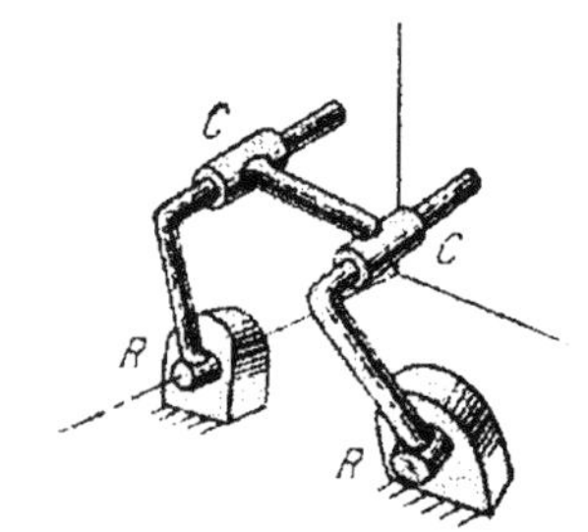

Fig. 2 The three mavericks.

These nine four-link mechanisms are the easiest to build because they contain only those joints that have area contact and are self-connecting. In the table, these joints are the five closed, lower pair types:

- *R* = Revolute joint, which permits rotation only
- *P* = Prism joint, which permits sliding motion only
- *H* = Helix or screw type of joint
- *C* = Cylinder joint, which permits both rotation and sliding (hence has two degrees of freedom)
- *S* = Sphere joint, which is the common ball joint permitting rotation in any direction (three degrees of freedom)

All these mechanisms can produce rotary or sliding output motion from a rotary input—the most common mechanical requirements for which linkage mechanisms are designed.

The type letters of the kinematic pairs in the table identify the mechanism by ordering the letter symbols consecutively around the closed kinematic chain. The first letter identifies the pair connecting the input link and the fixed link; the last letter identifies the output link, or last link, with the fixed link. Thus, a mechanism labeled *R-S-C-R* is a double-crank mechanism with a spherical pair between the input crank and the coupler, and a cylindrical pair between the coupler and the output crank.

The Mavericks

The Kutzbach criterion is inadequate for the job because it cannot predict the existence of such mechanisms as the Bennett *R-R-R-R* mechanism, the double-ball joint *R-S-S-R* mechanism, and the *R-C-C-R* mechanism (Fig. 2). These "special" mechanisms require special geometric conditions to have a single degree of freedom. The *R-R-R-R* mechanism requires a particular orientation of the revolute axes and a particular ratio of link lengths to function as a single degree of freedom space mechanism. The *R-S-S-R* configuration, when functioning as a single degree-of-freedom mechanism, will have a passive degree of freedom of its coupler link. When properly constructed, the configuration *R-C-C-R* will also have a passive degree-of-freedom of its coupler, and it will function as a single-degree space mechanism.

Of these three special four-link mechanisms, the *R-S-S-R* mechanism is seen as the outstanding choice. It is the most versatile and practical configuration for meeting double-crank motion requirements.

Classification of kinematic pairs

Degree of freedom	Type number*	Type of joint	
		Symbol	Name
1	100	R	Revolute
	010	P	Prism
	001	H	Helix
2	200	T	Torus
	110	C	Cylinder
	101	T_H	Torus-helix
	020	. .	. . . .
	011	. .	. . . .
3	300	S	Sphere
	210	S_S	Sphere-slotted cylinder
	201	S_{SH}	Sphere-slotted helix
	120	P_L	Plane
	021	. .	. . . .
	111	. .	. . . .
4	310	S_G	Sphere-groove
	301	S_{GH}	Sphere-grooved helix
	220	C_P	Cylinder-plane
	121	. .	. . . .
	211	. .	. . . .
5	320	S_P	Sphere-plane
	221	. .	. . . .
	311		. . . .

* Number of freedoms, give in the order of N_R, N_T, N_H.

SEVEN THREE-DIMENSIONAL LINKAGE DRIVES

The main advantage of three-dimensional drives is their ability to transmit motion between nonparallel shafts. They can also generate other types of helpful motion. This roundup includes descriptions of seven industrial applications for the drives.

Spherical Crank Drive

This type of drive is the basis for most three-dimensional linkages, much as the common four-bar linkage is the basis for the two-dimensional field. Both mechanisms operate on similar principles. (In the accompanying sketches, a is the input angle, and β the output angle. This notation has been used throughout this section.)

In the four-bar linkage, the rotary motion of driving crank *1* is transformed into an oscillating motion of output link 3. If the fixed link is made the shortest of all, then it is a double-crank mechanism; both the driving and driven members make full rotations.

The spherical crank drive, link *1* is the input, link *3* the output. The axes of rotation intersect at point *O;* the lines connecting *AB*, *BC*, *CD*, and *DA* can be considered to be parts of great circles of a sphere. The length of the link is best represented by angles a, b, c, and d.

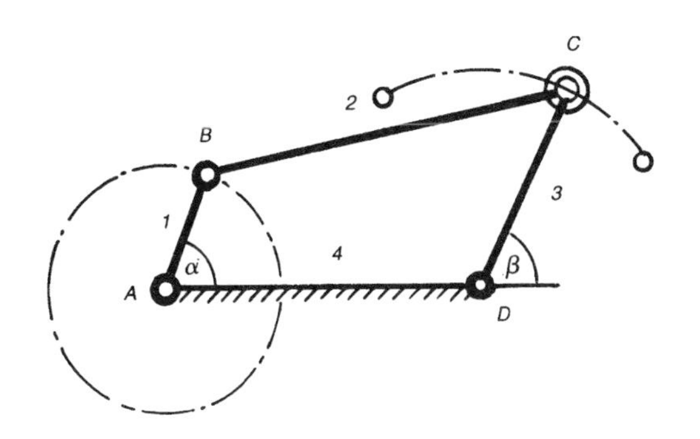

Four-bar linkage

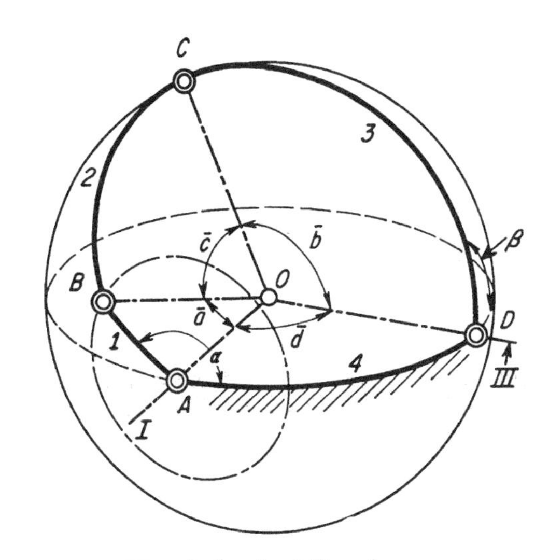

The Spherical Crank

Spherical-Slide Oscillator Drive

The two-dimensional slider crank is obtained from a four-bar linkage by making the oscillating arm infinitely long. By making an analogous change in the spherical crank, the spherical slider crank is obtained.

The uniform rotation of input shaft *I* is transferred into a nonuniform oscillating or rotating motion of output shaft *III*. These shafts intersect at an angle δ, corresponding to the frame link *4* of the spherical crank. Angle γ corresponds to the length of link *1*, and axis *II* is at right angle to axis *III*.

The output oscillates when γ is smaller than δ, but it rotates when γ is larger than δ.

The relation between input angle *a* and output angle β as designated in the skewed Hooke's joint is:

$$\tan\beta = \frac{(\tan\gamma)(\sin\alpha)}{\sin\delta + (\tan\gamma)(\cos\delta)(\cos\alpha)}$$

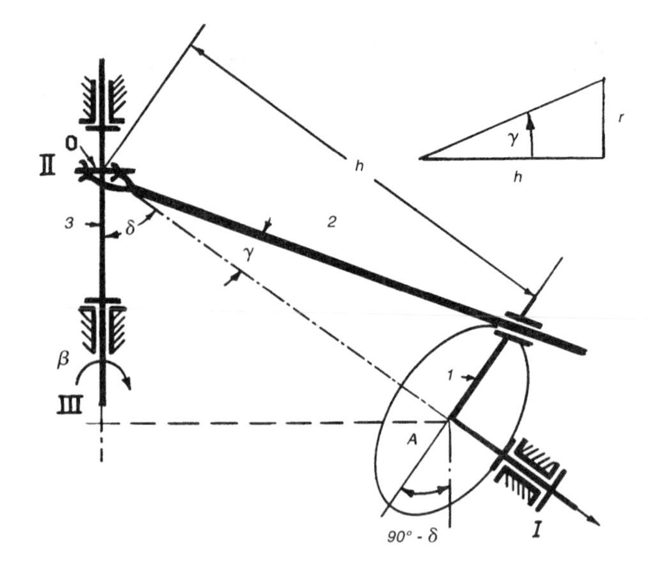

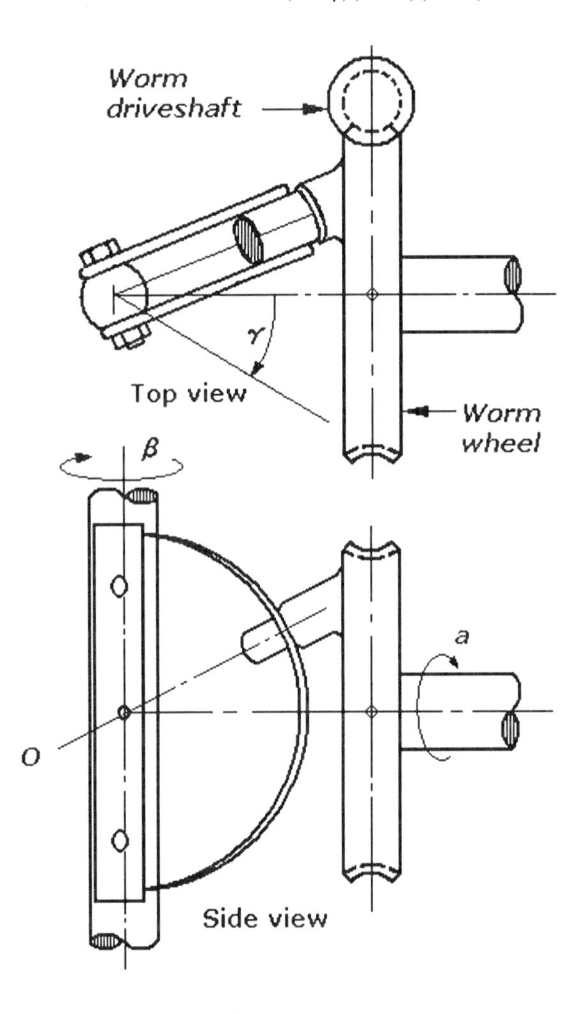

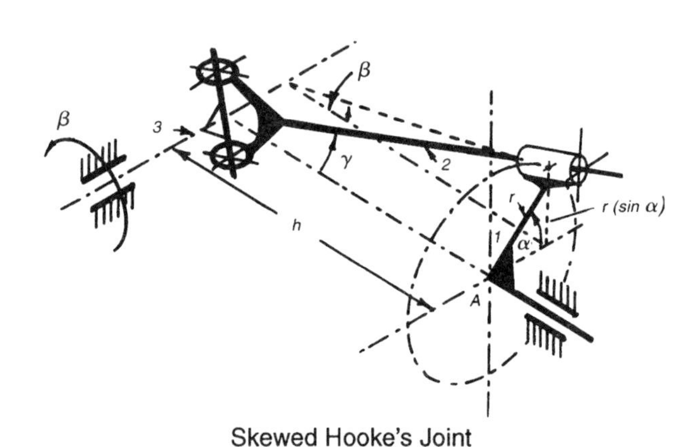

Skewed Hooke's Joint

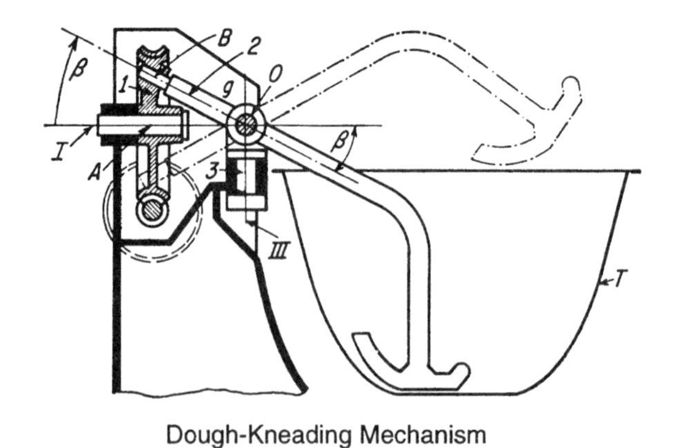

Dough-Kneading Mechanism

Skewed Hooke's Joint Drive

This variation of the spherical crank is specified where an almost linear relation is desired between the input and output angles for a large part of the motion cycle.

The equation defining the output in terms of the input can be obtained from the skewed Hooke's joint equation by making δ = 90°. Thus, sin δ = 1, cos δ = 0, and

$$\tan\beta = \tan\gamma\sin\alpha$$

The principle of the skewed Hooke's joint has been applied to the drive of a washing machine (see sketch).

Here, the driveshaft drives the worm wheel *1* which has a crank fashioned at an angle γ. The crank rides between two plates and causes the output shaft *III* to oscillate in accordance with the equation.

The dough-kneading drive is also based on the Hooke's joint, but it follows the path of link *2* to give a wobbling motion that kneads dough in the tank.

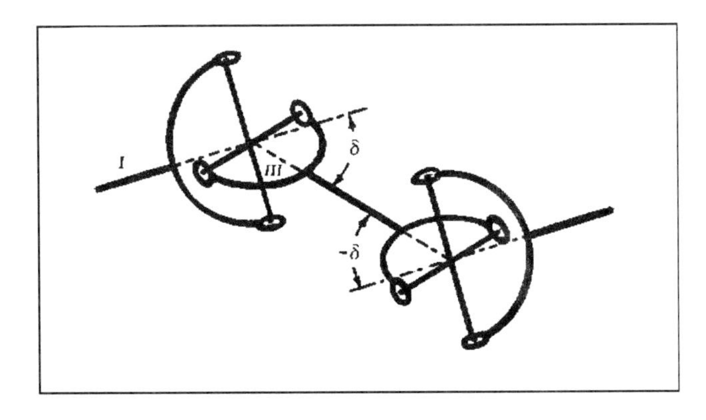

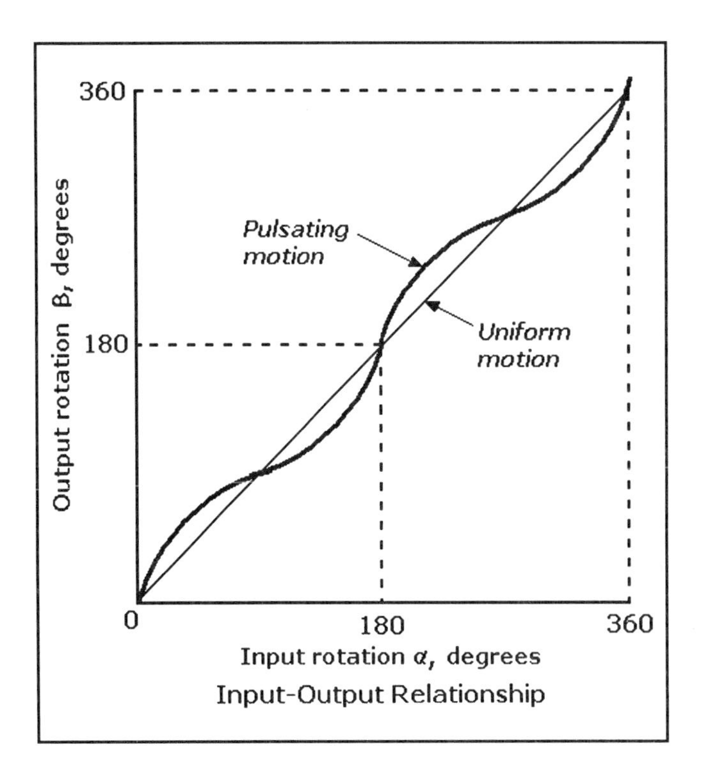

The Universal Joint Drive

The universal joint is a variation of the spherical-slide oscillator, but with angle $\gamma = 90°$. This drive provides a totally rotating output and can be operated as a pair, as shown in the diagram.

The equation relating input with output for a single universal joint, where γ is the angle between the connecting link and shaft *I*, is:

$$\tan \beta = \tan \alpha \cos \delta$$

The output motion is pulsating (see curve) unless the joints are operates as pairs to provide a uniform motion.

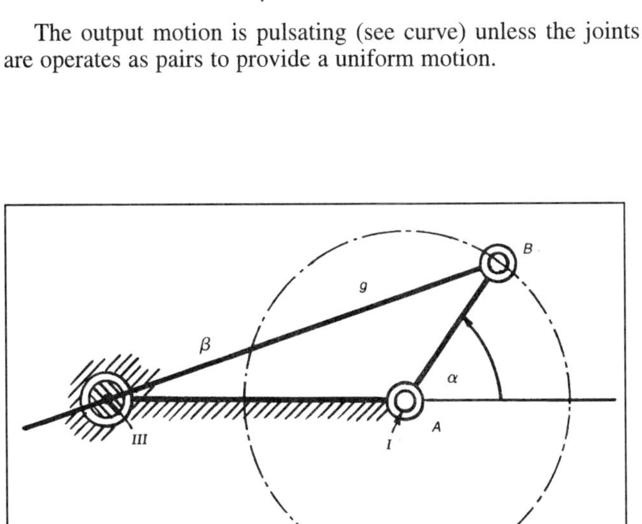

The 3-D Crank Slide Drive

The three-dimensional crank slide is a variation of a plane crank slide (see sketch), with a ball point through which link *g* always slides, while a point B on link *g* describes a circle. A 3-D crank is obtained from this mechanism by shifting output shaft *III* so that it is not normal to the plane of the circle; another way to accomplish this is to make shafts *I* and *III* nonparallel.

A practical variation of the 3-D crank slide is the agitator mechanism (see sketch). As input gear *I* rotates, link *g* swivels around (and also lifts) shaft *III*. Hence, the vertical link has both an oscillating rotary motion and a sinusoidal harmonic translation in the direction of its axis of rotation. The link performs what is essentially a twisting motion in each cycle.

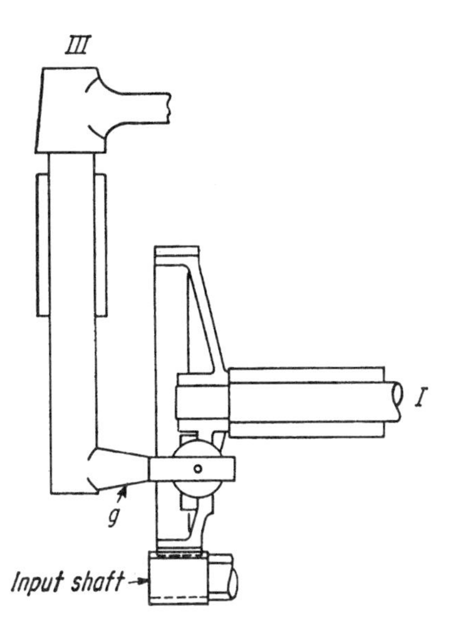

Agitator Mechanism

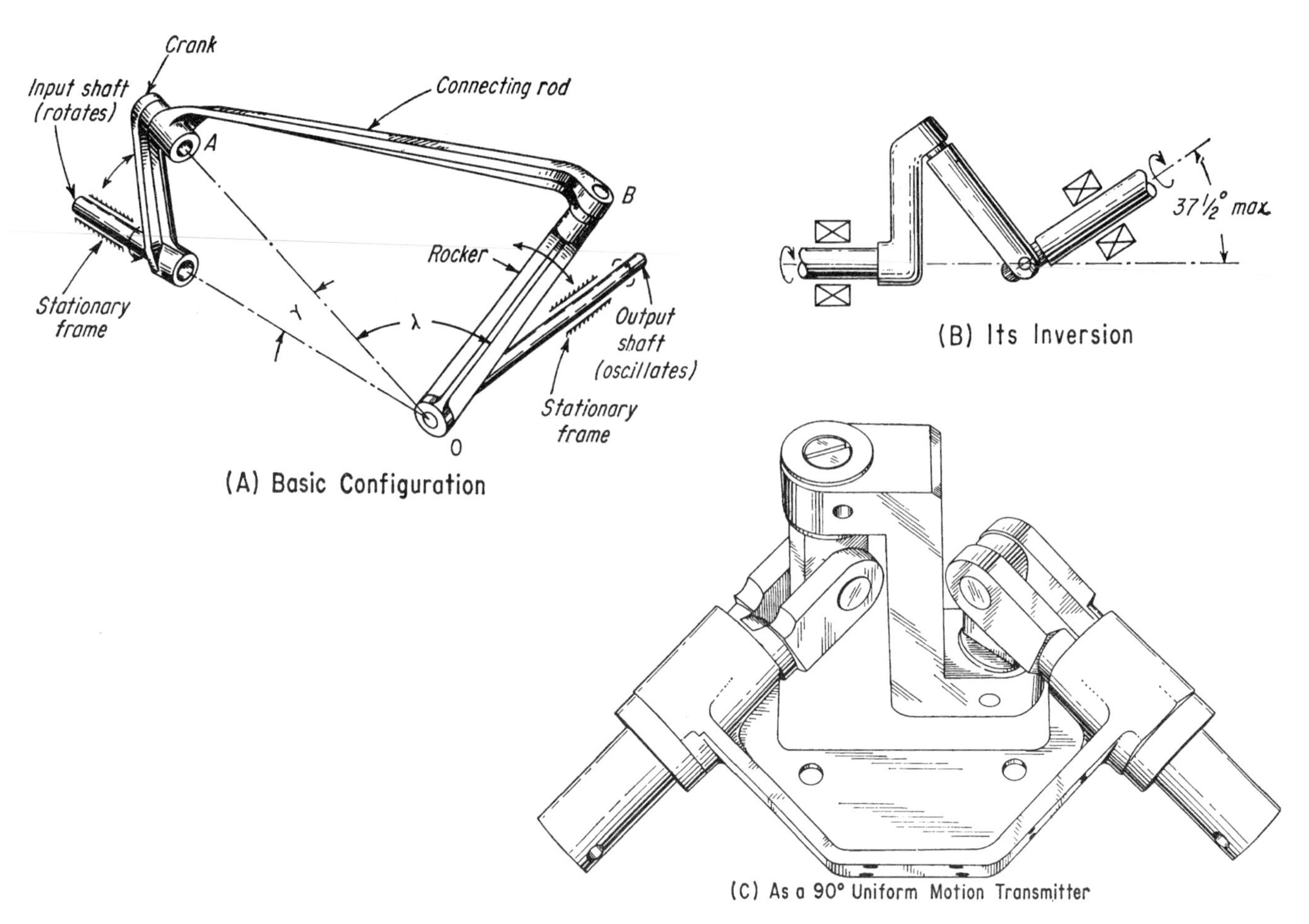

The Space Crank Drive

One of the more recent developments in 3-D linkages is the space crank shown in (A). It resembles the spherical crank, but has different output characteristics. The relationship between the input and output displacements is:

$$\cos \beta = (\tan \gamma)(\cos \alpha)(\sin \beta) - \frac{\cos \lambda}{\cos \gamma}$$

The velocity ratio is:

$$\frac{\omega_o}{\omega_i} = \frac{\tan \gamma \sin \alpha}{1 + \tan \gamma \cos \alpha \cot \beta}$$

where ω_0 is the output velocity and ω_i is the constant input velocity.

An inversion of the space crank is shown in (B). It can couple intersecting shafts, and permits either shaft to be driven with full rotations. Motion is transmitted up to $37^1/_2$° misalignment.

By combining two inversions (C), a method for transmitting an exact motion pattern around a 90° bend is obtained. This unit can also act as a coupler or, if the center link is replaced by a gear, it can drive two output shafts; in addition, it can transmit uniform motion around two bends.

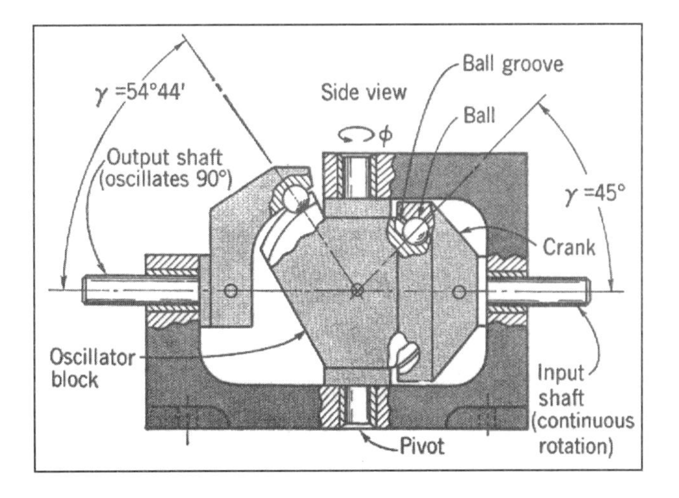

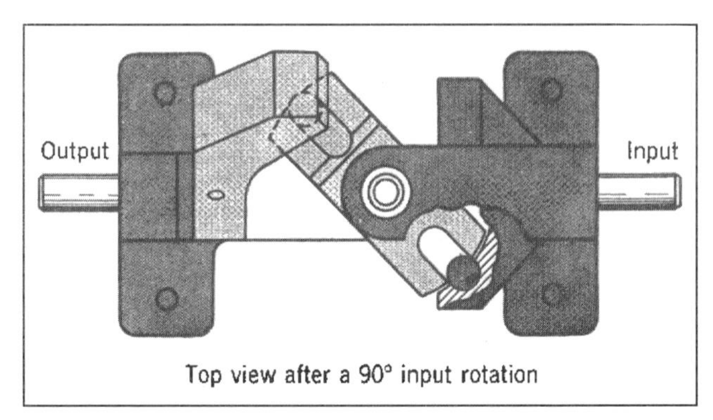

Steel balls riding within spherical grooves convert a continuous rotary input motion into an output that oscillates the shaft back and forth.

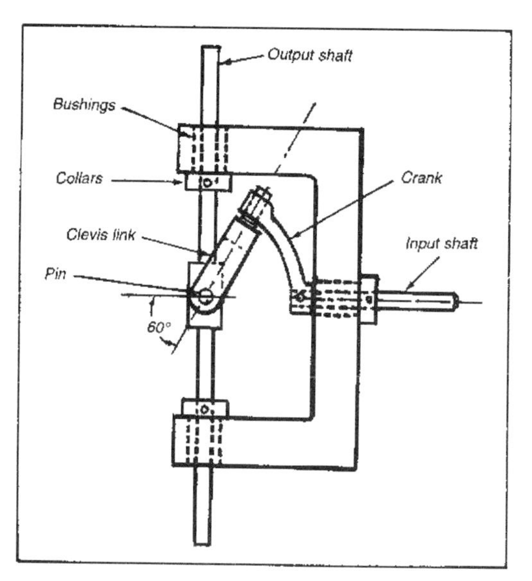

The oscillating motion is powered at right angles. The input shaft, in making full rotations, causes the output shaft to oscillate 120°.

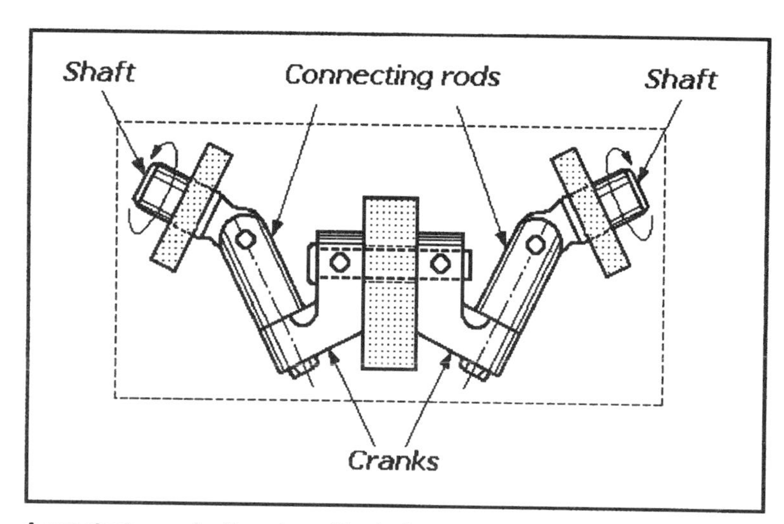

A constant-speed-ratio universal is obtained by placing two "inversions" back-to-back. Motion is transmitted up to a 75° misalignment.

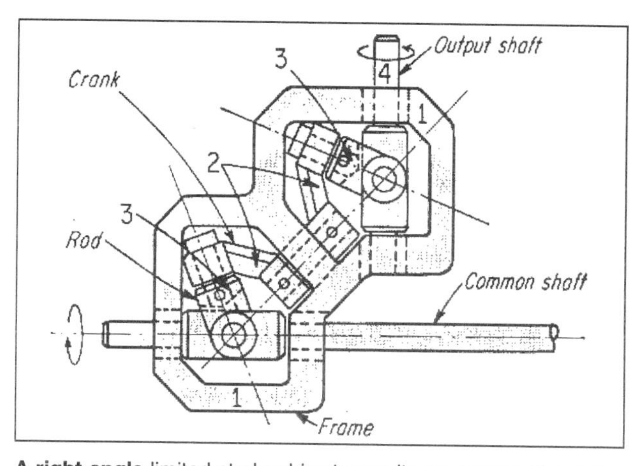

A right-angle limited-stroke drive transmits an exact motion pattern. A multiplicity of fittings can be operated from a common shaft.

The Elliptical Slide Drive

The output motion, β, of a spherical slide oscillator can be duplicated with a two-dimensional "elliptical slide." The mechanism has a link g that slides through a pivot point D and is fastened to a point P moving along an elliptical path. The ellipse can be generated by a Cardan drive, which is a planetary gear system whose planet gear has half the diameter of its internal gear. The center of the planet, point M, describes a circle; any point on its periphery describes a straight line, and any point in between, such as point P, describes an ellipse.

There are special relationships between the dimensions of the 3-D spherical slide and the 2-D elliptical slide: $\tan \gamma/\sin \delta = a/d$ and $\tan \gamma/\cot \delta = b/d$, where a is the major half-axis, b the minor half-axis of the ellipse, and d is the length of the fixed link DN. The minor axis lies along this link.

If point D is moved within the ellipse, a completely rotating output is obtained, corresponding to the rotating spherical crank slide.

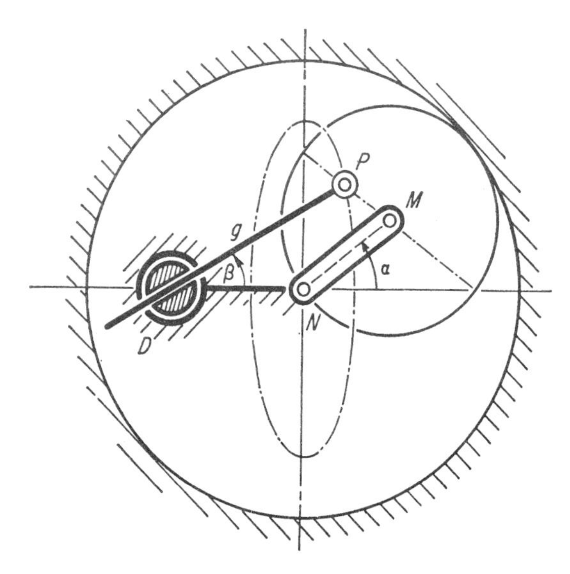

THIRTEEN DIFFERENT TOGGLE LINKAGE APPLICATIONS

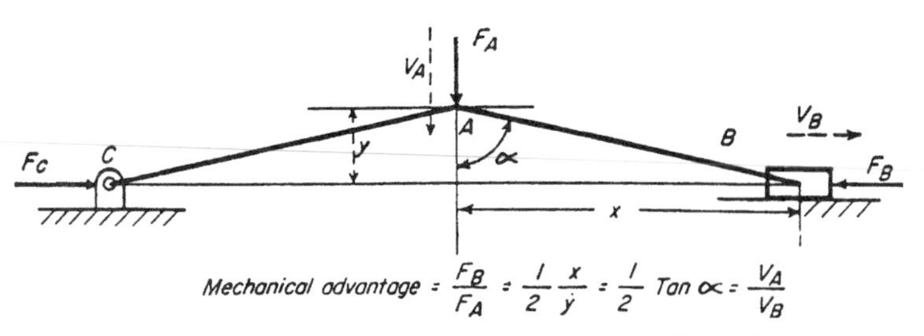

$$\text{Mechanical advantage} = \frac{F_B}{F_A} = \frac{1}{2}\frac{x}{y} = \frac{1}{2}\,\text{Tan}\,\alpha = \frac{V_A}{V_B}$$

Fig. 1 Many mechanical linkages are based on the simple toggle that consists of two links which tend to line up in a straight line at one point in their motion. The mechanical advantage is the velocity ratio of the input point A with respect to the outpoint point B: or V_A/V_B. As the angle is α approaches 90°, the links come into toggle, and the mechanical advantage and velocity ratio both approach infinity. However, frictional effects reduce the forces to much les than infinity, although they are still quite high.

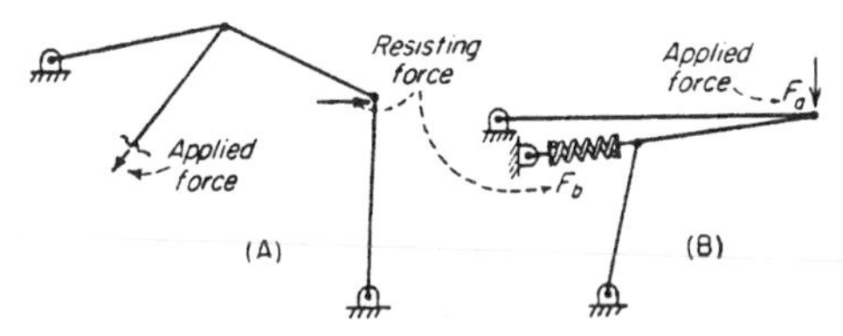

Fig. 2 Forces can be applied through other links, and need not be perpendicular to each other. (A) One toggle link can be attached to another link rather than to a fixed point or slider. (B) Two toggle links can come into toggle by lining up on top of each other rather than as an extension of each other. The resisting force can be a spring.

HIGH MECHANICAL ADVANTAGE

Fig. 3 In punch presses, large forces are needed at the lower end of the work stroke. However, little force is required during the remainder of the stroke. The crank and connecting rod come into toggle at the lower end of the punch stroke, giving a high mechanical advantage at exactly the time it is most needed.

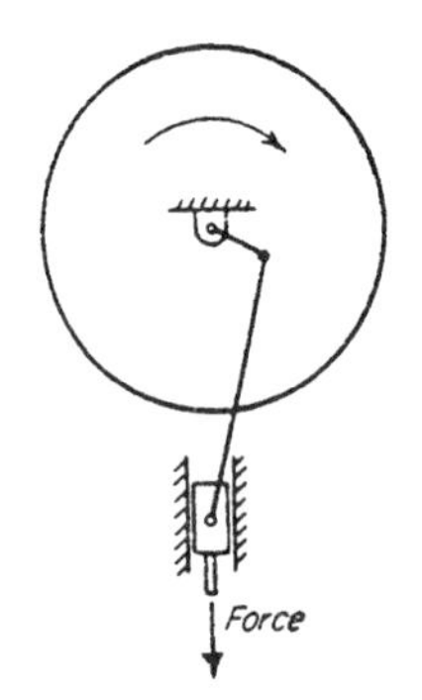

Fig. 4 A cold-heading rivet machine is designed to give each rivet two successive blows. Following the first blow (point 2) the hammer moves upward a short distance (to point 3). Following the second blow (at point 4), the hammer then moves upward a longer distance (to point 1) to provide clearance for moving the work-piece. Both strokes are produced by one revolution of the crank, and at the lowest point of each stroke (points 2 and 4) the links are in toggle.

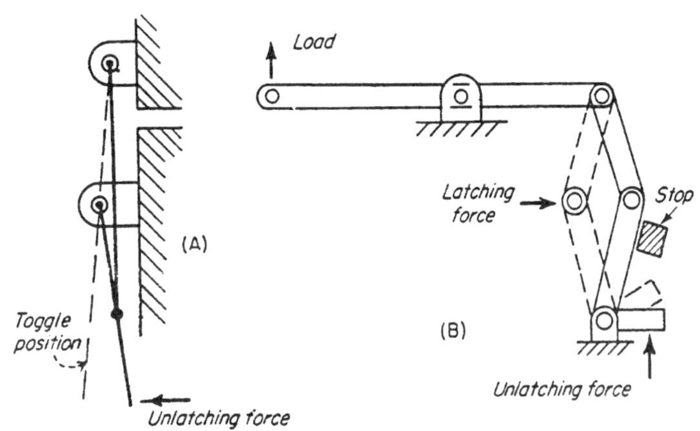

Fig. 5 Locking latches produce a high mechanical advantage when in the toggle portion of the stroke. A simple latch exerts a large force in the locked position (Fig. 5A). For positive locking, the closed position of latch is slightly beyond the toggle position. A small unlatching force opens the linkage (Fig. 5B).

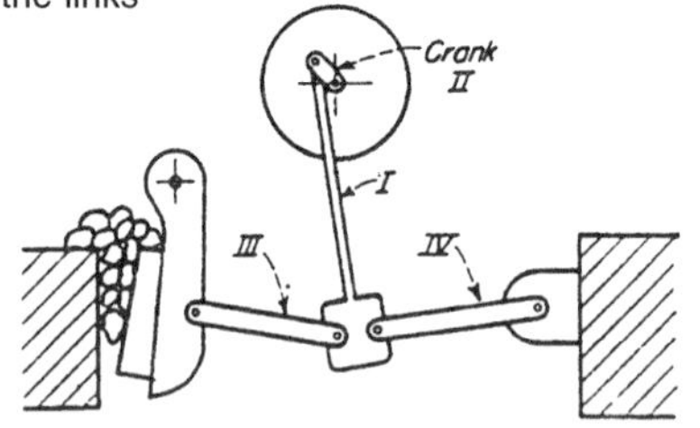

Fig. 6 A stone crusher has two toggle linkages in series to obtain a high mechanical advantage. When the vertical link I reaches the top of its stroke, it comes into toggle with the driving crank II; at the same time, link III comes into toggle and link IV. This multiplication results in a very large crushing force.

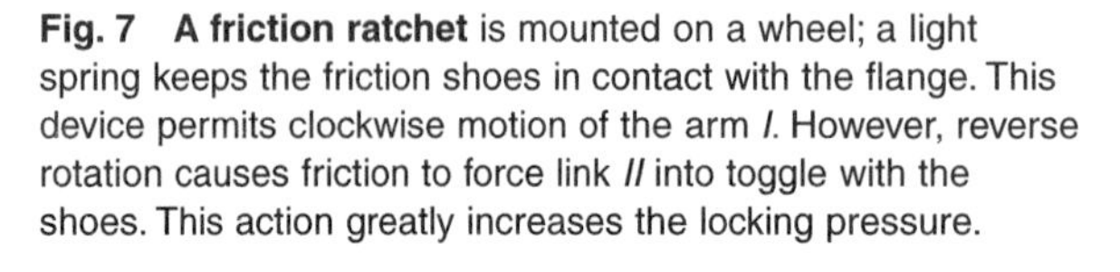

Fig. 7 A friction ratchet is mounted on a wheel; a light spring keeps the friction shoes in contact with the flange. This device permits clockwise motion of the arm I. However, reverse rotation causes friction to force link II into toggle with the shoes. This action greatly increases the locking pressure.

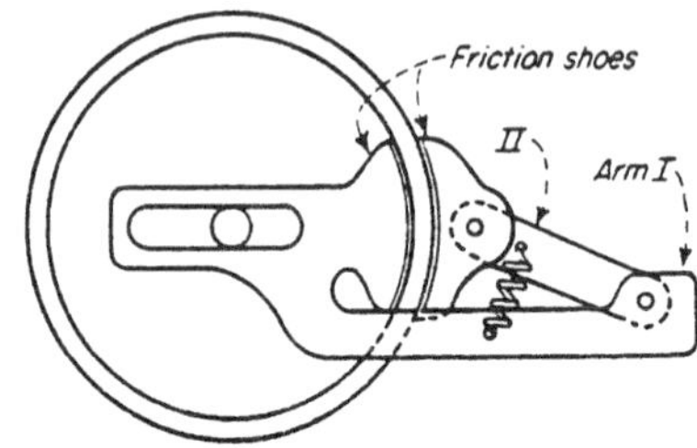

HIGH VELOCITY RATIO

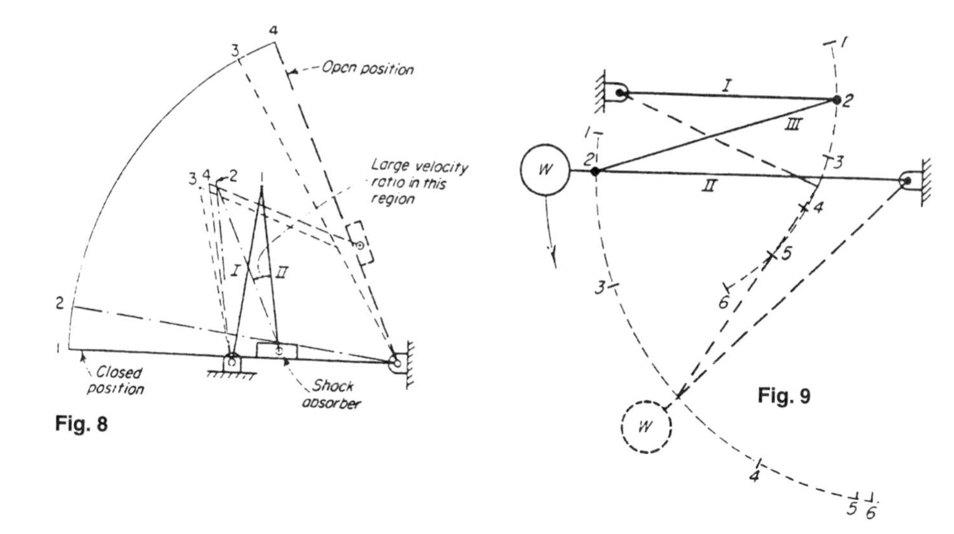

Fig. 8 Door check linkage gives a high velocity ratio during the stroke. As the door swings closed, connecting link *I* comes into toggle with the shock absorber arm *II*, giving it a large angular velocity. The shock absorber is more effective in retarding motion near the closed position.

Fig. 9 An impact reducer is on some large circuit breakers. Crank *I* rotates at constant velocity while the lower crank moves slowly at the beginning and end of the stroke. It moves rapidly at the mid-stroke when arm *II* and link *III* are in toggle. The accelerated weight absorbs energy and returns it to the system when it slows down.

VARIABLE MECHANICAL ADVANTAGE

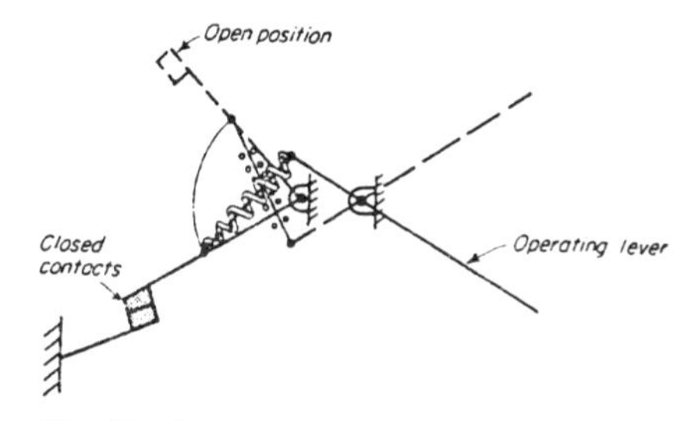

Fig. 10 A toaster switch has an increasing mechanical advantage to aid in compressing a spring. In the closed position, the spring holds the contacts closed and the operating lever in the down position. As the lever is moved upward, the spring is compressed and comes into toggle with both the contact arm and the lever. Little effort is required to move the links through the toggle position; beyond this point, the spring snaps the contacts open. A similar action occurs on closing.

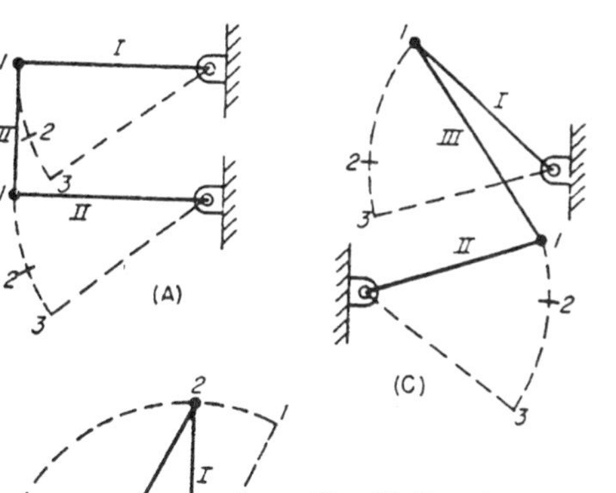

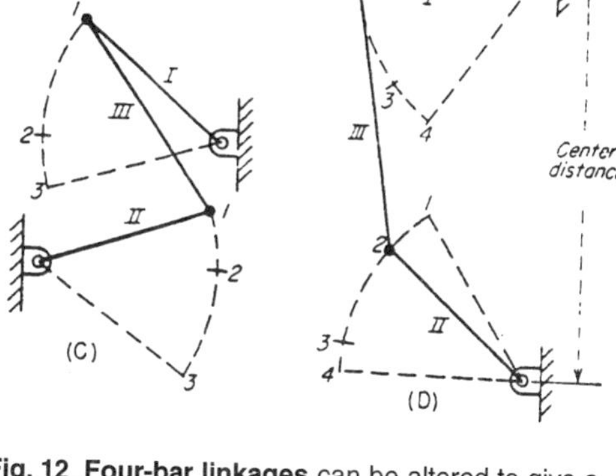

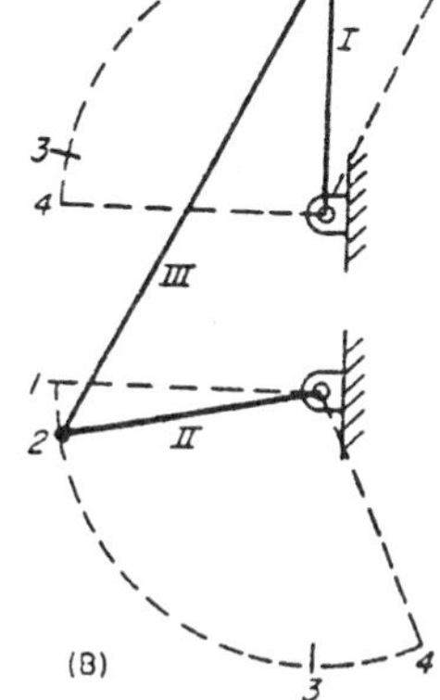

Fig. 12_Four-bar linkages can be altered to give a variable velocity ratio (or mechanical advantage). (Fig. 12A) Since the cranks *I* and *II* both come into toggle with the connecting link *III* at the same time, there is no variation in mechanical advantage. (Fig. 12B) increasing the length of link *III* gives an increased mechanical advantage between positions 1 and 2, because crank *I* and connecting link *III* are near toggle. (Fig. 12C) Placing one pivot at the left produces similar effects as in (Fig. 12B). (Fig. 12D) increasing the center distance puts crank *II* and link *III* near toggle at position 1; crank *I* and link *III* approach the toggle position at *4*.

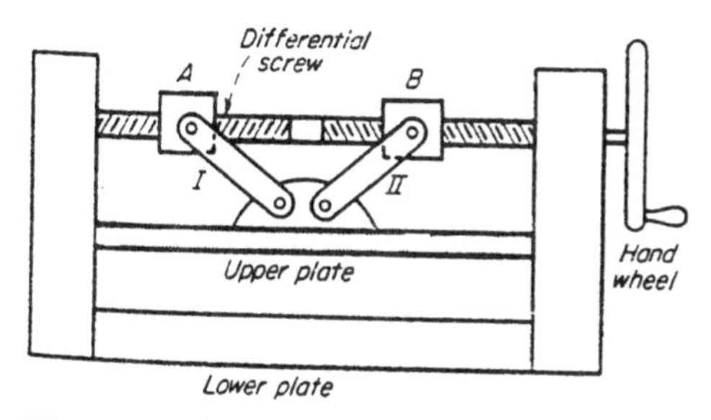

Fig. 11 A toggle press has an increasing mechanical advantage to counteract the resistance of the material being compressed. A rotating handwheel with a differential screw moves nuts *A* and *B* together, and links *I* and *II* are brought into toggle.

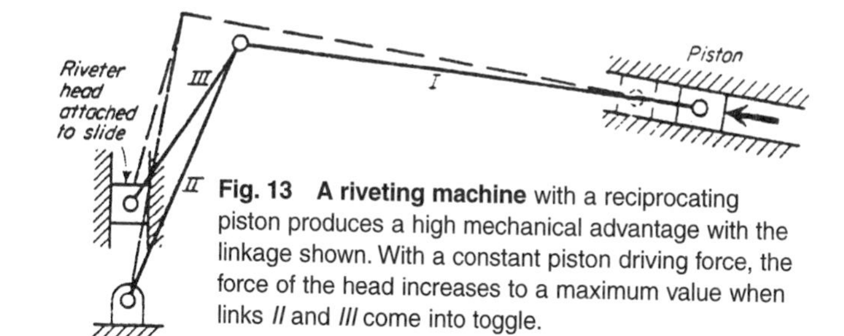

Fig. 13 A riveting machine with a reciprocating piston produces a high mechanical advantage with the linkage shown. With a constant piston driving force, the force of the head increases to a maximum value when links *II* and *III* come into toggle.

HINGED LINKS AND TORSION BUSHINGS SOFT-START DRIVES

Centrifugal force automatically draws up the linkage legs, while the torsional resistance of the bushings opposes the deflection forces.

A spidery linkage system combined with a rubber torsion bushing system formed a power-transmission coupling. Developed by a British company, Twiflex Couplings Ltd., Twickenham, England, the device (drawing below) provides ultra-soft starting characteristics. In addition to the torsion system, it also depends on centrifugal force to draw up the linkage legs automatically, thus providing additional soft coupling at high speeds to absorb and isolate any torsional vibrations arising from the prime mover.

The TL coupling has been installed to couple marine main engines to gearbox-propeller systems. Here the coupling reduces propeller vibrations to negligible proportions even at high critical speeds. Other applications are also foreseen, including their use in diesel drives, machine tools, and off-the-road construction equipment. The coupling's range is from 100 hp to 4000 rpm to 20,000 hp at 400 rpm.

Articulating links. The key factor in the TL coupling, an improvement over an earlier Twiflex design, is the circular grouping of hinged linkages connecting the driving and driven coupling flanges. The forked or tangential links have resilient precompressed bonded-rubber bushings at the outer flange attachments, while the other pivots ride on bearings.

When torque is applied to the coupling, the linkages deflect in a positive or negative direction from the neutral position (drawings, below). Deflection is opposed by the torsional resistance of the rubber bushings at the outer pins. When the coupling is rotating, the masses of the linkage give rise to centrifugal forces that further oppose coupling deflection. Therefore, the working position of the linkages depends both on the applied torque and on the speed of the coupling's rotation.

Tests of the coupling's torque/deflection characteristics under load have shown that the torsional stiffness of the coupling increases progressively with speed and with torque when deflected in the positive direction. Although the geometry of the coupling is asymmetrical the torsional characteristics are similar for both directions of drive in the normal working range. Either half of the coupling can act as the driver for either direction of rotation.

The linkage configuration permits the coupling to be tailored to meet the exact stiffness requirements of individual systems or to provide ultra-low torsional stiffness at values substantially softer than other positive-drive couplings. These characteristics enable the Twiflex coupling to perform several tasks:

- It detunes the fundamental mode of torsional vibration in a powertransmission system. The coupling is especially soft at low speeds, which permits complete detuning of the system.
- It decouples the driven machinery from engine-excited torsional vibration. In a typical geared system, the major machine modes driven by the gearboxes are not excited if the ratio of coupling stiffness to transmitted torque is less than about 7:1—a ratio easily provide by the Twiflex coupling.
- It protects the prime mover from impulsive torques generated by driven machinery. Generator short circuits and other causes of impulsive torques are frequently of sufficient duration to cause high response torques in the main shafting.

Using the example of the TL 2307G coupling design—which is suitable for 10,000 hp at 525 rpm—the torsional stiffness at working points is largely determined by coupling geometry and is, therefore, affected to a minor extent by the variations in the properties of the rubber bushings. Moreover, the coupling can provide torsional-stiffness values that are accurate within 5.0%.

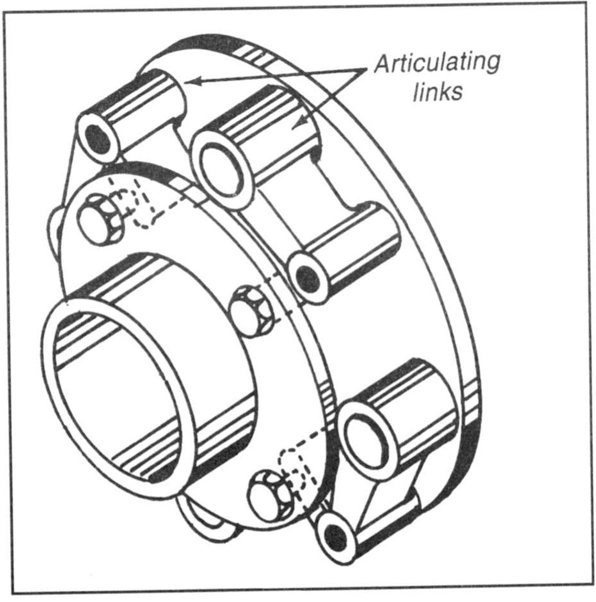

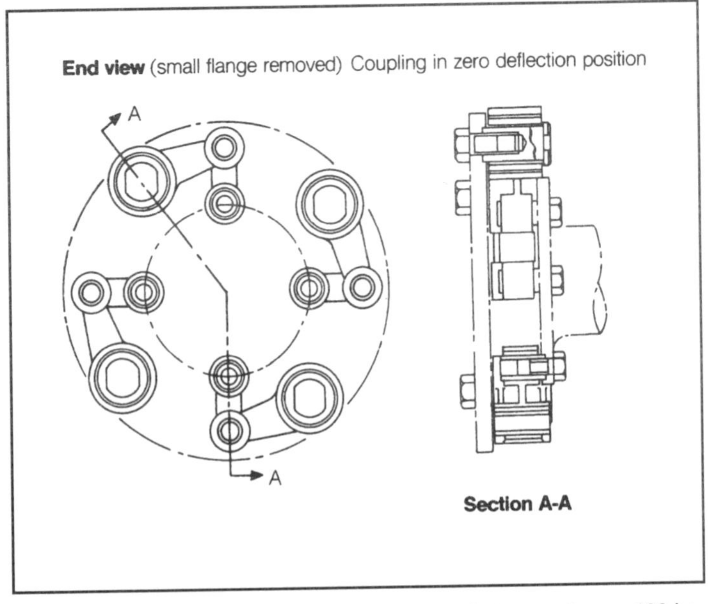

Articulating links of the new coupling (left) are arranged around the driving flanges. A four-link design (right) can handle torques from a 100-hp prime mover driving at 4000 rpm.

EIGHT LINKAGES FOR BAND CLUTCHES AND BRAKES

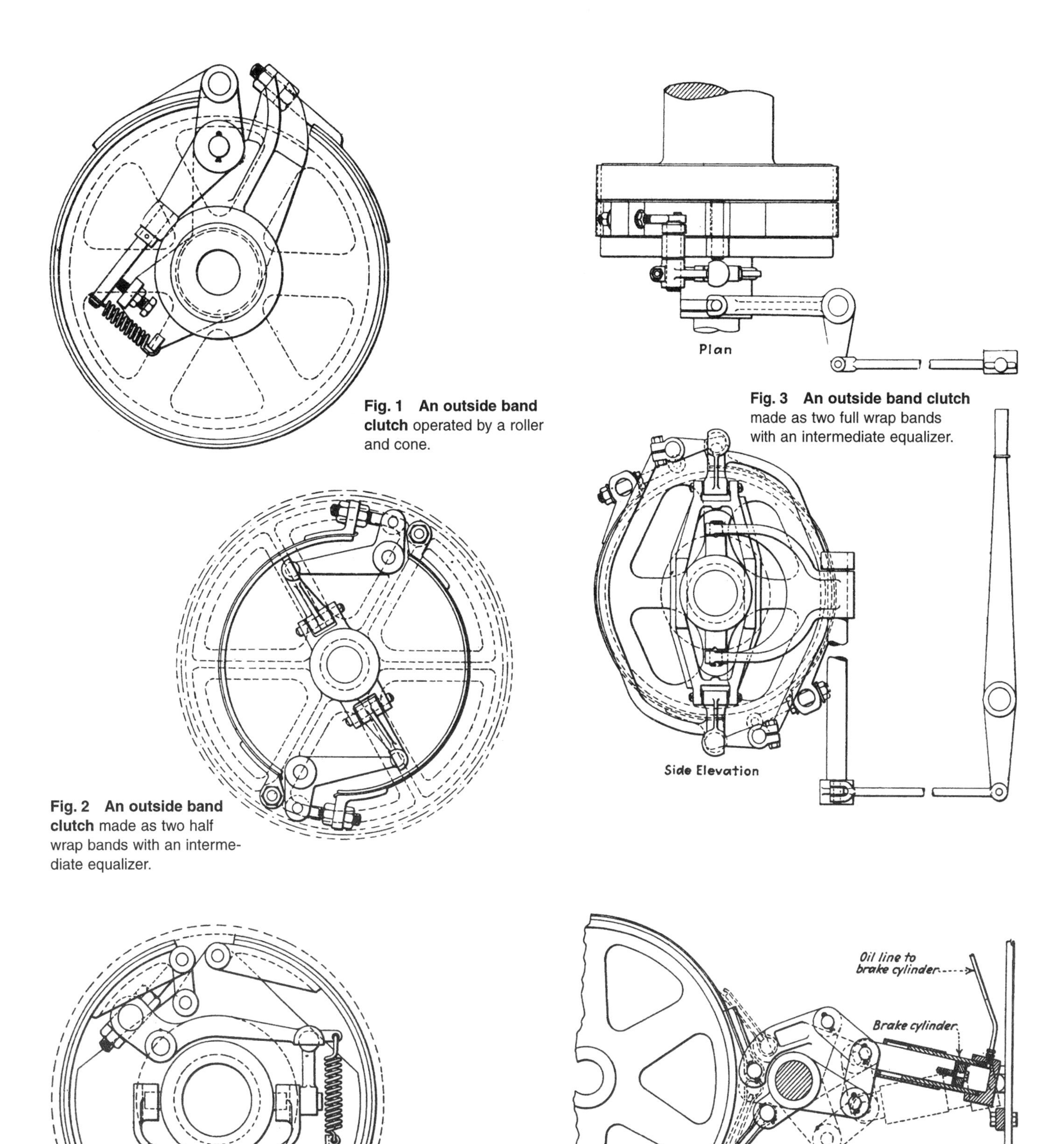

Fig. 1 An outside band clutch operated by a roller and cone.

Fig. 2 An outside band clutch made as two half wrap bands with an intermediate equalizer.

Fig. 3 An outside band clutch made as two full wrap bands with an intermediate equalizer.

Fig. 4 An inside band clutch operated by a yoke having movement along the shaft.

Fig. 5 A two-way acting band brake operated hydraulically.

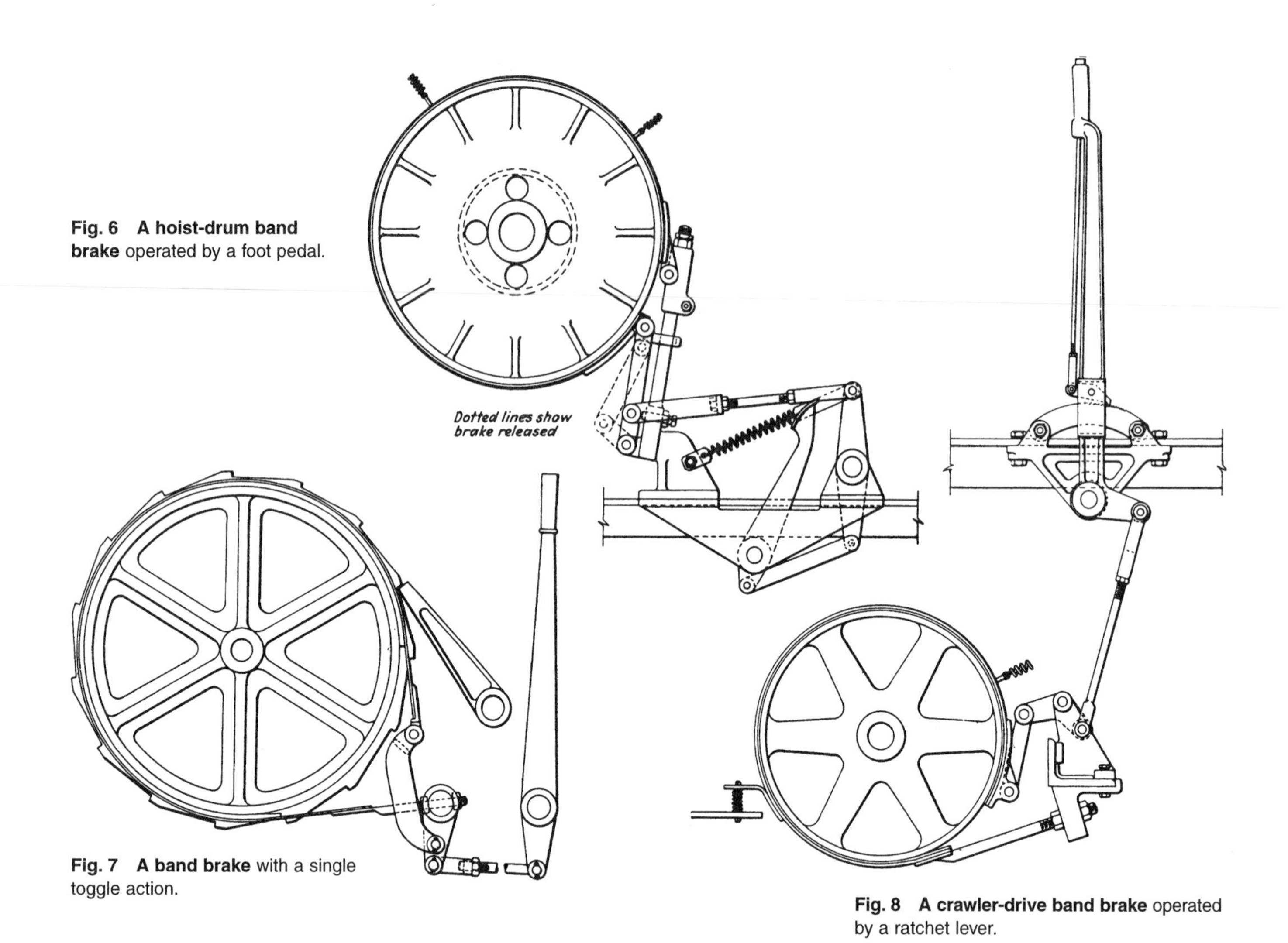

Fig. 6 A hoist-drum band brake operated by a foot pedal.

Fig. 7 A band brake with a single toggle action.

Fig. 8 A crawler-drive band brake operated by a ratchet lever.

DESIGN OF CRANK-AND-ROCKER LINKS FOR OPTIMUM FORCE TRANSMISSION

Four-bar linkages can be designed with a minimum of trial and error by a combination of tabular and iteration techniques.

The determination of optimum crank-and-rocker linkages has most effectively been performed on a computer because of the complexity of the equations and calculations involved. Thanks to the work done at Columbia University's Department of Mechanical and Nuclear Engineering, all you need now is a calculator and the computer-generated tables presented here. The computations were done by Mr. Meng-Sang Chew, at the university.

A crank-and-rocker linkage, *ABCD*, is shown in the first figure. The two extreme positions of the rocker are shown schematically in the second figure. Here ψ denotes the rocker swing angle and ϕ denotes the corresponding crank rotation, both measured counterclockwise from the extended dead-center position, AB_1C_1D.

The problem is to find the proportions of the crank-and-rocker linkage for a given rocker swing angle, ψ, a prescribed corresponding crank rotation, ϕ, and optimum force transmission. The latter is usually defined in terms of the transmission angle, m, the angle μ between coupler *BC* extended and rocker *CD*.

Considering static forces only, the closer the transmission angle is to 90°, the greater is the ratio of the driving component of the force exerted on the rocker to the component exerting bearing pressure on the rocker. The control of transmission-angle variation becomes especially important at high speeds and in heavy-duty applications.

How to find the optimum. The steps in the determination of crank-and-rocker proportions for a given rocker swing angle, corresponding crank rotation, and optimum transmission, are:

- Select (ψ, ϕ) within the following range:

  $0° < \psi < 180°$
  $(90° + 1/2\,\psi) < \phi < (270° + 1/2\,\psi)$

- Calculate: $t = \tan 1/2\,\phi$
  $u = \tan 1/2\,(\phi - \psi)$
  $v = \tan 1/2\,\psi$

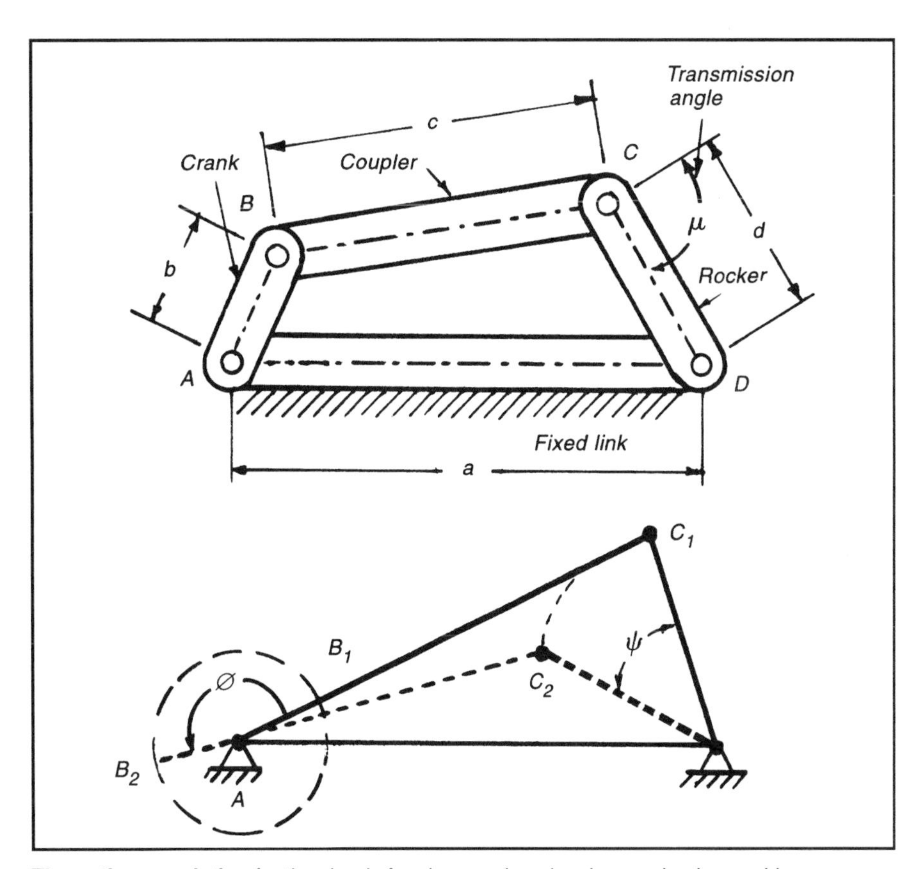

The optimum solution for the classic four-bar crank-and-rocker mechanism problem can now be obtained with only the accompanying table and a calculator.

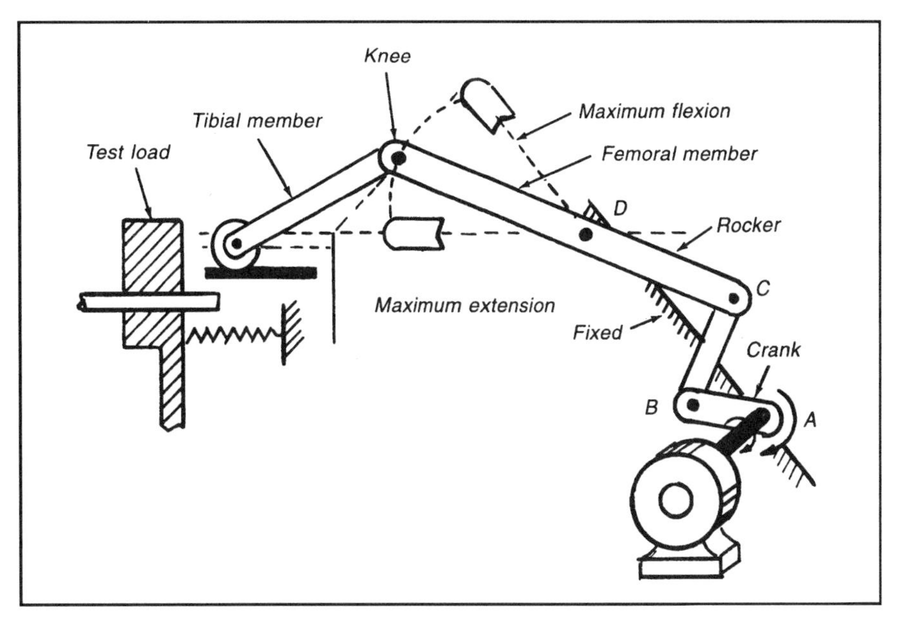

An example in this knee-joint tester designed and built by following the design and calculating procedures outlined in this article.

- Using the table, find the ratio λ_{opt} of coupler to crank length that minimizes the transmission-angle deviation from 90°. The most practical combinations of (ψ, ϕ) are included in the table. If the (ψ, ϕ) combination is not included, or if $\phi = 180°$, go to next steps (a,b,c):
- (a) If $\phi = 180°$ and (ψ, ϕ) fall outside the range given in the table, determine the arbitrary intermediate value Q from the equation: $Q^3 + 2Q^2 - t^2Q - (t^2/u^2)(1 + t^2) = 0$ where $(1/u^2 < Q < t^2)$.

This is conveniently accomplished by numerical iteration:

Set $$Q_1 = \frac{1}{2}\left(t^2 + \frac{1}{u^2}\right)$$

Calculate $Q_2, Q_3, \ldots$ from the recursion equation:

$$Q_{i+1} = \frac{2Q_i^2(Q_i + 1) + (t^2/u^2)(1 + t^2)}{Q_i(3Q_i + 4) - t^2}$$

Iterate until the ratio $[(Q_i + 1 - Q_i)/Q_i]$ is sufficiently small, so that you obtain the desired number of significant figures. Then:

$$\lambda_{opt} = t^2/Q$$

(b) If $\phi = 180°$ and the determination of λ_{opt} requires interpolation between two entries in the table, let $Q_1 = t^2\lambda^2$, where λ corresponds to the nearest entry in the table, and continue as in (a) above to determine Q and λ_{opt}. Usually one or two iterations will suffice.

(c) $\phi = 180°$. In this case, $a^2 + b^2 = c^2 + d^2$; $\psi = 2\sin^{-1}(b/d)$; and the maximum deviation, Δ, of the transmission angle from 90° is equal to $\sin^{-1}(ab/cd)$.

- Determine linkage proportions as follows:

$$(a')^2 = \frac{u^2 + \lambda_{opt}^2}{1 + u^2}$$

$$(b')^2 = \frac{v^2}{1 + v^2}$$

$$(c')^2 = \frac{\lambda_{opt}^2 v^2}{1 + v^2}$$

$$(d')^2 = \frac{t^2 + \lambda_{opt}^2}{1 + t^2}$$

Then: $a = ka'$; $b = kb'$; $c = kc'$; $d = kd'$ where k is a scale factor, such that the length of any one link, usually the crank, is equal to a design value. The

Optimum values of lambda ratio for given ø and ψ

ø deg	ψ, deg 160	162	164	166	168	170	172	174	176	178
10	2.3532	2.4743	2.6166	2.7873	2.9978	3.2669	3.6284	4.1517	5.0119	6.8642
12	2.3298	2.4491	2.5891	2.7570	2.9636	3.2272	3.5804	4.0899	4.9224	6.6967
14	2.3064	2.4239	2.5617	2.7266	2.9293	3.1874	3.5324	4.0283	4.8342	6.5367
16	2.2831	2.3988	2.5344	2.6964	2.8953	3.1479	3.4848	3.9675	4.7482	6.3853
18	2.2600	2.3740	2.5073	2.6664	2.8615	3.1089	3.4380	3.9080	4.6650	6.2427
20	2.2372	2.3494	2.4805	2.6368	2.8282	3.0704	3.3920	3.8499	4.5848	6.1087
22	2.2145	2.3250	2.4540	2.6076	2.7954	3.0327	3.3470	3.7935	4.5077	5.9826
24	2.1922	2.3010	2.4279	2.5789	2.7631	2.9956	3.3030	3.7388	4.4338	5.8641
26	2.1701	2.2773	2.4022	2.5505	2.7314	2.9594	3.2602	3.6857	4.3628	5.7524
28	2.1483	2.2539	2.3768	2.5227	2.7004	2.9239	3.2185	3.6344	4.2948	5.6469
30	2.1268	2.2309	2.3519	2.4954	2.6699	2.8893	3.1779	3.5847	4.2295	5.5472
32	2.1056	2.2082	2.3273	2.4685	2.6401	2.8554	3.1384	3.5367	4.1668	5.4526
34	2.0846	2.1858	2.3032	2.4421	2.6108	2.8223	3.0999	3.4901	4.1066	5.3628
36	2.0640	2.1637	2.2794	2.4162	2.5821	2.7899	3.0624	3.4449	4.0486	5.2773
38	2.0436	2.1420	2.2560	2.3908	2.5540	2.7583	3.0259	3.4012	3.9927	5.1957
40	2.0234	2.1205	2.2330	2.3657	2.5264	2.7274	2.9903	3.3587	3.9388	5.1177
42	2.0035	2.0994	2.2103	2.3411	2.4994	2.6971	2.9556	3.3175	3.8868	5.0430
44	1.9839	2.0785	2.1879	2.3169	2.4728	2.6675	2.9217	3.2773	3.8364	4.9712
46	1.9644	2.0579	2.1659	2.2931	2.4468	2.6384	2.8886	3.2383	3.7877	4.9023
48	1.9452	2.0375	2.1441	2.2696	2.4211	2.6100	2.8563	3.2003	3.7404	4.8358
50	1.9262	2.0174	2.1227	2.2465	2.3959	2.5820	2.8246	3.1632	3.6945	4.7717

max deviation, Δ, of the transmission angle from 90° is:

$$\sin\Delta = \frac{|(a \pm b)^2 - c^2 - d^2|}{2cd}$$

$0° \leq \Delta \leq 90°$
+ sign if $\phi < 180°$
– sign if $\phi < 180°$

An actual example. A simulator for testing artificial knee joints, built by the Department of Orthopedic Surgery, Columbia University, under the direction of Dr. N. Eftekhar, is shown schematically. The drive includes an adjustable crank-and-rocker, ABCD. The rocker swing angle ranges from a maximum of about 48° to a minimum of about one-third of this value. The crank is 4 in. long and rotates at 150 rpm. The swing angle adjustment is obtained by changing the length of the crank.

Find the proportions of the linkage, assuming optimum-transmission proportions for the maximum rocker swing angle, as this represents the most severe condition. For smaller swing angles, the maximum transmission-angle deviation from 90° will be less.

Crank rotation corresponding to 48° rocker swing is selected at approximately 170°. Using the table, find $\lambda_{opt} = 2.6100$. This gives $a' = 1.5382$, $b' = 0.40674$, $c' = 1.0616$, and $d' = 1.0218$.

For a 4 in. crank, $k = 4/0.40674 = 9.8343$ and $a = 15.127$ in., $b = 4$ in., $c = 10.440$ in., and $d = 10.049$ in., which is very close to the proportions used. The maximum deviation of the transmission angle from 90° is 47.98°.

This procedure applies not only for the transmission optimization of crank-and-rocker linkages, but also for other crank-and-rocker design. For example, if only the rocker swing angle and the corresponding crank rotation are prescribed, the ratio of coupler to crank length is arbitrary, and the equations can be used with any value of λ_2 within the range $(1, u^2t^2)$. The ratio λ can then be tailored to suit a variety of design requirements, such as size, bearing reactions, transmission-angle control, or combinations of these requirements.

The method also was used to design dead-center linkages for aircraft landing-gear retraction systems, and it can be applied to any four-bar linkage designs that meet the requirements discussed here.

Optimum values of lambda ratio for given ø and ψ

ø deg	ψ, deg 182	184	186	188	190	192	194	196	198	200
10	7.2086	5.3403	4.4560	3.9112	3.5318	3.2478	3.0245	2.8428	2.6911	2.5616
12	7.0369	5.2692	4.4227	3.8969	3.5282	3.2507	3.0317	2.8528	2.7030	2.5748
14	6.8646	5.1881	4.3795	3.8739	3.5174	3.2478	3.0341	2.8589	2.7117	2.5855
16	6.6971	5.1013	4.3287	3.8435	3.5000	3.2392	3.0317	2.8610	2.7171	2.5934
18	6.5371	5.0121	4.2726	3.8071	3.4768	3.2252	3.0245	2.8589	2.7189	2.5982
20	6.3857	4.9226	4.2131	3.7663	3.4487	3.2065	3.0129	2.8528	2.7171	2.5998
22	6.2431	4.8344	4.1518	3.7221	3.4167	3.1837	2.9972	2.8428	2.7117	2.5982
24	6.1090	4.7484	4.0900	3.6759	3.3818	3.1575	2.9780	2.8293	2.7030	2.5934
26	5.9830	4.6652	4.0284	3.6284	3.3447	3.1286	2.9558	2.8127	2.6911	2.5855
28	5.8644	4.5849	3.9676	3.5804	3.3062	3.0976	2.9311	2.7833	2.6763	2.5748
30	5.7527	4.5079	3.9080	3.5324	3.2669	3.0652	2.9045	2.7718	2.6592	2.5616
32	5.6472	4.4339	3.8500	3.4849	3.2272	3.0318	2.8764	2.7484	2.6399	2.5461
34	5.5475	4.3630	3.7936	3.4380	3.1875	2.9979	2.8473	2.7236	2.6190	2.5287
36	5.4529	4.2949	3.7388	3.3920	3.1480	2.9636	2.8175	2.6977	2.5967	2.5097
38	5.3631	4.2296	3.6858	3.3470	3.1089	2.9294	2.7873	2.6711	2.5734	2.4894
40	5.2776	4.1669	3.6345	3.3031	3.0705	2.8953	2.7570	2.6440	2.5492	2.4680
42	5.1960	4.1067	3.5848	3.2602	3.0327	2.8615	2.7266	2.6166	2.5246	2.4459
44	5.1180	4.0487	3.5367	3.2185	2.9956	2.8282	2.6964	2.5891	2.4996	2.4232
46	5.0432	3.9928	3.4901	3.1779	2.9594	2.7954	2.6665	2.5617	2.4744	2.4001
48	4.9715	3.9389	3.4450	3.1384	2.9239	2.7631	2.6369	2.5344	2.4491	2.3767
50	4.9025	3.8869	3.4012	3.0999	2.8893	2.7314	2.6076	2.5073	2.4239	2.3533

DESIGN OF FOUR-BAR LINKAGES FOR ANGULAR MOTION

How to use four-bar linkages to generate continuous or intermittent angular motions required by feeder mechanisms

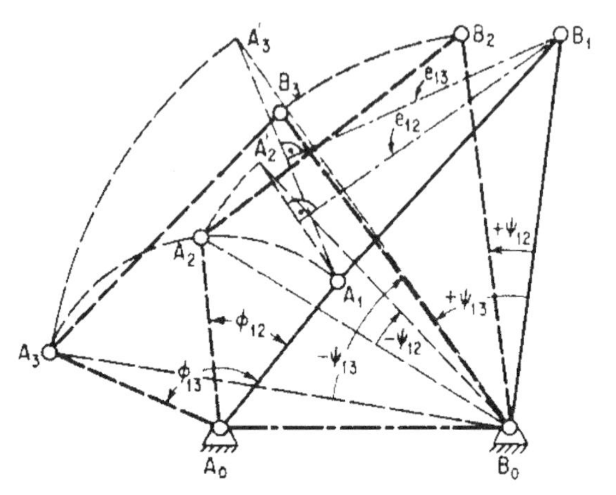

Fig. 1 Four-bar linkage synchronizes two angular movements, ϕ_{12} and ϕ_{13}, with ψ_{12}, and ψ_{14}.

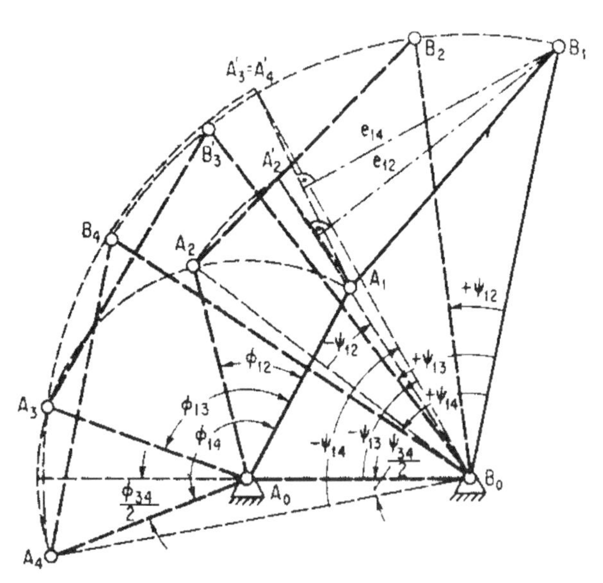

Fig. 2 Three angular positions, ϕ_{12}, ϕ_{13}, ϕ_{14}, are synchronized by four-bar linkage here with ψ_{12}, ψ_{13}, and ψ_{14}.

In putting feeder mechanisms to work, it is often necessary to synchronize two sets of angular motions. A four-bar linkage offers one way. For example, in Fig. 1 two angular motions, ϕ_{12} and ϕ_{13}, must be synchronized with two others, ψ_{12} and ψ_{13}, about the given pivot points A_0 and B_0 and the given crank length A_0A. This means that crank length B_0B. must be long enough so that the resulting four-bar linkage will coordinate angular motions ϕ_{12} and $\phi 13$ with ψ_{12} and ψ_{13}. The procedure is:

1. Obtain point A'_2 by revolving A_2 about B_0 through angle—ψ_{12} but in the opposite direction.
2. Obtain point A'_3 similarly by revolving A_3 about B_0 through angle – ψ_{13}.
3. Draw lines $A_1A'_2$ and $A_1A'_3$ and the perpendicular bisectors of the lines which intersect at desired point B_1.
4. The quadrilateral $A_0A_1B_1B_0$ represents the four-bar linkage that will produce the required relationship between the angles ϕ_{12}, ϕ_{13}, and ψ_{12}, ψ_{13}.

Three angles with four relative positions can be synchronized in a similar way. Figure 2 shows how to synchronize angles ψ_{12}, ψ_{13}, ψ_{14} with corresponding angles ψ_{12}, ψ_{13}, and ψ_{14}, using freely chosen pivot points A_0 and B_0. In this case, crank length A_0A as well as B_0B is to be determined, and the procedure is:

1. Locate pivot points A_0 and B_0 on a line those bisects angle A_3A_0A4, the length A_0B_0 being arbitrary.
2. Measure off $^1/_2$ of angle $B_3B_0B_4$ and with this angle draw B_0A_4 which establishes crank length A_0A at intersection of A_0A_4. This also establishes points A_3, A_2 and A_1.
3. With B_0 as center and B_0B_4 as radius mark off angles $-\psi_{14}$, $-\psi_{13}$, $-\psi_{12}$, the negative sign indicating they are in opposite sense to ψ_{14}, ψ_{13} and ψ_{12}. This establishes points A'_2, A'_3 and A'_4, but here A'_3 and A'_4 coincide because of symmetry of A_3 and A_4 about A_0B_0.
4. Draw lines $A_1A'_2$ and $A_1A'_4$, and the perpendicular bisectors of these lines, which intersect at the desired point B_1.
5. The quadrilateral $A_0A_1B_0B_1$ represents the four-bar linkage that will produce the required relationship between the angles ϕ_{12}, ϕ_{13}, ϕ_{14}, and ψ_{12}, ψ_{13}, ψ_{14}.

The illustrations show how these angles must be coordinated within the given space. In Fig. 3A, input angles of the crank must be coordinated with the output angles of the forked escapement. In Fig. 3B, input angles of the crank are coordinated with the output angles of the tilting hopper. In Fig. 3C, the input angles of the crank are coordinated with the output angles of the segment. In Fig. 3D, a box on a conveyor is tilted 90° by an output crank, which is actuated by an input crank through a coupler. Other mechanisms shown can also coordinate the input and output angles; some have dwell periods between the cycles, others give a linear output with dwell periods.

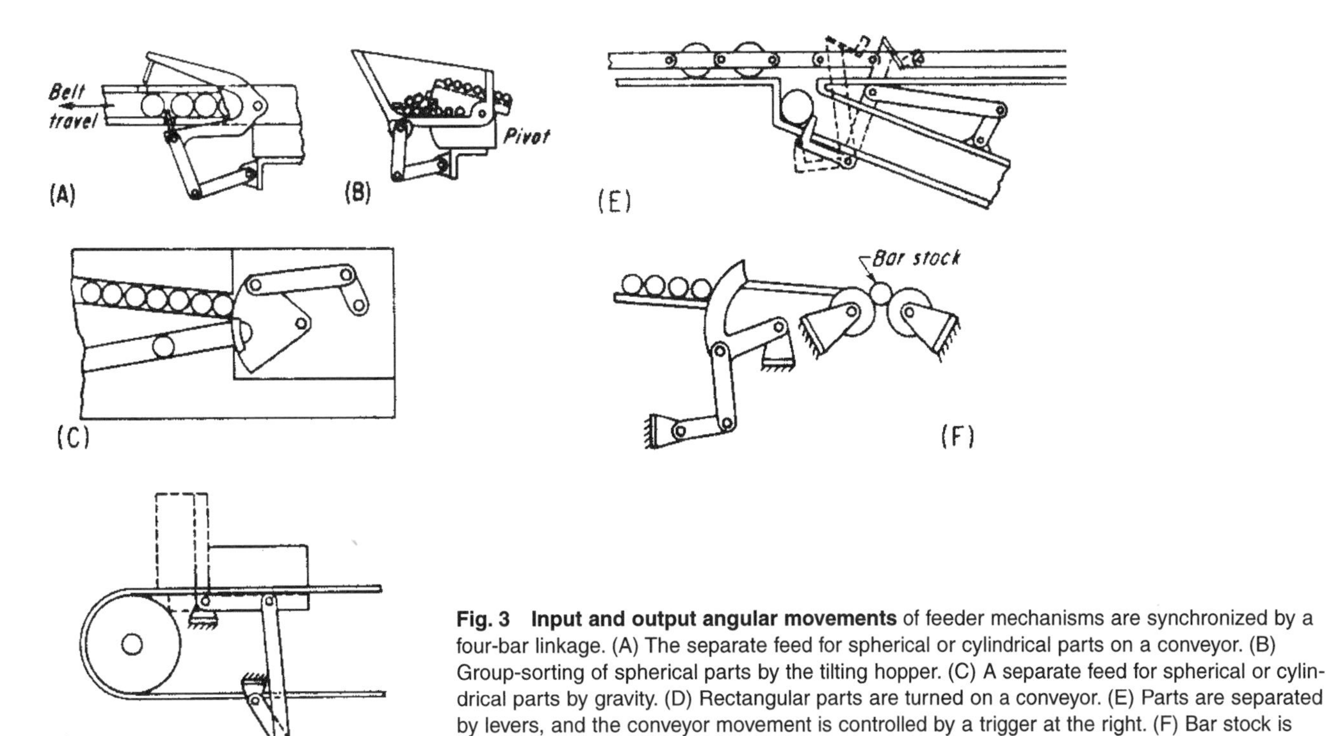

Fig. 3 Input and output angular movements of feeder mechanisms are synchronized by a four-bar linkage. (A) The separate feed for spherical or cylindrical parts on a conveyor. (B) Group-sorting of spherical parts by the tilting hopper. (C) A separate feed for spherical or cylindrical parts by gravity. (D) Rectangular parts are turned on a conveyor. (E) Parts are separated by levers, and the conveyor movement is controlled by a trigger at the right. (F) Bar stock is positioned by the angular oscillation of an output lever when the input crank is actuated.

MULTIBAR LINKAGES FOR CURVILINEAR MOTIONS

Four-bar linkages can be combined into six, eight, or more linkages for the feeder mechanisms in cameras, automatic lathes, farm machinery, and torch-cutting machines.

When feeder mechanisms require complex curvilinear motions, it might be necessary to use compound linkages rather than four links. However, four-bar linkages can be synthesized to produce curvilinear motions of various degrees of complexity, and all possibilities for four-bar linkages should be considered before selecting more complex linkages.

For example, a camera film-advancing mechanism, Fig. 1, has a simple four-bar linage with a coupler point d, which generates a curvilinear and straight-line motion a resembling a D. Another more complex curvilinear motion, Fig. 2, is also generated by a coupler point E of a four-bar linkage, which controls an automatic profile cutter. Four-bar linkages can generate many different curvilinear motions, as in Fig. 3. Here the points of the coupler prongs, g_1, g_2, and g_3 on coupler b, and g_4 and g_5 on coupler e, are chosen so that their motions result in the desired progressive feeding of straw into a press.

A similar feeding and elevating device is shown in Fig. 4. The rotating device crank a moves coupler b and swinging lever c, which actuates the guiding arm f through the link e. The bar h carries the prone fingers g_1 through g_7. They generate coupler curves a_1 through a_7.

As another practical example, consider the torch-cutting machine in Fig. 5A designed to cut sheet metal along a curvilinear path a. Here the points A_0 and B_0 are fixed in the machine, and the lever A_0A_1 has an adjustable length to suit the different curvilinear paths a desired.

The length B_1B_1 is also fixed. The challenge is to find the length of the levers A_1B_1 and E_1B_1 in the four-bar linkage to give the desired path a, which is to be traced by the coupler point E on which the cutting torch is mounted.

The graphical solution for this problem, as shown in Fig. 5B, requires the selection of the points A_1 and E_4 so that the distances A_1E_1 to A_8E_8 are equal and the points E_1 to E_8 lie on the desired coupler curved a. In this case, only the points E_4 to E_8 represent the desired profile to be cut. The correct selection of points A_1 and E_1 depends upon making the following triangles congruent:

$$\Delta E_2A_2B_{01} = \Delta E_1A_1B_{02}$$
$$\Delta E_3A_3B_{01} = \Delta E_1A_1B_{03}$$
$$\Delta E_8A_8B_{01} = \Delta E_1A_1B_{08}$$

and so on until $E_8A_8B_{01} = E_1A_1B_{08}$. At the same time, all points A_1 to A_8 must lie on the arc having B_1 as center.

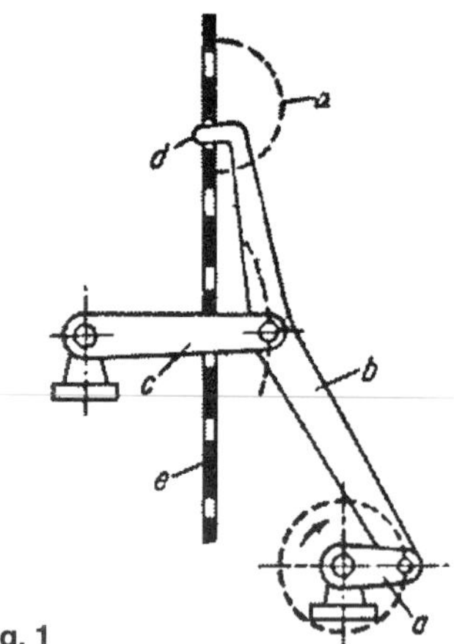

Fig. 1

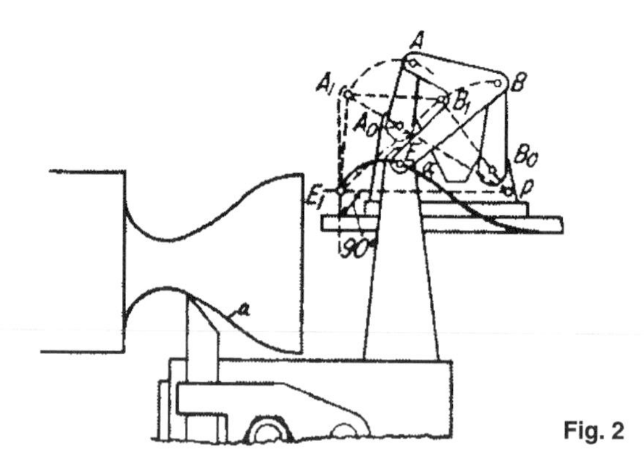

Fig. 2

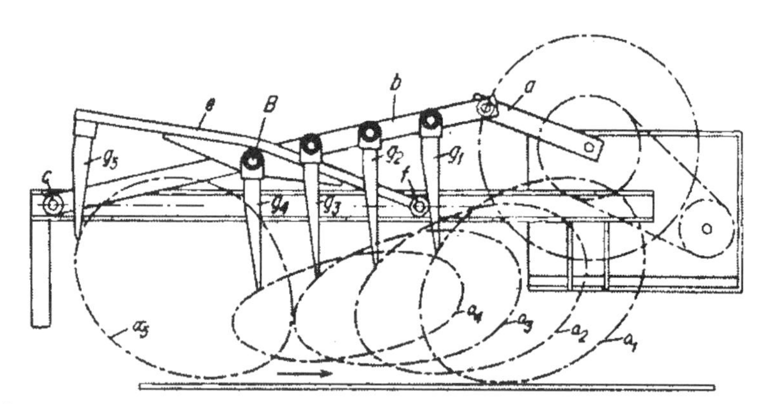

Fig. 3

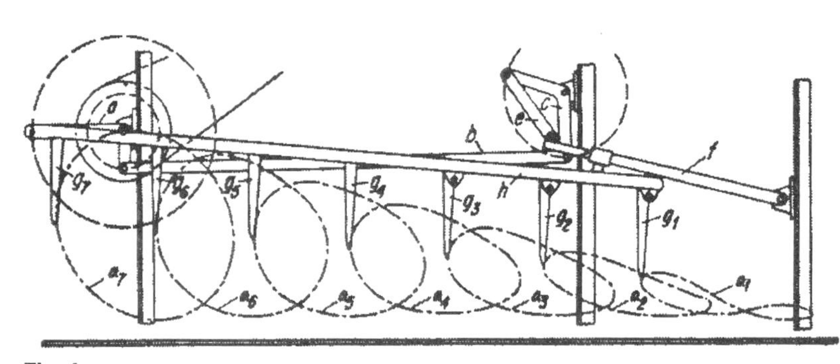

Fig. 4

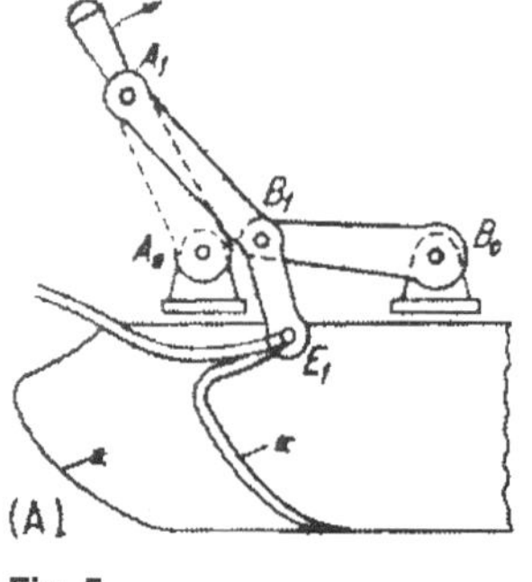

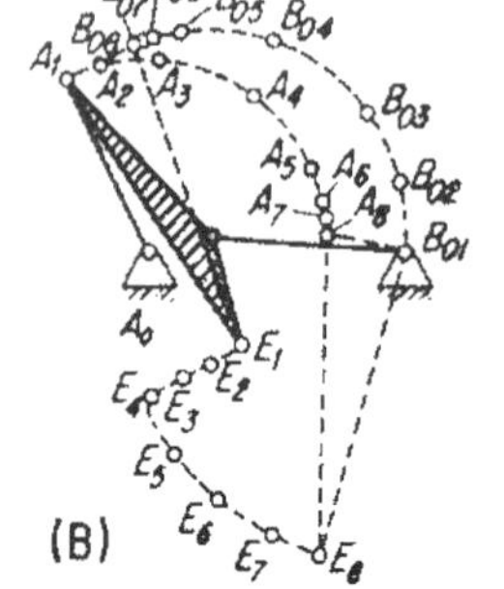

Fig. 5

Synthesis of an Eight-Bar Linkage

Design a linkage with eight precision points, as shown in Fig. 6. In this mechanism the curvilinear motion of one four-bar linkage is coordinated with the angular oscillation of a second four-bar linkage. The first four-bar linkage consists of $A\,A_0\,BB_0$ with coupler point E which generates γ with eight precision points E_1 through E_8 and drives a second four-bar linkage HH_0GG_0. Coupler point F generates curve δ with precision points F_1 through F_8. The coupler points F_2, F_4, F_6, F_8 are coincident because straight links GG_0 and GH are in line with one another in these coupler positions. This is what permits HH_0 to oscillate, despite the continuous motion of the coupler point F. The coupler points F_1 coincident with F_5, and F_3 coincident with F_7, have been chosen so that F_1 is the center of a circle k_1 and F_3 is the center of a circle k_3. These circles are tangent to coupler curve γ at E_1, and E_5, E_3, and E_7, and they indicate the limiting positions of the second four-bar linkage HH_0GG_0.

The limiting angular oscillation of HH_0, which is one of the requirements of this mechanism, is represented by positions H_0H_1 and H_0H_3. It oscillates four times for each revolution of the input crank AA_0, and the positions H_1 to H_8 correspond to input crank positions A_1 to A_8.

The synthesis of a compound linkage with dwell periods and coordinated intermittent motion is shown in Fig. 7. The four-bar linkage AA_0BB_0 generates an approximately triangular curve with coupler point E, which has six precision points E_1 through E_6. A linkage that will do this is not unusual and can be readily proportioned from known methods of four-bar linkage synthesis. However, the linkage incorporates dwell periods that produce coordinated intermittent motion with a second four-bar linkage FF_0HB_0. Here the tangent arcs k_{12}, k_{34} and k_{56} are drawn with EF as the radius from centers F_{12}, F_{34} and F_{56}.

These centers establish the circle with F_0 as the center and pivot point for the second four-bar linkage. Each tangent arc causes a dwell of the link FF_0, while AA_0 rotates continuously. Thus, the link FF_0, with three rest periods in one revolution, can produce intermittent curvilinear motion in the second four-bar linkage FF_0HB_0. In laying out the center, F_0 must be selected so that the angle EFF_0 deviates only slightly from 90° because this will minimize the required torque that is to be applied at E. The length of B_0H can be customized, and the rest periods at H_{34}, H_{12} and H_{56} will correspond to the crank angles ϕ_{34}, ϕ_{12} and ϕ_{56}.

A compound linkage can also produce a 360° oscillating motion with a dwell period, as in Fig. 8. The two four-bar linkages are AA_0BB_0 and BB_0FF_0, and the output coupler curve γ is traversed only through segment E_1, E_2. The oscillating motion is produced by lever HH_0, connected to the coupler point by EH. The fixed point H_0 is located within the loop of the coupler curve γ. The dwell occurs at point H_3, which is the center of circular arc k tangent to the coupler curve γ during the desired dwell period. In this example, the dwell is made to occur in the middle of the 360° oscillation. The coincident positions H_1 and H_2 indicate the limiting positions of the link HH_0, and they correspond to the positions E_1 and E_2 of the coupler point.

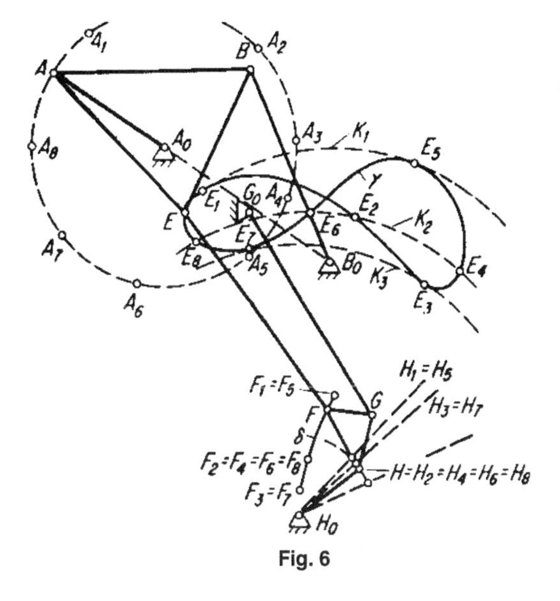

Fig. 6

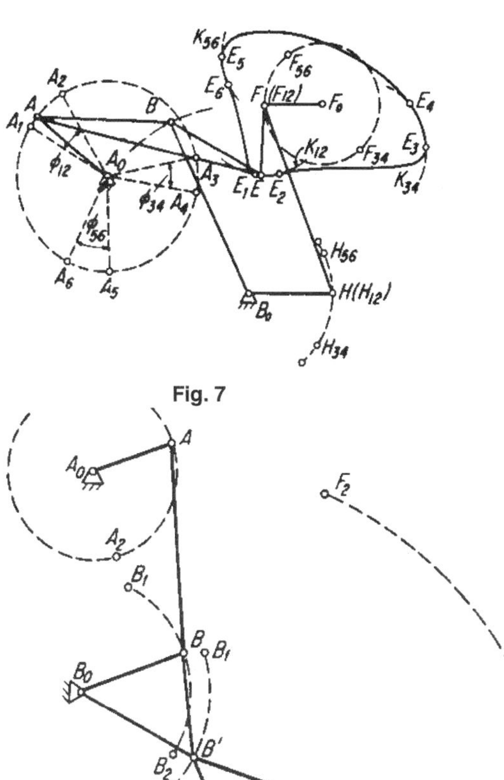

Fig. 7

Fig. 8

ROBERTS' LAW HELPS TO DESIGN ALTERNATE FOUR-BAR LINKAGES

The three linkage examples

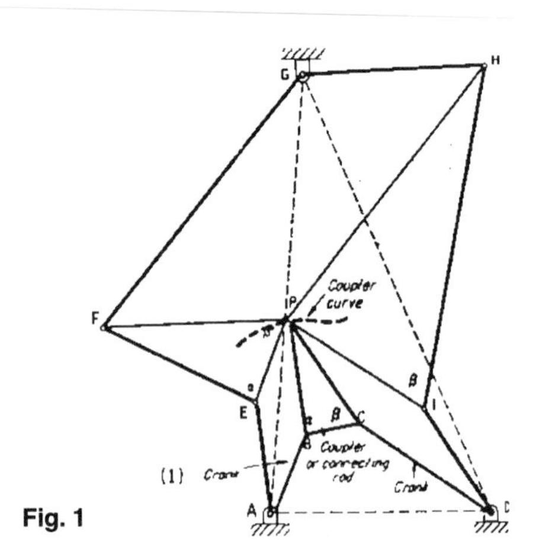

Fig. 1

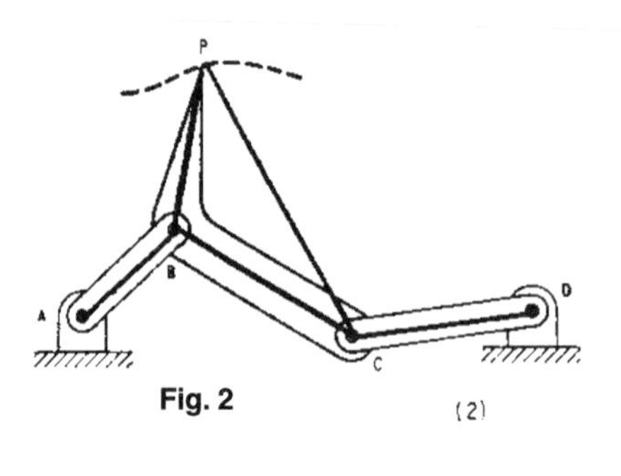

Fig. 2

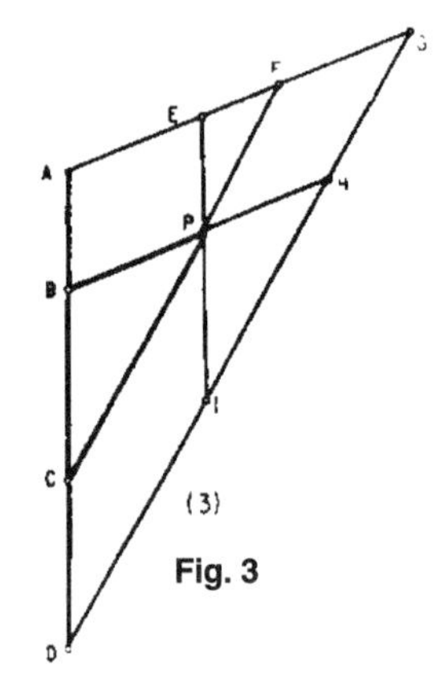

Fig. 3

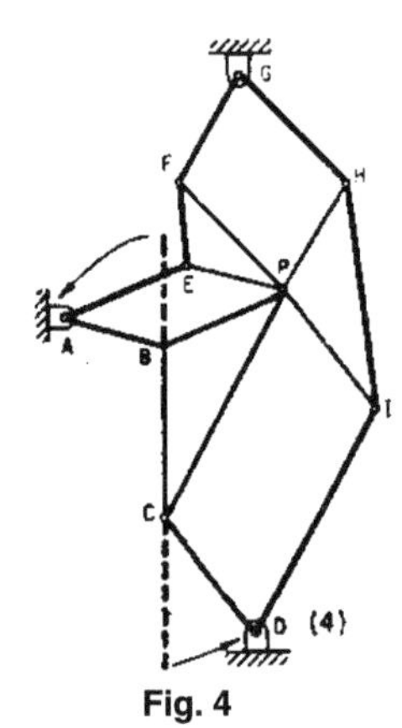

Fig. 4

When a four-bar linkage has been designed or selected from a catalog to produce a desired coupler curve, it is often found that one of the pivot points is inconveniently located or that the transmission angles are not suitable. (A coupler curve is produced by a point on the connecting rod joining the two cranks of the four-bar linkage). According to *Roberts' Law* there are at least two other four-bar linkages that will generate the same coupler curve. One of these linkages might be more suitable for the application.

Robert's Law states that the two alternate linkages are related to the first by a series of similar triangles. This leads to graphical solutions; three examples are shown. The first involves similar triangles, the second is a more convenient step-by-step method, and the third illustrates the solution of a special case where the coupler point lies along the connecting rod.

Method of Similar Triangles

Four-bar linkage *ABCD* in Fig. 1 uses point *P*, which is actually an extension of the connecting rod *BC*, to produce desired curve. Point *E* is found by constructing *EP* parallel to *AB*, and *EA* parallel to *PB*. Then triangle *EFP* is constructed similar to triangle *BPC*. This calls for laying out angles a and β.

Point *H* is found in a similar way, and point *G* is located by drawing *GH* parallel to *FP* and *GF* parallel to *HP*.

The two alternate linkages to *ABCD* are *GFEA* and *GHID*. All use point *P* to produce the desired curve, and given any one of the three, the other two can be determined.

The Step-by-Step Method

With the similar-triangle method just described, slight errors in constructing the proper angles lead to large errors in link dimensions. The construction of angles can be avoided by laying off the link lengths along a straight line.

Thus, linkage *ABCD* in Fig. 2 is laid off as a straight line from *A* to *D*in Fig. 3. Included in the transfers is point *P*. Points *EFGHI* are quickly found by either extending the original lines or constructing parallel lines. Fig. 3, which now has all the correct dimensions of all the links, is placed under a sheet of tracing paper and, with the aid of a compass, links *AB* and *CD* are rotated (see Fig. 4) so that linkage *ABCD* is identical to that in Fig. 2. Links *PEF* and *PHI* are rotated parallel to *AB* and *CD*, respectively. Completion of the parallelogram gives the two alternate linkages, *AEFG* and *GHID*.

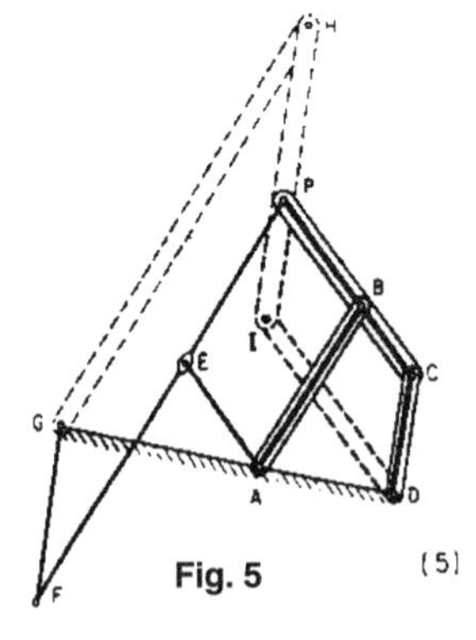

Fig. 5

Fig. 1 The method of similar triangles.
Fig. 2, 3, 4 A step-by-step method.
Fig. 5 This special case shows the simplicity of applying Roberts' Law.

Special Case

It is not uncommon for the coupler point *P* to lie on a line through *BC*, as in Fig. 5. Links *EA*, *EP* and *ID* can be found quickly by constructing the appropriate parallel lines. Point *G* is located by using the proportion: *CB*:*BP* = *DA*:*AG*. Points *H* and *F* are then located by drawing lines parallel to *AB* and *CD*.

SLIDER-CRANK MECHANISM

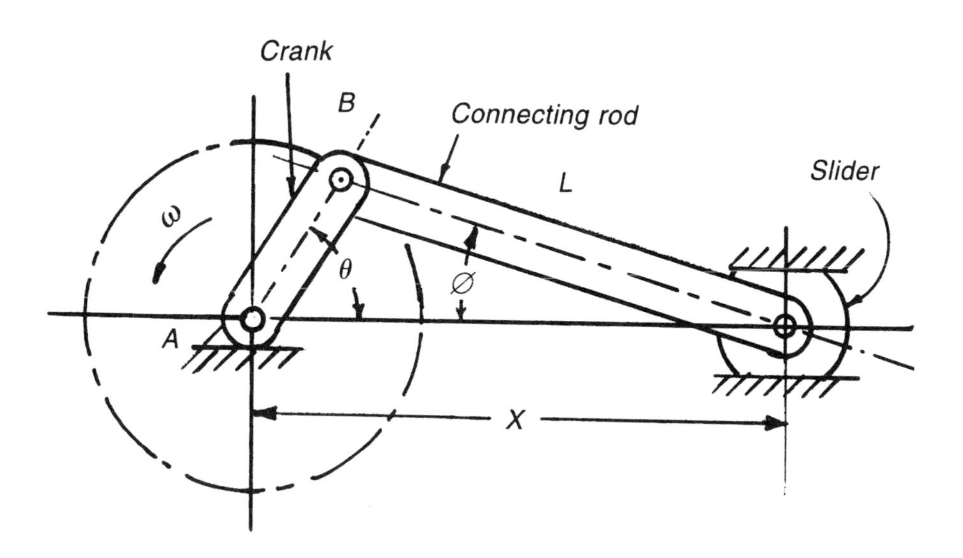

The slider crank, an efficient mechanism for changing reciprocating motion to rotary motion, is widely used in engines, pumps, automatic machinery, and machine tools.

The equations developed here for finding these factors are in a more simplified form than is generally found in text books.

SYMBOLS

L = length of connecting rod
R = crank length; radius of crank circle
x = distance from center of crankshaft A to wrist pin C
x' = slider velocity (linear velocity of point C)
x'' = slider acceleration
θ = crank angle measured from dead center (when slider is fully extended)
ϕ = angular position of connecting rod; $\phi = 0$ when $\theta = 0$
ϕ' = connecting-rod angular velocity = $d\phi/dt$
ϕ'' = connecting-rod angular acceleration = $d^2\phi/dl^2$
ω = constant crank angle velocity

Displacement of slider

$$x = L\cos\phi + R\cos\phi$$

Also:

$$\cos\phi = \left[1 - \left(\frac{R}{L}\right)^2 \sin^2\theta\right]^{1/2}$$

Angular velocity of the connecting rod

$$\phi' = \omega\left[\frac{(R/L)\cos\theta}{[1 - (R/L)^2 \sin{}^2\theta]^{1/2}}\right]$$

Linear velocity of the piston

$$\frac{x'}{L} = -\omega\left[1 + \frac{\phi'}{\omega}\right]\left(\frac{R}{L}\right)\sin\theta$$

Angular acceleration of the connecting rod

$$\phi'' = \frac{\omega^2(R/L)\sin\theta[(R/L)^2 - 1]}{[1 - (R/L)^2 \sin{}^2\theta]^{3/2}}$$

Slider acceleration

$$\frac{x''}{L} = -\omega^2\left(\frac{R}{L}\right)\left[\cos\theta + \frac{\phi''}{\omega^2}\sin\theta + \frac{\phi'}{\omega}\cos\theta\right]$$

CHAPTER 6

GEARS: DEVICES, DRIVES, AND MECHANISMS

GEARS AND ECCENTRIC DISK PROVIDE QUICK INDEXING

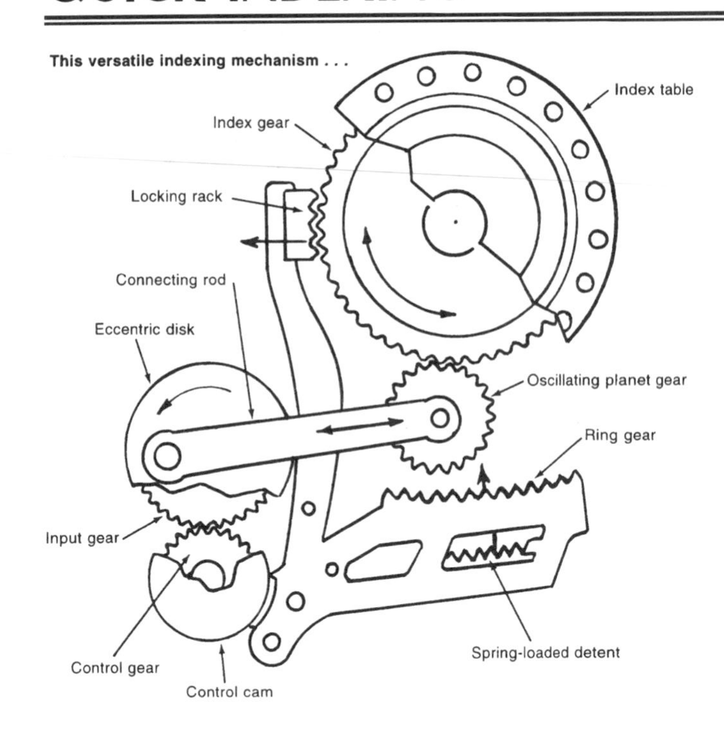

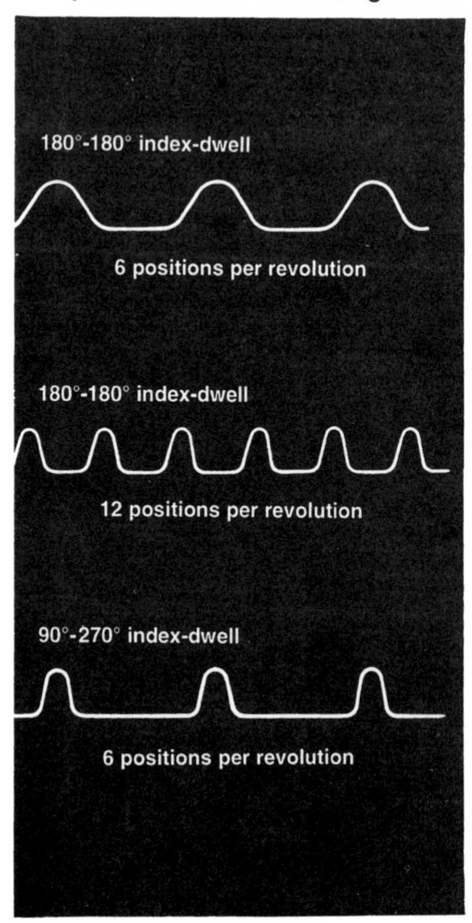

Both stops and dwell are adjustable.

An ingenious intermittent mechanism with its multiple gears, gear racks, and levers provides smoothness and flexibility in converting constant rotary motion into a start-and-stop type of indexing.

It works equally well for high-speed operations, as fast as 2 seconds per cycle, including index and dwell, or for slows-peed assembly functions.

The mechanism minimizes shock loads and offers more versatility than the indexing cams and genevas usually employed to convert rotary motion into start-stop indexing. The number of stations (stops) per revolution of the table can easily be changed, as can the period of dwell during each stop.

Advantages. This flexibility broadens the scope of such automatic machine operations as feeding, sorting, packaging, and weighing that the rotary table can perform. But the design offers other advantages, too:

- Gears instead of cams make the mechanism cheaper to manufacture, because gears are simpler to machine.
- The all-mechanical interlocked system achieves an absolute time relationship between motions.
- Gearing is arranged so that the machine automatically goes into a dwell when it is overloaded, preventing damage during jam-ups.
- Its built-in anti-backlash gear system averts rebound effects, play, and lost motion during stops.

How it works. Input from a single motor drives an eccentric disk and connecting rod. In the position shown in the drawing, the indexing gear and table are locked by the rack—the planet gear rides freely across the index gear without imparting any motion to it. Indexing of the table to its next position begins when the control cam simultaneously releases the locking rack from the index gear and causes the spring control ring gear to pivot into mesh with the planet.

This is a planetary gear system containing a stationary ring gear, a driving planet gear, and a "sun" index gear. As the crank keeps moving to the right, it begins to accelerate the index gear with harmonic motion—a desirable type of motion because of its low acceleration-deceleration characteristics—while it is imparting high-speed transfer to the table.

At the end of 180° rotation of the crank, the control cam pivots the ring-gear segment out of mesh and, simultaneously, engages the locking rack. As the connecting rod is drawn back, the planet gear rotates freely over the index gear, which is locked in place.

The cam control is so synchronized that all toothed elements are in full engagement briefly when the crank arm is in full toggle at both the beginning and end of index. The device can be operated just as easily in the other direction.

Overload protection. The ring gear segment includes a spring-load detent mechanism (simplified in the illustration) that will hold the gearing in full engagement under normal indexing forces. If rotation of the table is blocked at any point in index, the detent spring force is overcome and the ring gear pops out of engagement with the planet gear.

A detent roller (not shown) will then snap into a second detent position, which will keep the ring gear free during the remainder of the index portion of the cycle. After that, the detent will automatically reset itself.

Incomplete indexing is detected by an electrical system that stops the machine at the end of the index cycle.

Easy change of settings. To change indexes for a new job setup, the eccentric is simply replaced with one heaving a different crank radius, which gives the proper drive stroke for 6, 8, 12, 16, 24, 32, or 96 positions per table rotation.

Because indexing occurs during one-half revolution of the eccentric disk, the input gear must rotate at two or three times per cycle to accomplish indexing of $^1/_2$, $^1/_4$, or $^1/_{16}$ of the total cycle time (which is the equivalent to index-to-dwell cycles of 180/180°, 90/270° or 60/300°). To change the cycle time, it is only necessary to mount a difference set of change gears between input gear and control cam gear.

ODD-SHAPED PLANETARY GEARS SMOOTH STOP AND GO

This intermittent-motion mechanism for automatic processing machinery combines gears with lobes; some pitch curves are circular and some are noncircular.

This intermittent-motion mechanism combines circular gears with noncircular gears in a planetary arrangement, as shown in the drawing.

The mechanism was developed by Ferdinand Freudenstein, a professor of mechanical engineering at Columbia University. Continuous rotation applied to the input shaft produces a smooth, stop-and-go unidirectional rotation in the output shaft, even at high speeds.

This jar-free intermittent motion is sought in machines designed for packaging, production, automatic transfer, and processing.

Varying differential. The basis for Freudenstein's invention is the varying differential motion obtained between two sets of gears. One set has lobular pitch circles whose curves are partly circular and partly noncircular.

The circular portions of the pitch curves cooperate with the remainder of the mechanism to provide a dwell time or stationary phase, or phases, for the output member. The non-circular portions act with the remainder of the mechanism to provide a motion phase, or phases, for the output member.

Competing genevas. The main competitors to Freudenstein's "pulsating planetary" mechanism are external genevas and starwheels. These devices have a number of limitations that include:

- Need for a means, separate from the driving pin, for locking the output member during the dwell phase of the motion. Moreover, accurate manufacture and careful design are required to make a smooth transition from rest to motion and vice versa.
- Kinematic characteristics in the geneva that are not favorable for high-speed operation, except when the number of stations (i.e., the number of slots in the output member) is large. For example, there is a sudden change of acceleration of the output member at the beginning and end of each indexing operation.
- Relatively little flexibility in the design of the geneva mechanism. One factor alone (the number of slots

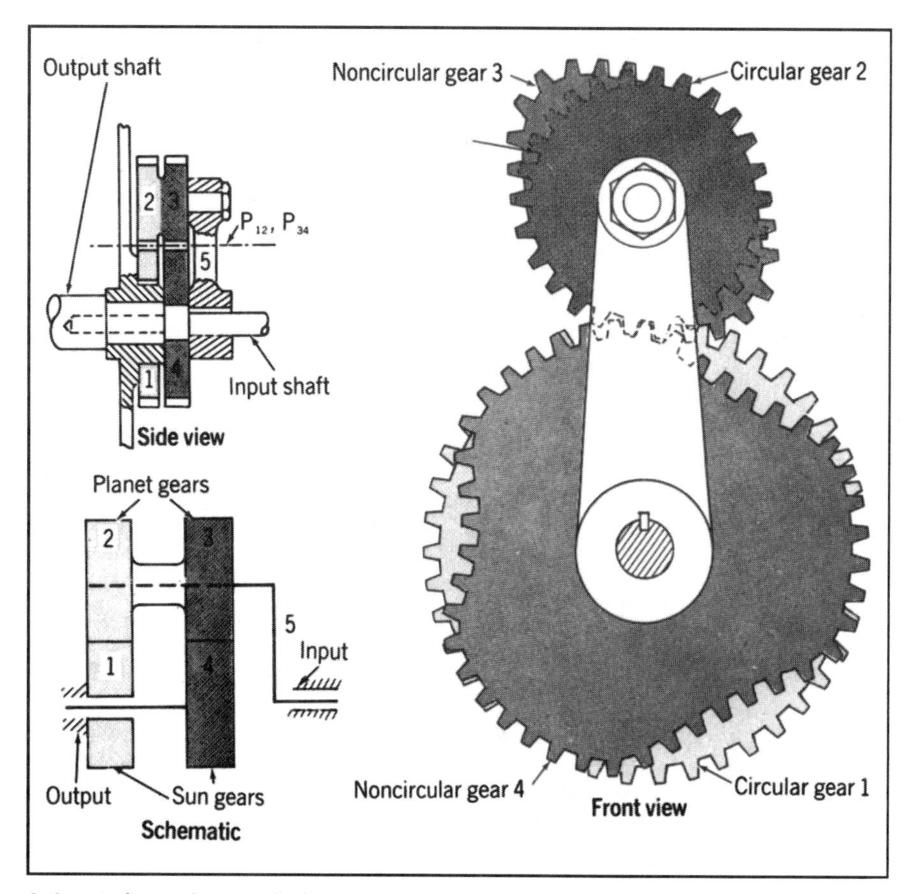

At heart of new planetary (in front view, circular set stacked behind noncircular set), two sets of gears when assembled (side view) resemble conventional unit (schematic).

in the output member) determines the characteristics of the motion. As a result, the ratio of the time of motion to the time of dwell cannot exceed one-half, the output motion cannot be uniform for any finite portion of the indexing cycle, and it is always opposite in sense to the sense of input rotation. The output shaft, moreover, must always be offset from the input shaft.

Many modifications of the standard external geneva have been proposed, including multiple and unequally spaced driving pins, double rollers, and separate entrance and exit slots. These proposals have, however, been only partly successful in overcoming these limitations.

Differential motion. In deriving the operating principle of his mechanism, Freudenstein first considered a conventional epicyclic (planetary) drive in which the input to the cage or arm causes a planet set with gears *2* and *3* to rotate the output "sun," gear *4*, while another sun, gear *1*, is kept fixed (see drawing).

Letting r_1, r_2, r_3, r_4, equal the pitch radii of the circular *1*, *2*, *3*, *4*, then the output ratio, defined as:

$$R = \frac{\text{angular velocity of output gear}}{\text{angular velocity of arm}}$$

is equal to: $R = 1 - \frac{r_1 r_3}{r_2 r_4}$

Now, if $r_1 = r_4$ and $r_2 = r_3$, there is no "differential motion" and the output remains stationary. Thus if one gear pair, say *3* and *4*, is made partly circular and partly noncircular, then where $r_2 = r_3$ and $r_1 = r_4$ for the circular portion, gear *4* dwells. Where $r_2 \neq r_3$ and $r_1 \neq r_4$ for the noncircular portion, gear *4* has motion. The magnitude of this motion depends on the difference in radii, in accordance with the previous equation. In this manner, gear *4* undergoes an intermittent motion (see graph).

Advantages. The pulsating planetary approach demonstrates some highly useful characteristics for intermittent-motion machines:

- The gear teeth serve to lock the output member during the dwell as well as to drive that member during motion.
- Superior high-speed characteristics are obtainable. The profiles of the pitch curves of the noncircular gears can be tailored to a wide variety of desired kinematic and dynamic characteristics. There need be no sudden terminal acceleration change of the driven member, so the transition from dwell to motion, and vice versa, will be smooth, with no jarring of machine or payload.
- The ratio of motion to dwell time is adjustable within wide limits. It can even exceed unity, if desired. The number of indexing operations per revolution of the input member also can exceed unity.
- The direction of rotation of the output member can be in the same or opposite sense relative to that of the input member, according to whether the pitch axis P_{34} for the noncircular portions of gears *3* and *4* lies wholly outside or wholly inside the pitch surface of the planetary sun gear *1*.
- Rotation of the output member is coaxial with the rotation of the input member.
- The velocity variation during motion is adjustable within wide limits. Uniform output velocity for part of the indexing cycle is obtainable; by varying the number and shape of the lobes, a variety of other desirable motion characteristics can be obtained.
- The mechanism is compact and has relatively few moving parts, which can be readily dynamically balanced.

Design hints. The design techniques work out surprisingly simply, said Freudenstein. First the designer must select the number of lobes L_3 and L_4 on the gears *3* and *4*. In the drawings, $L_3 = 2$ and $L_4 = 3$. Any two lobes on the two gears (i.e., any two lobes of which one is on one gear and the other on the other gear) that are to mesh together must have the same arc length. Thus, every lobe on gear *3* must mesh with every lobe on gear *4*, and $T_3/T_4 = L_3/L_4 = 2/3$, where T_3 and T_4 are the numbers of teeth on gears *3* and *4*. T_1 and T_2 will denote the numbers of teeth on gears *1* and *2*.

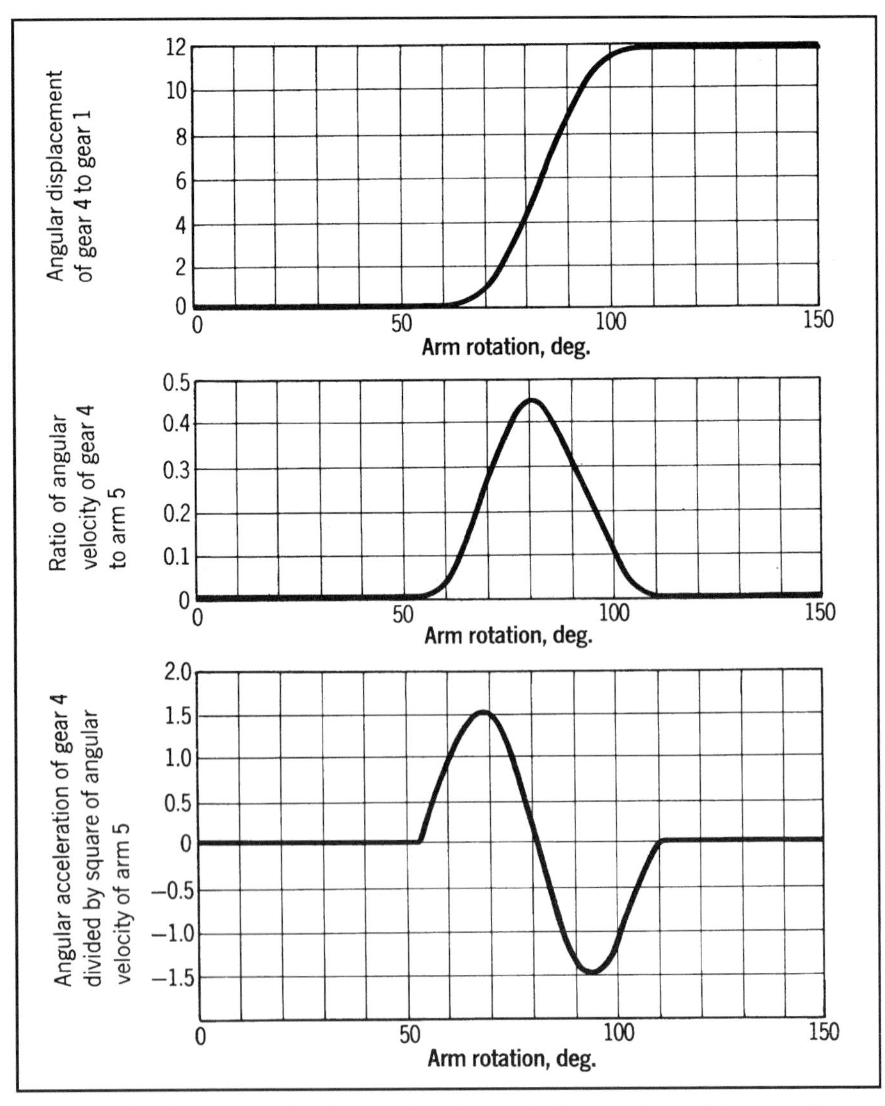

Output motion (upper curve) has long dwell periods; velocity curve (center) has smooth transition from zero to peak; acceleration at transition is zero (bottom).

Next, select the ratio S of the time of motion of gear *4* to its dwell time, assuming a uniform rotation of the arm *5*. For the gears shown, $S = 1$. From the geometry,

$$(\theta_{30} + \Delta\theta_{30})L_3 = 360°$$

and

$$S = \Delta\theta_{30}/\theta_{30}$$

Hence

$$\theta_{30}(1 + S)L_3 = 360°$$

For $S = 1$ and $L_3 + 2$,

$$\theta_{30} = 90°$$

and

$$\Delta\theta_3 = 90°$$

Now select a convenient profile for the noncircular portion of gear *3*. One profile (see the profile drawing) that Freudenstein found to have favorable high-speed characteristics for stop-and-go mechanisms is

$$r_3 = R_3$$

$$\left[1 + \frac{\lambda}{2}\left(1 - \cos\frac{2\pi(\theta_3 - \theta_{30})}{\Delta\theta_3}\right)\right]$$

The profile defined by this equation has, among other properties, the characteristic that, at transition from rest to motion and vice versa, gear *4* will have zero acceleration for the uniform rotation of arm *5*.

In the above equation, λ is the quantity which, when multiplied by R_3, gives the maximum or peak value of $r_3 - R_3$, differing by an amount h' from the radius R_3 of the circular portions of the gear. The noncircular portions of each lobe are, moreover, symmetrical about their midpoints, the midpoints of these portions being indicated by *m*.

To evaluate the quantity λ, Freudenstein worked out the equation:

$$\lambda = \frac{1 - \mu}{\mu} \times$$

$$\frac{[S + \alpha - (1 + \alpha)\mu][\alpha - S - (1 + \alpha)\mu]}{[\alpha - (1 + \alpha)\mu]^2}$$

where $R_3\lambda$ = height of lobe

$$\mu = \frac{R_3}{A} = R_3/(R_3 + R_4)$$

$$\alpha = S + (1 + S)L_3/L_4$$

To evaluate the equation, select a suitable value for μ that is a reasonably simple rational fraction, i.e., a fraction such as $^3/_8$ whose numerator and denominator are reasonably small integral numbers.

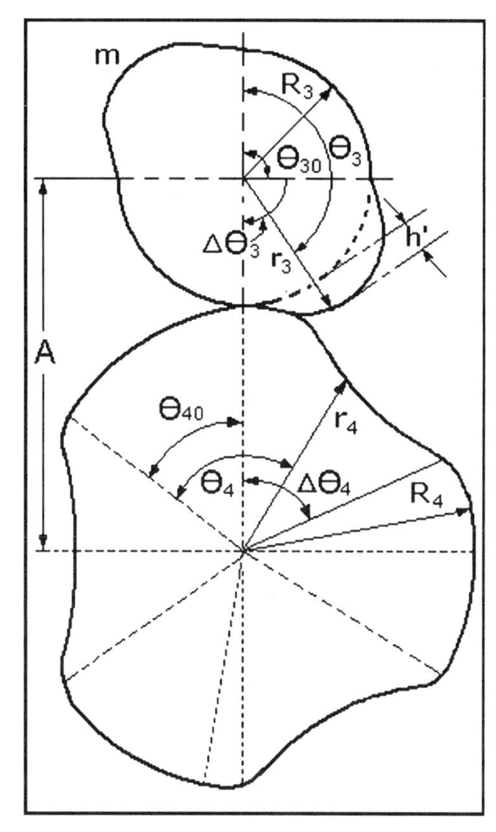

Profiles for noncircular gears are circular arcs blended to special cam curves.

Thus, without a computer or lengthy trial-and-error procedures, the designer can select the configuration that will achieve his objective of smooth intermittent motion.

CYCLOID GEAR MECHANISM CONTROLS PUMP STROKE

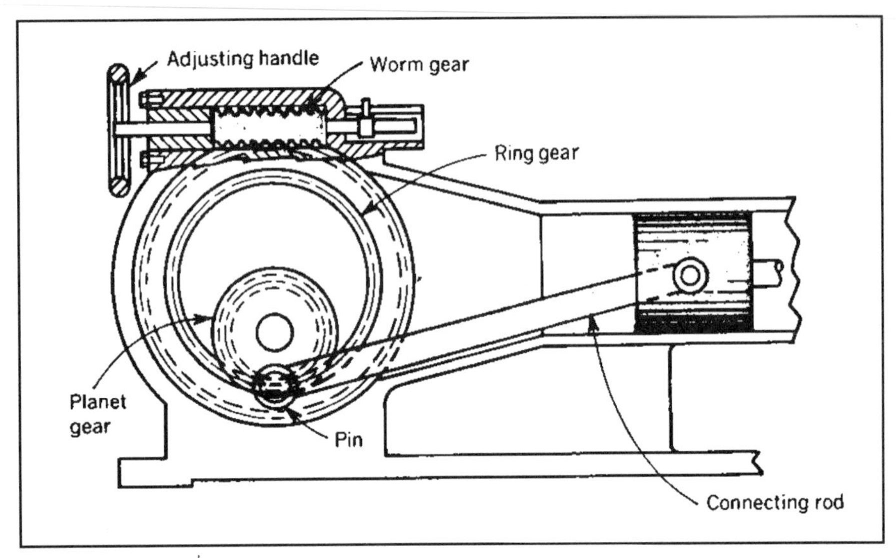

An adjustable ring gear meshes with a planet gear having half of its diameter to provide an infinitely variable stroke in a pump. The adjustment in the ring gear is made by engaging other teeth. In the design below, a yoke replaces the connecting rod.

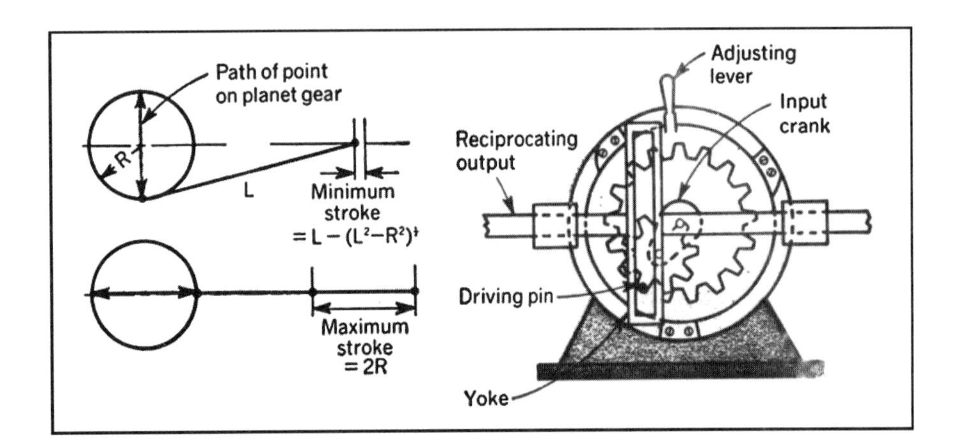

A metering pump for liquid or gas has an adjustable ring gear that meshes with a special-size planet gear to provide an infinitely variable stroke in the pump. The stroke can be set manually or automatically when driven by a servomotor. Flow control from 180 to 1200 liter/hr. (48 to 317 gal./hr.) is possible while the pump is at a standstill or running.

Straight-line motion is key. The mechanism makes use of a planet gear whose diameter is half that of the ring gear. As the planet is rotated to roll on the inside of the ring, a point on the pitch diameter of the planet will describe a straight line (instead of the usual hypocycloid curve). This line is a diameter of the ring gear. The left end of the connecting rod is pinned to the planet at this point.

The ring gear can be shifted if a second set of gear teeth is machined in its outer surface. This set can then be meshed with a worm gear for control. Shifting the ring gear alters the slope of the straight-line path. The two extreme positions are shown in the diagram. In the position of the mechanism shown, the pin will reciprocate vertically to produce the minimum stroke for the piston. Rotating the ring gear 90° will cause the pin to reciprocate horizontally to produce the maximum piston stroke.

The second diagram illustrates another version that has a yoke instead of a connecting rod. This permits the length of the stroke to be reduced to zero. Also, the length of the pump can be substantially reduced.

GEARS CONVERT ROTARY-TO-LINEAR MOTION

A compact gear system that provides linear motion from a rotating shaft was designed by Allen G. Ford of The Jet Propulsion Laboratory in California. It has a planetary gear system so that the end of an arm attached to the planet gear always moves in a linear path (drawing).

The gear system is set in motion by a motor attached to the base plate. Gear *A*, attached to the motor shaft, turns the case assembly, causing Gear *C* to rotate along Gear *B*, which is fixed.

The arm is the same length as the center distance between Gears *B* and *C*. Lines between the centers of Gear *C*, the end of the arm, and the case axle form an isosceles triangle, the base of which is always along the plane through the center of rotation. So the output motion of the arm attached to Gear *C* will be in a straight line.

When the end of travel is reached, a switch causes the motor to reverse, returning the arm to its original position.

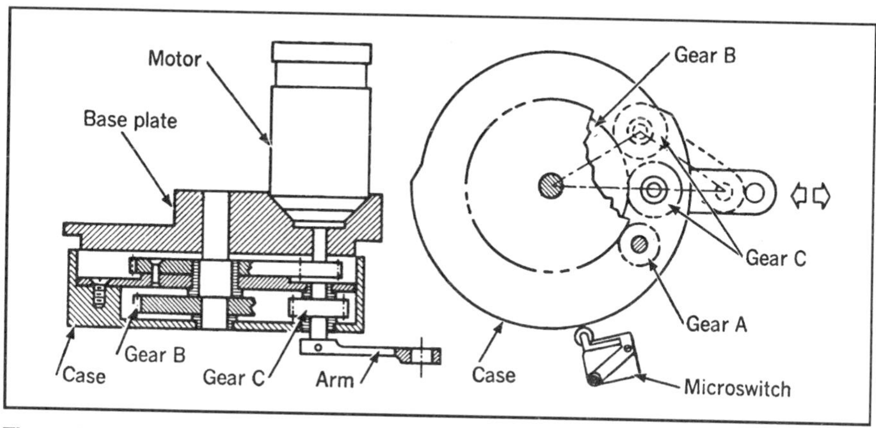

The end of arm moves in a straight line because of the triangle effect (right).

TWIN-MOTOR PLANETARY GEARS OFFER SAFETY AND DUAL-SPEED

Many operators and owners of hoists and cranes fear the possible catastrophic damage that can occur if the driving motor of a unit should fail for any reason. One solution to this problem is to feed the power of two motors of equal rating into a planetary gear drive.

Power supply. Each of the motors is selected to supply half the required output power to the hoisting gear (see diagram). One motor drives the ring gear, which has both external and internal teeth. The second motor drives the sun gear directly.

Both the ring gear and sun gear rotate in the same direction. If both gears rotate at the same speed, the planetary cage, which is coupled to the output, will also revolve at the same speed (and in the same direction). It is as if the entire inner works of the planetary were fused together. There would be no relative motion. Then, if one motor fails, the cage will revolve at half its original speed, and the other motor can still lift with undiminished capacity. The same principle holds true when the ring gear rotates more slowly than the sun gear.

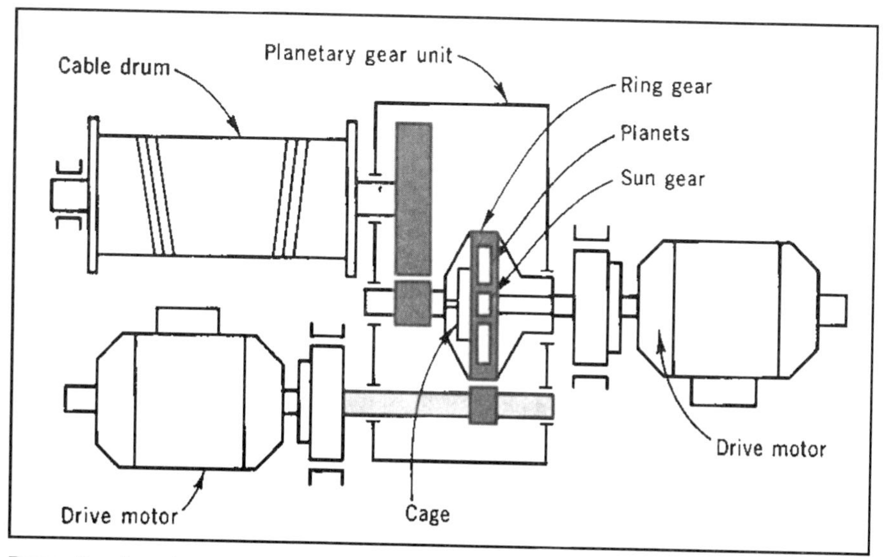

Power flow from two motors combine in a planetary that drives the cable drum.

No need to shift gears. Another advantage is that two working speeds are available as a result of a simple switching arrangement. This makes is unnecessary to shift gears to obtain either speed.

The diagram shows an installation for a steel mill crane.

ELEVEN CYCLOID GEAR MECHANISMS

The appeal of cycloidal mechanisms is that they can be tailored to provide one of these three common motions:

- **Intermittent**—with either short or long dwells
- **Rotary with progressive oscillation**—where the output undergoes a cycloidal motion during which the forward motion is greater than the return motion
- **Rotary**-to-linear with a dwell period

All the cycloidal mechanisms shown here are general. This results in compact positive mechanisms capable of operating at relatively high speeds with little backlash or "slop." These mechanisms can be classified into three groups:

Hypocycloid—the points tracing the cycloidal curves are located on an external gear rolling inside an internal ring gear. This ring gear is usually stationary and fixed to the frame.

Epicycloid—the tracing points are on an external gear that rolls in another external (stationary) gear.

Pericycloid—the tracing points are located on an internal gear that rolls on a stationary external gear.

Basic hypocycloid curves

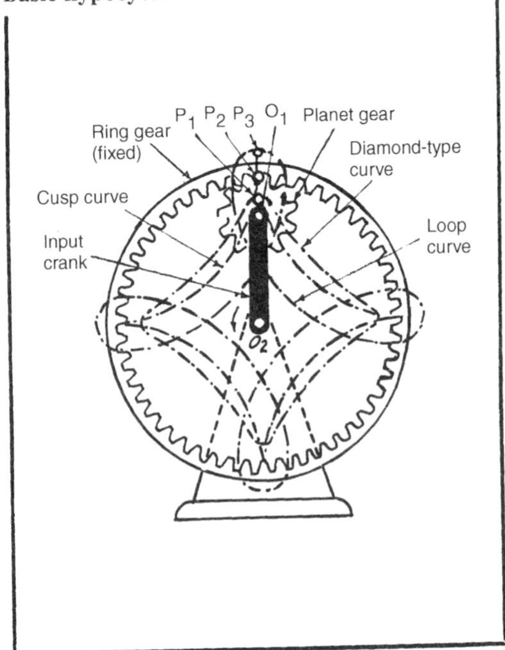

The input drives a planet in mesh with a stationary ring gear. Point P_1 on the planet gear describes a diamond-shape curve, point P_2 on the pitch line of the planet describes the familiar cusp curve, and point P_3, which is on an extension rod fixed to the planet gear, describes a loop-type curve. In one application, an end miller located at P_1 machined a diamond-shaped profile.

Double-dwell mechanism

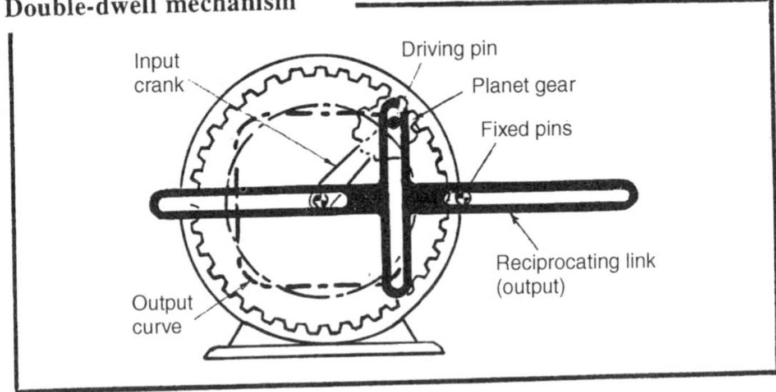

Coupling the output pin to a slotted member produces a prolonged dwell in each of the extreme positions. This is another application of the diamond-type hypocycloidal curve.

Long-dwell drive mechanism

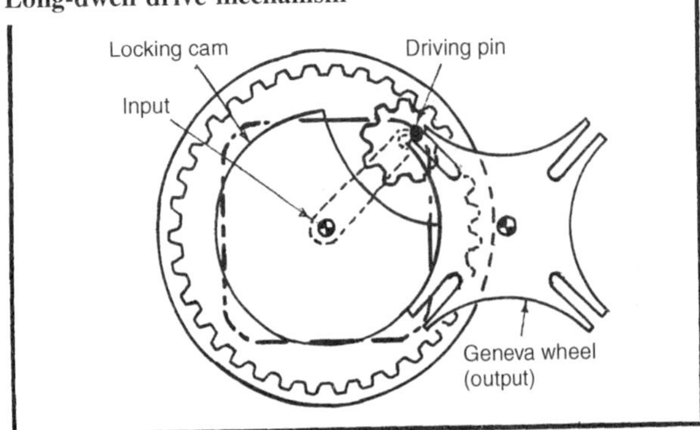

In common with standard, four-station genevas, each rotation of the input of this drive indexes the slotted geneva 90°. A pin fastened to the planet gear causes the drive to describe a rectangular-shaped cycloidal curve. This produces a smoother indexing motion because the driving pin moves on a noncircular path.

Internal drive mechanism

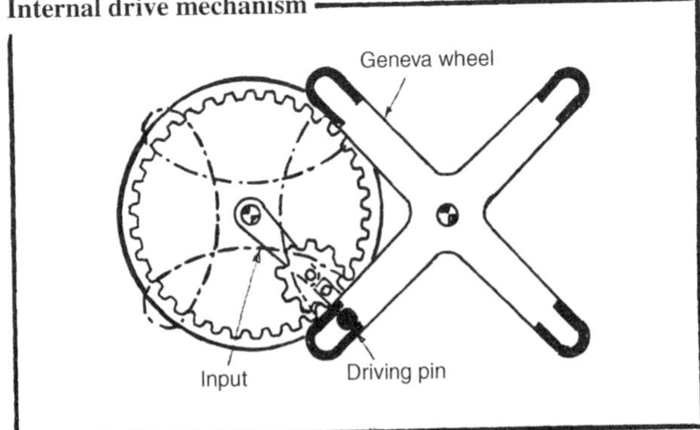

A loop-type curve permits the driving pin to enter the slot in a direction that is radially outward from the center. The pin then loops over to index the cross member rapidly. As with other genevas, the output rotates 90° before going into a long dwell period during each 270° rotation of the input element.

Cycloidal parallelogram mechanism

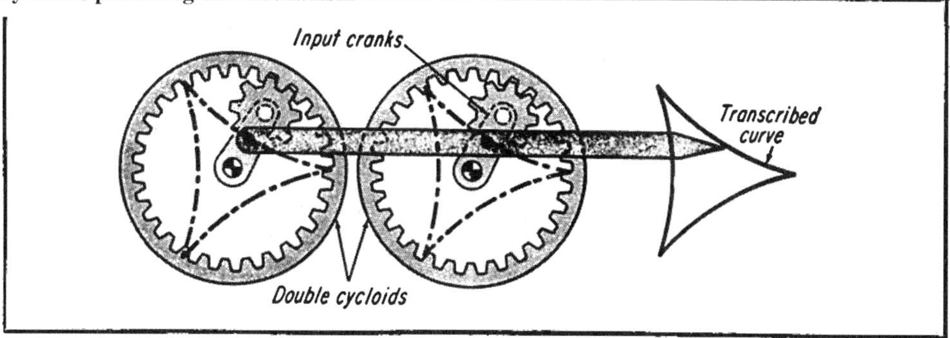

Two identical hypocycloid mechanisms guide the point of the bar along the triangularly shaped path. The mechanisms are useful where space is limited in the area where the curve must be described. These double-cycloid mechanisms can be designed to produce other curve shapes.

Cycloidal short-dwell rotary mechanism

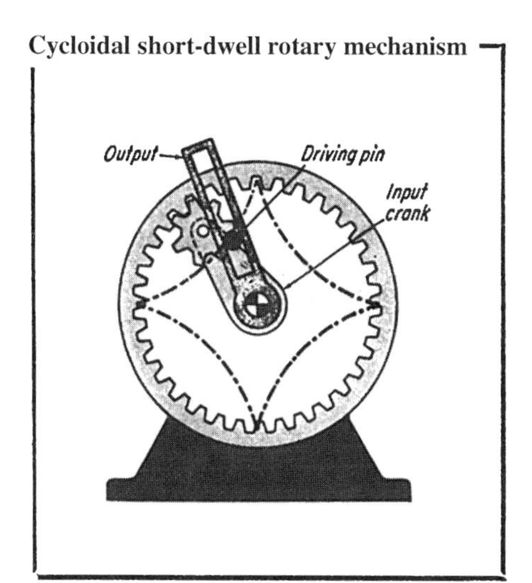

The pitch circle of this planet gear is exactly one-quarter that of the ring gear. A pin on the planet gear will cause the slotted output member to dwell four times during each revolution of the input shaft.

Cycloidal rocker mechanism

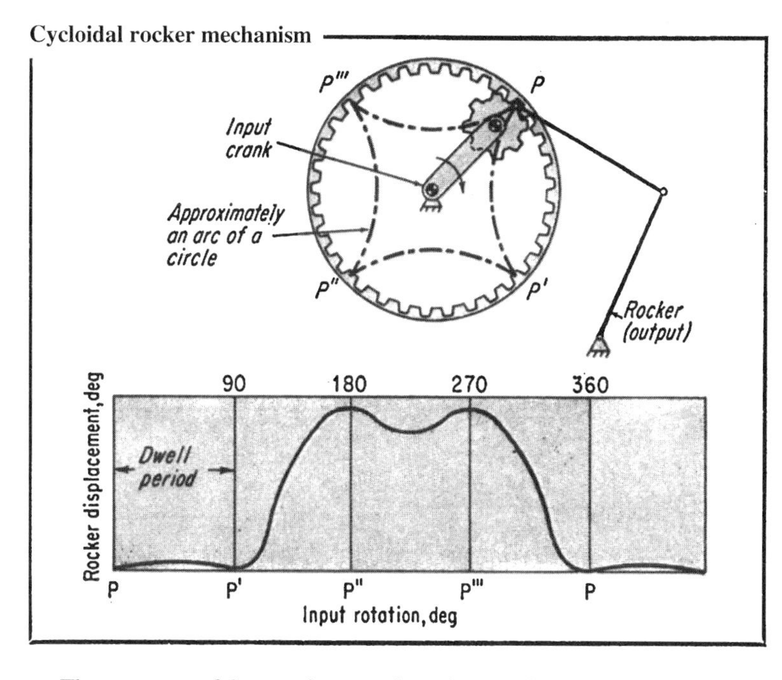

The curvature of the cusp is approximately that of an arc of a circle. Hence the rocker reaches a long dwell at the right extreme position while point P moves to P'. There is then a quick return from P' to P'', with a momentary dwell at the end of this phase. The rocker then undergoes a slight oscillation from point P'' to P''', as shown in the rocker displacement diagram.

Cycloidal reciprocator mechanism

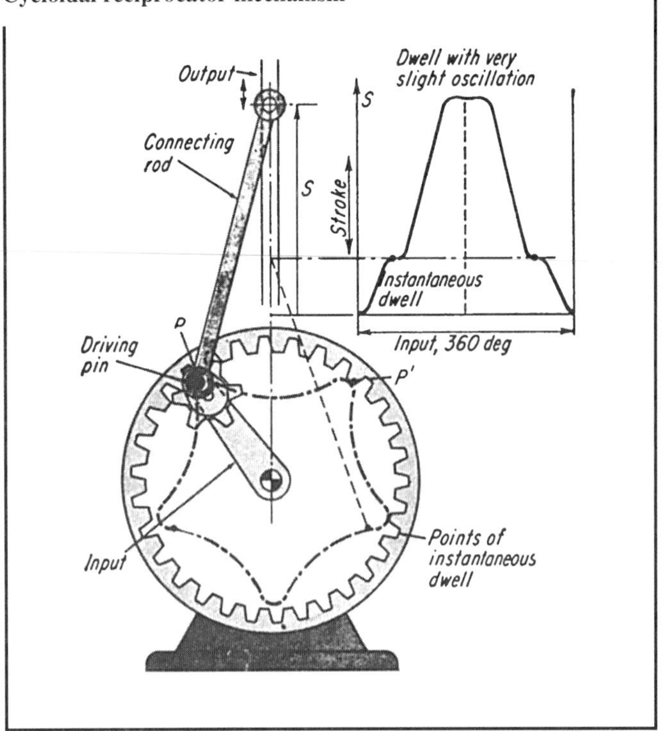

Part of curve *P-P′* produces a long dwell, but the five-lobe cycloidal curve avoids a marked oscillation at the end of the stroke. There are also two points of instantaneous dwell where the curve is perpendicular to the connecting rod.

Adjustable harmonic drive mechanism

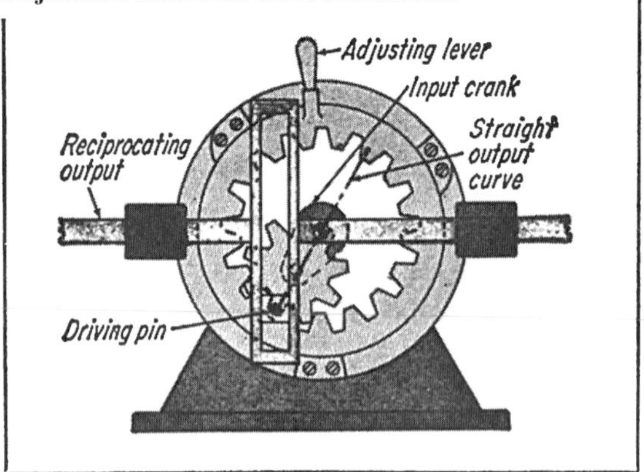

By making the planet gear diameter half that of the internal gear, a straight-line output curve can be produced by the driving pin which is fastened to the planet gear. The pin engages the slotted member to cause the output to reciprocate back and forth with harmonic (sinusoidal) motion. The position of the fixed ring gear can be changed by adjusting the lever, which in turn rotates the straight-line output curve. When the curve is horizontal, the stroke is at a maximum; when the curve is vertical, the stroke is zero.

Elliptical-motion drive mechanism

By making the pitch diameter of the planet gear equal to half that of the ring gear, every point on the planet gear (such as points P_2 and P_3) will describe elliptical curves which get flatter as the points are selected closer to the pitch circle. Point P_1, at the center of the planet, describes a circle; point P_4, at the pitch circle, describes a straight line. When a cutting tool is placed at P_3, it will cut almost-flat sections from round stock, as when milling flats on a bolt. The other two flats of the bolt can be cut by rotating the bolt or the cutting tool 90º.

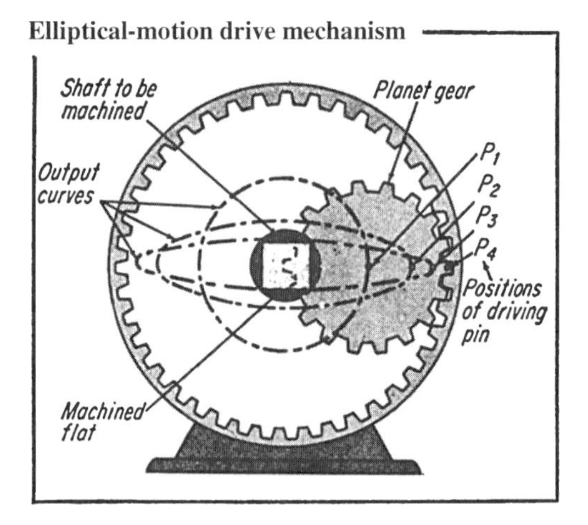

Epicycloid reciprocator mechanism

Here the sun gear is fixed and the planet is gear driven around it by the input link. There is no internal ring gear as with the hypocycloid mechanisms. Driving pin *P* on the planet describes the curve shown, which contains two almost-flat portions. If the pin rides in the slotted yoke, a short dwell is produced at both the extreme positions of the output member. The horizontal slots in the yoke ride the end-guides, as shown.

FIVE CARDAN-GEAR MECHANISMS

These gearing arrangements convert rotary into straight-line motion, without the need for slideways.

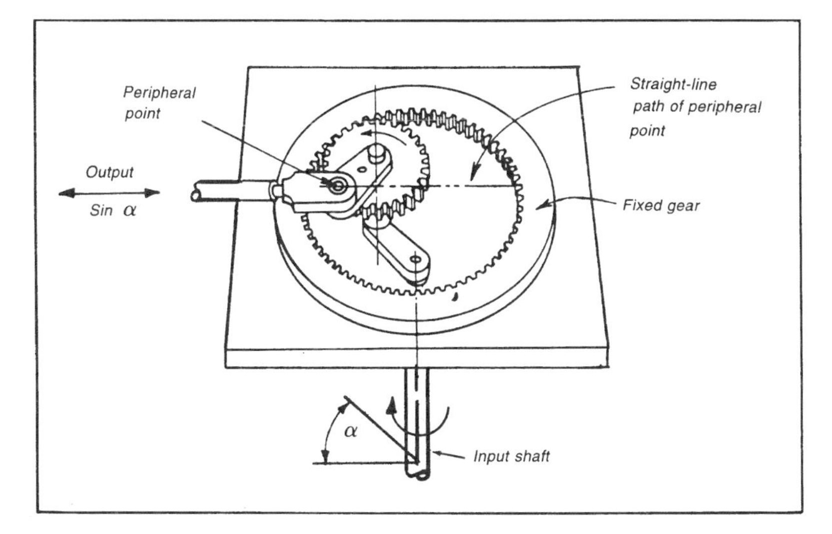

Fig. 1 Cardan gearing works on the principle that any point on the periphery of a circle rolling on the inside of another circle describes, in general, a hypocyloid. This curve degenerates into a true straight line (diameter of the larger circle) if the diameters of both circles are in the ratio of 1:2. The rotation of the input shaft causes a small gear to roll around the inside of the fixed gear. A pin located on the pitch circle of the small gear describes a straight line. Its linear displacement is proportional to the theoretically true sine or cosine of the angel through which the input shaft is rotated.

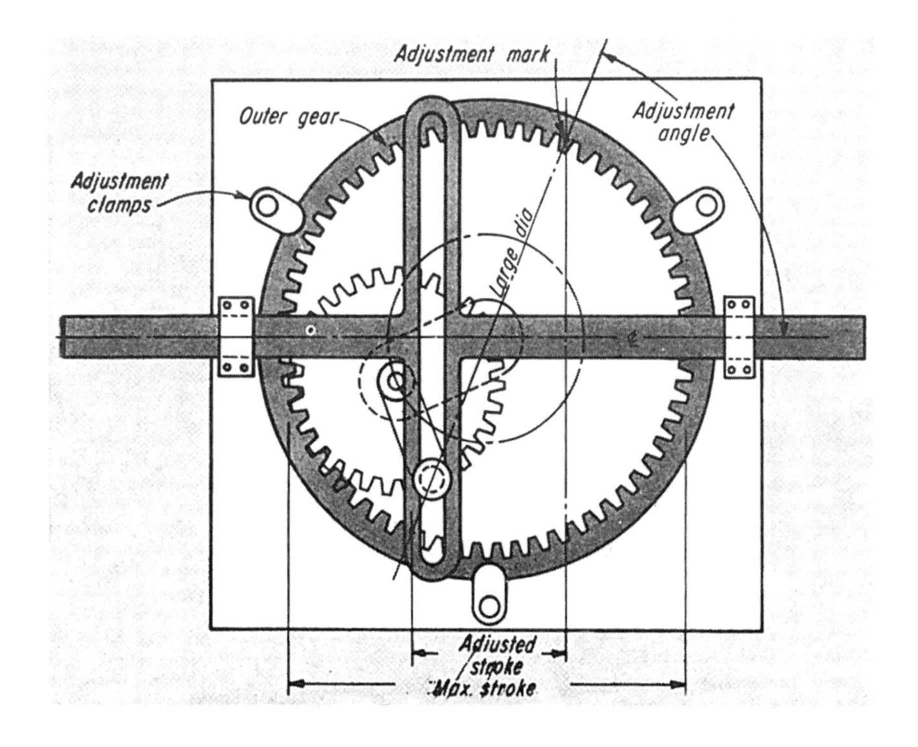

Fig. 2 Cardan gearing and a Scotch yoke in combination provide an adjustable stroke. The angular position of the outer gear is adjustable. The adjusted stroke equals the projection of the large diameter, along which the drive pin travels, on the Scotch-yoke's centerline. The yoke motion is simple harmonic.

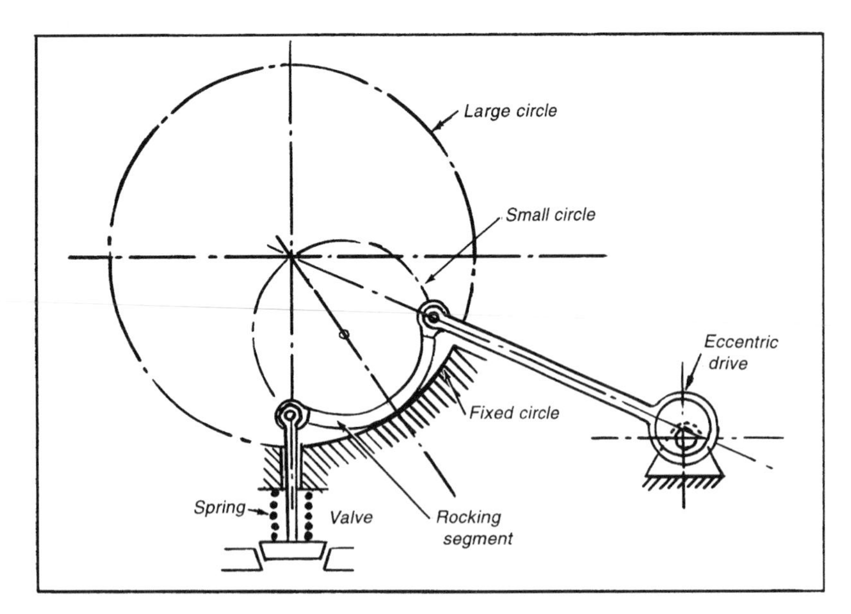

Fig. 3 A valve drive demonstrates how the Cardan principle can be applied. A segment of the smaller circle rocks back and forth on a circular segment whose radius is twice as large. The input and output rods are each attached to points on the small circle. Both these points describe straight lines. The guide of the valve rod prevents the rocking member from slipping.

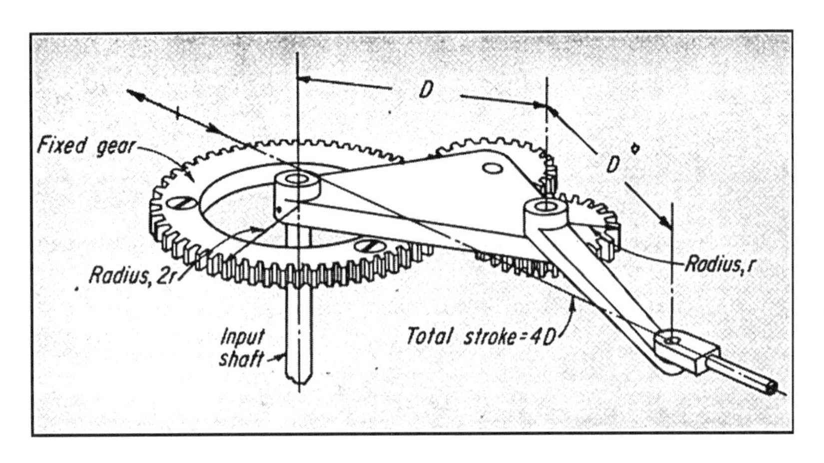

Fig. 4 A simplified Cardan mechanism eliminates the need for the relatively expensive internal gear. Here, only spur gears are used, and the basic requirements must be met, i.e., the 1:2 ratio and the proper direction of rotation. The rotation requirement is met by introducing an idler gear of appropriate size. This drive delivers a large stroke for the comparative size of its gears.

Fig. 5 A rearrangement of gearing in the simplified Cardan mechanism results in another useful motion. If the fixed sun gear and planet pinion are in the ratio of 1:1, an arm fixed to the planet shaft will stay parallel to itself during rotation, while any point on the arm describes a circle of radius *R*. When arranged in conjugate pairs, the mechanism can punch holes on moving webs of paper.

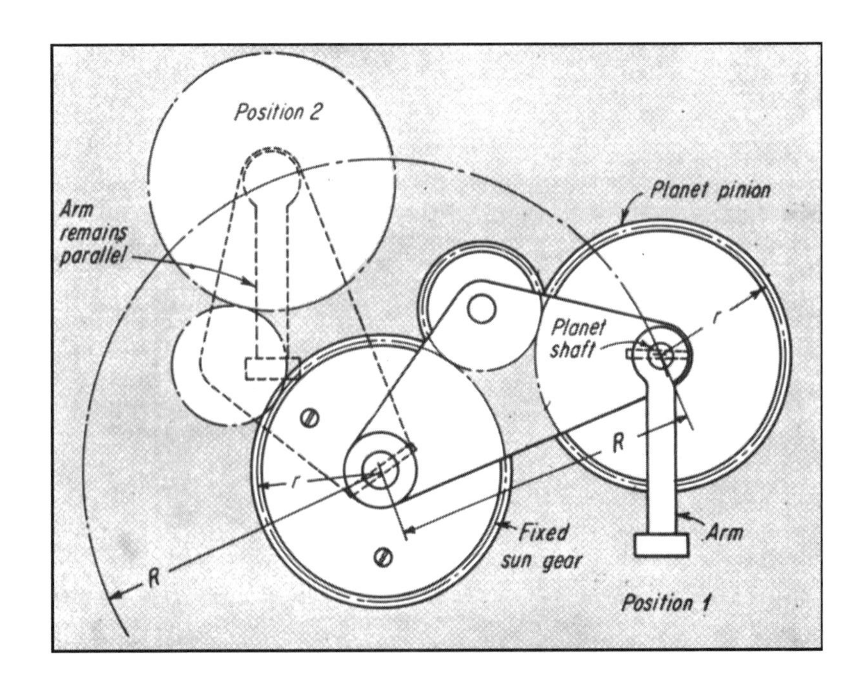

CONTROLLED DIFFERENTIAL GEAR DRIVES

By coupling a differential gear assembly to a variable speed drive, a drive's horsepower capacity can be increased at the expense of its speed range. Alternatively, the speed range can be increased at the expense of the horsepower range. Many combinations of these variables are possible. The features of the differential depend on the manufacturer. Some systems have bevel gears, others have planetary gears. Both single and double differentials are employed. Variable-speed drives with differential gears are available with ratings up to 30 hp.

Horsepower-increasing differential (Fig. 1). The differential is coupled so that the output of the motor is fed into one side and the output of the speed variator is fed into the other side. An additional gear pair is employed as shown in Fig. 1

Output speed

$$n_4 = \frac{1}{2}\left(n_1 + \frac{n_2}{R}\right)$$

Output torque

$$T_4 = 2T_3 = 2RT_2$$

Output hp

$$\text{hp} = \left(\frac{Rn_1 + n_2}{63{,}025}\right)T_2$$

hp increase

$$\Delta\text{hp} = \left(\frac{Rn_1}{63{,}025}\right)T_2$$

Speed variation

$$n_{4\,\max} - n_{4\,\min} = \frac{1}{2R}(n_{2\,\max} - n_{2\,\min})$$

Speed range increase differential (Fig. 2). This arrangement achieves a wide range of speed with the low limit at zero or in the reverse direction.

Fig. 1

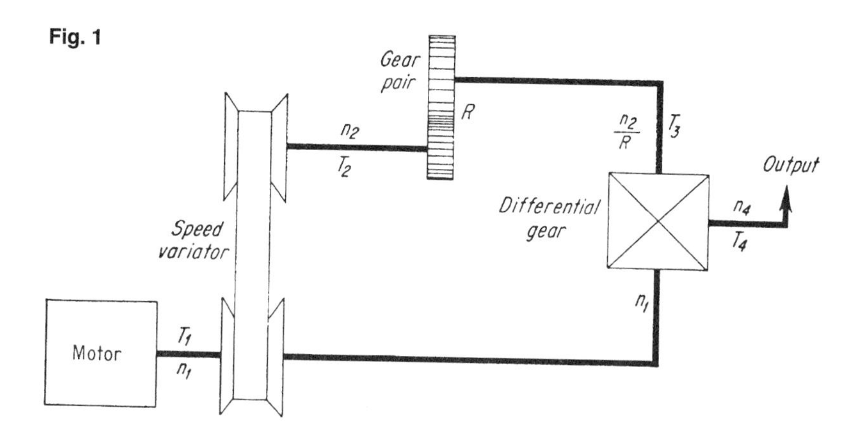

Fig. 2

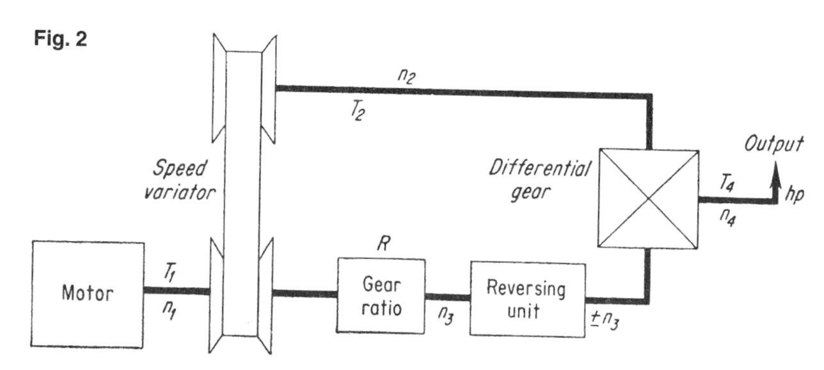

Fig. 3

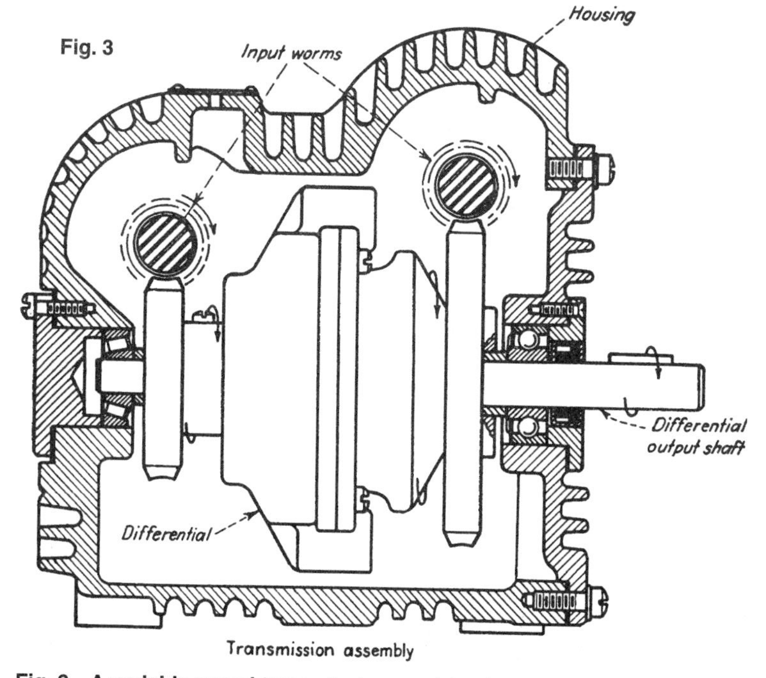

Fig. 3 A variable-speed transmission consists of two sets of worm gears feeding a differential mechanism. The output shaft speed depends on the difference in rpm between the two input worms. When the worm speeds are equal, output is zero. Each worm shaft carries a cone-shaped pulley. These pulley are mounted so that their tapers are in opposite directions. Shifting the position of the drive belt on these pulleys has a compound effect on their output speed.

FLEXIBLE FACE-GEARS ARE EFFICIENT HIGH-RATIO SPEED REDUCERS

A system of flexible face-gearing provides designers with a means for obtaining high-ratio speed reductions in compact trains with concentric input and output shafts.

With this approach, reduction ratios range from 10:1 to 200:1 for single-stage reducers, whereas ratios of millions to one are possible for multi-stage trains. Patents on the flexible face-gear reducers were held by Clarence Slaughter of Grand Rapids, Michigan.

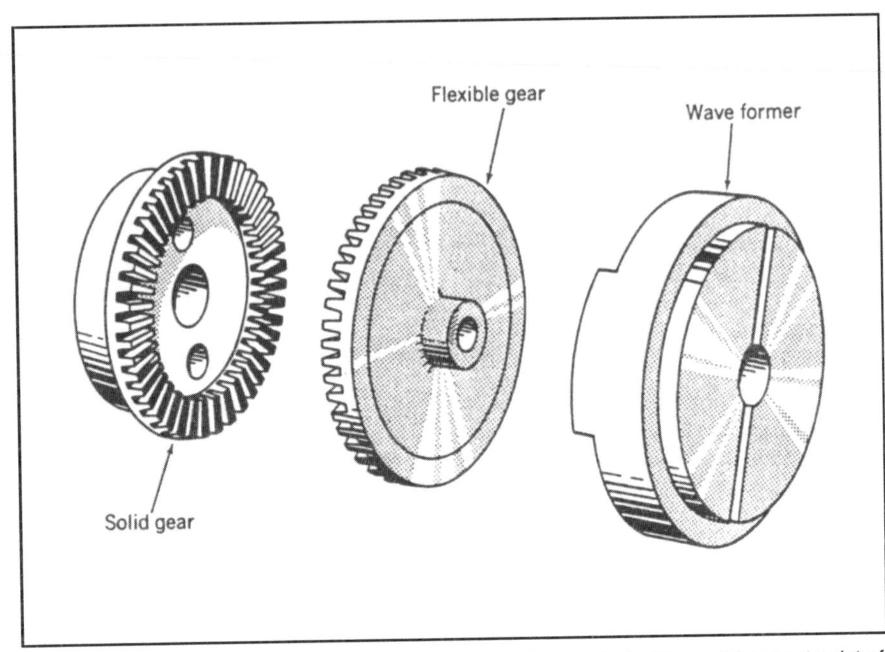

A flexible face-gear is flexed by a rotating wave former into contact with a solid gear at point of mesh. The two gears have slightly different numbers of teeth.

Building blocks. Single-stage gear reducers consist of three basic parts: a flexible face-gear made of plastic or thin metal; a solid, non-flexing face-gear; and a wave former with one or more sliders and rollers to force the flexible gear into mesh with the solid gear at points where the teeth are in phase.

The high-speed input to the system usually drives the wave former. Low-speed output can be derived from either the flexible or the solid face gear; the gear not connected to the output is fixed to the housing.

Teeth make the difference. Motion between the two gears depends on a slight difference in their number of teeth (usually one or two teeth). But drives with gears that have up to a difference of 10 teeth have been devised.

On each revolution of the wave former, there is a relative motion between the two gears that equals the difference in their numbers of teeth. The reduction ratio equals the number of teeth in the output gear divided by the difference in their numbers of teeth.

Two-stage and four-stage gear reducers are made by combining flexible and solid gears with multiple rows of teeth and driving the flexible gears with a common wave former.

Hermetic sealing is accomplished by making the flexible gear serve as a full seal and by taking output rotation from the solid gear.

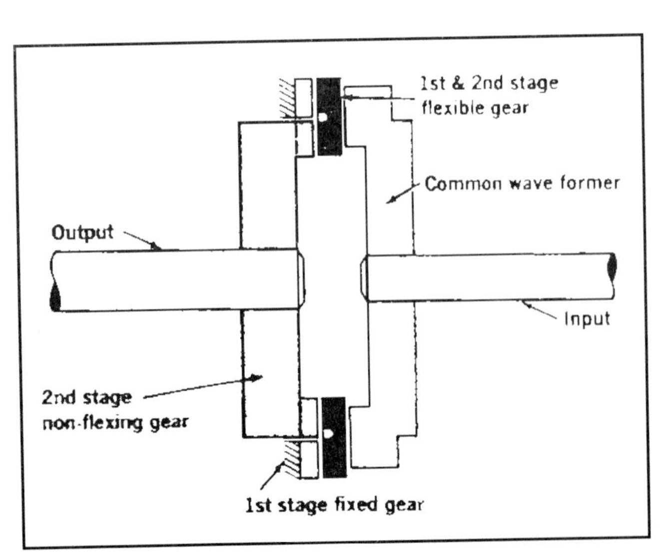

A two-stage speed reducer is driven by a common-wave former operating against an integral flexible gear for both stages.

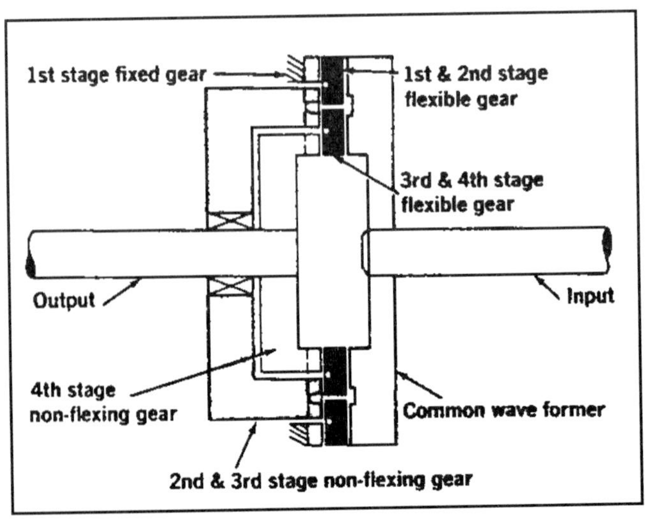

A four-stage speed reducer can, theoretically, attain reductions of millions to one. The train is both compact and simple.

ROTARY SEQUENCER GEARS TURN COAXIALLY

Two coaxial rotations, one clockwise and one counterclockwise, are derived from a single clockwise rotation.

A proposed rotary sequencer is assembled from a conventional planetary differential gear-set and a latching mechanism. Its single output and two rotary outputs (one clockwise and one counterclockwise) are coaxial, and the output torque is constant over the entire cycle. Housed in a lightweight, compact, cylindrical package, the sequencer requires no bulky ratchets, friction clutches, or cam-and-track followers. Among its possible applications are sequencing in automated production-line equipment, in home appliances, and in vehicles.

The sequencer is shown in Figure 1. A sun gear connects with four planetary gears that engage a ring gear. With the ring gear held stationary, clockwise rotation of the sun gear causes the entire planetary-gear carrier also to rotate clockwise. If the planetary-gear carrier is held fixed, the ring gear will rotate counterclockwise when the sun gear rotates clockwise.

Figure 2 shows the latch. It consists of a hook (the carrier hook) that is rigidly attached to the planetary-gear carrier, a rind that is rigidly attached to the ring gear, and a latch pivot arm with a pair of latch rollers attached to one end. The other end of the pivot arm rotates about a short shaft that extends from the fixed wall of the housing.

The sequencer cycle starts with the ring latch roller resting in a slot in the ring. This locks the ring and causes the planetary-gear carrier to rotate clockwise with the input shaft (Fig. 2a). When the carrier hook has rotated approximately three-quarters of a complete cycle, it begins to engage the planet-carrier latch roller (Fig. 2b), causing the latch pivot arm to rotate and the ring latch roller to slip out of its slot (Fig. 2c). This frees the ring and ring gear for counterclockwise motion, while locking the carrier. After a short interval of concurrent motion, the planetary-gear output shaft ceases its clockwise motion, and the ring-gear output shaft continues its clockwise motion.

When the ring reaches the position in Fig. 2d, the cycle is complete, and the input shaft is stopped. If required, the input can then be rotated counterclockwise, and the sequence will be reversed until the starting position (Fig. 2a) is reached again.

In a modified version of the sequencer, the latch pivot arm is shortened until its length equals the radii of the rollers. This does away with the short overlap of output rotations when both are in motion. For this design, the carrier motion ceases before the ring begins its rotation.

This work was done by Walter T. Appleberry of Rockwell International Corp. for **Johnson Space Center,** *Houston, Texas.*

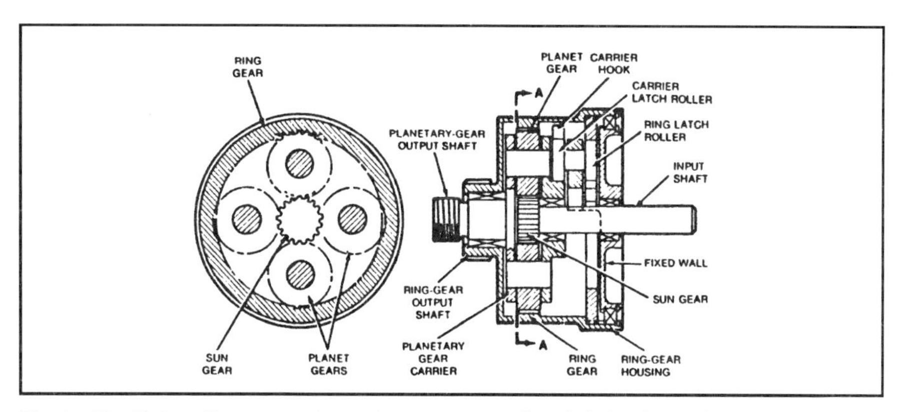

Fig. 1 The Rotary Sequencer has a ring-gear output (in color) that is coaxial with a planetary-gear output (in gray). Clockwise rotation of the input is converted to clockwise rotation of the planetary-gear output followed by counterclockwise rotation of the ring-gear output. The sequence is controlled by the latch action described in Fig. 2.

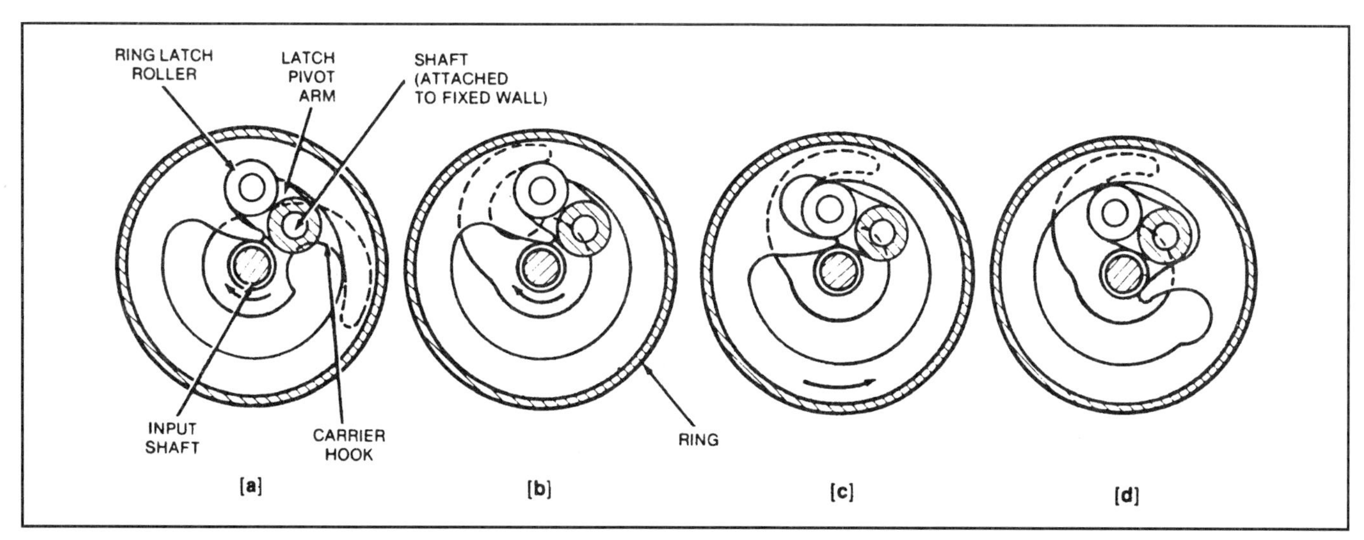

Fig. 2 The Latch Sequence is shown in four steps: (a) The input shaft rotates the carrier clockwise while the ring latch roller holds the ring gear stationary; (b) the carrier hook begins to engage the carrier latch roller; (c) the ring latch roller begins to move out of its slot, and the carrier motion ceases while the ring begins to move; and (d) the sequence has ended with the ring in its final position.

PLANETARY GEAR SYSTEMS

Designers keep finding new and useful planetaries. Forty-eight popular types are given here with their speed-ratio equations.

MISSILE SILO COVER DRIVE

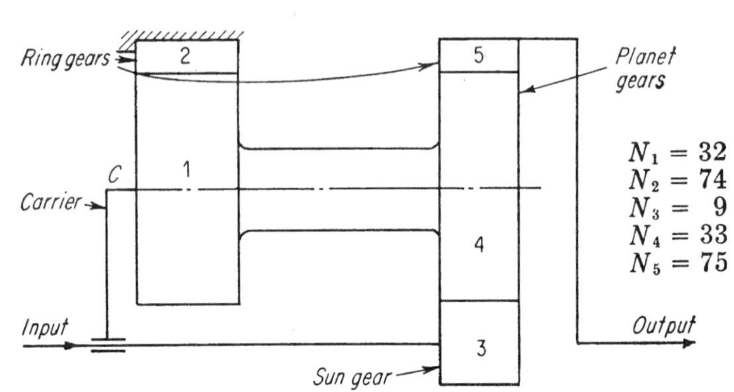

$N_1 = 32$
$N_2 = 74$
$N_3 = 9$
$N_4 = 33$
$N_5 = 75$

Ring gear 2 fixed; ring gear 5 output

Speed-ratio equation

$$R = \frac{1 + \dfrac{N_4 N_2}{N_3 N_1}}{1 - \dfrac{N_4 N_2}{N_5 N_1}} = \frac{1 + \dfrac{(33)(74)}{(9)(32)}}{1 - \dfrac{(33)(74)}{(75)(32)}} = -541\tfrac{2}{3}$$

Symbols

C = carrier (also called "spider")—a non-gear member of gear train whose rotation affects gear ratio

N = number of teeth

R = overall speed reduction ratio

1, 2, 3, etc. = gears in a train (corresponding to labels on schematic diagram)

DOUBLE-ECCENTRIC DRIVE

Input is through double-throw crank (carrier). **Gear 1**

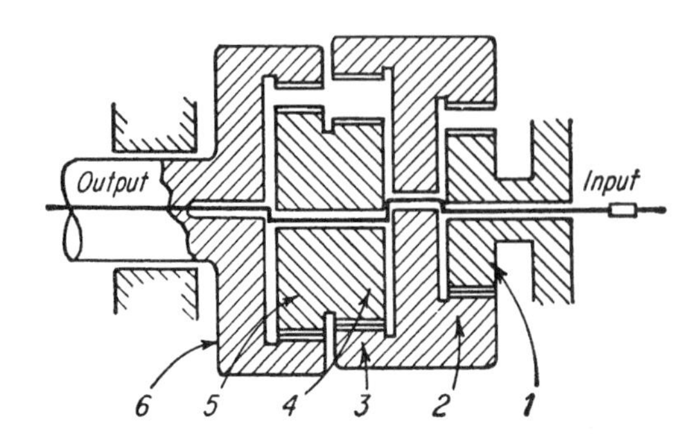

$$R = \frac{1}{1 - \dfrac{N_5 N_3 N_1}{N_6 N_4 N_2}}$$

When $N_1 = 103$, $N_2 = 110$, $N_3 = 109$, $N_4 = 100$, $N_5 = 94$, $N_6 = 96$

$$R = \frac{1}{1 - \dfrac{(94)(109)(103)}{(96)(100)(110)}} = 1505$$

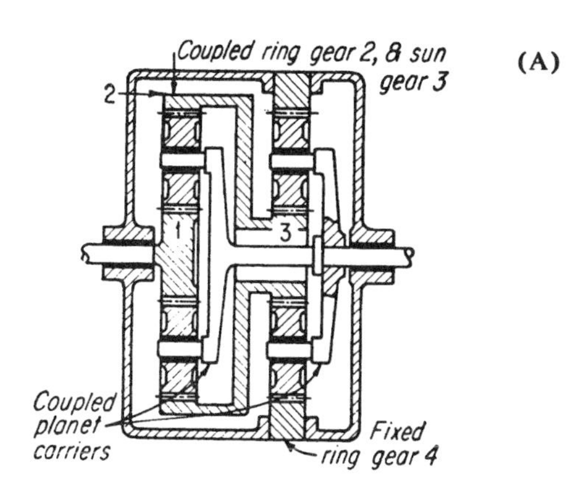

(A)

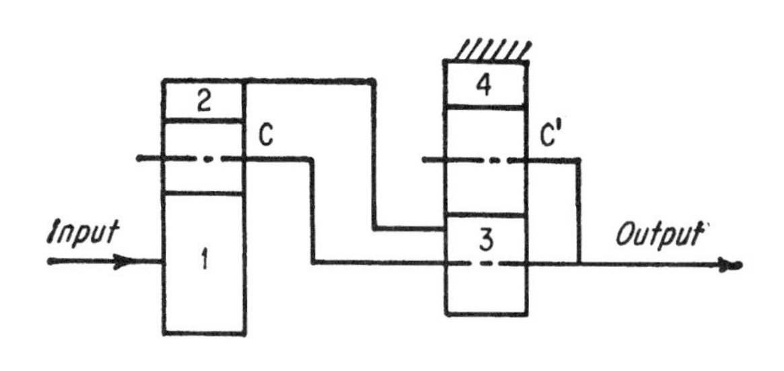

$$R = 1 - \frac{N_2 N_4}{N_1 N_3}$$

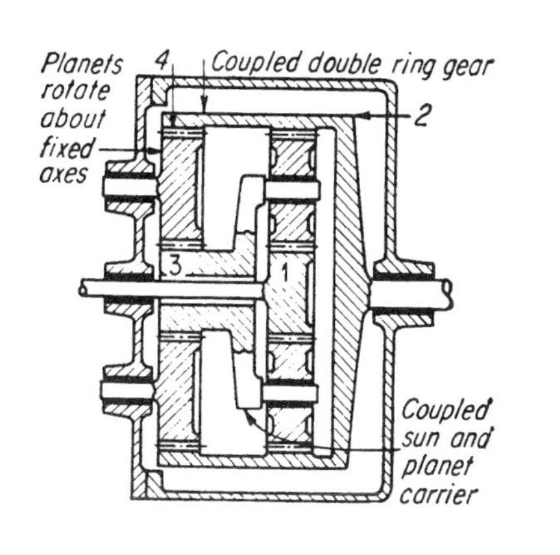

(B)

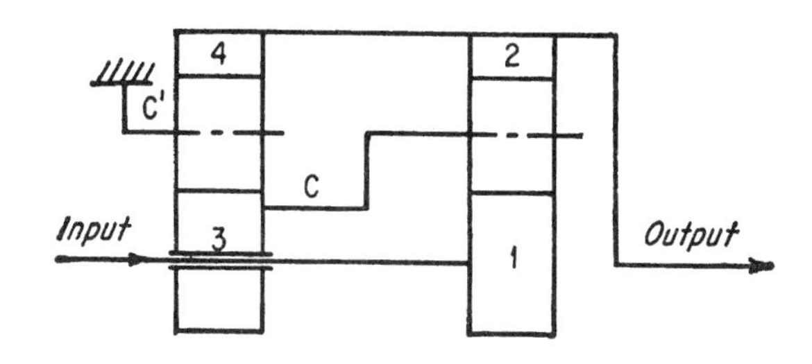

$$R = \left(1 + \frac{N_2}{N_1}\right)\left(-\frac{N_4}{N_3}\right) - \frac{N_2}{N_1}$$

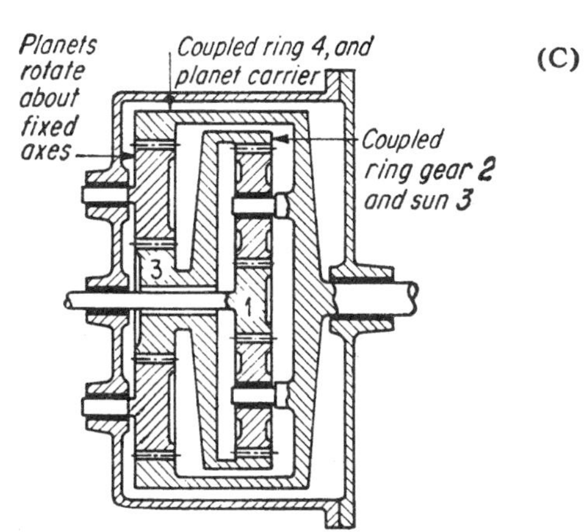

(C)

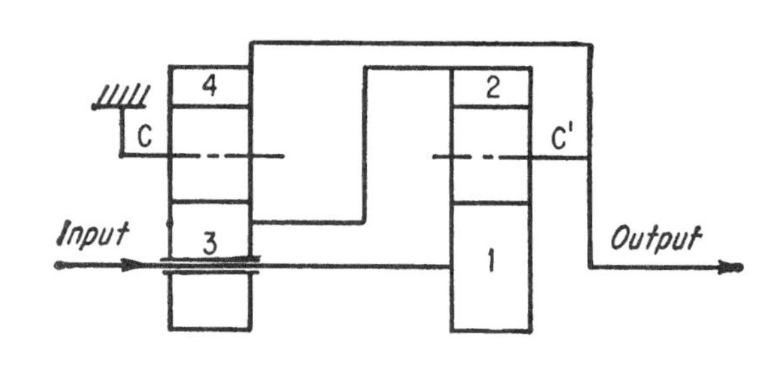

$$R = 1 + \frac{N_2}{N_1}\left(1 + \frac{N_4}{N_3}\right)$$

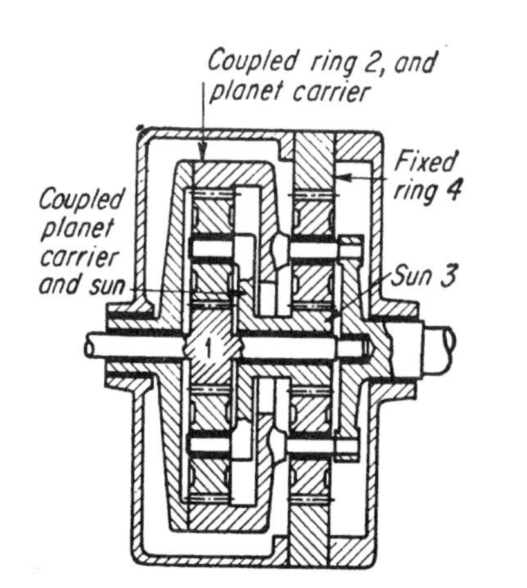

(D)

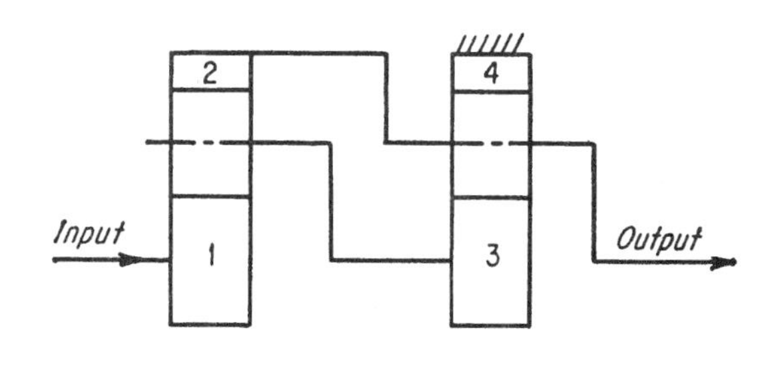

$$R = 1 + \frac{N_4}{N_3}\left(1 + \frac{N_2}{N_1}\right)$$

Output is difference between speeds of two parts leading to high reduction ratios

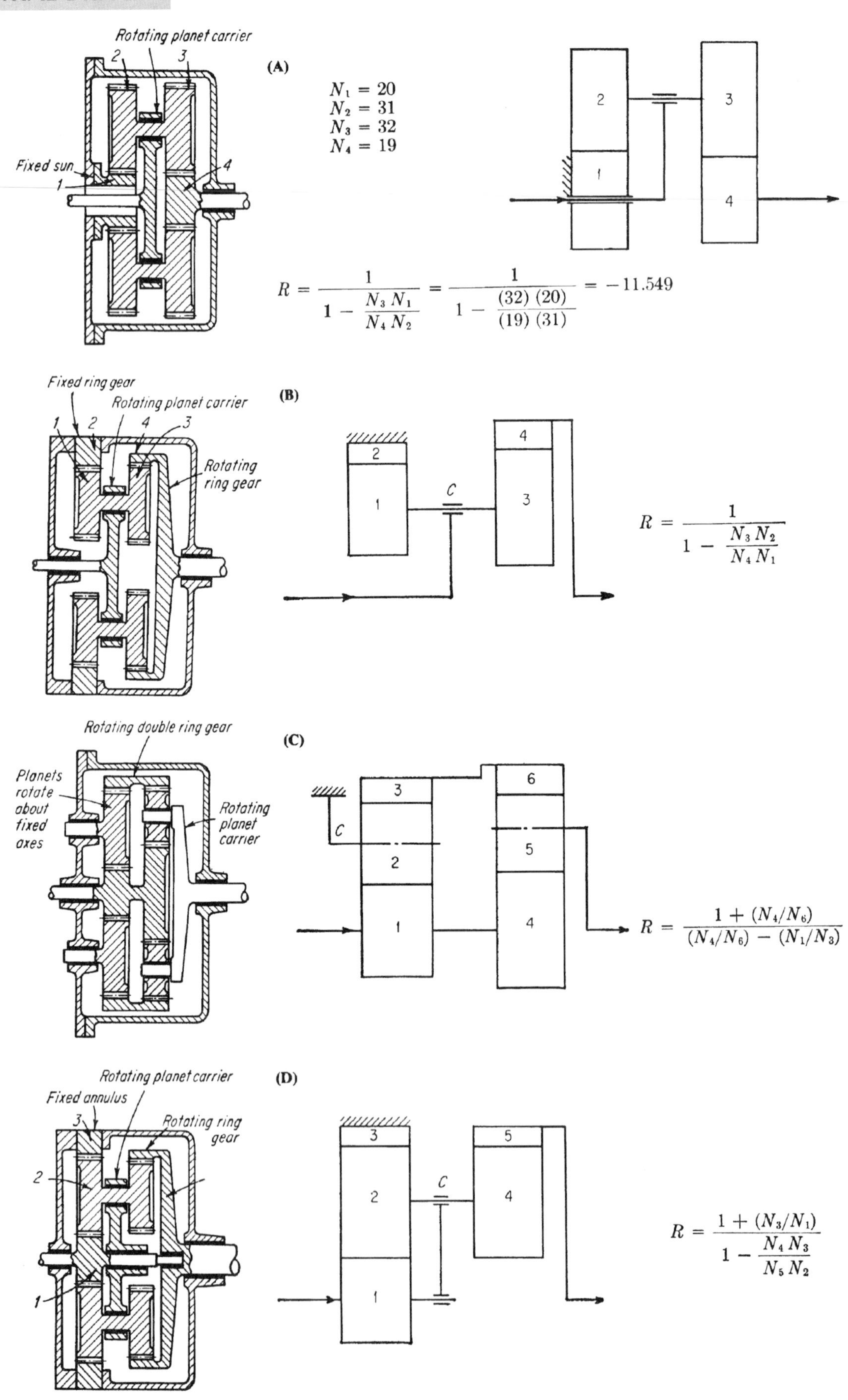

SIMPLE PLANETARIES AND INVERSIONS

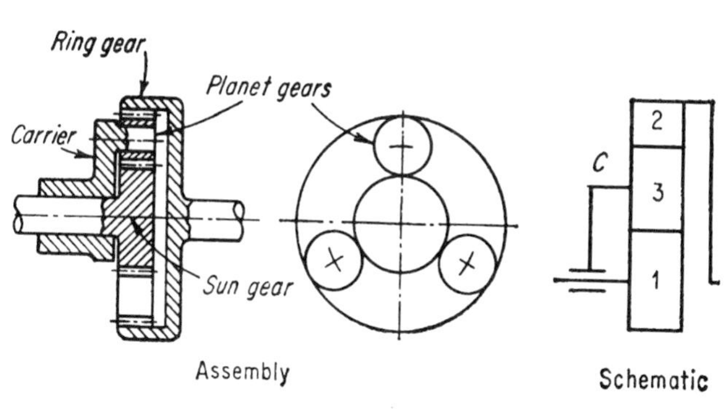

Input member	Fixed member	Output member	Speed-ratio equation
1	C	2	$R=-N_2/N_1$
2	C	1	$R=-N_1/N_2$
1	2	C	$R=1+(N_2/N_1)$
2	1	C	$R=1+(N_1/N_2)$
C	2	1	$R=\dfrac{1}{1+(N_2/N_1)}$
C	1	2	$R=\dfrac{1}{1+(N_1/N_2)}$

Input member	Fixed member	Output member	Speed-ratio equation
1	C	3	$R=\dfrac{N_2 N_3}{N_1 N_4}$
1	3	C	$R=1-\dfrac{N_2 N_3}{N_1 N_4}$
3	1	C	$R=1-\dfrac{N_1 N_4}{N_2 N_3}$
3	C	1	$R=\dfrac{N_4 N_1}{N_3 N_2}$
C	1	3	$R=1\Big/\left(1-\dfrac{N_1 N_4}{N_2 N_3}\right)$
C	3	1	$R=1\Big/\left(1-\dfrac{N_2 N_3}{N_1 N_4}\right)$

HUMPAGE'S BEVEL GEARS

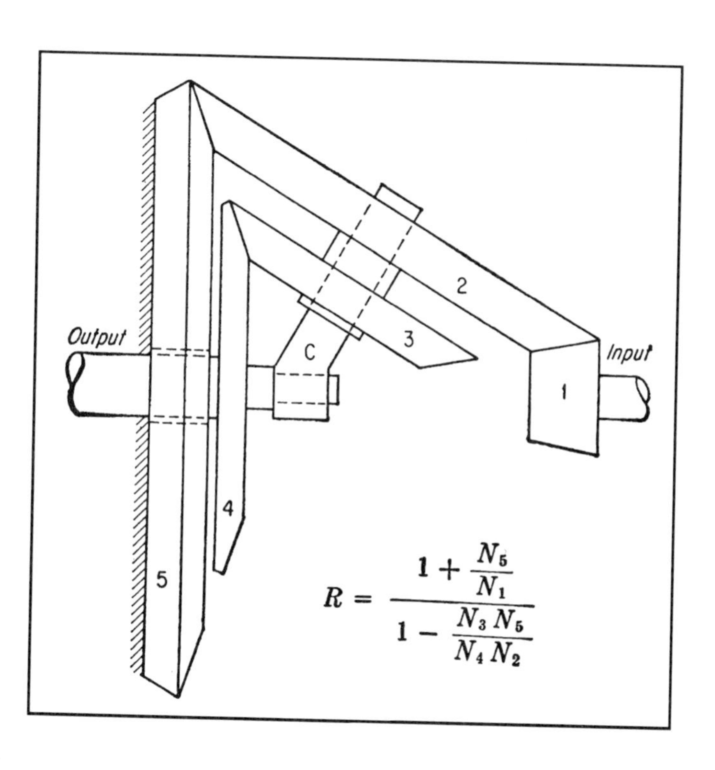

$$R=\frac{1+\dfrac{N_5}{N_1}}{1-\dfrac{N_3 N_5}{N_4 N_2}}$$

TWO-SPEED FORDOMATIC (Ford Motor Co.)

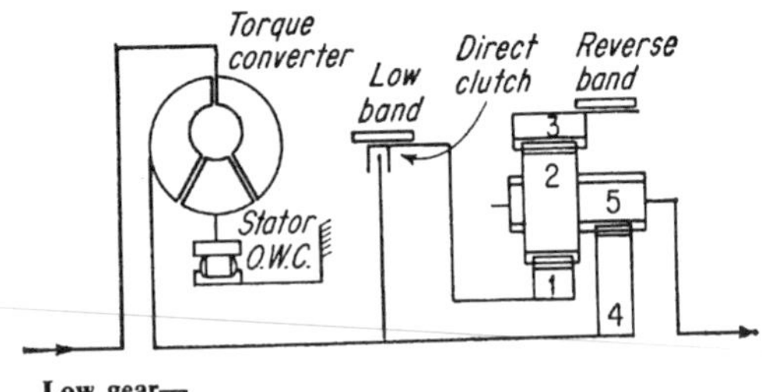

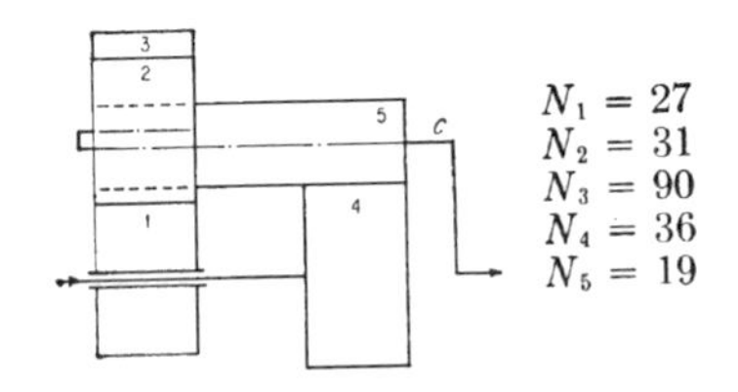

$N_1 = 27$
$N_2 = 31$
$N_3 = 90$
$N_4 = 36$
$N_5 = 19$

Low gear— gear 1 fixed

$$R = 1 + \frac{N_1}{N_4} = 1.75$$

Reverse gear— gear 3 fixed

$$R = 1 - \frac{N_3}{N_4} = -1.50$$

Note: Power-Glide Transmission is similar to above, but with $N_1 = 23$, $N_2 = 28$, $N_3 = 79$, $N_4 = 28$, $N_5 = 18$. This produces identical ratios in low and reverse.

$$R = 1 + \frac{23}{28} = 1.82 \qquad R = 1 - \frac{79}{28} = -1.82$$

CRUISE-O-MATIC 3-SPEED TRANSMISSION (Ford Motor Co.)

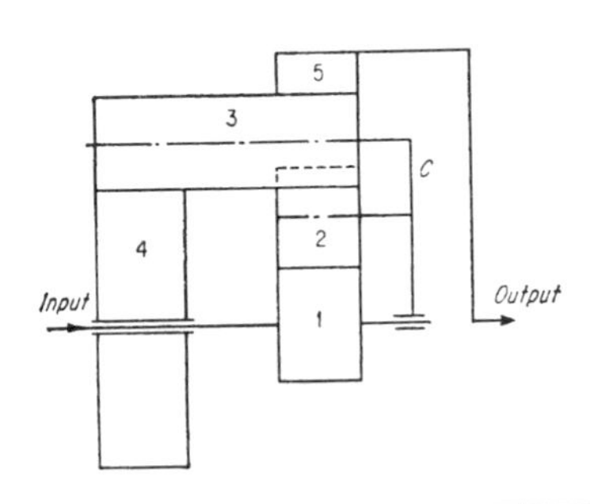

Long planet, $N_3 = 18$
Short planet, $N_2 = 18$
Sun gears, $N_4 = 36$, $N_1 = 30$
Ring gears, $N_5 = 72$

Low gear—Input to 1 C fixed

$$R = \frac{N_5}{N_1} = 2.4$$

Intermediate gear—
Input to 1, gear 4 fixed

$$R = \frac{1 + \frac{N_4}{N_1}}{1 + \frac{N_4}{N_5}} = 1.467$$

Reverse gear—
Input to 4, C fixed

$$R = \frac{N_5}{N_4} = -2.0$$

HYDRAMATIC 3-SPEED TRANSMISSION (General Motors)

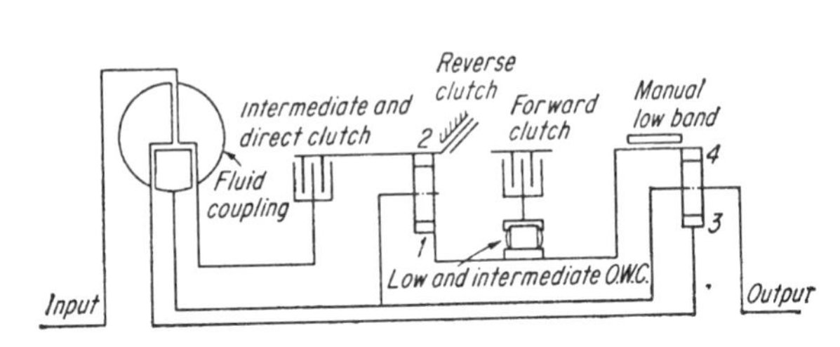

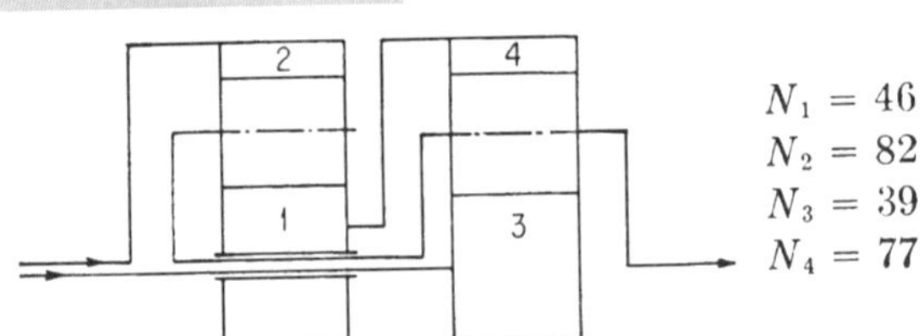

$N_1 = 46$
$N_2 = 82$
$N_3 = 39$
$N_4 = 77$

Low gear—
Input to 3, 4 fixed

$$R = 1 + \frac{N_4}{N_3} = 2.97$$

Intermediate gear— Input to 2, 1 fixed

$$= 1 + \frac{N_1}{N_2} = 1.56$$

Reverse gear— Input to 3, 2 fixed

$$R = 1 - \frac{N_4 N_2}{N_3 N_1} = -2.52$$

TRIPLE PLANETARY DRIVES

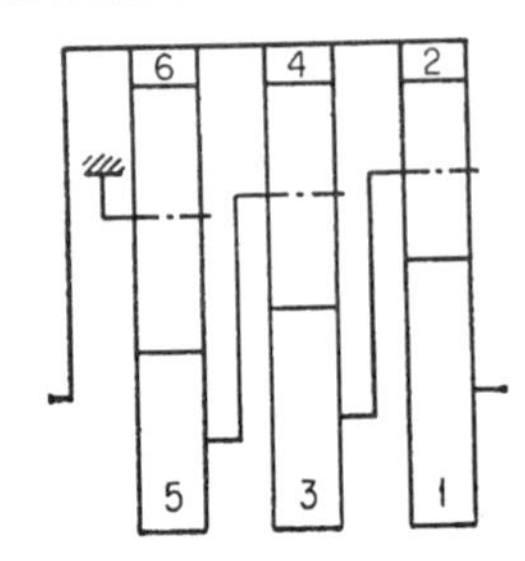

Input to gear 1, output from gear 6

$$R = \left(1 + \frac{N_2}{N_1}\right)\left[\left(1 + \frac{N_4}{N_3}\right)\left(-\frac{N_6}{N_5}\right) - \frac{N_4}{N_3}\right] - \frac{N_2}{N_1}$$

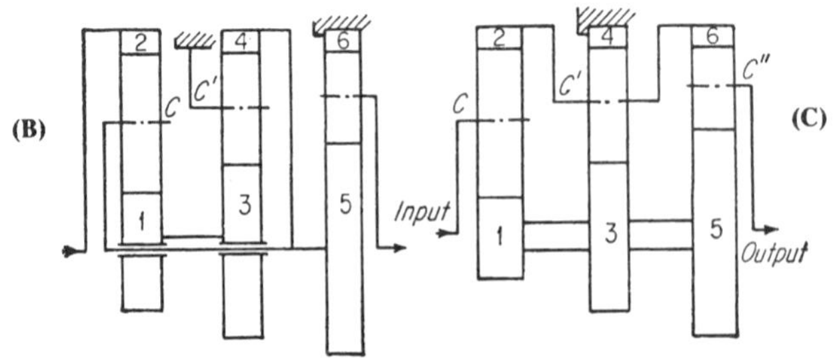

(B) $$R = \left[1 + \frac{N_1}{N_2}\left(1 + \frac{N_4}{N_3}\right)\right]\left(1 + \frac{N_6}{N_5}\right)$$

(C) $$R = \left[1 + \frac{N_4/N_3}{1 + (N_2/N_1)}\right] \Big/ \left[1 + \frac{N_4/N_3}{1 + (N_6/N_5)}\right]$$

FORD TRACTOR DRIVES

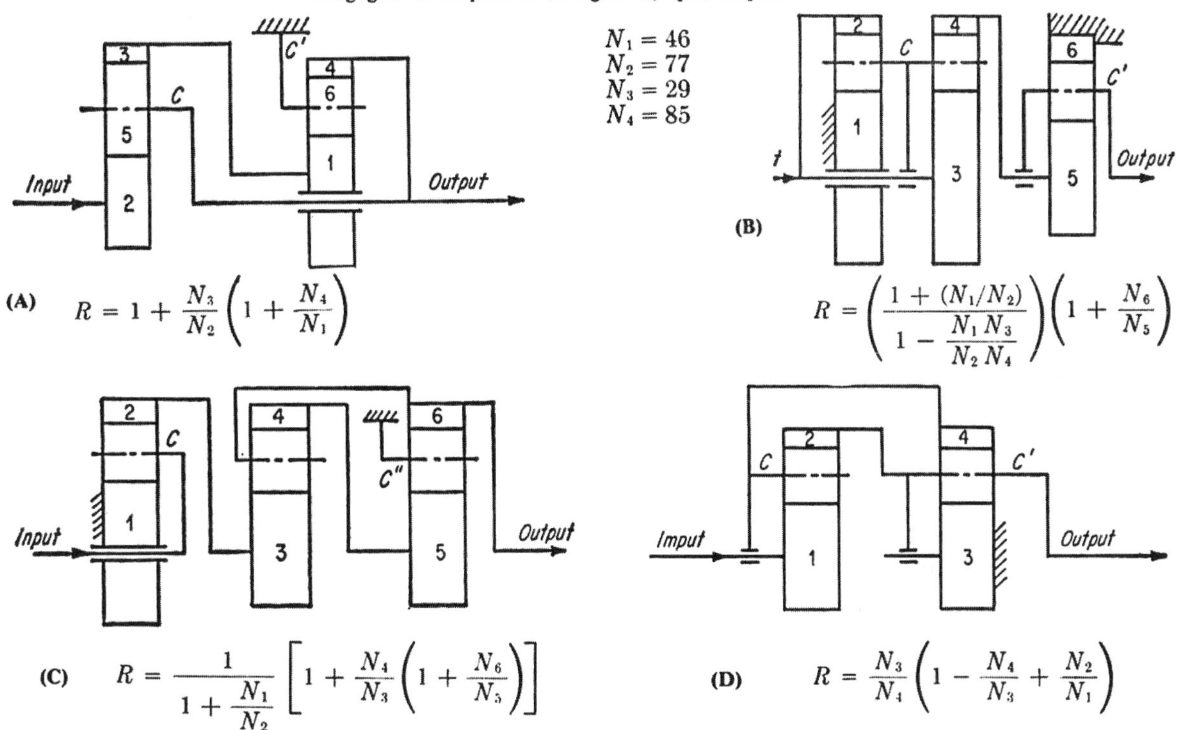

Ring gear 3 coupled to sun gear 1; split output.

$N_1 = 46$
$N_2 = 77$
$N_3 = 29$
$N_4 = 85$

(A) $$R = 1 + \frac{N_3}{N_2}\left(1 + \frac{N_4}{N_1}\right)$$

(B) $$R = \left(\frac{1 + (N_1/N_2)}{1 - \dfrac{N_1 N_3}{N_2 N_4}}\right)\left(1 + \frac{N_6}{N_5}\right)$$

(C) $$R = \frac{1}{1 + \dfrac{N_1}{N_2}}\left[1 + \frac{N_4}{N_3}\left(1 + \frac{N_6}{N_5}\right)\right]$$

(D) $$R = \frac{N_3}{N_4}\left(1 - \frac{N_4}{N_3} + \frac{N_2}{N_1}\right)$$

LYCOMING TURBINE DRIVE

$$R = \left(1 + \frac{N_3}{N_2}\right) \times \left(1 + \frac{N_4}{N_1}\right)$$

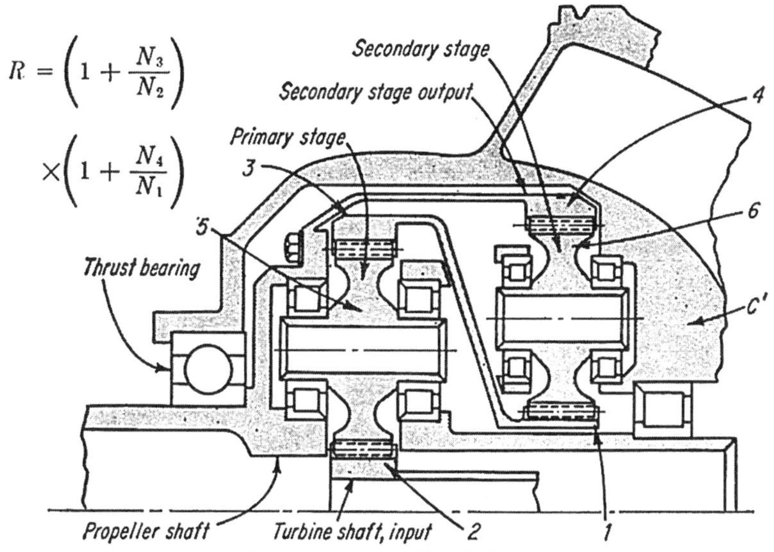

Input to sun gear 2, output to propeller shaft.

Basically same system as the Ford tractor drive, (gears are numbered the same way) and will have the same speed-ratio.

COMPOUND SPUR-BEVEL GEAR DRIVE

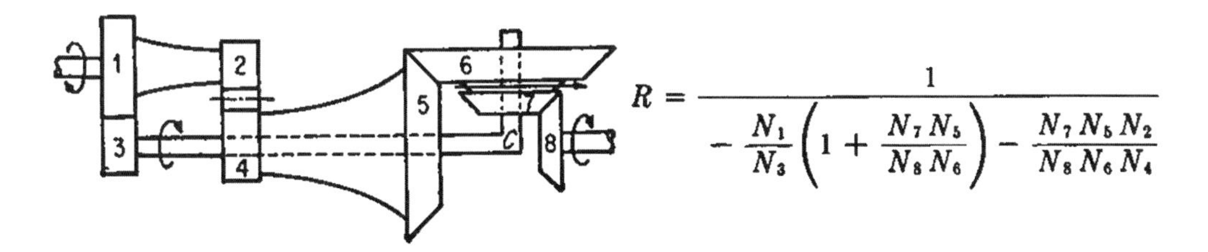

$$R = \frac{1}{-\dfrac{N_1}{N_3}\left(1 + \dfrac{N_7 N_5}{N_8 N_6}\right) - \dfrac{N_7 N_5 N_2}{N_8 N_6 N_4}}$$

TWO-GEAR PLANETARY DRIVES

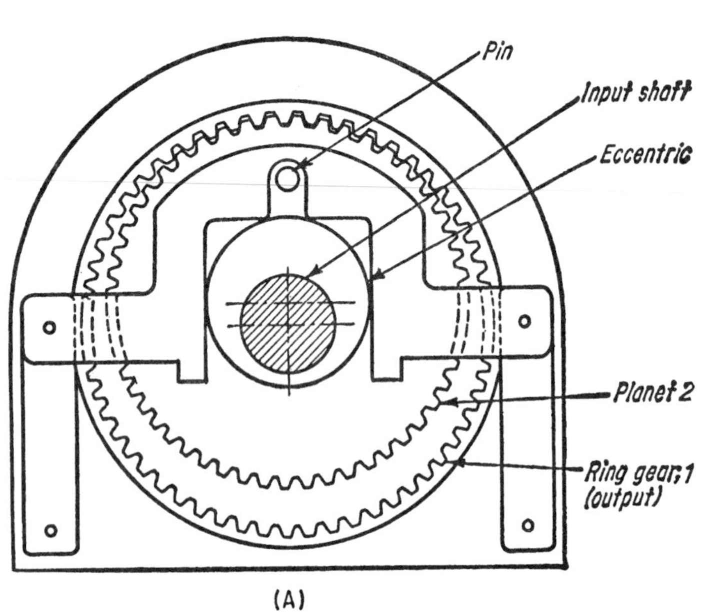

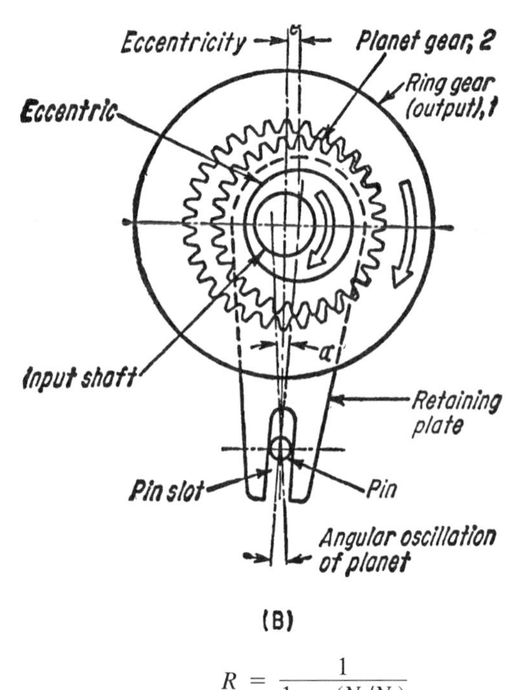

The pin in (A) is fixed to the planet which is mounted on the eccentric hub of the input shaft. The ring gear is the output gear. The system in (B) is simplified, but it produces slight pulsations in output.

$$R = \frac{1}{1 - (N_1/N_2)}$$

PLANOCENTRIC DRIVE

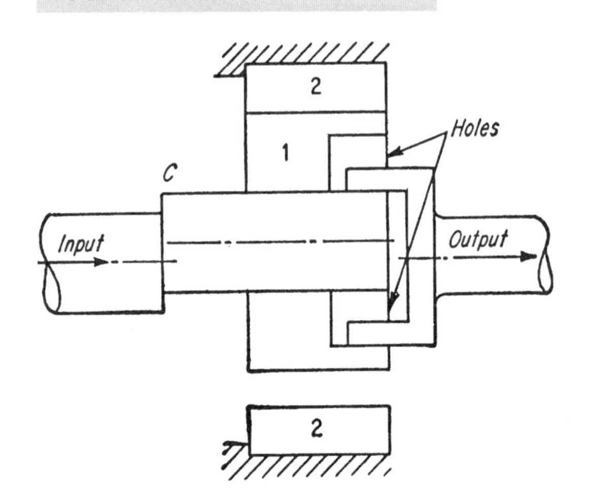

$N_2 = 65$
$N_2 = 64$

The planet gear 1 is eccentrically mounted to the input gear (plant 1 is not rigidly connected to the eccentric). The output is driven by holes.

$$R = \frac{N_1}{N_1 - N_2} = \frac{64}{64 - 65} = -64$$

WOBBLE-GEAR DRIVE

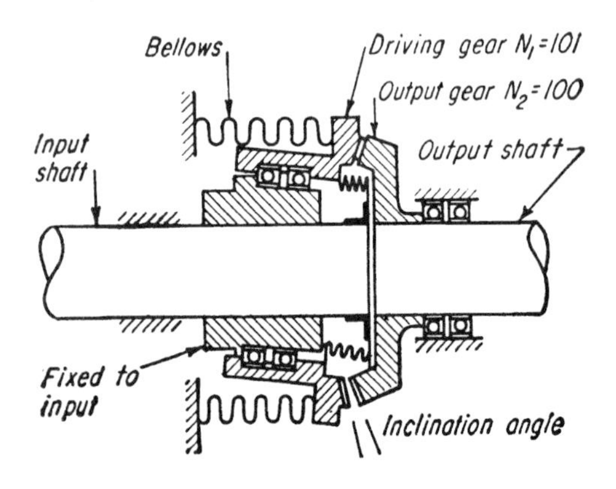

This drive is a close relative of the harmonic drive. The bevel "wobble" gear mesh at only one point on the circumference because of the slight angle of inclination of the driving gear, N_1, which has one tooth more than output gear, N_2. The driving gear, N_1, does not rotate: it yaws and pitches only.

$$R = R_i = \frac{1}{1 - m_{or}}$$

$$R = \frac{1}{1 - \dfrac{N_1}{N_2}} = \frac{1}{1 - \dfrac{101}{100}} = -100$$

NONCIRCULAR GEARS ARE BALANCED FOR SPEED

Noncircular gears generally cost more than competitive components such as linkages and cams. But with the development of modern production methods, such as the computer-controlled gear shaper, cost has gone down considerably. Also, in comparison with linkages, noncircular gears are more compact and balanced —and they can be more easily balanced. These are important considerations in high-speed machinery. Furthermore, the gears can produce continuous, unidirectional cyclic motion—a point in their favor when compared with cams. The disadvantage of cams is that they offer only reciprocating motion.

Applications can be classified into two groups:

- Where only an over-all change in angular velocity of the driven member is required, as in quick-return drives, intermittent mechanisms in such machines as printing presses, planers, shears, winding machines, and automatic-feed machines.
- Where precise, nonlinear functions must be generated, as in mechanical computing machines for extracting roots of numbers, raising numbers to any power, or generating trigonometric and logarithmic functions.

Noncircular Gears

It is always possible to design a specially shaped gear to roll and mesh properly with a gear of any shape. The sole requirement is that the distance between the two axes must be constant. However, the pitch line of the mating gear might turn out to be an open curve, and the gears can be rotated only for a portion of a revolution—as with two logarithmic-spiral gears (illustrated in Fig. 1).

True elliptical gears can only be made to mesh properly if they are twins, and if they are rotated about their focal points. However, gears resembling ellipses can be generated from a basic ellipse. These "higher-order" ellipses (see Fig. 2) can be meshed in various interesting combinations to rotate about centers *A, B, C,* or *D*. For example, two second-order elliptical gears can be meshed to rotate about their geometric center; however, they will produce two complete speed cycles per revolution. The difference in contour between a basic ellipse and a second-order ellipse is usually very slight. Note also that the fourth-order "ellipses" resemble square gears (this explains why the square gears, sometimes found as ornaments on tie clasps, illustrated in Fig. 3, actually work).

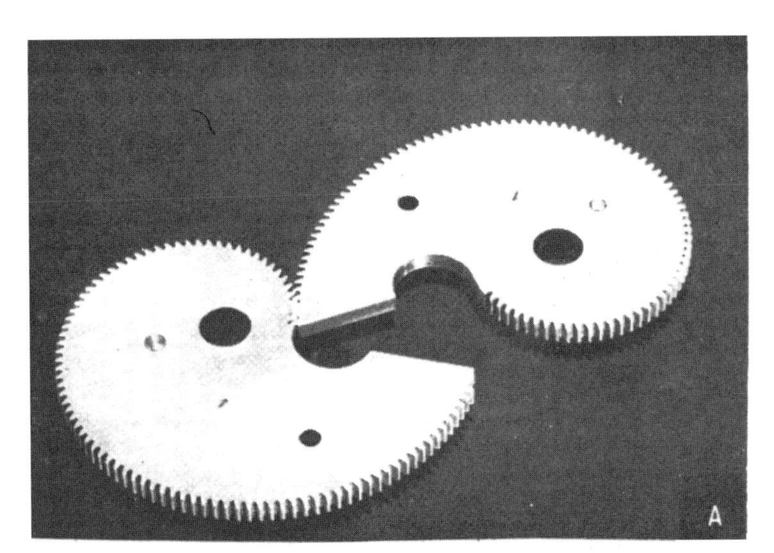

Fig. 1 The logarithmic spiral gears shown in (A), are open-curved. They are usually components in computing devices. The elliptically shaped gears, shown in (B), are closed curved. They are components in automatic machinery. The specially shaped gears, shown in (C), offer a wider range of velocity and acceleration characteristics.

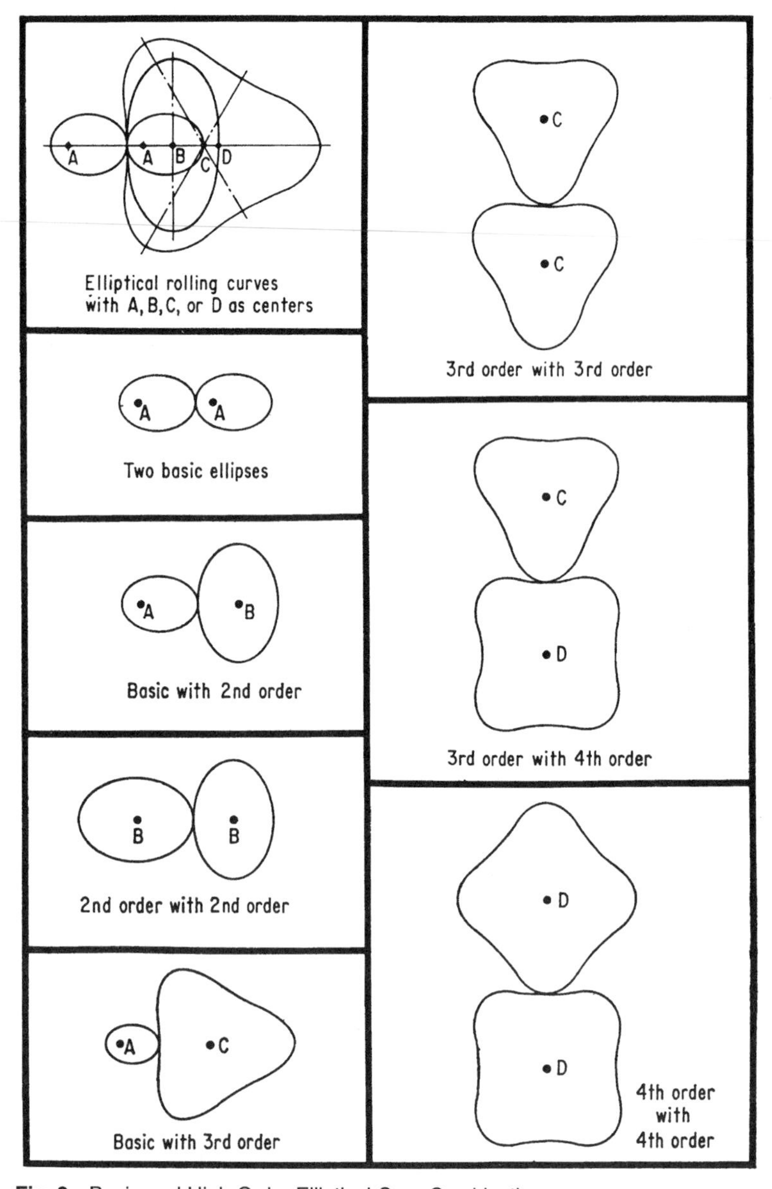

Fig. 2 Basic and High-Order Elliptical Gear Combinations.

A circular gear, mounted eccentrically, can roll properly only with specially derived curves (shown in Fig. 4). One of the curves, however, closely resembles an ellipse. For proper mesh, it must have twice as many teeth as the eccentric gear. When the radiis *r,* and eccentricity, *e,* are known, the major semiaxis of the elliptically shaped gear becomes $2r + e$, and the minor $2r - e$. Note also that one of the gears in this group must have internal teeth to roll with the eccentric gear. Actually, it is possible to generate internal-tooth shapes to rotate with noncircular gears of any shape (but, again, the curves can be of the open type).

Noncircular gears can also be designed to roll with specially shaped racks (shown in Fig. 5). Combinations include: an elliptical gear and a sinusoid-like rack. A third-order ellipse is illustrated, but any of the elliptical rolling curves can be used in its place. The main advantage of those curves is that when the ellipse rolls, its axis of rotation moves along a straight line; other combinations include a logarithmic spiral and straight rack. The rack, however, must be inclined to its direction of motion by the angle of the spiral.

DESIGN EQUATIONS

Equations for noncircular gears are given here in functional form for three common design requirements. They are valid for any noncircular gear pair. Symbols are defined in the box.

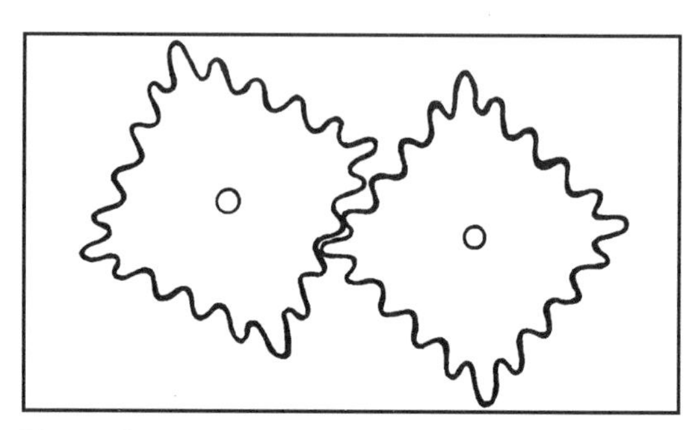

Fig. 3 Square gears seem to defy basic kinematic laws, but they are a takeoff on a pair of fourth-order ellipses.

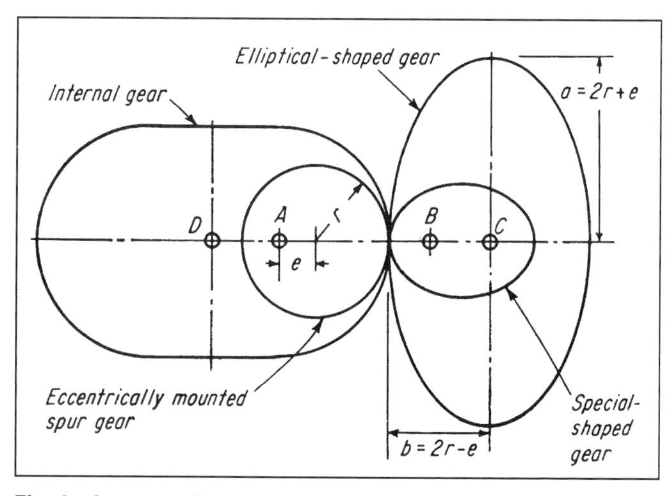

Fig. 4 An eccentric spur gear rotating about point **A**, will mesh properly with any of the three gears shown whose centers are at points B, C and D.

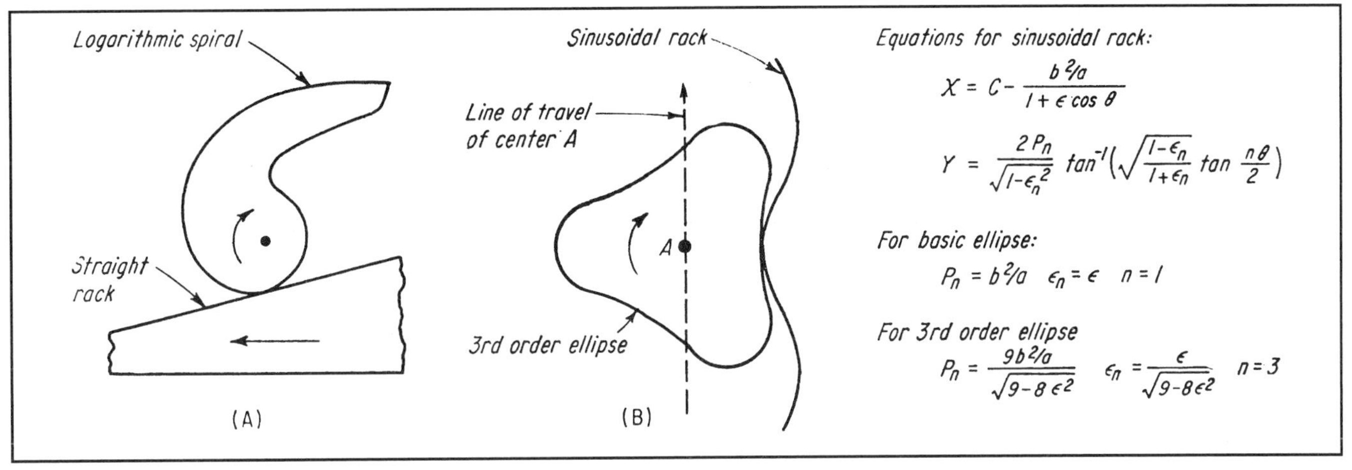

Fig. 5 Rack and gear combinations are possible with noncircular gears. The straight rack for the logarithmic spiral (A) must move obliquely; the center of third-order ellipse (B) follows a straight line.

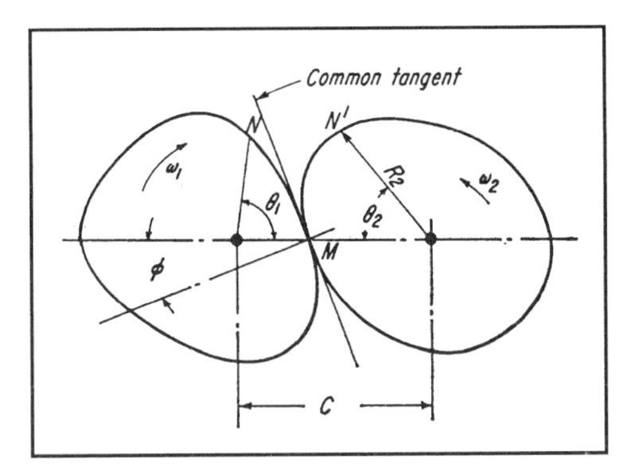

Symbols

- a = semi-major axis of ellipse
- b = semi-minor axis of ellipse
- C = center distance (see above sketch)
- ϵ = eccentricity of an ellipse = $\sqrt{1-(b/a)^2}$
- e = eccentricity of an eccentrically mounted spur gear
- N = number of teeth
- P = diametral pitch
- r_c = radius of curvature
- R = active pitch radius
- S = length of periphery of pitch circle
- X, Y = rectangular coordinates
- θ = polar angle to R
- ϕ = angle of obliquity
- ω = angular velocity
- $f(\theta)$, $F(\theta)$, $G(\theta)$ = various functions of θ
- $f'(\theta)$, $F'(\theta)$, $G'(\theta)$ = first derivatives of functions of θ

CASE I Polar equation of one curve and center distance are known; to find the polar equation of the mating gear:

$$R_1 = f(\theta_1)$$

$$R_2 = C - f(\theta_1)$$

$$\theta_2 = -\theta_1 + C\int \frac{d\theta_1}{C - f(\theta_1)}$$

CASE II The relationship between angular rotation of the two members and the center distance are known; to find the polar equations of both members:

$$\theta_2 = F(\theta_1)$$

$$R_1 = \frac{CF'(\theta_1)}{1 + F'(\theta_1)}$$

$$R_2 = C - R_1 = \frac{C}{1 + F'(\theta_1)}$$

CASE III The relationship between angular velocities of the two members and the center distance are known; to find the polar equations of both members:

$$\omega_2 = \omega_1 G(\theta_1)$$

$$R_1 = \frac{CG(\theta_1)}{1 + G(\theta_1)}$$

$$R_2 = C - R_1$$

$$\theta_1 = \int G(\theta_1) d\theta_1$$

Velocity equations and the characteristics of five types of noncircular gears are listed in the table.

CHECKING FOR CLOSED CURVES

Gears can be quickly analyzed to determine whether their pitch curves are open or close with the following equations:

In case I, if $R = f(\theta) = f(\theta + 2N_\pi)$, the pitch curve is closed.

In case II, if $\theta_1 = F(\theta_2)$ and $F(\theta_0) = 0$, the curve is closed with the equation $F(\theta_0 + 2_\pi/N_1) = 2_\pi/N_2$ can be satisfied by substituting integers or rational fractions for N_1 and N_2. If fractions must be used to solve this equation, the curve will have double points (intersect itself), which is, or course, an undesirable condition.

In case III, if $\theta_2 = \int G(\theta_1)d\theta_1$, let $G(\theta_1)d\,\theta_1 = F(\theta_1)$, and use the same method as for Case II, with the subscripts reversed.

With some gear sets, the mating gear will be a closed curve only if the correct center distance is employed. This distance can be found from the equation:

$$4\pi = \int_0^{2\pi} \frac{d\theta_1}{C - f(\theta_1)}$$

Characteristics of Five Noncircular Gear Systems

Type	Comments	Basic equations	Velocity equations ω_1 = constant
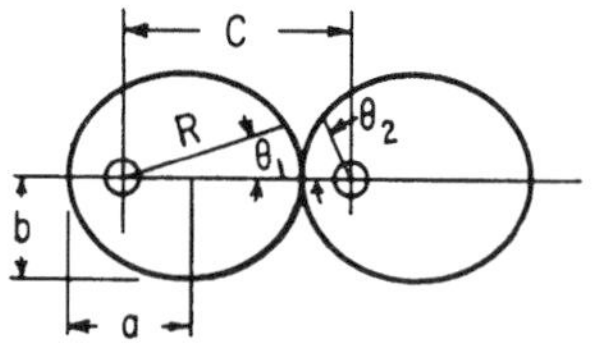 Two ellipses rotating about foci	Gears are identical. Comparatively easy to manufacture. Used for quick-return mechanisms, printing presses, automatic machinery	$R = \frac{b^2}{a[1+\epsilon\cos\theta]}$ ϵ = eccentricity $= \sqrt{1-\left(\frac{b}{a}\right)^2}$ $a = \frac{1}{2}$ major axis $b = \frac{1}{2}$ minor axis	$\omega_2 = \omega_1\left[\frac{r^2+1+(r^2-1)\cos\theta_2}{2r}\right]$ where $r = \frac{R\ max}{R\ min}$ 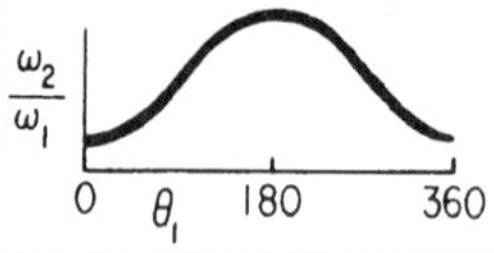
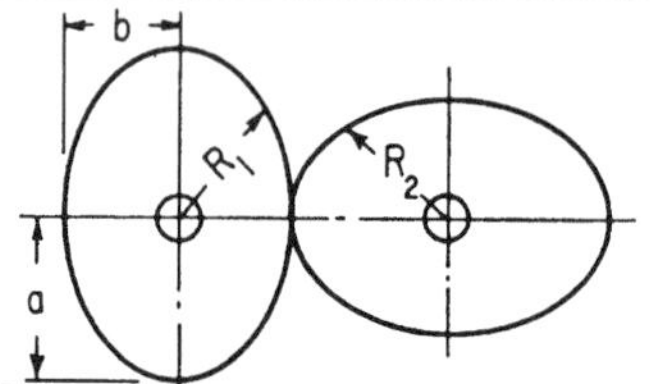 2nd Order elliptical gears rotating about their geometric centers	Gears are identical. Geometric properties well known. Better balanced than true elliptical gears. Used where two complete speed cycles are required for one revolution	$R = \frac{2ab}{(a+b)-(a-b)\cos 2\theta}$ $C = a + b$ a = maximum radius b = minimum radius	$\omega_2 = \omega_1\left[\frac{r^2+1-(r^2-1)\cos 2\theta_2}{2r}\right]$ where $r = \frac{a}{b}$ 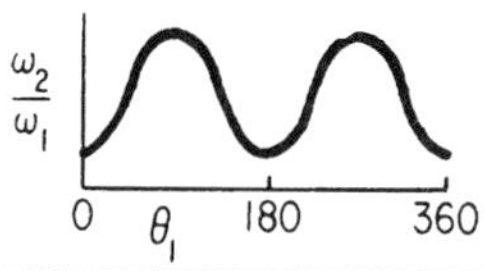
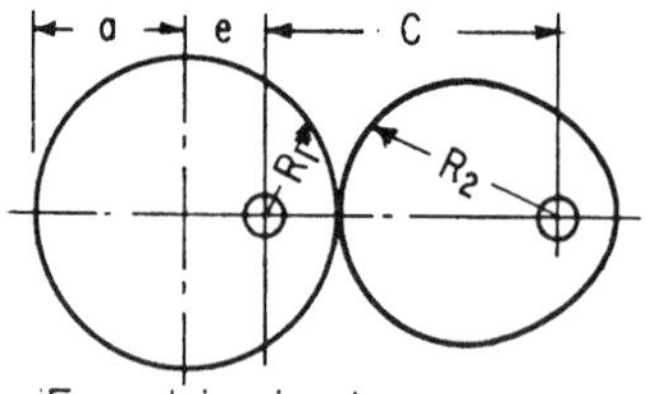 Eccentric circular gear rotating with its conjugate	Standard spur gear can be employed as the eccentric. Mating gear has special shape	$R_1 = \sqrt{a^2+e^2+2ae\cos\theta_1}$ $\theta_2 = \theta_1 + C\int\frac{d\theta_1}{C-R_1}$ $C = R_1 + R_2$	$\frac{\omega_2}{\omega_1} = \frac{\sqrt{a^2+e^2+2ae\cos\theta_1}}{C-\sqrt{a^2+e^2+2ae\cos\theta_1}}$ 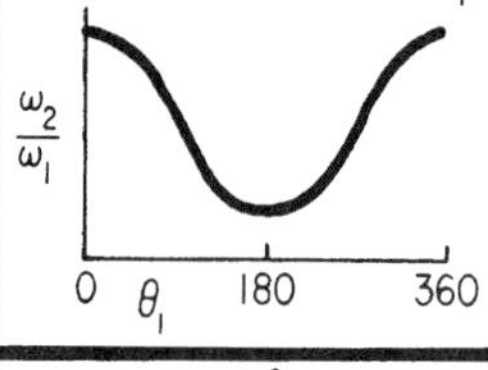
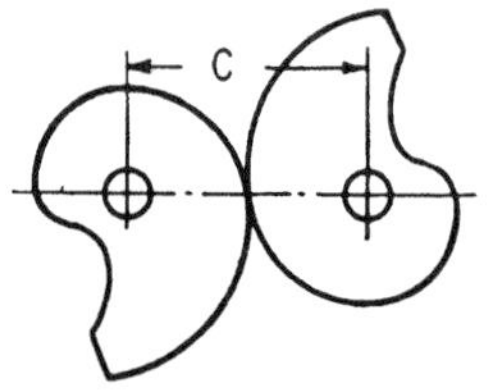 Logarithmic spiral gears	Gears can be identical although can be used in conbinations to give variety of functions. Must be open gears	$R_1 = Ae^{k\theta_1}$ $R_2 = C - R_1$ $= Ae^{k\theta_2}$ $\theta_2 = \frac{1}{k}\log(C-Ae)^{k\theta_1}$ e = natural log base	$\frac{\omega_2}{\omega_1} = \frac{Ae^{k\theta_1}}{C-Ae^{k\theta_1}}$ 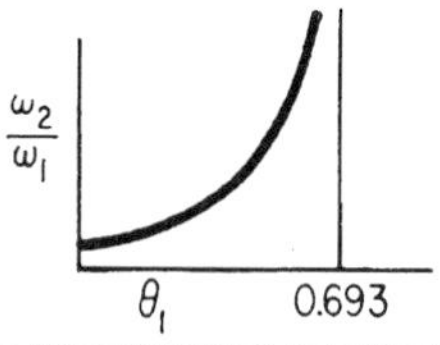
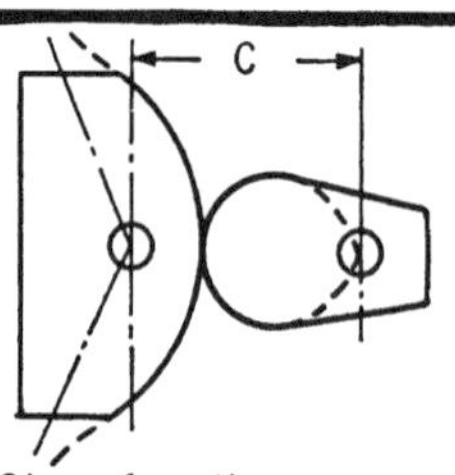 Sine-function gears	For producing angular displacement proportional to sine of input angle. Must be open gears	$\theta_2 = \sin^{-1}(k\theta_1)$ $R_2 = \frac{C}{1+k\cos\theta_1}$ $R_1 = C - R_2$ $= \frac{Ck\cos\theta_1}{1+k\cos\theta_1}$	$\frac{\omega_2}{\omega_1} = k\cos\theta_1$ 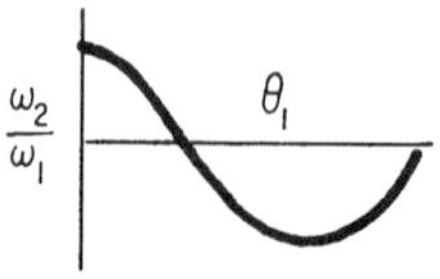

SHEET-METAL GEARS, SPROCKETS, WORMS, AND RATCHETS FOR LIGHT LOADS

When a specified motion must be transmitted at intervals rather than continuously, and the loads are light, these mechanisms are ideal because of their low cost and adaptability to mass production.

Although not generally considered precision parts, ratchets and gears can be stamped to tolerances of ± 0.007 in, and if necessary, shaved to close dimensions.

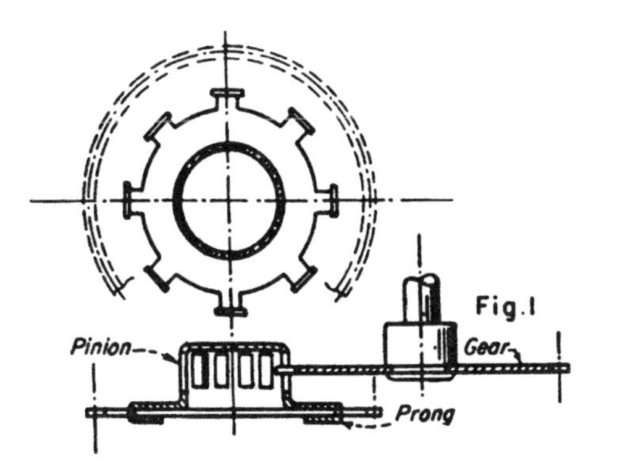

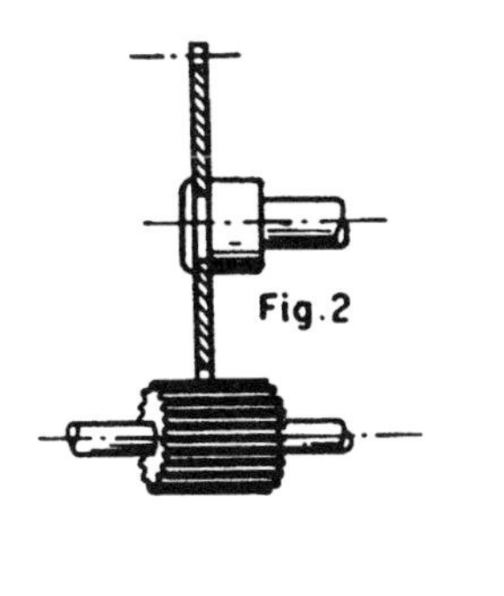

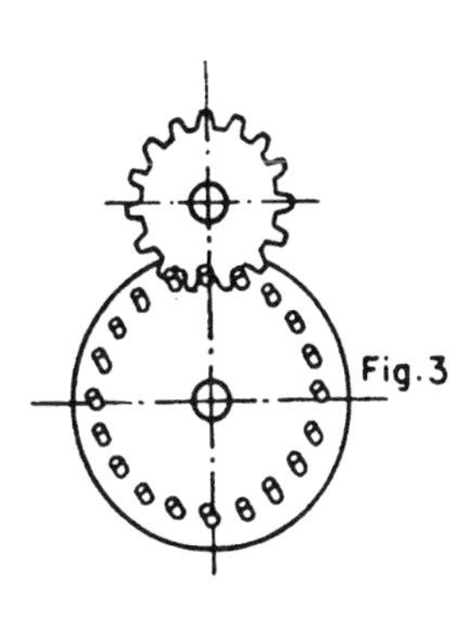

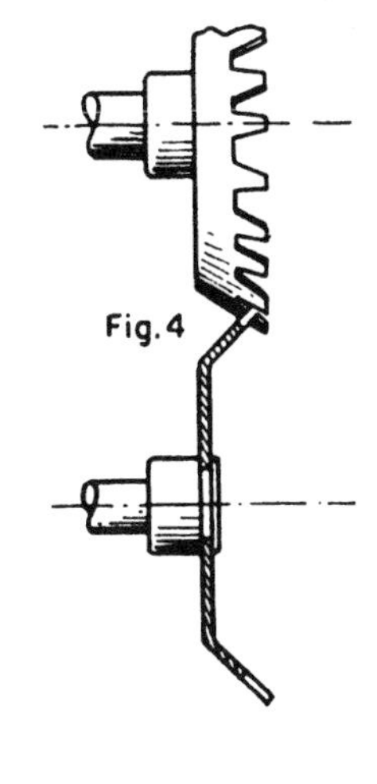

Fig. 1 The pinion is a sheet metal cup with rectangular holes serving as teeth. The meshing gear is sheet metal, blanked with specially formed teeth. The pinion can be attached to another sheet metal wheel by prongs, as shown, to form a gear train.

Fig. 2 The sheet-metal wheel gear meshes with a wide-face pinion, which is either extruded or machined. The wheel is blanked with teeth of conventional form.

Fig. 3 The pinion mates with round pins on a circular disk made of metal, plastic or wood. The pins can be attached by staking or with threaded fasteners.

Fig. 4 Two blanked gears, conically formed after blanking, become bevel gears meshing on a parallel axis. Both have specially formed teeth.

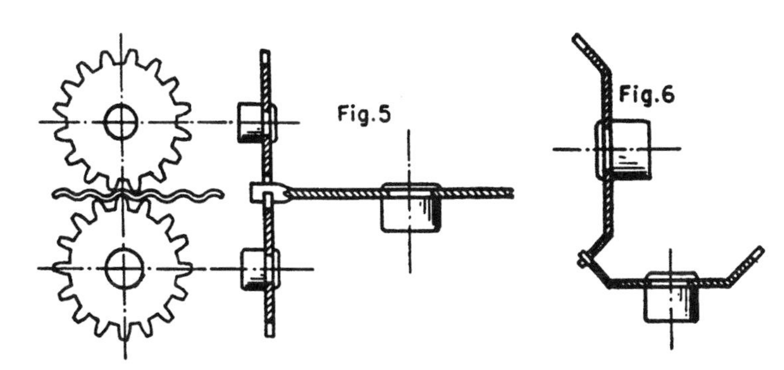

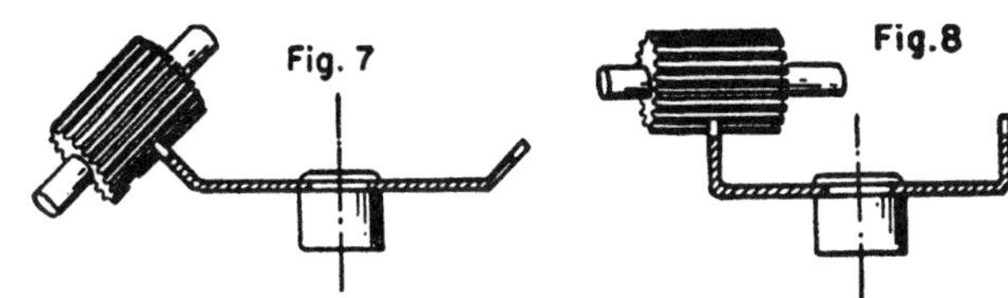

Fig. 7 The blanked and formed bevel-type gear meshes with a machined or extruded pinion. Conventional teeth can be used on both the gear and pinion.

Fig. 8 The blanked, cup-shaped wheel meshes with a solid pinion on 90° intersecting axes.

Fig. 5 The horizontal wheel with waves on its out rim replacing teeth, meshes with either one or two sheet-metal pinions. They have specially formed teeth and are mounted on intersecting axes.

Fig. 6 Two bevel-type gears, with specially formed teeth, are mounted on 90° intersecting axes. They can be attached by staking them to hubs.

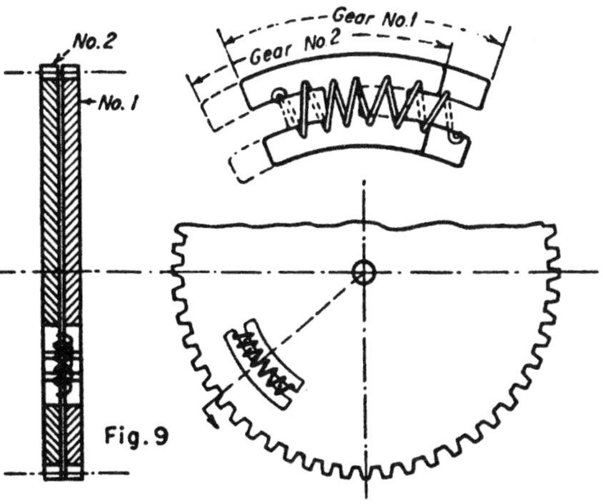

Fig. 9 Backlash can be eliminated from stamped gears by stacking two identical gears and displacing them by one tooth. The spring then bears one projection on each gear, taking up lost motion.

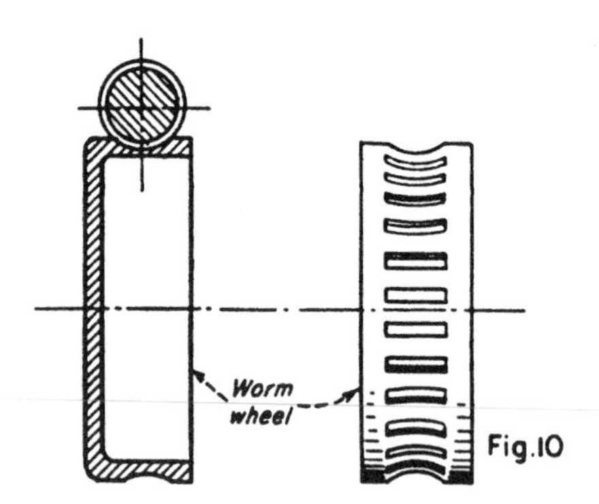

Fig. 10 A sheet metal cup with indentations replacing worm-wheel teeth, meshes with a standard coarse-thread screw.

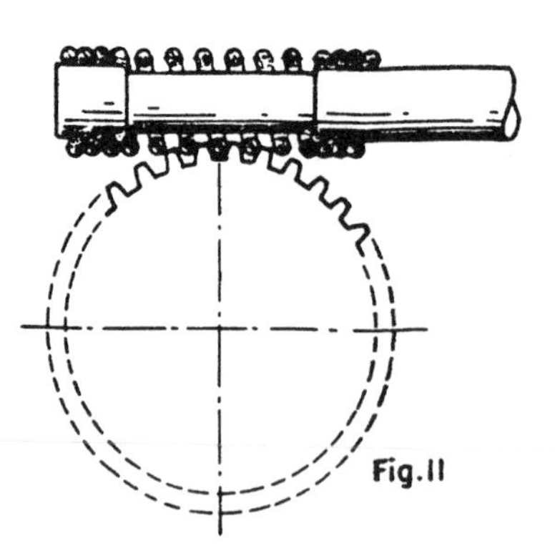

Fig. 11 A blanked wheel, with specially formed teeth, meshes with a helical spring mounted on a shaft, which serves as the worm.

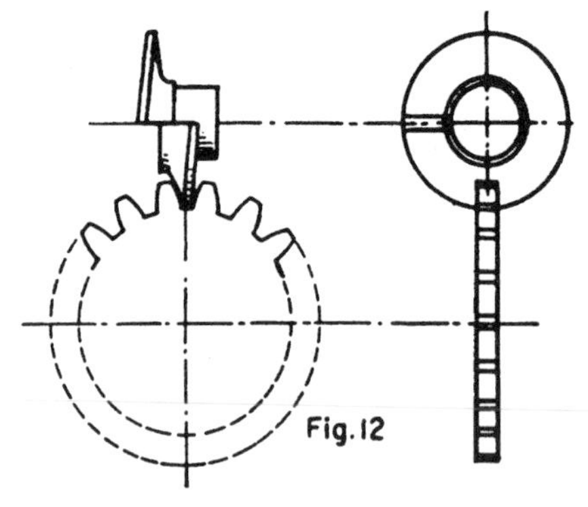

Fig. 12 This worm wheel is blanked from sheet metal with specially formed teeth. The worm is a sheet-metal disk that was split and helically formed.

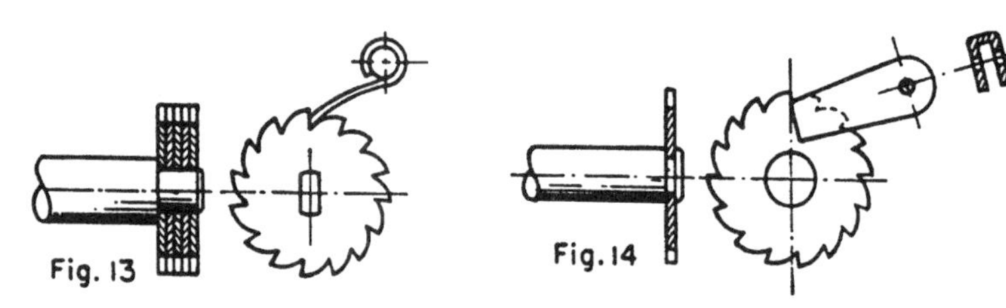

Fig. 13 Blanked ratchets with one-sided teeth are stacked to fit a wide-sheet-metal finger when single thickness is inadequate. The ratchet gears can be spot-welded.

Fig. 14 To avoid stacking, a single ratchet is used with a U-shaped finger, also made of sheet metal.

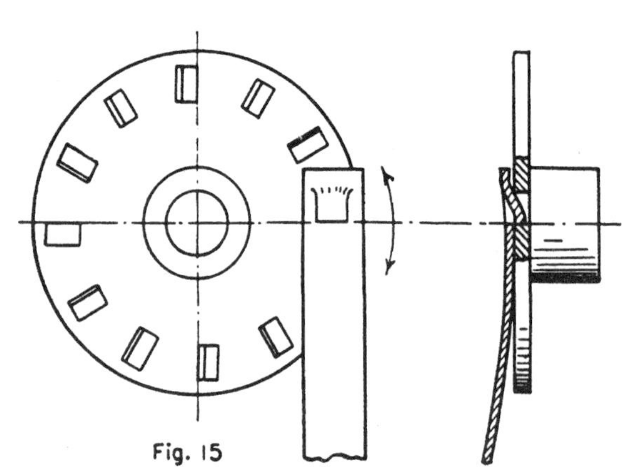

Fig. 15 This wheel is a punched disk with square-punched holes serving as teeth. The pawl is spring steel.

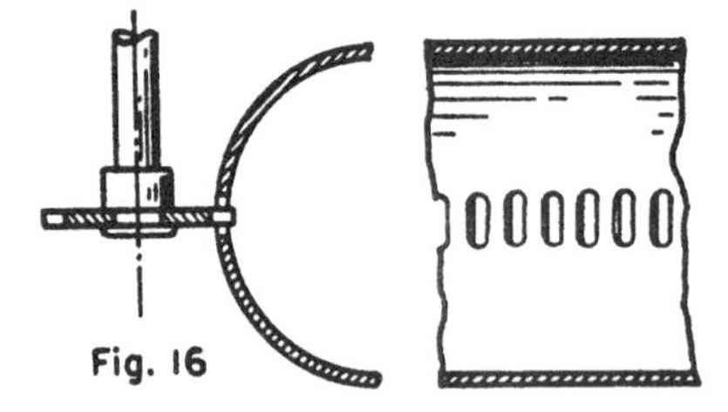

Fig. 16 This sheet-metal blanked pinion, with specially formed teeth, meshes with windows blanked in a sheet metal cylinder. They form a pinion-and-rack assembly.

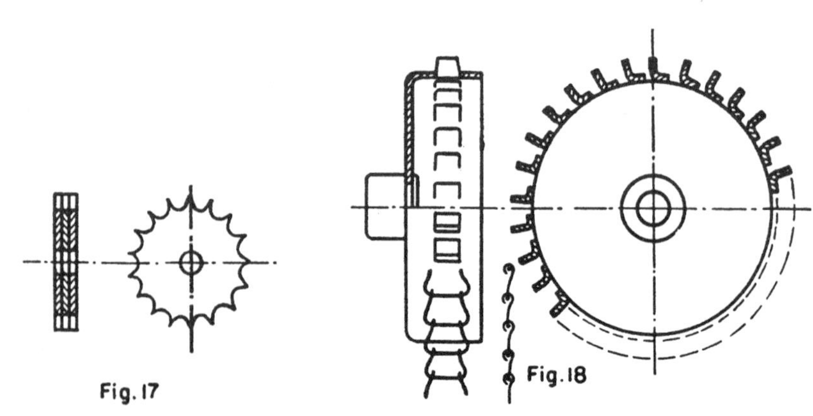

Fig. 17 This sprocket, like that in Fig. 13, can be fabricated from separate stampings.

Fig. 18 For a wire chain as shown, the sprocket is made by bending out punched teeth on a drawn cup.

THIRTEEN WAYS GEARS AND CLUTCHES CAN CHANGE SPEED RATIOS

13 ways of arranging gears and clutches to obtain changes in speed ratios

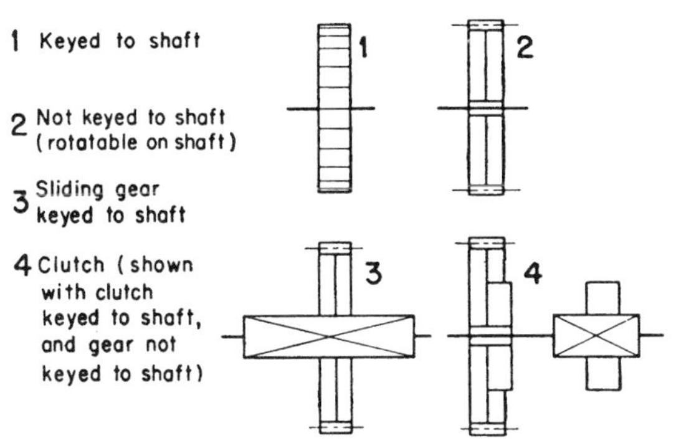

Fig. 1 The schematic symbols used in the following illustrations to represent gears and clutches.

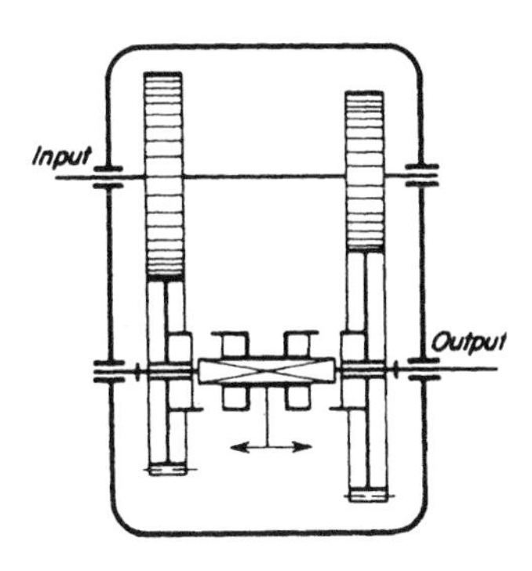

Fig. 2 Double-clutch drive. Two pairs of gears are permanently in mesh. Pair I or II transmits motion to the output shaft depending on the position of the coupling; the other pair idles. The coupling is shown in a neutral position with both gear pairs idle. Herring-bone gears are recommended for quieter running.

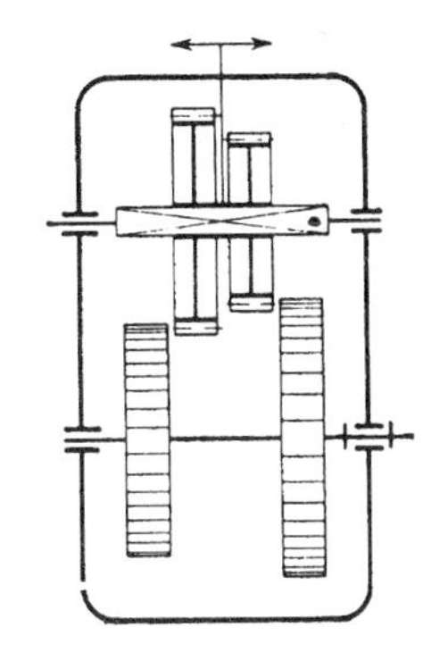

Fig. 3 Sliding-change drive. Gears are meshed by lateral sliding. Up to three gears can be mounted on a sliding sleeve. Only one pair is in mesh in any operating position. This drive is simpler, cheaper, and more extensively used than the drive of Fig. 2. Chamfering the sides of the teeth eases their engagement.

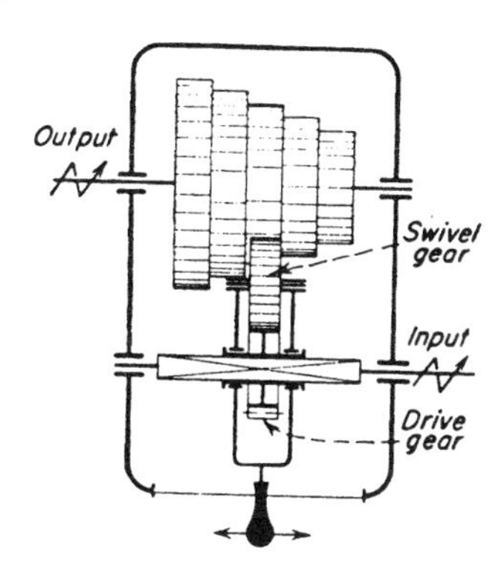

Fig. 4 Swivel-gear drive. Output gears are fastened to the shaft. A handle is pushed down, then shifted laterally to obtain transmission through any output gear. This drive is not suitable for the transmission of large torques because the swivel gear tends to vibrate. Its overall ratio should not exceed 1:3.

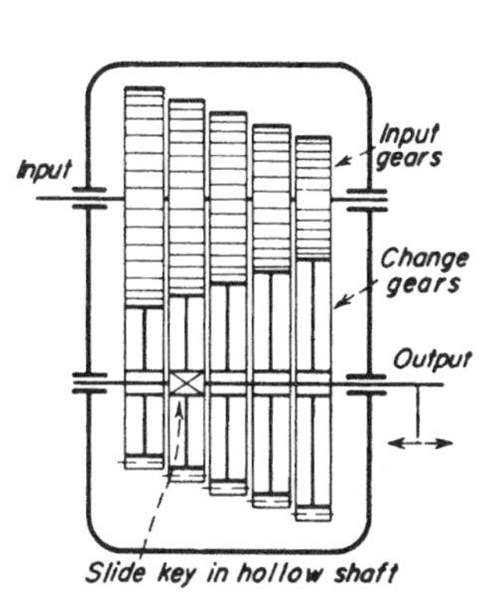

Fig. 5 Slide-key drive. A spring-loaded slide key rides inside a hollow output shaft. The slide key snaps out of the shaft when it is in position to lock a specific change gear to the output shaft. No central position is shown.

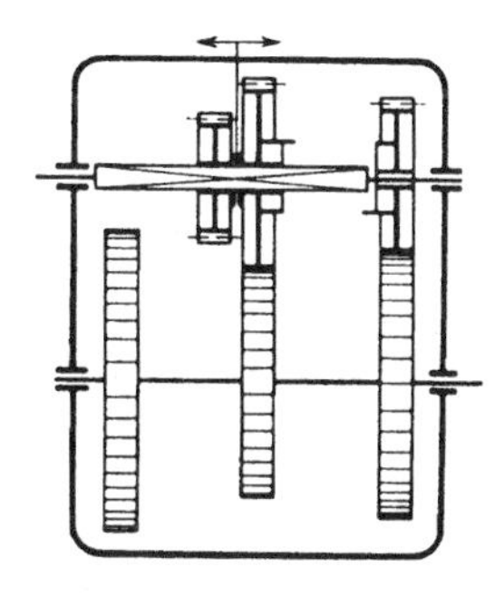

Fig. 6 This is a combination coupling and slide gears. It has three ratios: a direct mesh for ratios I and II; a third ratio is transmitted through gears II and III, which couple together.

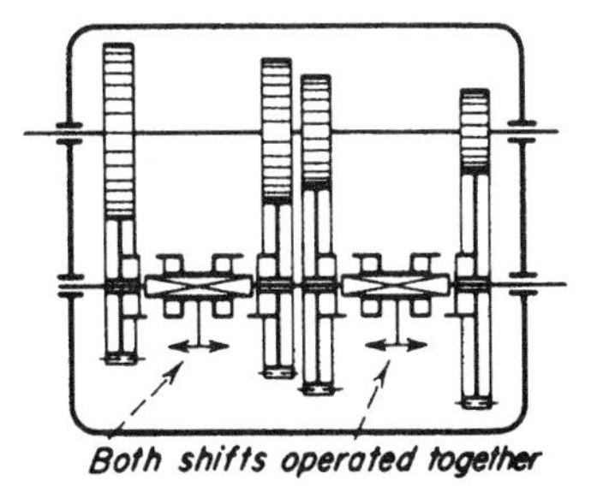

Fig. 7 Double-shift drive. One shift must always be in a neutral position. That might require both levers to be shifted when making a change. However, only two shafts are used to achieve four ratios.

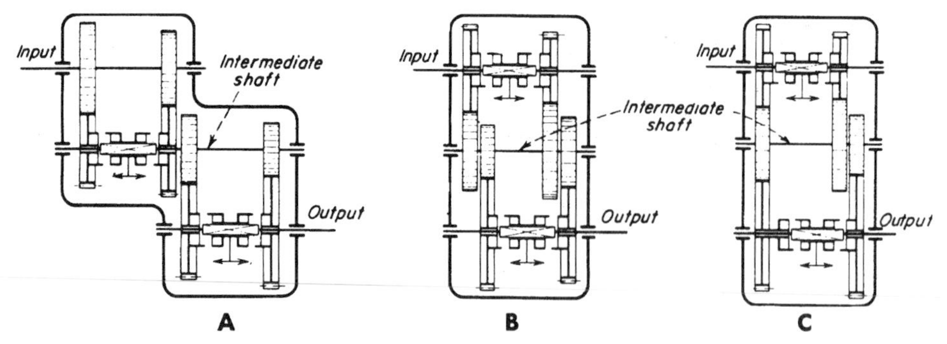

Fig. 8 A triple shaft drive gives four ratios. (A) The output of the first drive serves as the input for the second. The presence of an intermediate shaft eliminates the requirement for ensuring that one shift is always in the neutral position. A wrong shift-lever position cannot cause damage. (B) A space-saving modification; the coupling is on shaft *A* instead of the intermediate shaft (C) Still more space is saved if one gear replaces a pair on the intermediate shaft. Ratios can be calculated to allow this.

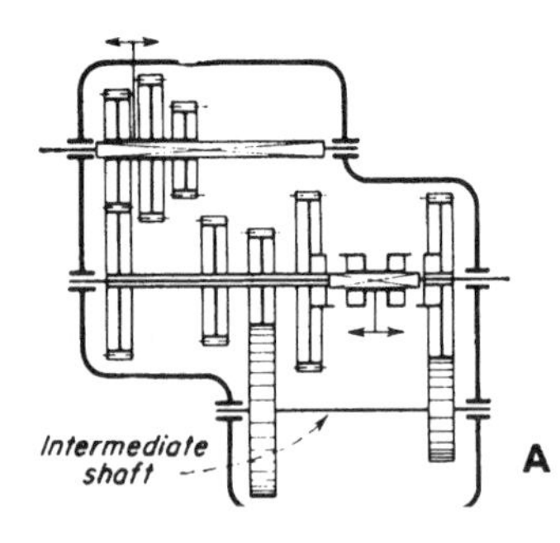

Fig. 9 Six ratios are available with two couplings and (A) ten gears, (B) eight gears. Up to six gears can be in permanent mesh. It is not necessary to ensure that one shift is in neutral.

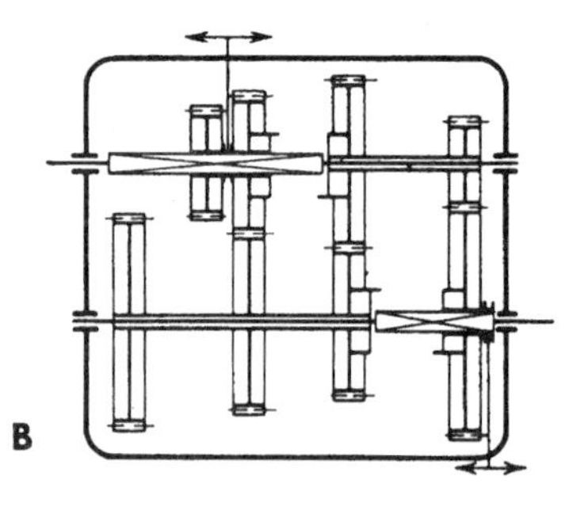

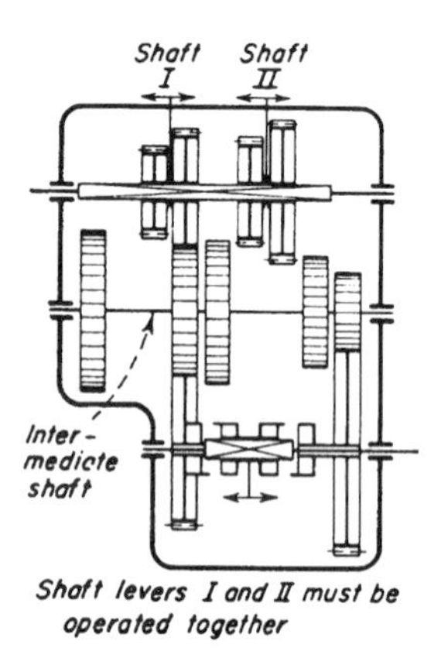

Fig. 10 This eight-ratio drive has two slide gears and a coupling. This arrangement reduces the number of parts and meshes. The position of shifts I and II are interdependent. One shift must be in neutral if the other is in mesh.

Fig. 11 This drive has eight ratios; a coupled gear drive and slide-key drive are in series. Comparatively low strength of the slide key limits the drive to small torque.

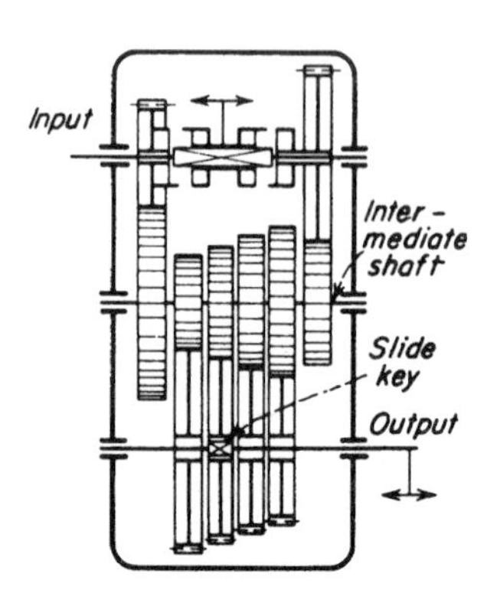

GEAR AND CLUTCH SHIFTING MECHANISMS

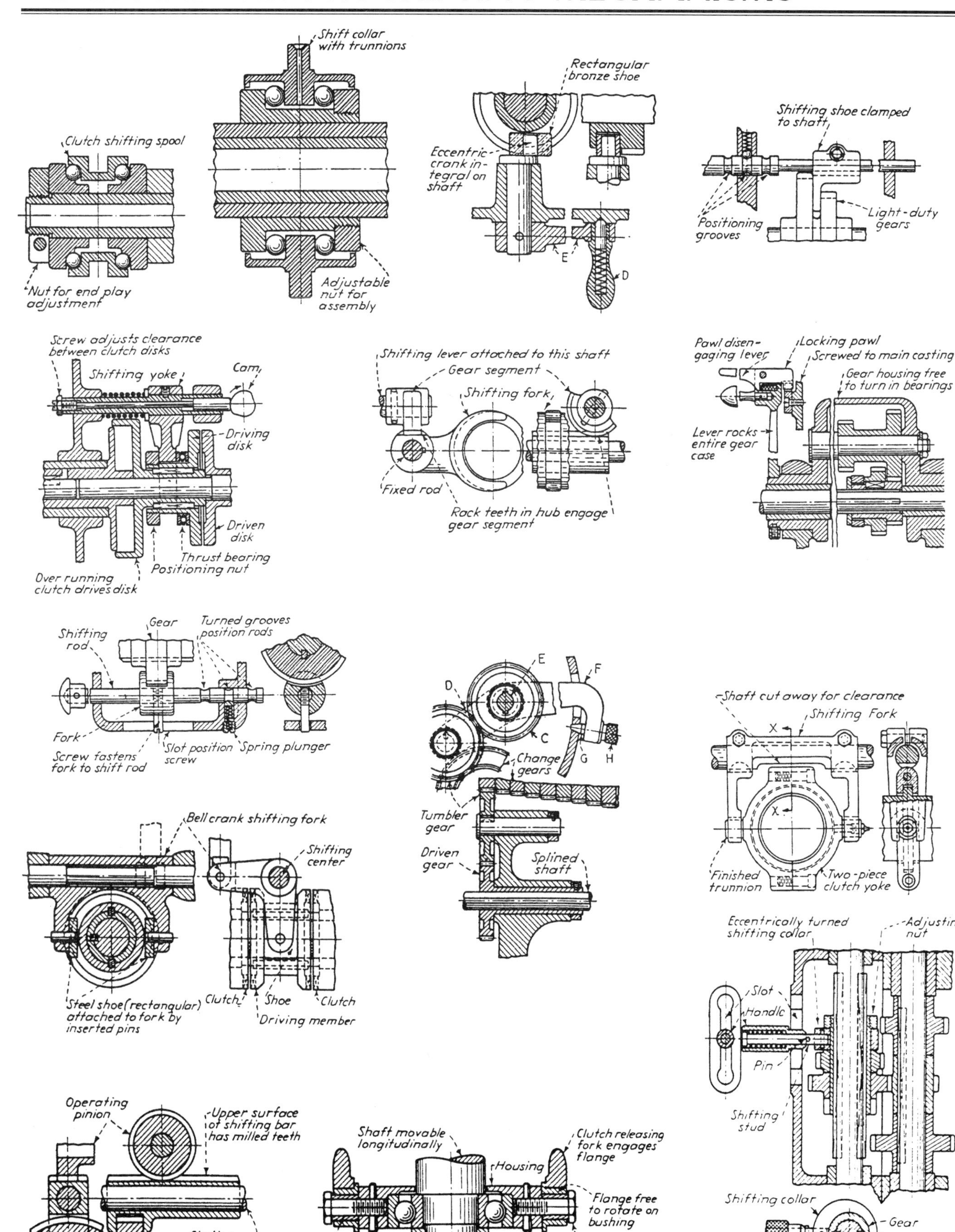

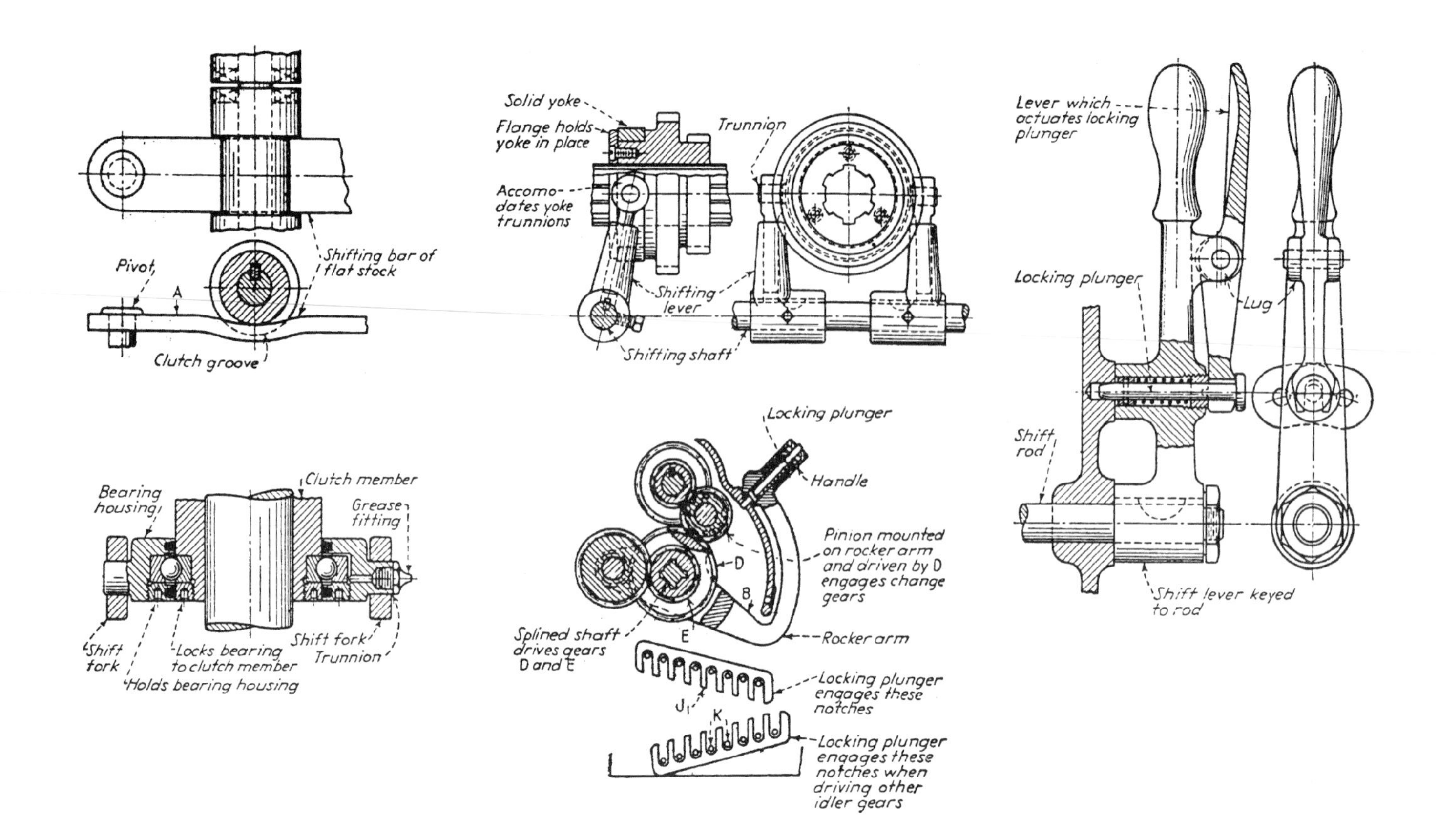

Spiderless differential

If you've ever been unable to drive your car out of a ditch because one wheel spun uselessly while the other sat torqueless and immobile, you'll thank the inventors (Seliger and Hegar) of the limited-slip differential shown here.

In straight running, it performs as a drive axle, driven by the driveshaft pinion through the ring gear. The differential action occurs only when one wheel loses traction, travels along a different arc, or otherwise attempts to turn at a speed that is different from that of the other. Then the wedge-type, two-way, over-running clutch (second figure) disengages, freeing the wheel to spin without drag.

Variations. Each clutch has three positions: forward drive, idle, and reverse drive. Thus, there are many combinations of drive-idle, depending on road conditions and turn direction. US Patent 3,124,972 describes a few:

- For left turns, the left wheel is driving, and the right wheel is forced to turn faster—thus over-running and disengaging the clutch. A friction ring built into each clutch assembly does the shifting. Wear is negligible.
- If power should be removed from the driveshaft during the left turn, the friction rings will shift each clutch and cause the left wheel to run free and the right wheel to drag in full coupling with the car's driveshaft.
- If your car is on the straightaway, under power and one wheel is lifted out of contact with the road, the other immediately transmits full torque. (The conventional spider differential performs in the opposite manner.)

On or off. Note one limitation, however: There is no gradual division of power. A wheel is either clutched in and turning at exactly the same speed as its opposite, or it is clutched out. It is not the same kind of mechanism as the conventional spider differential, which divides the driving load variably at any ratio of speeds.

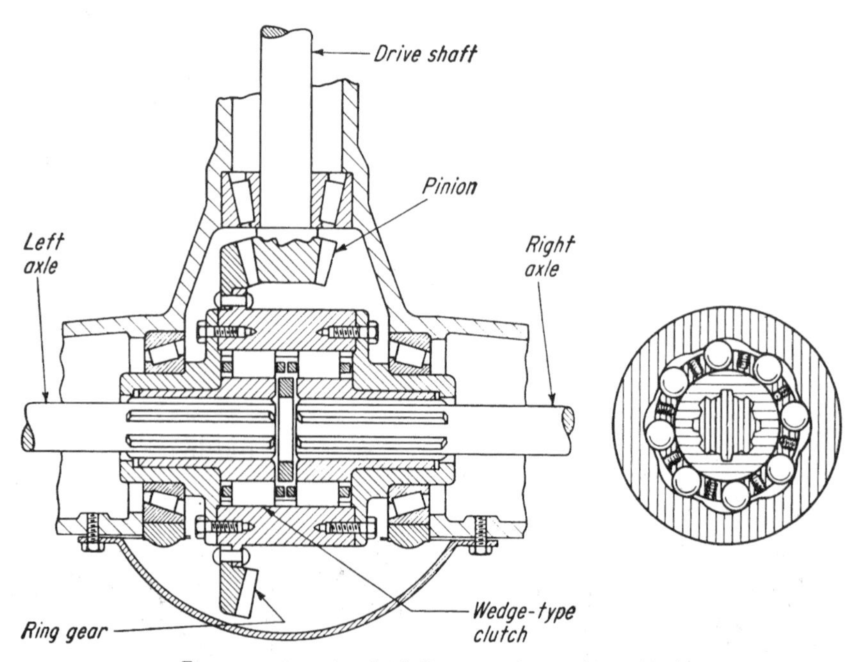

Two-way over-running clutch disengages the non-driving wheel

TWINWORM GEAR DRIVE OFFERS BIDIRECTIONAL OUTPUT

The term "self-locking" as applied to gear systems denotes a drive that gives the input gear freedom to rotate the output gear in either direction. But the output gear locks with the input when an outside torque attempts to rotate the output in either direction. This characteristic is often sought by designers who want to be sure that loads on the output side of the system cannot affect the position of the gears. Worm gears are one of the few gear systems that can be made self-locking, but at the expense of efficiency. It seldom exceeds 40% when the gears are self-locking.

An Israeli engineer, B. Popper, invented a simple dual-worm gear system that not only provided self-locking with over 90% efficiency, but exhibited a phenomenon which the inventor calls "deceleration-locking."

The "Twinworm" drive has been employed in Israel-designed counters and computers for years with marked success.

The Twinworm drive is simply constructed. Two threaded rods, or "worm" screws, are meshed together. Each worm is wound in a different direction and has a different pitch angle. For proper mesh, the worm axes are not parallel, but slightly skewed. (If both worms had the same pitch angle, a normal, reversible drive would result—similar to helical gears.) But y selecting proper, and different, pitch angles, the drive will exhibit either self-locking, or a combination of self-locking and deceleration-locking characteristics, as desired. Deceleration-locking is a completely new property best described in this way.

When the input gear decelerates (for example, when the power source is shut off, or when an outside force is applied to the output gear in a direction that tends to help the output gear), the entire transmission immediately locks up and comes to an abrupt stop, moderated only by any elastic "stretch" in the system.

Almost any type of thread will work with the new drive—standard threads, 60° screw threads, Acme threads, or any arbitrary shallow-profile thread. Hence, the worms can be manufactured on standard machine-shop equipment.

JOBS FOR THE NEW DRIVE

Applications for Twinworm can be divided into two groups:

(1) Those employing self-locking characteristics to prevent the load from affecting the system.

(2) Those employing deceleration-locking characteristics to brake the system to an abrupt stop if the input decelerates.

Self-locking occurs as soon as tan ϕ_1 is equal to or smaller than μ, or when

$$\tan \phi_1 = \frac{\mu}{S_1}$$

Angles ϕ_1 and ϕ_2 represent the respective pitch angles of the two worms, and $\phi_2 - \phi_1$ is the angle between the two worm shafts ϕ angle of misalignment). Angle ϕ_1 is quite small (usually in the order of 2° to 5°).

Here, S_1 represents a "safety factor" (selected by the designer). It must be somewhat greater than one to make sure that self-locking is maintained, even if μ should fall below an assumed value. Neither ϕ_2 nor the angle $(\phi_2 - \phi_1)$ affects the self-locking characteristic.

Deceleration-locking occurs as soon as tan ϕ_2 is also equal to or smaller than μ; or, if a second safety factor S_2 is employed (where $S_2 > 1$), when

$$\tan \phi_2 = \frac{\mu}{S_2}$$

For the equations to hold true, ϕ_2 must always be made greater than ϕ_1. Also, μ refers to the idealized case where the worm threads are square. If the threads are inclined (as with Acme-threads or V-threads) then a modified value of μ must be employed, where

$$\mu_{modified} = \frac{\mu_{true}}{\cos \theta}$$

A relationship between the input and output forces during rotation is:

$$\frac{p_1}{P_2} = \frac{\sin \phi_1 + \mu \cos \phi_1}{\sin \phi_2 + \mu \cos \phi_2}$$

Efficiency is determined from the equation:

$$\eta = \frac{1 + \mu/\tan \phi_2}{1 + \mu/\tan \phi_1}$$

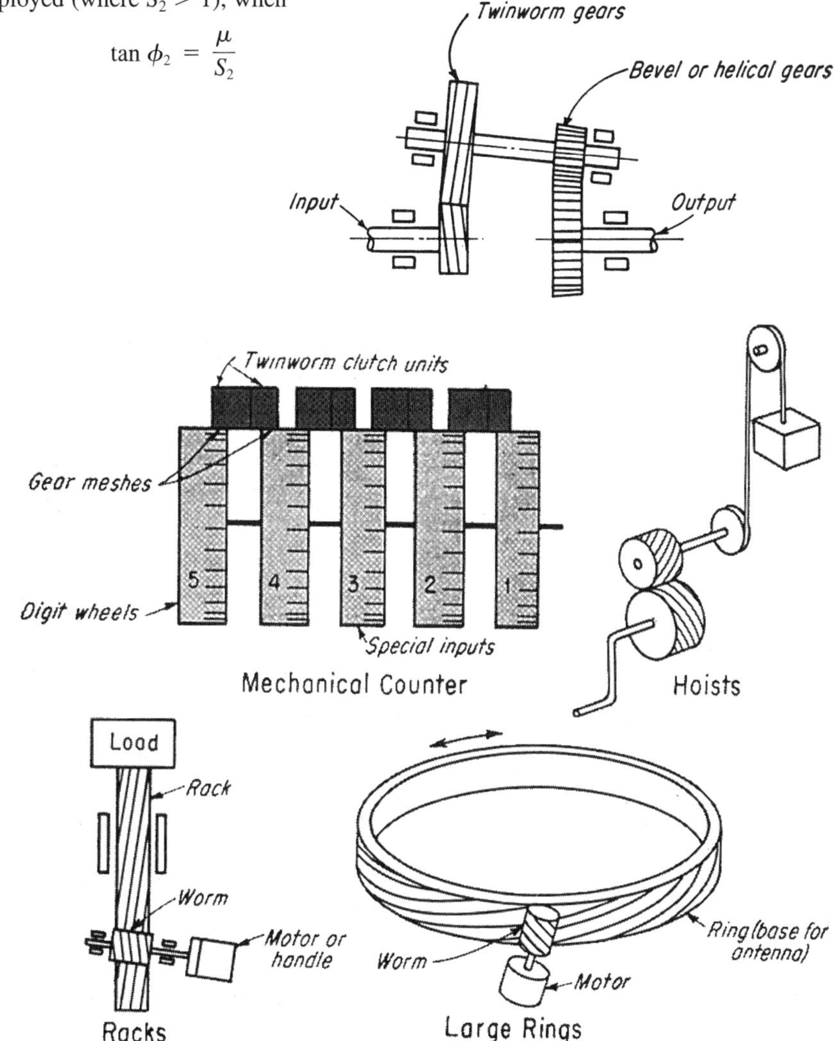

ELASTOMERIC GEAR BEARINGS EQUALIZE TORQUE LOADS

Elastomeric bearings make torque loads more nearly equal.
Lewis Research Center, Cleveland, Ohio

A set of elastomeric bearings constitutes a springy coupling between a spur gear and a drive shaft. The gear, bearings, and shaft are parts of a split-drive (redundant) mechanical transmission, and the compliance of the coupling helps to distribute torque nearly equally along the load paths of the split drive. Compliance is necessary because without it, even slight deviations in the dimensions of the redundant gears can cause grossly unequal sharing of loads. Indeed, in the absence of compliant coupling, the gears along one load path can assume the entire load while those along another load path can freewheel. Thus, the advantage of reduced loads on gear teeth is lost.

The figure illustrates one version of the shaft/bearing/gear assembly. An inner, concentric elastomeric bearing lies between a central drive shaft and an extension of a ring spur gear. A set of padlike outer elastomeric bearings joins outward protrusions on an extension of the drive shaft with facing inward protrusions on the ring spur gear.

The inner elastomeric bearing has high radial stiffness and low circumferential stiffness. This bearing centers the ring spur gear on the axis of the drive shaft and provides compliance in a circumferential direction. In a representative design of a redundant helicopter transmission, it should be at least 0.5 in. (1.27 cm) thick so that it transmits little torque.

The outer elastomeric bearings, in contrast, have low radial stiffness and high circumferential stiffness. They thus transmit torque effectively between facing protrusions. Nevertheless, they are sufficiently compliant circumferentially to accommodate the desired amount of circumferential displacement [up to $^1/_{16}$ in. (1.6 mm) in the helicopter transmission application].

The process of assembling the compliant gearing begins with the pressing of the inner elastomeric bearing onto the drive shaft. Then, with the help of an alignment tool, the ring spur gear is pressed onto the inner elastomeric bearing. The outer elastomeric bearings are ground to fit the spaces between the protrusions and bonded in place on the protrusions. As an alternative to bonding, the entire assembly can be potted in a soft matrix that holds the outer bearings in place but allows rotation with little restraint.

This work was done by C. Isabelle and J. Kish of United Technologies Corp. for **Lewis Research Center.**

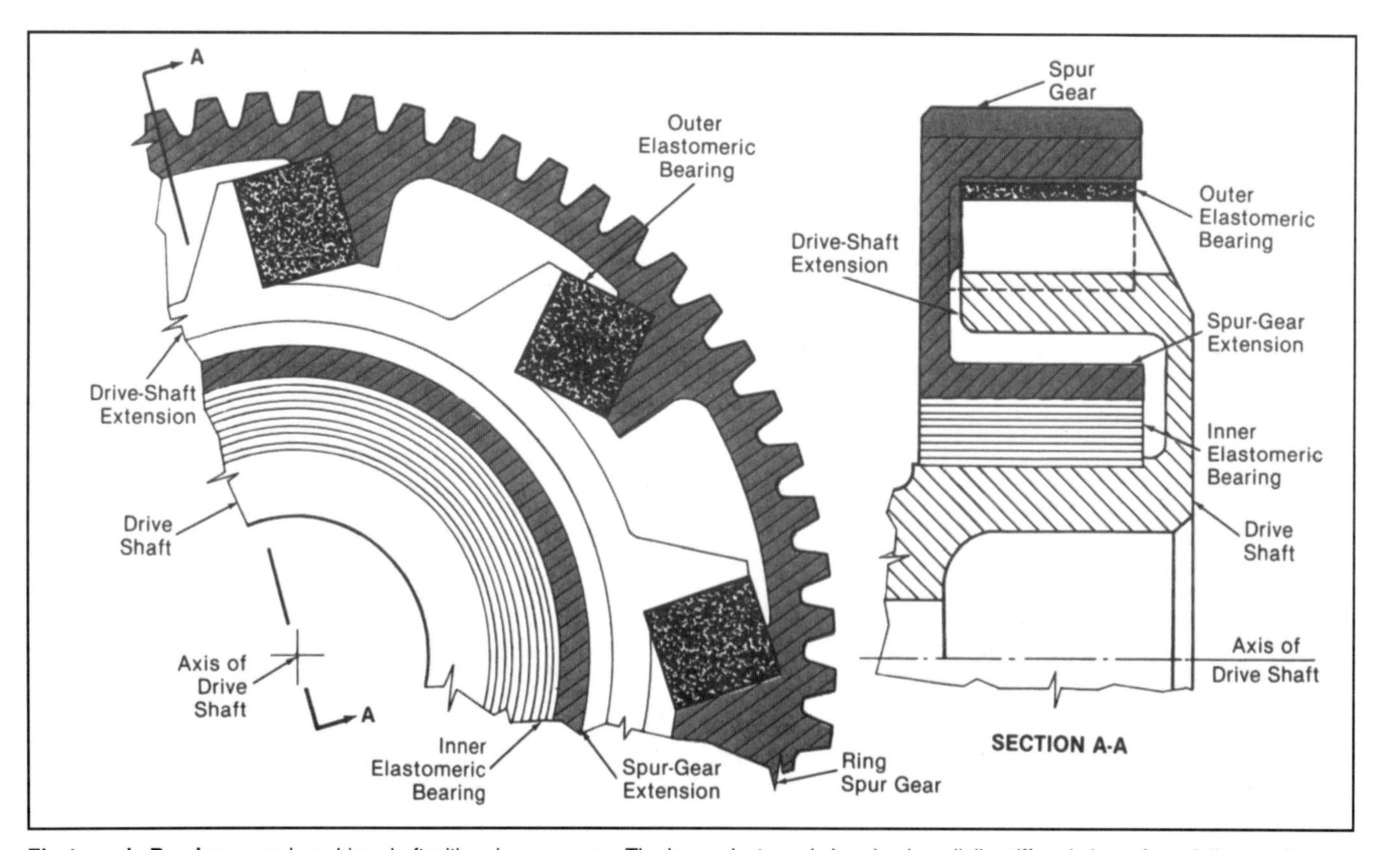

Elastomeric Bearings couple a drive shaft with a ring spur gear. The inner elastomeric bearing is radially stiff and circumferentially compliant, while the outer elastomeric bearings are circumferentially stiff and radially compliant. The combination accommodates minor variations in the dimensions and placements of gears, shafts, and other components.

REDUNDANT GEARING IN HELICOPTER TRANSMITS TORQUE

Redundant gearing transmits torque through an angle or angles. *Lewis Research Center, Cleveland, Chio*

An improved gear system intended primarily for use in a helicopter transmits torque from the horizontal or nearly horizontal shafts of two engines to the vertical output shaft that supports the rotor. The system apportions torques equally along multiple, redundant drive paths, thereby reducing the stresses on individual gear teeth, and it enables one engine to continue to turn the rotor when the other engine fails. The underlying design concept could also be applied to couple two airplane engines to a set of propellers in such a way that both propellers turn as long as at least one engine operates.

The system exploits the special advantages of the geometry of the meshing of a spur-gear-type pinion with a face gear. In comparison with other gear geometries that have been used in helicopter transmissions, this one is much more forgiving of (1) errors in manufacturing and alignment and (2) thermal and vibrational changes in the sizes and positions of the meshing components. One of the benefits is a reduction of gear-tooth-contact noise and vibration. Another benefit is the possibility of achieving a high (> 4) speed-reduction ratio in a single, efficient mesh, and the consequent possibility of reducing the number of parts, the size, the cost, and the weight of the gear system. Of course, the reduction of the number of parts confers yet another benefit by increasing the reliability of the system.

The system is shown schematically in the figure. The output of each engine is coupled by a pinion shaft to a spur-gear-type pinion. Each pinion engages an upper and a lower face gear, and each face gear is coupled by a face-gear shaft to an upper spur gear. The upper spur gears feed torque into a large combining gear. The pinion end of each pinion shaft is lightly spring-loaded in a nominal lateral position and is free to shift laterally through a small distance to take up slack, compensate for misalignments, and apportion torques equally to the two face gears with which it is engaged.

The combining gear is splined to a shaft that flares outwardly to a sun gear. The sun gear operates in conjunction with planetary gears and a stationary outer ring gear. The torque is coupled from the sun gear through the planetary gears to the planet-carrier ring, which is mounted on the output shaft.

This work was done by Robert B. Bossler, Jr., of Lucas Western, Inc., for **Lewis Research Center.**

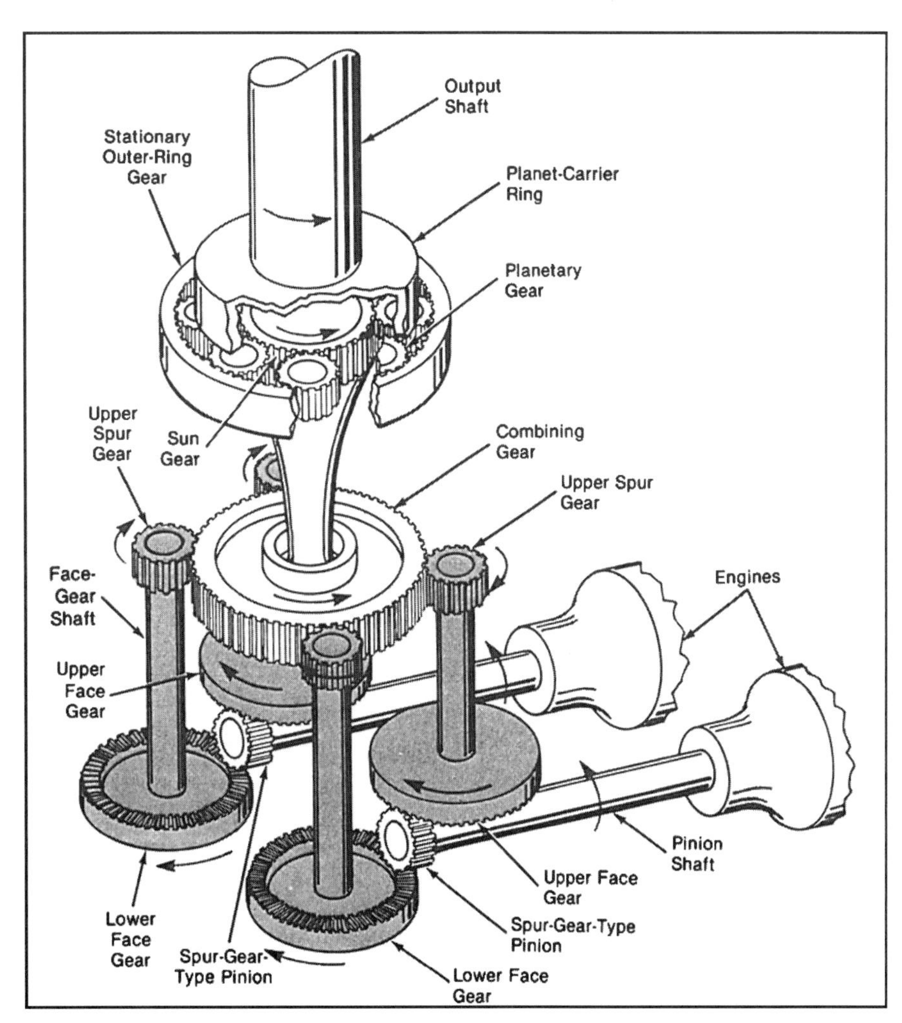

Torque from each engine is split and transmitted to the combining gear along two redundant paths. Should one engine fail, the other engine could still turn the output shaft.

WORM GEAR FRICTION REDUCED BY OIL PRESSURE

Friction would be reduced greatly.
Lewis Research Center, Cleveland, Ohio

In a proposed worm-gear transmission, oil would be pumped at high pressure through the meshes between the teeth of the gear and the worm coil (see Figure 1). The pressure in the oil would separate the meshing surfaces slightly, and the oil would reduce the friction between these surfaces. Each of the separating forces in the several meshes would contribute to the torque on the gear and to an axial force on the worm. To counteract this axial force and to reduce the friction that it would otherwise cause, oil would also be pumped under pressure into a counterforce hydrostatic bearing at one end of the worm shaft.

This type of worm-gear transmission was conceived for use in the drive train between the gas-turbine engine and the rotor of a helicopter and might be useful in other applications in which weight is critical. Worm gear is attractive for such weight-critical applications because (1) it can transmit torque from a horizontal engine (or other input) shaft to a vertical rotor (or other perpendicular output) shaft, reducing the speed by the desired ratio in one stage, and (2) in principle, a one-stage design can be implemented in a gearbox that weighs less than does a conventional helicopter gearbox.

Heretofore, the high sliding friction between the worm coils and the gear teeth of worm-gear transmissions has reduced efficiency so much that such transmissions could not be used in helicopters. The efficiency of the proposed worm-gear transmission with hydrostatic engagement would depend partly on the remaining friction in the hydrostatic meshes and on the power required to pump the oil. Preliminary calculations show that the efficiency of the proposed transmission could be the same as that of a conventional helicopter gear train.

Figure 2 shows an apparatus that is being used to gather experimental data pertaining to the efficiency of a worm gear with hydrostatic engagement. Two stationary disk sectors with oil pockets represent the gear teeth and are installed in a caliper frame. A disk that represents the worm coil is placed between the disk sectors in the caliper and is rotated rapidly by a motor and gearbox. Oil is pumped at high pressure through the clearances between the rotating disk and the stationary disk sectors. The apparatus is instrumented to measure the frictional force of meshing and the load force.

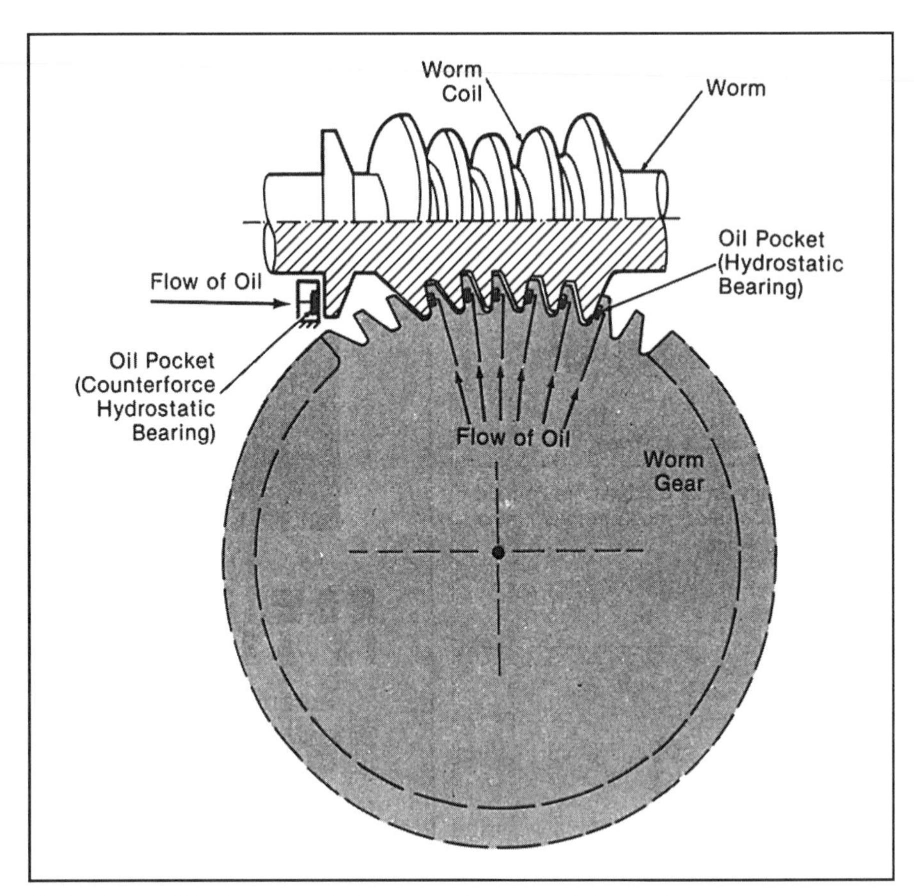

Fig.1 Oil would be injected at high pressure to reduce friction in critical areas of contact.

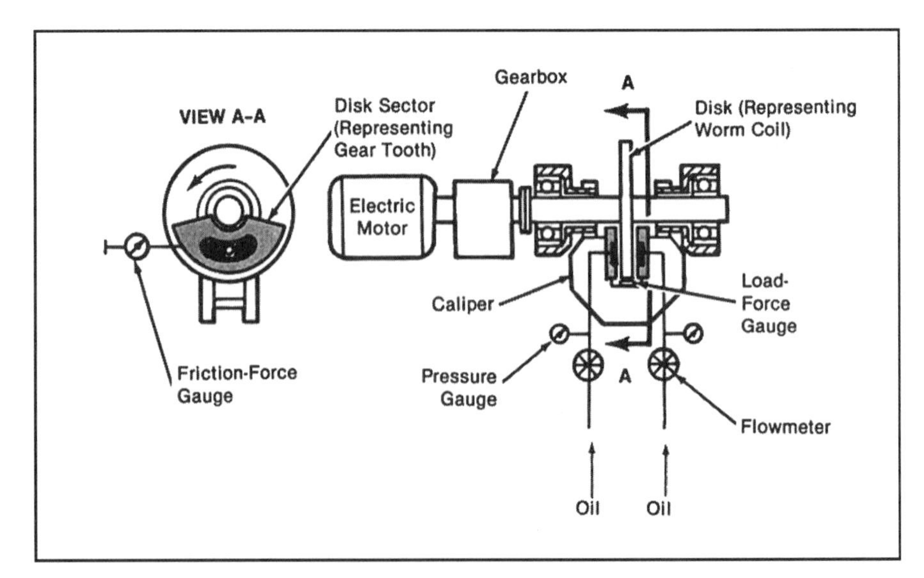

Fig. 2 This test apparatus simulates and measures some of the loading conditions of the proposed worm gear with hydrostatic engagement. The test data will be used to design efficient worm-gear transmissions.

The stationary disk sectors can be installed with various clearances and at various angles to the rotating disk. The stationary disk sectors can be made in various shapes and with oil pockets at various positions. A flowmeter and pressure gauge will measure the pump power. Oils of various viscosities can be used. The results of the tests are expected to show the experimental dependences of the efficiency of transmission on these factors.

It has been estimated that future research and development will make it possible to make worm-gear helicopter transmission that weigh half as much as conventional helicopter transmissions do. In addition, the new hydrostatic meshes would offer longer service life and less noise. It might even be possible to make the meshing worms and gears, or at least parts of them, out of such lightweight materials as titanium, aluminum, and composites.

This work was done by Lev. I. Chalko of the U.S. Army Propulsion Directorate (AVSCOM) for **Lewis Research Center.**

BEVEL AND HYPOID GEAR DESIGN PREVENTS UNDERCUTTING

Lengths and radii of shafts can be chosen to prevent undercutting.
Lewis Research Center, Cleveland, Ohio

A computer-assisted method of analysis of straddle designs for spiral bevel and hypoid gears helps to prevent undercutting of gear shafts during cutting of the gear teeth. Figure 1 illustrates a spiral bevel gear or straddle design, in which the shaft extends from both ends of the toothed surface to provide double bearing support. One major problem in such a design is to choose the length and radius of the shaft at the narrow end (equivalently, the radial coordinate r and axial coordinate u) such that the head cutter that generates the gear teeth does not collide with, and thereby undercut, the shaft.

The analytical method and computer program are based on the equations for the surface traced out by the motion of the head cutter, the equation for the cylindrical surface of the shaft, and the equations that express the relationships among the coordinate systems fixed to the various components of the gear-cutting machine tool and to the gear. The location of a collision between the shaft and the cutter is defined as the vector that simultaneously satisfies the equations for head-cutter-traced and shaft surfaces. The solution of these equations yields the u and r coordinates of the point of collision.

Given input parameters in the form of the basic machine-tool settings for cutting the gear, the computer program finds numerical values of r and u at a representative large number of points along the path of the cutter. These computations yield a family of closed curves (see Fig. 2) that are the loci of collision points. The region below the curves is free of collisions: thus, it contains the values of r and u that can be chosen by the designer to avoid collisions between the shaft and the head cutter.

This work was done by Robert F. Handschuh of the U.S. Army Aviation Systems Command; Faydor L. Litvin, Chihping Kuan, and Jonathan Kieffer of the University of Illinois at Chicago; and Robert Bossler of Lucas Western, Inc., for **Lewis Research Center.**

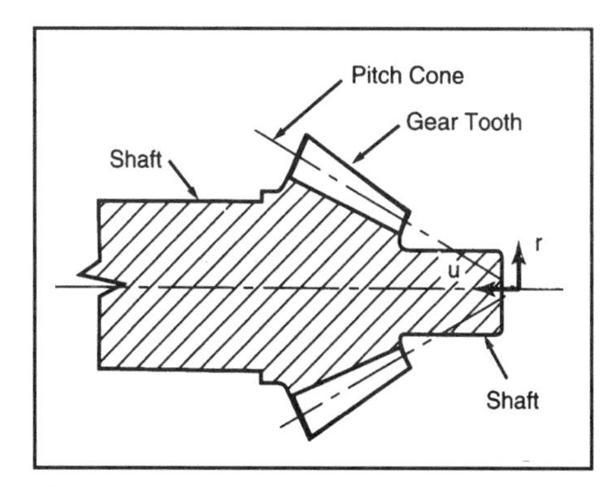

Fig. 1 A straddle-design spiral bevel gear includes two integral shaft extensions. One of these could terminate near or even beyond the apex of the pitch cone.

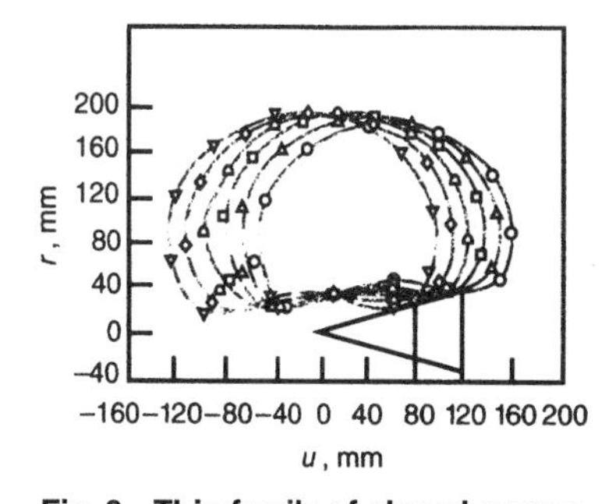

Fig. 2 This family of closed curves applies to a typical hypoid gear. It will help in the selection of the length and radius of the shaft at the narrow end. The region below the curves is free of collisions between the head cutter and the shaft.

GEARED ELECTROMECHANICAL ROTARY JOINT

Springy planetary gears provide low-noise electrical contact.

The figure illustrates a geared rotary joint that provides low-noise ac or dc electrical contact between electrical subsystems that rotate relative to each other. This joint is designed to overcome some of the disadvantages of older electromechanical interfaces—especially the intermittency (and, consequently, the electrical noise) of sliding-contact and rolling-contact electromechanical joints.

The firs electrical subsystem is mounted on, or at least rotates with, the shaft and the two inner gears attached to the shaft. The inner gears are separated axially by an electrically insulating disk. Each inner gear constitutes one of two electrical terminals through which electrical power is fed to or from the first electrical subsystem.

The second electrical subsystem is mounted on, or at least rotates with, the outer (ring) gears. As was done to the inner gears, the ring gears are separated axially by an electrically insulating annular disk. The ring gears act as the electrical terminals through which power is fed from or to the second electrical subsystem.

Electrical contact between the inner and outer (ring) gears is provided by multiple, equally spaced, flexible planetary gears formed as hollow cylinders with thin, fluted walls. These gears mesh with the inner and outer (ring) gears. Those gears are slightly oversize with respect to the gaps between the inner and outer gears, but their flexibility makes it possible to compress them slightly to install them in the gaps. After installation, meshing of the gears maintains the even angular interval between the planetary gears at all rotational speeds.

The planetary gears are made of beryllium copper, which is preferred for electrical contacts because it is a self-cleaning material that exhibits excellent current-carrying characteristics. Atypical flexible planetary gear has 13 teeth. Both have an axial length and an average diameter of 0.25 in. (6.35 mm), and a wall thickness of 0.004 in. (0.10 mm). Because each planetary gear is independently sprung into a cylinder-in-socket configuration with respect to the inner and outer gears, it maintains continuous electrical contact between them. The reliability and continuity of the electrical contact is further ensured by the redundancy of the multiple planetary gears. The multiplicity of the contacts also ensures low electrical resistance and large current-carrying capability.

The springiness of the planetary gears automatically compensates for thermal expansion, thermal contraction, and wear; moreover, wear is expected to be minimal. Finally, the springiness of the planetary gears provides an antibacklash capability in a gear system that is simpler and more compact in comparison with conventional antibacklash gear systems.

This work was done by John M. Vranish of **Goddard Space Flight Center.**

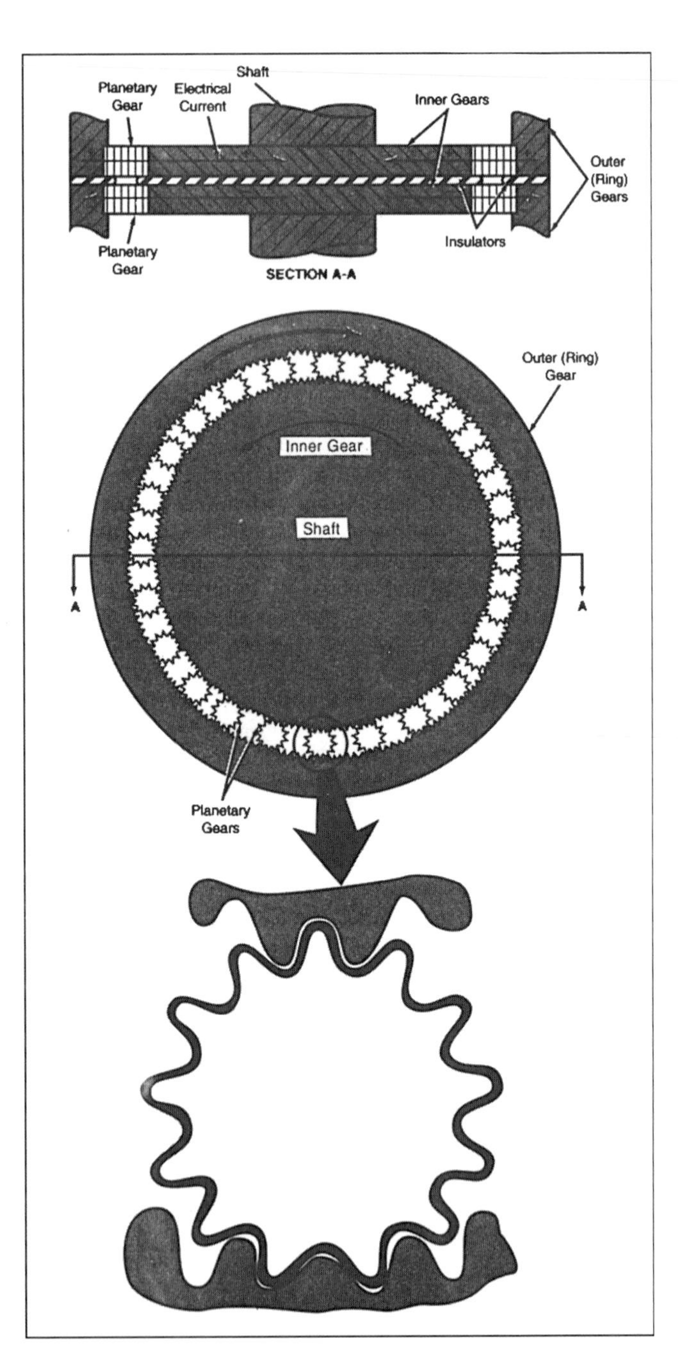

Hollow, springy, planetary gears provide continuous, redundant, low-noise electrical contact between the ginner and outer gears.

GEARED SPEED REDUCERS OFFER ONE-WAY OUTPUT

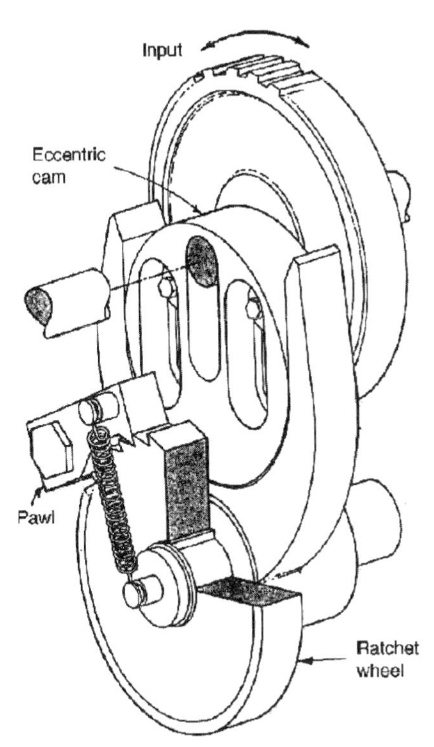

This eccentric cam adjusts over a range of high reduction ratios, but unbalance limits it to low speeds. When its direction of input changes, thee is no lag in output rotation. The output shaft moves in steps because of a ratchet drive through a pawl which is attached to a U follower.

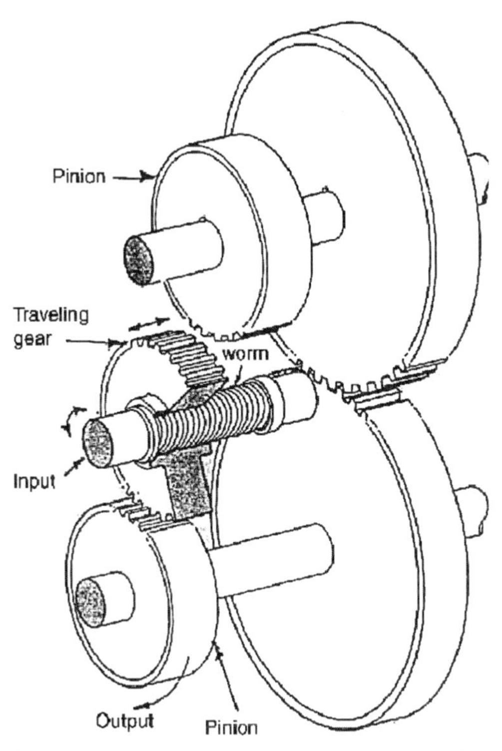

A traveling gear moves along a worm and transfers drive torque to the other pinion when the input rotation changes direction. To ease the gear engagement, the gear teeth are tapered at their ends. Output rotation is smooth, but there is a lag after direction changes as the gear shifts. The gear cannot be wider than the axial offset between pinions or there will be destructive interference.

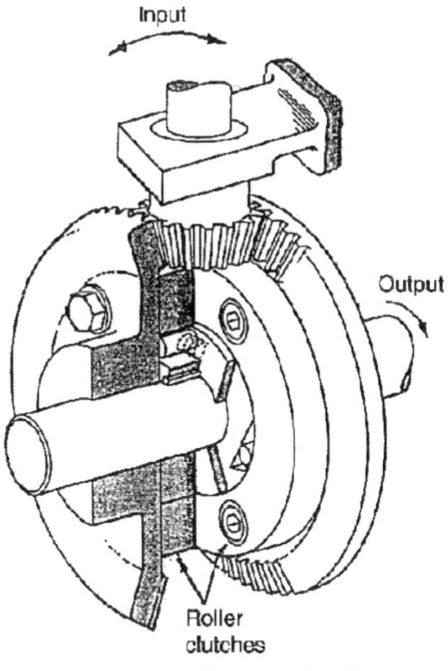

Two bevel gears drive through roller clutches. One clutch catches in one direction and the other catches in the opposite direction. There is little or no interruption of smooth output rotation when the input direction changes.

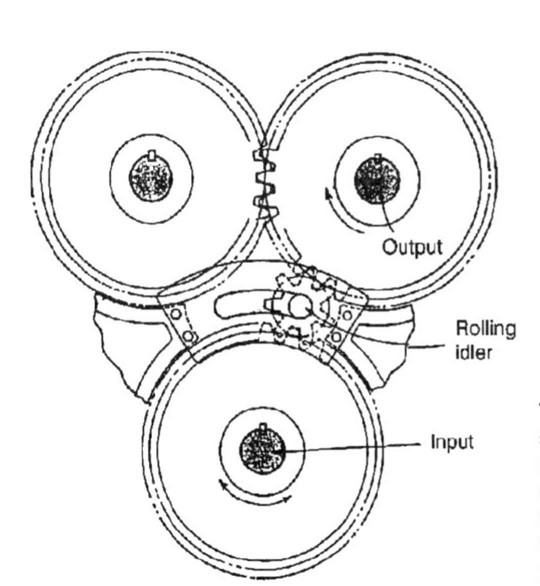

This rolling idler also provides a smooth output and a slight lag after its input direction changes. A small drag on the idler is necessary so that it will transfer smoothly into engagement with the other gear and not remain spinning between the gears.

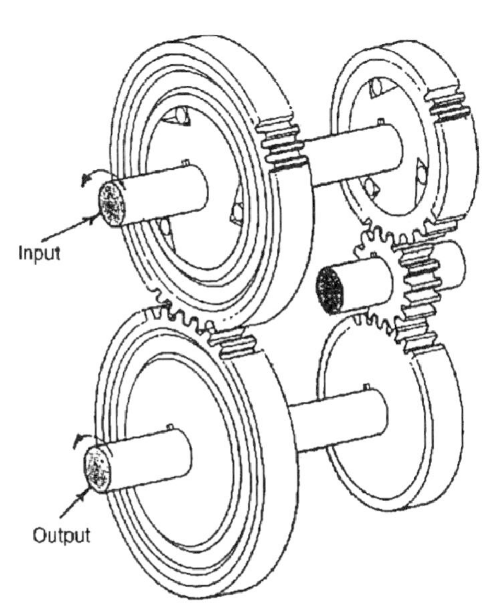

Roller clutches are on the input gears in this drive. These also give smooth output speed and little output lag as the direction changes.

DESIGN OF GEARED FIVE-BAR MECHANISMS

Geared five-bar mechanisms offer excellent force-transmission characteristics and can produce more complex output motions—including dwells—than conventional four-bar mechanisms.

It is often necessary to design a mechanism that will convert uniform input rotational motion into nonuniform output rotation or reciprocation. Mechanisms designed for such purposes are usually based on four-bar linkages. Those linkages produce a sinusoidal output that can be modified to yield a variety of motions.

Four-bar linkages have their limitations, however. Because they cannot produce dwells of useful duration, the designer might have to include a cam when a dwell is desired, and he might have to accept the inherent speed restrictions and vibration associated with cams. A further limitation of four-bar linkages is that only a few kinds have efficient force-transmission capabilities.

One way to increase the variety of output motions of a four-bar linkage, and obtain longer dwells and better force transmissions, is to add a link. The resulting five-bar linkage would become impractical, however, because it would then have only two degrees of freedom and would, consequently, require two inputs to control the output.

Simply constraining two adjacent links would not solve the problem. The five-bar chain would then function effectively only as a four-bar linkage. If, on the other hand, any two nonadjacent links are constrained so as to remove only one degree of freedom, the five-bar chain becomes a functionally useful mechanism.

Gearing provides solution. There are several ways to constrain two nonadjacent links in a five-bar chain. Some possibilities include the use of gears, slot-and-pin joints, or nonlinear band mechanisms. Of these three possibilities, gearing is the most attractive. Some practical gearing systems (Fig. 1) included paired external gears, planet gears revolving within an external ring gear, and planet gears driving slotted cranks.

In one successful system (Fig. 1A) each of the two external gears has a fixed crank that is connected to a crossbar by a rod. The system has been successful in high-speed machines where it transforms rotary motion into high-impact linear motion. The Stirling engine includes a similar system (Fig. 1B).

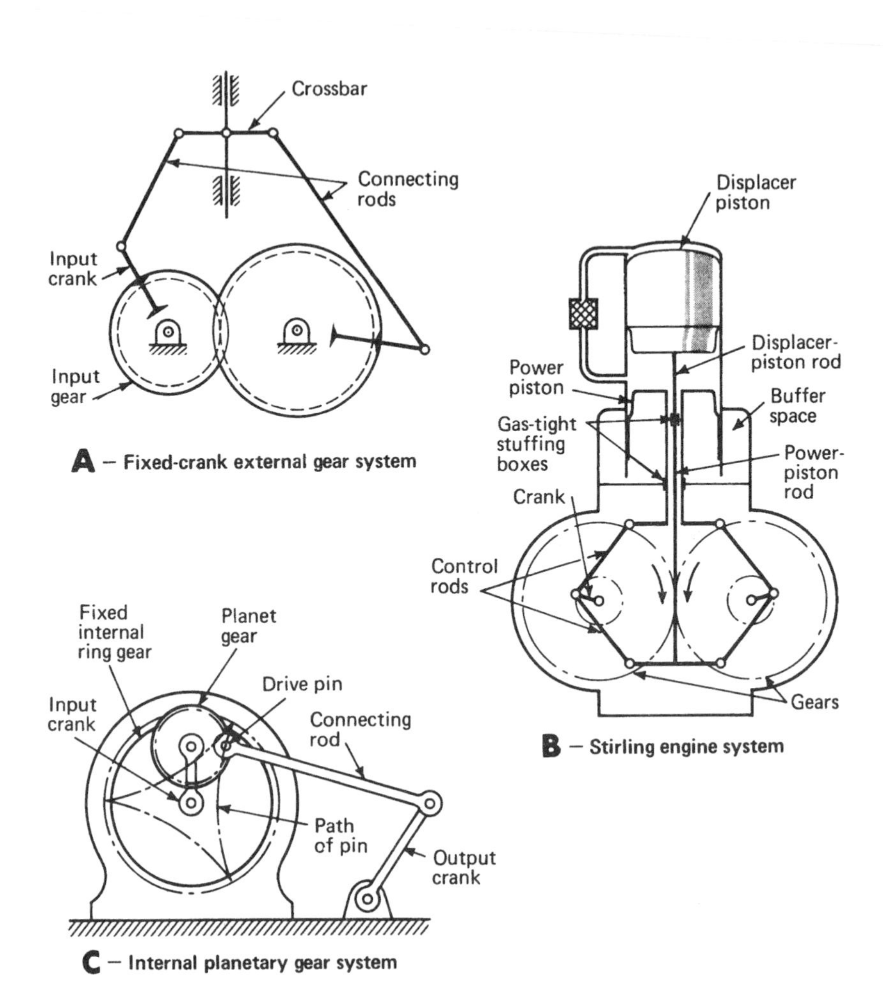

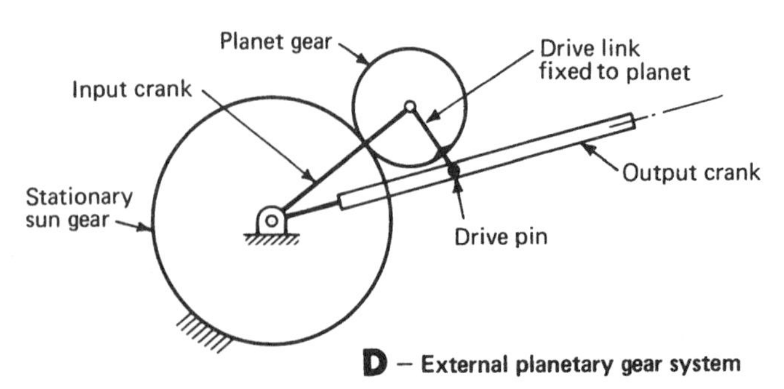

Fig. 1 Five-bar mechanism designs can be based on paired external gears or planetary gears. They convert simple input motions into complex outputs.

In a different system (Fig. 1C) a pin on a planet gear traces an epicyclic, three-lobe curve to drive an output crank back and forth with a long dwell at the extreme right-hand position. A slotted output crank (Fig. 1D) will provide a similar output.

Two professors of mechanical engineering, Daniel H. Suchora of Youngstown State University, Youngstown, Ohio, and Michael Savage of the University of Akron, Akron, Ohio, studied a variation of this mechanism in detail.

Five kinematic inversions of this form (Fig. 2) were established by the two researchers. As an aid in distinguishing between the five, each type is named according to the link which acts as the fixed link. The study showed that the Type 5 mechanism would have the greatest practical value.

In the Type 5 mechanism (Fig. 3A), the gear that is stationary acts as a sun gear. The input shaft at Point E drives the input crank which, in turn, causes the planet gear to revolve around the sun gear. Link a_2, fixed to the planet, then drives the output crank, Link a_4, by means of the connecting link, Link a_3. At any input position, the third and fourth links can be assembled in either of two distinct positions or "phases" (Fig. 3B).

Variety of outputs. The different kinds of output motions that can be obtained from a Type 5 mechanism are based on the different epicyclic curves traced by link joint B. The variables that control the shape of a "B-curve" are the gear ratio GR ($GR = N_2/N_5$), the link ratio a_2/a_1 and the initial position of the gear set, defined by the initial positions of θ_1 and θ_2, designated as θ_{10} and θ_{20}, respectively.

Typical B-curve shapes (Fig. 4) include ovals, cusps, and loops. When the B-curve is oval (Fig. 4B) or semioval (Fig. 4C), the resulting B-curve is similar to the true-circle B-curve produced by a four-bar linkage. The resulting output motion of Link a_4 will be a sinusoidal type of oscillation, similar to that produced by a four-bar linkage.

When the B-curve is cusped (Fig. 4A), dwells are obtained. When the B-curve is looped (Figs. 4D and 4E), a double oscillation is obtained.

In the case of the cusped B-curve (Fig. 4A), dwells are obtained. When the B-curve is looped (Figs. 4D and 4E), a double oscillation is obtained.

In the case of the cusped B-curve (Fig. 4A), by selecting a_2 to be equal to the pitch radius of the planet gear r_2, link joint B becomes located at the pitch circle of the planet gear. The gear ratio in all the cases illustrated is unity ($GR = 1$).

Professors Suchora and Savage analyzed the different output motions produced by the geared five-bar mechanisms by plotting the angular position θ_4 of the

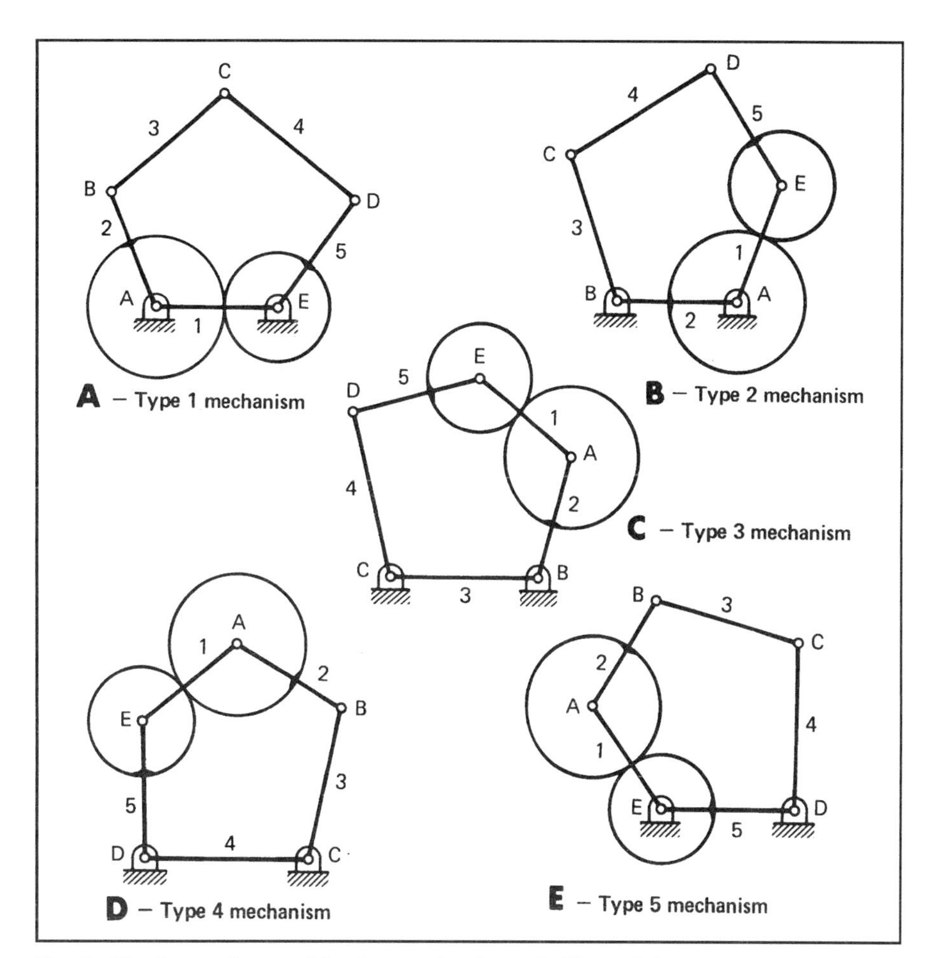

Fig. 2 Five types of geared five-bar mechanisms. A different link acts as the fixed link in each example. Type 5 might be the most useful for machine design.

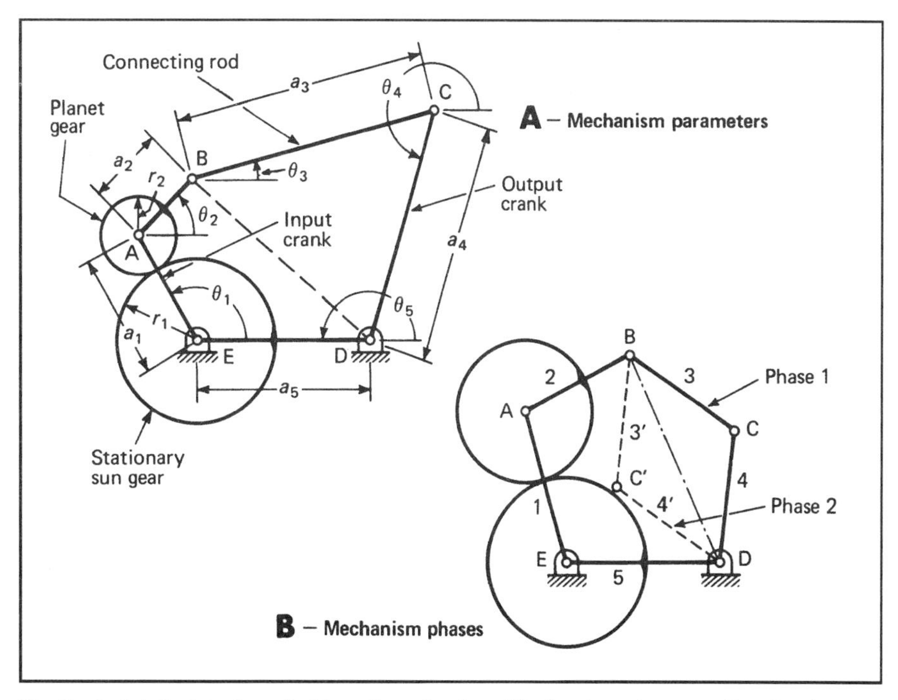

Fig. 3 A detailed design of a Type-5 mechanism. The input crank causes the planet gear to revolve around the sun gear, which is always stationary.

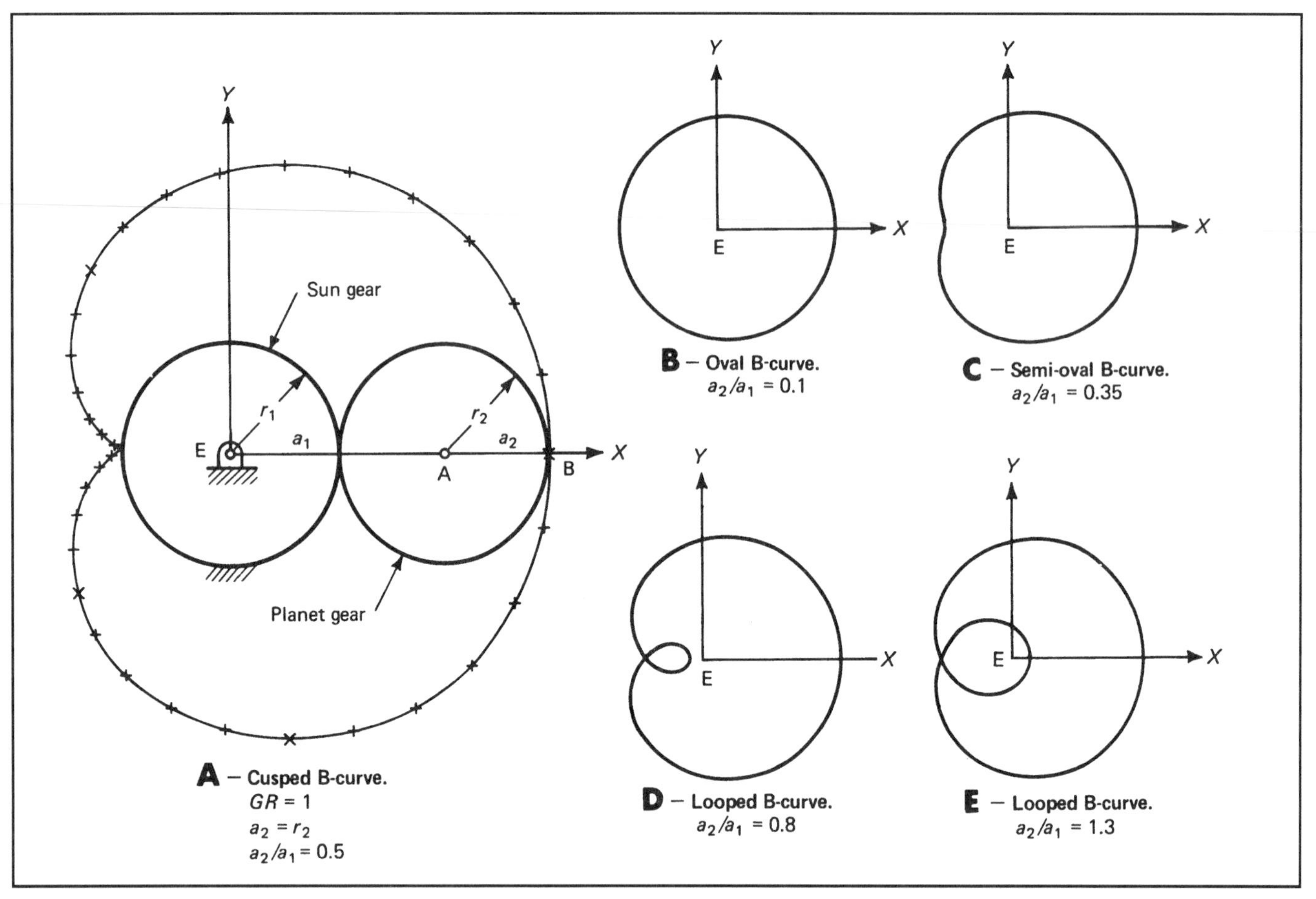

Fig. 4 Typical B-curve shapes obtained from various Type-5 geared five-bar mechanisms. The shape of the epicyclic curved is changed by the link ratio a_2/a_1 and other parameters, as described in the text.

Calculating displacement, velocity and acceleration

Displacement θ_4 can be found from the following equation:

$$\theta_4 = 2\tan^{-1}\left(\frac{I \pm \sqrt{I^2 + H^2 - J^2}}{H + J}\right)$$

where $H = a_1 \cos\theta_1 + a_2 \cos\theta_2 - a_5$; $I = a_1 \sin\theta_1 + a_2 \sin\theta_2$; and $J = 1/2a_4\,(a^2_3 - a^2_4 - H^2 - I^2)$; and where $\theta_2 = \theta_{20} + (1 + 1/GR)(\theta_1 - \theta_{10})$, where θ_{10} and θ_{20} are the initial values of the angles θ_1 and θ_2, respectively.

For layout purposes, once θ_4 is determined, θ_3 can be found from:

$$\theta_3 = \tan^{-1}\left(\frac{a_4 \sin\theta_4 + I}{a_4 \cos\theta_4 + H}\right)$$

To find velocities θ'_4 and θ'_3, use these equations:

$$\theta'_4 = \frac{a_1 \sin(\theta_3 - \theta_1) + a_2 \sin(\theta_3 - \theta_2)\,\theta'_2}{a_4 \sin(\theta_4 - \theta_3)}$$

$$\theta'_3 = \frac{a_1 \sin(\theta_1 - \theta_4) + a_2 \sin(\theta_2 - \theta_4)\,\theta'_2}{a_4 \sin(\theta_4 - \theta_3)}$$

where $\theta'_2 = (1 + 1/GR)$.

Use these equations to determine accelerations θ''_4 and θ''_3:

$$\theta''_4 = \frac{L}{a_3\, a_4 \sin(\theta_4 - \theta_3)}$$

$$\theta''_3 = \frac{K}{a_3\, a_4 \sin(\theta_4 - \theta_3)}$$

where $K = a_3a_4\cos(\theta_3 - \theta_4)\,\theta'^2_3 + a_4^2\theta'^2_4 + a_1a_2\cos(\theta_1 - \theta_4) + a_2a_4\cos(\theta_2 - \theta_4)\theta'^2_2$ and $L = a_3^2\theta'^2_3 - a_3a_4\cos(\theta_3 - \theta_4)\theta'^2_4 - a_1a_3\cos(\theta_3 - \theta_1) + a_2a_3\cos(\theta_3 - \theta_2)\theta'^2_2$.

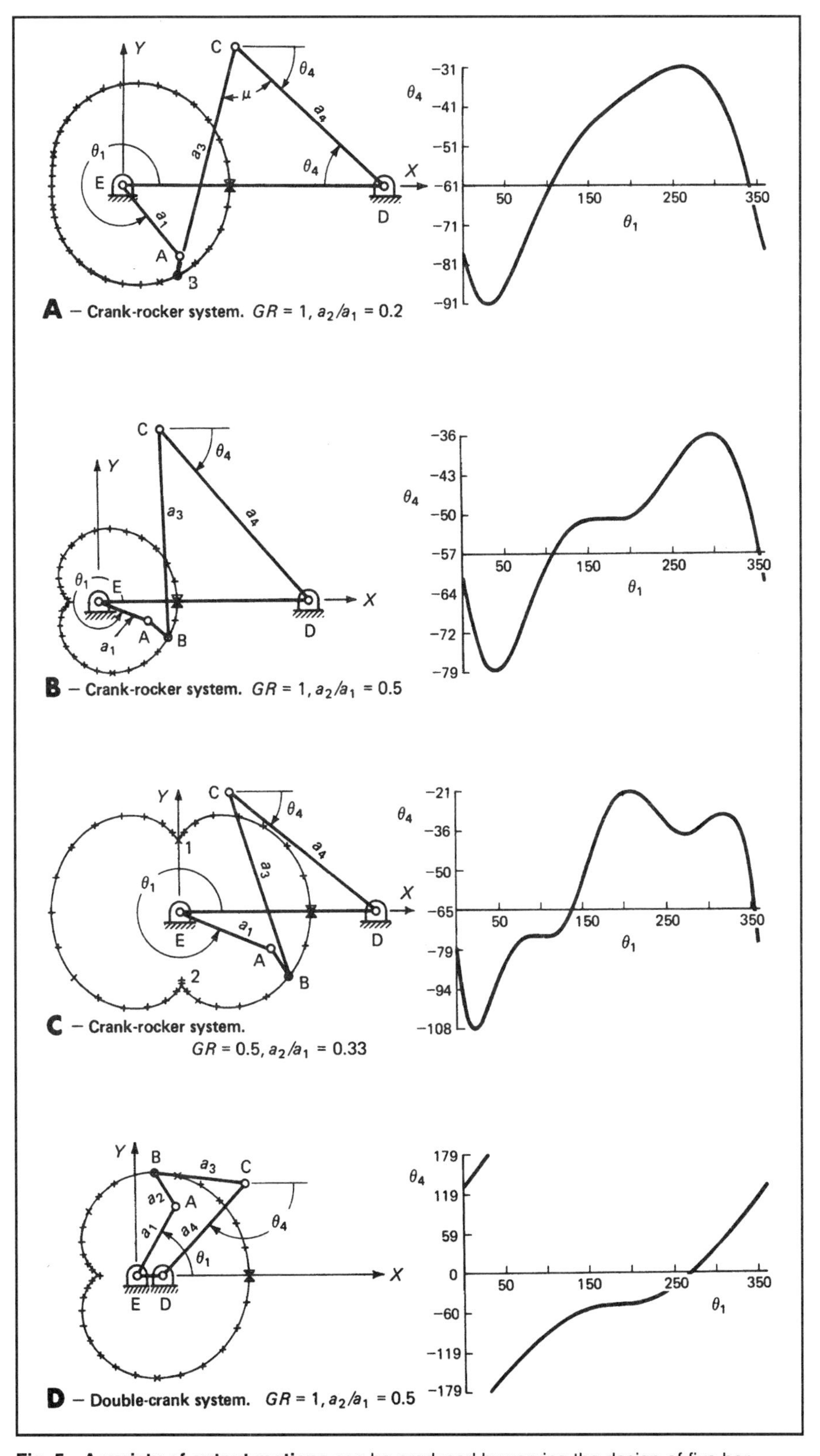

Fig. 5 A variety of output motions can be produced by varying the design of five-bar geared mechanisms. Dwells are obtainable with proper design. Force transmission is excellent. In these diagrams, the angular position of the output link is plotted against the angular position of the input link for various five-bar mechanism designs.

output link a_4 of the output link a_4 against the angular position of the input link θ_1 for a variety of mechanism configurations (Fig. 5).

In three of the four cases illustrated, $GR = 1$, although the gear pairs are not shown. Thus, one input rotation generates the entire path of the B-curve. Each mechanism configuration produces a different output.

One configuration (Fig. 5A) produces an approximately sinusoidal reciprocating output motion that typically has better force-transmission capabilities than equivalent four-bar outputs. The transmission angle μ should be within 45 to 135° during the entire rotation for best results.

Another configuration (Fig. 5B) produces a horizontal or almost-horizontal portion of the output curve. The output link, link, a_4, is virtually stationary during this period of input rotation—from about 150 to 200° of input rotation θ_1 in the case illustrated. Dwells of longer duration can be designed.

By changing the gear ratio to 0.5 (Fig. 5C), a complex motion is obtained; two intermediate dwells occur at cusps 1 and 2 in the path of the B-curve. One dwell, from $\theta_1 = 80$ to 110°, is of good quality. The dwell from 240 to 330° is actually a small oscillation.

Dwell quality is affected by the location of Point D with respect to the cusp, and by the lengths of links a_3 and a_4. It is possible to design this form of mechanism so it will produce two usable dwells per rotation of input.

In a double-crank version of the geared five-bar mechanism (Fig. 5D), the output link makes full rotations. The output motion is approximately linear, with a usable intermediate dwell caused by the cusp in the path of the B-curve.

From this discussion, it's apparent that the Type 5 geared mechanism with $GR = 1$ offers many useful motions for machine designers. Professors Suchora and Savage have derived the necessary displacement, velocity, and acceleration equations (see the "Calculating displacement, velocity, and acceleration" box).

EQUATIONS FOR DESIGNING GEARED CYCLOID MECHANISMS

Fig. 1 Equation for epicycloid drives.

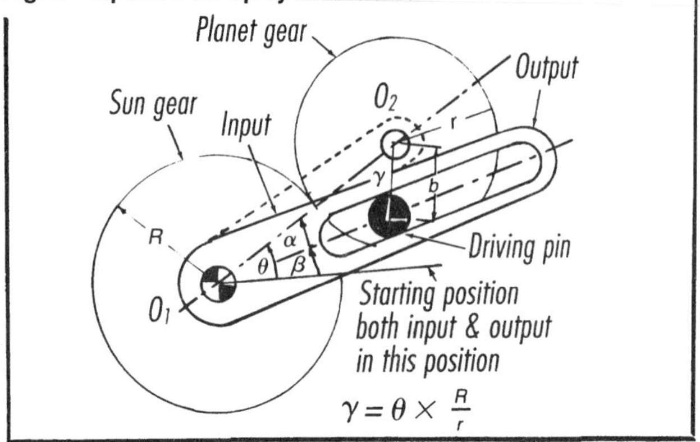

Angular displacement

$$\tan \beta = \frac{(R + r) \sin \theta - b \sin (\theta + \gamma)}{(R + r) \cos \theta - b \cos (\theta + \gamma)} \qquad (1)$$

Angular velocity

$$V = \omega \frac{1 + \dfrac{b^2}{r(R + r)} - \left(\dfrac{2r + R}{r}\right)\left(\dfrac{b}{R + r}\right)\left(\cos \dfrac{R}{r} \theta\right)}{1 + \left(\dfrac{b^2}{R + r}\right)^2 - \left(\dfrac{2b}{R + r}\right)\left(\cos \dfrac{R}{r} \theta\right)} \qquad (2)$$

Angular acceleration

$$V = \omega^2 \frac{\left(1 - \dfrac{b^2}{(R + r)^2}\right)\left(\dfrac{R^2}{r^2}\right)\left(\dfrac{b}{R + r}\right)\left(\sin \dfrac{R}{r} \theta\right)}{\left[1 + \dfrac{b^2}{(R + r)^2} - \left(\dfrac{2b}{R + r}\right)\left(\cos \dfrac{R}{r}\theta\right)\right]^2} \qquad (3)$$

Symbols

- A = angular acceleration of output, degrees per second^2
- b = radius of driving pin from center of planet gear
- r = pitch radius of planet gear
- R = pitch radius of fixed sun gear
- V = angular velocity of output, degrees per second
- β = angular displacement of output, degree
- γ = $\theta R/r$
- θ = input displacement, degree
- ω = angular velocity of input, degrees per second

The equations for angular displacement, velocity, and acceleration for a basic epicyclic drive are given below.

$$\tan\beta = \frac{\sin \theta - \left(\dfrac{b}{R - r}\right)\left(\sin \dfrac{R - r}{r} \theta\right)}{\cos \theta + \left(\dfrac{b}{R - r}\right)\left(\cos \dfrac{R - r}{r} \theta\right)} \qquad (4)$$

$$V = \omega \frac{1 - \left(\dfrac{R - r}{r}\right)\left(\dfrac{b^2}{(R - r)^2}\right) + \left(\dfrac{2r - R}{r}\right)\left(\dfrac{b}{R - r}\right)\left(\cos \dfrac{R}{r} \theta\right)}{1 + \dfrac{b^2}{(R - r)^2} + \left(\dfrac{2b}{R - r}\right)\left(\cos \dfrac{R}{r} \theta\right)} \qquad (5)$$

$$A = \omega^2 \frac{\left(1 - \dfrac{b^2}{(R - r)^2}\right)\left(\dfrac{b}{R - r}\right)\left(\dfrac{R^2}{r^2}\right)\left(\sin \dfrac{R}{r} \theta\right)}{\left[1 + \dfrac{b^2}{(R - r)^2} + \left(\dfrac{2b}{R - r}\right)\left(\cos \dfrac{R}{r} \theta\right)\right]^2} \qquad (6)$$

Fig. 2 Equations for geared cycloid mechanisms.

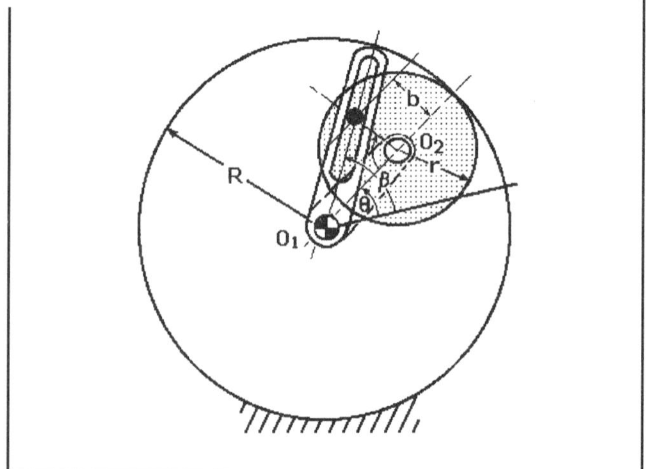

Fig. 3 A gear rolling on a gear flattens curves.

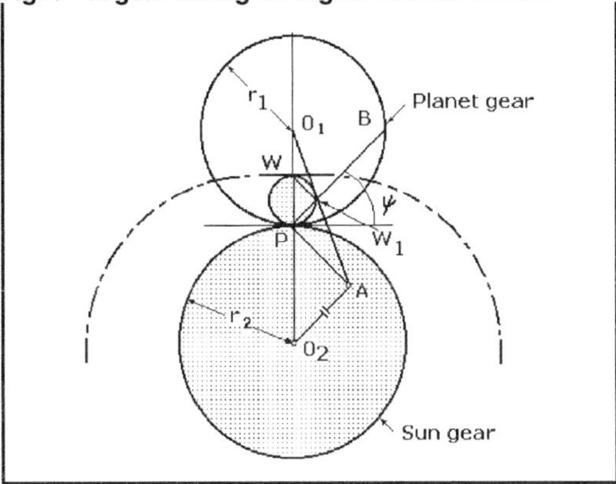

It is frequently desirable to find points on the planet gear that will describe approximately straight lines for portions of the output curves. These points will yield dwell mechanisms. Construction is as follows (see drawing):

1. Draw an arbitrary line *PB*.
2. Draw its parallel O_2A.
3. Draw its perpendicular *PA* at *P*. Locate point *A*.
4. Draw O_1A. Locate W_1.
5. Draw perpendicular to PW_1 at W_1 to locate *W*.
6. Draw a circular with *PW* as the diameter.

Fig. 4 A gear rolling on a rack describes vee curves.

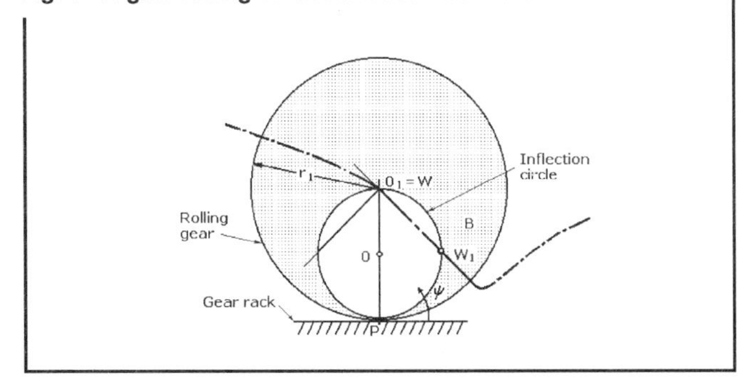

All points on this circle describe curves with portions that are approximately straight. This circle is also called the **inflection circle** because all points describe curves that have a point of inflection at the position illustrated. (The curve passing through point *W* is shown.)

This is a special case. Draw a circle with a diameter half that of the gear (diameter O_1P). This is the inflection circle. Any point, such as point W_1, will describe a curve that is almost straight in the vicinity selected. Tangents to the curves will always pass through the center of the gear, O_1 (as shown).

To find the inflection circle for a gear rolling inside a gear:

1. Draw arbitrary line *PB* from the contact point *P*.
2. Draw its parallel O_2A, and its perpendicular, *PA*. Locate *A*.
3. Draw line AO_1 through the center of the rolling gear. Locate W_1.
4. Draw a perpendicular through W_1. Obtain *W*. Line *WP* is the diameter of the inflection circle. Point W_1, which is an arbitrary point on the circle, will trace a curve of repeated almost-straight lines, as shown.

Fig. 5 A gear rolling inside a gear describes a zig-zag.

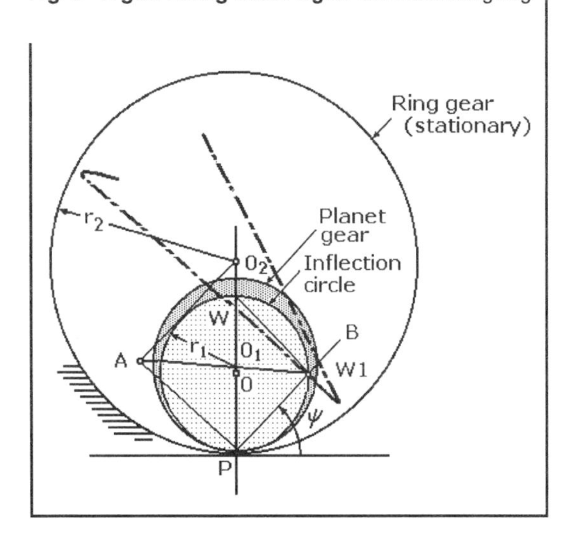

DESIGNING FOR DWELLS

Fig. 6 The center of curvature: a gear rolling on gear

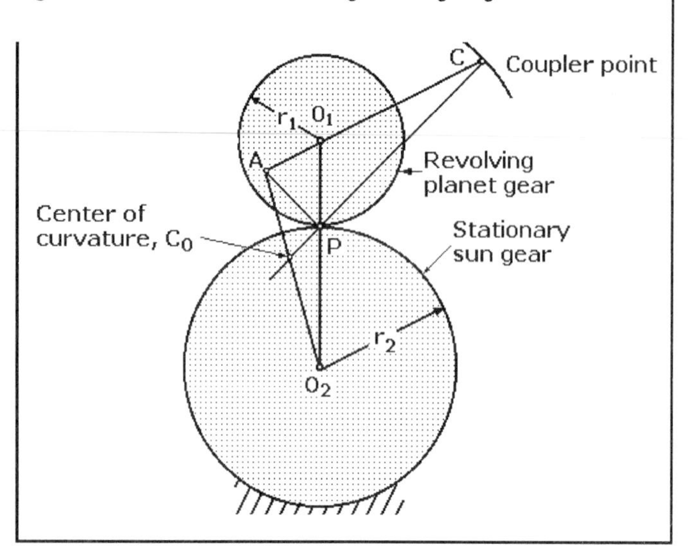

By locating the centers of curvature at various points, one can determine the length of the rocking or reciprocating arm to provide long dwells.

1. Draw a line through points C and P.
2. Draw a line through points C and O_1.
3. Draw a perpendicular to CP at P. This locates point A.
4. Draw line AO_2, to locate C_0, the center of curvature.

Fig. 7 The center of curvature: a gear rolling on a rack

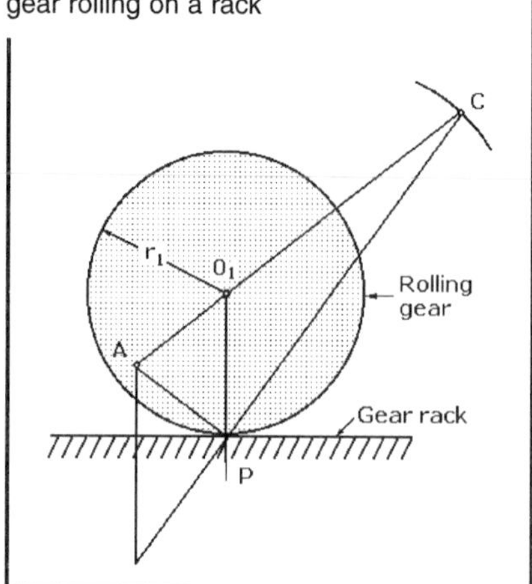

Construction is similar to that of the previous case.

1. Draw an extension of line CP.
2. Draw a perpendicular at P to locate A.
3. Draw a perpendicular from A to the straight surface to locate C.

Fig. 9 Analytical solutions.

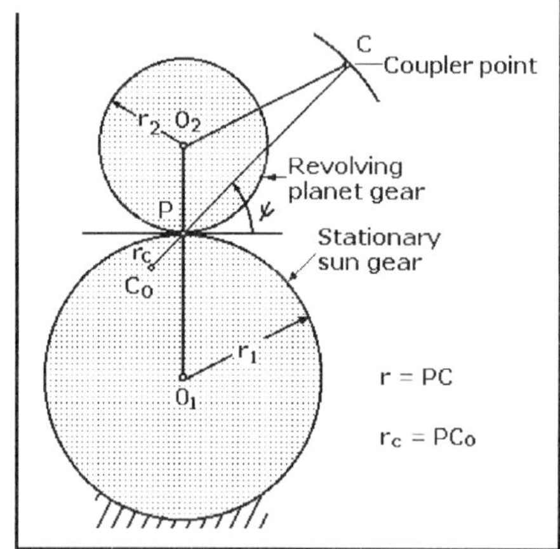

Fig. 8 The center of curvature: a gear rolling iside a gear.

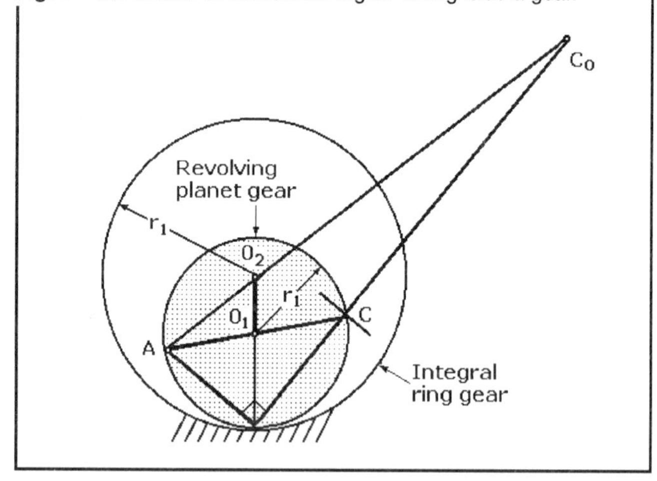

1. Draw extensions of CP and CO_1.
2. Draw a perpendicular of PC at P to locate A.
3. Draw AO_2 to locate C_0.

The center of curvature of a gear rolling on an external gear can be computed directly from the Euler-Savary equation:

$$\left(\frac{1}{r} - \frac{1}{r_c}\right)\sin\psi = \text{constant} \quad (7)$$

where angle ψ and r locate the position of C.

By applying this equation twice, specifically to point O_1 and O_2, which have their own centers of rotation, the following equation is obtained:

$$\left(\frac{1}{r_2} - \frac{1}{r_1}\right)\sin 90^\circ = \left(\frac{1}{r} + \frac{1}{r_c}\right)\sin\psi$$

or

$$\frac{1}{r_2} + \frac{1}{r_1} = \left(\frac{1}{r} + \frac{1}{r_c}\right)\sin\psi$$

This is the final design equation. All factors except r_c are known; hence, solving for r_c leads to the location of C_0.

For a gear rolling inside an internal gear, the Euler-Savary equation is:

$$\left(\frac{1}{r} + \frac{1}{r_c}\right)\sin\psi = \text{constant}$$

which leads to:

$$\frac{1}{r_2} - \frac{1}{r_1} = \left(\frac{1}{r} - \frac{1}{r_c}\right)\sin\psi$$

DESIGN CURVES AND EQUATIONS FOR GEAR-SLIDER MECHANISMS

What is a gear-slider mechanism? It is little more than a crank-and-slider with two gears meshed in line with the crank (Fig. 1). But, because one of the gears (planet gear, 3) is prevented from rotating because it is attached to the connecting rod, the output is taken from the sun gear, not the slider. This produces a variety of cyclic output motions, depending on the proportions of the members.

In his investigation of the capabilities of the mechanism, Professor Preben Jensen of Bridgeport, Connecticut derived the equations defining its motion and acceleration characteristics. He then devised some variations of his own (Figs. 5 through 8). These, he believes, will outperform the parent type. Jensen illustrated how the output of one of the new mechanisms, Fig. 8, can come to dead stop during each cycle, or progressively oscillate to new positions around the clock. Amachine designer, therefore, can obtain a wide variety of intermittent motions from the arrangement and, by combining two of these units, he can tailor the dwell period of the mechanism to fit the automatic feed requirements of a machine.

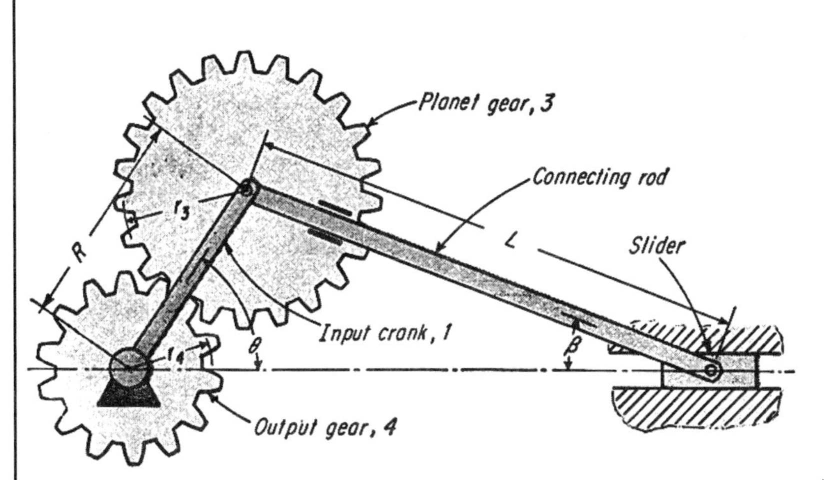

Fig. 1 A basic gear-slider mechanism. It differs from the better known three-gear drive because a slider restricts the motion of the planet gear. The output is taken from the gear, which is concentric with the input shaft, and not from the slider.

Symbols

L = Length of connecting rod, in.
r_3 = radius of gear fixed to connecting rod, in.
r_4 = radius of output gear, in.
R = length of crank, in.
α = angular acceleration of the input crank, rad/sec^2
β = connecting rod displacement, deg
γ = output rotation, deg
θ = input rotation, deg
θ_o = crank angle rotation during which the output gear reverses its motion, deg
ϕ = angle through which the output gear rotates back
ω = angular velocity of input crank, rad/sec

A single prime mark denotes angular velocity, rad/sec; double prime marks denote angular acceleration, rad/sec^2.

The Basic Form

The input motion is to crank 1, and the output motion is from gear 4. As the crank rotates, say counterclockwise, it causes planet gear 3 to oscillate while following a satellite path around gear 4. This imparts a varying output motion to gear 4, which rotates twice in the counterclockwise direction (when $r_3 = r_4$) for every revolution of the input.

Jensen's equations for angular displacement, velocity, and acceleration of gear 4, when driven at a speed of ω by crank 1, are as follows:

Angular Displacement

$$\gamma = \theta + \frac{r_3}{r_4}(\theta + \beta) \qquad (1)$$

where β is computed from the following relationship (see the list of symbols in this article):

$$\sin \beta = \frac{R}{L} \sin \theta \qquad (2)$$

Angular Velocity

$$\gamma' = \omega + \frac{r_3}{r_4}(\omega + \beta') \qquad (3)$$

where

$$\frac{\beta'}{\omega} = \frac{R}{L} \frac{\cos \theta}{\left[1 - \left(\frac{R}{L}\right)^2 \sin^2 \theta\right]^{1/2}} \qquad (4)$$

Angular Acceleration

$$\gamma'' = \alpha + \frac{r_3}{r_4}(\alpha + \beta'') \qquad (5)$$

where

$$\frac{\beta''}{\omega^2} = \frac{R}{L} \frac{\sin \theta\left[\left(\frac{R}{L}\right)^2 - 1\right]}{\left[1 - \left(\frac{R}{L}\right)^2 \sin^2 \theta\right]^{3/2}} \qquad (6)$$

For a constant angular velocity, Eq. 5 becomes

$$\gamma'' = \frac{r_3}{r_4}\beta'' \qquad (7)$$

Design Charts

The equations were solved by Professor Jensen for various L/R ratios and positions of the crank angle θ to obtain the design charts in Figs. 2, 3, and 4. Thus, for a mechanism with

$L = 12$ in. $r_3 = 2.5$
$R = 4$ in. $r_4 = 1.5$
$\omega = 1000$ per second
= radians per second

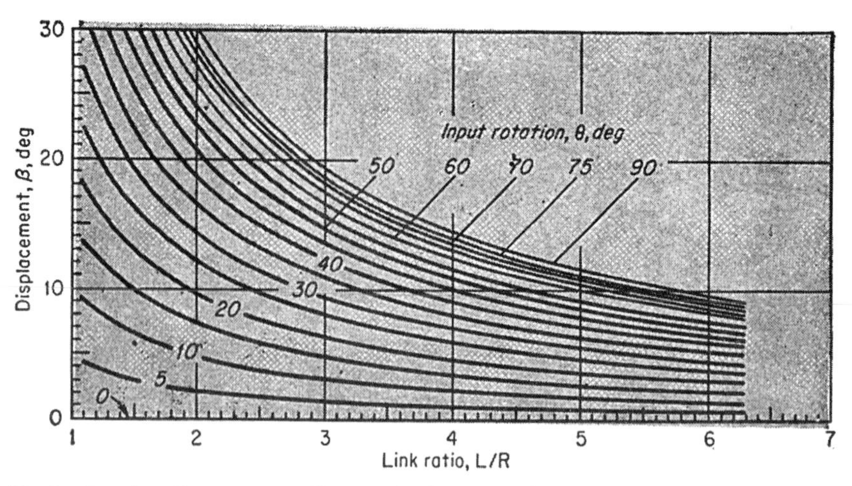

Fig. 2 Angular displacement diagram for the connecting rod.

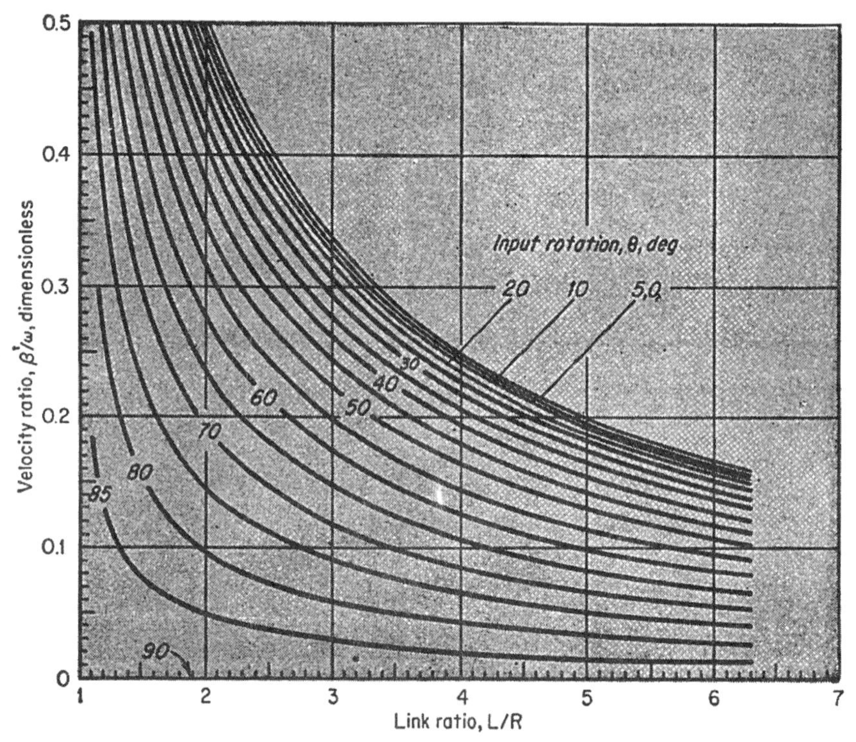

Fig. 3 Angular velocity curves for various crank angles.

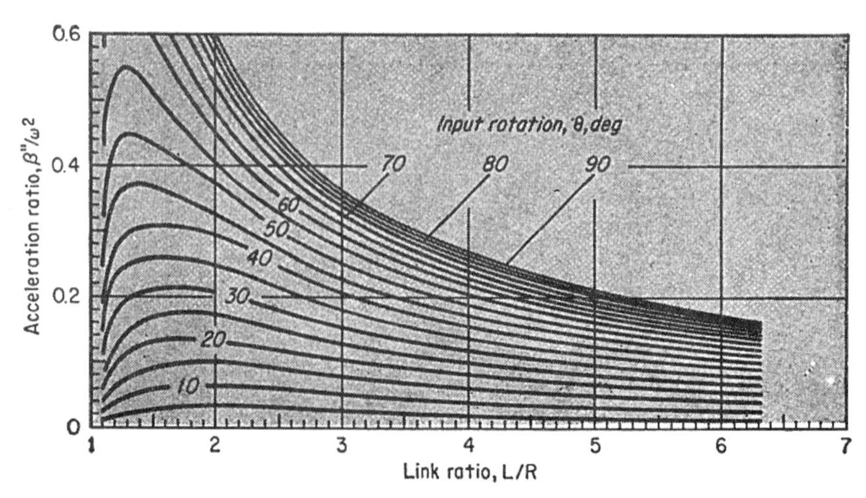

Fig. 4 Angular acceleration curves for various crank angles.

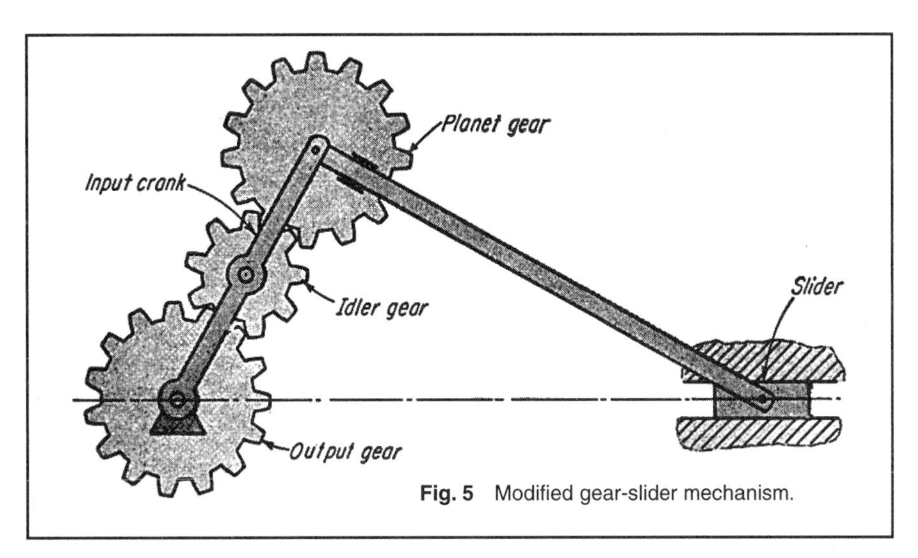

Fig. 5 Modified gear-slider mechanism.

the output velocity at crank angle $\theta = 60°$ can be computed as follows:

$$L/R = 12/4 = 3$$

From Fig. 3 $\beta'/\omega = 0.175$

$$\beta' = 0.175(1000)$$
$$= 175 \text{ radians per second}$$

From Eq. 3

$$\gamma = 2960 \text{ radians per second}$$

Three-Gear Variation

One interesting variation, shown in Fig. 5, is obtained by adding idler gear 5 to the drive. If gears 3 and 4 are then made equal in side, output gear 4 will then oscillate with exactly the same motion as connecting rod 2.

One use for this linkage, Jensen said, is in machinery where a sleeve is to ride concentrically over an input shaft, and yet must oscillate to provide a reciprocating motion. The shaft can drive the sleeve with this mechanism by making the sleeve part of the output gear.

Internal-Gear Variations

By replacing one of the external gears of Fig. 1 with an internal one, two mechanisms are obtained (Figs. 6 and 7) which have wider variable output abilities. But it is the mechanism in Fig. 7 that interested Jensen. This could be proportioned to give either a dwell or a progressive oscillation, that is, one in which the output rotates forward, say 360°, turns back to 30°, moves forward 30°, and then proceeds to repeat the cycle by moving forward again for 360°.

In this mechanism, the crank drives the large ring gear 3 which is fixed to the connecting rod 2. Output is from gear 4. Jensen derived the following equations:

Output Motion

$$\omega_4 = -\left(\frac{L - R - r_4}{Lr_4}\right)R\omega_1 \qquad (8)$$

When $r_4 = L - R$, then $(_4 = 0$ from Eq. 8, and the mechanism is proportioned to give instantaneous dwell. To obtain a progressive oscillation, r_4 must be greater than $L - R$, as shown in Jensen's model (Fig. 8).

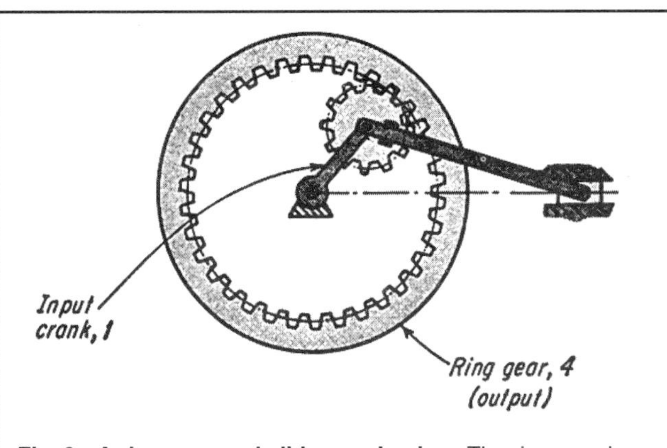

Fig. 6 A ring-gear and slider mechanism. The ring gear is the output and it replaces the center gear in Fig. 1.

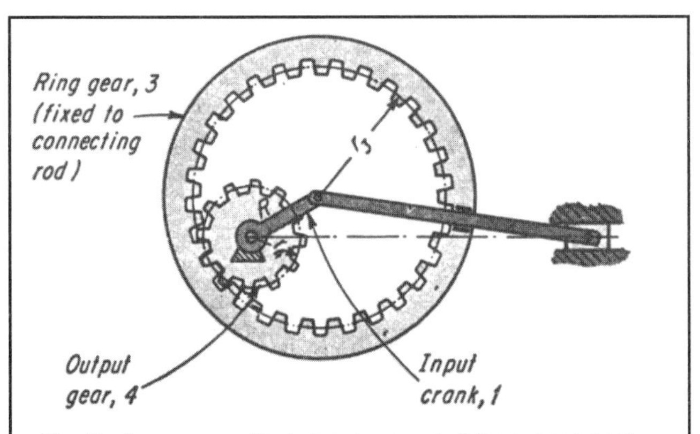

Fig. 7 A more practical ring-gear and slider arrangement. The output is now from the smaller gear.

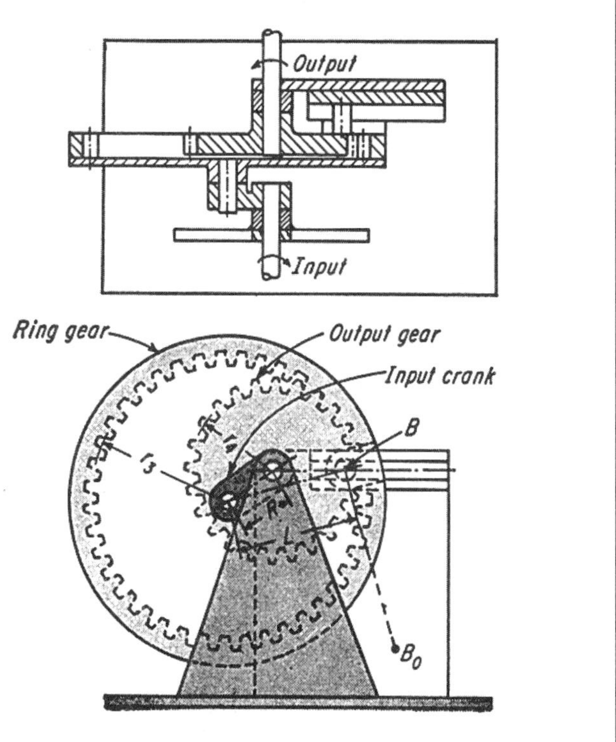

Fig. 8 Jensen's model of the ring-gear and slider mechanism shown in Fig. 7. A progressive oscillation motion is obtained by making r_4 greater than *L-R*.

If gear 4 turns back and then starts moving forward again, there must be two positions where the motion of gear 4 is zero. Those two mechanisms are symmetrical with respect to A_0B. If θ_0 equals the crank-angle rotation (of input), during which the output gear reverses its motion, and ϕ equals the angle through which gear 4 rotates back, then

$$\cos\frac{\theta_0}{2} = \left[\frac{L^2 - R^2}{r_4(2R + r_4)}\right]^{1/2} \quad (9)$$

and

$$\gamma = \theta_0 - \frac{r_3}{r_4}(\theta_0 - \beta_0) \quad (10)$$

where

$$\sin\beta_0 = \frac{R}{L}\sin\frac{\theta_0}{2} \quad (11)$$

Chart for Proportioning

The chart in Fig. 9 helps proportion the mechanism of Fig. 8 to provide a specific kind of progressive oscillation. It is set up for R equals 1 in. For other values of R, convert the chart values for r_4 proportionally, as shown below.

For example, assume that the output gear, during each cycle, is to rotate back 9.2°. Thus ϕ = 9.2°. Also given is R = 0.75 in. and L = 1.5 in. Thus L/R = 2.

From the right side of the chart, go to the ϕ-curve for L = 2, then upward to the θ_0-curve for L = 2 in. Read θ_0 = 82° at the left ordinate.

Now return to the second intersection point and proceed upward to read on the abscissa scale for L = 2, a value of r_4 = 1.5. Since R = 0.75 in., and the chart is for R 1, convert r_4 as follows: r_4 = 0.75 (1.5) = 1.13 in.

Thus, if the mechanism is built with an output gear of radius r_4 = 1.13 in., then during 82° rotation of the crank, the output gear 4 will go back 9.2°. Of course, during the next 83°, gear 4 will have reversed back to its initial position—and then will keep going forward for the remaining 194° of the crank rotation.

Future Modifications

The mechanism in Fig. 8 is designed to permit changing the output motion easily from progressive oscillation to instantaneous dwell or nonuniform CW or CCW rotation. This is accomplished by shifting the position of the pin which acts as the sliding piece of the centric slider crank. It is also possible to use an eccentric slider crank, a four-bar linkage, or a sliding-block linkage as the basic mechanism.

Two mechanisms in series will give an output with either a prolonged dwell or two separate dwells. The angle between the separated dwells can be adjusted during its operation by interposing a gear differential so that the position of the output shaft of the first mechanism can be changed relative to the position of the input shaft of the second mechanism.

The mechanism can also be improved by introducing an additional link, B-B_0, to guide pin B along a circular arc instead of a linear track. This would result in a slight improvement in the performance of the mechanism.

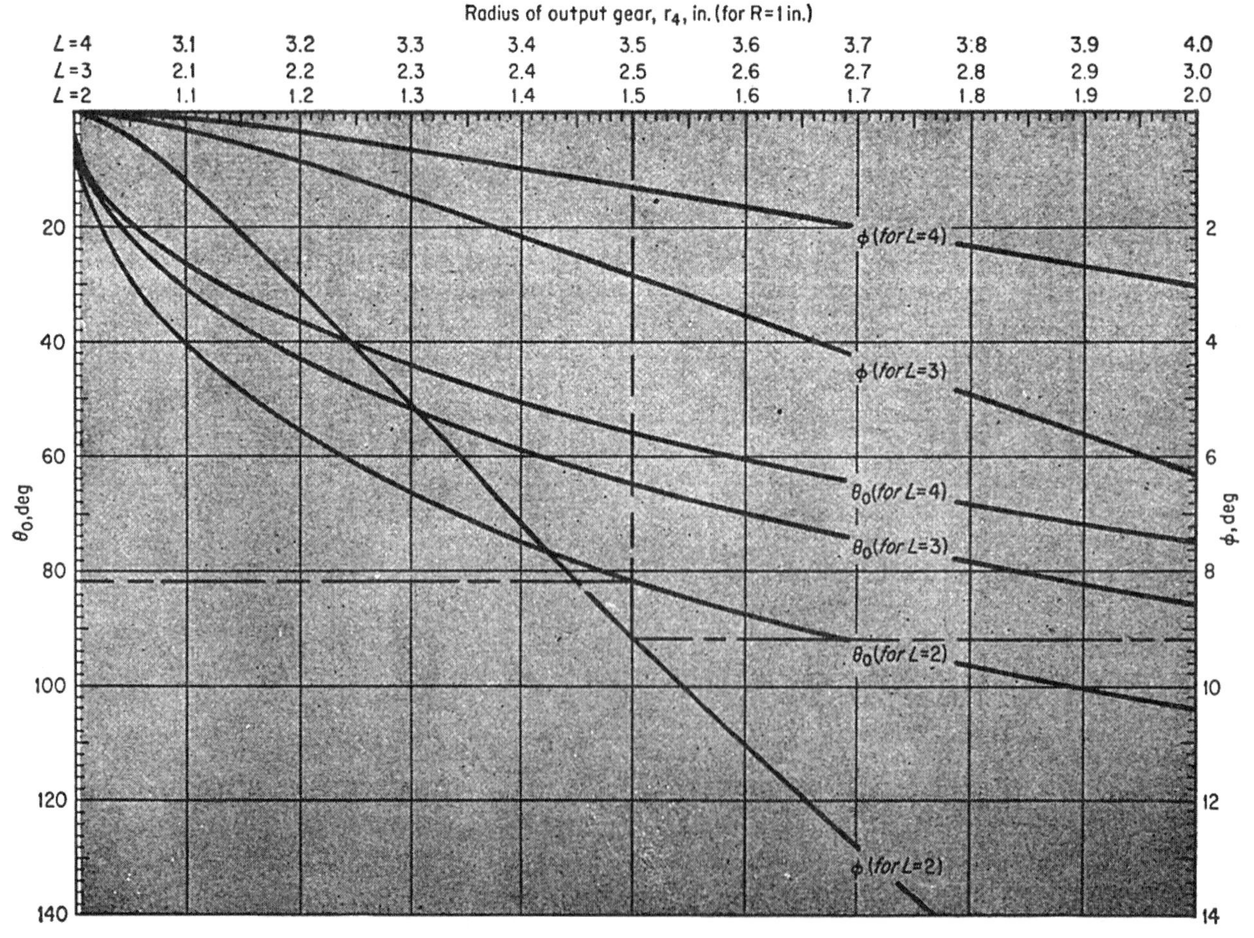

Fig. 9 **A chart** for proportioning a ring-gear and slider mechanism.

CHAPTER 7

CAM, GENEVA, AND RATCHET DRIVES AND MECHANISMS

CAM-CONTROLLED PLANETARY GEAR SYSTEM

By incorporating a grooved cam a novel mechanism can produce a wide variety of output motions.

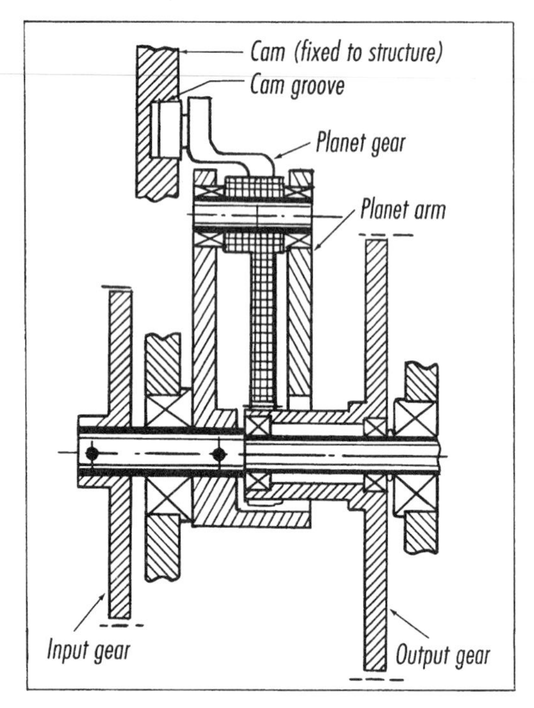

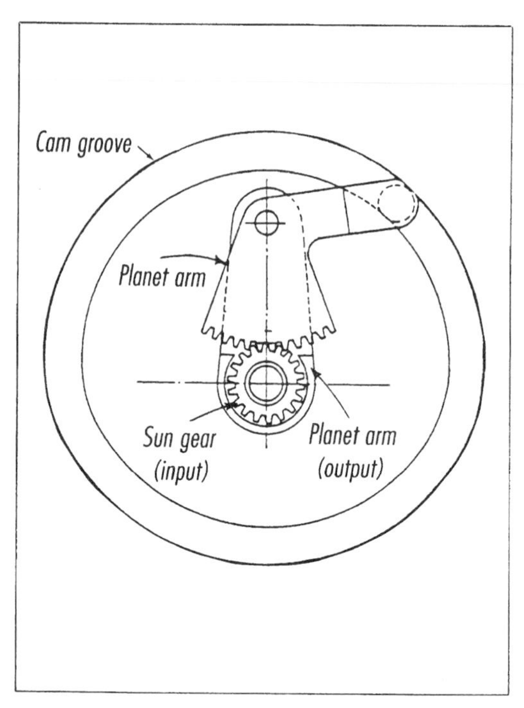

Construction details of a cam-planetary mechanism used in a film drive.

Do you want more variety in the kinds of output motion given by a planetary gear system? You can have it by controlling the planet with a grooved cam. The method gives the mechanism these additional features:

- Intermittent motion, with long dwells and minimum acceleration and deceleration.
- Cyclic variations in velocity.
- Two levels, or more, of constant speed during each cycle of the input.

The design is not simple because of need to synchronize the output of the planetary system with the cam contour. However, such mechanisms are now at work in film drives and should prove useful in many automatic machines. Here are equations, tables, and a step-by-step sequence that will make the procedure easier.

How the Mechanism Works

The planet gear need not be cut in full—a gear sector will do because the planet is never permitted to make a full revolution. The sun gear is integral with the output gear. The planet arm is fixed to the input shaft, which is coaxial with the output shaft. Attached to the planet is a follower roller which rides in a cam groove. The cam is fixed to the frame.

The planet arm (input) rotates at constant velocity and makes one revolution with each cycle. Sun gear (output) also makes one revolution during each cycle. Its motion is modified, however, by the oscillatory motion of the planet gear relative to the planet arm. It is this motion that is controlled by the cam (a constant-radius cam would not affect the output, and the drive would give only a constant one-to-one ratio).

Comparison with Other Devices

A main feature of this cam-planetary mechanism is its ability to produce a wide range of nonhomogeneous functions. These functions can be defined by no less than two mathematical expressions, each valid for a discrete portion of the range. This feature is not shared by the more widely known intermittent mechanisms: the external and internal genevas, the three-gear drive, and the cardioid drive.

Either three-gear or cardioid can provide a dwell period—but only for a comparatively short period of the cycle. With the cam-planetary, one can obtain over 180° of dwell during a 360° cycle by employing a 4-to-1 gear ratio between planet and sun.

And what about a cam doing the job by itself? This has the disadvantage of producing reciprocating motion. In other words, the output will always reverse during the cycle—a condition unacceptable in many applications.

Design Procedure

The basic equation for an epicyclic gear train is:

$$d\theta_S = d\theta_A - n\,d\theta_{P\text{-}A}$$

where: $d\theta_S$ = **rotation of sun gear (output), deg**
$d\theta_A$ = **rotation of planet arm (input), deg**
$d\theta_{P\text{-}A}$ = **rotation of planet gear with respect to arm, deg**
n = **ratio of planet to sun gear.**

The required output of the system is usually specified in the form of kinematic curves. Design procedure then is to:

- Select the proper planet-sun gear ratio
- Develop the equations of the planet motion (which also functions as a cam follower)
- Compute the proper cam contour

FIVE CAM-STROKE-AMPLIFYING MECHANISMS

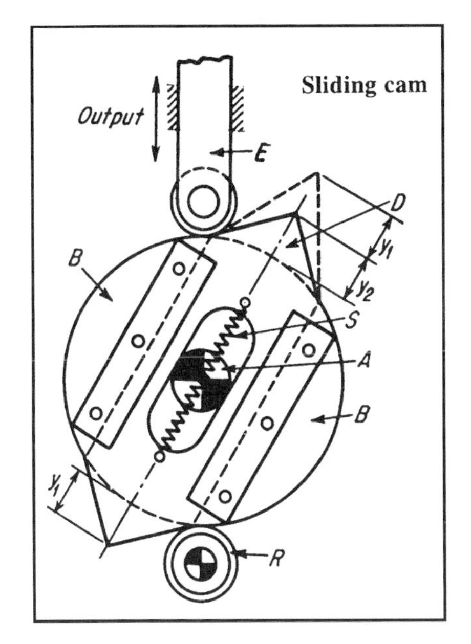

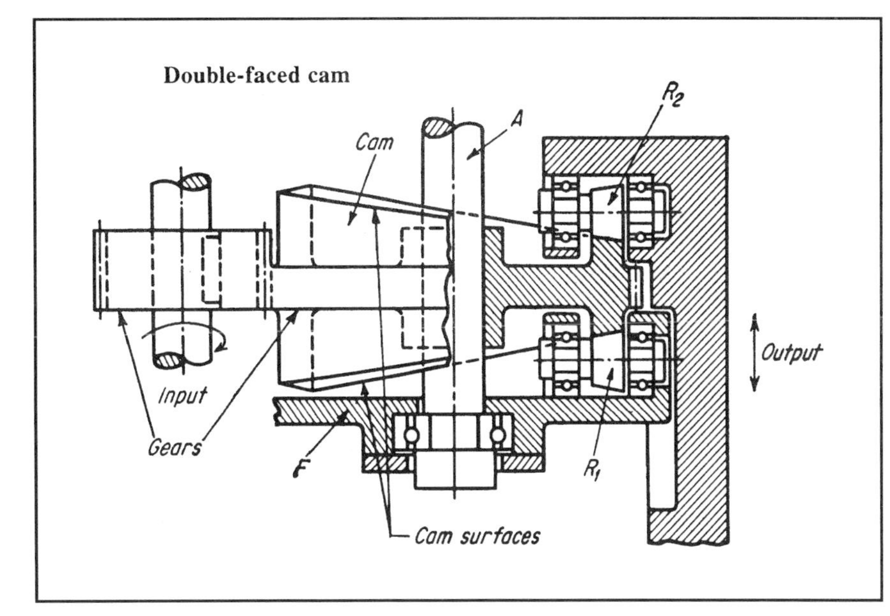

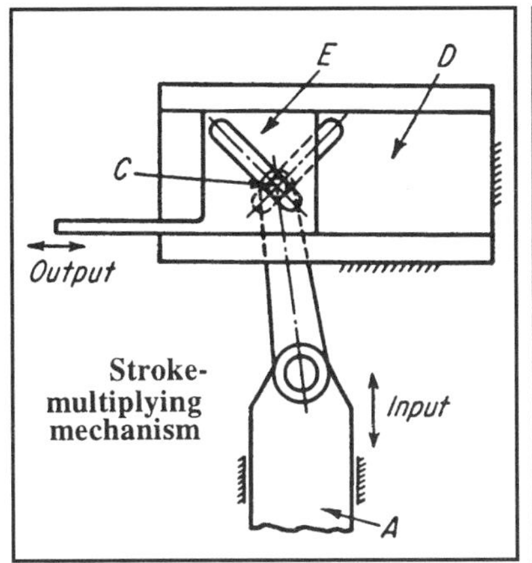

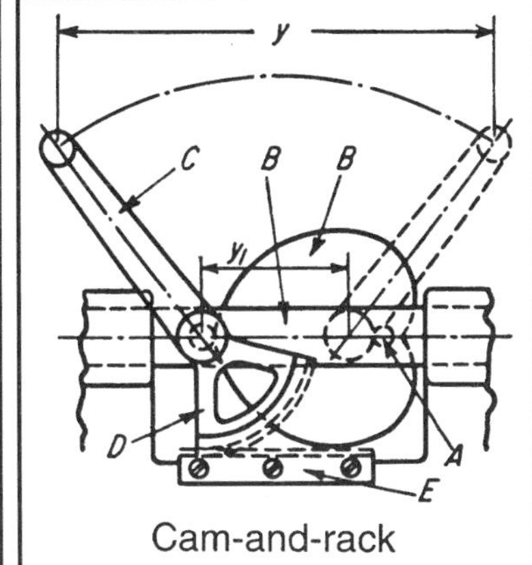

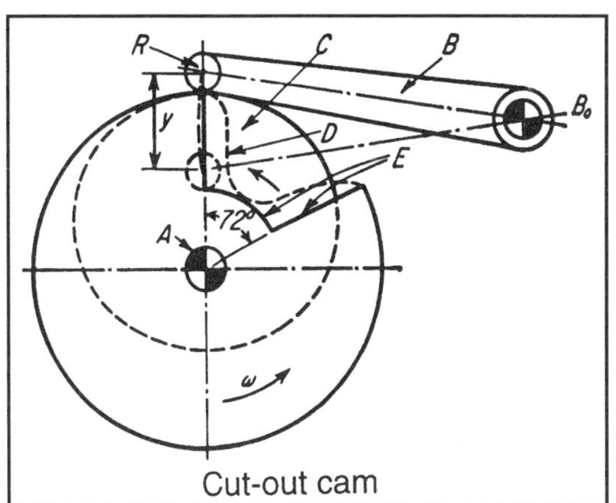

When the pressure angles of stroke-amplifying mechanisms are too high to satisfy the design requirements, and it is undesirable to enlarge the cam size, certain devices can be installed to reduce the pressure angles:

Sliding cam—This mechanism is used on a wire-forming machine. Cam *D* has a pointed shape because of the special motion required for twisting wires. The machine operates at slow speeds, but the principle employed here is also applicable to high-speed cams.

The original stroke desired was ($y_1 + y_2$) but this results in a large pressure angle. The stroke therefore is reduced to y_2 on one side of the cam, and a rise of y_1 is added to the other side. Flanges *B* are attached to cam shaft *A*. Cam *D*, a rectangle with the two cam ends (shaded), is shifted upward as it cams off stationary roller *R* when the cam follower *E* is being cammed upward by the other end of cam *D*.

Stroke-multiplying mechanisms—This mechanism is used in power presses. The opposing slots, the first in a fixed member *D*, and the second in the movable slide *E*, multiply the motion of the input slide *A* driven by the cam. As *A* moves upward, *E* moves rapidly to the right.

Double-faced cam—This mechanism doubles the stroke, hence reduces the pressure angles to one-half of their original values. Roller R_1 is stationary. When the cam rotates, its bottom surface lifts itself on R_1, while its top surface adds an additional motion to the movable roller R_2. The output is driven linearly by roller R_2 and thus is approximately the sum of the rise of both of these surfaces.

Cam-and-rack—This mechanism increases the throw of a lever. Cam *B* rotates around *A*. The roller follower travels at distances y_1; during this time, gear segment *D* rolls on rack *E*. Thus the output stroke of lever *C* is the sum of transmission and rotation, giving the magnified stoke *y*.

Cut-out cam—A rapid rise and fall within 72° was desired. This originally called for the cam contour, *D*, but produced severe pressure angles. The condition was improved by providing an additional cam *C*. This cam also rotates around the cam center *A*, but at five times the speed of cam *D* because of a 5:1 gearing arrangement (not shown). The original cam was then completely cut away for the 72° (see surfaces *E*). The desired motion, expanded over 360° (because $72° \times 5 = 360°$), is now designed into cam *C*. This results in the same pressure angle as would occur if the original cam rise occurred over 360° instead of 72°.

CAM-CURVE-GENERATING MECHANISMS

It usually doesn't pay to design a complex cam curve if it can't be easily machined—so check these mechanisms before starting your cam design.

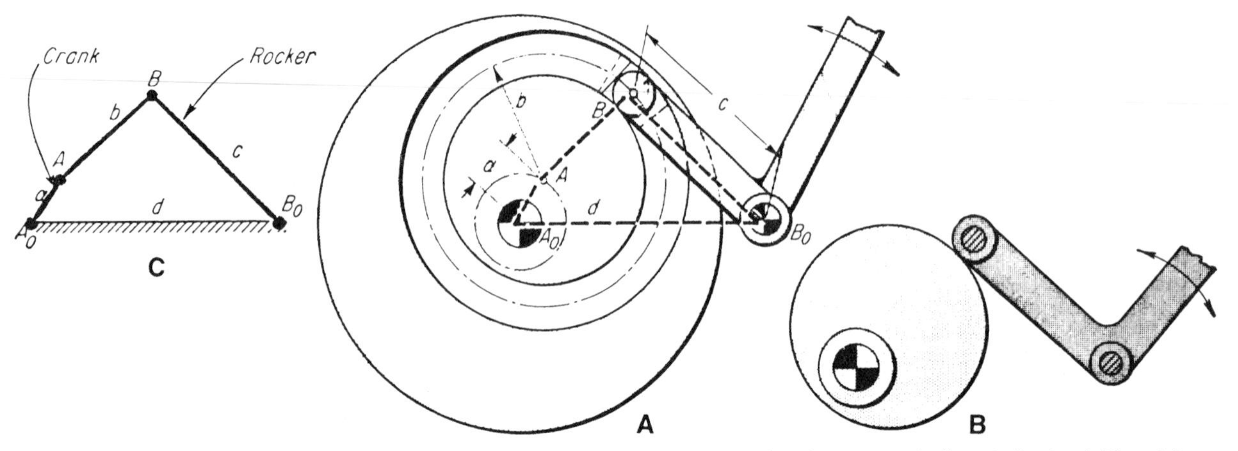

Fig. 1 A circular cam groove is easily machined on a turret lathe by mounting the plate eccentrically onto the truck. The plate cam in **(B)** with a spring-load follower produces the same output motion. Many designers are unaware that this type of cam has the same output motion as four-bar linkage **(C)** with the indicated equivalent link lengths. Thus, it's the easiest curve to pick when substituting a cam for an existing linkage.

If you have to machine a cam curve into the metal blank without a master cam, how accurate can you expect it to be? That depends primarily on how precisely the mechanism you use can feed the cutter into the cam blank. The mechanisms described here have been carefully selected for their practicability. They can be employed directly to machine the cams, or to make master cams for producing other cams.

The cam curves are those frequently employed in automatic-feed mechanisms and screw machines They are the circular, constant-velocity, simple-harmonic, cycloidal, modified cycloidal, and circular-arc cam curve, presented in that order.

Circular Cams

This is popular among machinists because of the ease in cutting the groove. The cam (Fig. 1A) has a circular groove whose center, A, is displaced a distance a from the cam-plate center, A_0, can simply be a plate cam with a spring-loaded follower (Fig. 1B).

Interestingly, with this cam you can easily duplicate the motion of a four-bar linkage (Fig. 1C). Rocker BB_0 in Fig. 1C, therefore, is equivalent to the motion of the swinging follower shown in Fig. 1A.

The cam is machined by mounting the plate eccentrically on a lathe. Consequently, a circular groove can be cut to close tolerances with an excellent surface finish.

If the cam is to operate at low speeds, you can replace the roller with an arc-formed slide. This permits the transmission of high forces. The optimum design of these "power cams" usually requires time-consuming computations.

The disadvantages (or sometimes, the advantage) of the circular-arc cam is that, when traveling from one given point, its follower reaches higher-speed accelerations than with other equivalent cam curves.

Constant-Velocity Cams

A constant-velocity cam profile can be generated by rotating the cam plate and feeding the cutter linearly, both with uniform velocity, along the path the translating roller follower will travel later (Fig. 2A). In the example of a swinging follower, the tracer (cutter) point is placed on an arm whose length is equal to the length of the swinging roller follower, and the arm is rotated with uniform velocity (Fig. 2B).

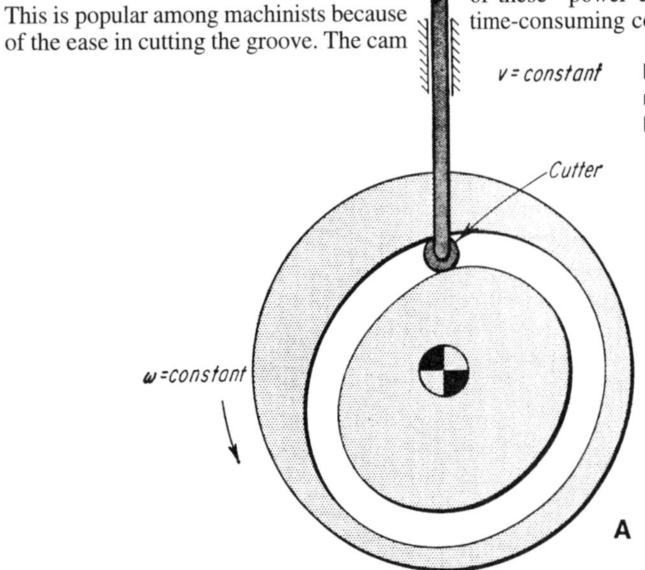

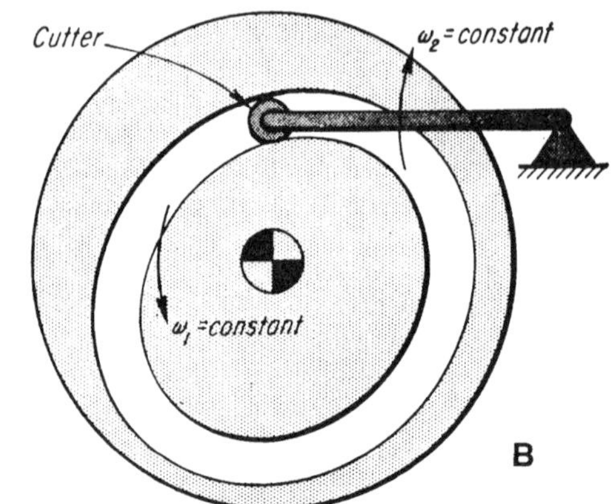

Fig. 2 A constant-velocity cam is machined by feeding the cutter and rotating the cam at constant velocity. The cutter is fed linearly **(A)** or circularly **(B),** depending on the type of follower.

Fig. 3 **For producing simple harmonic curves:** **(A)** a scotch yoke device feeds the cutter while the gearing arrangement rotates the cam; **(B)** a truncated-cylinder slider for a cylindrical cam; **(C)** a scotch-yoke inversion linkage for avoiding gearing; **(D)** an increase in acceleration when a translating follower is replaced by a swinging follower.

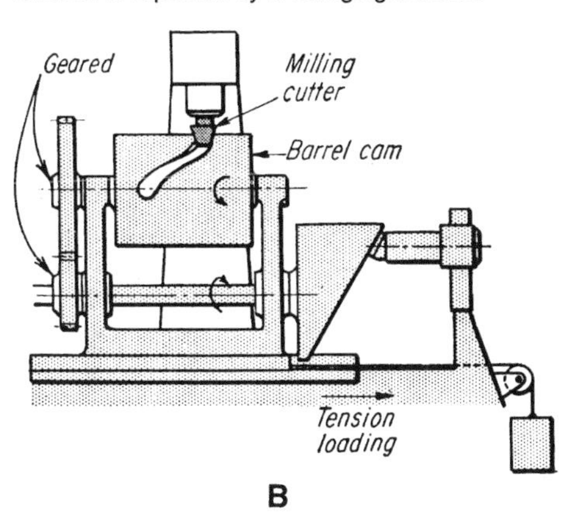

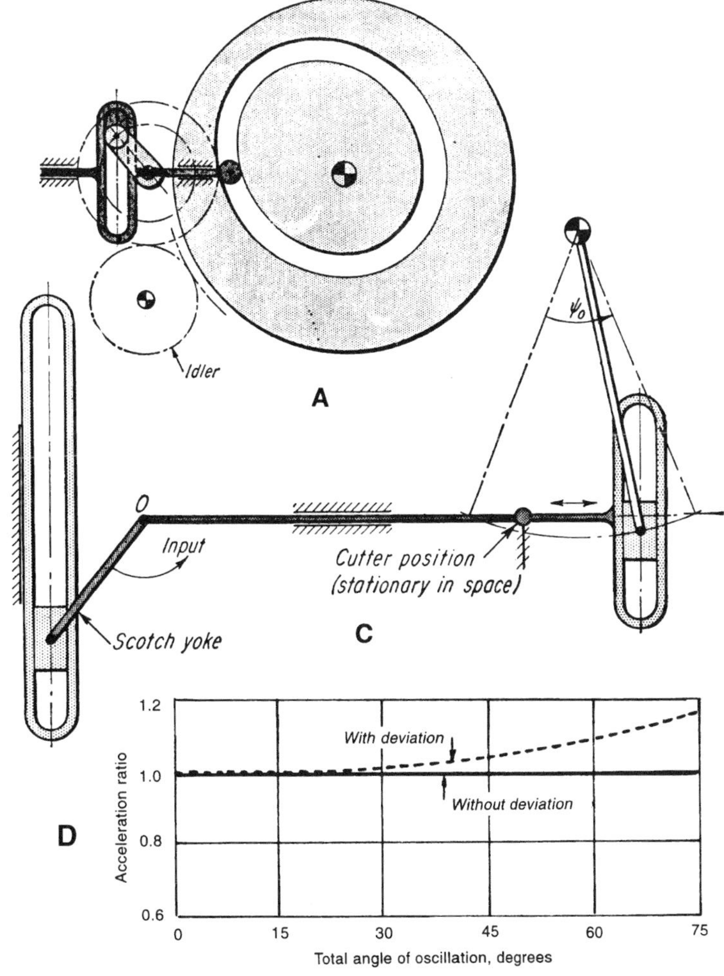

Simple-Harmonic Cams

The cam is generated by rotating it with uniform velocity and moving the cutter with a scotch yoke geared to the rotary motion of the cam. Fig. 3A shows the principle for a radial translating follower; the same principle is applicable for offset translating and the swinging roller follower. The gear ratios and length of the crank working in the scotch yoke control the pressures angles (the angles for the rise or return strokes).

For barrel cams with harmonic motion, the jig in Fig. 3B can easily be set up to do the machining. Here, the barrel cam is shifted axially by the rotating, weight-loaded (or spring-loaded) truncated cylinder.

The scotch-yoke inversion linkage (Fig. 3C) replaces the gearing called for in Fig. 3A. It will cut an approximate simple-harmonic motion curve when the cam has a swinging roller follower, and an exact curve when the cam has a radial or offset translating roller follower. The slotted member is fixed to the machine frame *1*. Crank 2 is driven around the center *0*. This causes link *4* to oscillate back and forward in simple harmonic motion. The sliding piece *5* carries the cam to be cut, and the cam is rotated around the center of *5* with uniform velocity. The length of arm *6* is made equal to the length of the swinging roller follower of the actual am mechanism and the device adjusted so that the extreme position of the center of *5* lie on the center line of *4*.

The cutter is placed in a stationary spot somewhere along the centerline of member *4*. If a radial or offset translating roller follower is used, sliding piece *5* is fastened to *4*.

The deviation from simple harmonic motion, when the cam has a swinging follower, causes an increase in acceleration ranging from 0 to 18% (Fig. 3D), which depends on the total angle of oscillation of the follower. Note that for a typical total oscillating angle of 45° the increase in acceleration is about 5%.

Cycloidal Motion

This curve is perhaps the most desirable from a designer's viewpoint because of its excellent acceleration characteristic. Luckily, this curve is comparatively easy to generate. Before selecting the mechanism, it is worth looking at the underlying theory of cycloids because it is possible to generate not only cycloidal motion but a whole family of similar curves.

The cycloids are based on an offset sinusoidal wave (Fig. 4). Because the

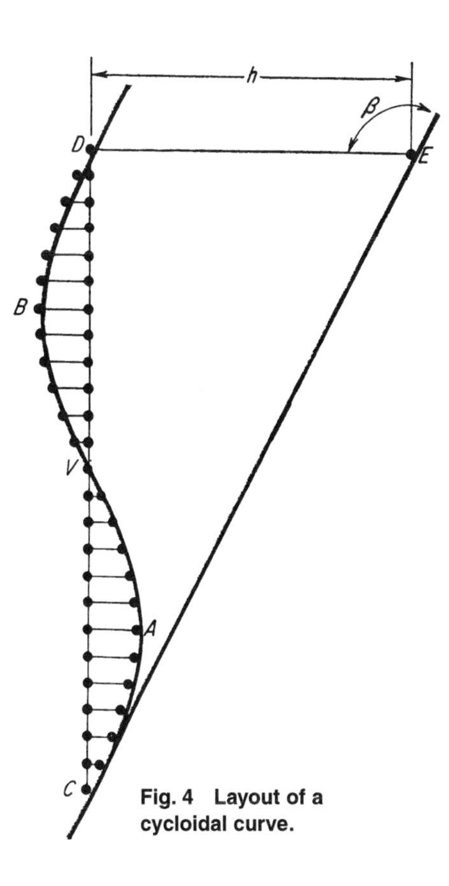

Fig. 4 Layout of a cycloidal curve.

radii of curvatures in points C, V, and D are infinite (the curve is "flat" at these points), if this curve was a cam groove and moved in the direction of line CVD, a translating roller follower, actuated by this cam, would have zero acceleration at points C, V, and D no matter in what direction the follower is pointed.

Now, if the cam is moved in the direction of CE and the direction of motion of the translating follower is lined up perpendicular to CE, the acceleration of the follower in points, C, V, and D would still be zero. This has now become the basic cycloidal curve, and it can be considered as a sinusoidal curve of a certain amplitude (with the amplitude measured perpendicular to the straight line) superimposed on a straight (constant-velocity) line.

The cycloidal is considered to be the best standard cam contour because of its low dynamic loads and low shock and vibration characteristics. One reason for these outstanding attributes is that sudden changes in acceleration are avoided during the cam cycle. But improved performance is obtainable with certain modified cycloidals.

Modified Cycloids

To modify the cycloid, only the direction and magnitude of the amplitude need to be changed, while keeping the radius of curvature infinite at points C, V, and D.

Comparisons are made in Fig. 5 of some of the modified curves used in industry. The true cycloidal is shown in the cam diagram of Fig. 5A. Note that the sine amplitudes to be added to the constant-velocity line are perpendicular to the base. In the Alt modification shown in Fig. 5B (named after Hermann Alt, a German kinematician who first analyzed it), the sine amplitudes are perpendicular to the constant-velocity line. This results in improved (lower) velocity characteristics (Fig. 5D), but higher acceleration magnitudes (Fig. 5E).

The Wildt modified cycloidal (after Paul Wildt) is constructed by selecting a point w which is 0.57 the distance $T/2$, and then drawing line wp through yp which is midway along OP. The base of the sine curve is then constructed perpendicular to yw. This modification results in a maximum acceleration of $5.88\ h/T^2$. By contrasts, the standard cycloidal curve has a maximum acceleration of $6.28\ h/T^2$. This is a 6.8 reduction in acceleration.

(It's a complex task to construct a cycloidal curve to go through a particular point P—where P might be anywhere within the limits of the box in Fig. 5C—and with a specific scope at P. There is a growing demand for this kind of cycloidal modification.

Generating Modified Cycloidals

One of the few methods capable of generating the family of modified cycloidals consists of a double carriage and rack arrangement (Fig. 6A).

The cam blank can pivot around the spindle, which in turn is on the movable carriage I. The cutter center is stationary. If the carriage is now driven at constant speed by the leadscrew in the direction of the arrow, steel bands 1 and 2 will also cause the cam blank to rotate. This rotation-and-translation motion of the cam will cut a spiral groove.

For the modified cycloidals, a second motion must be imposed on the cam to compensate for the deviations from the true cycloidal. This is done by a second

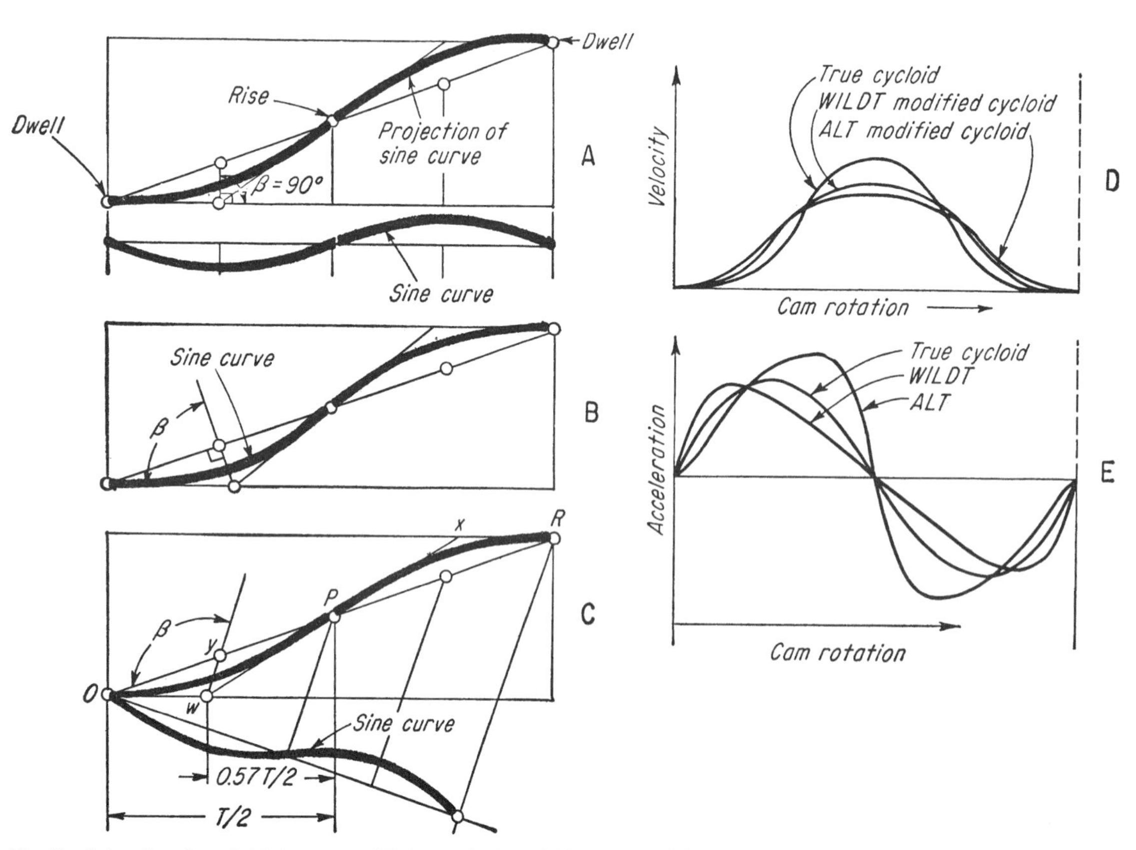

Fig. 5 A family of cycloidal curves: (A) A standard cycloidal motion; **(B)** A modification according to H. Alt; **(C)** A modification according to P. Wildt; **(D)** A comparison of velocity characteristics; **(E)** A comparison of acceleration curves.

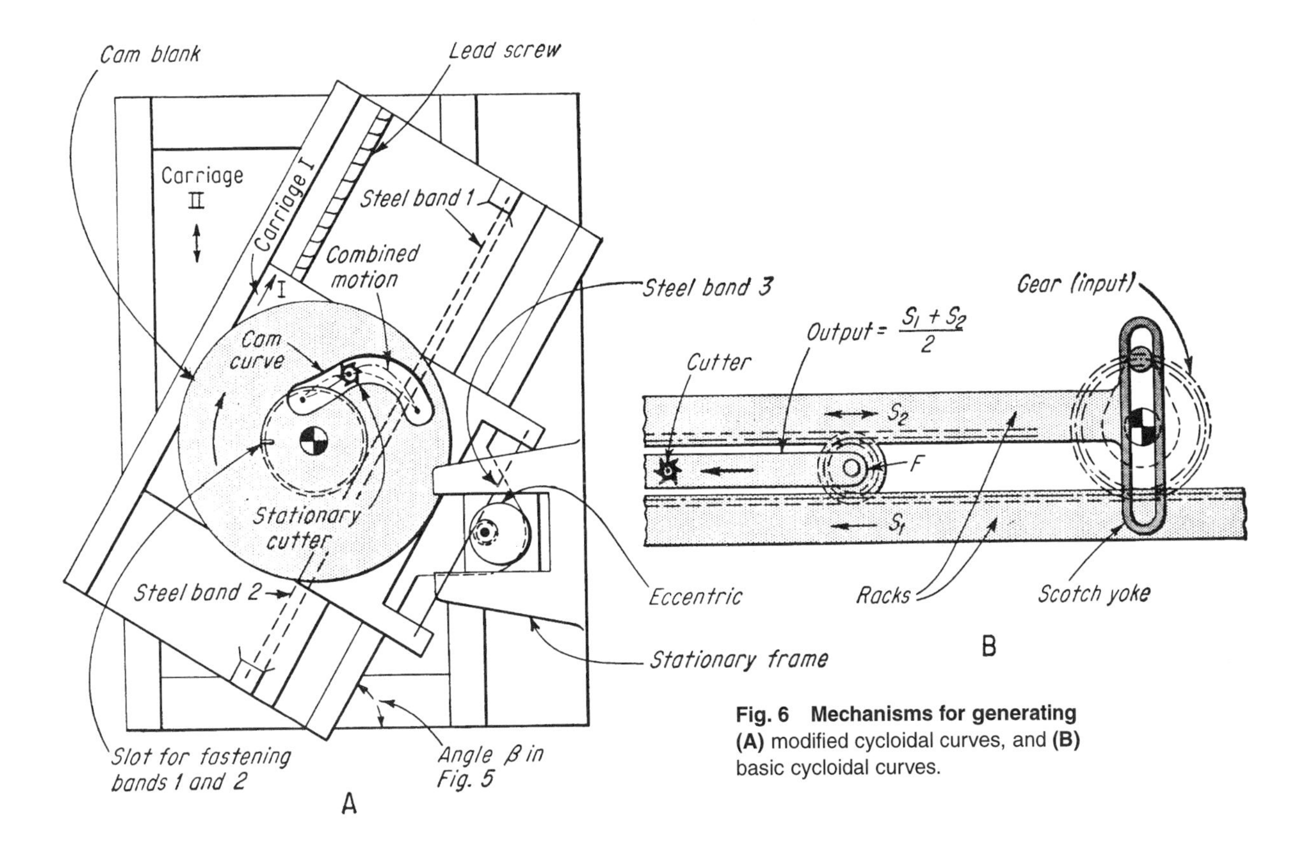

Fig. 6 Mechanisms for generating (A) modified cycloidal curves, and **(B)** basic cycloidal curves.

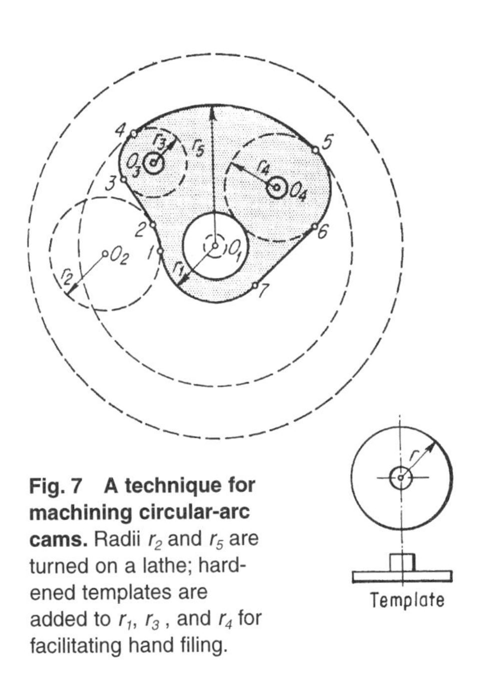

Fig. 7 A technique for machining circular-arc cams. Radii r_2 and r_5 are turned on a lathe; hardened templates are added to r_1, r_3, and r_4 for facilitating hand filing.

steel-band arrangement. As carriage I moves, bands 3 and 4 cause the eccentric to rotate. Because of the stationary frame, the slide surrounding the eccentric is actuated horizontally. This slide is part of carriage II. As a result, a sinusoidal motion is imposed on the cam.

Carriage I can be set at various angles β to match angle β in Fig. 5B and C. The mechanism can also be modified to cut cams with swinging followers.

Circular-Arc Cams

In recent years it has become customary to turn to the cycloidal and other similar curves even when speeds are low. However, there are still many applications for circular-arc cams. Those cams are composed of circular arcs, or circular arc and straight lines. For comparatively small cams, the cutting technique illustrated in Fig. 7 produces accurate results.

Assume that the contour is composed of circular arc *1-2* with center at 0_2, arc *3*-4 with center at 0_3, arc *4*-5 with center at 0_1 , arc 5-*6* with center at 0_4 , arc *7-1* with center at 0_1, and the straight lines *2-3* and *6-7*. The method calls for a combination of drilling, lathe turning, and template filing.

First, small holes about 0.1 in. in diameter are drilled at 0_1, 0_3, and 0_4. Then a hole drilled with the center at 0_2, and radius of r_2 . Next the cam is fixed in a turret lathe with the center of rotation at 0_1, and the steel plate is cut until it has a diameter of $2r_5$. This completes the larger convex radius. The straight lines *6-7* and *2-3* are then milled on a milling machine.

Finally, for the smaller convex arcs, hardened pieces are turned with radii r_1, r_3, and r_4. One such piece is shown in Fig. 7. The templates have hubs that fit into the drilled holes at 0_1, 0_3, and 0_4 . Next the arcs *7-1*, *3-4*, and *5-6* are filed with the hardened templates as a guide. The final operation is to drill the enlarged hole at 0_1 to a size that will permit a hub to be fastened to the cam.

This method is usually better than copying from a drawing or filing the scallops from a cam on which a large number of points have been calculated to determine the cam profile.

Compensating for Dwells

One disadvantage with the previous generating machines is that, with the exception of the circular cam, they cannot include a dwell period within the rise-and-fall cam cycle. The mechanisms must be disengaged at the end of the rise, and the cam must be rotated the exact number of degrees to the point where the

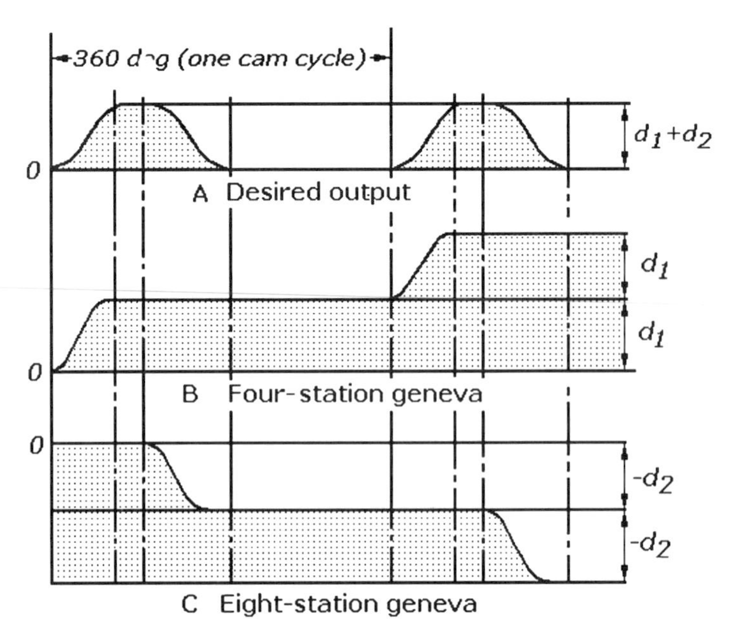

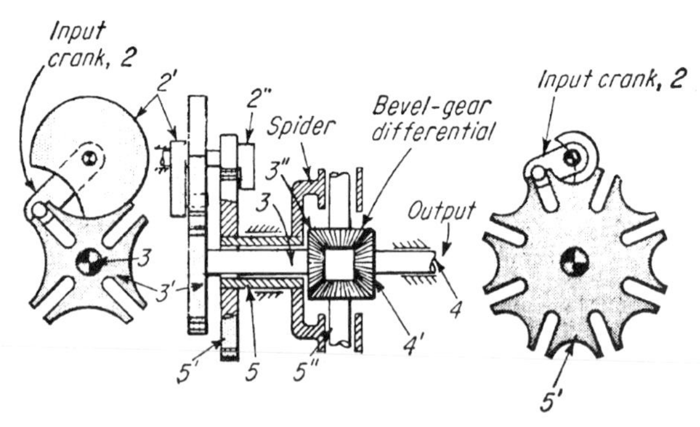

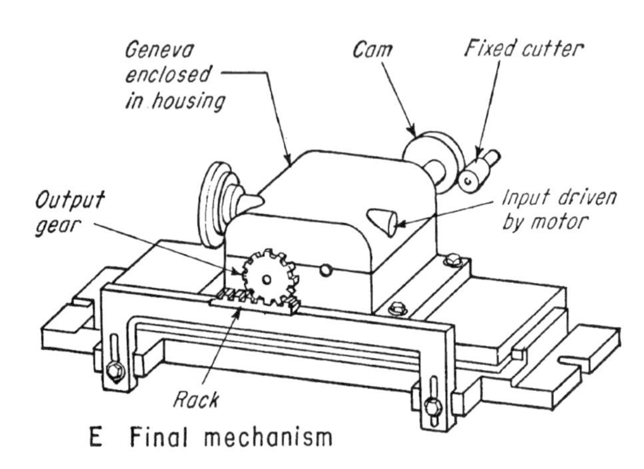

Fig. 8 Double genevas with differentials for obtaining long dwells. The desired output characteristic **(A)** of the cam is obtained by adding the motion **(B)** of a four-station geneva to that of **(C)** an eight-station geneva. The mechanical arrangement of genevas with a differential is shown in **(D)**; the actual device is shown in **(E)**. A wide variety of output dwells **(F)** are obtained by varying the angle between the driving cranks of the genevas.

fall cycle begins. This increases the possibility of inaccuracies and slows down production.

There are two mechanisms, however, that permit automatic cam machining through a specific dwell period: the double-geneva drive and the double eccentric mechanism.

Double-Genevas with Differential

Assume that the desired output contains dells (of specific duration) at both the rise and fall portions, as shown in Fig. 8A. The output of a geneva that is being rotated clockwise will produce an intermittent motion similar to the one shown in Fig. 8B—a rise-dwell-rise-dwell motion. These rise portions are distorted simple-harmonic curves, but are sufficiently close to the pure harmonic to warrant their use in many applications.

If the motion of another geneva, rotating counterclockwise as shown in (Fig. 8C), is added to that of the clockwise geneva by a differential (Fig. 8D), then the sum will be the desired output shown in (Fig. 8A).

The dwell period of this mechanism is varied by shifting the relative positions between the two input cranks of the genevas.

The mechanical arrangement of the mechanism is shown in Fig. 8D. The two driving shafts are driven by gearing (not shown). Input from the four-star geneva to the differential is through shaft *3*; input from the eight-station geneva is through the spider. The output from the differential, which adds the two inputs, is through shaft *4*.

The actual mechanism is shown in Fig. 8E. The cutter is fixed in space. Output is from the gear segment that rides on a fixed rack. The cam is driven by the motor, which also drives the enclosed genevas. Thus, the entire device reciprocates back and forth on the slide to feed the cam properly into the cutter.

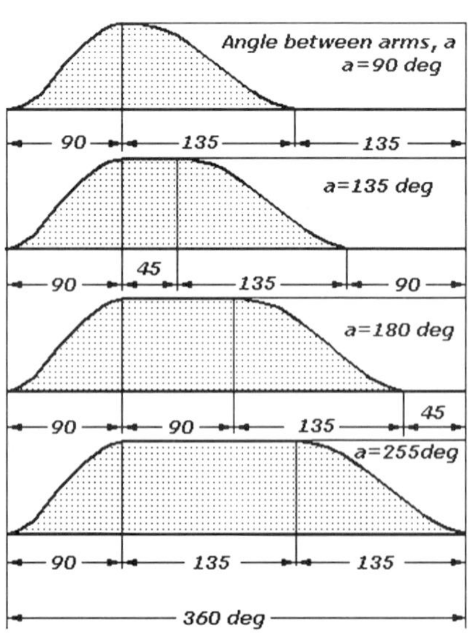

F Various dwell resultants

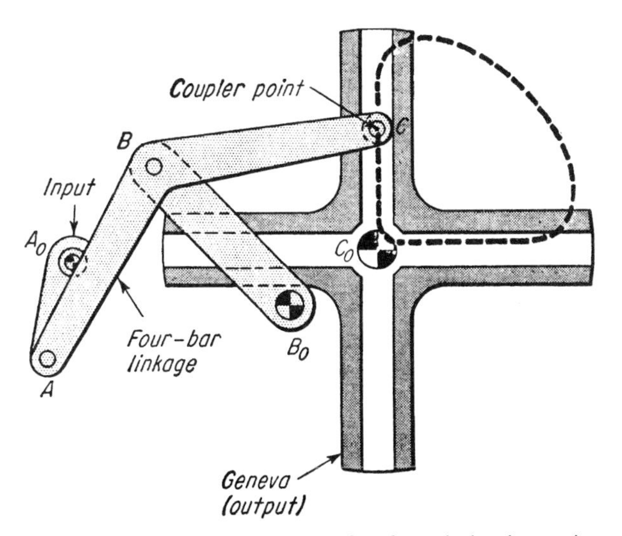

Fig. 9 A four-bar coupler mechanism for replacing the cranks in genevas to obtain smoother acceleration characteristics.

Genevas Driven by Couplers

When a geneva is driven by a constant-speed crank, as shown in Fig. 8D, it has a sudden change in acceleration at the beginning and end of the indexing cycle (as the crank enters or leaves a slot). These abrupt changes can be avoided by employing a four-bar linkage with a coupler in place of the crank. The motion of the coupler point *C* (Fig. 9) permits its smooth entry into the geneva slot.

Double Eccentric Drive

This is another machine for automatically cutting cams with dwells. The rotation of crank *A* (Fig. 10) imparts an oscillating motion to the rocker *C* with a prolonged dwell at both extreme positions. The cam, mounted on the rocker, is rotated by the chain drive and then is fed into the cutter with the proper motion. During the dwells of the rocker, for example, a dwell is cut into the cam.

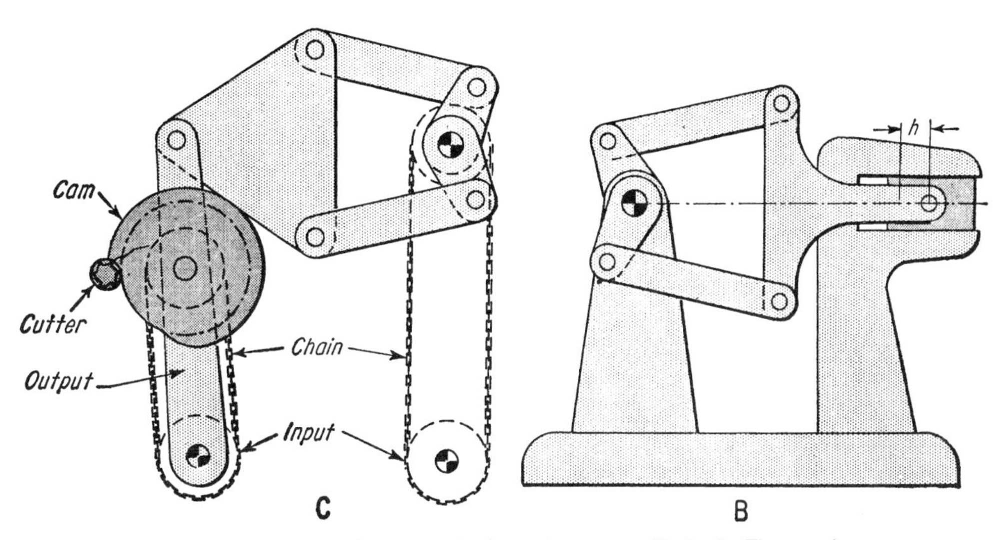

Fig. 10 A double eccentric drive for automatically cutting cams with dwells. The cam is rotated and oscillated, with dwell periods at extreme ends of oscillation corresponding to desired dwell periods in the cam.

FIFTEEN DIFFERENT CAM MECHANISMS

This assortment of devices reflects the variety of ways in which cams can be put to work.

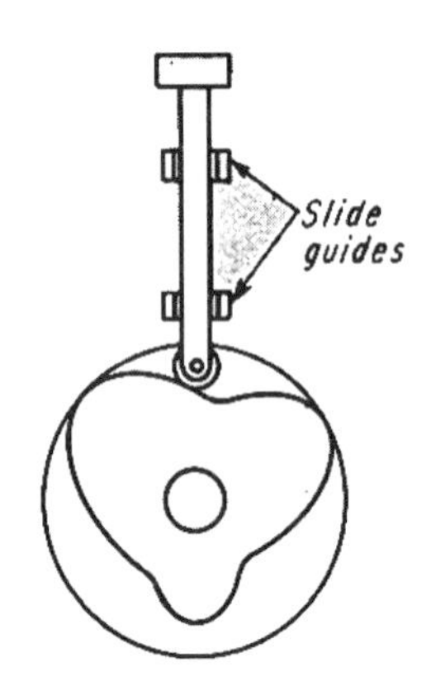

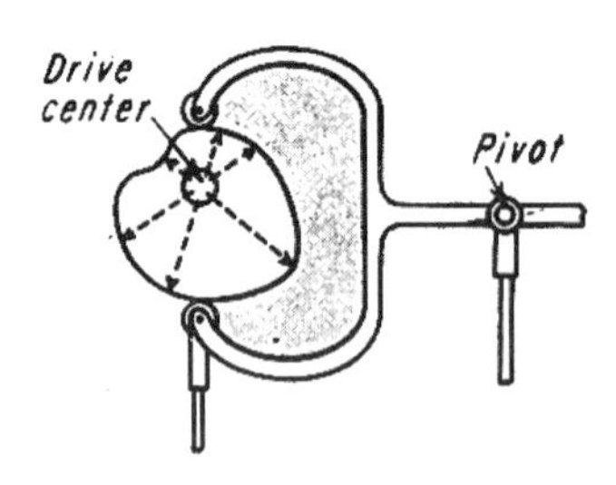

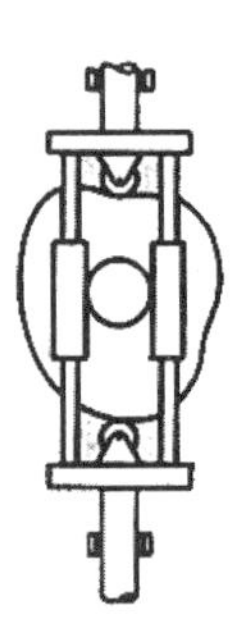

Figs. 1, 2, and 3 A constant-speed rotary motion is converted into a variable, reciprocating motion (Fig. 1); rocking or vibratory motion of a simple forked follower (Fig. 2); or a more robust follower (Fig. 3), which can provide valve-moving mechanisms for steam engines. Vibratory-motion cams must be designed so that their opposite edges are everywhere equidistant when they are measured through their drive-shaft centers.

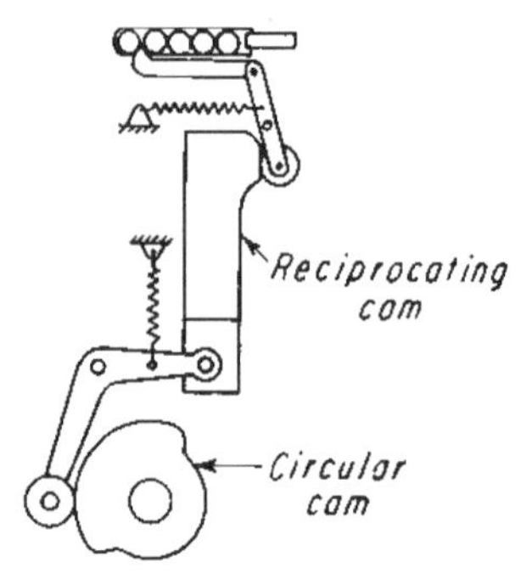

Fig. 4 An automatic feed for automatic machines. There are two cams, one with circular motion, the other with reciprocating motion. This combination eliminates any trouble caused by the irregularity of feeding and lack of positive control over stock feed.

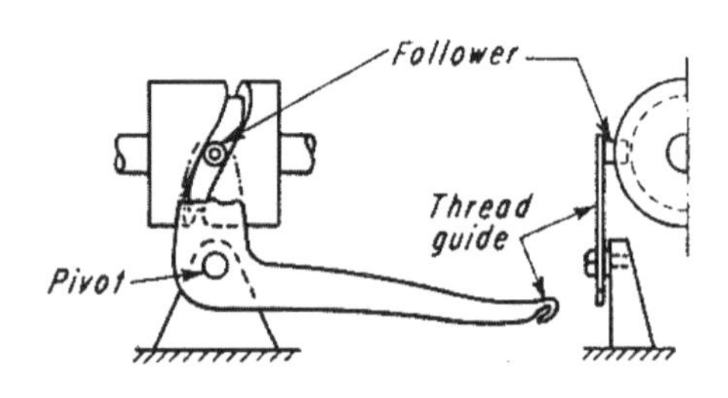

Fig. 5 A barrel cam with milled grooves is used in sewing machines to guide thread. This kind of cam is also used extensively in textile manufacturing machines such as looms and other intricate fabric-making machines.

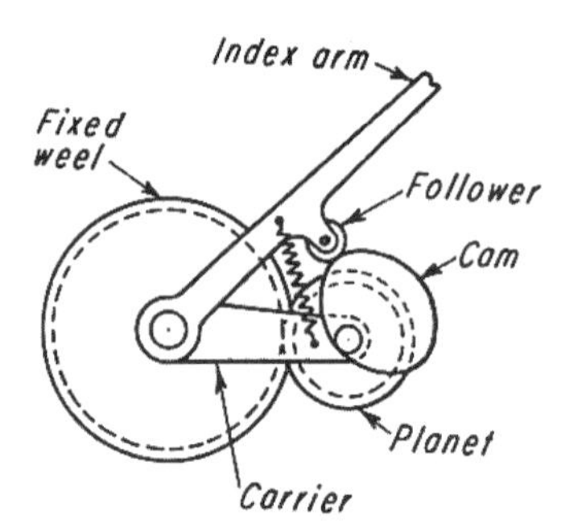

Fig. 6 This indexing mechanism combines an epicyclic gear and cam. A planetary wheel and cam are fixed relative to one another; the carrier is rotated at uniform speed around the fixed wheel. The index arm has a nonuniform motion with dwell periods.

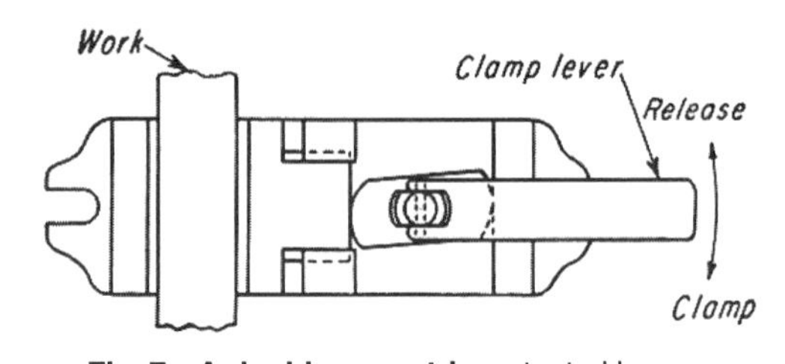

Fig. 7 A double eccentric, actuated by a suitable handle, provides powerful clamping action for a machine-tool holding fixture.

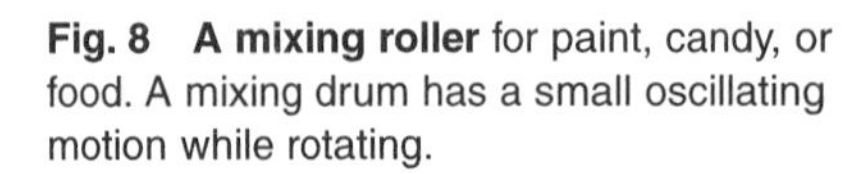

Fig. 8 A mixing roller for paint, candy, or food. A mixing drum has a small oscillating motion while rotating.

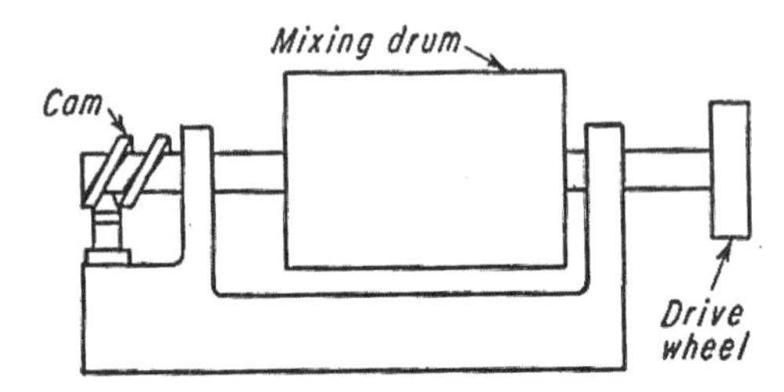

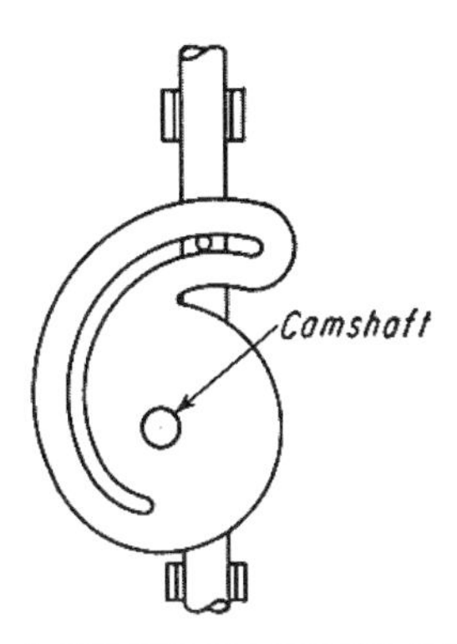

Fig. 9 A slot cam converts the oscillating motion of a camshaft to a variable but straight-line motion of a rod. According to slot shape, rod motion can be made to suit specific design requirements, such as straight-line and logarithmic motion.

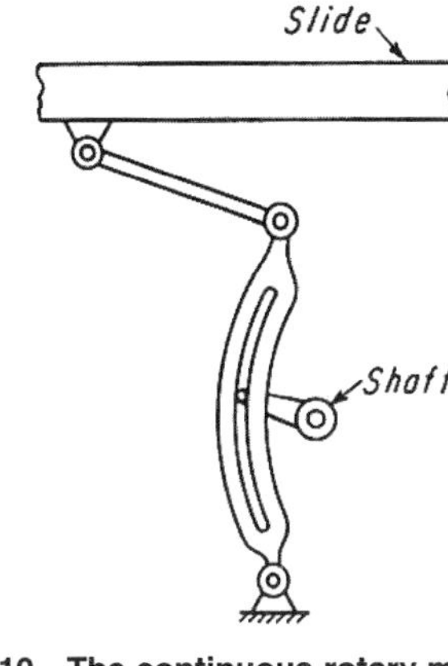

Fig. 10 The continuous rotary motion of a shaft is converted into the reciprocating motion of a slide. This device is used on sewing machines and printing presses.

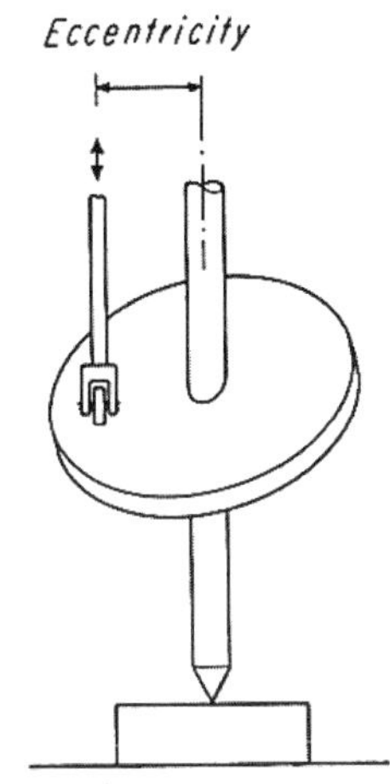

Fig. 11 Swash-plate cams are feasible for light loads only, such as in a pump. The cam's eccentricity produces forces that cause excessive loads. Multiple followers can ride on a plate, thereby providing smooth pumping action for a multipiston pump.

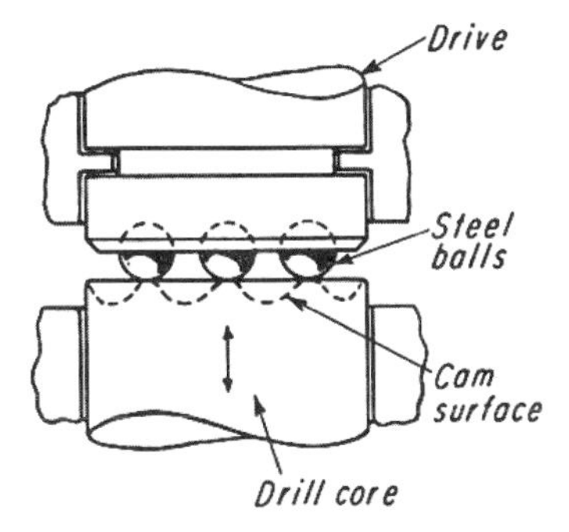

Fig. 12 This steel-ball cam can convert the high-speed rotary motion of an electric drill into high-frequency vibrations that power the drill core for use as a rotary hammer for cutting masonry, and concrete. This attachment can also be designed to fit hand drills.

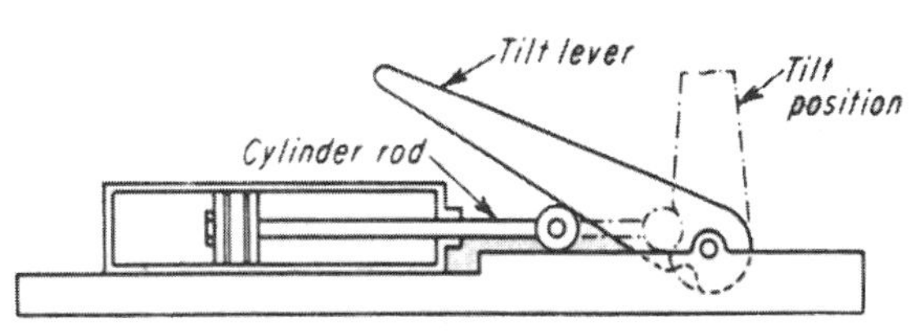

Fig. 13 This tilting device can be designed so that a lever remains in a tilted position when the cylinder rod is withdrawn, or it can be spring-loaded to return with a cylinder rod.

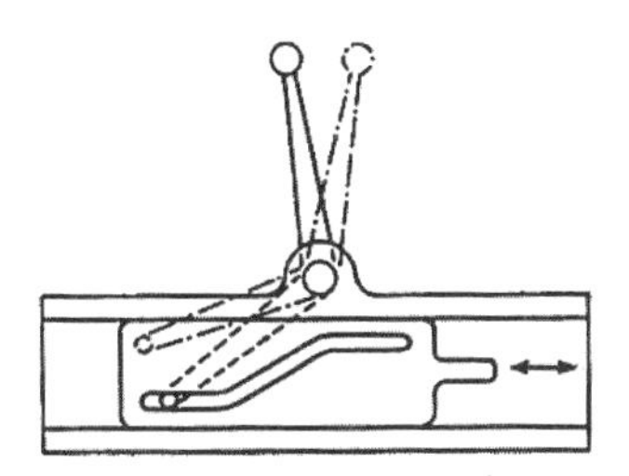

Fig. 14 This sliding cam in a remote control can shift gears in a position that is otherwise inaccessible on most machines.

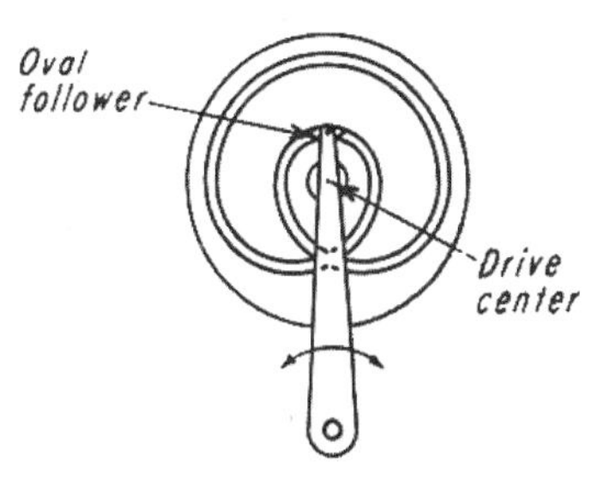

Fig. 15 A groove and oval follower form a device that requires two revolutions of a cam for one complete follower cycle.

TEN SPECIAL-FUNCTION CAMS

Fig. 1—A quick drop of the follower is obtained by permitting the cam to be pushed out of the way by the follower itself as it reaches the edge of the cam. Lugs *C* and *C′* are fixed to the camshaft. The cam is free to turn (float) on the camshaft, limited by lug *C* and the adjusting screw. With the cam rotating clockwise, lug C drives the cam through lug *B*. At the position shown, the roller will drop off the edge of the cam, which is then accelerated clockwise until its cam lug *B* strikes the adjusting screw of lug *C′*.

Fig. 2—Instantaneous drop is obtained by the use of two integral cams and followers. The roller follower rides on cam *1*. Continued rotation will transfer contact to the flat-faced follower, which drops suddenly off the edge of cam *2*. After the desired dwell, the follower is restored to its initial position by cam *1*.

Fig. 3—The dwell period of the cam can be varied by changing the distance between the two rollers in the slot.

Fig. 4—A reciprocating pin (not shown) causes the barrel cam to rotate intermittently. The cam is stationary while a pin moves from *1* to *2*. Groove *2-3* is at a lower level; thus, as the pin retracts, it cams the barrel cam; then it climbs the incline from *2* to the new position of *1*.

Fig. 5—A double-groove cam makes two revolutions for one complete movement of the follower. The cam has movable switches, *A* and *B*, which direct the follower alternately in each groove. At the instant shown, *B* is ready to guide the roller follower from slot *1* to slot *2*.

Figs. 6 and 7—Increased stroke is obtained by permitting the cam to shift on the input shaft. Total displacement of the follower is therefore the sum of the cam displacement on the fixed roller plus the follower displacement relative to the cam.

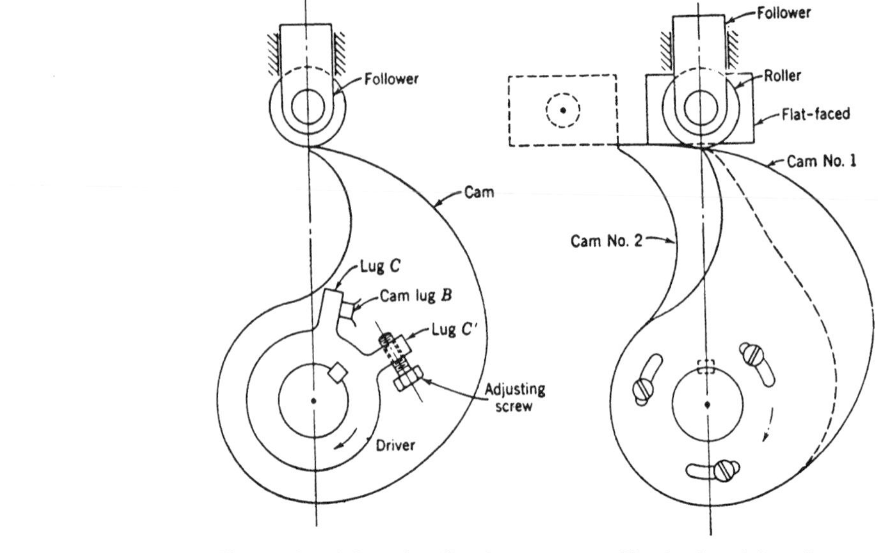

Fig. 1 A quick-acting floating cam.

Fig. 2 A quick-acting dwell cams.

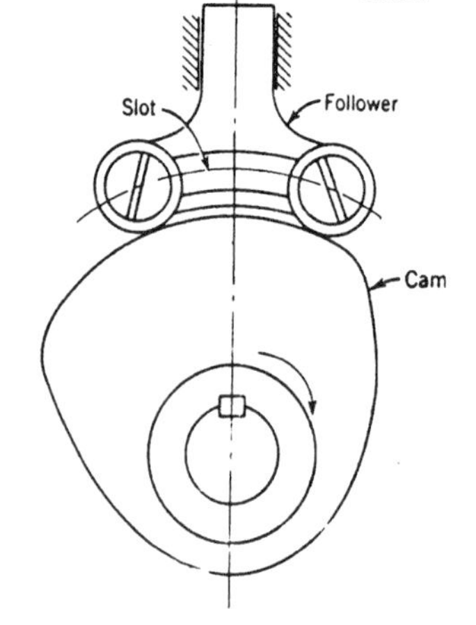

Fig. 3 An adjustable-dwell cam.

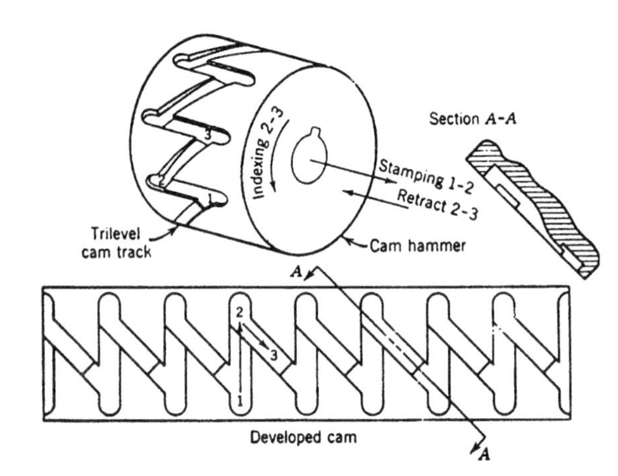

Fig. 4 An indexing cam.

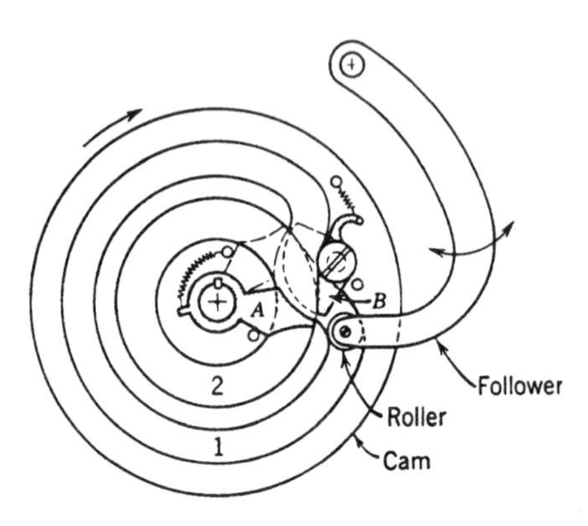

Fig. 5 A double-revolution cam.

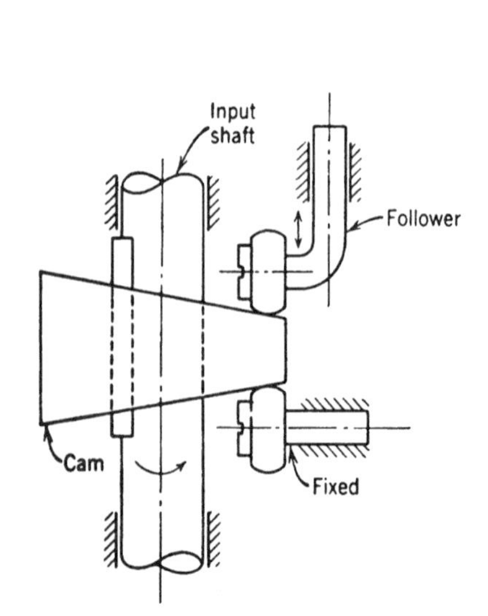

Fig. 6 An increased-stroke barrel cam.

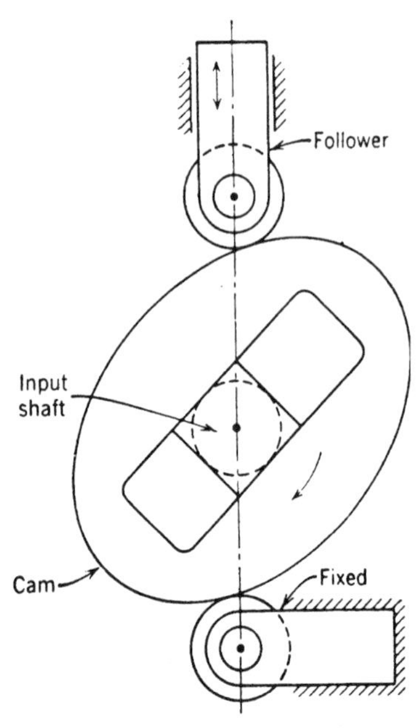

Fig. 7 An increased-stroke plate cam.

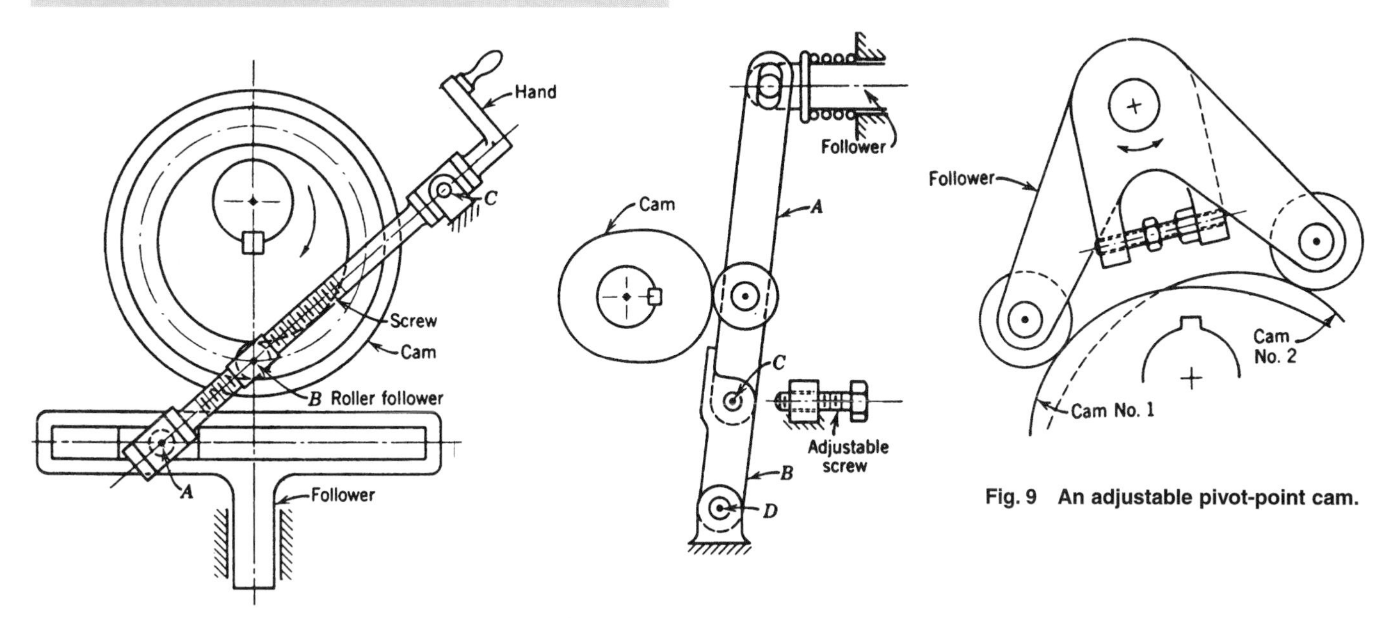

Fig. 8 An adjustable roller-position cam.

Fig. 9 An adjustable pivot-point cam.

Fig. 8—The stroke of the follower is adjusted by turning the screw handle which changes distance *AB*.

Fig. 9—The pivot point of the connecting link to the follower is changed from point *D* to point *C* by adjusting the screw.

Fig. 10—Adjustable dwell is obtained by having the main cam, with lug *A*, pinned to the revolving shaft. Lug *A* forces the plunger up into the position shown, and allows the latch to hook over the catch, thus holding the plunger in the up position. The plunger is unlatched by lug *B*. The circular slots in the cam plate permit the shifting of lug *B*, thereby varying the time that the plunger is held in the latched position.

REFERENCE: Rothbart, H. A. *Cams—Design, Dynamics, and Accuracy*, John Wiley and Sons, Inc., New York.

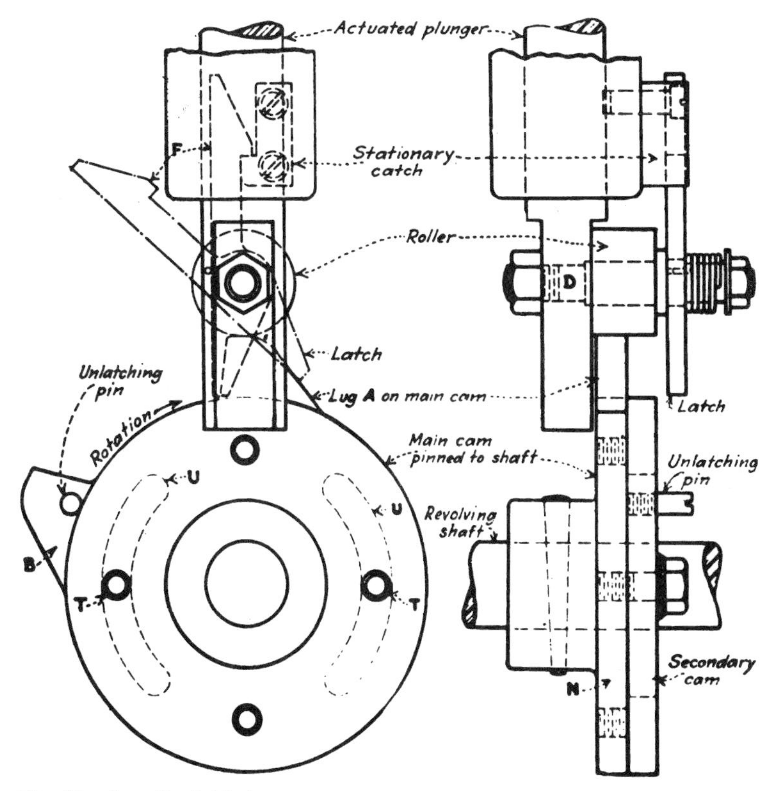

Fig. 10 An adjustable lug cam.

TWENTY GENEVA DRIVES

Locking-arm geneva drive

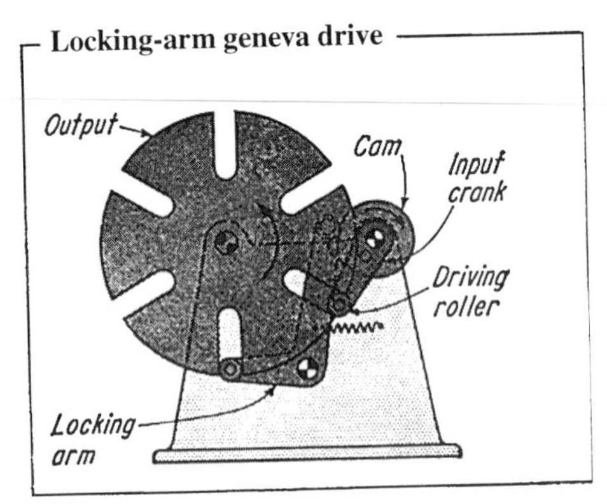

The driving follower on the rotating input crank of this geneva enters a slot and rapidly indexes the output. In this version, the roller of the locking-arm (shown leaving the slot) enters the slot to prevent the geneva from shifting when it is not indexing.

Twin geneva drive

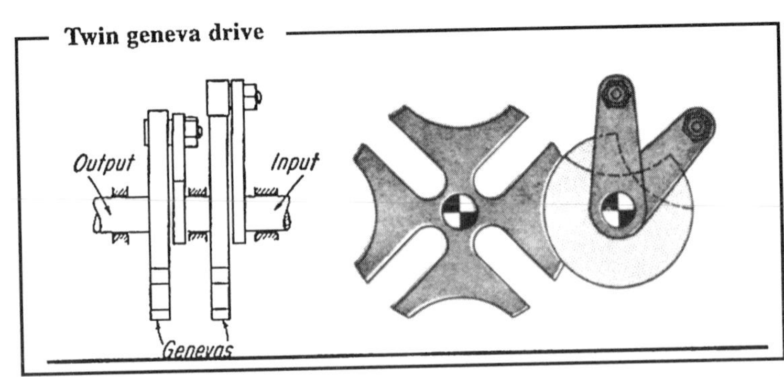

The driven member of the first geneva acts as the driver for the second geneva. This produces a wide variety of output motions including very long dwells between rapid indexes.

Groove cam geneva drive

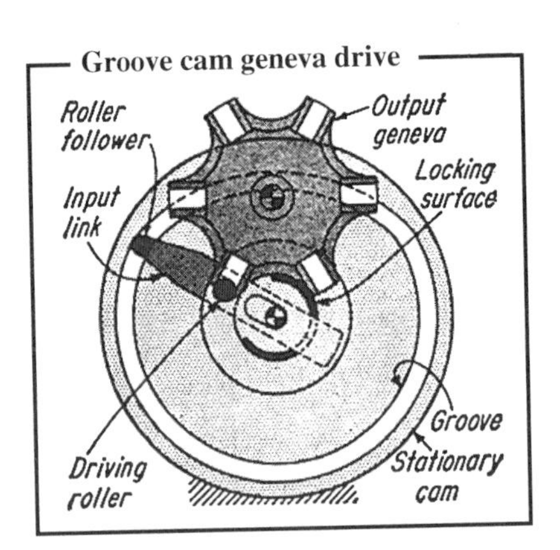

When a geneva is driven by a roller rotating at a constant speed, it tends to have very high acceleration and deceleration characteristics. In this modification, the input link, which contains the driving roller, can move radially while being rotated by the groove cam. Thus, as the driving roller enters the geneva slot, it moves radially inward. This action reduces the geneva acceleration force.

Planetary gear geneva drive

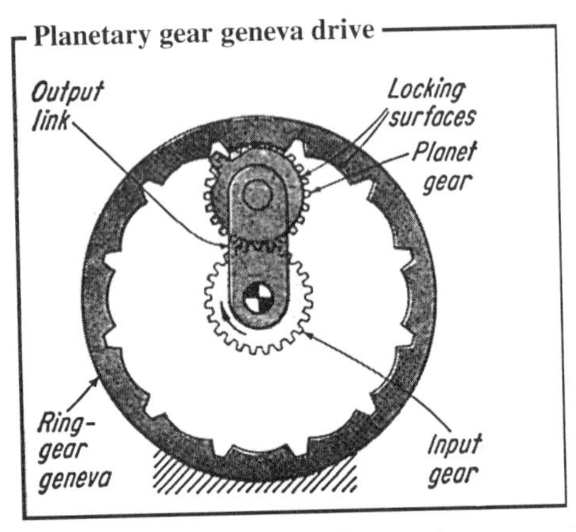

The output link remains stationary while the input gear drives the planet gear with single tooth on the locking disk. The disk is part of the planet gear, and it meshes with the ring-gear geneva to index the output link one position.

Locking-slide geneva drive

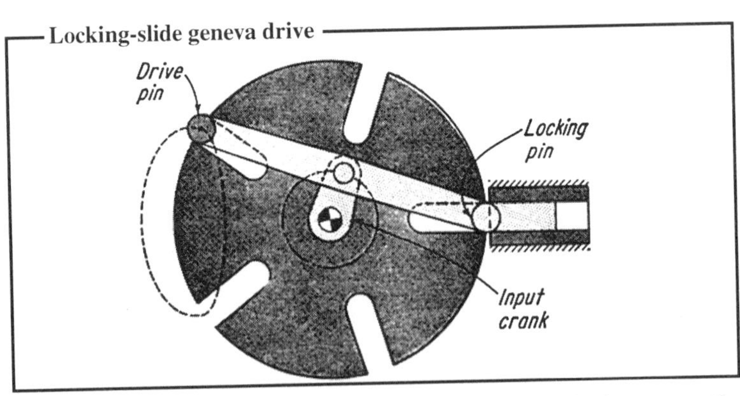

One pin locks and unlocks the geneva; the second pin rotates the geneva during the unlocked phase. In the position shown, the drive pin is about to enter the slot to index the geneva. Simultaneously, the locking pin is just clearing the slot.

Four-bar geneva drive

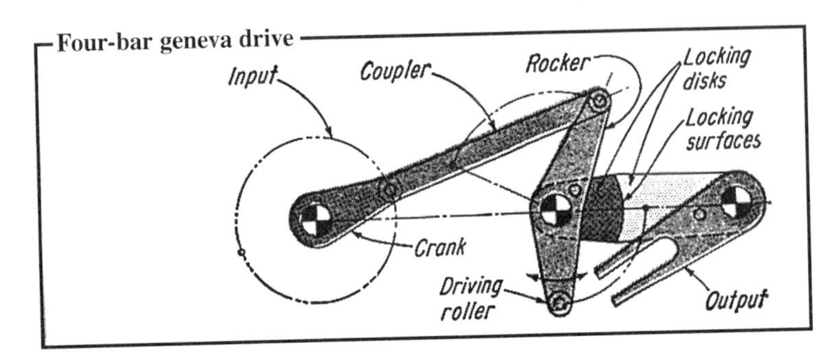

A four-bar geneva produces a long-dwell motion from an oscillating output. The rotation of the input wheel causes a driving roller to reciprocate in and out of the slot of the output link. The two disk surfaces keep the output in the position shown during the dwell period.

Rapid-transfer geneva drive

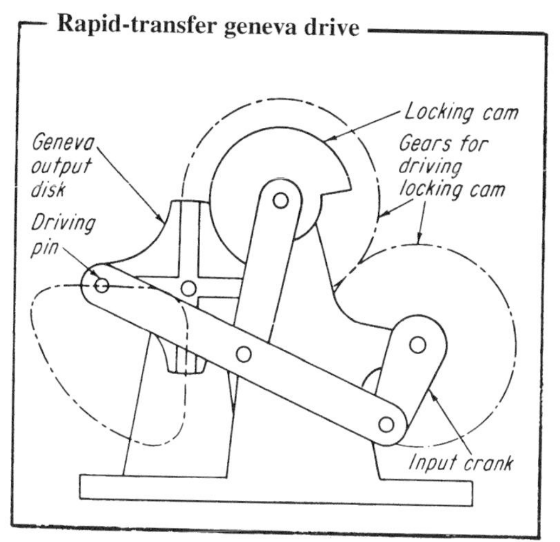

The coupler point at the extension of the connecting link of the four-bar mechanism describes a curve with two approximately straight lines, 90° apart. This provides a favorable entry situation because there is no motion in the geneva while the driving pin moves deeply into the slot. Then there is an extremely rapid index. A locking cam, which prevents the geneva from shifting when it is not indexing, is connected to the input shaft through gears.

Long-dwell geneva drive

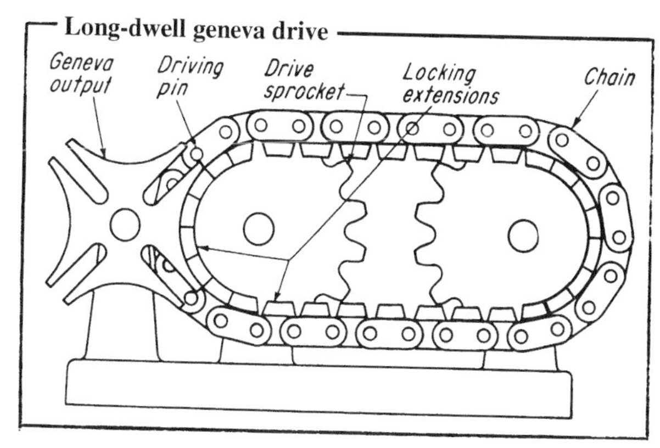

This geneva arrangement has a chain with an extended pin in combination with a standard geneva. This permits a long dwell between each 90° shift in the position of the geneva. The spacing between the sprockets determines the length of dwell. Some of the links have special extensions to lock the geneva in place between stations.

Modified motion geneva drive

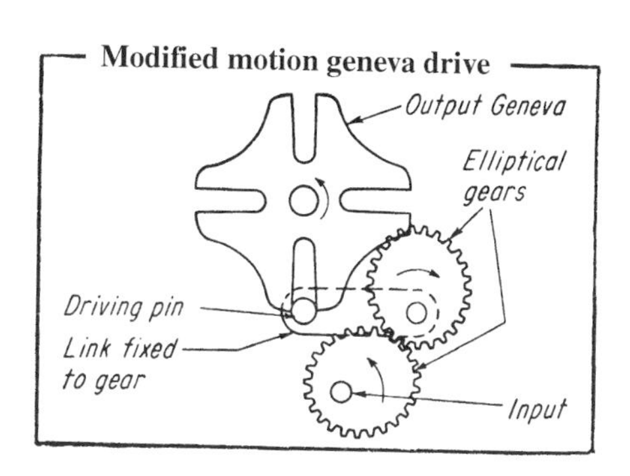

The input link of a normal geneva drive rotates at constant velocity, which restricts flexibility in design. That is, for given dimensions and number of stations, the dwell period is determined by the speed of the input shaft. Elliptical gears produce a varying crank rotation that permits either extending or reducing the dwell period.

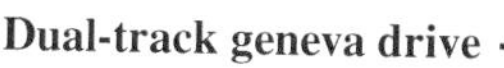

Dual-track geneva drive

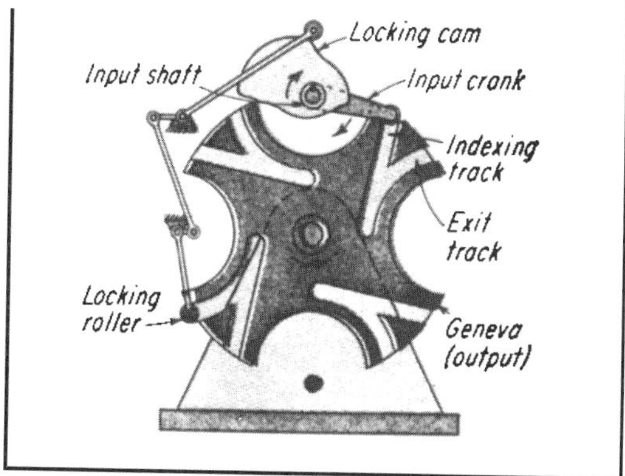

The key consideration in the design of genevas is to have the input roller enter and leave the geneva slots tangentially (as the crank rapidly indexes the output). This is accomplished in the novel mechanism shown with two tracks. The roller enters one track, indexes the geneva 90° (in a four-stage geneva), and then automatically follows the exit slot to leave the geneva.

The associated linkage mechanism locks the geneva when it is not indexing. In the position shown, the locking roller is just about to exit from the geneva.

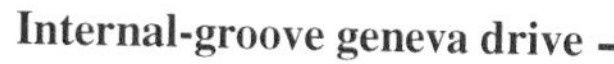

Internal-groove geneva drive

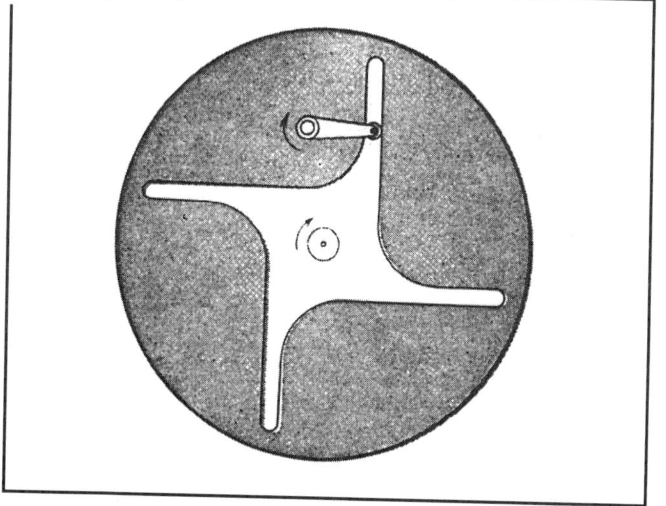

This arrangement permits the roller to exit and enter the driving slots tangentially. In the position shown, the driving roller has just completed indexing the geneva, and it is about to coast for 90° as it goes around the curve. (During this time, a separate locking device might be necessary to prevent an external torque from reversing the geneva.)

Controlled-output escapement

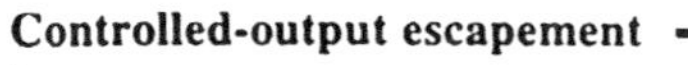

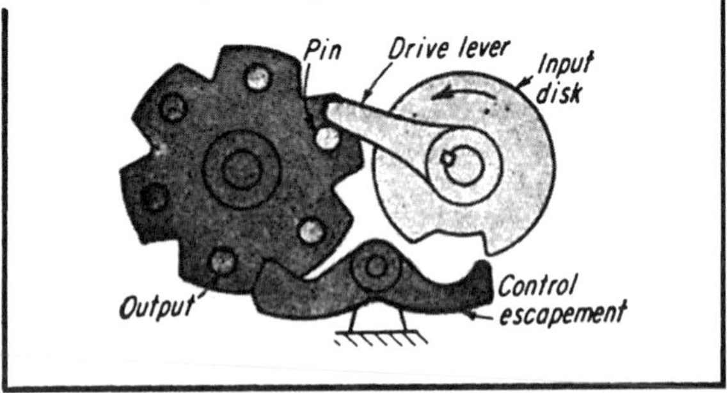

The output in this simple mechanism is prevented from turning in either direction—unless it is actuated by the input motion. In operation, the drive lever indexes the output disk by bearing on the pin. The escapement is cammed out of the way during indexing because the slot in the input disk is positioned to permit the escapement tip to enter it. But as the lever leaves the pin, the input disk forces the escapement tip out of its slot and into the notch. That locks the output in both directions.

Progressive oscillating drive

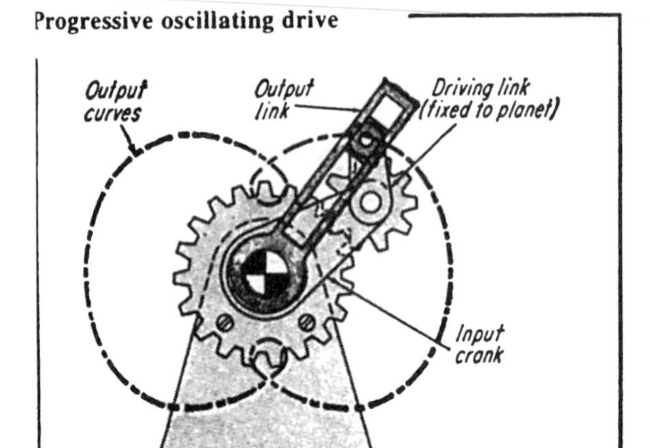

A crank attached to the planet gear can make point *P* describe the double loop curve illustrated. The slotted output crank oscillates briefly at the vertical positions.

Sinusoidal reciprocator drive

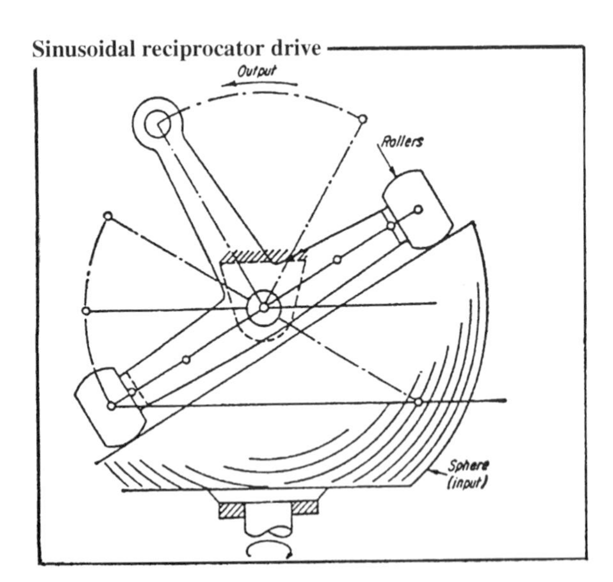

This reciprocator transforms rotary motion into a reciprocating motion in which the oscillating output member is in the same plane as the input shaft. The output member has two arms with rollers which contact the surface of the truncated sphere. The rotation of the sphere causes the output to oscillate.

Parallel-guidance drive

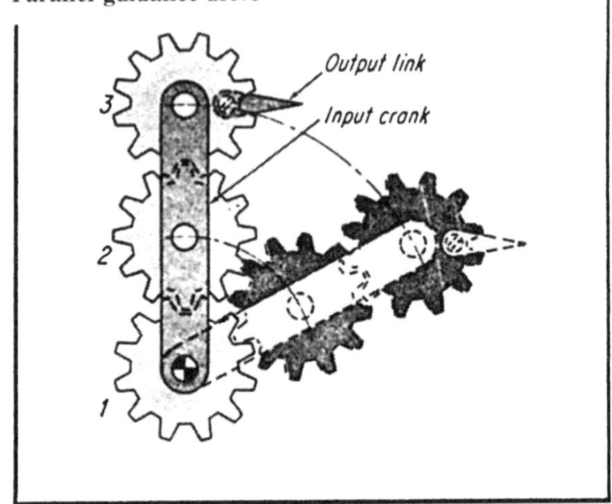

The input crank contains two planet gears. The center sun gear is fixed. By making the three gears equal in diameter and having gear *2* serve as an idler, any member fixed to gear *3* will remain parallel to its previous posi-tions throughout the rotation of the input ring crank.

Rotating-cam reciprocator drive

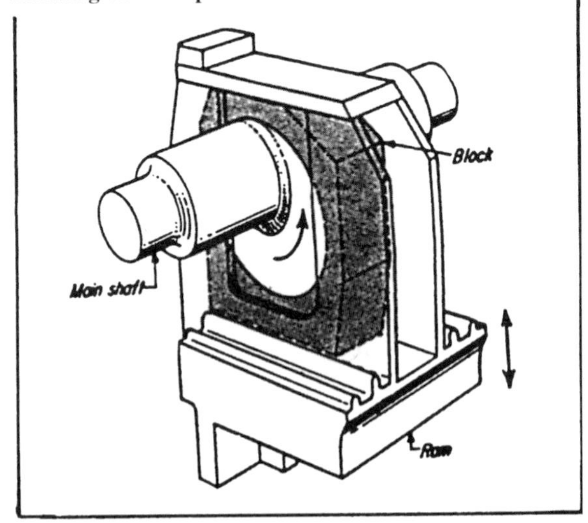

The high-volume 2500-ton press is designed to shape such parts as connecting rods, tractor track links, and wheel hubs. A simple automatic-feed mechanism makes it possible to produce 2400 forgings per hour.

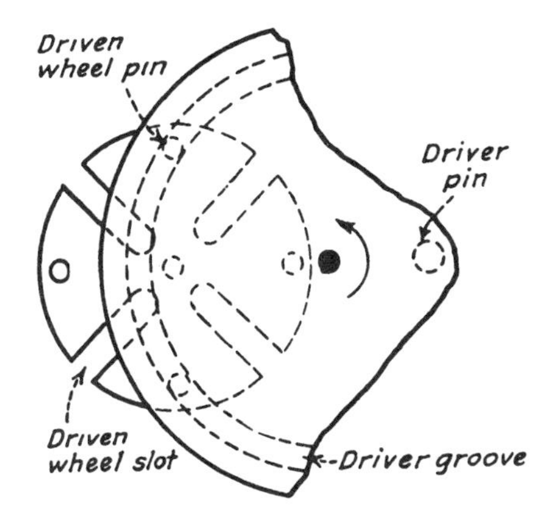

An external geneva drive. The driver grooves lock the driven wheel pins during dwell. During movement, the driver pin mates with the driven-wheel slot.

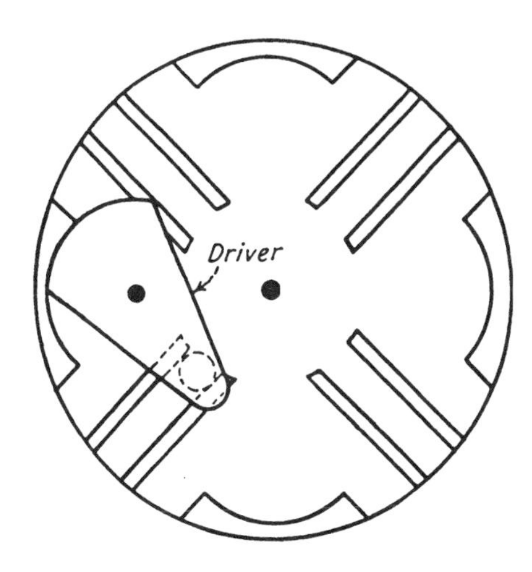

An internal geneva drive. The driver and driven wheel rotate in same direction. The duration of dwell is more than 180º of driver rotation.

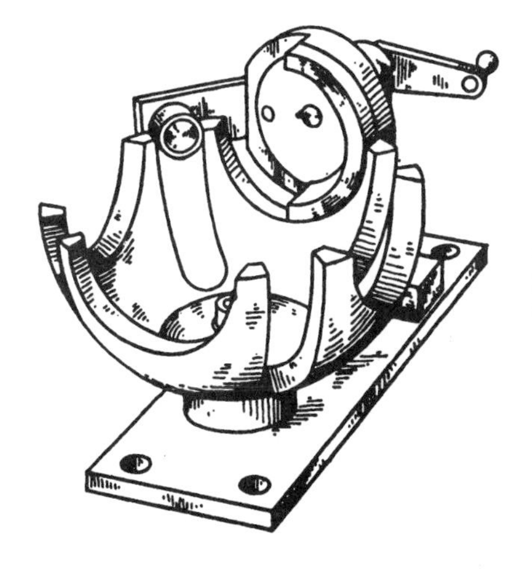

A spherical geneva drive. The driver and driven wheel are on perpendicular shafts. The duration of dwell is exactly 180° of driver rotation.

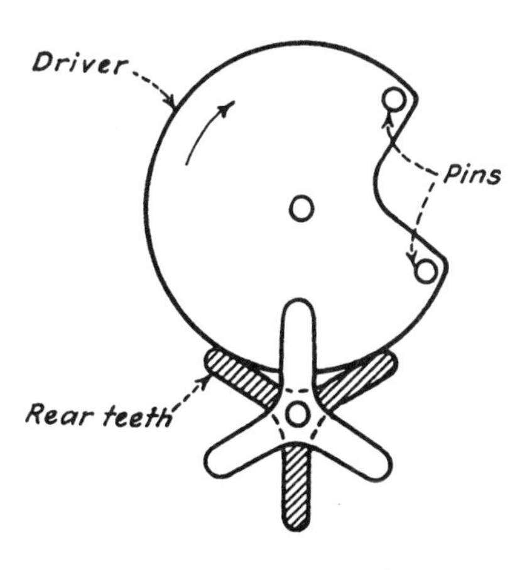

An intermittent counter drive. One revolution of the driver advances the driven wheel 120°. The driven-wheel rear teeth are locked on the cam surface during dwell.

SIX MODIFIED GENEVA DRIVES

The mechanisms shown here add a varying velocity component to conventional geneva motion.

Fig. 1 With a conventional external geneva drive, a constant-velocity input produces an output consisting of a varying velocity period plus a dwell. The motion period of the modified geneva shown has a constant-velocity interval which can be varied within limits. When spring-loaded driving roller *a* enters the fixed cam *b*, the output-shaft velocity is zero. As the roller travels along the cam path, the output velocity rises to some constant value, which is less than the maximum output of an unmodified geneva with the same number of slots. The duration of constant-velocity output is arbitrary within limits. When the roller leaves the cam, the output velocity is zero. Then the output shaft dwells until the roller re-enters the cam. The spring produces a variable radial distance of the driving roller from the input shaft, which accounts for the described motions. The locus of the roller's path during the constant-velocity output is based on the velocity-ratio desired.

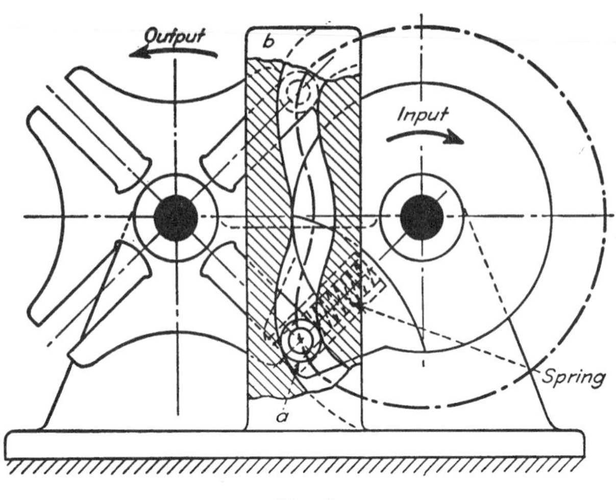

Fig. 1

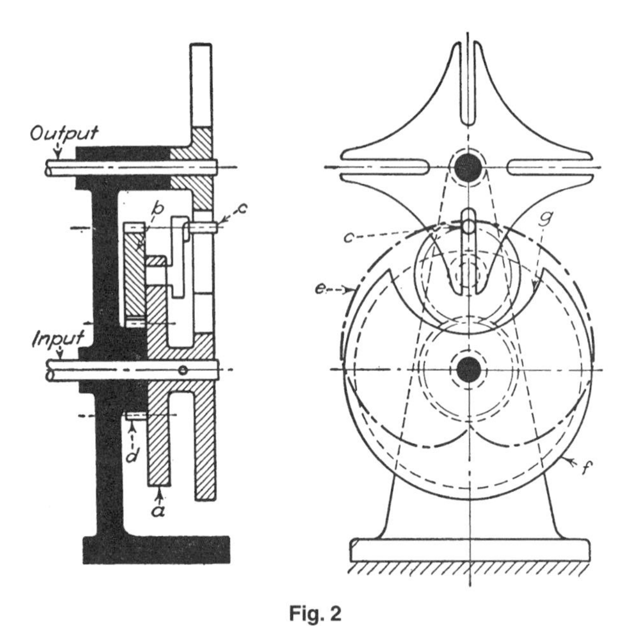

Fig. 2

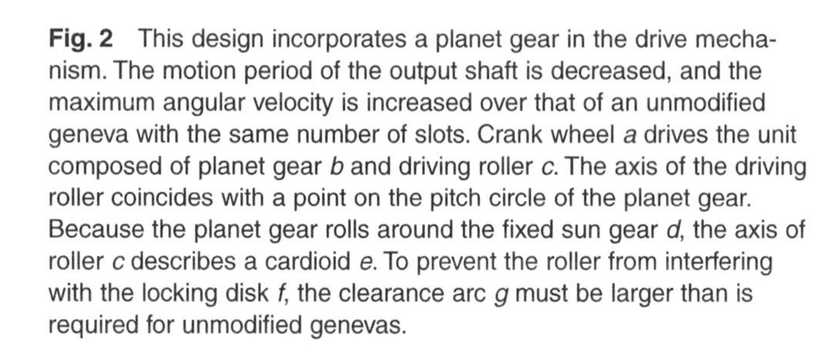
Fig. 2 This design incorporates a planet gear in the drive mechanism. The motion period of the output shaft is decreased, and the maximum angular velocity is increased over that of an unmodified geneva with the same number of slots. Crank wheel *a* drives the unit composed of planet gear *b* and driving roller *c*. The axis of the driving roller coincides with a point on the pitch circle of the planet gear. Because the planet gear rolls around the fixed sun gear *d*, the axis of roller *c* describes a cardioid *e*. To prevent the roller from interfering with the locking disk *f*, the clearance arc *g* must be larger than is required for unmodified genevas.

Fig. 3 A motion curve similar to that of Fig. 2 can be derived by driving a geneva wheel with a two-crank linkage. Input crank *a* drives crank *b* through link *c*. The variable angular velocity of driving roller *d*, mounted on *b*, depends on the center distance *L*, and on the radii *M* and *N* of the crank arms. This velocity is about equivalent to what would be produced if the input shaft were driven by elliptical gears.

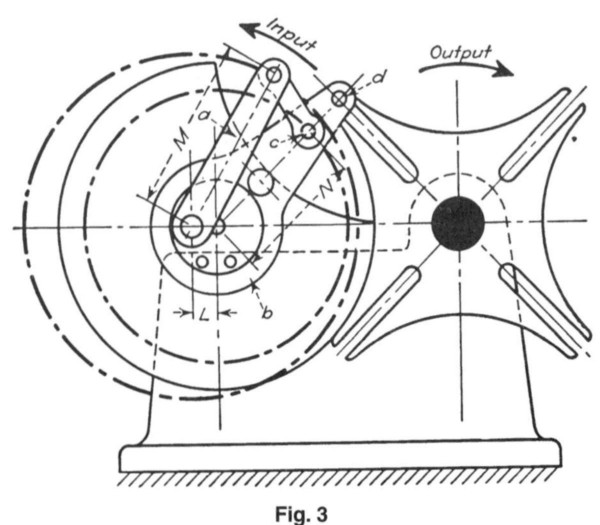

Fig. 3

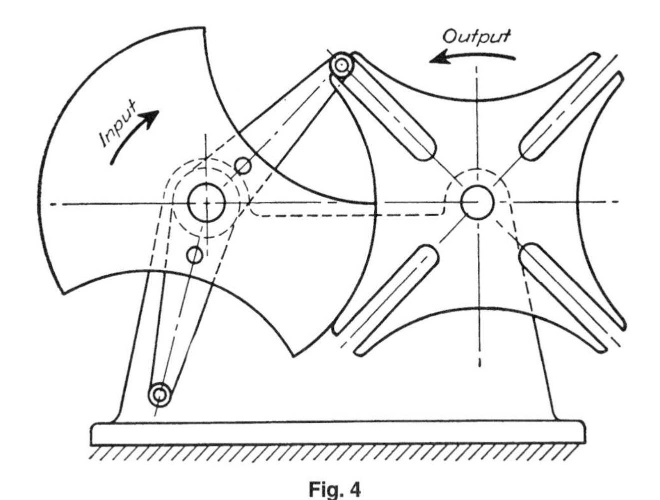

Fig. 4

Fig. 4 The duration of the dwell periods is changed by arranging the driving rollers unsymmetrically around the input shaft. This does not affect the duration of the motion periods. If unequal motion periods and unequal dwell periods are desired, the roller crank-arms must be unequal in length and the star must be suitably modified. This mechanism is called an irregular geneva drive.

Fig. 5 In this intermittent drive, the two rollers drive the output shaft and lock it during dwell periods. For each revolution of the input shaft, the output shaft has two motion periods. The output displacement ϕ is determined by the number of teeth. The driving angle, ψ, can be chosen within limits. Gear *a* is driven intermittently by two driving rollers mounted on input wheel *b*, which is bearing-mounted on frame *c*. During the dwell period the rollers circle around the top of a tooth. During the motion period, a roller's path *d*, relative to the driven gear, is a straight line inclined towards the output shaft. The tooth profile is a curve parallel to path *d*. The top land of a tooth becomes the arc of a circle of radius *R*, and the arc approximates part of the path of a roller.

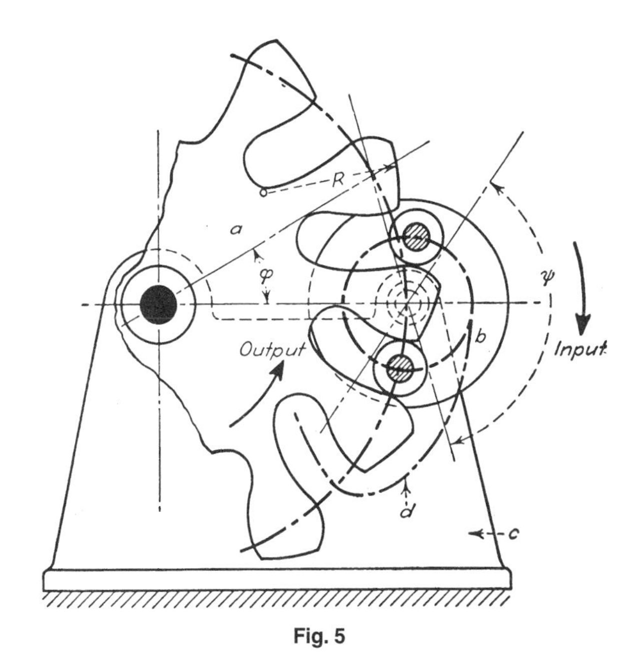

Fig. 5

Driven

Driver

Fig. 6

Fig. 6 An intermittent drive with a cylindrical lock. Shortly before and after the engagement of two teeth with driving pin *d* at the end of the dwell period, the inner cylinder *f* is unable to cause positive locking of the driven gear. Consequently, a concentric auxiliary cylinder *e* is added. Only two segments are necessary to obtain positive locking. Their length is determined by the circular pitch of the driven gear.

KINEMATICS OF EXTERNAL GENEVA WHEELS

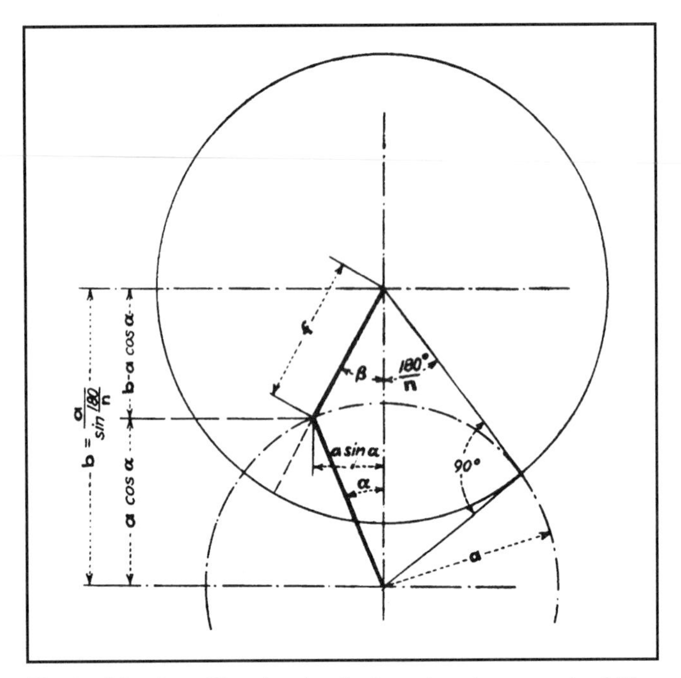

Fig. 1 **A basic outline drawing** for the external geneva wheel. The symbols are identified for application in the basic equations.

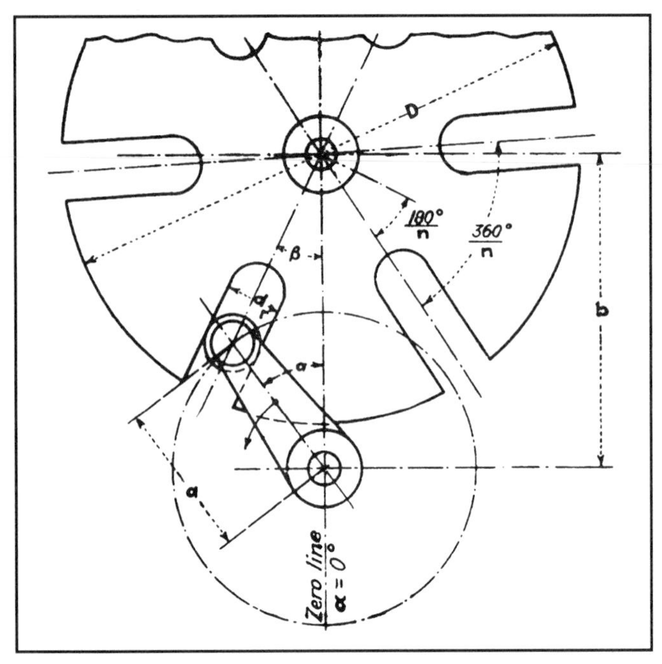

Fig. 2 **A schematic drawing** of a six-slot geneva wheel. Roller diameter, d_r, must be considered when determining D.

Table I—Notation and Formulas for the External Geneva Wheel

Assumed or given: a, n, d and p

a = crank radius of driving member

n = number of slots

d_r = roller diameter

p = constant velocity of driving crank in rpm

$$m = \frac{1}{\sin \frac{180}{n}}$$

b = center distance = am

$$D = \text{diameter of driven member} = 2\sqrt{\frac{d_r^2}{4} + a^2 \cot^2 \frac{180}{n}}$$

$$\omega = \text{constant angular velocity of driving crank} = \frac{p\pi}{30} \text{ radians per sec}$$

α = angular position of driving crank at any time

β = angular displacement of driven member corresponding to crank angle α

$$\cos\beta = \frac{m - \cos\alpha}{\sqrt{1 + m^2 - 2m\cos\alpha}}$$

$$\text{Angular Velocity of driven member} = \frac{d\beta}{dt} = \omega\left(\frac{m\cos\alpha - 1}{1 + m^2 - 2m\cos\alpha}\right)$$

$$\text{Angular Acceleration of driven member} = \frac{d^2\beta}{dt^2} = \omega^2\left(\frac{m\sin\alpha\,(1 - m^2)}{(1 + m^2 - 2m\cos\alpha)^2}\right)$$

Maximum Angular Acceleration occurs when $\cos\alpha =$

$$\sqrt{\left(\frac{1 + m^2}{4m}\right)^2 + 2} - \left(\frac{1 + m^2}{4m}\right)$$

Maximum Angular Velocity occurs at $\alpha = 0$ deg, and equals

$$\frac{\omega}{m - 1} \text{ radians per sec}$$

One of the most commonly applied mechanisms for producing intermittent rotary motion from a uniform input speed is the external geneva wheel.

The driven member, or star wheel, contains many slots into which the roller of the driving crank fits. The number of slots determines the ratio between dwell and motion period of the driven shaft. The lowest possible number of slots is three, while the highest number is theoretically unlimited. In practice, the three-slot geneva is seldom used because of the extremely high acceleration values encountered. Genevas with more than 18 slots are also infrequently used because they require wheels with comparatively large diameters.

In external genevas of any number of slots, the dwell period always exceeds the motion period. The opposite is true of the internal geneva. However, for the spherical geneva, both dwell and motion periods are 180°.

For the proper operation of the external geneva, the roller must enter the slot tangentially. In other words, the centerline of the slot and the line connecting the roller center and crank rotation center must form a right angle when the roller enters or leaves the slot.

The calculations given here are based on the conditions stated here.

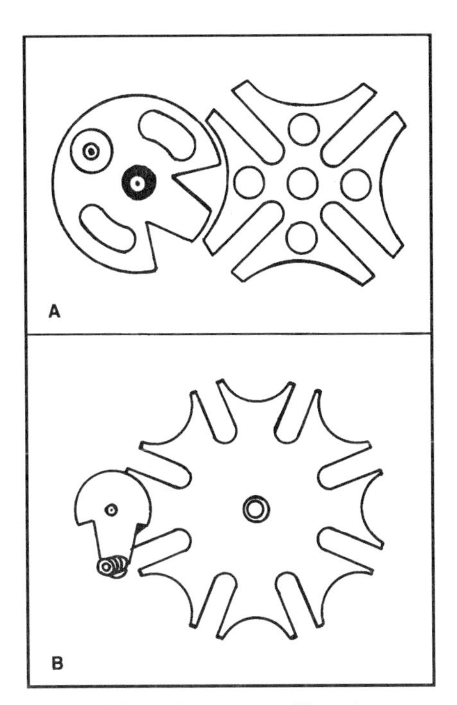

Fig. 3 A four-slot geneva (A) and an eight-slot geneva (B). Both have locking devices.

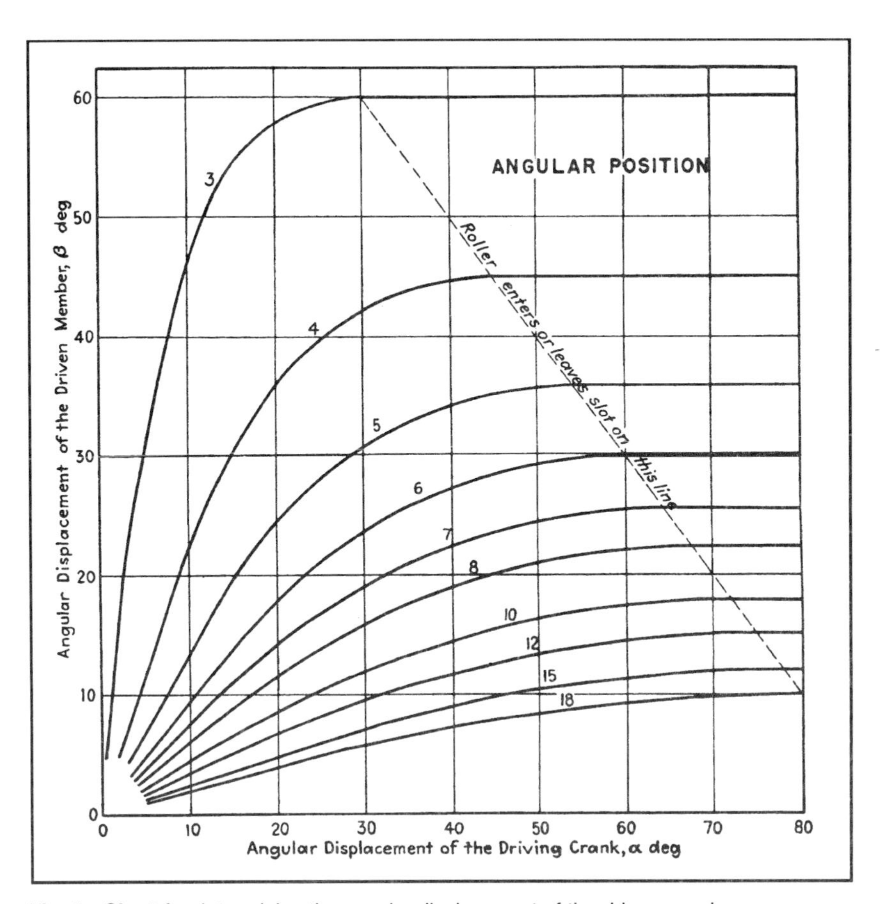

Fig. 4 Chart for determining the angular displacement of the driven member.

Consider an external geneva wheel, shown in Fig. 1, in which
n = number of slots
a = crank radius
From

Fig. 1, b = center distance $= \dfrac{a}{\sin \dfrac{180}{n}}$

Let $\dfrac{1}{\sin \dfrac{180}{n}} = m$

Then $b = am$

It will simplify the development of the equations of motion to designate the connecting line of the wheel and crank centers as the zero line. This is contrary to the practice of assigning the zero value of α, representing the angular position of the driving crank, to that position of the crank where the roller enters the slot.

Thus, from Fig. 1, the driven crank radius f at any angle is:

$$f = \sqrt{(am - a\cos\alpha)^2 + \alpha^2 \sin^2\alpha}$$
$$= \alpha\sqrt{1 + m^2 - 2m\cos\alpha} \quad (1)$$

and the angular displacement β can be found from:

$$\cos\beta = \frac{m - \cos a}{\sqrt{1 + m^2 - 2m\cos\alpha}} \quad (2)$$

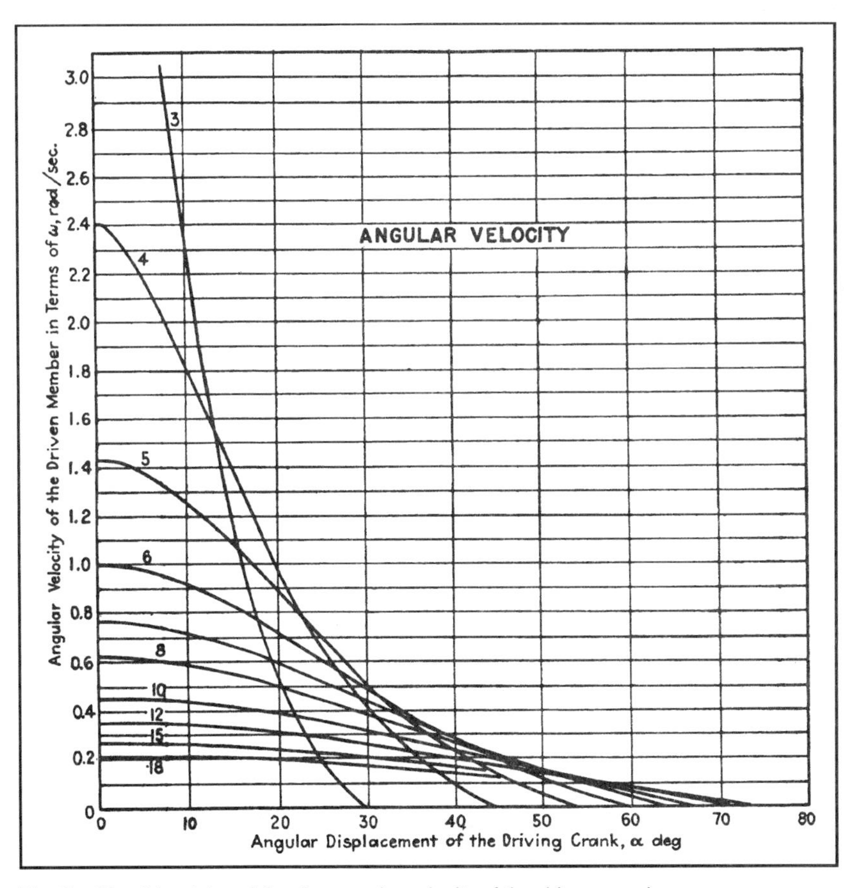

Fig. 5 Chart for determining the angular velocity of the driven member.

Table II—Principal Kinematic Data for External Geneva Wheel

No. of Slots	$\frac{360°}{n}$	Dwell period	Motion period	*m* and center-distance for $a = 1$	Maximum angular velocity of driven member, radians per sec. equals ω multiplied by values tabulated. Crank at 0° position	Angular acceleration of driven member when roller enters slot, radians² per sec², equals ω^2 multiplied by values tabulated.			Maximum angular Acceleration of driven member, radians² per sec², equals ω^2 multiplied by values tabulated		
						α	β	Multiplier	α	β	Multiplier
3	120°	300°	60°	1.155	6.458	30°	60°	1.729	4°	27° 58′	29.10
4	90°	270°	90°	1.414	2.407	45°	45°	1.000	11° 28′	25° 11′	5.314
5	72°	252°	108°	1.701	1.425	54°	36°	0.727	17° 31′	21° 53′	2.310
6	60°	240°	120°	2.000	1.000	60°	30°	0.577	22° 55′	19° 51′	1.349
7	51° 25′ 43″	231° 30′	128° 30′	2.305	0.766	64° 17′ 8″	25° 42′ 52″	0.481	27° 41′	18° 11′	0.928
8	45°	225°	135°	2.613	0.620	67° 30′	22° 30′	0.414	31° 38′	16° 32′	0.700
9	40°	220°	140°	2.924	0.520	70°	20°	0.364	35° 16′	15° 15′	0.559
10	36°	216°	144°	3.236	0.447	72°	18°	0.325	38° 30′	14° 16′	0.465
11	32° 43′ 38″	212° 45′	147° 15′	3.549	0.392	73° 38′ 11″	16° 21′ 49″	0.294	41° 22′	13° 16′	0.398
12	30°	210°	150°	3.864	0.349	75°	15°	0.268	44°	12° 26′	0.348
13	27° 41′ 32″	207° 45′	152° 15′	4.179	0.315	76° 9′ 14″	13° 50′ 46″	0.246	46° 23′	11° 44′	0.309
14	25° 42′ 52″	205° 45′	154° 15′	4.494	0.286	77° 8′ 34″	21° 51′ 26″	0.228	48° 32′	11° 3′	0.278
15	24°	204°	156°	4.810	0.263	78°	12°	0.213	50° 30′	10° 27′	0.253
16	22° 30′	202° 30′	157° 30′	5.126	0.242	78° 45′	11° 15′	0.199	52° 24′	9° 57′	0.232
17	21° 10′ 35″	201°	159°	5.442	0.225	79° 24′ 43″	10° 35′ 17″	0.187	53° 58′	9° 26′	0.215
18	20°	200°	160°	5.759	0.210	80°	10°	0.176	55° 30′	8° 59′	0.200

A six-slot geneva is shown schematically in Fig. 2. The outside diameter *D* of the wheel (when accounting for the effect of the roller diameter *d*) is found to be:

$$D = 2\sqrt{\frac{d_r^2}{4} + a^2 \cot^2 \frac{180}{n}} \qquad (3)$$

Differentiating Eq. (2) and dividing by the differential of time, *dt*, the angular velocity of the driven member is:

$$\frac{d\beta}{dt} = \omega\left(\frac{m \cos \alpha - 1}{1 + m^2 - 2m \cos \alpha}\right) \qquad (4)$$

where ω represents the constant angular velocity of the crank.

By differentiation of Eq. (4) the acceleration of the driven member is found to be:

$$\frac{d^2\beta}{dt^2} = \omega^2\left(\frac{m \sin \alpha(1 - m^2)}{(1 + m^2 - 2m \cos \alpha)^2}\right) \qquad (5)$$

All notations and principal formulas are given in Table I for easy reference. Table II contains all the data of principal interest for external geneva wheels having from 3 to 18 slots. All other data can be read from the charts: Fig. 4 for angular position, Fig. 5 for angular velocity, and Fig. 6 for angular acceleration.

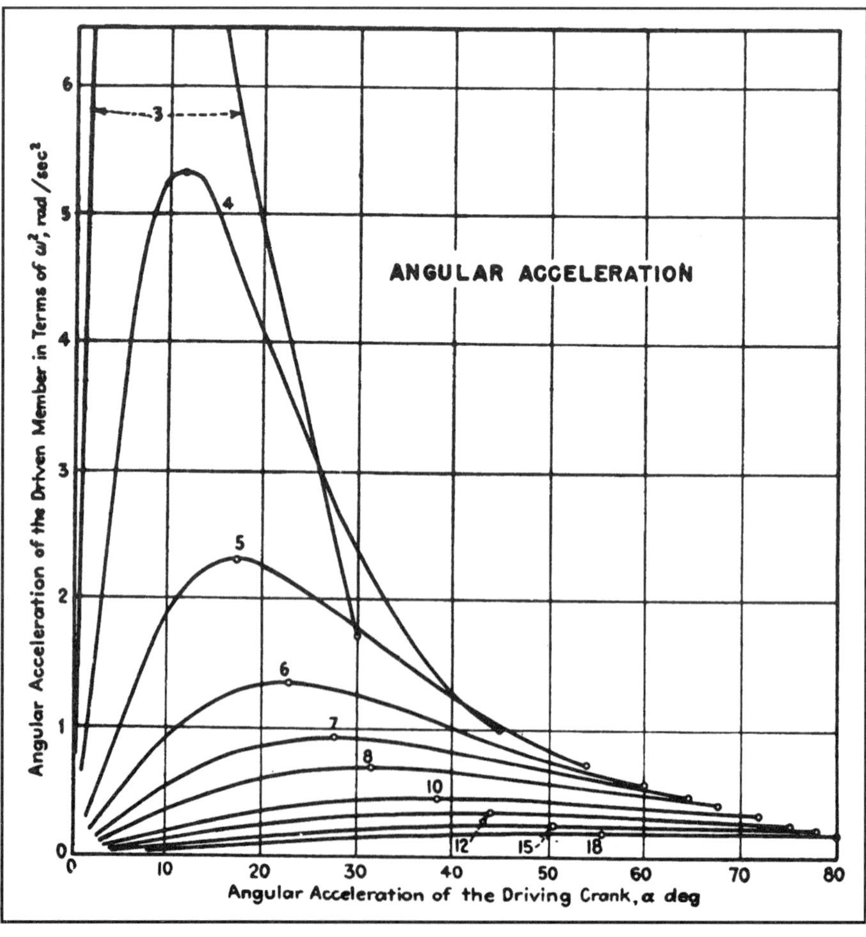

Fig. 6 Chart for determining the angular acceleration of the driven member.

KINEMATICS OF INTERNAL GENEVA WHEELS

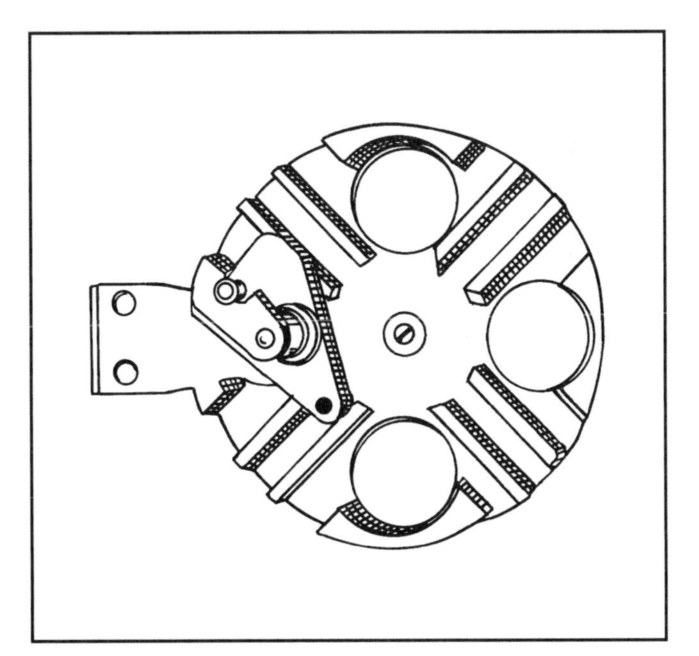

Fig. 1 A four-slot internal geneva wheel incorporating a locking mechanism. The basic sketch is shown in Fig. 3.

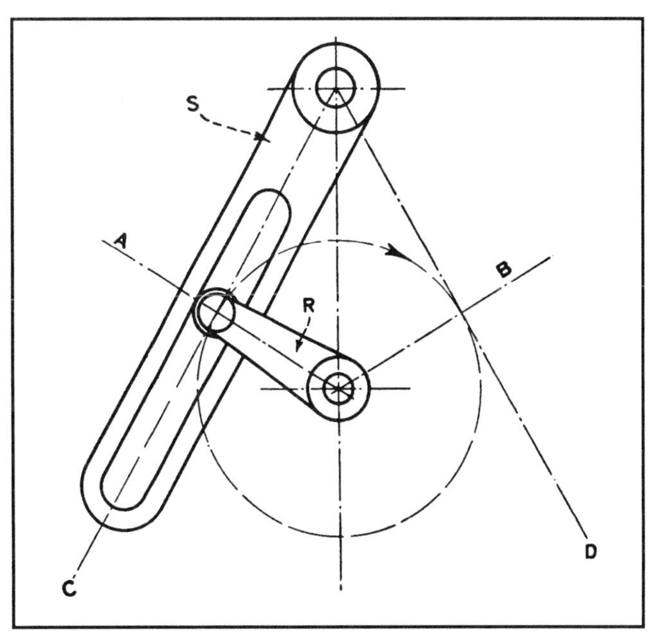

Fig. 2 Slot-crank motion from *A* to *B* represents external geneva action; from *B* to *A* represents internal geneva motion.

Where intermittent drives must provide dwell periods of more than 180°, the external geneva wheel design is satisfactory and is generally the standard device employed. But where the dwell period must be less than 180°, other intermittent drive mechanisms must be used. The internal geneva wheel is one way of obtaining this kind of motion.

The dwell period of all internal genevas is always smaller than 180°. Thus, more time is left for the star wheel to reach maximum velocity, and acceleration is lower. The highest value of angular acceleration occurs when the roller enters or leaves the slot. However, the acceleration occurs when the roller enters or leaves the slot. However, the acceleration curve does not reach a peak within the range of motion of the driven wheel. The geometrical maximum would occur in the continuation of the curve. But this continuation has no significance because the driven member will have entered the dwell phase associated with the high angular displacement of the driving member.

The geometrical maximum lies in the continuation of the curve, falling into the region representing the motion of the external geneva wheel. This can be seen by the following considerations of a crank and slot drive, drawn in Fig. 2.

When the roller crank *R* rotates, slot link *S* will perform an oscillating movement, for which the displacement, angular velocity,

Table I—Notation and Formulas for the Internal Geneva Wheel

Assumed or given: *a*, *n*, *d* and *p*

a = crank radius of driving member
n = number of slots
d = roller diameter
p = constant velocity of driving crank in rpm

$$m = \frac{1}{\sin \frac{180°}{n}}$$

b = center distance = $a\,m$

$$D = \text{inside diameter of driven member} = 2\sqrt{\frac{d^2}{4} + a^2 \cot^2 \frac{180°}{n}}$$

ω = constant angular velocity of driving crank in radians per sec = $\frac{p\pi}{30}$ radians per sec

α = angular position of driving crank at any time
β = angular displacement of driven member corresponding to crank angle α

$$\cos\beta = \frac{m + \cos\alpha}{\sqrt{1 + m^2 + 2m\cos\alpha}}$$

$$\text{Angular velocity of driven member} = \frac{d\beta}{dt} = \omega\left(\frac{1 + m\cos\alpha}{1 + m^2 + 2m\cos\alpha}\right)$$

$$\text{Angular acceleration of driven member} = \frac{d^2\beta}{dt^2} = \omega^2\left[\frac{m\sin\alpha\,(1 - m^2)}{(1 + m^2 + 2m\cos\alpha)^2}\right]$$

Maximum angular velocity occurs at $\alpha = 0°$ and equals = $\frac{\omega}{1 + m}$ radians per sec

Maximum angular acceleration occurs when roller enters slot and equals = $\frac{\omega^2}{\sqrt{m^2 - 1}}$ radians² per sec²

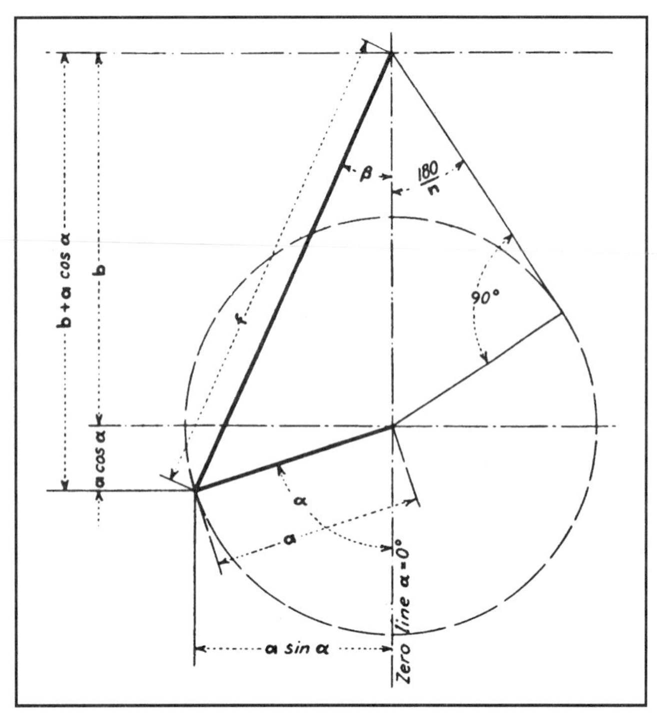

Fig. 3 A basic outline for developing the equations of the internal geneva wheel, based on the notations shown.

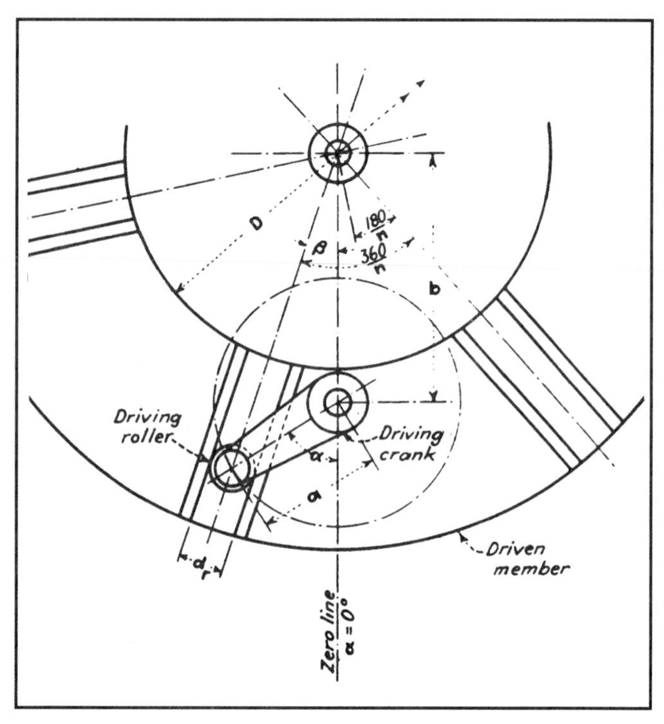

Fig. 4 A drawing of a six-slot internal geneva wheel. The symbols are identified, and the motion equations are given in Table I.

and acceleration can be given in continuous curves.

When the crank *R* rotates from *A* to *B*, then the slot link *S* will move from *C* to *D*, exactly reproducing all moving conditions of an external geneva of equal slot angle. When crank *R* continues its movement from *B* back to *A*, then the slot link *S* will move from *D* back to *C*, this time reproducing exactly (though in a mirror picture with the direction of motion being reversed) the moving conditions of an internal geneva.

Therefore, the characteristic curves of this motion contain both the external and internal geneva wheel conditions; the region of the external geneva lies between *A* and *B*, the region of the internal geneva lies between *B* and *A*.

The geometrical maxima of the acceleration curves lie only in the region between *A* and *B*, representing that portion of the curves which belongs to the external geneva.

The principal advantage of the internal geneva, other than its smooth operation, is it sharply defined dwell period. A disadvantage is the relatively large size of the driven member, which increases the force resisting acceleration. Another feature, which is sometimes a disadvantage, is the cantilever arrangement of the roller crank shaft. This shaft cannot be a through shaft because the crank must be fastened to the overhanging end of the input shaft.

To simplify the equations, the connecting line of the wheel and crank centers is taken as the zero line. The angular

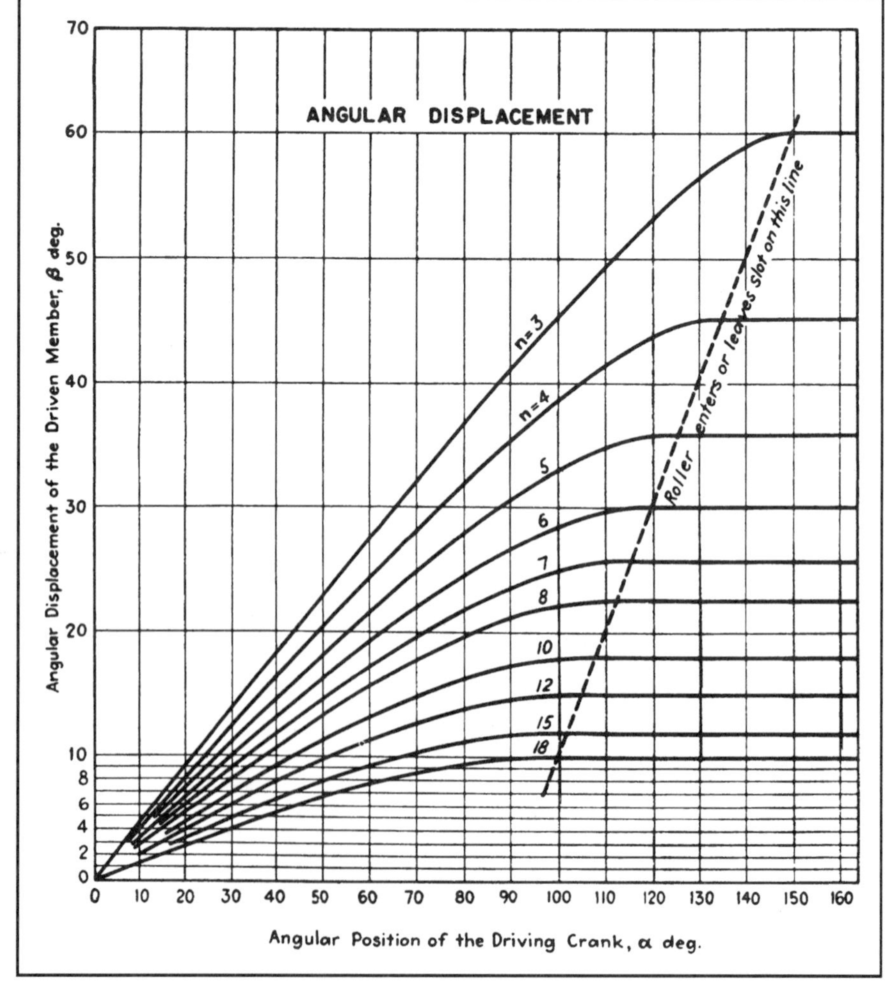

Fig. 5 Angular displacement of the driven member can be determined from this chart.

position of the driving crank α is zero when it is on this line. Then the follow-ing relations are developed, based on Fig. 3.

n = number of slots

a = crank radius

b = center distance $= \dfrac{a}{\sin \dfrac{180°}{n}}$

Let

$$\frac{1}{\sin \dfrac{180°}{n}} = m,$$

then; $b = am$

To find the angular displacement β of the driven member, the driven crank radius f is first calculated from:

$$f = \sqrt{a^2 \sin^2 \alpha + (am + a \cos \alpha)^2}$$
$$= \alpha\sqrt{1 + m^2 + 2m \cos \alpha} \quad (1)$$

and because

$$\cos \beta = \frac{m + \cos \alpha}{f}$$

it follows:

$$\cos \beta = \frac{m + \cos \alpha}{\sqrt{1 + m^2 - 2m \cos \alpha}} \quad (2)$$

From this formula, β, the angular displacement, can be calculated for any angle α, the angle of the mechanism's driving member.

The first derivative of Eq. (2) gives the angular velocity as:

$$\frac{d\beta}{dt} = \omega\left(\frac{1 + m \cos \alpha}{1 + m^2 + 2m \cos \alpha}\right) \quad (3)$$

where ω designates the uniform speed of the driving crank shaft, namely:

$$\omega = \frac{p\pi}{30}$$

if p equals its number of revolutions per minute.

Differentiating Eq. (3) once more develops the equation for the angular acceleration:

$$\frac{d^2\beta}{dt^2} = \omega^2\left(\frac{m \sin \alpha(1 - m^2)}{(1 + m^2 + 2m \cos \alpha)^2}\right) \quad (4)$$

The maximum angular velocity occurs, obviously, at $\alpha = 0°$. Its value is found by substituting 0° for α in Eq. (3). It is:

$$\frac{d\beta}{dt_{\max}} = \frac{\omega}{1 + m} \quad (5)$$

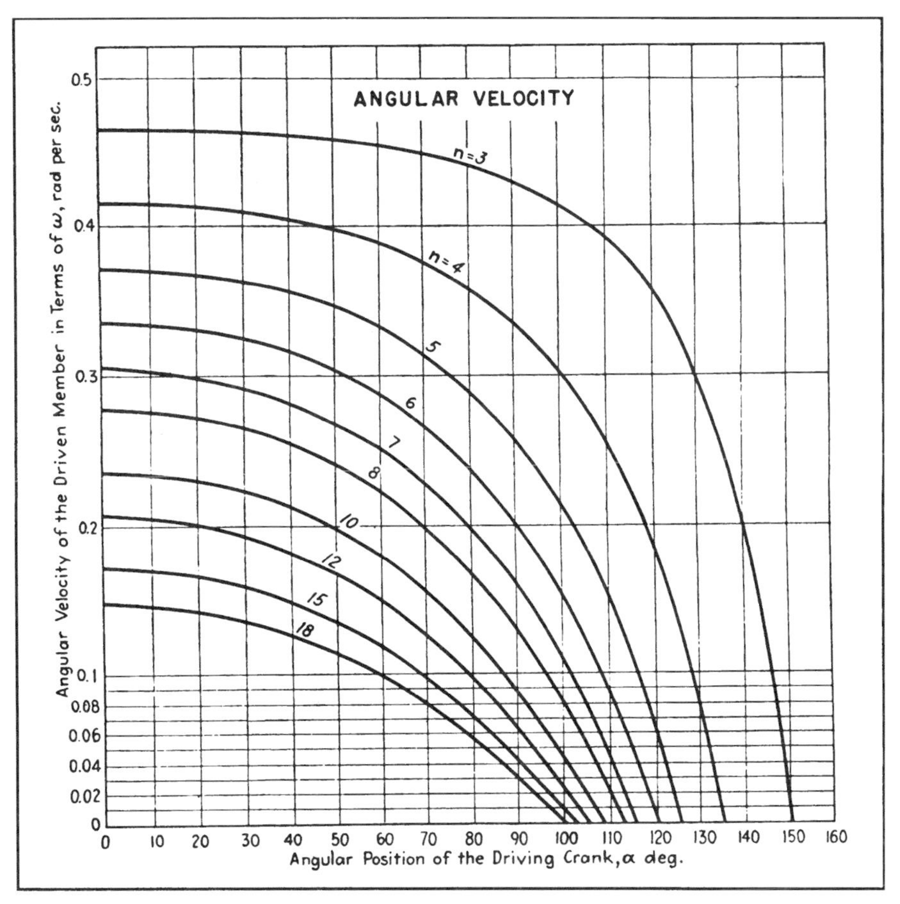

Fig. 6 Angular velocity of the driven member can be determined from this chart.

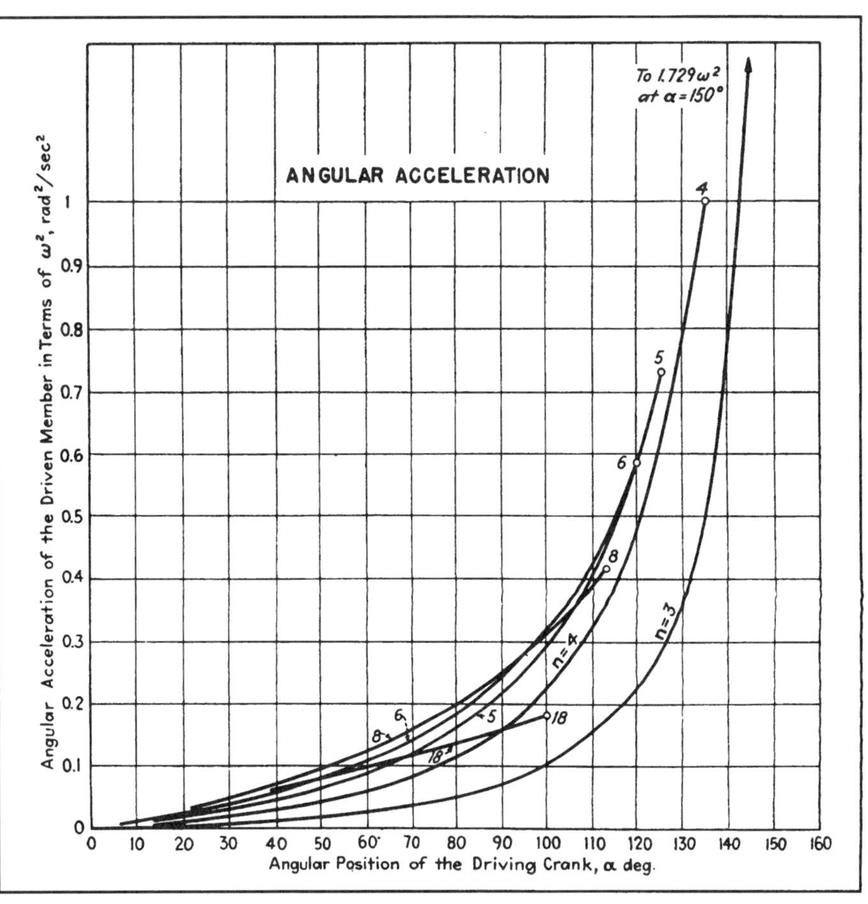

Fig. 7 Angular acceleration of the driven member can be determined from this chart.

Table II—Kinematic Data For the Internal Geneva Wheel

Number of slots, n	$\frac{360°}{n}$	Dwell period	Motion period	m and center-distance for $a = 1$	Maximum angular velocity of driven member equals ω radians per sec. multiplied by values tabulated. Both α and β in 0° position	Angular acceleration of driven member when roller enters slot equals ω^2 radians² per sec² multiplied by values tabulated		
						α	β	Multiplier
3	120°	60°	300°	1.155	0.464	150°	60°	1.729
4	90°	90°	270°	1.414	0.414	135°	45°	1.000
5	72°	108°	252°	1.701	0.370	126°	36°	0.727
6	60°	120°	240°	2.000	0.333	120°	30°	0.577
7	51° 25′ 43″	128° 30′	231° 30′	2.305	0.303	115° 42′ 52″	25° 42′ 52″	0.481
8	45°	135°	225°	2.613	0.277	112° 30′	22° 30′	0.414
9	40°	140°	220°	2.924	0.255	110°	20°	0.364
10	36°	144°	216°	3.236	0.236	108°	18°	0.325
11	32° 43′ 38″	147° 15′	212° 45′	3.549	0.220	106° 21′ 49″	16° 21′ 49″	0.294
12	30°	150°	210°	3.864	0.206	105°	15°	0.268

The highest value of the acceleration is found by substituting $180/n + 980$ for α in Eq. (4):

$$\frac{d^2\beta}{dt^2_{\max}} = \frac{\omega^2}{\sqrt{m^2 - 1}} \qquad (6)$$

A layout drawing for a six-slot internal geneva wheel is shown in Fig. 4. All the symbols in this drawing and throughout the text are compiled in Table I for easy reference.

Table II contains all the data of principal interest on the performance of internal geneva wheels that have from 3 to 18 slots. Other data can be read from the charts: Fig. 5 for angular position, Fig. 6 for angular velocity, and Fig. 7 for angular acceleration.

STAR WHEELS CHALLENGE GENEVA DRIVES FOR INDEXING

Star wheels with circular-arc slots can be analyzed mathematically and manufactured easily.

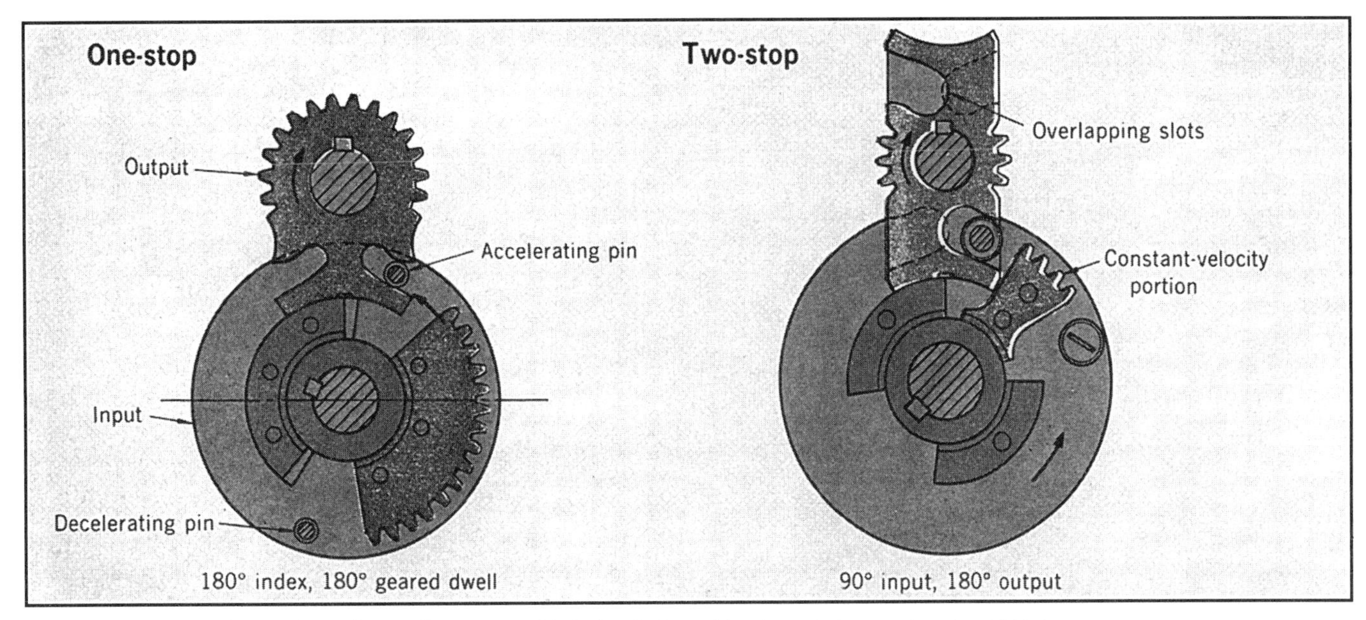

Star Wheels vary in shape, depending on the degree of indexing that must be done during one input revolution.

A family of star wheels with circular instead of the usual epicyclic slots (see drawings) can produce fast start-and-stop indexing with relatively low acceleration forces.

This rapid, jar-free cycling is important in a wide variety of production machines and automatic assembly lines that move parts from one station to another for drilling, cutting, milling, and other processes.

The circular-slot star wheels were invented by Martin Zugel of Cleveland, Ohio.

The motion of older star wheels with epicyclic slots is difficult to analyze and predict, and the wheels are hard to make.

The star wheels with their circular-arc slots are easy to fabricate, and because the slots are true circular arcs, they can be visualized for mathematical analysis as four-bar linkages during the entire period of pin-slot engagement.

Strong points. With this approach, changes in the radius of the slot can be analyzed and the acceleration curve varied to provide inertia loads below those of the genevas for any practical design requirement.

Another advantage of the star wheels is that they can index a full 360° in a relatively short period (180°). Such one-stop operation is not possible with genevas. In fact, genevas cannot do two-stop operations, and they have difficulty producing three stops per index. Most two-stop indexing devices available are cam-operated, which means they require greater input angles for indexing.

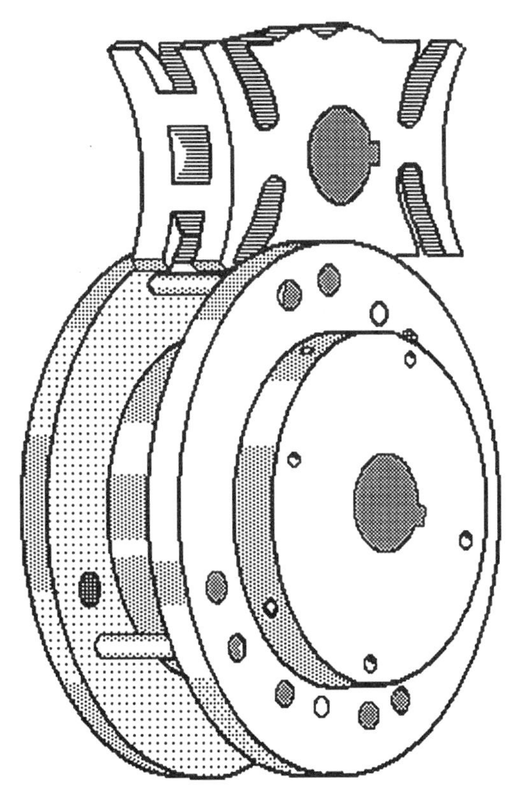

Geared star sector indexes smoothly a full 360° during a 180° rotation of the wheel, then it pauses during the other 180° to allow the wheel to catch up.

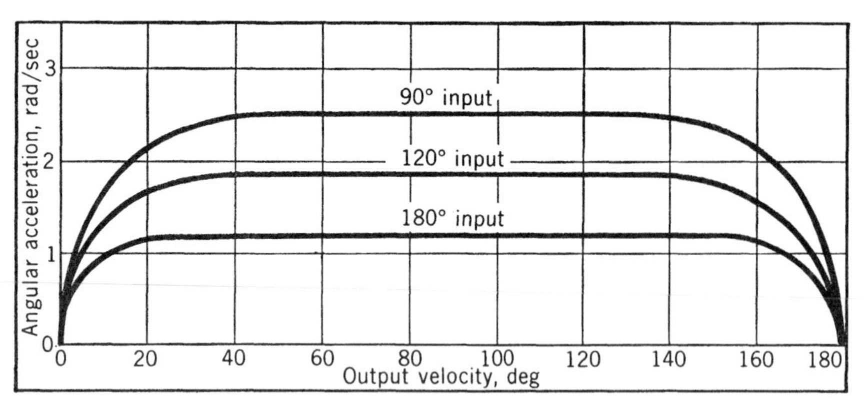

The one-stop index motion of the unit can be designed to take longer to complete its indexing, thus reducing its index velocity.

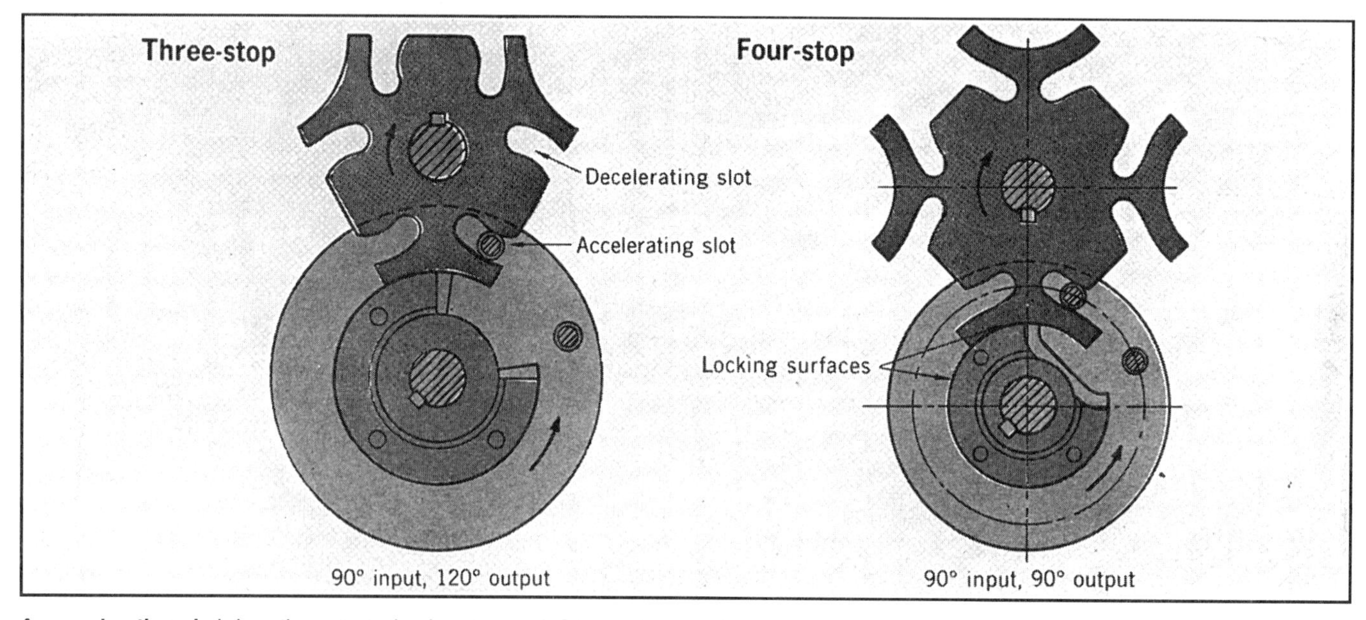

An accelerating pin brings the output wheel up to speed. Gear sectors mesh to keep the output rotating beyond 180°.

Operating sequence. In operation, the input wheel rotates continuously. A sequence starts (see drawing) when the accelerating pin engages the curved slot to start indexing the output wheel clockwise. Simultaneously, the locking surface clears the right side of the output wheel to permit the indexing.

Pin C in the drawings continues to accelerate the output wheel past the midpoint, where a geneva wheel would start deceleration. Not until the pins are symmetrical (see drawing) does the acceleration end and the deceleration begin. Pin D then takes the brunt of the deceleration force.

Adaptable. The angular velocity of the output wheel, at this stage of exit of the acceleration roller from Slot 1, can be varied to suit design requirements. At this point, for example, it is possible either to engage the deceleration roller as described or to start the engagement of a constant-velocity portion of the cycle. Many more degrees of output index can be obtained by interposing gear-element segments between the acceleration and deceleration rollers.

The star wheel at left will stop and start four times in making one revolution, while the input turns four times in the same period. In the starting position, the output link has zero angular velocity, which is a prerequisite condition for any star wheel intended to work at speeds above a near standstill.

In the disengaged position, the angular velocity ratio between the output and input shafts (the "gear" ratio) is entirely dependent upon the design angles α and β and independent of the slot radius, r.

Design comparisons. The slot radius, however, plays an important role in the mode of the acceleration forces. A four-stop geneva provides a good basis for comparison with a four-stage "Cyclo-Index" system.

Assume, for example, that $\alpha = \beta = 22.5°$. Application of trigonometry yields:

$$R = A\left[\frac{\sin \beta}{\sin (\alpha + \beta)}\right]$$

which yields $R = 0.541A$. The only restriction on r is that it be large enough to allow the wheel to pass through its mid-position. This is satisfied if:

$$r > \frac{RA(1 - \cos \alpha)}{A - 2R - A \cos \alpha} \approx 0.1A$$

There is no upper limit on r, so that slot can be straight.

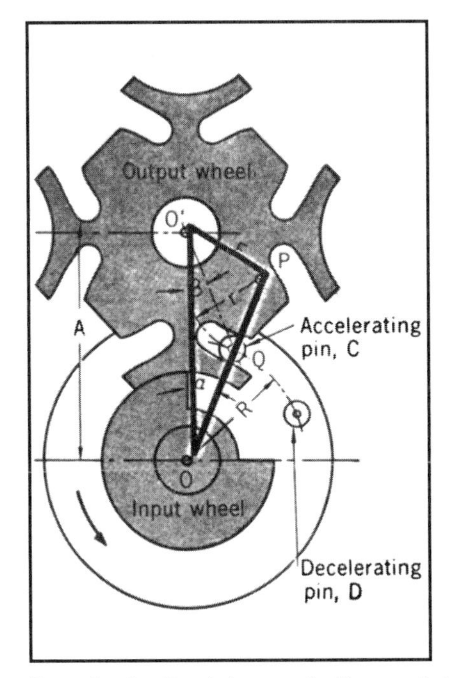

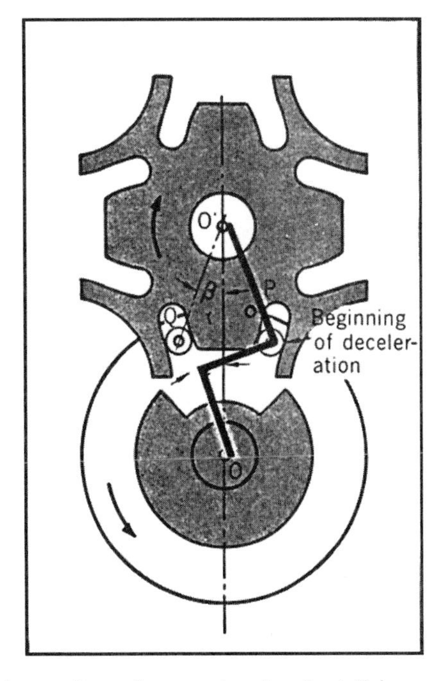

Star-wheel action is improved with curved slots over the radius r, centered on the initial-contact line OP. The units then act as four-bar linkages, 00^1PQ.

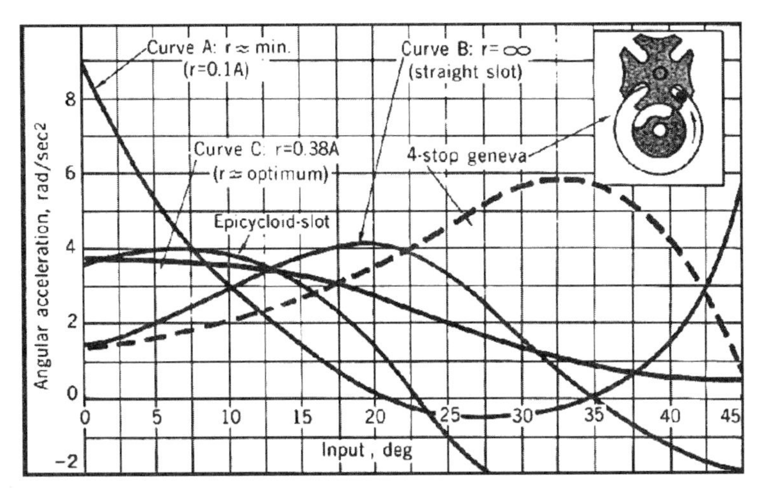

The accelerating force of star wheels (curves A, B, C) varies with input rotation. With an optimum slot (curve C), it is lower than for a four-stop geneva.

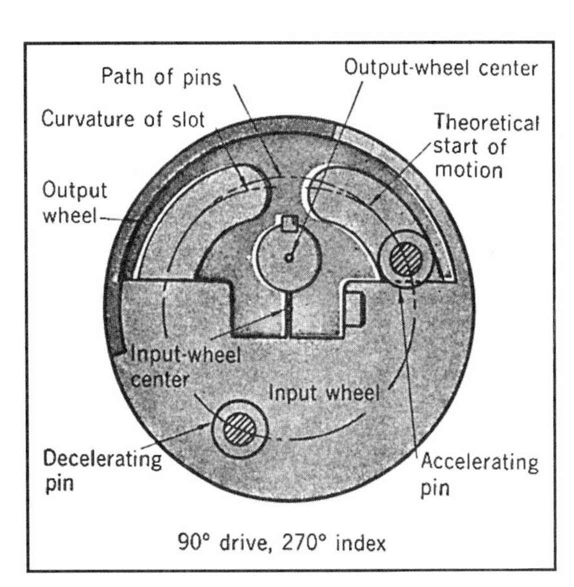

This internal star wheel has a radius difference to cushion the indexing shock.

RATCHET-TOOTH SPEED-CHANGE DRIVE

An in-line shaft drive, with reduction ratios of 1:1 and 1:16 or 1:28, combined in a single element, was designed by Telefunken of Germany. It consists basically of friction wheels that drip each other elastically.

Crown wheel with a gear ratio of 1:1 provide the coarse adjustment, and friction spur gearing, with a ratio of 1:16 or 1:28, provides the fine or vernier adjustment.

A spring (see diagram) applies pressure to the fine-adjustment pinion, preventing backlash while the coarse adjustment is in use. It uncouples the coarse adjustment when the vernier is brought into play by forward movement of the front shaft. The spring also ensures that the front shaft is always in gear.

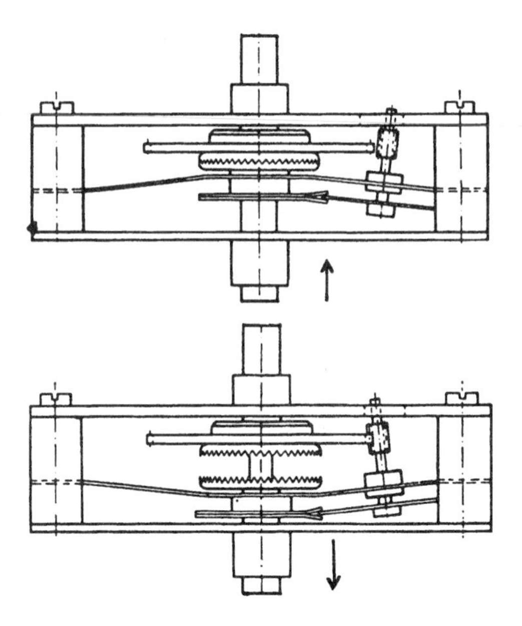

MODIFIED RATCHET DRIVE

A ratchet drive was designed to assure movement, one tooth at a time, in only one direction, without overriding. The key element is a small stub that moves along from the bottom of one tooth well, across the top of the tooth, and into an adjacent tooth well, while the pawl remains at the bottom of another tooth well.

The locking link, which carries the stub along with the spring, comprises a system that tends to hold the link and pawl against the outside circumference of the wheel and to push the stub and pawl point toward each other and into differently spaced wells between the teeth. A biasing element, which might be another linkage or solenoid, is provided to move the anchor arm from one side to the other, between the stops, as shown by the double arrow. The pawl will move from one tooth well to the next tooth well only when the stub is at the bottom of a tooth well and is in a position to prevent counter-rotation.

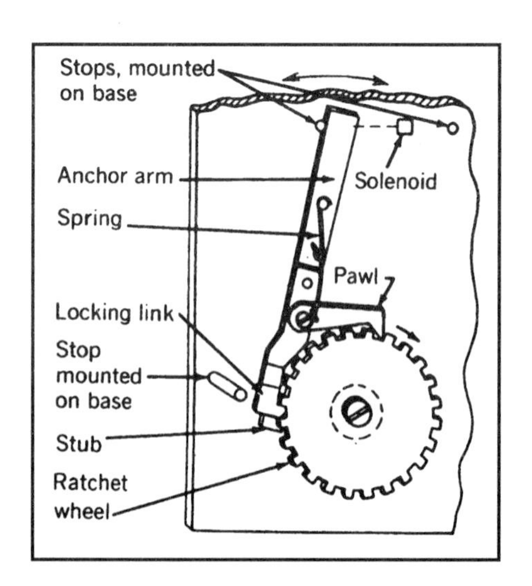

EIGHT TOOTHLESS RATCHETS

Ratchets with springs, rollers, and other devices keep motion going one way.

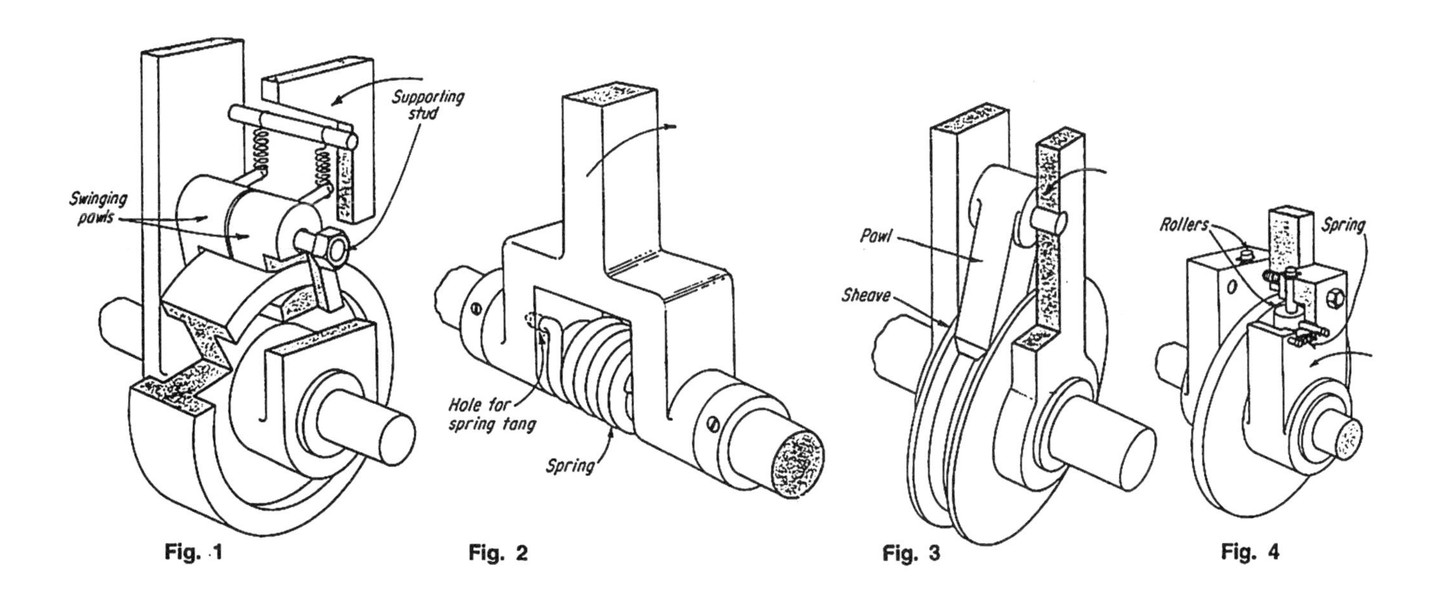

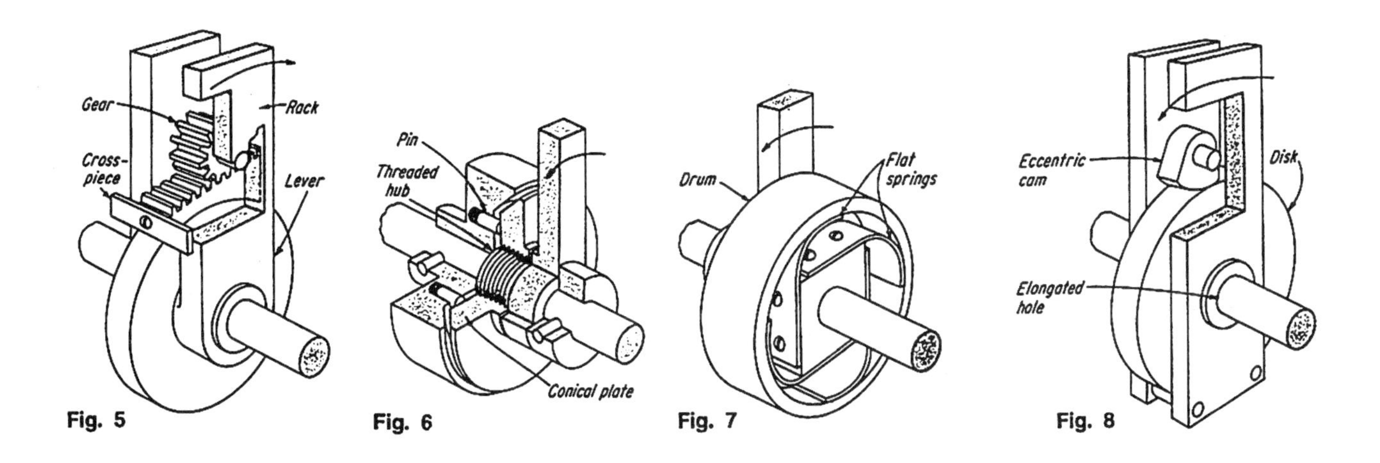

Fig. 1 Swinging pawls lock on the rim when the lever swings forward, and release on the return stroke. Oversize holes for the supporting stud make sure that both the top and bottom surfaces of the pawls make contact.

Fig. 2 A helical spring grips the shaft because its inner diameter is smaller than the outer diameter of shaft. During the forward stroke, the spring winds tighter; during the return stroke, it expands.

Fig. 3 A V-belt sheave is pushed around when pawl wedges in the groove. For a snug fit, the bottom of the pawl is tapered like a V-belt.

Fig. 4 Eccentric rollers squeeze a disk on its forward stroke. On the return stroke, rollers rotate backwards and release their grip. Springs keep the rollers in contact with the disk.

Fig. 5 A rack is wedge-shaped so that it jams between the rolling gear and the disk, pushing the shaft forward. When the driving lever makes its return stroke, it carries along the unattached rack by the cross-piece.

Fig. 6 A conical plate moves like a nut back and forth along the threaded center hub of the lever. The light friction of spring-loaded pins keeps the plate from rotating with the hub.

Fig. 7 Flat springs expand against the inside of a drum when a lever moves one way, but they drag loosely when the lever turns the drum in the opposite direction.

Fig. 8 An eccentric cam jams against the disk during the motion half of a cycle. Elongated holes in the levers allow the cam to wedge itself more tightly in place.

ANALYSIS OF RATCHET WHEELS

The ratchet wheel is widely used in machinery, mainly to transmit intermittent motion or to allow shaft rotation in one direction only. Ratchet-wheel teeth can be either on the perimeter of a disk or on the inner edge of a ring.

The pawl, which engages the ratchet teeth, is a beam pivoted at one end; the other end is shaped to fit the ratchet-tooth flank. Usually, a spring or counterweight maintains constant contact between wheel and pawl.

It is desirable, in most designs, to keep the spring force low. It should be just large enough to overcome the separation forces—inertia, weight, and pivot friction. Excess spring force should not be considered for engaging the pawl and holding it against the load.

To ensure that the pawl is automatically pulled in and kept in engagement independently of the spring, a properly drawn tooth flank is necessary.

The requirement for self-engagement is:

$$Pc + M > \mu Pb + P\sqrt{(1 + \mu^2)}\mu_1 r_1$$

Neglecting weight and pivot friction:

$$Pc > \mu Pb$$

but $c/b = r/a = \tan \phi$, and because $\tan \phi$ is approximately equal to $\sin \phi$:

$$c/b = r/R$$

Substituting in term (1)

$$rR > \mu$$

For steel on steel, dry, $\mu = 0.15$ Therefore, using

$$r/R = 0.20 \text{ to } 0.25$$

the margin of safety is large; the pawl will slide into engagement easily. For internal teeth with ϕ of 30°, c/b is tan 30° or 0.577, which is larger than μ, and the teeth are therefore self-engaging.

When laying out the ratchet wheel and pawl, locate points O, A and O_1 on the same circle. AO and AO_1 will then be perpendicular to one another; this will ensure that the smallest forces are acting on the system.

Ratchet and pawl dimensions are governed by design sizes and stress. If the tooth, and thus pitch, must be larger than required to be strong enough, a multiple pawl arrangement can be used. The pawls can be arranged so that one of them will engage the ratchet after a rotation of less than the pitch.

A fine feed can be obtained by placing many pawls side by side, with the corresponding ratchet wheels uniformly displaced and interconnected.

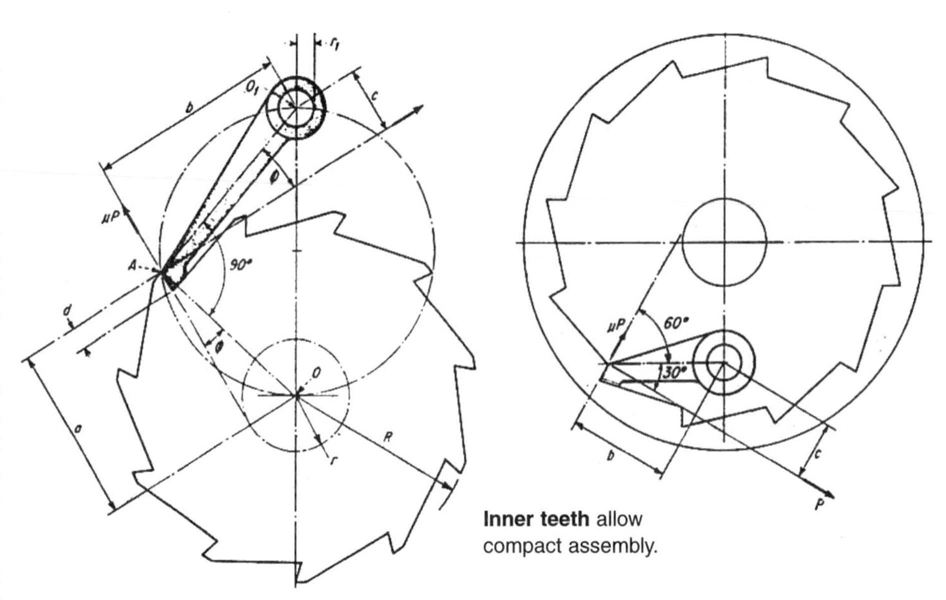

Pawl in compression has tooth pressure P and weight of pawl producing a moment that tends to engage pawl. Friction-force μP and pivot friction tend to oppose pawl engagement.

Inner teeth allow compact assembly.

a = moment arm of wheel torque

M = moment about O_1 caused by weight of pawl

$O_1 - O_2$ = ratchet and pawl pivot centers respectively

P = tooth pressure = wheel torque/a

$P\sqrt{(1 + \mu^2)}$ = load on pivot pin

μ, μ_1 = friction coefficients

Other symbols as defined in diagrams.

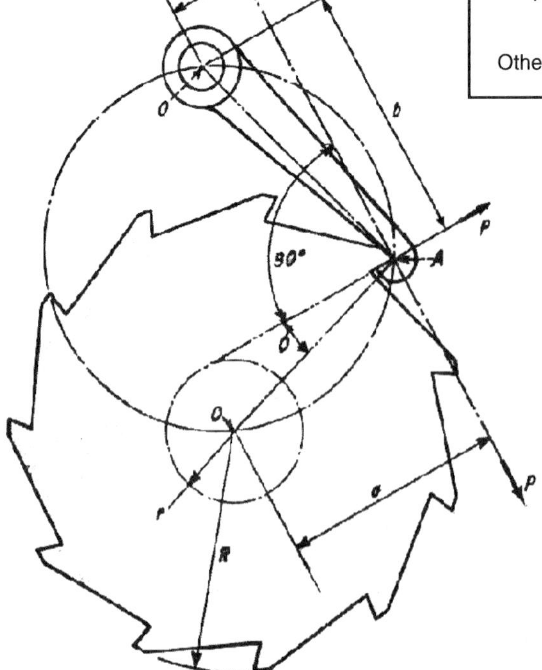

Pawl in tension has the same forces acting on the unit as other arrangements. The same layout principles apply.

CHAPTER 8
CLUTCHES AND BRAKES

TWELVE CLUTCHES WITH EXTERNAL OR INTERNAL CONTROL

Both friction and positive clutches are illustrated here. Figures 1 to 7 show externally controlled clutches, and Figures 8 to 12 show internally controlled clutches which are further divided into overload relief, overriding, and centrifugal versions.

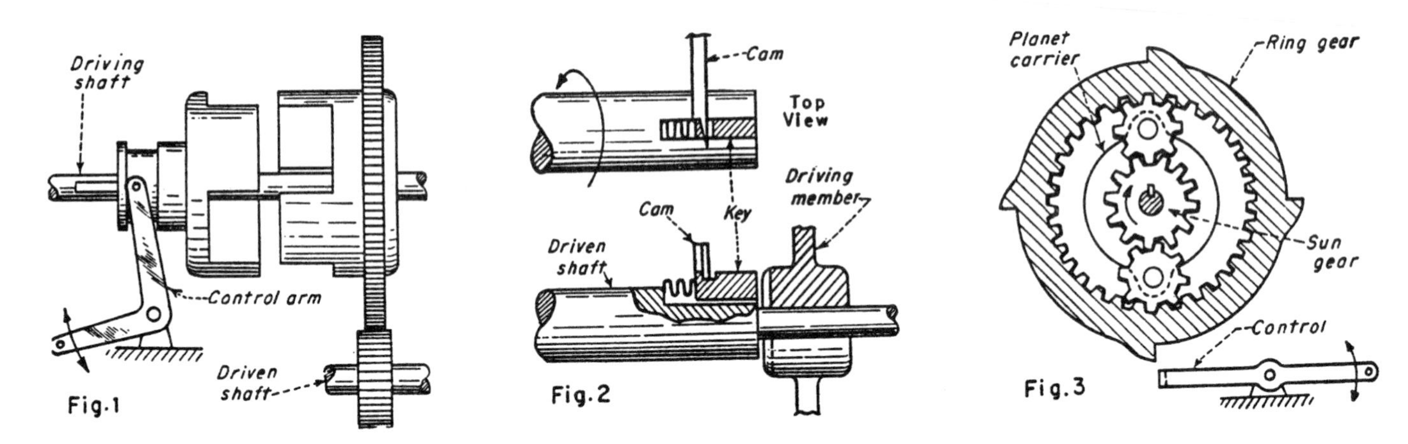

Fig. 1 Jaw Clutch: The left sliding half of this clutch is feathered to the driving shaft while the right half rotates freely. The control arm activates the sliding half to engage or disengage the drive. However, this simple, strong clutch is subject to high shock during engagement and the sliding half exhibits high inertia. Moreover, engagement requires long axial motion.

Fig. 2 Sliding Key Clutch: The driven shaft with a keyway carries the freely rotating member with radial slots along its hub. The sliding key is spring-loaded but is restrained from the engaging slots by the control cam. To engage the clutch, the control cam is raised and the key enters one of the slots. To disengage it, the cam is lowered into the path of the key and the rotation of the driven shaft forces the key out of the slot in the driving member. The step on the control cam limits the axial movement of the key.

Fig. 3 Planetary Transmission Clutch: In the disengaged position shown, the driving sun gear causes the free-wheeling ring gear to idle counter-clockwise while the driven planet carrier remains motionless. If the control arm blocks ring gear motion, a positive clockwise drive to the driven planet carrier is established.

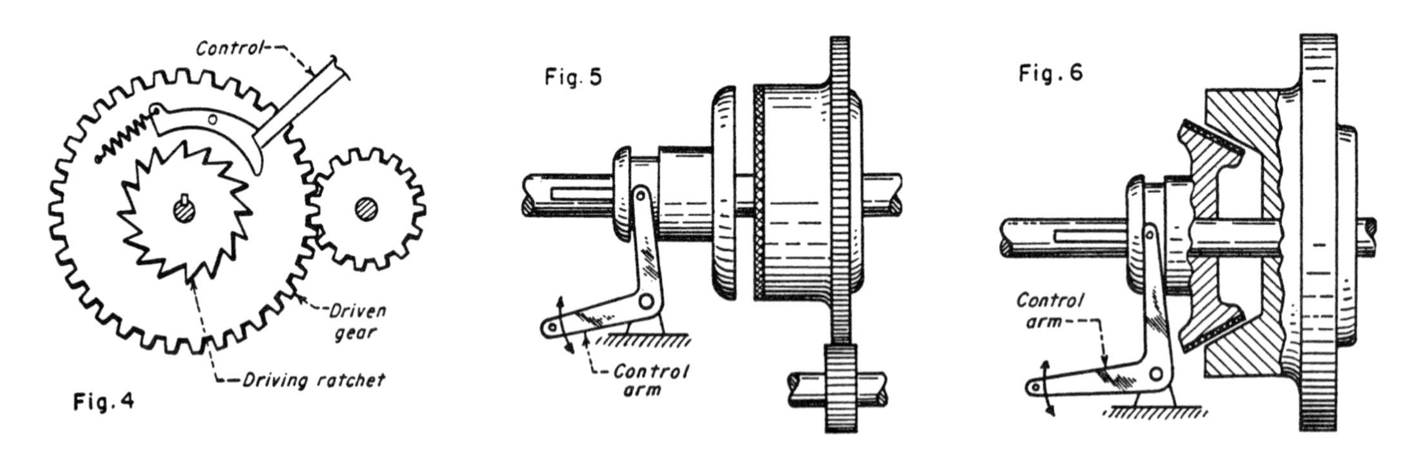

Fig. 4 Pawl and Ratchet Clutch: (External Control) The driving ratchet of this clutch is keyed to the driving shaft, and the pawl is pinned to the driven gear which can rotate freely on the driving shaft. When the control arm is raised, the spring pulls in the pawl to engage the ratchet and drive the gear. To disengage the clutch the control arm is lowered so that driven gear motion will disengage the pawl and stop the driven assembly against the control member.

Fig. 5 Plate Clutch: The plate clutch transmits power through the friction developed between the mating plate faces. The left sliding plate is fitted with a feather key, and the right plate member is free to rotate on the shaft. Clutch torque capacity depends on the axial force exerted by the control half when it engages the sliding half.

Fig. 6 Cone Clutch: The cone clutch, like the plate clutch, requires axial movement for engagement, but less axial force is required because of the increased friction between mating cones. Friction material is usually applied to only one of the mating conical surfaces. The free member is mounted to resist axial thrust.

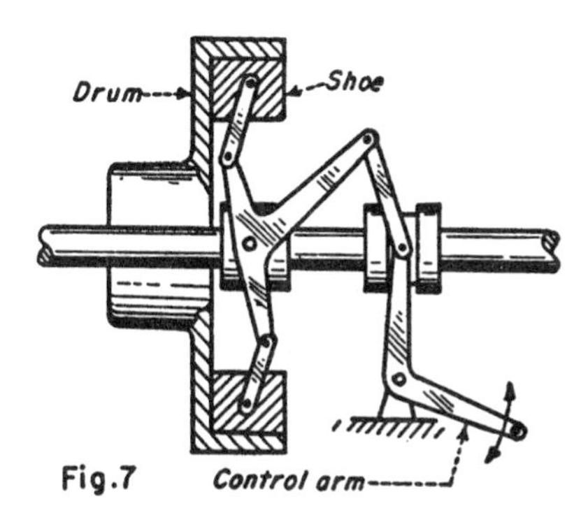

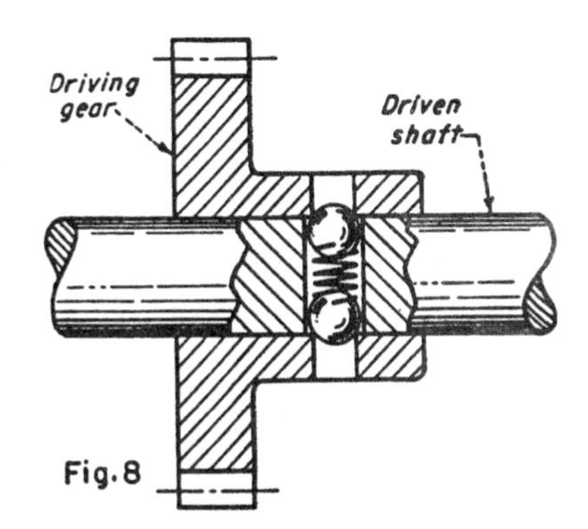

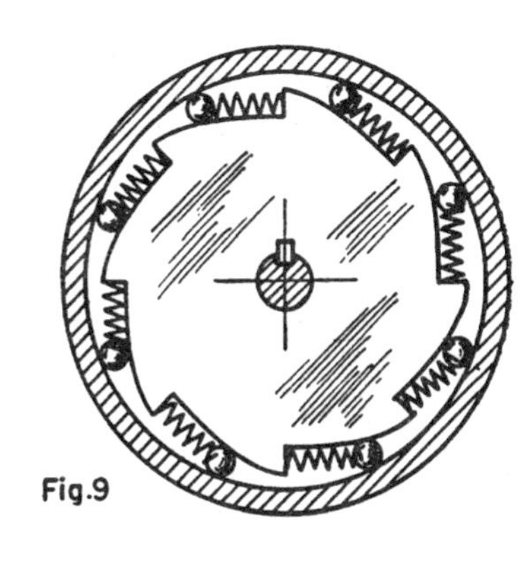

Fig. 7 Expanding Shoe Clutch: This clutch is engaged by the motion of the control arm. It operates linkages that force the friction shoes radially outwards so that they contact the inside surface of the drum.

Fig. 8 Spring and Ball Radial Detent Clutch: This clutch will hold the driving gear and driven gear in a set timing relationship until the torque becomes excessive. At that time the balls will be forced inward against their springs and out of engagement with the holes in the hub. As a result the driving gear will continue rotating while the drive shaft is stationary.

Fig. 9 Cam and Roller Clutch: This over-running clutch is better suited for higher-speed free-wheeling than a pawl-and-ratchet clutch. The inner driving member has cam surfaces on its outer rim that hold light springs that force the rollers to wedge between the cam surfaces and the inner cylindrical face of the driven member. While driving, friction rather than springs force the rollers to wedge tightly between the members to provide positive clockwise drive. The springs ensure fast clutching action. If the driven member should begin to run ahead of the driver, friction will force the rollers out of their tightly wedged positions and the clutch will slip.

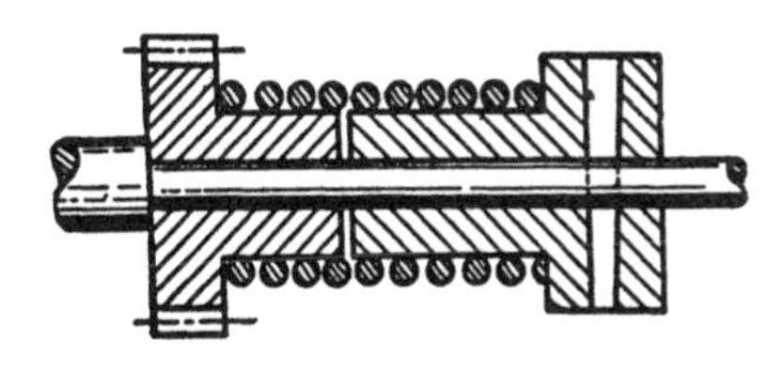

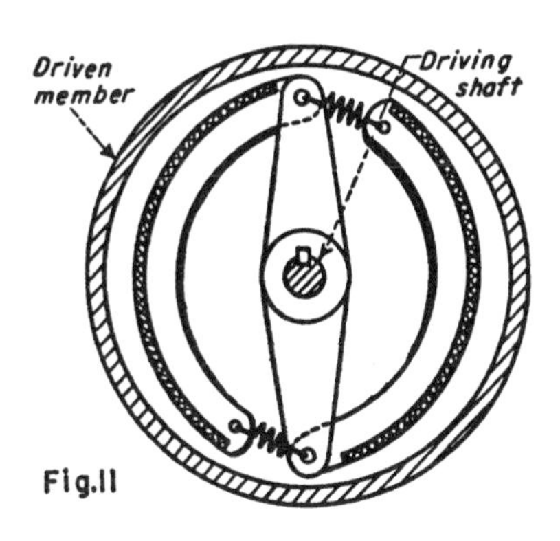

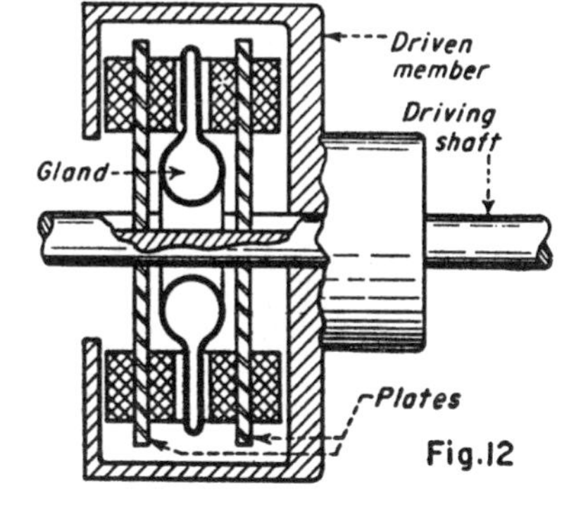

Fig. 10 Wrapped Spring Clutch: This simple unidirectional clutch consists of two rotating hubs connected by a coil spring that is press-fit over both hubs. In the driving direction the spring tightens around the hubs increasing the friction grip, but if driven in the opposite direction the spring unwinds causing the clutch to slip.

Fig. 11 Expanding Shoe Centrifugal Clutch: This clutch performs in a similar manner to the clutch shown in Fig. 7 except that there is no external control. Two friction shoes, attached to the driving member, are held inward by springs until they reach the "clutch-in" speed. At that speed centrifugal force drives the shoes outward into contact with the drum. As the drive shaft rotates faster, pressure between the shoes against the drum increases, thus increasing clutch torque.

Fig. 12 Mercury Gland Clutch: This clutch contains two friction plates and a mercury-filled rubber bladder. At rest, mercury fills a ring-shaped cavity around the shaft, but when rotated at a sufficiently high speed, the mercury is forced outward by centrifugal force. The mercury then spreads the rubber bladder axially, forcing the friction plates into contact with the opposing faces of the housing to drive it.

SPRING-WRAPPED CLUTCH SLIPS AT PRESET TORQUE

The simple spring clutch becomes even more useful when designed to slip at a predetermined torque. Unaffected by temperature extremes or variations in friction, these clutches are simple—they can even be "homemade." Information is provided here on two dual-spring, slip-type clutches. Two of the dual-spring clutches are in the tape drive shown.

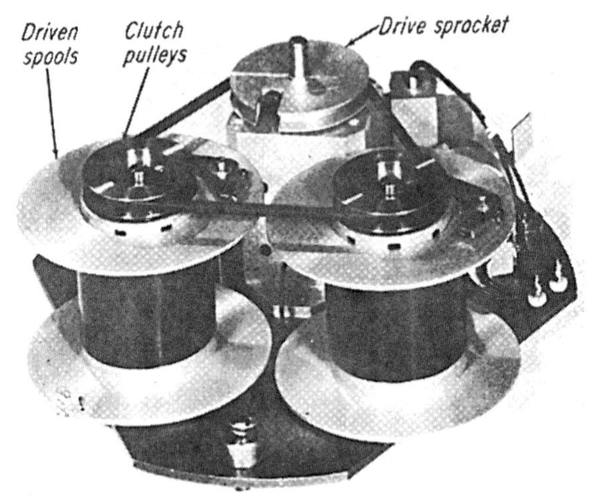

Fig. 1 Two dual-spring clutches are in this tape drive.

Spring clutches are devices for driving a load in one direction and uncoupling it when the output is overdriven or the direction of the input rotation is reversed. A spring clutch was modified to give a predetermined slip in either direction—hence the designation of this type as a "slip clutch." A stepped helical spring was employed to accomplish that modification. Later it was developed further by introducing an intermediate clutch member between two helical springs. This dual-spring innovation was preferred where more output torque accuracy was required.

Most designs employ either a friction-disk clutch or a shoe clutch to obtain a predetermined slip (in which the input drives output without slippage until a certain torque level is reached—then a drag-slippage occurs). But the torque capacity (or slip torque) for friction-disk clutches is the same for both directions of rotation.

By contrast, the stepped-spring slip clutch, pictured on the next page, can be designed to have either the same or different torque capacities for each direction of rotation. Torque levels where slippage occurs are independent of each other, thus providing wide latitude of design.

The element producing slip is the stepped spring. The outside diameter of the large step of the spring is assembled tightly in the bore of the output gear. The inside diameter of the smaller step fits tightly over the shaft. Rotation of the shaft in one direction causes the coils in contact with the shaft to grip tightly, and the coils inside the bore to contract and produce slip. Rotation in the opposite direction reverses the action of the spring parts, and slip is effected on the shaft.

Dual-Spring Slip Clutch

This innovation also permits bi-directional slip and independent torque capacities for the two directions of rotation. It requires two springs, one right-handed and one left-handed, for coupling the input, intermediate and output members. These members are coaxial, with the intermediate and input free to rotate on the output shaft. The rotation of input in one direction causes the spring, which couples the input and intermediate member, to grip tightly. The second spring, which couples the intermediate and output members, is oppositely wound, tends to expand and slip. The rotation in the opposite direction reverses the action of the two springs so that the spring between the input and intermediate members provides the slip. Because this design permits greater independence in the juggling of dimensions, it is preferred where more accurate slip-torque values are required.

Repeatable Performance

Spring-wrapped slip clutches and brakes have remarkably repeatable slip-torque characteristics which do not change with service temperature. Torque capacity remains constant with or without lubrication, and is unaffected by variations in the coefficient of friction. Thus, break-away torque capacity is equal to the sliding torque capacity. This stability makes it unnecessary to overdesign slip members to obtain reliable operation. These advantages are absent in most slip clutches.

Brake and Clutch Combinations

An interesting example of how slip brakes and clutches worked together to maintain proper tension in a tape drive, in either direction of operation, is pictured above and shown schematically on the opposite page. A brake here is simply a slip clutch with one side fastened to the frame of the unit. Stepped-spring clutches and brakes are shown for simplicity although, in the actual drive, dual-spring units were installed.

The sprocket wheel drives both the tape and belt. This allows the linear speed of the tape to be constant (one of the requirements). The angular speed of the spools, however, will vary as they wind or unwind. The task here is to maintain proper tension in the tape at all times and in either direction. This is done with a brake-clutch combination. In a counterclockwise direction, for example, the brake might become a "low-torque brake" that resists with a 0.1 in.-lb. Torque. The clutch in this direction is a "high-torque clutch"—it will provide a 1-in.-lb torque. Thus, the clutch overrides the brake with a net torque of 0.9 in.-lb.

When the drive is reversed, the same brake might now act as a high-torque brake, resisting with a 1 in.-lb torque, while the clutch acts as a low-torque clutch, resisting with 0.1 in.-lb. Thus, in the first direction the clutch drives the spool, in the other direction,

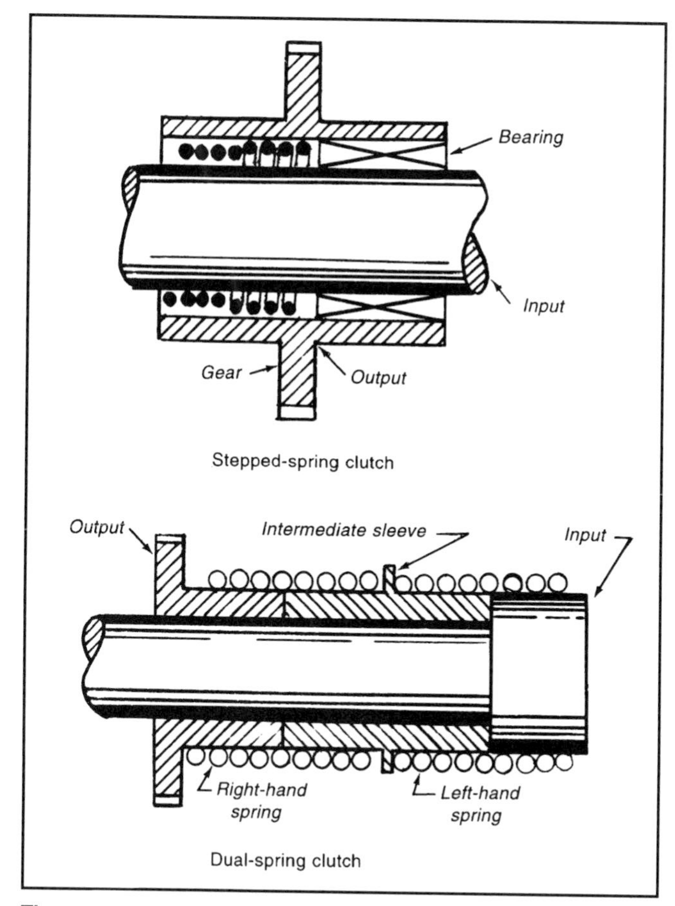

These two modifications of spring clutches offer independent slip characteristics in either direction of rotation.

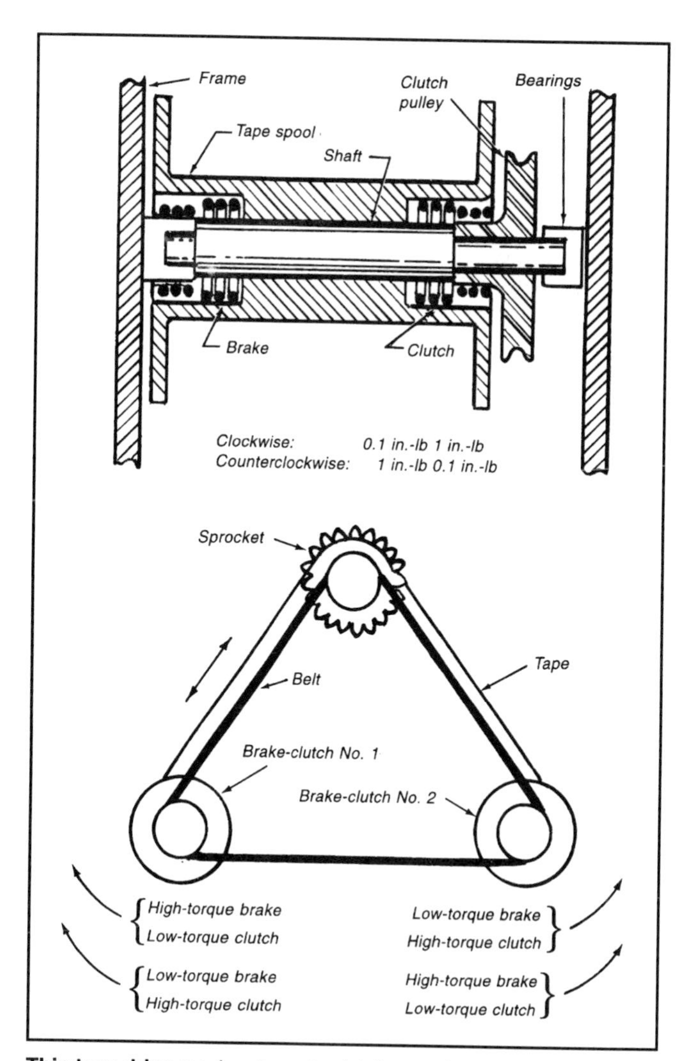

This tape drive requires two slip clutches and two brakes to ensure proper tension for bidirectional rotation. The detail of the spool (above) shows a clutch and brake unit.

the brake overcomes the clutch and provides a steady resisting force to provide tension in the tape. Of course, the clutch also permits the pulley that is driven by the belt to overdrive.

Two brake-clutch units are required. The second unit will provide opposing torque values—as listed in the diagram. The drive necessary to advance the tape only in a clockwise direction would be the slip clutch in unit 2 and the brake in unit 1. Advancing the tape in the other direction calls for use of the clutch in unit 1 and the brake in unit 2.

For all practical purposes, the low torque values in the brakes and clutches can be made negligible by specifying minimum interference between the spring and the bore or shaft. The low torque is amplified in the spring clutch at the level necessary to drive the tensioning torques of the brake and slip clutches.

Action thus produced by the simple arrangement of directional slip clutches and brakes cannot otherwise be duplicated without resorting to more complex designs.

Torque capacities of spring-wrapped slip clutches and brakes with round, rectangular, and square wire are, respectively:

$$T = \frac{\pi E d^4 \delta}{32D^2};\ T = \frac{Ebt^3\delta}{6D^2};\ T = \frac{Et^4\delta}{6D^2}$$

where E = modules of elasticity, psi; d = wire diameter, inches; D = diameter of shaft or bore, inches; ε = diametral interference between spring and shaft, or spring and bore, inches; t = wire thickness, inches; b = width of rectangular wire, inches; and T = slip torque capacity, pound-inches.

Minimum interference moment (on the spring gripping lightly) required to drive the slipping spring is:

$$M = \frac{T}{e^{\mu\theta} - 1}$$

where e = natural logarithmic base (e = 2.716; θ = angle of wrap of spring per shaft, radians, μ = coefficient of friction, M = interference moment between spring and shaft, pound-inches.

Design Example

Required: to design a tape drive similar to the one shown above. The torque requirements for the slip clutches and brakes for the two directions of rotation are:

(1) Slip clutch in normal takeup capacity (active function) is 0.5 to 0.8 in.-lb.

(2) Slip clutch in override direction (passive function) is 0.1 in.-lb (maximum).

(3) Brake in normal supply capacity (active function) is 0.7 to 1.0 in.-lb.

(4) Brake in override direction (passive function) is 0.1 in.-lb (maximum).

Assume that the dual-spring design shown previously is to include 0.750-in. drum diameters. Also available is an axial length for each spring, equivalent to 12 coils which are divided equally between the bridged shafts. Assuming round wire, calculate the wire diameter of the springs if 0.025 in. is maximum diametral interference desired for the active functions. For the passive functions use round wire that produces a spring index not more than 25.

Slip clutch, active spring:

$$d = \sqrt[4]{\frac{32D^2T}{\pi E\delta}} = \sqrt[4]{\frac{(32)(0.750)^2(0.8)}{\pi(30 \times 10^6)(0.025)}} = 0.050 \text{ in.}$$

The minimum diametral interference is (0.025) (0.5)/0.8 = 0.016 in. Consequently, the ID of the spring will vary from 0.725 to 0.734 in.

Slip clutch, passive spring:

$$\text{Wire dia.} = \frac{\text{drum dia.}}{\text{spring index}} = \frac{0.750}{25} = 0.030 \text{ in.}$$

Diametral interference:

$$\delta = \frac{32D^2T}{\pi E d^4} = \frac{(32)(0.750^2)(0.1)}{\pi(30 \times 10^6)(0.030)^4} = 0.023 \text{ in.}$$

Assuming a minimum coefficient of friction of 0.1, determine the minimum diametral interference for a spring clutch that will drive the maximum slip clutch torque of 0.8 lb-in.

Minimum diametral interference:

$$M = \frac{T}{e^{\mu\theta} - 1} = \frac{0.8}{e^{(0.1\pi)(6)} - 1}$$

ID of the spring is therefore 0.727 to 0.745 in.

$$\text{min.} = 0.023 \times \frac{0.019}{0.1} = 0.0044 \text{ in.}$$

Brake springs

By similar computations the wire diameter of the active brake spring is 0.053 in., with an ID that varies from 0.725 and 0.733 in.; wire diameter of the passive brake spring is 0.030 in., with its ID varying from 0.727 to 0.744 in.

CONTROLLED-SLIP EXPANDS SPRING CLUTCH APPLICATIONS

A remarkably simple change in spring clutches is solving a persistent problem in tape and film drives—how to keep drag tension on the tape constant, as its spool winds or unwinds. Shaft torque has to be varied directly with the tape diameter so many designers resort to adding electrical control systems, but that calls for additional components; an extra motor makes this an expensive solution. The self-adjusting spring brake (Fig. 1) developed by Joseph Kaplan, Farmingdale, NY, gives a constant drag torque ("slip" torque) that is easily and automatically varied by a simple lever arrangement actuated by the tape spool diameter (Fig. 2). The new brake is also being employed to test the output of motors and solenoids by providing levels of accurate slip torque.

Kaplan used his "controlled-slip" concept in two other products. In the controlled-torque screwdriver (Fig. 3) a stepped spring provides a 11.4-in.-lb slip when turned in either direction. It avoids overtightening machine screws in delicate instrument assemblies. A stepped spring is also the basis for the go/no-go torque gage that permits production inspection of output torques to within 1%.

Interfering spring. The three products were the latest in a series of slip clutches, drag brakes, and slip couplings developed by Kaplan for instrument brake drives. All are actually outgrowths of the spring clutch. The spring in this clutch is normally prevented from gripping the shaft by a detent response. Upon release of the detent, the spring will grip the shaft. If the shaft is turning in the proper direction, it is self-energizing. In the other direction, the spring simply overrides. Thus, the spring clutch is a "one-way" clutch.

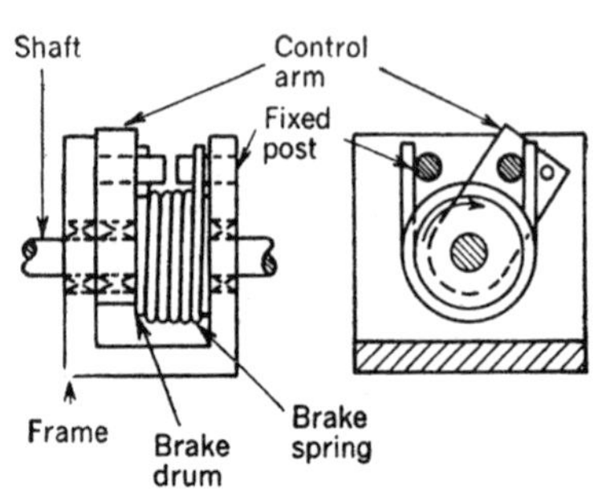

Fig. 1 Variable-torque drag brake . . .

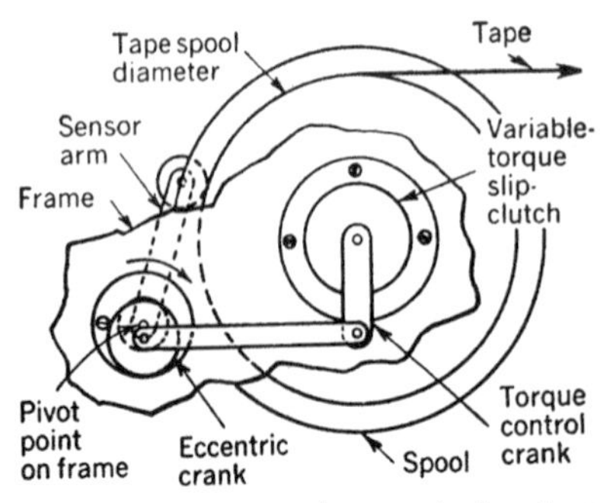

Fig. 2 . . . holds tension constant on tape

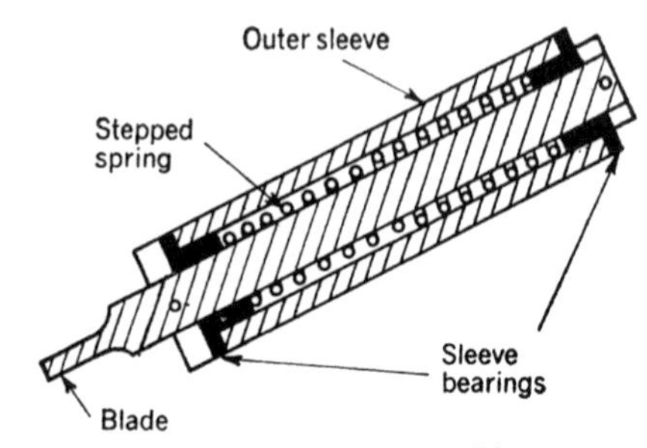

Fig. 3 Constant-torque screwdriver

SPRING BANDS IMPROVE OVERRUNNING CLUTCH

An overrunning clutch that takes up only half the space of most clutches has a series of spiral-wound bands instead of conventional rollers or sprags to transmit high torques. The design (see drawing) also simplifies the assembly, cutting costs as much as 40% by eliminating more than half the parts in conventional clutches.

The key to the savings in cost and space is the clutches' freedom from the need for a hardened outer race. Rollers and sprags must have hardened races because they transmit power by a wedging action between the inner and outer races.

Role of spring bands. Overrunning clutches, including the spiral-band type, slip and overrun when reversed (see drawing). This occurs when the outer member is rotated clockwise and the inner ring is the driven member.

The clutch, developed by National Standard Co., Niles, Michigan, contains a set of high-carbon spring-steel bands (six in the design illustrated) that grip the inner member when the clutch is driving.

The outer member simply serves to retain the spring anchors and to play a part in actuating the clutch. Because it isn't subject to wedging action, it can be made of almost any material, and this accounts for much of the cost saving. For example, in the automotive torque converter in the drawing at right, the bands fit into the aluminum die-cast reactor.

Reduced wear. The bands are spring-loaded over the inner member of the clutch, but they are held and rotated by the outer member. The centrifugal force on the bands then releases much of the force on the inner member and considerably decreases the overrunning torque. Wear is consequently greatly reduced.

The inner portion of the bands fits into a V-groove in the inner member. When the outer member is reversed, the bands wrap, creating a wedging action in this V-groove. This action is similar to that of a spring clutch with a helical-coil spring, but the spiral-band type has very little unwind before it overruns, compared with the coil type. Thus, it responds faster.

Edges of the clutch bands carry the entire load, and there is also a compound action of one band upon another. As the torque builds up, each band pushes down on the band beneath it, so each tip is forced more firmly into the V-groove. The bands are rated for torque capacities from 85 to 400 ft.-lb. Applications include their use in auto transmissions, starters, and industrial machinery.

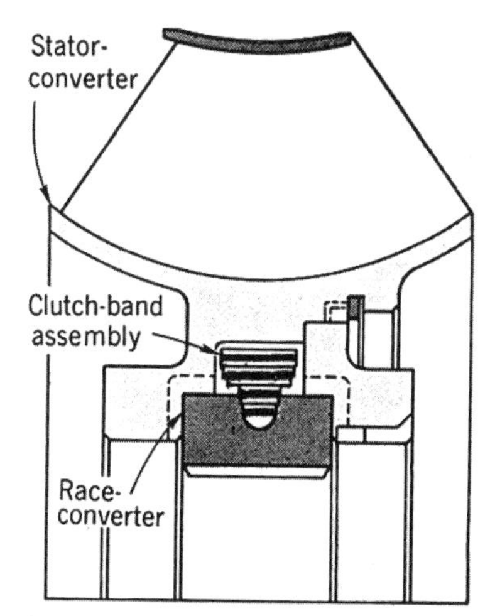

Spiral clutch bands can be purchased separately to fit the user's assembly.

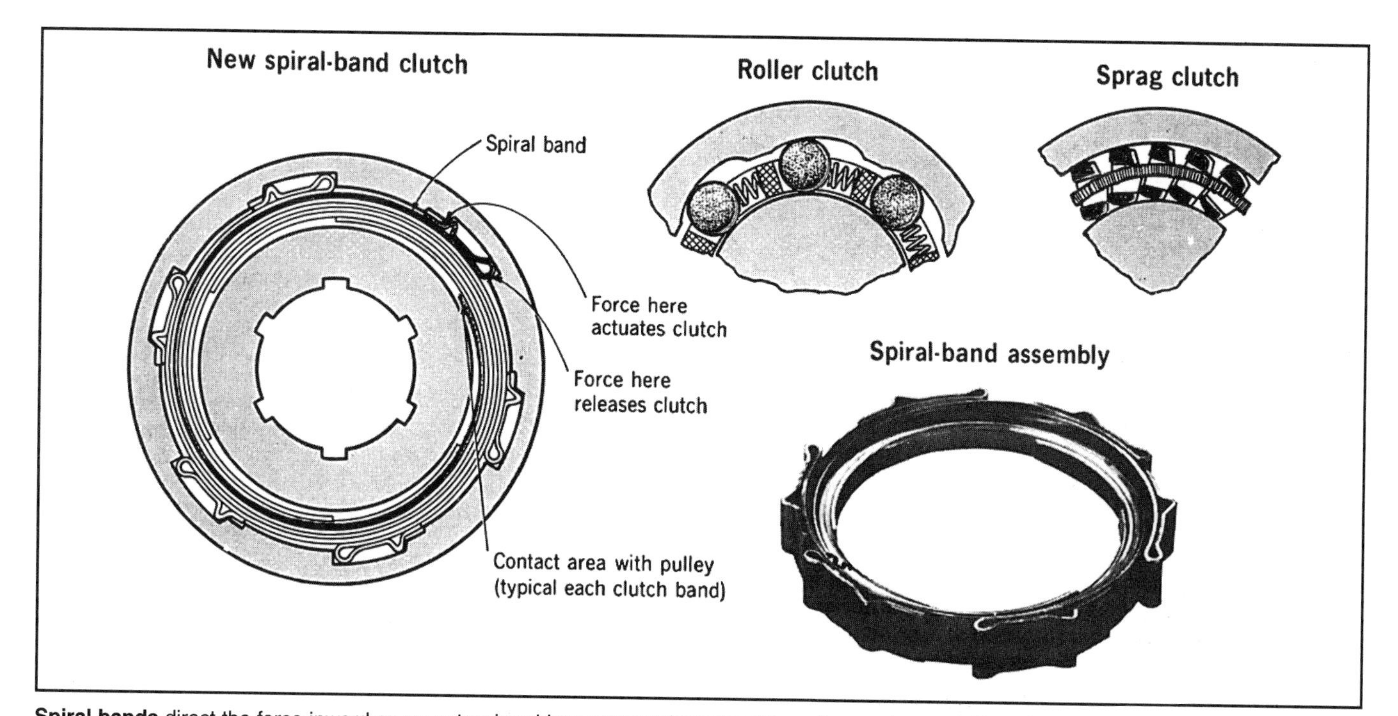

Spiral bands direct the force inward as an outer ring drives counterclockwise. The rollers and sprags direct the force outward.

SLIP AND BIDIRECTIONAL CLUTCHES COMBINE TO CONTROL TORQUE

A torque-limiting knob includes a dual set of miniature clutches—a detent slip clutch in series with a novel bidirectional-locking clutch—to prevent the driven member from backturning the knob. The bi-directional clutch in the knob locks the shaft from backlash torque originating within the panel, and the slip clutch limits the torque transmitted from outside the panel. The clutch was invented by Ted Chanoux, of Medford, N.Y.

The clutch (see drawing) is the result of an attempt to solve a problem that often plagues design engineers. A mechanism behind a panel such as a precision potentiometer or switch must be operated by a shaft that protrudes from the panel. The mechanism, however, must not be able to turn the shaft. Only the operator in front of the knob can turn the shaft, and he must limit the amount of torque he applies.

Solving design problem. This problem showed up in the design of a navigational system for aircraft.

The counter gave a longitudinal or latitudinal readout. When the aircraft was ready to take off, the navigator or pilot set a counter to some nominal figure, depending on the location of his starting point, and he energized the system. The computer then accepts the directional information from the gyro, the air speed from instruments in the wings, plus other data, and feeds a readout at the counter.

The entire mechanism was subjected to vibration, acceleration and deceleration, shock, and other high-torque loads, all of which could feed back through the system and might move the counter. The new knob device positively locks the mechanism shaft against the vibration, shock loads, and accidental turning, and it also limits the input torque to the system to a preset value.

Operation. To turn the shaft, the operator depresses the knob $^1/_{16}$ in. and turns it in the desired direction. When it is released, the knob retracts, and the shaft immediately and automatically locks to the panel or frame with zero backlash. Should the shaft torque exceed the preset value because of hitting a mechanical stop after several turns, or should the knob turn in the retracted position, the knob will slip to protect the system mechanism.

Internally, pushing in the knob turns both the detent clutch and the bidirectional-clutch release cage via the keyway. The fingers of the cage extend between the clutch rollers so that the rotation of the cage cams out the rollers, which are usually kept jammed between the clutch cam and the outer race with the roller springs. This action permits rotation of the cam and instrument shaft both clockwise and counterclockwise, but it locks the shaft securely against inside torque up to 30 oz.-in.

Applications. The detent clutch can be adjusted to limit the input torque to the desired values without removing the knob from the shaft. The outside diameter of the shaft is only 0.900 in., and the total length is 0.940 in. The exterior material of the knob is anodized aluminum, black or gray, and all other parts are stainless steel. The device is designed to meet the military requirements of MIL-E-5400, class 3 and MILK-3926 specifications.

Applications were seen in counter and reset switches and controls for machines and machine tools, radar systems, and precision potentiometers.

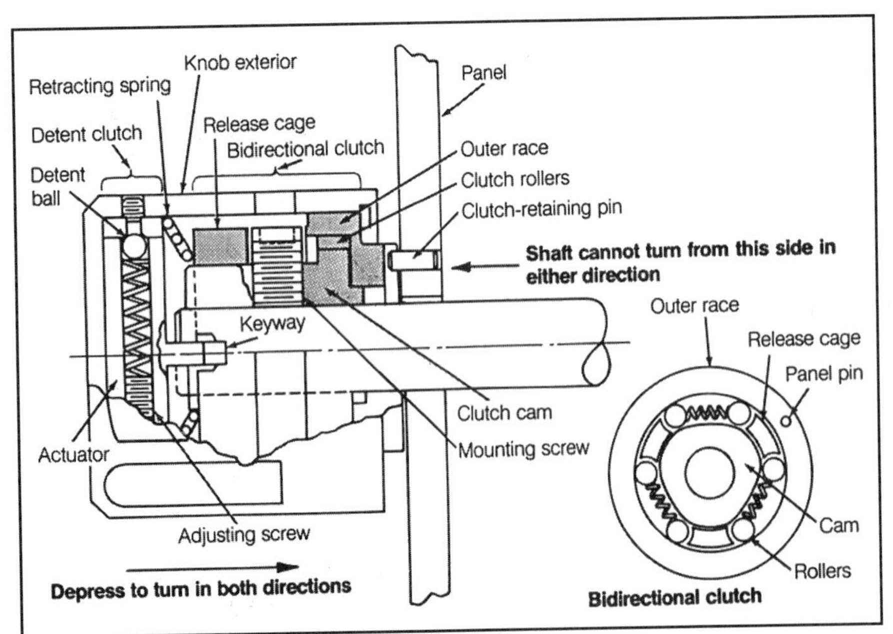

Miniature knob is easily operated from outside the panel by pushing it in and turning it in the desired direction. When released, the bi-directional clutch automatically locks the shaft against all conditions of shock and vibration.

Eight-Joint Coupler

A novel coupler combines two parallel linkage systems in a three-dimensional arrangement to provide wide angular and lateral off-set movements in pipe joints. By including a bellows between the connecting pipes, the connector can join high-pressure and high-temperature piping such as is found in refineries, steam plants, and stationary power plants.

The key components in the coupler are four pivot levers (drawing) mounted

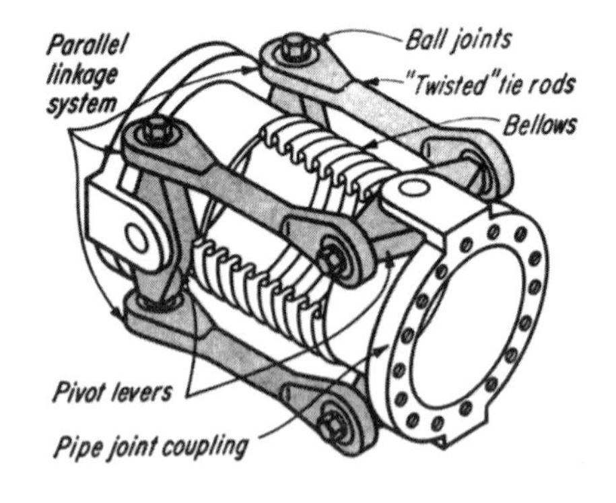

in two planes. Each pivot lever has provisions for a ball joint at each end. "Twisted" tie rods, with holes in different planes, connect the pivot levers to complete the system. The arrangement permits each pipe face to twist through an appreciable arc and also to shift orthogonally with respect to the other.

Longer tie rods can be formed by joining several bellows together with center tubes.

The connector was developed by Ralph Kuhm Jr. of El Segundo, California.

WALKING PRESSURE PLATE DELIVERS CONSTANT TORQUE

This automatic clutch causes the driving plate to move around the surface of the driven plate to prevent the clutch plates from overheating if the load gets too high. The "walking" action enables the clutch to transmit full engine torque for hours without serious damage to the clutch plates or the engine.

The automatic centrifugal clutch, manufactured by K-M Clutch Co., Van Nuys, California, combines the principles of a governor and a wedge to transmit torque from the engine to the drive shaft (see drawing).

How it works. As the engine builds up speed, the weights attached to the levers have a tendency to move towards the rim of the clutch plate, but they are stopped by retaining springs. When the shaft speed reaches 1600 rpm, however, centrifugal force overcomes the resistance of the springs, and the weights move outward. Simultaneously, the tapered end of the lever wedges itself in a slot in pin E, which is attached to the driving clutch plate. The wedging action forces both the pin and the clutch plate to move into contact with the driven plate.

A pulse of energy is transmitted to the clutch each time a cylinder fires. With every pulse, the lever arm moves outward, and there is an increase in pressure between the faces of the clutch. Before the next cylinder fires, both the lever arm and the driving plate return to their original positions. This pressure fluctuation between the two faces is repeated throughout the firing sequence of the engine.

Plate walks. If the load torque exceeds the engine torque, the clutch immediately slips, but full torque transfer is maintained without serious overheating. The pressure plate then momentarily disengages from the driven plate. However, as the plate rotates and builds up torque, it again comes in contact with the driven plate. In effect, the pressure plate "walks" around the contact surface of the driven plate, enabling

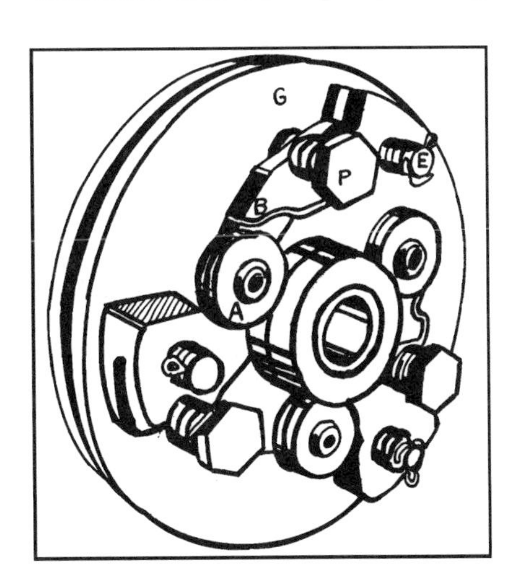

the clutch to continuously transmit full engine torque.

Applications. The clutch has undergone hundreds of hours of development testing on 4-stroke engines that ranged from 5 to 9 hp. According to the K-M Clutch Co., the clutch enables designers to use smaller motors than they previously could because of its no-load starting characteristics.

The clutch also acts as a brake to hold engine speeds within safe limits. For example, if the throttle accidentally opens when the driving wheels or driven mechanisms are locked, the clutch will stop.

The clutch can be fitted with sprockets, sheaves, or a stub shaft. It operates in any position, and can be driven in both directions. The clutch could be installed in ships so that the applied torque would come from the direction of the driven plate.

The pressure plate was made of cast iron, and the driven-plate casting was made of magnesium. To prevent too much wear, the steel fly weights and fly levers were pre-hardened.

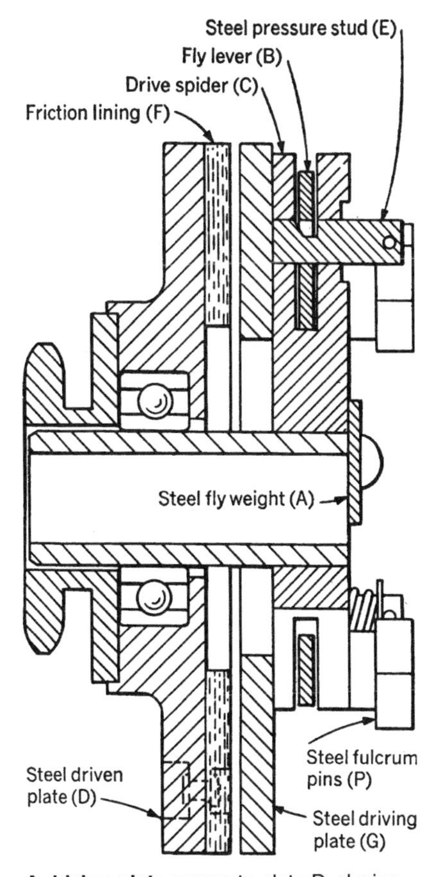

A driving plate moves to plate D, closing the gap, when speed reaches 1600 rpm.

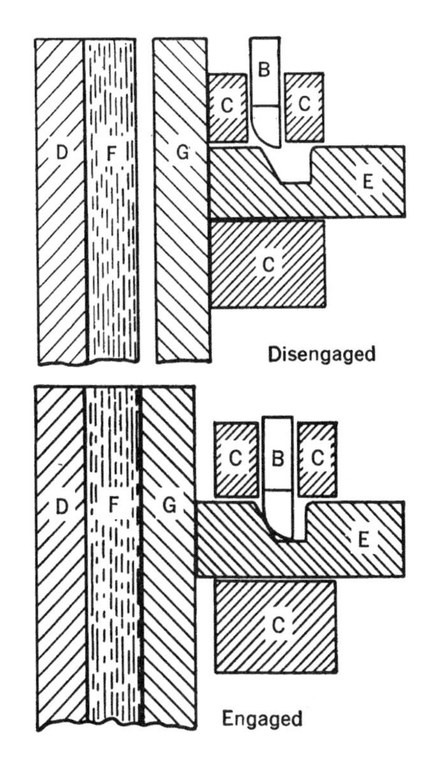

When a centrifugal force overcomes the resistance of the spring force, the lever action forces the plates together.

SEVEN OVERRUNNING CLUTCHES

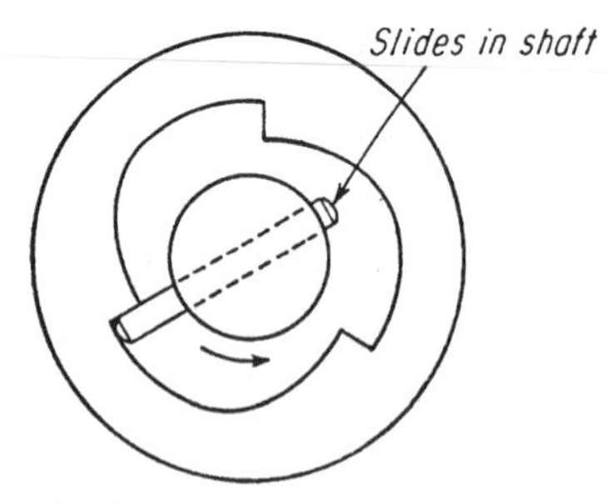

Fig. 1 A lawnmower clutch.

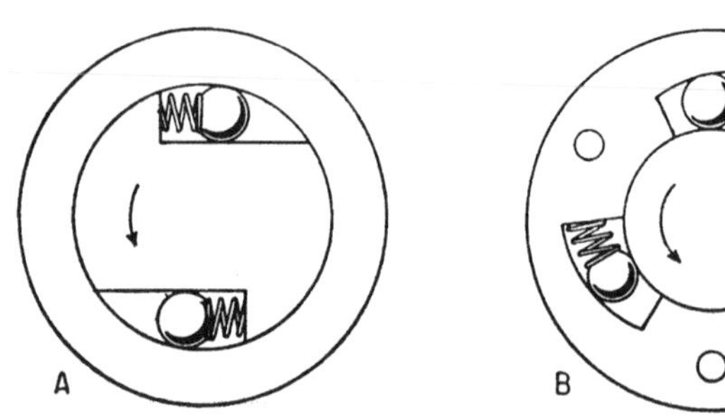

Fig. 2 Wedging balls or rollers: internal (A); external (B) clutches.

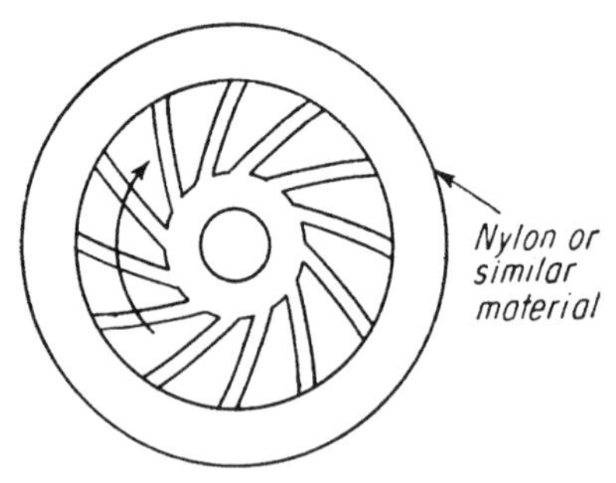

Fig. 3 Molded sprags (for light duty).

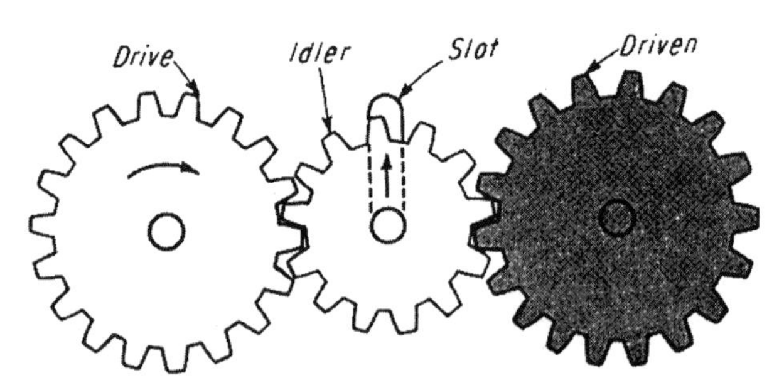

Fig. 4 A disengaging idler rises in a slot when the drive direction is reversed.

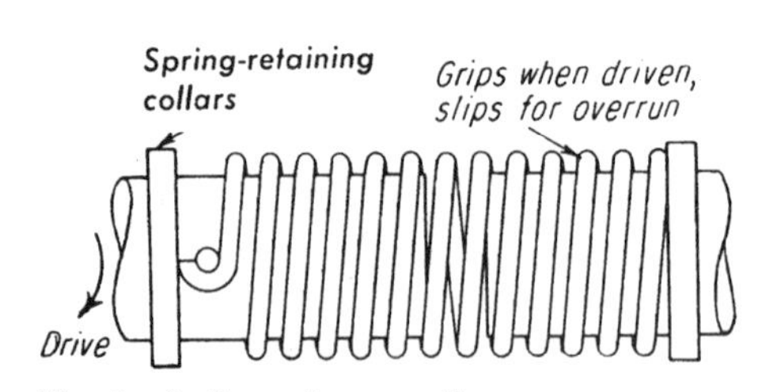

Fig. 5 A slip-spring coupling.

Fig. 6 An internal ratchet and spring-loaded pawls.

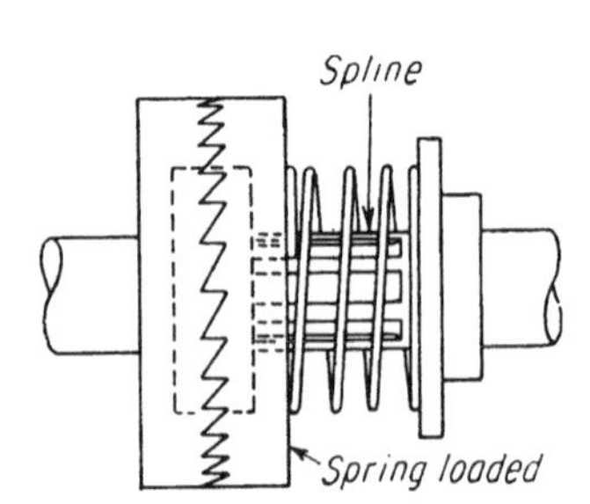

Fig. 7 A one-way dog clutch.

ONE-WAY CLUTCH HAS SPRING-LOADED PINS AND SPRAGS

Sprags combined with cylindrical rollers in a bearing assembly can provide a simple, low-cost method for meeting the torque and bearing requirements of most machine applications. Designed and built by Est. Nicot of Paris, this unit gives one-direction-only torque transmission in an overrunning clutch. In addition, it also serves as a roller bearing.

The torque rating of the clutch depends on the number of sprags. A minimum of three, equally spaced around the circumference of the races, is generally necessary to get acceptable distribution of tangential forces on the races.

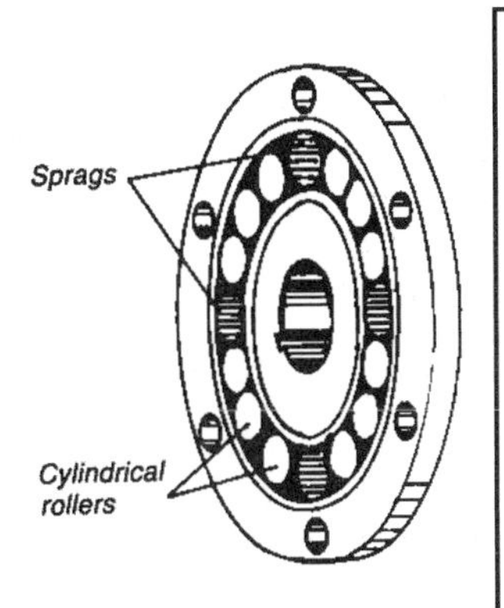

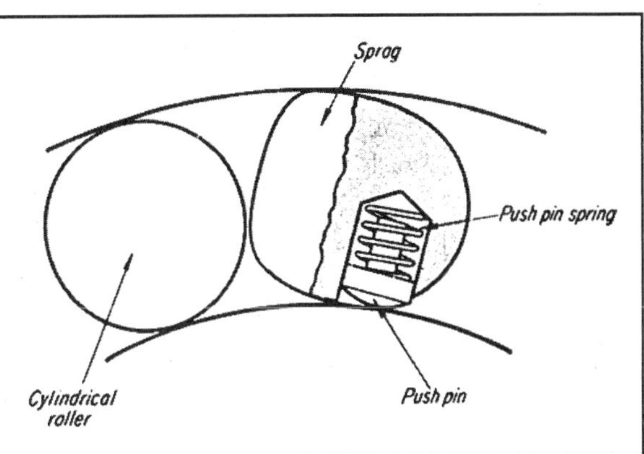

Races are concentric; a locking ramp is provided by the sprag profile, which is composed of two nonconcentric curves of different radius. A spring-loaded pin holds the sprag in the locked position until the torque is applied in the running direction. A stock roller bearing cannot be converted because the hard-steel races of the bearing are too brittle to handle the locking impact of the sprag. The sprags and rollers can be mixed to give any desired torque value.

ROLLER CLUTCH PROVIDES TWO OUTPUT SPEEDS

This clutch can be adapted for either electrical or mechanical actuation, and will control $^1/_2$ hp at 1500 rpm with only 7 W of power in the solenoid. The rollers are positioned by a cage (integral with the toothed control wheel —see diagram) between the ID of the driving housing and the cammed hub (integral with the output gear).

When the pawl is disengaged, the drag of the housing on the friction spring rotates the cage and wedges the rollers into engagement. This permits the housing to drive the gear through the cam.

When the pawl engages the control wheel while the housing is rotating, the friction spring slips inside the housing and the rollers are kicked back, out of engagement. Power is therefore interrupted.

According to the manufacturer, Tiltman Langley Ltd, Surrey, England, the unit operated over the full temperature range of –40° to 200°F.

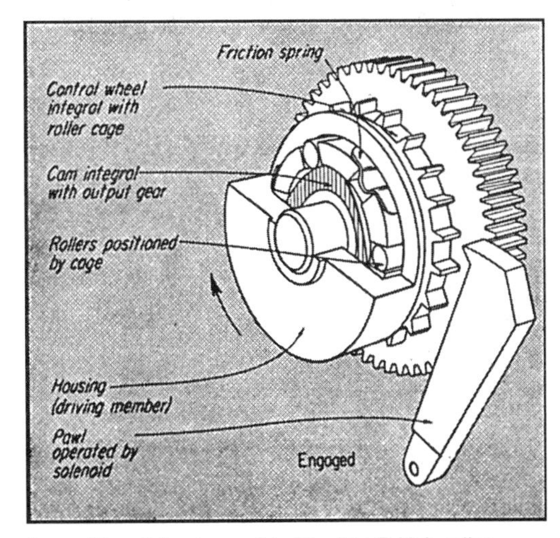

A positive drive is provided by this British roller clutch.

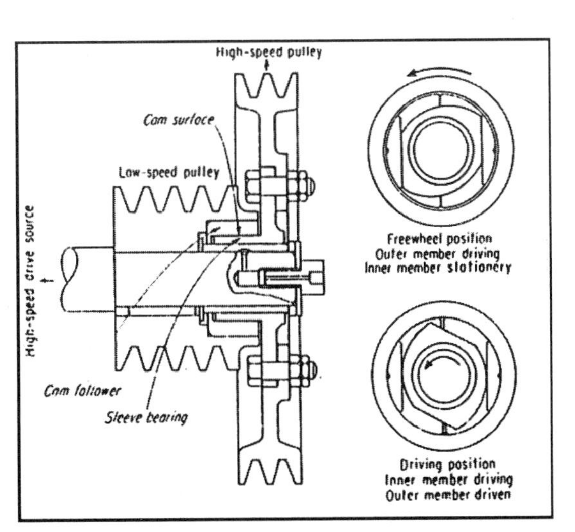

Two-speed operation is provided by the new cam clutch

This clutch consists of two rotary members (see diagrams), arranged so that the outer (follower) member acts on its pulley only when the inner member is driving. When the outer member is driving, the inner member idles. One application was in a dry-cleaning machine. The clutch functions as an intermediary between an ordinary and a high-speed motor to provide two output speeds that are used alternately.

SEVEN OVERRIDING CLUTCHES

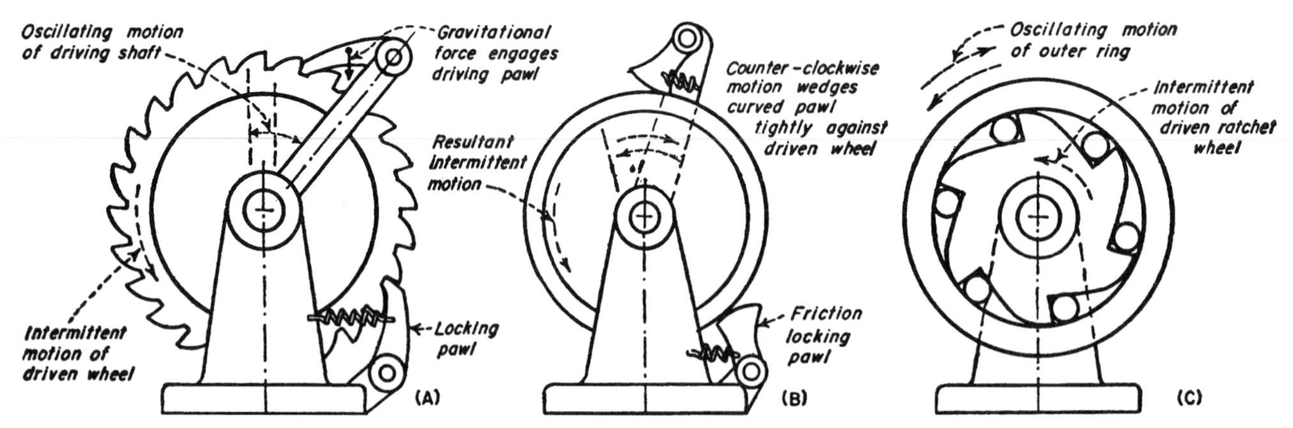

Fig. 1 Elementary overriding clutches: (A) A ratchet and pawl mechanism converts reciprocating or oscillating movement to intermittent rotary motion. This motion is positive but limited to a multiple of the tooth pitch. (B) A friction-type clutch is quieter, but it requires a spring device to keep the eccentric pawl in constant engagement. (C) Balls or rollers replace the pawls in this device. Motion of the outer race wedges the rollers against the inclined surfaces of the ratchet wheel.

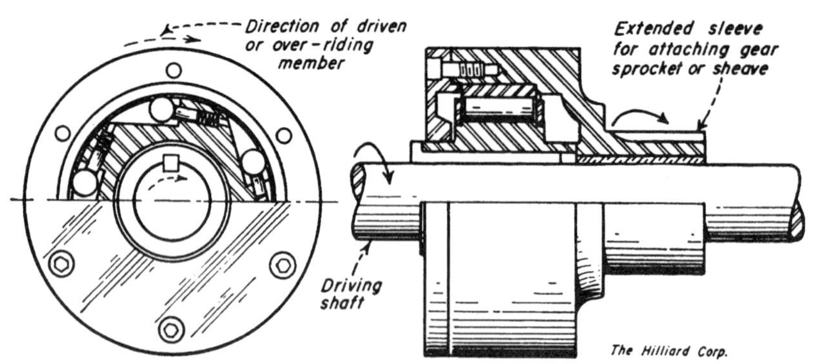

Fig. 2 A commercial overriding clutch has springs that hold the rollers in continuous contact between the cam surfaces and the outer race; thus, there is no backlash or lost motion. This simple design is positive and quiet. For operation in the opposite direction, the roller mechanism can easily be reversed in the housing.

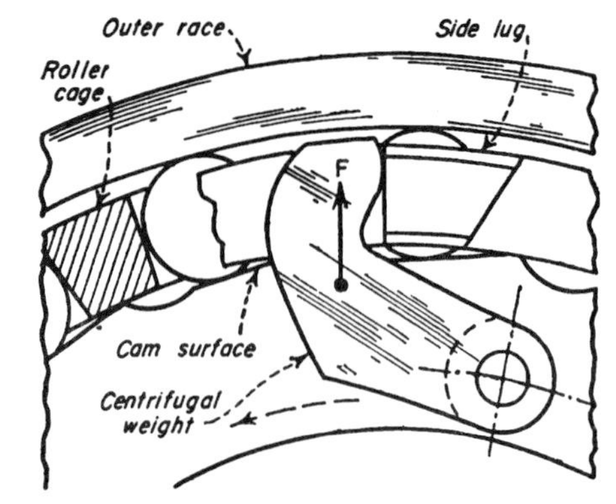

Fig. 3 A centrifugal force can hold the rollers in contact with the cam and outer race. A force is exerted on the lugs of the cage that controls the position of the rollers.

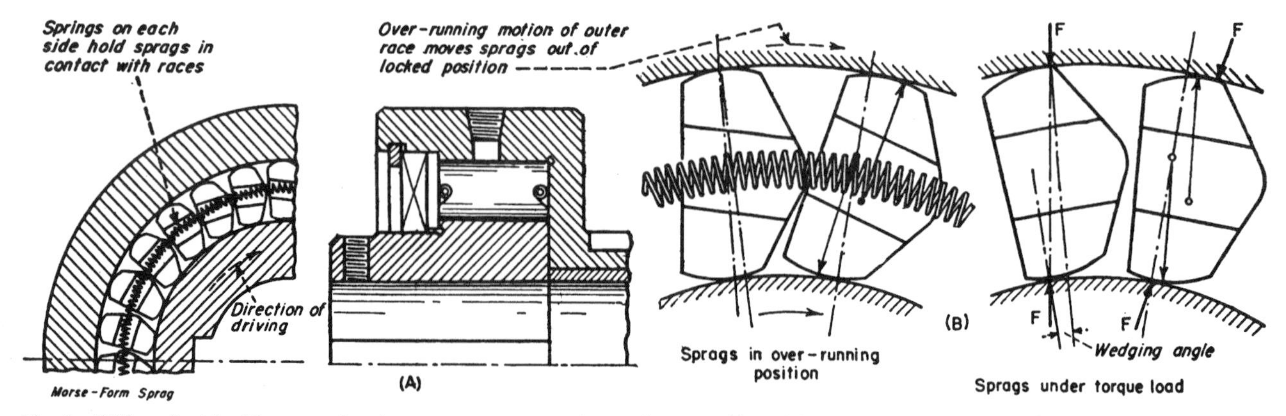

Fig. 4 With cylindrical inner and outer races, sprags can transmit torque. Energizing springs serve as a cage to hold the sprags. (A) Compared to rollers, the shape of a sprag permits a greater number within a limited space; thus higher torque loads are possible. Special cam surfaces are not required, so this version can be installed inside gear or wheel hubs. (B) Rolling action wedges the sprags tightly between the driving and driven members. A relatively large wedging angle ensures positive engagement.

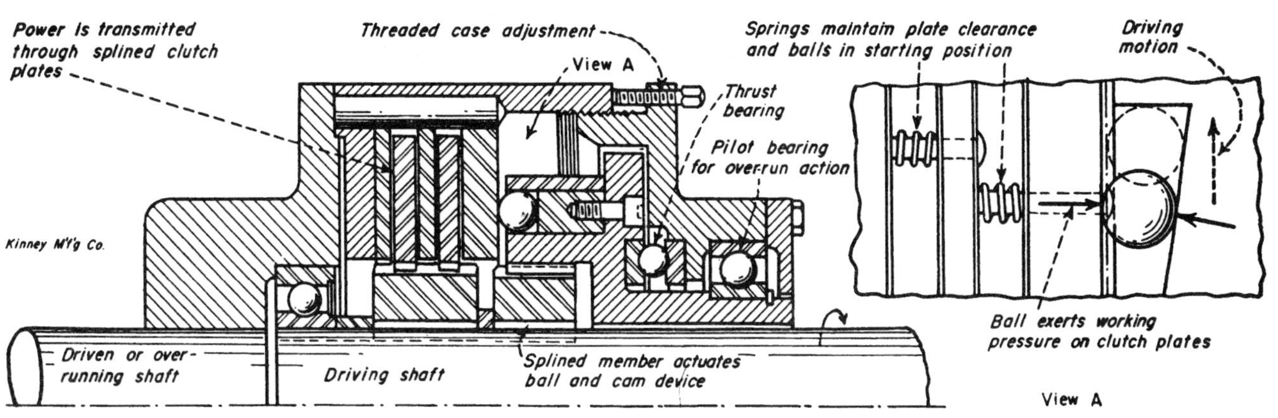

Fig. 5 A multidisk clutch is driven by several sintered-bronze friction surfaces. Pressure is exerted by a cam-actuating device that forces a series of balls against a disk plate. A small part of the transmitted torque is carried by the actuating member, so capacity is not limited by the localized deformation of the contacting balls. The slip of the friction surfaces determines the capacity and prevents rapid shock loads. The slight pressure of disk springs ensures uniform engagement.

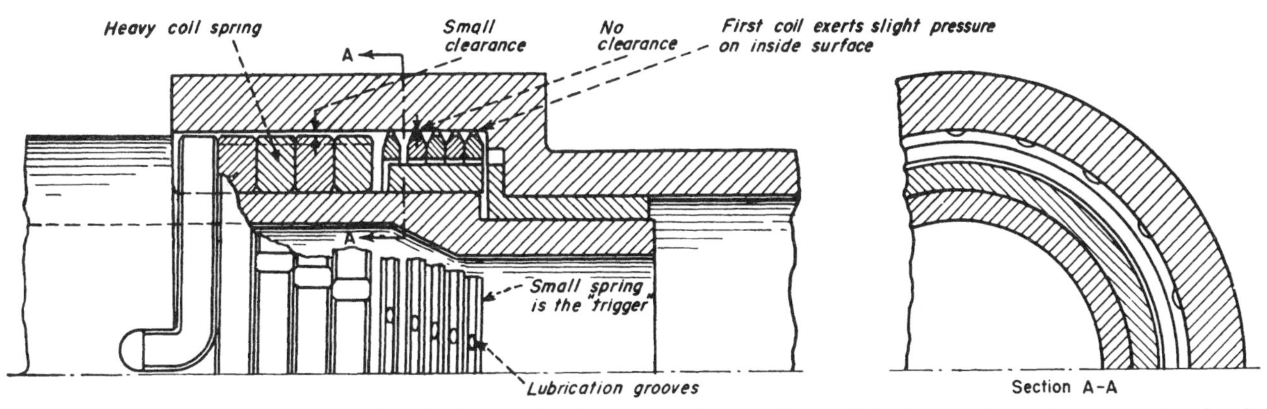

Fig. 6 An engaging device consists of a helical spring that is made up of two sections: a light trigger spring and a heavy coil spring. It is attached to and driven by the inner shaft. The relative motion of the outer member rubbing on the trigger causes this spring to wind up. This action expands the spring diameter, which takes up the small clearance and exerts pressure against the inside surface until the entire spring is tightly engaged. The helix angle of the spring can be changed to reverse the overriding direction.

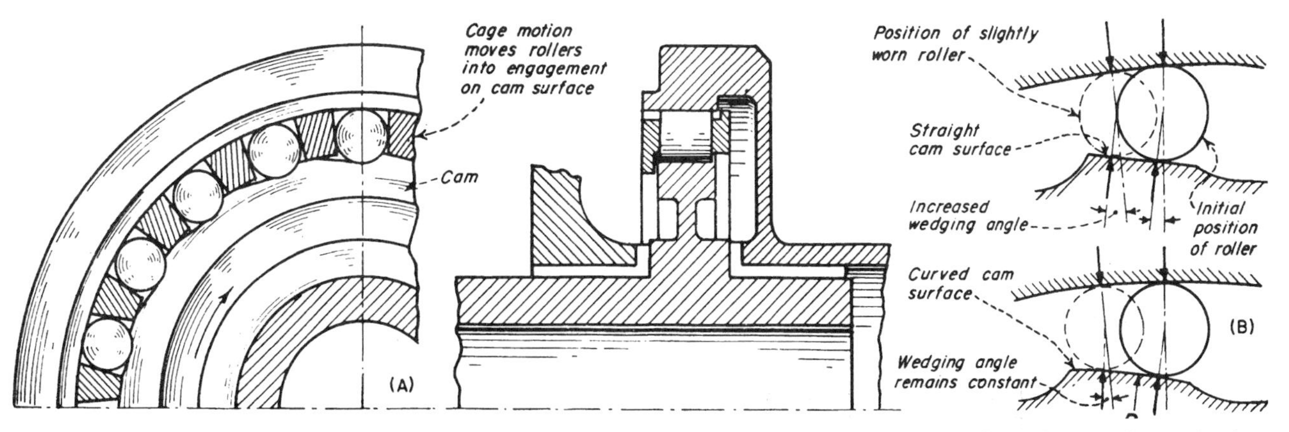

Fig. 7 A free-wheeling clutch widely used in power transmission has a series of straight-sided cam surfaces. An engaging angle of about 3° is used; smaller angles tend to become locked and are difficult to disengage while larger ones are not as effective. (A) The inertia of a floating cage wedges the rollers between the cam and outer race. (B) Continual operation causes the wear of surfaces; 0.001 in. wear alters the angle to 8.5° on straight-sided cams. Curved cam surfaces maintain a constant angle.

TEN APPLICATIONS FOR OVERRUNNING CLUTCHES

These clutches allow freewheeling, indexing, and backstopping; they will solve many design problems. Here are examples.

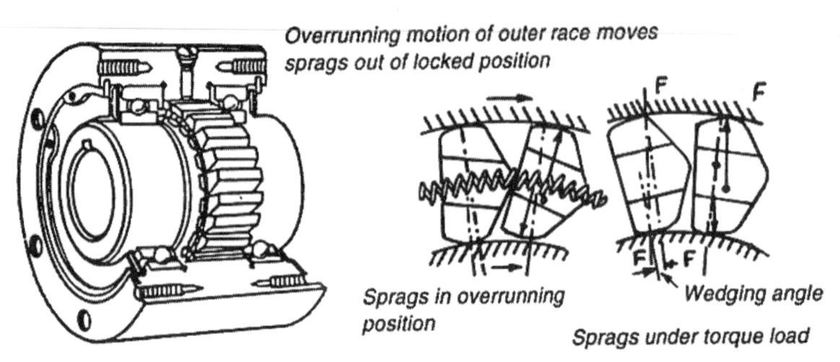

Fig. 1 Precision sprags act as wedges and are made of hardened alloy steel. In the formsprag clutch, torque is transmitted from one race to another by the wedging action of sprags between the races in one direction; in the other direction the clutch freewheels.

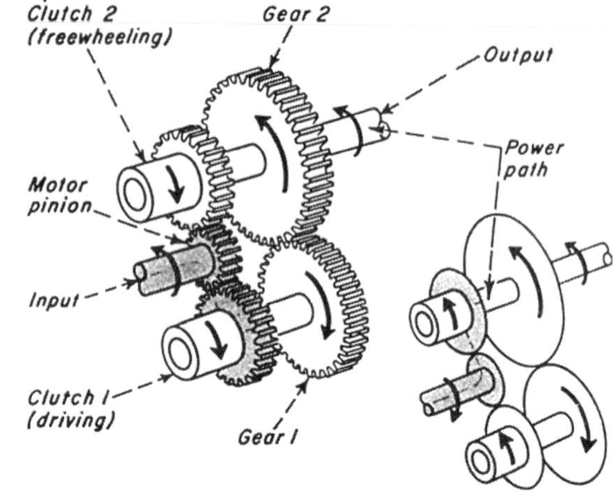

Fig. 2 This speed drive requires input rotation to be reversible. Counterclockwise input (as shown in the diagram) drives gear 1 through clutch 1; the output is counterclockwise; clutch 2 overruns. Clockwise input (schematic) drives gear 2 through clutch 2; the output is still counterclockwise; clutch 1 overruns.

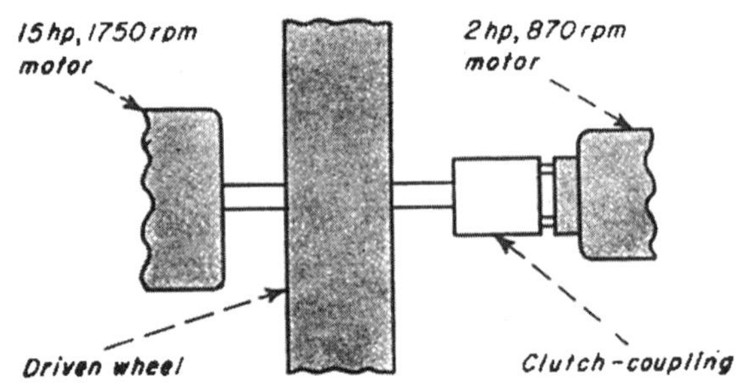

Fig. 3 This speed drive for a grinding wheel can be a simple, in-line assembly if the overrunning clutch couples two motors. The outer race of the clutch is driven by a gearmotor; the inner race is keyed to a grinding-wheel shaft. When the gearmotor drives, the clutch is engaged; when the larger motor drives, the inner race overruns.

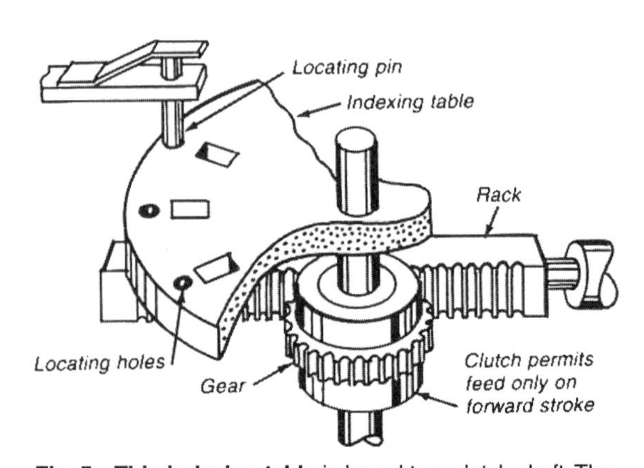

Fig. 5 This indexing table is keyed to a clutch shaft. The table is rotated by the forward stroke of the rack; power is transmitted through the clutch by its outer-ring gear only during this forward stroke. Indexing is slightly short of the position required. The exact position is then located by a spring-loaded pin that draws the table forward to its final positioning. The pin now holds the table until the next power stroke of the hydraulic cylinder.

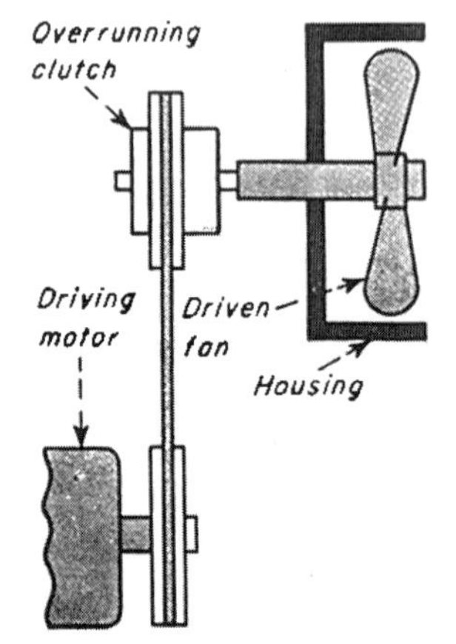

Fig. 4 This fan freewheels when driving power is shut off. Without an overrunning clutch, fan momentum can cause belt breakage. If the driving source is a gearmotor, excessive gear stress can also occur by feedback of kinetic energy from the fan.

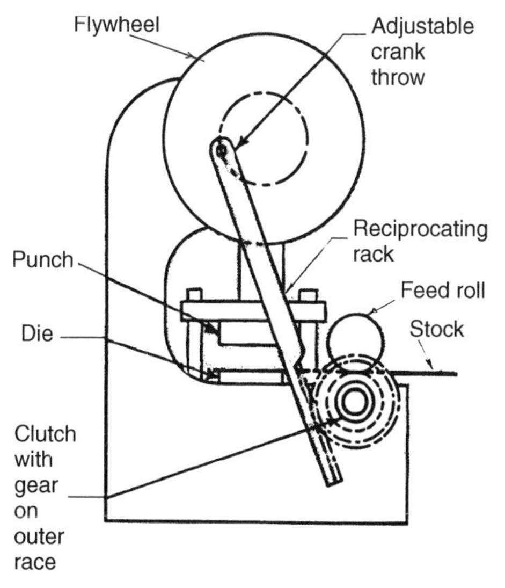

Fig. 6 This punch press feed is arranged so that the strip is stationary on the downstroke of the punch (clutch freewheels); feed occurs during the upstroke when the clutch transmits torque. The feed mechanism can easily be adjusted to vary the feed amount.

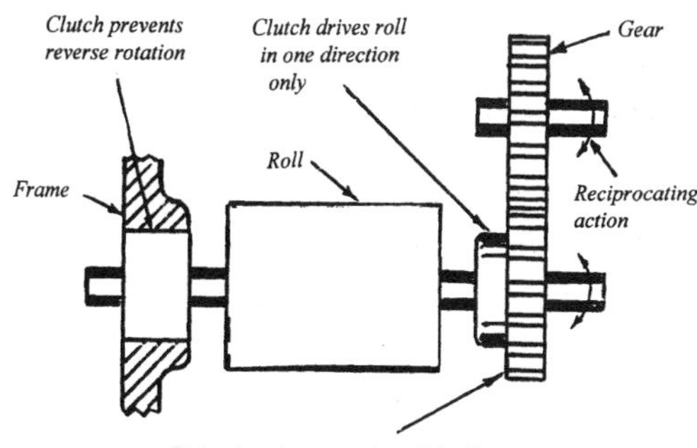

Fig. 7 Indexing and backstopping is done with two clutches arranged so that one drives while the other freewheels. The application shown here is for a capsuling machine; gelatin is fed by the roll and stopped intermittently so the blade can precisely shear the material to form capsules.

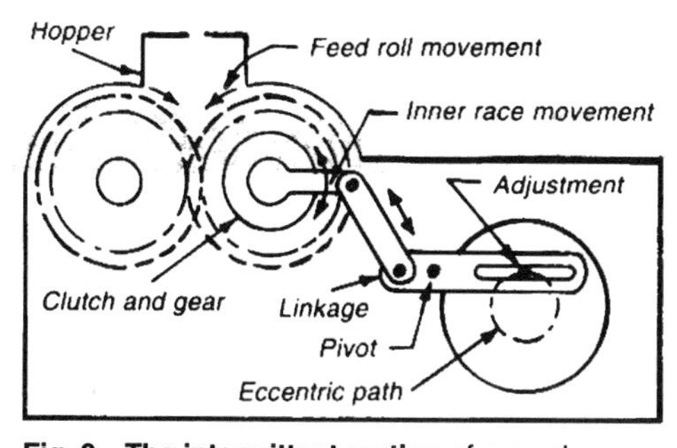

Fig. 8 The intermittent motion of a candy machine is adjustable. The clutch ratchets the feed rolls around. This keeps the material in the hopper agitated.

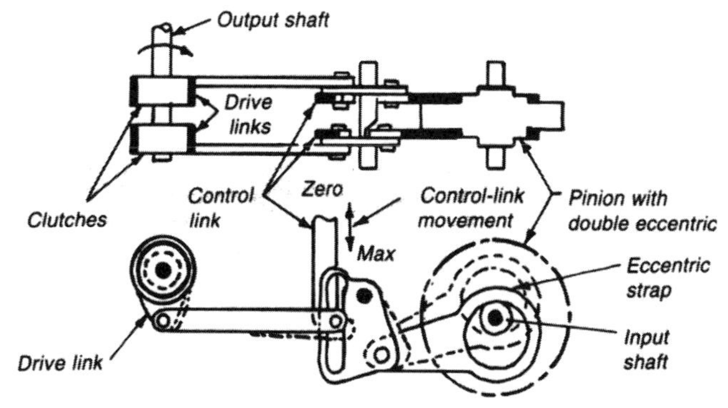

Fig. 9 This double-impulse drive has double eccentrics and drive clutches. Each clutch is indexed 180° out of phase with the other. One revolution of the eccentric produces two drive strokes. Stroke length, and thus the output rotation, can be adjusted from zero to maximum by the control link.

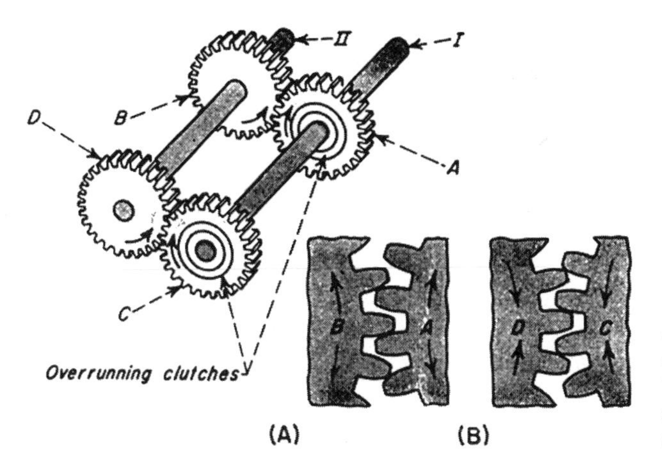

Fig. 10 This anti-backlash device depends on overrunning clutches to ensure that no backlash is left in the unit. Gear *A* drives *B* and shaft *II* with the gear mesh and backlash, as shown in (A). The overrunning clutch in gear *C* permits gear *D* (driven by shaft *II*) to drive gear *C* and results in the mesh and backlash shown in (B). The overrunning clutches never actually overrun. They provide flexible connections (something like split and sprung gears) between shaft *I* and gears *A* and *C* to allow absorption of all backlash.

EIGHT SPRAG CLUTCH APPLICATIONS

Overrunning sprag clutches transmit torque in one direction and reduce speed, rest, hold, or free-wheel in the reverse direction. Applications include overrunning, backstopping, and indexing. Their selection—similar to other mechanical devices—requires a review of the torque to be transmitted, overrunning speed, type of lubrication, mounting characteristics, environmental conditions, and shock conditions that might be encountered.

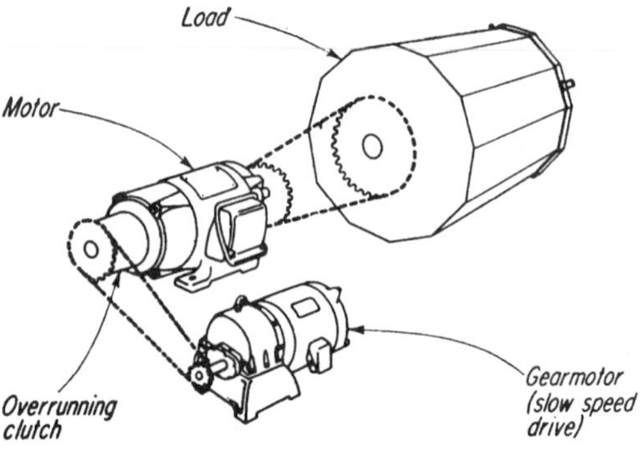

Fig. 1 Overrunning permits torque transmission in one direction and free wheels or overruns in the opposite direction. For example, the gar motor drives the load by transmitting torque through the overrunning clutch and the high-speed shaft. Energizing the high-speed motor causes the inner member to rotate at the rpm of the high-speed motor. The gear motor continues to drive the inner member, but the clutch is freewheeling.

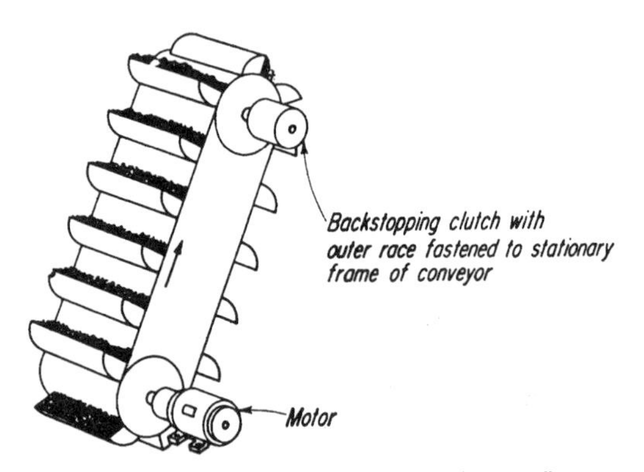

Fig. 2 Backstopping permits rotation in one direction only. The clutch serves as a counter-rotation holding device. An example is a clutch mounted on the headshaft of a conveyor. The outer race is restrained by torque-arming the stationary frame of the conveyor. If, for any reason, power to the conveyor is interrupted, the back-stopping clutch will prevent the buckets from running backwards and dumping the load.

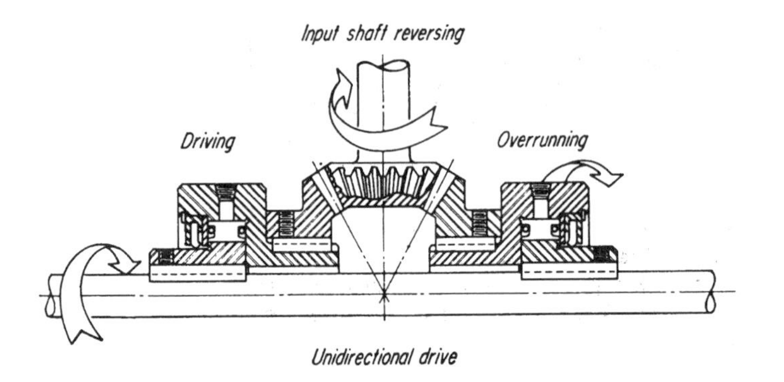

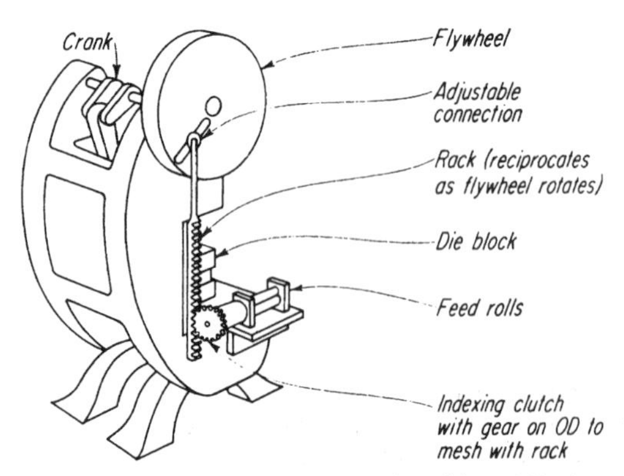

Fig. 3 Indexing is the transmission of intermittent rotary motion in one direction; an example is the feed rolls of a punch press. On each stroke of the press crankshaft, a feed stroke on the feed roll is accomplished by the rack-and-pinion system. The system feeds the material into the dies of the punch press.

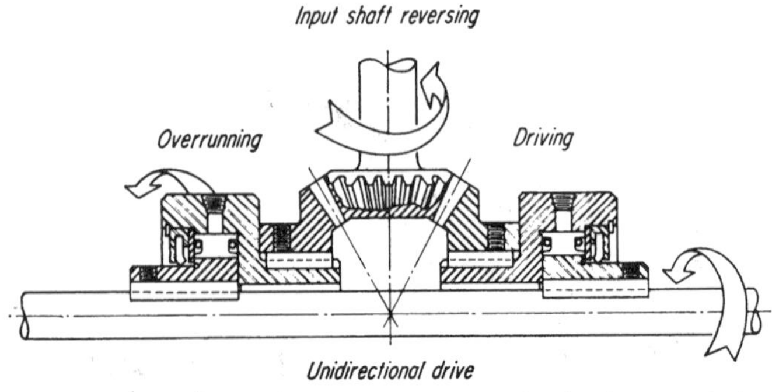

Fig. 4 Unidirectional drives with reverse mechanism incorporate two overrunning clutches into the gears, sheaves, or sprockets. Here, a 1:1 ratio right-angle drive is shown with a reversing input shaft. The output shaft rotates clockwise, regardless of the input shaft direction. By changing gear sizes, combinations of continuous or intermittent unidirectional output relative to the input can be obtained.

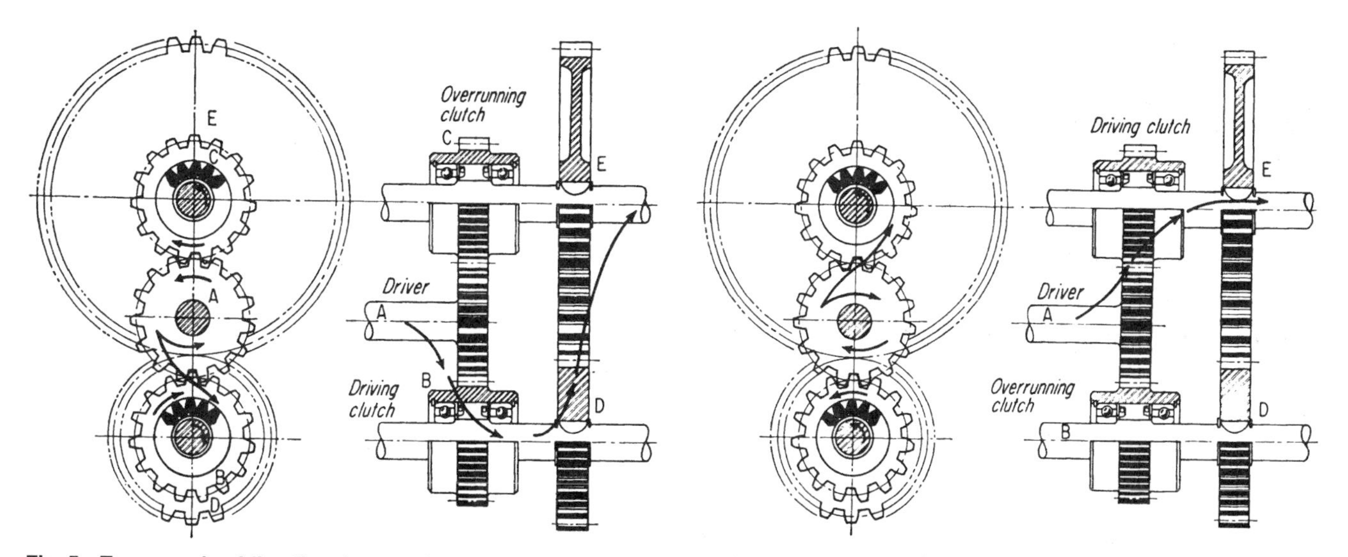

Fig. 5 Two-speed unidirectional output is made possible by using spur gears and reversing the direction of the input shaft. The rotation of shaft A transfers the power of gears B, D, and E to the output. Counterclockwise rotation engages the lower clutch, freewheeling the upper clutch because gear C is traveling at a faster rate than the shaft. This is caused by the reduction between gears B and E. Clockwise rotation of A engages the upper clutch, while the lower clutch freewheels because of the speed increase between gears D and E.

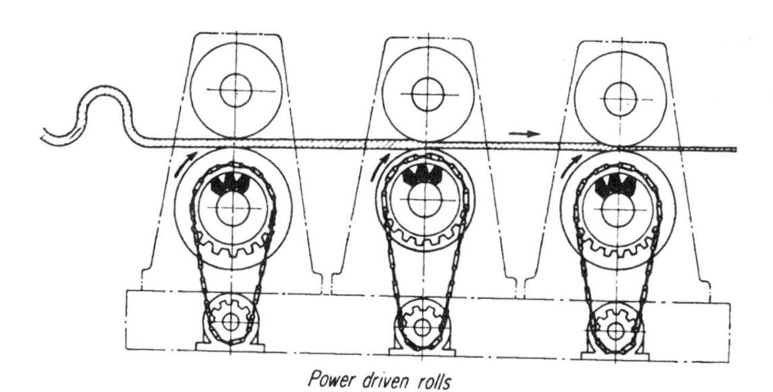

Fig. 6 A speed-differential or compensation is required where a different speed range for a function is desired, while retaining the same basic speed for all other functions. A series of individually driven power rolls can have different surface speeds because of drive or diameter variations of the rolls. An overrunning clutch permits the rolls with slower peripheral speed to overspeed and adjust to the material speed.

Fig. 7 A speed differential application permits the operation of engine accessories within a narrow speed range while the engine operates over a wide range. Pulley No. 2 contains the overrunning clutch. When the friction or electric clutch is disengaged, the driver pulley drives pulley No. 2 through the overrunning clutch, rotating the driven shaft. The engagement of the friction or electric clutch causes high-speed driven shaft rotation. This causes an overrun condition in the clutch at pulley No. 2.

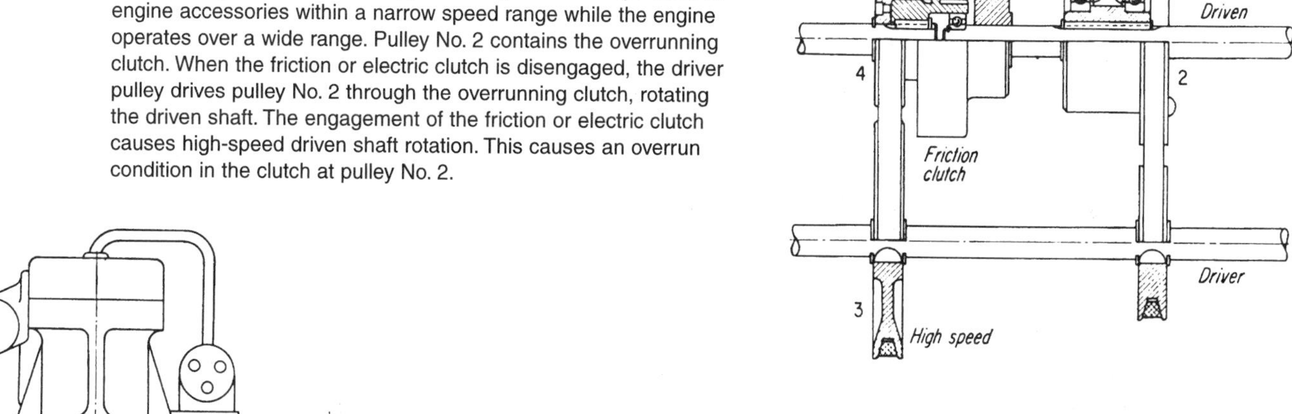

Engine

Generator

Fig. 8 High inertia dissipation avoids driving back through a power system. In machines with high resistances, it prevents power train damage. If the engine is shut down while the generator is under a no-load condition, it would have a tendency to twist off the generator shaft. The overrunning clutch allows generator deceleration at a slower rate than the engine deceleration.

SIX SMALL CLUTCHES PERFORM PRECISE TASKS

Clutches for small machines must have: (1) Quick response—lightweight moving parts; (2) Flexibility—permit multiple members to control operation; (3) Compactness—for equivalent capacity positive clutches are smaller than friction; (4) Dependability; and (5) Durability.

Fig. 1 A pawl and ratchet, single-cycle Dennis clutch. The primary parts of this clutch are the driving ratchet B, the driven cam plate C, and the connecting pawl D, which is carried by the cam plate. The pawl is normally held disengaged by the lower tooth of clutch arm A. When activated, arm A rocks counterclockwise until it is out of the path of rim F on cam plate C.

This permits pawl D, under the effect of spring E, to engage with ratchet B. Cam plate C then turns clockwise until, near the end of one cycle, pin G on the plate strikes the upper part of arm A, camming it clockwise back to its normal position. The lower part of A then performs two functions: (1) it cams pawl D out of engagement with the driving ratchet B, and (2) it blocks the further motion of rim F and the cam plate.

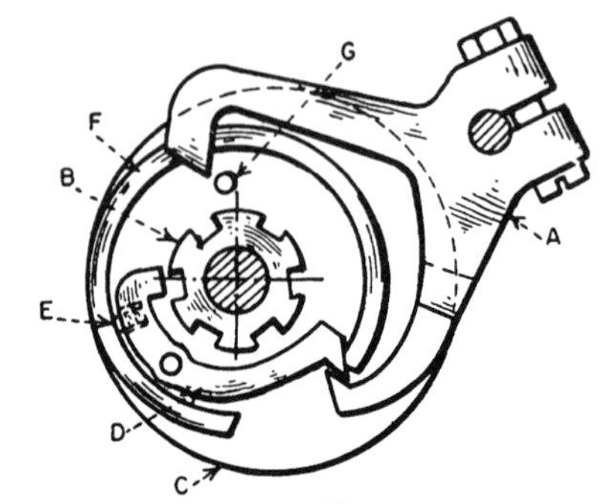

Fig. 1

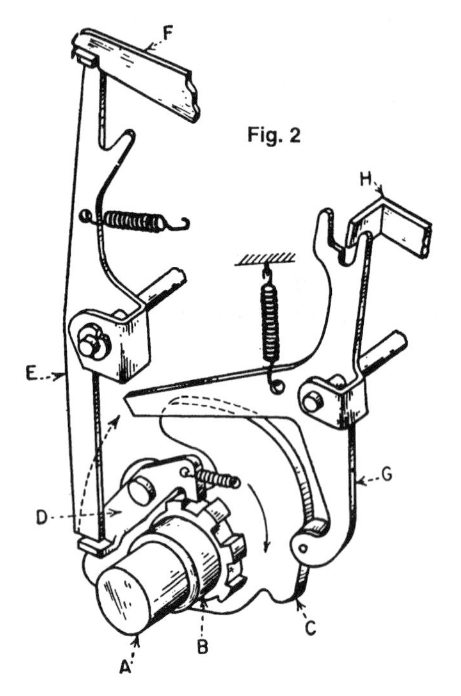

Fig. 2 A pawl and ratchet, single-cycle, dual-control clutch. The principal parts of this clutch are driving ratchet B, driven crank C, and spring-loaded ratchet pawl D. Driving ratchet B is directly connected to the motor and free to rotate on rod A. Driven crank C is directly connected to the main shaft of the machine and is also free to move on rod A. Spring-loaded ratchet pawl D is carried by crank C, which is normally held disengaged by latch E.

To activate the clutch, arm F is raised, permitting latch E to trip and pawl D to engage with ratchet B. The left arm of clutch latch G, which is in the path of the lug on pawl D, is normally permitted to move out of the way by the rotation of the camming edge of crank C. For certain operations, block H is temporarily lowered. This prevents the motion of latch G, resulting in the disengagement of the clutch after part of the cycle. It remains disengaged until the subsequent raising of block H permits the motion of latch G and the resumption of the cycle.

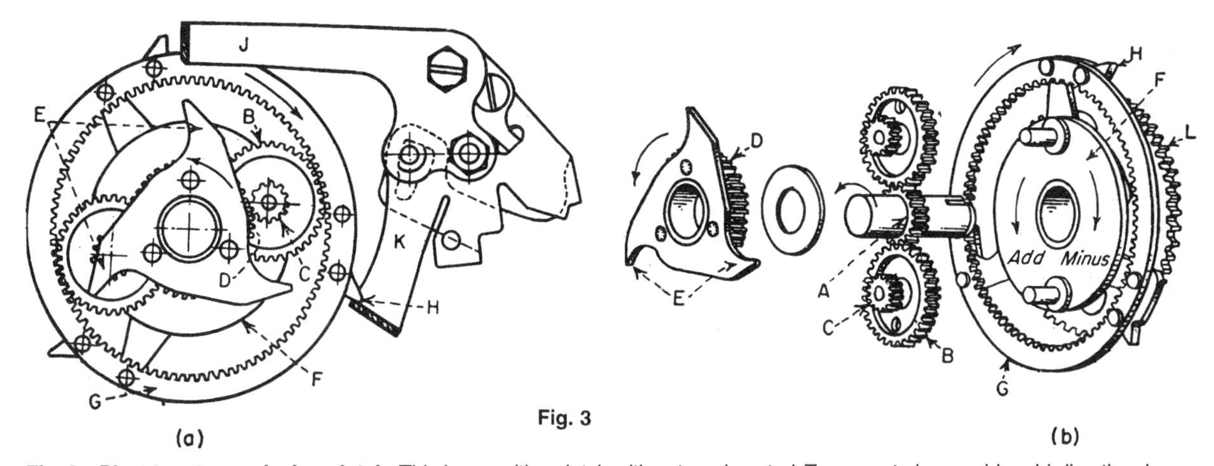

Fig. 3 Planetary transmission clutch. This is a positive clutch with external control. Two gear trains provide a bi-directional drive to a calculator for cycling the machine and shifting the carriage. Gear A is the driver; gear L, the driven member, is directly connected to the planet carrier F. The planet consists of integral gears B and C. Gear B meshes with free-wheeling gear D. Gears D and G carry projecting lugs E and H, respectively. Those lugs can contact formings on arms J and K of the control yoke.

When the machine is at rest, the yoke is centrally positioned so that arms J and K are out of the path of the projecting lugs, permitting both D and G to free-wheel. To engage the drive, the yoke rocks clockwise, as shown, until the forming on arm K engages lug H, blocking further motion of ring gear G. A solid gear train is thereby established, driving F and L in the same direction as the drive A. At the same time, the gear train alters the speed of D as it continues counterclockwise. A reversing signal rotates the yoke counterclockwise until arm J encounters lug E, blocking further motion of D. This actuates the other gear train with the same ratio.

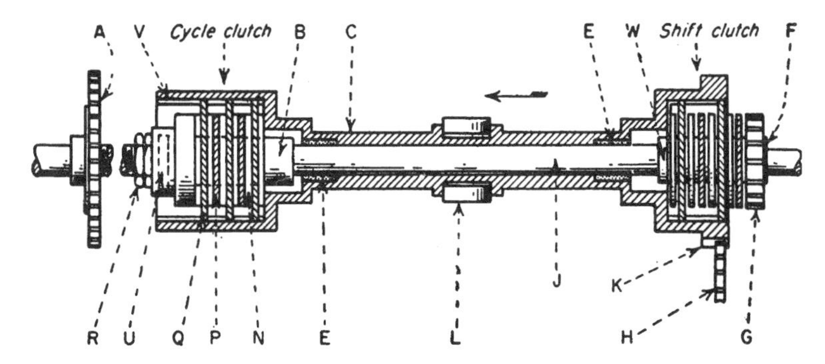

Fig. 4

Fig. 4 A multiple-disk friction clutch. Two multiple-disk friction clutches are combined in a single, two-position unit that is shown shifted to the left. A stepped cylindrical housing, C, encloses both clutches. Internal self-lubricated bearings support the housing on coaxial shaft J that is driven by transmission gear H, meshing with housing gear teeth K. At the other end, the housing carries multiple metal disks Q that engage keyways V and can make frictional contact with phenolic laminate disks N. They, in turn, can contact a set of metal disks P that have slotted openings for couplings with flats located on sleeves B and W.

In the position shown, pressure is exerted through rollers L, forcing the housing to the left, making the left clutch compress against adjusting nuts R. Those nuts drive gear A through sleeve B, which is connected to jack shaft J by pin U. When the carriage is to be shifted, rollers L force the housing to the right. However, it first relieves the pressure between the adjoining disks on the left clutch. Then they pass through a neutral position in which both clutches are disengaged, and they finally compress the right clutch against thrust bearing F. That action drives gear G through sleeve W, which rotates freely on the jack shaft.

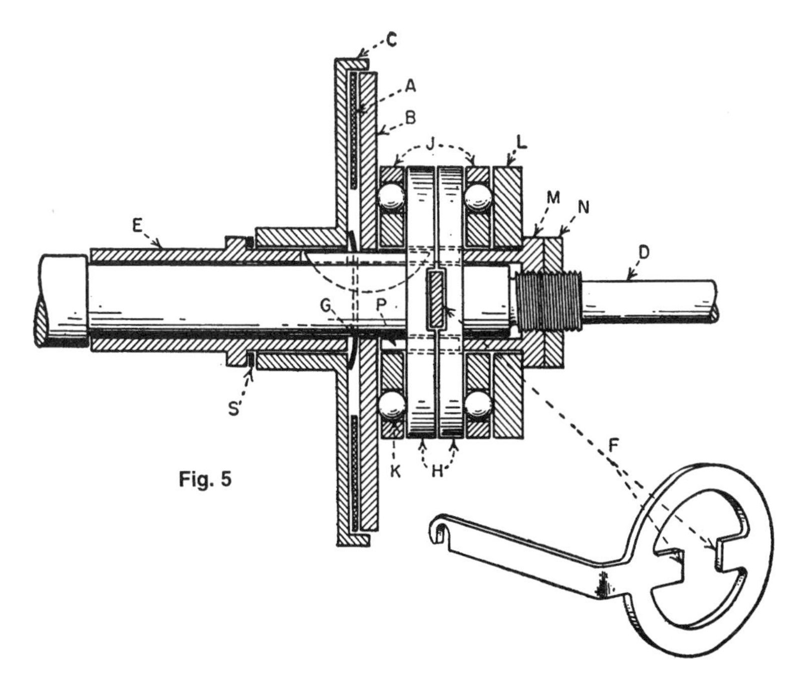

Fig. 5

Fig. 5 A single-plate friction clutch. The basic parts of this clutch are the phenolic laminate clutch disk A, steel disk B, and drum C. They are normally kept separated by spring washer G. To engage the drive, the left end of a control arm is raised, causing ears F, which are located in slots in plates H, to rock clockwise. This action spreads the plates axially along sleeve P. Sleeves E and P and plate B are keyed to the drive shaft; all other members can rotate freely.

The axial motion loads the assembly to the right through the thrust ball bearings K against plate L and adjusting nut M. It also loads them to the left through friction surfaces on A, B, and C to thrust washer S, sleeve E, and against a shoulder on shaft D. This response then permits phemolic laminate disk A to drive drum C.

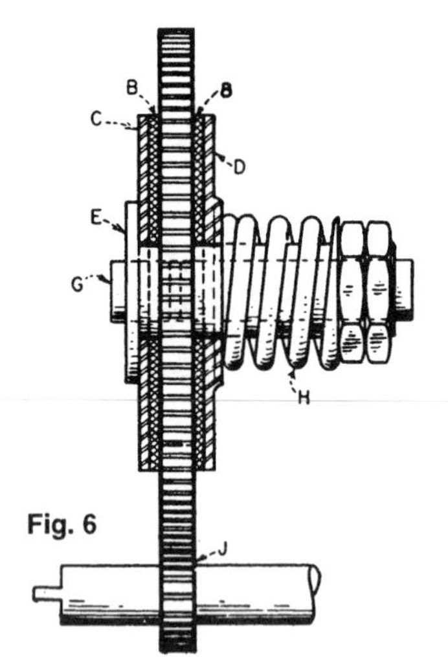

Fig. 6

Fig. 6 An overload relief clutch. This is a simple, double-plate, friction coupling with spring loading. Shaft G drives collar E, which drives slotted plates C and D faced with phenolic laminate disks B. Spring H is held in compression by the two adjusting nuts on the threaded end of collar E. These maintain the unit under axial pressure against the shoulder at the left end of the collar.

This enables the phenolic laminate disks B to drive through friction against both faces of the gar, which is free to turn o the collar. This motion of the gear causes output pinion J to rotate. If the machine to which the clutch is attached should jam and pinion J is prevented from turning, the motor can continue to run without overloading. However, slippage can occur between the phenolic laminate clutch plates B and the large gear.

TWELVE DIFFERENT STATION CLUTCHES

Innumerable variations of these station clutches can be designed for starting and stopping machines at selected points in their operation cycles.

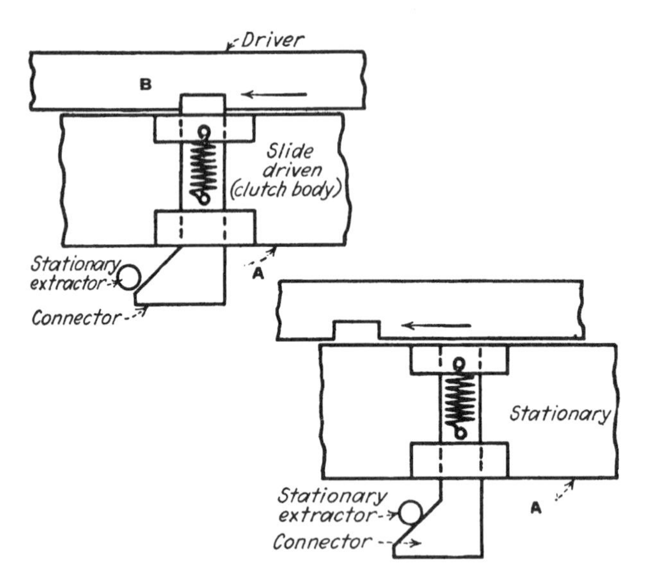

Fig. 1(*a*) The driver and clutch body move in unison with the connector just coming in contact with the extractor. In **Fig. 1(*b*)** continued movement withdraws the connector.

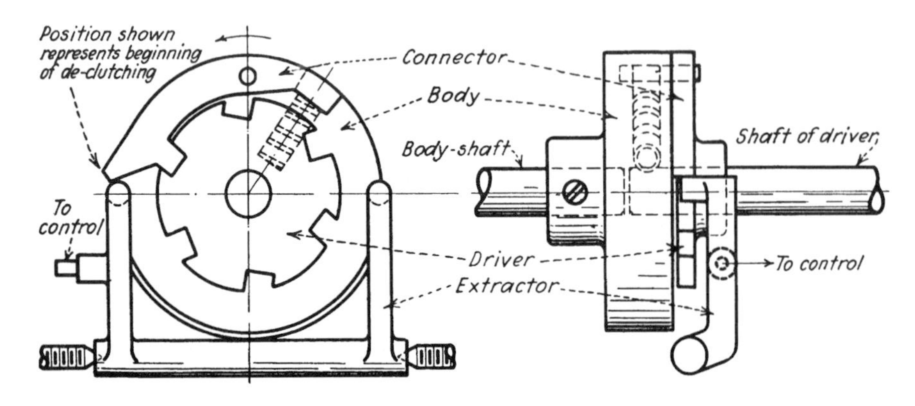

Fig. 2 A two-station clutch whose stations are 180° apart. Because it has only one extractor arm, this mechanism can function as a one-station clutch.

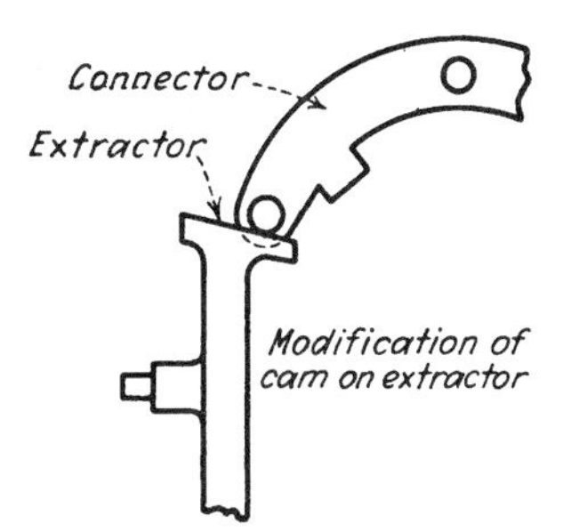

Fig. 3 A modification of the cam extractor shown in Fig. 2.

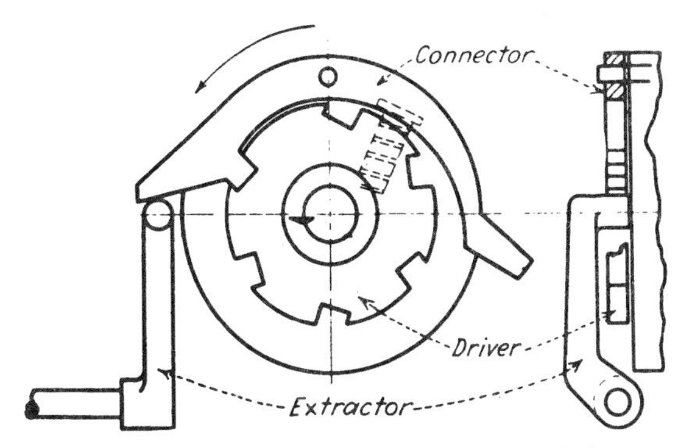

Fig. 4 A single extractor two-station clutch with the stations that are 180° apart. Only one extractor is required because the connector has two cams.

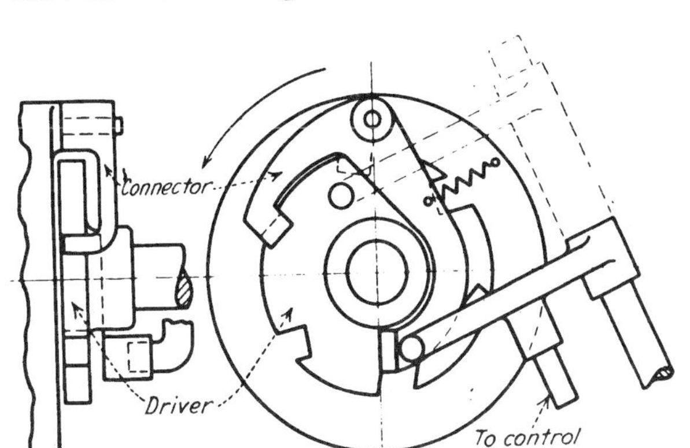

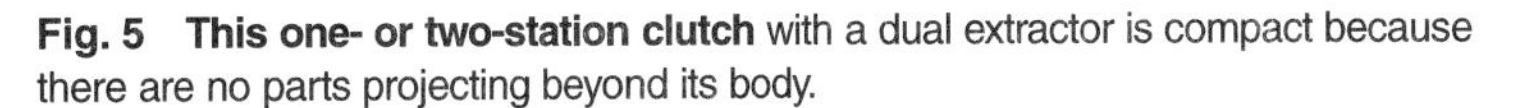

Fig. 5 This one- or two-station clutch with a dual extractor is compact because there are no parts projecting beyond its body.

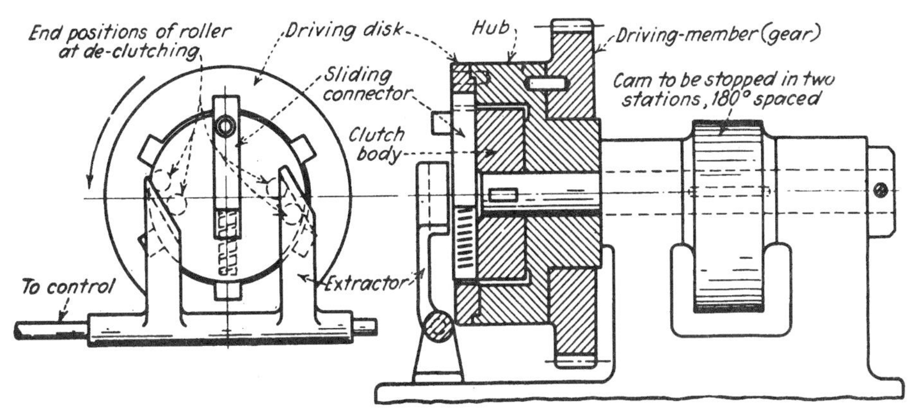

Fig. 6 The end and longitudinal section of a station clutch with internal driving recesses.

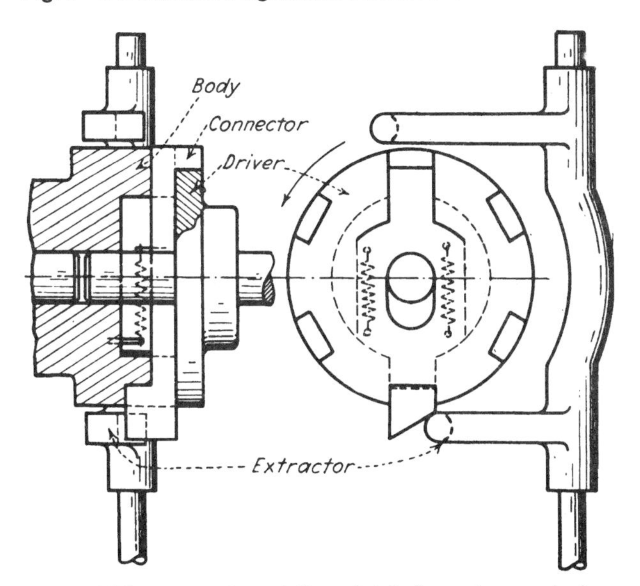

Fig. 7 This one- or two-station clutch depends on a single or a dual extractor. Its stations are spaced 180° apart.

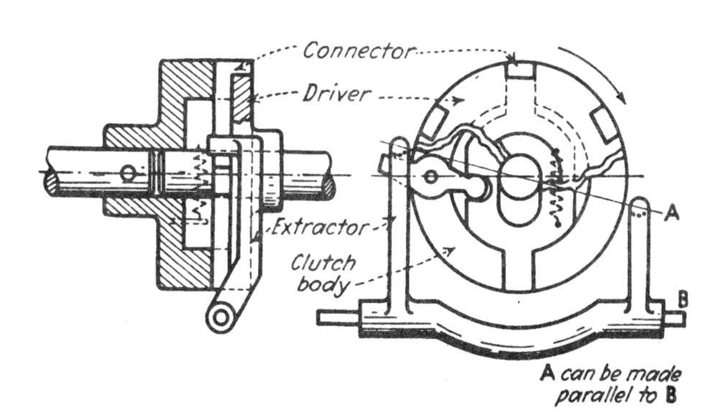

Fig. 8 This is another one- or two-station clutch. It has a single or dual extractor with stations spaced 180° apart.

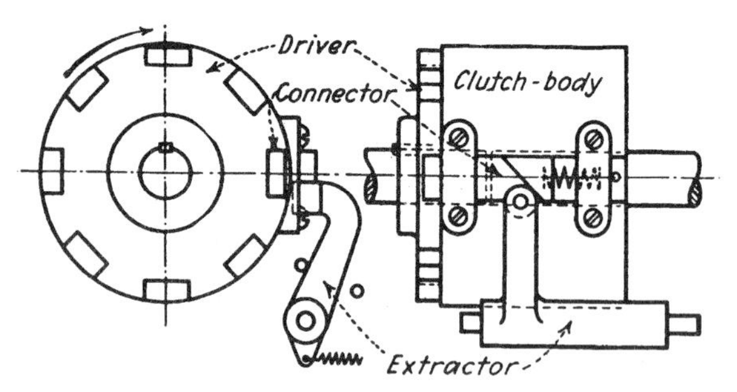

Fig. 9 A one-station axial connector clutch.

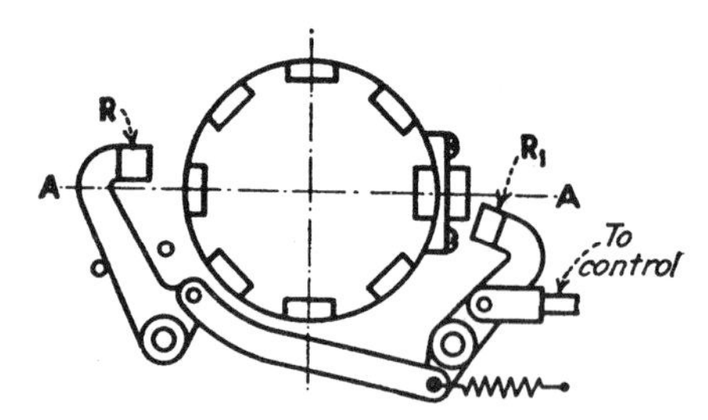

Fig. 10 A two-station clutch. The rollers R and R1 of the extractor can also be arranged on the center-line A-A.

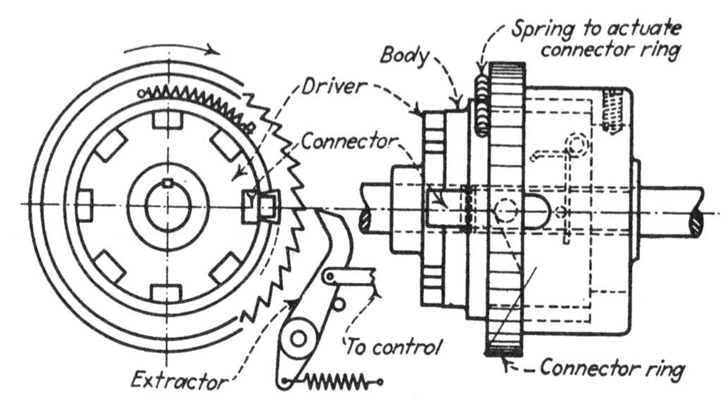

Fig. 11 A nonselective multistation clutch for instantaneous stopping in any position.

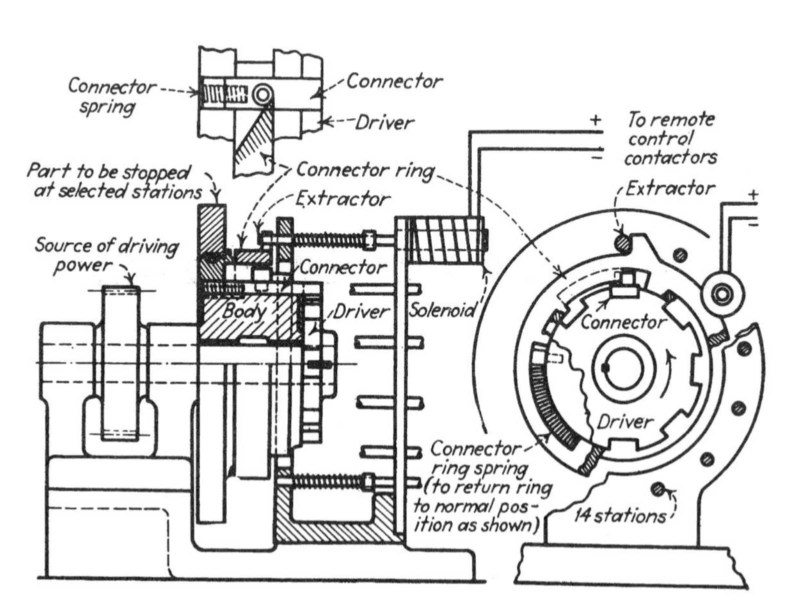

Fig. 12 A multistation clutch with remote control. The extractor pins are actuated by solenoids that either hold the extractor pin in position against spring pressure or release the pin.

TWELVE APPLICATIONS FOR ELECTROMAGNETIC CLUTCHES AND BRAKES

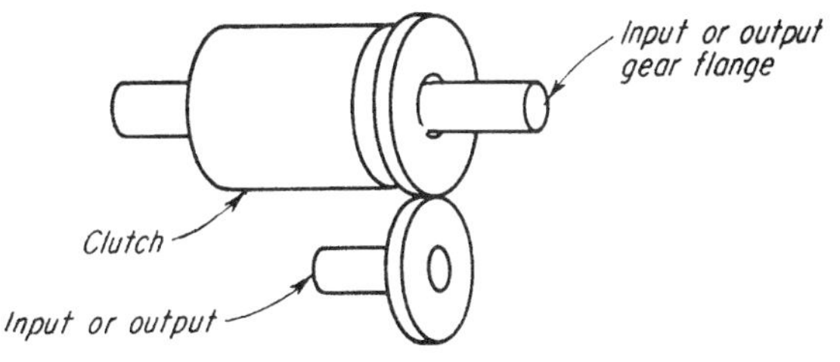

Fig. 1 Coupling or uncoupling power or sensing device.

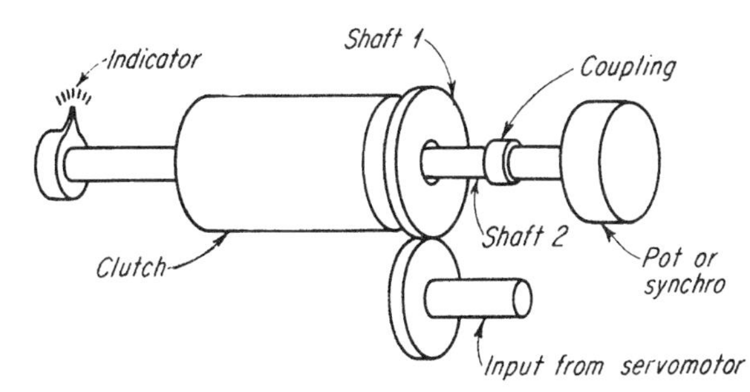

Fig. 2 Calibration protection (energize to adjust).

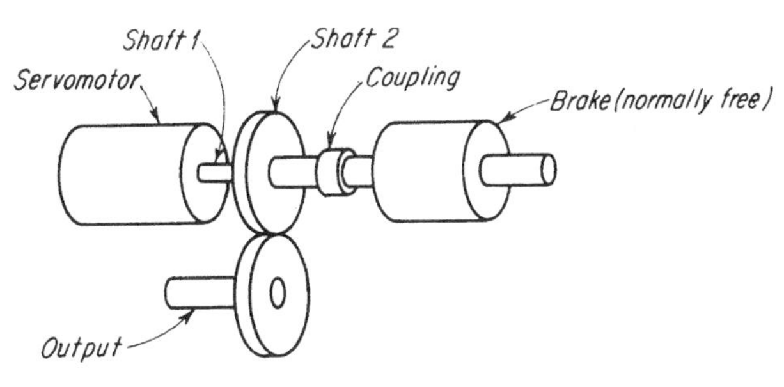

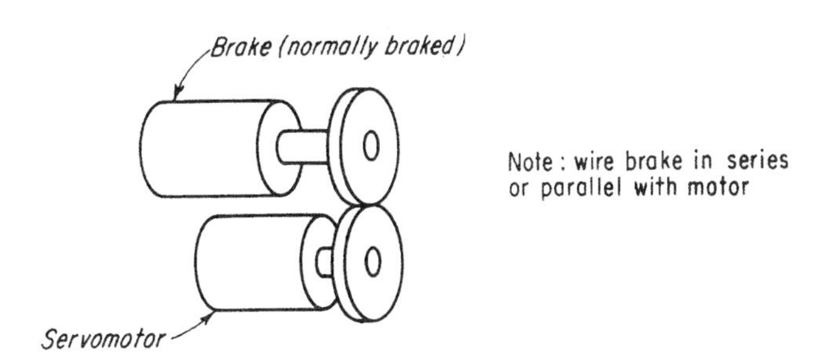

Figs. 3 & 4 Simple servomotor brakes.

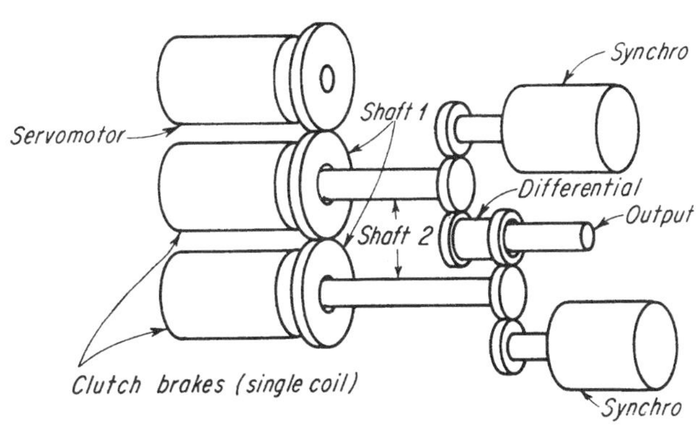

Fig. 5 Adding or subtracting two inputs.

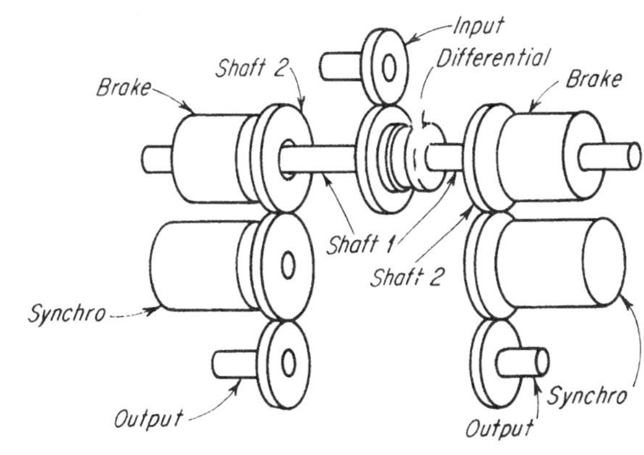

Fig. 6 Controlling output from a differential.

Magnetic Friction Clutches

The simplest and most adaptable electromagnetic control clutch is the magnetic friction clutch. It works on the same principle as a simple solenoid-operated electric relay with a spring return to normal. Like the relay, it is a straightforward automatic switch for controlling the flow of power (in this cases, torque) through a circuit.

Rotating or Fixed Field?

This is a question primarily of magnetic design. Rotating-field clutches include a rotating coil, energized through brushes and slip rings. Fixed-field units have a stationary coil. Rotating-field units are still more common, but there has been a marked trend toward the fixed-field versions.

Generally speaking, a *rotating-field clutch* is a two-member unit, with the coil carried in the driving (input) member. It can be mounted directly on a motor or speed-reducer shaft without loading down the driving motor. In the smaller sizes, it offers a better ratio of size to rated output than the fixed-field type, although the rotating coil increases inertia in the larger models.

A *fixed-field clutch,* on the other hand, is a three-member unit with rotating input and output members and a stationary coil housing. It eliminates the need for brushes and slip rings, but it demands additional bearing supports, and it can require close tolerances in mounting.

Purely Magnetic Clutches

Probably less familiar than the friction types are *hysteresis* and *eddy-current clutches.* They operate on straight magnetic principles and do not depend on mechanical contact between their members. The two styles are almost identical in construction, but the magnetic segments of the hysteresis clutch are electrically isolated, and those of the eddy-current clutch are interconnected.

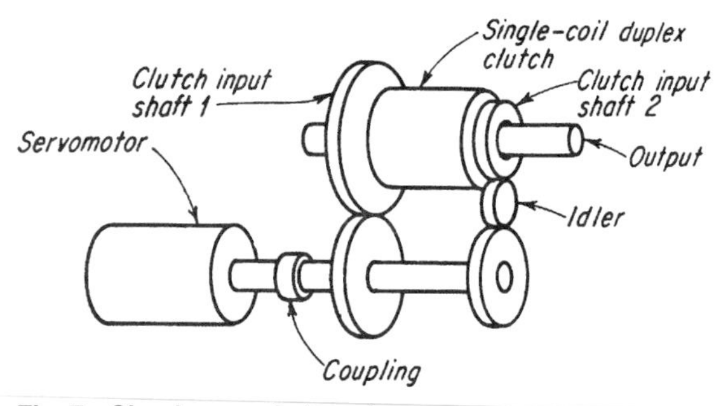

Fig. 7 Simple-speed changing.

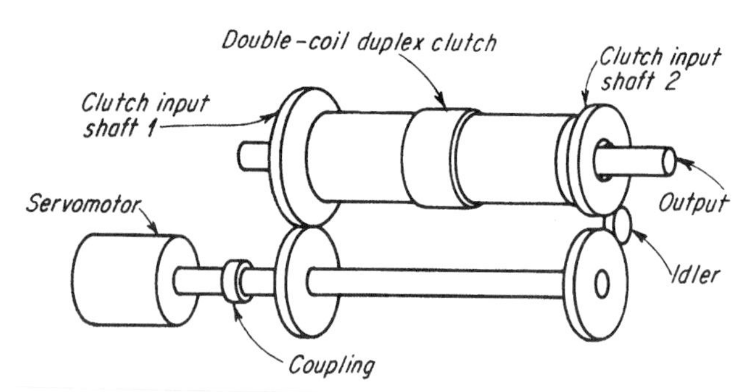

Fig. 8 Speed-changing and uncoupling.

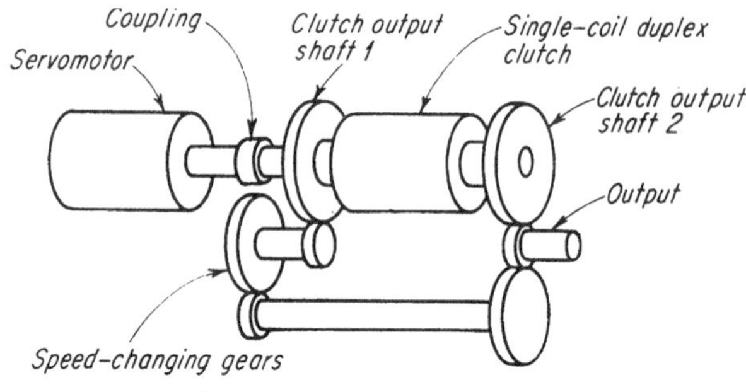

Fig. 9 Simple direction-changing.

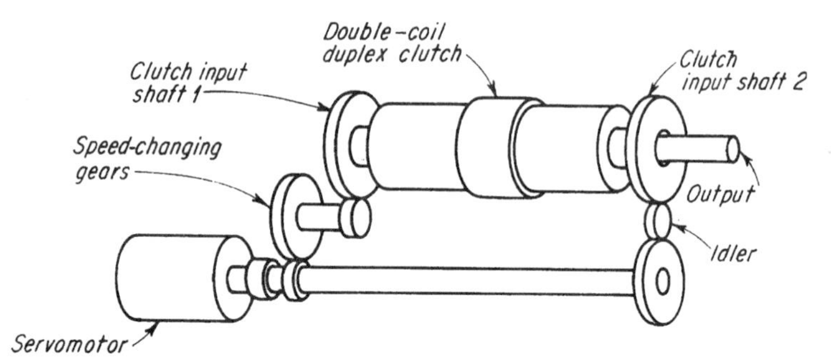

Fig. 10 Direction-changing and uncoupling.

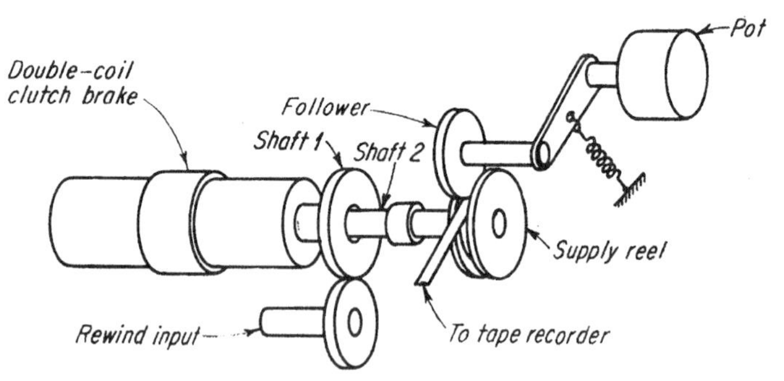

Fig. 11 Constant tensioning.

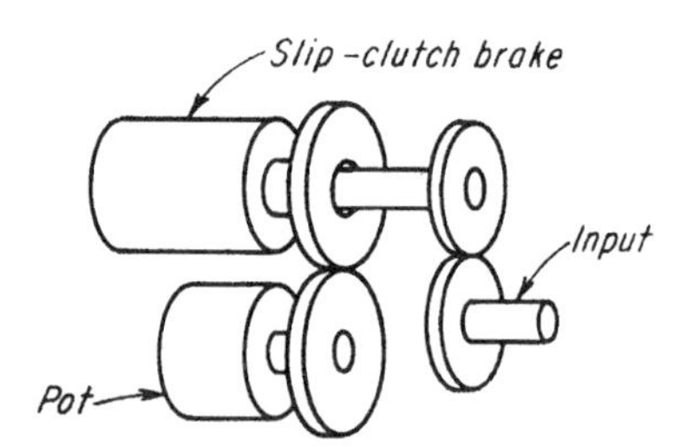

Normal : input drives pot, with slip protection
Energized: input disconnected and pot locked

Fig. 12 Potentiometer control.

The magnetic analogy of both styles is similar in that the flux is passed between the two clutch members.

Hysteresis Clutches

The hysteresis clutch is a proportional-torque control device. As its name implies, it exploits the hysteresis effect in a permanent-magnet rotor ring to produce a substantially constant torque that is almost completely independent of speed (except for slight, unavoidable secondary eddy-current torques—which do not seriously reduce performance). It is capable of synchronous driving or continuous slip, with almost no torque variation at any slip differential for a given control current. Its control-power requirement can be met by a transistor drive. Typical applications include wire or tape tensioning, servo-control actuation, and torque control in dynamometers.

Eddy-Current Clutches

Eddy-current clutches on the other hand, are inherently speed-sensitive devices. They exhibit virtually no hysteresis, and develop torque by dissipating eddy currents through the electrical resistance of the rotor ring. This torque is almost a linear function of slip speed. These clutches perform best in speed-control applications, and as oscillation dampers.

Particle and Fluid Magnetic Clutches

There is no real difference between *magnetic-particle* and *magnetic-field clutches.* However, the magnetic medium in the particle clutch is a dry powder; in the fluid clutch it is a powder suspended in oil. In both clutches the ferromagnetic medium is introduced into the airgap between the input and output faces, which do not actually contact one another. When the clutch coil is energized, the particles are excited in the magnetic field between the faces; as they shear against each other, they produce a drag torque between the clutch members.

Theoretically, those clutches can approach the proportional control characteristics of a hysteresis clutch within the small weight and size limits of a comparably rated miniature friction clutch. But in practice, the service life of miniature magnetic-particle clutches has so far been too short for industrial service.

Other Magnetic Clutches

Two sophisticated concepts—neither of them yet developed to the point of practical application—might be of interest to anyone researching this field.

Electrostatic clutches depend on high voltages instead of a magnetic field to create force-producing suspensions.

Magnetostrictive clutches depend on a magnetic force to change the dimensions of a crystal or metal bar poised between two extremely precise facts.

ROLLER LOCKING MECHANISM CONTAINS TWO OVERRUNNING CLUTCHES

The figure is a simplified cross-sectional view of an electromagnetically releasable roller-locking mechanism that functions as a brake or clutch in clockwise or counterclockwise rotation. In essence, the mechanism contains two back-to-back overrunning clutches such as those that are commonly used in industry to roll freely in one direction and lock against rolling in the opposite direction. In addition to bidirectionality, the novel design of this mechanism offers advantages of efficiency and controllability over older clutches and brakes.

As in other roller-locking mechanisms, lock is achieved in this mechanism by jamming rollers between a precise surface on one rotating or stationary subassembly (in this case, the inner surface of a reaction ring in a housing) and a precise surface on another rotating or stationary subassembly (in this case, the outer surface of a disk integral with a drive shaft). There are two sets of rollers: CW and CCW locking. They feature cam surfaces that jam against the disk and reaction ring in the event of clockwise and counterclockwise rotation, respectively. The mechanism is called a "trip roller clutch" because of the manner in which the rollers are unjammed or tripped to allow rotation, as explained later.

The rollers are arranged in pairs around the disk and the reaction ring. Each pair contains one CW and one CCW locking roller. A tripping anvil fixed to the reaction ring is located between the rollers in each pair. Each roller is spring-loaded to translate toward a prescribed small distance from the tripping anvil and to rotate toward the incipient-jamming position. In the absence of any tripping or releasing action, the clutch remains in lock; that is, any attempt at clockwise or counterclockwise rotation of the drive shaft result sin jamming of the CW or CCW rollers, respectively.

Release is effected by energizing the electromagnet coil in the housing. The resulting magnetic force pulls a segmented striker disk upward against spring bias. Attached to each segment of the striker disk is a tripper, which slides toward a CW or CCW roller on precisely angled surfaces in the tripping anvil. Each tripper then pushes against its associated CW or CCW locking roller with a small blocking force. But, in blocking the locking roller, the locking cam angles are effectively increased and slipping (followed by release) occurs. Thus, the clutch is "tripped" out of lock into release.

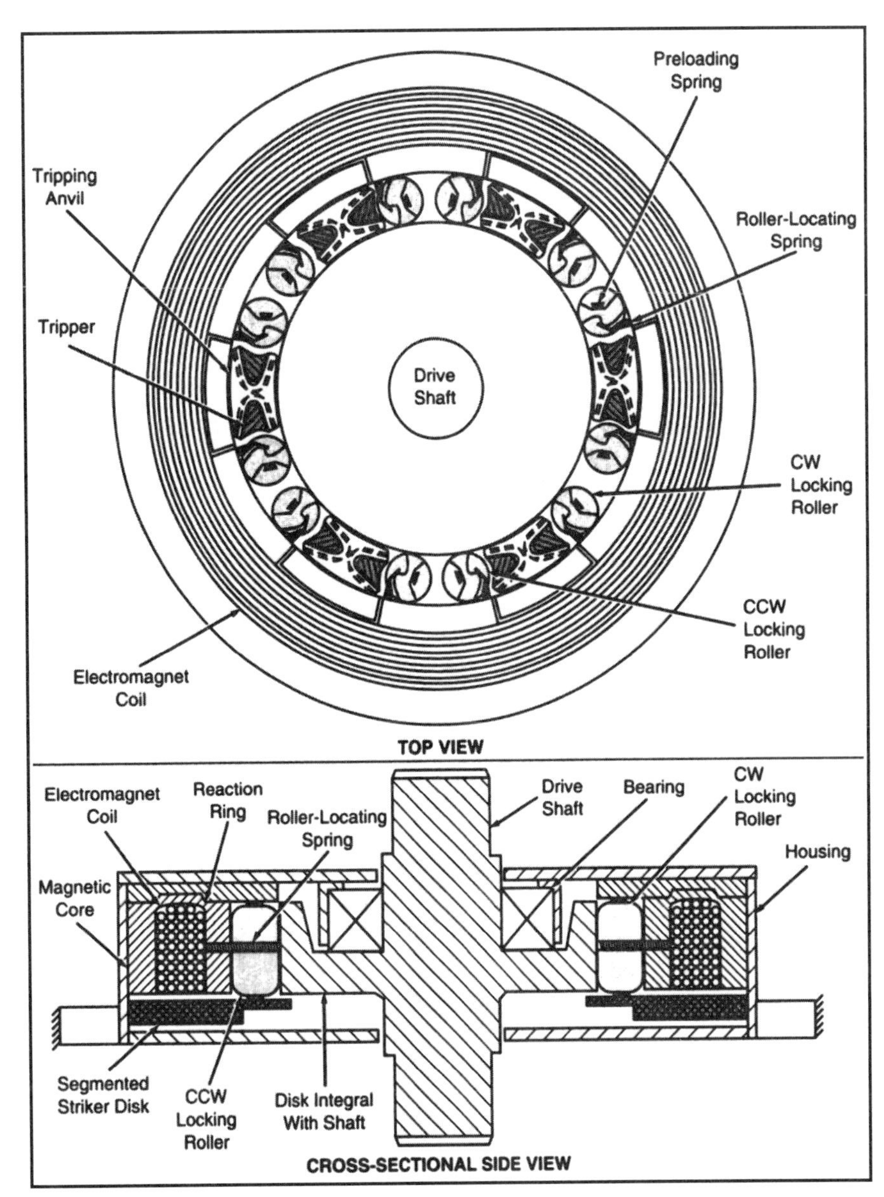

The trip roller clutch contains back-to-back roller-locking, overrunning clutches that can be released (tripped) with small magnetic forces.

Very little force is needed for this releasing action, even though the forces in lock can be very large. Because the gap between the striker plate and the magnetic core is zero or very small during release, very little magnetic force is needed to maintain release. Thus, the electromagnet coil and the power is consumes can be made smaller than in comparable prior mechanisms, with a corresponding gain in power efficiency and decrease in size and in weight. To lock the clutch, one simply turns off the electromagnet, allowing the springs to retract the trippers and restore the rollers to the incipient-jamming position.

The excellent frequency response and high mechanical efficiency, inherent in roller locking, enable the trip roller clutch to be lockable and releasable precisely at a desired torque under sensory interactive computer control. For the same reasons, the trip roller clutch can be opened and closed repeatedly in a pulsating manner to maintain precise torque(s).

This work was done by John M. Vranish of **Goddard Space Flight Center** *and supported by Honeybee Robotics, NY.*

CHAPTER 9

LATCHING, FASTENING, AND CLAMPING DEVICES AND MECHANISMS

SIXTEEN LATCH, TOGGLE, AND TRIGGER DEVICES

Diagrams of basic latching and quick-release mechanisms.

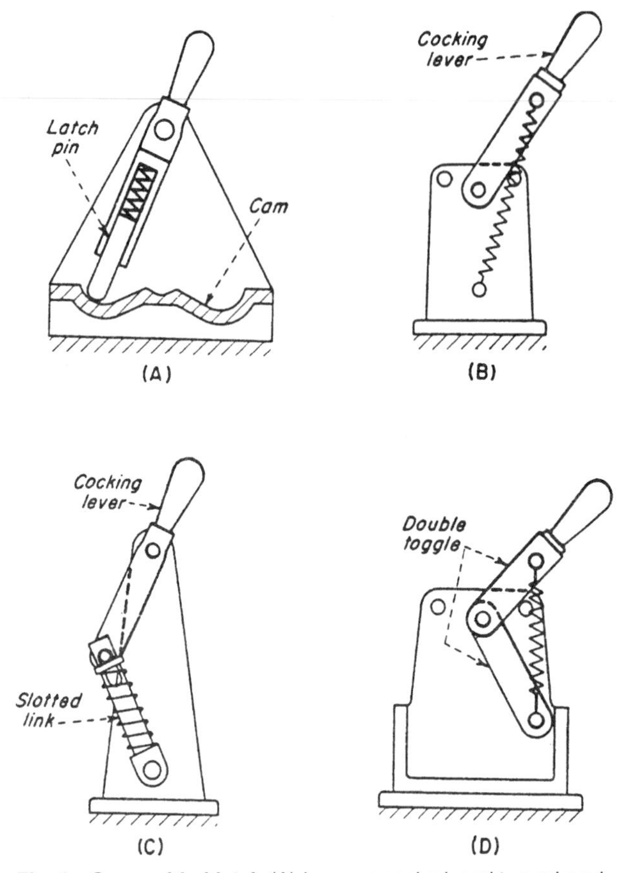

Fig. 1 Cam-guided latch (A) has one cocked, and two relaxed positions, (B) Simple overcenter toggle action. (C) An overcenter toggle with a slotted link. (D) A double toggle action often used in electrical switches.

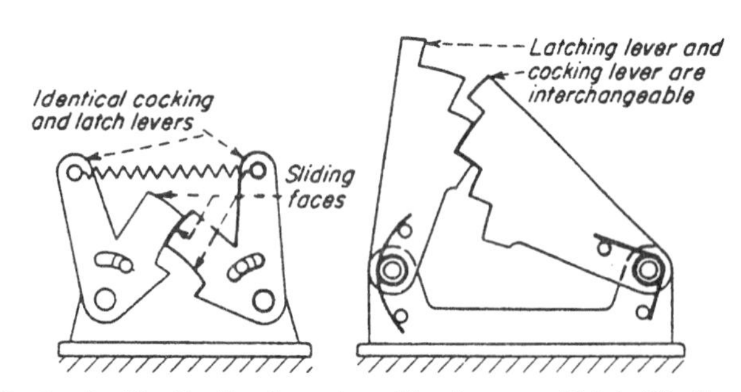

Fig. 2 An identically shaped cocking lever and latch (A) allow their functions to be interchangeable. The radii of the sliding faces must be dimensioned for a mating fit. The stepped latch (B) offers a choice of several locking positions.

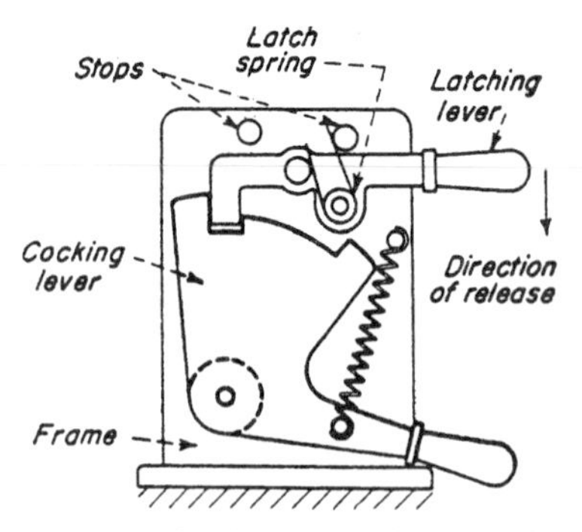

Fig. 3 A latch and cocking lever is spring-loaded so that latch movement releases the cocking lever. The cocked position can be held indefinitely. Studs in the frame provide stops, pivots, or mounts for the springs.

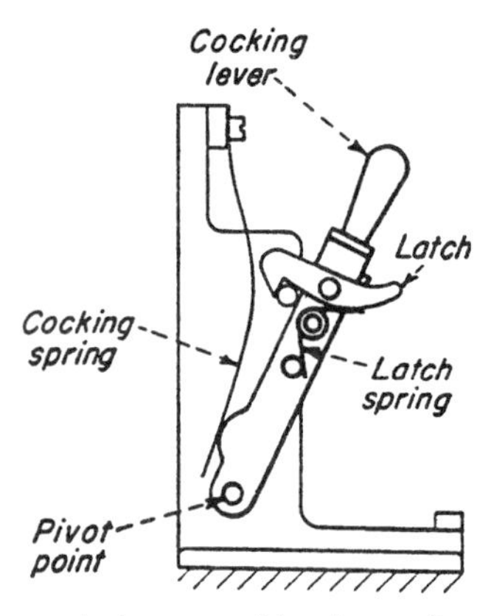

Fig. 4 A latch mounted on a cocking lever allows both levers to be reached at the same time with one hand. After release, the cocking spring initiates clockwise lever movement; then gravity takes over.

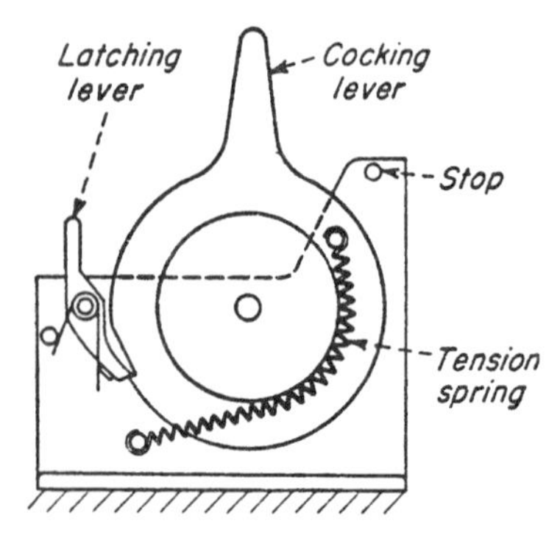

Fig. 5 A disk-shaped cocking has a tension spring resting against the cylindrical hub. Spring force always acts at a constant radius from the lever pivot point.

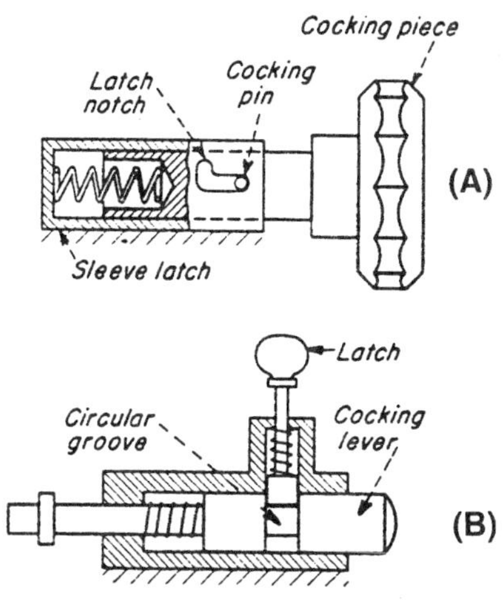

Fig. 6 A sleeve latch (A) as an L-shaped notch. A pin in the shaft rides in a notch. Cocking requires a simple push and twist action. (B) The Latch and plunger depend on axial movement for setting and release. A circular groove is needed if the plunger is to rotate.

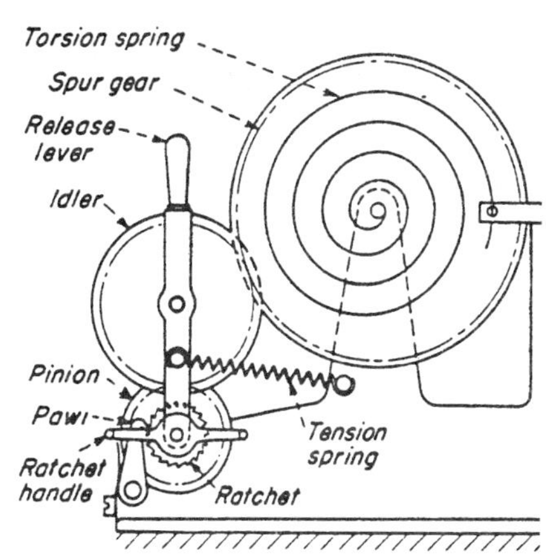

Fig. 7 A geared cocking device has a ratchet fixed to a pinion. A torsion spring exerts clockwise force on the spur gear; a tension spring holds the gar in mesh. The device is wound by turning the ratchet handle counterclockwise, which in turn winds the torsion spring. Moving the release-lever permits the spur gear to unwind to its original position without affecting the ratchet handle.

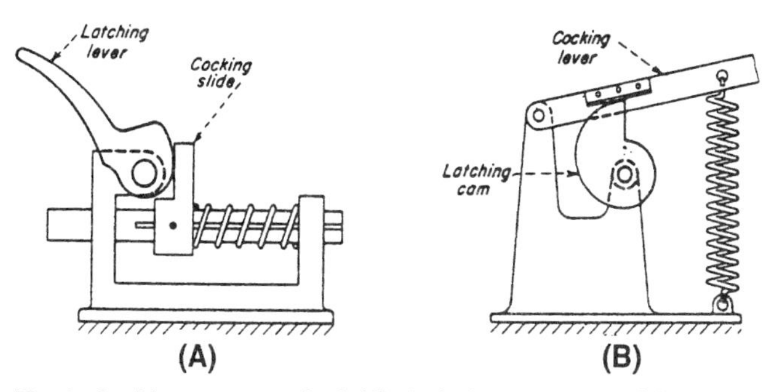

Fig. 8 In this overcenter lock (A) clockwise movement of the latching lever cocks and locks the slide. A counterclockwise movement is required to release the slide. (B) A latching-cam cocks and releases the cocking lever with the same counterclockwise movement as (A).

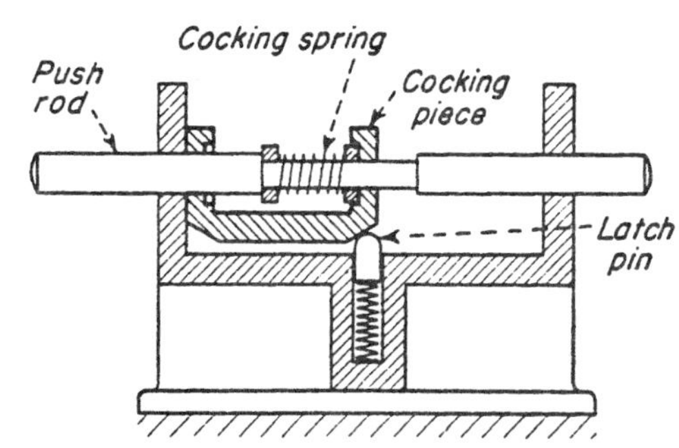

Fig. 9 A spring-loaded cocking piece has chamfered corners. Axial movement of the push-rod forces the cocking piece against a spring-loaded ball or pin set in a frame. When cocking builds up enough force to overcome the latch-spring, the cocking piece snaps over to the right. The action can be repeated in either direction.

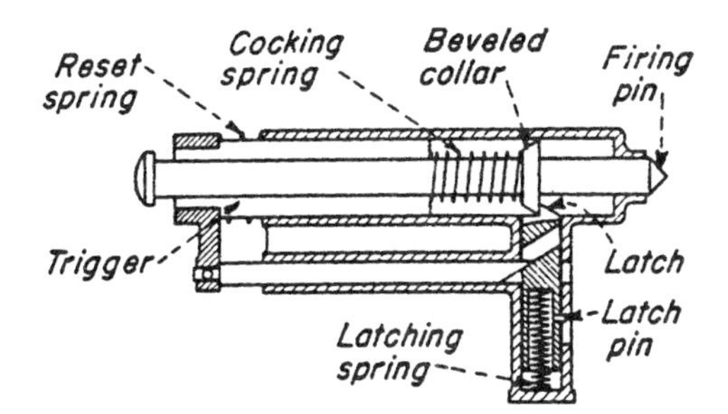

Fig. 10 A firing-pin mechanism has a beveled collar on a pin. Pressure on the trigger forces the latch down until it releases the collar when the pin snaps out, under the force of cocking the spring. A reset spring pulls the trigger and pin back. The latch is forced down by a beveled collar on a pin until it snaps back, after overcoming the force of the latch spring. (A latch pin retains the latch if the trigger and firing pin are removed.)

FOURTEEN SNAP-ACTION DEVICES

These diagrams show fourteen ways to produce mechanical snap action.

Mechanical snap action results when a force is applied to a device over a period of time; buildup of this force to a critical level causes a sudden motion to occur. The ideal snap device would have no motion until the force reached a critical level. This, however, is not possible, and the way in which the mechanism approaches this ideal is a measure of its efficiency as a snap device. Some of the designs shown here approach the ideal closely; others do not, but they have other compensating good features.

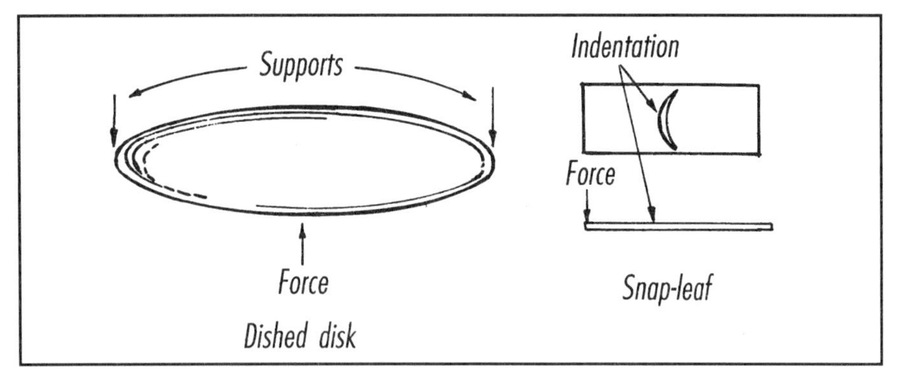

Fig. 1 A dished disk is a simple, common method for producing snap action. A snap leaf made from spring material can have various-shaped impressions stamped at the point where the overcentering action occurs. A "Frog clacker" is, of course, a typical applications. A bimetal element made in this way will reverse itself at a predetermined temperature.

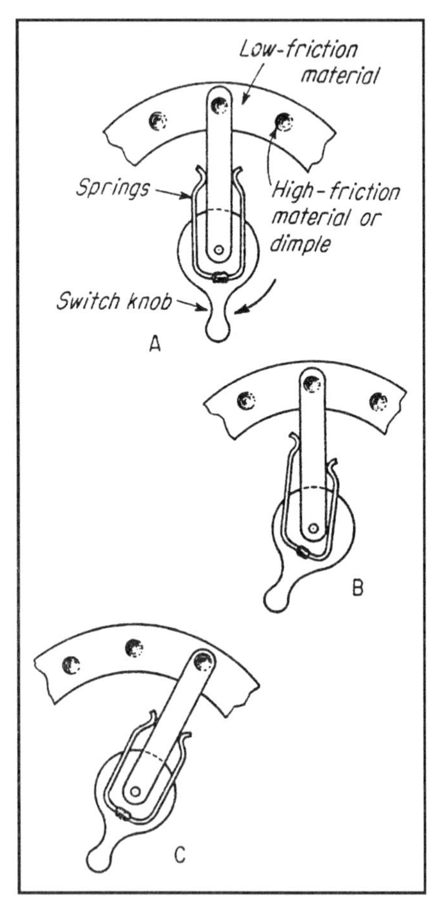

Fig. 2 Friction override can hold against an increasing load until friction is suddenly overcome. This is a useful action for small sensitive devices where large forces and movements are undesirable. This is the way we snap our fingers. That action is probably the original snap mechanism.

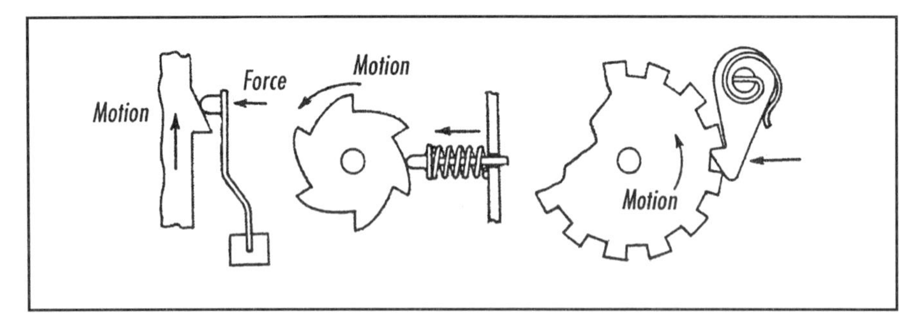

Fig. 3 A ratchet-and-pawl combination is probably the most widely used form of snap mechanism. Its many variations are an essential feature in practically every complicated mechanical device. By definition, however, this movement is not true snap-action.

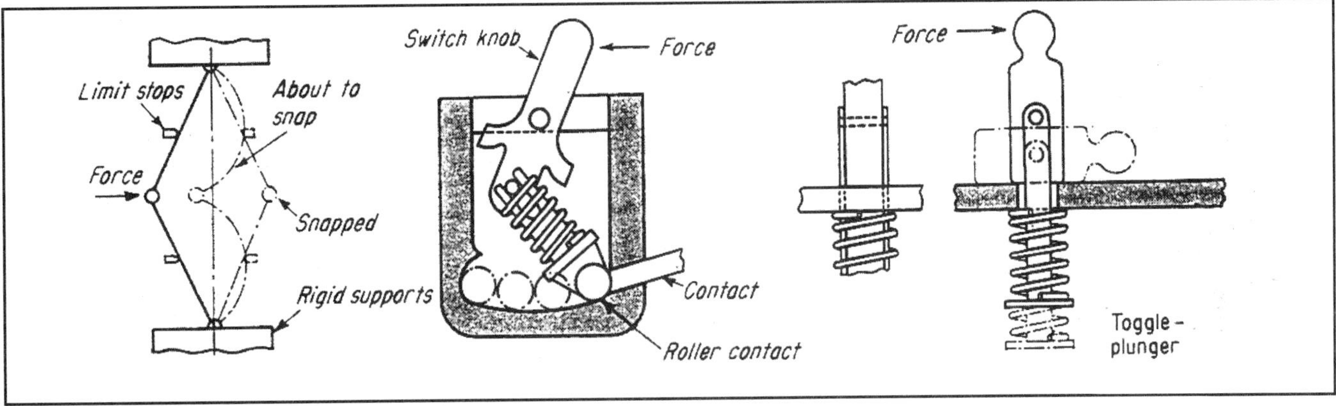

Fig. 4 Over-centering mechanisms find many applications in electrical switches. Considerable design ingenuity has been applied to fit this principle into many different mechanisms. It is the basis of most snap-action devices.

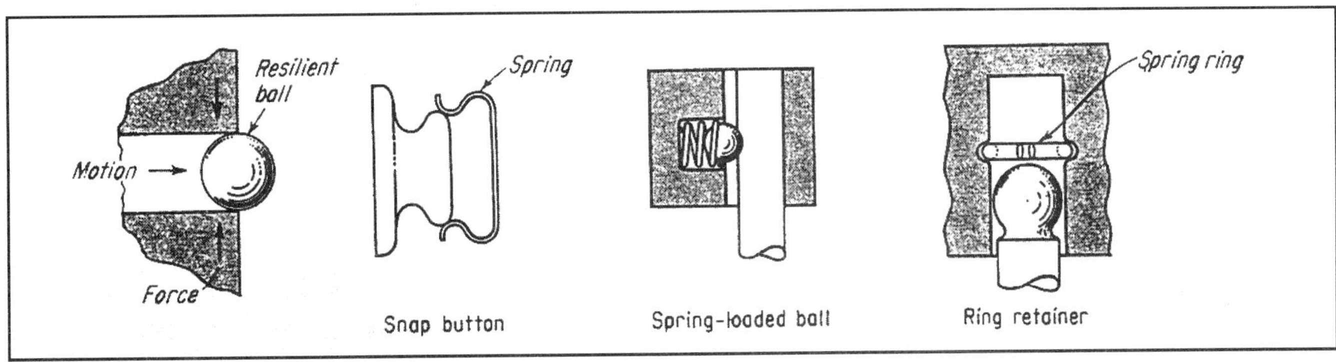

Fig. 5 The sphere ejection principle is based on snap buttons, spring-loaded balls and catches, and retaining-rings for fastening that must withstand repeated use. Their action can be designed to provide either easy or difficult removal. Wear can change the force required.

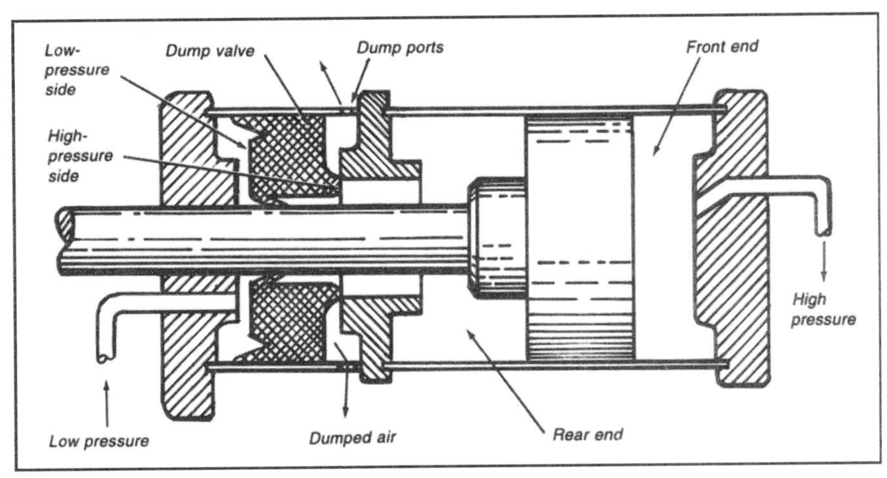

Fig. 6 A pneumatic dump valve produces snap action by preventing piston movement until air pressure has built up in the front end of the cylinder to a relatively high pressure. Dump-valve area in the low-pressure end is six times larger than its area on the high-pressure side. Thus the pressure required on the high-pressure side to dislodge the dump valve from its seat is six times that required on the low-pressure side to keep the valve properly seated.

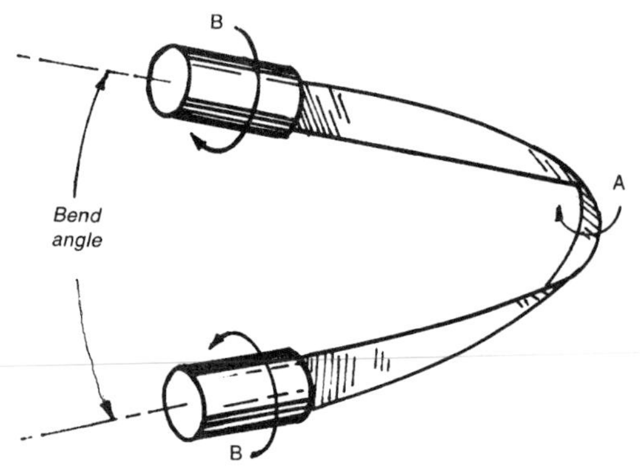

Fig. 1 A torsion ribbon bent as shown will turn "inside out" at A with a snap action when twisted at B. Design factors are ribbon width, thickness, and bend angle.

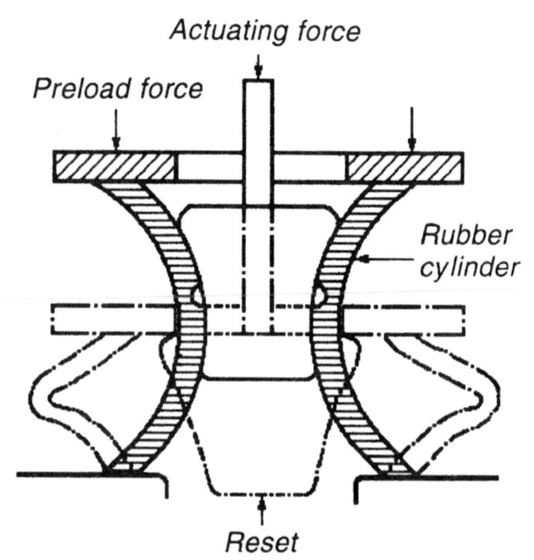

Fig. 2 A collapsing cylinder has elastic walls that can be deformed gradually until their stress changes from compressive to bending, with the resulting collapse of the cylinder.

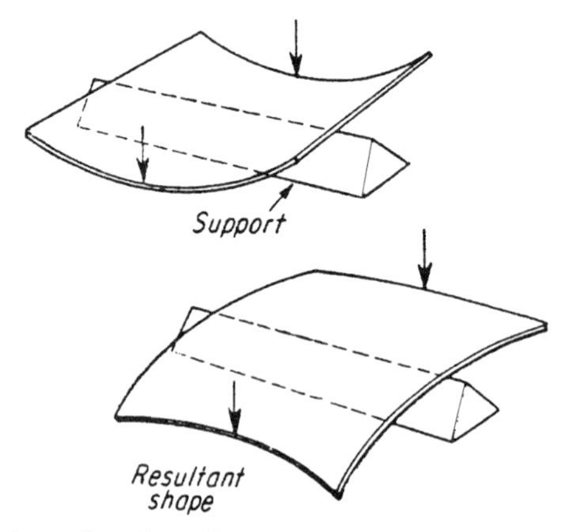

Fig. 3 A bowed spring will collapse into a new shape when it is loaded as shown A. A "push-pull" steel measuring tape illustrates this action; the curved material stiffens the tape so that it can be held out as a cantilever until excessive weight causes it to collapse suddenly.

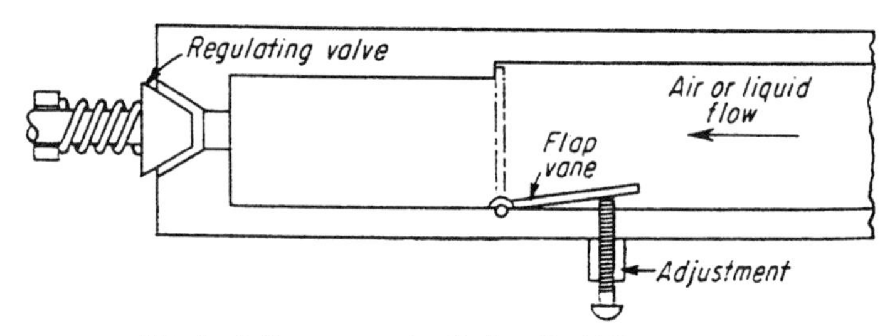

Fig. 4 A flap vane cuts off air or liquid flow at a limiting velocity. With a regulating valve, the vane will snap shut (because of increased velocity) when pressure is reduced below a design value.

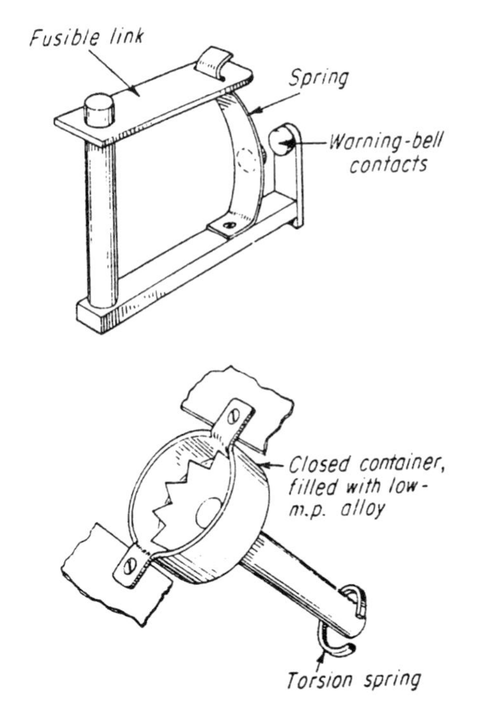

Fig. 5 A sacrificing link is useful where high temperature or corrosive chemicals would be hazardous. If the temperature becomes too high, or atmosphere too corrosive, the link will yield at design conditions. The device usually is required to act only once, although a device like the lower one can be quickly reset. However, it is restricted to temperature control.

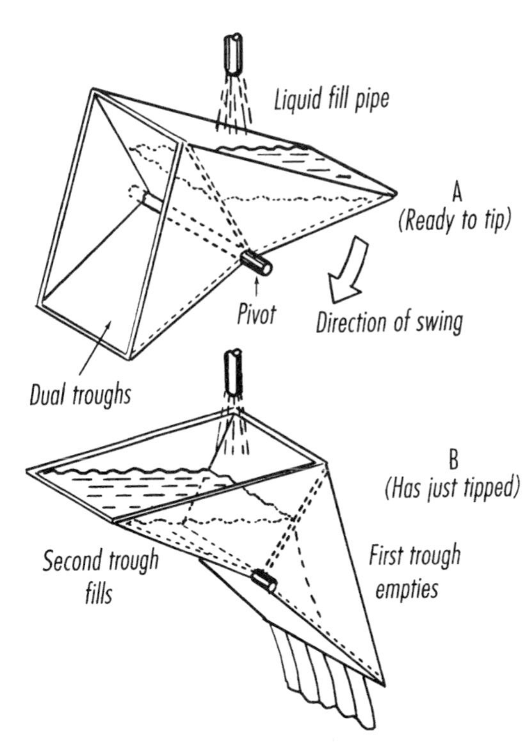

Fig. 6 Gravity-tips, although slower acting than most snap mechanisms, can be called snap mechanisms because they require an accumulation of energy to trigger an automatic release. A tripping trough that spreads sewerage is one example. As shown in A, it is ready to trip. When overbalanced, it trips rapidly, as in B.

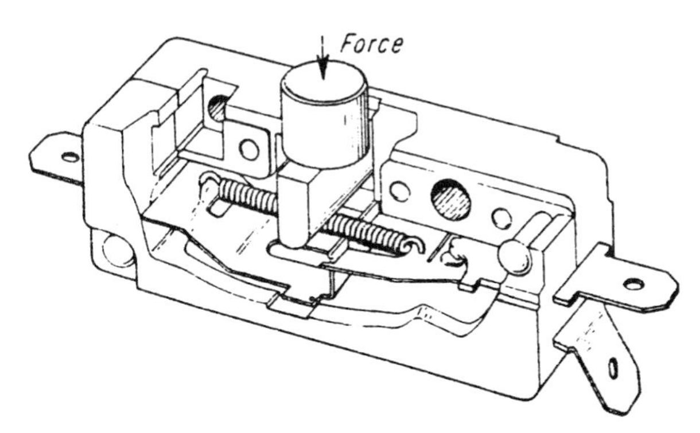

Fig. 7 An overcentering tension spring combined with a pivoted contact-strip is one arrangement used in switches. The example shown here is unusual because the actuating force bears on the spring itself.

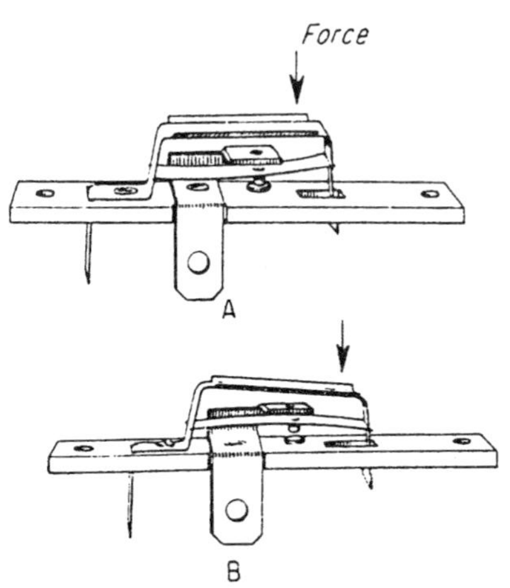

Fig. 8 An overcentering leaf-spring action is also the basis for many ingenious snap-action switches for electrical control. Sometimes spring action is combined with the thermostatic action of a bimetal strip to make the switch respond to heat or cold, either for control purposes or as a safety feature.

REMOTE CONTROLLED LATCH

This simple mechanism engages and disengages parallel plates carrying couplings and connectors.

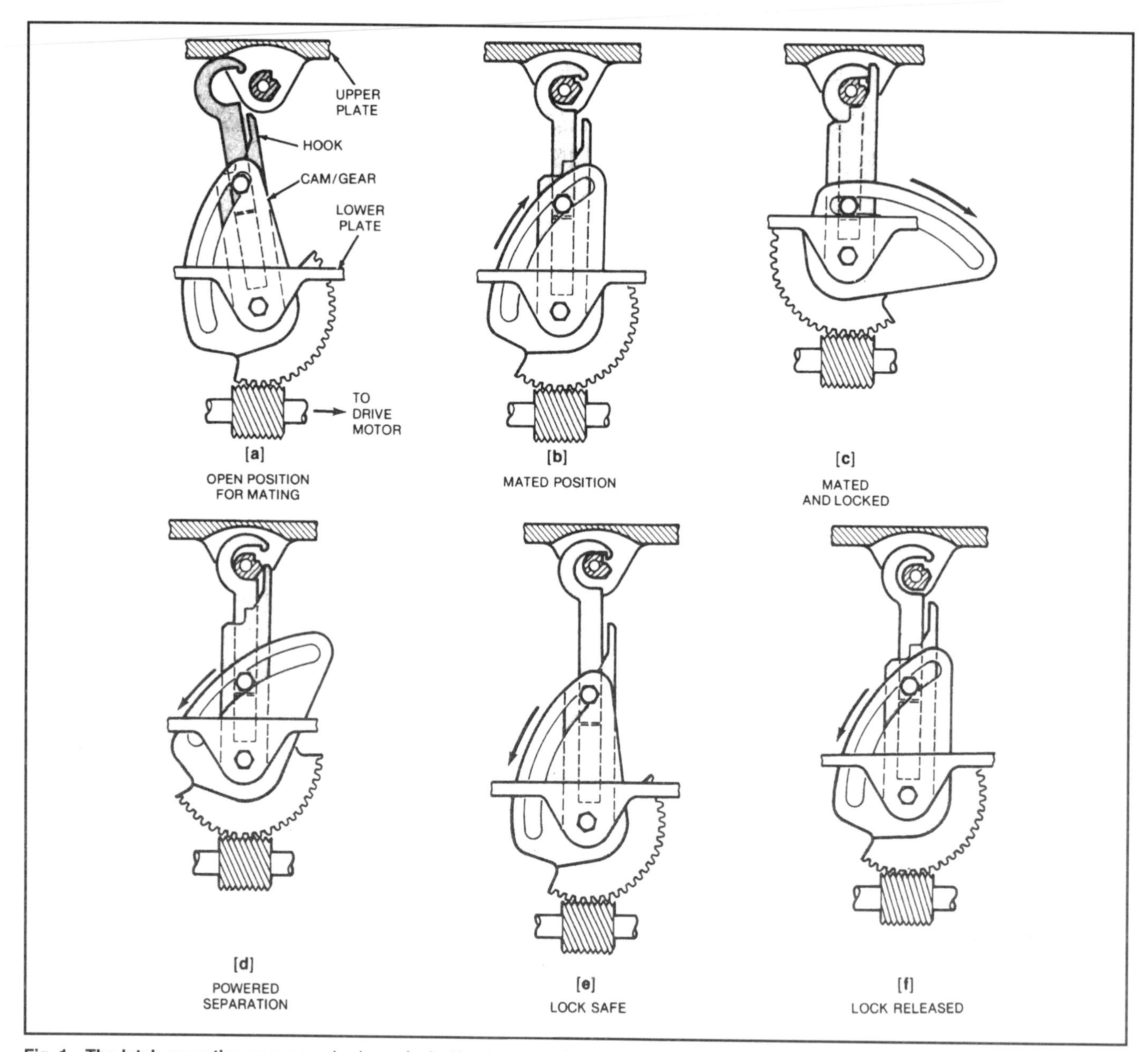

Fig. 1 The latch operation sequence is shown for locking in steps (a) through (c) and for unlocking in steps (d) through (f).

A new latch mates two parallel plates in one continuous motion (see Fig. 1). On the Space Shuttle, the latch connects (and disconnects) plates carrying 20 fluid couplings and electrical connectors. (The coupling/connector receptacles are one plate, and mating plugs are on the other plate). Designed to lock items in place for handling, storage, or processing under remote control, the mechanism also has a fail-safe feature: It does not allow the plates to separate completely unless both are supported. Thus, plates cannot fall apart and injure people or damage equipment.

The mechanism employs four cam/gear assemblies, one at each corner of the lower plate. The gears on each side of the plate face inward to balance the loading and help align the plates. Worm gears on the cam-gear assemblies are connected to a common drive motor.

Figure 1 illustrates the sequence of movements as a pair of plates is latched and unlatched. Initially, the hook is extended and tilted out. The two plates are brought together, and when they are 4.7 in. (11.9 cm) apart, the drive motor is started (a). The worm gear rotates the hook until it closes on a pin on the opposite plate (b). Further rotation of the worm gear shortens the hook extension and raises the lower plate (c). At that point, the couplings and connectors on the two plates are fully engaged and locked.

To disconnect the plates, the worm gear is turned in the opposite direction. This motion lowers the bottom plate and pulls the couplings apart (d). However, if the bottom plate is unsupported, the latch safety feature operates. The hook cannot clear the pin if the lower plate hangs freely (e). If the bottom plate is supported, the hook extension lifts the hook clear of the pin (f) so that the plates are completely separated.

This work was done by Clifford J. Barnett, Paul Castiglione, and Leo R. Coda of Rockwell International Corp. for **Johnson Space Center.**

TOGGLE FASTENER INSERTS, LOCKS, AND RELEASES EASILY

A pin-type toggle fastener, invented by C.C. Kubokawa at NASA's Ames Research Center, can be used to fasten plates together, fasten things to walls or decks, or fasten units with surfaces of different curvatured, such as a concave shape to a convex surface.

With actuator pin. The cylindrical body of the fastener has a tapered end for easy entry into the hole; the head is threaded to receive a winged locknut and, if desired, a ring for pulling the fastener out again after release. Slots in the body hold two or more toggle wings that respond to an actuator pin. These wings are extended except when the spring-loaded pin is depressed.

For installation, the actuator pin is depressed, retracting the toggle wings. When the fastener is in place, the pin is released, and the unit is then tightened by screwing the locknut down firmly. This exerts a compressive force on the now expanded toggle wings. For removal, the locknut is loosened and the pin is again depressed to retract the toggle wings. Meanwhile, the threaded outer end of the cylindrical body functions as a stud to which a suitable pull ring can be screwed to facilitate removal of the fastener.

This invention has been patented by NASA (U.S. Patent No. 3,534,650).

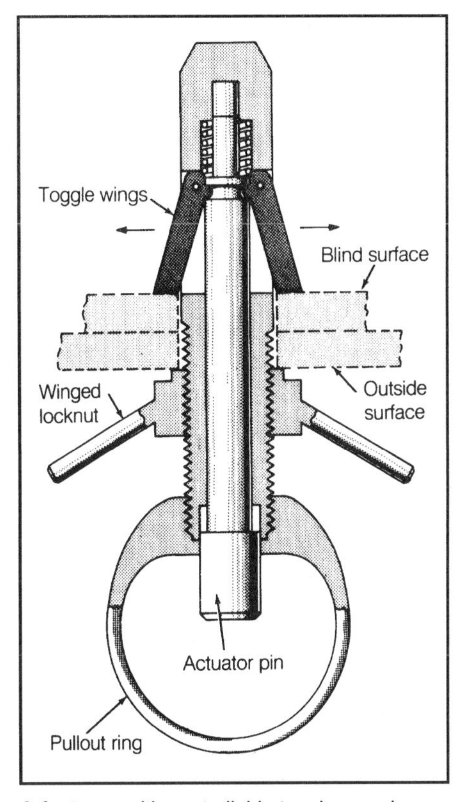

A fastener with controllable toggles can be inserted and locked from only one side.

GRAPPLE FREES LOADS AUTOMATICALLY

A simple grapple mechanism, designed at Argonne National Laboratory in Illinois, engages and releases loads from overhead cranes automatically. This self-releasing mechanism was developed to remove fuel rods from nuclear reactors. It can perform tasks where human intervention is hazardous or inefficient, such as lowering and releasing loads from helicopters.

The mechanism (see drawing) consists of two pieces: a lift knob secured to the load and a grapple member attached to the crane. The sliding latch-release collar under the lift knob is the design's key feature.

Spring magic. The grapple housing, which has a cylindrical inner surface, contains a machined groove fitted with a garter spring and three metal latches. When the grapple is lowered over the lift knob, these latches recede into the groove as their edges come into contact with the knob. After passing the knob, they spring forward again, locking the grapple to the knob. Now the load can be lifted.

When the load is lowered to the ground again, gravity pull or pressure from above forces the grapple housing down until the latches come into contact with a double cone-shaped release collar. The latches move back into the groove as they pass over the upper cone's surface and move forward again when they slide over the lower cone.

The grapple is then lifted so that the release collar moves up the cylindrical rod until it is housed in a recess in the lift knob. Because the collar can move no farther, the latches are forced by the upward pull to recede again into the groove—allowing the grapple to be lifted free.

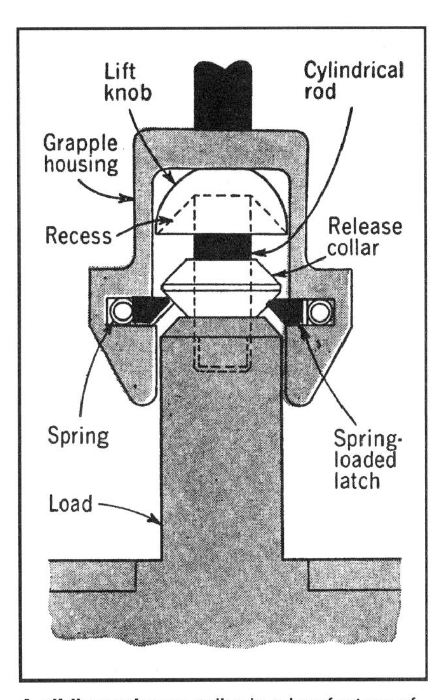

A sliding release collar is a key feature of this automatic grapple.

QUICK-RELEASE LOCK PIN HAS A BALL DETENT

A novel quick-release locking pin has been developed that can be withdrawn to separate the linked members only when stresses on the joint are negligible.

The pin may be the answer to the increasing demand for locking pins and fasteners that will pull out quickly and easily when desired, yet will stay securely in place without chance of unintentional release.

The key to this foolproof pin is a group of detent balls and a matching grooved. The ball must be in the groove whenever the pin is either installed or pulled out of the assembly. This is easy to do during installation, but during removal the load must be off the pin to get the balls to drop into the groove.

How it works. The locking pin was developed by T.E. Othman, E.P. Nelson, and L.J. Zmuda under contract to NASA's Marshall Space Flight Center. It consists of a forward-pointing sleeve with a spring-loaded sliding handle as its rear end, housing a sliding plunger that is pushed backward (to its locking position) by a spring within the handle.

To some extend the plunger can slide forward against the plunger spring, and the handle can slide backward against the handle spring. A groove near the front end of the plunger accommodates the detent balls when the plunger is pushed forward by the compression of its spring. When the plunger is released backward, the balls are forced outward into holes in the sleeve, preventing withdrawal of the pin.

To install the pin, the plunger is pressed forward so that the balls fall into their groove and the pin is pushed into the hole. When the plunger is released, the balls lock the sleeve against accidental withdrawal.

To withdraw the pin, the plunger is pressed forward to accommodate the locking balls, and at the same time the handle is pulled backward. If the loading on the pin is negligible, the pin is withdrawn from the joint; if it is considerable, the handle spring is compressed and the plunger is forced backward by the handle so the balls will return to their locking position.

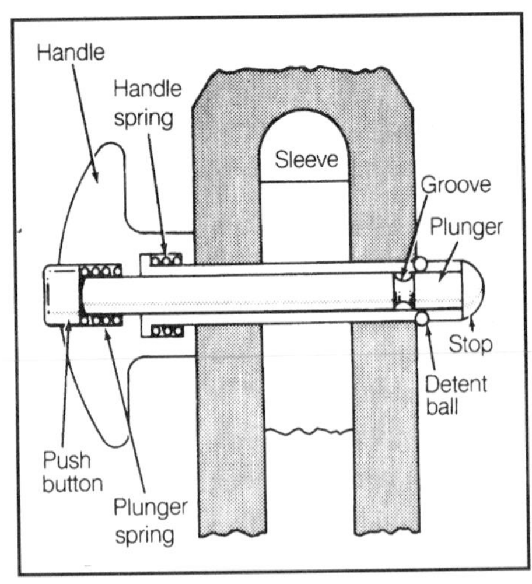

A foolproof locking pin releases quickly when the stress on the joint is negligible.

The allowable amount of stress on the joint that will permit its removal can be varied by adjusting the pressure required for compressing the handle spring. If the stresses on the joint are too great or the pin to be withdrawn in the normal manner, hammering on the forward end of the plunger simply ensures that the plunger remains in its rearward position, with the locking balls preventing the withdrawal of the pin. A stop on its forward end prevents the plunger from being driven backward.

AUTOMATIC BRAKE LOCKS HOIST WHEN DRIVING TORQUE CEASES

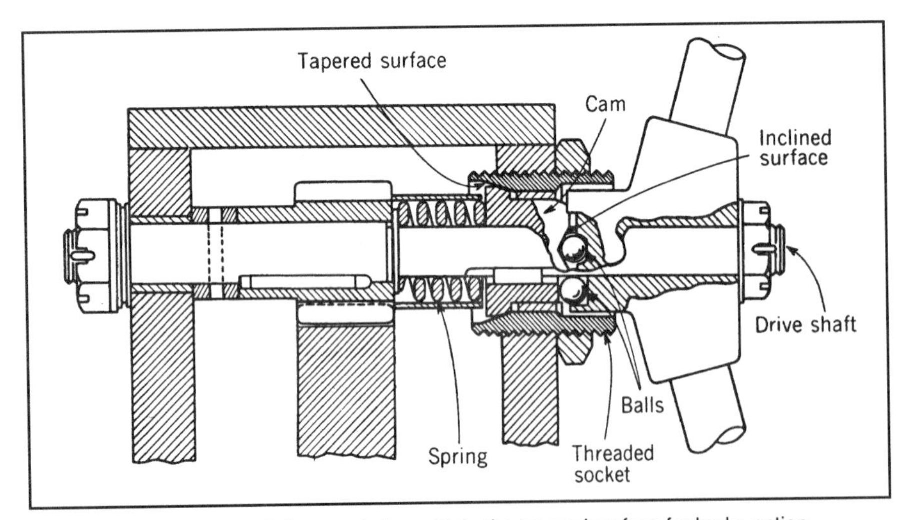

When torque is removed, the cam is forced into the tapered surface for brake action.

A brake mechanism attached to a chain hoist is helping engineers lift and align equipment accurately by automatically locking it in position when the driving torque is removed from the hoist.

According to the designer, Joseph Pizzo, the brake could also be used on wheeled equipment operating on slopes, to act as an auxiliary brake system.

How it works. When torque is applied to the driveshaft (as shown in the figure), four steel balls try to move up the inclined surfaces of the cam. Although called a cam by the designer, it is really a concentric collar with a cam-like surface on one of its end faces. Because the balls are contained by four cups in the hub, the cam is forced to move forward axially to the left. Because the cam moves away from the tapered surface, the cam and the driveshaft that is keyed to it are now free to rotate.

If the torque is removed, a spring resting against the cam and the driveshaft gear forces the cam back into the tapered surface of the threaded socket for instant braking.

Although this brake mechanism (which can rotate in either direction) was designed for manual operation, the principle can be applied to powered systems.

LIFT-TONG MECHANISM FIRMLY GRIPS OBJECTS

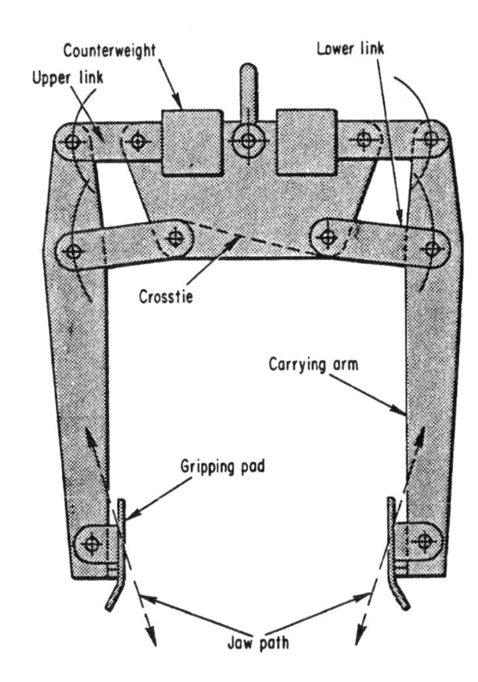

Twin four-bar linkages are the key components in this long mechanism that can grip with a constant weight-to-grip force ratio any object that fits within its grip range. The long mechanism relies on a cross-tie between the two sets of linkages to produce equal and opposite linkage movement. The vertical links have extensions with grip pads mounted at their ends, while the horizontal links are so proportioned that their pads move in an inclined straight-line path. The weight of the load being lifted, therefore, wedges the pads against the load with a force that is proportional to the object's weight and independent of its size.

PERPENDICULAR-FORCE LATCH

The installation and removal of equipment modules are simplified.

A latching mechanism simultaneously applies force in two perpendicular directions to install or remove electronic-equipment modules. The mechanism (see Fig. 1) requires only the simple motion of a handle to push or pull an avionic module to insert or withdraw connectors on its rear face into or from spring-loaded mating connectors on a panel and to force the box downward onto or release the box from a mating cold plate that is part of the panel assembly. The concept is also adaptable to hydraulic, pneumatic, and mechanical systems. Mechanisms of this type can simplify the manual installation and removal of modular equipment where a technician's movement is restricted by protective clothing, as in hazardous environments, or where the installation and removal are to be performed by robots or remote manipulators.

Figure 2 sows an installation sequence. In step 1, the handle has been installed on the handle cam and turned downward. In step 2,

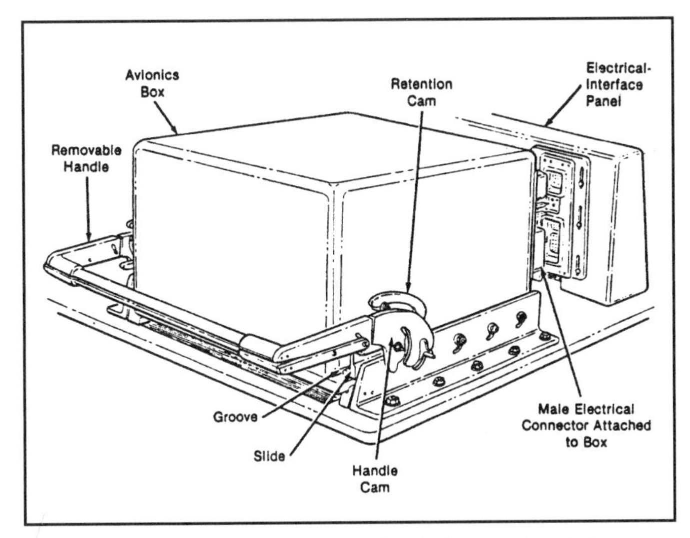

Fig. 1 An avionics box mates with electrical connectors in the rear and is locked in position on the cold plate when it is installed with the latching mechanism.

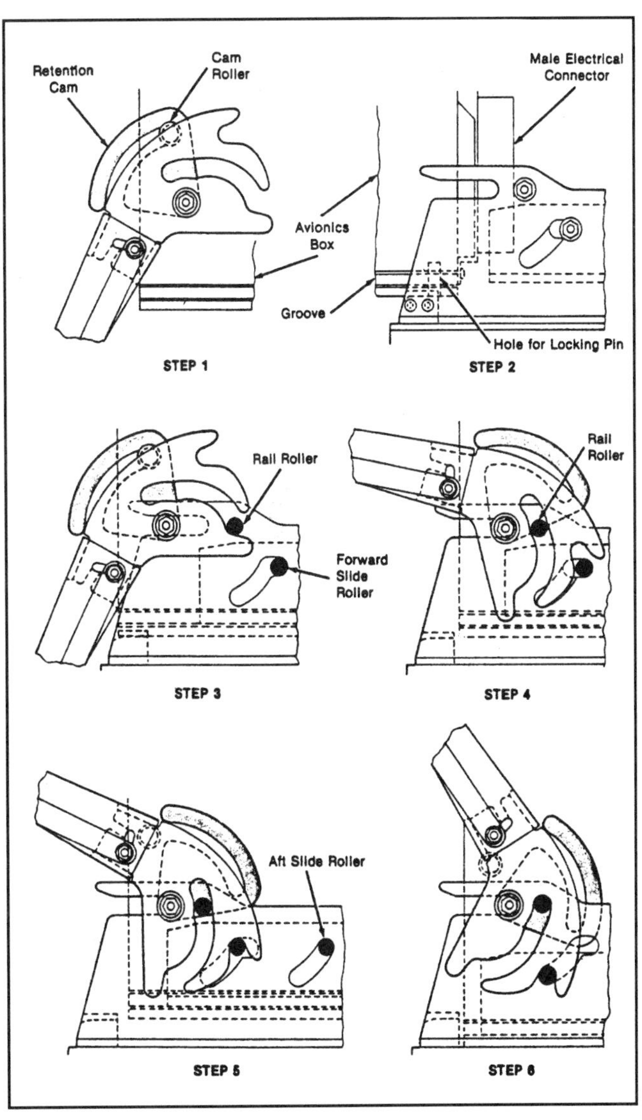

Fig. 2 This installation sequence shows the positions of the handle and retention cams as the box is moved rearward and downward.

Perpendicular-Force Latch (*continued*)

the technician or robot pushes the box rearward as slides attached to the rails enter grooves near the bottom of the box. In step 3, as the box continues to move to the rear, the handle cam automatically aligns with the slot in the rail and engages the rail roller.

In step 4, the handle is rotated upward 75°, forcing the box rearward to mate with the electrical connectors. In step 5, the handle is pushed upward an additional 15°, locking the handle cam and the slide. In step 6, the handle is rotated an additional 30°, forcing the box and the mating spring-loaded electrical connectors downward so that the box engages the locking pin and becomes clamped to the cold plate. The sequence for removal is identical except that the motions are reversed.

TWO QUICK-RELEASE MECHANISMS

QUICK-RELEASE MECHANISM

Quick release mechanisms have many applications. Although the design shown here operates as a tripping device for a quick-release hook, the mechanical principles involved have many other applications. Fundamentally, it is a toggle-type mechanism with the characteristic that the greater the load the more effective the toggle.

The hook is suspended from the shackle, and the load or work is supported by the latch, which is machined to fit the fingers C. The fingers C are pivoted about a pin. Assembled to the fingers are the arms E, pinned at one end and joined at the other by the sliding pin G. Enclosing the entire unit are the side plates H, containing the slot J for guiding the pin G in a vertical movement when the hook is released. The helical spring returns the arms to the bottom position after they have been released.

To trip the hook, the tripping lever is pulled by the cable M until the arms E pass their horizontal centerline. The toggle effect is then broken, releasing the load.

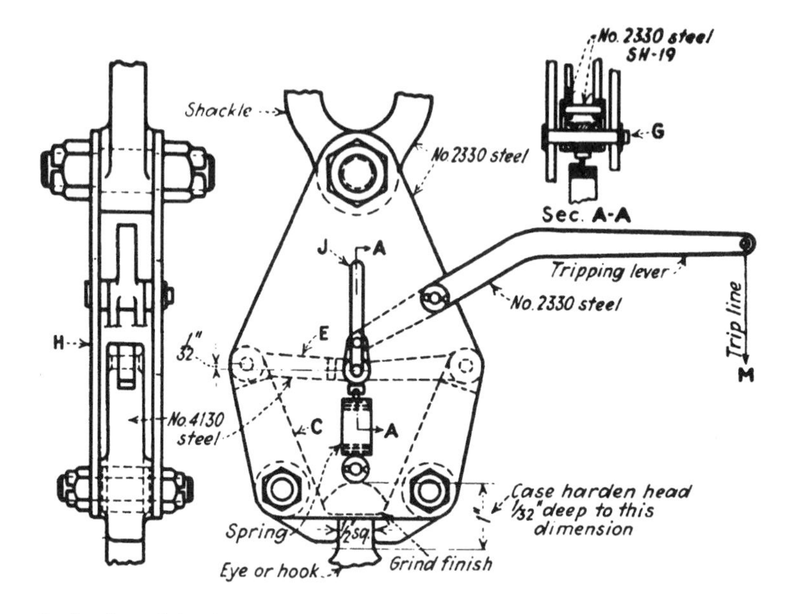

A simple quick-release toggle mechanism was designed for tripping a lifting hook.

POSITIVE LOCKING AND QUICK-RELEASE MECHANISM

The object here was to design a simple device that would hold two objects together securely and quickly release them on demand.

One object, such as a plate, is held to another object, such as a vehicle, by a spring-loaded slotted bolt, which is locked in position by two retainer arm. The retainer arms are constrained from movement by a locking cylinder. To release the plate, a detent is actuated to lift the locking cylinder and rotate the retainer arms free from contact with the slotted bolt head. As a result of this action, the spring-loaded bolt is ejected, and the plate is released from the vehicle.

The actuation of the slidable detent can be initiated by a squib, a fluid-pressure device, or a solenoid. The principle of this mechanism can be applied wherever a positive engagement that can be quickly released on demand is required. Some suggested applications for this mechanism are in coupling devices for load-carrying carts or trucks, hooks or pick-up attachments for cranes, and quick-release mechanisms for remotely controlled manipulators.

Detent
Retainer Arm
Locking Cylinder
Slot
Plate to be released
Vehicle
Bolt
Spring

This quick-release mechanism is shown locking a vehicle and plate.

RING SPRINGS CLAMP PLATFORM ELEVATOR INTO POSITION

A simple yet effective technique keeps a platform elevator locked safely in position without an external clamping force. The platform (see drawing) contains special ring assemblies that grip the four column-shafts with a strong force by the simple physical interaction of two tapered rings.

Thus, unlike conventional platform elevators, no outside power supply is required to hold the platform in position. Conventional jacking power is employed, however, in raising the platform from one position to another.

How the rings work. The ring assemblies are larger versions of the ring springs sometimes installed for shock absorption. In this version, the assembly is made up of an inner nonmetallic ring tapering upward and an outer steel ring tapered downward (see drawing).

The outside ring is linked to the platform, and the inside ring is positioned against the circumference of the column shaft. When the platform is raised to the designed height, the jack force is removed, and the full weight of the platform bears downward on the outside ring with a force that, through a wedging action, is transferred into a horizontal inward force of the inside ring.

Thus, the column shaft is gripped tightly by the inside ring; the heavier the platform the larger the gripping force produced.

The advantage of the technique is that the shafts do not need notches or threads, and cost is reduced. Moreover, the shafts can be made of reinforced concrete.

Outside steel ring
Inside nonmetallic ring
Column shaft
Platform rising
Ring-springs clamping

Ring springs unclamp the column as the platform is raised (upper). As soon as the jack power is removed (lower), the column is gripped by the inner ring.

CAMMED JAWS IN HYDRAULIC CYLINDER GRIP SHEET METAL

A single, double-acting hydraulic cylinder in each work holder clamps and unclamps the work and retracts or advances the jaws as required. With the piston rod fully withdrawn into the hydraulic cylinder (A), the jaws of the holder are retracted and open. When the control valve atop the work holder is actuated, the piston rod moves forward a total of 12 in. The first 10 in. of movement (B) brings the sheet-locater bumper into contact with the work. The cammed surface on the rod extension starts to move the trip block upward, and the locking pin starts to drop into position. The next $^{3}/_{4}$ in. of piston-rod travel (C) fully engages the work-holder locking pin and brings the lower jaw of the clamp up to the bottom of the work. The work holder slide is now locked between the forward stop and the locking pin. The last $1^{1}/_{4}$ in. of piston travel (D) clamps the workpiece between the jaws with a pressure of 2500 lbs. No adjustment for work thickness is necessary. A jaws-open limit switch clamps the work holder in position (C) for loading and unloading operations.

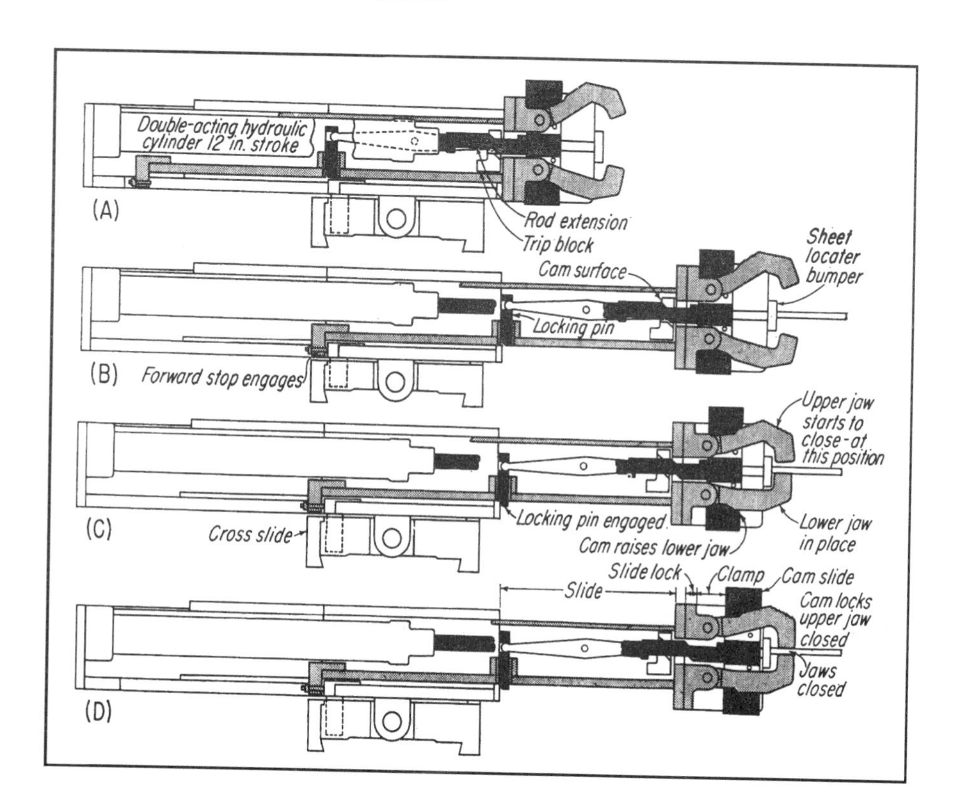

QUICK-ACTING CLAMPS FOR MACHINES AND FIXTURES

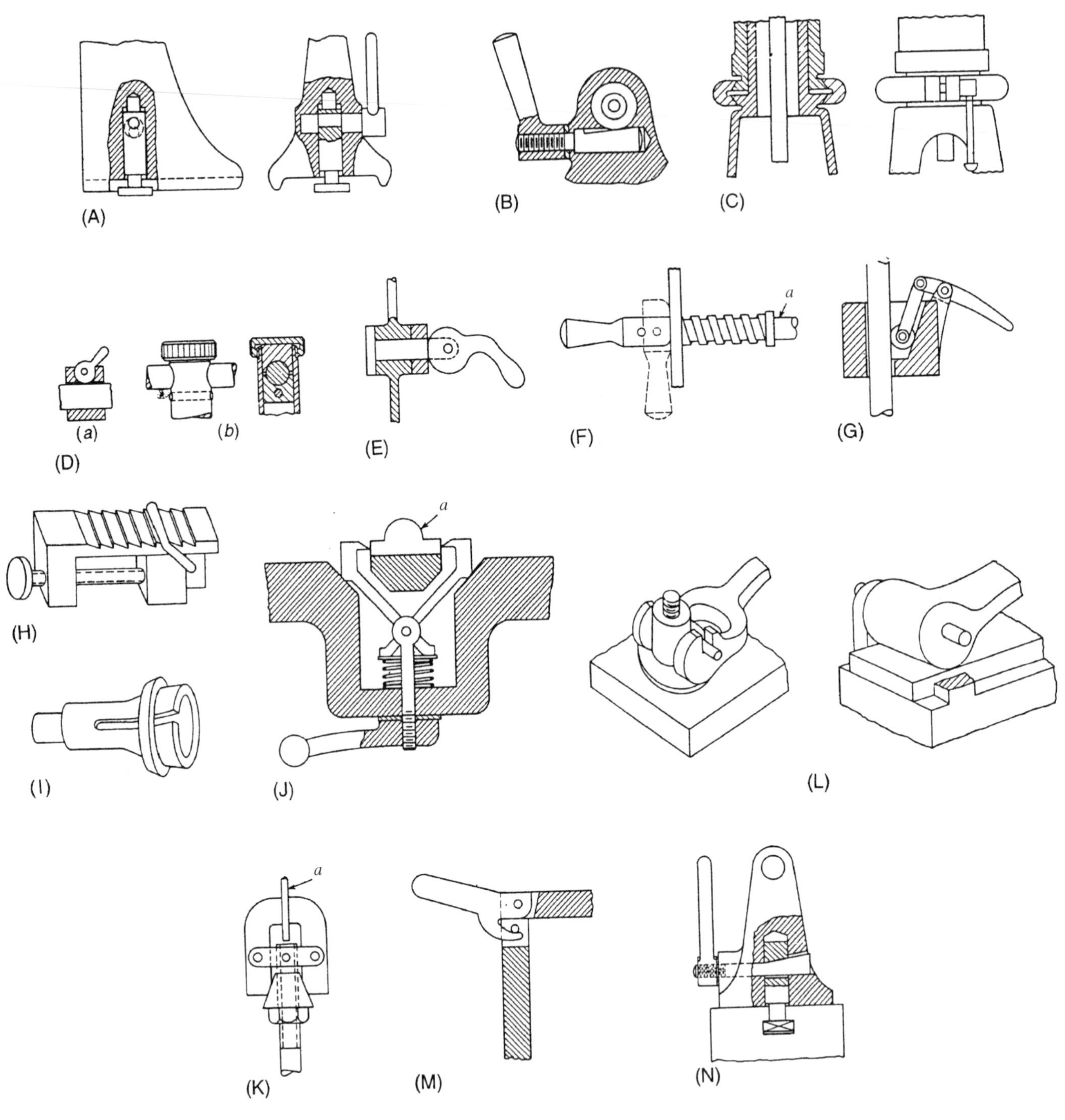

(A) An eccentric clamp. (B) A spindle-clamping bolt. (C) A method for clamping a hollow column to a structure. It permits quick rotary adjustment of the column. (D) (a) A cam catch for clamping a rod or rope. (b) A method for fastening a small cylindrical member to a structure with a thumb nut and clamp jaws. It permits quick longitudinal adjustment of a shaft in the structure. (E) A cam catch can lock a wheel or spindle. (F) A spring handle. Movement of the handle in the vertical or horizontal position provides movement at a. (G) A roller and inclined slot for locking a rod or rope. (H) A method for clamping a light member to a structure. The serrated edge on the structure permits the rapid accommodation of members with different thicknesses. (I) A spring taper holder with a sliding ring. (J) A special clamp for holding member a. (K) The cone, nut, and levers grip member a. The grip can have two or more jaws. With only two jaws, the device serves as a small vise. (L) Two different kinds of cam clamps. (M) A cam cover catch. Movement of the handle downward locks the cover tightly. (N) The sliding member is clamped to the slotted structure with a wedge bolt. This permits the rapid adjustment of a member on the structure.

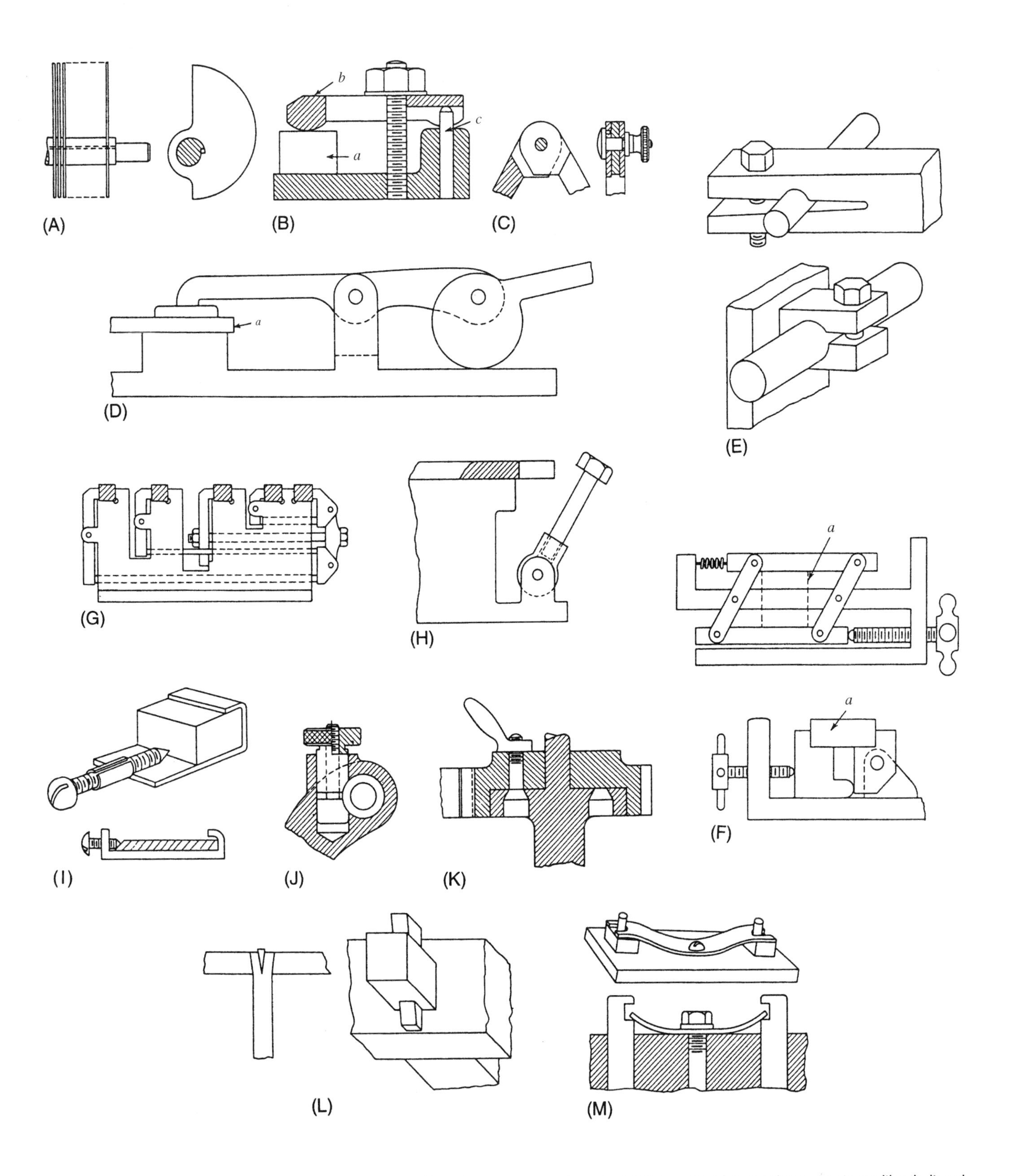

(A) A method for fastening capacitor plates to a structure with a circular wedge. Rotation of the plates in a clockwise direction locks the plates to the structure. (B) A method for clamping member *a* with a special clamp. Detail *b* pivots on pin *c*. (C) A method for clamping two movable parts so that they can be held in any angular position with a clamping screw. (D) A cam clamp for clamping member *a*. (E) Two methods for clamping a cylindrical member. (F) Two methods for clamping member *a* with a special clamp. (G) A special clamping device that permits the parallel clamping of five parts by the tightening of one bolt. (H) A method for securing a structure with a bolt and a movable detail that provides a quick method for fastening the cover. (I) A method for quickly securing, adjusting, or releasing the center member. (J) A method for securing a bushing in a structure with a clamp screw and thumb nut. (K) A method for securing an attachment to a structure with a bolt and hand lever used as a nut. (L) A method for fastening a member to a structure with a wedge. (M) Two methods for fastening two members to a structure with a spring and one screw. The members can be removed without loosening the screw.

NINE FRICTION CLAMPING DEVICES

Many different devices for gaining mechanical advantage have been used in the design of friction clamps. These clamps can grip moderately large loads with comparatively small smooth surfaces, and the loads can be tightened or released with simple controls. The clamps illustrated here can be tightened or released with screws, levers, toggles, wedges, and combinations of them.

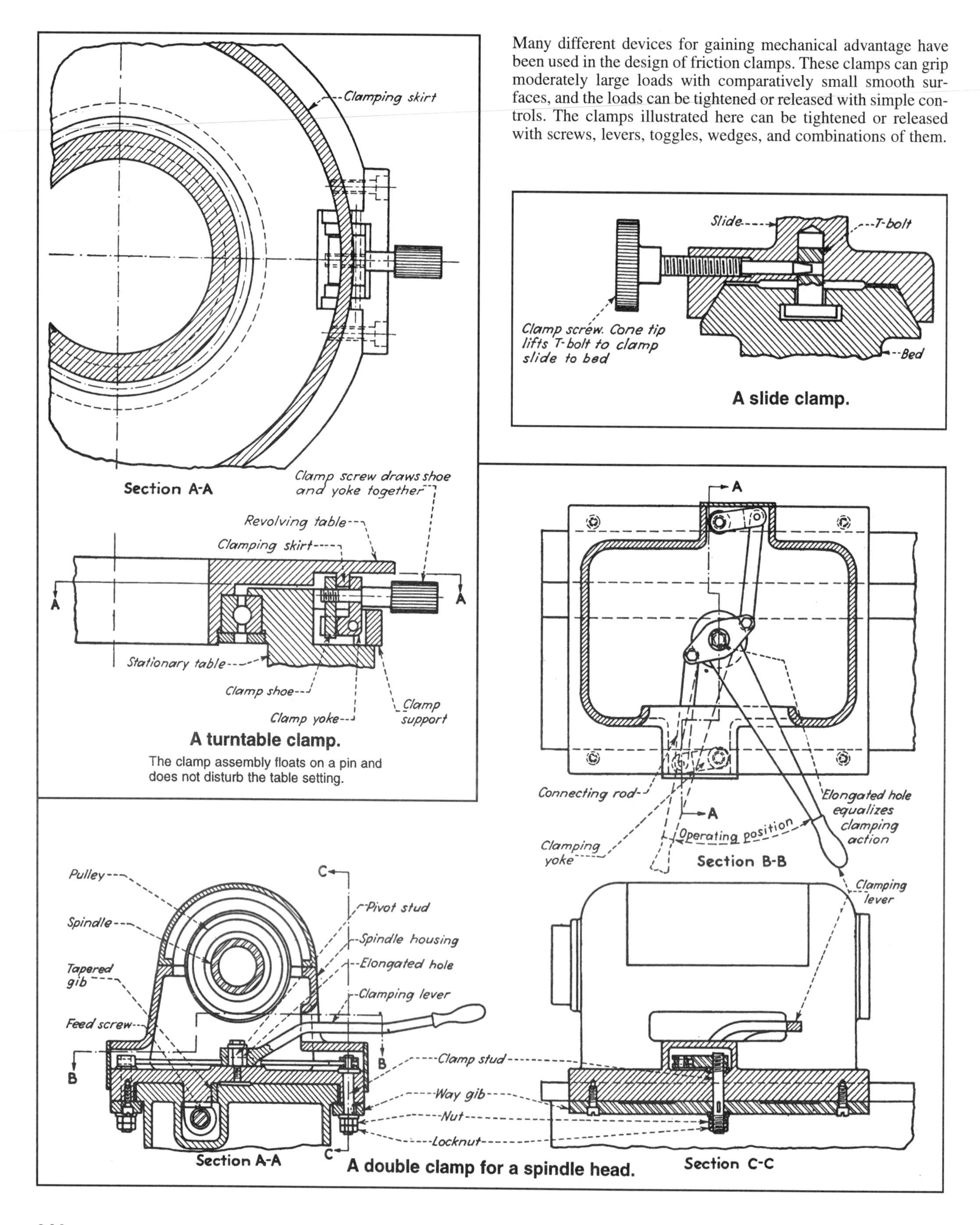

A turntable clamp.

The clamp assembly floats on a pin and does not disturb the table setting.

A slide clamp.

A double clamp for a spindle head.

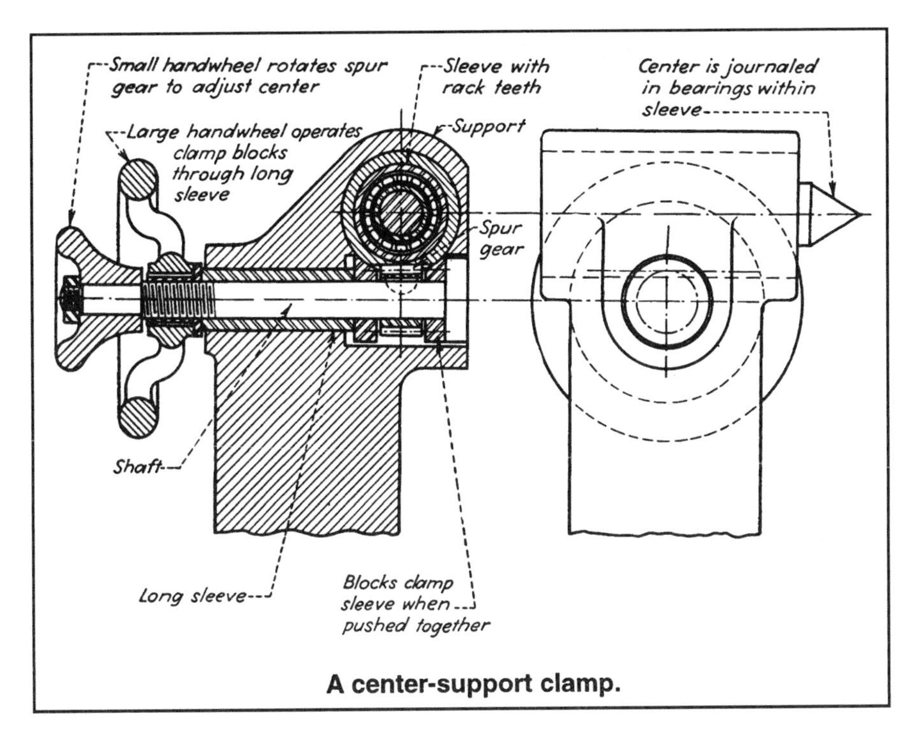

A center-support clamp.

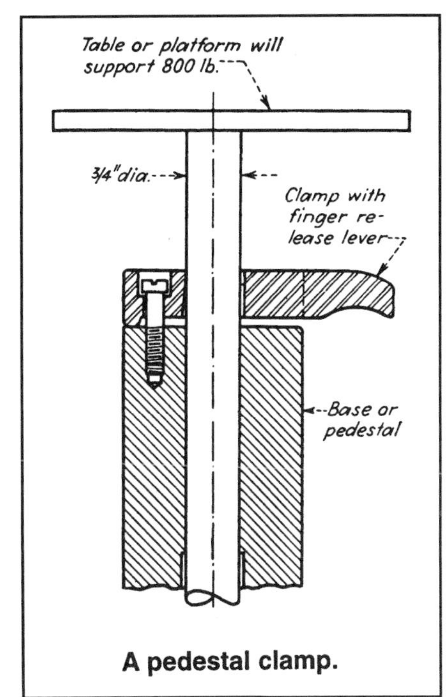

A pedestal clamp.

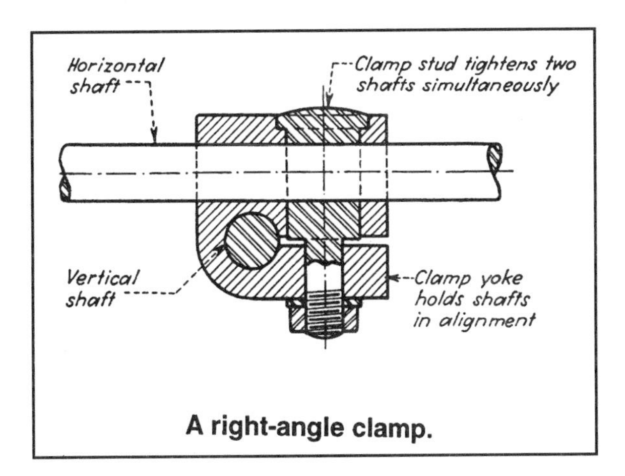

A right-angle clamp.

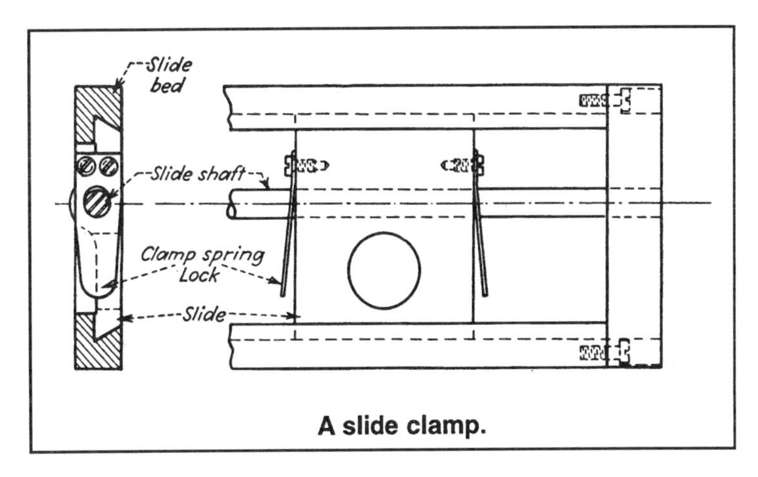

A slide clamp.

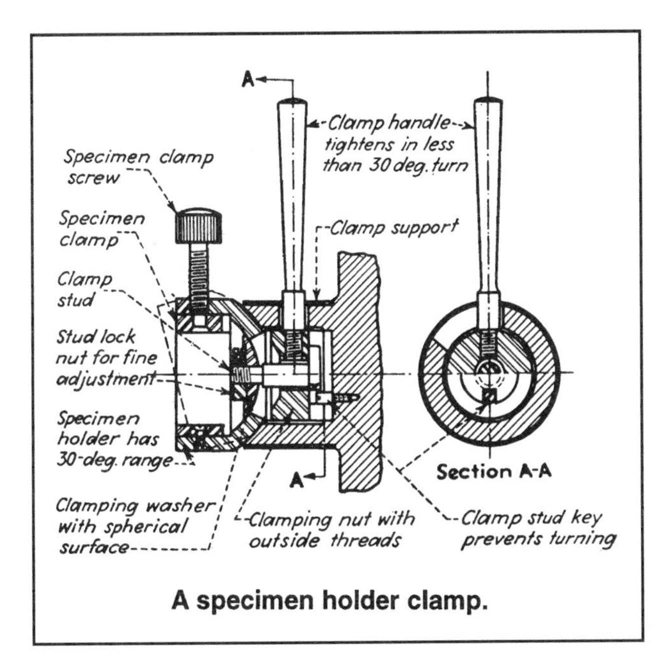

A specimen holder clamp.

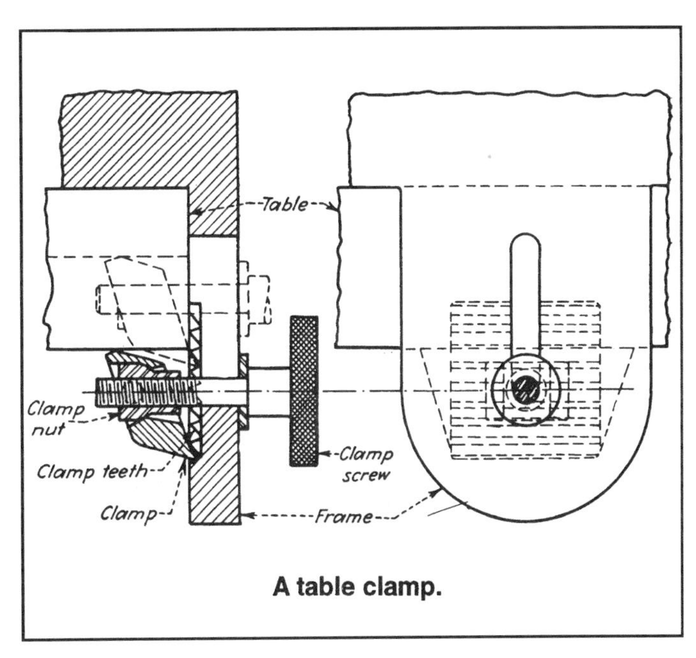

A table clamp.

DETENTS FOR STOPPING MECHANICAL MOVEMENTS

Some of the more robust and practical devices for stopping mechanical movements are illustrated here.

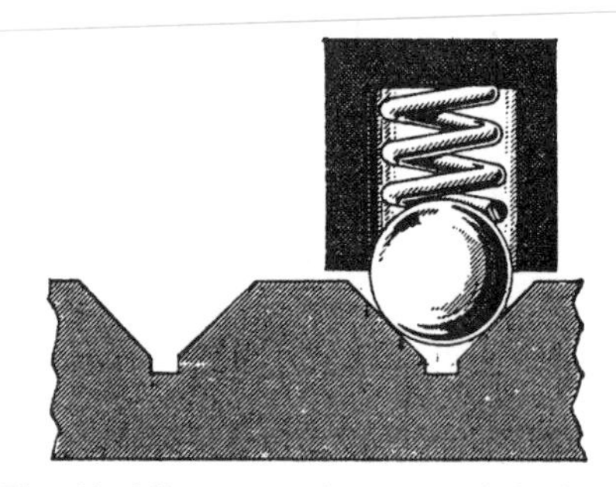

Fixed holding power is constant in both directions.

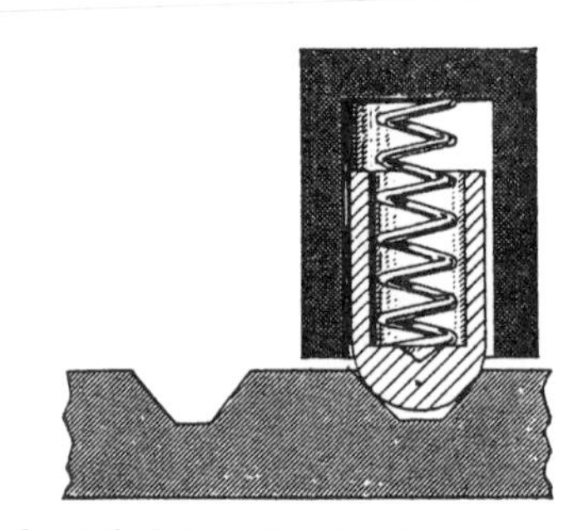

A domed plunger has long life.

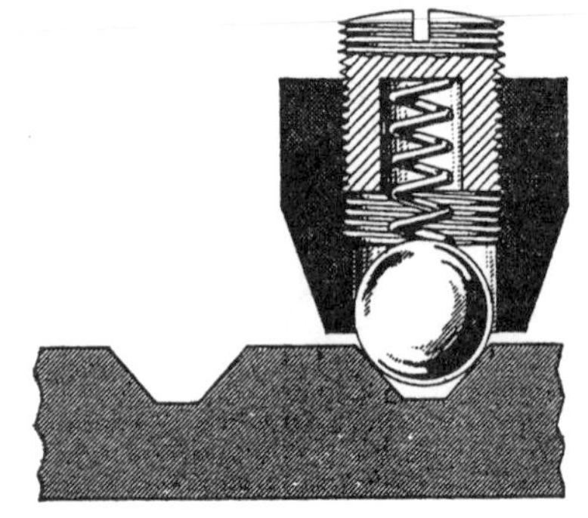

The screw provides adjustable holding.

Wedge action locks the movement in the direction of the arrow.

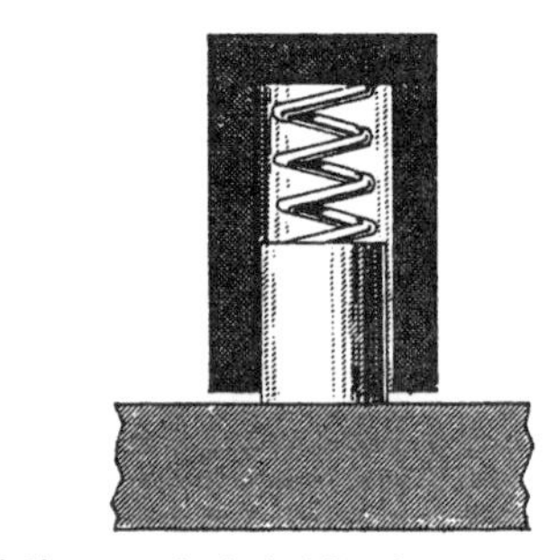

Friction results in holding force.

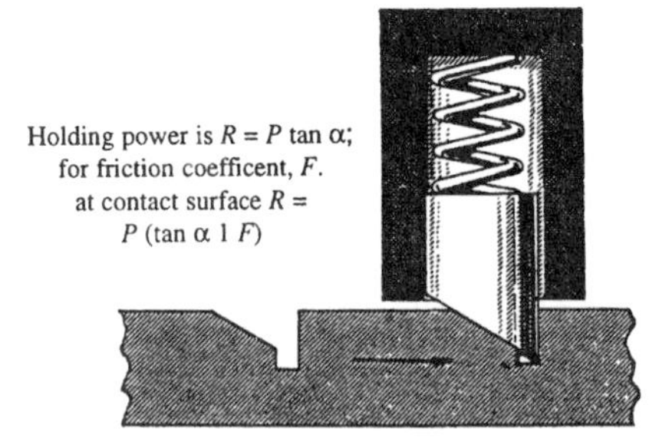

A notch shape dictates the direction of rod motion.

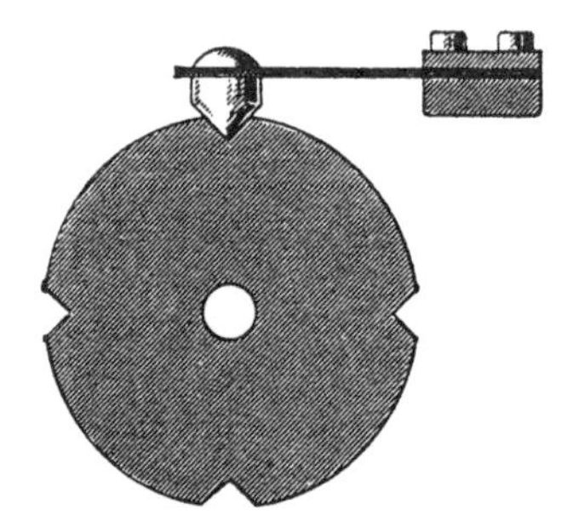

A leaf spring provides limited holding power.

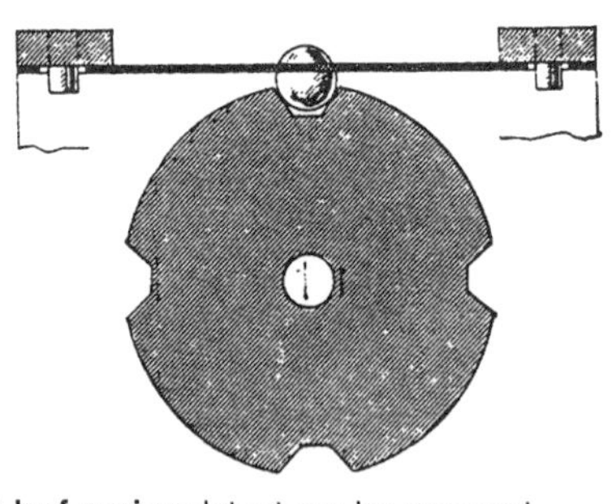

A leaf-spring detent can be removed quickly.

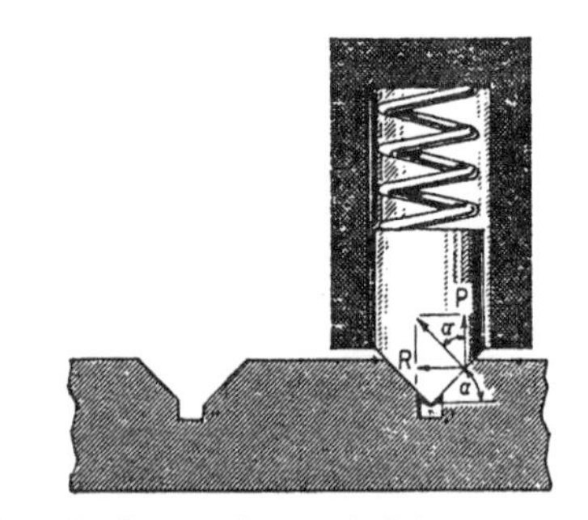

A conical or wedge-ended detent.

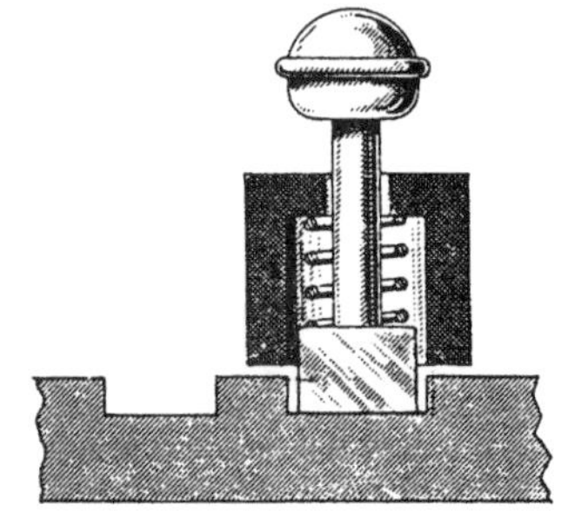

A positive detent has a manual release.

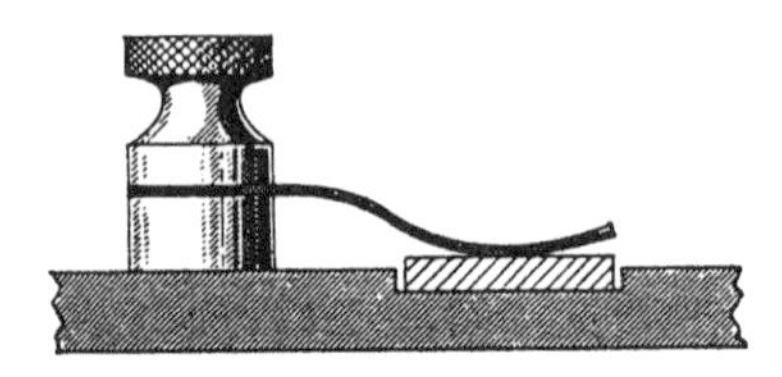

.**A leaf spring** for holding flat pieces.

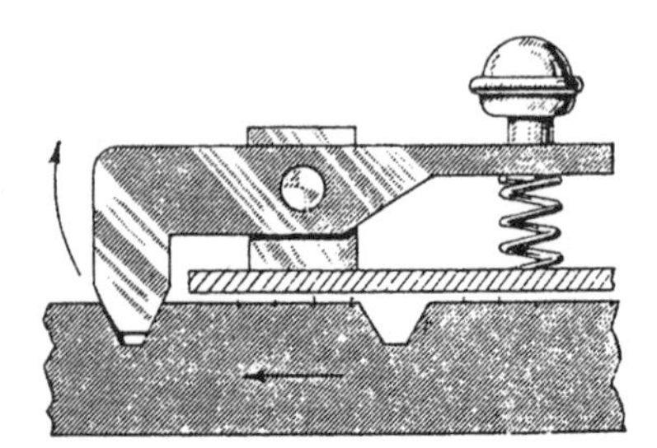

An automatic release occurs in one direction; manual release is needed in the other direction.

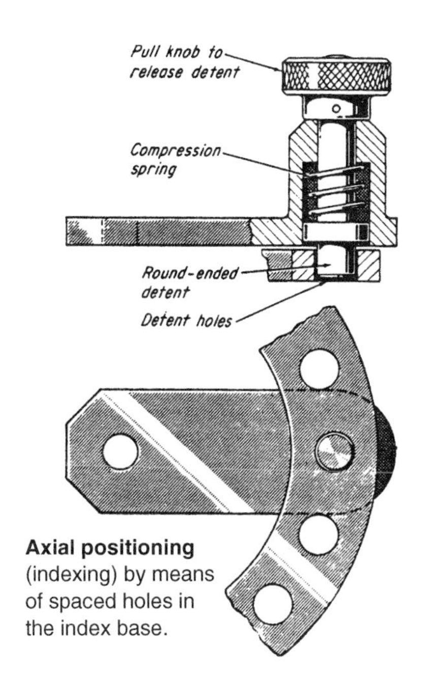

Axial positioning (indexing) by means of spaced holes in the index base.

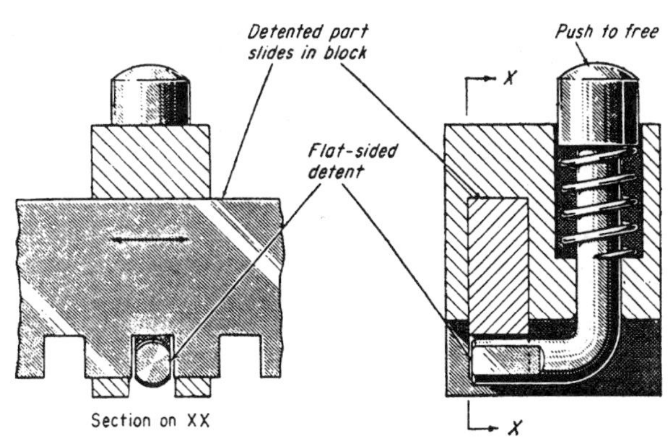

A positive detent has a push-button release for straight rods.

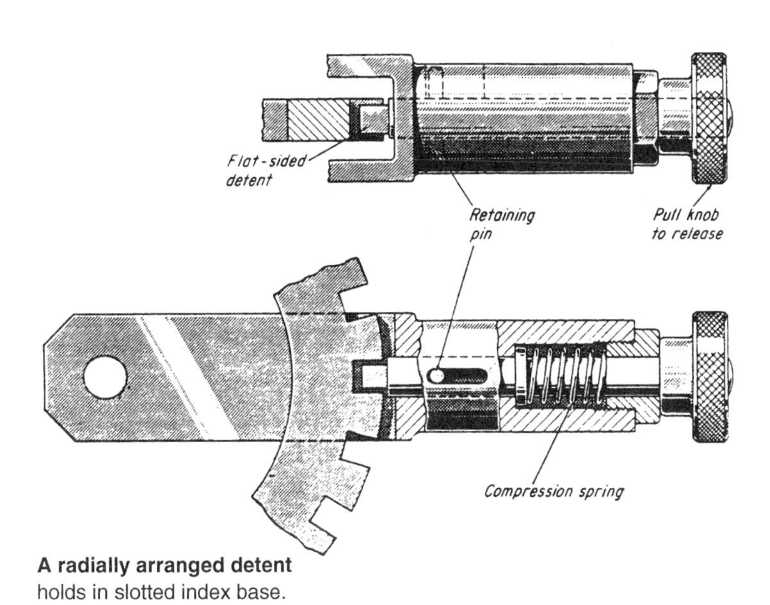

A radially arranged detent holds in slotted index base.

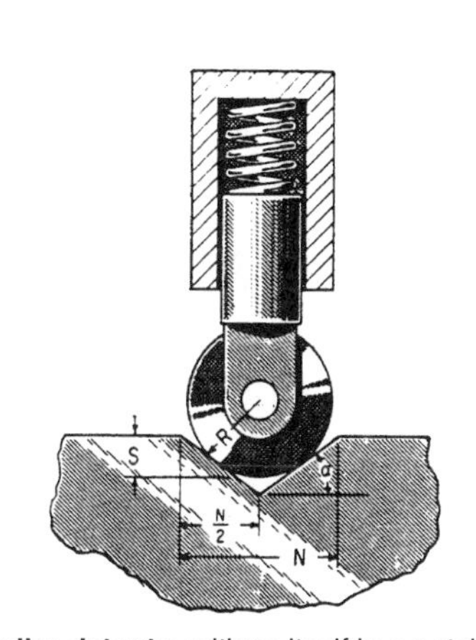

A roller detent positions itself in a notch.

$$\text{Rise, } S = \frac{N \tan a}{2} - R \times \frac{1 - \cos}{\cos a}$$

Roller Radius, $R =$

$$\left(\frac{N \tan a}{2} - S\right)\left(\frac{\cos a}{1 - \cos a}\right)$$

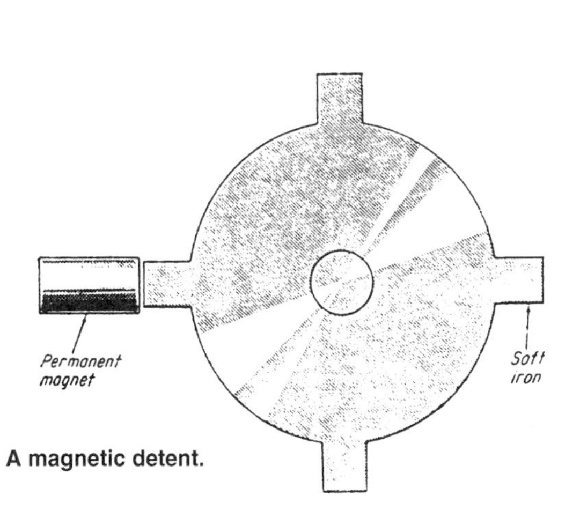

A magnetic detent.

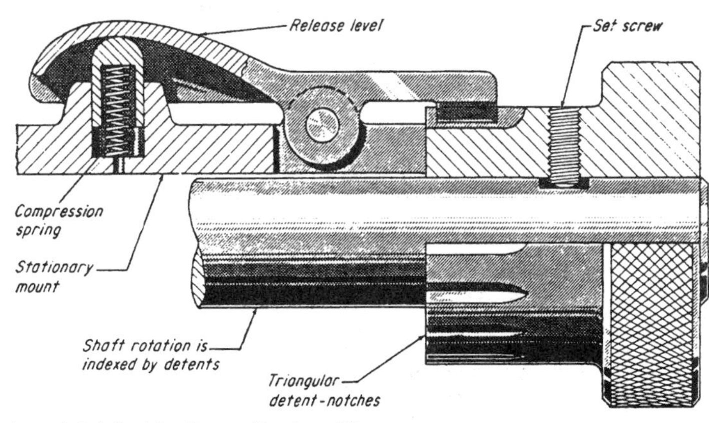

An axial detent for the positioning of the adjustment knob with a manual release.

TWELVE CLAMPING METHODS FOR ALIGNING ADJUSTABLE PARTS

Methods for clamping parts that must be readily movable are as numerous and as varied as the requirements. In many instances, a clamp of any design is satisfactory, provided it has sufficient strength to hold the parts immovable when tightened. However, it is sometimes necessary that the movable part be clamped to maintain accurate alignment with some fixed part. Examples of these clamps are described and illustrated.

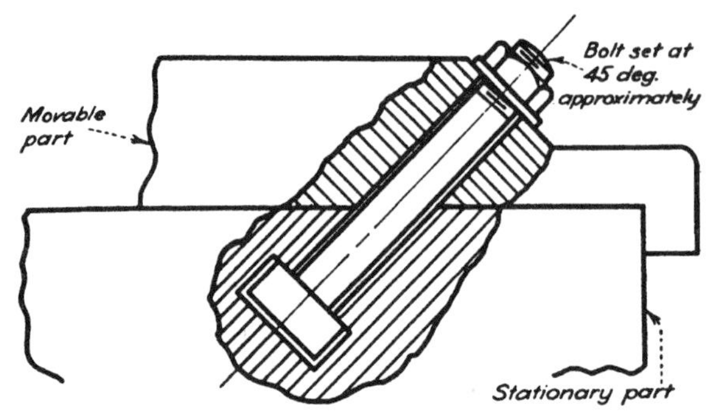

Fig. 1 When a nut is tightened, the flange on the edge of the movable part is drawn against the machined edge of the stationary part. This method is effective, but the removal of the clamped part can be difficult if it is heavy or unbalanced.

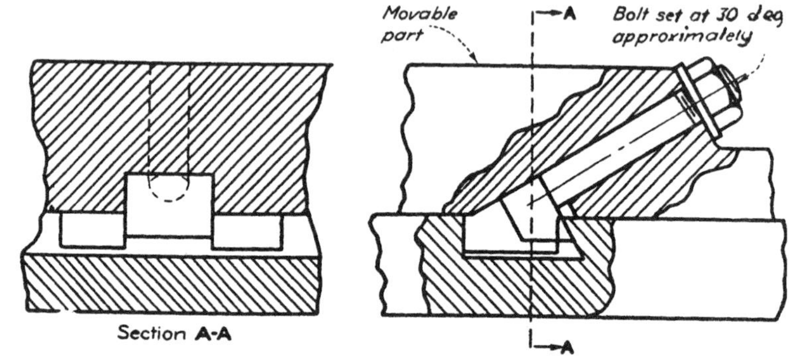

Fig. 2 The lower edged of the bolt head contacts the angular side of locating groove, causing the keys to be held tightly against the opposite side of the groove. This design permits easy removal of the clamped part, but it is effective only if the working pressure is directly downward or in a direction against the perpendicular side of the slot.

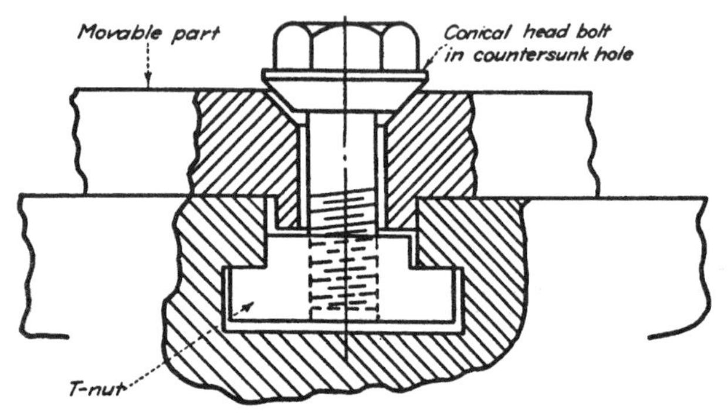

Fig. 3 The movable part is held against one side of the groove while the T-nut is forced against the other side. The removal of the screw permits easy removal of the clamped part. Heavy pressure toward the side of the key out of contact with the slot can permit slight movement due to the springing of the screw.

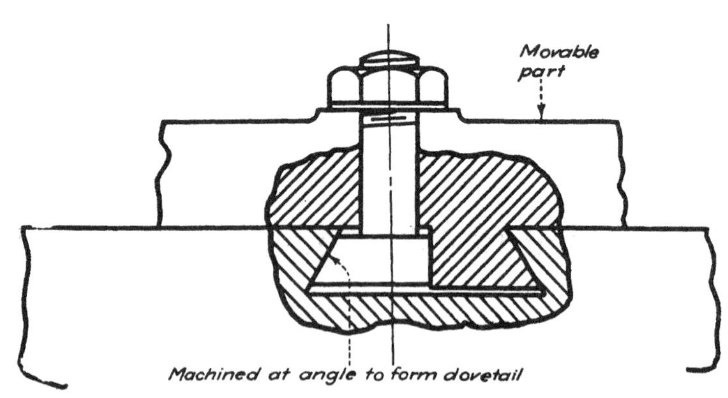

Fig. 4 One side of the bolt is machined at an angle to form a side of the dovetail, which tightens in the groove as the nut is drawn tight. The part must be slid the entire length of the slot for removal.

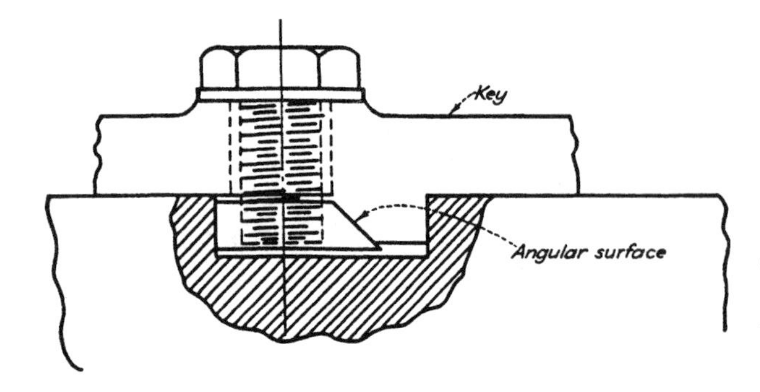

Fig. 5 The angular surface of the nut contacts the angular side of the key, and causes it to move outward against the side of the groove. This exerts a downward pull on the clamped part due to the friction of the nut against the side of a groove as the nut is drawn upward by the screw.

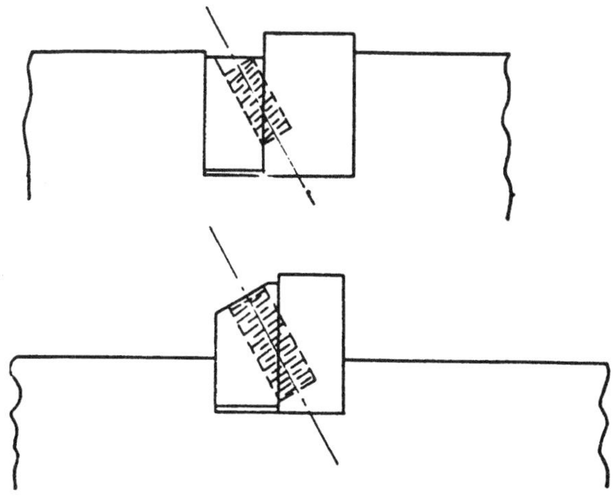

Fig. 6 and 7 These designs differ only in the depth of the grooves. They cannot withstand heavy pressure in an upward direction but have the advantage of being applicable to narrow grooves.

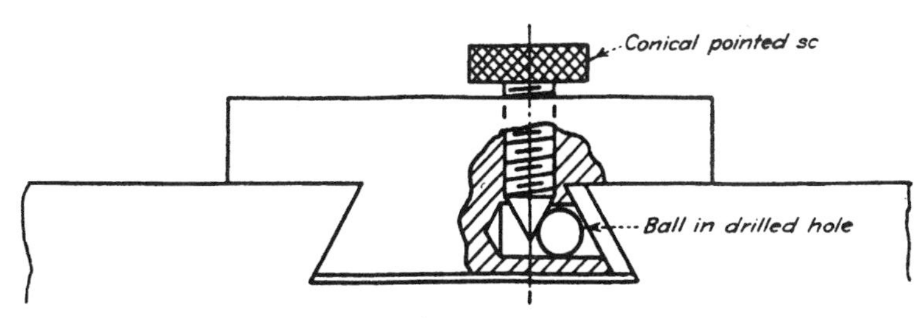

Fig. 8 Screw contact causes the ball to exert an outward pressure against the gib. The gib is loosely pinned to the movable part. This slide can be applied to broad surfaces where it would be impractical to apply adjusting screws through the stationary part.

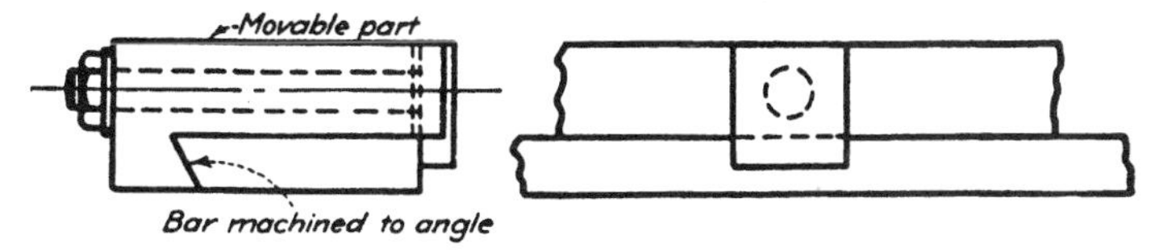

Fig. 10 One edge of a bar is machined at an angle which fits into mating surfaces on the movable part. When the bolt, which passes through the movable part, is drawn tight, the two parts are clamped firmly together.

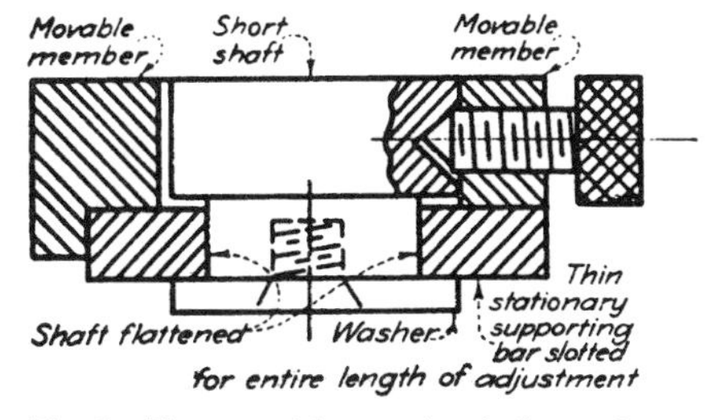

Fig. 9 The movable member is flanged on one side and carries a conical pointed screw on the other side. A short shaft passes through both members and carries a detent slightly out of alignment with the point of the screw. This shaft is flattened on opposite sides where it passes through the stationary member to prevent it from turning when the movable member is removed. A heavy washer is screwed to the under side of the shaft. When the knurled screw is turned inward, the shaft is drawn upward while the movable member is drawn downward and backward against the flange. The shaft is forced forward against the edge of the slot. The upper member can thus be moved and locked in any position. Withdrawing the point of the screw from the detent in the shaft permits the removal of the upper member.

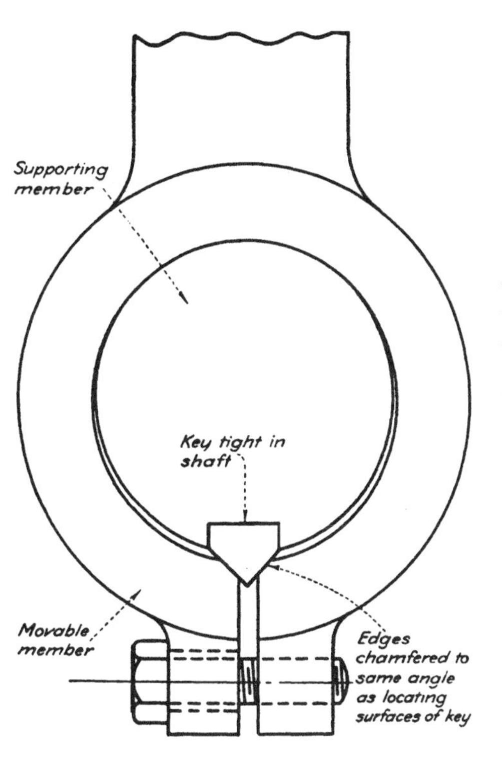

Fig. 11 As the screw is tightened, the chamfered edges of the cut tend to ride outward on the angular surfaces of the key. This draws the movable member tightly against the opposite side of the shaft.

Fig. 12 As the screw is turned, it causes the movable side, which forms one side of the dovetail groove, to move until it clamps tightly on the movable member. The movable side should be as narrow as possible, because there is a tendency for this part to ride up on the angular surface of the clamped part.

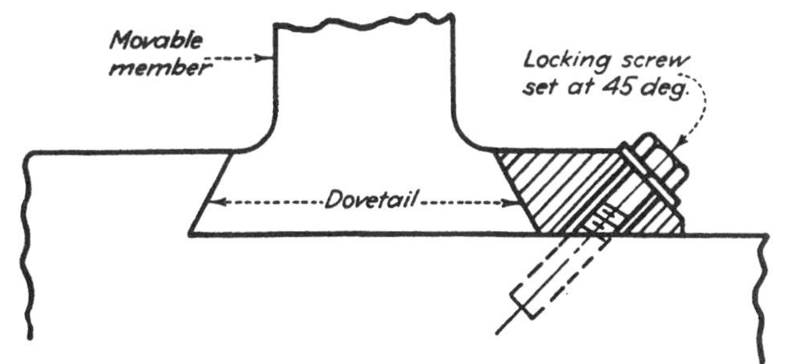

SPRING-LOADED CHUCKS AND HOLDING FIXTURES

Spring-loaded fixtures for holding work can be preferable to other fixtures. Their advantages are shorter setup time and quick workpiece change. Work distortion is reduced because the spring force can be easily and accurately adjusted.

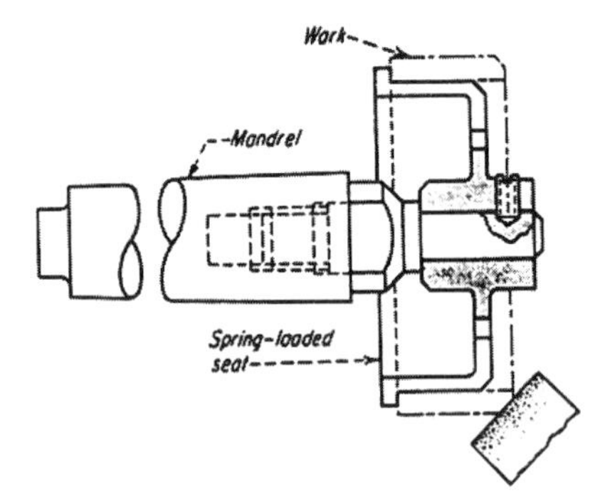

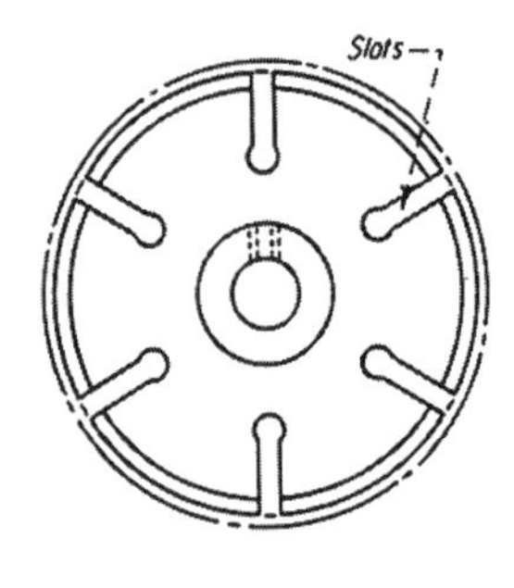

A spring-loaded nest has radial slots extending into its face. These ensure an even grip on the work, which is pushed over the rim. A slight lead on the rim makes mounting work easier. The principal application of this fixture is for ball-bearing race grinding where only light cutting forces are applied.

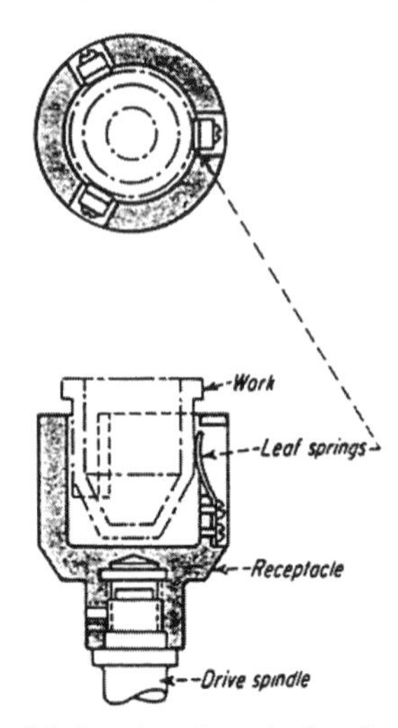

A cupped fixture has three leaf-springs equally spaced in a wall. The work, usually to be lacquered, is inserted into the cup during its rotation. Because the work is placed in the fixture by hand, the spindle is usually friction-driven for safety.

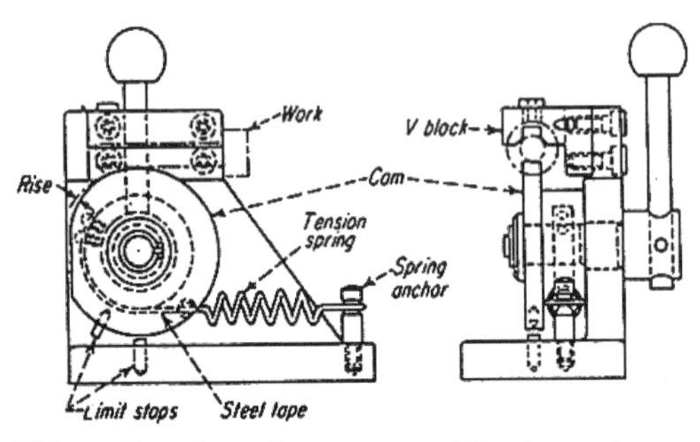

This spring clamp has a cam-and-tension spring that applies a clamping force. A tension spring activates the cam through a steel band. When the handle is released, the cam clamps the work against the V-bar. Two stop-pins limit travel when there is no work in the fixture.

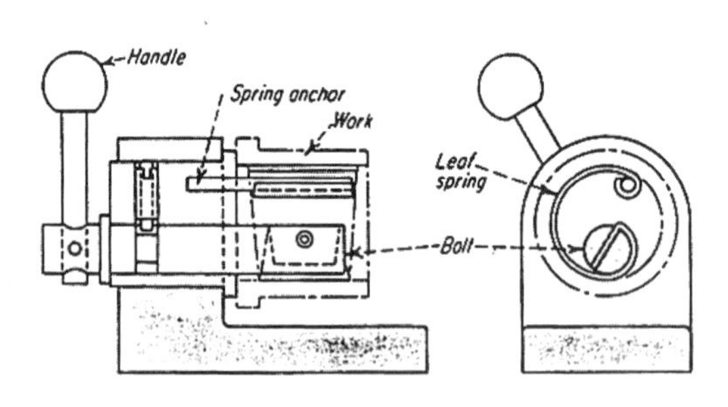

A leaf-spring gripper is used mainly to hold work during assembly. One end of a flat coil-spring is anchored in the housing; the other end is held in a bolt. When the bolt is turned, the spring is tightened, and its outside diameter is decreased. After the work is slid over the spring, the bolt handle is released. The spring then presses against the work, holding it tight.

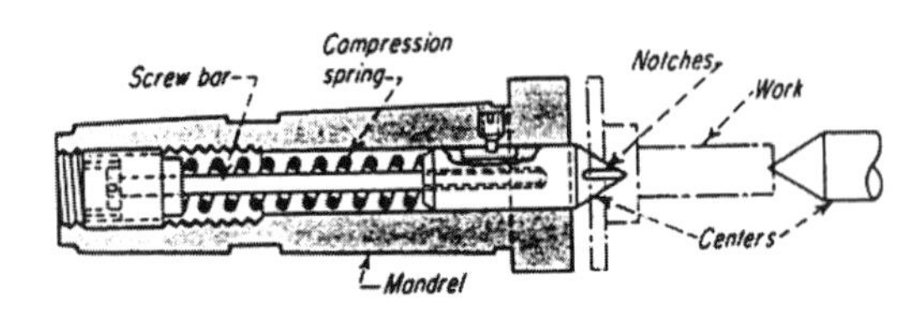

This lathe center is spring loaded and holds the work with spring pressure alone. Eight sharp-edged notches on the conical surface of the driving center bite into the work and drive it. Its spring tension is adjustable.

CHAPTER 10

CHAIN AND BELT DEVICES AND MECHANISMS

TWELVE VARIABLE-SPEED BELT AND CHAIN DRIVES

Variable-speed drives provide an infinite number of speed ratios within a specific range. They differ from the stepped-pulley drives in that the stepped drives offer only a discrete number of velocity ratios.

Mechanical "all-metal" drives employ friction or preloaded cones, disks, rings, and spheres, which undergo a certain amount of slippage. Belt drives, on the other hand, have little slippage or frictional losses, and chain has none—it maintains a fixed phase relationship between the input and output shafts.

Belt Drives

Belt drives offer high efficiency and are relatively low in price. Most use V-belts, reinforced by steel wires to 3 inches in width.

Speed adjustment in belt drives is obtained through one of the four basic arrangements shown below.

Variable-distance system (Fig. 1). A variable-pitch sheave on the input shaft opposes a solid (fixed-pitch) sheave on the output shaft. To vary the speed, the center distance is varied, usually by an adjustable base, tilting or sliding motor (Fig. 6).

Speed variations up to 4:1 are easily achieved, but torque and horsepower characteristics depend on the location of the variable-diameter sheave.

Fixed-distance system (Fig. 2). Variable-pitch sheaves on both input and output shafts maintain a constant center distance between shafts. The sheaves are controlled by linkage. Either the pitch diameter of one sheave is positively controlled and the disks of the other sheave under spring tension, adjust automatically or the pitch diameters of both sheaves are positively controlled by the linkage system (Fig. 5). Pratt & Whittney has applied the system in Fig. 5 to the spindle drive of numerically controlled machines.

Speed variations up to 11:1 are obtained, which means that with a 1200-rpm motor, the maximum output speed will be $1200\sqrt{11} = 3984$ rpm, and the minimum output speed = 3984/11 = 362 rpm.

Double-reduction system (Fig. 3). Solid sheaves are on both the input and output shafts, but both sheaves on the intermediate shaft are of variable-pitch type. The center distance between input and output is constant.

Coaxial shaft system (Fig. 4). The intermediate shaft in this arrangement permits the output shaft to be coaxial with that of the input shaft. To maintain a fixed center distance, all four sheaves must be of the variable-pitch type and controlled by linkage, similar to the system in Fig. 6. Speed variation up to 16:1 is available.

Packaged belt units (Fig. 7). These combine the motor and variable-pitch transmissions as an integral unit. The belts are usually ribbed, and speed ratios can be dialed by a handle.

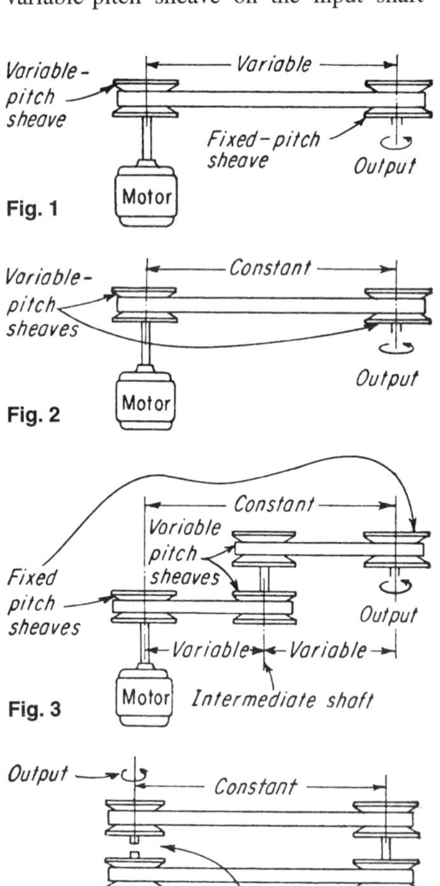

Four basic belt arrangements for varying output speed.

Fig. 5 Linkage controlled pulleys.

Fig. 6 Tandem arrangement employs dual belt-system to produce high speed-reduction.

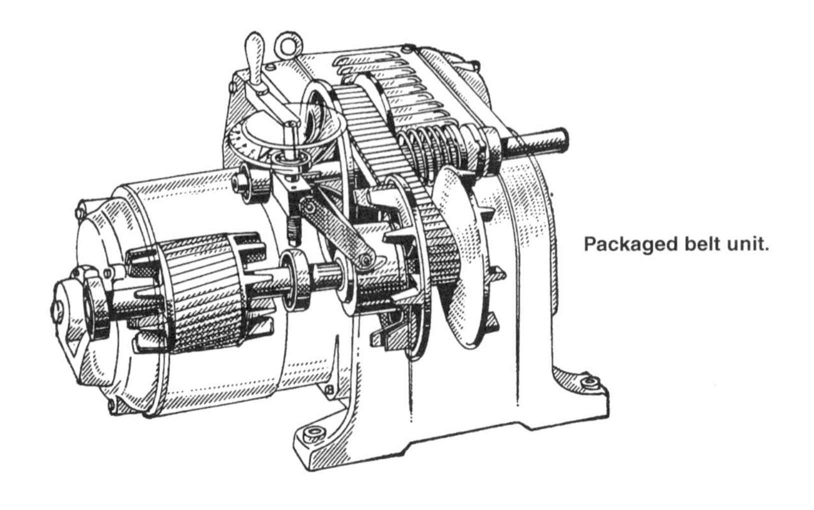

Packaged belt unit.

Sheave Drives

The axial shifting of variable-pitch sheaves is controlled by one of four methods:

Linkage actuation. The sheave assemblies in Fig. 5 are directly controlled by linkages which, in turn, are manually adjusted.

Spring pressure. The cons of the sheaves in Figs. 2 and 4 are axially loaded by spring force. A typical pulley of this type is illustrated in Fig. 8. These pulleys are used in conjunction with directly controlled sheaves, or with variable center-distance arrangements.

Cam-controlled sheave. The cones of this sheave (Fig. 9) are mounted on a floating sheave, free to rotate on the pulley spindle. Belt force rotates the cones, whose surfaces are cammed by the inclined plane of the spring. The camming action wedges the cones against the belt, thus providing sufficient pressure to prevent slippage at the higher speeds, as shown in the curve.

Centrifugal-force actuator. In this unique sheave arrangement (Fig. 10) the pitch diameter of the driving sheave is controlled by the centrifugal force of steel balls. Another variable-pitch pulley mounted on the driven shaft is responsive to the torque. As the drive speed increases, the centrifugal force of the balls forces the sides of the driving sheave together. With a change in load, the movable flange of the driven sheave rotates in relation to the fixed flange. The differential rotation of the sheave flanges cams them together and forces the V-belt to the outer edge of the driven sheave, which has a lower transmission ratio. The driving sheave is also shifted as the load rises with decreasing speed. With a stall load, it is moved to the idling position. When the torque-responsive sheave is the driving member, any increase" the flanges of the centrifugal member, thus maintaining a constant output speed. The drive has performed well in transmissions with ratings ranging from 2 to 12 hp.

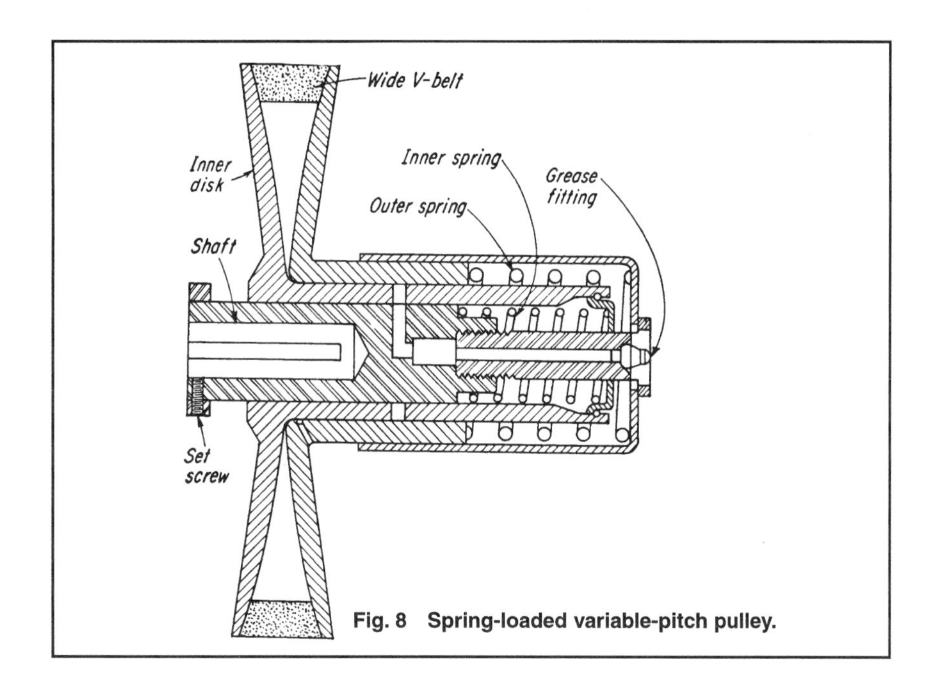

Fig. 8 Spring-loaded variable-pitch pulley.

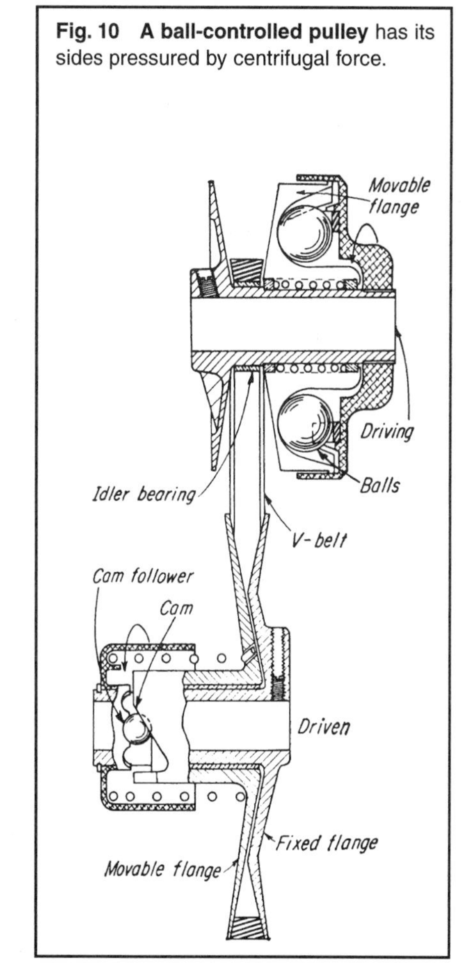

Fig. 10 A ball-controlled pulley has its sides pressured by centrifugal force.

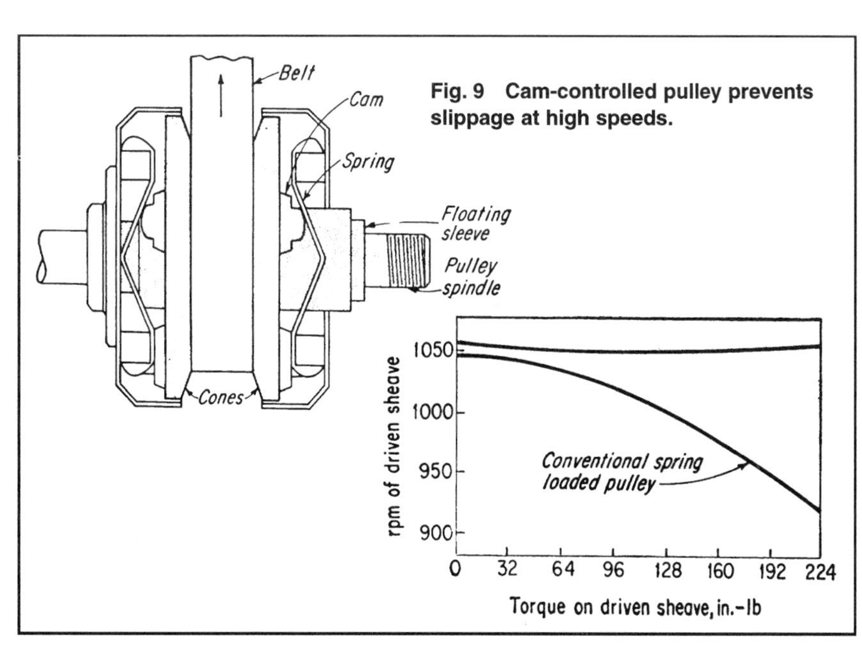

Fig. 9 Cam-controlled pulley prevents slippage at high speeds.

Chain Drives

PIV drive (Fig. 11). This chain drive (positive, infinitely variable) eliminates any slippage between the self-forming laminated chain teeth and the chain sheaves. The individual laminations are free to slide laterally to take up the full width of the sheave. The chain runs in radially grooved faces of conical surface sheaves which are located on the input and output shafts. The faces are not straight cones, but have a slight convex curve to maintain proper chain tension at all positions. The pitch diameters of both sheaves are positively controlled by the linkage. Booth action is positive throughout operating range. It is rated to 25 hp with speed variation of 6:1.

Double-roller chain drive (Fig. 12). This specially developed chain is built for capacities to 22 hp. The hardened rollers are wedged between the hardened conical sides of the variable-pitch sheaves. Radial rolling friction results in smooth chain engagement.

Single-roller chain drive (Fig. 13). The double strand of this chain boosts the capacity to 50 hp. The scissor-lever control system maintains the proper proportion of forces at each pair of sheave faces throughout the range.

Fig. 11 A PIV drive chain grips radially grooved faces of a variable-pitch sheave to prevent slippage.

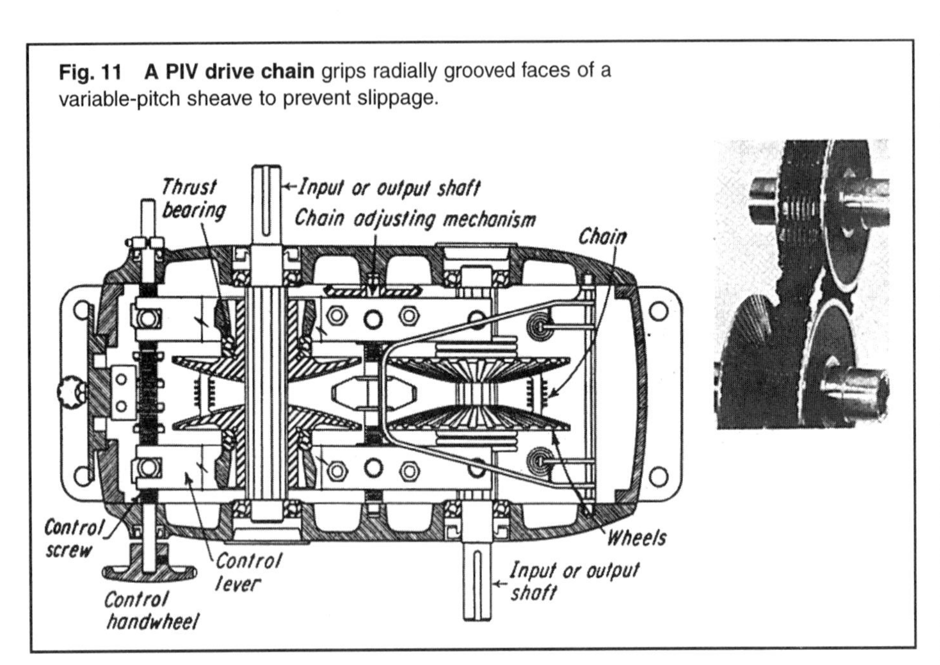

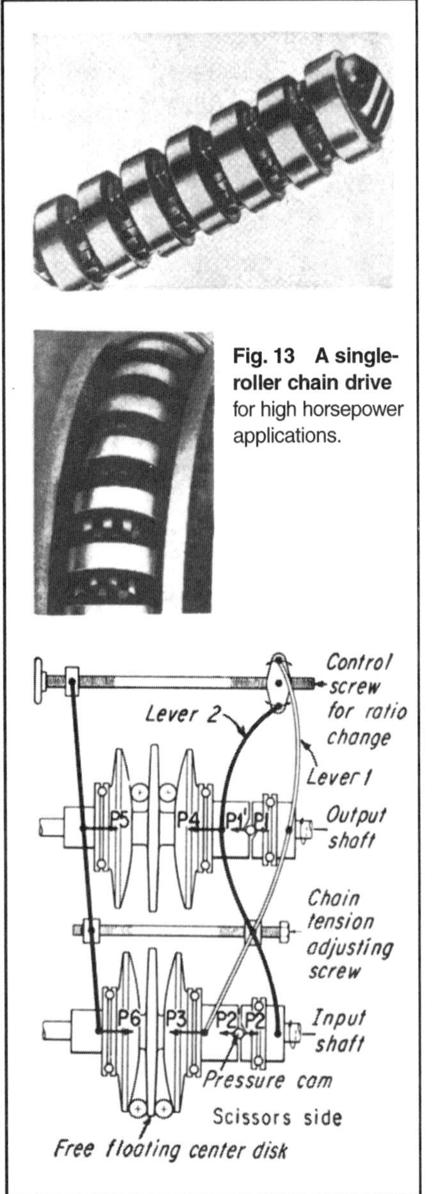

Fig. 13 A single-roller chain drive for high horsepower applications.

Fig. 12 A double-roller chain drive combines strength with ease in changing speed.

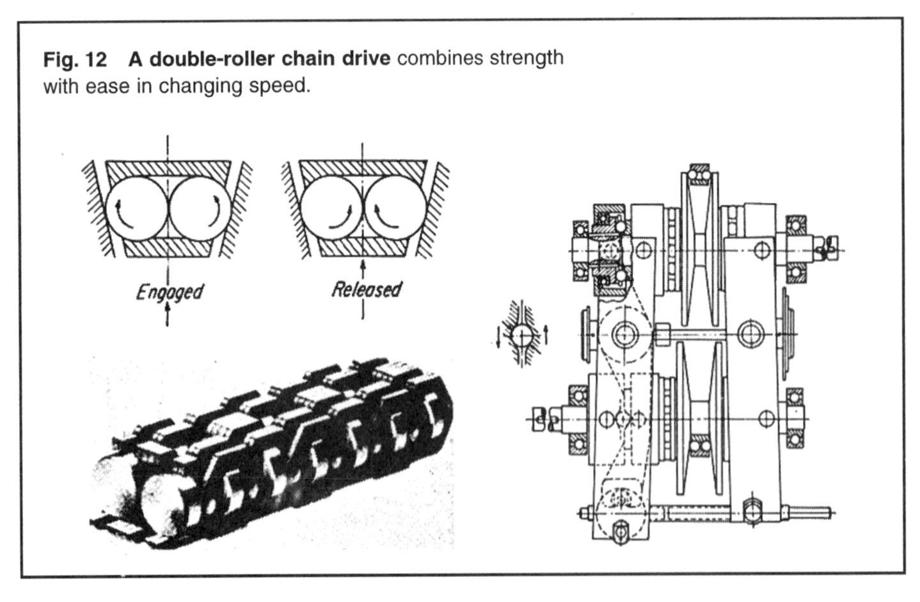

BELTS AND CHAINS ARE AVAILABLE IN MANYDIFFERENT FORMS

Imaginative fusions of belts, cables, gears and chains are expanding the horizons for light-duty synchronous drives.

Belts have long been used for the transfer of mechanical power. Today's familiar flat belts and V-belts are relatively light, quiet, inexpensive, and tolerant of alignment errors. They transmit power solely through frictional contacts. However, they function best at moderate speeds (4000 to 6000 fpm) under static loads. Their efficiencies drop slightly at low speeds, and centrifugal effects limit their capacities at high speeds. Moreover, they are inclined to sip under shock loads or when starting and braking. Even under constant rotation, standard belts tend to creep. Thus, these drives must be kept under tension to function properly, increasing loads on pulley shaft bearings.

Gears and chains, on the other hand, transmit power through bearing forces between positively engaged surfaces. They do not slip or creep, as measured by the relative motions of the driving and driven shafts. But the contacts themselves can slip significantly as the chain rollers and gear teeth move in and out of mesh.

Positive drives are also very sensitive to the geometries of the mating surfaces. A gear's load is borne by one or two teeth, thus magnifying small tooth-to-tooth errors. A chain's load is more widely distributed, but chordal variations in the driving wheel's effective radius produce small oscillations in the chain's velocity.

To withstand these stresses, chains and gears must be carefully made from hard materials and must then be lubricated in operations. Nevertheless, their operating noise betrays sharp impacts and friction between mating surfaces.

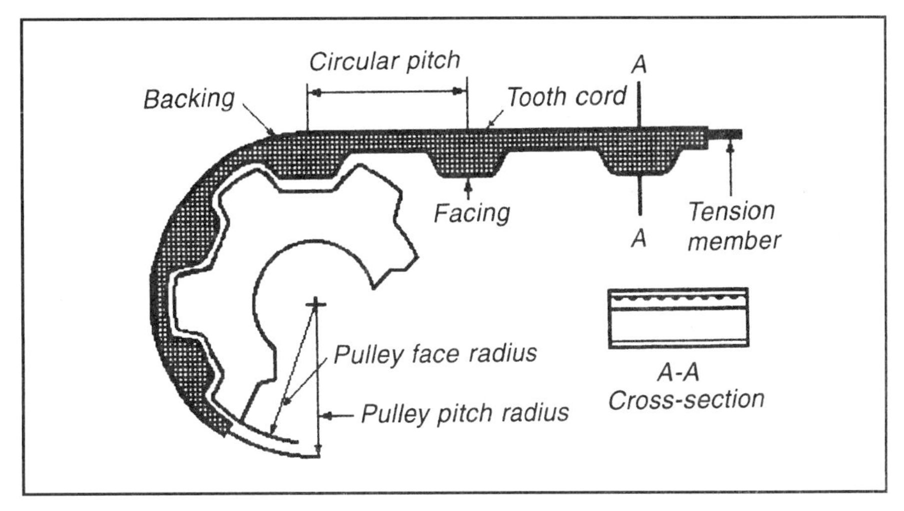

Fig. 1 Conventional timing belts have fiberglass or polyester tension members, bodies of neoprene or polyurethane, and trapezoidal tooth profiles.

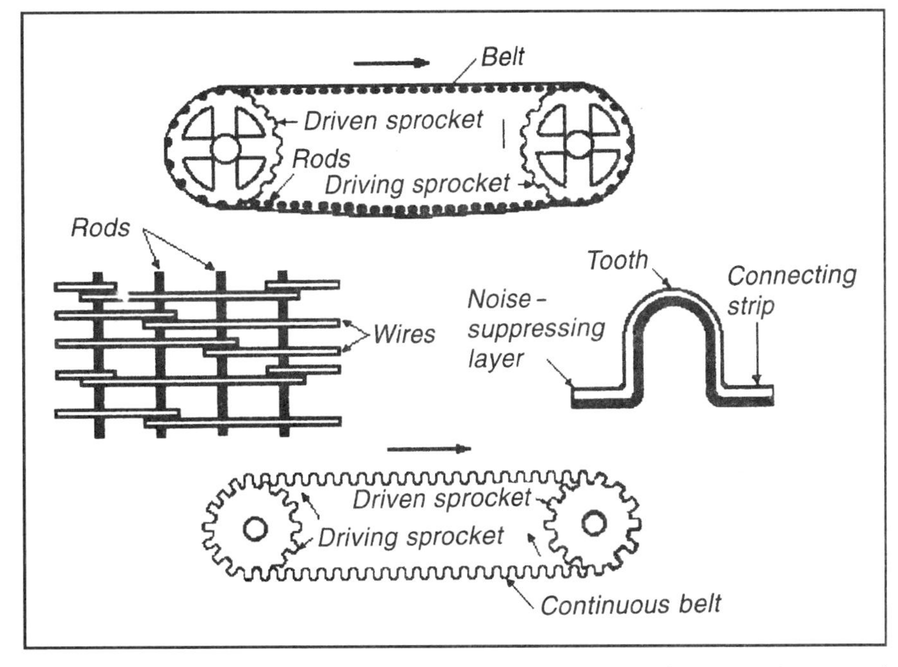

Fig. 2 NASA metal timing belts exploit stainless steel's strength and flexibility, and are coated with sound-and friction-reducing plastic.

The cogged timing belt, with its trapezoidal teeth (Fig. 1), is the best-known fusion of belt, gear, and chain. Though these well-established timing belts can handle high powers (up to 800 hp), many of the newer ideas in synchronous belting have been incorporated into low and fractional horsepower drives for instruments and business machines.

Steel Belts for Reliability

Researchers at NASA's Goddard Space Flight Center (Greenbelt, MD) turned to steel in the construction of long-lived toothed transmission belts for spacecraft instrument drives.

The NASA engineers looked for a belt design that would retain its strength and hold together for long periods of sustained or intermittent operation in hostile environments, including extremes of heat and cold.

Two steel designs emerged. In the more chain-like version (Fig. 2A), wires running along the length of the belt are wrapped at intervals around heavier rods running across the belt. The rods do double duty, serving as link pins and as teeth that mesh with cylindrical recesses cut into the sprocket. The assembled belt is coated with plastic to reduce noise and wear.

In the second design (Fig. 2B), a strip of steel is bent into a series of U-shaped teeth. The steel is supple enough to flex as it runs around the sprocket with its protruding transverse ridges, but the material resists stretching. This belt, too, is plastic-coated to reduce wear and noise.

The V-belt is best formed from a continuous strip of stainless steel "not much thicker than a razor blade," according to the agency, but a variation can be made by welding several segments together.

NASA has patented both belts, which are now available for commercial licensing. Researchers predict that they will be particularly useful in machines that must be dismantled to uncover the belt pulleys, in permanently encased machines, and in machines installed in remote places. In addition, stainless-steel belts might find a place in high-precision instrument drives because they neither stretch nor slip.

Though plastic-and-cable belts don't have the strength or durability of the NASA steel belts, they do offer versatility and production-line economy. One of the least expensive and most adaptable is the modern version of the bead chain, now common only in key chains and light-switch pull-cords.

The modern bead chain—if chain is the proper word—has no links. It has, instead, a continuous cable of stainless steel or aramid fiber which is covered with polyurethane. At controlled intervals, the plastic coating is molded into a bead (Fig. 3A). The length of the pitches thus formed can be controlled to within 0.001 in.

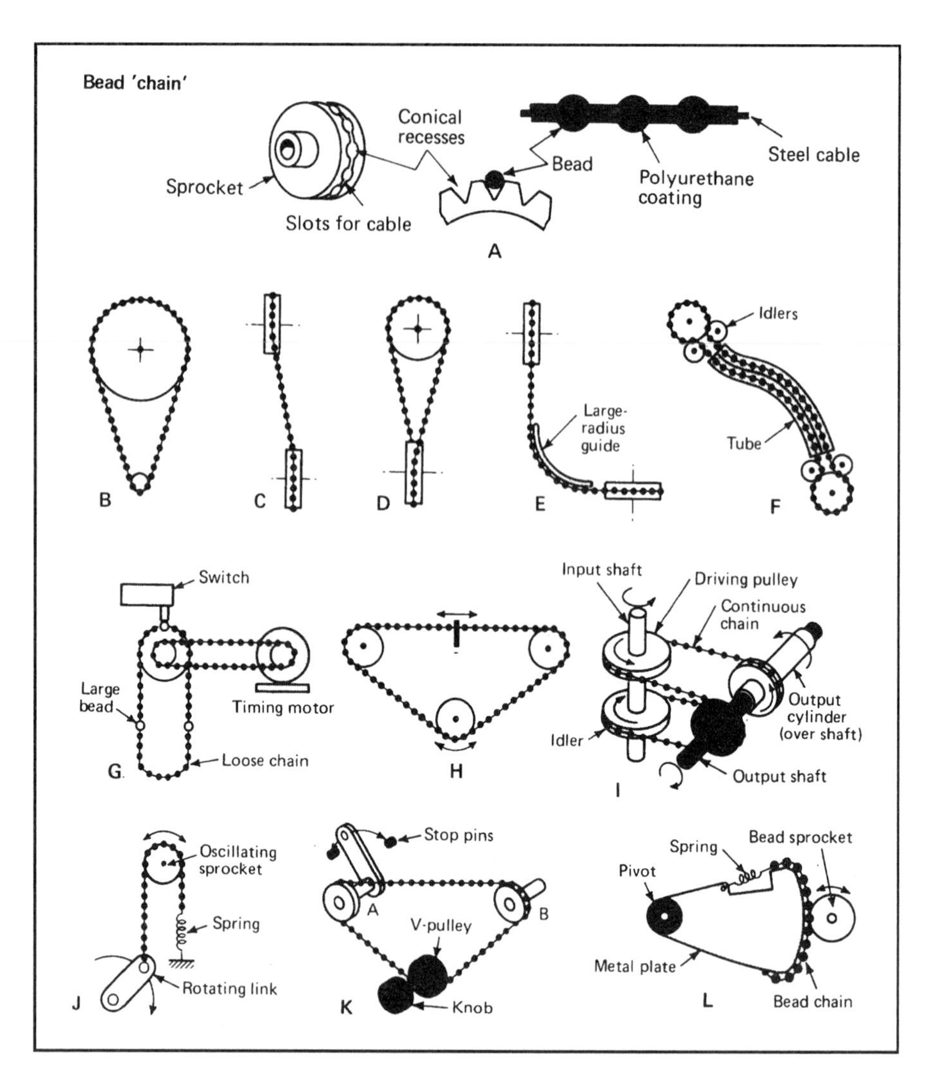

Fig. 3 Polyurethane-coated steel-cable "chains"—both beaded and 4-pinned—can cope with conditions unsuitable for most conventional belts and chains.

Table 1. Conventional Timing Belts

Type	Circular pitch, in.	Wkg. tension lb/in. width	Centr. loss const., K_c
Standard (Fig 1)			
MXL	0.080	32	10×10^{-9}
XL	0.200	41	27×10^{-9}
L	0.375	55	38×10^{-9}
H	0.500	140	53×10^{-9}
40DP	0.0816	13	—
High-torque (Fig 8)			
3 mm	0.1181	60	15×10^{-9}
5 mm	0.1968	100	21×10^{-9}
8 mm	0.3150	138	34×10^{-9}

Courtesy Stock Drive Products

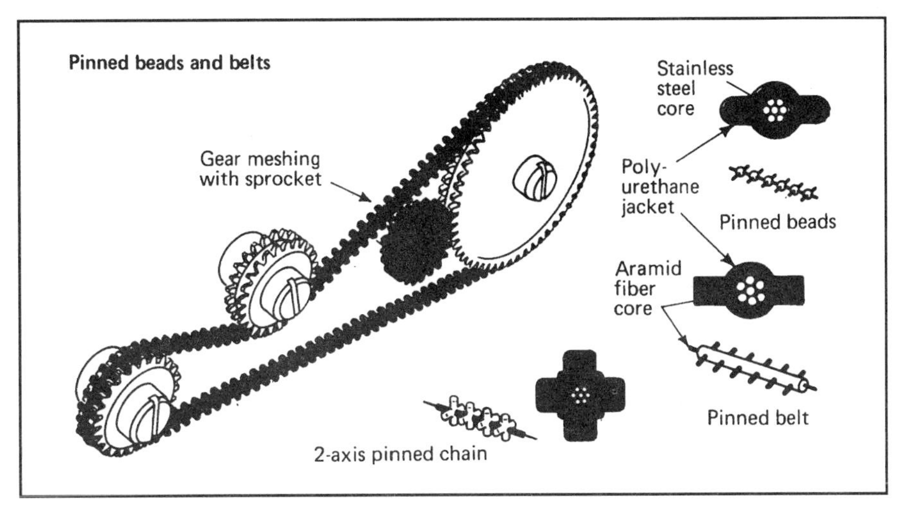

Fig. 4 Plastic pins eliminate the bead chain's tendency to cam out of pulley recesses, and permit greater precision in angular transmission.

In operation, the cable runs in a grooved pulley; the beads seat in conical recesses in the pulley face. The flexibility, axial symmetry, and positive drive of bead chain suit a number of applications, both common and uncommon:

- An inexpensive, high-ratio drive that resists slipping and requires no lubrication (Fig. 3B). As with other chains and belts, the bead chain's capacity is limited by its total tensile strength (typically 40 to 80 lb for a single-strand steel-cable chain), by the speed-change ratio, and by the radii of the sprockets or pulleys.
- Connecting misaligned sprockets. If there is play in the sprockets, or if the sprockets are parallel but lie in different planes, the bead chain can compensate for up to 20º of misalignment (Fig. 3C).
- Skewed shafts, up to 90º out of phase (Fig. 3D).
- Right-angle and remote drives using guides or tubes (Figs. 3E and 3F). These methods are suitable only for low-speed, low-torque applications. Otherwise, frictional losses between the guide and the chain are unacceptable.
- Mechanical timing, using oversize beads at intervals to trip a microswitch (Fig. 3G). The chain can be altered or exchanged to give different timing schemes.
- Accurate rotary-to-linear motion conversion (Fig. 3H).
- Driving two counter-rotating outputs from a single input, using just a single belt (Fig. 3I).
- Rotary-to-oscillatory motion conversion (Fig. 3J).
- Clutched adjustment (Fig. 3K). A regular V-belt pulley without recesses permits the chain to slip when it reaches a pre-set limit. At the same time, bead-pulleys keep the output shafts synchronized. Similarly, a pulley or sprocket with shallow recesses permits the chain to slip one bead at a time when overloaded.
- Inexpensive "gears" or gear segments fashioned by wrapping a bead chain round the perimeter of a disk or solid arc of sheet metal (Fig. 3L). The sprocket then acts as a pinion. (Other designs are better for gear fabrication.)

Fig. 5 A plastic-and-cable ladder chain in an impact-printer drive. In extreme conditions, such hybrids can serve many times longer than steel.

A More Stable Approach

Unfortunately, bead chains tend to cam out of deep sprocket recesses under high loads. In its first evolutionary step, the simple spherical bead grew limbs—two pins projecting at right angles to the cable axis (Fig. 4). The pulley or sprocket looks like a spur gear grooved to accommodate the belt; in fact, the pulley can mesh with a conventional spur gear of proper pitch.

Versions of the belt are also available with two sets of pins, one projecting vertically and the other horizontally. This arrangement permits the device to drive a series of perpendicular shafts without twisting the cable, like a bead chain but without the bead chain's load limitations. Reducing twist increases the transmission's lifetime and reliability.

These belt-cable-chain hybrids can be sized and connected in the field, using metal crimp-collars. However, nonfactory splices generally reduce the cable's tensile strength by half.

Parallel-Cable Drives

Another species of positive-drive belt uses parallel cables, sacrificing some flexibility for improved stability and greater strength. Here, the cables are connected by rungs molded into the plastic coating, giving the appearance of a ladder (Fig. 6). This "ladder chain" also meshes with toothed pulleys, which need not be grooved.

A cable-and-plastic ladder chain is the basis for the differential drive system in a Hewlett-Packard impact printer (Fig. 5). When the motors rotate in the same direction at the same speed, the carriage moves to the right or left. When they rotate in opposite directions, but at the same speed, the carriage remains stationary and the print-disk rotates. A differential motion of the motors produces a combined translation and rotation of the print-disk.

The hybrid ladder chain is also well suited to laboratory of large spur gears from metal plates or pulleys (Fig. 6). Such a "gear" can run quietly in mesh with a pulley or a standard gear pinion of the proper pitch.

Another type of parallel-cable "chain," which mimics the standard chain, weighs just 1.2 oz/ft, requires no lubrication, and runs almost silently.

A Traditional Note

A new high-capacity tooth profile has been tested on conventional cogged belts. It has a standard cord and elastic body construction, but instead of the usual trapezoid, it has curved teeth (Fig. 7). Both 3-mm and 5-mm pitch versions have been introduced.

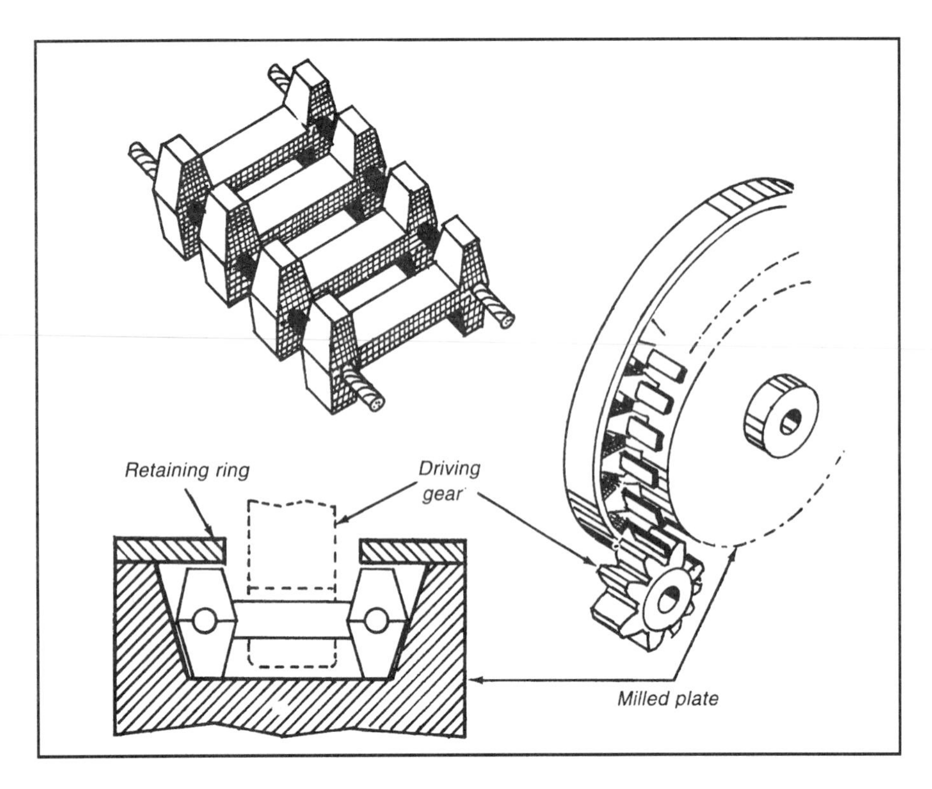

Fig. 6 A gear chain can function as a ladder chain, as a wide V-belt, or, as here, a gear surrogate meshing with a standard pinion.

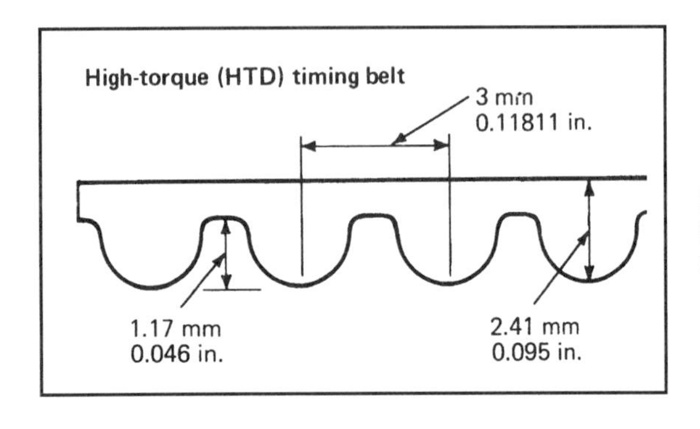

Fig. 7 Curved high-torque tooth profiles (just introduced in 3-mm and 5-mm pitches) increase load capacity of fine-pitch neoprene belts.

CHANGE CENTER DISTANCE WITHOUT ALTERING SPEED RATIO

Increasing the gap between the roller and knife changes chain lengths from *F* to *E*. Because the idler moves with the roller sprocket, length *G* changes to *H*. The changes in chain length are similar in value but opposite in direction. Chain lengths *E* minus *F* closely approximate *G* minus *H*. Variations in required chain length occur because the chains do not run parallel. Sprocket offset is required to avoid interference. Slack produced is too minute to affect the drive because it is proportional to changes in the cosine of a small angle (2° to 5°). For the 72-in. chain, variation is 0.020 in.

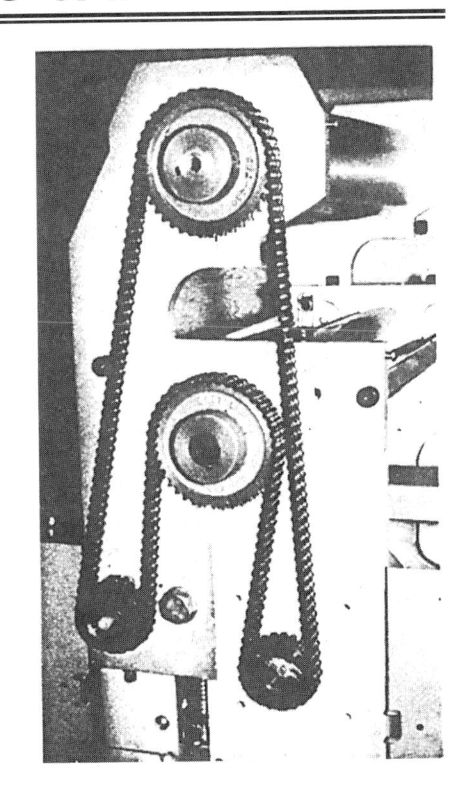

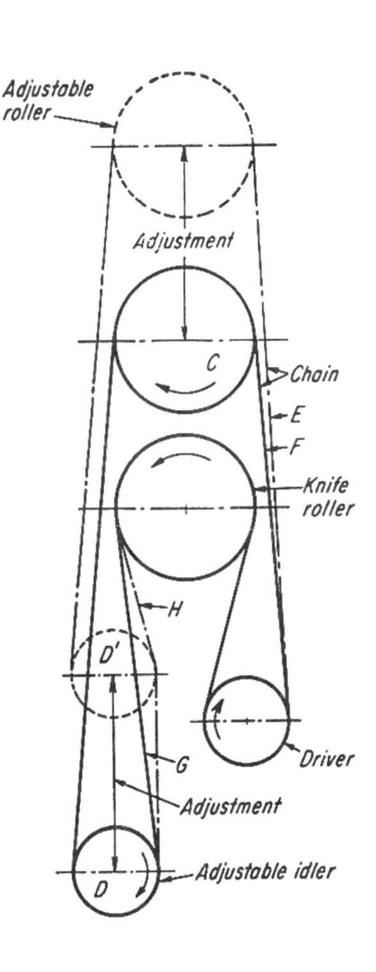

MOTOR MOUNT PIVOTS TO CONTROL BELT TENSION

Belt tensioning proportional to load

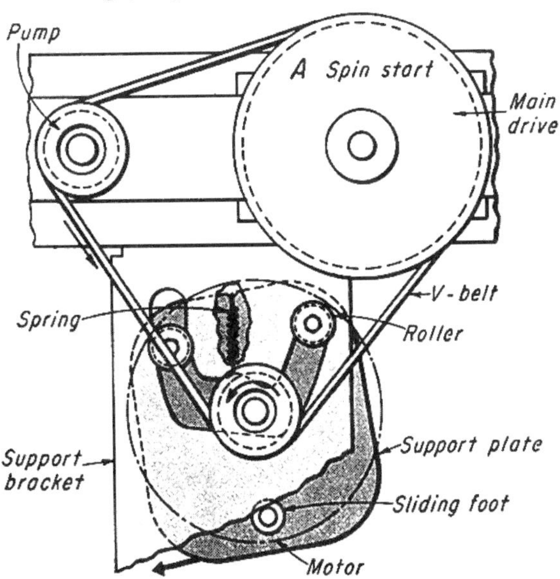

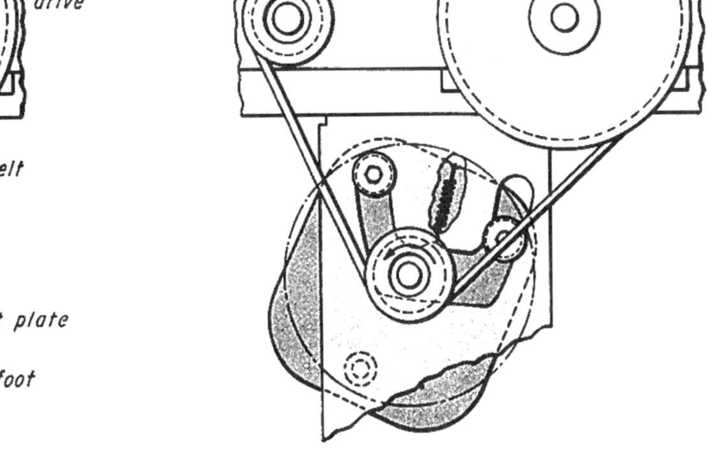

When the agitation cycle is completed, the motor is momentarily idle with the right roller bottomed in the right-hand slot. When spin-dry starts, (A) the starting torque produces a reaction at the stator, pivoting the motor on the bottomed roller. The motor pivots until the opposite roller bottoms in the left-hand slot. The motor now swings out until restrained by the V-belt, which drives the pump and basket.

The motor, momentarily at zero rpm, develops maximum torque and begins to accelerate the load of basket, water, and wash. The motor pivots (B) about the left roller increasing belt tension in proportion to the output torque. When the basket reaches maximum speed, the load is reduced and belt tension relaxes. The agitation cycle produces an identical reaction in the reverse direction.

TEN ROLLER CHAINS AND THEIR ADAPTATIONS

Various roller, side-plate and pin configurations for power transmissions, conveying, and elevating.

STANDARD ROLLER CHAIN—FOR POWER TRANSMISSION AND CONVEYING

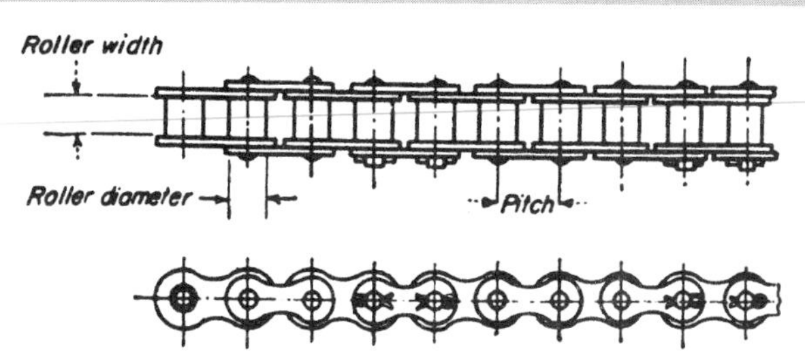

SINGLE WIDTH—Sizes $^5/_8$ in and smaller have a spring-clip connecting link; those $^3/_4$ in and larger have a cotter pin.

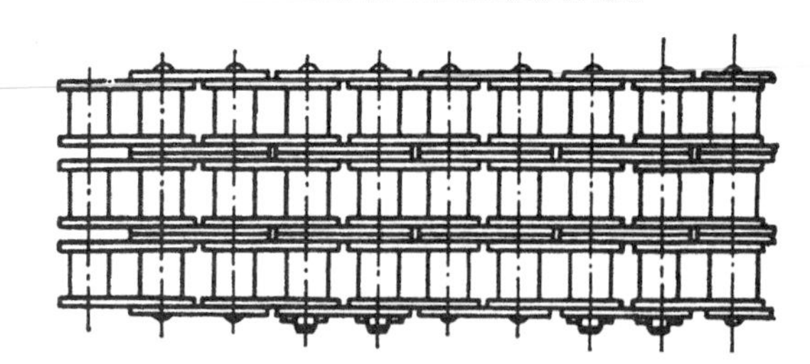

MULTIPLE WIDTH—Similar to single-width chain. It is made in widths up to 12 strands.

EXTENDED PITCH CHAIN—FOR CONVEYING

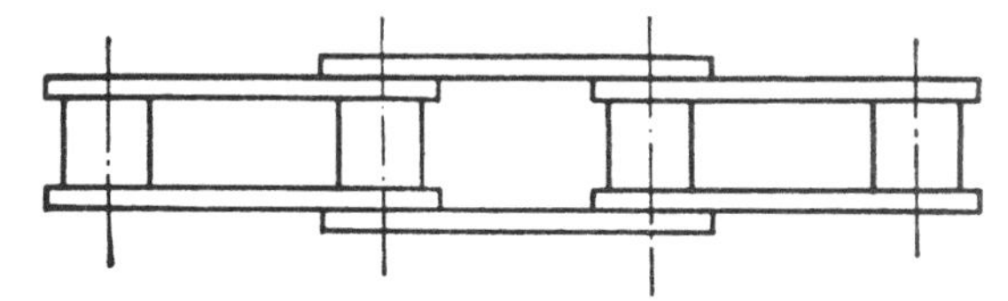

STANDARD ROLLER DIAMETER—made with 1 to 4 in pitch and cotter-pin-type connecting links.

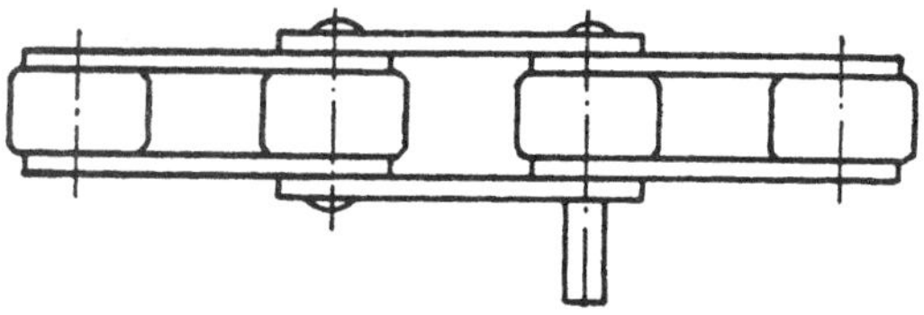

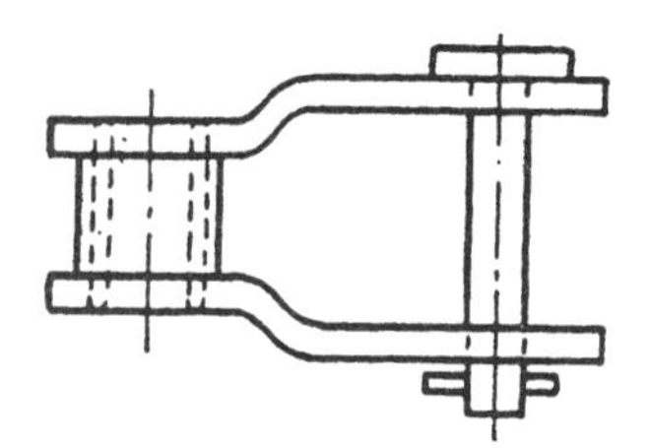

OVERSIZED ROLLER DIAMETER—Same base chain as standard roller type but not made in multiple widths.

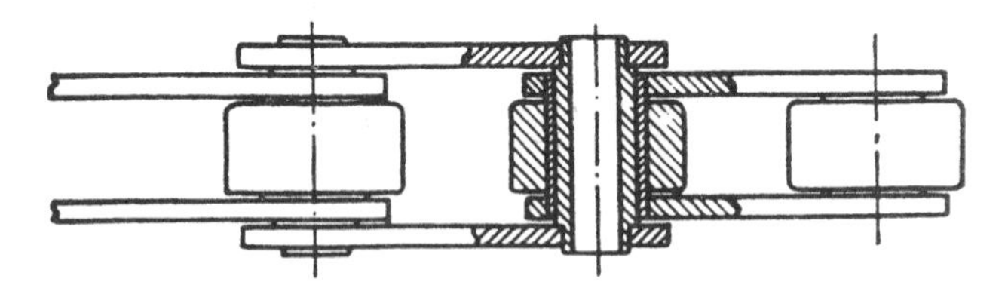

HOLLOW PIN—Made with $1^1/_4$ to 15 in pitch. It is adaptable to a variety of bolted attachments.

OFFSET LINK—Used when length requires an odd number of pitches and to shorten and lengthen a chain by one pitch.

STANDARD PITCH ADAPTATIONS

STRAIGHT LUG—Lugs on one or both sides can be spaced as desired. A standard roller is shown.

BENT LUG—Similar to straight-lug type for adaptations. A standard roller is shown.

EXTENDED PITCH ADAPTATIONS

STRAIGHT LUG—An oversized diameter roller is shown.

BENT LUG—An oversized diameter roller is shown.

HOLLOW PIN

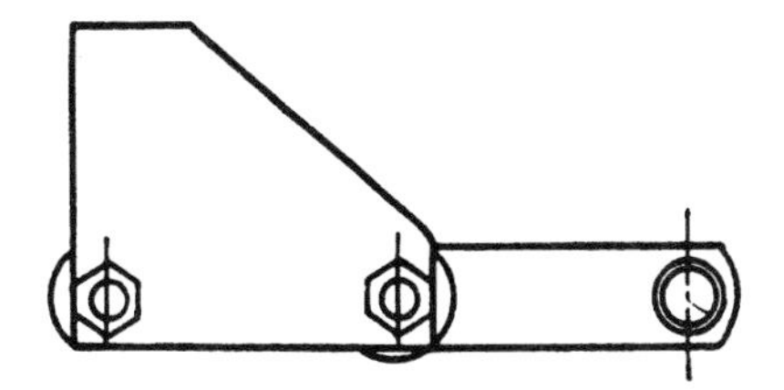

STRAIGHT LUG—Lugs are detachable for field adaptation.

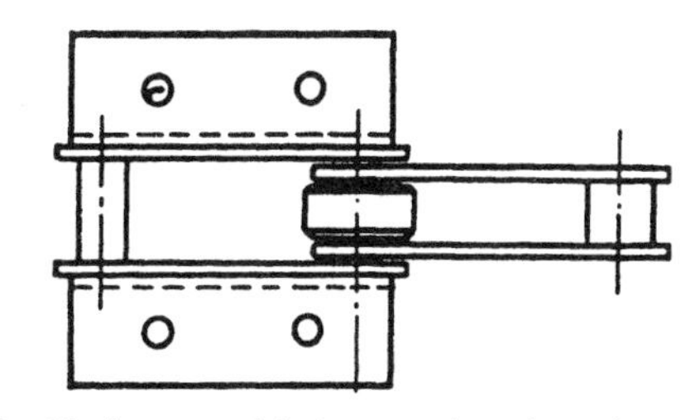

BENT LUG—Similar to straight lug type for adaptations.

EXTENDED PIN CHAINS

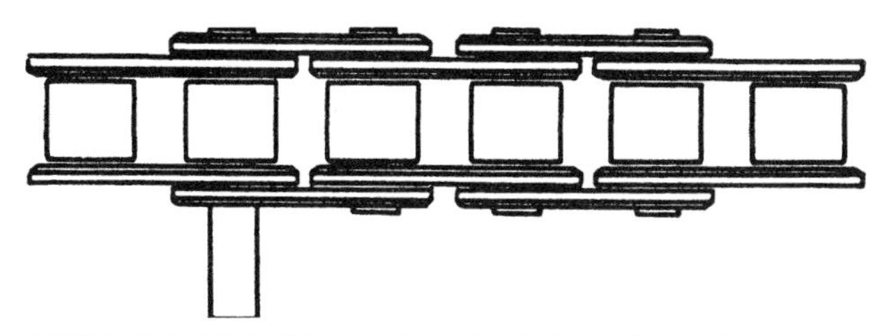

STANDARD PITCH—Pins can be extended on either side.

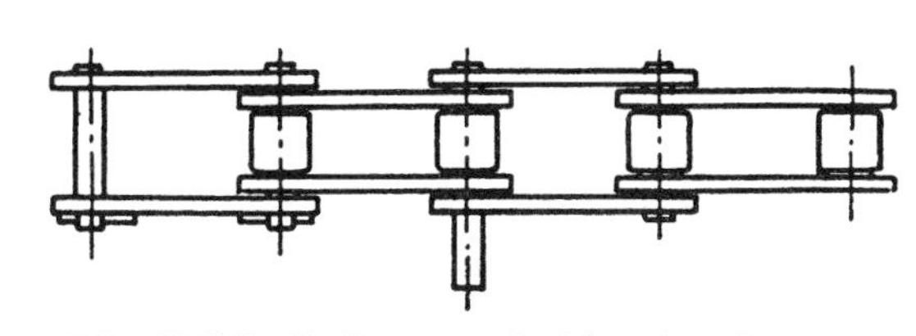

EXTENDED PITCH—Similar to standard for adaptations.

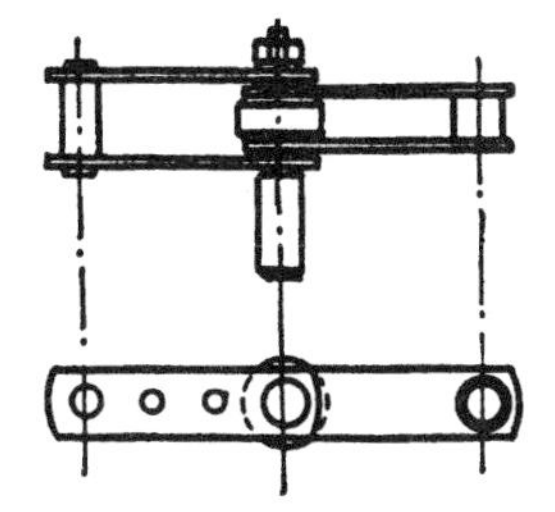

HOLLOW PIN—Pins are designed for field adaptation.

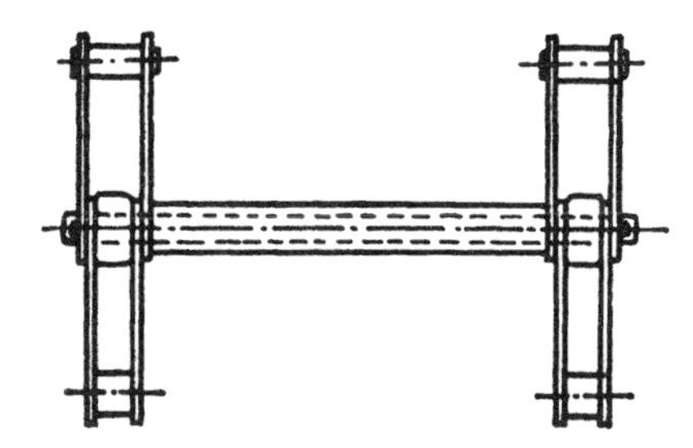

CROSS ROD—The rod can be removed from the hollow pins.

SPECIAL ADAPTATIONS

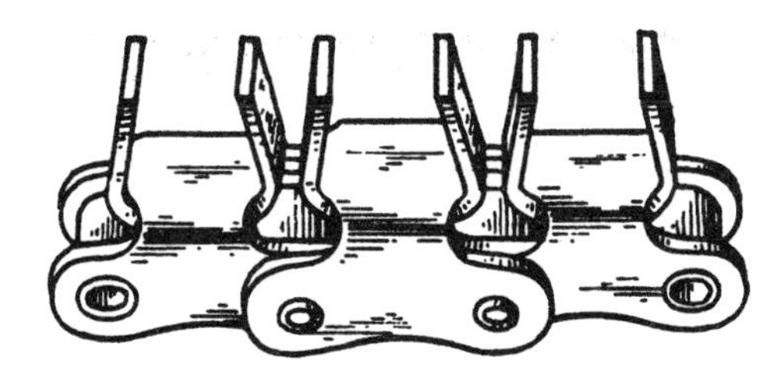

Used for holding conveyed objects.

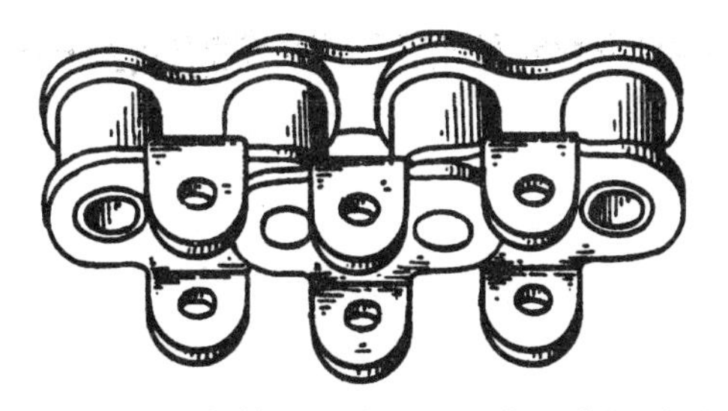

Used to keep conveyed object on the center-line of the chain.

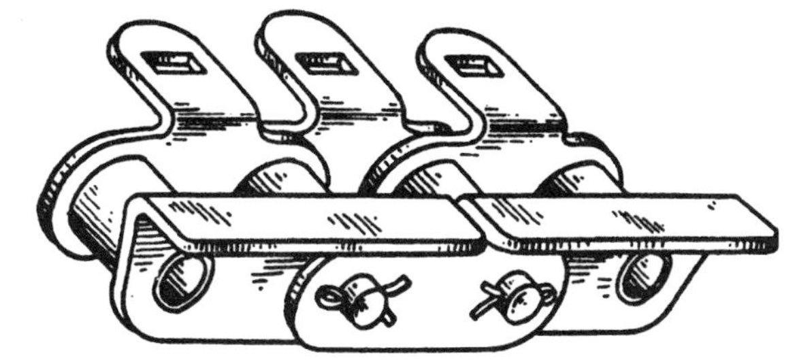

Used when flexing is desired in one direction only.

Used for supporting concentrated loads.

TWELVE APPLICATIONS FOR ROLLER CHAIN

This low-cost industrial chain can be applied in a variety of ways to perform tasks other than simply transmitting power.

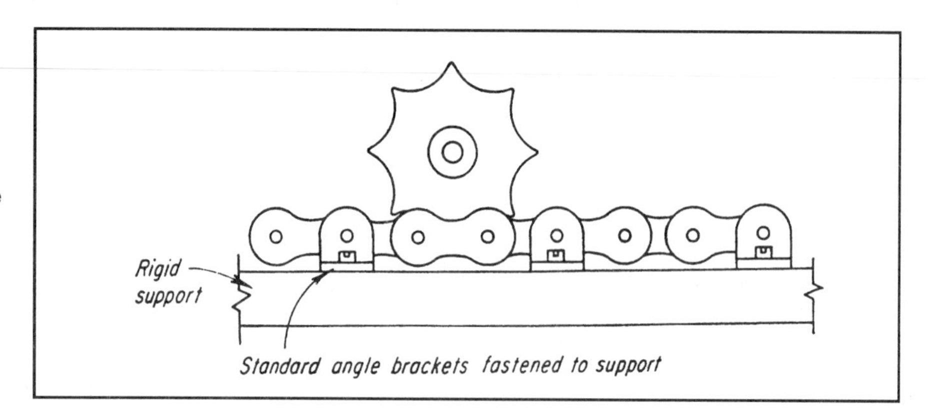

Fig. 1 This low-cost rack-and-pinion device is easily assembled from standard parts.

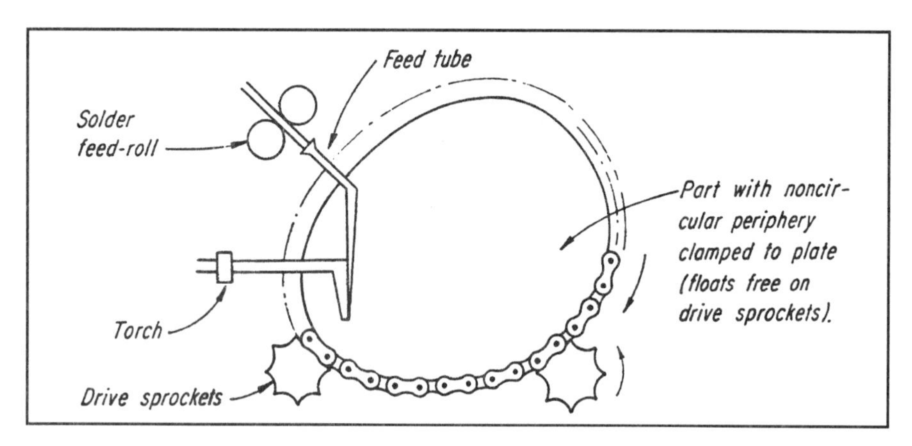

Fig. 2 An extension of the rack-and-pinion principle—This is a soldering fixture for non-circular shells. Positive-action cams can be similarly designed. Standard angle brackets attach the chain to a cam or fixture plate.

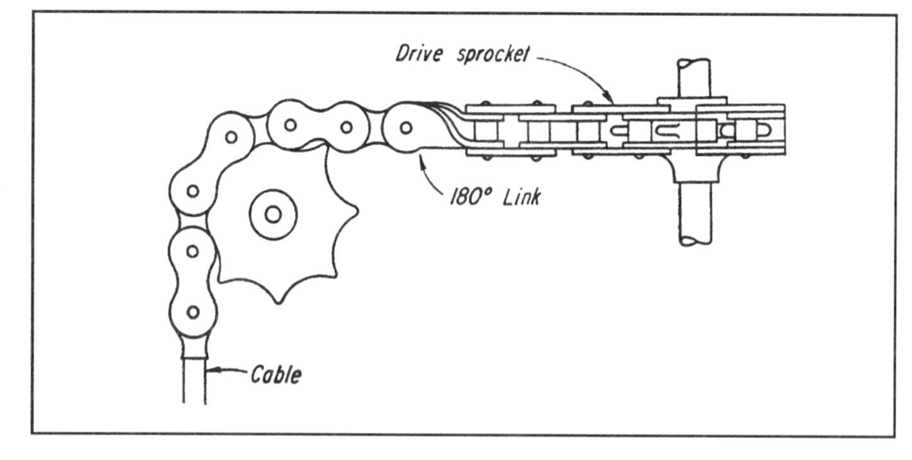

Fig. 3 This control-cable direction-changer is extensively used in aircraft.

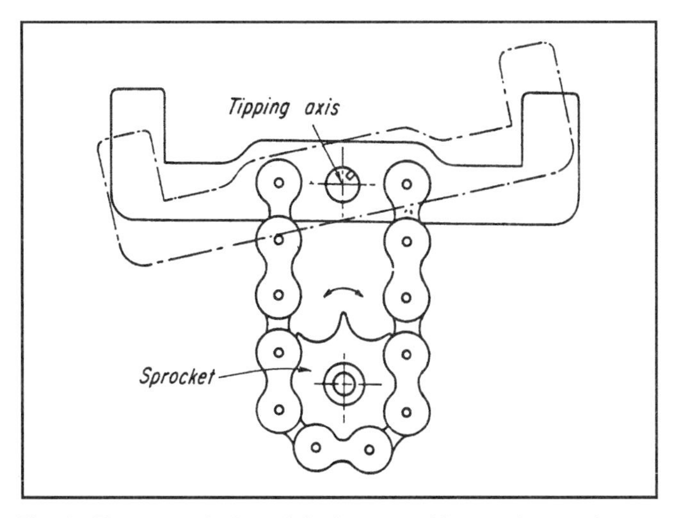

Fig. 4 The transmission of tipping or rocking motion can be combined with the previous example (Fig. 3) to transmit this kind of motion to a remote location and around obstructions. The tipping angle should not exceed 40°.

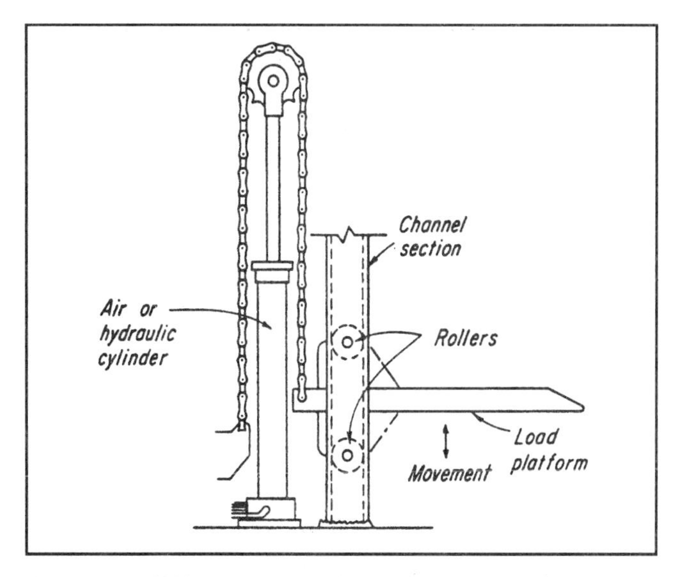

Fig. 5 This lifting device is simplified by roller chain.

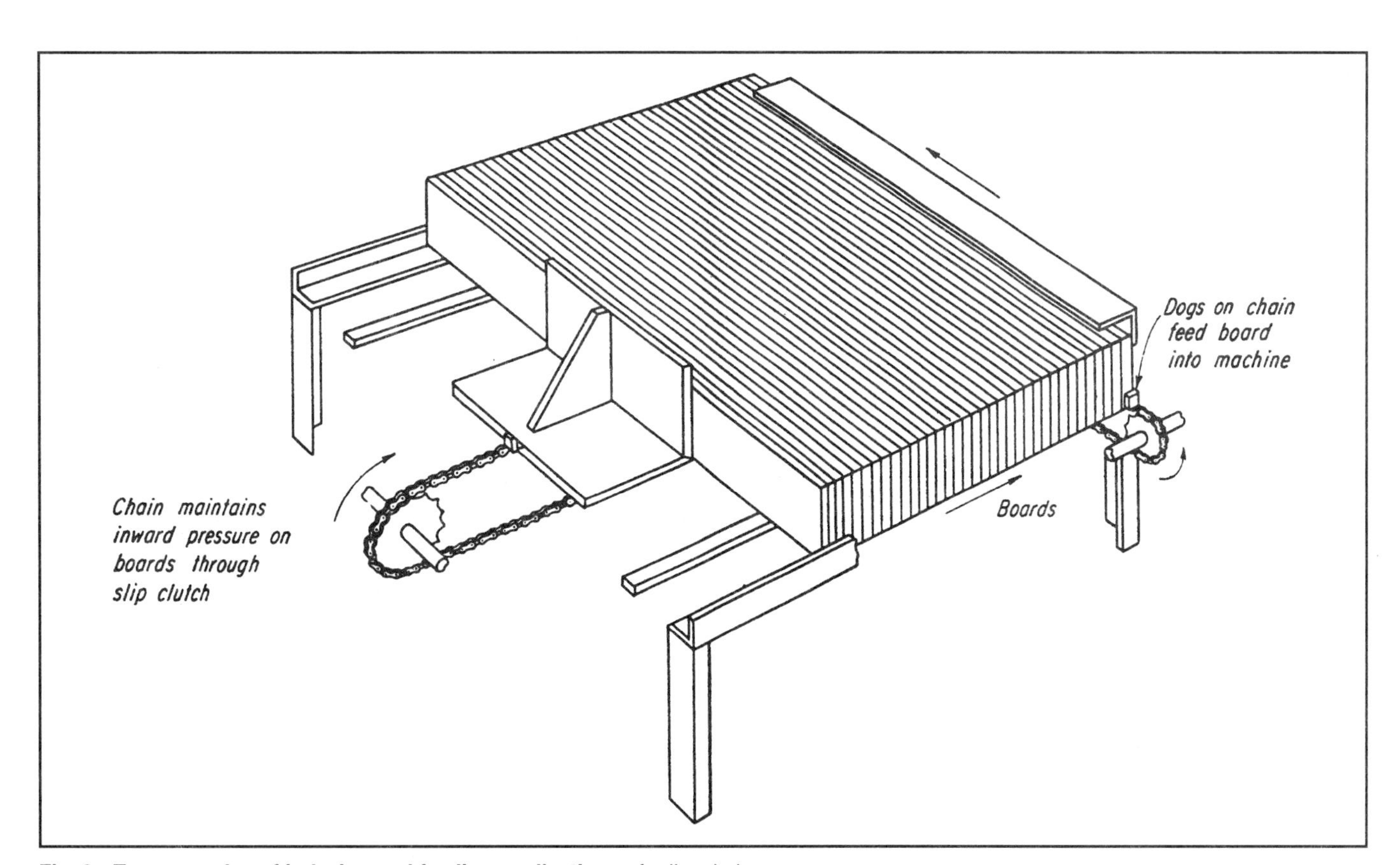

Fig. 6 Two examples of indexing and feeding applications of roller chain are shown here. This setup feeds plywood strips into a machine. The advantages of roller chain as used here are its flexibility and long feed.

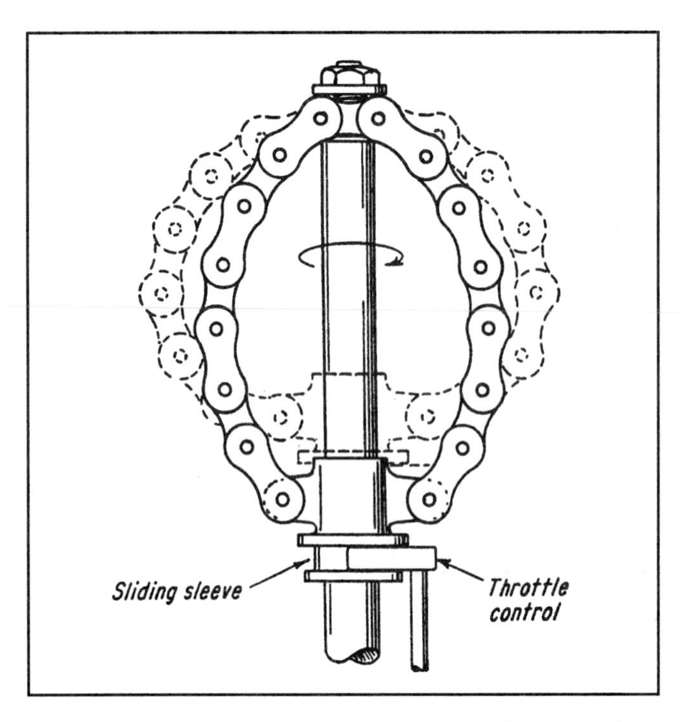

Fig. 7 Simple governor weights can be attached by means of standard brackets to increase response force when rotation speed is slow.

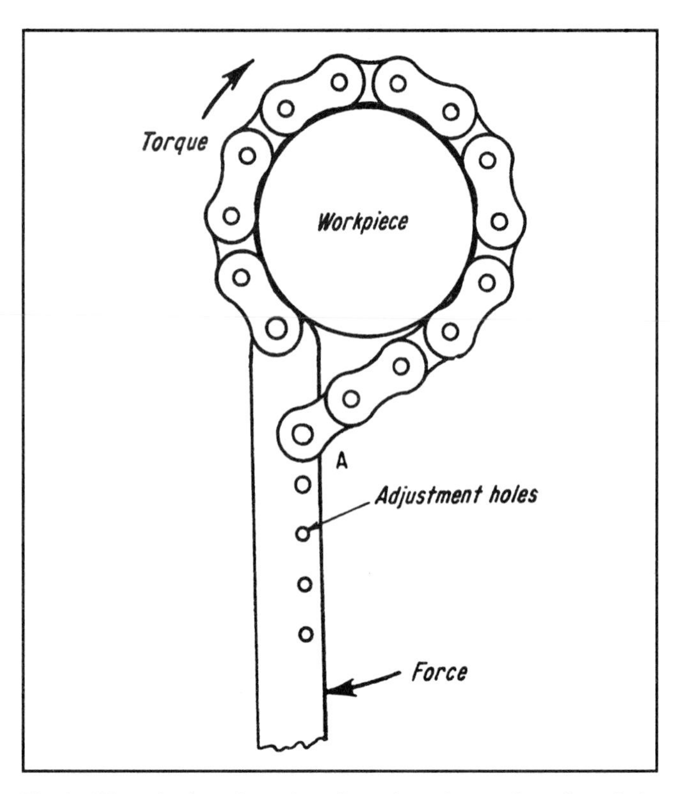

Fig. 8 Wrench pivot A can be adjusted to grip a variety of regularly or irregularly shaped objects.

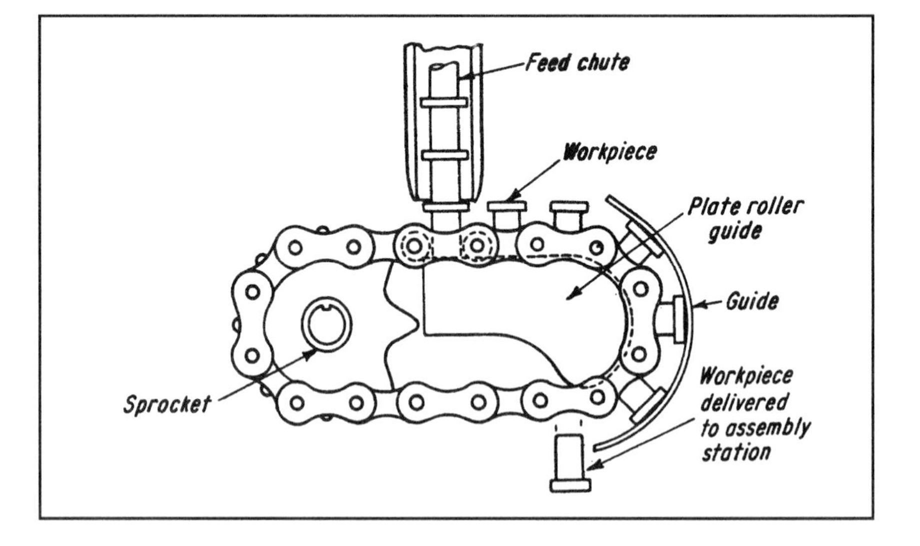

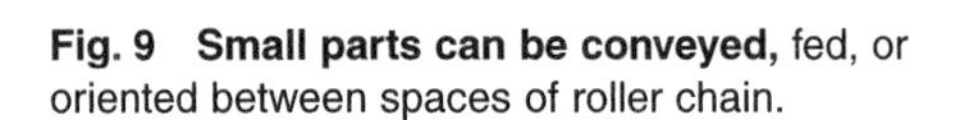

Fig. 9 Small parts can be conveyed, fed, or oriented between spaces of roller chain.

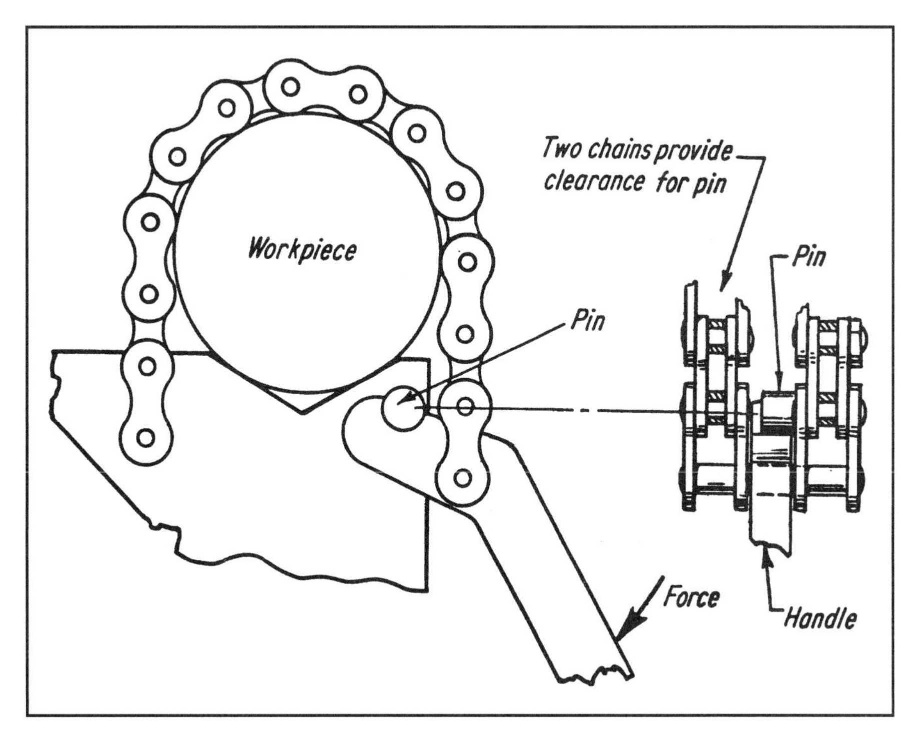

Fig. 10 Clamp toggle action is supplied by two chains, thus clearing pin at fulcrum.

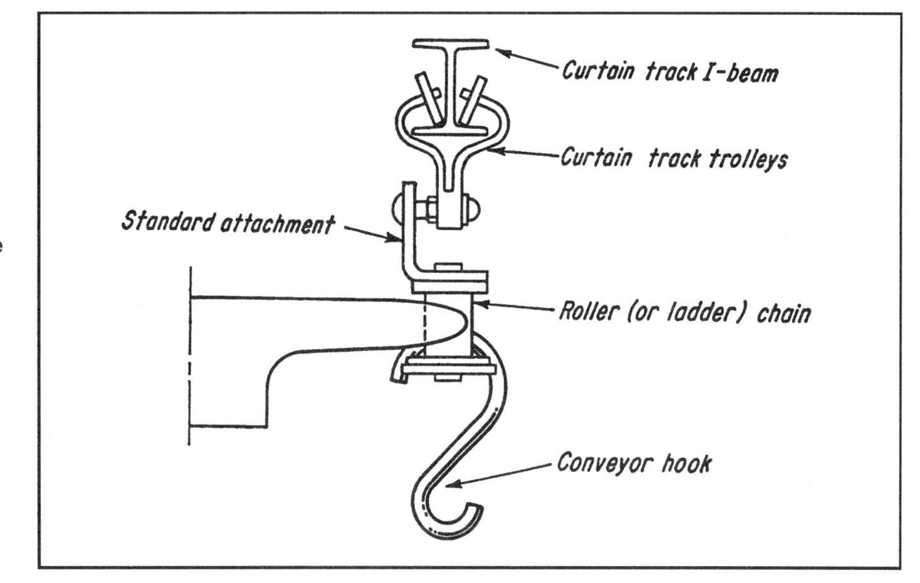

Fig. 11 Light-duty trolley conveyors can be made by combining standard roller-chain components with standard curtain-track components. Small gearmotors are used to drive the conveyor.

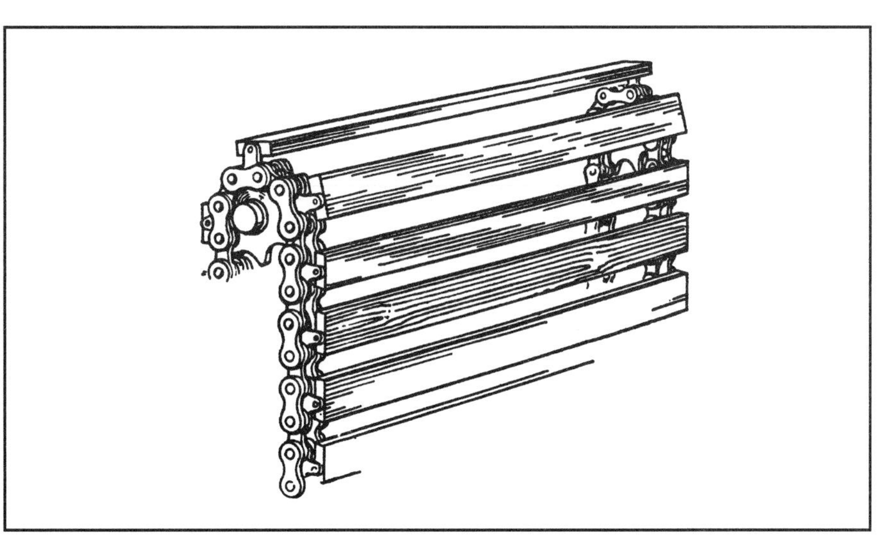

Fig. 12 Slatted belt, made by attaching wood, plastic, or metal slats, can serve as adjustable safety guard, conveyor belt, fast-acting security-wicket window.

SIX MECHANISMS FOR REDUCING PULSATIONS IN CHAIN DRIVES

Pulsations in chain motion created by the chordal action of chain and sprockets can be minimized or avoided by introducing a compensating cyclic motion in the driving sprocket. Mechanisms for reducing fluctuating dynamic loads in chain drives and the pulsations resulting from them include noncircular gears, eccentric gears, and cam-activated intermediate shafts.

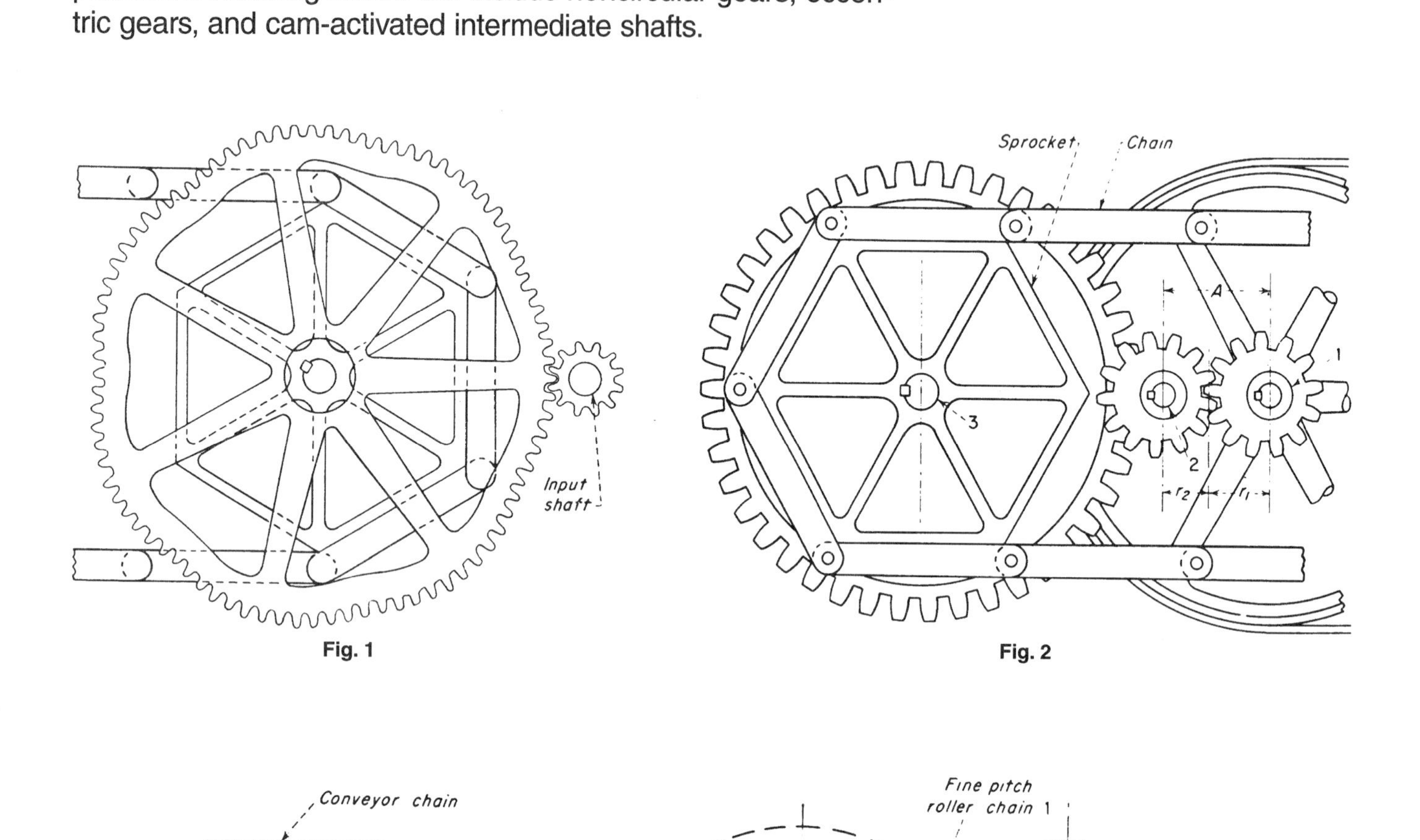

Fig. 1

Fig. 2

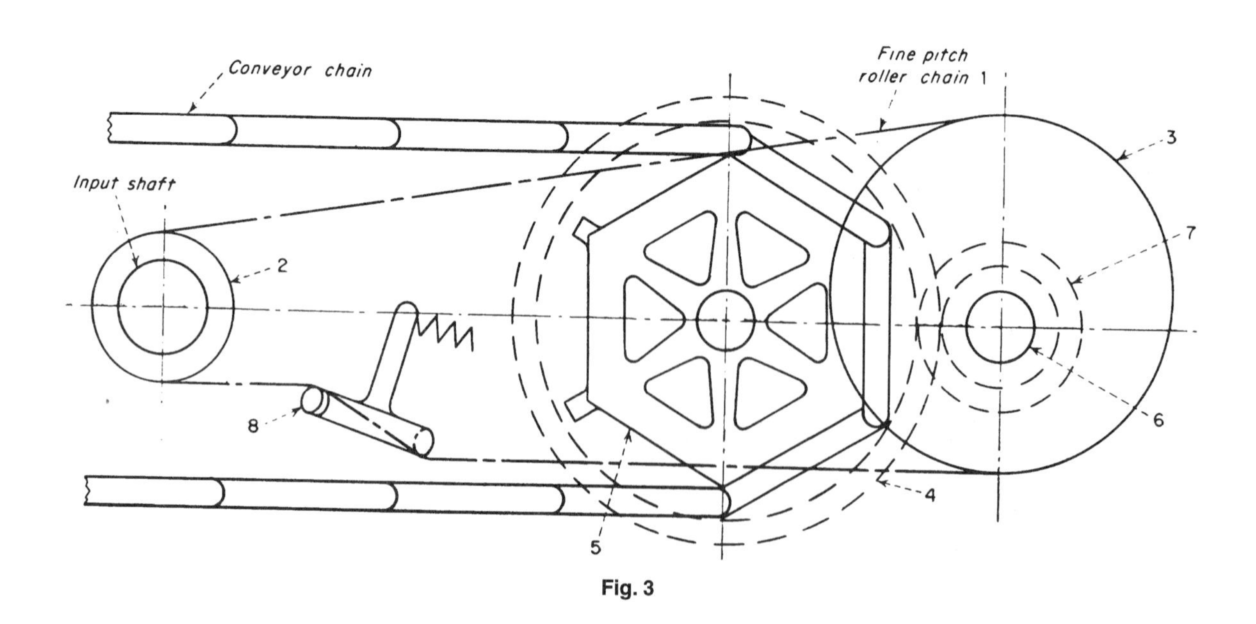

Fig. 3

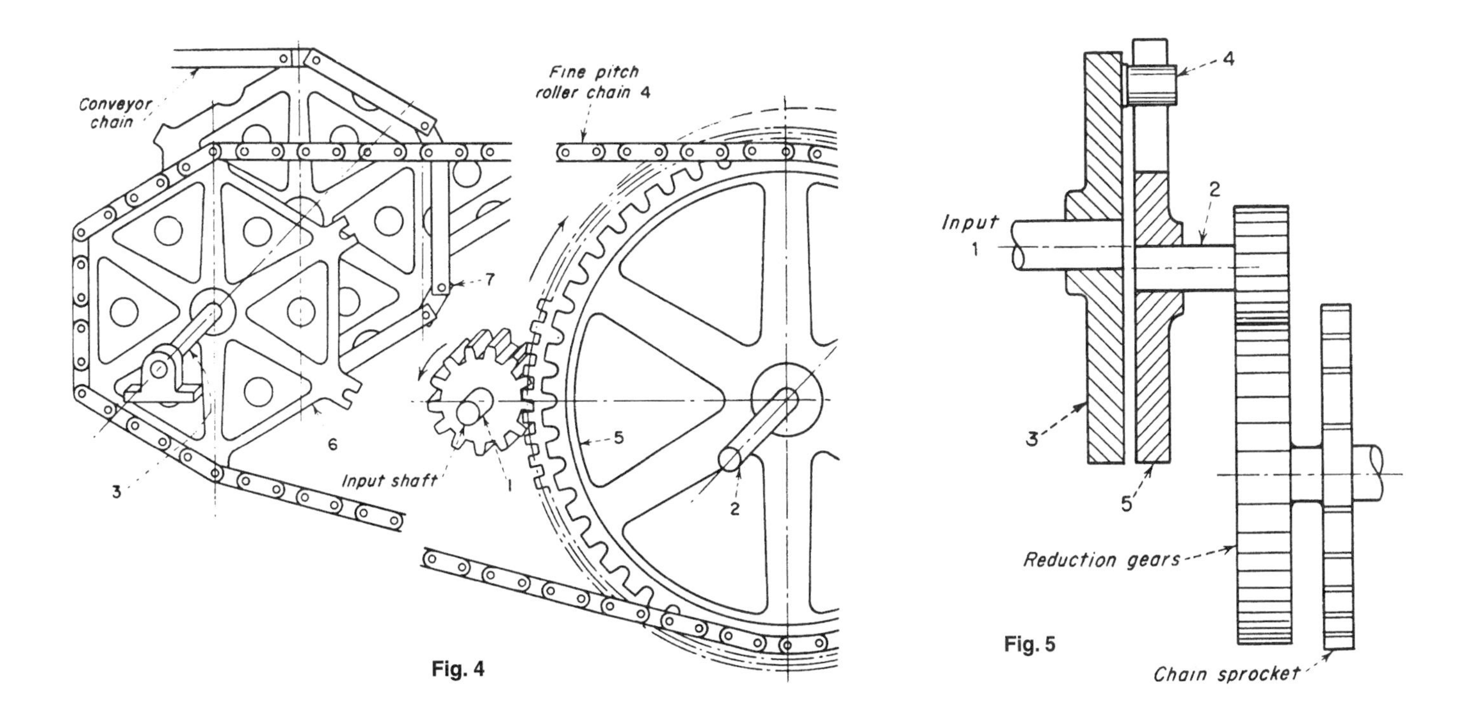

Fig. 1 The large cast-tooth, noncircular gear, mounted on the chain sprocket shaft, has a wavy outline in which the number of waves equals the number of teeth on a sprocket. The pinion has a corresponding noncircular shape. Although requiring special-shaped gears, the drive completely equalizes the chain pulsations.

Fig. 2 This drive has two eccentrically mounted spur pinions (1 and 2). Input power is through the belt pulley keyed to the same shaft as pinion 1. Pinion 3 (not shown), keyed to the shaft of pinion 2, drives the large gear and sprocket. However, the mechanism does not completely equalize chain velocity unless the pitch lines of pinions 1 and 2 are noncircular instead of eccentric.

Fig. 3 An additional sprocket 2 drives the noncircular sprocket 3 through a fine-pitch chain 1. This imparts pulsating velocity to shaft 6 and to the long-pitch conveyor sprocket 5 through pinion 7 and gear 4. The ratio of the gear pair is made the same as the number of teeth of sprocket 5. Spring-actuated lever and rollers 8 take up the slack. Conveyor motion is equalized, but the mechanism has limited power capacity because the pitch of chain 1 must be kept small. Capacity can be increased by using multiple strands of fine-pitch chain.

Fig. 4 Power is transmitted from shaft 2 to sprocket 6 through chain 4, thus imparting a variable velocity to shaft 3, and through it, to the conveyor sprocket 7. Because chain 4 has a small pitch and sprocket 5 is relatively large, the velocity of 4 is almost constant. This induces an almost constant conveyor velocity. The mechanism requires the rollers to tighten the slack side of the chain, and it has limited power capacity.

Fig. 5 Variable motion to the sprocket is produced by disk 3. It supports pin and roller 4, as well as disk 5, which has a radial slot and is eccentrically mounted on shaft 2. The ratio of rpm of shaft 2 to the sprocket equals the number of teeth in the sprocket. Chain velocity is not completely equalized.

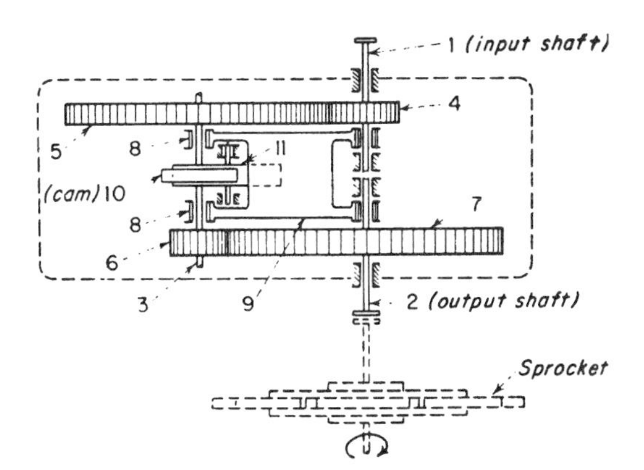

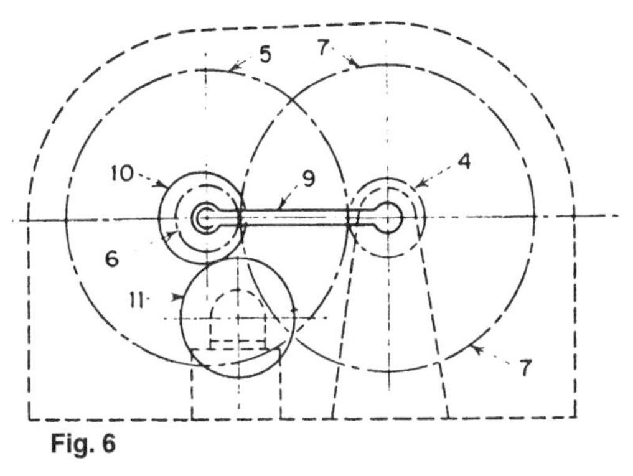

Fig. 6

Fig. 6 The integrated "planetary gear" system (gears 4, 5, 6 and 7) is activated by cam 10, and it transmits a variable velocity to the sprocket synchronized with chain pulsations through shaft 2, thus completely equalizing chain velocity. Cam 10 rides on a circular idler roller 11. Because of the equilibrium of the forces, the cam maintains positive contact with the roller. The unit has standard gears, acts simultaneously as a speed reducer, and can transmit high horsepower.

CHAPTER 11

SPRING AND SCREW DEVICES AND MECHANISMS

FLAT SPRINGS IN MECHANISMS

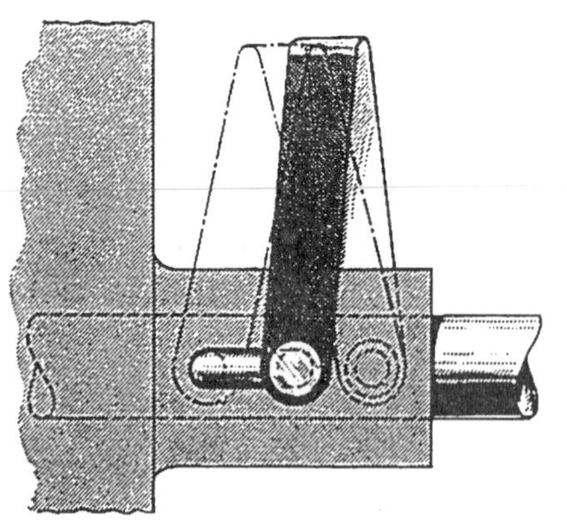

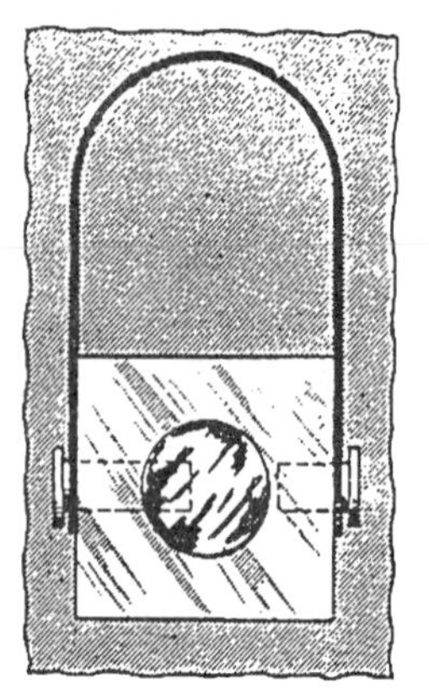

Constant force is approached because of the length of this U-spring. Don't align the studs or the spring will fall.

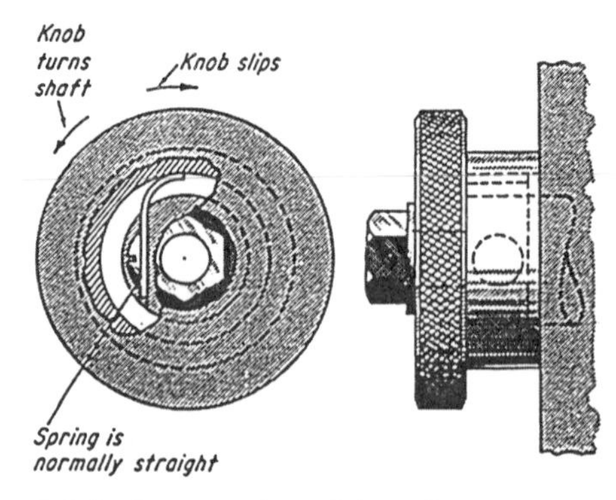

A flat-wire sprag is straight until the knob is assembled: thus tension helps the sprag to grip for one-way clutching.

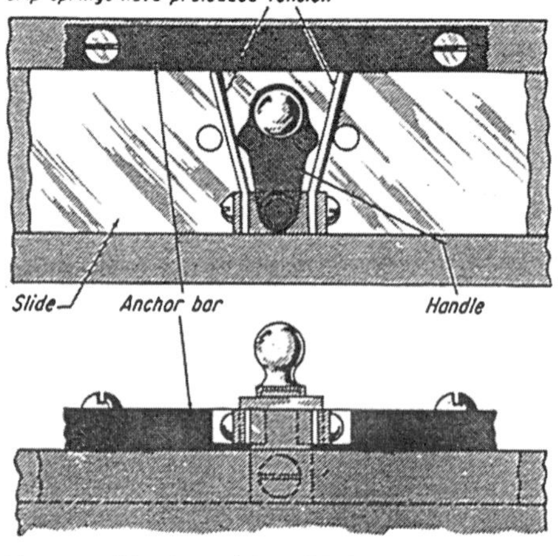

Easy positioning of the slide is possible when the handle pins move a grip spring out of contact with the anchor bar.

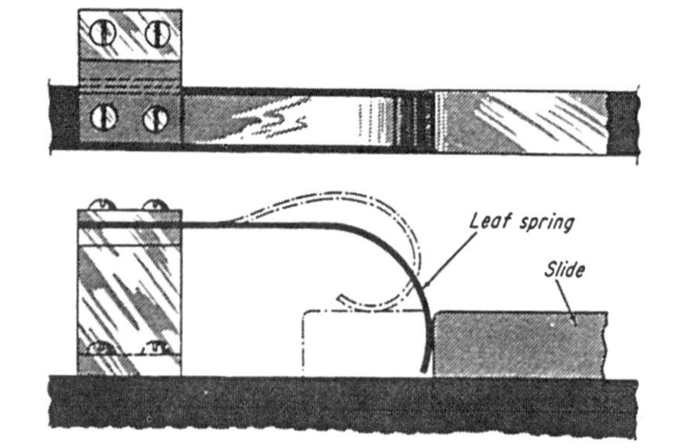

A spring-loaded slide will always return to its original position unless it is pushed until the spring kicks out.

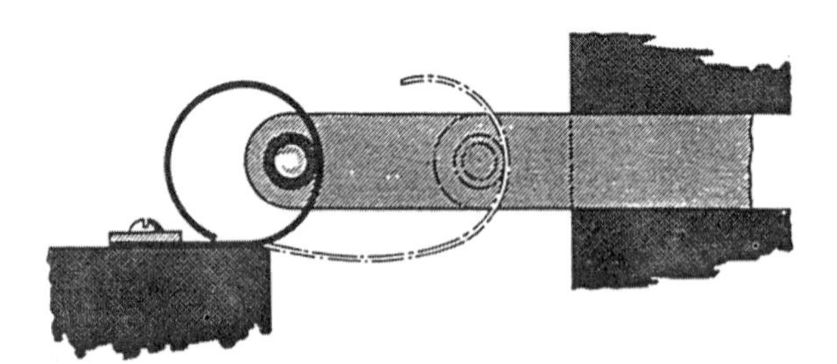

Nearly constant tension in the spring, as well as the force to activate the slide, is provided by this single coil.

Increasing support area as the load increases on both upper and lower platens is provided by a circular spring.

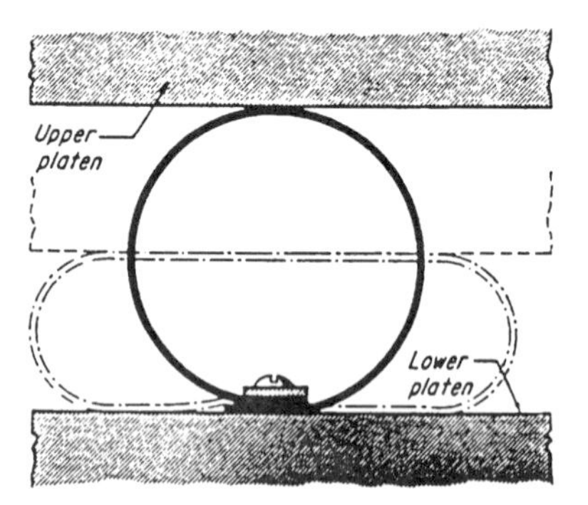

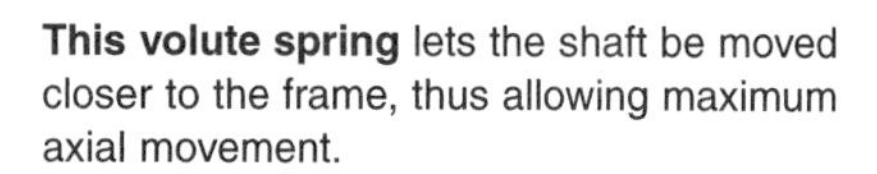

This volute spring lets the shaft be moved closer to the frame, thus allowing maximum axial movement.

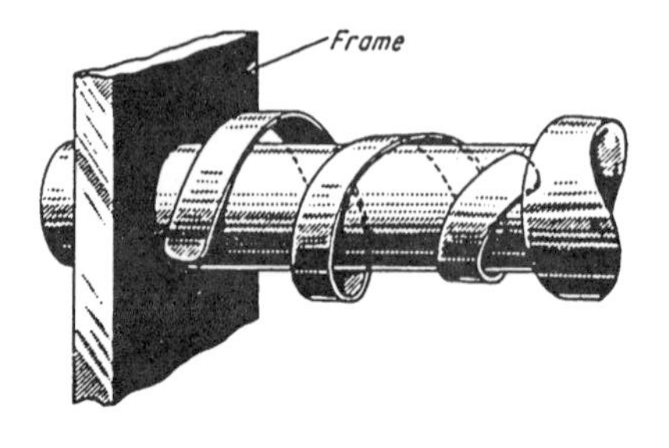

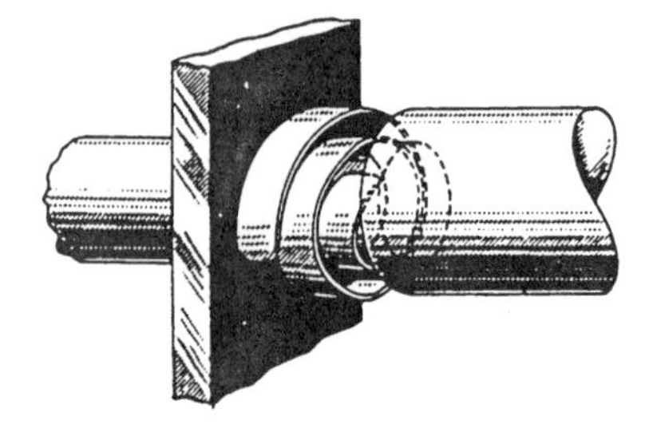

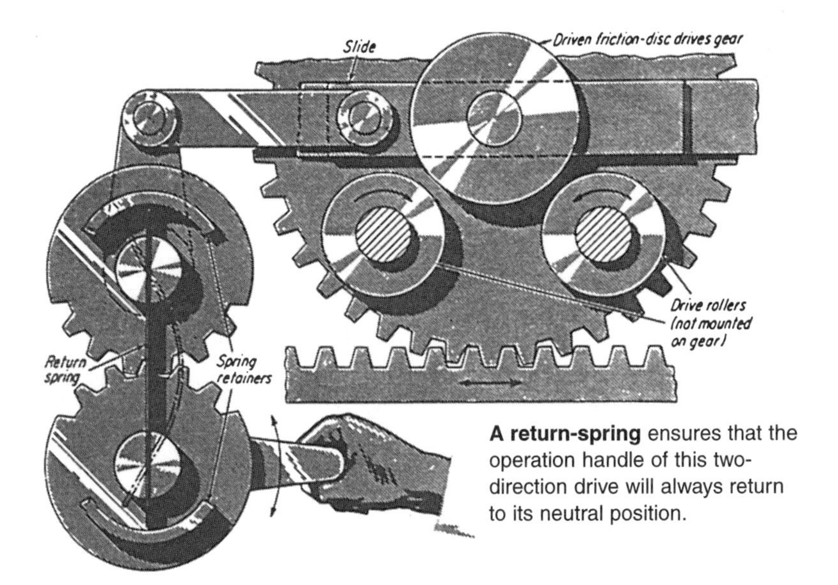

A return-spring ensures that the operation handle of this two-direction drive will always return to its neutral position.

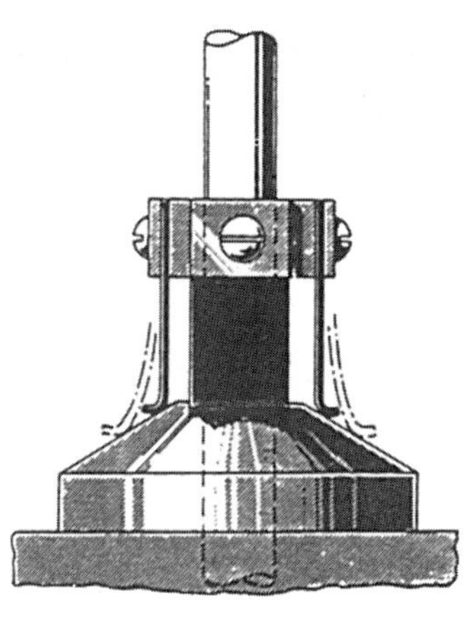

This cushioning device imparts rapid increase of spring tension because of the small pyramid angle. Its rebound is minimum.

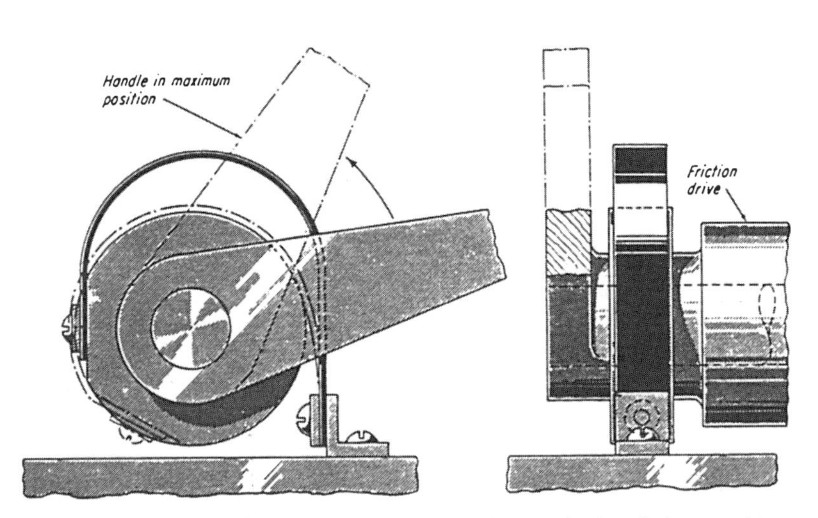

This spring-mounted disk changes its center position as the handle is rotated to move the friction drive. It also acts as a built-in limit stop.

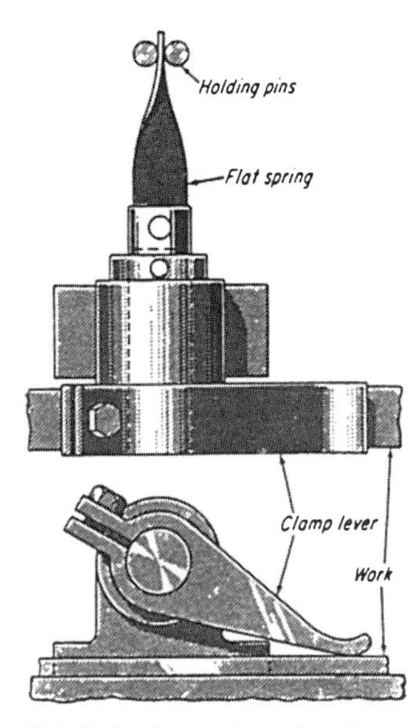

This hold-down clamp has its flat spring assembled with an initial twist to provide a clamping force for thin material.

Indexing is accomplished simply, efficiently, and at low cost by flat-spring arrangement shown here.

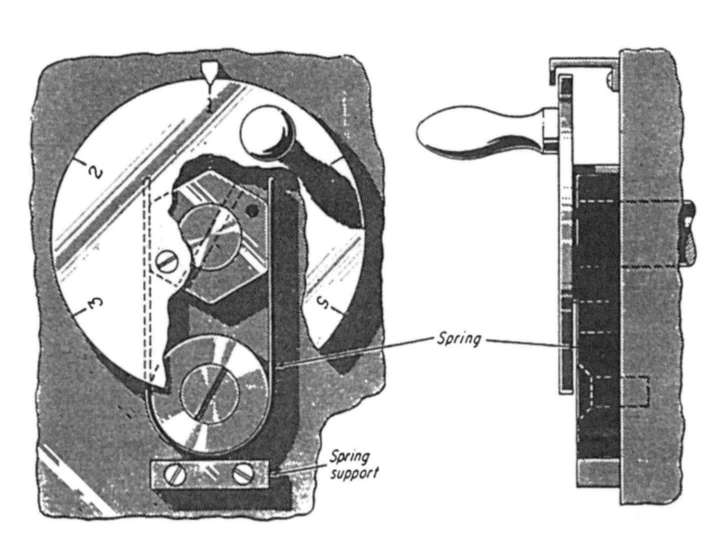

TWELVE WAYS TO USE METAL SPRINGS

Variable-rate arrangements, roller positioning, space saving, and other ingenious ways to get the most from springs.

This setup provides a **variable rate** with a sudden change from a light load to a heavy load by limiting the low-rate extension with a spring.

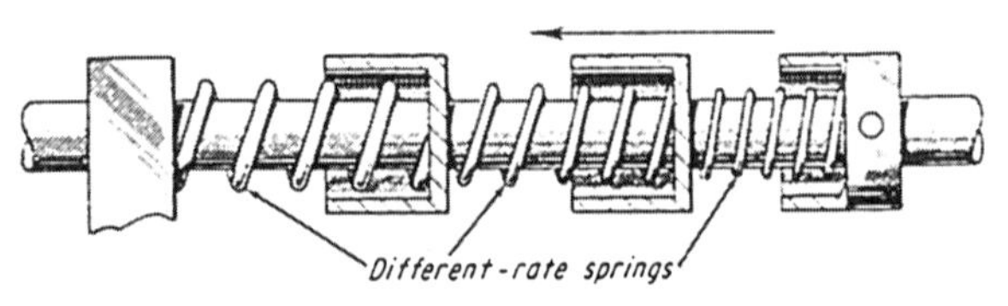

This mechanism provides a **three-step rate** change at predetermined positions. The lighter springs will always compress first, regardless of their position.

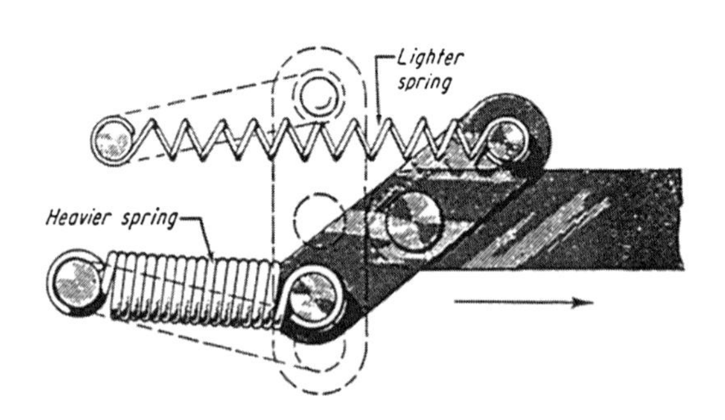

This differential-rate linkage sets the actuator stroke under light tension at the start, then allows a gradual transition to heavier tension.

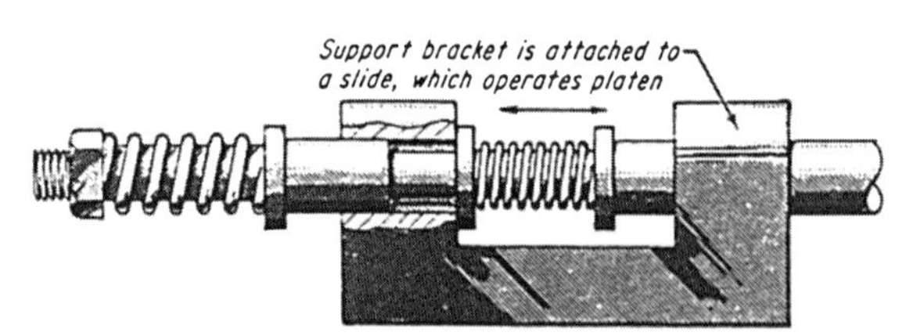

This compressing mechanism has a dual rate for double-action compacting. In one direction pressure is high, but in the reverse direction pressure is low.

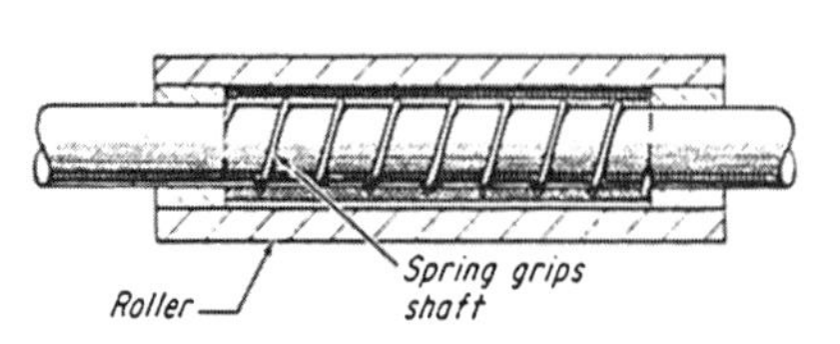

Roller positioning by a tightly wound spring on the shaft is provided by this assembly. The roller will slide under excess end thrust.

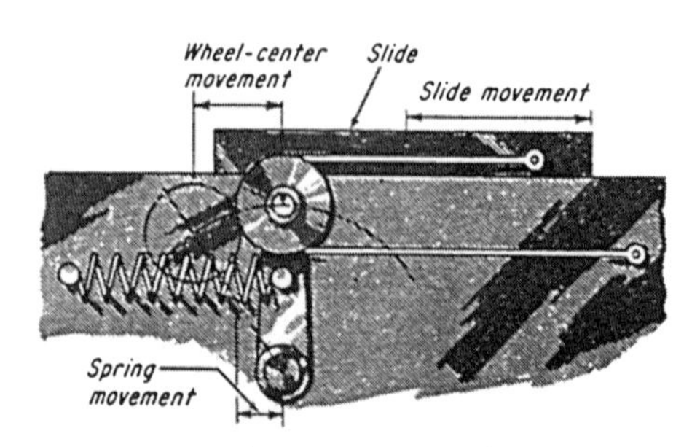

A short extension of the spring for a long movement of the slide keeps the tension change between maximum and minimum low.

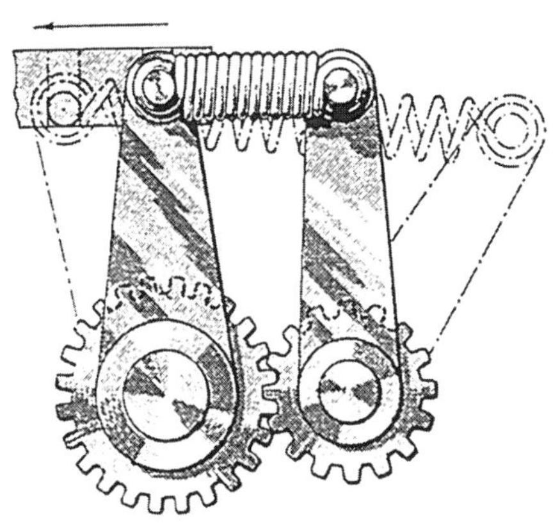

Increased tension for the same movement is gained by providing a movable spring mount and gearing it to the other movable lever.

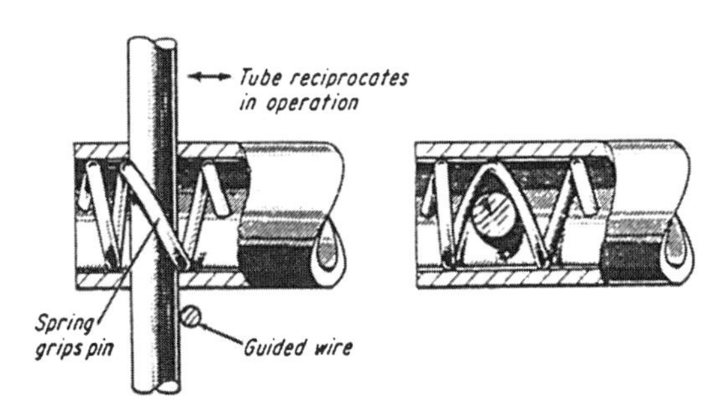

This pin grip is a spring that holds a pin by friction against end movement or rotation, but lets the pin be repositioned without tools.

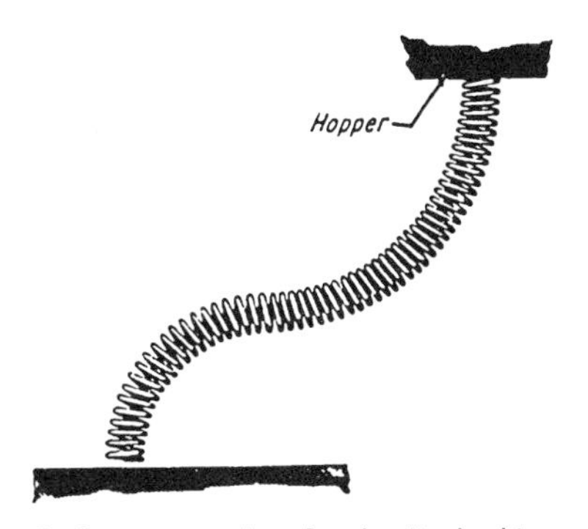

A close-wound spring is attached to a hopper, and it will not buckle when it is used as a movable feed-duct for nongranular material.

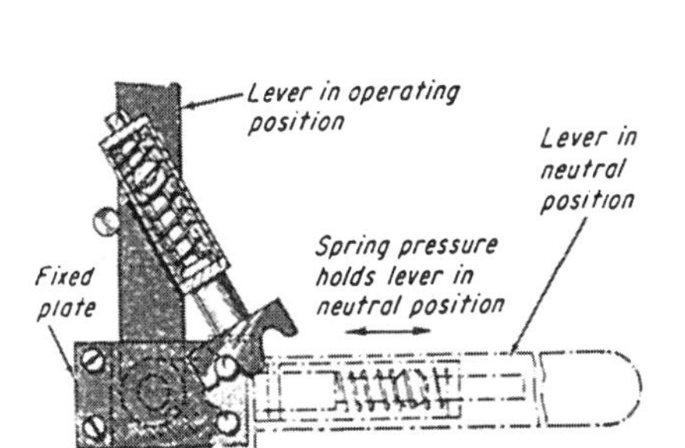

Toggle action here ensures that the gear-shift lever will not inadvertently be thrown past its neutral position.

Tension varies at a different rate when the brake-applying lever reaches the position shown. The rate is reduced when the tilting lever tilts.

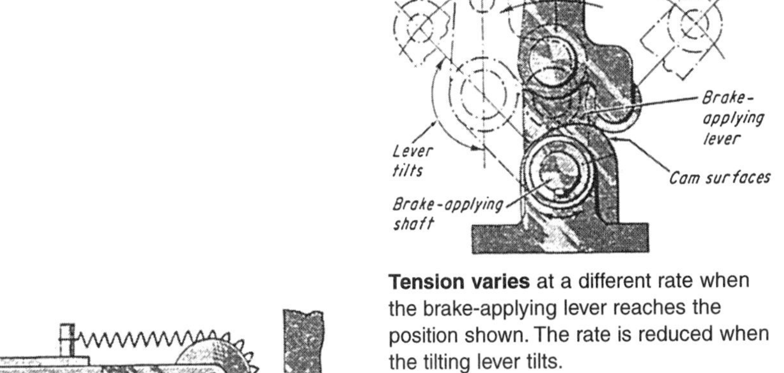

The spring wheel helps to distribute deflection over more coils that if the spring rested on the corner. The result is less fatigue and longer life.

SEVEN OVERRIDING SPRING MECHANISMS FOR LOW-TORQUE DRIVES

Overriding spring mechanisms are widely used in the design of instruments and controls. All of the arrangements illustrated allow an incoming motion to override the outgoing motion whose limit has been reached. In an instrument, for example, the spring mechanism can be placed between the sensing and indicating elements to provide overrange protection. The dial pointer is driven positively up to its limit before it stops while the input shaft is free to continue its travel. Six of the mechanisms described here are for rotary motion of varying amounts. The last is for small linear movements.

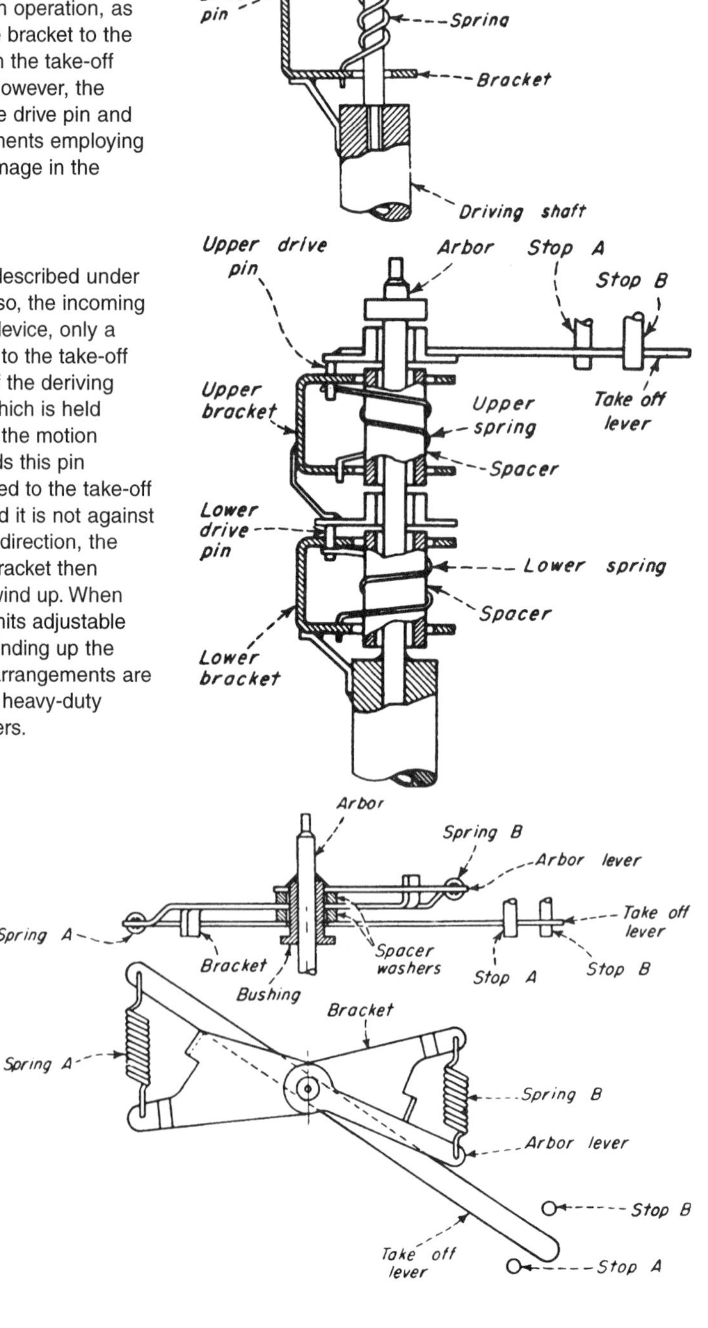

Fig. 1 Unidirectional override. The take-off lever of this mechanism can rotate nearly 360°. Its movement is limited only by one stop pin. In one direction, motion of the driving shaft is also impeded by the stop pin. But in the reverse direction the driving shaft is capable or rotating approximately 270° past the stop pin. In operation, as the driving shaft is turned clockwise, motion is transmitted through the bracket to the take-off lever. The spring holds the bracket against the drive pin. When the take-off lever has traveled the desired limit, it strikes the adjustable stop pin. However, the drive pin can continue its rotation by moving the bracket away from the drive pin and winding up the spring. An overriding mechanism is essential in instruments employing powerful driving elements, such as bimetallic elements, to prevent damage in the overrange regions.

Fig. 2 Two-directional override. This mechanism is similar to that described under Fig. 1, except that two stop pins limit the travel of the take-off lever. Also, the incoming motion can override the outgoing motion in either direction. With this device, only a small part of the total rotation of the driving shaft need be transmitted to the take-off lever, and this small part can be anywhere in the range. The motion of the deriving shaft is transmitted through the lower bracket to the lower drive pin, which is held against the bracket by the spring. In turn, the lower drive pin transfers the motion through the upper bracket to the upper drive pin. A second spring holds this pin against the upper drive bracket. Because the upper drive pin is attached to the take-off lever, any rotation of the drive shaft is transmitted to the lever, provided it is not against either stop *A* or *B*. When the driving shaft turns in a counterclockwise direction, the take-off lever finally strikes against the adjustable stop *A*. The upper bracket then moves away from the upper drive pin, and the upper spring starts to wind up. When the driving shaft is rotated in a clock-wise direction, the take-off lever hits adjustable stop *B*, and the lower bracket moves away from the lower drive pin, winding up the other spring. Although the principal applications for overriding spring arrangements are in instrumentation, it is feasible to apply these devices in the drives of heavy-duty machines by strengthening the springs and other load-bearing members.

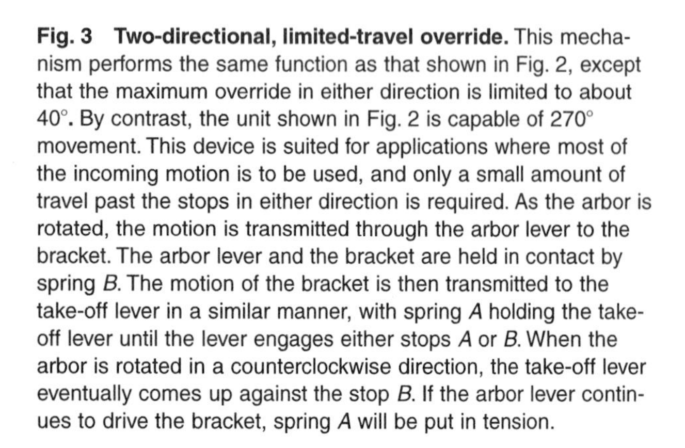

Fig. 3 Two-directional, limited-travel override. This mechanism performs the same function as that shown in Fig. 2, except that the maximum override in either direction is limited to about 40°. By contrast, the unit shown in Fig. 2 is capable of 270° movement. This device is suited for applications where most of the incoming motion is to be used, and only a small amount of travel past the stops in either direction is required. As the arbor is rotated, the motion is transmitted through the arbor lever to the bracket. The arbor lever and the bracket are held in contact by spring *B*. The motion of the bracket is then transmitted to the take-off lever in a similar manner, with spring *A* holding the take-off lever until the lever engages either stops *A* or *B*. When the arbor is rotated in a counterclockwise direction, the take-off lever eventually comes up against the stop *B*. If the arbor lever continues to drive the bracket, spring *A* will be put in tension.

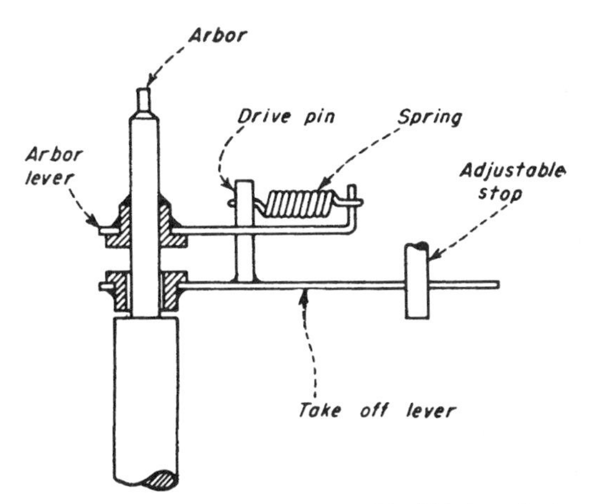

Fig. 4 Unidirectional, 90° override. This is a single overriding unit that allows a maximum travel of 90° past its stop. The unit, as shown, is arranged for overtravel in a clockwise direction, but it can also be made for a counterclockwise override. The arbor lever, which is secured to the arbor, transmits the rotation of the arbor to the take-off lever. The spring holds the drive pin against the arbor lever until the take-off lever hits the adjustable stop. Then, if the arbor lever continues to rotate, the spring will be placed in tension. In the counterclockwise direction, the drive pin is in direct contact with the arbor lever so that no overriding is possible.

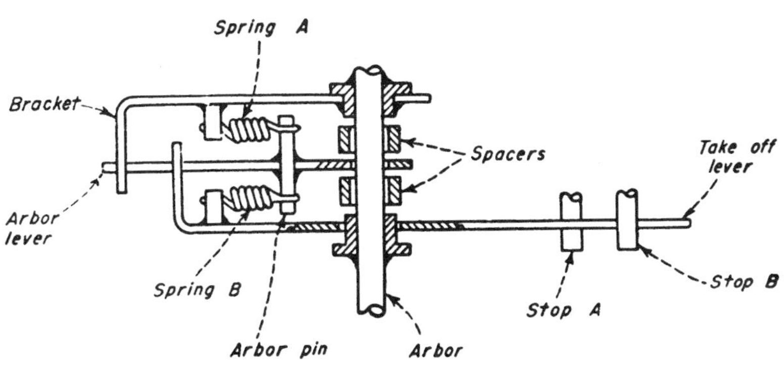

Fig. 5 Two-directional, 90° override. This double-overriding mechanism allows a maximum overtravel of 90° in either direction. As the arbor turns, the motion is carried from the bracket to the arbor lever, then to the take-off lever. Both the bracket and the take-off lever are held against the arbor lever by spring *A* and *B*. When the arbor is rotated counterclockwise, the takeoff lever hits stop *A*. The arbor lever is held stationary in contact with the take-off lever. The bracket, which is fastened to the arbor, rotates away from the arbor lever, putting spring *A* in tension. When the arbor is rotated n a clockwise direction, the take-off lever comes against stop *B*, and the bracket picks up the arbor lever, putting spring *B* in tension.

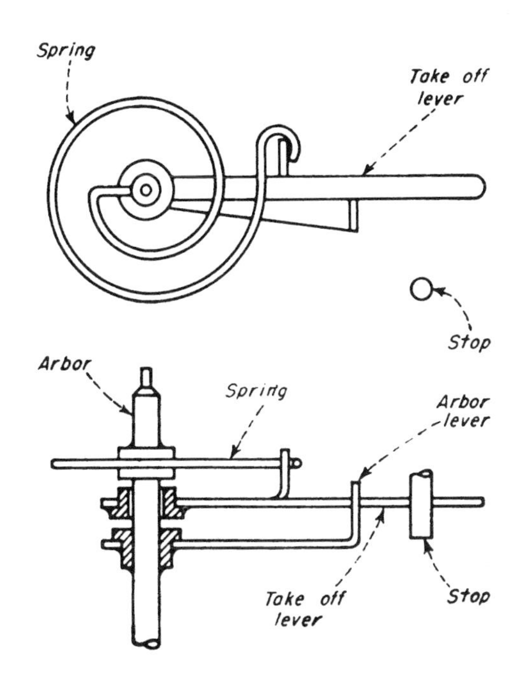

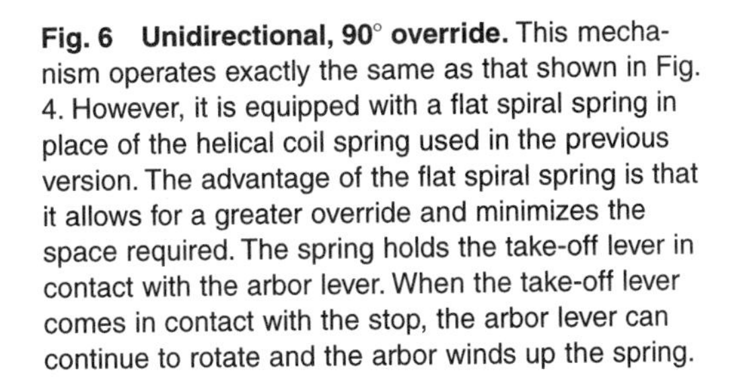

Fig. 6 Unidirectional, 90° override. This mechanism operates exactly the same as that shown in Fig. 4. However, it is equipped with a flat spiral spring in place of the helical coil spring used in the previous version. The advantage of the flat spiral spring is that it allows for a greater override and minimizes the space required. The spring holds the take-off lever in contact with the arbor lever. When the take-off lever comes in contact with the stop, the arbor lever can continue to rotate and the arbor winds up the spring.

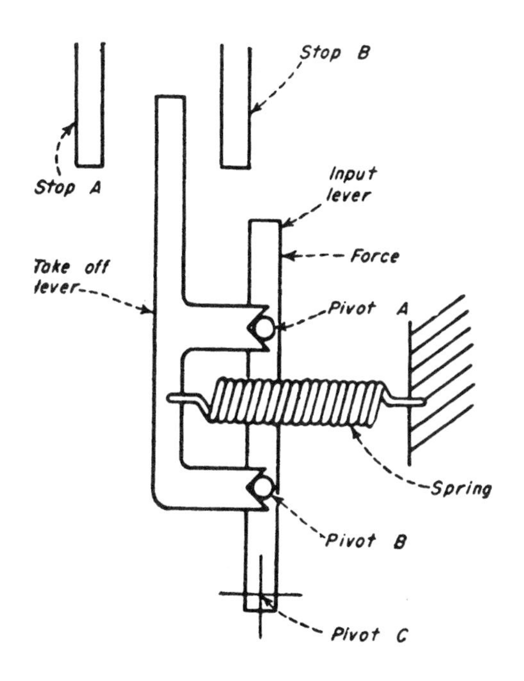

Fig. 7 Two-directional override, linear motion. The previous mechanisms were overrides for rotary motion. The device in Fig. 7 is primarily a double override for small linear travel, although it could be used on rotary motion. When a force is applied to the input lever, which pivots about point *C*, the motion is transmitted directly to the take-off lever through the two pivot posts, *A* and *B*. The take-off lever is held against these posts by the spring. When the travel causes the take-off lever to hit the adjustable stop *A*, the take-off lever revolves about pivot post *A*, pulling away from pivot post *B*, and putting additional tension in the spring. When the force is diminished, the input lever moves in the opposite direction until the take-off lever contacts the stop *B*. This causes the take-off lever to rotate about pivot post B, and pivot post *A* is moved away from the take-off lever.

SIX SPRING MOTORS AND ASSOCIATED MECHANISMS

Many applications of spring motors in clocks, motion picture cameras, game machines, and other mechanisms offer practical ideas for adaptation to any mechanism that is intended to operate for an appreciable length of time. While spring motors are usually limited to comparatively small power application where other sources of power are unavailable or impracticable, they might also be useful for intermittent operation requiring comparatively high torque or high speed, using a low-power electric motor or other means for building up energy.

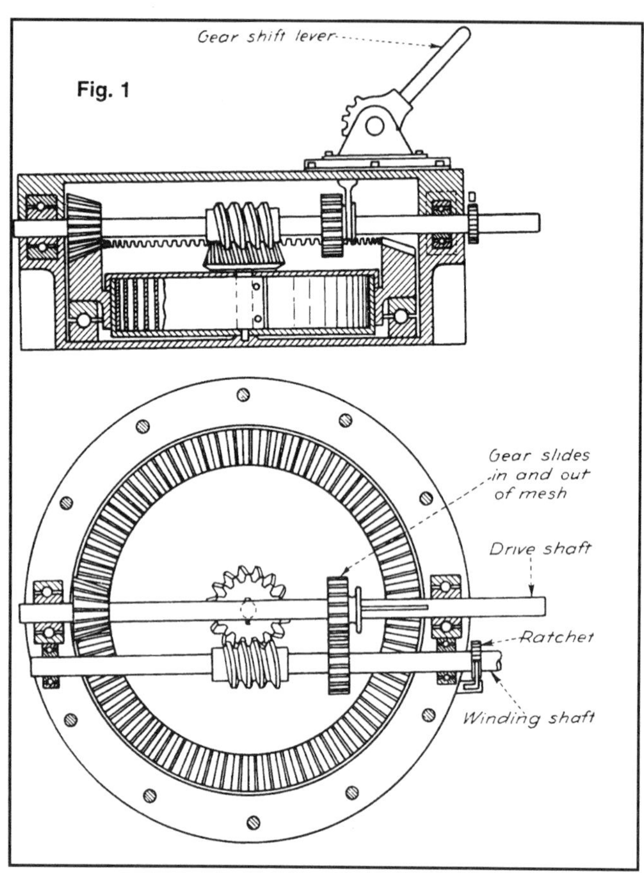

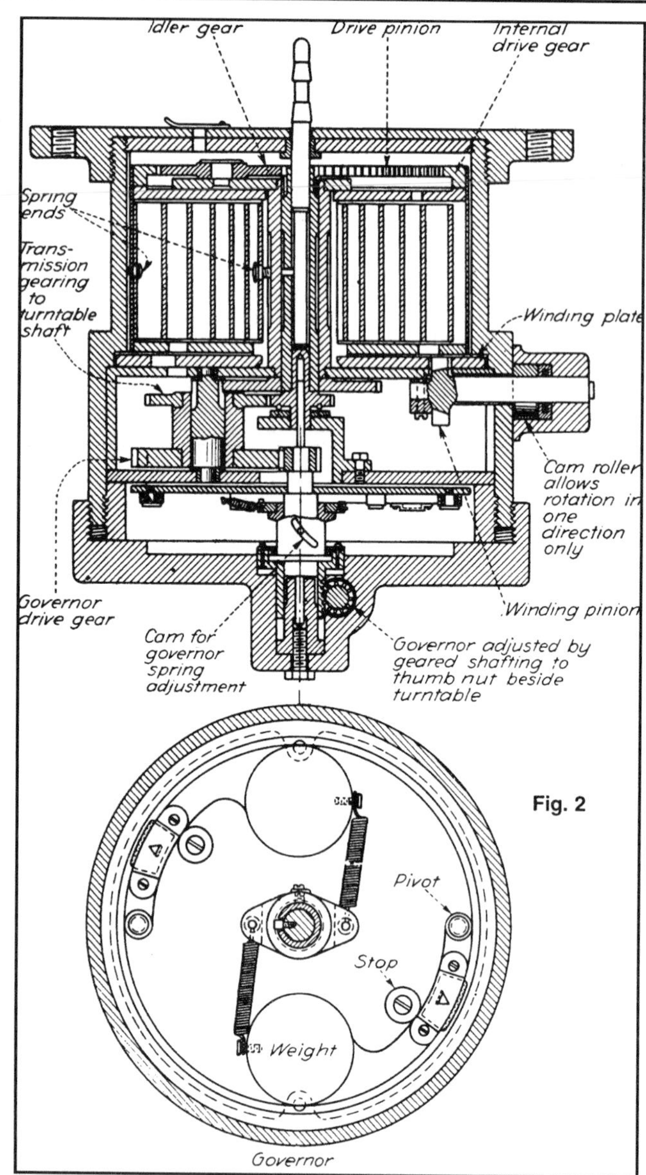

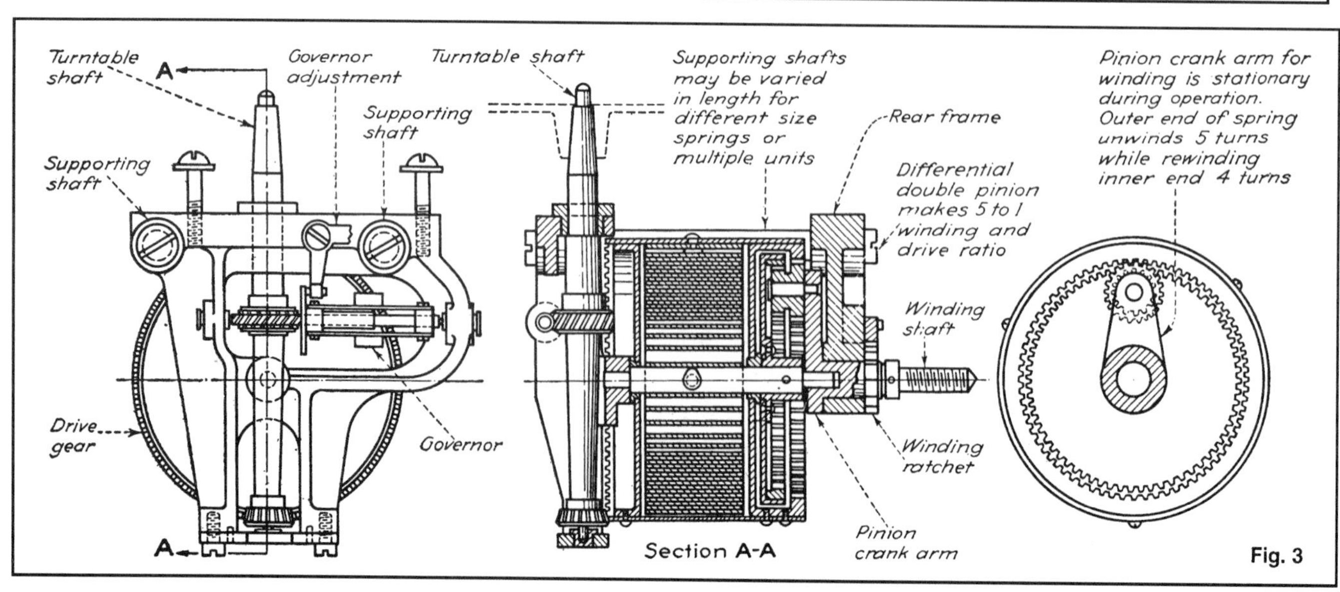

The accompanying patented spring motor designs show various methods for the transmission and control of spring-motor power. Flat-coil springs, confined in drums, are most widely used because they are compact, produce torque directly, and permit long angular displacement. Gear trains and feedback mechanisms reduce excess power drain so that power can be applied for a longer time. Governors are commonly used to regulate speed.

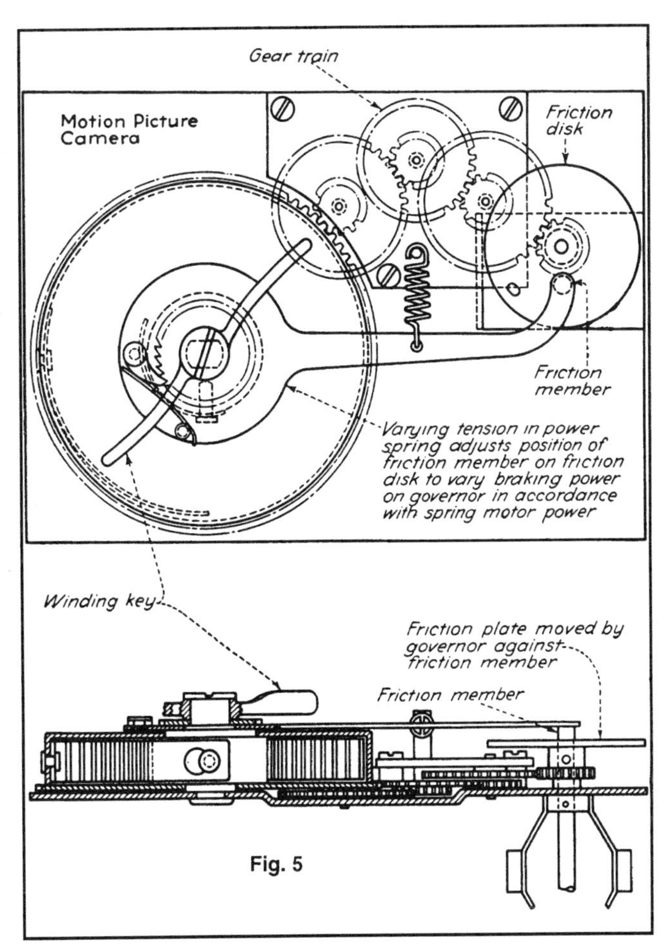

Fig. 5

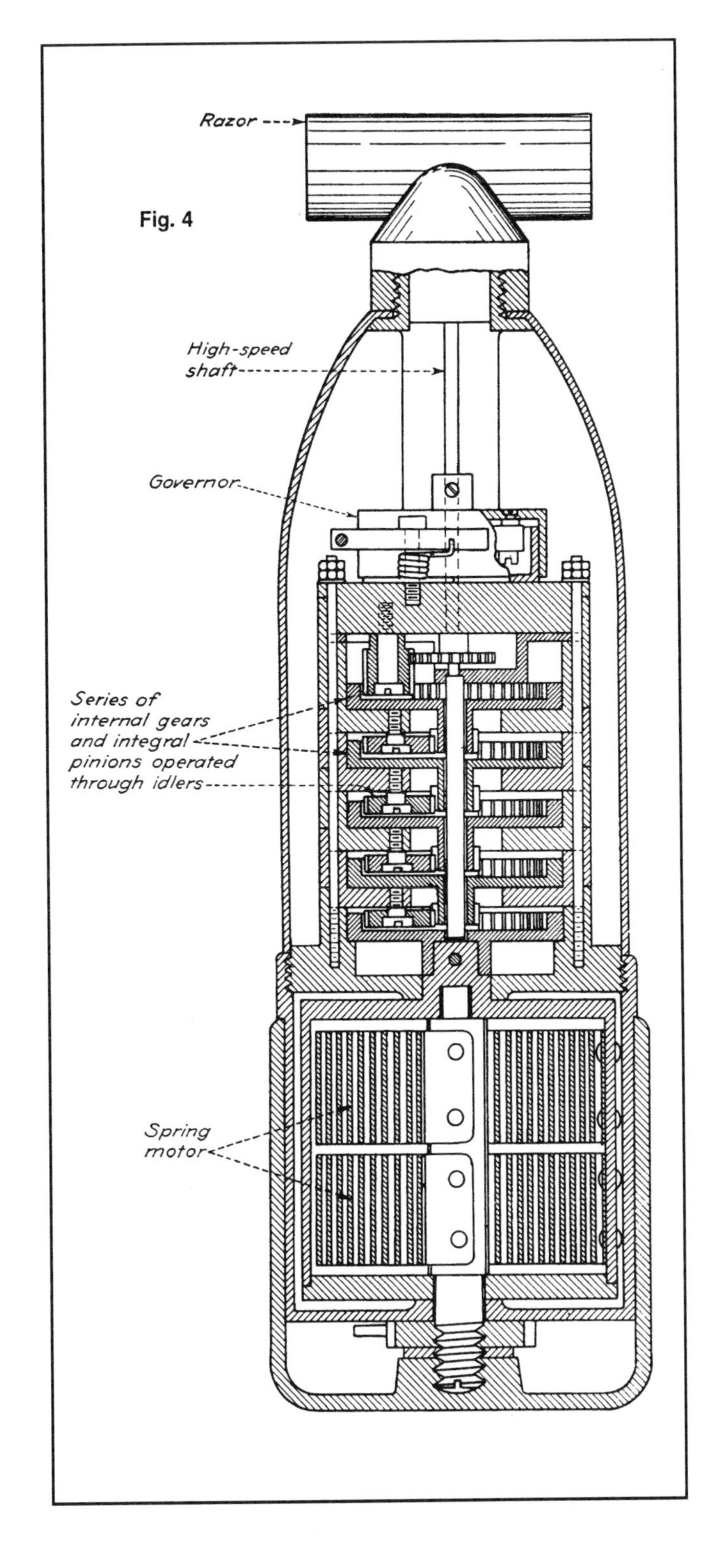

Fig. 4

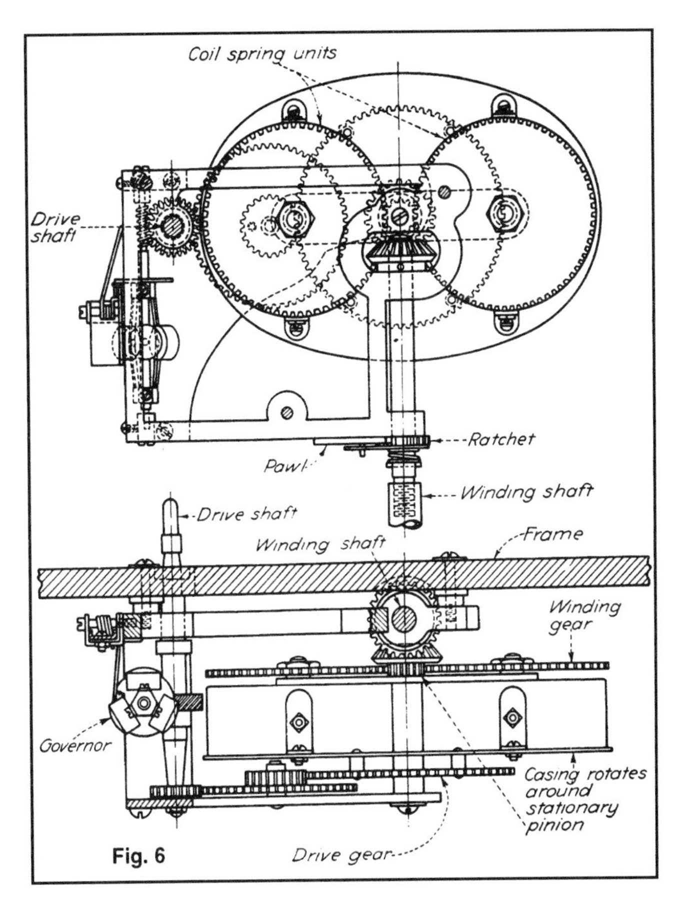

Fig. 6

TWELVE AIR SPRING APPLICATIONS

EIGHT WAYS TO ACTUATE MECHANISMS WITH AIR SPRINGS

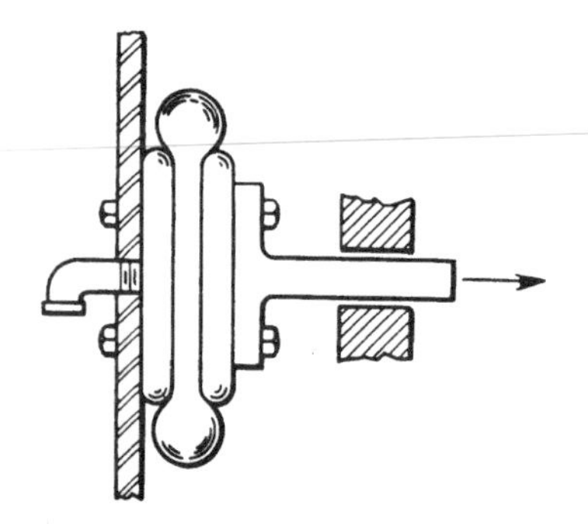

Linear force link: A one- or two-convolution air spring drives the guide rod. The rod is returned by gravity, opposing force, metal spring or, at times, internal stiffness of an air spring.

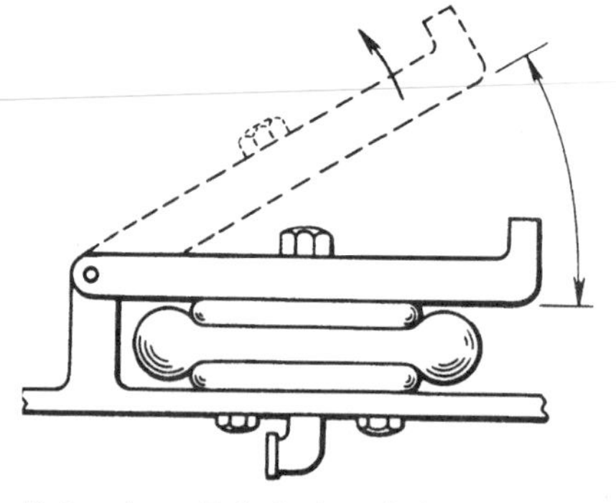

Rotary force link: A pivoted plate can be driven by a one-convolution or two-convolution spring to 30° of rotation. The limitation on the angle is based on permissible spring misalignment.

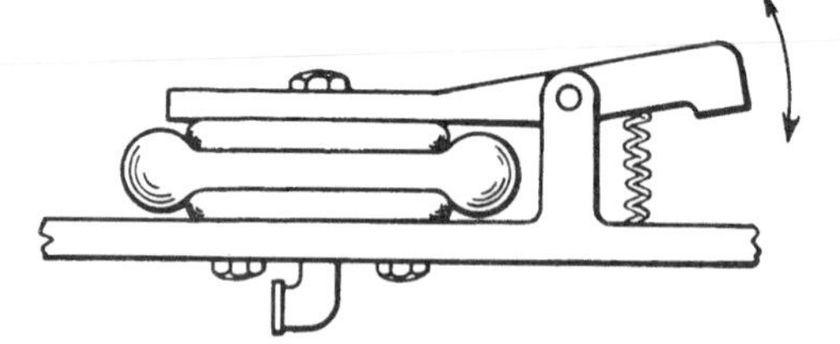

Clamp: A jaw is normally held open by a metal spring. Actuation of the air spring then closes the clamp. The amount of opening in the jaws of the clamp can be up to 30° of arc.

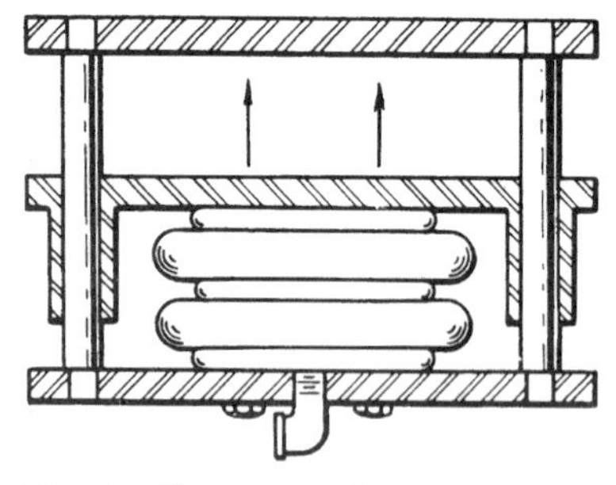

Direct-acting press: One-, two-, or three-convolution air springs are assembled singly or in gangs. They are naturally stable when used in groups. Gravity returns the platform to its starting position.

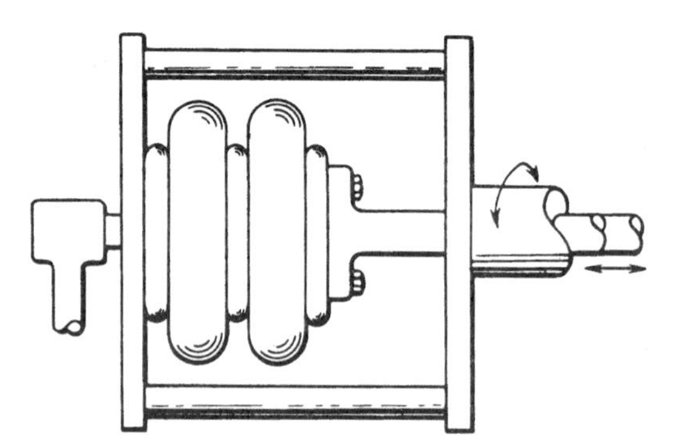

Rotary shaft actuator: The activator shifts the shaft longitudinally while the shaft is rotating. Air springs with one, two, or three convolutions can be used. A standard rotating air fitting is required.

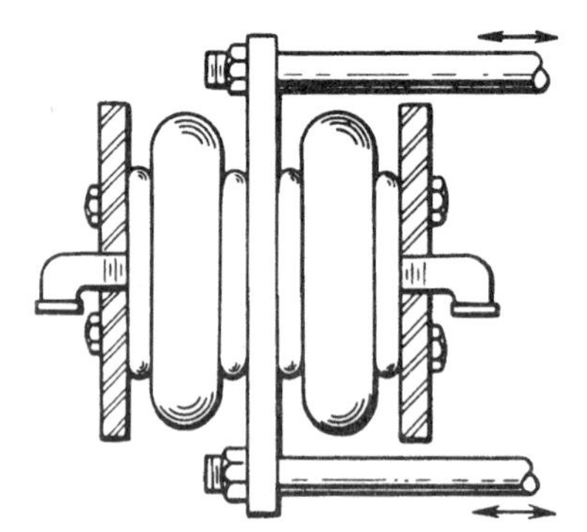

Reciprocating linear force link: It reciprocates with one-, two-, or three-convolution air springs in a back-to-back arrangement. Two- and three-convolution springs might need guides for their force rods.

POPULAR TYPES OF AIR SPRINGS

Air is an ideal load-carrying medium. It is highly elastic, its spring rate can be easily varied, and it is not subject to permanent set.

Air springs are elastic devices that employ compressed air as the spring element. They maintain a soft ride and a constant vehicle height under varying load. In industrial applications they control vibration (isolate or amplify it) and actuate linkages to provide either rotary or linear movement. Three kinds of air springs (bellows, rolling sleeve, and rolling diaphragm) are illustrated.

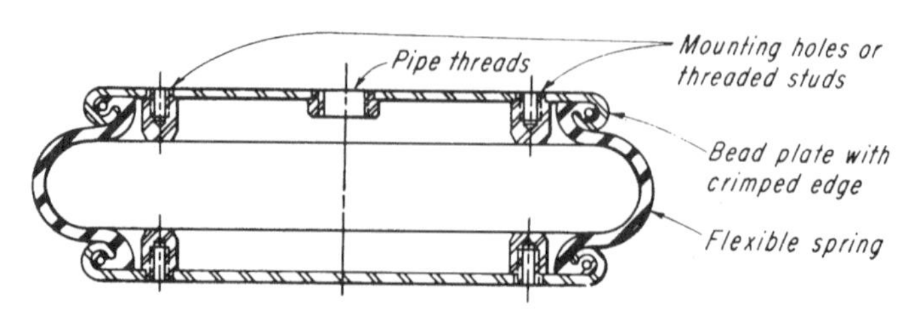

ONE-CONVOLUTION BELLOW

Bellows Type

A single-convolution spring looks like a tire lying on its side. It has a limited stroke and a relatively high spring rate. Its natural frequency is about 150 cpm without auxiliary volume for most sizes, and as high as 240 cpm for the smallest size. Lateral stiffness is high (about half the vertical rate); therefore the spring is quite stable laterally when used for industrial vibration isolation. It can be filled manually or kept inflated to a constant

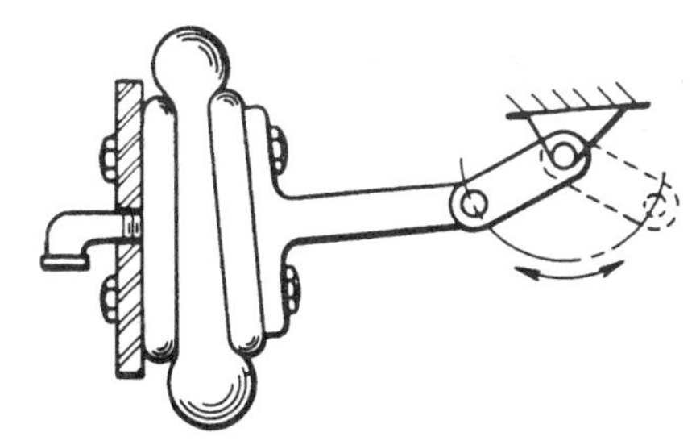

Pivot mechanism: It rotates a rod through 145° of rotation. It can accept a 30° misalignment because of the circular path of its connecting-link pin. A metal spring or opposing force retracts the link.

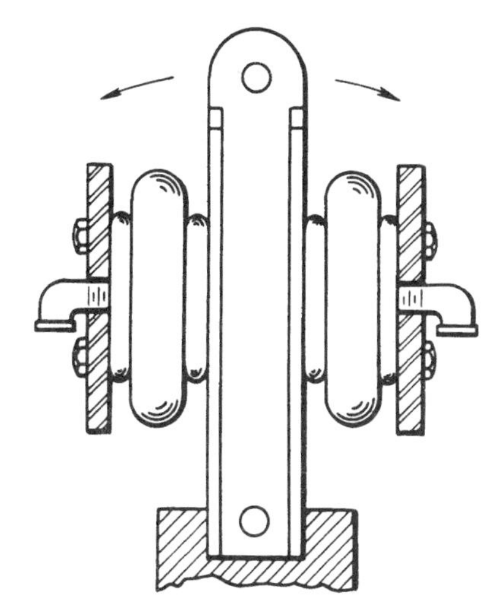

Reciprocating rotary motion with one-convolution and two-convolution springs. An arc up to 30° is possible. It can pair a large air spring with a smaller one or a lengthen lever.

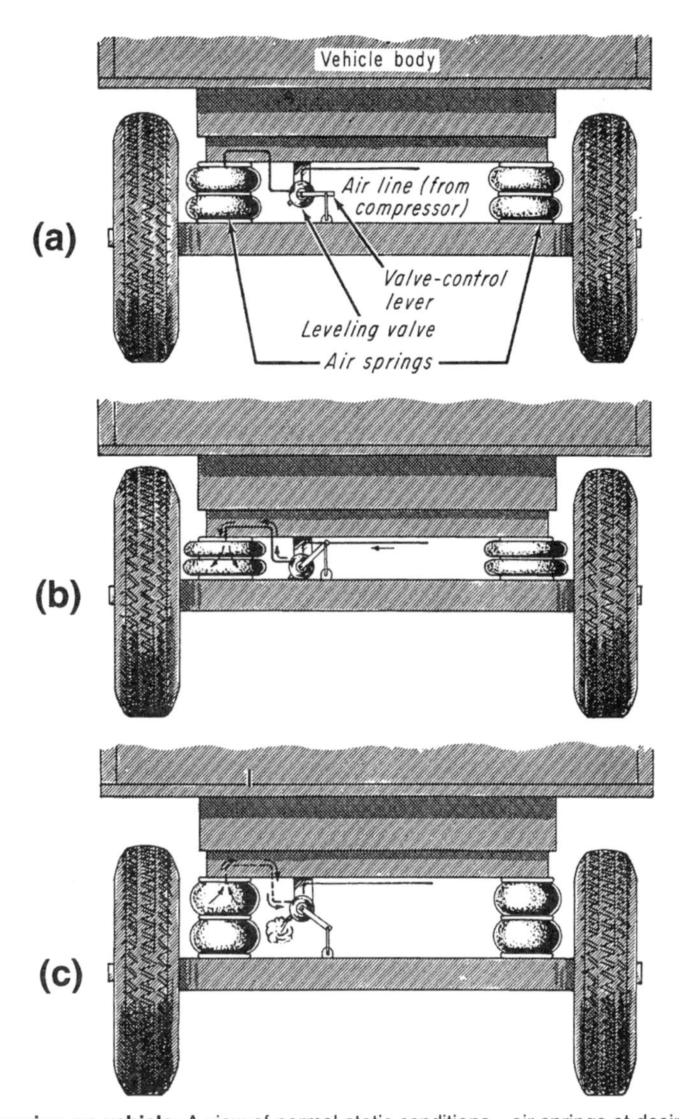

Air suspension on vehicle: A view of normal static conditions—air springs at desired height and height-control valve closed (a). When a load is added to the vehicle—the valve opens to admit air to the springs and restore height, but at higher pressure (b). With load removed from the vehicle—valve permits bleeding off excess air pressure to atmosphere and restores its design height (c).

height if is connected to factory air supply through a pressure regulator. This spring will also actuate linkages where short axial length is desirable. It is seldom used in vehicle suspension systems.

Rolling-Sleeve Type

This spring is sometimes called the reversible-sleeve or rolling-lobe type. It has a telescoping action—the lobe at the bottom of the air spring rolls up and down along the piston. The spring is used primarily in vehicle suspensions because lateral stiffness is almost zero.

Rolling-Diaphragm Type

These are laterally stable and can be used as vibration isolators, actuators, or constant-force spring. But because of their negative effective-area curve, their pressure is not generally maintained by pressure regulators.

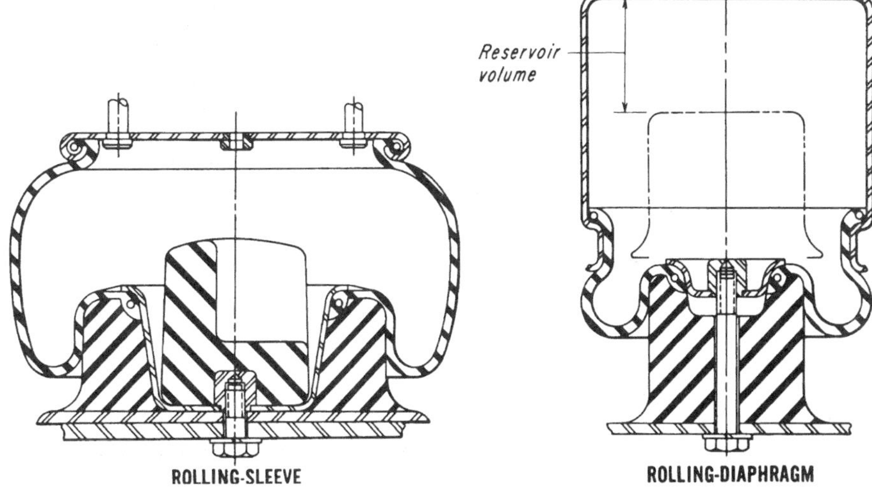

NOVEL APPLICATIONS FOR DIFFERENT SPRINGS

How stops, cams, linkages, and other arrangements can vary the load/deflection ratio during extension or compression

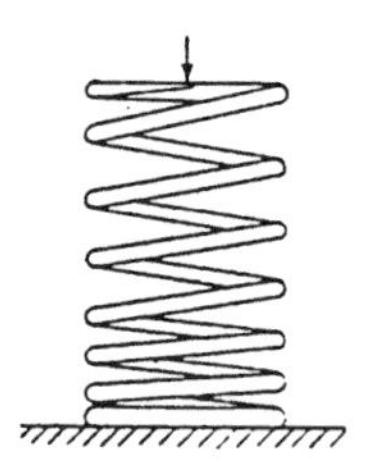

With tapered-pitch spring the number of effective coils changes with deflection—the coils "bottom" progressively.

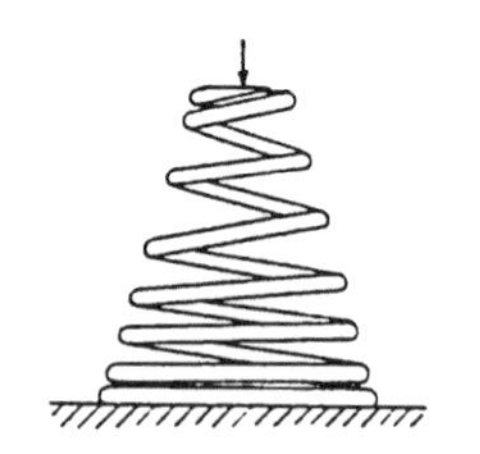

A tapered outside diameter and pitch combine to produce a similar effect except that the spring with tapered O.D. will have a shorter solid height.

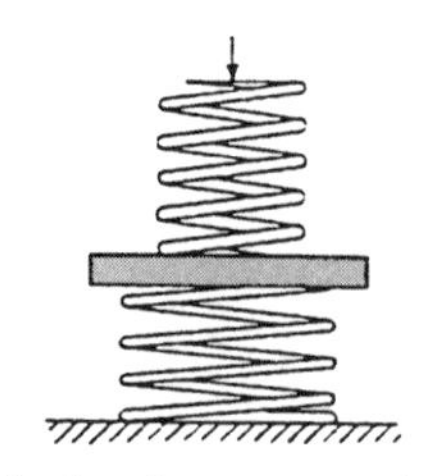

In dual springs, one spring closes completely before the other.

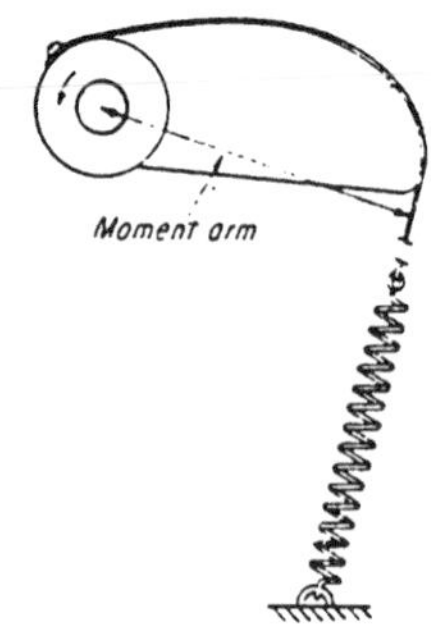

A cam-and-spring device causes the torque relationship to vary during rotation as the moment arm changes.

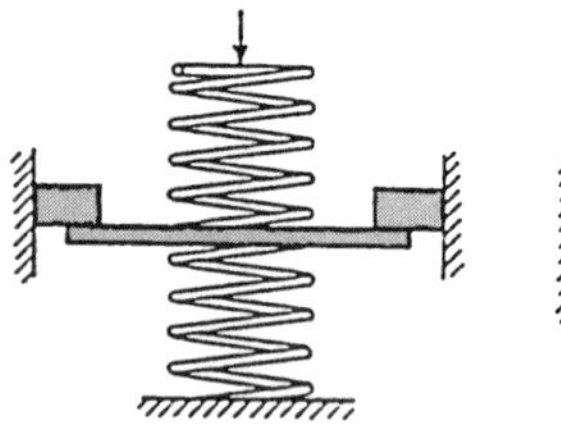

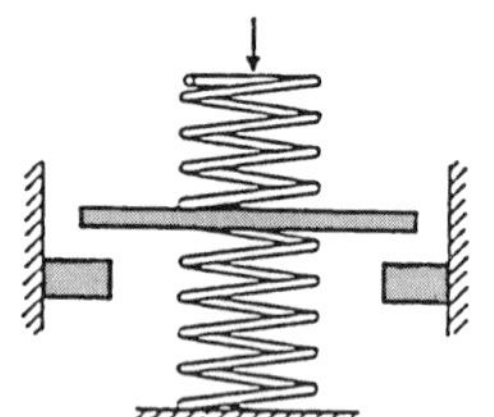

Stops can be used with either compression or extension springs.

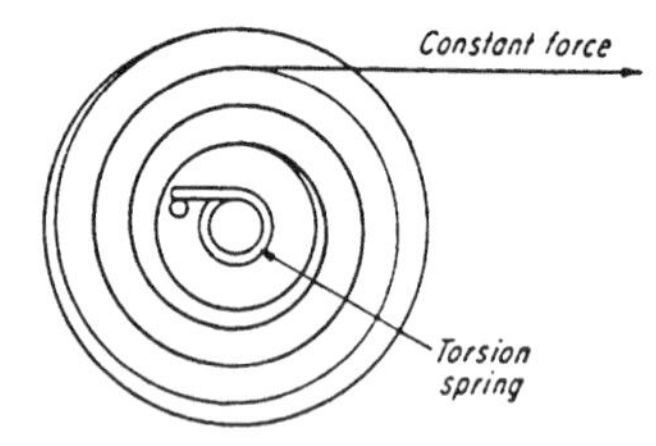

Torsion spring combined with a variable-radius pulley gives a constant force.

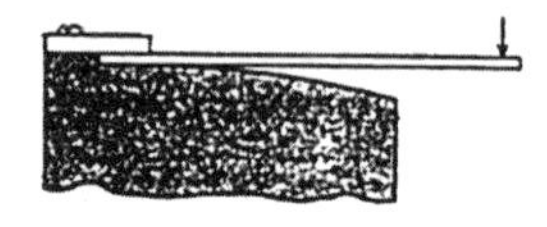

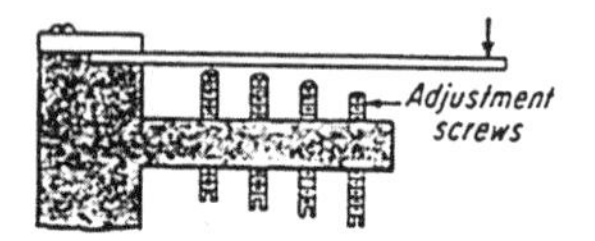

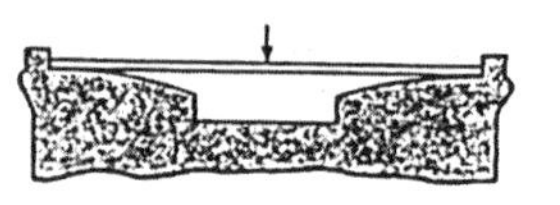

Leaf springs can be arranges so that their effective lengths change with deflection.

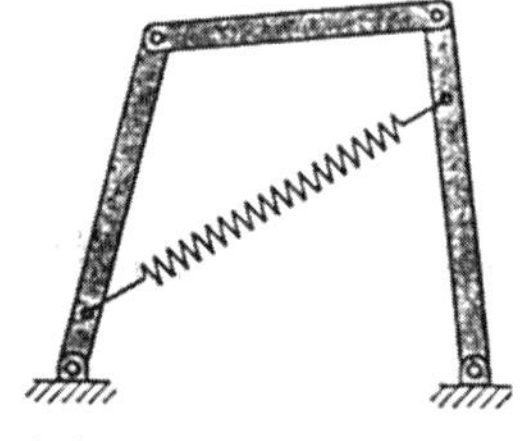

A four-bar mechanism in conjunction with a spring has a wide variety of load/deflection characteristics.

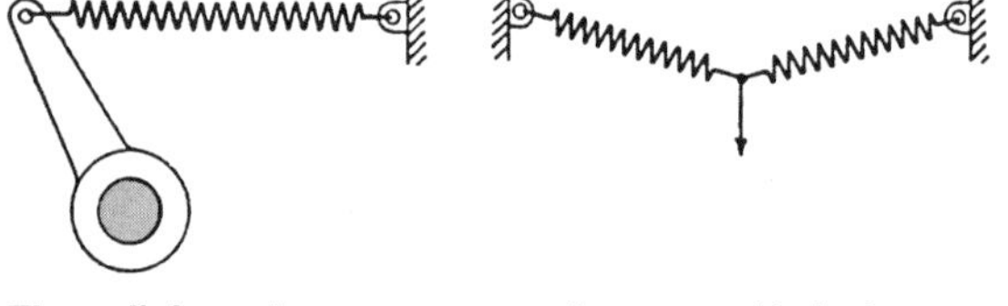

These linkage-type arrangements are used in instruments where torque control or anti-vibration suspension is required.

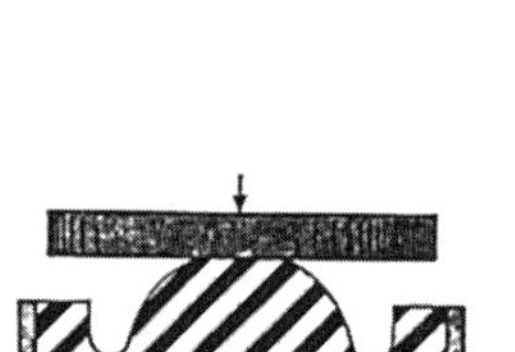

A molded-rubber spring has deflection characteristics that vary with its shape.

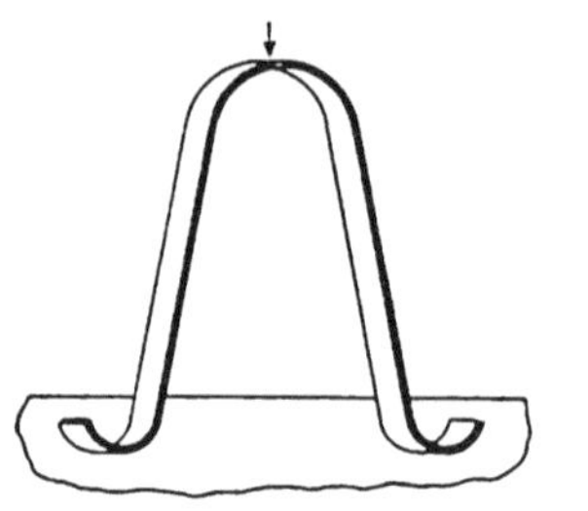

An arched leaf-spring gives an almost constant force when it is shaped like the one illustrated.

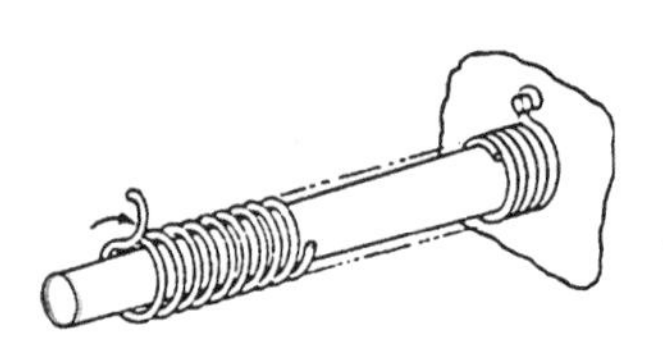

With a tapered mandrel and torsion spring the effective number or coils decreases with torsional deflection.

APPLICATIONS FOR BELLEVILLE SPRINGS

Belleville springs are low-profile conical rings with differing height (h) to thickness (t) ratios, as shown in Fig. 1. Four way to stack them are shown in Fig. 2.

Belleville springs lend themselves to a wide variety; of applications:

For height to spring ratios of about 0.4—A linear spring rate and high load resistance with small deflections.

For height to spring ratios between 0.8 and 1.0—An almost linear spring rate for fasteners and bearing and in stacks.

For rations of around 1.6—A constant (flat) spring rate starting at about 60% of the deflection (relative to the fully compressed flat position) and proceeding to the flat position and, if desired, on to the flipped side to a deflection of about 140%. In most applications, the flat position is the limit of travel, and for deflections beyond the flat, the contact elements must be allowed unrestricted travel.

One application of bellevilles with constant spring rate is on live spindles on the tailpiece of a lathe. The work can be loaded on the lathe, and as the piece heats up and begins to expand, the belleville will absorb this change in length without adding any appreciable load.

For high height to spring ratios exceeding about 2.5—The spring is stiff, and as the stability point (high point on the curve) is passed the spring rate becomes negative causing resistance to drop rapidly. If allowed, the belleville will snap through the flat position. In other words, it will turn itself inside out.

Working in groups. Belleville washers stacked in the parallel arrangement have been used successfully in a variety of applications.

One is a pistol or rifle buffer mechanism (Fig. 3) designed to absorb repeated, high-energy shock loads. A preload nut predeflects the washers to stiffen their resistance. The stacked washers are guided by a central shaft, an outside guide cylinder, guide rings, or a combination of these.

A windup starter mechanism for diesel engines (shown in Fig. 4) replaces a heavy-duty electric starter or auxiliary gas engine. To turn over the engine, energy is manually stored in a stack of bellevilles compressed by a hand crank. When released, the expanding spring pack rotates a pinion meshed with the flywheel ring gear to start the engine.

Figure 5 shows a belleville as a loading spring for a clutch.

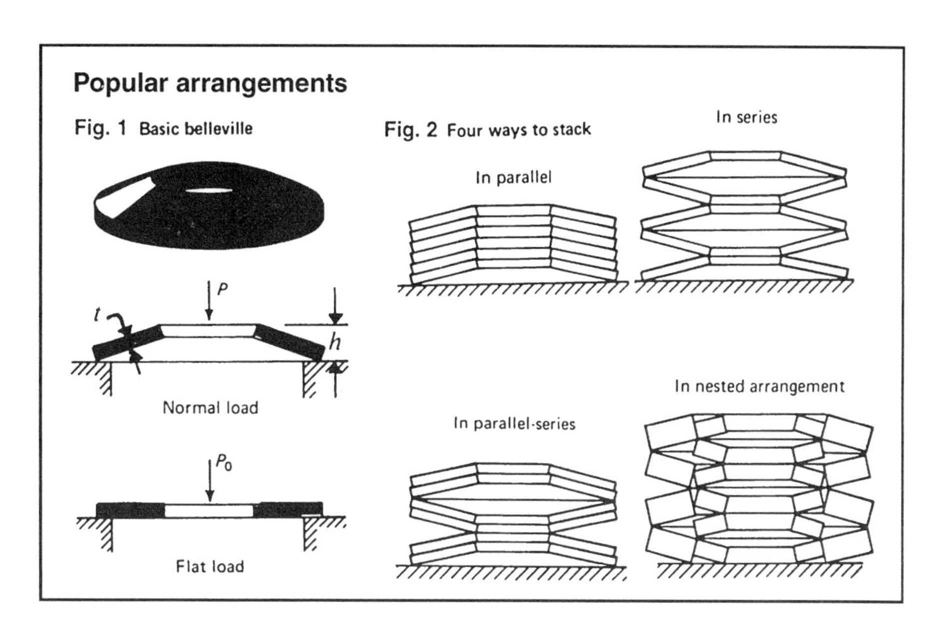

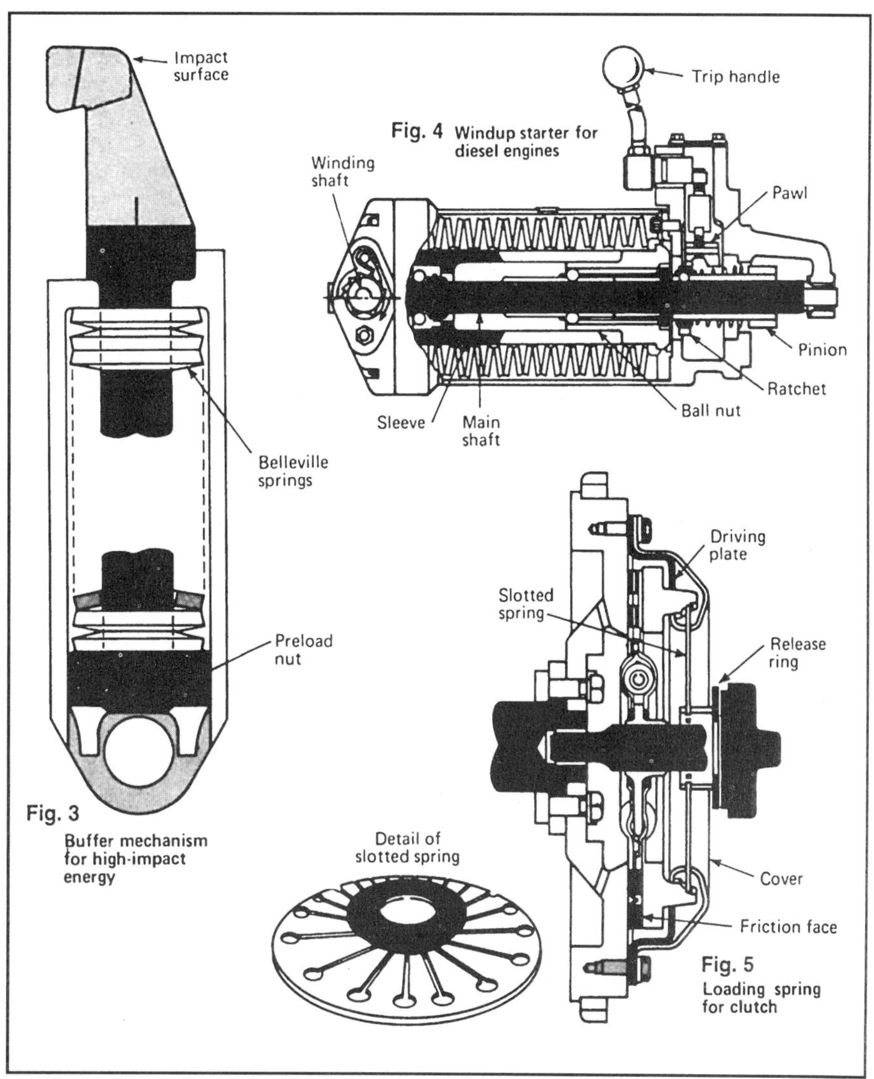

Fig. 3 Buffer mechanism for high-impact energy

Fig. 5 Loading spring for clutch

VIBRATION CONTROL WITH SPRING LINKAGE

Do you need a buffer between vibrating machinery and the surrounding structure? These isolators, like capable fighters, absorb the light jabs and stand firm against the forces that inflict powerful haymakers

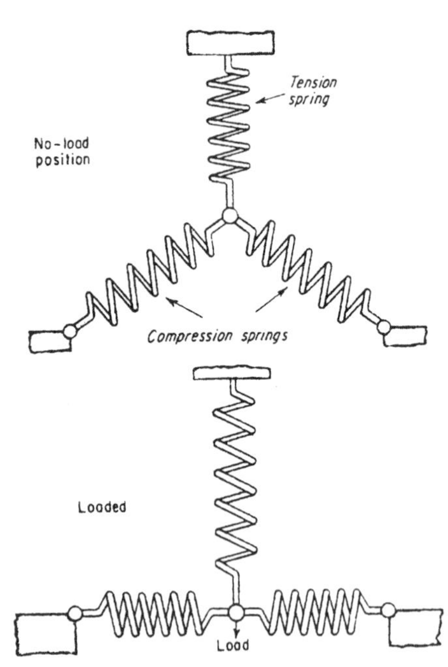

Fig. 1 This basic spring arrangement has zero stiffness, and is as "soft as a cloud" when compression springs are in line, as illustrated in the loaded position. But change the weight or compression-spring alignment, and stiffness increases greatly. This support is adequate for vibration isolation because zero stiffness give a greater range or movement than the vibration amplitude generally in the hundredths-of-an-inch range.

Arrangements shown here are highly absorbent when required, yet provide a firm support when large force changes occur. By contrast, isolators that depend upon very "soft" springs, such as the sine spring, are unsatisfactory in many applications; they allow a large movement of the supported load with any slight weight change or large-amplitude displacing force.

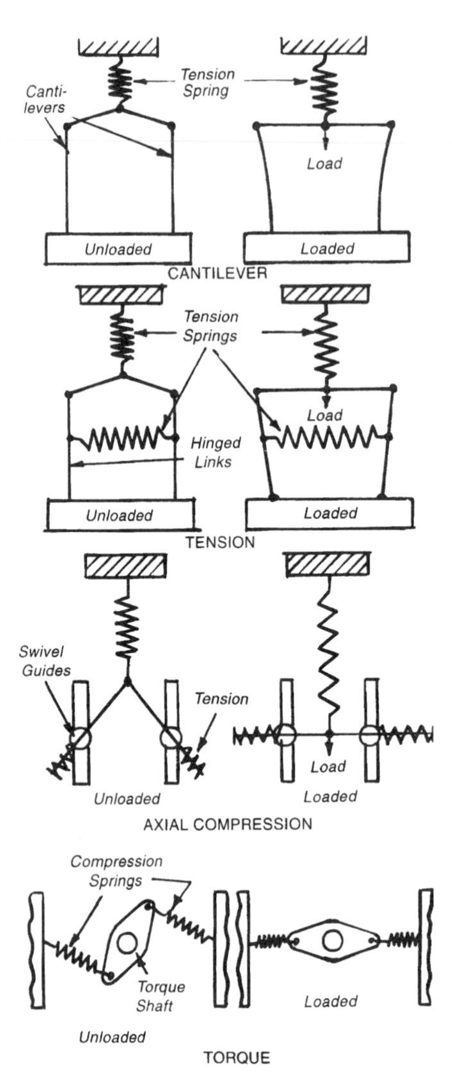

Fig. 2 Alternative arrangements illustrate adaptability of basic design. Here, instead of the inclined, helical compression springs, wither tension or cantilever springs can serve. Similarly, different type of springs can replace the axial, tension spring. Zero torsional stiffness can also be provided.

Various applications of the principle of vibration isolation show how versatile the design is. Coil spring (Fig. 4) as well as cantilever and torsion-bar suspension of automobiles can all be reduced in stiffness by adding an inclined spring; stiffness of the tractor seat (Fig. 5) and, consequently, transmitted shocks can be similarly reduced. Mechanical tension meter (Fig. 6) provides a sensitive indication of small variations in tension. A weighing scale, for example, could detect small variations in nominally identical objects. A nonlinear torque mete (Fig. 7) provides a sensitive indication of torque variations about a predetermined level.

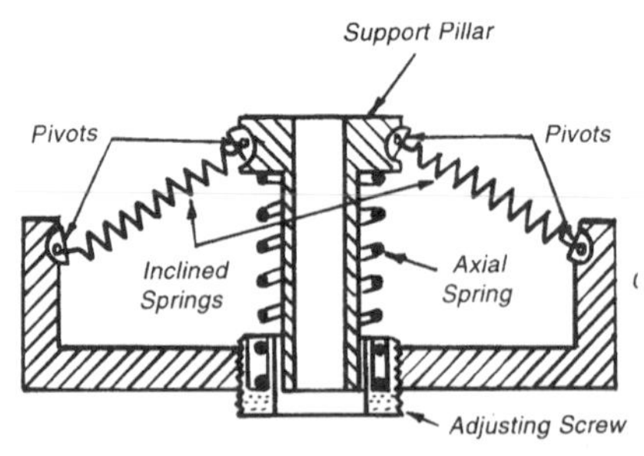

Fig. 3 A general-purpose support is based on basic spring arrangement, except that an axial compression spring is substituted for a tension spring. Inclined compression springs, spaced around a central pillar, carry the component to be isolated. When a load is applied, adjustment might be necessary to bring the inclined springs to zero inclination. Load range that can be supported with zero stiffness on a specific support is determined by the adjustment range and physical limitations of the axial spring.

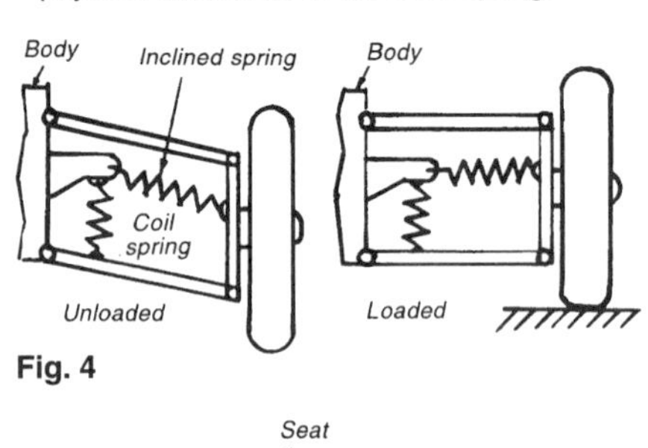

Fig. 4

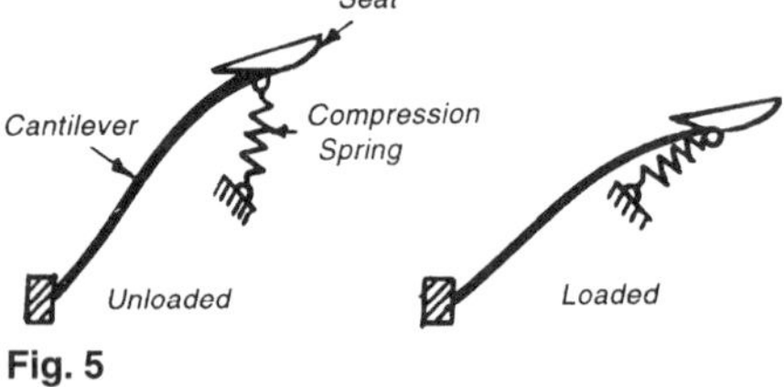

Fig. 5

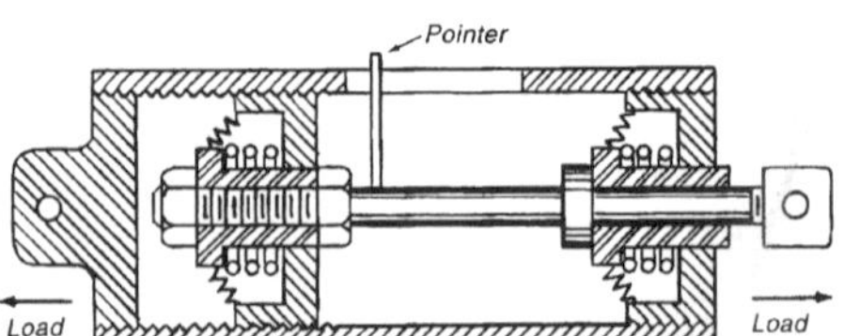

Fig. 6

Loaded
Unloaded
Torsion Spring
Compression Spring

Fig. 7

TWENTY SCREW DEVICES

A threaded shaft and a nut plus some way to make one of these members rotate without translating and the other to translate without rotating are about all you need to do practically all of the adjusting, setting, or locking in a machine design.

Most of these applications have low-precision requirements. That's why the thread might be a coiled wire or a twisted strip; the nut might be a notched ear on a shaft or a slotted disk. Standard screws and nuts from hardware store shelves can often serve at very low cost.

Here are the basic motion transformations possible with screw threads (Fig. 1):

- Transform rotation into linear motion or reverse (A),
- Transform helical motion into linear motion or reverse (B),
- Transform rotation into helical motion or reverse (C).

Of course the screw thread can be combined with other components: in a four-bar linkage (Fig. 2), or with multiple screw elements for force or motion amplification.

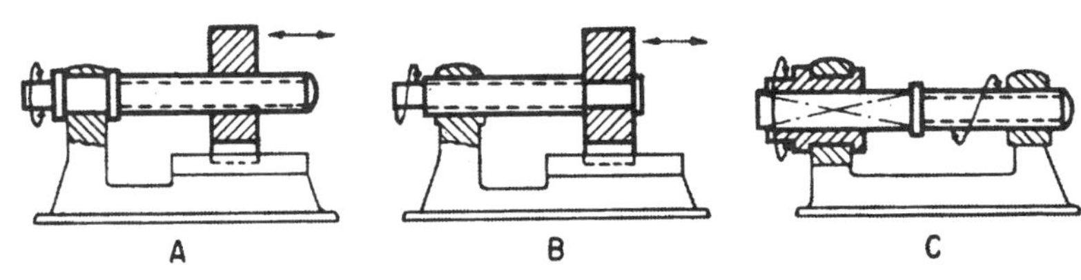

Fig. 1 Motion transformations of a screw thread include: rotation to translation (A), helical to translation (B), rotation to helical (C). These are reversible if the thread is not self-locking. (The thread is reversible when its efficiency is over 50%.)

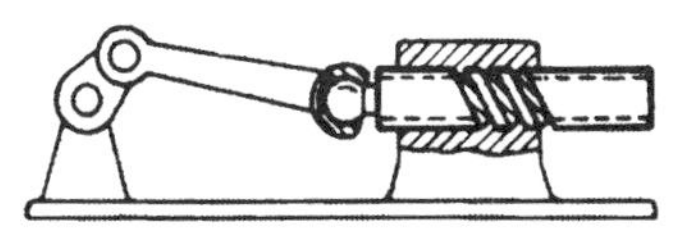

Fig. 2 Standard four-bar linkage has a screw thread substituted for a slider. The output is helical rather than linear.

ROTATION TO TRANSLATION

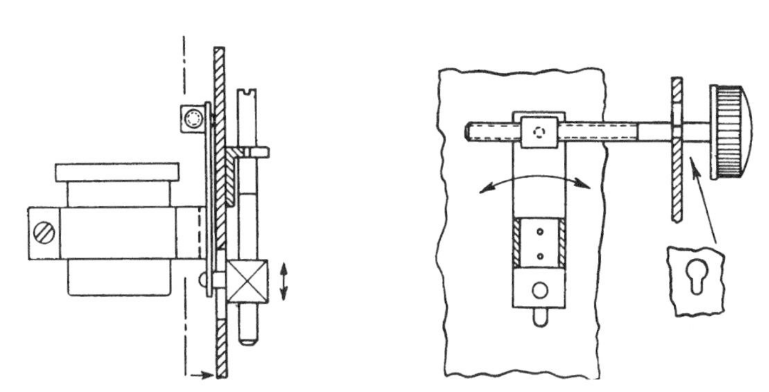

Fig. 3 A two-directional lamp adjustment with screwdriver will move a lamp up and down. A knob adjust (right) rotates the lamp about a pivot.

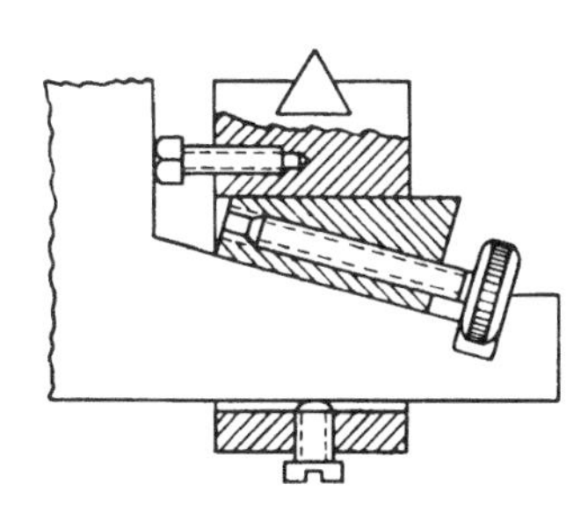

Fig. 4 A knife-edge bearing is raised or lowered by a screw-driven wedge. Two additional screws position the knife edge laterally and lock it.

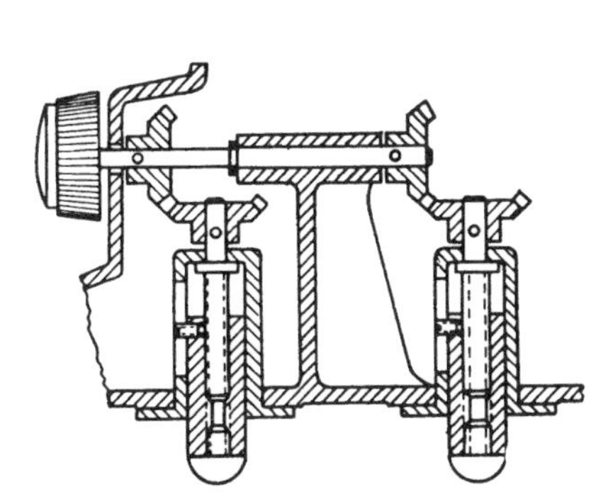

Fig. 5 A parallel arrangement of tandem screw threads raises the projector evenly.

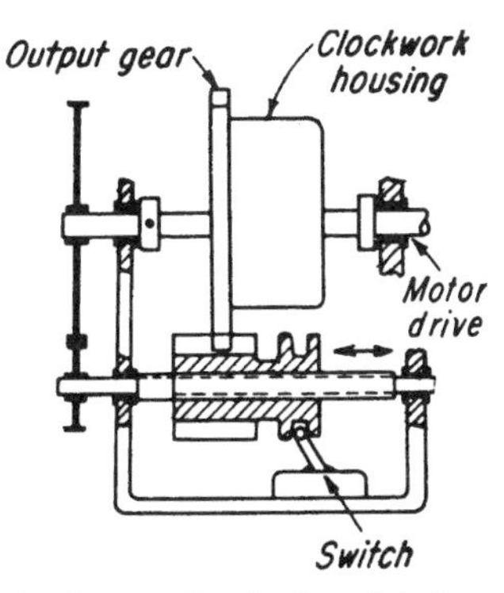

Fig. 6 Automatic clockwork is kept would taut by an electric motor turned on and off by a screw thread and nut. The motor drive must be self-locking or it will permit the clock to unwind as soon as the switch is turned off.

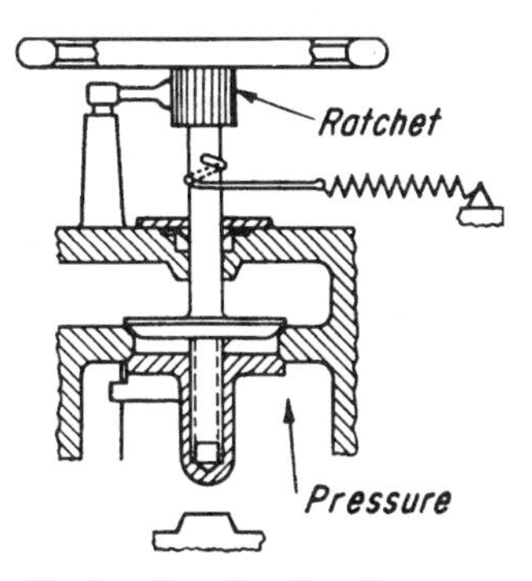

Fig. 7 A valve stem has two oppositely moving valve cones. When opening the upper cone moves up first, until it contacts its stop. Further turning of the valve wheel forces the lower cone out of its seat. The spring is wound up at the same time. When the ratchet is released, the spring pulls both cones into their seats.

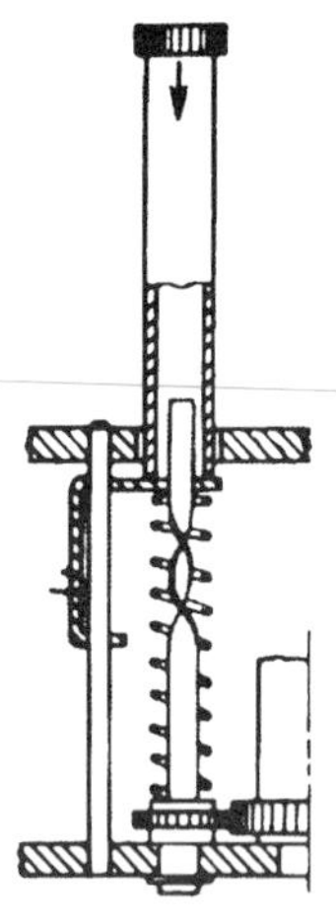

Fig. 8 A metal strip or square rod can be twisted to make a long-lead thread. It is ideal for transforming linear into rotary motion. Here a pushbutton mechanism winds a camera. The number of turns or dwell of the output gear is easily altered by changing (or even reversing) the twist of the strip.

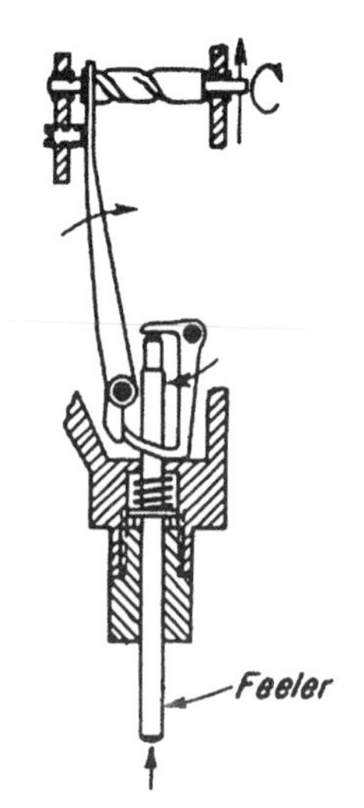

Fig. 9 A feeler gage has its motion amplified through a double linkage and then transformed to rotation for moving a dial needle.

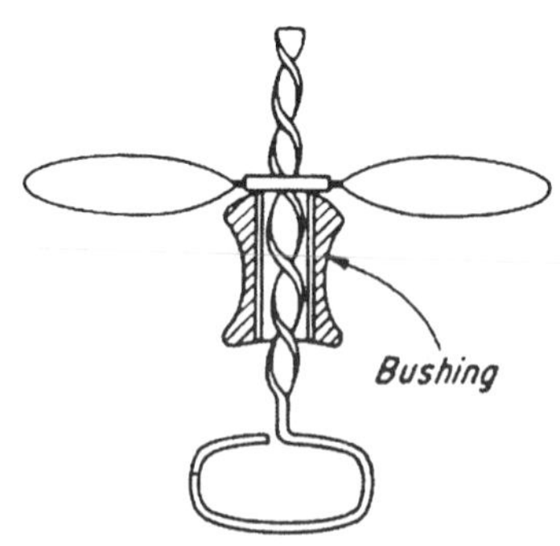

Fig. 10 The familiar flying propeller-toy is operated by pushing the bushing straight up and off the thread.

SELF-LOCKING

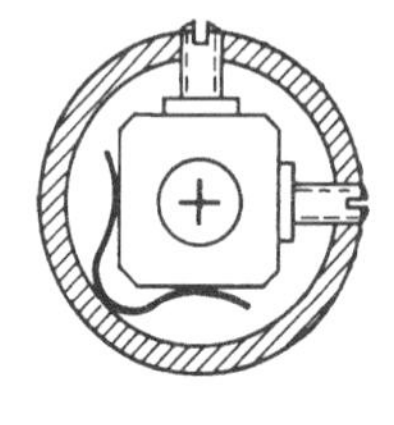

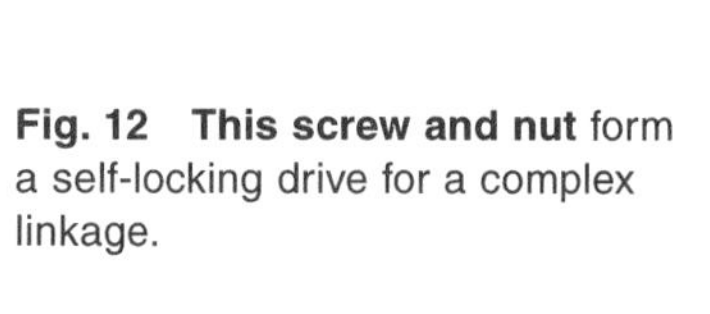

Fig. 11 A hairline adjustment for a telescope with two alternative methods for drive and spring return.

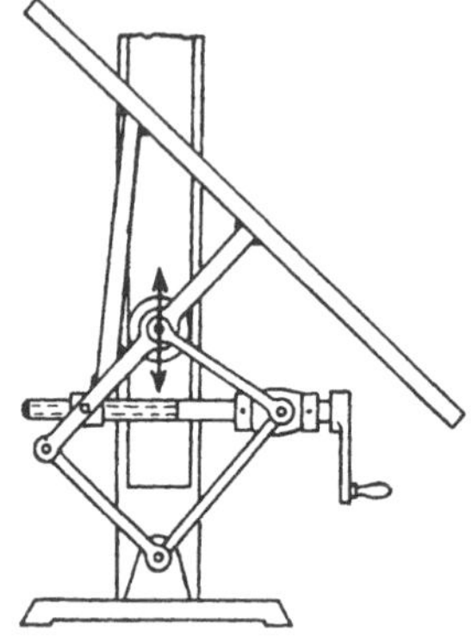

Fig. 12 This screw and nut form a self-locking drive for a complex linkage.

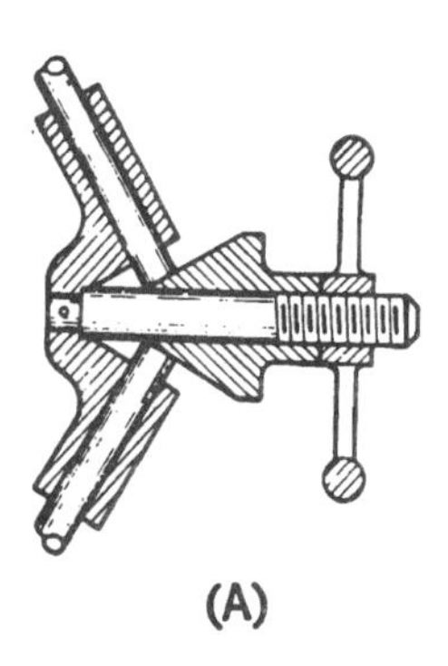

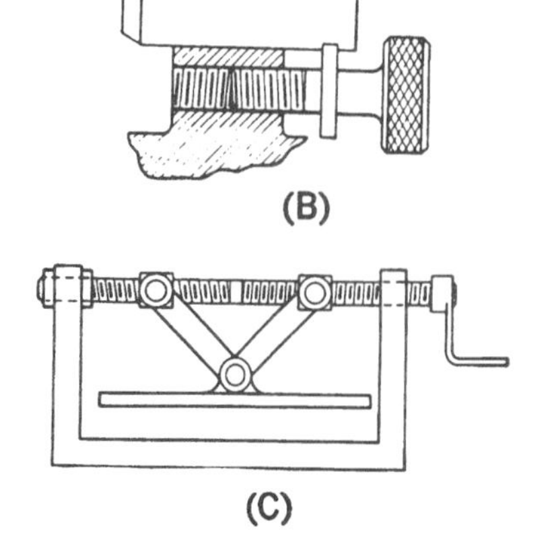

Fig. 13 Force translation. The threaded handle in (A) drives a coned bushing that thrusts rods outwardly for balanced pressure. The screw in (B) retains and drives a dowel pin for locking purposes. A right- and left-handed shaft (C) actuates a press.

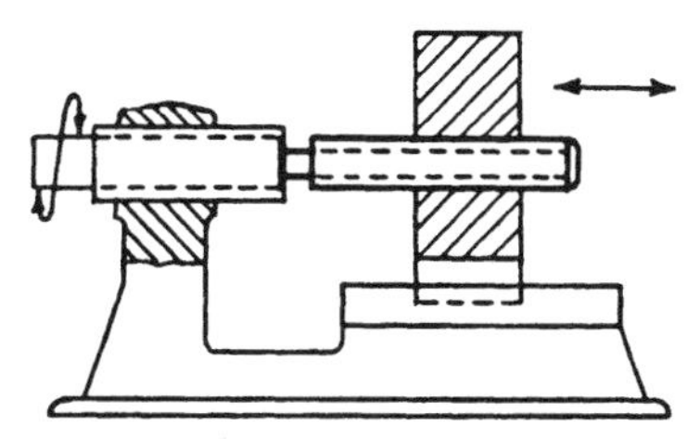

Fig. 14 Double-threaded screws, when used as differentials, permit very fine adjustment for precision equipment at relatively low cost.

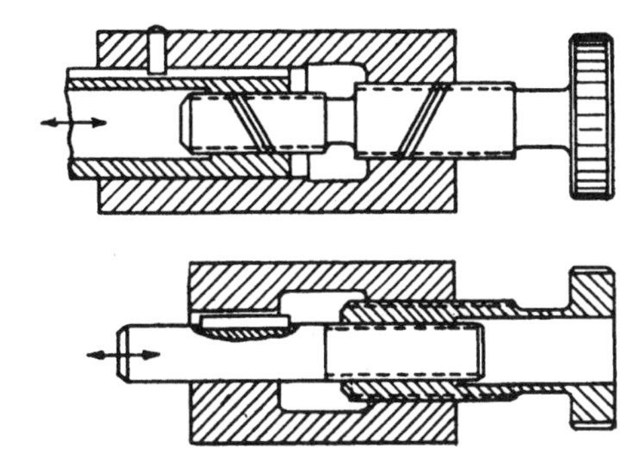

Fig. 15 Differential screws can be made in dozens of forms. Here are two methods: in the upper figure, two opposite-hand threads on a single shaft; in the lower figure, same-hand threads on independent shafts.

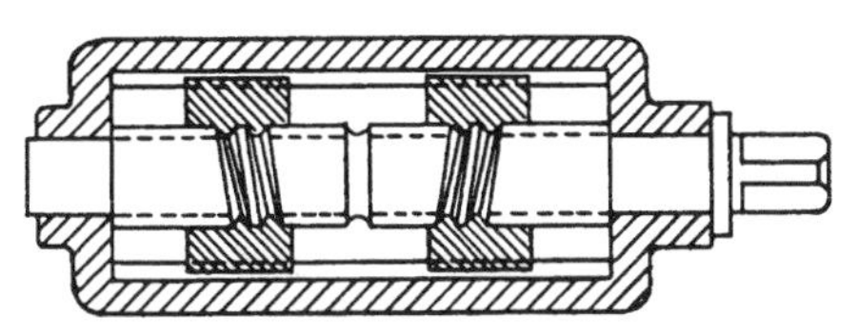

Fig. 16 Opposite-hand threads make a high-speed centering clamp out of two moving nuts.

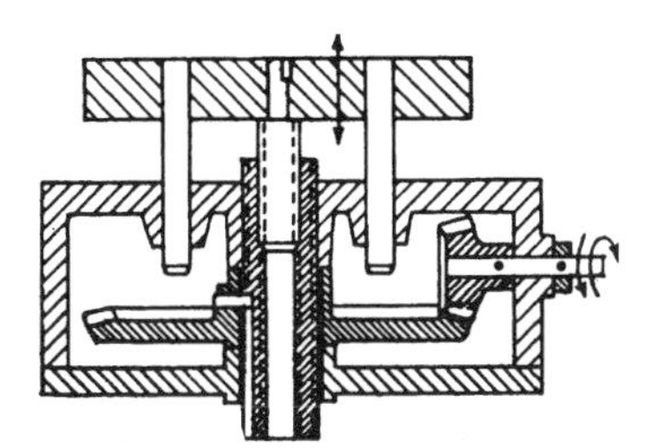

Fig. 17 A measuring table rises very slowly for many turns of the input bevel gear. If the two threads are $1^1/_2$ to 12 and $^3/_4$ to 16, in the fine-thread series, the table will rise approximately 0.004 in. per input-gear revolution.

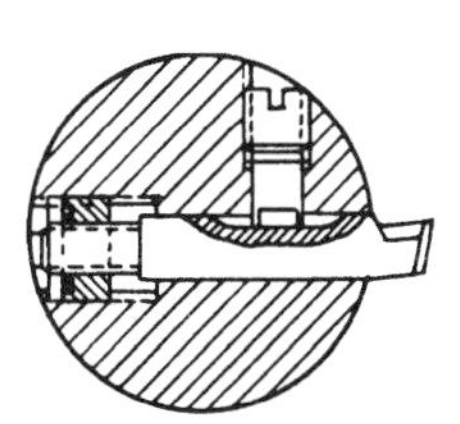

Fig. 18 A lathe turning tool in a drill rod is adjusted by a differential screw. A special double-pin wrench turns the intermediate nut, advancing the nut and retracting the threaded tool simultaneously. The tool is then clamped by a setscrew.

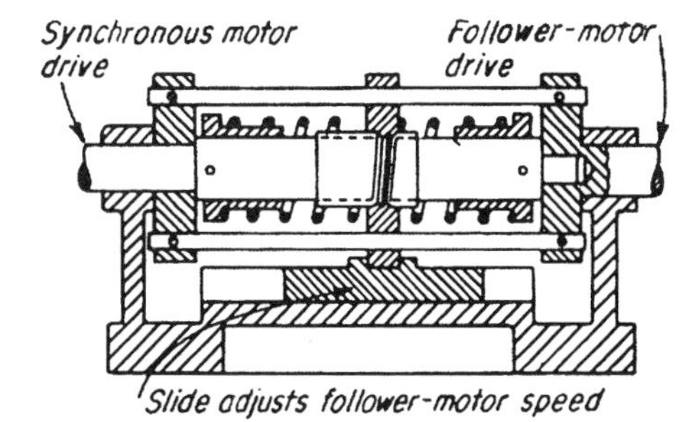

Fig. 19 Any variable-speed motor can be made to follow a small synchronous motor by connecting them to the two shafts of this differential screw. Differences in the number of revolutions between the two motors appear as motion of the traveling nut and slide, thus providing electrical speed compensation.

Fig. 20 A wire fork is the nut in this simple tube-and-screw device.

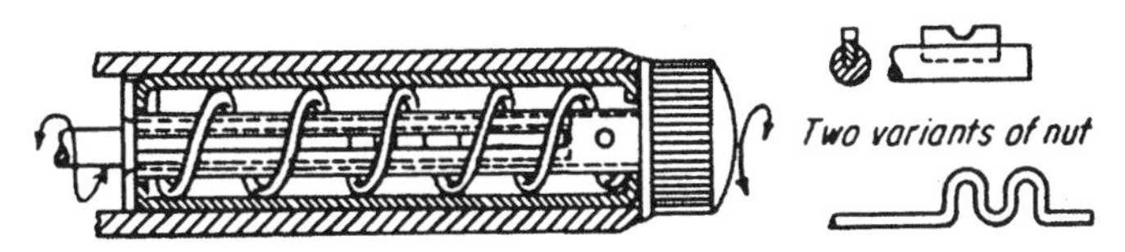

Fig. 21 A mechanical pencil includes a spring as the screw thread and a notched ear or a bent wire as the nut.

TEN APPLICATIONS FOR SCREW MECHANISMS

Three basic components of screw mechanisms are: actuating member (knob, wheel, handle), threaded device (screw-nut set), and sliding device (plunger-guide set).

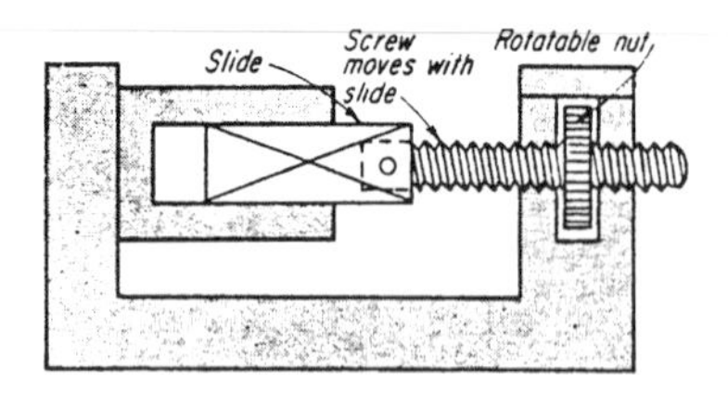

A nut can rotate but will not move longitudinally. Typical applications: screw jacks, heavy vertically moved doors; floodgates, opera-glass focusing, vernier gages, and Stillson wrenches.

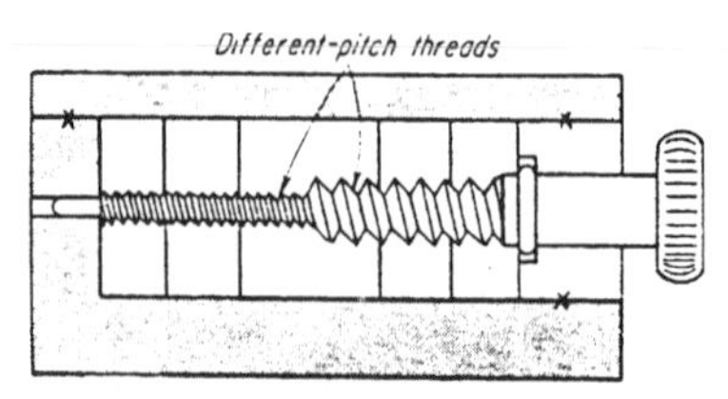

A differential movement is given by threads of different pitch. When the screw is rotated, the nuts move in the same direction but at different speeds.

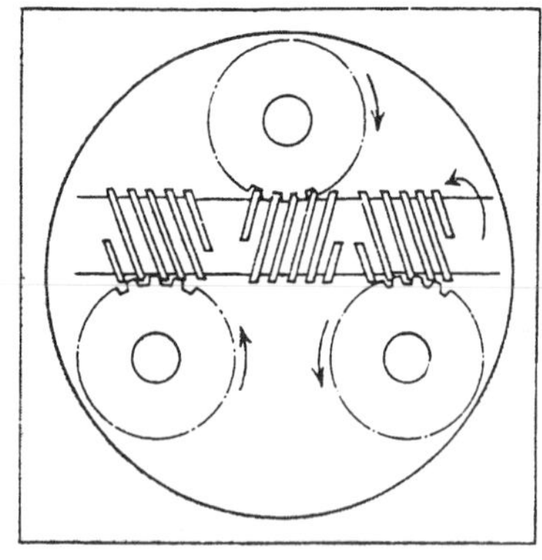

One screw actuates three gears simultaneously. The axes of gears are at right angles to that of the screw. This mechanism can replace more expensive gear setups there speed reduction and multiple output from a single input is required.

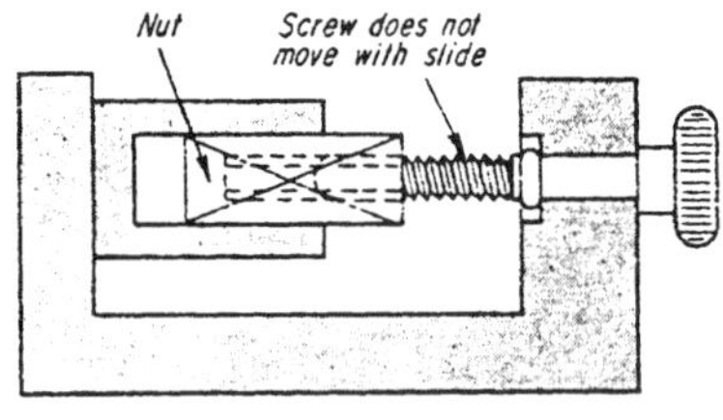

A screw can rotate but only the nut moves longitudinally. Typical applications: lathe tailstock feed, vises, lathe apron.

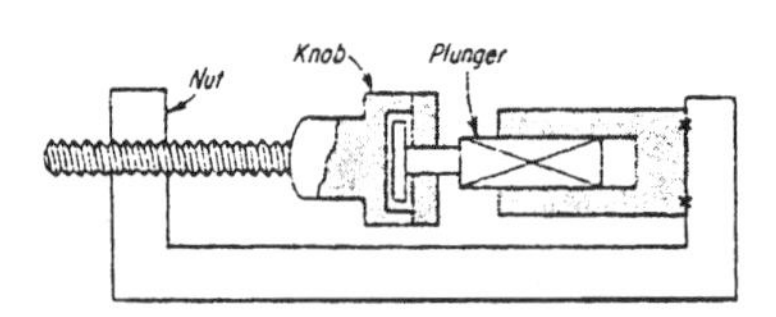

A screw and plunger are attached to a knob. The nut and guide are stationary. It is used on: screw presses, lathe steady-rest jaws for adjustment, and shaper slide regulation.

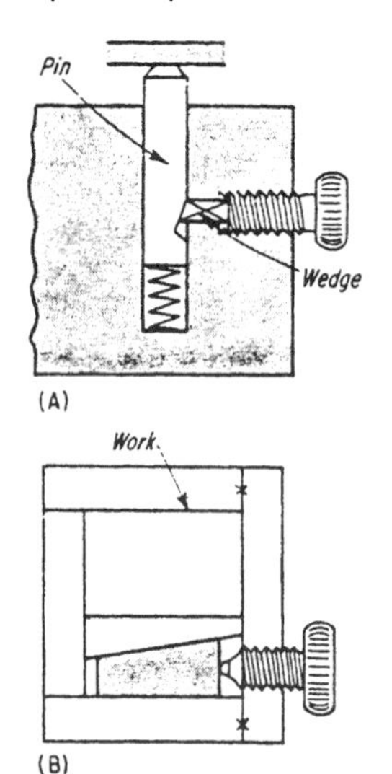

Screw-actuated wedges lock locating pin A and hold the work in fixture (B). These are just two of the many tool and diemaking applications for these screw actions.

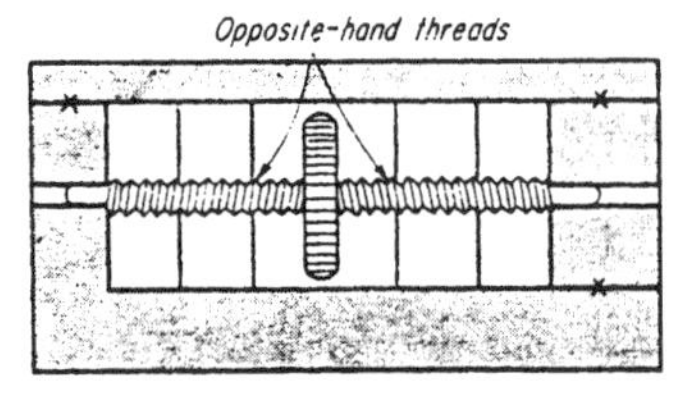

Opposing movement of lateral slides; adjusting members or other screw-actuated parts can be achieved with opposite-hand threads.

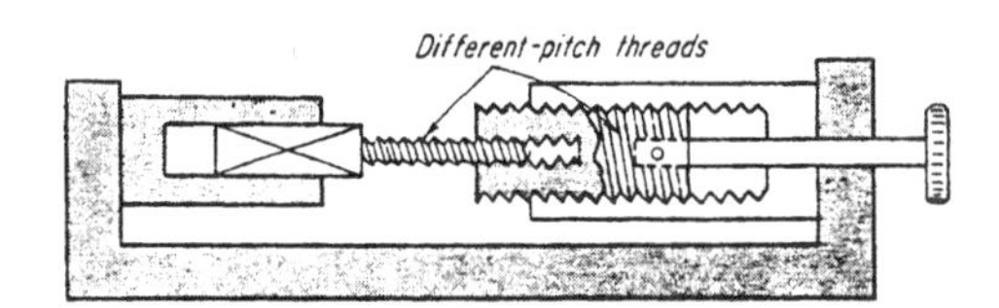

Concentric threading also gives differential movement. Such movements are useful wherever rotary mechanical action is required. A typical example is a gas-bottle valve, where slow opening is combined with easy control.

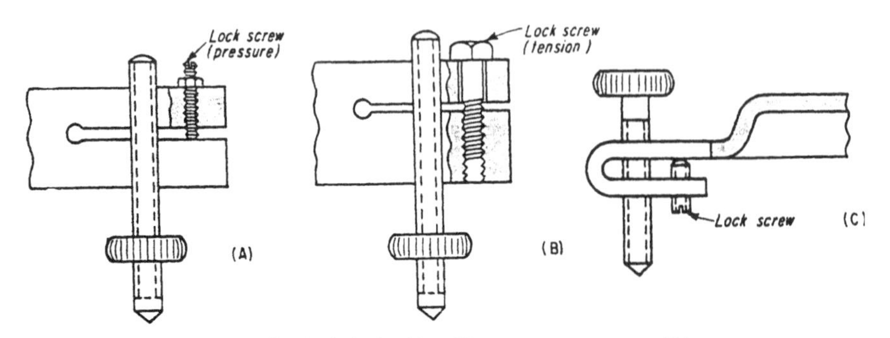

Adjustment screws are effectively locked by either a pressure screw (A) or tension screw (B). If the adjusting screw is threaded into a formed sheet-metal component (C), a setscrew can be used to lock the adjustment.

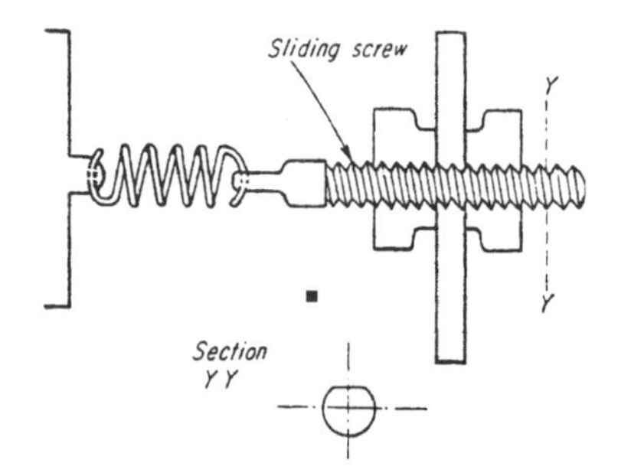

Locking nuts can be placed on opposite sides of a panel to prevent axial screw movement and simultaneously lock against vibrations. Drill-press depth stops and adjustable stops for shearing and cutoff dies are some examples.

SEVEN SPECIAL SCREW ARRANGEMENTS

Differential, duplex, and other types of screws can provide slow and fast feeds, minute adjustments, and strong clamping action.

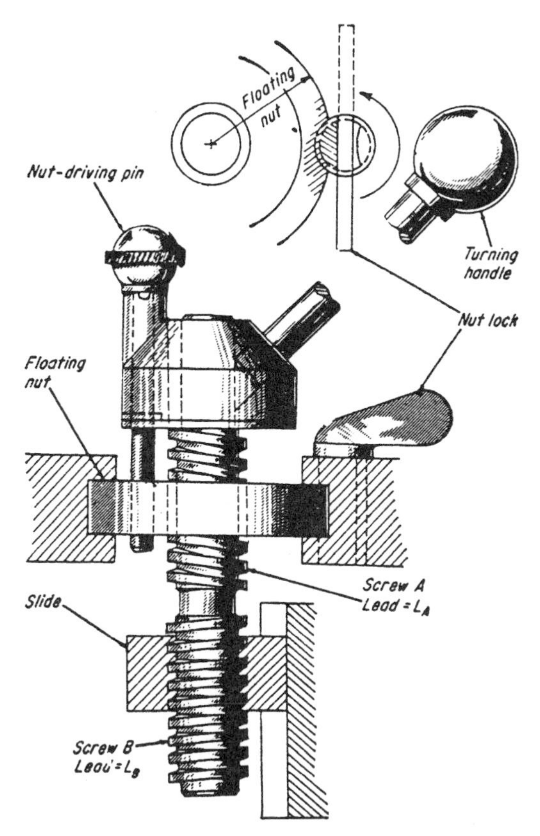

Rapid and slow feed. With left- and right-hand threads, slide motion with the nut locked equals L_A plus L_B per turn; with the nut floating, slide motion per turn equals L_B. Extremely fine feed with a rapid return motion is obtained when the threads are differential.

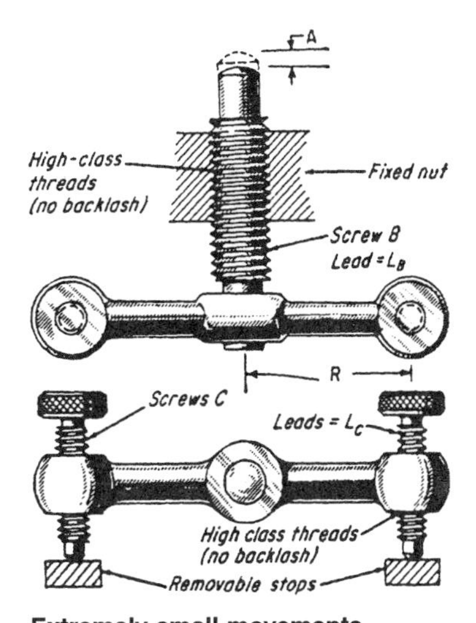

Extremely small movements. Microscopic measurements, for example, are characteristic of this arrangement. Movement A is equal to $N(L_B \times L_t)$ $12\pi R$, where N equals the number of turns of screw C.

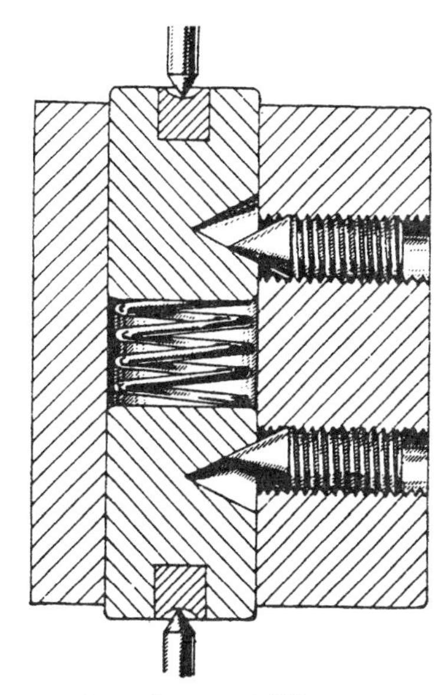

Bearing adjustment. This screw arrangement is a handy way for providing bearing adjustment and overload protection.

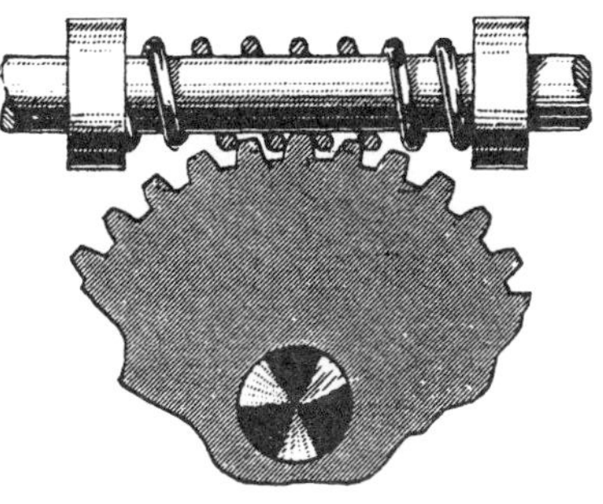

Shock absorbent screw. When the springs coiled as shown are used as worm urives for light loads, they have the advantage of being able to absorb heavy shocks.

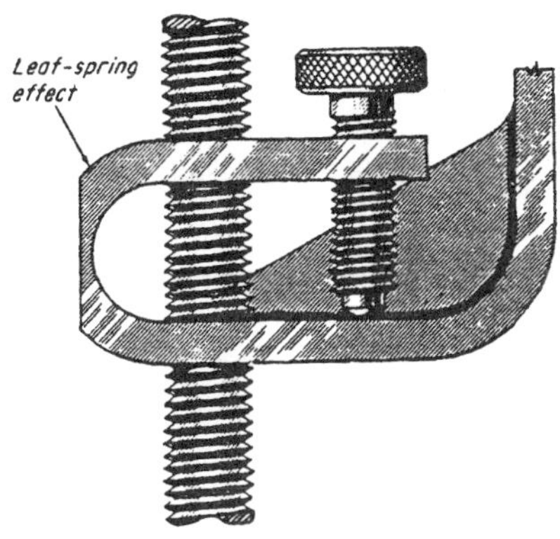

Backlash elimination. The large screw is locked and all backlash is eliminated when the knurled screw is tightened; finger torque is sufficient.

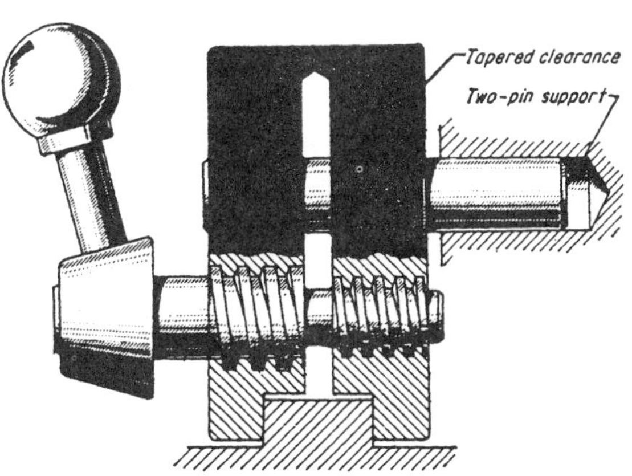

Differential clamp. This method of using a differential screw to tighten clamp jaws combines rugged threads with high clamping power. Clamping pressure, $P = Te$ $[R(\tan \phi + \tan \alpha]$, where T = torque at handle, R = mean radius of screw threads, ϕ = angle of friction (approx. 0.1), α = mean pitch angle or screw, and e = efficiency of screw generally about 0.8).

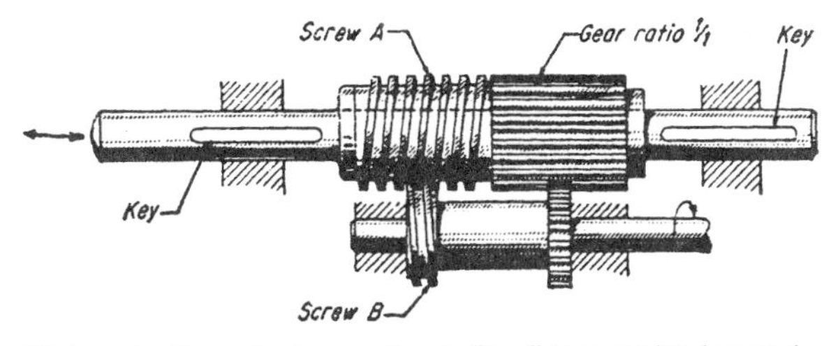

High reduction of rotary motion to fine linear motion is possible here. This arrangement is for low forces. Screws are left and right hand. $L_A = L_B$ plus or minus a small increment. When $L_B = 1/10$ and $L_A = 1/10.5$, the linear motion f screw A will be 0.05 in. per turn. When screws are the same hand, linear motion equals $L_A + L_B$.

FOURTEEN SPRING AND SCREW ADJUSTING DEVICES

Here is a selection of some basic devices that provide and hold mechanical adjustment.

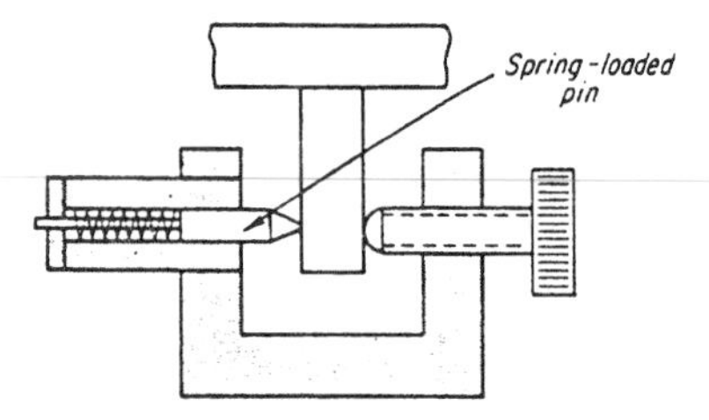

Fig. 1 A spring-loaded pin supplies a counterforce against which an adjustment force must always act. A leveling foot would work against gravity, but for most other setups a spring is needed to give a counter-force.

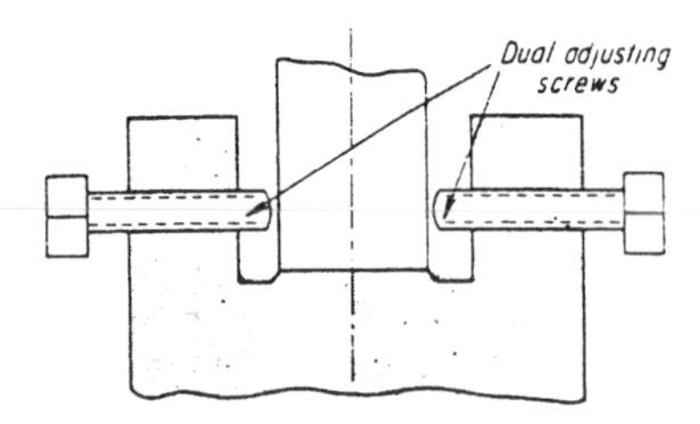

Fig. 2 Dual screws provide an inelastic counterforce. Backing-off one screw and tightening the other allows extremely small adjustments to be made. Also, once adjusted, the position remains solid against any forces tending to move the device out of adjustment

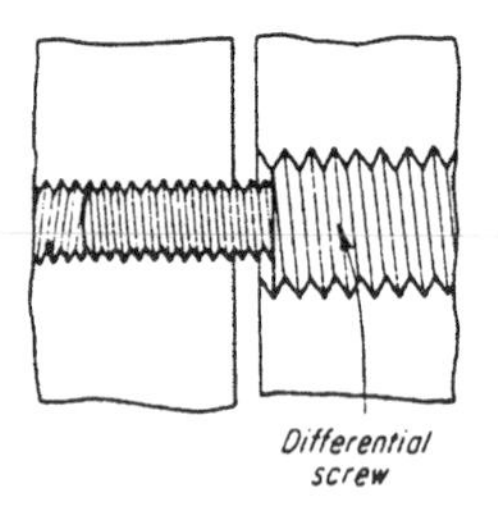

Fig. 3 A differential screw has same-hand threads but with different pitches. The relative distance between the two components can be adjusted with high precision by differential screws.

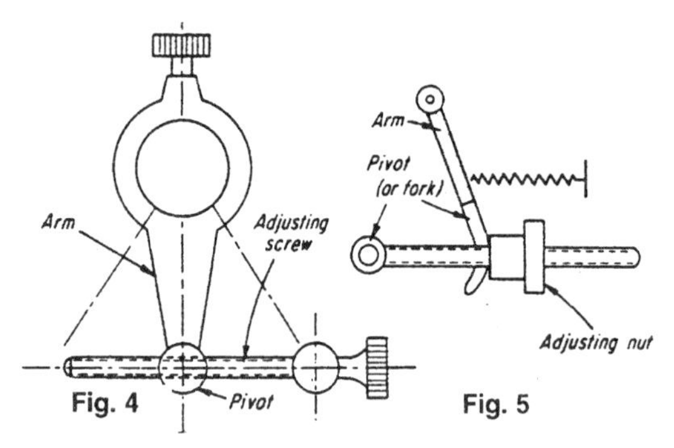

Figs. 4 and 5 Swivel motion is necessary in (Fig. 4) between the adjusting screw and arm because of a circular locus of female threads in the actuated member. Similar action (Fig. 5) requires either the screw to be pivoted or the arm to be forked.

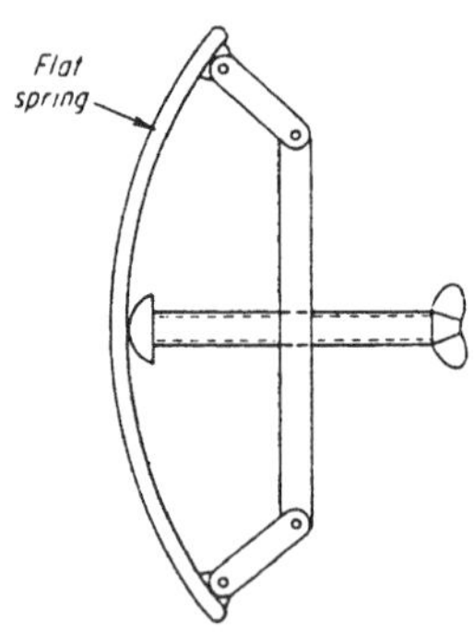

Fig. 6 This arc-drafting guide is an example of an adjusting device. One of its components, the flat spring, both supplies the counter-force and performs the mechanism's main function—guiding the pencil.

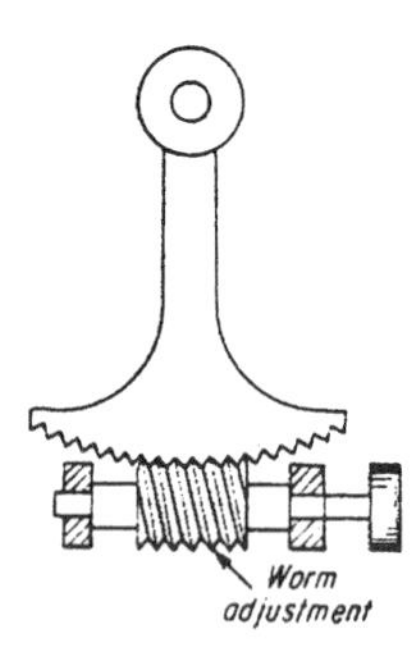

Fig. 7 The worm adjustment shown here is in a device for varying the position of an arm. Measuring instruments, and other tools requiring fine adjustments, include this adjusting device.

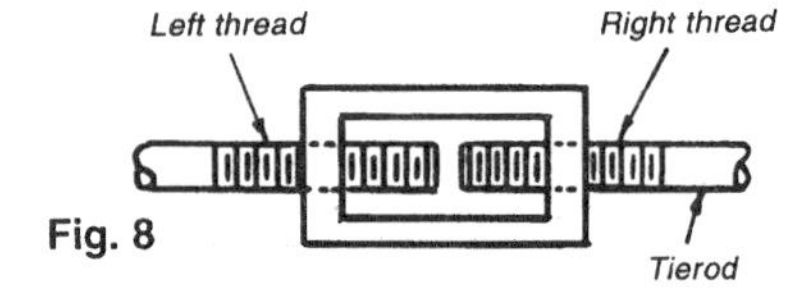

Figs. 8 and 9 Tierods with opposite-hand threads at their ends (Fig. 8) require only a similarly threaded nut to provide simple, axial adjustment. Flats on the rod ends (Fig. 9) make it unnecessary to restrain both the rods against rotation when the adjusting screw is turned; restraining one rod is enough.

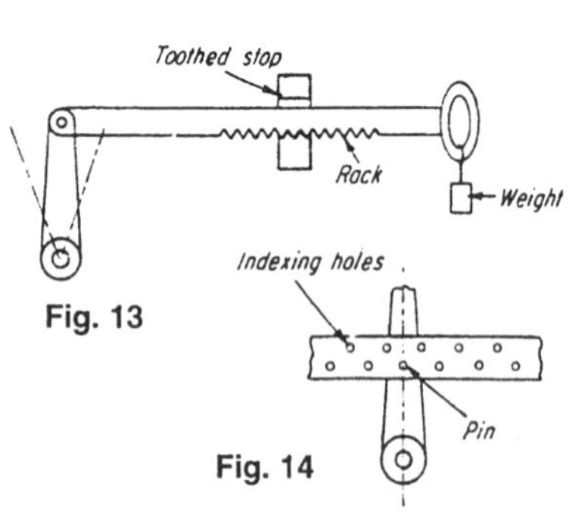

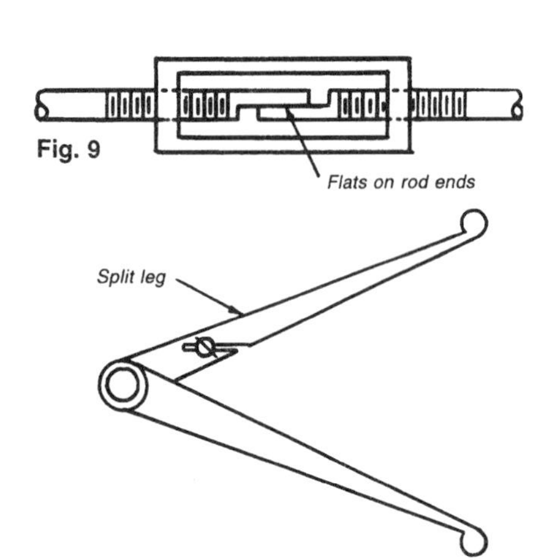

Fig. 10 A split-leg caliper is an example of a simple but highly efficient adjusting device. A tapered screw forces the split leg part, thus enlarging the opening between the two legs.

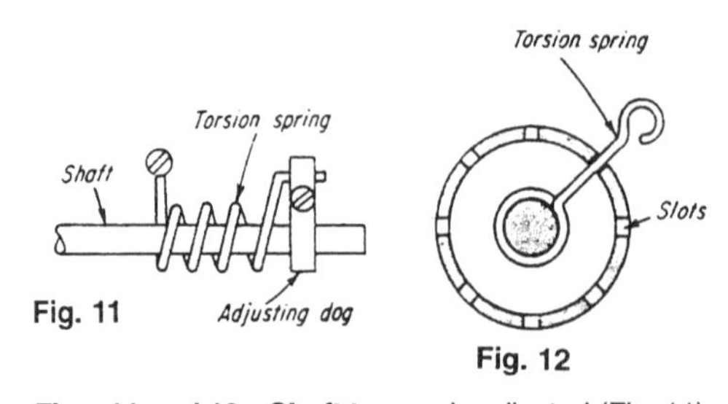

Figs. 11 and 12 Shaft torque is adjusted (Fig. 11) by rotating the spring-holding collar relative to the shaft, and locking the collar at a position of desired torque. Adjusting slots (Fig. 12) accommodate the torsion-spring arm after the spring is wound to the desired torque.

Figs. 13 and 14. Rack and toothed stops (Fig. 13) are frequently used to adjust heavy louvers, boiler doors and similar equipment. The adjustment is not continuous; it depends on the rack pitch. Large counter-adjustment forces might require a weighted rack to prevent tooth disengagement. Indexing holes (Fig. 14) provide a similar adjustment to the rack. The pin locks the members together.

CHAPTER 12
SHAFT COUPLINGS AND CONNECTIONS

FOUR COUPLINGS FOR PARALLEL SHAFTS

Fig. 1 One method of coupling shafts makes use of gears that can replace chains, pulleys, and friction drives. Its major limitation is the need for adequate center distance. However, an idler can be used for close centers, as shown. This can be a plain pinion or an internal gear. Transmission is at a constant velocity and there is axial freedom.

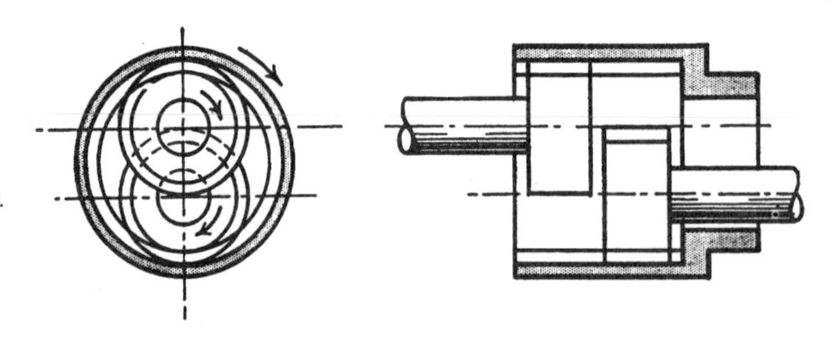

Fig. 2 This coupling consists of two universal joints and a short shaft. Velocity transmission is constant between the input and output shafts if the shafts remain parallel and if the end yokes are arranged symmetrically. The velocity of the central shaft fluctuates during rotation, but high speed and wide angles can cause vibration. The shaft offset can be varied, but axial freedom requires that one shaft be spline mounted.

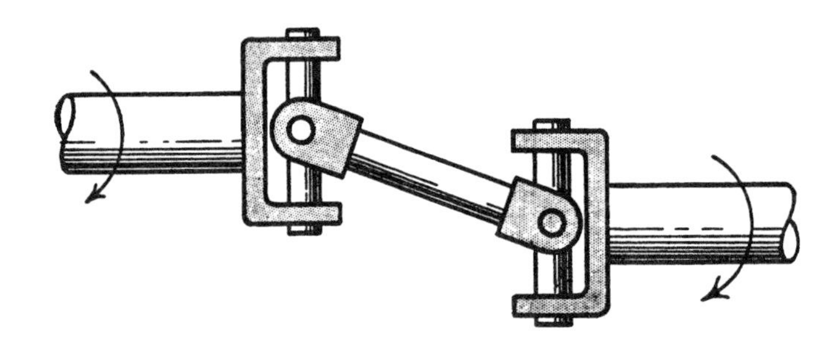

Fig. 3 This crossed-axis yoke coupling is a variation of the mechanism shown in Fig. 2. Each shaft has a yoke connected so that it can slide along the arms of a rigid cross member. Transmission is at a constant velocity, but the shafts must remain parallel, although the offset can vary. There is no axial freedom. The central cross member describes a circle and is thus subjected to centrifugal loads.

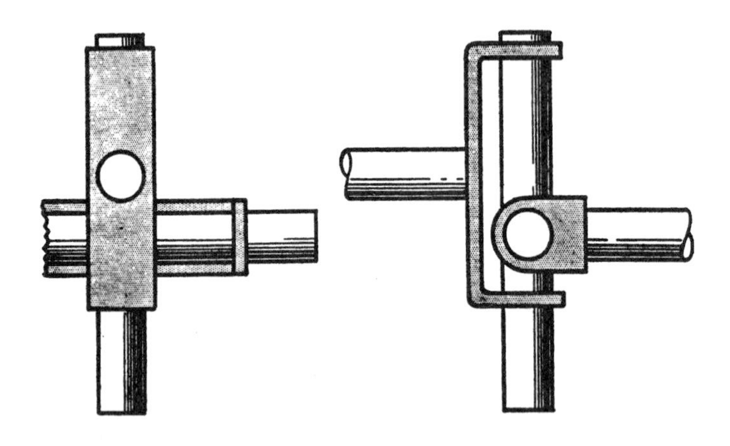

Fig. 4 This Oldham coupling provides motion at a constant velocity as its central member describes a circle. The shaft offset can vary, but the shafts must remain parallel. A small amount of axial freedom is possible. A tilt in the central member can occur because of the offset of the slots. This can be eliminated by enlarging its diameter and milling the slots in the same transverse plane.

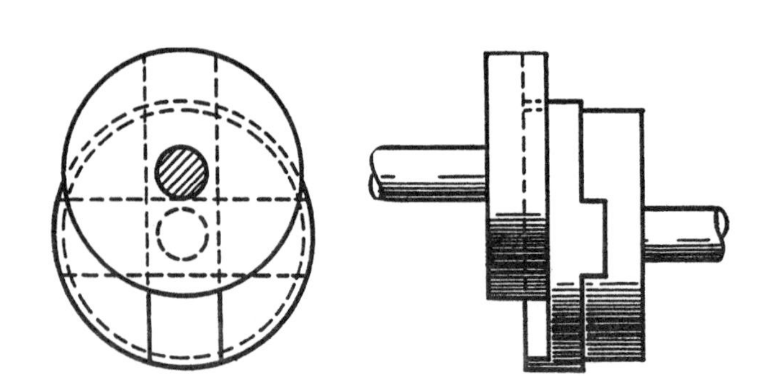

LINKS AND DISKS COUPLE OFFSET SHAFTS

An unorthodox yet remarkably simple arrangement of links and disks forms the basis of a versatile parallel-shaft coupling. This coupling—essentially three disks rotating in unison and interconnected in series by six links (se drawing)—can adapt to wide variations in axial displacement while it is running under load.

Changes in radial displacement do not affect the constant-velocity relationship between the input and output shafts, nor do they affect initial radial reaction forces that might cause imbalance in the system. Those features open up unusual applications for it in automotive, marine, machine-tool, and rolling-mill machinery (see drawings).

How it works. The inventor of the coupling, Richard Schmidt of Madison, Alabama, said that a similar link arrangement had been known to some German engineers for years. But those engineers were discouraged from applying the theory because they erroneously assumed that the center disk had to be retained by its own bearing. Actually, Schmidt found that the center disk is free to assume its own center of rotation. In operation, all three disks rotate with equal velocity.

The bearing-mounted connections of links to disks are equally spaced at 120° on pitch circles of the same diameter. The distance between shafts can be varied steplessly between zero (when the shafts are in line) and a maximum that is twice the length of the links (see drawings.) There is no phase shift between shafts while the coupling is undulating.

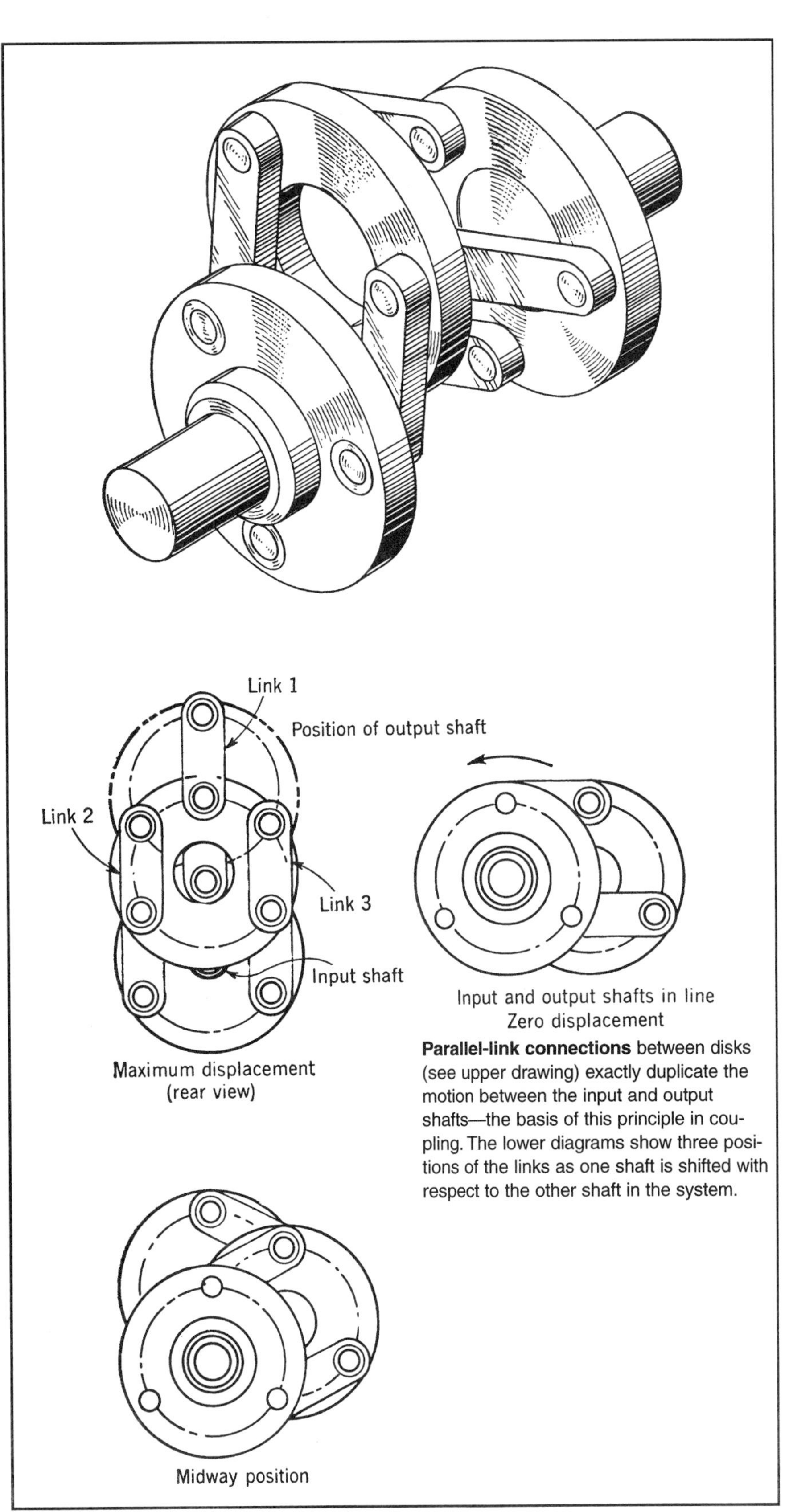

Parallel-link connections between disks (see upper drawing) exactly duplicate the motion between the input and output shafts—the basis of this principle in coupling. The lower diagrams show three positions of the links as one shaft is shifted with respect to the other shaft in the system.

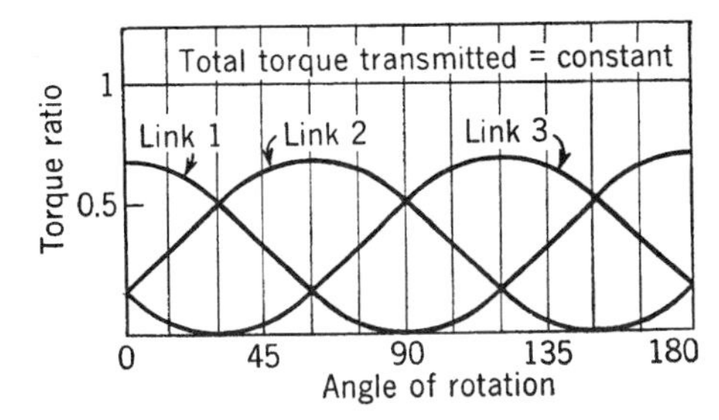

Torque transmitted by three links in the group adds up to a constant value, regardless of the angle of rotation.

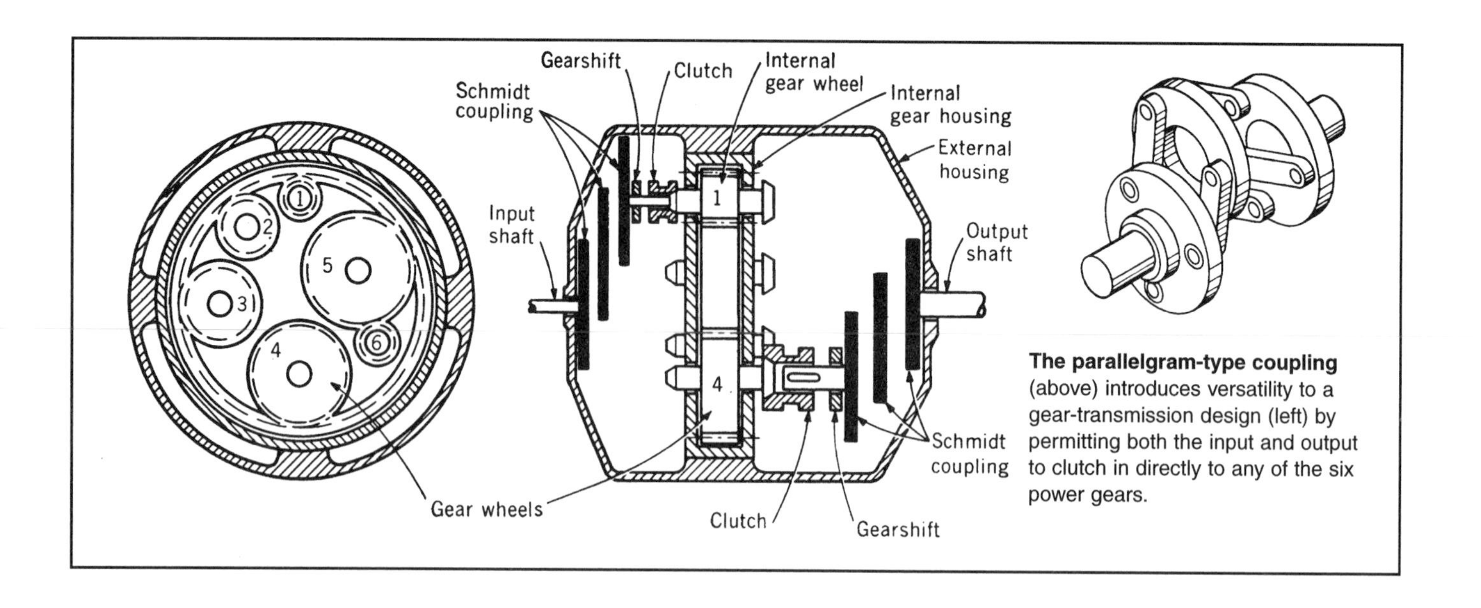

The parallelgram-type coupling (above) introduces versatility to a gear-transmission design (left) by permitting both the input and output to clutch in directly to any of the six power gears.

DISK-AND-LINK COUPLINGS SIMPLIFY TORQUE TRANSMISSION

A unique disk-and-link coupling that can handle large axial displacement between shafts, while the shafts are running under load, has opened up new approaches to transmission design. It was developed by Richard Schmidt of Madison, Alabama.

The coupling (drawing, upper right) maintains a constant transmission ratio between input and output shafts while the shafts undergo axial shifts in their relative positions. This permits gear-and-belt transmissions to be designed that need fewer gears and pulleys.

Half as many gears. In the internal-gear transmission shown, a Schmidt coupling on the input side permits the input to be plugged in directly to any one of six gears, all of which are in mesh with the internal gear wheel.

On the output side, after the power flows through the gear wheel, a second Schmidt coupling permits a direct power takeoff from any of the same six gears. Thus, any one of 6 × 6 minus 5 or 31 different speed ratios can be selected while the unit is running. A more orthodox design would require almost twice as many gears.

Powerful pump. In the worm-type pump (bottom left), as the input shaft rotates clockwise, the worm rotor is forced to roll around the inside of the gear housing, which has a helical groove running from end to end. Thus, the rotor center-line will rotate counterclockwise to produce a powerful pumping action for moving heavy liquids.

In the belt drive (bottom right), the Schmidt coupling permits the belt to be shifted to a different bottom pulley while remaining on the same top pulley. Normally, because of the constant belt length, the top pulley would have to be shifted too, to provide a choice of only three output speeds. With this arrangement, nine different output speeds can be obtained.

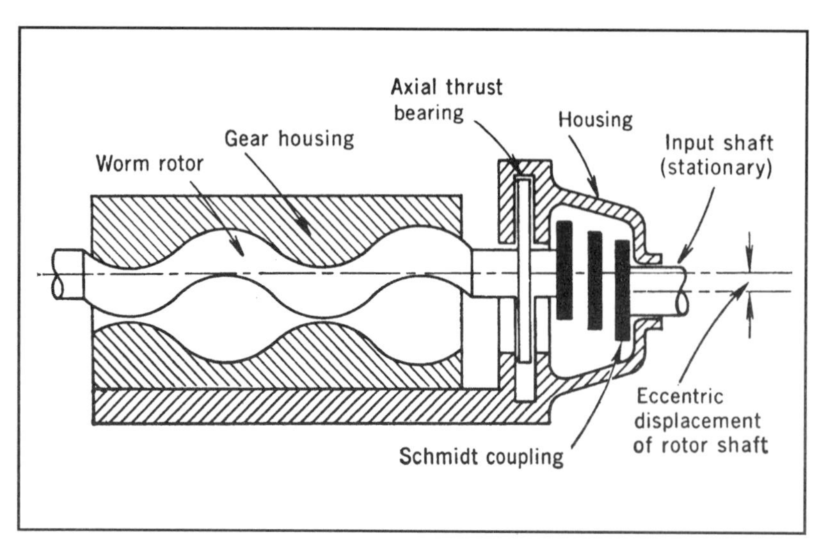

The coupling allows a helically-shaped rotor to oscillate for pumping purposes.

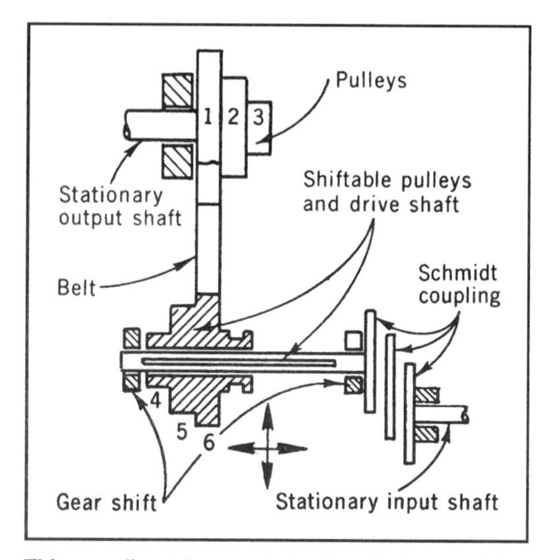

This coupling takes up slack when the bottom shifts.

INTERLOCKING SPACE-FRAMES FLEX AS THEY TRANSMIT SHAFT TORQUE

This coupling tolerates unusually high degrees of misalignment, with no variation in the high torque that's being taken from the shaft.

A concept in flexible drive-shaft couplings permits unusually large degrees of misalignment and axial motion during the transmission of high amounts of torque. Moreover, the rotational velocity of the driven member remains constant during transmission at angular misalignments; in other words, cyclic pulsations are not induced as they would be if, say, a universal coupling or a Hooke's joint were employed.

The coupling consists essentially of a series of square space-frames, each bent to provide offsets at the diagonals and each bolted to adjacent members at alternate diagonals. The concept was invented by Robert B. Bossler, Jr. He was granted U.S. Patent No. 3,177,684.

Couplings accommodate the inevitable misalignments between rotating shafts in a driven train. These misalignments are caused by imperfect parts, dimensional variations, temperature changes, and deflections of the supporting structures. The couplings accommodate misalignment either with moving contacts or by flexing.

Most couplings, however, have parts with moving contacts that require lubrication and maintenance. The rubbing parts also absorb power. Moreover, the lubricant and the seals limit the coupling environment and coupling life. Parts wear out, and the coupling can develop a large resistance to movement as the parts deteriorate. Then, too, in many designs, the coupling does not provide true constant velocity.

For flexibility. Bossler studied the various types of couplings n the market and first developed a new one with a moving contact. After exhaustive tests, he became convinced that if there were to be the improvements he wanted, he had to design a coupling that flexed without any sliding or rubbing.

Flexible-coupling behavior, however, is not without design problems. Any flexible coupling can be proportioned with strong, thick, stiff members that easily transmit a design torque and provide the stiffness to operate at design speed. However, misalignment requires flexing of these members. The flexing produces alternating stresses that can limit coupling life. The greater the strength and stiffness of a member, the higher the alternating stress from a given misalignment. Therefore, strength and stiffness provisions that transmit torque at speed will be detrimental to misalignment accommodation capability.

The design problem is to proportion the flexible coupling to accomplish torque transmission and overcome misalignment for the lowest system cost. Bossler looked at a drive shaft, a good example of power transmission—and wondered how he could convert it into one with flexibility.

He began to evolve it by following basic principles. How does a drive shaft transmit torque? By tension and compression. He began paring it down to the important struts that could transmit torque and found that they are curved beams. But a curved beam in tension and compression is not as strong as a straight beam. He ended up with the beams straight in a square space-frame with what might be called a *double helix arrangement.* One helix contained elements in compression; the other helix contained elements in tension.

Flattening the helix. The total number of plates should be an even number to obtain constant velocity characteristics during misalignment. But even with an odd number, the cyclic speed variations are minute, not nearly the magnitude of those in a Hooke's joint.

Although the analysis and resulting equations developed by Bossler are based on a square-shaped unit, he concluded that the perfect square is not the ideal for the coupling, because of the position of the mounting holes. The flatter the helix—in other words the smaller the distance *S*—the more misalignment the coupling will tolerate.

Hence, Bossler began making the space-frames slightly rectangular instead of square. In this design, the bolt-heads that fasten the plates together are offset from adjoining pairs, providing enough clearance for the design of a "flatter" helix. The difference in stresses between a coupling with square-shaped plates and one with slightly rectangular plates is so insignificant that the square-shape equations can be employed with confidence.

Design equations. By making a few key assumptions and approximations, Bossler boiled the complex analytical relationships down to a series of straightforward design equations and charts. The derivation of the equations and the resulting verification from tests are given in the NASA report *The Bossler Coupling,* CR-1241.

Torque capacity. The ultimate torque capacity of the coupling before buckling that might occur in one of the space-frame struts under compression is given by Eq. 1. The designer usually knows or establishes the maximum continuous torque that the coupling must transmit. Then he must allow for possible shock loads and overloads. Thus, the clutch should be designed to have an ultimate torque capacity that is at least twice as much, and perhaps three times as much, as the expected continuous torque, according to Bossler.

Induced stress. At first glance, Eq. 1 seems to allow a lot of leeway in selecting the clutch size. The torque capacity is easily boosted, for example, by picking a smaller bolt-circle diameter, *d,* which

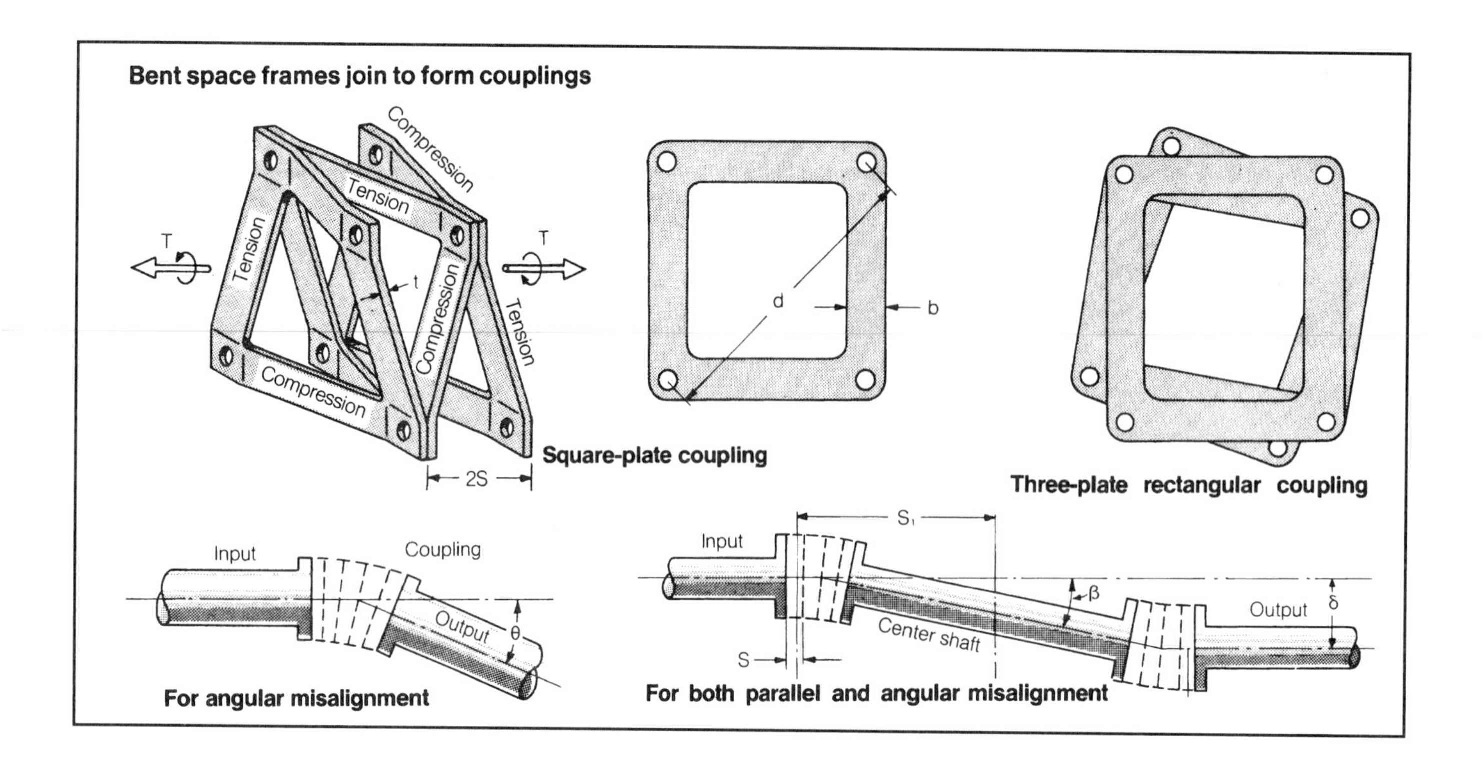

Design equations for the Bossler coupling

Ultimate torque capacity

(1) $$T = 11.62\frac{Ebt^3}{dn^{0.9}}$$

Maximum stress per degree of misalignment.

(2) $$\sigma_{max} = 0.0276\ Et/L$$

Minimum thickness to meet required torque strength

(3) $$t = 0.4415\left(\frac{dT}{bE}\right)^{1/3} n^{0.3}$$

Weight of coupling with minimum-thickness plates

(4) $$W = 1.249\ w\left(\frac{T}{E}\right)^{1/3} d^{4/3}\, b^{2/3}\, n^{1.3}$$

Maximum permissible misalignment

(5) $$\theta_{max} = 54.7\left[\frac{bd^2}{TE^2}\right]^{1/3} \sigma_c\, n^{0.7}$$

Maximum permissible misalignment (simplified)

(6) $$\theta/d = 10.9\frac{n^{0.7}}{T^{1/3}}$$

Maximum permissible offset-angle

(7) $$\beta = 54.7\left[\frac{bd^2}{TE^2}\right]^{1/3} \frac{\sigma_e C}{n^{0.3}}$$

where: $$\sum_{x=1}^{x=n}\left[1 - (x - 1)\frac{S}{S_1}\right]^2$$

Maximum permissible offset-angle (simplified)

(8) $$\beta/d = \frac{10.9C}{T^{1/3} n^{0.3}}$$

Critical speed frequency

(9) $$f = \frac{60}{2\pi}\left(\frac{k}{M}\right)^{1/2}$$

where: $k = \dfrac{24(El)_e}{(nS)^3}$ and $(El)_e = 0.886Ebt^3S/L$

List of symbols

b = Width of an element
d = Diameter at the bold circle
E = Modulus of elasticity
f = First critical speed, rpm
l = Flatwise moment of inertia of an element = $bt^3/12$
k = Spring constant for single degree of freedom
L = Effective length of an element. This concept is required because joint details tend to stiffen the ends of the elements. L = 0.667 d is recommended
M = Mass of center shaft plus mass of one coupling with fasteners
n = Number of plates in each coupling
S = Offset distance by which a plate out of plane
t = Thickness of an element
T = Torque applied to coupling, useful ultimate, usually taken as lowest critical buckling torque
w = Weight per unit volume
W = Total weight of plates in a coupling
$(El)_e$ = Flexural stiffness, the moment that causes one radian of flexural angle change per unit length of coupling
β = Equivalent angle change at each coupling during parallel offset misalignment, deg
υ = Total angular misalignment, deg
σ_c = Characteristic that limits stress for the material: yield stress for static performance, endurance limit stress for fatigue performance

makes the clutch smaller, or by making the plates thicker. But either solution would also make the clutch stiffer, hence would restrict the misalignment permitted before the clutch becomes over- stressed. The stress-misalignment relationship is given in Eq. 2, which shows the maximum flat-wise bending stress produced when a plate is misaligned 1° and is then rotated to transmit torque.

Plate thickness. For optimum misalignment capability, the plates should be selected with the least thickness that will provide the required torque strength. To determine the minimum thickness, Bossler found it expedient to rearrange Eq. 1 into the form shown in Eq. 3. The weight of any coupling designed in accordance to the minimum-thickness equation can be determined from Eq. 4.

Maximum misalignment. Angular misalignment occurs when the centerlines of the input and output shafts intersect at some angle—the angle of misalignment. When the characteristic limiting stress is known for the material selected—and for the coupling's dimensions—the maximum allowable angle of misalignment can be computed from Eq. 5.

If this allowance is not satisfactory, the designer might have to juggle the size factors by, say, adding more plates to the unit. To simplify eq. 5, Bossler made some assumptions in the ratio of endurance limit to modulus and in the ratio of *dsb* to obtain Eq. 6.

Parallel offset. This condition exists when the input and output shafts remain parallel but are displaced laterally. As with Eq. 6, Eq. 7 is a performance equation and can be reduced to design curves. Bossler obtained Eq. 8 by making the same assumptions as in the previous case.

Critical speed. Because of the noncircular configurations of the coupling, it is important that the operating speed of the unit be higher than its critical speed. It should not only be higher but also should avoid an integer relationship.

Bossler worked out a handy relationship for critical speed (Eq. 9) that employs a somewhat idealized value for the spring constant k.

Bossler also made other recommendations where weight reduction is vital:

- **Size of plates.** Use the largest d consistent with envelope and centrifugal force loading. Usually, centrifugal force loading will not be a problem below 300 ft/s tip speed.
- **Number of plates.** Pick the least n consistent with the required performance.
- **Thickness of plates.** Select the smallest t consistent with the required ultimate torque.
- **Joint details.** Be conservative; use high-strength tension fasteners with high preload. Provide fretting protection. Make element centerlines and bolt centerlines intersect at a point.
- **Offset distance.** Use the smallest S consistent with clearance.

COUPLING WITH OFF-CENTER PINS CONNECTS MISALIGNED SHAFTS

Two Hungarian engineers developed an all-metal coupling (see drawing) for connecting shafts where alignment is not exact—that is, where the degree of misalignment does not exceed the magnitude of the shaft radius.

The coupling is applied to shafts that are being connected for either high-torque or high-speed operation and that must operate at maximum efficiency. Knuckle joints are too expensive, and they have too much play; elastic joints are too vulnerable to the influences of high loads and vibrations.

How it's made. In essence, the coupling consists of two disks, each keyed to a splined shaft. One disk bears four fixed-mounted steel studs at equal spacing; the other disk has large-diameter holes drilled at points facing the studs.

Each large hole is fitted with a bearing that rotates freely inside it on rollers or needles. The bore of the bearings, however, is off-center. The amount of eccentricity of the bearing bore is identical to the deviation of the two shaft center lines.

In operation, input and output shafts can be misaligned, yet they still rotate with the same angular relationship they would have if perfectly aligned.

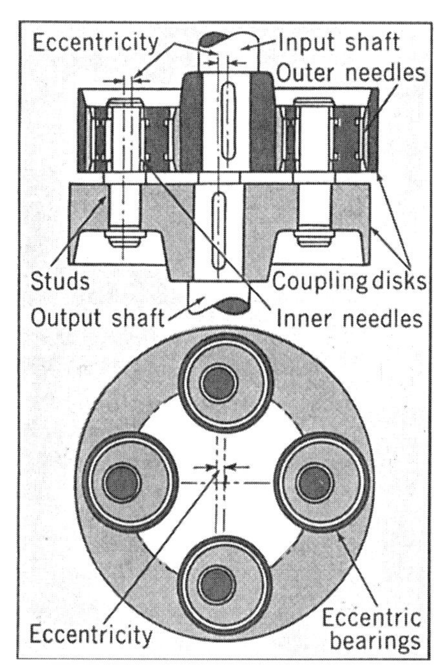

Eccentrically bored bearings rotate to make up for misalignment between shafts.

UNIVERSAL JOINT TRANSMITS TORQUE 45° AT CONSTANT SPEED

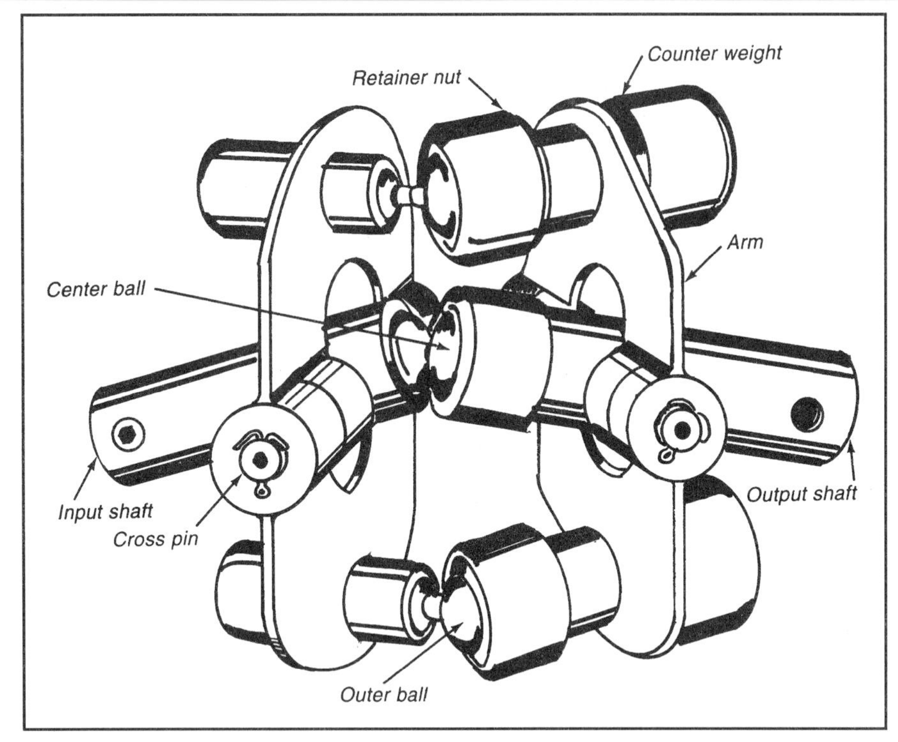

A novel arrangement of pivots and ball-socket joints transmits uniform motion.

A universal joint that transmits power at constant speeds through angles up to 45° was designed by Malton Miller of Minnesota.

Models of the true-speed drive that can transmit up to 20 hp have been developed.

It had not been possible to transmit power at constant speeds with only one universal joint. Engineers had to specify an intermediate shaft between two Hooke's joints or use a Rzappa-type joint to get the desired effect.

Ball-and-socket. Basically, the True-Speed joint is a system of ball-and-socket connections with large contact areas (low unit pressure) to transmit torsional forces across the joint. This arrangement minimizes problems when high bearing pressures build up against running surfaces. The low-friction bearings also increase efficiency. The joint is balanced to keep vibration at high speeds to a minimum.

The joint consists of driving and driven halves. Each half has a coupling sleeve at its end of the driveshaft, a pair of driving arms opposite each other and pivoted on a cross pin that extends through the coupling sleeve, and a ball-and-socket coupling at the end of each driving arm.

As the joint rotates, angular flexure in one plane of the joint is accommodated by the swiveling of the all-and-socket couplings and, in the 90° plane, by the oscillation of the driving arms about the transverse pin. As rotation occurs, torsion is transmitted from one half of the joint to the other half through the swiveling ball-and-socket couplings and the oscillating driving arms.

Balancing. Each half of the joint, in effect, rotates about its own center shaft, so each half is considered separate for balancing. The center ball-and-socket coupling serves only to align and secure the intersection point of the two shafts. It does not transmit any forces to the entire drive unit.

Balancing for rotation is achieved by equalizing the weight of the two driving arms of each half of the joint. Balancing the acceleration forces due to the oscillation of the ball-and-socket couplings, which are offset from their swiveling axes, is achieved by the use of counterweights extending from the opposite side of each driving arm.

The outer ball-and-socket couplings work in two planes of motion, swiveling widely in the plane perpendicular to the main shaft and swiveling slightly about the transverse pin in the plane parallel to the main shaft. In this coupling configuration, the angular displacement of the driving shaft is exactly duplicated in the driven shaft, providing constant rotational velocity and constant torque at all shaft intersection angles.

Bearings. The only bearing parts are the ball-and-socket couplings and the driving arms on the transverse pins. Needle bearings support the driving arms on the transverse pin, which is hardened and ground. A high-pressure grease lubricant coats the bearing surfaces of the ball-and-socket couplings. Under maximum rated loadings of 600 psi on the ball-and-socket surfaces, there is no appreciable heating or power loss due to friction.

Capabilities. Units have been laboratory-tested at all rated angles of drive under dynamometer loadings. Although the first available units were for smaller capacities, a unit designed for 20 hp at 550 rpm, suitable for tractor power take-off drive, has been tested.

Similar couplings have been designed as pump couplings. But the True-Speed drive differs in that the speed and transfer elements are positive. With the pump coupling, on the other hand, the speed might fluctuate because of spring bounce.

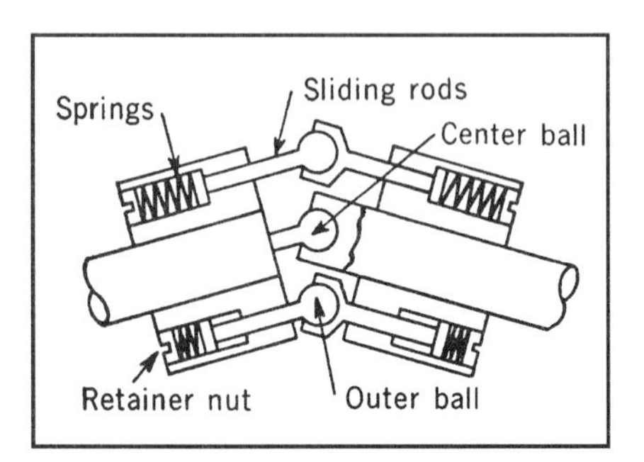

An earlier version for angled shafts required spring-loaded sliding rods.

TEN UNIVERSAL SHAFT COUPLINGS

Hooke's Joints

The commonest form of a universal coupling is a *Hooke's joint.* It can transmit torque efficiently up to a maximum shaft alignment angle of about 36°. At slow speeds, on hand-operated mechanisms, the permissible angle can reach 45°. The simplest arrangement for a Hooke's joint is two forked shaft-ends coupled by a cross-shaped piece. There are many variations and a few of them are included here.

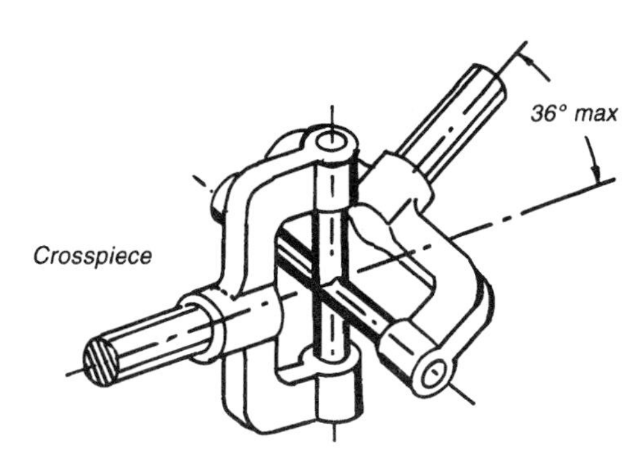

Fig. 1 The Hooke's joint can transmit heavy loads. Anti-friction bearings are a refinement often used.

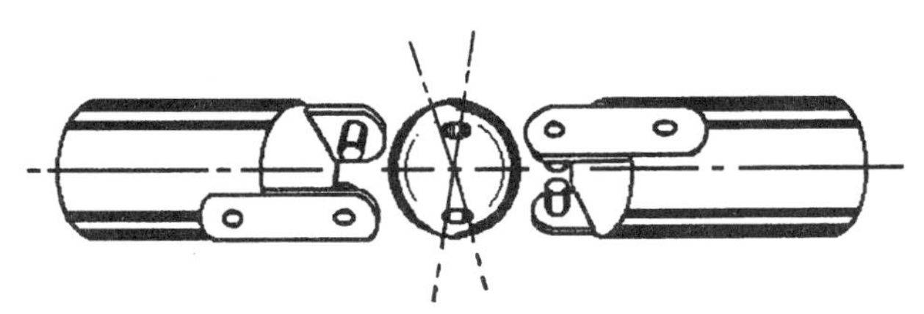

Fig. 2 A pinned sphere shaft coupling replaces a cross-piece. The result is a more compact joint.

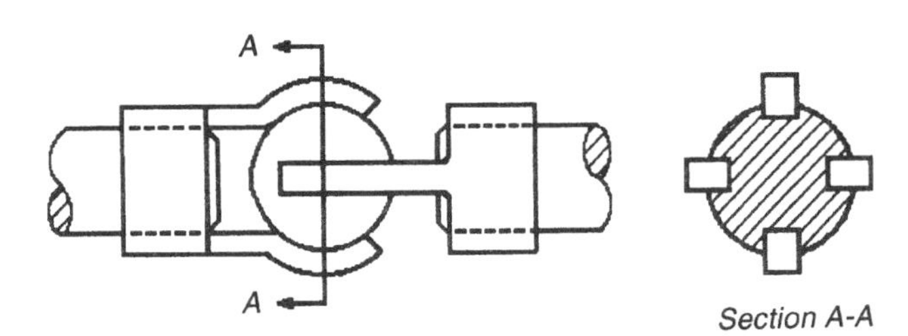

Fig. 3 A grooved-sphere joint is a modification of a pinned sphere. Torques on fastening sleeves are bent over the sphere on the assembly. Greater sliding contact of the torques in grooves makes simple lubrication essential at high torques and alignment angles.

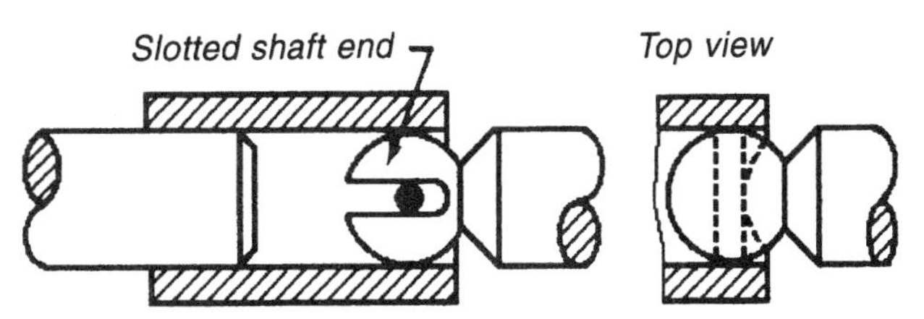

Fig. 4 A pinned-sleeve shaft-coupling is fastened to one shaft that engages the forked, spherical end on the other shaft to provide a joint which also allows for axial shaft movement. In this example, however, the angle between shafts must be small. Also, the joint is only suitable for low torques.

Constant-Velocity Couplings

The disadvantages of a single Hooke's joint is that the velocity of the driven shaft varies. Its maximum velocity can be found by multiplying driving-shaft speed by the secant of the shaft angle; for minimum speed, multiply by the cosine. An example of speed variation: a driving shaft rotates at 100 rpm; the angle between the shafts is 20°. The minimum output is 100 × 0.9397, which equals 93.9 rpm; the maximum output is 1.0642 × 100, or 106.4 rpm. Thus, the difference is 12.43 rpm. When output speed is high, output torque is low, and vice versa. This is an objectionable feature in some mechanisms. However, two universal joints connected by an intermediate shaft solve this speed-torque objection.

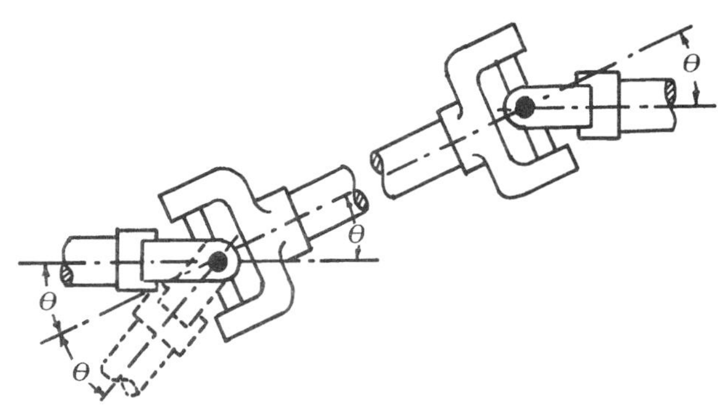

Fig. 5 A constant-velocity joint is made by coupling two Hooke's joints. They must have equal input and output angles to work correctly. Also, the forks must be assembled so that they will always be in the same plane. The shaft-alignment angle can be double that for a single joint.

This single constant-velocity coupling is based on the principle (Fig. 6) that the contact point of the two members must always lie on the homokinetic plane. Their rotation speed will then always be equal because the radius to the contact point of each member will always be equal. Such simple couplings are ideal for toys, instruments, and other light-duty mechanisms. For heavy duty, such as the front-wheel drives of military vehicles, a more complex coupling is shown diagrammatically in Fig. 7A. It has two joints close-coupled with a sliding member between them. The exploded view (Fig. 7B) shows these members. There are other designs for heavy-duty universal couplings; one, known as the *Rzeppa,* consists of a cage that keeps six balls in the homokinetic plane at all times. Another constant-velocity joint, the *Bendix-Weiss,* also incorporates balls.

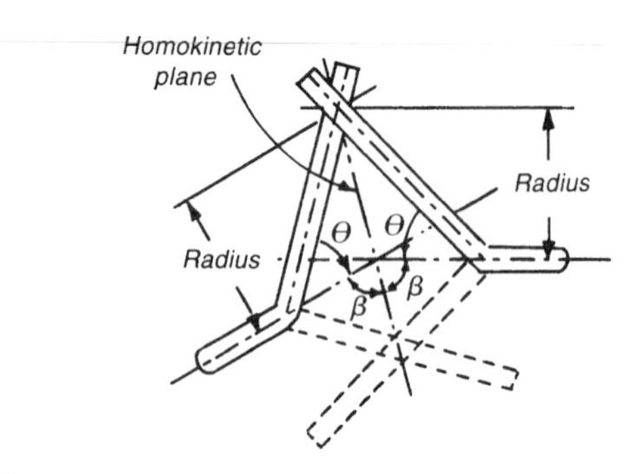

Fig. 6

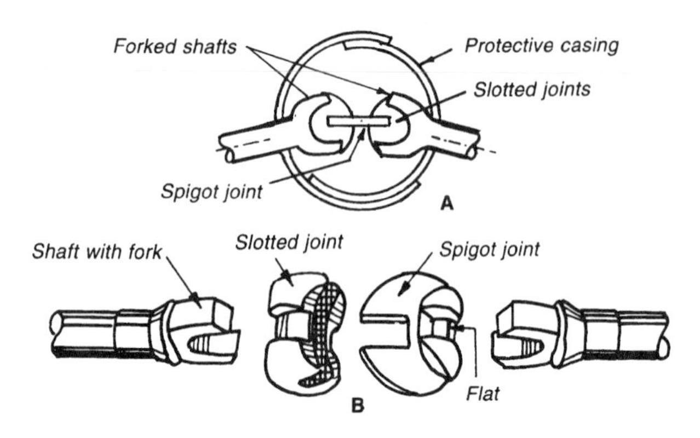

Fig. 7

Fig. 8 This flexible shaft permits any shaft angle. These shafts, if long, should be supported to prevent backlash and coiling.

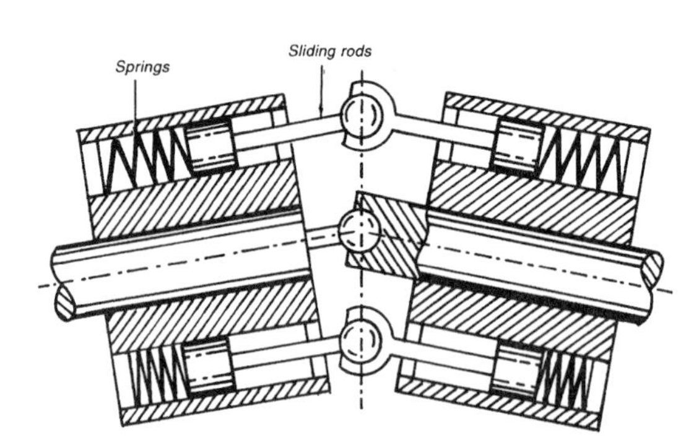

Fig. 9 This pump-type coupling has the reciprocating action of sliding rods that can drive pistons in cylinders.

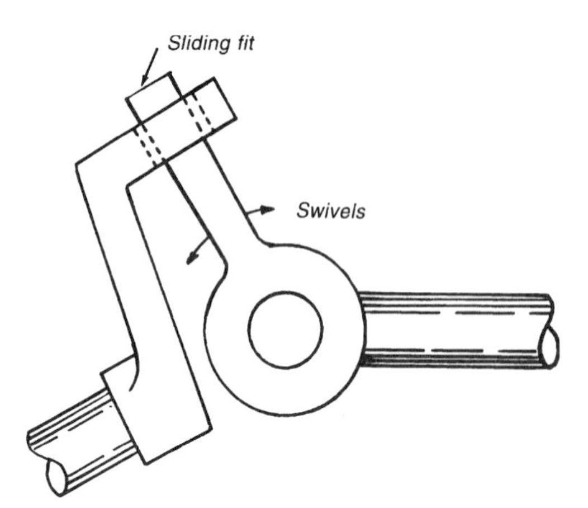

Fig. 10 This light-duty coupling is ideal for many simple, low-cost mechanisms. The sliding swivel-rod must be kept well lubricated at all times.

NINETEEN METHODS FOR COUPLING ROTATING SHAFTS

Methods for coupling rotating shafts vary from simple bolted flange assembles to complex spring and synthetic rubber assembles. Those including chain belts, splines, bands, and rollers are shown here.

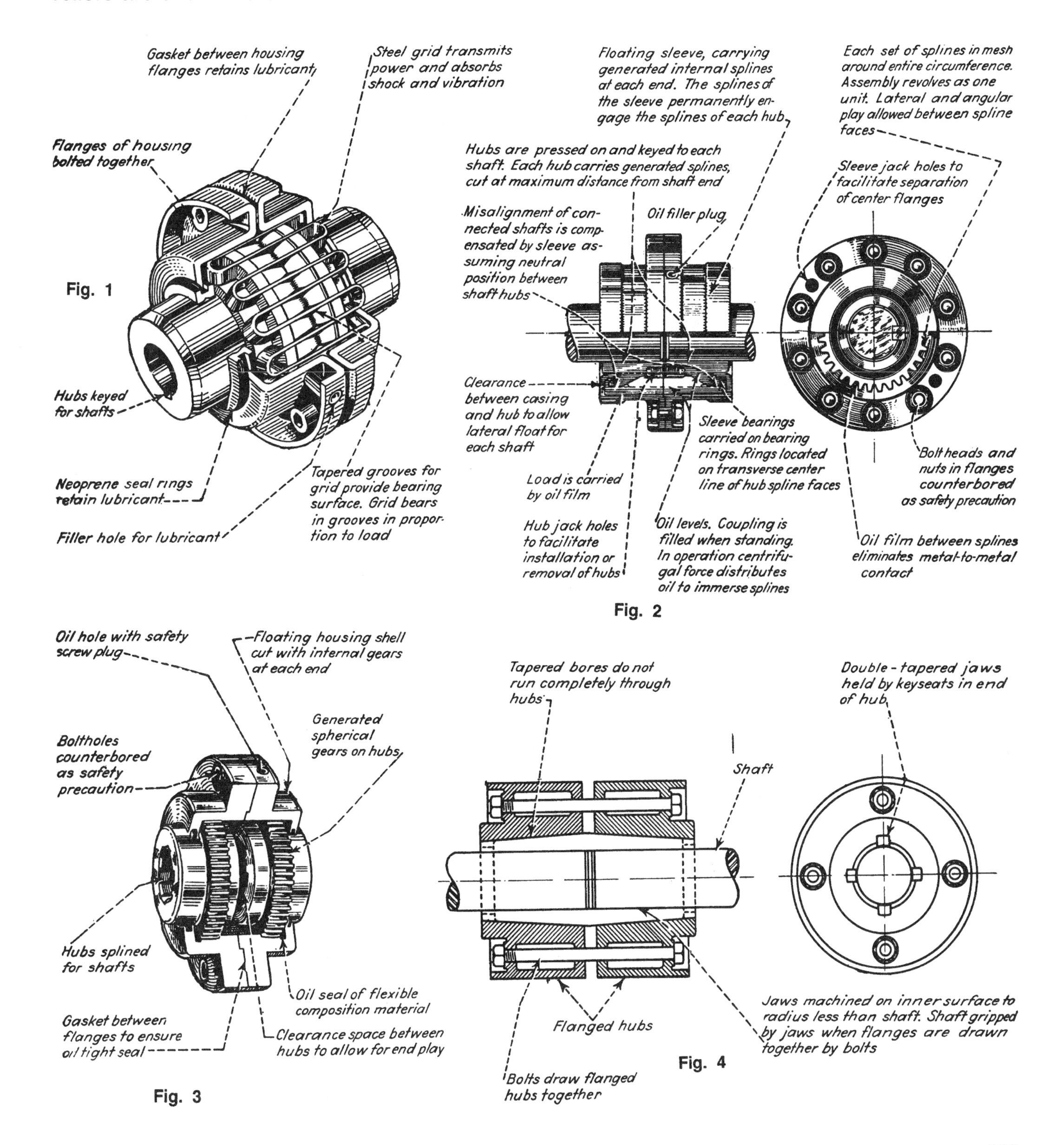

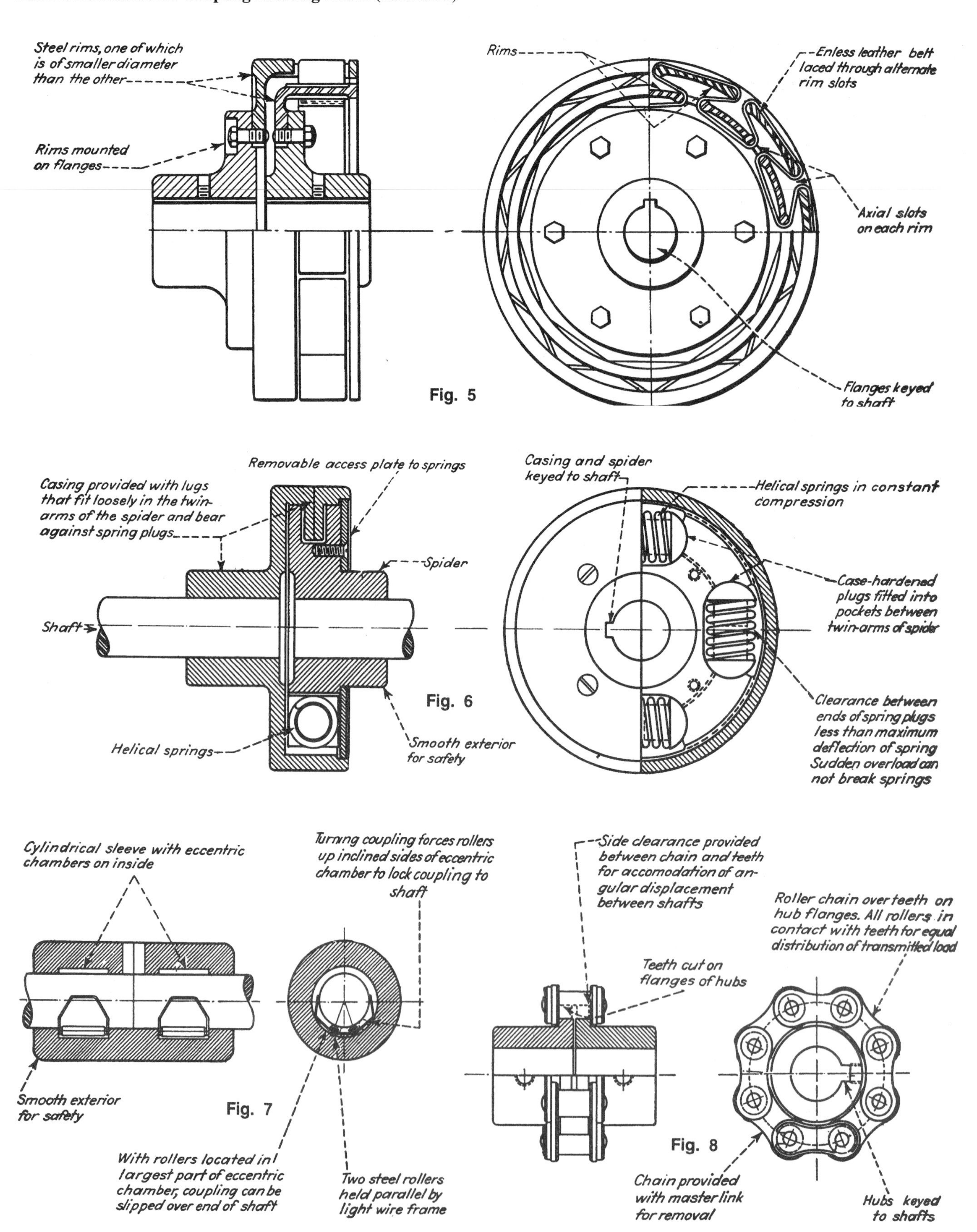

Steel rims, one of which is of smaller diameter than the other
Rims mounted on flanges
Rims
Enless leather belt laced through alternate rim slots
Axial slots on each rim
Flanges keyed to shaft
Fig. 5
Removable access plate to springs
Casing provided with lugs that fit loosely in the twin-arms of the spider and bear against spring plugs
Spider
Shaft
Helical springs
Smooth exterior for safety
Fig. 6
Casing and spider keyed to shaft
Helical springs in constant compression
Case-hardened plugs fitted into pockets between twin-arms of spider
Clearance between ends of spring plugs less than maximum deflection of spring Sudden overload can not break springs
Cylindrical sleeve with eccentric chambers on inside
Turning coupling forces rollers up inclined sides of eccentric chamber to lock coupling to shaft
Smooth exterior for safety
Fig. 7
With rollers located in largest part of eccentric chamber, coupling can be slipped over end of shaft
Two steel rollers held parallel by light wire frame
Side clearance provided between chain and teeth for accomodation of angular displacement between shafts
Roller chain over teeth on hub flanges. All rollers in contact with teeth for equal distribution of transmitted load
Teeth cut on flanges of hubs
Fig. 8
Chain provided with master link for removal
Hubs keyed to shafts

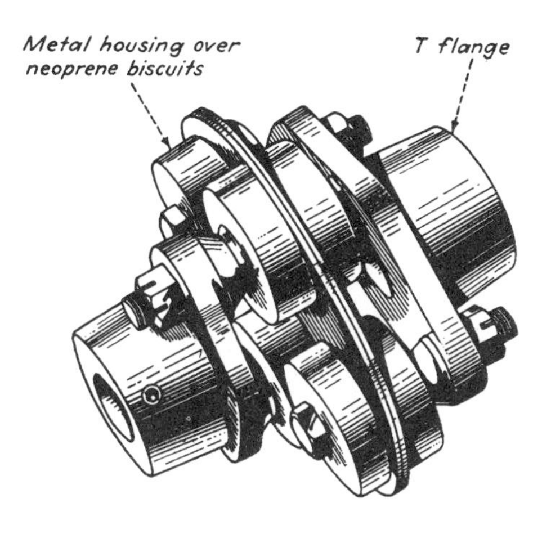

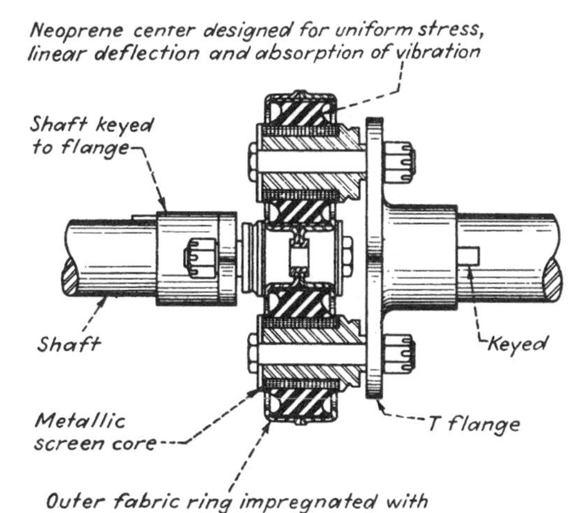

Fig. 9

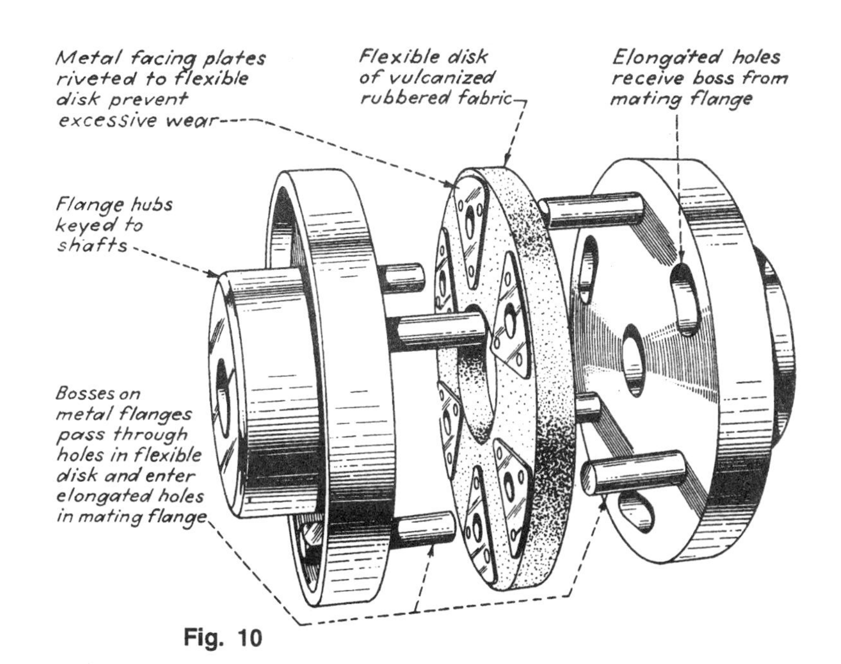

Fig. 10

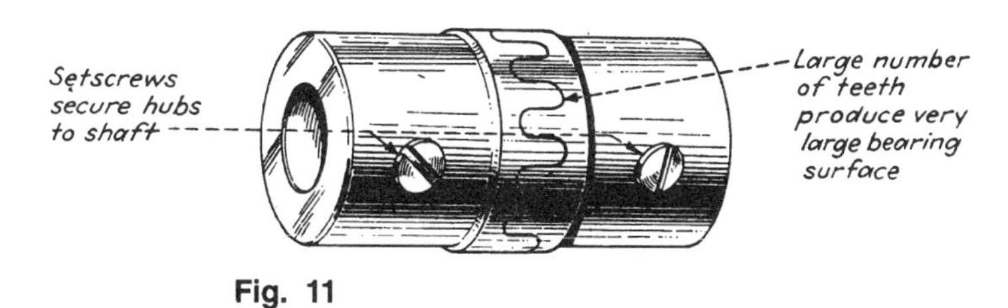

Fig. 11

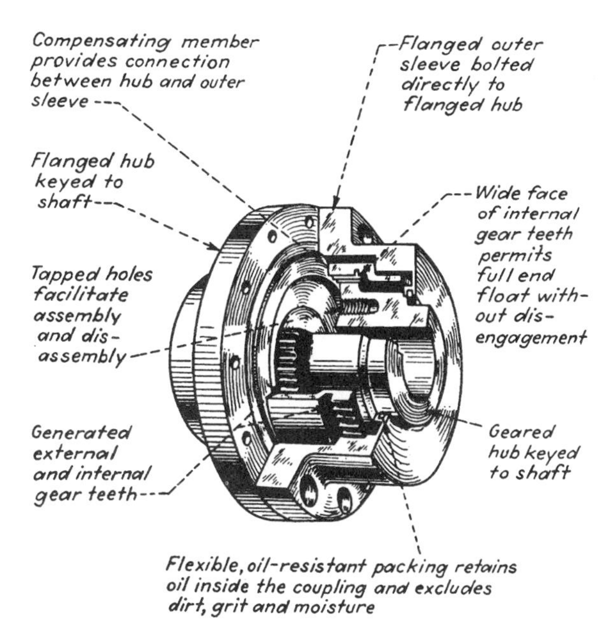

Fig. 12

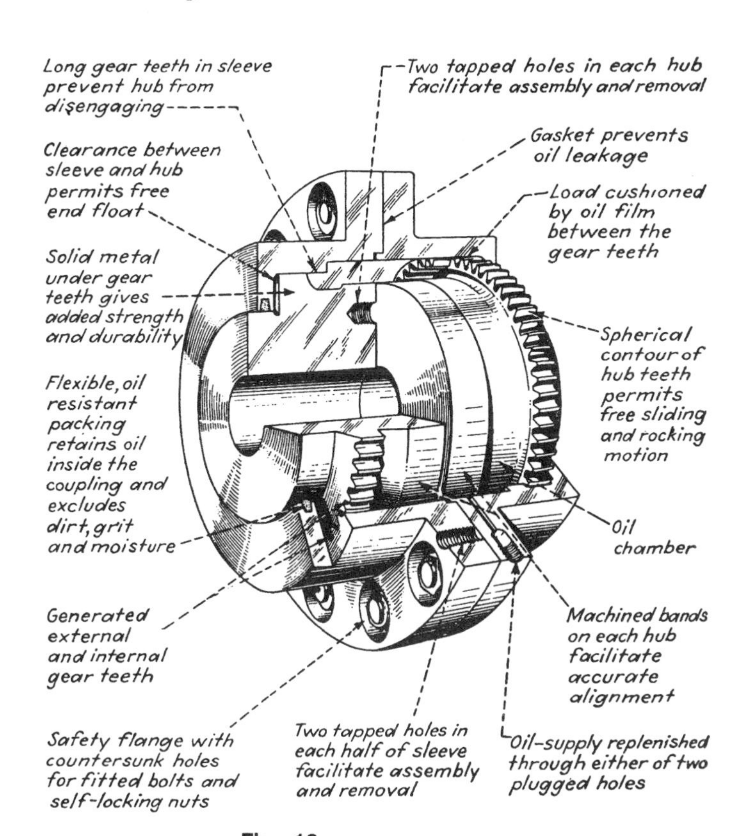

Fig. 13

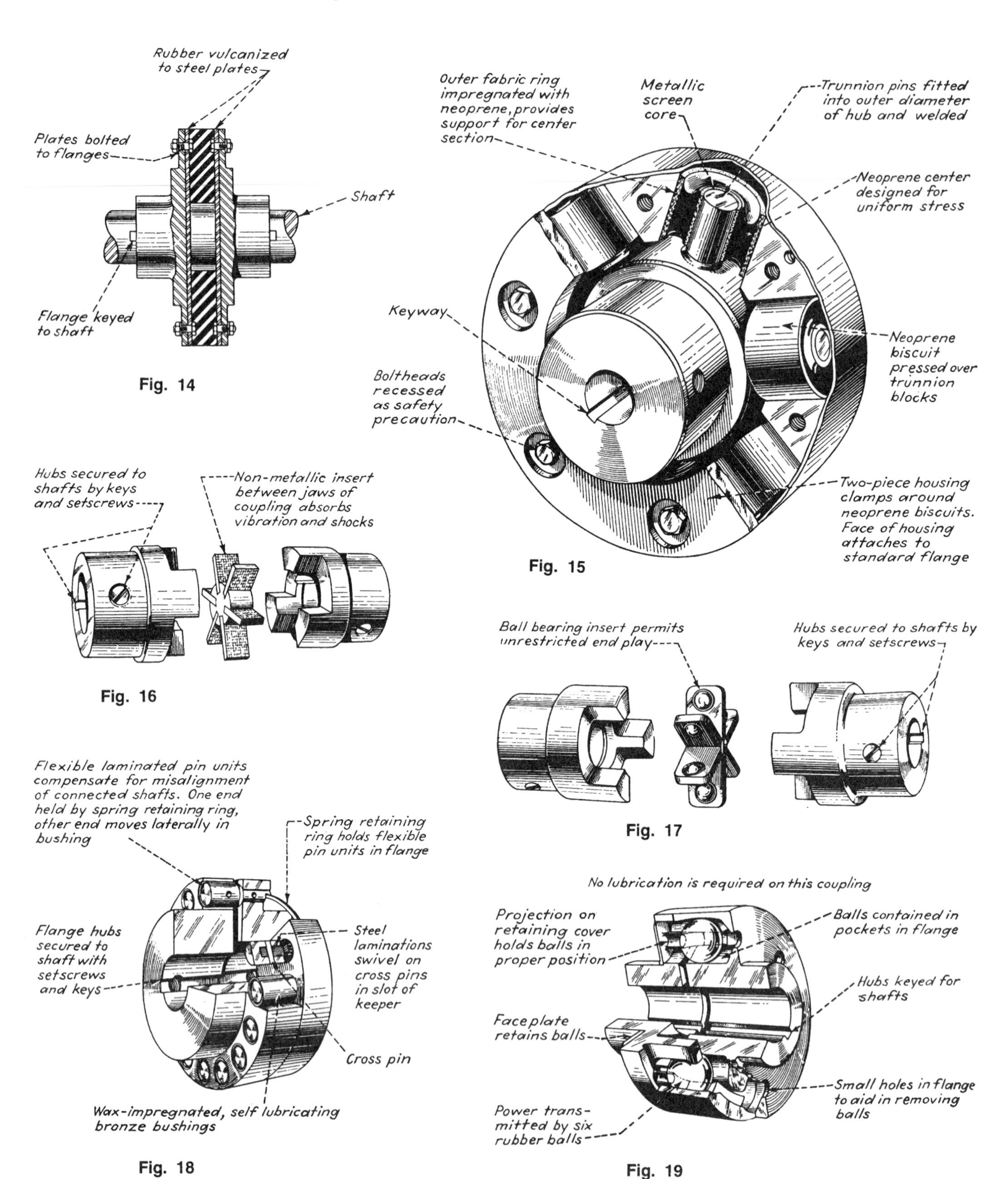

Fig. 14

Fig. 15

Fig. 16

Fig. 17

Fig. 18

Fig. 19

FIVE DIFFERENT PIN-AND-LINK COUPLINGS

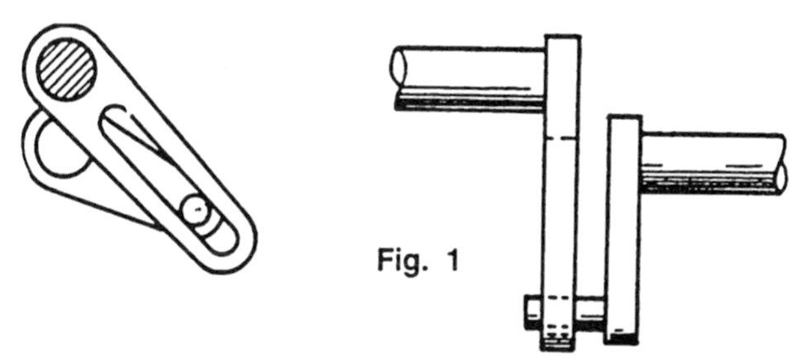

Fig. 1 If constant velocity is not required, a pin and slot coupling can be used. Velocity transmission is irregular because the effective radius of operation is continually changing. The shafts must remain parallel unless a ball joint is placed between the slot and pin. Axial freedom is possible, but any change in the shaft offset will further affect the fluctuation of velocity transmission.

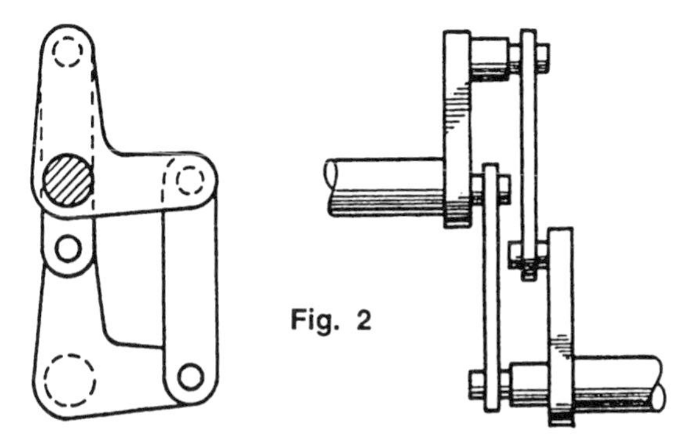

Fig. 2 This parallel-crank coupling drives the overhead camshaft on engines. Each shaft has at lest two cranks connected by links. Each must have full symmetry for constant velocity action and to avoid dead points. By attaching ball joints at the ends of the links, displacement between the crank assembles is possible.

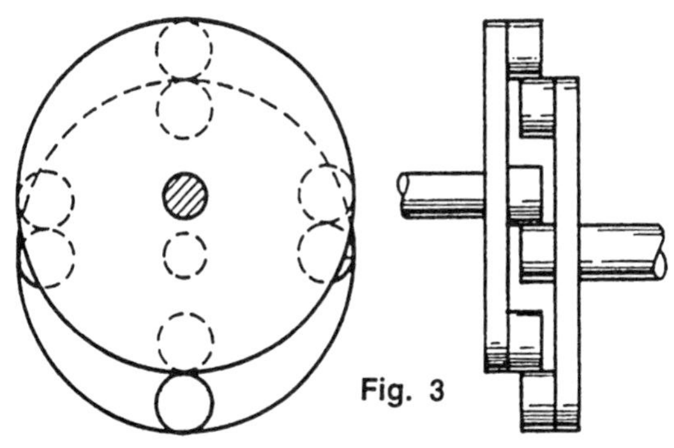

Fig. 3 This coupling is kinematically equivalent to Fig. 2. It can be made by substituting two circular and contacting pins for each link. Each shaft has a disk carrying three or more projecting pins. The sum of the radii of the pins is equal to the eccentricity of offset of the shafts. The center lines between each pair of pins remain parallel as the coupling rotates. The pins need not have equal diameters. Transmission is at a constant velocity, and axial freedom is possible.

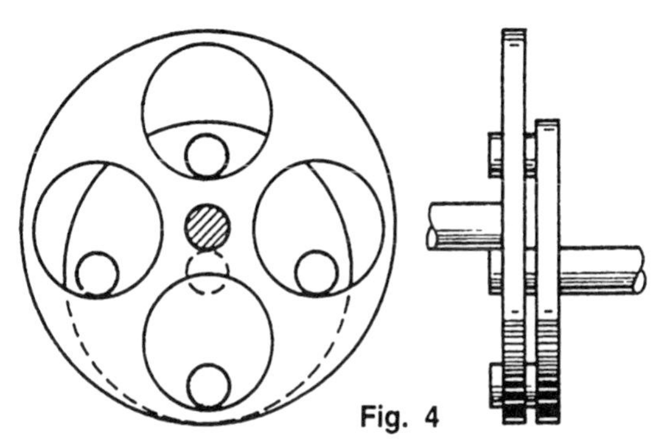

Fig. 4 This coupling is similar to the coupling shown in Fig. 3. However, holes replace one set of pins. The difference in radii is equal to the eccentricity or offset. Velocity transmission is constant; axial freedom is possible, but as in Fig. 3, the shaft axes must remain fixed. This type of coupling can be installed in epicyclic reduction gear boxes.

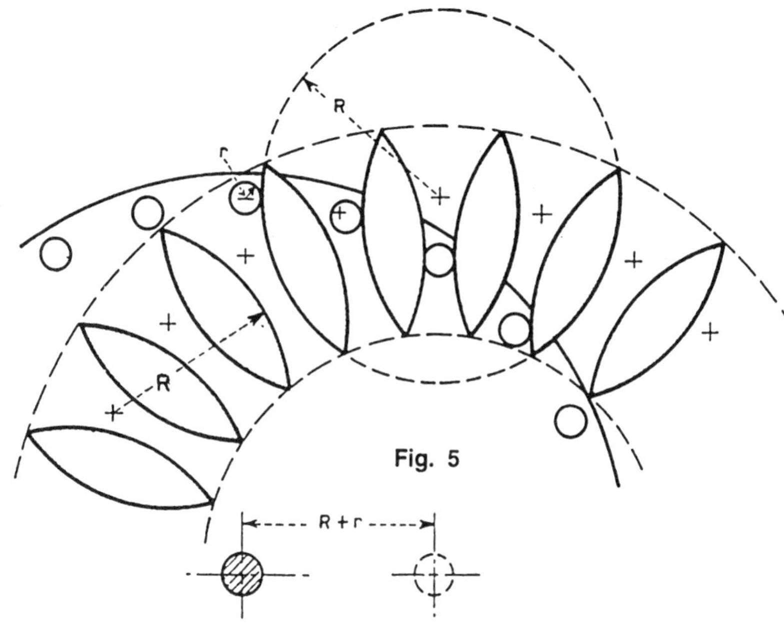

Fig. 5 An unusual development in pin coupling is shown. A large number of pins engages the lenticular or shield-shaped sections formed from segments of theoretical large pins. The axes forming the lenticular sections are struck from the pith points of the coupling, and the distance $R + r$ is equal to the eccentricity between the shaft centers. Velocity transmission is constant; axial freedom is possible, but the shafts must remain parallel.

TEN DIFFERENT SPLINED CONNECTIONS

CYLINDRICAL SPLINES

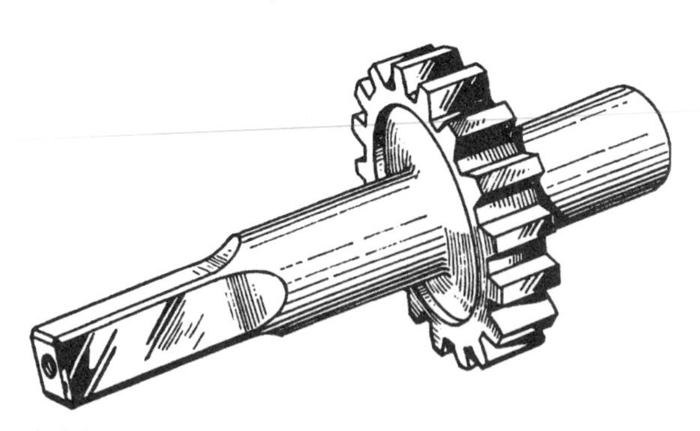

1. SQUARE SPLINES make simple connections. They are used mainly for transmitting light loads, where accurate positioning is not critical. This spline is commonly used on machine tools; a cap screw is required to hold the enveloping member.

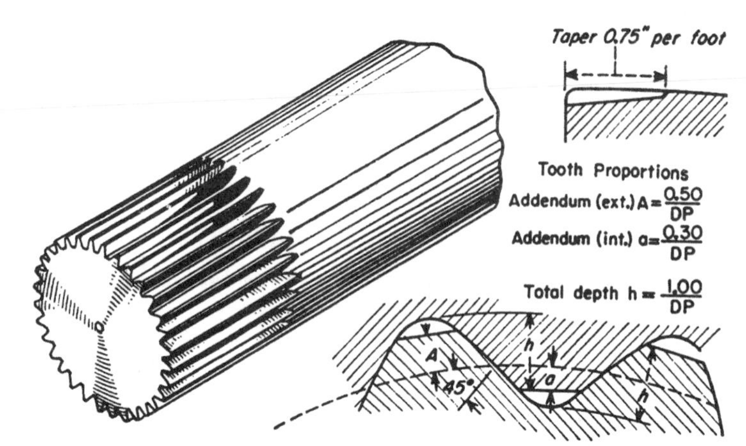

2. SERRATIONS of small size are used mostly for transmitting light loads. This shaft forced into a hole of softer material makes an inexpensive connection. Originally straight-sided and limited to small pitches, 45° serrations have been standardized (SAE) with large pitches up to 10 in. dia. For tight fits, the serrations are tapered.

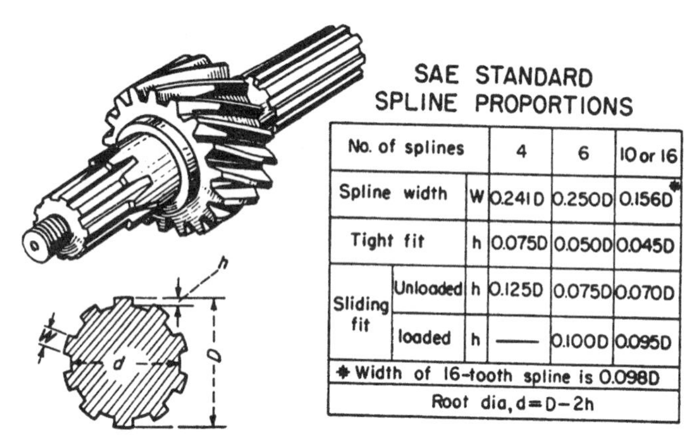

SAE STANDARD SPLINE PROPORTIONS

No. of splines			4	6	10 or 16
Spline width		W	0.241D	0.250D	0.156D*
Tight fit		h	0.075D	0.050D	0.045D
Sliding fit	Unloaded	h	0.125D	0.075D	0.070D
	loaded	h	—	0.100D	0.095D
* Width of 16-tooth spline is 0.098D					
Root dia, d = D − 2h					

3. STRAIGHT-SIDED splines have been widely used in the automotive field. Such splines are often used for sliding members. The sharp corner at the root limits the torque capacity to pressures of approximately 1,000 psi on the spline projected area. For different applications, tooth height is altered, as shown in the table above.

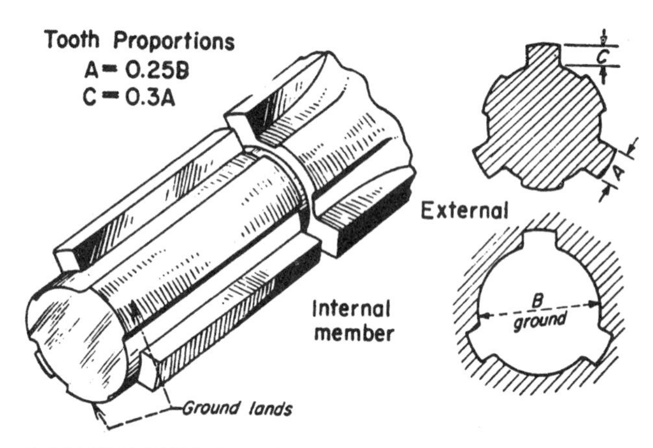

4. MACHINE-TOOL splines have wide gaps between splines to permit accurate cylindrical grinding of the lands—for precise positioning. Internal parts can be ground readily so that they will fit closely with the lands of the external member.

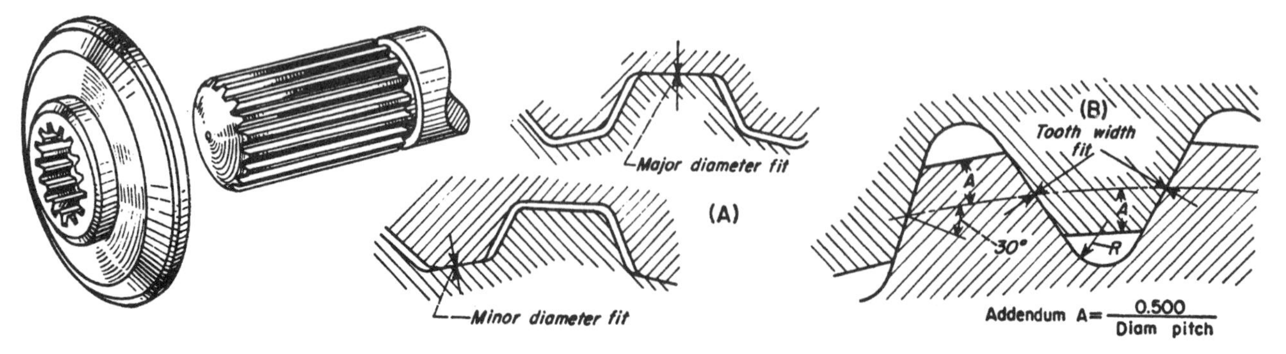

5. INVOLUTE-FORM splines are used where high loads are to be transmitted. Tooth proportions are based on a 30° stub tooth form. (A) Splined members can be positioned either by close fitting major or minor diameters. (B) Use of the tooth width or side positioning has the advantage of a full fillet radius at the roots. Splines can be parallel or helical. Contact stresses of 4,000 psi are used for accurate, hardened splines. The diametral pitch shown is the ratio of teeth to the pitch diameter.

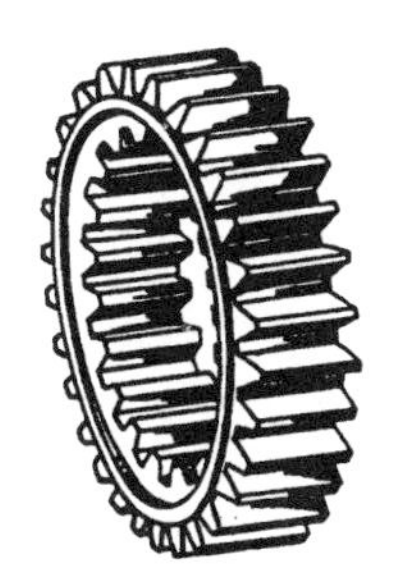

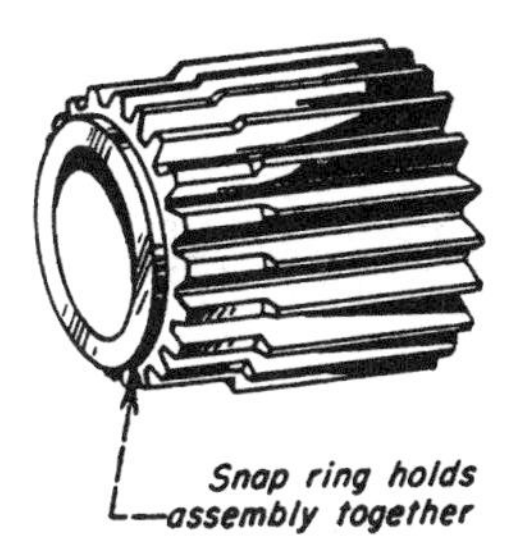

6. SPECIAL INVOLUTE splines are made by using gear tooth proportions. With full depth teeth, greater contact area is possible. A compound pinion is shown made by cropping the smaller pinion teeth and internally splining the larger pinion.

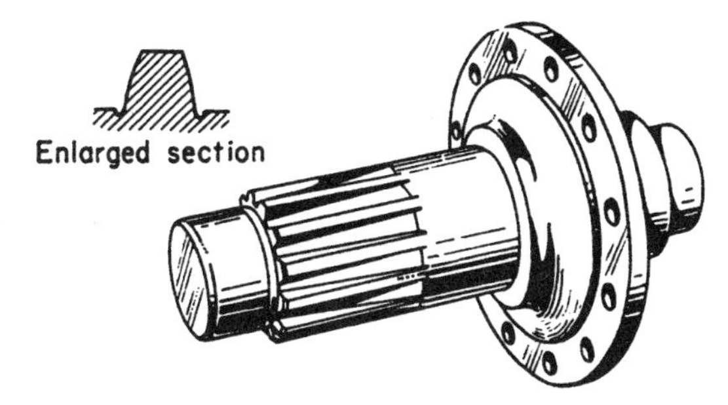

7. TAPER-ROOT splines are for drivers that require positive positioning. This method holds mating parts securely. With a 30° involute stub tooth, this type is stronger than parallel root splines and can be hobbed with a range of tapers.

FACE SPLINES

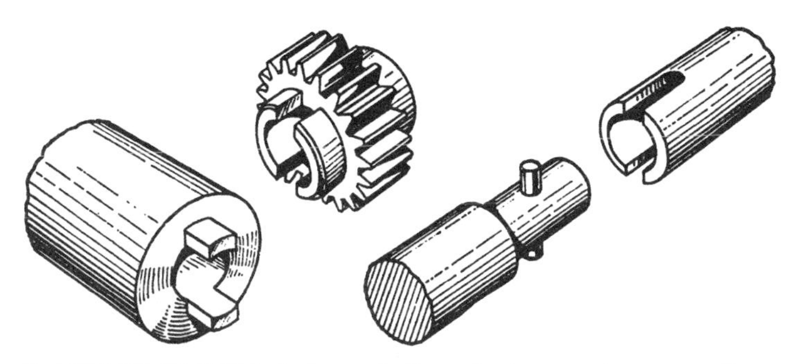

8. MILLED SLOTS in hubs or shafts make inexpensive connections. This spline is limited to moderate loads and requires a locking device to maintain positive engagement. Apin and sleeve method is used for light torques and where accurate positioning is not required.

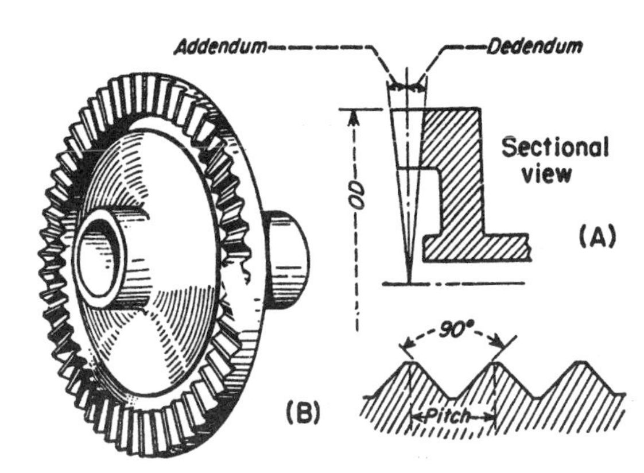

9. RADIAL SERRATIONS made by milling or shaping the teeth form simple connections. (A) Tooth proportions decrease radially. (B) Teeth can be straight-sided (castellated) or inclined; a 90° angle is common.

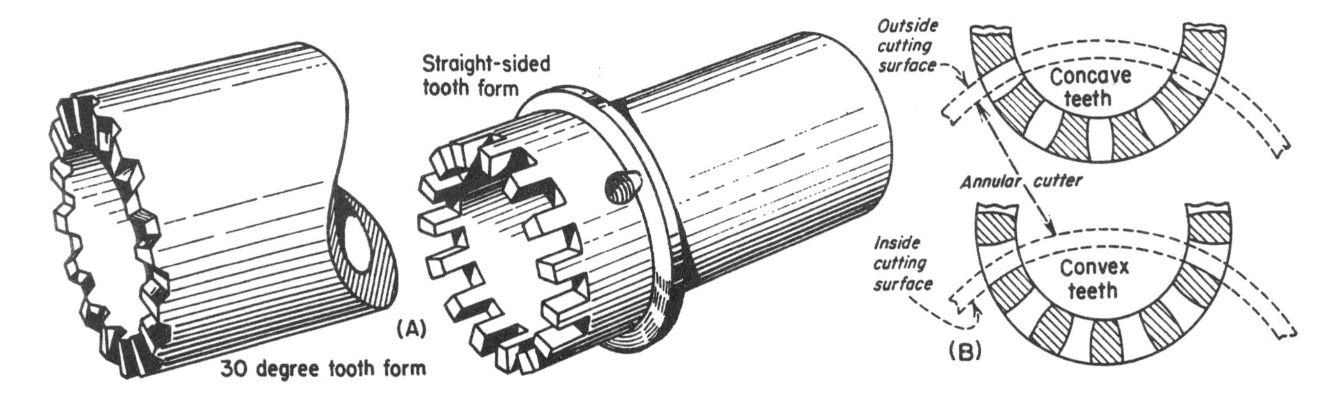

10. CURVIC COUPLING teeth are machined by a face-mill cutter. When hardened parts are used that require accurate positioning, the teeth can be ground. (A) This process produces teeth with uniform depth. They can be cut at any pressure angle, although 30° is most common. (B) Due to the cutting action, the shape of the teeth will be concave (hour-glass) on one member and convex on the other—the member with which it will be assembled.

FOURTEEN WAYS TO FASTEN HUBS TO SHAFTS

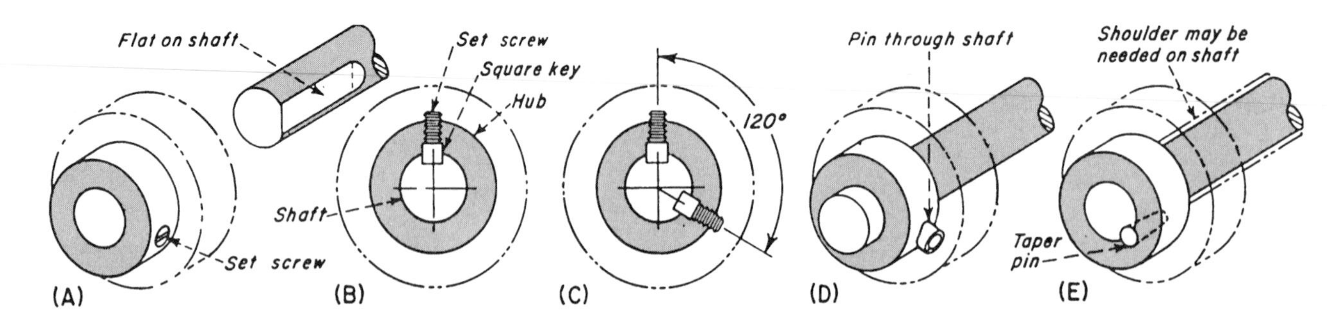

Fig. 1 A cup-point setscrew in hub (A) bears against a flat on a shaft. This fastening is suitable for fractional horsepower drives with low shock loads but is unsuitable when frequent removal and assembly are necessary. The key with setscrew (B) prevents shaft marring from frequent removal and assembly. It can withstand high shock loads. Two keys 120° apart (C) transmit extra heavy loads. Straight or tapered pin (D) prevents end play. For experimental setups an expanding pin is suitable yet easy to remove. Taper pin (E) parallel to shaft might require a shoulder on the shaft. It can be used when a gear or pulley has no hub.

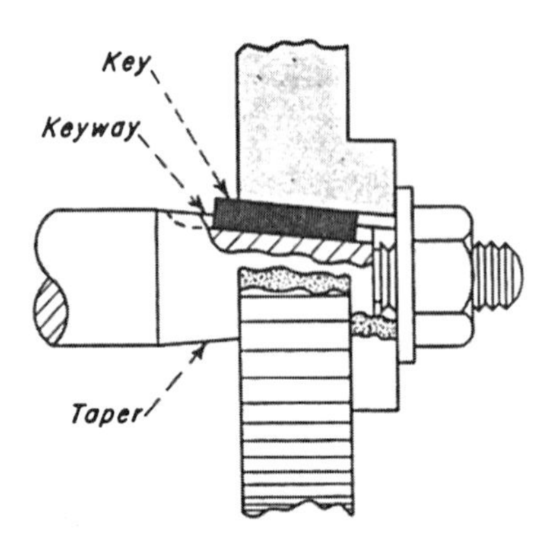

Fig. 2 A tapered shaft with a key and threaded end is a rigid concentric assembly. It is suitable for heavy-duty applications, yet it can be easily disassembled.

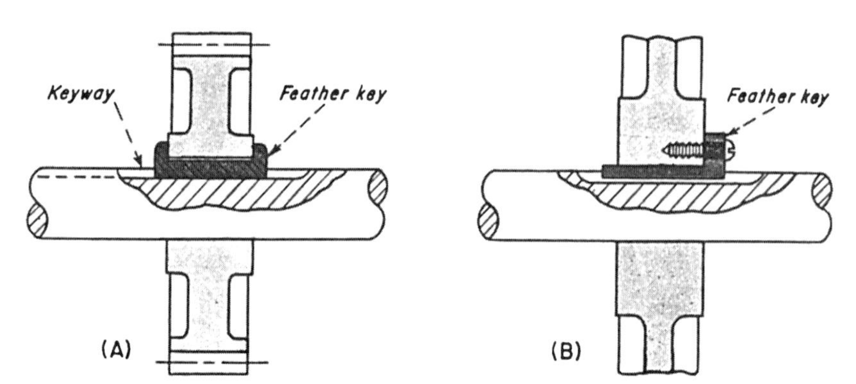

Fig. 3 A feather key (A) allows axial gear movement. A keyway must be milled to the end of the shaft. For a blind keyway (B) the hub and key must be drilled and tapped, but the design allows the gear to be mounted anywhere on the shaft with only a short keyway.

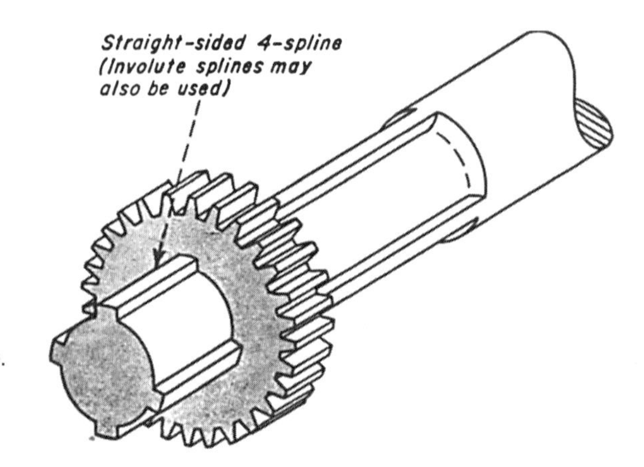

Fig. 4 Splined shafts are frequently used when a gear must slide. Square splines can be ground to close minor diameter gaps, but involute splines are self-centering and stronger. Non-sliding gears can be pinned to the shaft if it is provided with a hub.

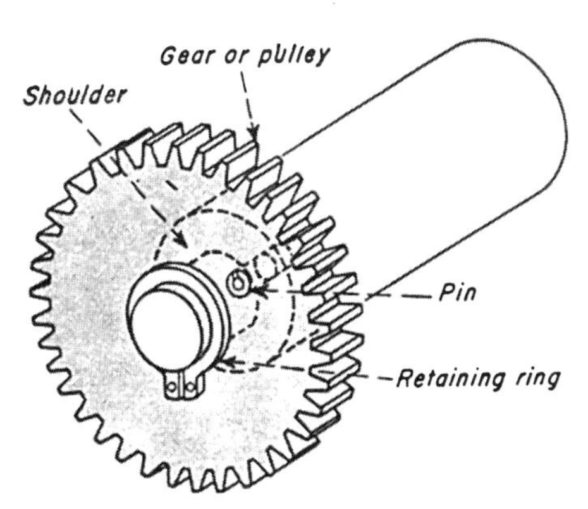

Fig. 5 A retaining ring allows quick gear removal in light-load applications. A shoulder on the shaft is necessary. A shear pin can secure the gear to the shaft if protection against an excessive load is required.

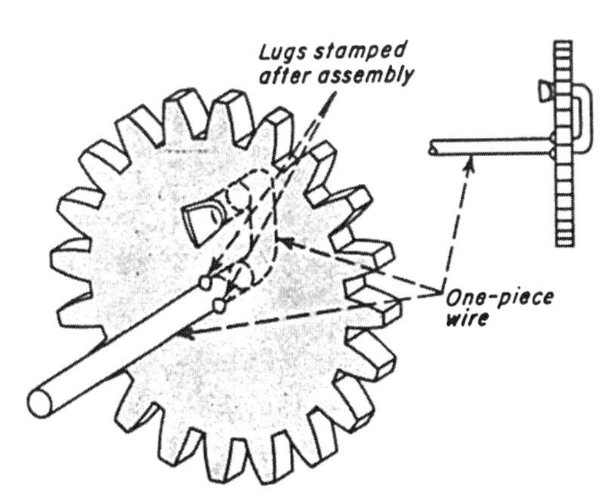

Fig. 6 A stamped gear and formed wire shaft can be used in light-duty application. Lugs stamped on both legs of the wire prevent disassembly. The bend radii of the shaft should be small enough to allow the gear to seat.

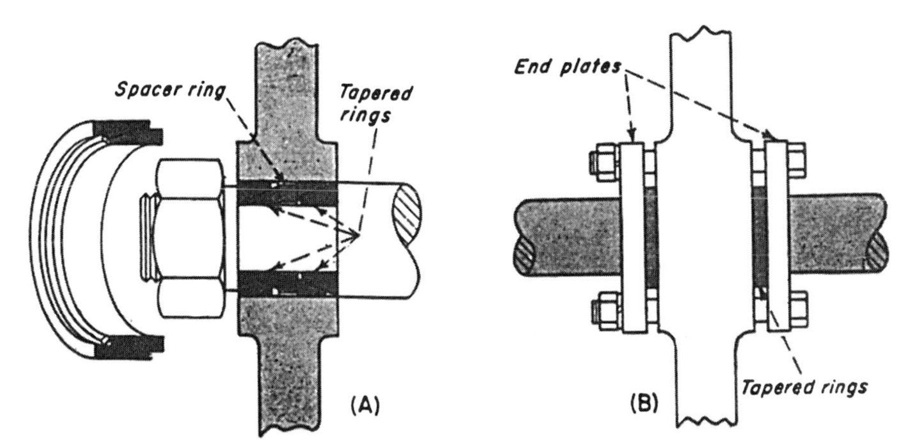

Fig. 7 Interlocking tapered rings hold the hub tightly to the shaft when the nut is tightened. Coarse machining of the hub and shaft does not affect concentricity as in pinned and keyed assemblies. A shoulder is required (A) for end-of-shaft mounting. End plates and four bolts (B) allow the hub to be mounted anywhere on the shaft.

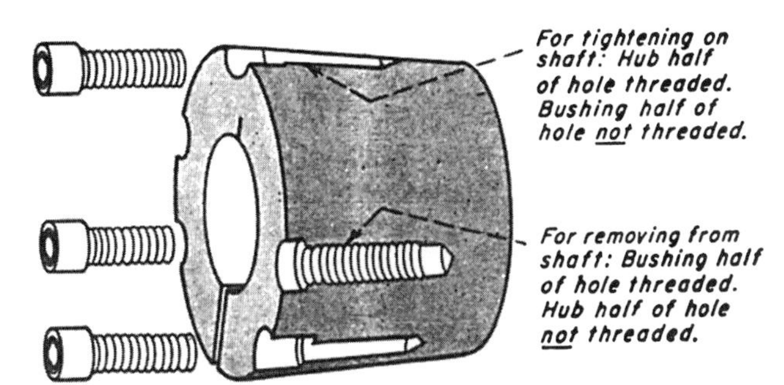

Fig. 8 This split bushing has a tapered outer diameter. Split holes in the bushing align with split holes in the hub. For tightening, the hub half of the hole is tapped, and the bushing half is un-tapped. A screw pulls the bushing into the hub as it is tightened, and it is removed by reversing the procedure.

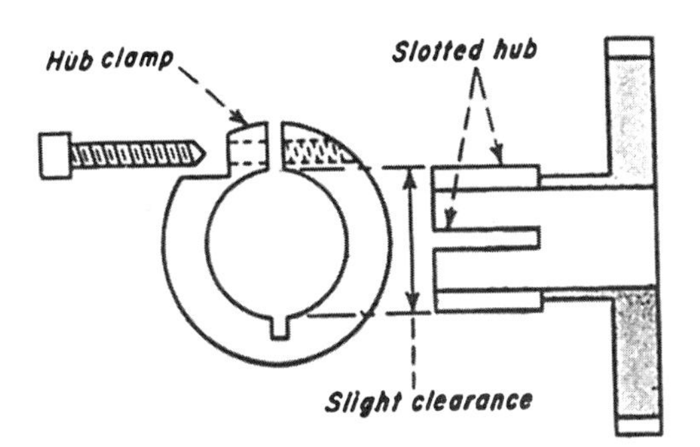

Fig. 9 The split hub of a stock precision gear is clamped onto a shaft with a separate hub clamp. Manufacturers list correctly dimensioned hubs and clamps so that they can be efficiently fastened to a precision-ground shaft.

CHAPTER 13

MOTION-SPECIFIC DEVICES, MECHANISMS, AND MACHINES

TIMING BELTS, FOUR-BAR LINKAGE TEAM UP FOR SMOOTH INDEXING

A class of intermittent mechanisms based on timing belts, pulleys, and linkages (see drawing) instead of the usual genevas or cams is capable of cyclic start-and-stop motions with smooth acceleration and deceleration.

Developed by Eric S. Buhayar and Eugene E. Brown of the Engineering Research Division, Scott Paper Co. (Philadelphia), the mechanisms are employed in automatic assembly lines.

These mechanisms, moreover, can function as phase adjusters in which the rotational position of the input shaft can be shifted as desired in relation to the output shaft. Such phase adjusters have been used in the textile and printing industries to change the "register" of one roll with that of another, when both rolls are driven by the same input.

Outgrowth from chains. Intermittent-motion mechanisms typically have ingenious shapes and configurations. They have been used in watches and in production machines for many years. There has been interest in the chain type of intermittent mechanism (see drawing), which ingeniously routes a chain around four sprockets to produce a dwell-and-index output.

The input shaft of such a device has a sprocket eccentrically fixed to it. The input also drives another shaft through one-to-one gearing. This second shaft mounts a similar eccentric sprocket that is, however, free to rotate. The chain passes first around an idler pulley and then around a second pulley, which is the output.

As the input gear rotates, it also pulls the chain around with it, producing a modulated output rotation. Two spring-loaded shoes,

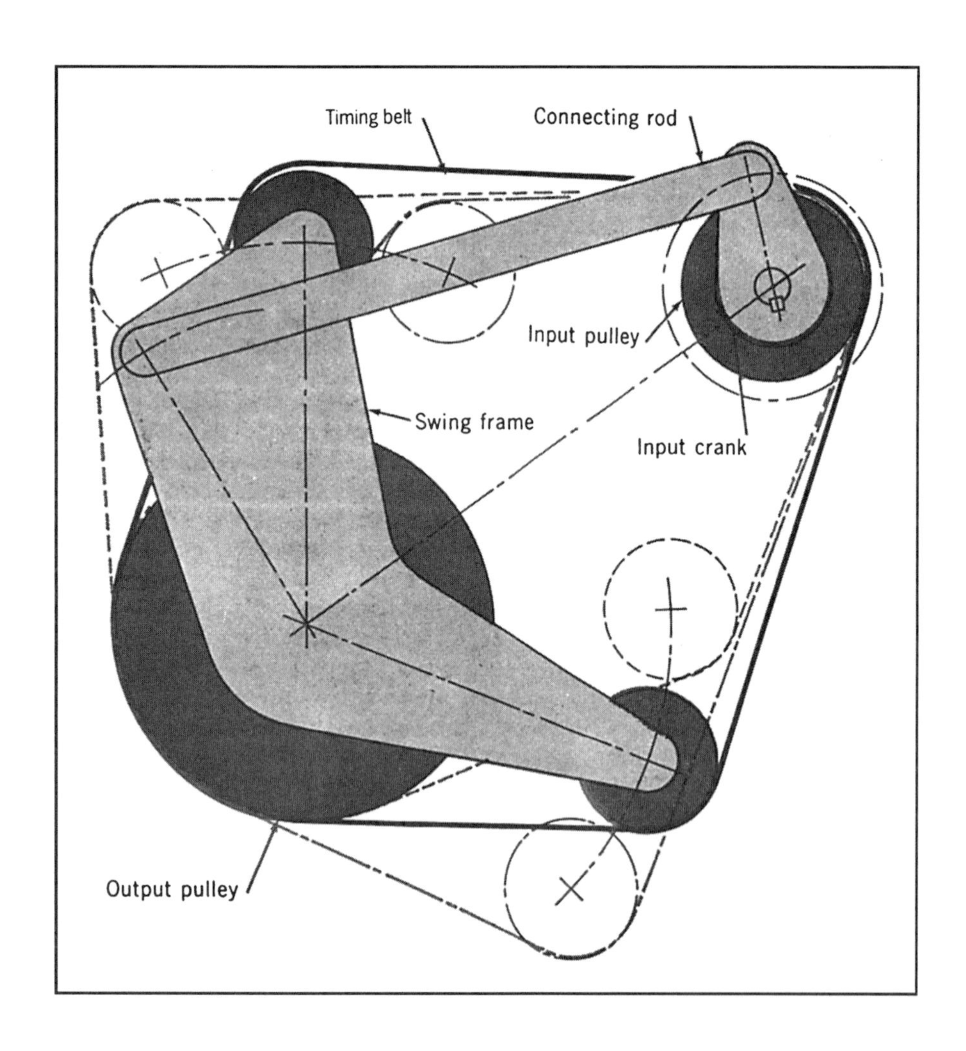

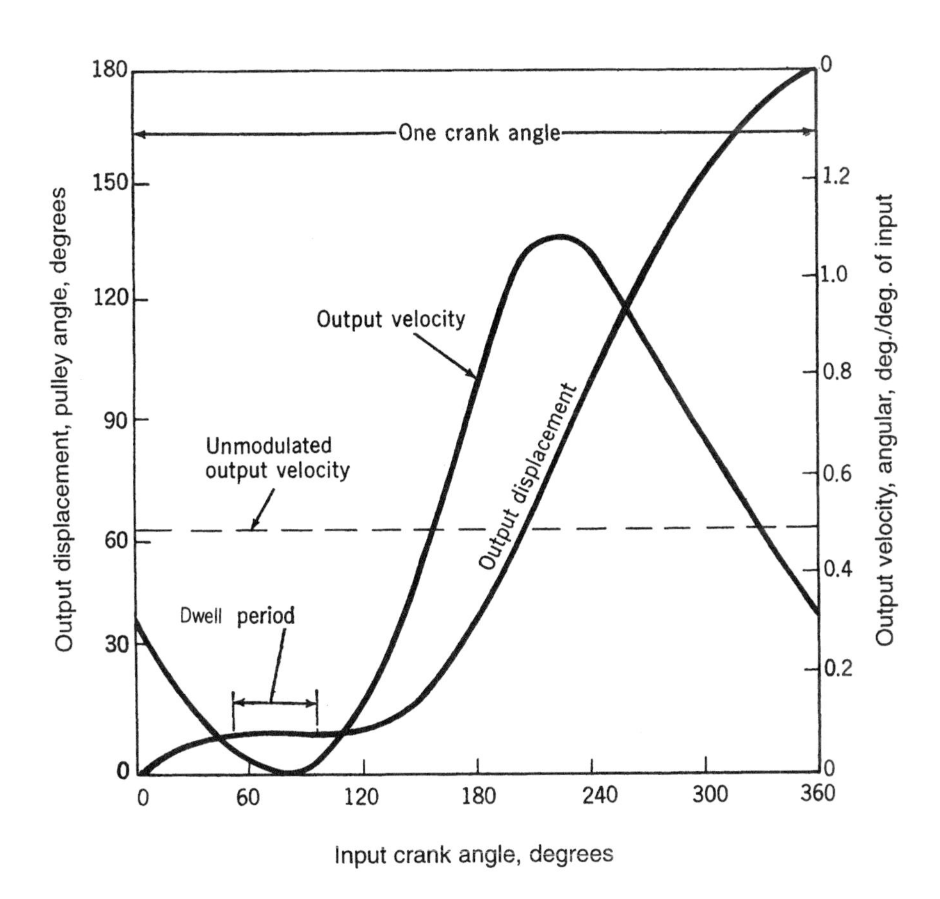

however, must be employed because the perimeter of the pulleys is not a constant figure, so the drive has varying slack built into it.

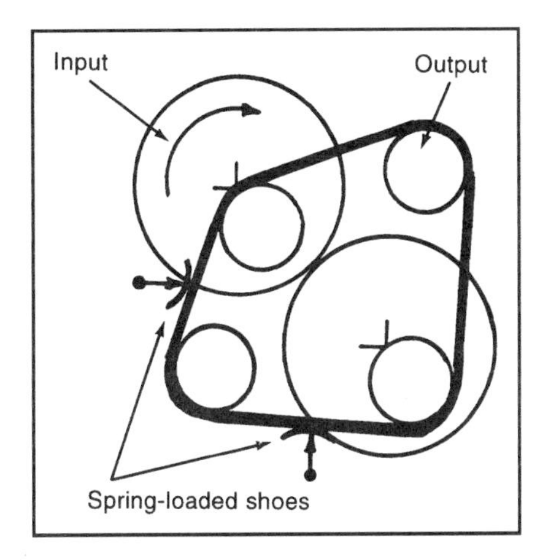

Commercial type. A chain also links the elements of a commercial phase-adjuster drive. A handle is moved to change the phase between the input and output shafts. The theoretical chain length is constant.

In trying to improve this chain device, Scott engineers decided to keep the input and output pulleys at fixed positions and maintain the two idlers on a swing frame. The variation in wraparound length turned out to be surprisingly little, enabling them to install a timing belt without spring-loaded tensioners instead of a chain.

If the swing frame is held in one position, the intermittent mechanism produces a constant-speed output. Shifting the swing frame to a new position automatically shifts the phase relationship between the input and output.

Computer consulted. To obtain intermittent motion, a four-bar linkage is superimposed on the mechanism by adding a crank to the input shaft and a connecting rod to the swing frame. The developers chose an iterative program on a computer to optimize certain variables of the four-bar version.

In the design of one two-stop drive, a dwell period of approximately 50º is obtained. The output displacement moves slowly at first, coming to a "pseudo dwell," in which it is virtually stationary. The output then picks up speed smoothly until almost two-thirds of the input rotation has elapsed (240º). After the input crank completes a full circle of rotation, it continues at a slower rate and begins to repeat its slowdown—dwell—speed-up cycle.

TEN INDEXING AND INTERMITTENT MECHANISMS

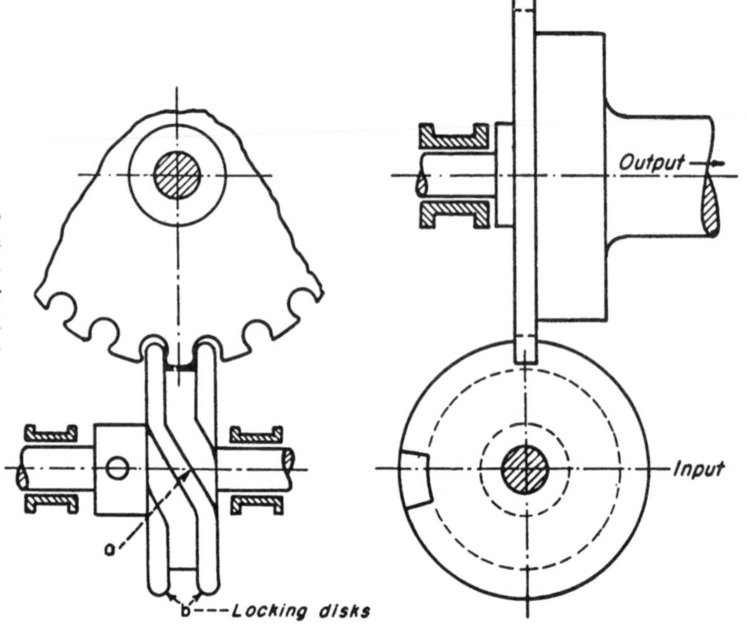

This mechanism transmits intermittent motion between two skewed shafts. The shafts need not be at right angles to one another. Angular displacement of the output shaft per revolution of input shaft equals the circular pitch of the output gear wheel divided by its pitch radius. The duration of the motion period depends on the length of the angular joint *a* of the locking disks *b*.

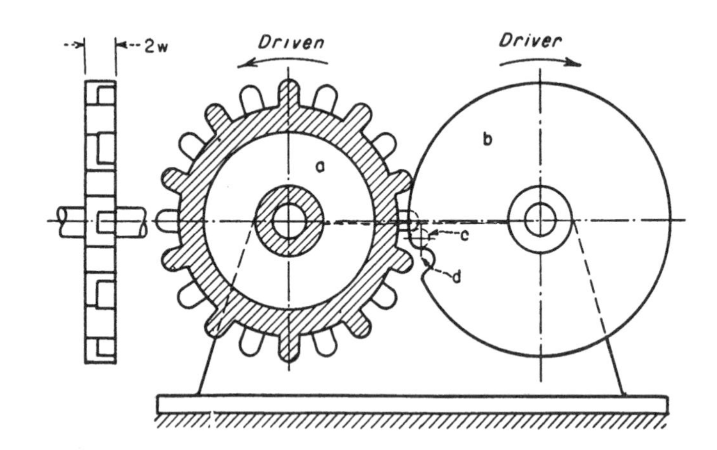

A "mutilated tooth" intermittent drive mechanism. Driver *b* is a circular disk of width *w* with a cutout *d* on its circumference. It carries a pin *c* close to the cutout. The driven gear, *a*, of width 2*w* has an even number of standard spur gear teeth. They alternately have full and half-width (mutilated) teeth. During the dwell period, two full-width teeth are in contact with the circumference of the driving disk, thus locking it. The mutilated tooth between them is behind the driver. AT the end of the dwell period, pin *c* contacts the mutilated tooth and turns the driven gear one circular pitch. Then, the full-width tooth engages the cutout *d*, and the driven gear moves one more pitch. Then the dwell period starts again and the cycle is repeated.

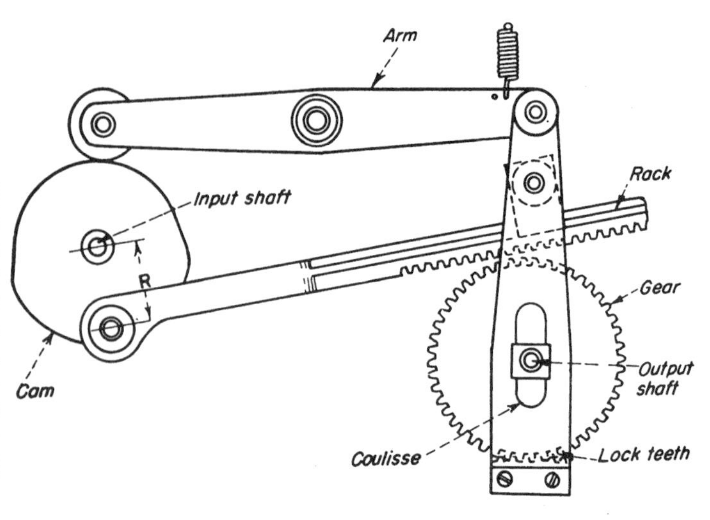

An operating cycle of 180° motion and 180° dwell is produced by this mechanism. The input shaft drives the rack, which is engaged with the output shaft gear during half the cycle. When the rack engages, the lock teeth at the lower end of the coulisse are disengaged and, conversely, when the rack is disengaged, the coulisse teeth are engaged. This action locks the output shaft positively. The changeover points occur at the dead-center positions, so that the motion of the gear is continuously and positively governed. By varying the radius *R* and the diameter of the gear, the number of revolutions made by the output shaft during the operating half of the cycle can be varied to suit many differing requirements.

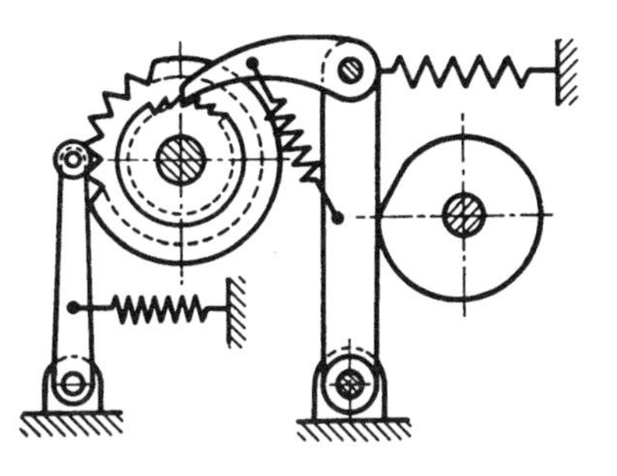

A cam-driven ratchet.

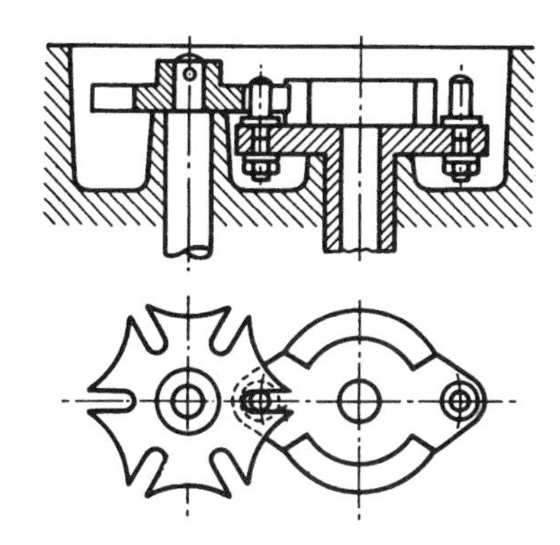

A six-sided Maltese cross and double driver give a 3:1 ratio.

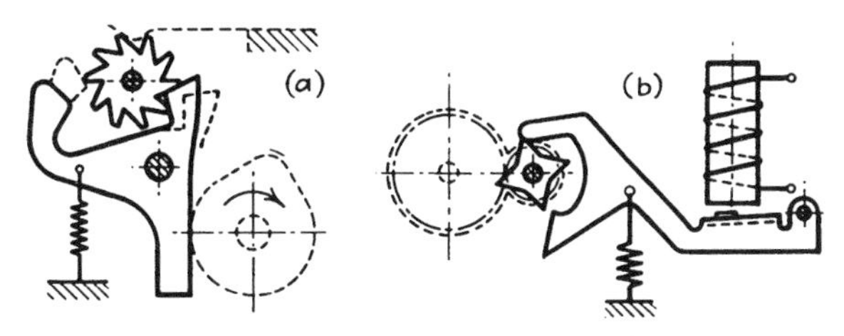

A cam operated escapement on a taximeter (a). A solenoid-operated escapement (b).

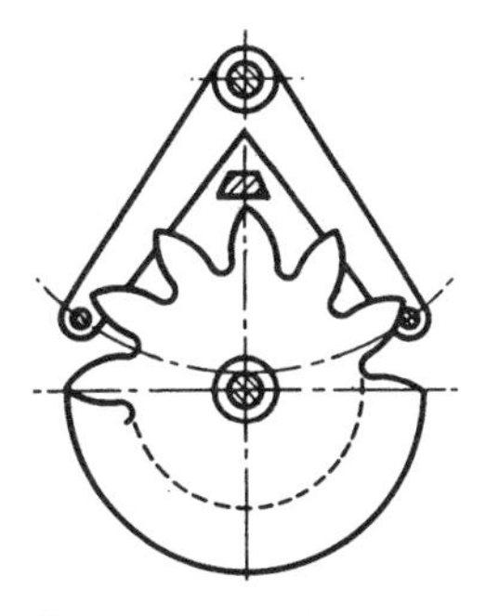

An escapement on an electric meter.

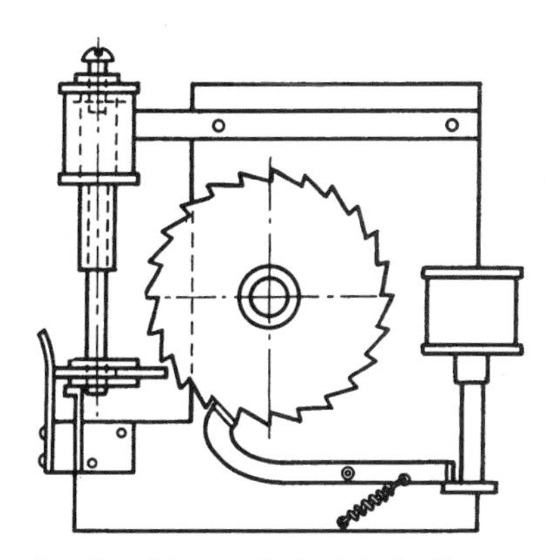

A solenoid-operated ratchet with a solenoid-resetting mechanism A sliding washer engages the teeth.

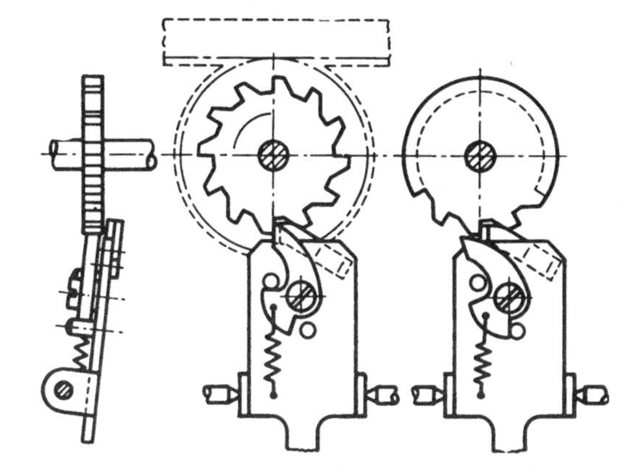

A plate oscillating across the plane of a ratchet-gear escapement carries stationary and spring-held pawls.

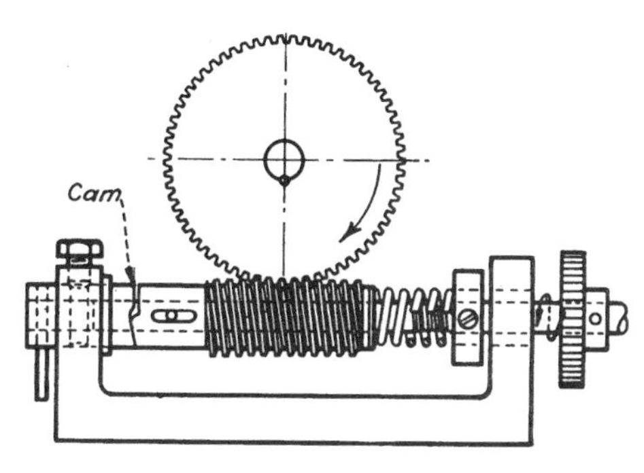

A worm drive, compensated by a cam on a work shaft, produces intermittent motion of the gear.

TWENTY-SEVEN ROTARY-TO-RECIPROCATING MOTION AND DWELL MECHANISMS

Four-bar slider mechanism

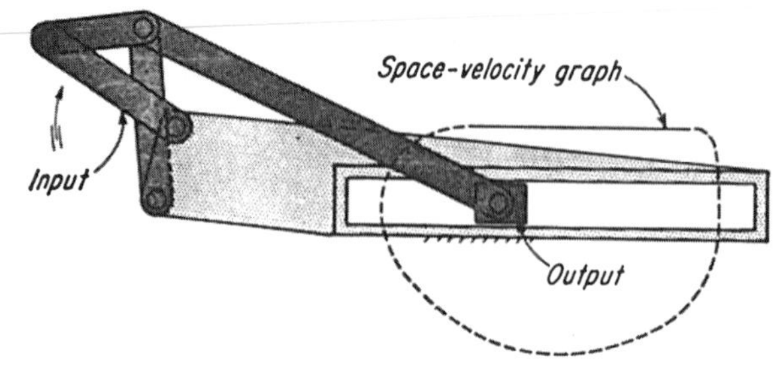

With proper dimensions, the rotation of the input link can impart an almost-constant velocity motion to the slider within the slot.

Oscillating-chain mechanism

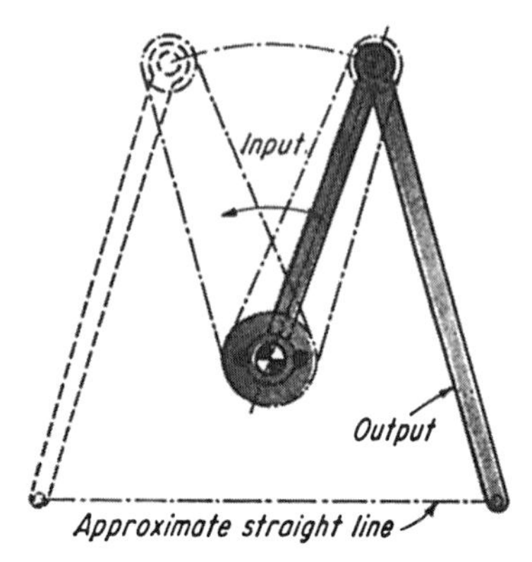

The rotary motion of the input arm is translated into linear motion of the linkage end. The linkage is fixed to the smaller sprocket, and the larger sprocket is fixed to the frame.

Three-gear stroke multiplier mechanism

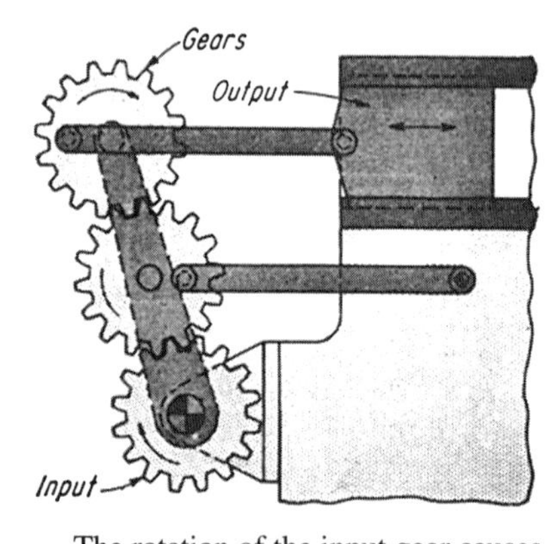

The rotation of the input gear causes the connecting link, attached to the machine frame, to oscillate. This action produces a large-stroke reciprocating motion in the output slider.

Rack and gear sector mechanism

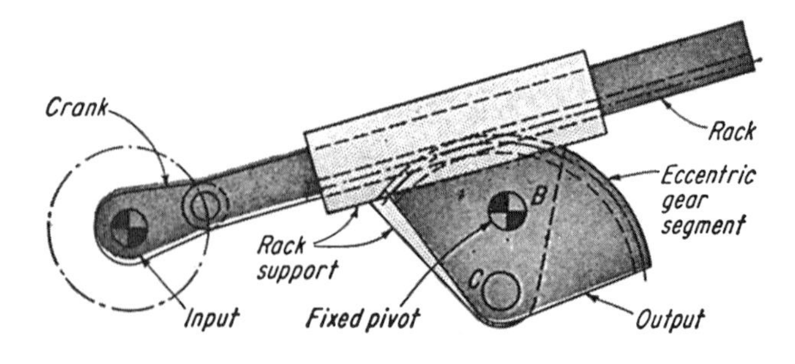

The rotary motion of the input shaft is translated into an oscillating motion of the output gear segment. The rack support and gear sector are pinned at *C* but the gear itself oscillates around *B*.

Linear reciprocator mechanism

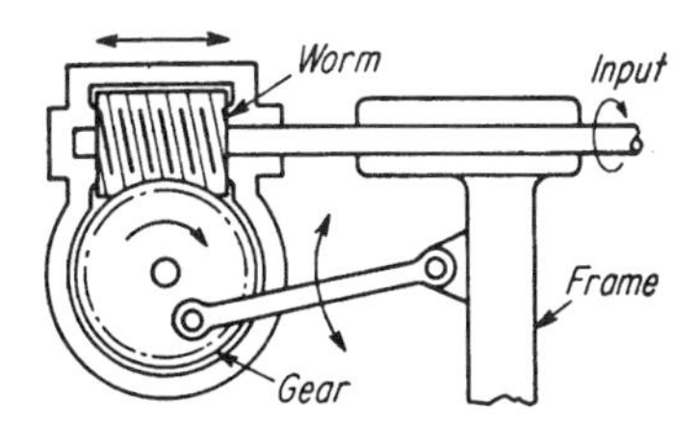

This linear reciprocator converts a rotary motion into a reciprocating motion that is *in line* with the input shaft. Rotation of the shaft drives the worm gear which is attached to the machine frame with a rod. Thus input rotation causes the worm gear to draw itself (and the worm) to the right—thus providing a reciprocating motion.

Disk and roller drive mechanism

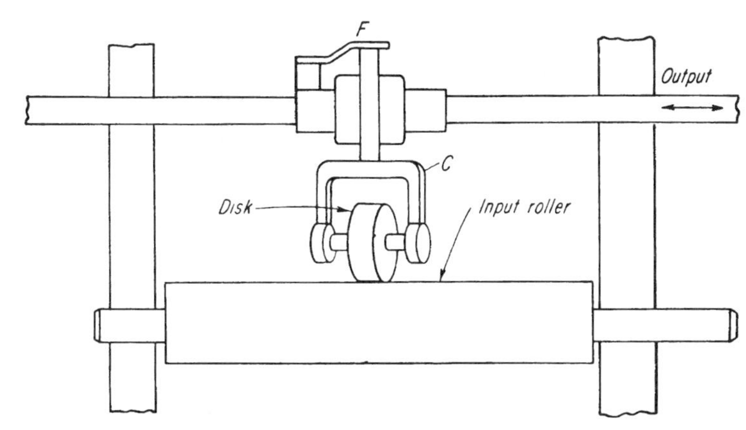

A hardened disk in this drive, riding at an angle to the axis of an input roller, transforms the rotary motion into linear motion parallel to the axis of the input. The roller is pressed against the input shaft by flat spring *F*. The feed rate is easily varied by changing the angle of the disk. This arrangement can produce an extremely slow feed with a built-in safety factor in case of possible jamming.

Bearing and roller drive mechanism

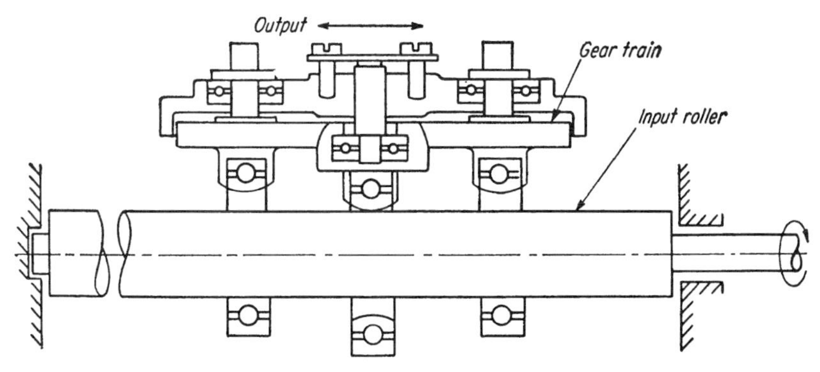

This drive arrangement avoids large Hertzian stresses between the disk and roller by including three ball bearings in place of the single disk. The inner races of the bearings make contact on one side or the other. Hence a gearing arrangement is required to alternate the angle of the bearings. This arrangement also reduces the bending moment on the shaft.

Reciprocating space crank mechanism

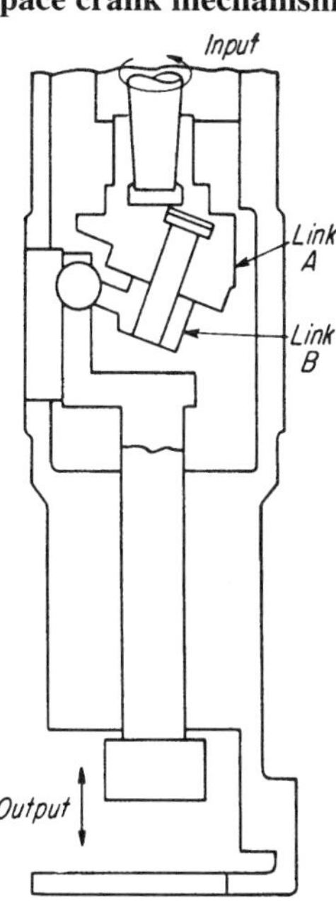

The rotary input of this crank causes the bottom surface of link *A* to wobble with respect to the center link. Link *B* is free of link *A*, but it is restrained from rotating by the slot. This causes the output member to reciprocate linearly.

Oscillating crank and planetary drive mechanism

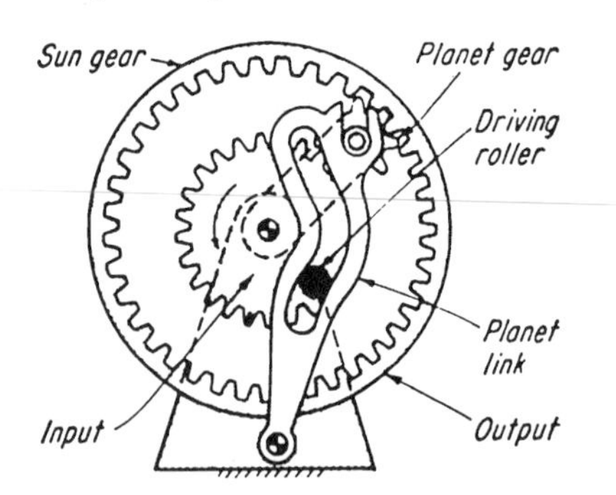

The planet gear is driven with a stop-and-go motion. The driving roller is shown entering the circular-arc slot on the planet link. The link and the planet remain stationary while the roller travels along this section of the slot. As a result, the output sun gear has a rotating output motion with a progressive oscillation.

Chain-slider drive mechanism

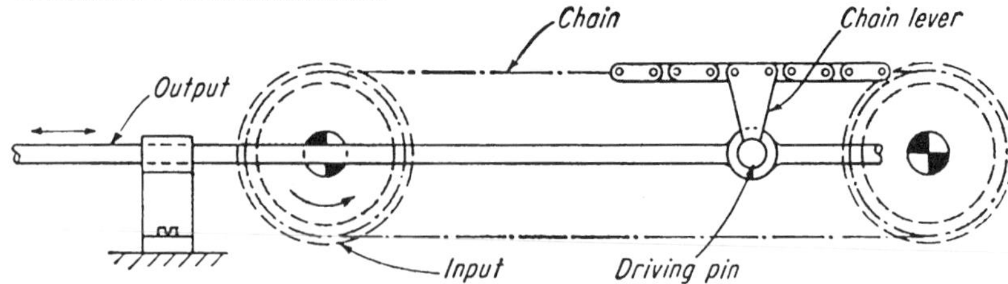

The chain link drives a lever that oscillates. A slowdown-dwell occurs when the chain pin passes around the left sprocket.

The output shaft reciprocates with a constant velocity, but it reaches a long dwell at both ends as the chain lever, whose length is equal to the radius of the sprockets, goes around both sprockets.

Chain-oscillating drive mechanism

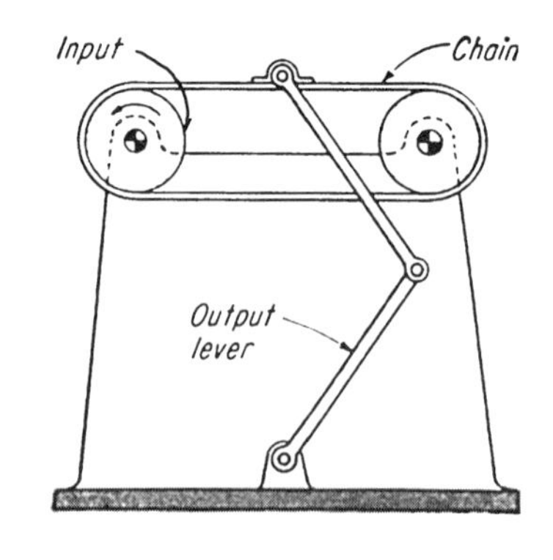

Chain and slider dirve mechanism

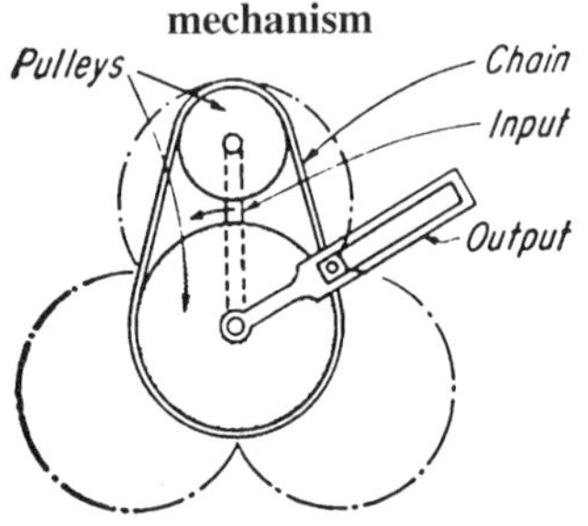

The input crank causes the small pulley to orbit around the stationary larger pulley. A pivot point attached to the chain slides inside the slot of the output link. In the position shown, the output is about to start a long dwell period of about 120°.

Epicyclic dwell mechanism

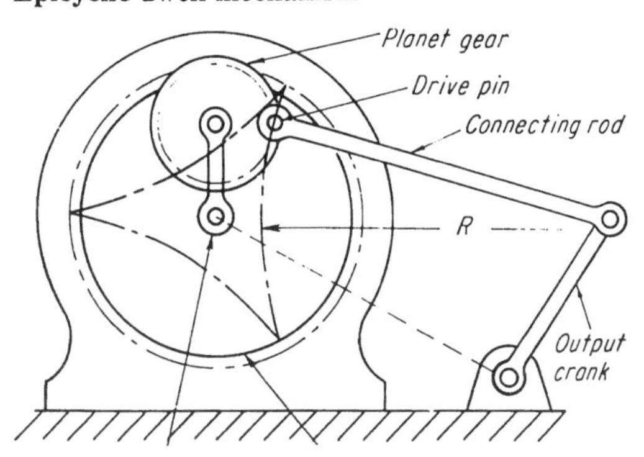

The output crank pulsates back and forth with a long dwell at its extreme right position. The input shaft rotates the planet gear with a crank. The pin on the planet gear traces the epicyclic three-lobe curve shown. The right side of the curve is a near circular arc of radius R. If the connecting rod length equals R, the output crank reaches a virtual standstill during a third of the total rotation of the input crank. The crank then reverses, stops at its left position, reverses, and repeats its dwell.

Cam-worm dwell mechanism

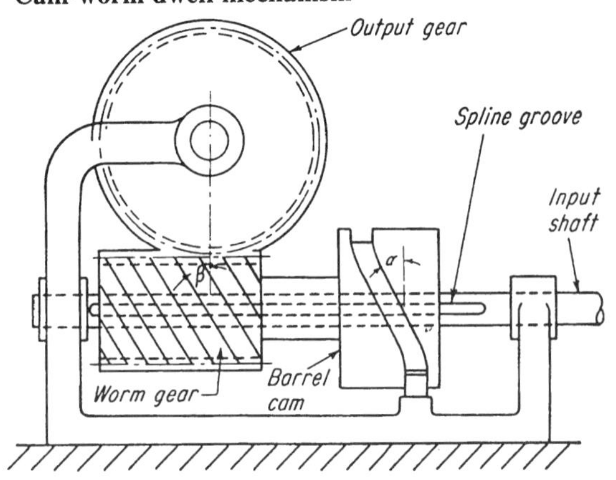

Without the barrel cam, the input shaft would drive the output gear by the worm gear at constant speed. The worm and the barrel cam, however, can slide linearly on the input shaft. The rotation of the input shaft now causes the worm gear to be cammed back and forth, thus adding or subtracting motion to the output. If barrel cam angle α is equal to the worm angle β, the output stops during the limits of rotation shown. It then speeds up to make up for lost time.

Cam-helical dwell mechanism

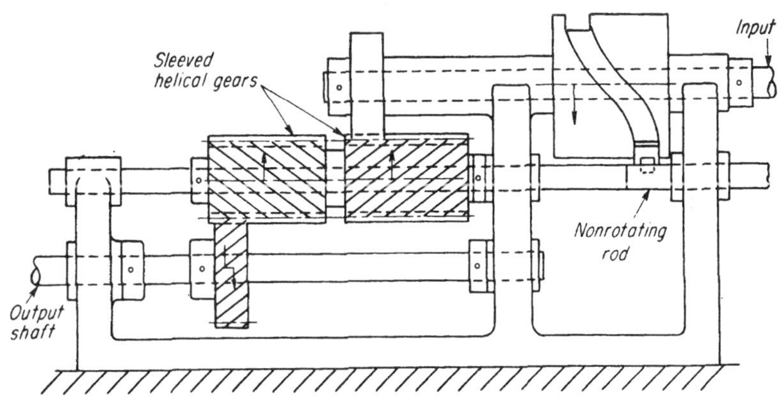

When one helical gear is shifted linearly (but prevented from rotating) it will impart rotary motion to the mating gear because of the helix angle. This principle is applied in the mechanism illustrated. The rotation of the input shaft causes the intermediate shaft to shift to the left, which in turn adds or subtracts from the rotation of the output shaft.

Six-bar dwell mechanism

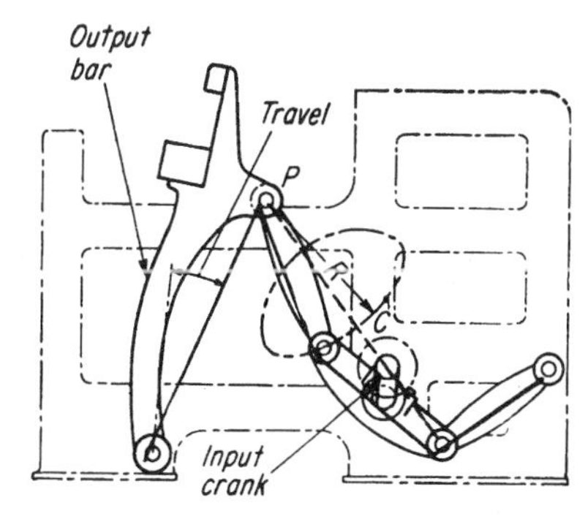

The rotation of the input crank causes the output bar to oscillate with a long dwell at its extreme right position. This occurs because point C describes a curve that is approximately a circular arc (from C to C' with its center at P. The output is almost stationary during that part of the curve.

Cam-roller dwell mechanism

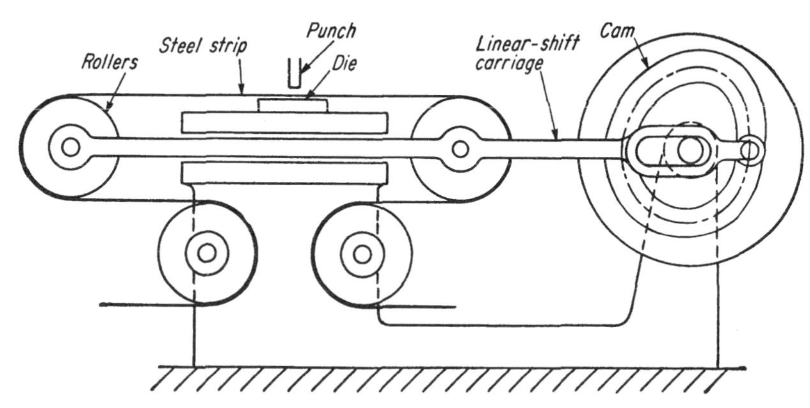

A steel strip is fed at constant linear velocity in this mechanism. But at the die station (illustrated), it is desired to stop the strip so that the punching operation can be performed. The strip passes over movable rollers which, when shifted to the right, cause the strip to move to the right. Since the strip is normally fed to the left, proper design of the cam can nullify the linear feed rates so that the strip stops, and then speeds to catch up to the normal rate.

Three-gear drive mechanism

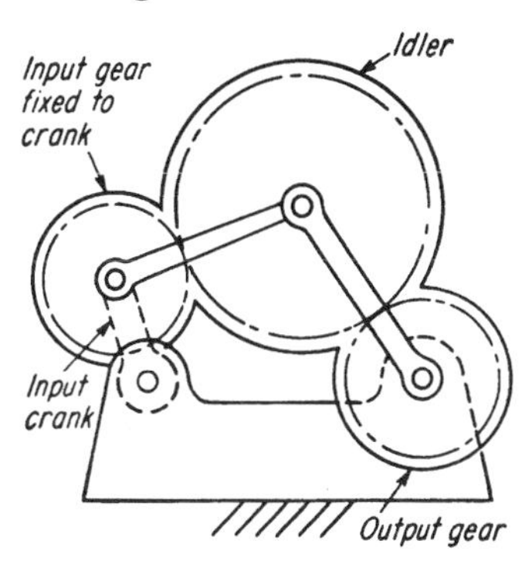

This is actually a four-bar linkage combined with three gears. As the input crank rotates, it turns the input gear which drives the output gear through the idler. Various output motions are possible. Depending on the relative diameters of the gears, the output gear can pulsate, reach a short dwell, or even reverse itself briefly.

Double-crank dwell mechanism

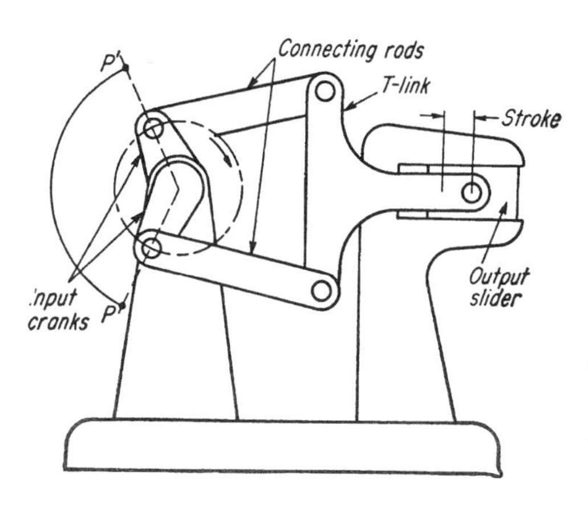

Both cranks are connected to a common shaft which also acts as the input shaft. Thus the cranks always remain a constant distance apart from each other. There are only two frame points—the center of the input shaft and the guide for the output slider. As the output slider reaches the end of its stroke (to the right), it remains at a virtual standstill while one crank rotates through angle PP'.

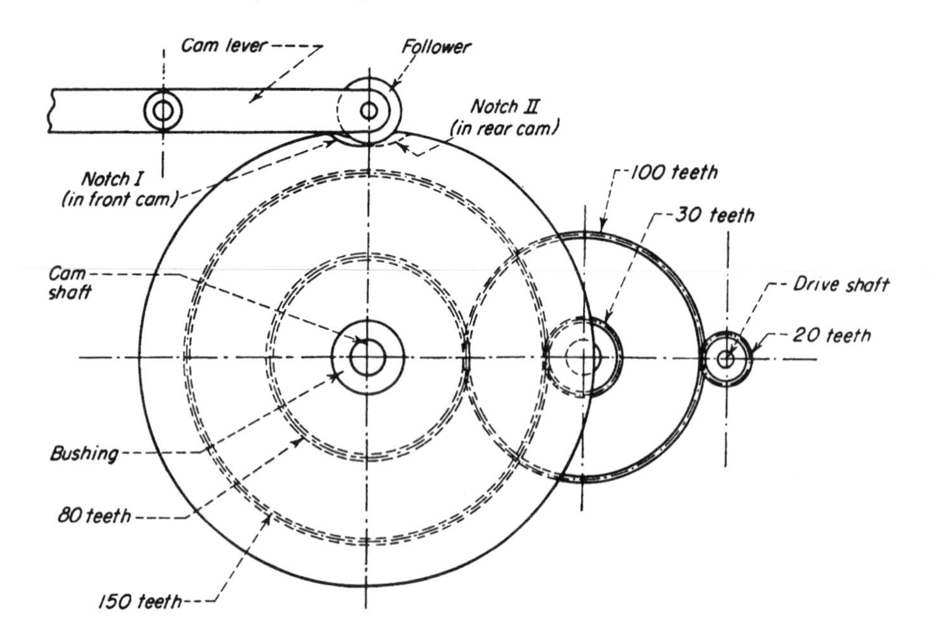

Fast Cam-Follower Motion

Fast cam action every n cycles (where n is a relatively large number) can be obtained with this manifold cam and gear mechanism. A single notched cam geared $1/n$ to a shaft turning once per cycle moves relatively slowly under the follower. The double notched-cam arrangement shown is designed to operate the lever once in 100 cycles, imparting a rapid movement to it. One of the two identical cams and the 150-tooth gear are keyed to the bushing which turns freely around the cam shaft. The cam shaft carries the second cam and the 80-tooth gear. The 30- and 100-tooth gears are integral, while the 20-tooth gear is attached to the one-cycle drive shaft. One of the cams turns in the ratio of 20/80 or 1/4; the other turns in the ratio 20/100 times 30/150 or 1/25. The notches therefore coincide once every 100 cycles (4 × 25). Lever movement is the equivalent of a cam turning in a ratio of 1 to 4 in relation to the drive shaft. To obtain fast cam action, n must be reduced to prime factors. For example, if 100 were factored into 5 and 20, the notches would coincide after every 20 cycles.

Intermittent Motion

This mechanism can be adapted to produce a stop, a variable speed without stop, or a variable speed with momentary reverse motion. A uniformly rotating input shaft drives the chain around the sprocket and idler. The arm serves as a link between the chain and the end of the output shaft crank. The sprocket drive must be in the ratio N/n with the cycle of the machine, where n is the number of teeth on the sprocket and N the number of links in the chain. When point P travels around the sprocket from point A to position B, the crank rotates uniformly. Between B and C, P decelerates; between C and A it accelerates; and at C there is a momentary dwell By changing the size and position of the idler, or the lengths of the arm and crank, a variety of motions can be obtained. If the length of the crank is shortened, a brief reverse period will occur in the vicinity of C; if the crank is lengthened, the output velocity will vary between a maximum and minimum without reaching zero.

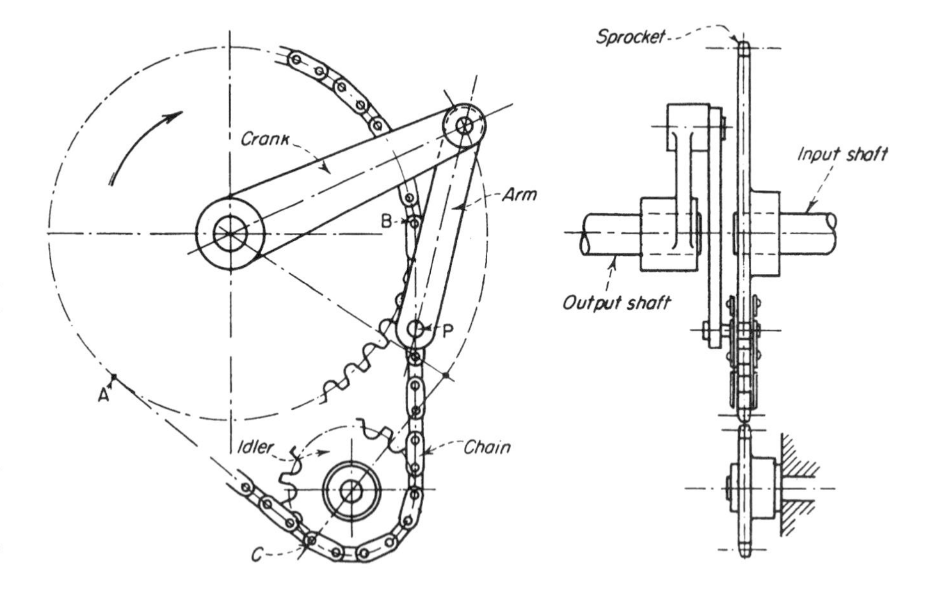

Gear-slider crank mechanism

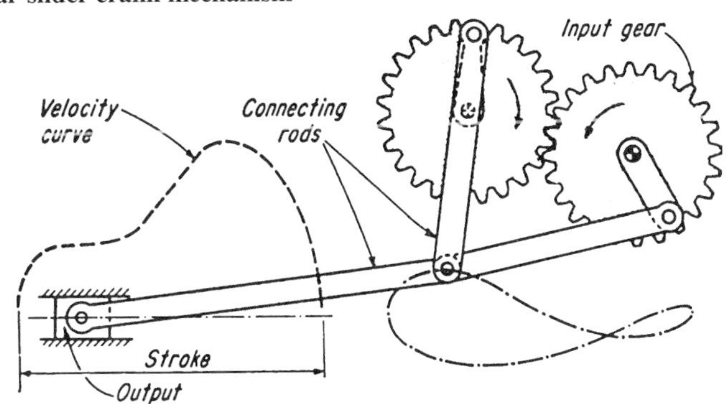

The input shaft drives both gears which, in turn, drive the connecting rods to produce the velocity curve shown. The piston moves with a low constant velocity.

Gear oscillating crank mechanism

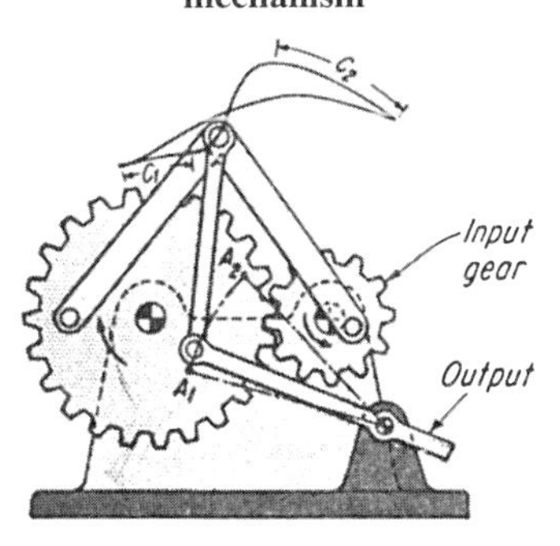

In this arrangement, the curve described by the pin connection has two parts, C_1 and C_2, which are very close to circular arc with its centers at A_1 and A_2. Consequently the driven link will have a dwell at both of its extreme positions.

Curve-slider drive mechanism

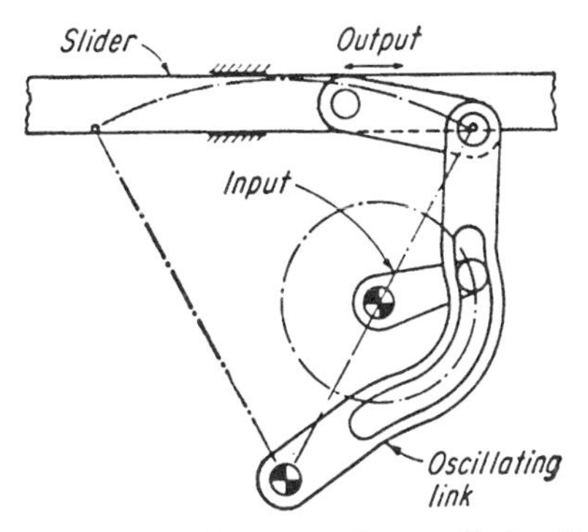

The circular arc on the oscillating link permits the link to reach a dwell during the right position of the output slider.

Triple-harmonic drive mechanism

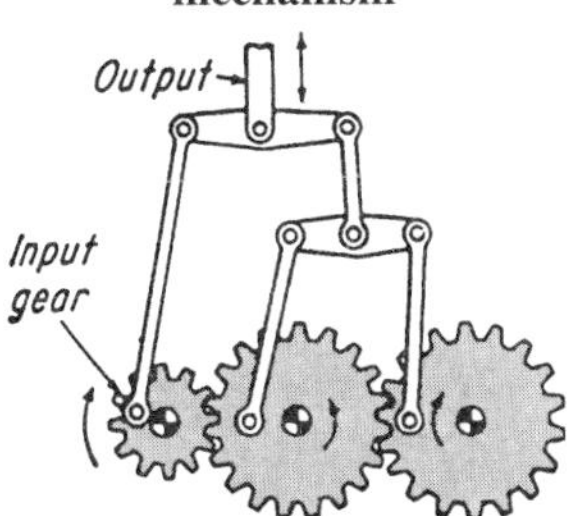

The input shaft drives three gears with connecting rods. A wide variety of reciprocating output motions can be obtained by selecting different lengths for the linkages. In addition, one to several dwells can be obtained per cycle.

Whitworth quick-return drive mechanism

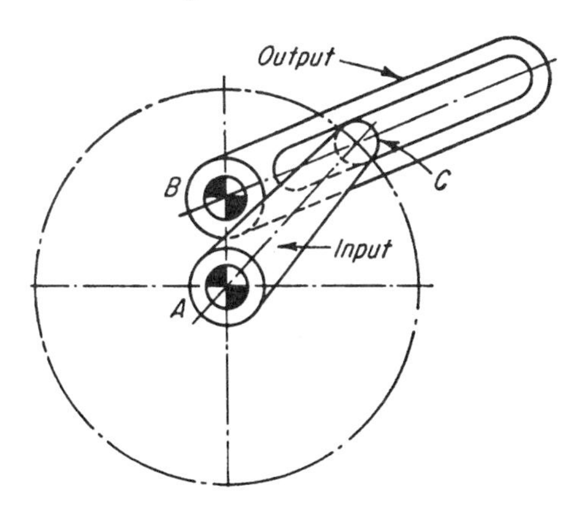

Varying motion can be imparted simply to output shaft B. However, the axes, A and B, are not colinear.

Wheel and slider drive mechanism

For each revolution of the input disk, the slider moves in to engage the wheel and index it one tooth width. A flat spring keeps the wheel locked while it is stationary.

FIVE FRICTION MECHANISMS FOR INTERMITTENT ROTARY MOTION

Friction mechanisms do not have the disadvantages inherent in conventional pawl and ratchet drives such as: (1) noisy operation; (2) backlash needed for pawl engagement; (3) load concentrated on one tooth of the ratchet; and (4) pawl engagement dependent on an external spring. Each of the five mechanisms presented here converts the reciprocating motion of a connecting rod into an intermittent rotary motion. The connecting rod stroke to the left drives a shaft counterclockwise and that shaft is uncoupled. It remains stationary during the return stroke of the connecting rod to the right.

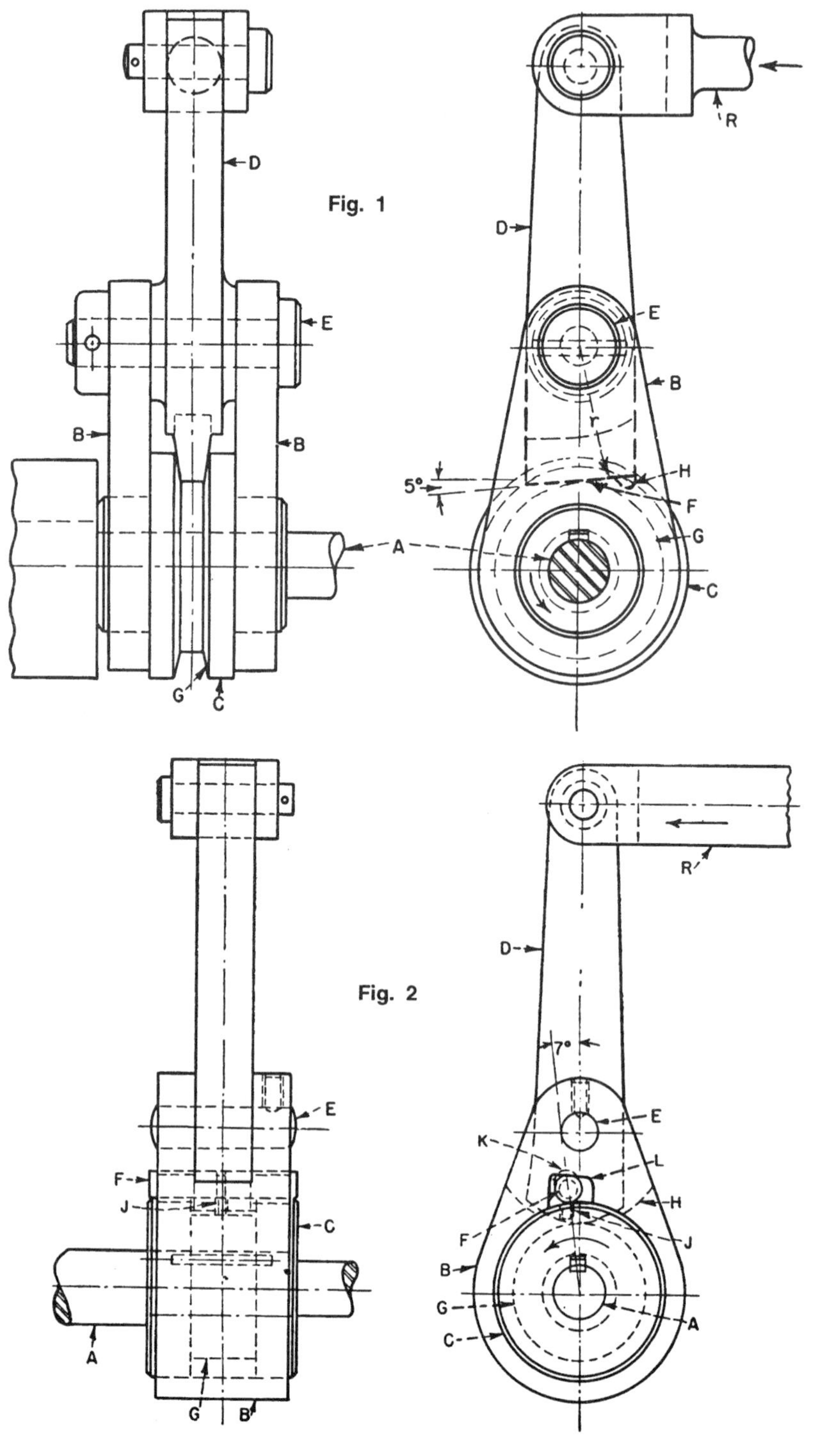

Fig. 1 The wedge and disk mechanism consists of shaft *A* supported in bearing block *J*; ring *C* is keyed to *A* and it contains an annular groove *G*; body *B*, which can pivot around the shoulders of *C*; lever *D*, which can pivot about *E*; and connecting rod *R*, which is driven by an eccentic (not shown). Lever *D* is rotated counter-clockwise about *E* by the connecting rod moving to the left until surface *F* wedges into groove *G*. Continued rotation of *D* causes *A*, *B*, and *D* to rotate counterclockwise as a unit about *A*. The reversal of input motion instantly swivels *F* out of *G*, thus unlocking the shaft, which remains stationary during its return stroke because of friction induced by its load. As *D* continues to rotate clockwise about *E*, node *H*, which is hardened and polished to reduce friction, bears against the bottom of *G* to restrain further swiveling. Lever *D* now rotates with *B* around *A* until the end of the stroke.

Fig. 2 The pin and disk mechanism: Lever *D*, which pivots around *E*, contains pin *F* in an elongated hole *K*. The hole permits slight vertical movement of the pin, but set screw *J* prevents horizontal movement. Body *B* can rotate freely about shaft *A*. Cut-outs *L* and *H* in body *B* allow clearances for pin *F* and lever *D*, respectively. Ring *C*, which is keyed to shaft *A*, has an annular groove *G* to permit clearance for the tip of lever *D*. Counterclockwise motion of lever *D*, actuated by the connecting rod, jams a pin between *C* and the top of cut-out *L*. This occurs about 7º from the vertical axis. *A*, *B*, and *D* are now locked together and rotate about *A*. The return stroke of *R* pivots *D* clockwise around *E* and unwedges the pin until it strikes the side of *L*. Continued motion of *R* to the right rotates *B* and *D* clockwise around *A*, while the uncoupled shaft remains stationary because of its load.

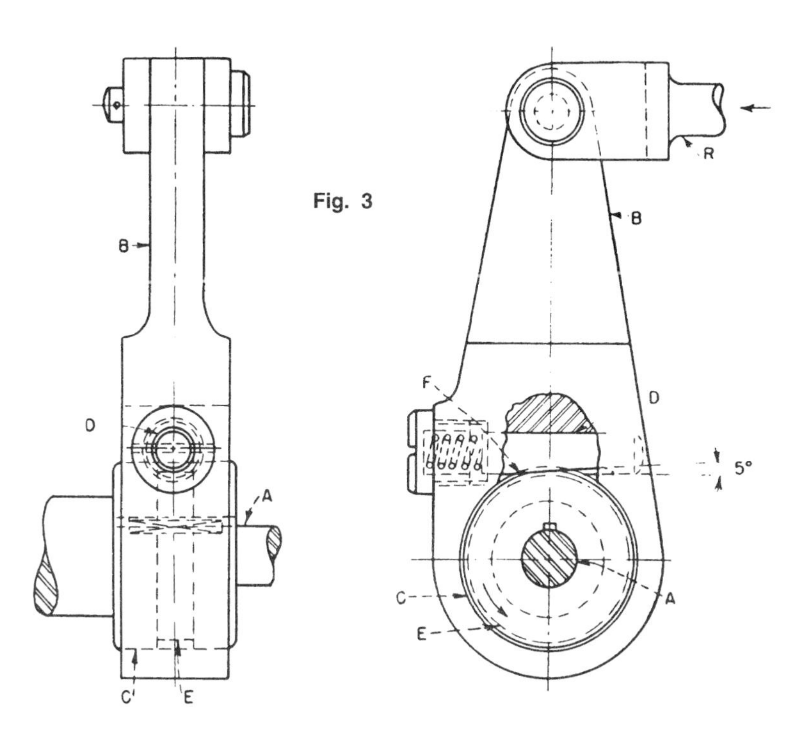

Fig. 3 The sliding pin and disk mechanism: The counterclockwise movement of body *B* about shaft *A* draws pin *D* to the right with respect to body *B*, aided by spring pressure, until the flat bottom *F* of the pin is wedged against the annular groove *E* of ring *C*. The bottom of the pin is inclined about 5º for optimum wedging action. Ring *C* is keyed to *A*, and parts, *A*, *C*, *D* and *B* now rotate counterclockwise as a unit until the end of the connecting rod's stroke. The reversal of *B* draws the pin out of engagement so that *A* remains stationary while the body completes its clockwise rotation.

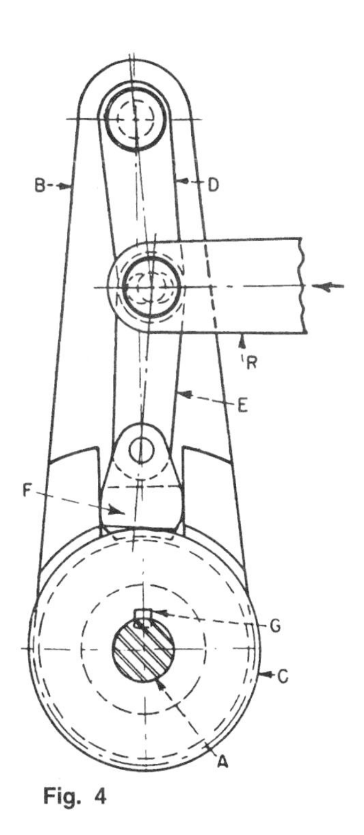

Fig. 4

Fig. 4 The toggle link and disk mechanism: The input stroke of connecting rod *R* (to the left) wedges block *F* in groove *G* by straightening toggle links *D* and *E*. Body *B*, toggle links, and ring *C*, which is keyed to shaft *A*, rotate counterclockwise together about *A* until the end of the stroke. The reversal of connecting rod motion lifts the block, thus uncoupling the shaft, while body *B* continues clockwise rotation until the end of stroke.

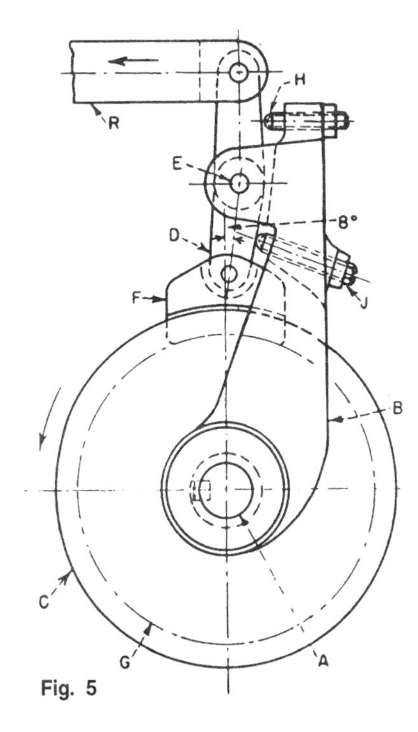

Fig. 5

Fig. 5 The rocker arm and disk mechanism: Lever *D*, activated by the reciprocating bar *R* moving to the left, rotates counterclockwise on pivot *E*, thus wedging block *F* into groove *G* of disk *C*. Shaft *A* is keyed to *C* and rotates counterclockwise as a unit with body *B* and lever *D*. The return stroke of *R* to the right pivots *D* clockwise about *E* and withdraws the block from the groove so that shaft is uncoupled while *D*, striking adjusting screw *H*, travels with *B* about *A* until the completion of stroke. Adjusting screw *J* prevents wedging block *F* from jamming in the groove.

NINE DIFFERENT BALL SLIDES FOR LINEAR MOTION

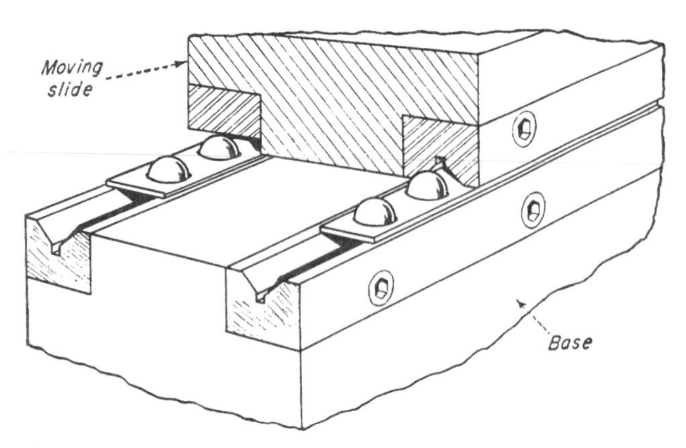

Fig. 1 V-grooves and flat surface make a simple horizontal ball slide for reciprocating motion where no side forces are present and a heavy slide is required to keep the balls in continuous contact. The ball cage ensures the proper spacing of the balls and its contacting surfaces are hardened and lapped.

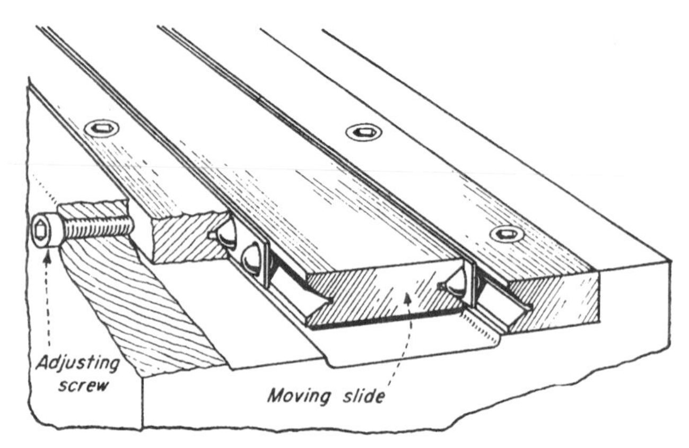

Fig. 2 Double V grooves are necessary where the slide is in a vertical position or when transverse loads are present. Screw adjustment or spring force is required to minimize any looseness in the slide. Metal-to-metal contact between the balls and grooves ensure accurate motion.

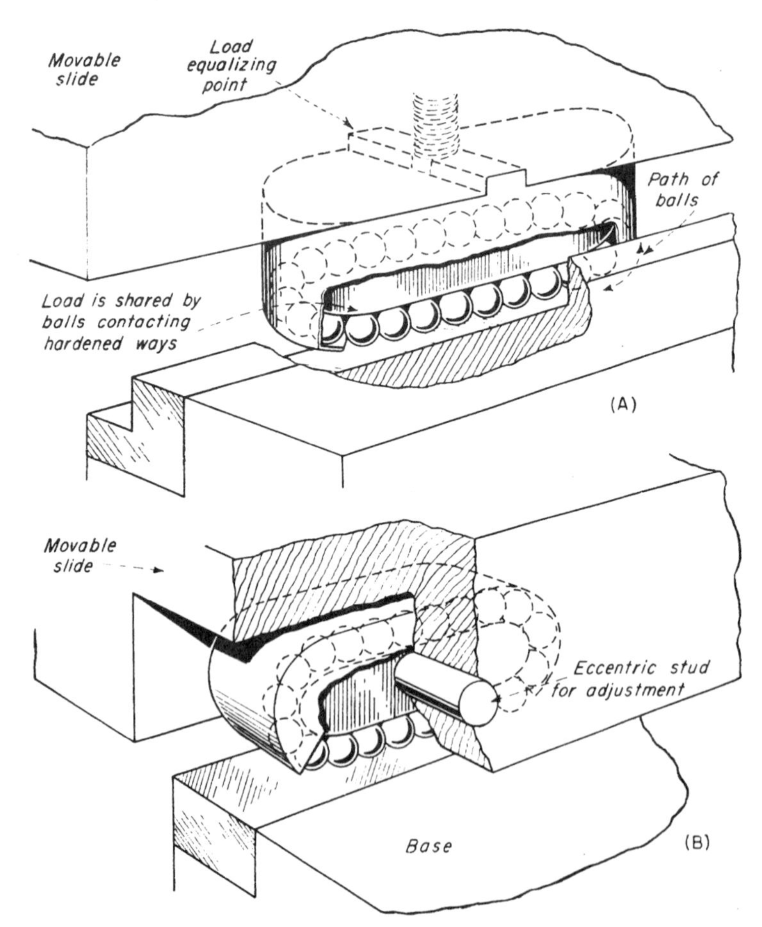

Fig. 3 The ball cartridge has the advantage of unlimited travel because the balls are free to recirculate. Cartridges are best suited for vertical loads. (A) Where lateral restraint is also required, this type is used with a side preload. (B) For flat surfaces the cartridge is easily adjusted.

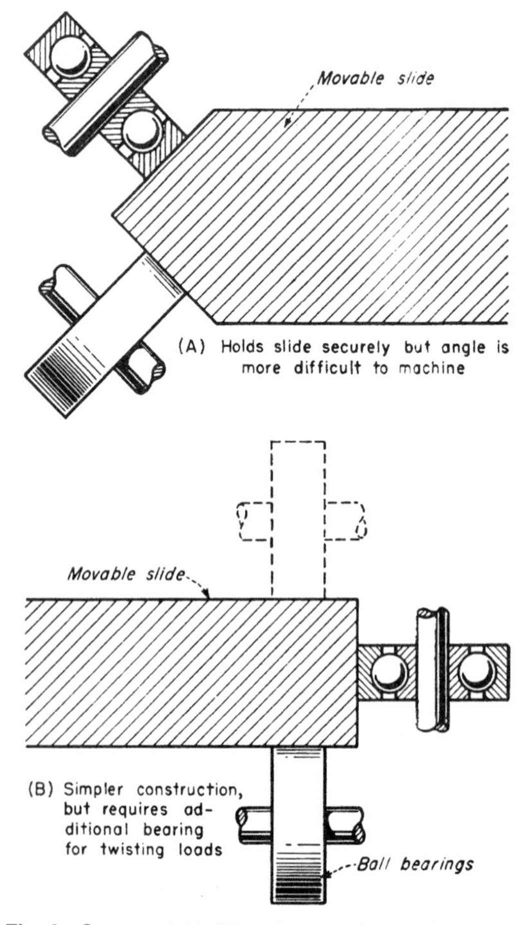

Fig. 4 Commercial ball bearings can be used to make a reciprocating slide. Adjustments are necessary to prevent looseness of the slide. (A) Slide with beveled ends, (B) Rectangular-shaped slide.

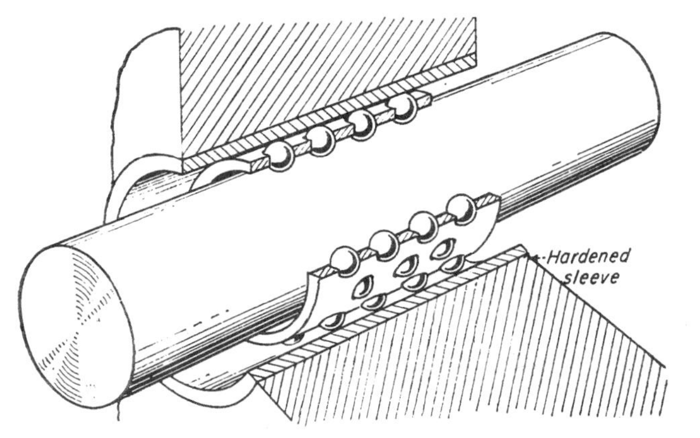

Fig. 5 This sleeve bearing, consisting of a hardened sleeve, balls, and retainer, can be used for reciprocating as well as oscillating motion. Travel is limited in a way similar to that of Fig. 6. This bearing can withstand transverse loads in any direction.

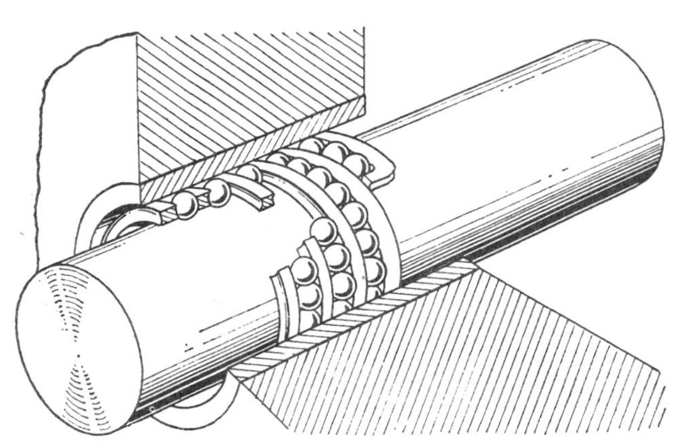

Fig. 6 This ball reciprocating bearing is designed for rotating, reciprocating or oscillating motion. A formed-wire retainer holds the balls in a helical path. The stroke is about equal to twice the difference between the outer sleeve and the retainer length.

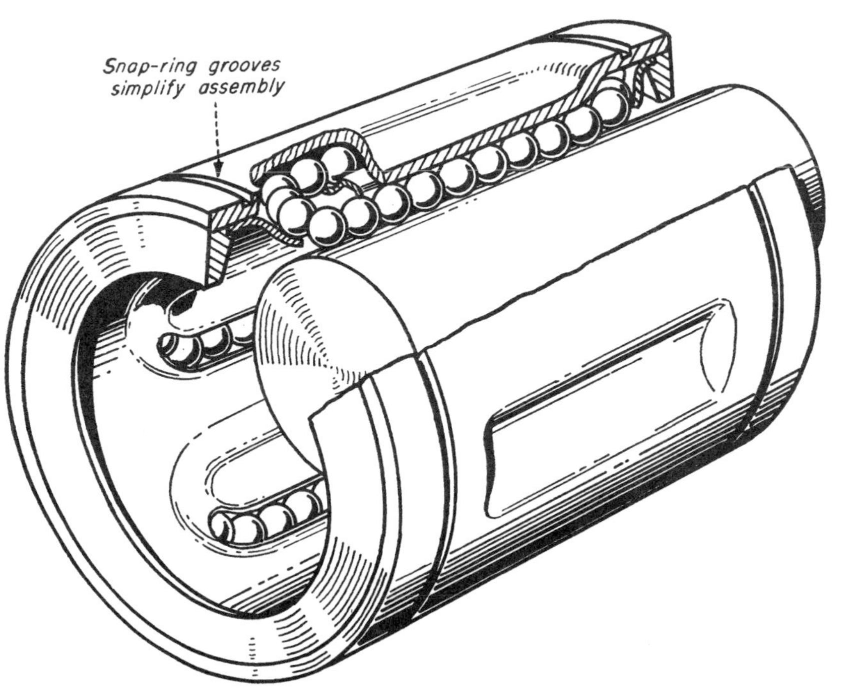

Fig. 7 This ball bushing has several recirculating systems of balls that permit unlimited linear travel. Very compact, this bushing requires only a bored hole for installation. For maximum load capacity, a hardened shaft should be used.

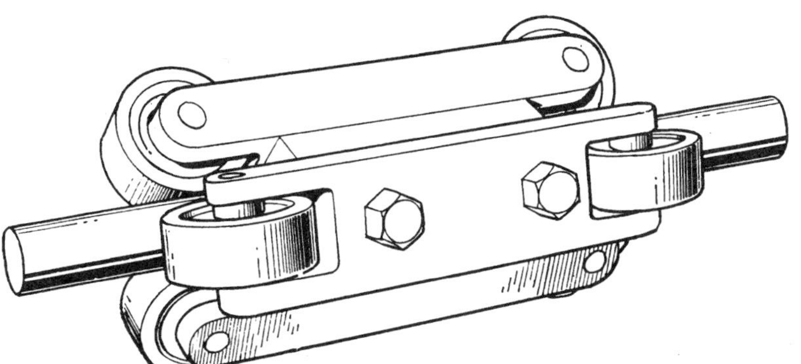

Fig. 8 Cylindrical shafts can be held by commercial ball bearings that are assembled to make a guide. These bearings must be held tightly against the shaft to prevent any looseness.

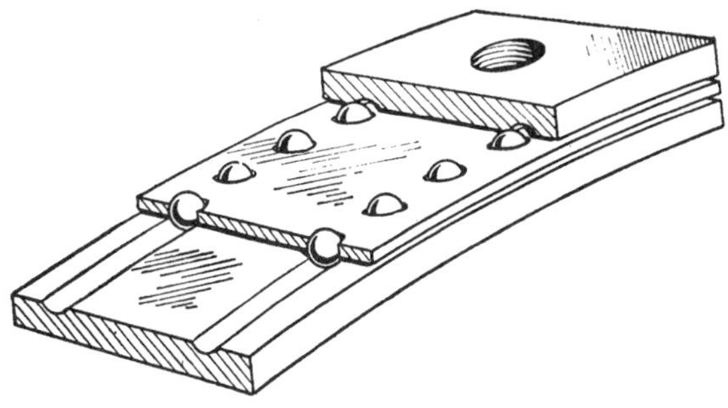

Fig. 9 Curvilinear motion in a plane is possible with this device when the radius of curvature is large. However, uniform spacing between its grooves is important. Circular-sectioned grooves decrease contact stresses.

BALL-BEARING SCREWS CONVERT ROTARY TO LINEAR MOTION

This cartridge-operated rotary actuator quickly retracts the webbing to separate a pilot forcibly from his seat as the seat is ejected in emergencies. It eliminates the tendency of both pilot and seat to tumble together after ejection, preventing the opening of the chute. Gas pressure from the ejection device fires the cartridge in the actuator to force the ball-bearing screw to move axially. The linear motion of the screw is translated into the rotary motion of a ball nut. This motion rapidly rolls up the webbing (stretching it as shown) so that the pilot is snapped out of his seat.

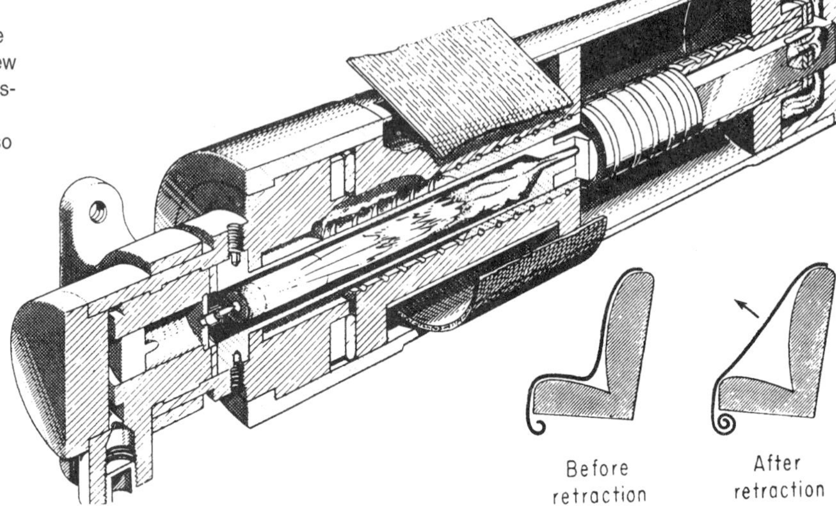

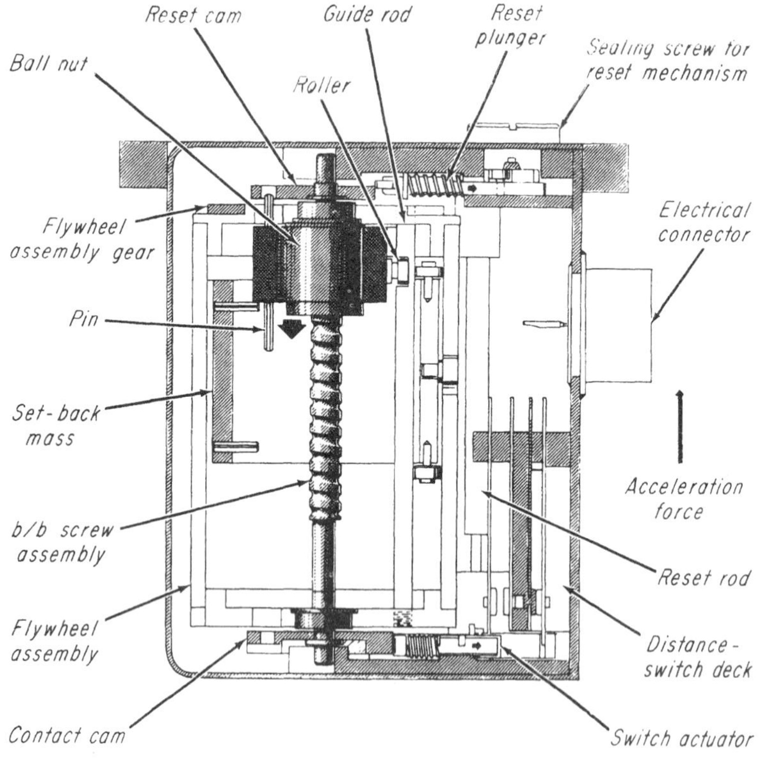

This time-delay switching device integrates a time function with a missile's linear travel. Its purpose is to arm the warhead safely. A strict "minimum G-time" system might arm a slow missile too soon for the adequate protection of friendly forces because a fast missile might arrive before the warhead is fused. The weight of the nut plus the inertia under acceleration will rotate the ball-bearing screw which has a flywheel on its end. The screw pitch is selected so that the revolutions of the flywheel represent the distance the missile has traveled.

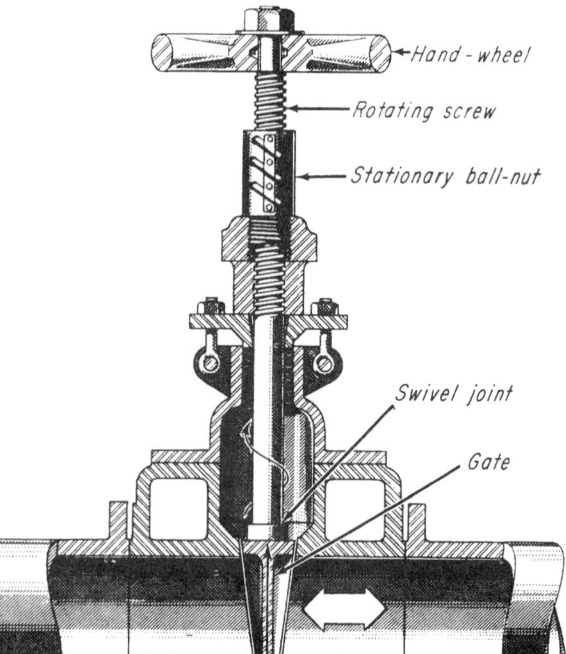

Fast, easy, and accurate control of fluid flow through a valve is obtained by the rotary motion of a screw in the stationary ball nut. The screw produces linear movement of the gate. The swivel joint eliminates rotary motion between the screw and the gate.

NINETEEN ARRANGEMENTS FOR CHANGING LINEAR MOTION

These arrangements of linkages, slides, friction drives, gears, cams, pistons, and solenoids permit linear motion to be changed.

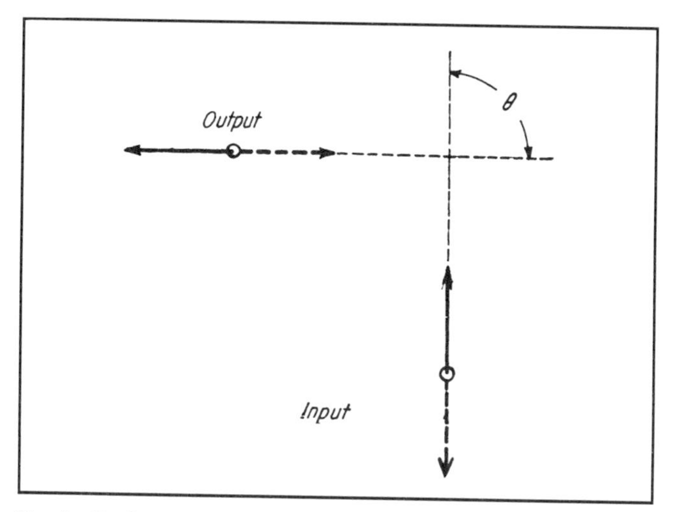

Fig. 1 Basic problem (θ is generally close to 90°).

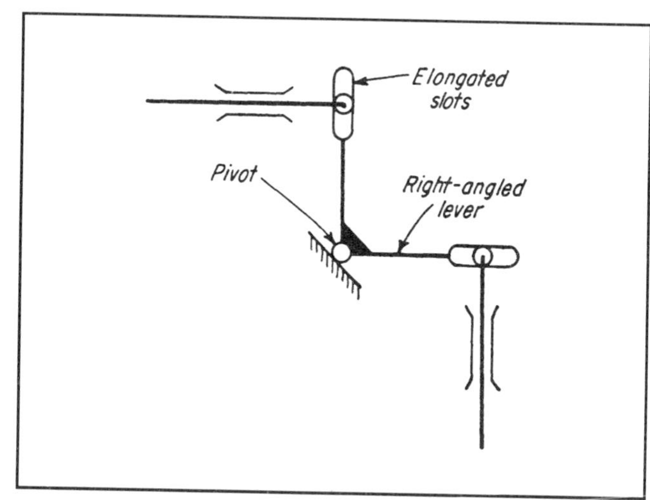

Fig. 2 Slotted lever.

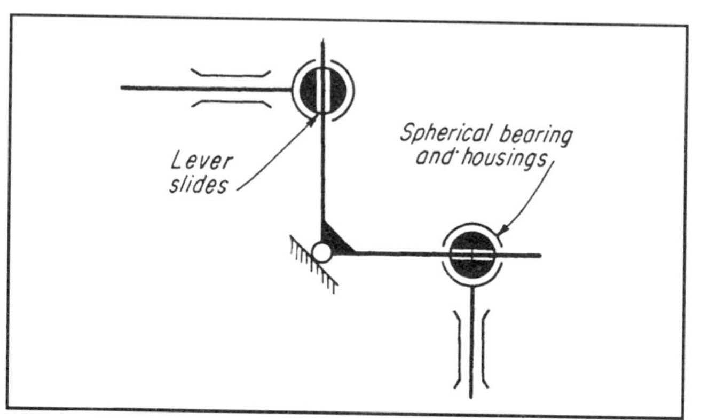

Fig. 3 Spherical bearings.

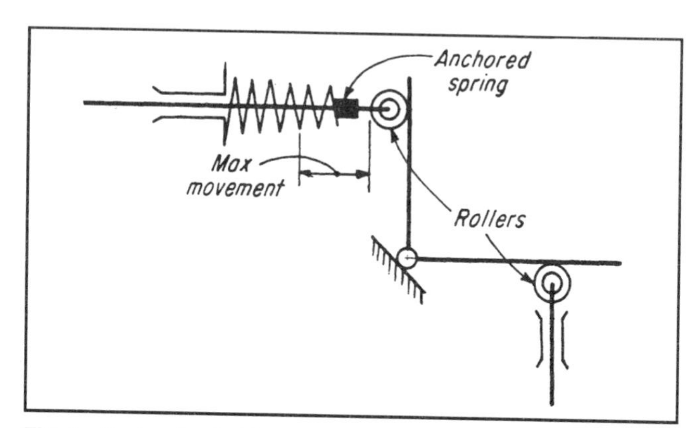

Fig. 4 Spring-loaded lever.

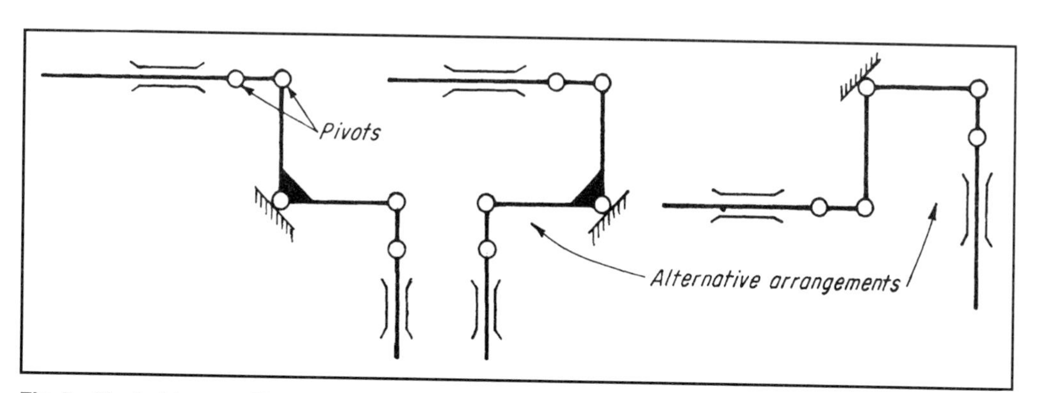

Fig. 5 Pivoted levers with alternative arrangements.

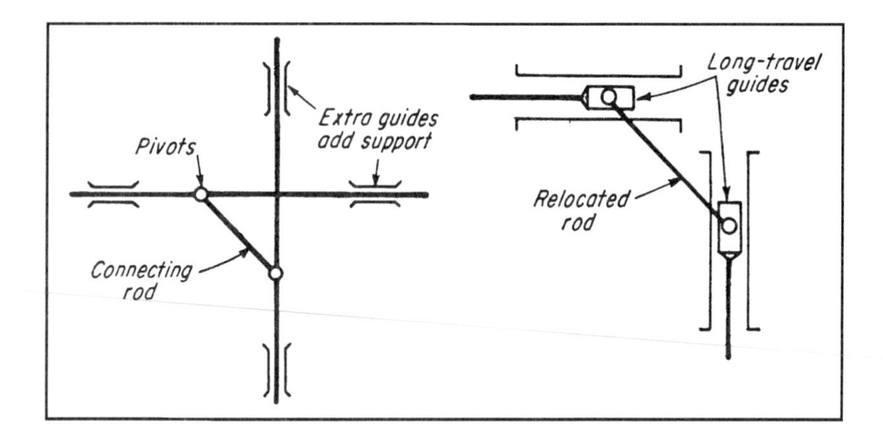

Fig. 6 Single connecting rod (left) is relocated (right) to eliminate the need for extra guides.

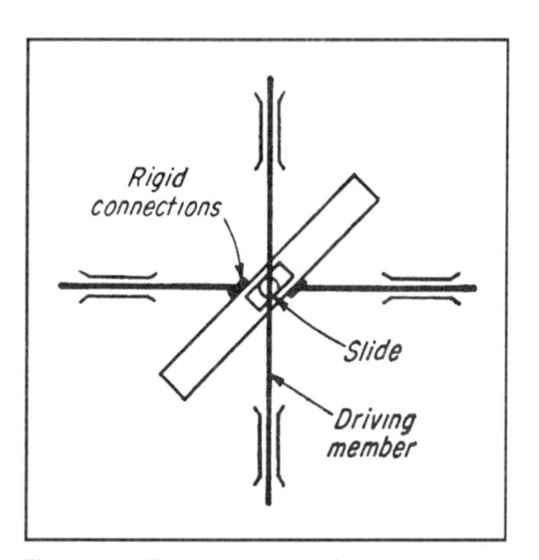

Fig. 7 Inclined bearing-guide.

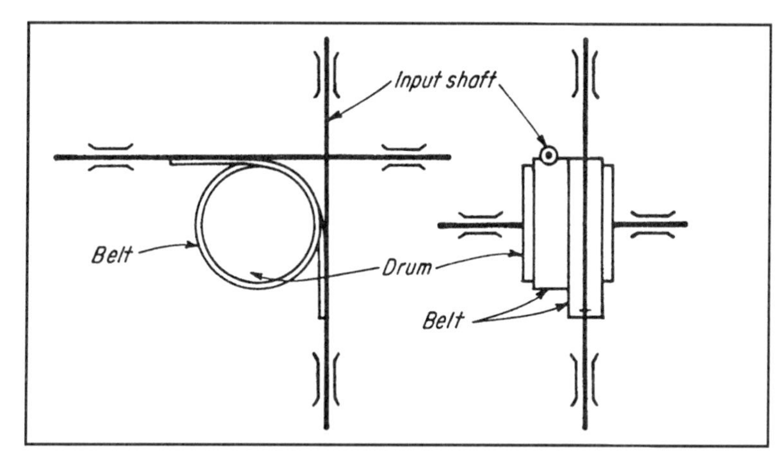

Fig. 8 A belt, steel band, or rope around the drum is fastened to the driving and driven members; sprocket-wheels and chain can replace the drum and belt.

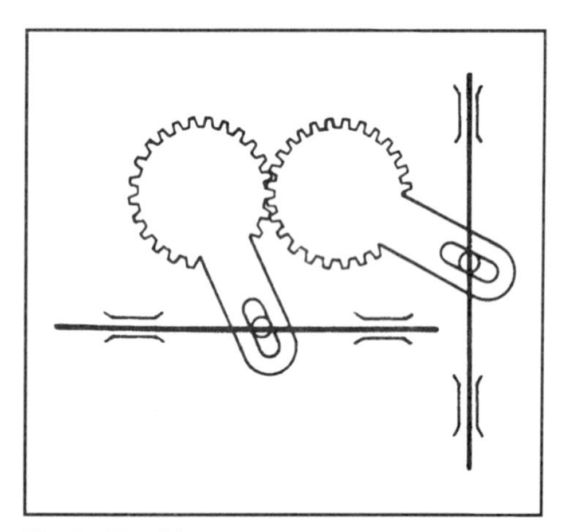

Fig. 9 Matching gear-segments.

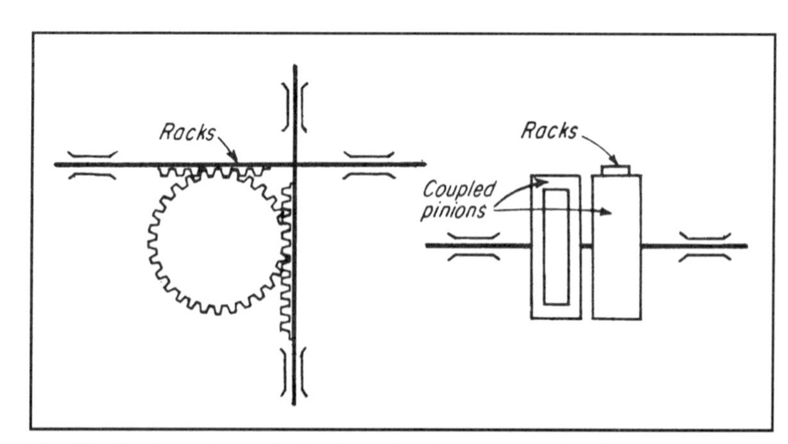

Fig. 10 Racks and coupled pinions (can be substituted as friction surfaces for a low-cost setup).

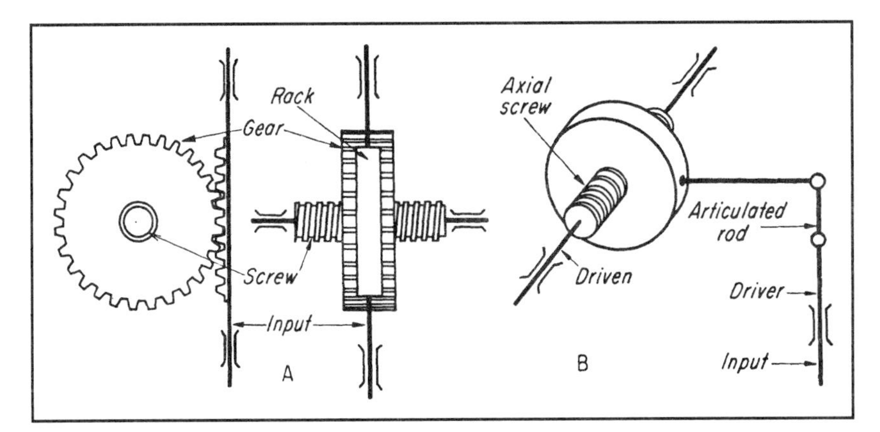

Fig. 11 An axial screw with a rack-actuated gear (A) and an articulated driving rod (B) are both irreversible movements, i.e., the driver must always drive.

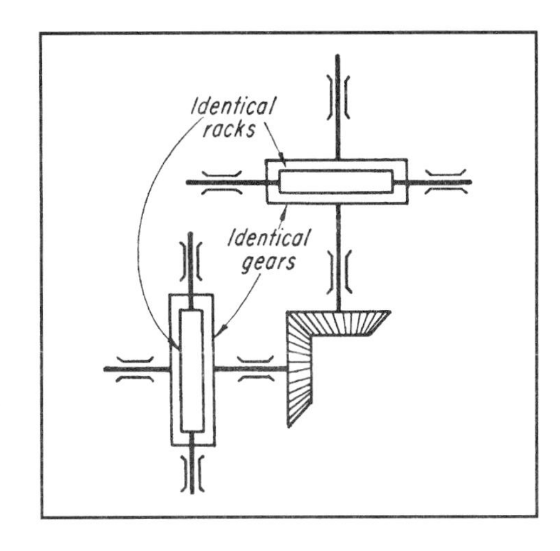

Fig. 12 A rack-actuated gear with associated bevel gears is reversible.

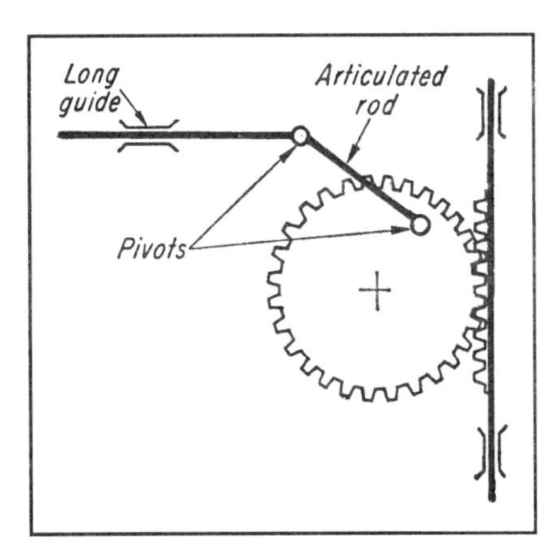

Fig. 13 An articulated rod on a crank-type gear with a rack driver. Its action is restricted to comparatively short movements.

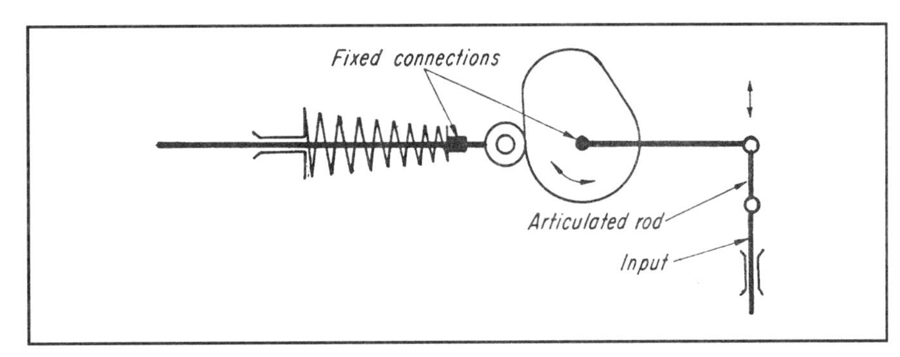

Fig. 14 A cam and spring-loaded follower allows an input/output ratio to be varied according to cam rise. The movement is usually irreversible.

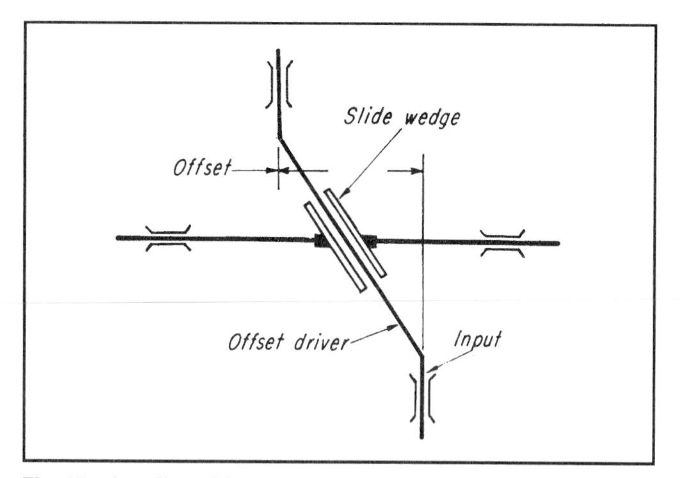

Fig. 15 An offset driver actuates a driven member by wedge action. Lubrication and materials with a low coefficient of friction permit the offset to be maximized.

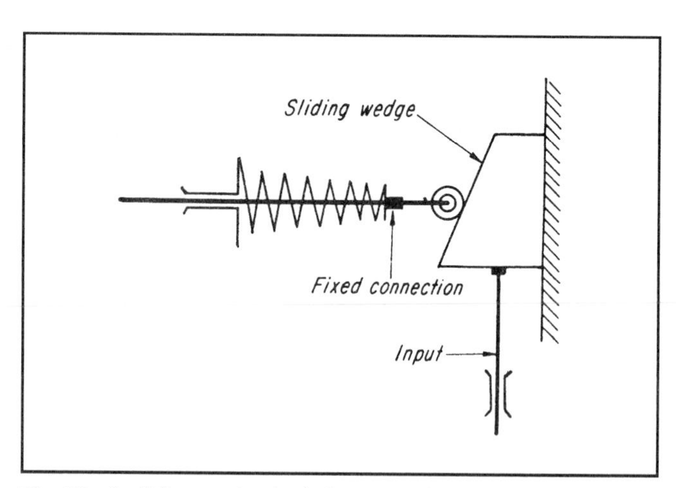

Fig. 16 A sliding wedge is similar to an offset driver but it requires a spring-loaded follower; also, low friction is less critical with a roller follower.

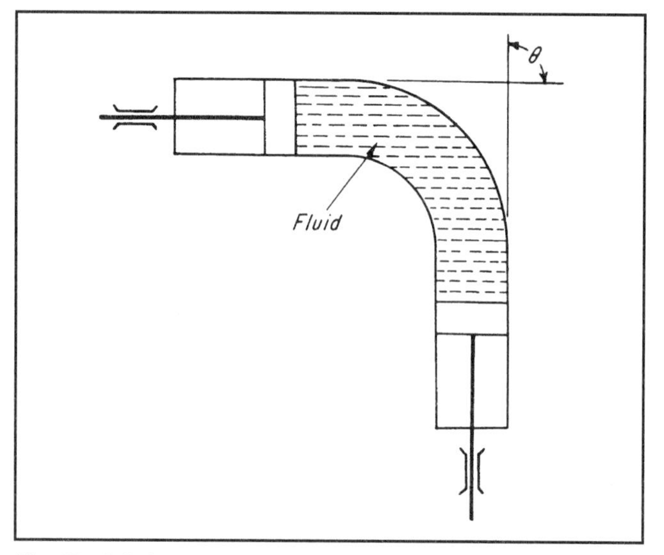

Fig. 17 A fluid coupling allows motion to be transmitted through any angle. Leak problems and accurate piston-fitting can make this method more expensive than it appears to be. Also, although the action is reversible, it must always be compressive for the best results.

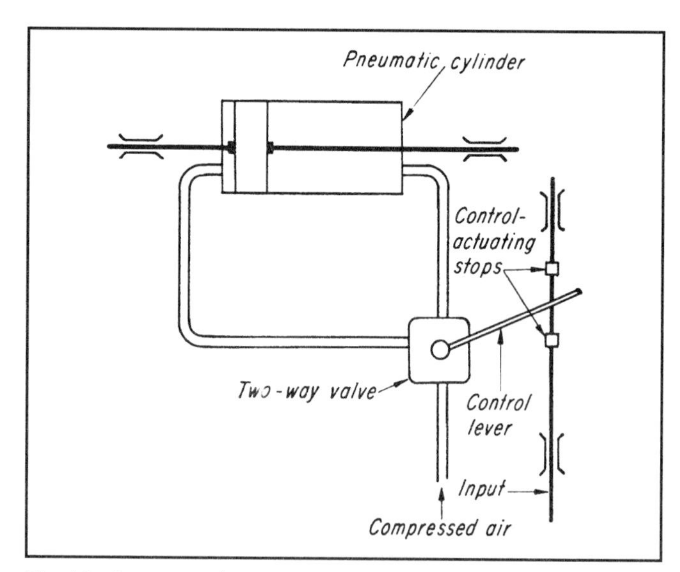

Fig. 18 A pneumatic system with a two-way valve is ideal when only two extreme positions are required. The action is irreversible. The speed of a driven member can be adjusted by controlling the input of air to the cylinder.

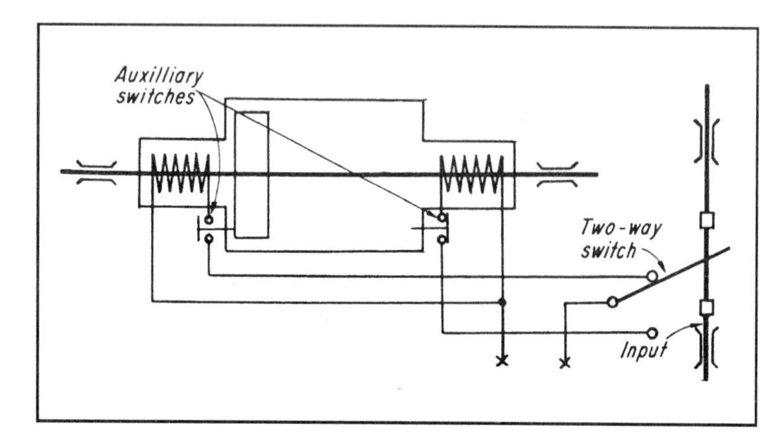

Fig. 19 Solenoids and a two-way switch are organized as an analogy of a pneumatic system. Contact with the energized solenoid is broken at the end of each stroke. The action is irreversible.

FIVE ADJUSTABLE-OUTPUT MECHANISMS

Linkage-motion adjuster

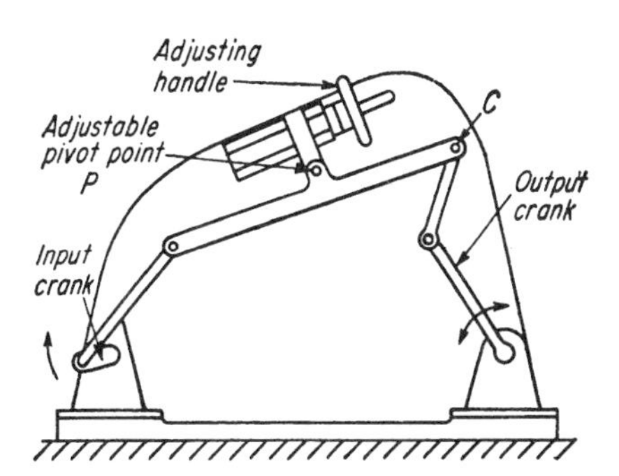

Here the motion and timing of the output link can be varied during its operation by shifting the pivot point of the intermediate link of the six-bar linkage illustrated. Rotation of the input crank causes point *C* to oscillate around the pivot point *P*. This, in turn, imparts an oscillating motion to the output crank. A screw device shifts point *P*.

Cam-motion adjuster

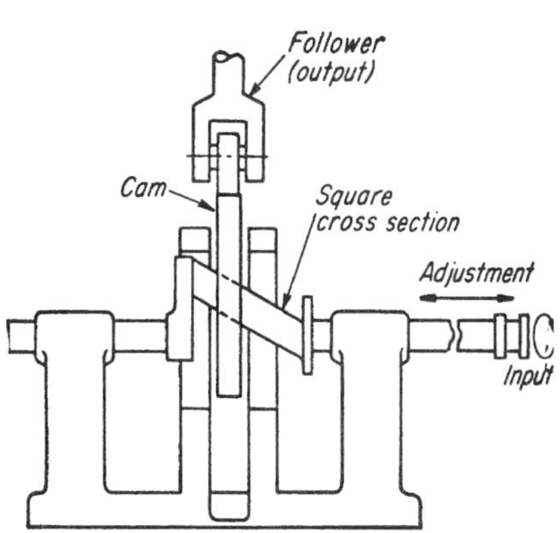

The output motion of the cam follower is varied by linearly shifting the input shaft to the right or left during its operation. The cam has a square hole which fits over the square cross section of the crank shaft. Rotation of the input shaft causes eccentric motion in the cam. Shifting the input shaft to the right, for instance, causes the cam to move radially outward, thus increasing the stroke of the follower.

Double-cam adjuster

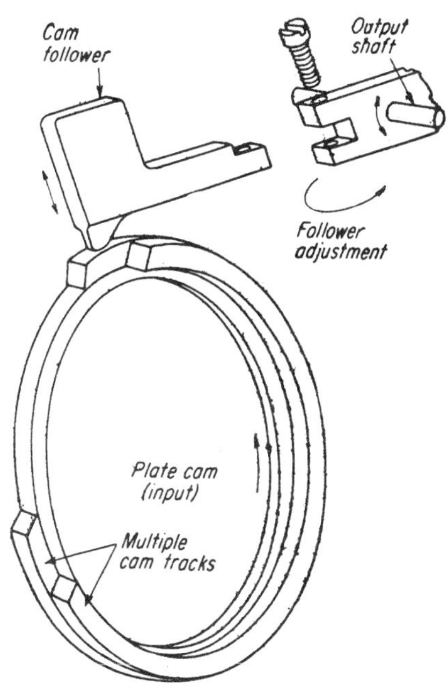

This is a simple but effective mechanism for changing the timing of a cam. The follower can be adjusted in the horizontal plane, but it is restricted in the vertical plane. The plate cam contains two or more cam tracks.

Valve-stroke adjuster

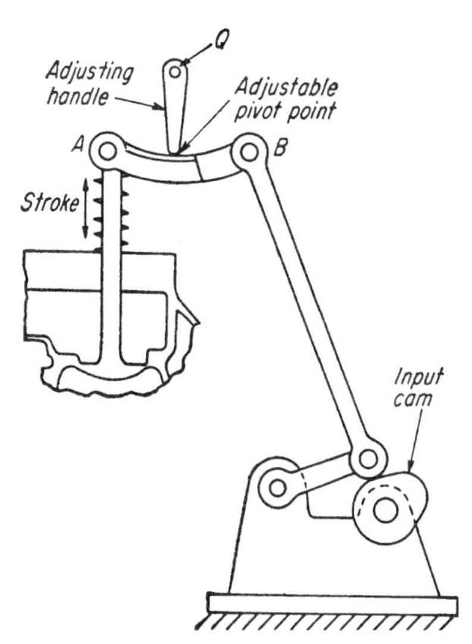

This mechanism adjusts the stroke of valves of combustion engines. One link has a curved surface and pivots around an adjustable pivot point. Rotating the adjusting link changes the proportion of strokes or points *A* and *B* and hence of the valve. The center of curvature of the curve link is at point *Q*.

Three-dimensional adjuster

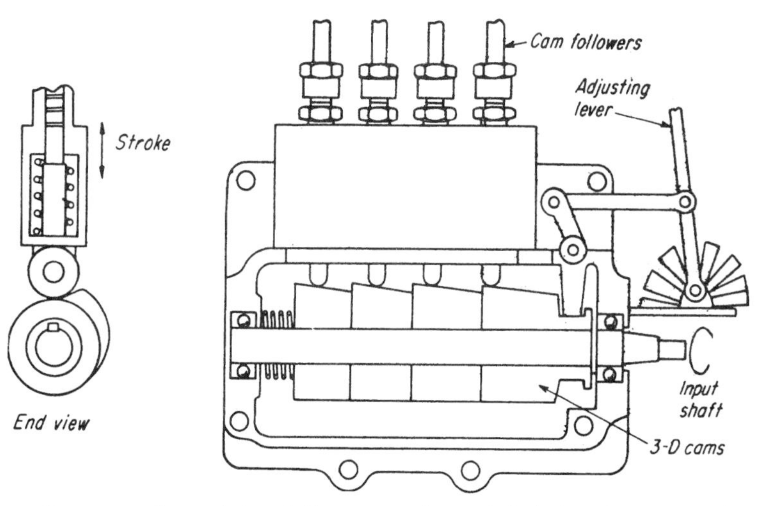

Output motions of four followers can be varied during the rotation by shifting the quadruple 3-D cam to the right or left. A linear shift can be made with the adjustment lever, which can be released in any of the six positions.

Piston-stroke adjuster

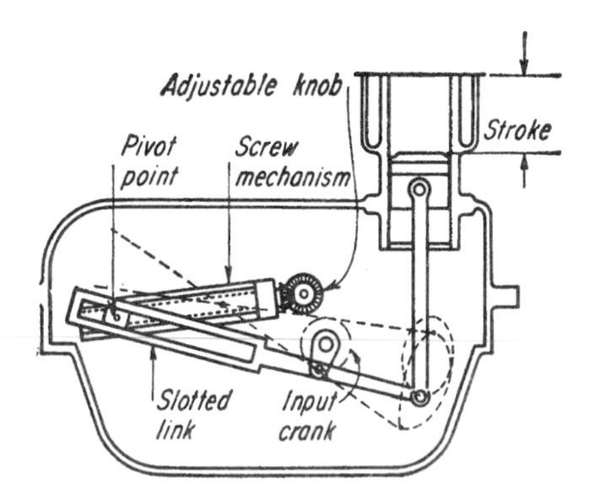

The input crank oscillates the slotted link to drive the piston up and down. The position of the pivot point can be adjusted with the screw mechanism even when the piston is under full load.

Shaft-synchronizer adjuster

The actual position of the adjusting shaft is normally kept constant. The input then drives the output with the bevel gears. Rotating the adjusting shaft in a plane at right angles to the input-output line changes the relative radial position of the input and output shafts. They introduce a torque into the system while running, synchronizing the input and output shafts, or changing the timing of a cam on the output shaft.

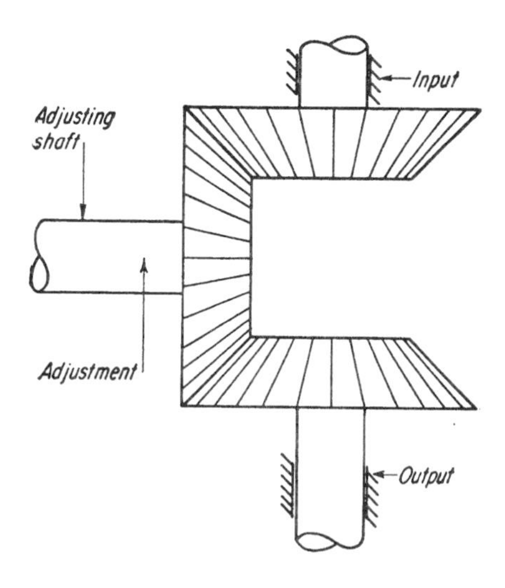

Eccentric pivot-point adjuster

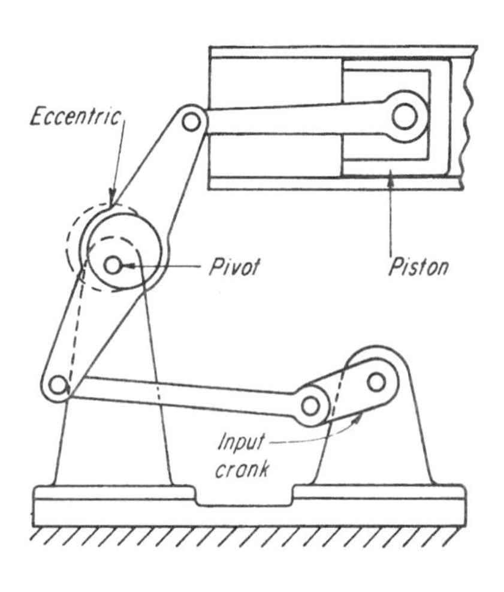

Rotation of the input crank causes the piston to reciprocate. The stroke length depends on the position of the pivot point which is easily adjusted, even during rotation, by rotating the eccentric shaft.

FOUR DIFFERENT REVERSING MECHANISMS

Double-link reverser

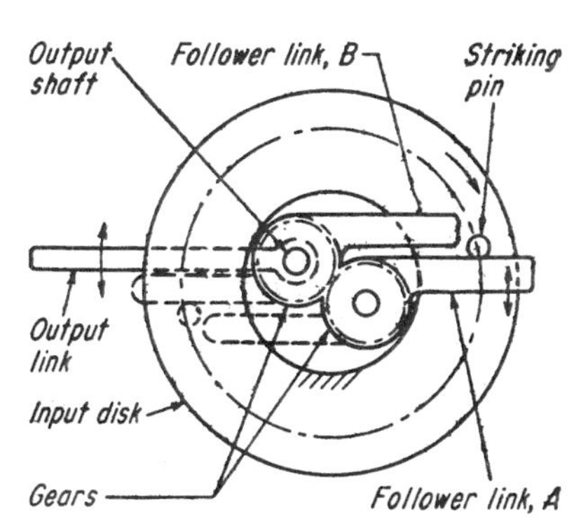

This mechanism automatically reverses the output drive for every 180° rotation of the input. The input disk has a press-fit pin which strikes link *A* to drive it clockwise. Link *A* in turn drives link *B* counterclockwise with its gear segments (or gears pinned to the links). The output shaft and the output link (which can be the working member) are connected to link *B*.

After approximately 180° of rotation, the pin slides past link *A* to strike link *B* coming to meet it—and thus reverses the direction of link *B* (and of the output). Then after another 180° rotation the pin slips past link *B* to strike link *A* and starts the cycle over again.

Toggle-link reverser

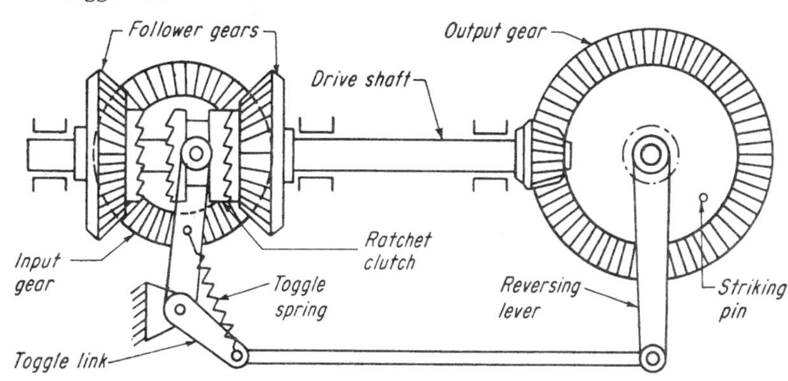

This mechanism also employs a striking pin—but here the pin is on the output member. The input bevel gear drives two follower bevels which are free to rotate on their common shaft. The ratchet clutch, however, is spline-connected to the shaft—although free to slide linearly. As shown, it is the right follower gear that is locked to the drive shaft. Hence the output gear rotates clockwise until the pin strikes the reversing level to shift the toggle to the left. Once past its center, the toggle spring snaps the ratchet to the left to engage the left follower gear. This instantly reverses the output, which now rotates counterclockwise until the pin again strikes the reversing level. Thus the mechanism reverses itself for every 360° rotation of the input.

Modified-Watt's reverser

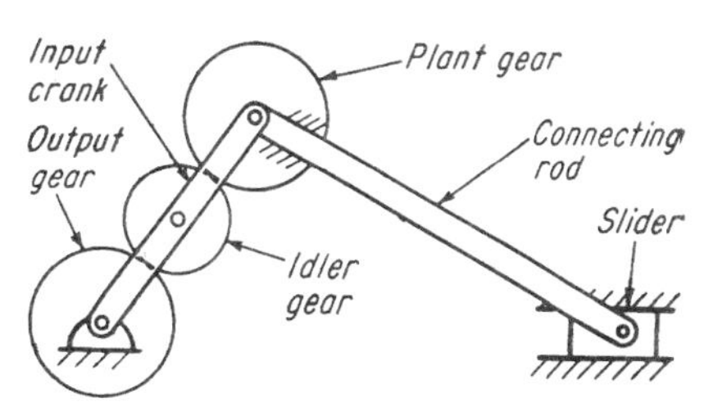

This is a modification of the well-known Watt crank mechanism. The input crank causes the planet gear to revolve around the output gear. But because the planet gear is fixed to the connecting rod, it causes the output gear to continually reverse itself. If the radii of the two gears are equal, each full rotation of the input link will cause the output gear to oscillate through the same angle as the rod.

Automatically switching from one pivot point to another in midstroke.

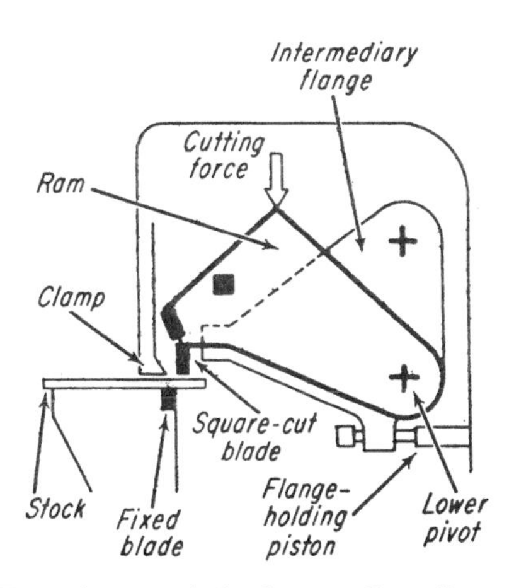

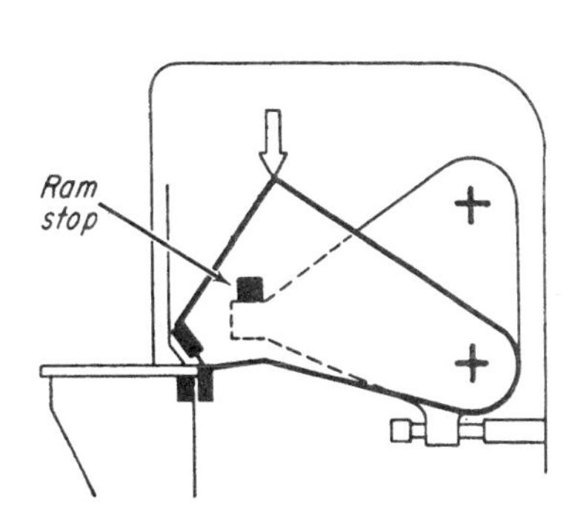

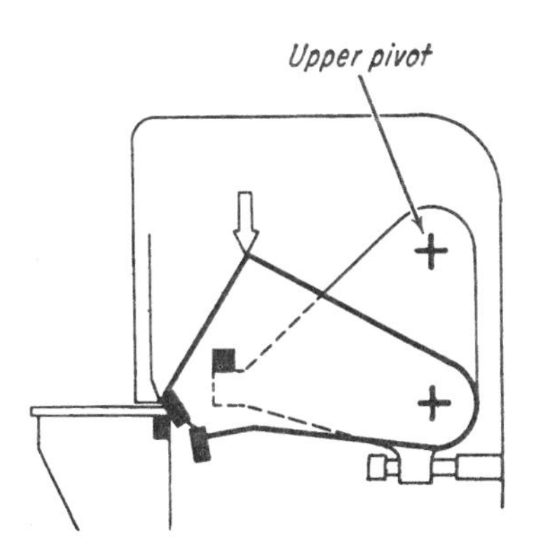

Two pivots and the intermediary flange govern the cutting sequence. The flange is connected to the press frame at the upper pivot, and the cutting ram is connected to the flange at the lower pivot. In the first part of the cycle, the ram turns around the lower pivot and shears the plate with the square-cut blade; the motion of the intermediary flange is restrained by the flange-holding piston.

After the shearing cut, the ram stop bottoms on the flange. This overcomes the restraining force of the flange-holding piston, and the ram turns around the upper pivot. This brings the beveling blade into contact with the plate for the bevel cut.

TEN MECHANICAL COMPUTING MECHANISMS

Analog computing mechanisms are capable of almost instantaneous response to minute variations in input. Basic units, similar to the examples shown, are combined to form the final mechanism. These mechanisms add, subtract, resolve vectors, or solve special or trigonometric functions.

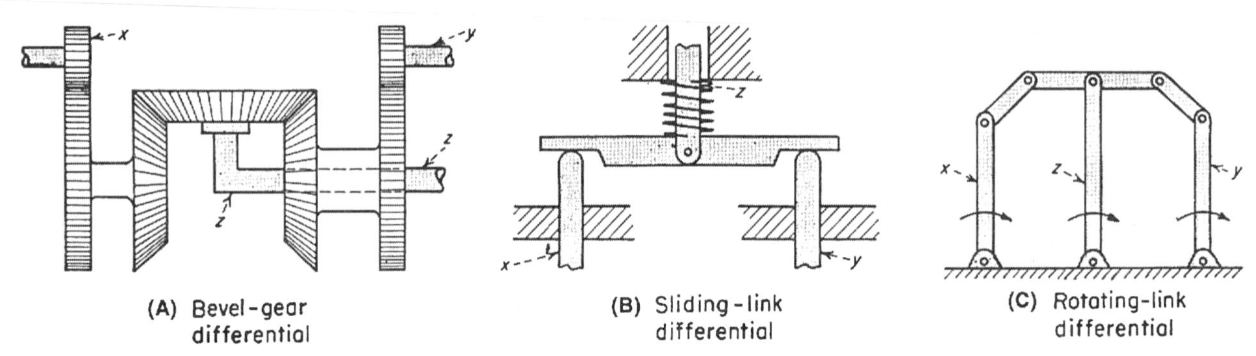

Fig. 1 Addition and subtraction is usually based on the differential principle; variations depend on whether inputs: (A) rotate shafts, (B) translate links, or (C) angularly displace links. Mechanisms can solve the equation: $z = c(x \pm y)$, where c is the scale factor, x and y are inputs, and z is the output. The motion of x and y in the same direction performs addition; in the opposite direction it performs subtraction.

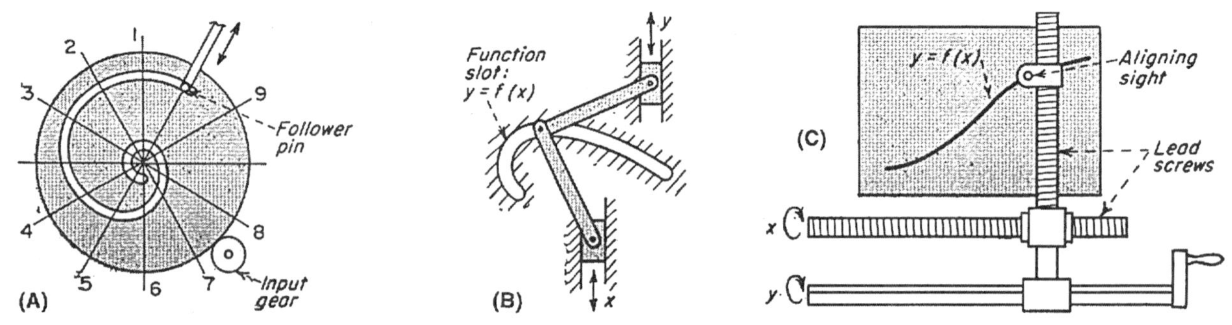

Fig. 2 Functional generators mechanize specific equations. (A) A reciprocal cam converts a number into its reciprocal. This simplifies division by permitting simple multiplication between a numerator and its denominator. The cam is rotated to a position corresponding to the denominator. The distance between the center of the cam to the center of the follower pin corresponds to a reciprocal. (B) A function-slot cam is ideal for performing complex functions involving one variable. (C) A function is plotted on a large sheet attached to a table. The x lead screw is turned at a constant speed by an analyzer. An operator or photoelectric follower turns the y output to keep the aligning sight on the curve.

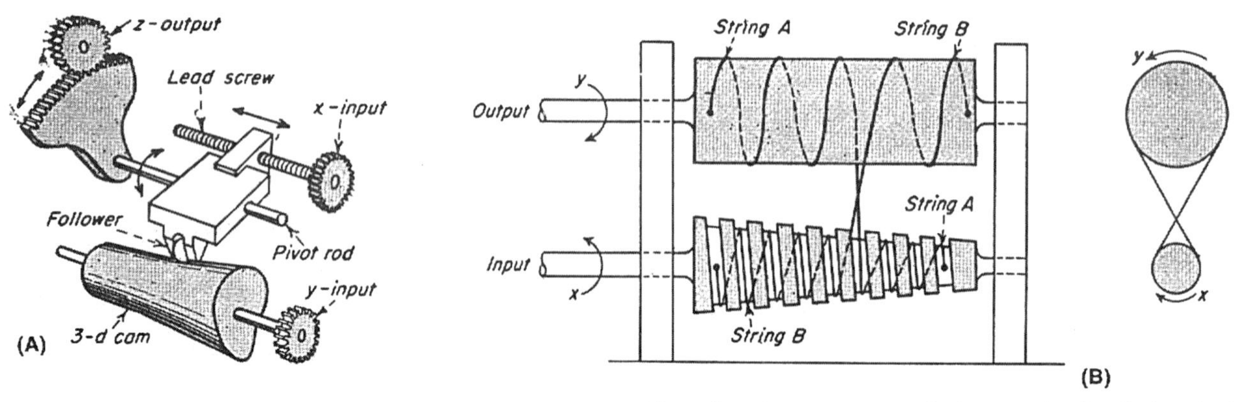

Fig. 3 (A) **A three-dimensional cam** generates functions with two variables: $z = f(x, y)$. A cam is rotated by the y-input; the x-input shifts a follower along a pivot rod. The contour of the cam causes a follower to rotate, giving angular displacement to the z-output gear. (B) **A conical cam** for squaring positive or negative inputs: $y = c(\pm x)^2$. The radius of a cone at any point is proportional to the length of string to the right of the point; therefore, cylinder rotation is proportional to the square of cone rotation. The output is fed through a gear differential to convert it to a positive number.

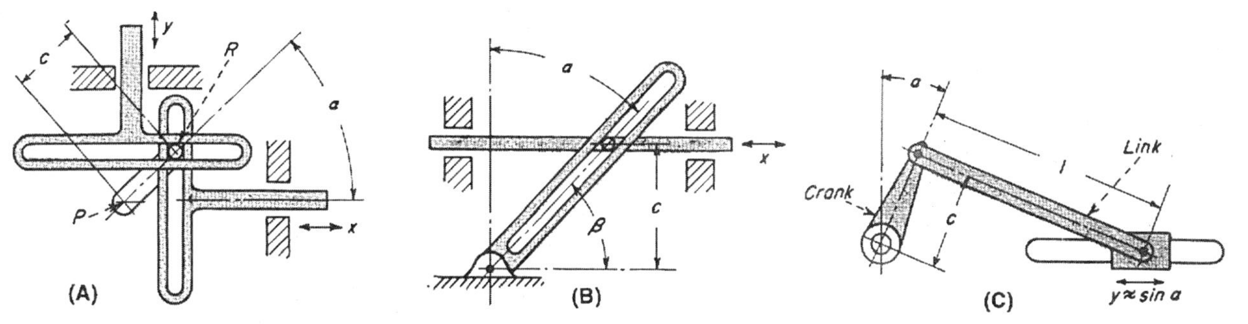

Mechanical Computing Mechanisms *(continued)*

Fig. 4 Trigonometric functions. (A) A Scotch-yoke mechanism for sine and cosine functions. A crank rotates about fixed point *P*, generating angle *a* and giving motion to the arms: $y = c \sin a$; $x = c \cos a$. (B) A tangent-cotangent mechanism generates $x = c \tan a$ or $x = c \cot \beta$. (C) The eccentric and follower is easily manufactured, but sine and cosine functions are approximate. The maximum error is zero at 90° and 270°; *l* is the length of the link, and *c* is the length of the crank.

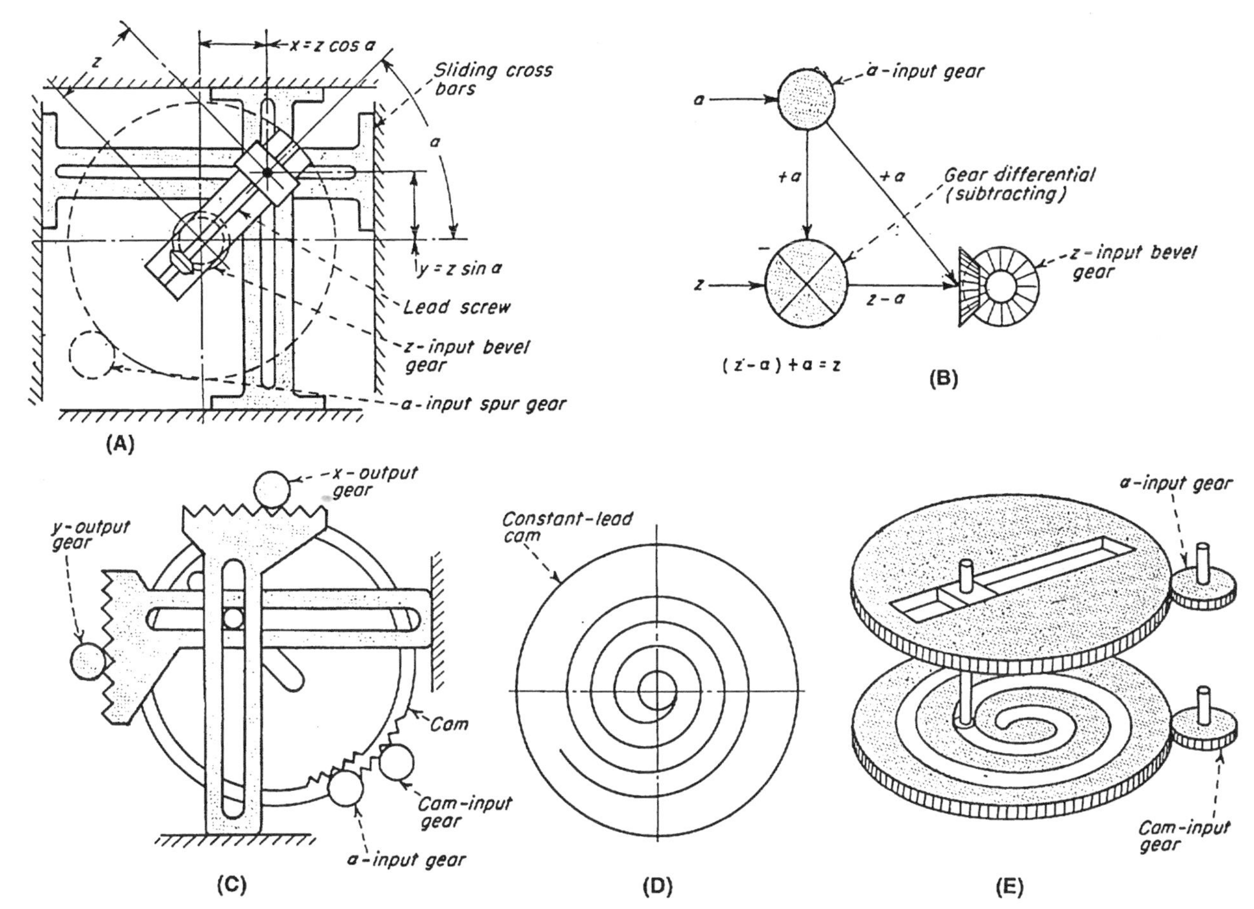

Fig. 5 Component resolvers determine *x* and *y* components of vectors that are continuously changing in both angle and magnitude. Equations are $x = z \cos a$, $y = z \sin a$, where *z* is magnitude of vector, and *a* is vector angle. Mechanisms can also combine components to obtain a resultant. Inputs in (A) are through bevel gears and lead screws for *z*-input, and through spur gears for *a*-input. Compensating gear differential (B) prevents the *a*-input from affecting the *z*-input. This problem is solved in (C) with constant-lead cams (D) and (E).

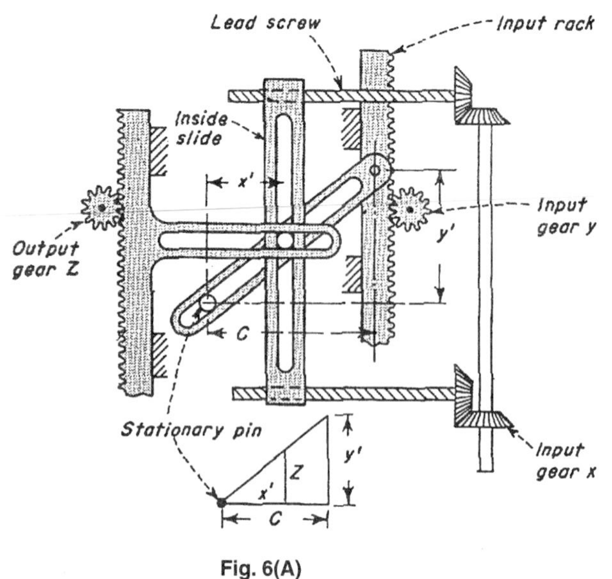

Fig. 6(A)

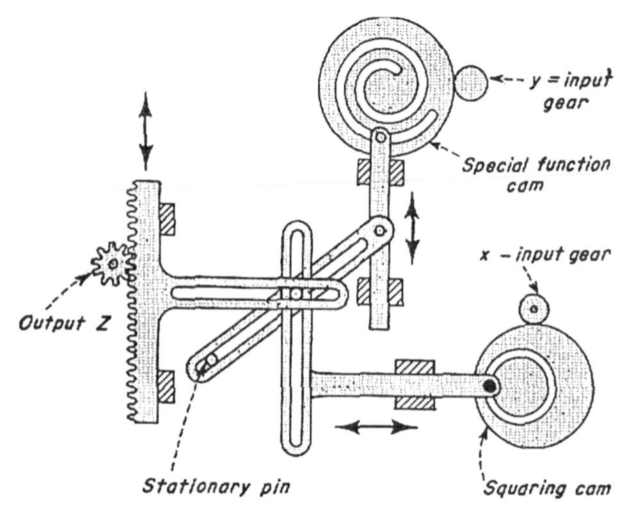

Fig. 7(A)

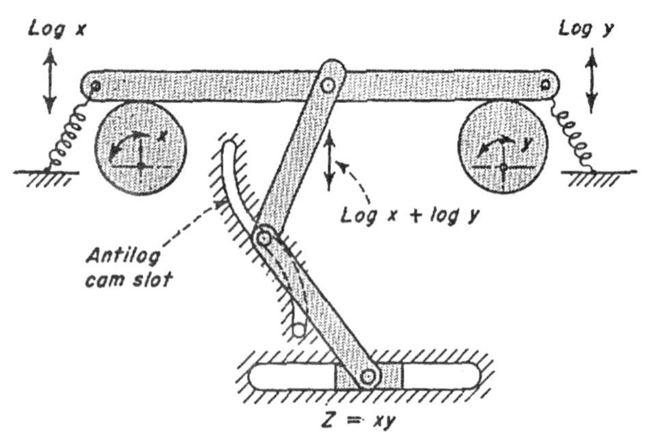

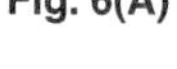

Fig. 6(B)

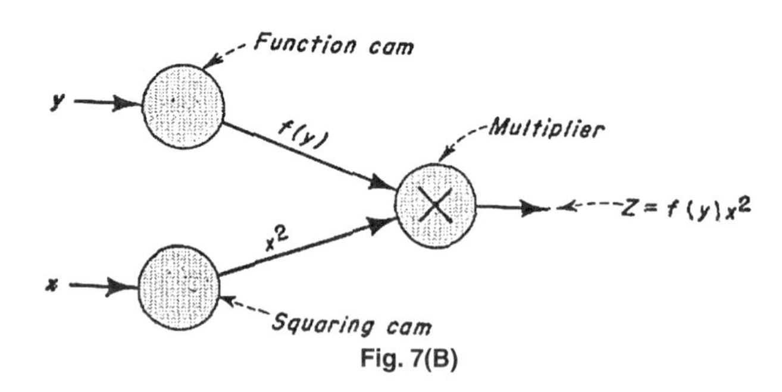

Fig. 7(B)

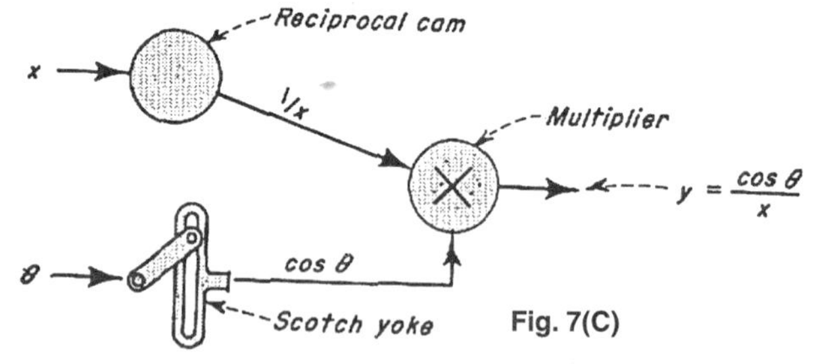

Fig. 7(C)

Fig. 6 The multiplication of two tables, *x* and *y*, can usually be solved by either: (A) The similar triangle method, or (B) the logarithmic method. In (A), lengths x' and y' are proportional to the rotation of input gears *x* and *y*. Distance *c* is constant. By similar triangles: $z/x = y/c$ or $z = xy/c$, where *z* is vertical displacement of output rack. The mechanism can be modified to accept negative variables. In (B), the input variables are fed through logarithmic cams to give linear displacements of log *x* and log *y*. The functions are then added by a differential link giving $z = \log x + \log xy$ (neglecting scale factors). The result is fed through the antilog cam so that the motion of the follower represents $z = xy$.

Fig. 7 Multiplication of complex functions can be accomplished by substituting cams in place of input slides and racks of the mechanism in Fig. 6. The principle of similar triangles still applies. The mechanism in (A) solves the equation: $z = f(y)x^2$. The schematic is shown in (B). Division of two variables can be done by feeding one of the variables through a reciprocal cam and then multiplying it by the other. The schematic in (C) shows the solution of $y = \cos\theta/x$.

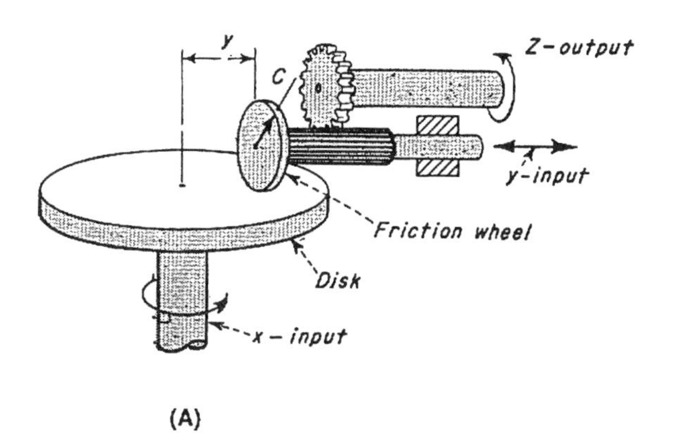

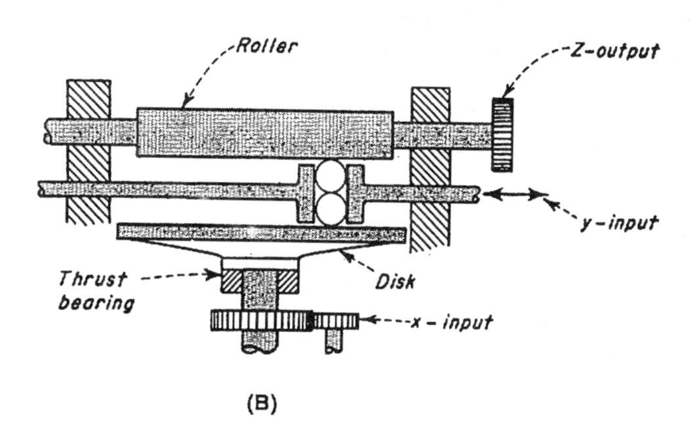

Fig. 8 Integrators are essentially variable-speed drives. The *x*-input shaft in Fig. 8 (A) rotates the disk which, in turn, rotates the friction wheel on the *y*-input shaft which is perpendicular to the *x*-input shaft. As the friction wheel turns, it rotates a spline on the movable *y*-input shaft. The gear on the end of the parallel *z*-output shaft drives that shaft.

Moving the *y*-input shaft along the radius dimension of the disk changes the rotational speed of the friction wheel from zero at the center of the disk to a maximum at the periphery. The *z*-axis output is thus a function of the rotational speed of the *x*-input, the diameter of the friction wheel, and *y*, the radius distance of the wheel on the disk.

In the integrator shown in Fig. 8 (B), two balls replace the friction wheel and spline of the *y*-input axis, and a roller replaces the gear on the *z*-output shaft to provide a variable-speed output as the *y*-input shaft is moved across the entire diameter of the disk.

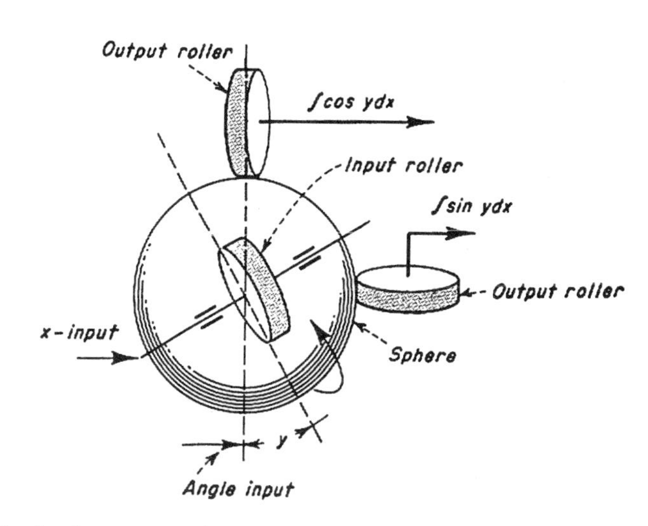

Fig. 9 A component integrator has three disks to obtain the *x* and *y* components of a differential equation. The input roller on the *x*-input shaft spins the sphere, and the *y*-input lever arm changes the angle of the roller with respect to the sphere. The sine and cosine output rollers provide integrals of components that parallel the *x* and *y* axes.

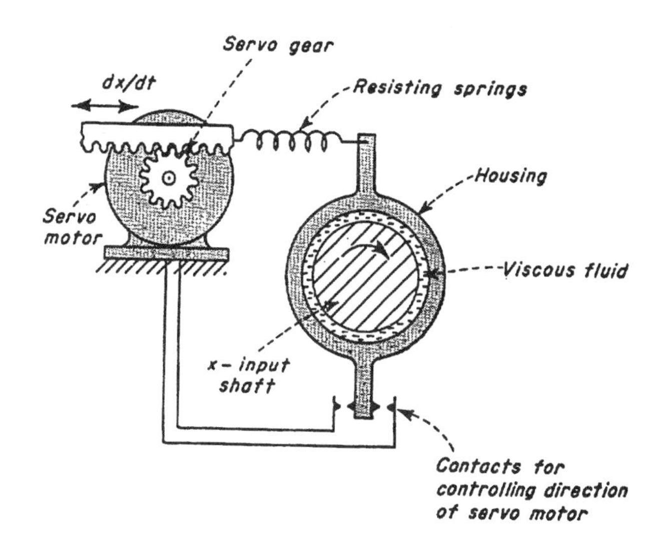

Fig. 10 This differentiator is based on the principle that a viscous drag force in a thin layer of fluid is proportional to the velocity of a rotating x-input shaft. The drag force is counteracted by resisting springs in tension. Spring length is regulated by a servomotor controlled by electrical contacts at the base of the housing. A change in shaft velocity causes a change in viscous torque. A shift in the housing closes one set of electrical contacts, causing the motor shaft to turn. This repositions a rack which adjusts the spring tension and balances the system. The total rotation of the servomotor gear is proportional to *dx*/*dt*.

SEVEN DIFFERENT MECHANICAL POWER AMPLIFIERS

Precise positioning and movement of heavy loads are two basic jobs for this all-mechanical torque booster.

This mechanical power amplifier has a fast response. Power from its continuously rotating drums is instantaneously available. When used for position-control applications, pneumatic, hydraulic, and electrical systems—even those with continuously running power sources—require transducers to change signals from one energy form to another. The mechanical power amplifier, on the other hand, permits direct sensing of the controlled motion.

Four major advantages of this all-mechanical device are:

1. Kinetic energy of the power source is continuously available for rapid response.
2. Motion can be duplicated and power amplified without converting energy forms.
3. Position and rate feedback are inherent design characteristics.
4. Zero slip between input and output eliminates the possibility of cumulative error.

One other important advantage is the ease with which this device can be adapted to perform special functions—jobs for which other types of systems would require the addition of more costly and perhaps less reliable components. The six applications which follow illustrate how those advantages have been put to work in solving widely divergent problems.

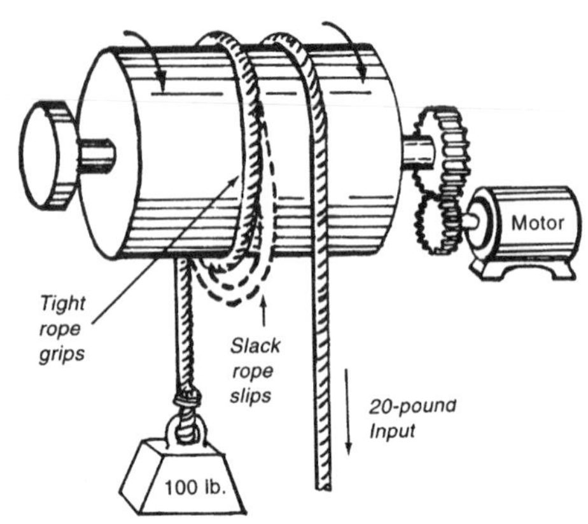

The capstan principle is the basis for the mechanical power amplifier described here that combines two counterrotating drums. The drums are continuously rotating but only transmit torque when the input shaft is rotated to tighten the band on drum A. Overrun of output is stopped by drum B, when overrun tightens the band on this drum.

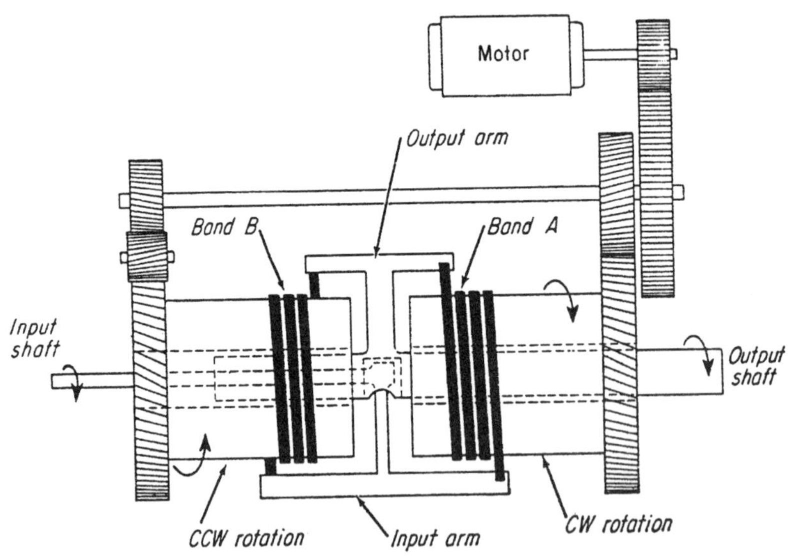

A capstan is a simple mechanical amplifier—rope wound on a motor-driven drum slips until slack is taken up on the free end. The force needed on the free end to lift the load depends on the coefficient of friction and the number of turns of rope. By connecting bands A and B to an input shaft and arm, the power amplifier provides an output in both directions, plus accurate angular positioning. When the input shaft is turned clockwise, the input arm takes up the slack on band A, locking it to its drum. Because the load end of locked band A is connected to the output arm, it transmits the CW motion of the driven drum on which it is wound to the output shaft. B and B therefore slacks off and slips on its drum. When the CW motion of the input shaft stops, tension on band A is released and it slips on its own drum. If the output shaft tries to overrun, the output arm will apply tension to band B, causing it to tighten on the CCW rotating rum and stop the shaft.

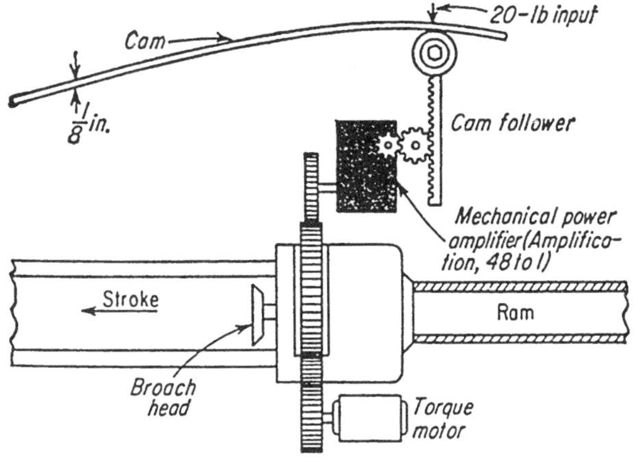

1. Nonlinear Broaching

Problem: In broaching large-fore rifles, the twist given to the lands and grooves represents a nonlinear function of barrel length. Development work on such rifles usually requires some experimentation with this function. At present, rotation of the broaching head is performed by a purely mechanical arrangement consisting of a long, heavy wedge-type cam and appropriate gearing. For steep twist angles, however, the forces acting on this mechanism become extremely high.

Solution: A suitable mechanical power amplifier, with its inherent position feedback, was added to the existing mechanical arrangement, as shown in Fig. 1. The cam and follower, instead of having to drive the broaching head, simply furnish enough torque to position the input shaft of the amplifier.

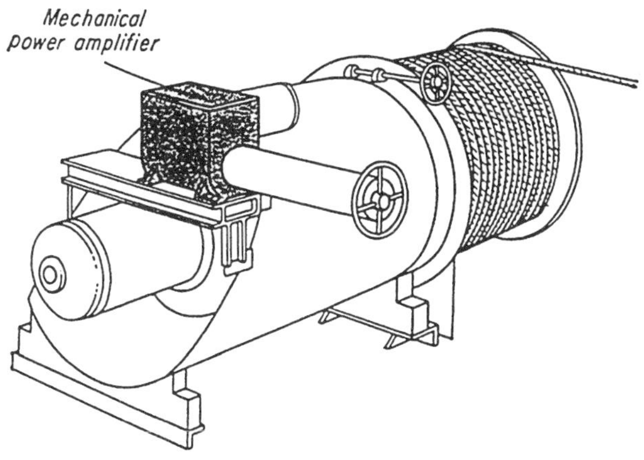

2. Hydraulic Winch Control

Problem: Hydraulic pump-motor systems are excellent for controlling position and motion at high power levels. In the 10- to 150-hp range, for example, the usual approach is to vary the output of a positive displacement pump in a closed-loop hydraulic circuit. In many of the systems that might be able to control this displacement, however, a force feedback proportional to system pressure can lead to serious errors or even oscillations.

Solution: Figure 2 shows an external view of the complete package. The output shaft of the mechanical power amplifier controls pump displacement, while its input is controlled by hand. In a more recent development requiring remote manual control, a servomotor replaces this local handwheel. Approximately 10 lb-in. torque drives a 600 lb-in. load. If this system had to transmit 600 lb-in., the equipment would be more expensive and more dangerous to operate.

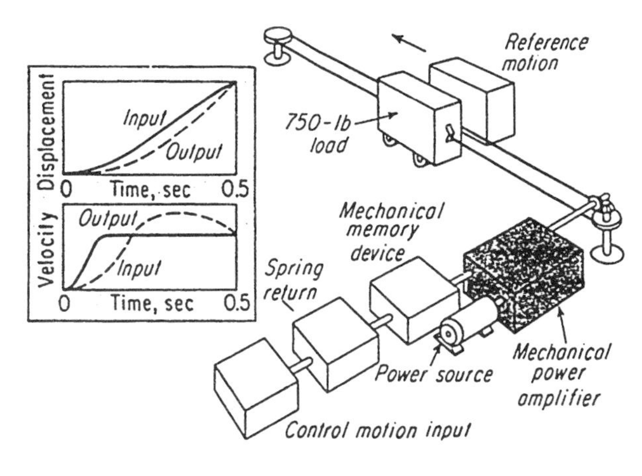

3. Load Positioning

Problem: It was necessary for a 750-lb load to be accelerated from standstill in 0.5 s and brought into speed and position synchronization with a reference linear motion. It was also necessary that the source of control motion be permitted to accelerate more rapidly than the load itself. Torque applied to the load could not be limited by any kind of slipping device.

Solution: A system with a single mechanical power amplifier provided the solution (Fig. 3). A mechanical memory device, preloaded for either rotation, drives the input shaft of the amplifier. This permits the input source to accelerate as rapidly as desired. The total control input travel minus the input travel of the amplifier shaft is temporarily stored. After 0.5 seconds, the load reaches proper speed, and the memory device transmits position information in exact synchronization with the input.

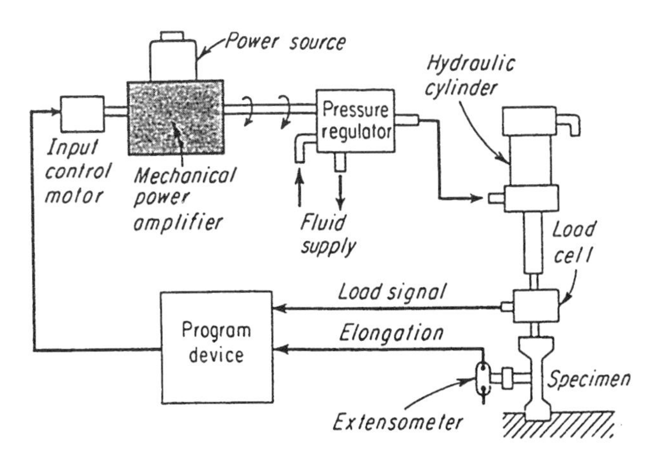

4. Tensile Testing Machine

Problem: On a hydraulic tensile testing machine, the stroke of the power cylinder had to be controlled as a function of two variables: tension in, and extension of, the test specimen. A programming device, designed to provide a control signal proportional to these variables, had an output power level of about 0.001 hp—too low to drive the pressure regulator controlling the flow to the cylinder.

Solution: An analysis of the problem revealed three requirements: the output of the programmer had to be amplified about 60 times, position accuracy had to be within 2°, and acceleration had to be held at a very low value. A mechanical power amplifier satisfied all three requirements. Figure 4 illustrates the completed system. Its design is based principally on steady-state characteristics.

5. Remote Metering and Counting

Problem: For a remote, liquid-metering job, synchro systems had been used to transmit remote meter readings to a central station and repeat this information on local indicating counters. The operation involved a large number of meters and indicators. As new equipment (e.g. ticket printers) was added, the torque requirement also grew.

Solution: Mechanical power amplifiers in the central station indicators not only supplied extra output torque but also made it possible to specify synchros that were even smaller than those originally selected to drive the indicators alone (see Fig. 5).
The synchro transmitters selected operate at a maximum speed of 600 rpm and produce only about 3 oz-in. of torque. The mechanical power amplifiers furnish up to 100 lb-in. of torque, and are designed to fit in the bottom of the registers shown in Fig. 5. Total accuracy is within 0.25 gallon, and error is noncumulative.

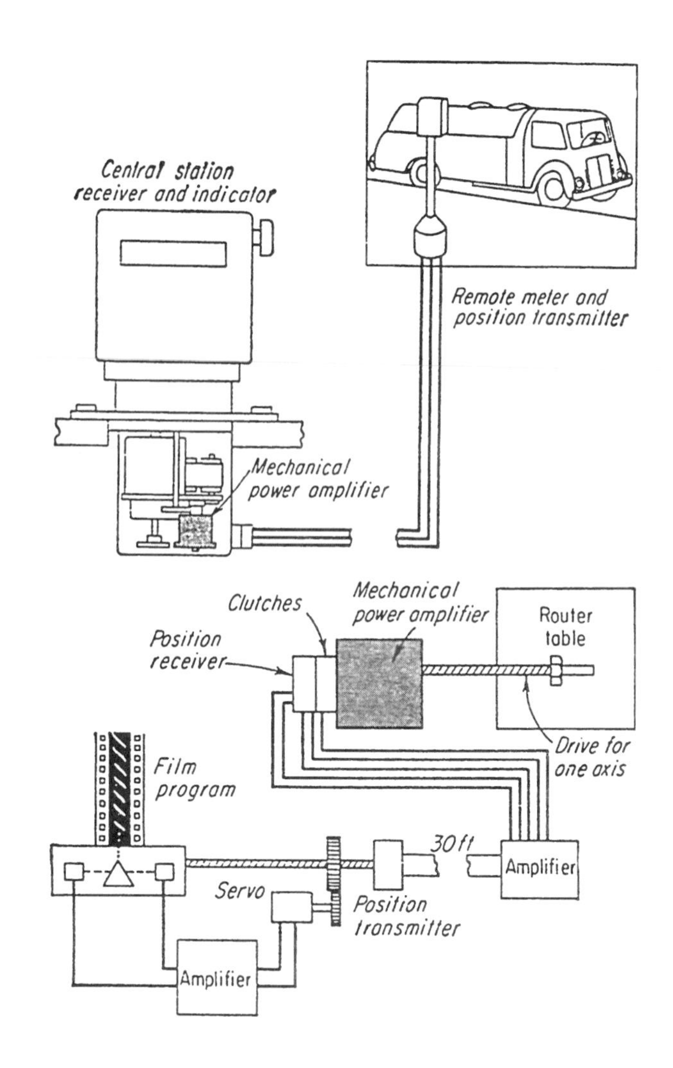

6. Irregular Routing

Problem: To control remotely the table position of a routing machine from information stored on a film strip. The servoloop developed to interpret this information produced only about 1 oz.-in. of torque. About 20 lb.-ft was required at the table feedscrew.

Solution: Figure 6 shows how a mechanical power amplifier supplied the necessary torque at the remote table location. A position transmitter converts the rotary motion output of the servoloop to a proportional electrical signal and sends it to a differential amplifier at the machine location. A position receiver, geared to the output shaft, provides a signal proportional to table position. The differential amplifier compares these, amplifies the difference, and sends a signal t either counterrotating electromagnetic clutch, which drives the input shaft of the mechanical power amplifier.

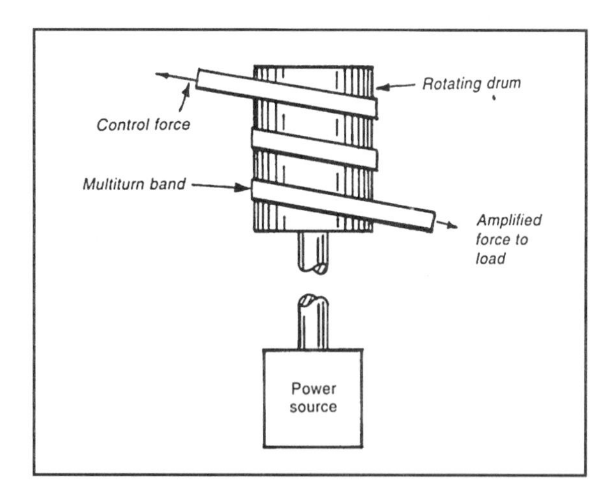

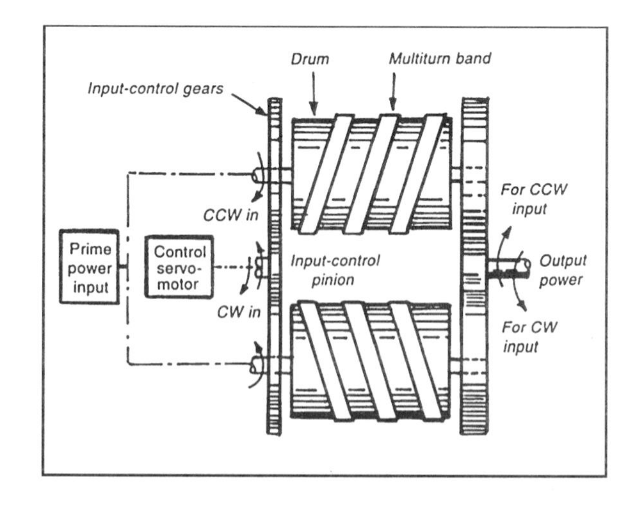

A mechanical power amplifier that drives a crossfeed slide is based on the principle of the windlass. By varying the control force, all or any part of power to the drum can be used.

Two drums mounted back to back supply the bi-directional power needed in servo systems. Replacing the operator with a two-phase induction servomotor permits electronic or magnetic signal amplification. A rotating input avoids a linear input and output of the simple windlass. Control and output ends of the multiturn bands are both connected to gears mounted concentrically with the drum axis.

When the servomotor rotates the control gear, it locks the band-drum combination, forcing output gear to rotate with it. Clockwise rotation of the servomotor produces CW power output while the second drum idles. Varying the servo speed, by changing servo voltage, varies output speed.

FORTY-THREE VARIABLE-SPEED DRIVES AND TRANSMISSIONS

CONE DRIVES

The simpler cone drives in this group have a cone or tapered roller in combination with a wheel or belt (Fig. 1). They have evolved from the stepped-pulley system. Even the more sophisticated designs are capable of only a limited (although infinite) speed range, and generally must be spring-loaded to reduce slippage.

Adjustable-cone drive (Fig. 1A). This is perhaps the oldest variable-speed friction system, and is usually custom built. Power from the motor-driven cone is transferred to the output shaft by the friction wheel, which is adjustable along the cone side to change the output speed. The speed depends upon the ratio of diameters at point of contact.

Two-cone drive (Fig. 1B). The adjustable wheel is the power transfer element, but this drive is difficult to preload because both input and output shafts would have to be spring loaded. The second cone, however, doubles the speed reduction range.

Cone-belt drives (Fig. 1C and D). In Fig. 1C the belt envelopes both cones; in Fig. 1D a long-loop endless belt runs between the cones. Stepless speed adjustment is obtained by shifting the belt along the cones. The cross section of the belt must be large enough to transmit the rated force, but the width must be kept to a minimum to avoid a large speed differential over the belt width.

Electrically coupled cones (Fig. 2). This drive is composed of thin laminates of paramagnetic material. The laminates are separated with semidielectric materials which also localize the effect of the inductive field. There is a field generating device within the driving cone. Adjacent to the cone is a positioning motor for the field generating device. The field created in a particular section of the driving cone induces a magnetic effect in the surrounding lamination. This causes the laminate and its opposing lamination to couple and rotate with the drive shaft. The ratio of diameters of the cones, at the point selected by positioning the field-generating component, determines the speed ratio.

Fig. 1

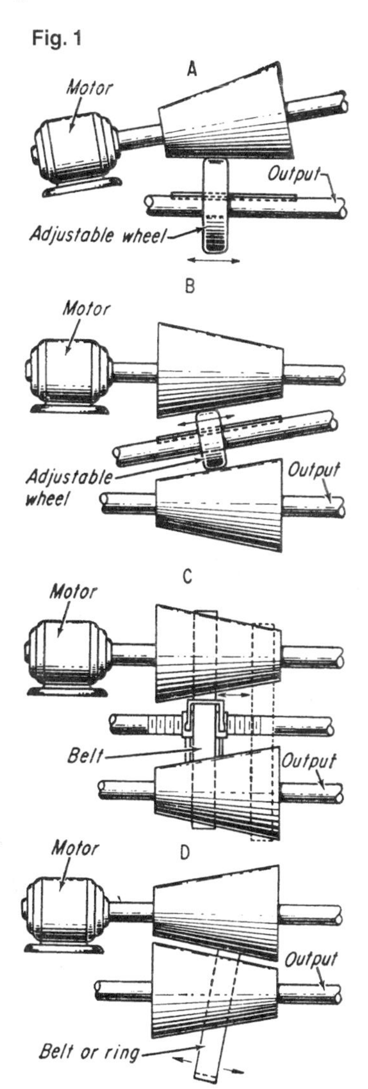

Fig. 2

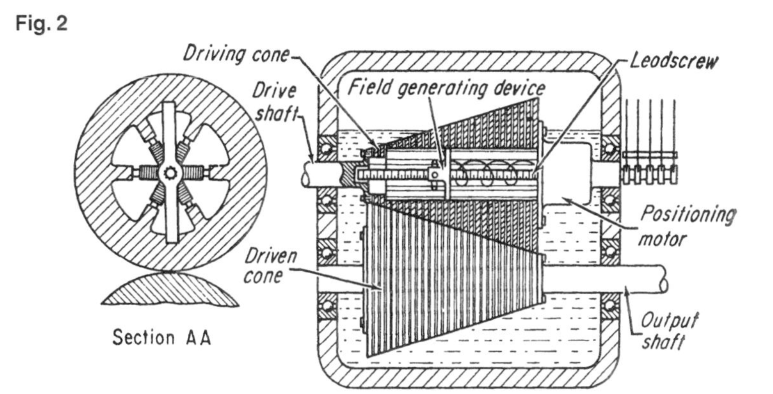

Fig. 3

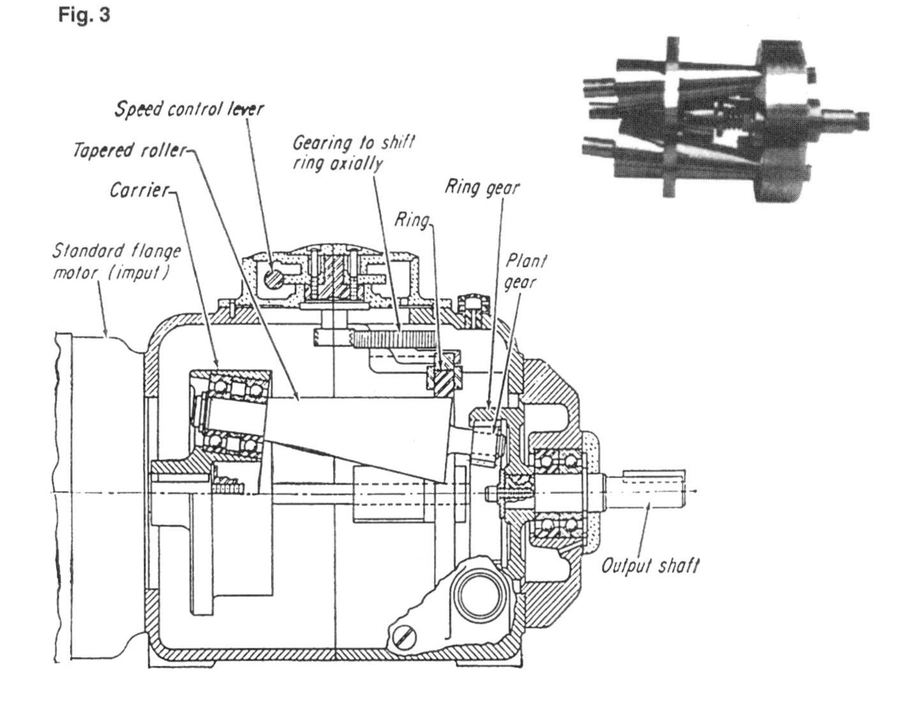

Graham drive (Fig. 3). This commercial unit combines a planetary-gear set and three tapered rollers (only one of which is shown). The ring is positioned axially by a cam and gear arrangement. The drive shaft rotates the carrier with the tapered rollers, which are inclined at an angle equal to their taper so that their outer edges are parallel to the centerline of the assembly. Traction pressure between the rollers and ring is created by centrifugal force, or spring loading of the rollers. At the end of each roller a pinion meshes with a ring gear. The ring gear is part of the planetary gear system and is coupled to the output shaft.

The speed ratio depends on the ratio of the diameter of the fixed ring to the effective diameter of the roller at the point of contact, and is set by the axial position of the ring. The output speed, even at its maximum, is always reduced to about one-third of input speed because of the differential feature. When the angular speed of the driving motor equals the angular speed of the centers of the tapered rollers around their mutual centerline (which is set by the axial position of the nonrotating friction ring), the output speed is zero. This drive is manufactured in ratings up to 3 hp; efficiency reaches 85%.

Cone-and-ring drive (Fig. 4). Here, two cones are encircled by a preloaded ring. Shifting the ring axially varies the output speed. This principle is similar to that of the cone-and-belt drive (Fig. 1C). In this case, however, the contact pressure between ring and cones increases with load to limit slippage.

Planetary-cone drive (Fig. 5). This is basically a planetary gear system but with cones in place of gears. The planet cones are rotated by the sun cone which, in turn, is driven by the motor. The planet cones are pressed between an outer non-rotating rind and the planet hold. Axial adjustment of the ring varies the rotational speed of the cones around their mutual axis. This varies the speed of the planet holder and the output shaft. Thus, the mechanism resembles that of the Graham drive (Fig. 3).

The speed adjustment range of the unit illustrated if from 4:1 to 24:1. The system is built in Japan in ratings up to 2 hp.

Fig. 4

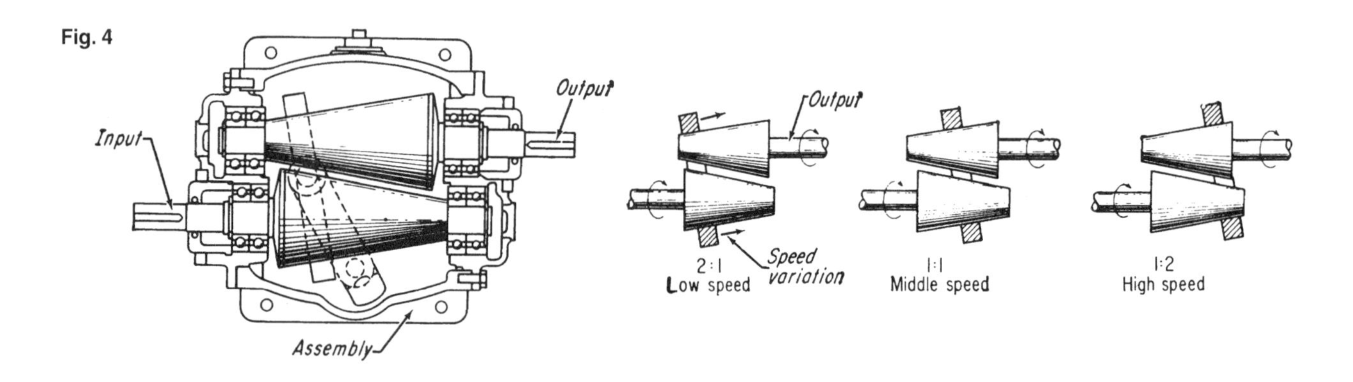

Fig. 5

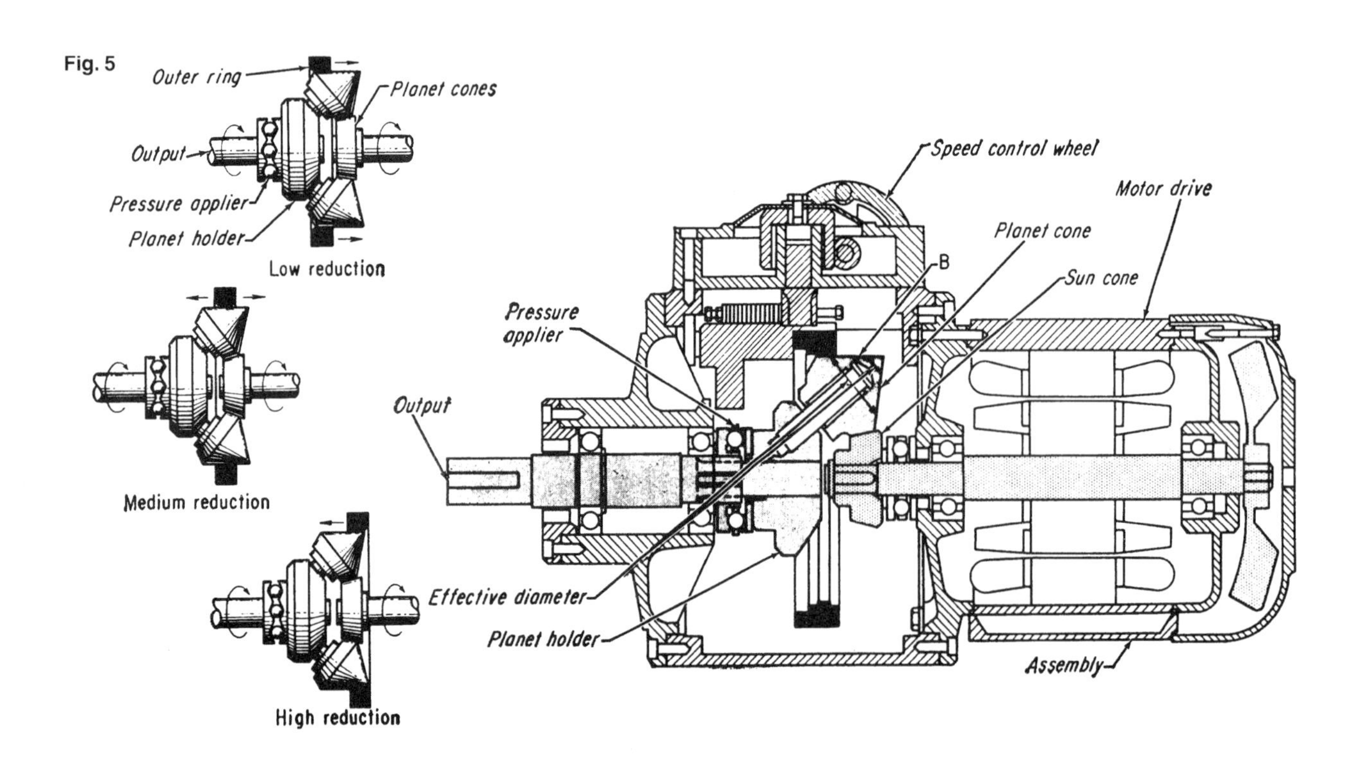

Adjustable disk drives (Figs. 6A and 6B). The output shaft in Fig. 7A is perpendicular to the input shaft. If the driving power, the friction force, and the efficiency stay constant, the output torque decreases in proportion to increasing output speed. The wheel is made of a high-friction material, and the disk is made of steel. Because of relatively high slippage, only small torques can be transmitted. The wheel can move over the center of the disk because this system has infinite speed adjustment.

To increase the speed, a second disk can be added. This arrangement (Fig. 6B) also makes the input and output shafts parallel.

Spring-loaded disk drive (Fig. 7). To reduce slippage, the contact force between the rolls and disks in this commercial drive is increased with the spring assembly in the output shaft. Speed adjustments are made by rotating the leadscrew to shift the cone roller in the vertical direction. The drive illustrated has a 4-hp capacity. Drives rated up to 20 hp can have a double assembly of rollers. Efficiency can be as high as 92%. Standard speed range is 6:1, but units of 10:1 have been build. The power transferring components, which are made hardened steel, operate in an oil mist, thus minimizing wear.

Planetary disk drive (Fig. 8). Four planet disks replace planet gears in this friction drive. Planets are mounted on levers which control radial position and therefore control the orbit. Ring and sun disks are spring-loaded.

Fig. 6

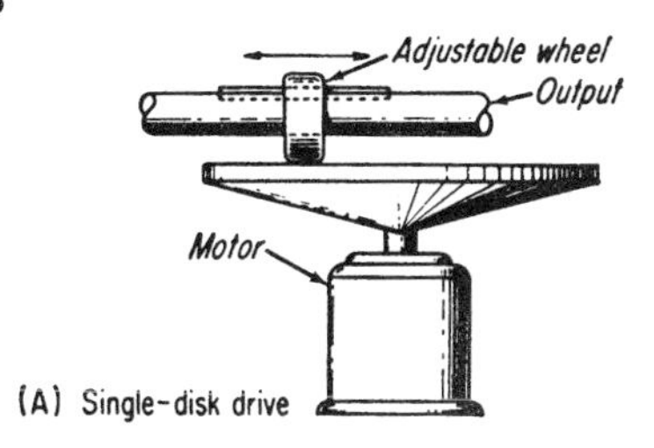

(A) Single-disk drive

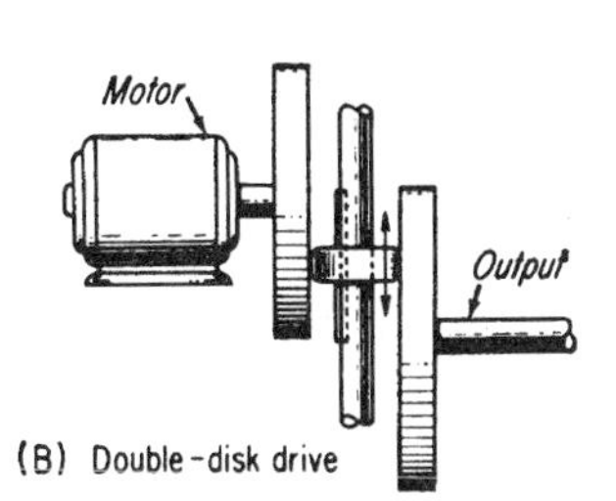

(B) Double-disk drive

Fig. 7

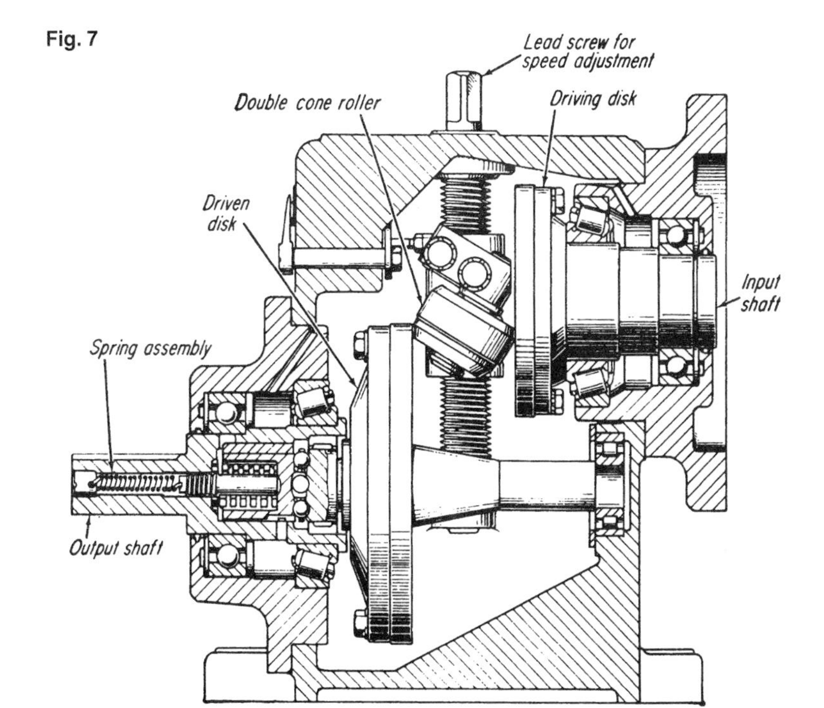

Fig. 8

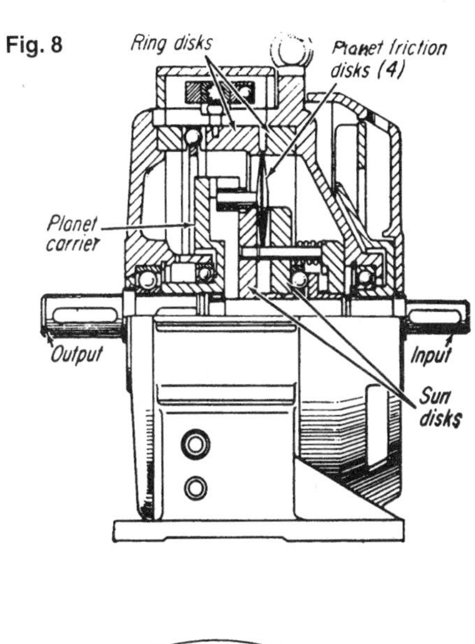

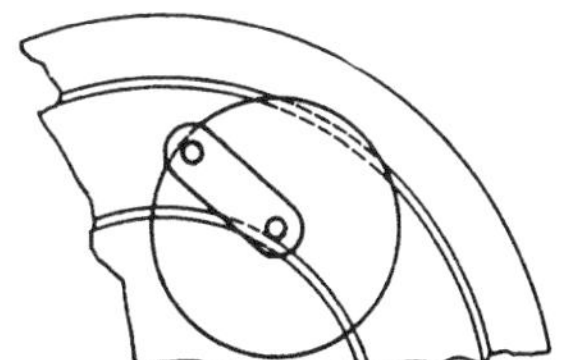

Ring-and-pulley drive (Fig. 9). A thick steel ring in this drive encircles two variable-pitch (actually variable-width) pulleys. A novel gear-and-linkage system simultaneously changes the width of both pulleys (see Fig. 9B). For example, when the top pulley opens, the sides of the bottom pulley close up. This reduces the effective pitch diameter of the top pulley and increases that of the bottom pulley, thus varying the output speed.

Normally, the ring engages the pulleys at points *A* and *B*. However, under load, the driven pulley resists rotation and the contact point moves from *B* to *D* because of the very small elastic deformation of the ring. The original circular shape of the ring is changed to a slightly oval form, and the distance between points of contact decreases. This wedges the ring between the pulley cones and increases the contact pressure between ring and pulleys in proportion to the load applied, so that constant horsepower at all speeds is obtained. The drive can have up to 3-hp capacity; speed variations can be 16:1, with a practical range of about 8:1.

Some manufacturers install rings with unusual cross sections (Fig. 10) formed by inverting one of the sets of sheaves.

Double-ring drive (Fig. 11). Power transmission is through two steel traction rings that engage two sets of disks mounted on separate shafts. This drive requires that the outer disks be under a compression load by a spring system (not illustrated). The rings are hardened and convex-ground to reduce wear. Speed is changed by tilting the ring support cage, forcing the rings to move to the desired position.

Fig. 9

a

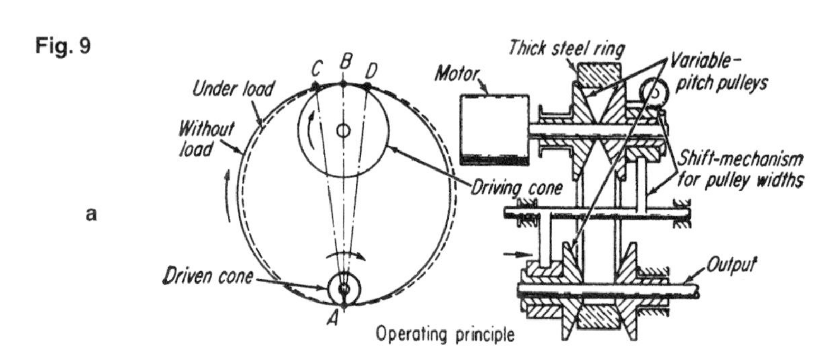

b

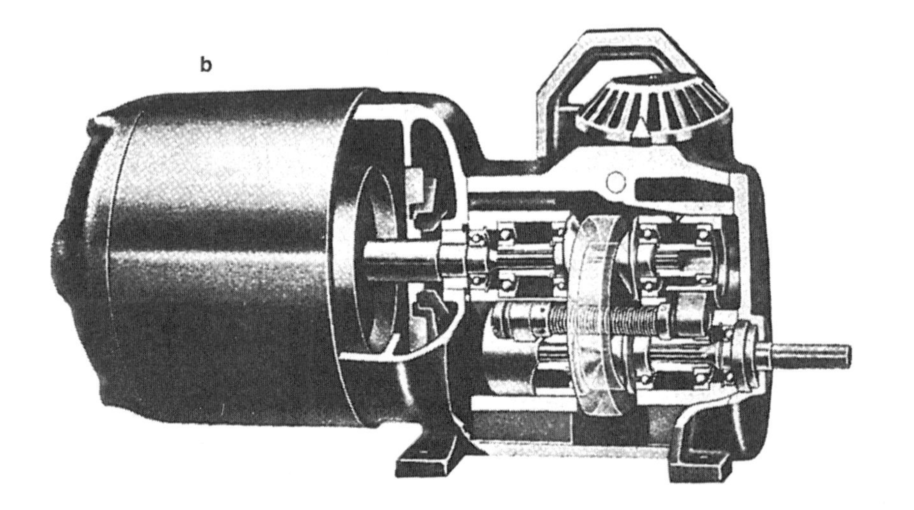

Fig. 10

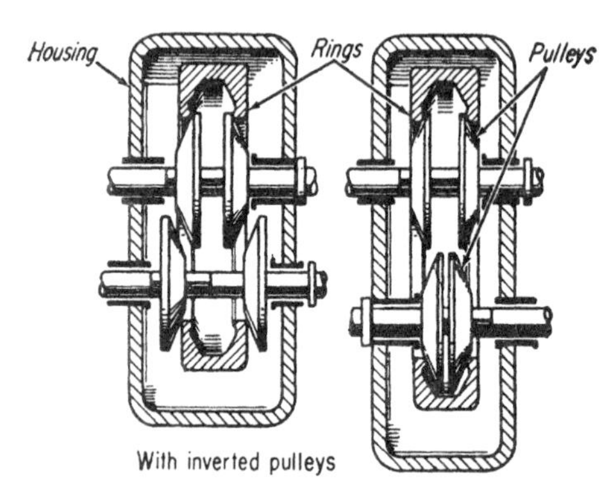

Fig. 11

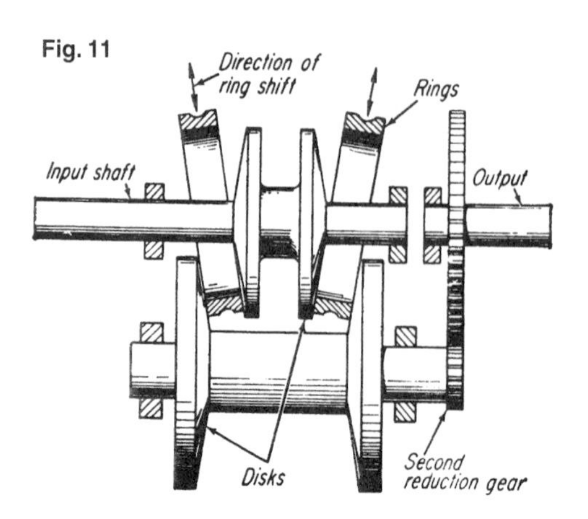

Sphere-and-disk drives (Figs. 12 and 13). The speed variations in the drive shown in Fig. 12 are obtained by changing the angle that the rollers make in contacting spherical disks. As illustrated, the left spherical disk is keyed to the driving shaft and the right disk contains the output gear. The sheaves are loaded together by a helical spring.

One commercial unit, shown in Fig. 13, is a coaxial input and output shaft-version of the Fig. 12 arrangement. The rollers are free to rotate on bearings and can be adjusted to any speed between the limits of 6:1 and 10:1. An automatic device regulates the contact pressure of the rollers, maintaining the pressure exactly in proportion to the imposed torque load.

Double-sphere drive (Fig. 14). Higher speed reductions are obtained by grouping a second set of spherical disks and rollers. This also reduces operating stresses and wear. The input shaft runs through the unit and carries two opposing spherical disks. The disks drive the double-sided output disk through two sets of three rollers. To change the ratio, the angle of the rollers is varied. The disks are axially loaded by hydraulic pressure.

Tilting-ball drive (Fig. 15). Power is transmitted between disks by steel balls whose rotational axes can be tilted to change the relative lengths of the two contact paths around the balls, and hence the output speed. The ball axes can be tilted uniformly in either direction; the effective rolling radii of balls and disks produce speed variations up to 3:1 increase, or 1:3 decrease, with the total up to 9:1 variation in output speed.

Tilt is controlled by a cam plate through which all ball axes project. To prevent slippage under starting or shock load, torque responsive mechanisms are located on the input and output sides of the drive. The axial pressure created is proportional to the applied torque. A worm drive positions the plate. The drives have been manufactured with capacities to 15-hp. The drive's efficiency is plotted in the chart.

Sphere and roller drive (Fig. 16). The roller, with spherical end surfaces, is

Fig. 12

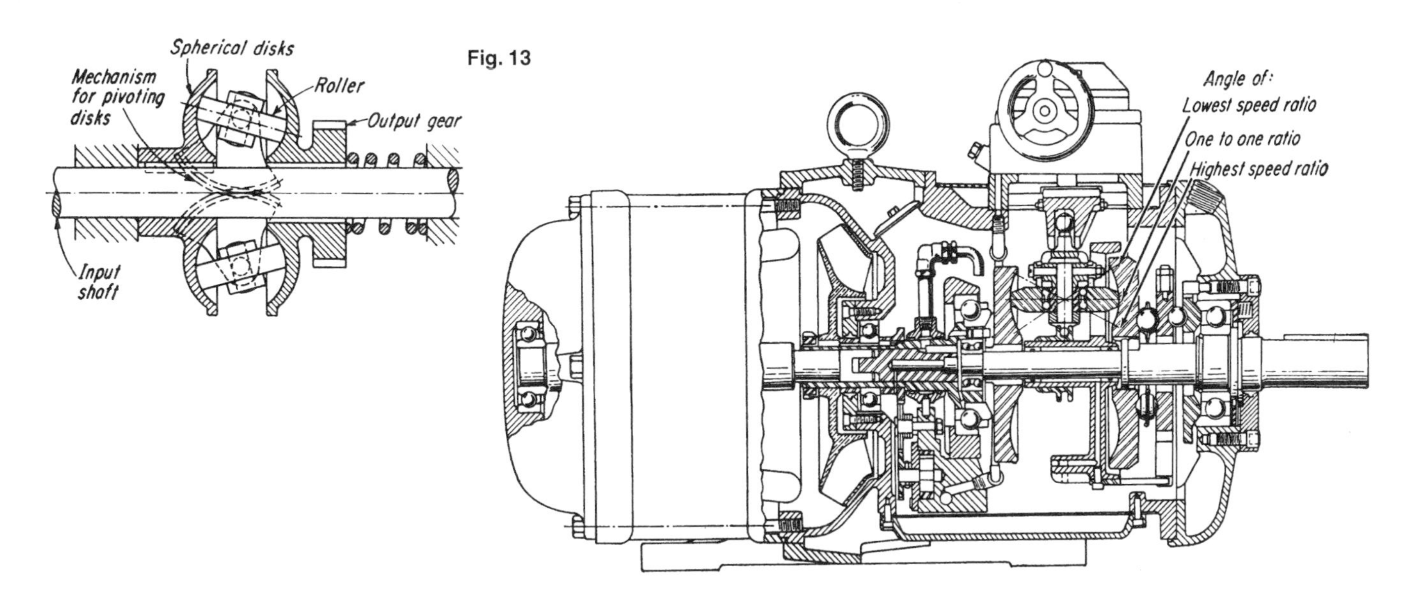

Fig. 13

Fig. 14

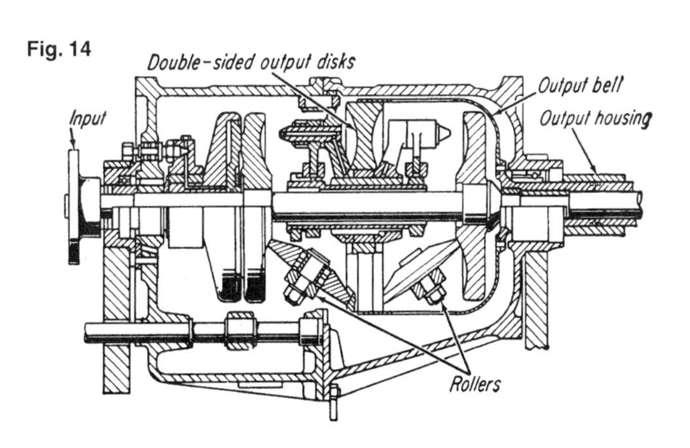

Fig. 15

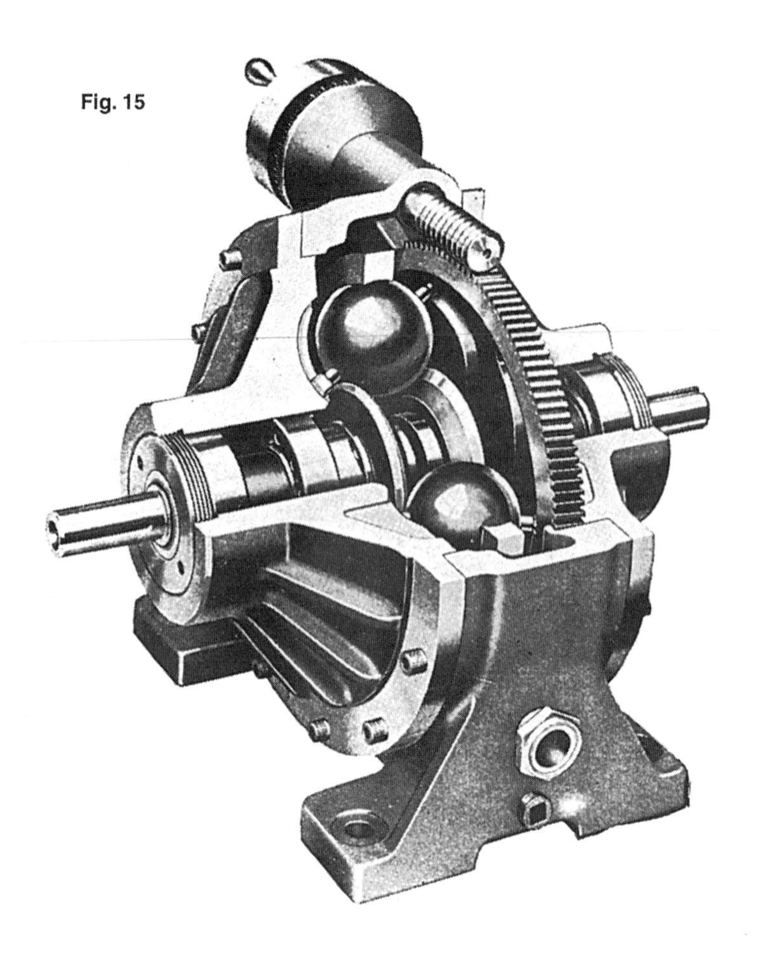

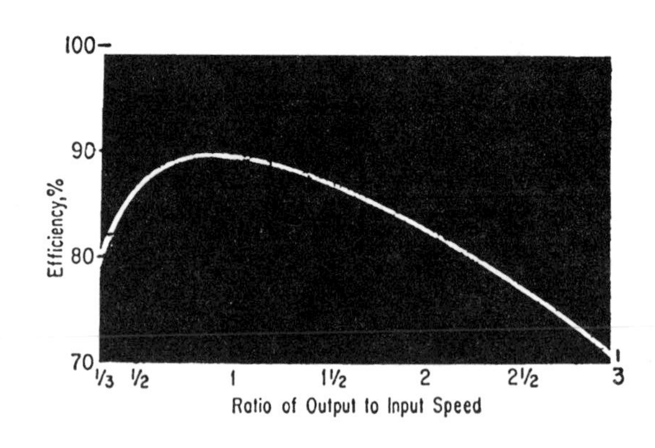

Efficiency of tilting-ball drive

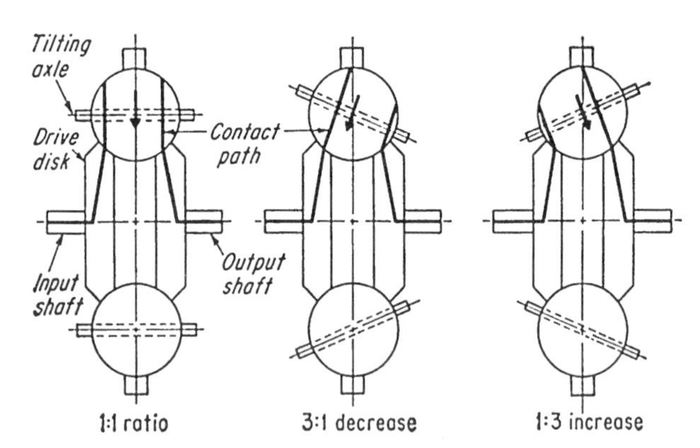

eccentrically mounted between the coaxial input and output spherical disks. Changes in speed ratio are made by changing the angular position of the roller.

The output disk rotates at the same speed as the input disk when the roller centerline is parallel to the disk center-line, as in Fig. 16A. When the contact point is nearer the centerline on the output disk and further from the centerline on the input disk, as in Fig. 16B, the output speed exceeds that of the input. Conversely, when the roller contacts the output disk at a large radius, as in Fig. 16C, the output speed is reduced.

A loading cam maintains the necessary contact force between the disks and power roller. The speed range reaches 9 to 1; efficiency is close to 90%.

Ball-and-cone drive (Fig. 17). In this simple drive the input and output shafts are offset. Two opposing cones with 90º internal vertex angles are fixed to each shaft. The shafts are preloaded against each other. Speed variation is obtained by positioning the ball that contacts the cones. The ball can shift laterally in relation to the ball plate. The conical cavities, as well as the ball, have hardened surfaces, and the drive operates in an oil bath.

Fig. 16

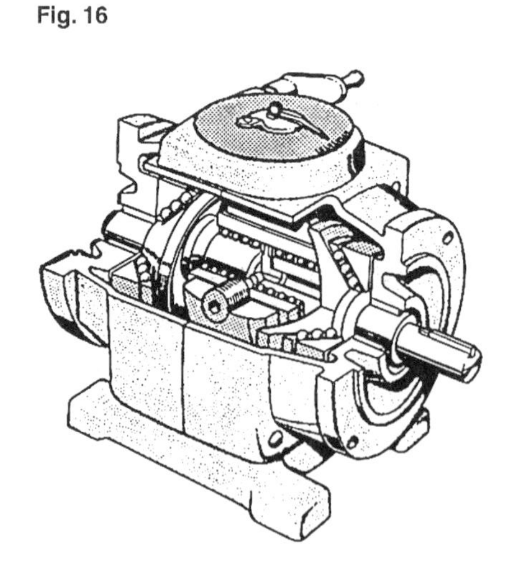

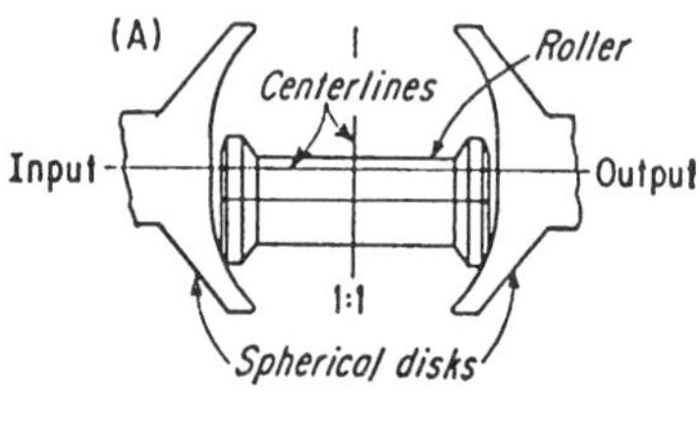

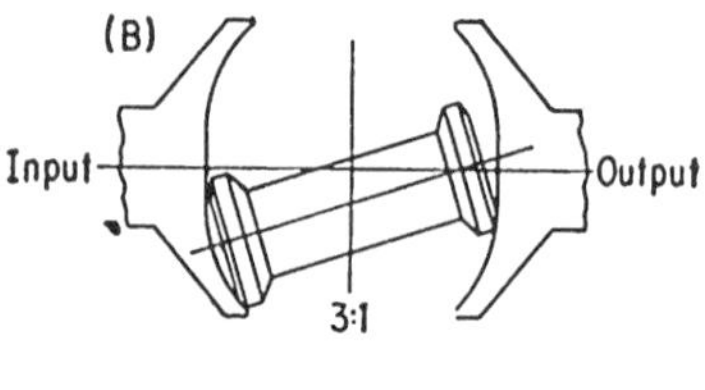

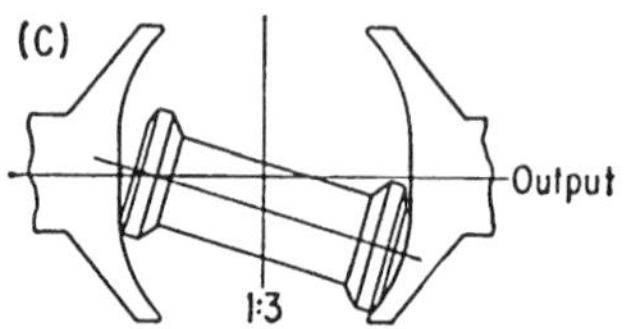

Fig. 17

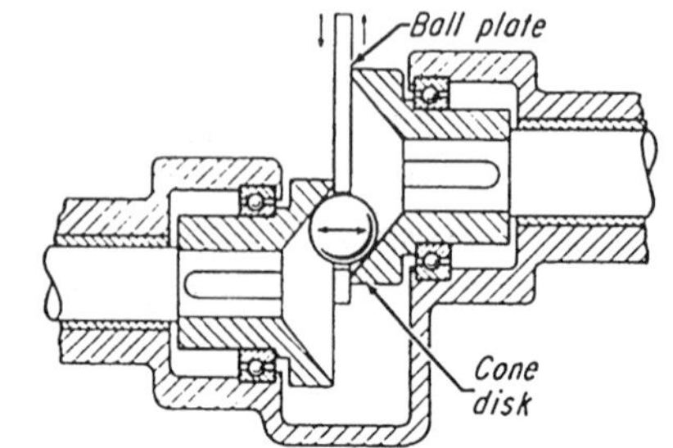

Ball-and-disk drive (Fig. 18). Friction disks are mounted on splined shafts to allow axial movement. The steel balls carried by swing arms rotate on guide rollers, and are in contact with driving and driven disks. Belleville springs provide the loading force between the balls and the disks. The position of the balls controls the ratio of contact radii, and thus the speed.

Only one pair of disks is required to provide the desired speed ratio; the multiple disks increase the torque capacity. If the load changes, a centrifugal loading device increases or decreases the axial pressure in proportion to the speed. The helical gears permit the output shaft to be coaxial with respect to the input shaft. Output to input speed ratios are from 1 to 1 to 1 to 5, and the drive's efficiency can reach 92%. Small ball and disk drives are rated to 9 hp, and large ball and disk drives are rated to 38 hp.

Oil-coated disks (Fig. 19). Power is transmitted without metal-to-metal contact at 85% efficiency. The interleaved disk sets are coated with oil when operating. At their points of contact, axial pressure applied by the rim disks compresses the oil film, increasing its viscosity. The cone disks transmit motion to the rim disks by shearing the molecules of the high-viscosity oil film.

Three stacks of cone disks (only one stack is shown) surround the central rim stack. Speed is changed by moving the cones radially toward the rim disks (output speed increases) or away from the rim disks (output speed decreases). A spring and cam on the output shaft maintain the pressure of the disks at all times.

Drives with ratings in excess of 60 hp have been built. The small drives are cooled, but water cooling is required for the larger units.

Under normal conditions, the drive can transmit its rated power with a 1% slip at high speeds and 3% slip at low speeds.

Fig. 18

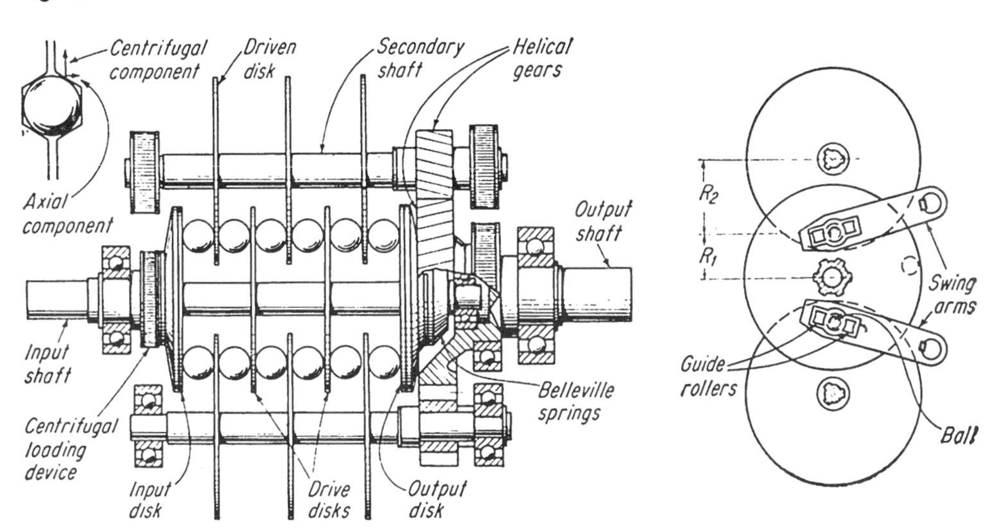

Fig. 19

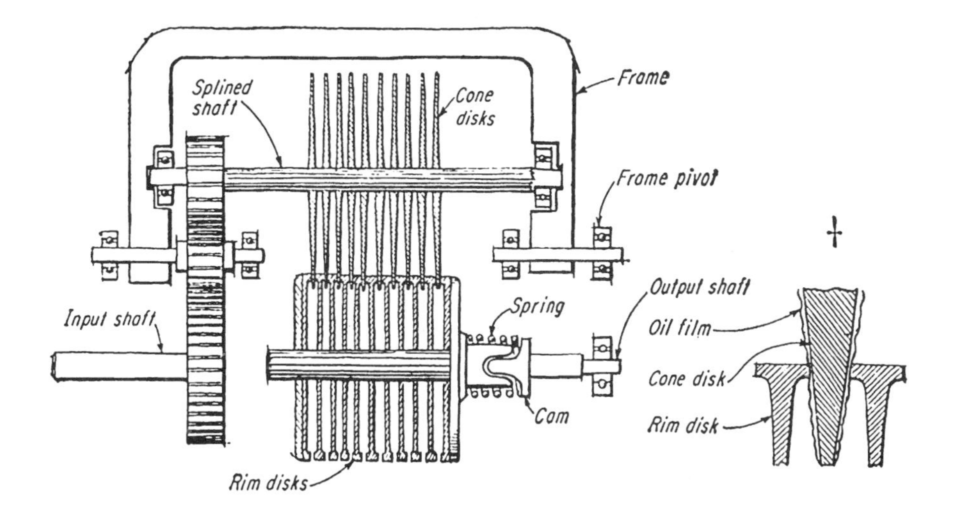

Variable-stroke drive (Fig. 20). This drive is a combination of a four-bar linkage with a one-way clutch or ratchet. The driving member rotates the eccentric that, through the linkage, causes the output link to rotate a fixed amount. On the return stroke, the output link overrides the output shaft. Thus a pulsating motion is transmitted to the output shaft, which in many applications such as feeders and mixers, is a distinct advantage. Shifting the adjustable pivot varies the speed ratio. By adding eccentrics, cranks, and clutches in the system, the frequency of pulsations per revolution can be increased to produce a smoother drive.

Morse drive (Fig. 21). The oscillating motion of the eccentric on the output shaft imparts motion to the input link, which in turn rotates the output gears. The travel of the input link is regulated by the control link that oscillates around its pivot and carries the roller, which rides in the eccentric cam track. Usually, three linkage systems and gear assemblies overlap the motions: two linkages on return, while the third is driving. Turning the handle repositions the control link and changes the oscillation angles of the input link, intermediate gear, and input gear. This is a constant-torque drive with limited range. The maximum torque output is 175 ft-lb at the maximum input speed of 180 rpm. Speed can be varied between 4.5 to 1 and 120 to 1.

Zero-Max drive (Fig. 22). This drive is also based on the variable-stroke principle. With an 1800-rpm input, it will deliver 7200 or more impulses per minute to the output shaft at all speed ratings above zero. The pulsations of this drive are damped by several parallel sets of mechanisms between the input and output shafts. (Figure 22 shows only one of these sets.)

At zero input speed, the eccentric on the input shaft moves the connecting rod up and down through an arc. The main link has no reciprocating motion. To set the output speed, the pivot is moved (upward in the figure), thus changing the direction of the connecting rod motion and imparting an oscillatory motion to the main link. The one-way clutch mounted on the output shaft provides the ratchet action. Reversing the input shaft rotation does not reverse the output. However, the drive can be reversed in two ways: (1) with a special reversible clutch, or (2) with a bellcrank mechanism in gearhead models.

This drive is classified as an infinite-speed range drive because its output speed passes through zero. Its maximum speed is 2000rpm, and its speed range is from zero to one-quarter of its input speed. It has a maximum rated capacity of $^{3}/_{4}$ hp.

Fig. 20

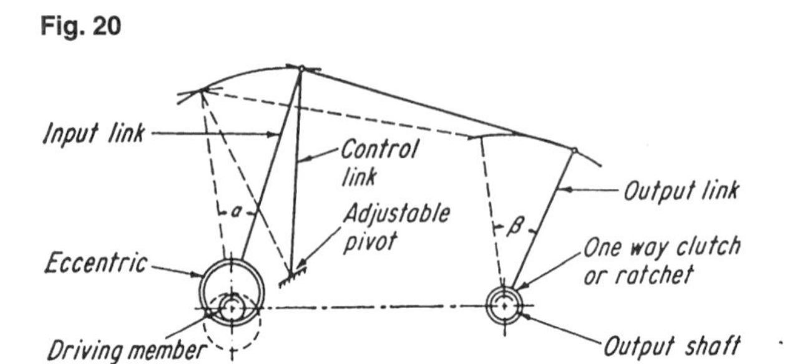

Fig. 21

Output gears
Cam (eccentric)
Input link
Control link
Pivot plate
Output shaft
One-way clutches

Fig. 22

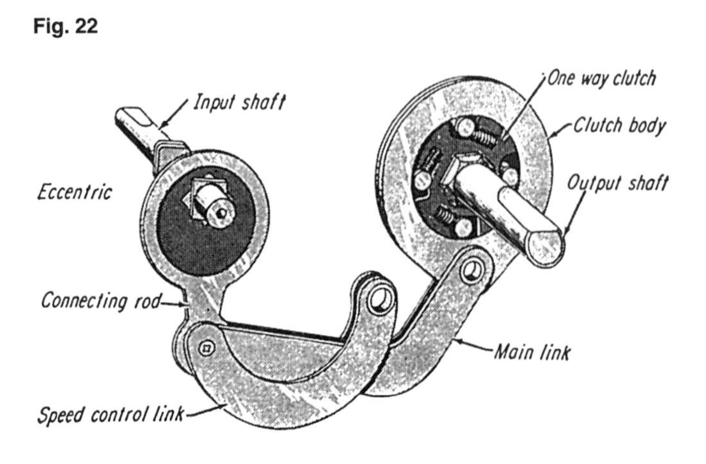

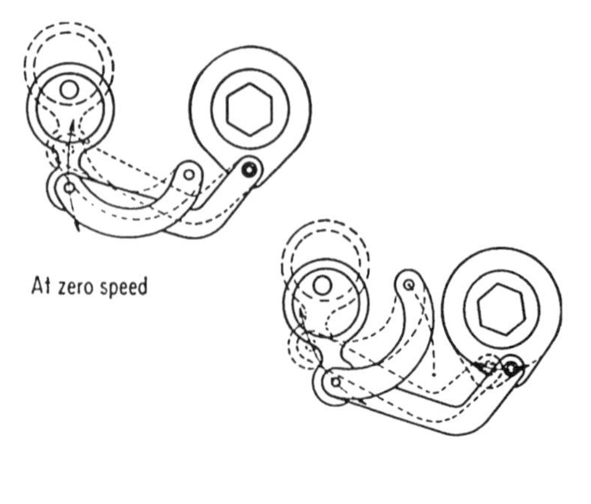

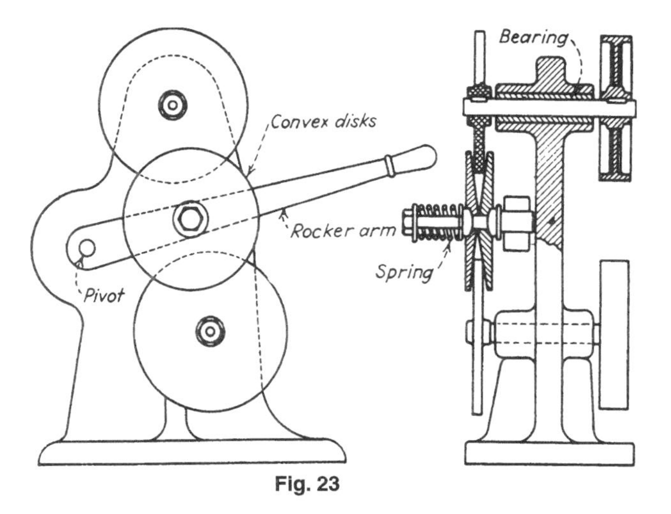

Fig. 23

Fig. 23 The Sellers' disks consist of a mechanism for transmitting power between fixed parallel shafts. Convex disks are mounted freely on a rocker arm, and they are pressed firmly against the flanges of the shaft wheels by a coiled spring to form the intermediate sheave. The speed ratio is changed by moving the rocker lever. No reverse is possible, but the driven shaft can rotate above or below the driver speed. The convex disk must be mounted on self-aligning bearings to ensure good contact in all positions.

Fig. 24 A curved disk device is formed by attaching a motor that is swung on its pivot so that it changes the effective diameters of the contact circles. This forms a compact drive for a small drill press.

Fig. 25 This is another motorized modification of the older mechanism shown in Fig. 2. It works on the principle that is similar that of Fig. 2, but it has only two shafts. Its ratio is changed by sliding the motor in vee guides.

Fig. 26 Two cones mounted close together and making contact through a squeezed belt permit the speed ratio to be changed by shifting the belt longitudinally. The taper on the cones must be moderate to avoid excessive wear on the sides of the belt.

Fig. 27 These cones are mounted at any convenient distance apart. They are connected by a belt whose outside edges consist of an envelope of tough, flexible rubberized fabric that is wear-resistant. It will withstand the wear caused by the belt edge traveling at a slightly different velocity that that part of the cone it actually contacts. The mechanism's speed ratio is changed by sliding the belt longitudinally

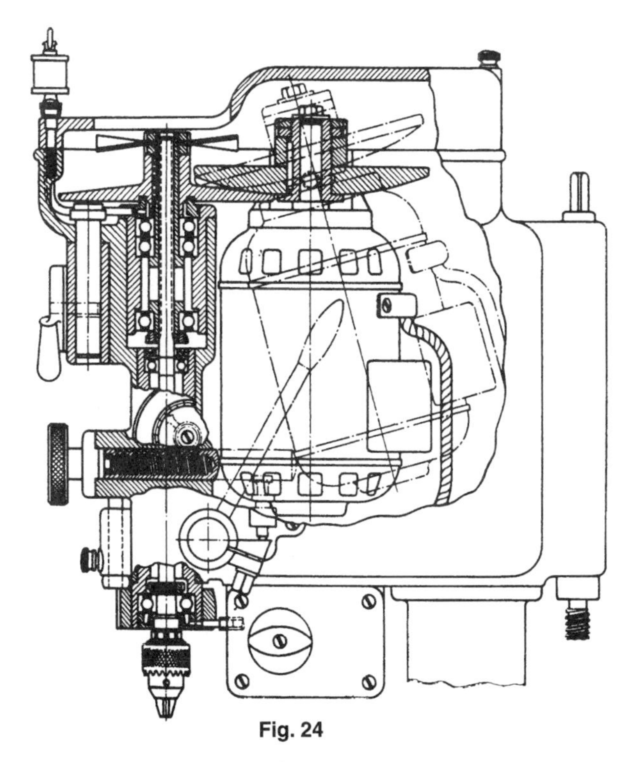
Fig. 24

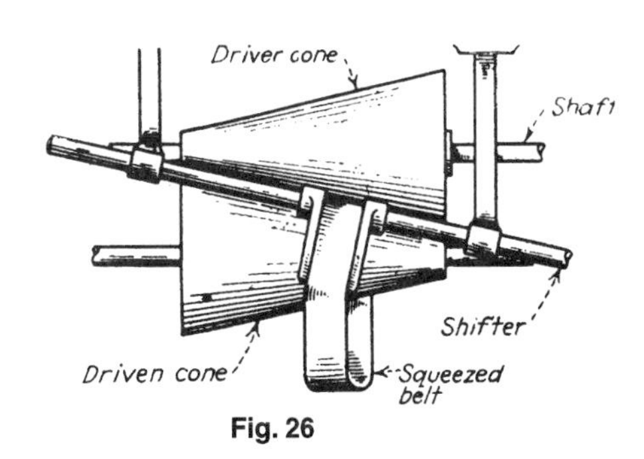

Fig. 26

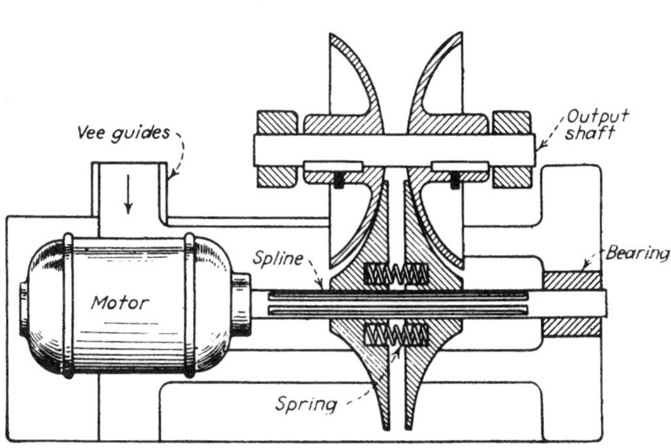

Fig. 25

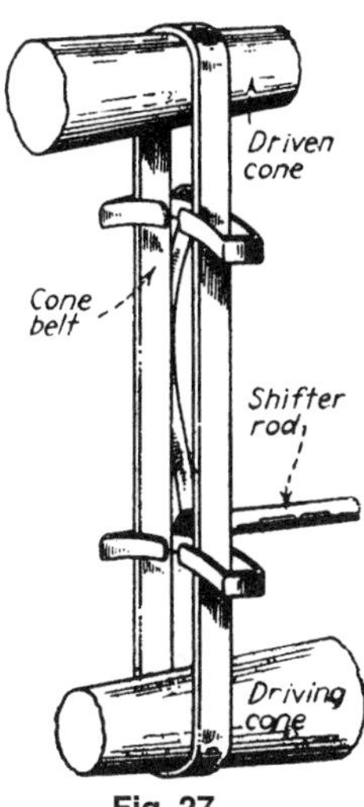

Fig. 27

Fig. 28 This drive avoids belt "creep" and wear in speed-cone transmissions. The inner bands are tapered on the inside, and they present a flat or crowned contact surface for the belt in all positions. The speed ratio is changed by moving the inner bands rather than the main belts.

Fig. 29 This drive avoids belt wear when the drive has speed cones. However, the creeping action of the belt is not eliminated, and the universal joints present ongoing maintenance problems.

Fig. 30 This drive is a modification of the drive shown in Fig. 7. A roller is substituted for the belt, reducing the overall size of the drive.

Fig. 31 The main component of this drive is a hollow internal cone driven by a conical pulley on the motor shaft. Its speed ratio can be changed by sliding the motor and pulley up or down in the vee slide. When the conical pulley on the motor shaft is moved to the center of the driving cone, the motor and cone run at the same speed. This feature makes the system attractive in applications where heavy torque requirements are met at the motor's rated speed and it is useful to have lower speeds for light preliminary operations.

Fig. 32 In this transmission, the driving pulley cone and driven cone are mounted on the same shaft with their small diameters directed toward each other. The driving pulley (at right) is keyed to the common shaft, and the driven cone (at left) is mounted on a sleeve. Power is transmitted by a series of rocking shafts with rollers mounted on their ends. The shafts are free to slide while they are pivoted within sleeves within a disk that is perpendicular to the driven-cone mounting sleeve. The speed ratio can be changed by pivoting the rocking shafts and allowing them to slide across the conical surfaces of the driving pulley and driven cone.

Fig. 33 This transmission has curved surfaces on its planetary rollers and races. The cone shaped inner races revolve with the drive shaft, but are free to slide longitudinally on sliding keys. Strong compression springs keeps the races in firm contact with the three planetary rollers.

Fig. 34 This Graham transmission has only five major parts. Three tapered rollers are carried by a spider fastened to the drive shaft. Each roller has a pinion that meshes with a ring gear connected to the output shaft. The speed of the rollers as well as the speed of the output shaft is varied by moving the contact ring longitudinally. This movement changes the ratio of the contacting diameters.

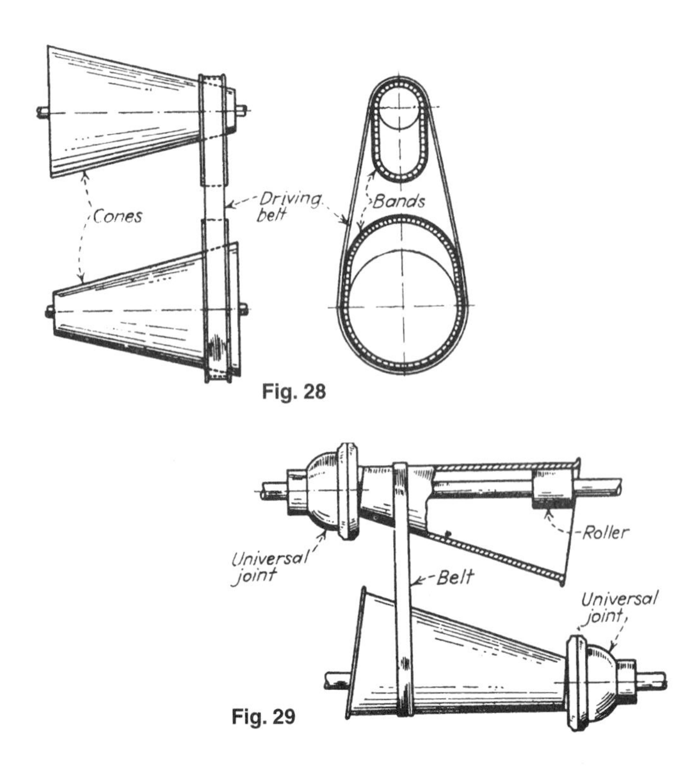

Fig. 28

Fig. 29

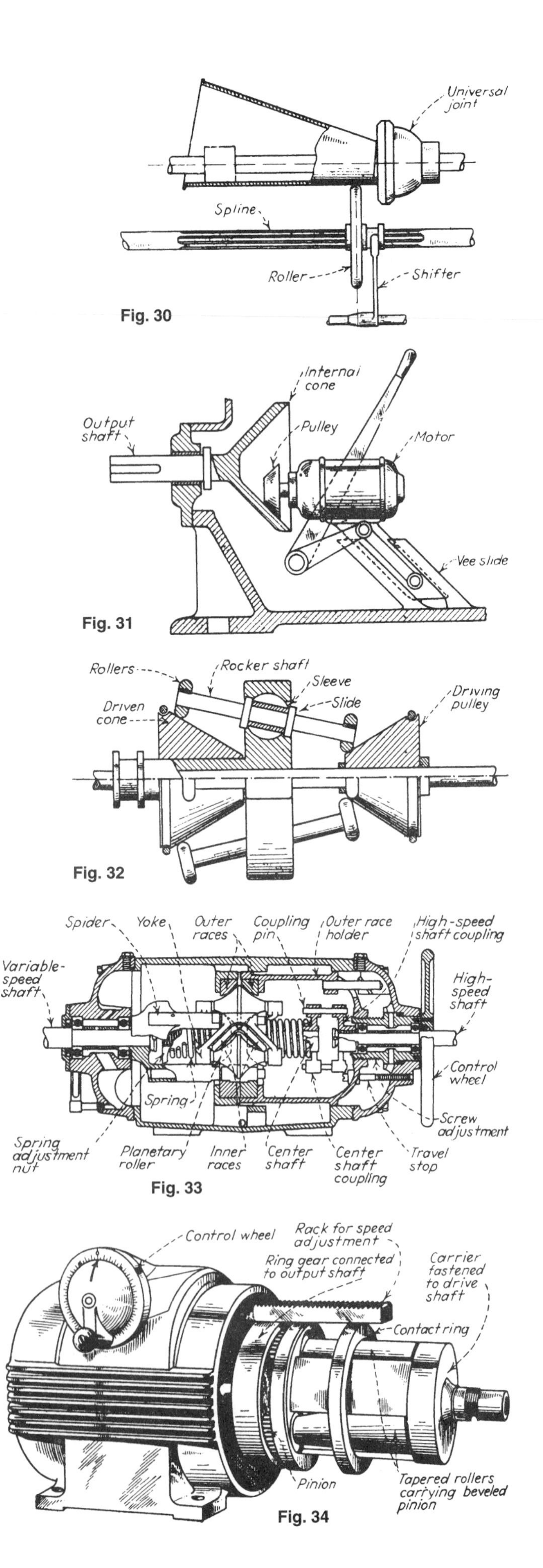

Fig. 30

Fig. 31

Fig. 32

Fig. 33

Fig. 34

These ratchet and inertial drives provide variable-speed driving of heavy and light loads.

Fig. 35 This variable-speed drive is suitable only for very light duty in a laboratory or for experimental work. The drive rod receives motion from the drive shaft and it rocks the lever. A friction clutch is formed in a lathe by winding wire around a drill rod whose diameter is slightly larger than the diameter of the driven shaft. The speed ratio can be changed when the drive is stationary by varying the length of the rods or the throw of the eccentric.

Fig. 36 This Torrington lubricator drive illustrates the general principles of ratchet transmission drives. Reciprocating motion from a convenient sliding part, or from an eccentric, rocks the ratchet lever. That motion gives the variable-speed shaft an intermittent unidirectional motion. The speed ratio can be changed only when the unit is stationary. The throw of the ratchet lever can be varied by placing a fork of the driving rod in a different hole.

Fig. 37 This drive is an extension of the principle illustrated in Fig. 2. The Lenney transmission replaces the ratchet with an over-running clutch. The speed of the driven shaft can be varied while the unit is in motion by changing the position of the connecting-lever fulcrum.

Fig. 38 This transmission is based on the principle shown in Fig. 3. The crank disk imparts motion to the connecting rod. The crosshead moves toggle levers which, in turn, give unidirectional motion to the clutch wheel when the friction pawls engage in a groove. The speed ratio is changed by varying the throw of the crank with the aid of a rack and pinion.

Fig. 39 This is a variable speed transmission for gasoline-powered railroad section cars. The connecting rod from the crank, mounted on a constant-speed shaft, rocks the oscillating lever and actuates the over-running clutch. This gives intermittent but unidirectional motion to the variable-speed shaft. The toggle link keeps the oscillating lever within the prescribed path. The speed ratio is changed by swinging the bell crank toward the position shown in the dotted lines, around the pivot attached to the frame. This varies the movement of the over-running clutch. Several units must be out-of-phase with each other for continuous shaft motion.

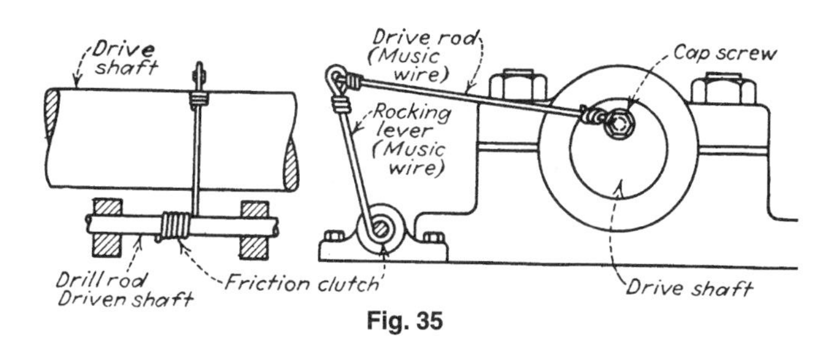

Fig. 35

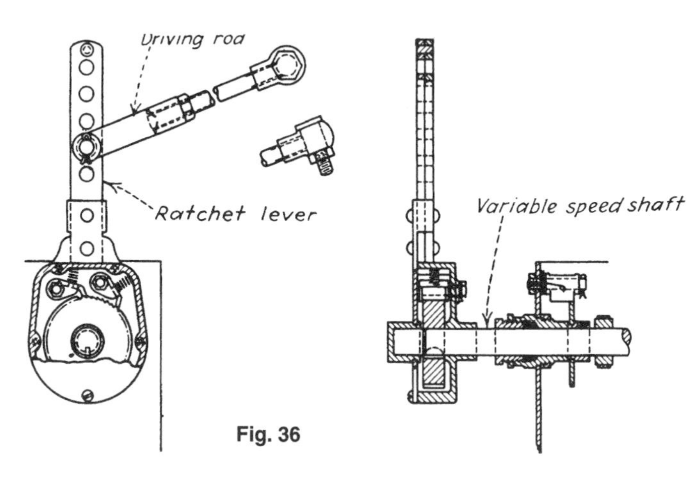

Fig. 36

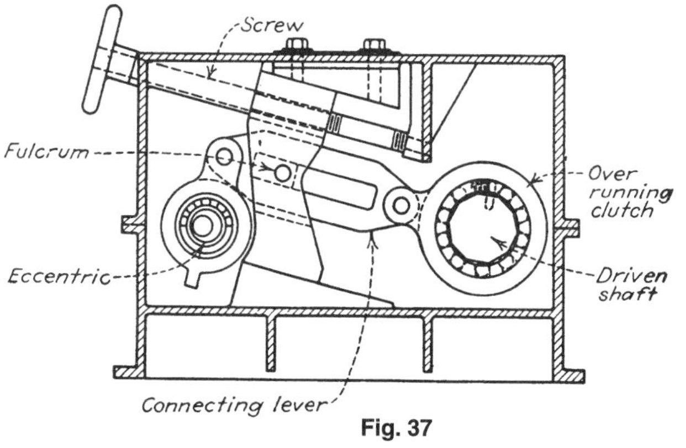

Fig. 37

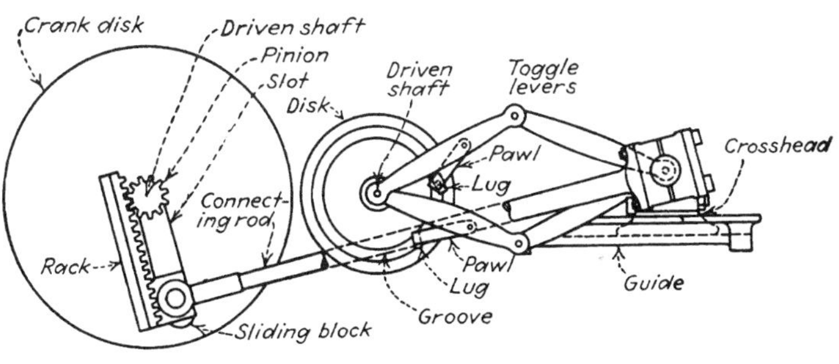

Fig. 38

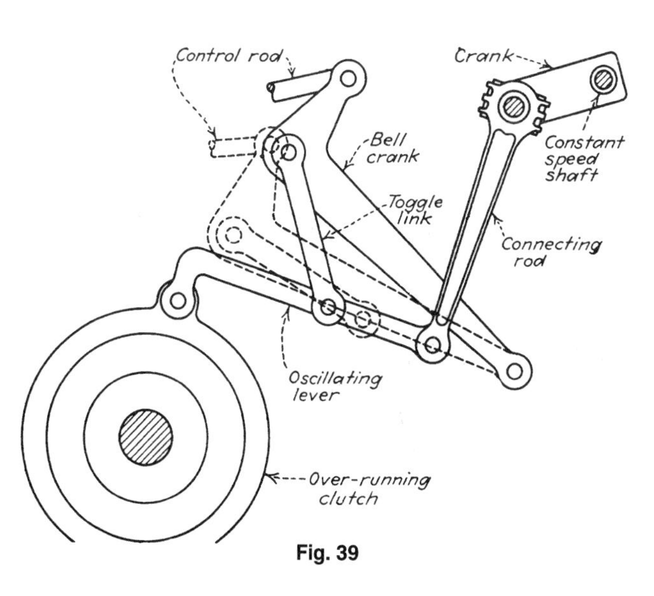

Fig. 39

Fig. 40 This Thomas transmission is an integral part of an automobile engine whose piston motion is transferred by a conventional connecting rod to the long arm of the bellcrank lever oscillating about a fixed fulcrum. A horizontal connecting rod, which rotates the crankshaft, is attached to the short arm of the bellcrank. Crankshaft motion is steadily and continuously maintained by a flywheel. However, no power other than that required to drive auxiliaries is taken from this shaft. The main power output is transferred from the bellcrank lever to the over-running clutch by a third connecting rod. The speed ratio is changed by sliding the top end of the third connecting rod within the bellcrank lever with a crosshead and guide mechanism. The highest ratio is obtained when the crosshead is farthest from the fulcrum, and movement of the crosshead toward the fulcrum reduces the ratio until a "neutral" position is reached. That occurs when the center line of the connecting rod coincides with the fulcrum.

Fig. 41 This Constantino torque converter is another automotive transmission system designed and built as part of the engine. It features an inherently automatic change of speed ratio that tracks the speed and load on the engine. The constant-speed shaft rotates a crank which, in turn, drives two oscillating levers with inertia weights at their ends. The other ends are attached by links to the rocking levers. These rocking levers include over-running clutches. At low engine speeds, the inertia weights oscillate through a wide angle. As a result, the reaction of the inertia force on the other end of the lever is very slight, and the link imparts no motion to the rocker lever. Engine speed increases cause the inertia weight reaction to increase. This rocks the small end of the oscillating lever as the crank rotates. The resulting motion rocks the rocking lever through the link, and the variable shaft is driven in one direction.

Fig. 42 This transmission has a differential gear with an adjustable escapement. This arrangement bypasses a variable portion of the drive-shaft revolutions. A constant-speed shaft rotates a freely mounted worm wheel that carries two pinion shafts. The firmly fixed pinions on these shafts, in turn, rotate the sun gear that meshes with other planetary gears. This mechanism rotates the small worm gear attached to the variable-speed output shaft.

Fig. 43 This Morse transmission has an eccentric cam integral with its constant-speed input shaft. It rocks three ratchet clutches through a series of linkage systems containing three rollers that run in a circular groove cut in the cam face. Unidirectional motion is transmitted to the output shaft from the clutches by planetary gearing. The speed ratio is changed by rotating an anchor ring containing a fulcrum of links, thus varying the stroke of the levers.

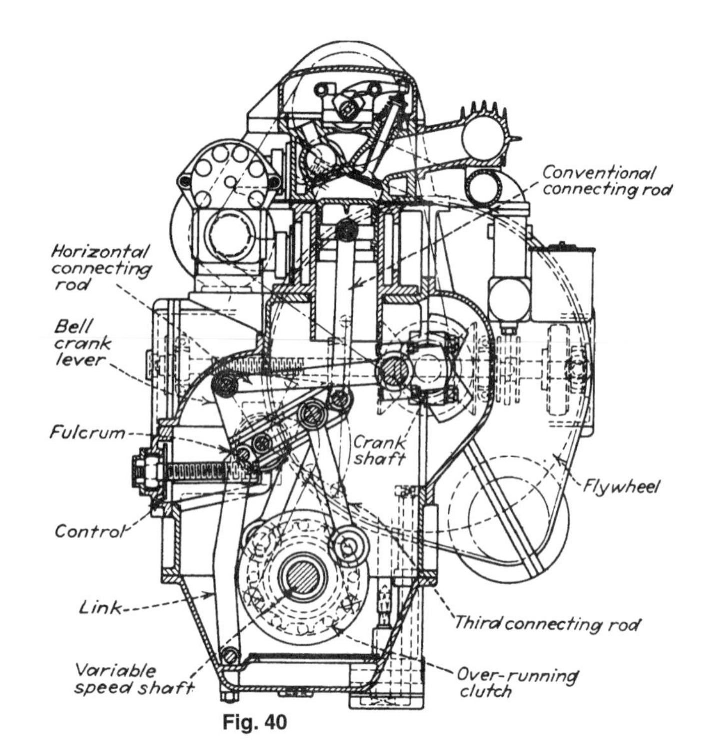

Fig. 40

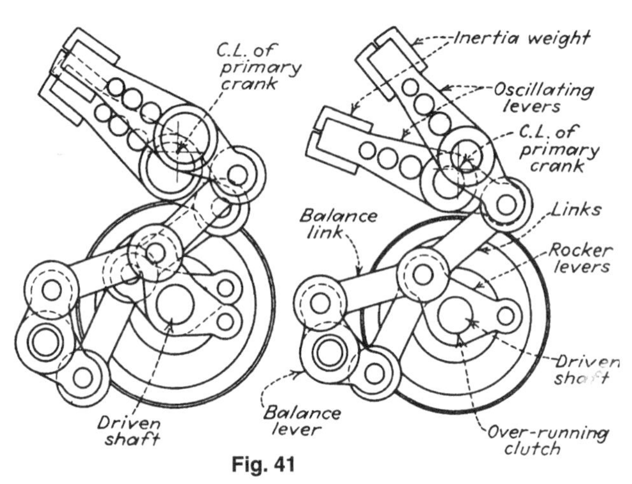

Fig. 41

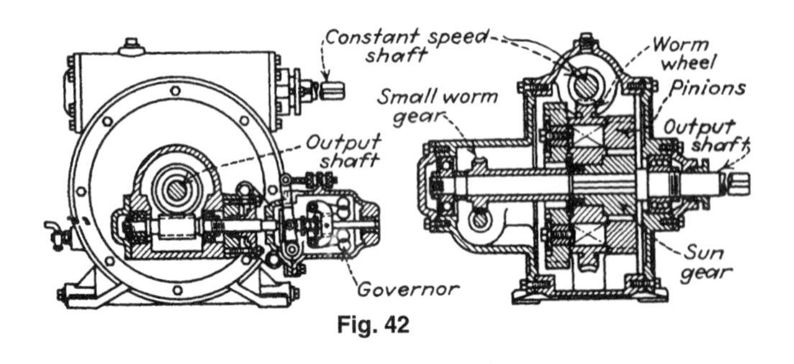

Fig. 42

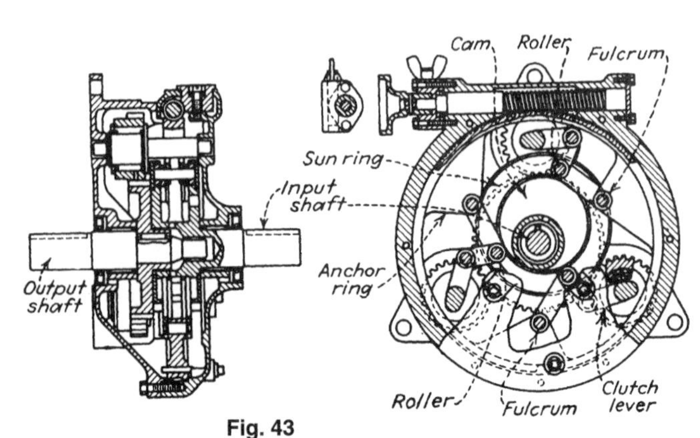

Fig. 43

TEN VARIABLE-SPEED FRICTION DRIVES

These drives can be used to transmit both high torque, as on industrial machines, and low torque, as in laboratory instruments. All perform best if they are used to reduce and not to increase speed. All friction drives have a certain amount of slip due to imperfect rolling of the friction members, but with effective design this slip can be held constant, resulting in constant speed of the driven member. Compensation for variations in load can be achieved by placing inertia masses on the driven end. Springs or similar elastic members can be used to keep the friction parts in constant contact and exert the force necessary to create the friction. In some cases, gravity will take the place of such members. Custom-made friction materials are generally recommended, but neoprene or rubber can be satisfactory. Normally only one of the friction members is made or lined with this material, while the other is metal.

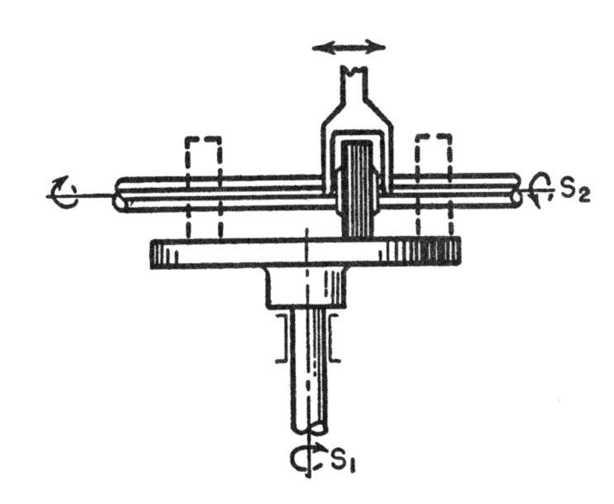

Fig. 1 A disk and roller drive. The roller is moved radially on the disk. Its speed ratio depends upon the operating diameter of the disk. The direction of relative rotation of the shafts is reversed when the roller is moved past the center of the disk, as indicated by dotted lines.

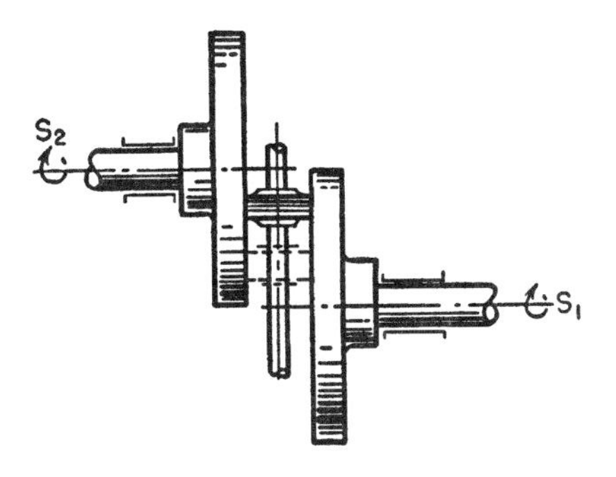

Fig. 2 Two disks have a free-spinning, movable roller between them. This drive can change speed rapidly because the operating diameters of the disks change in an inverse ratio.

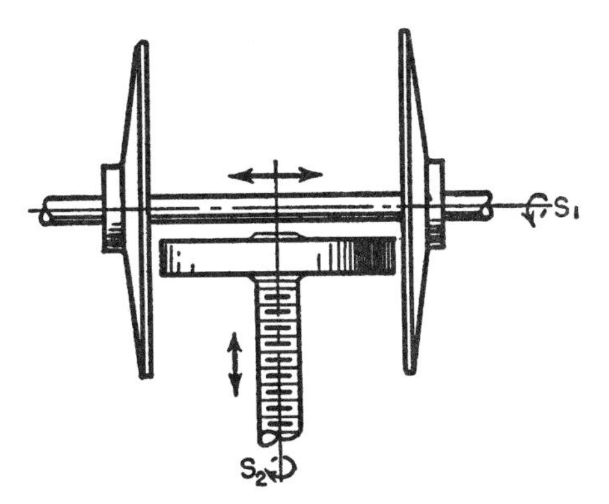

Fig. 3 Two disks are mounted on the same shaft and a roller is mounted on a threaded spindle. Roller contact can be changed from one disk to the other to change the direction of rotation. Rotation can be accelerated or decelerated by moving the screw.

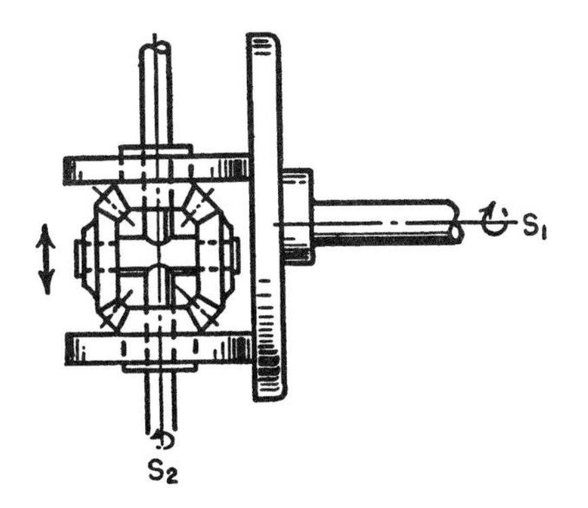

Fig. 4 A disk contacts two differential rollers. The rollers and their bevel gears are free to rotate on shaft S_2. The other two bevel gears are free to rotate on pins connected by S_2. This drive is suitable for the accurate adjustment of speed. S_2 will have the differential speed of the two rollers. The differential assembly is movable across the face of the disk.

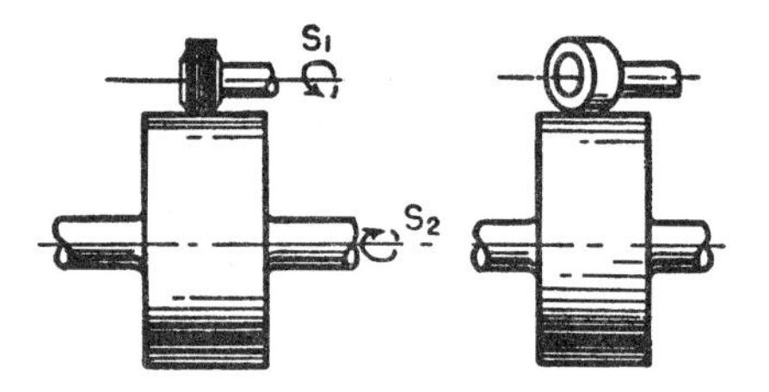

Fig. 5 This drive is a drum and roller. A change of speed is performed by skewing the roller relative to the drum.

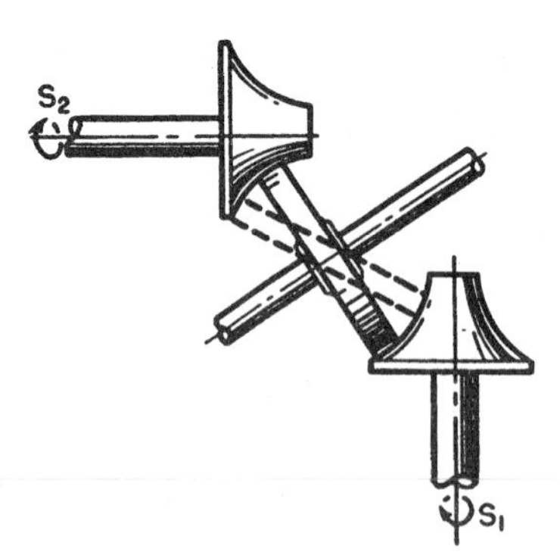

Fig. 6 This drive consists of two spherical cones on intersecting shafts and a free roller.

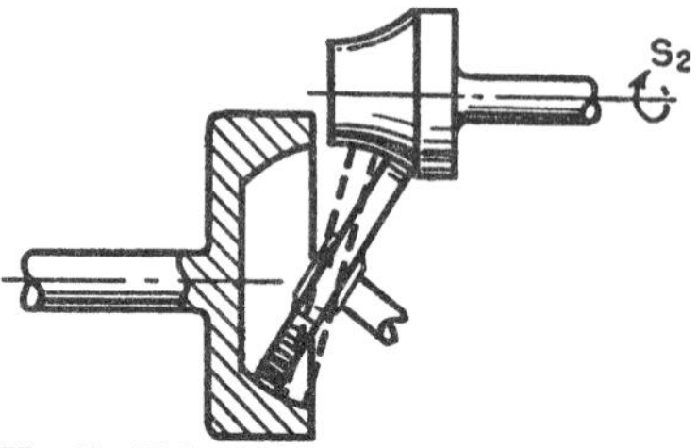

Fig. 7 This drive consists of a spherical cone and groove with a roller. It can be used for small adjustments in speed.

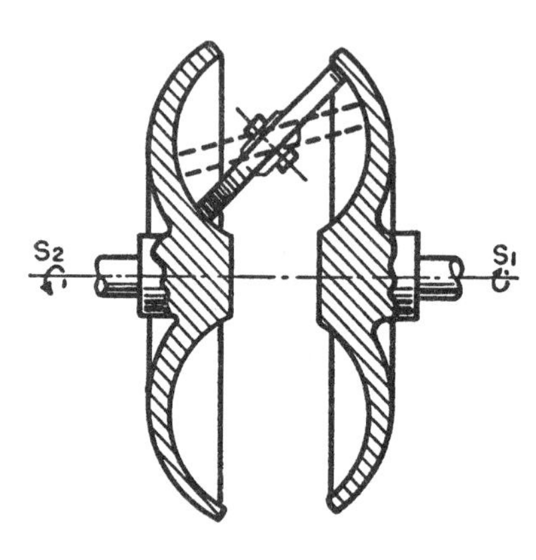

Fig. 8 This drive consists of two disks with torus contours and a free rotating roller.

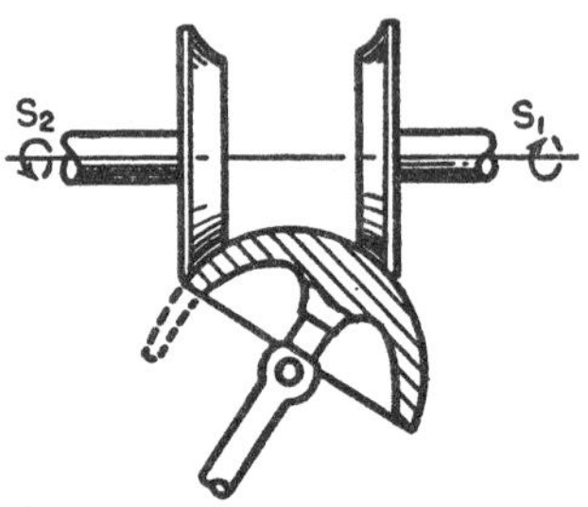

Fig. 9 This drive consists of two disks with a spherical free rotating roller.

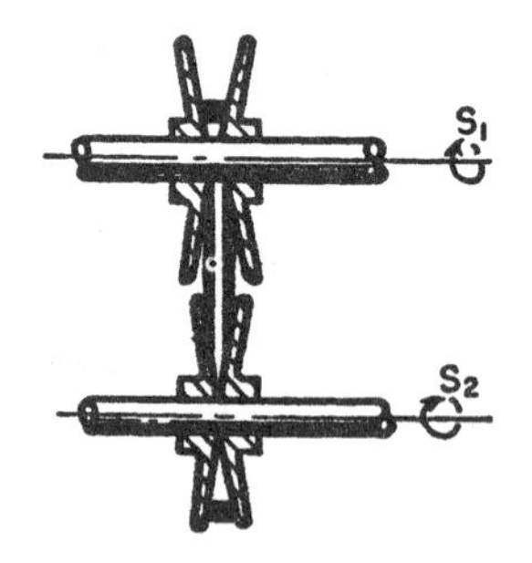

Fig. 10 This drive has split pulleys for V belts. The effective diameter of the belt grip can be adjusted by controlling the distance between the two parts of he pulley.

FOUR DRIVES CONVERT OSCILLATING MOTION TO ONE-WAY ROTATION

These four drives change oscillating motion into one-way rotation to perform feeding tasks and counting.

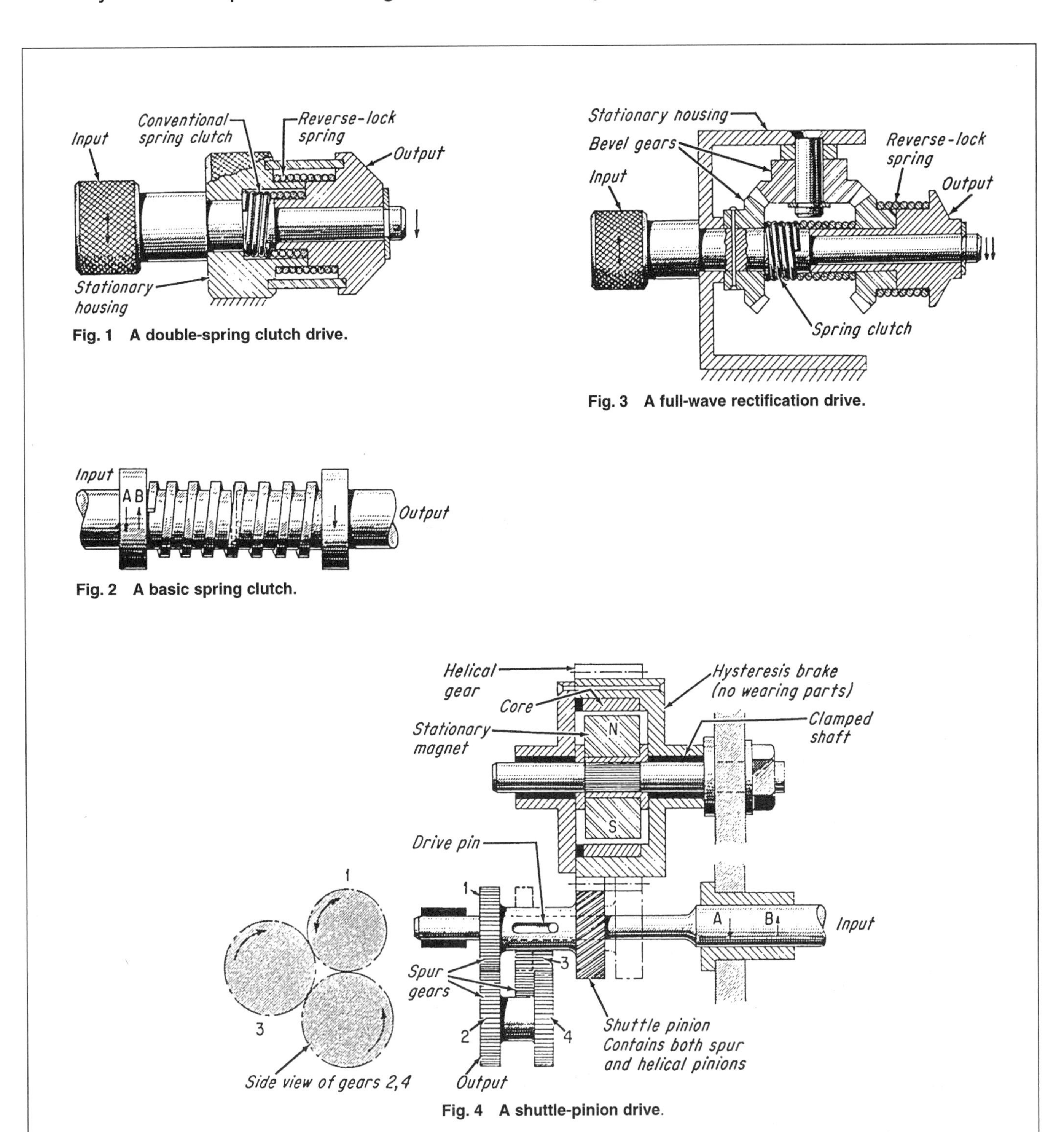

Fig. 1 A double-spring clutch drive.

Fig. 3 A full-wave rectification drive.

Fig. 2 A basic spring clutch.

Fig. 4 A shuttle-pinion drive.

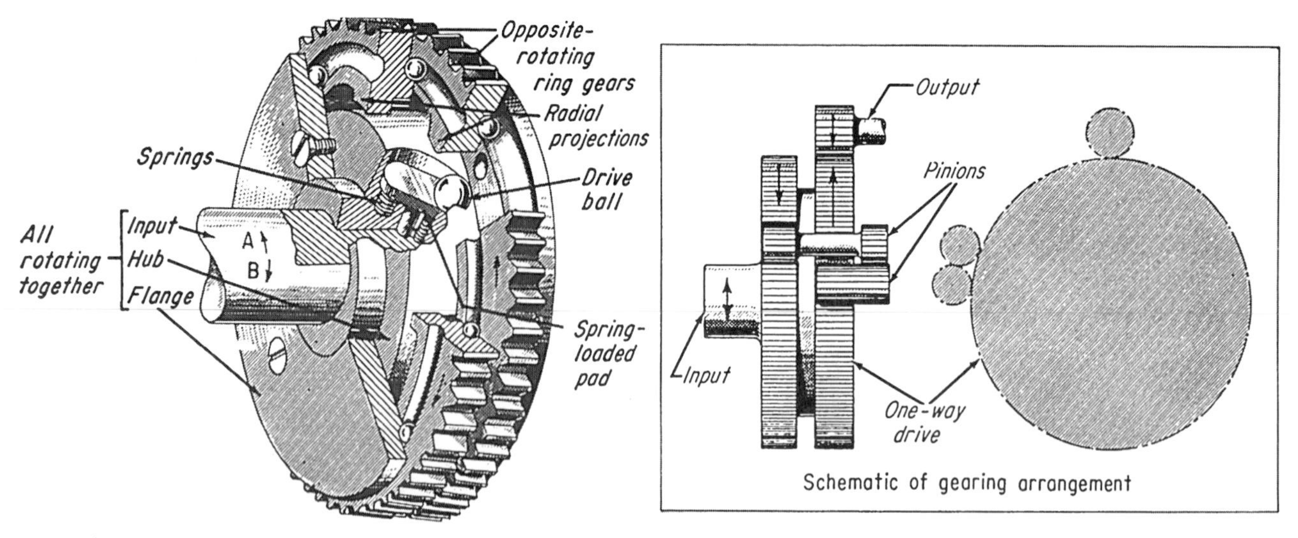

Fig. 5 A reciprocating-ball drive.

The one-way drive, shown in Fig. 1, was invented as a byproduct of the design of a money-order imprinter.

The task was to convert the oscillating motion of the input crank (20° in this case) into a one-way motion to advance an inking ribbon. One of the simplest known devices was used to obtain the one-way drive—a spring clutch which is a helical spring joining two co-linear butting shafts (Fig. 2). The spring is usually made of square or rectangular cross-section wire.

This clutch transmits torque in one direction only because it overrides when it is reversed. The helical spring, which bridges both shafts, need not be fastened at either end; a slight interference fit is acceptable. Rotating the input shaft in the direction tending to wind the spring (direction A in Fig. 2) causes the spring to grip both shafts and then transmit motion from the input to the output shaft. Reversing the input unwinds the spring, and it overrides the output shaft with a drag—but this drag, slight as it was, caused a problem in operation.

Double-Clutch Drive

The spring clutch (Fig. 2) did not provide enough friction in the tape drive to allow the spring clutch to slip on the shafts on the return stroke. Thus the output moved in sympathy with the input, and the desired one-way drive was not achieved.

At first, an attempt was made to add friction artificially to the output, but this resulted in an awkward design. Finally the problem was elegantly solved (Fig. 1) by installing a second helical spring, slightly larger than the first that served exactly the same purpose: transmission of motion in one direction only. This spring, however joined the output shaft and a stationary cylinder. In this way, with the two springs of the same hand, the undesirable return motion of the ribbon drive was immediately arrested, and a positive one-way drive was obtained quite simply.

This compact drive can be considered to be a mechanical *half-wave rectifier* in that it transmits motion in one direction only while it suppresses motion in the reverse direction.

Full-Wave Rectifier

The principles described will also produce a mechanical *full-wave rectifier* by introducing some reversing gears, Fig. 3. In this application the input drive in one direction is directly transmitted to the output, as before, but on the reverse stroke the input is passed through reversing gears so that the output appears in the opposite sense. In other words, the original sense of the output is maintained. Thus, the output moves forward twice for each back-and-forth movement of the input.

Shuttle-Gear Drive

Earlier, a one-way drive was developed that harnessed the axial thrust of a pair of helical gears to shift a pinion, Fig. 4. Although at first glance, it might look somewhat complicated, the drive is inexpensive to make and has been operating successfully with little wear.

When the input rotates in direction *A*, it drives the output through spur gears *1* and *2*. The shuttle pinion is also driving the helical gear whose rotation is resisted by the magnetic flux built up between the stationary permanent magnet and the rotating core. This magnet-core arrangement is actually a hysteresis brake, and its constant resisting torque produces an axial thrust in mesh of the helical pinion acting to the left. Reversing the input reverses the direction of thrust, which shifts the shuttle pinion to the right. The drive then operates through gears *1, 3,* and *4,* which nullifies the reversion to produce output in the same direction.

Reciprocating-Ball Drive

When the input rotates in direction A, Fig. 5, the drive ball trails to the right, and its upper half engages one of the radial projections in the right ring gear to drive it in the same direction as the input. The slot for the ball is milled at 45° to the shaft axes and extends to the flanges on each side.

When the input is reversed, the ball extends to the flanges on each side, trails to the left and deflects to permit the ball to ride over to the left ring gear, and engage its radial projection to drive the gear in the direction of the input.

Each gear, however, is constantly in mesh with a pinion, which in turn is in mesh with the other gear. Thus, regardless of the direction the input is turned, the ball positions itself under one or another ring gear, and the gears will maintain their respective sense of rotation (the rotation shown in Fig. 5). Hence, an output gear in mesh with one of the ring gears will rotate in one direction only.

OPERATING PRINCIPLES OF LIQUID, SEMISOLID, AND VACUUM PUMPS

These pumps are used to transfer liquids and supply hydraulic power.

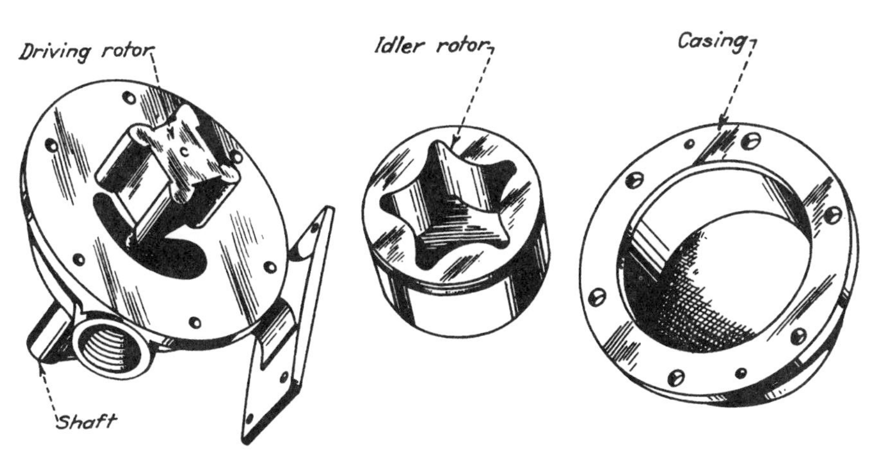

1. WITH BUT TWO MOVING PARTS, the rotors that turn in the same direction, this rotary pump has reduced friction to a minimum. The rotors rotate against flexible synthetic rubber cushions that allow sand, grit and other abrasives to flow freely through the pump without damage. It is a positive displacement pump that develops a constant pressure and will deliver a uniform flow at any given speed. The pump is reversible and can be driven by a gasoline engine or electric motor. The rubber cushions withstand the action of oil, kerosene, and gasoline, and the pump operates at any angle. It has been used in circulating water systems, cutting tool coolant oil systems and general applications.

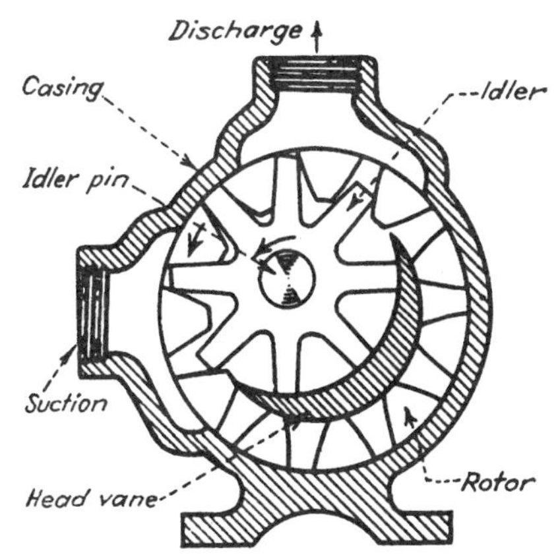

2. PUMPING ACTION is produced by the meshing of the idler and rotor teeth in this rotary pump. The idler is pin-mounted to the head and the rotor operates in either direction. This pump will not splash, foam, churn or cause pounding. Liquids of any viscosity that do not contain grit can be transferred by this pump which is made of iron and bronze.

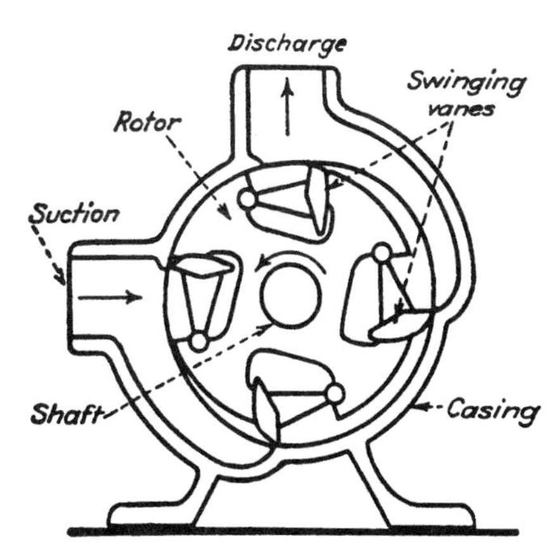

3. BASED on the swinging vane principle, this pump maintains its volumetric efficiency automatically. The action of the buckets, fitted loosely into recesses in the rotor, compensates for wear. In operation, the tip of the bucket is in light contact with the casing wall. Liquids are moved by sucking and pushing actions and are not churned or foamed.

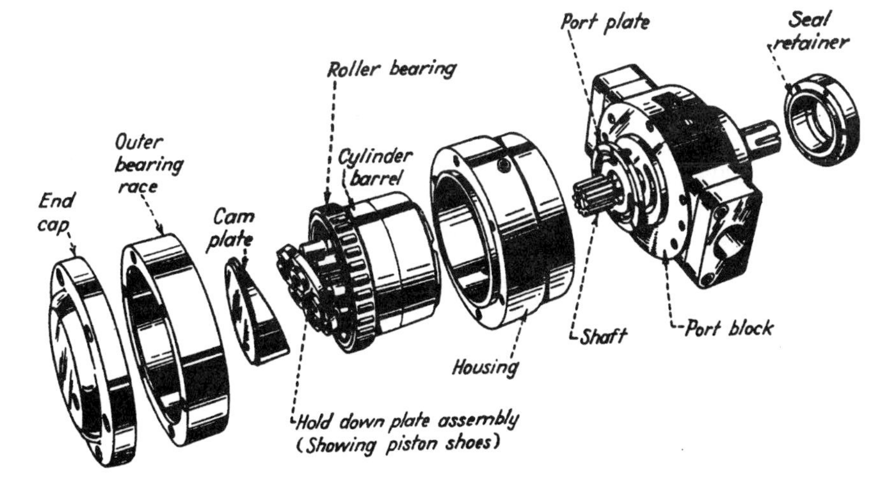

4. HIGH-PRESSURE, high-volume pumps of the axial piston, constant displacement type are rated at 3,500 psi for continuous duty operation; higher pressure is permissible for intermittent operation. A pressure-balanced piston shoe lubricates the cam plate and prevents direct contact between the shoe and cam plate. The use of the pressure balanced system removes the need for thrust bearings. The two-piece shaft absorbs deflection and minimizes bearing wear. The pump and electric driving motor are connected by a flexible coupling. The revolving cylinder barrel causes the axial reciprocation of the pumping pistons. These pumps only pump hydraulic fluids.

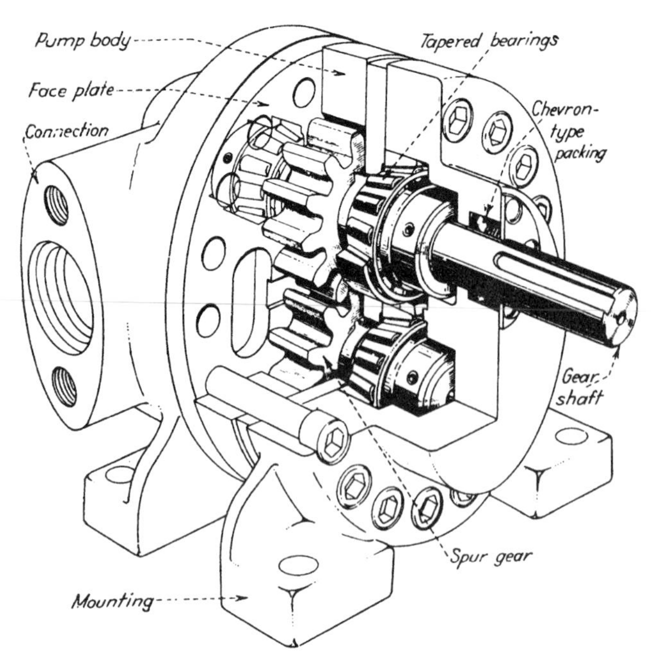

5. THE GEAR SHAFTS of this hydraulic gear pump are mounted on tapered roller bearings that accurately position the gears, decrease end play, and reduce wear to a minimum. This heavy-duty gear pump can be used at pressures up to 1,000 psi. These pumps were made with either single- or double-end shafts and can be foot- or flange-mounted. The drive shaft entrance packing is made from oil-resistant material, and the gear shafts are made from hardened molybdenum steel.

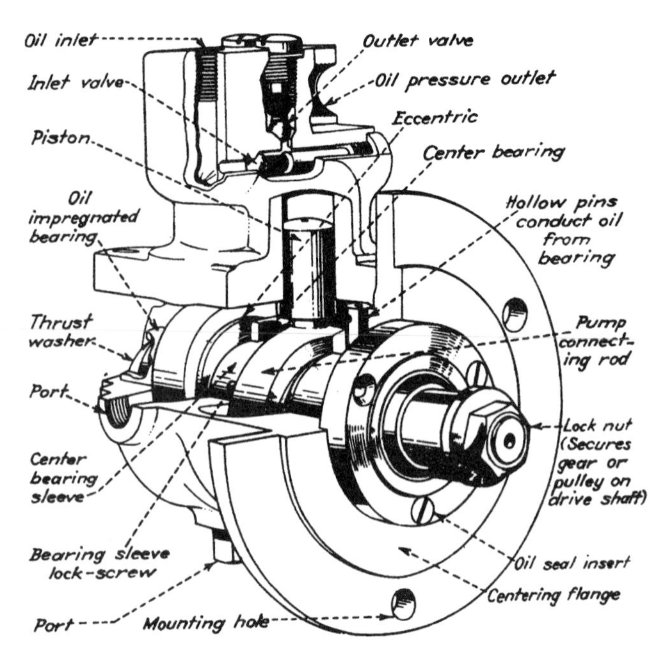

6. THIS HIGH-PRESSURE hydraulic pump has twin pistons that build pressures from 100 to 4,000 psi at speeds from 600 to 1,200 rpm. This pump can be operated continuously at 900 rpm and 2,500 psi with 1.37 gpm delivered. Because it can be mounted at any angle, and because it is used with small oil lines, small diameter rams and compact valves, the pump is suitable for installation in new equipment. This pump contains a pressure adjusting valve that is factory set to bypass at a predetermined pressure.

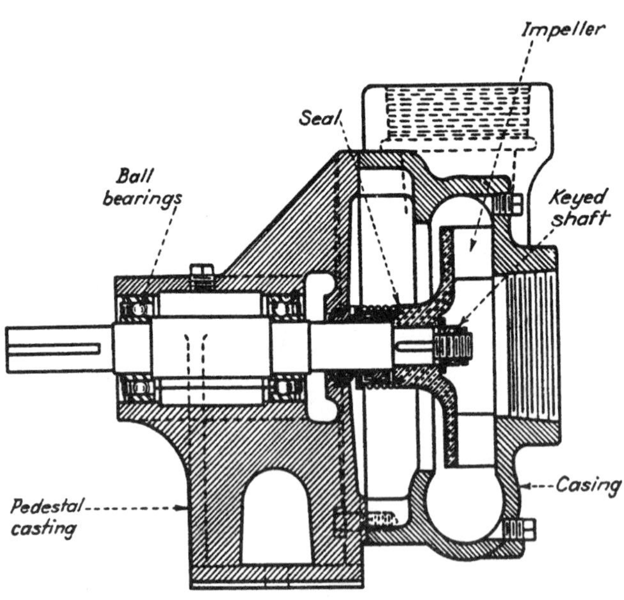

7. This pump is characterized by its pedestal mounting. The only non-critical fit is between the pedestal casting and the casing. Positive alignment is obtained because the sealed ball bearings and the shaft are supported in the single casting. The five-vaned, open, bronze impeller will move liquids that contain a high volume of solids. The pump is not for use with corrosive liquids. The five models of this pump, with ratings up to 500 gpm, are identical except for impeller and casing sizes.

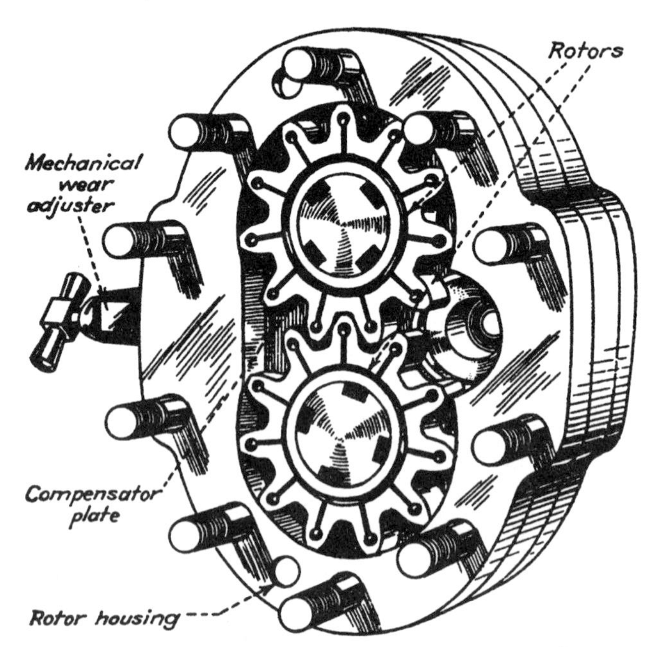

8. USED TO TRANSFER, meter, or proportion liquids of high or low viscosity, this pump is a positive displacement gear pump. It is made of stainless steel with a stainless steel armored, automatic take-up, shaft seal of the single-gland type. Automatic wear control compensates for normal wear and maintains volumetric efficiency. This pump will handle 5 to 300 gph without churning or foaming. It needs no lubrication and operates against high or low pressure.

These pumps are used to transfer liquids and semisolids, pump vacuums, and boost oil pressure

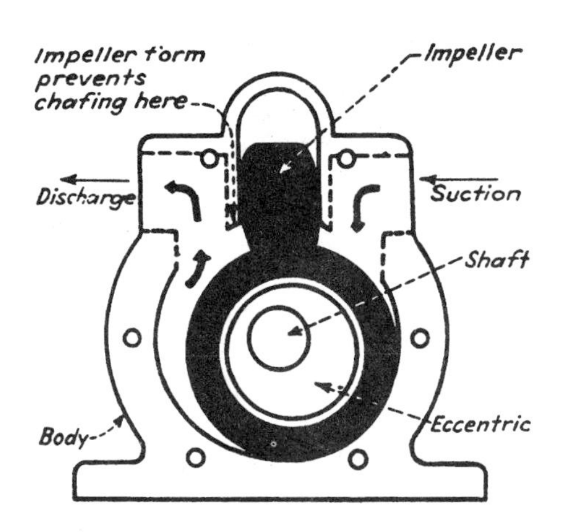

9. ORIGINALLY DESIGNED for use in the marine field, this gearless pump was made from stainless steel, monel, and bronze for handling acids, oils, and solvents. The impeller is made of pressure-vulcanized laminated layers of Hycar, 85 to 90 percent hard. Sand, grit, scale and fibrous materials will pass through. With capacities from 1 to 12 gpm and speeds from 200 to 3,500 rpm, these pumps will deliver against pressures up to 60 psi. Not self-priming, it can be installed with a reservoir. It operates in either direction and is self-lubricating.

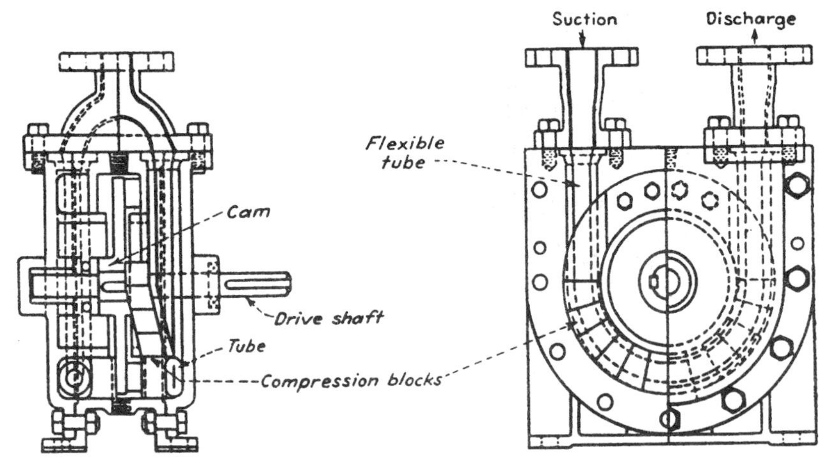

10. THE SQUEEGEE PUMP consists of a U-shaped flexible tube made of rubber, neoprene, or other flexible material. Acids and corrosive liquids or gases pass through the tube and do not contact working parts or lubricating oil. This prevents contamination of the liquid and avoids corrosion of metal parts. In operation, the tube is compressed progressively from the intake side to the discharge side by cams mounted on a driven shaft. Compression blocks move against the tube, closing the tube gradually and firmly from block to block, which forces the liquid out. As the cam passes the compression blocks, the tube returns to its original diameter. This creates a high vacuum on the intake side and causes the tube to be filled rapidly. The pump can be driven clockwise or counter-clockwise. The tube is completely encased and cannot expand beyond its original diameter. The standard pump is made of bronze and will handle volumes to 15 gpm. The Squeegee develops a vacuum of 25 in. of mercury and will work against pressures of 50 lb/in^2.

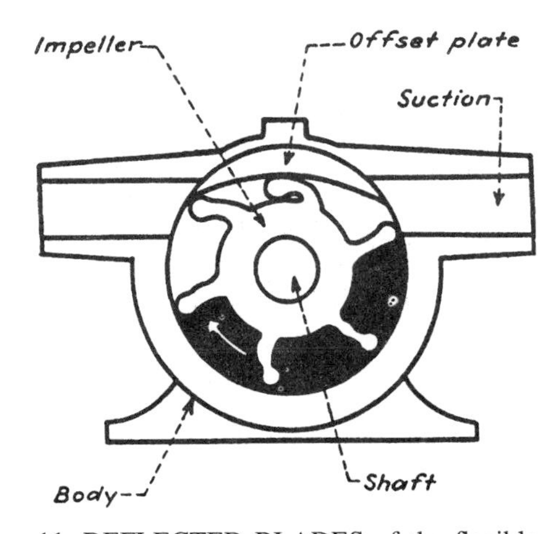

11. DEFLECTED BLADES of the flexible neoprene impeller straighten as they leave contact with offset plate. The high suction created draws fluid into pump, filling space between the blades. It handles animal, vegetable, and mineral oils but not napthas, gasoline, ordinary cleaning solvents, or paint thinners. The pump operates in either direction and can be mounted at any angle. It runs at 100 to 2,000 rpm, can deliver up to 55 gpm, and will operate against 25 psi. It operates at temperatures between 0 and 160 F.

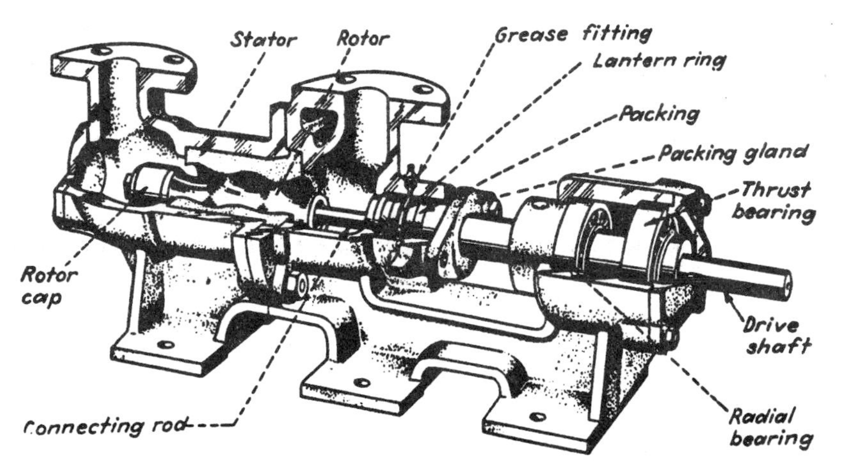

12. THIS PUMP CAN TRANSFER free-flowing liquids, non-pourable pastes, clean or contaminated with abrasives, chemically inert or active, homogeneous or containing solids in suspension. It is a positive displacement pump that delivers continuous, uniform flow. The one moving part, a low-alloy or tool-steel rotor, is a single helix, and the Hycar or natural rubber stator is a double internal helix. Pumping action is similar to that of a piston moving through a cylinder of infinite length. Containing no valves, this pump will self-prime at 28 ft of suction lift. The head developed is independent of speed, and capacity is proportional to speed. Slippage is a function of viscosity and pressure, and is predictable for any operating condition. The pump passes particles up to $^7/_8$ in. diameter through its largest pump. Pumping action can be in either direction. The largest standard pump, with two continuous seal lines, handles 150 gpm up to 200 psi.

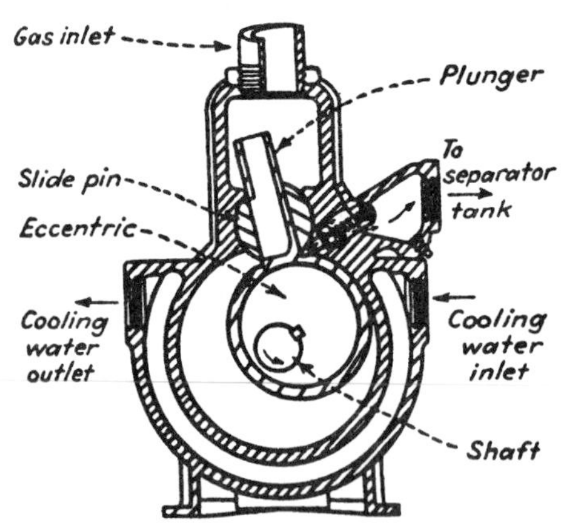

13. HIGH-VACUUM PUMPS operate with the rotating plunger action of liquid pumps. Sealing oil lubricates the three moving parts. Parts are accessible without disturbing connected piping. These pumps are used to rough pump a vacuum before connecting a diffusion pump; to evacute light bulbs and electronic tubes, and to vacuum dry and distill. Single pumps draw vacuums from 2 to 5 microns; in series to 0.5 micron, and compound pumps draw to 0.1 micron. They can be run in reverse for transferring liquids. Diagonal cored slots, closed by a slide pin, form the passageway and inlet valve. Popper or feature outlet valves are used.

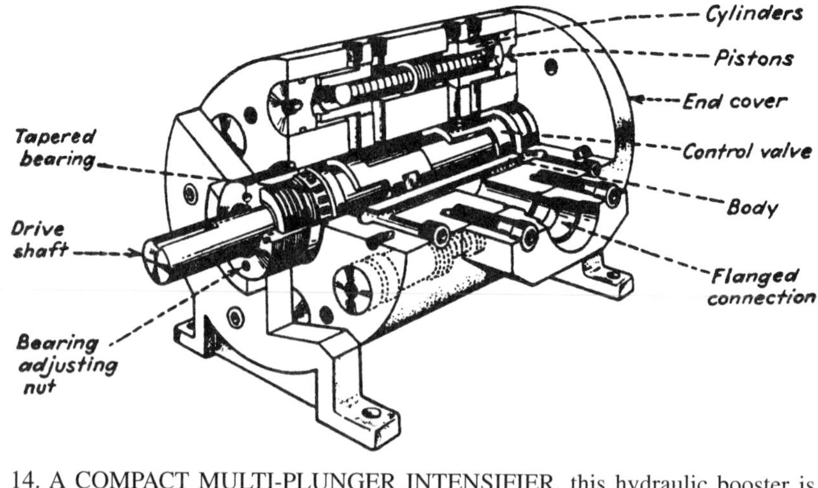

14. A COMPACT MULTI-PLUNGER INTENSIFIER, this hydraulic booster is designed to convert low pressure to high pressure in any oil-hydraulic circuit. No additional pumps are required. Because of its six plungers, the pressure flow from the booster is both smooth and uninterrupted. High-pressure pumps are not required, and no operating valves are needed to control the high-pressure system. Small cylinder and ram assemblies can be used on operating equipment because the pressure is high. Operating costs can be low because of the efficient use of connected horsepower. The inertia effects of the small operating rams are low, so high speed operations can be attained. These boosters were built in two standard sizes, each of which was available in two pressure ranges: 2 to 1 and 3 to 1. Volumetric output is in inverse proportion to the pressure ratio. All units have a maximum 7,500 psi discharge pressure. Pistons are double-acting, and the central valve admits oil to pistons in sequence and is always hydraulically balanced.

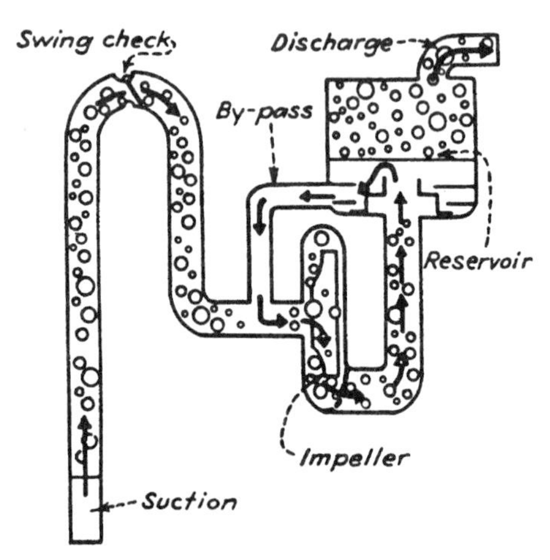

15. THIS SELF-PRIMING PUMP gives a rapid and smooth transition from priming cycle to centrifugal pumping. The pump starts with its priming chamber full. Liquid is recirculated through the impeller until the pump is primed. As priming liquid circulates, air is drawn through impeller and expelled through the discharge. When all air is evacuated, discharge velocity closes the priming valve completely. These pumps can have open or closed impellers. Solids up to 1 in. can be passed through a 3 in. size pump with an open impeller.

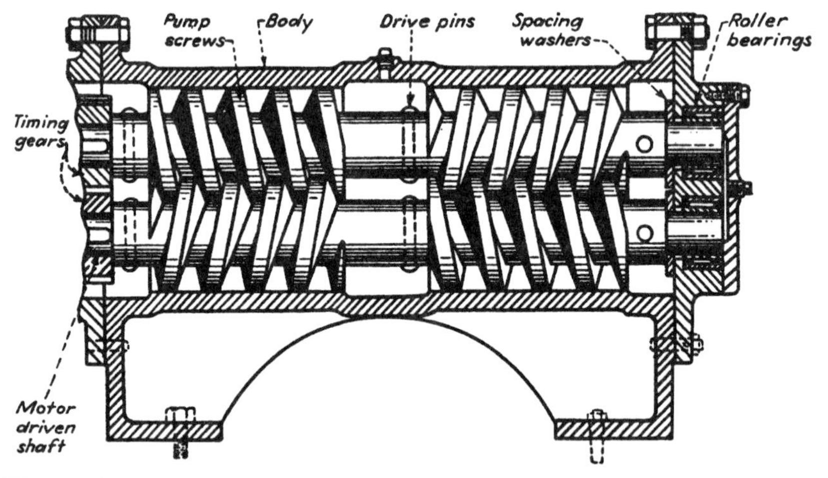

16. INTERNAL SCREW PUMPS can easily transfer high-viscosity petroleum products. They can be used as boiler fuel pumps because they deliver a pulseless flow of oil. For marine or stationary systems, the characteristic low vibration of screw pumps has allowed them to be mounted on light foundations. The absence of vibration and pulsing flow reduces strain on pipes, hose and fittings. The pumping screws are mounted on shafts and take in liquid at both ends of the pump body and move it to the center for discharge. This balanced pumping action makes it unnecessary to use thrust bearings except in installations where the pump is mounted at a high angle. The pumps can be used at any angle up to vertical. Where thrust bearings are needed, antifriction bearings capable of supporting the load of the shaft and screws are used. The intermeshing pumping screws are timed by a pair of precision-cut herringbone gears. These are self-centering, and do not allow the side wear of the screws while they are pumping. The pump is most efficient when driven less than 1,200 rpm by an electric motor and 1,300 rpm by a steam turbine.

TWELVE DIFFERENT ROTARY-PUMP ACTIONS

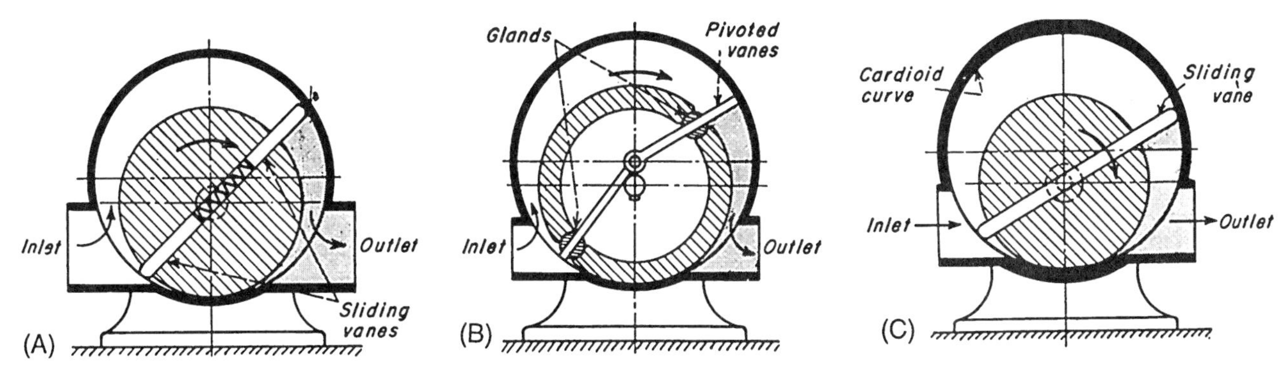

Fig. 1 (A) **A Ramelli pump** with spring-loaded vanes to ensure contact with the wall; vane ends are rounded for line contact. (B) Two vanes pivot in the housing and are driven by an eccentrically mounted disk; vanes slide in glands and are always radial to the housing, thus providing surface contact. (C) A housing with a cardioid curve allows the use of a single vane because opposing points on the housing in line with the disk center are equidistant.

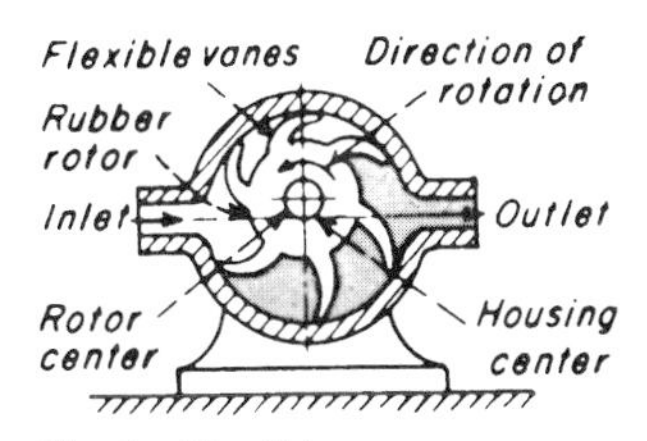

Fig. 2 Flexible vanes on an eccentric rubber rotor displace liquid as in sliding-vane pumps. Instead of the vanes sliding in and out, they bend against the casing to perform pumping.

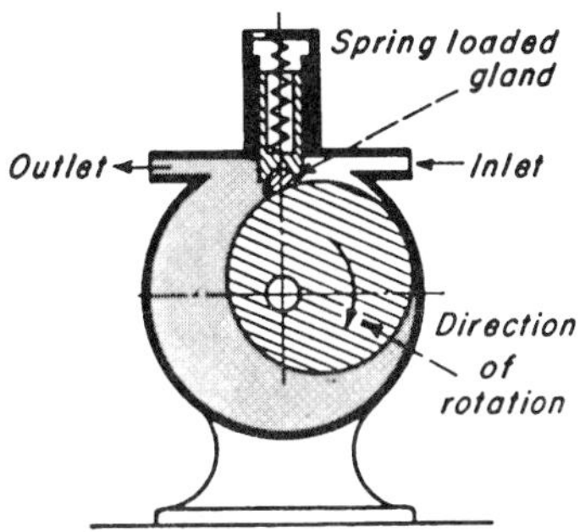
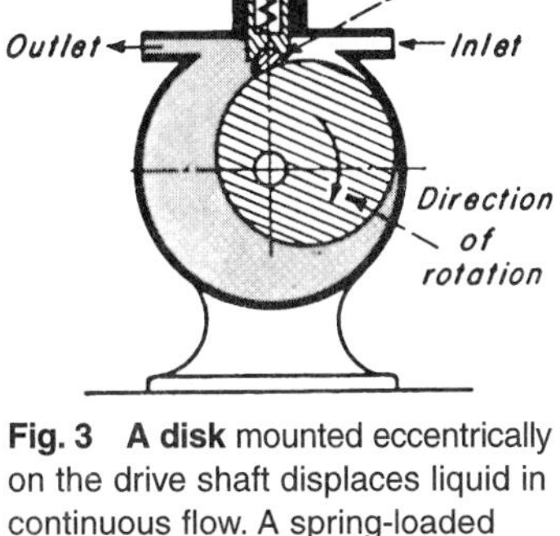

Fig. 3 A disk mounted eccentrically on the drive shaft displaces liquid in continuous flow. A spring-loaded gland separates the inlet from the outlet except when the disk is at the top of stroke.

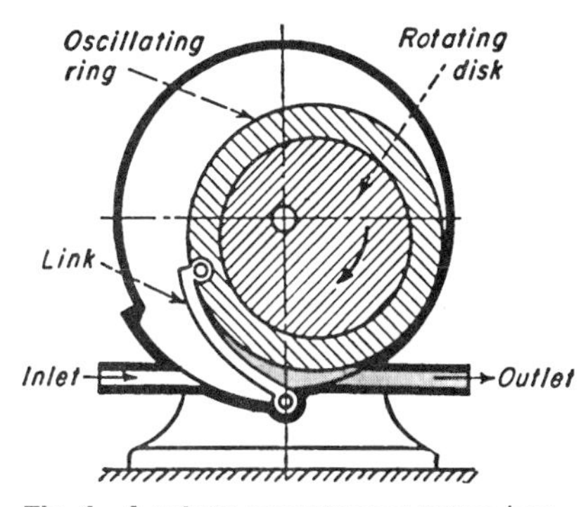

Fig. 4 A rotary compressor pump has a link separating its suction and compression sides. The link is hinged to a ring which oscillates while it is driven by the disk. Oscillating action pumps the liquid in a continuous flow.

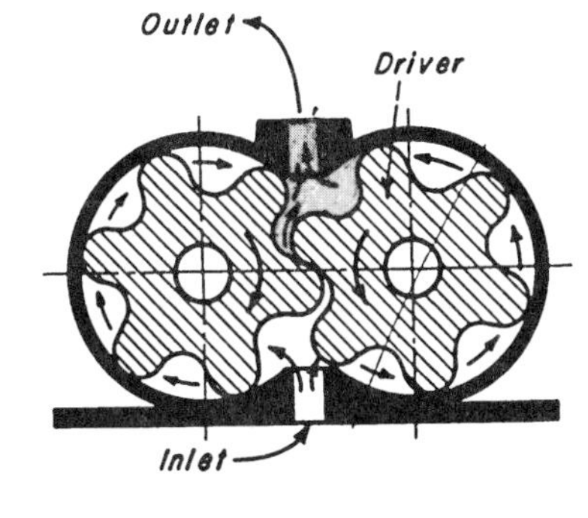

Fig. 5 A gear pump transports liquid between the tooth spaces and the housing wall. A circular tooth shape ha sonly one tooth making contact, and it is more efficient than an involute shape which might enclose a pocket between two adjoining teeth, recirculating part of the liquid. The pump has helical teeth.

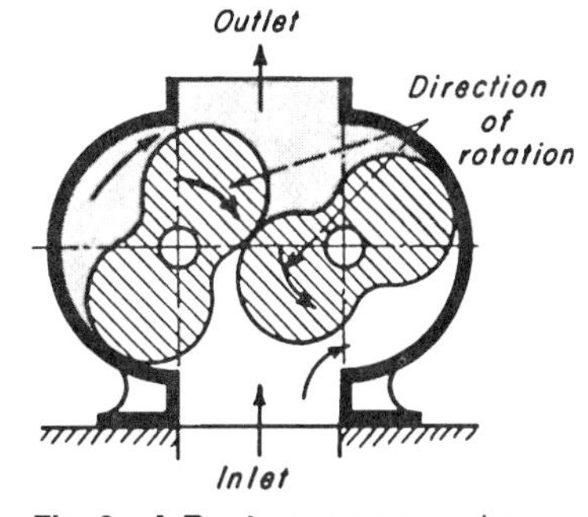

Fig. 6 A Roots compressor has two identical impellers with specially shaped teeth. The shafts are connected by external gearing to ensure constant contact between the impellers.

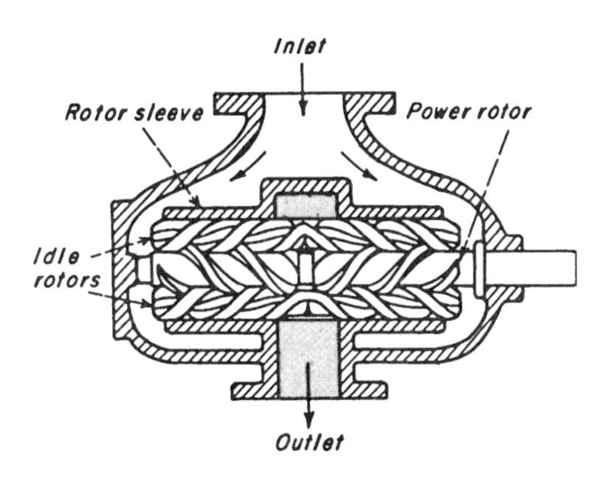

Fig. 7 A three-screw pump drives liquid between the screw threads along the axis of the screws. The idle rotors are driven by fluid pressure, not by metallic contact with the power rotor.

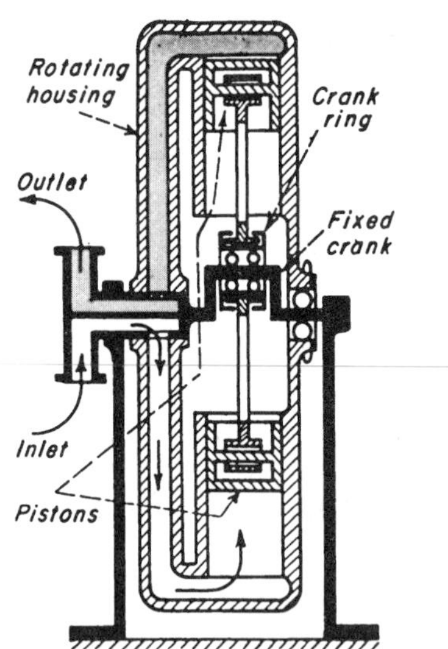

Fig. 8 The housing of the Hele-Shaw-Beachum pump rotates the round-cranked shaft. Connecting rods attached to the crank ring cause the pistons to oscillate as the housing rotates. No valves are necessary because the fixed hollow shaft, divided by a wall, has suction and compression sides that are always in correct register with the inlet and outlet ports.

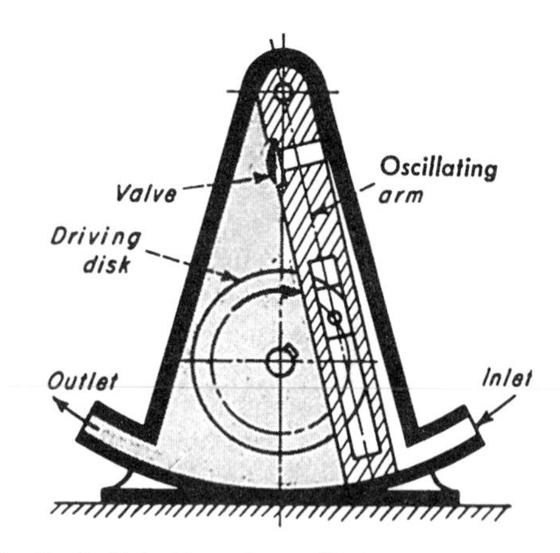

Fig. 9 A disk drives the oscillating arm which acts as piston. The velocity of the arm varies because of its quick-return mechanism. Liquid is slowly drawn in and expelled during the clockwise rotation of the arm; the return stroke transfers the liquid rapidly.

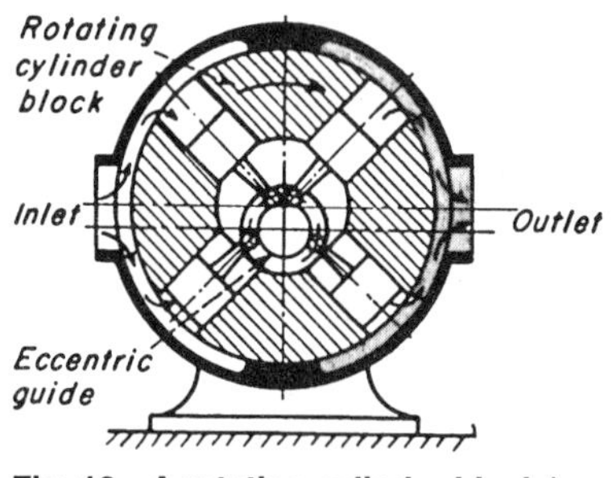

Fig. 10 A rotating cylinder block is mounted concentrically in a housing. Connecting-rod ends slide around an eccentric guide as the cylinders rotated and cause the pistons to reciprocate. The housing is divided into suction and compression compartments.

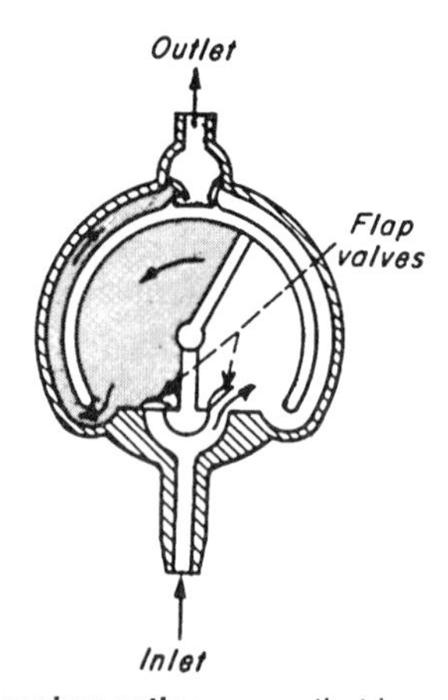

Fig. 11 A rotary-reciprocating pump that is normally operated manually to pump high-viscosity liquids such as oil.

OFFSET PLANETARY GEARS INDUCE ROTARY-PUMP ACTION

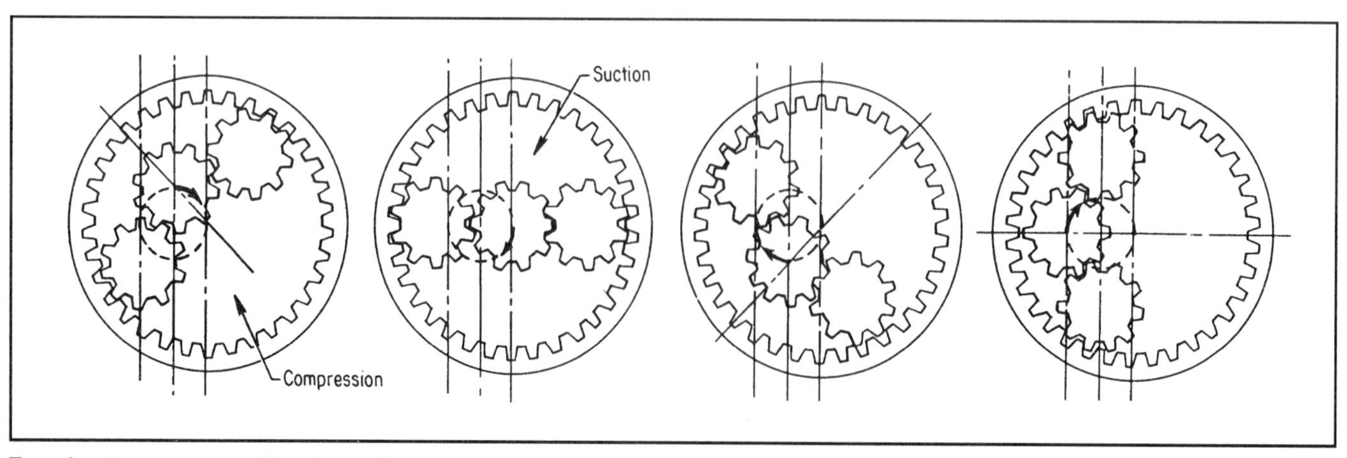

Two planetary gears are driven by an offset sun gear to provide the pumping action in this positive-displacement pump. A successively increasing/decreasing (suction/compression) is formed on either side of the sun and planet gears.

CHAPTER 14

PACKAGING, CONVEYING, HANDLING, AND SAFETY MECHANISMS AND MACHINES

FIFTEEN DEVICES THAT SORT, FEED, OR WEIGH

ORIENTING DEVICES

Orienting short, tubular parts

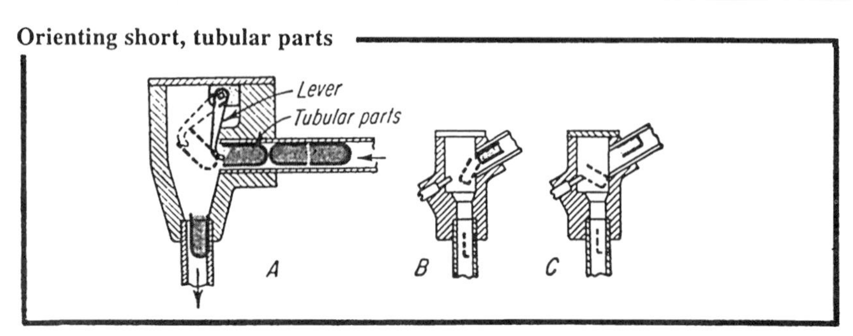

Here's a common problem; Parts arrive in either open-end or closed-end first; you need a device that will orient all the parts so they feed out facing the same way. In Fig. A. when a part comes in open-end first, it is pivoted by the swinging lever so that the open end is up. When it comes in closed-end first, the part brushes away the lever to flip over headfirst. Fig. B and C show a simpler arrangement with pin in place of lever.

Orienting dish-like parts

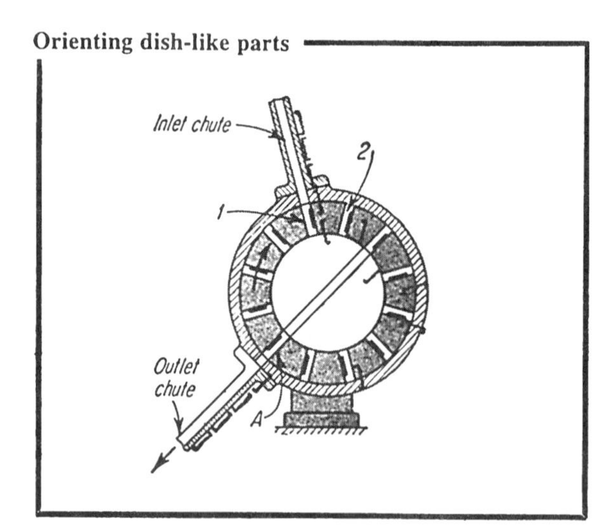

A part with its open-end facing to the right (part 1) falls on a matching projection as the indexing wheel begins to rotate clockwise. The projection retains the part for 230° to point A where it falls away from the projection to slide down the outlet chute, open-end up. An incoming part facing the other way (2) is not retained by the projection, hence it slides *through* the indexing wheel so that it too, passes through the outlet with its open-end up.

Orienting pointed-end parts

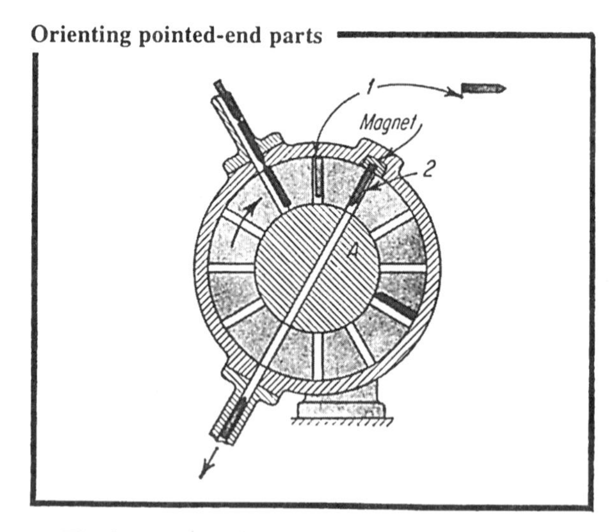

The important point here is that the built-in magnet cannot hold on to a part as it passes by if the part has its pointed end facing the magnet. Such a correctly oriented part (part 1) will fall through the chute as the wheel indexes to a stop. An incorrectly oriented part (part 2) is briefly held by the magnet until the indexing wheel continues on past the magnet position. The wheel and the core with the slot must be made from some nonmagnetic material.

Orienting U-shaped parts

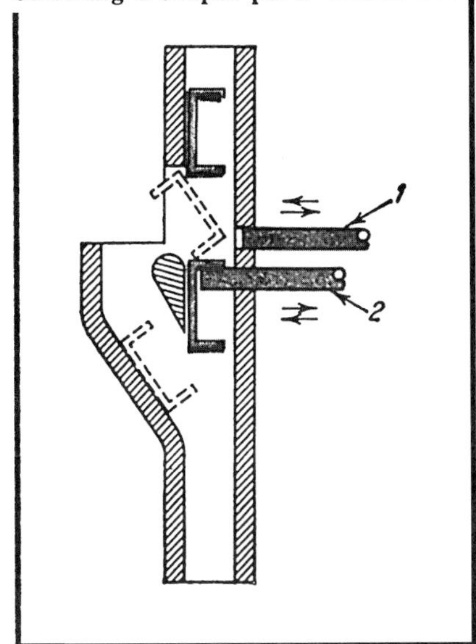

The key to this device is two pins that reciprocate one after another in the horizontal direction. The parts come down the chute with the bottom of the "U" facing either to the right or left. All pieces first strike and rest on pin 2. Pin 1 now moves into the passage way, and if the bottom of the "U" is facing to the right, the pin would kick over the part as shown by the dotted lines. If, on the other hand, the bottom of the "U" had been to the left, the motion of pin 1 would have no effect, and as pin 2 withdrew to the right, the part would be allowed to pass down through the main chute.

Orienting cone-shaped parts

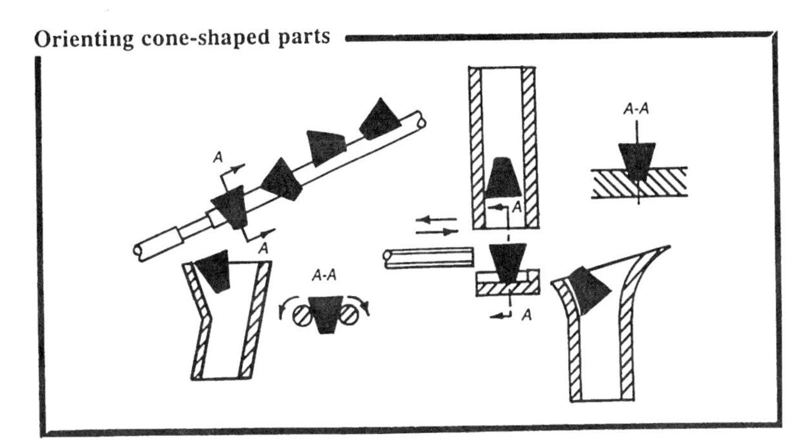

Regardless of which end of the cone faces forward as the cones slide down the cylindrical rods, the fact that both rods rotate in opposite directions causes the cones to assume the position shown in section *A-A* (above). When the cones reach the thinned-down section of the rods, they fall down into the chute, as illustrated.

In the second method of orienting cone-shaped parts (left), if the part comes down small end first, it will fit into the recess. The reciprocating rod, moving to the right, will then kick the cone over into the exit chute. But if the cone comes down with its large end first, it sits on top of the plate (instead of inside the recess), and the rod simply pushes it into the chute without turning it over.

Orienting stepped-disk parts

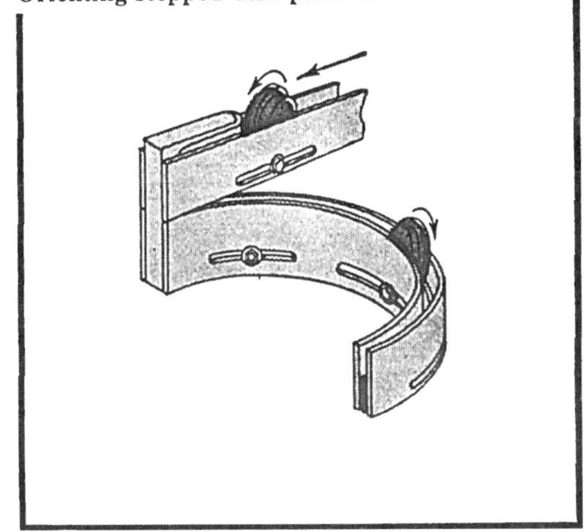

Parts rolling down the top rail to the left drop to the next rail which has a circular segment. The part, therefore, continue to roll on in the original direction, but their faces have now been rotated 180°. The idea of dropping one level might seem oversimplified, but it avoids the cam-based mechanisms more commonly used for accomplishing this job.

SIMPLE FEEDING DEVICES

Feeding a fixed number of parts

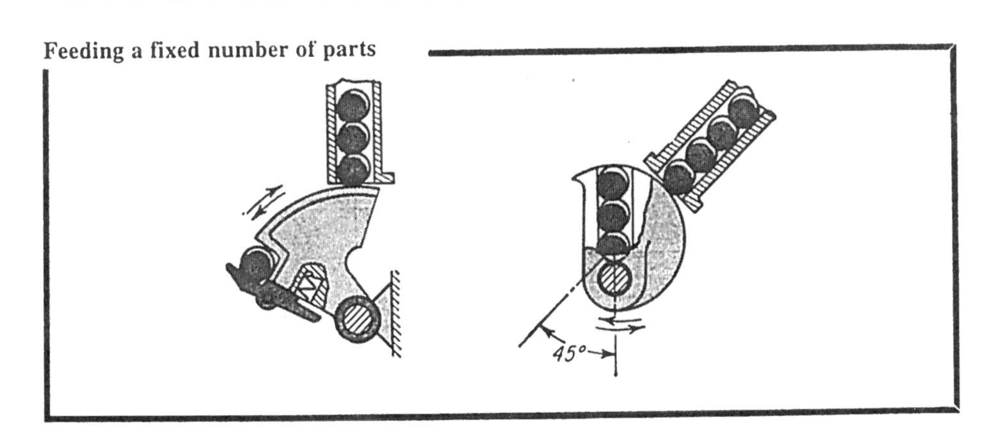

The oscillating sector picks up the desired number of parts, left diagram, and feeds them by pivoting the required number of degrees. The device for oscillating the sector must be able to produce dwells at both ends of the stroke to allow sufficient time for the parts to fall in and out of the sector.

The circular parts feed down the chute by gravity, and they are separated by the reciprocating rod. The parts first roll to station 3 during the downward stroke of the reciprocator, then to station 1 during the upward stroke; hence the time span between parts is almost equivalent to the time it takes for the reciprocator to make one complot oscillation.

The device in Fig. B is similar to the one in Fig. A, except that the reciprocator is replaced by an oscillating member.

One-by-one separating device

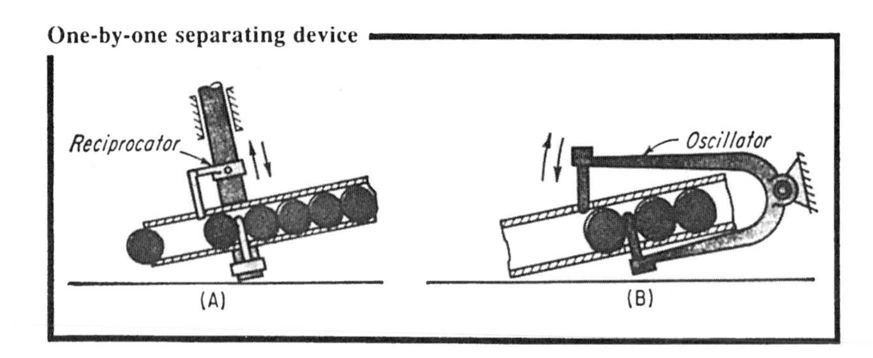

Mixing different parts together

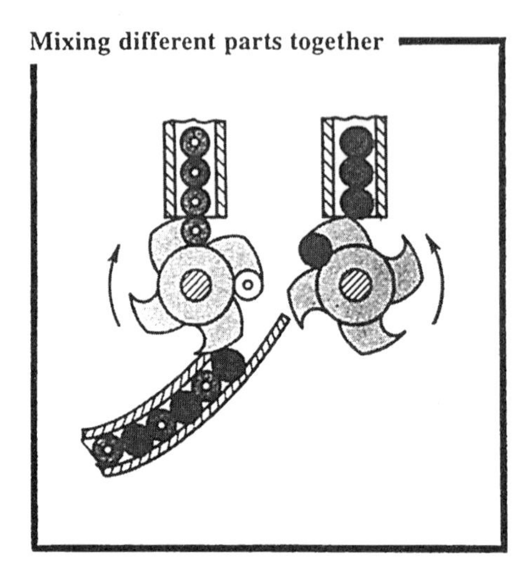

Two counter rotating wheels form a simple device for alternating the feed of two different workpieces.

Pausing until actuated

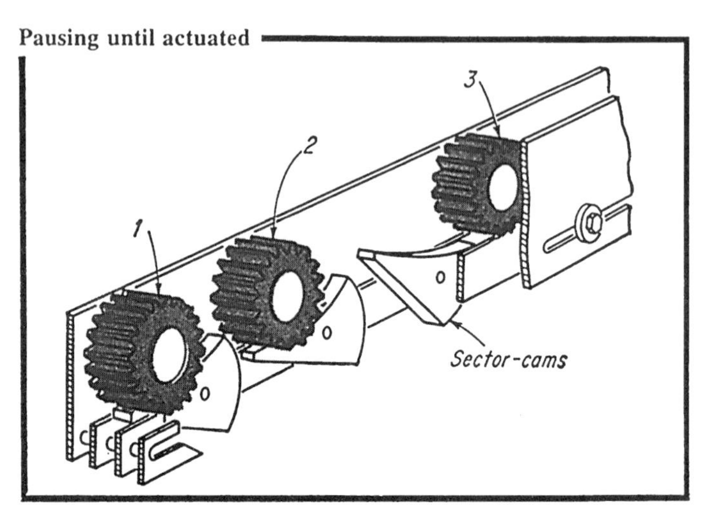

Each gear in this device is held up by a pivotable cam sector until the gear ahead of it moves forward. Thus, gear 3, rolling down the chute, kicks down its sector cam but is held up by the previous cam. When gear 1 is picked off (either manually, or mechanically), its sector cam pivots clockwise because of its own weight. This permits gear 2 to move into place of gear 1—and frees cam 2 to pivot clockwise. Thus, all gears in the row move forward one station.

SORTING DEVICES

In the simple device (A) the balls run down two inclined and slightly divergent rails. The smallest balls, therefore, will fall into the left chamber, the medium-size ones into the middle-size chamber, and the largest ones into the right chamber.

In the more complicated arrangement (B), the balls come down the hopper and must pass a gate which also acts as a latch for the trapdoor. The proper-size balls pass through without touching (actuating) the gate. Larger balls, however, brush against the gate which releases the catch on the bottom of the trapdoor, and fall through into the special trough for the rejects.

Sorting balls according to size

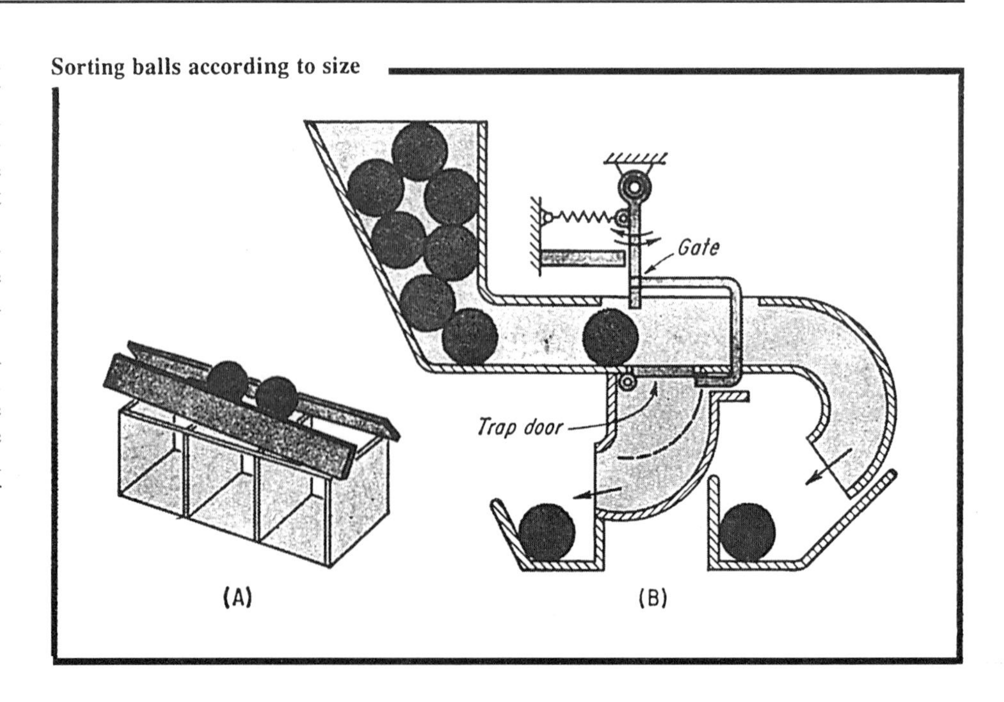

Sorting according to height

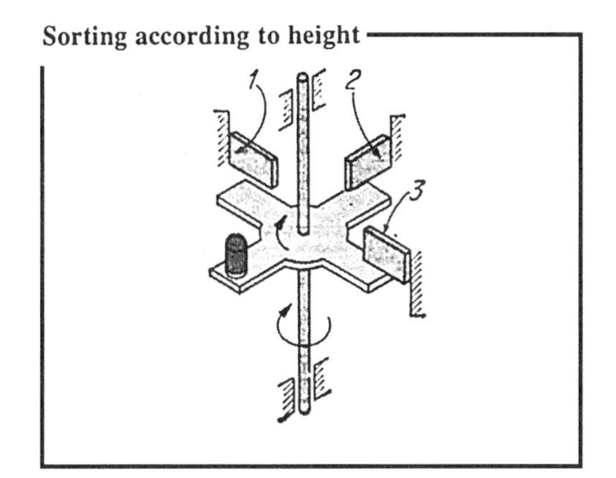

Workpieces of varying heights are placed on this slowly rotating cross-platform. Bars 1, 2, and 3 have been set at decreasing heights beginning with the highest bar (bar 1), down to the lowest bar (bar 3). The workpiece is therefore knocked off the platform at either station 1, 2, or 3, depending on its height.

WEIGHT-REGULATING ARRANGEMENTS

By varying the vibration amplitudes

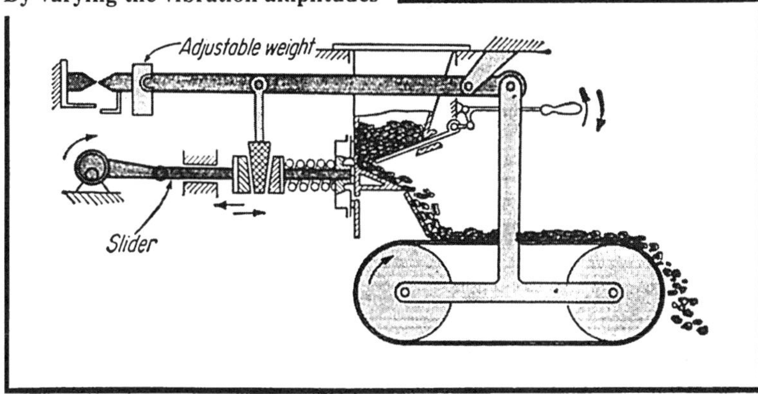

The material in the hopper is fed to a conveyor by the vibration of the reciprocating slider. The pulsating force of the slider is transmitted through the rubber wedge and on to the actuating rod. The amplitude of this force can be varied by moving the wedge up or down. This is done automatically by making the conveyor pivot around a central point. As the conveyor becomes overloaded, it pivots clockwise to raise the wedge, which reduces the amplitude of the force and slows the feed rate of the material.

Further adjustments in feed rate can be made by shifting the adjustable weight or by changing the speed of the conveyor belt.

By linkage arrangement

The loose material falls down the hopper and is fed to the right by the conveyor system which can pivot about the center point. The frame of the conveyor system also actuates the hopper gate so that if the amount of material on the belt exceeds the required amount, the conveyor pivots clockwise and closes the gate. The position of the counterweight on a frame determines the feed rate of the system.

By electric-eye and balancer

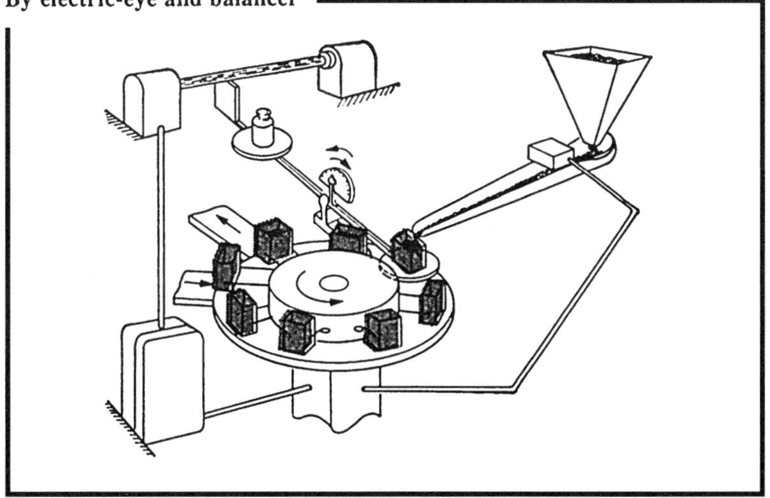

The indexing table automatically stops at the feed station. As the material drops into the container, its weight pivots the screen upward to cut off the light beam to the photocell relay. This in turn shuts the feed gate. The reactuation of the indexing table can be automatic after a time delay or by the cutoff response of the electric eye.

SEVEN CUTTING MECHANISMS

Clamping and cutting device

By pressing down on the foot pedal of this mechanism, the top knife and the clamp will be moved downward. However, when the clamp presses on the material, both it and link *EDO* will be unable to move further. Link *AC* will now begin to pivot around point *B*, drawing the lower knife up to begin the cutting action.

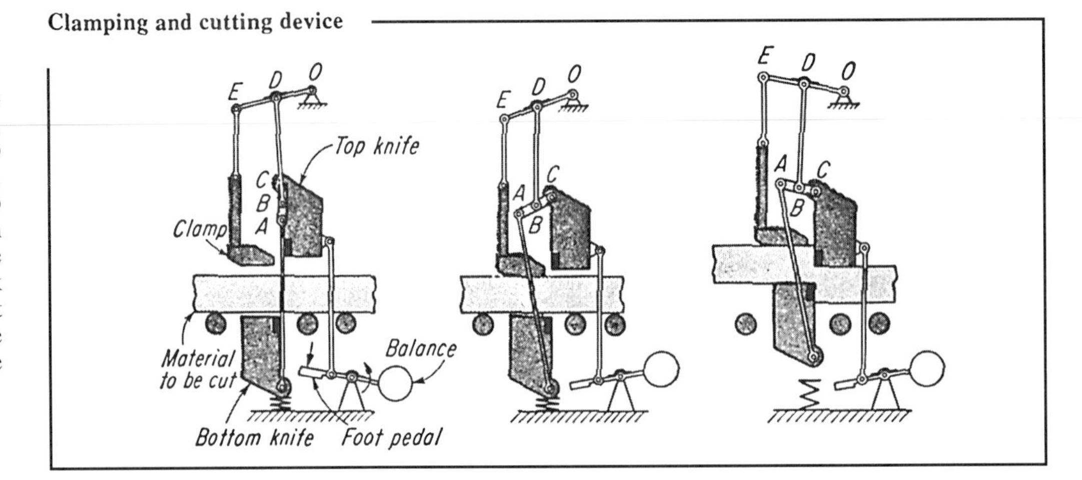

Four-bar mechanism

These 3 four-bar cutters provide a stable, strong, cutting action by coupling two sets of links to chain four-bar arrangements.

Parallel cutter mechanisms

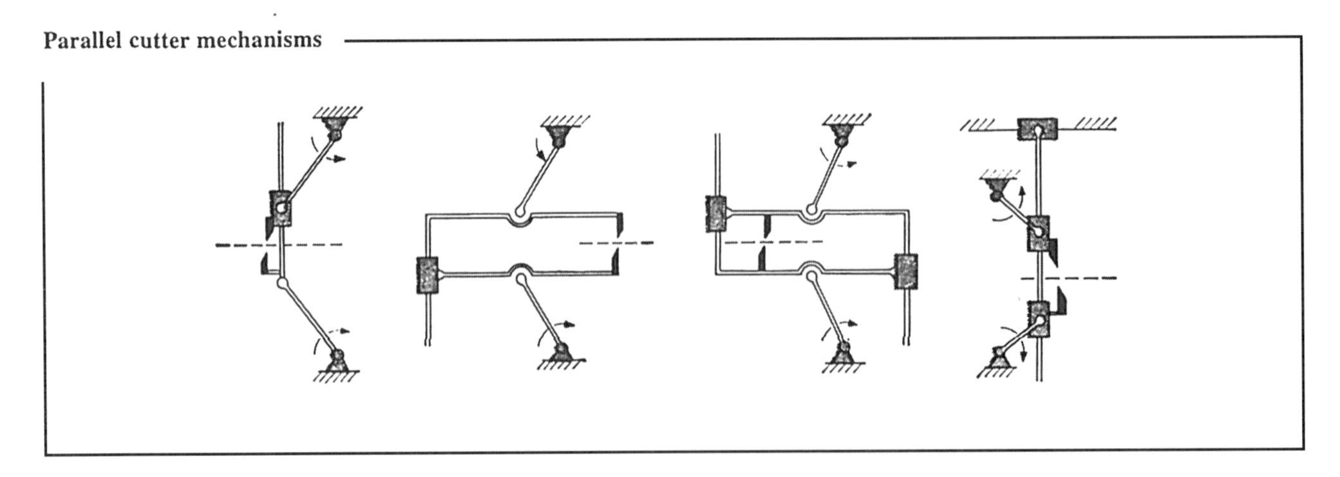

The cutting edges of the knives in the four mechanisms move parallel to each other, and they also remain vertical at all times to cut the material while it is in motion. The two cranks are rotated with constant velocity by a 1 to 1 gear system (not shown), which also feeds the material through the mechanism.

Curved-motion cutter mechanism

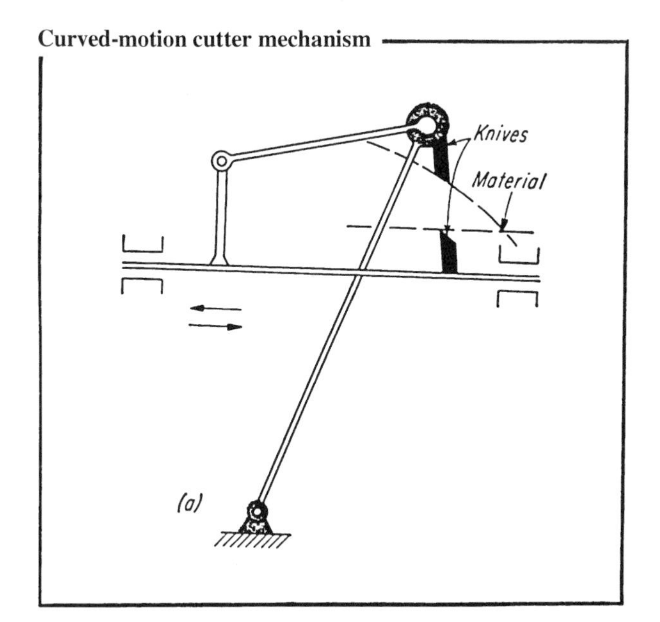

The material is cut while in motion by the reciprocating action of the horizontal bar. As the bar with the bottom knife moves to the right, the top knife will arc downward to perform the cutting operation.

Vertical cutter motion mechanism

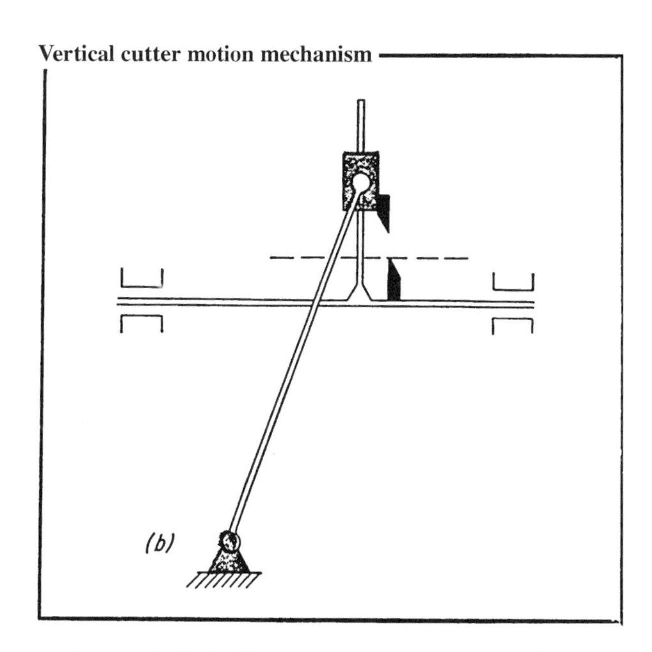

The top knife in this arrangement remains parallel to the bottom knife at all times during cutting to provide a true scissor-like action, but friction in the sliding member can limit the cutting force.

Slicing mechanism

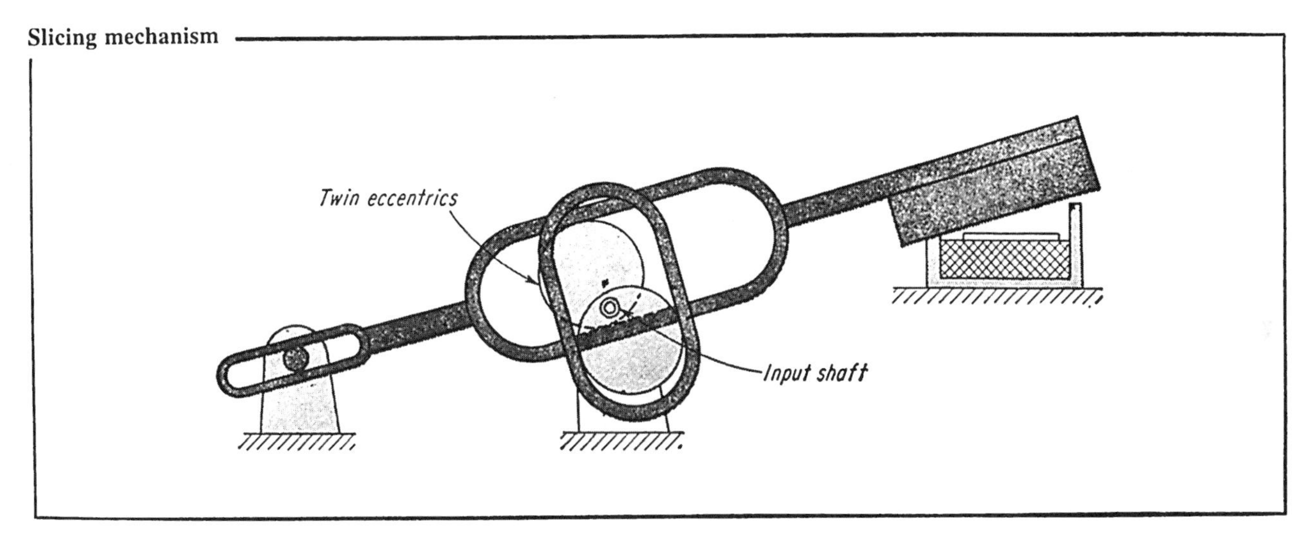

Slicing motion is obtained from the synchronized effort of two eccentric disks. The two looped rings actuated by the disks are welded together. In the position shown, the bottom eccentric disk provides the horizontal cutting movement, and the top disk provides the up-and-down force necessary for the cutting action.

Web-cutting mechanism

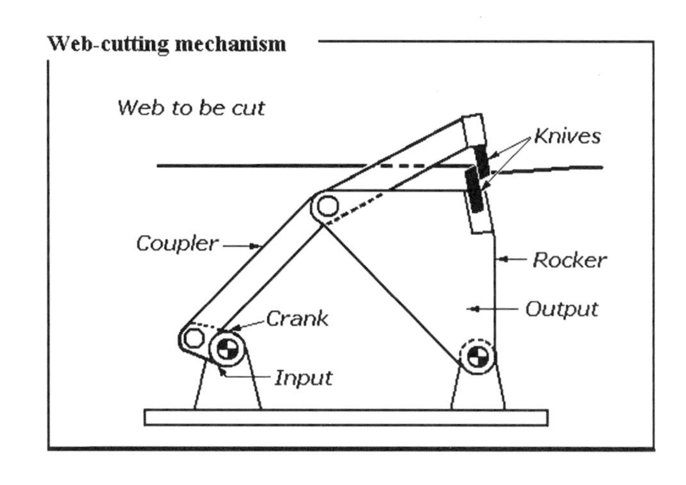

This four-bar linkage with an extended coupler can cut a web on the run at high speeds. The four-bar linkage shown is dimensioned to give the knife a velocity during the cutting operation that is equal to the linear velocity of the web.

TWO FLIPPING MECHANISMS

This mechanism can turn over a flat piece by driving two four-bar linkages from one double crank. The two flippers are actually extensions of the fourth members of the four-bar linkages. Link proportions are selected so that both flippers rise up at the same time to meet a line slightly off the vertical to transfer the piece from one flipper to the other by the momentum of the piece.

Turnover mechanism

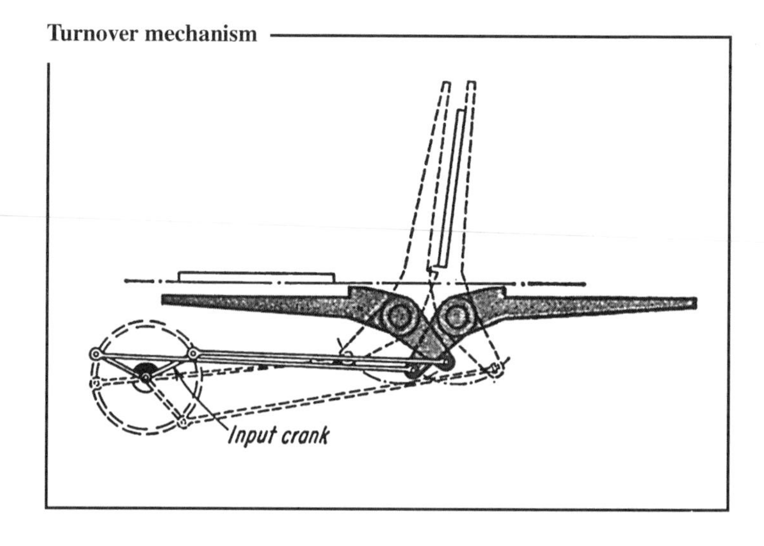

Upside-down flipper mechanism

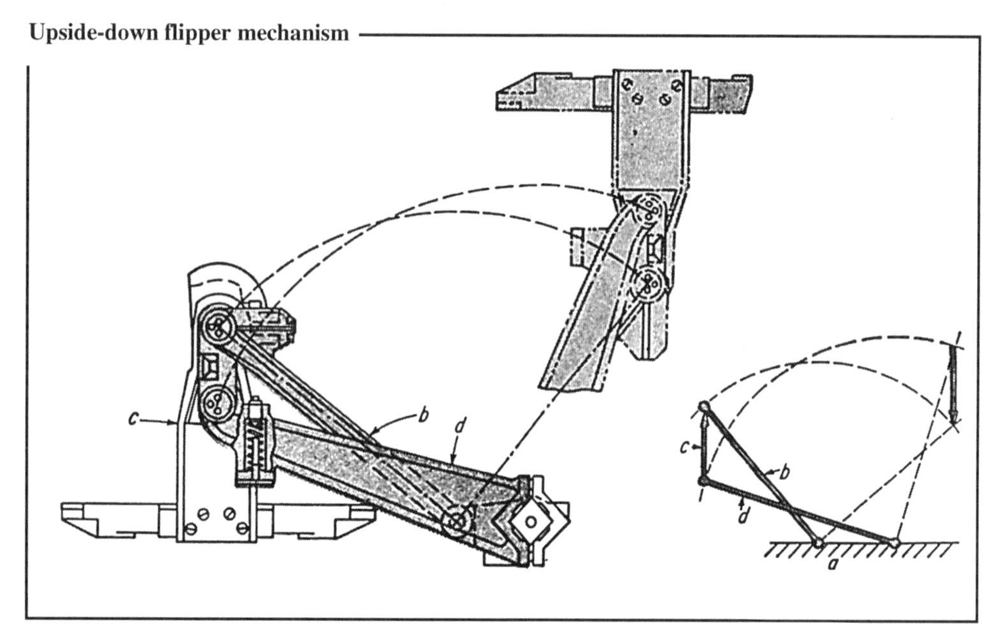

This is a four-bar linkage (links *a*, *b*, *c*, *d*) in which the part to be turned over is coupler *c* of the linkage. For the proportions shown, the 180° rotation of link *c* is accomplished during the 90° rotation of the input link.

ONE VIBRATING MECHANISM

Vibrating mechanism

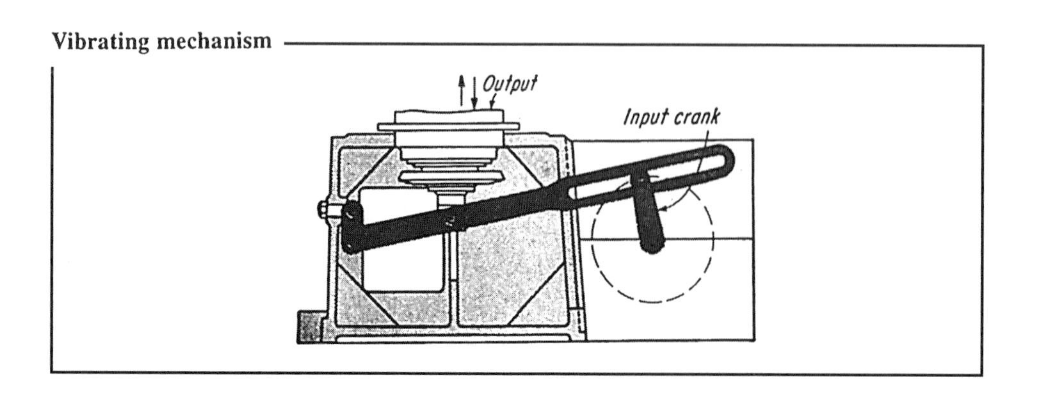

As the input crank rotates, the slotted link, which is fastened to the frame with an intermediate link, oscillates to vibrate the output table up and down.

SEVEN BASIC PARTS SELECTORS

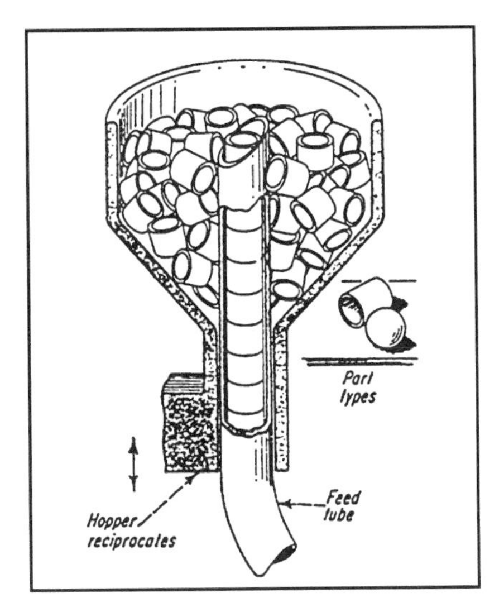

A reciprocating feed for spheres or short cyclinders is one of the simplest feed mehanisms. Either the hopper or the tube reciprocates. The hopper must be kept topped-up with parts unless the tube can be adjusted to the parts level.

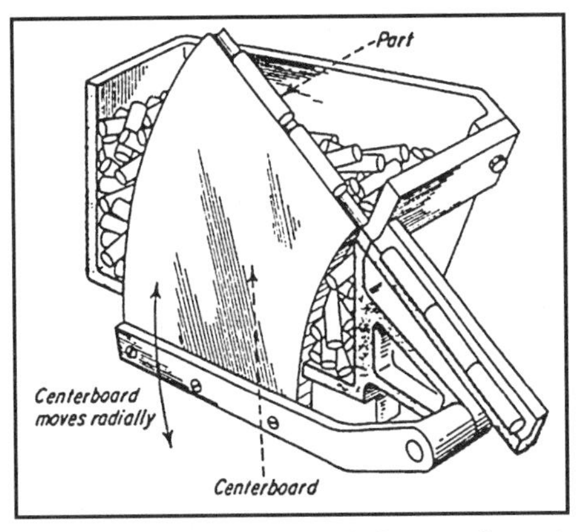

A centerboard selector is similar to reciprocating feed. The centerboard top can be milled to various section shapes to pick up moderately complex parts. I works best, however, with cylinders that are too long to be led with the reciprocating hopper. The feed can be continuous or as required.

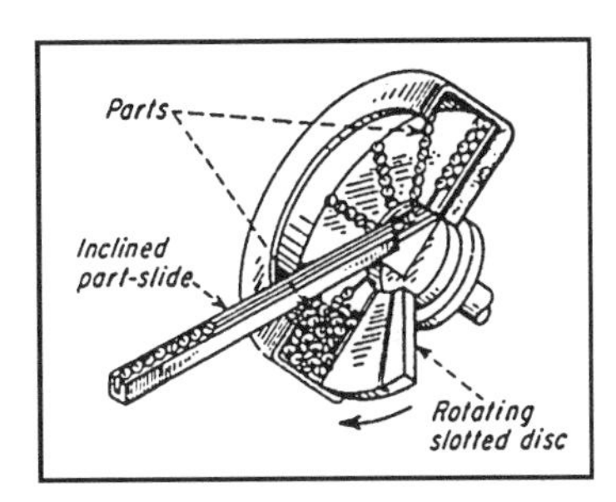

A rotary screw-feed handles screws, headed pings, shouldered shafts, and similar parts in most hopper feeds, random selection of chance-oriented parts calls for additional machinery if the parts must be fed in only one specific position. Here, however, all screws are fed in the same orientation (except for slot position) without separate machinery.

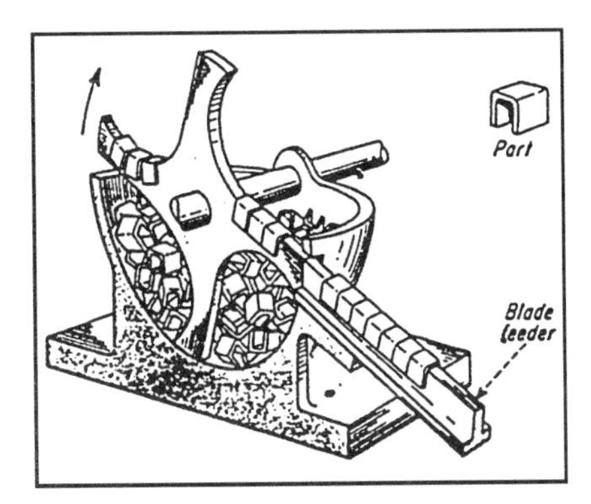

Rotary centerblades catch small U-shaped parts effectively if their legs are not too long. The parts must also be resilient enough to resist permanent set from displacement forces as the blades cut through a pile of parts. The feed is usual continuous.

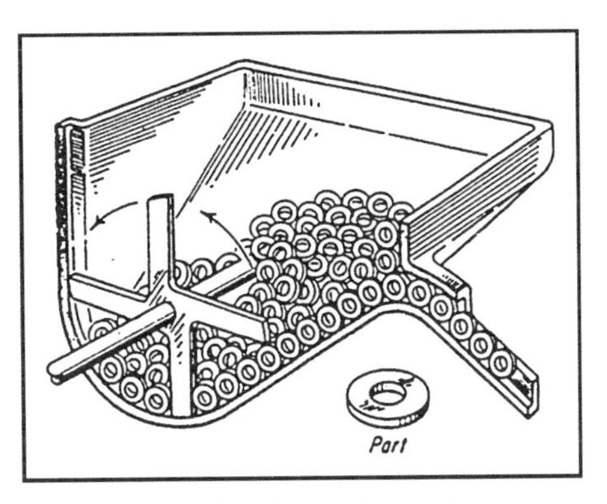

A paddle wheel is effective for feeding disk-shaped parts if they are stable enough. Thin, weak parts would bend and jam. Avoid these designs, if possible—Especially if automatic assembly methods will be employed.

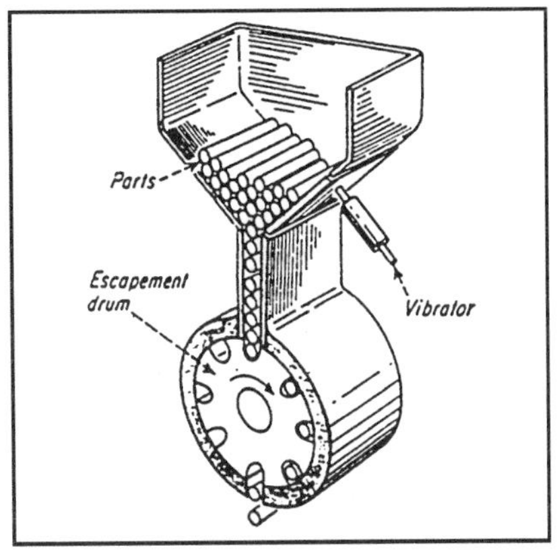

A long-cylinder feeder is a variation of the first two hoppers. If the cylinders have similar ends, the parts can be fed without proposition, thus assisting automatic assembly. A cylinder with differently shaped ends requires extra machinery to orientated the part before it can be assembled.

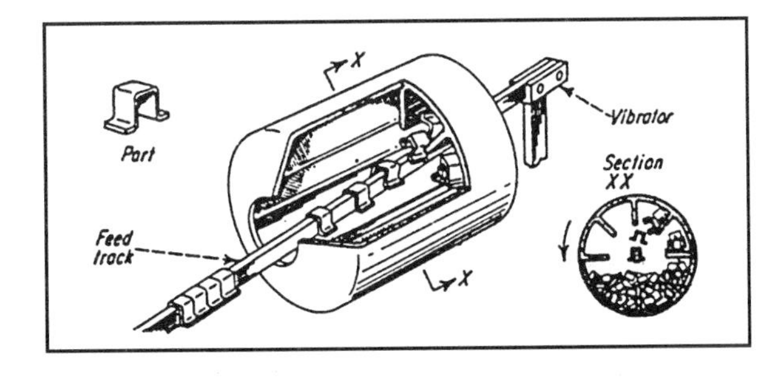

A barrel hopper is useful if parts lend to become entangled. The parts drop free of the rotating-barrel sides. By chance selection, some of them fall onto the vibrating rack and are fed out of the barrel. The parts should be stiff enough to resist excessive bending because the tumbling action can subject them to relatively severe loads. The tumbling can help to remove sharp burrs.

ELEVEN PARTS-HANDLING MECHANISMS

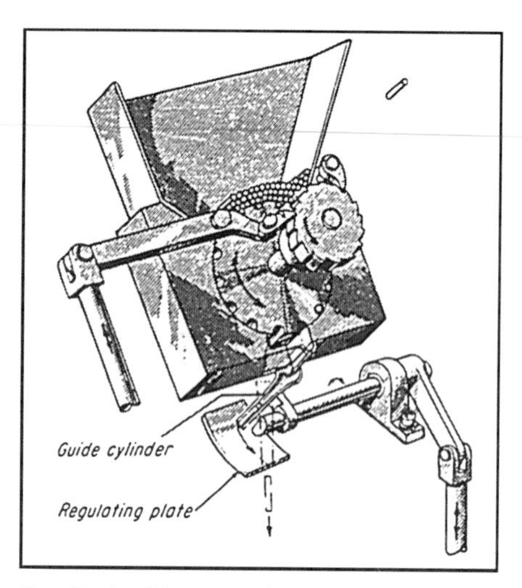

Gravity feed for rods. Single rods of a given length are transferred from the hopper to the lower guide cylinder by means of an intermittently rotating disk with a notched circumference. The guide cylinder, moved by a lever, delivers the rod when the outlet moves free of the regulating plate.

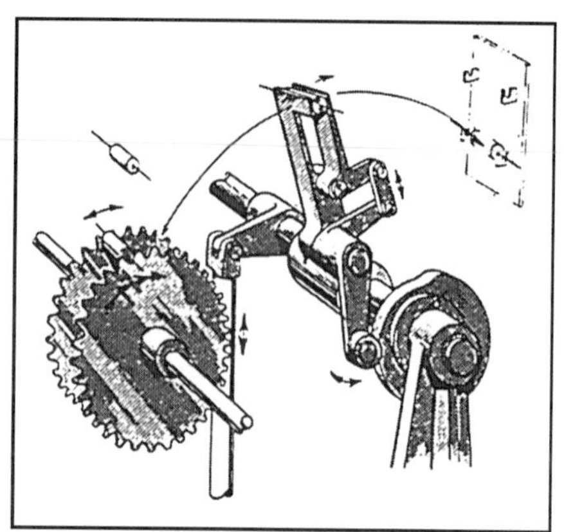

Feeding electronic components. Capacitors, for example, can be delivered by a pair of intermittently rotating gearlike disks with notched circumferences. Then a pick-up arm lifts the capacitor and it is carried to the required position by the action of a cam and follower.

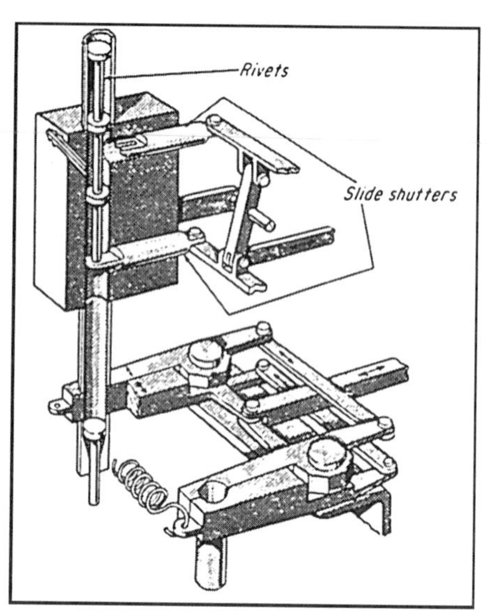

Feeding headed rivets. Headed rivets, correctly oriented, are supplied from a parts-feeder in a given direction. They are dropped, one by one, by the relative movement of a pair of slide shutters. Then the rivet falls through a guide cylinder to a clamp. Clamp pairs drop two rivets into corresponding holes.

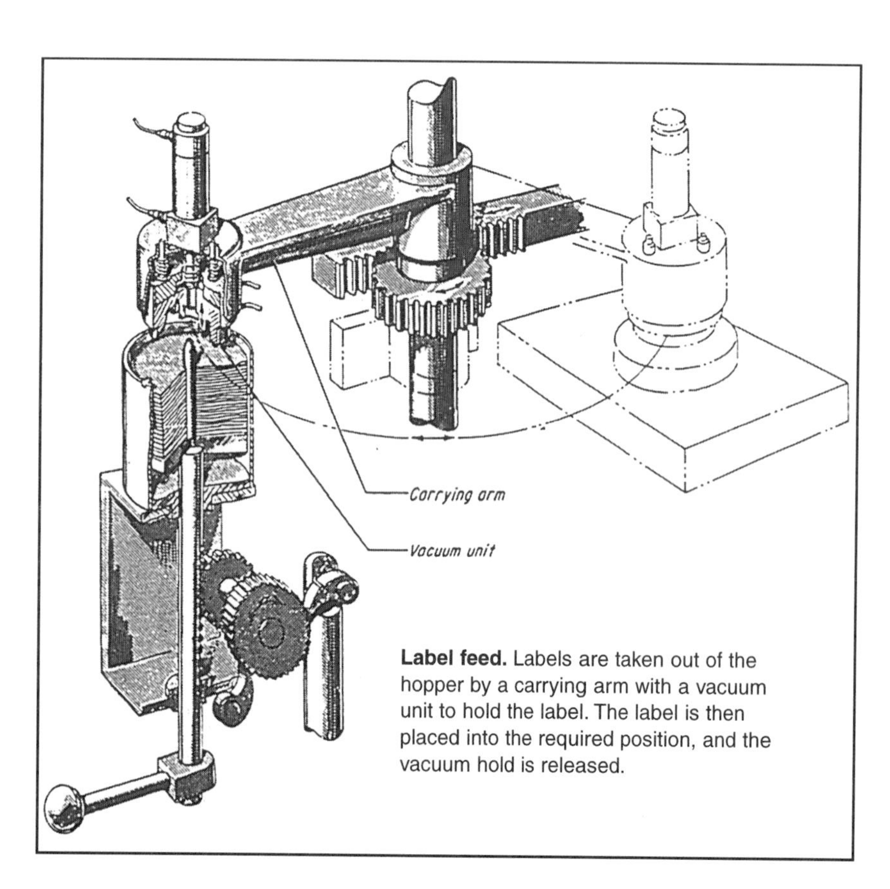

Label feed. Labels are taken out of the hopper by a carrying arm with a vacuum unit to hold the label. The label is then placed into the required position, and the vacuum hold is released.

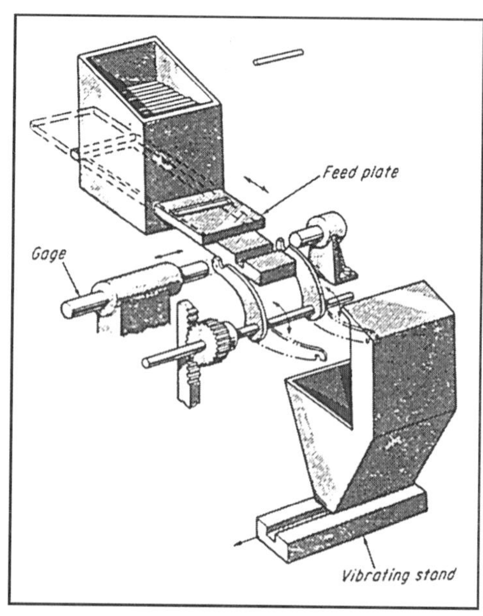

Horizontal feed for fixed-length rods. Single rods of a given length are brought from the hopper to the slot of a fixed plate by a moving plate. After being gauged in the notched portion of the fixed plate, each rod is moved to the chute by means of a lever, and is removed from the chute by a vibrating table.

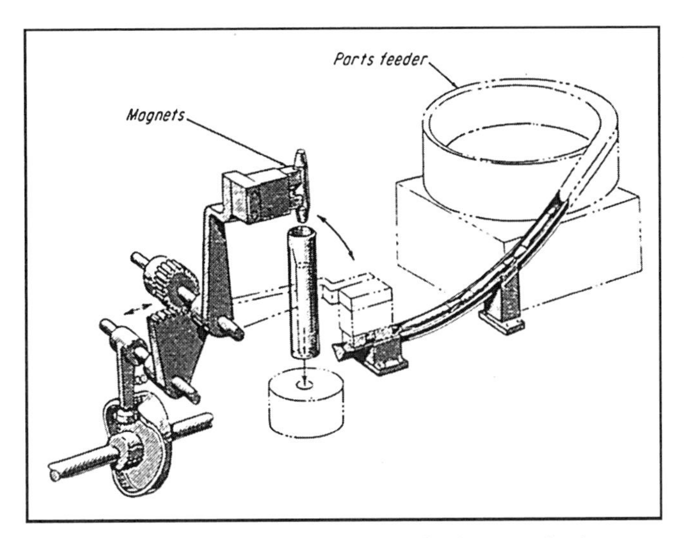

Pin inserter. Pins, supplied from the parts-feeder, are raised to a vertical position by a magnet arm. The pin drops through a guide cylinder when the electromagnet is turned off.

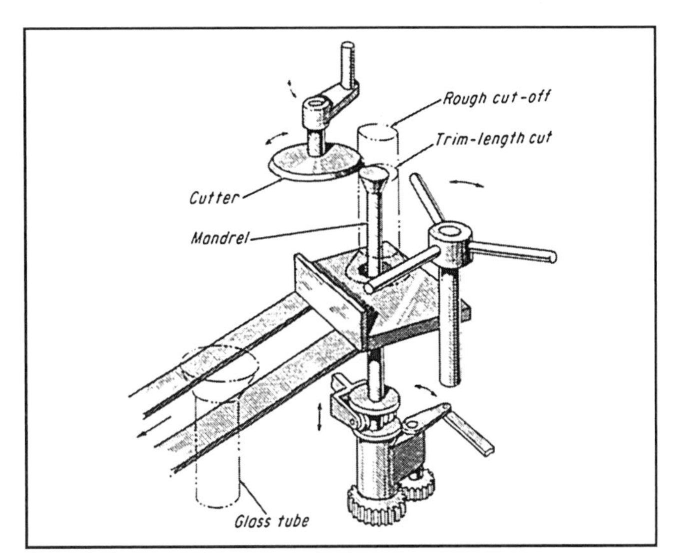

Cutoff and transfer devices for glass tubes. The upper part of a rotating glass tube is held by a chuck (not shown). When the cutter cuts the tube to a given length, the mandrel comes down and a spring member (not shown) drops the tube on the chute.

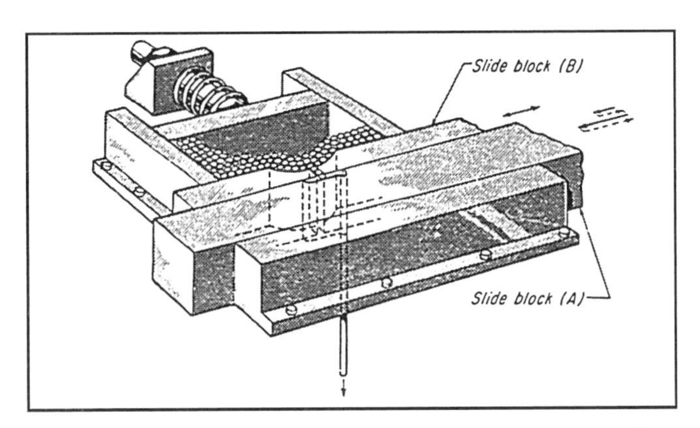

Vertical feed for wires. Wires of fixed length are stacked vertically, as illustrated. They are removed, one by one, as blocks A and B are slid by a cam and lever (not shown) while the wires are pressed into the hopper by a spring.

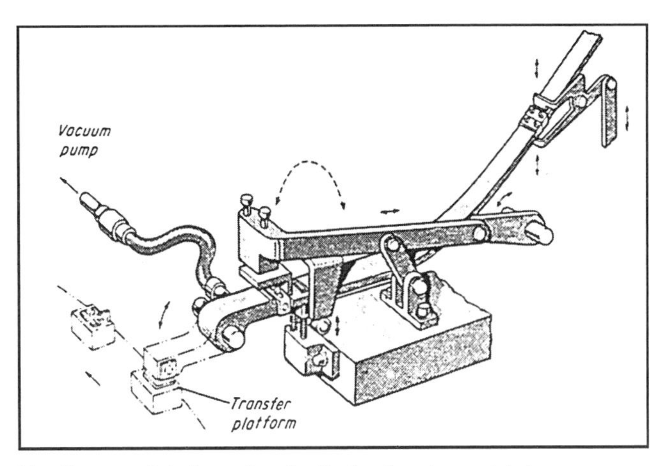

Feeding special-shaped parts. Parts of such special shapes as shown are removed, one by one, in a given direction, and are then moved individually into the corresponding indents on transfer platforms.

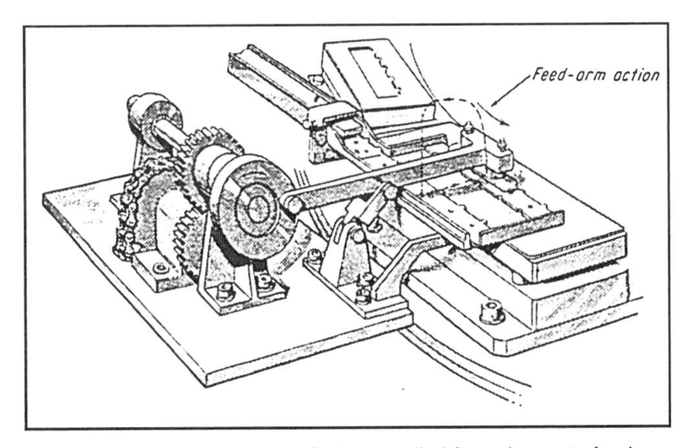

Lateral feed for plain strips. Strips supplied from the parts-feeder are put into the required position, one by one, by an arm that is part of a D-drive linkage.

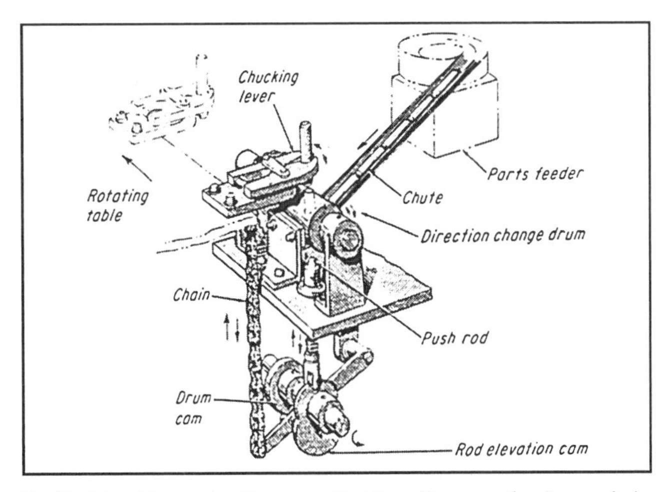

Vertical feed for rods. Rods supplied from the parts-feeder are fed vertically by a direction drum and a pushing bar. The rod is then drawn away by a chucking lever.

SEVEN AUTOMATIC-FEED MECHANISMS

The design of feed mechanisms for automatic or semiautomatic machines depends largely upon such factors as size, shape, and character of the materials or parts that are to be fed into a machine, and upon the kinds of operation to be performed. Feed mechanisms can be simple conveyors that give positive guidance, or they might include secure holding devices if the parts are subjected to processing operations while being fed through a machine. One of the functions of a feed mechanism is to extract single pieces from a stack or unassorted supply of stock. If the stock is a continuous strip of metal, roll of paper, long bar, or tube, the mechanism must maintain intermittent motion between processing operations. These conditions are illustrated in the accompanying drawings of feed mechanisms.

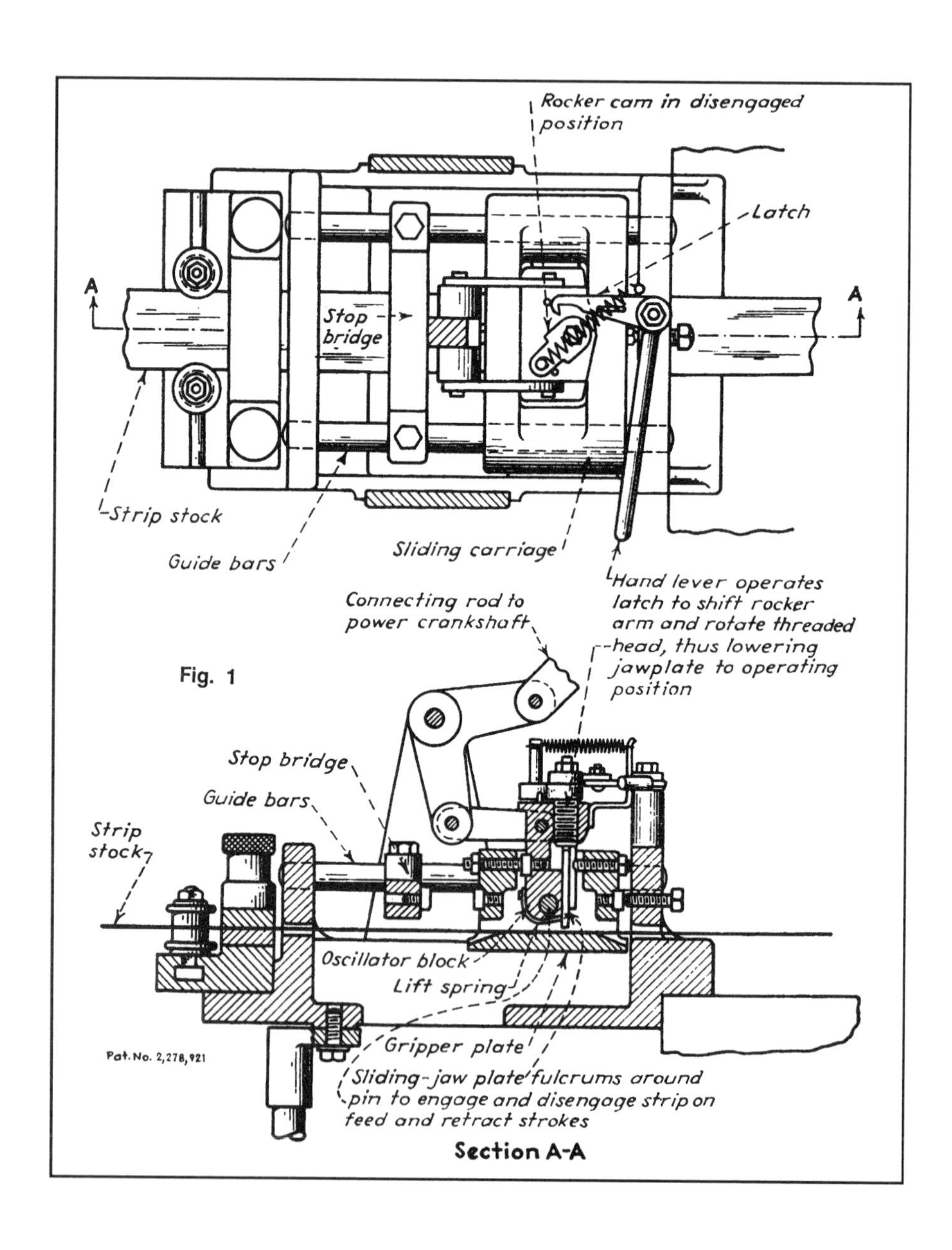

Fig. 1

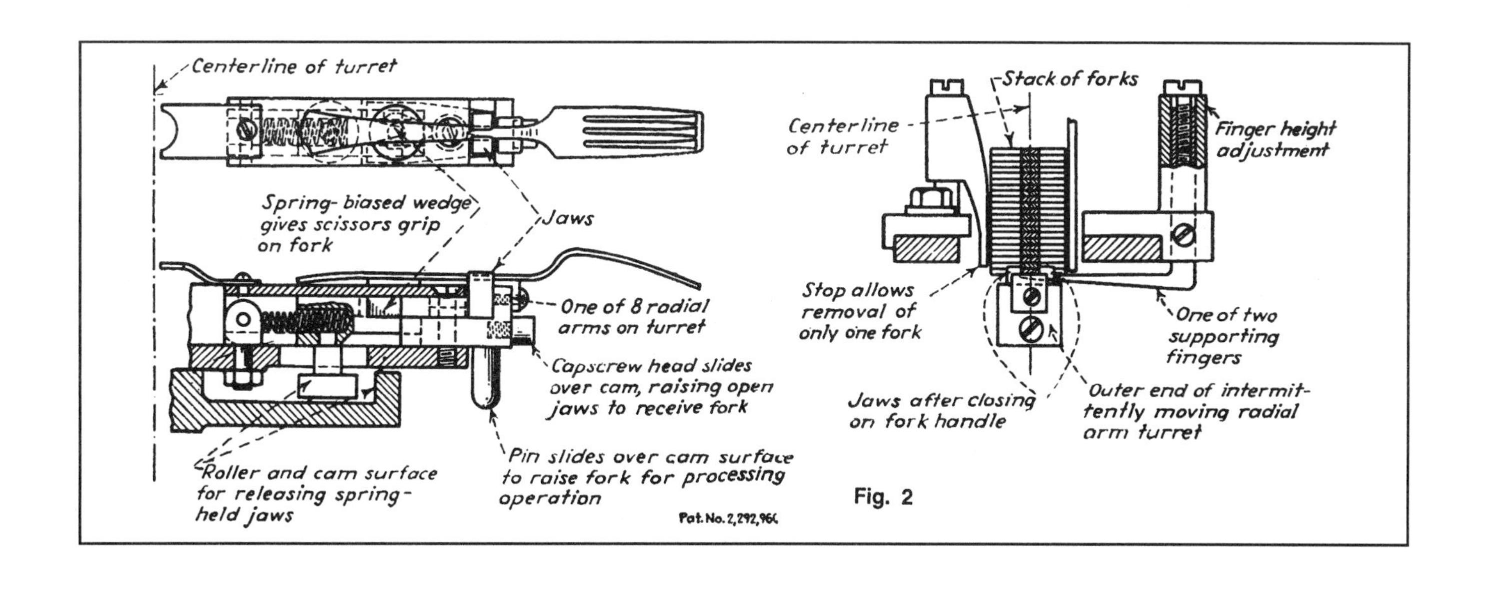
Centerline of turret
Spring- biased wedge gives scissors grip on fork
Jaws
One of 8 radial arms on turret
Capscrew head slides over cam, raising open jaws to receive fork
Roller and cam surface for releasing spring-held jaws
Pin slides over cam surface to raise fork for processing operation
Pat. No. 2,292,964
Stack of forks
Centerline of turret
Finger height adjustment
Stop allows removal of only one fork
One of two supporting fingers
Jaws after closing on fork handle
Outer end of intermittently moving radial arm turret

Fig. 2

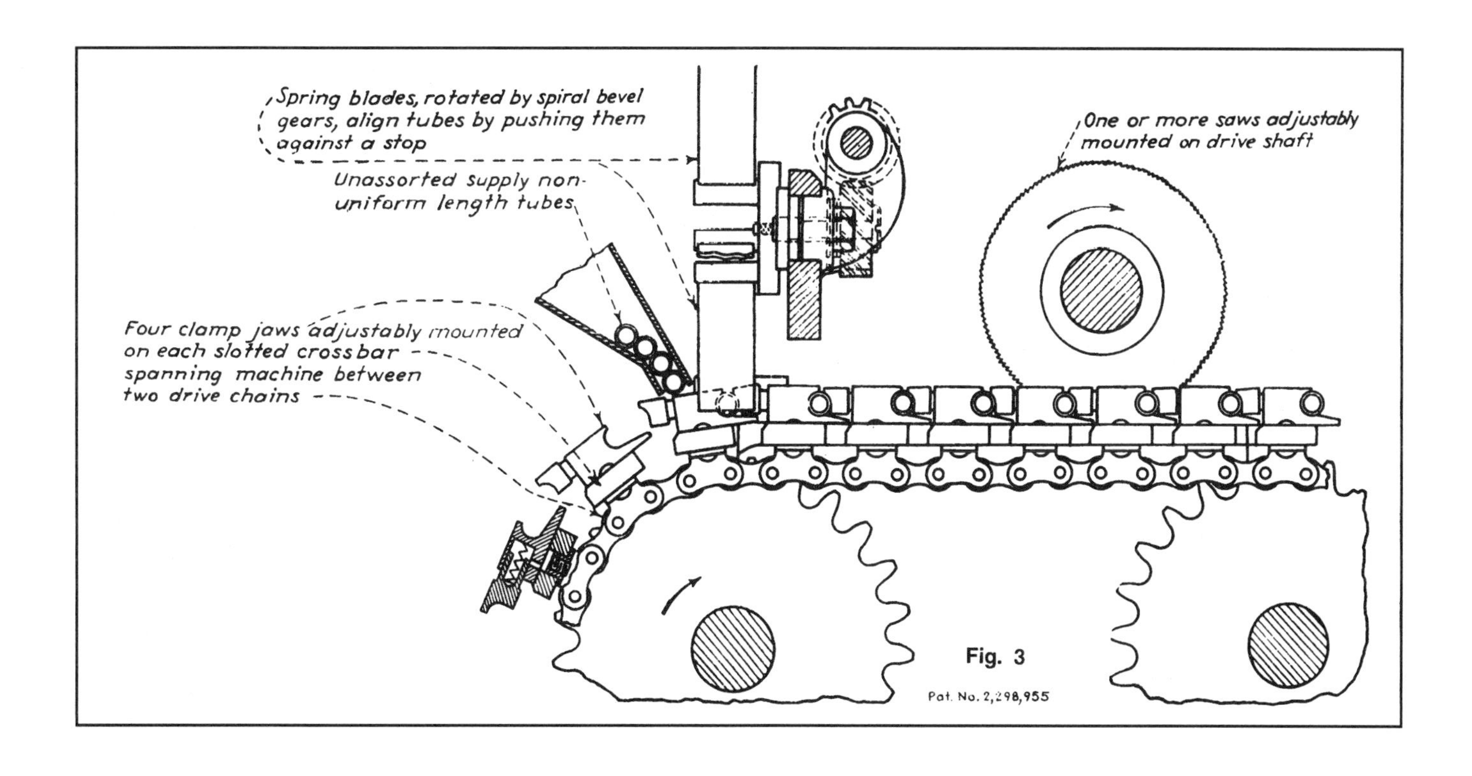
Spring blades, rotated by spiral bevel gears, align tubes by pushing them against a stop
One or more saws adjustably mounted on drive shaft
Unassorted supply non-uniform length tubes
Four clamp jaws adjustably mounted on each slotted crossbar spanning machine between two drive chains
Pat. No. 2,298,955

Fig. 3

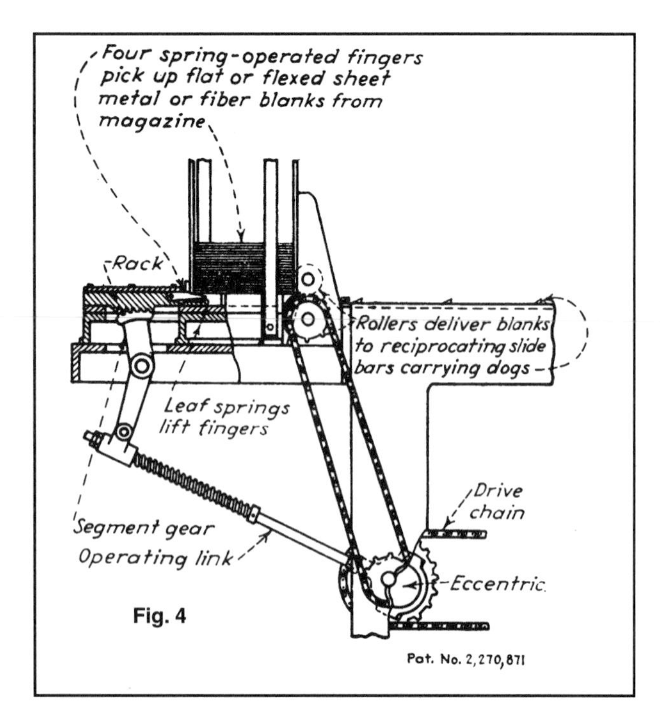

Fig. 4

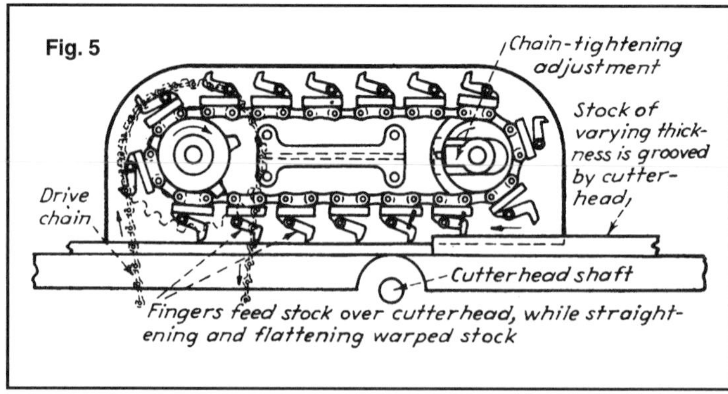

Fig. 5

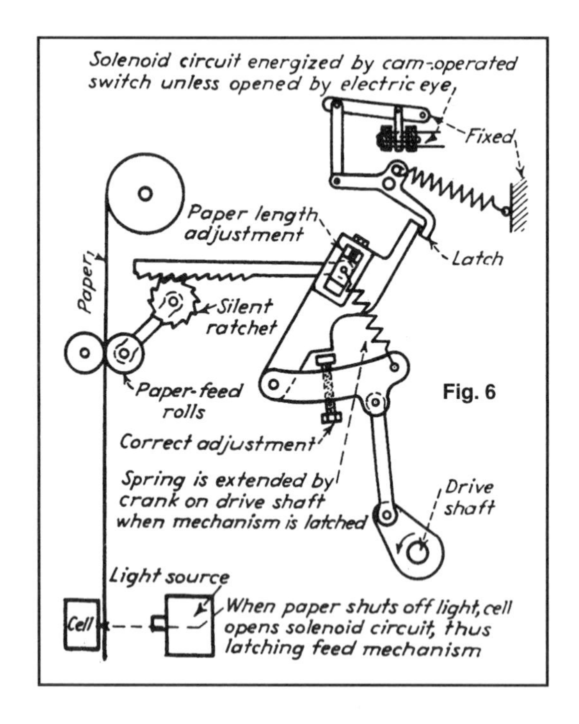

Fig. 6

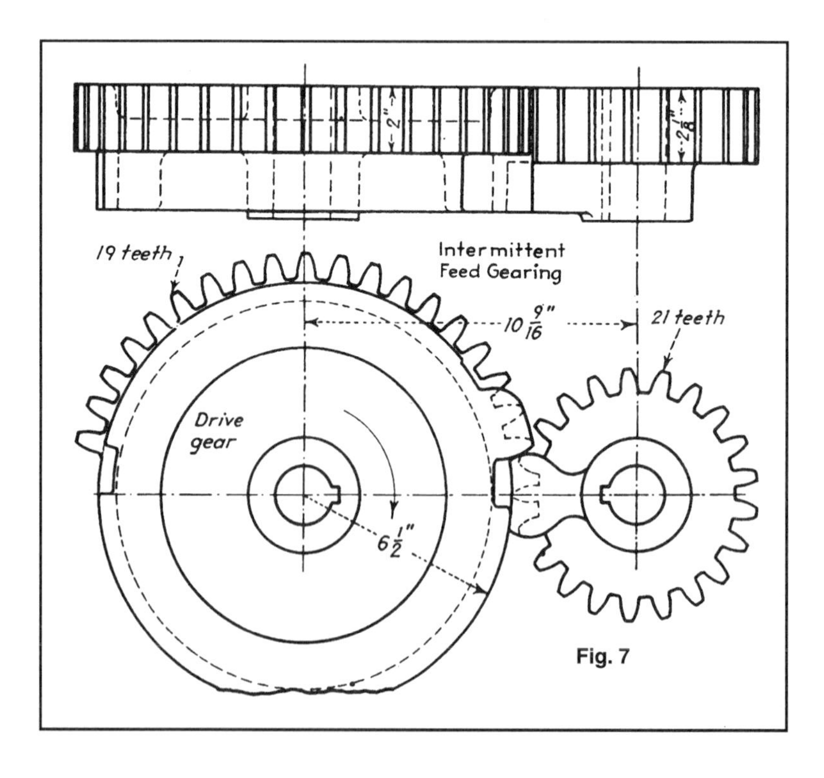

Fig. 7

FIFTEEN CONVEYOR SYSTEMS FOR PRODUCTION MACHINES

Conveyor systems can be divided into two classes: those that are a part of a machine for processing a product, and those that move products in various stages of fabrication. The movement might be from one worker to another or from one part of a plant to another. Most of the conveyors shown here are components in processing machines. Both continuous and intermittently moving equipment are illustrated.

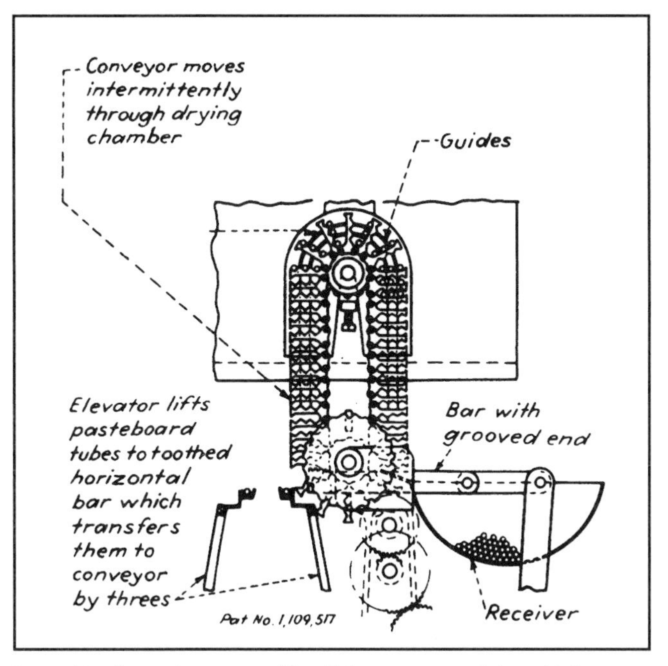

Intermittently moving grooved bar links convey pasteboard tubes through a drying chamber.

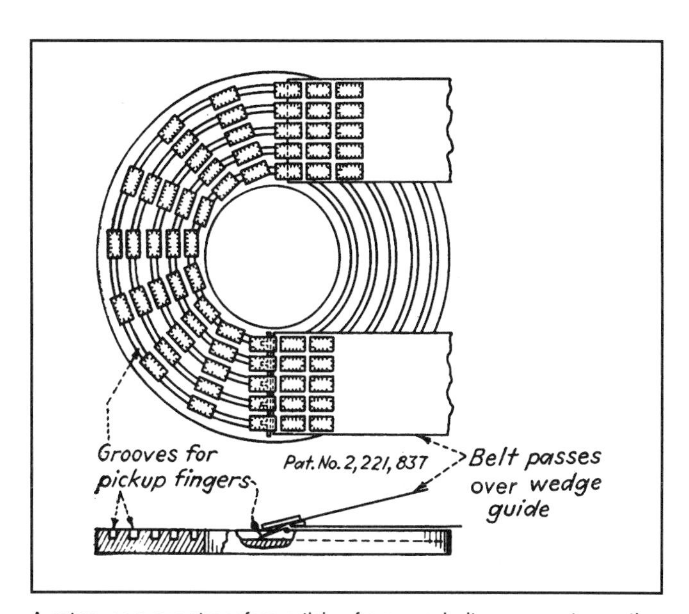

A rotary conveyor transfers articles from one belt conveyor to another without disturbing their relative positions.

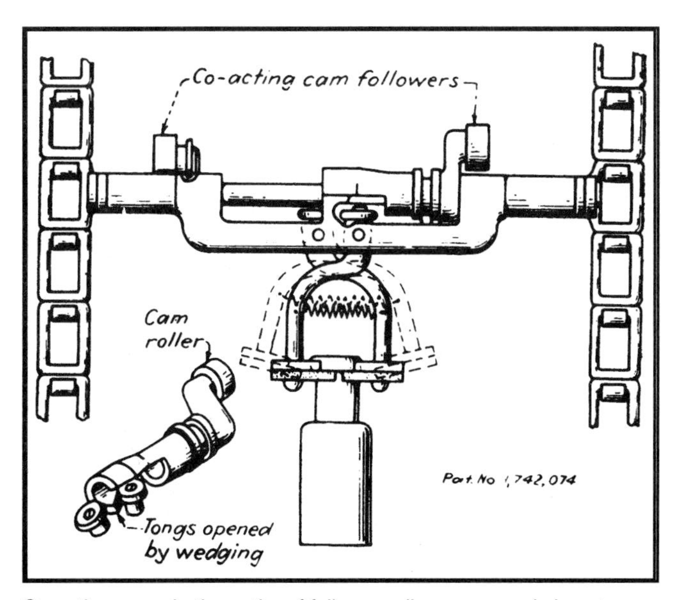

Co-acting cams in the paths of follower rollers open and close tongs over bottlenecks by a wedging action.

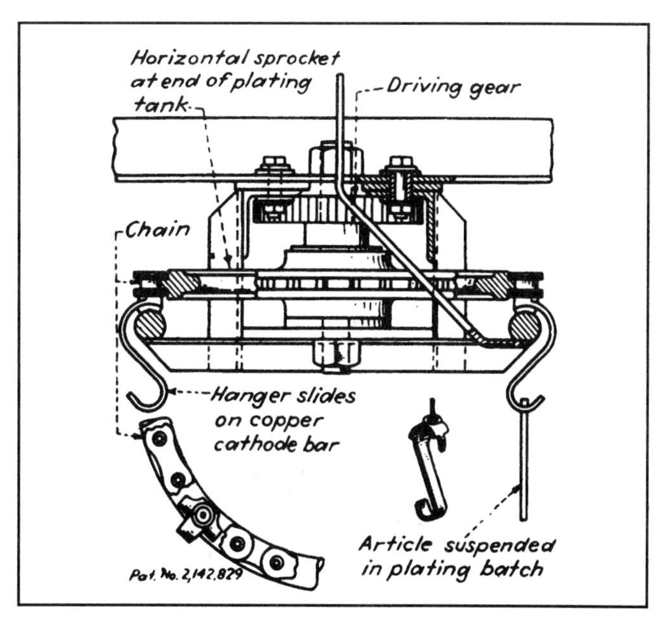

Hooks on a chain-driven conveyor move articles through a plating bath.

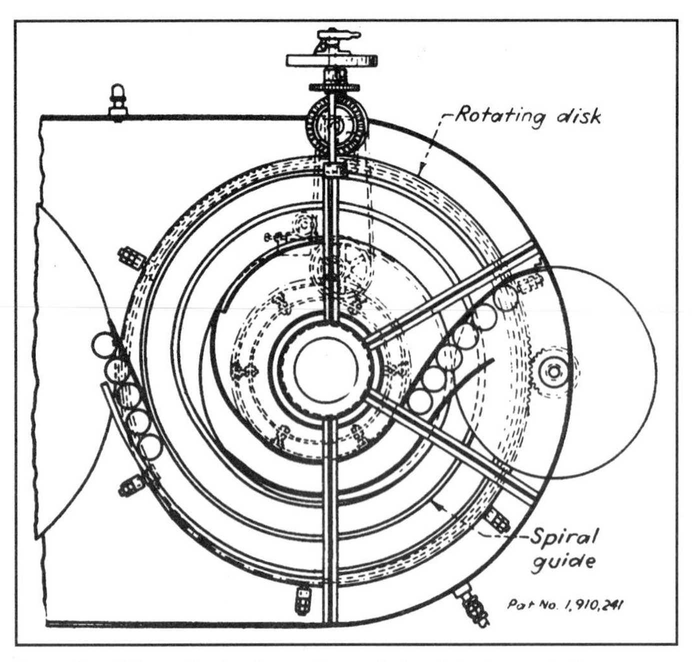

A rotating disk carries food cans in a spiral path between stationary guides for presealing heat treatment.

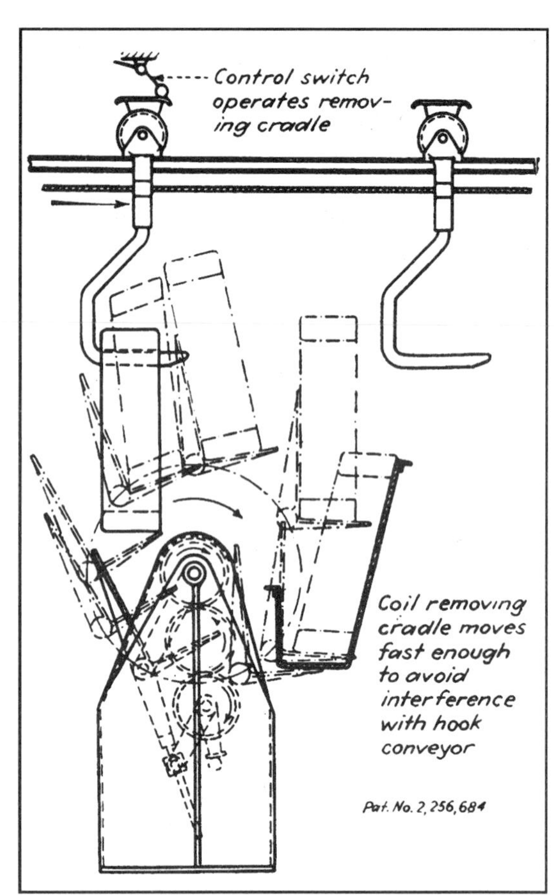

Hooks on a cable-driven conveyor and an automatic cradle for removing coils.

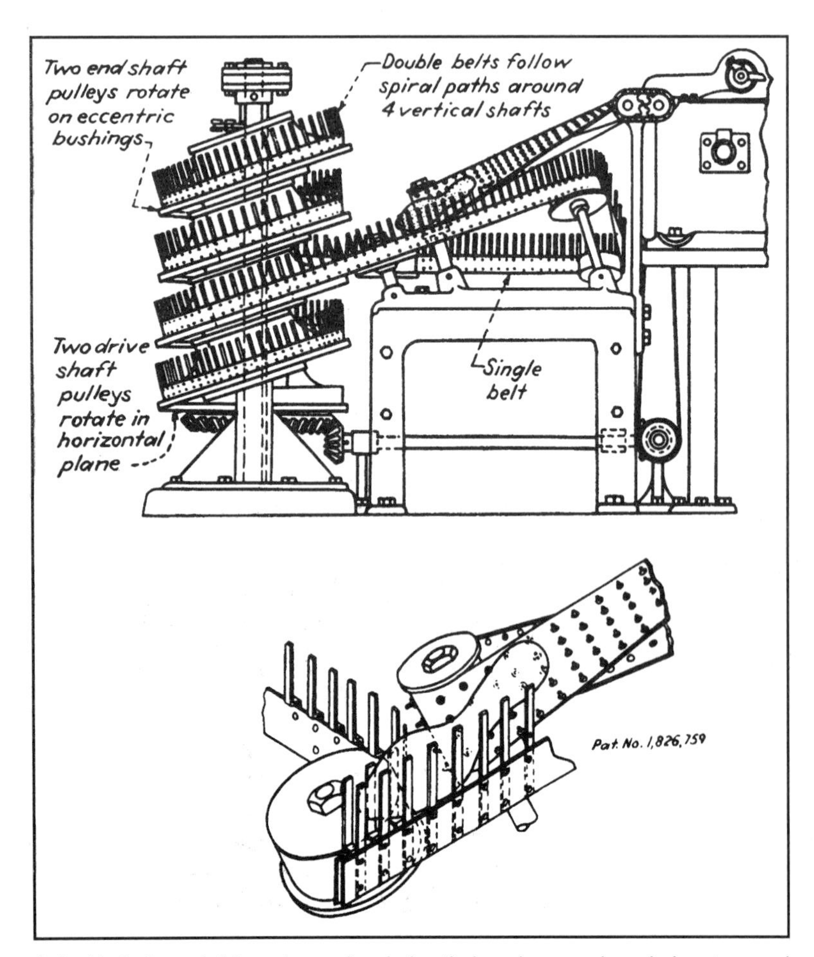

A double belt sandwiches shoe soles during their cycle around a spiral system and then separates to discharge the soles.

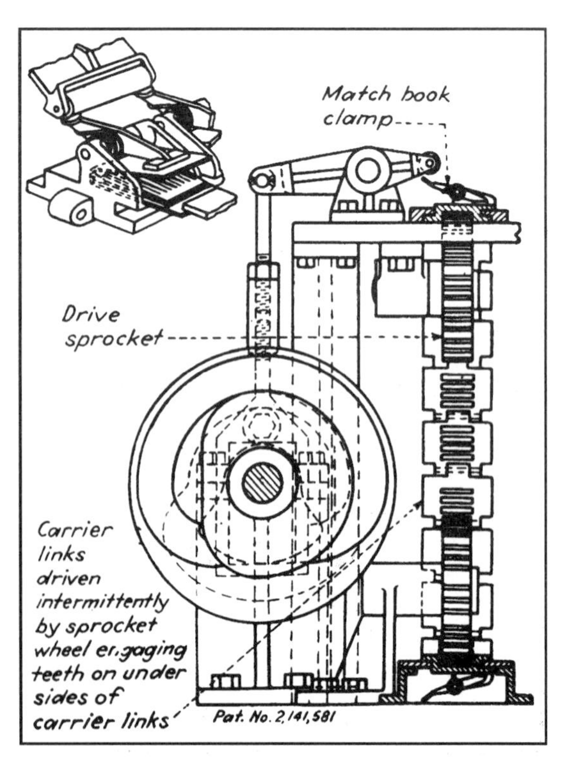

A matchbook carrier links with holding clips that are moved intermittently by sprockets.

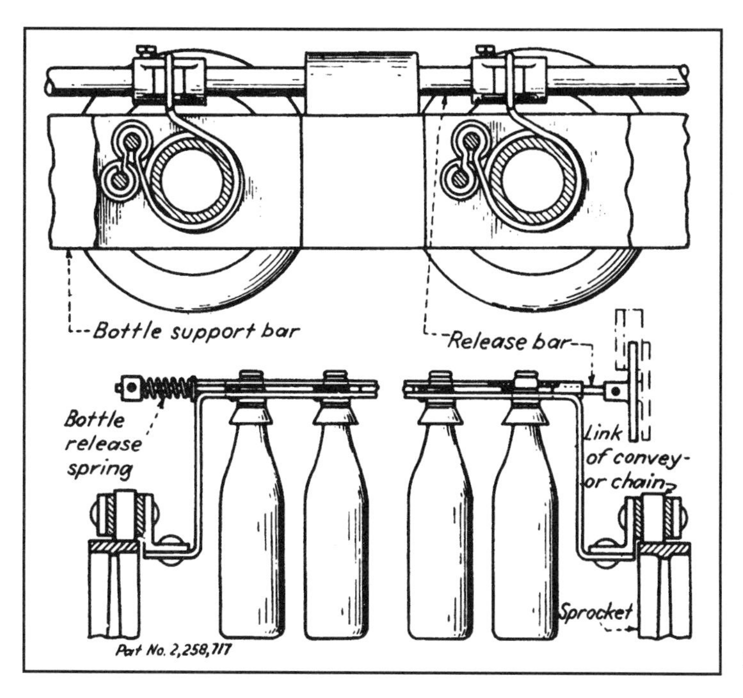

Bottle clips with release bars for automatic operation.

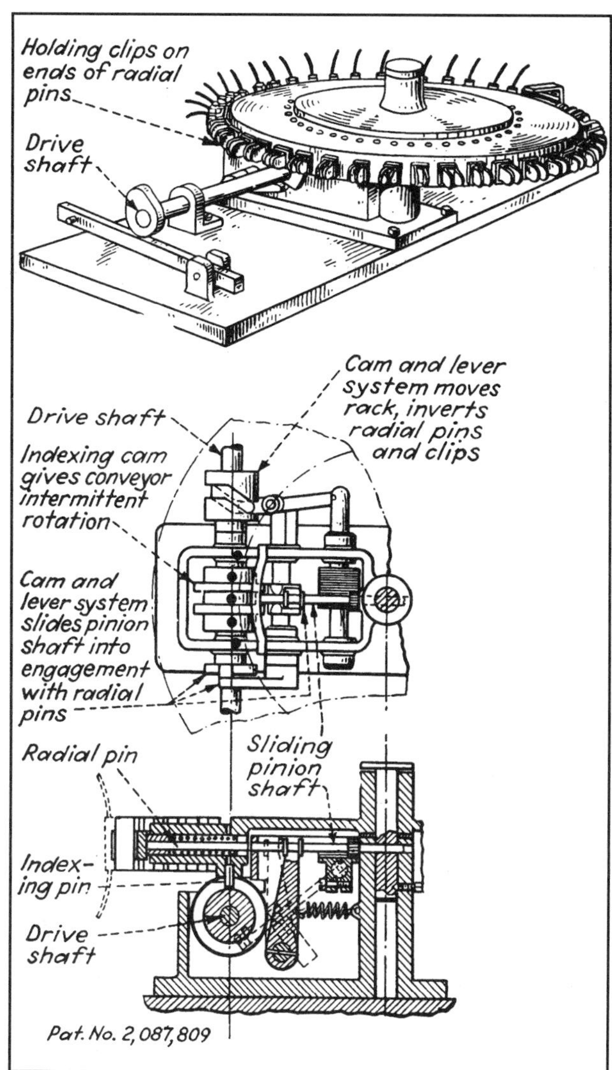

An intermittent rotary conveyor inverts electrical capacitors that are to be sealed at both ends by engaging radial pins which have holding clips attached.

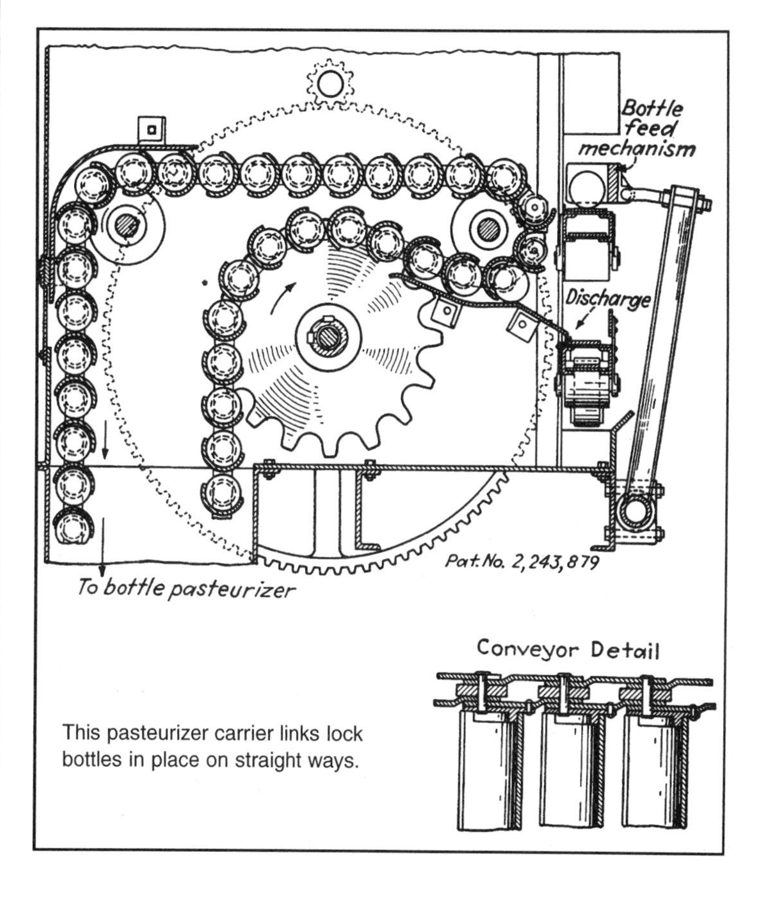

This pasteurizer carrier links lock bottles in place on straight ways.

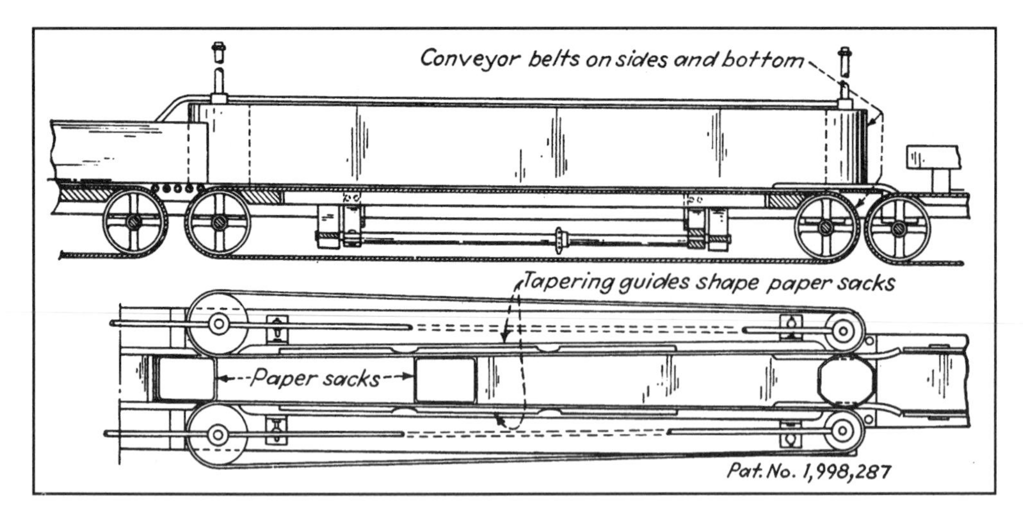

The wedging action of the side belts shapes paper sacks for wrapping an packing.

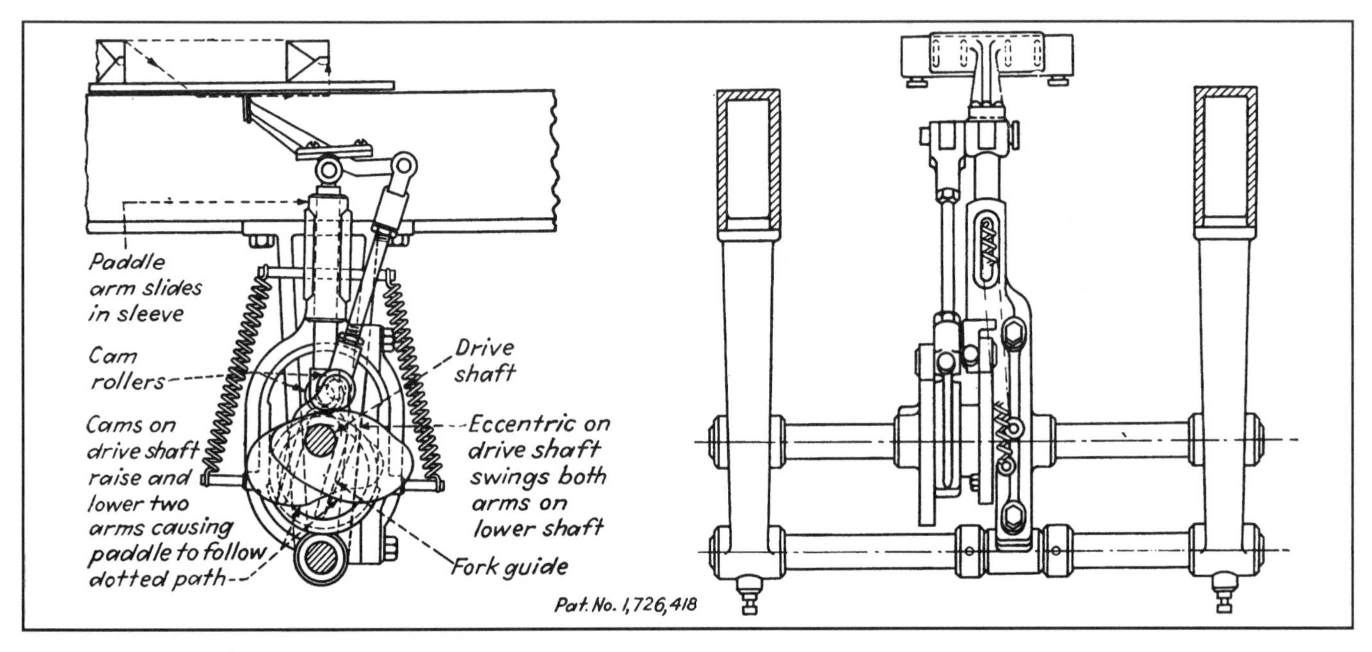

A reciprocating pusher plate is activated by an eccentric disk and two cams on a drive shaft.

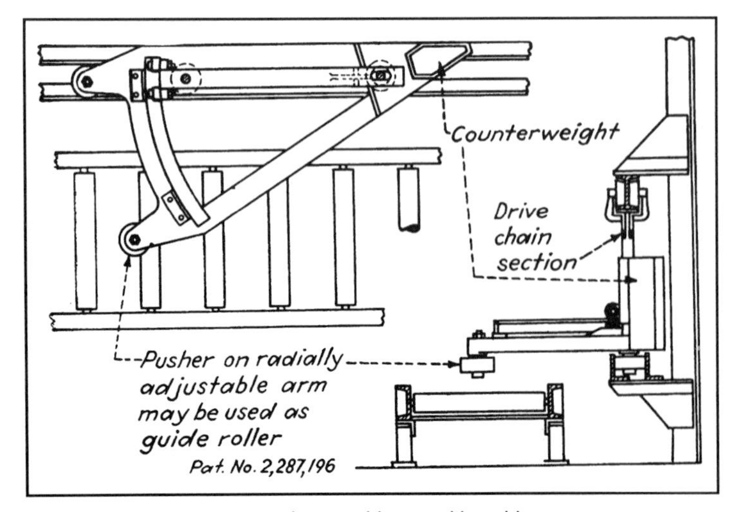

A pusher-type conveyor can have a drive on either side.

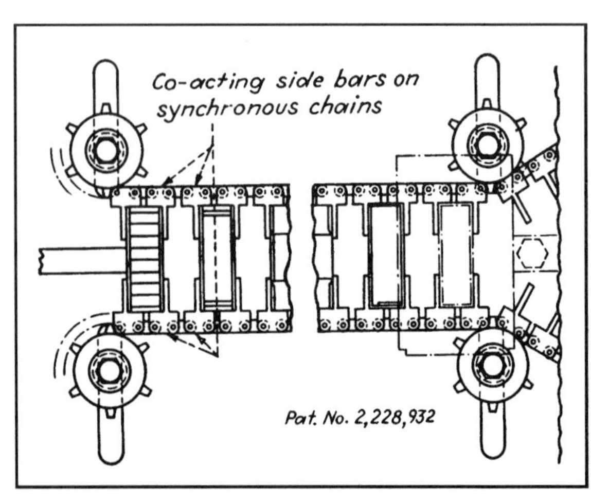

Synchronous chains with side arms grasp and move packages.

SEVEN TRAVERSING MECHANISMS FOR WINDING MACHINES

The seven mechanisms shown are parts of different yarn- and coil-winding machines. Their fundamentals, however, might be applicable to other machines that require similar changes of motion. Except for the leadscrews found on lathes, these seven represent the operating principles of all well-known, mechanical traversing devices.

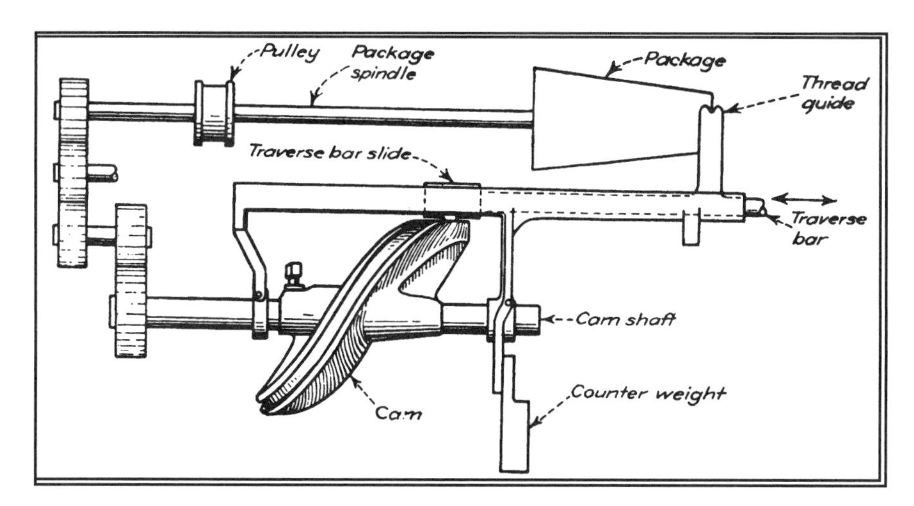

Fig. 1 A package is mounted on a belt-driven shaft on this precision winding mechanism. A camshaft imparts reciprocating motion to a traverse bar with a cam roll that runs in a cam groove. Gears determine the speed ratio between the cam and package. A thread guide is attached to the traverse bar, and a counter-weight keeps the thread guide against the package.

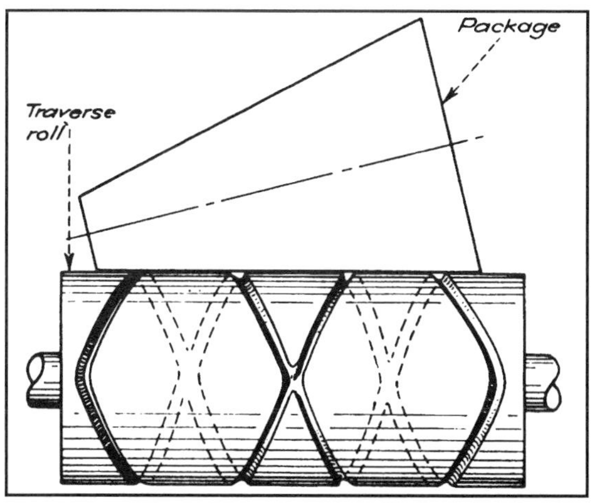

Fig. 2 A package is friction-driven from a traverse roll. Yarn is drawn from the supply source by traverse roll and is transferred to a package from the continuous groove in the roll. Different winds are obtained by varying the grooved path.

Fig. 3 Reversing bevel gears that are driven by a common bevel gear drive the shaft carrying the traverse screw. A traverse nut mates with this screw and is connected to the yarn guide. The guide slides along the reversing rod. When the nut reaches the end of its travel, the thread guide compresses the spring that actuates the pawl and the reversing lever. This action engages the clutch that rotates the traverse screw in the opposite direction. As indicated by the large pitch on the screw, this mechanism is limited to low speeds, but it permits longer lengths of traverse than most of the others shown.

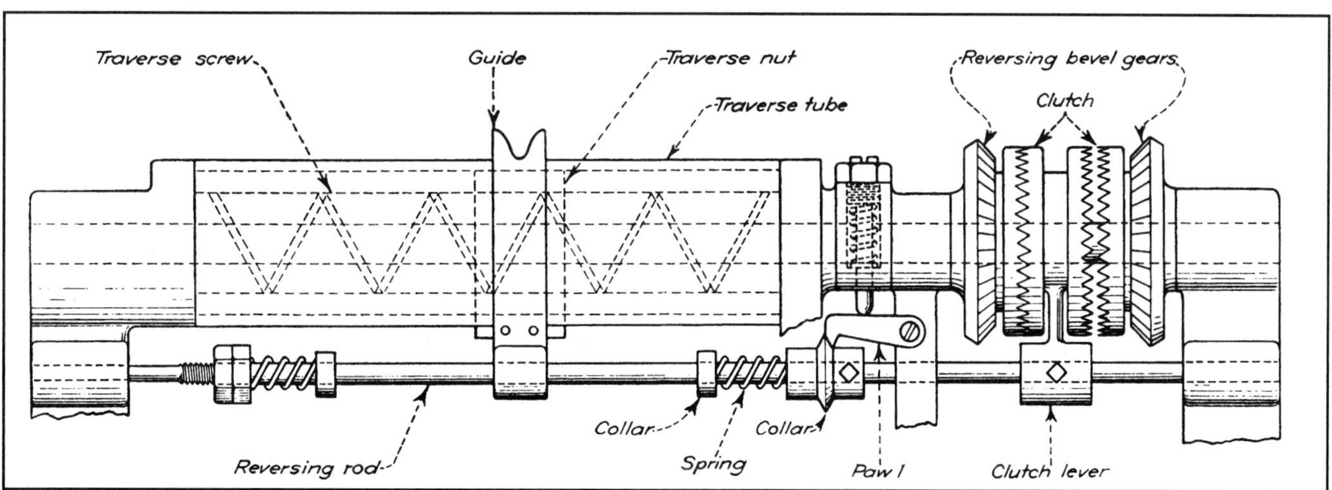

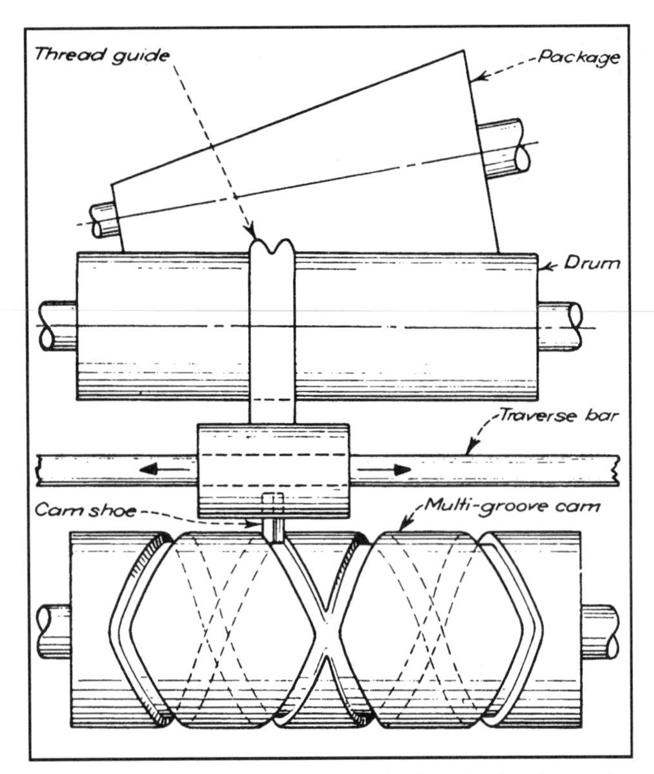

Fig. 4 A drum drives the package by friction. A pointed cam shoe, which pivots in the bottom side of the thread guide assembly, rides in cam grooves and produces a reciprocating motion of the thread guide assembly on the traverse bar. Plastic cams have proved to be satisfactory even with fast traverse speeds. Interchangeable cams permit a wide variety of winding methods.

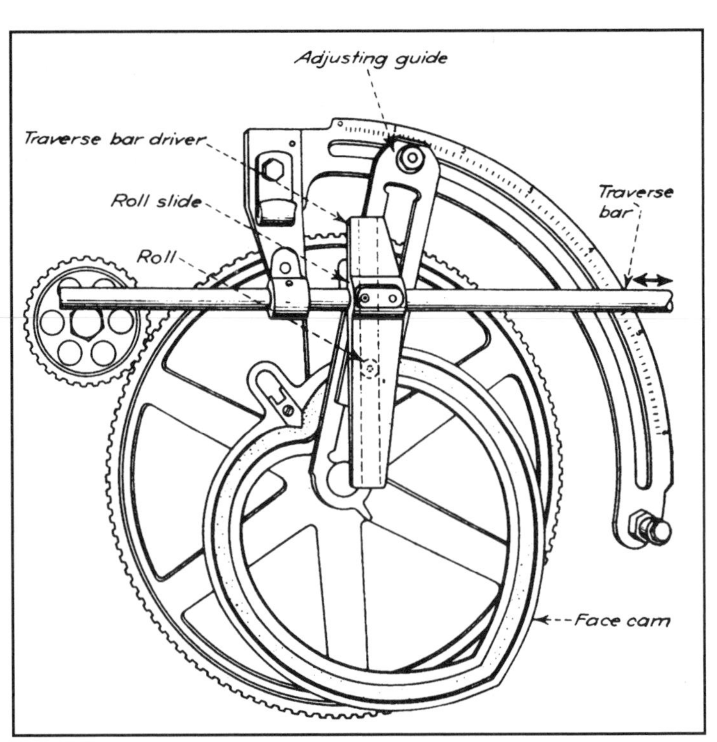

Fig. 5 A roll that rides in a heart-shaped cam groove engages a slot in a traverse bar driver which is attached to the traverse bar. Maximum traverse is obtained when the adjusting guide is perpendicular to the driver. As the angle between the guide and driver is decreased, traverse decreases proportionately. Inertia effects limit this mechanism to slow speeds.

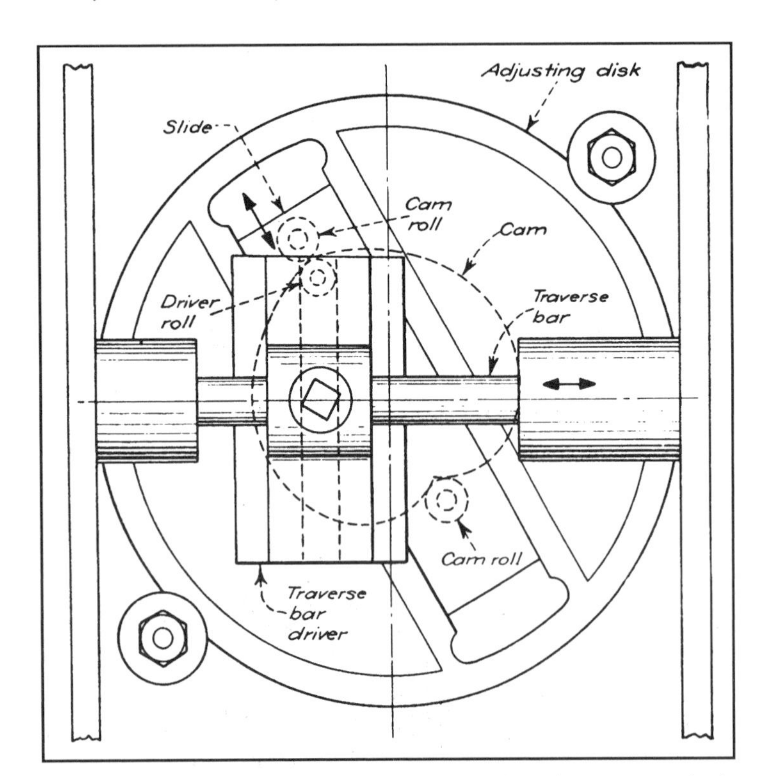

Fig. 6 The two cam rolls that engage this heart-shaped cam are attached to the slide. The slide has a driver roll that engages a slot in the traverse bar driver. Maximum traverse (to the capacity of the cam) occurs when the adjusting disk is set so the slide is parallel to the traverse bar. As the angle between the traverse bar and slide increases, traverse decreases. At 90° traverse is zero.

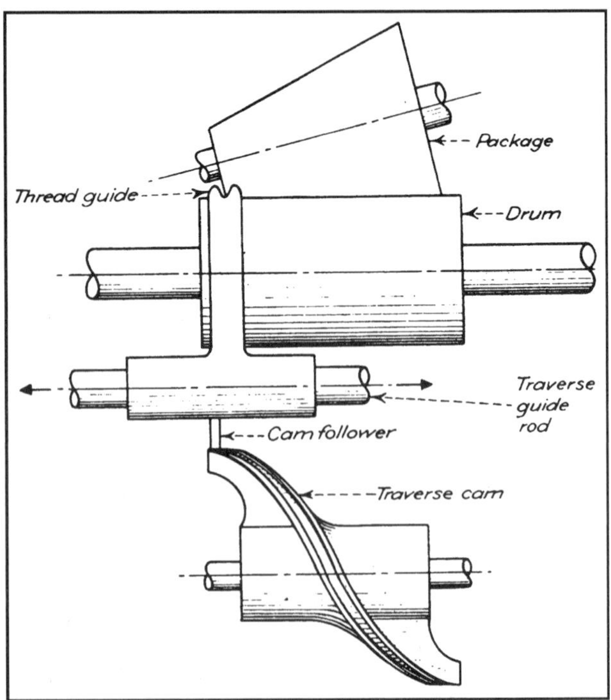

Fig. 7 A traverse cam imparts reciprocating motion to a cam follower that drives thread guides on traverse guide rods. The package is friction driven from the drum. Yarn is drawn from the supply source through a thread guide and transferred to the drum-driven package. The speed of this mechanism is determined by the weight of its reciprocating parts.

VACUUM PICKUP FOR POSITIONING PILLS

This pickup carries tablet cores to moving dies, places cores accurately in coating granulation, and prevents the formation of tablets without cores.

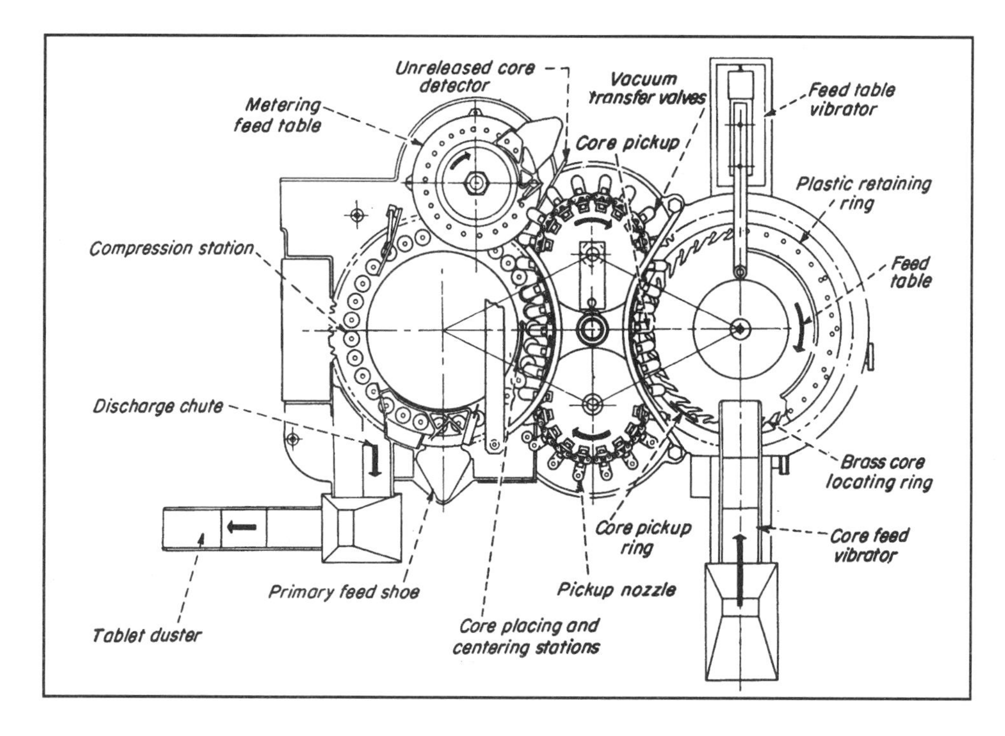

Cores are hopper fed to a rotating feeder disk through a tablet duster. This disk is vibrated clockwise under a slotted pick-up ring which rotates counter-clockwise. Each slot in the pickup ring holds two cores and lets broken tablets fall through to an area under the feeder table. Cores are picked from ring slots, carried to tablet press dies, and deposited in dies by vacuum nozzles fastened to a chain driven by the press die table. This chain also drives the pickup ring to synchronize the motion of ring slots and pickup nozzles. Coating granulation is fed into the dies ahead of and after the station where a vacuum pickup deposits a core in each die. Compressing rolls are at the left side of the machine. The principal design objective here was to develop a machine to apply dry coatings at speeds that lowered costs below those of liquid coating techniques.

MACHINE APPLIES LABELS FROM STACKS OR ROLLERS

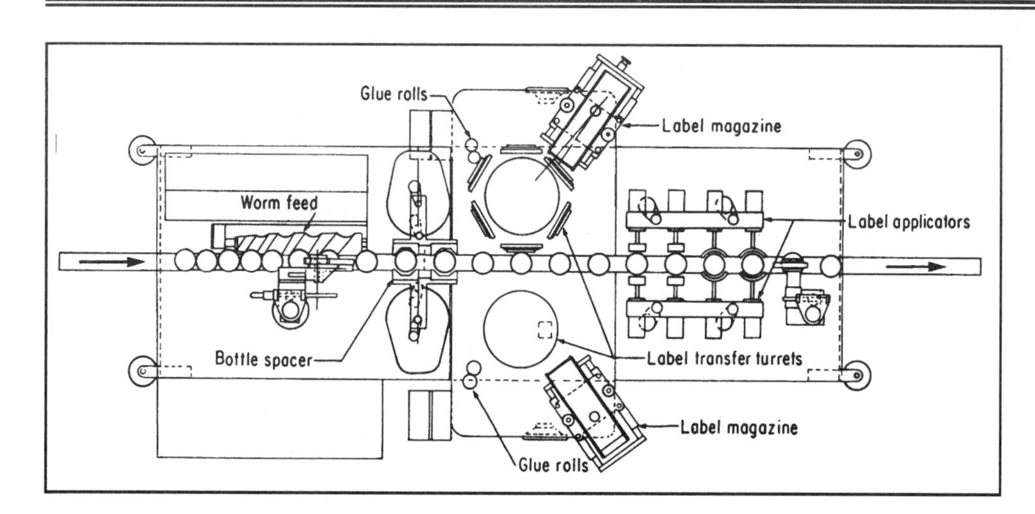

The flow of containers through this labeler is shown by the top-view drawing of the machine. Bottle spacers ensure that containers remain $7^1/_2$ in. apart on the conveyor. Dual label-transfer turrets allow for the simultaneous application of front and back labels.

This labeling machine can perform either conventional glue-label application or it can heat-seal labels in cut or roll form. The machine labels the front and back of round or odd-shaped containers at speeds of 60 to 160 containers per minute. The containers handled range from 1 in. diameter or thickness to $4\ ^1/_4$ in. diameter by $5\ ^1/_2$ in. wide. Container height can vary from 2 to 14 inches. The unit handles labels ranging from $^7/_8$ to $5^1/_2$ in. wide and $^7/_8$ to $6^1/_2$ in. high. The label hopper is designed for labels that are generally rectangular in shape, although it can be modified to handle irregular shapes. Provision has been made in design of the unit, according to the manufacturer, to allow labels to be placed at varying heights on the containers. The unit's cut-and-stacked label capacity is 4,500. An electric eye is provided for cutting labels in web-roll form.

TWENTY HIGH-SPEED MACHINES FOR APPLYING ADHESIVES

Viscous liquid adhesives are used to glue fabrics and paper, apply paper labels, make cardboard and wooden boxes and shoes, and bind books. Specially designed machines are required if the application of adhesives with different characteristics is to be satisfactorily controlled. The methods and machines shown here have been adapted to the application of adhesives in mass production. They might also work well for the application of liquid finishes such as primers, paint, and lacquer.

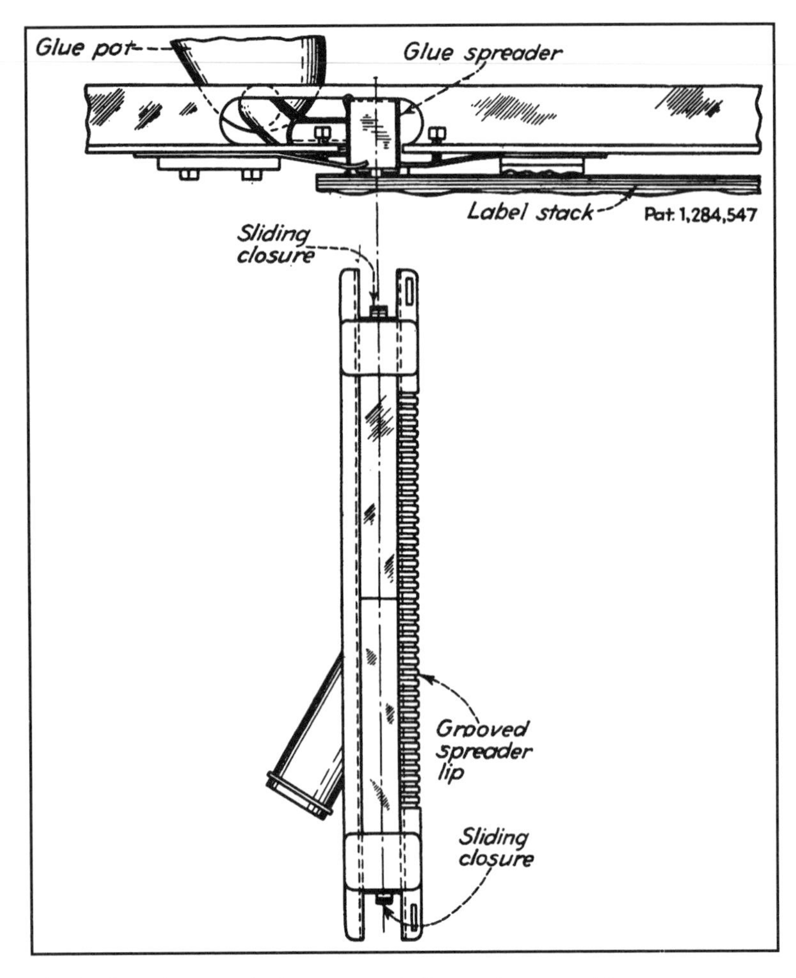

Fig. 1 A gravity spreader has an open bottom and a grooved lip.

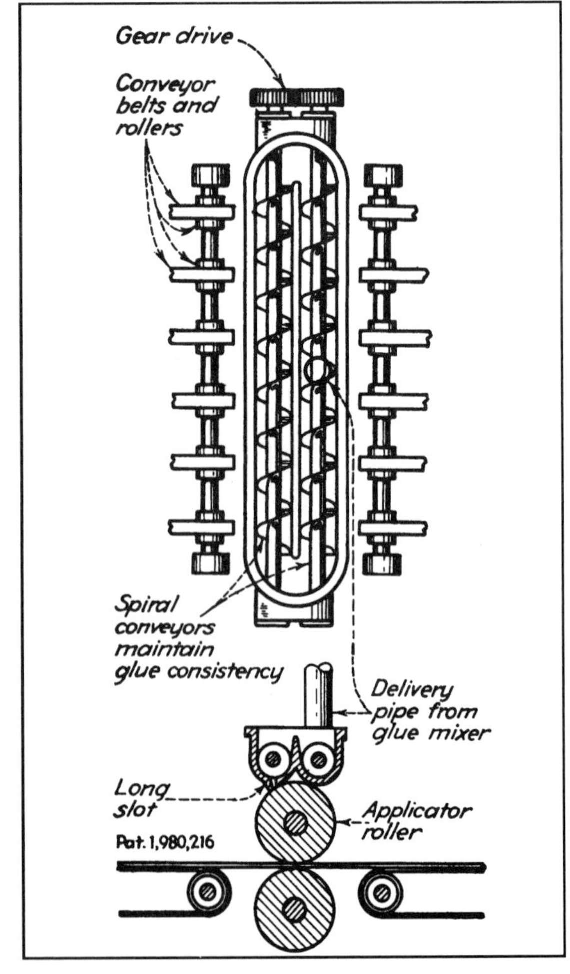

Fig. 2 Spiral conveyors feed the applicator roller by the force or gravity.

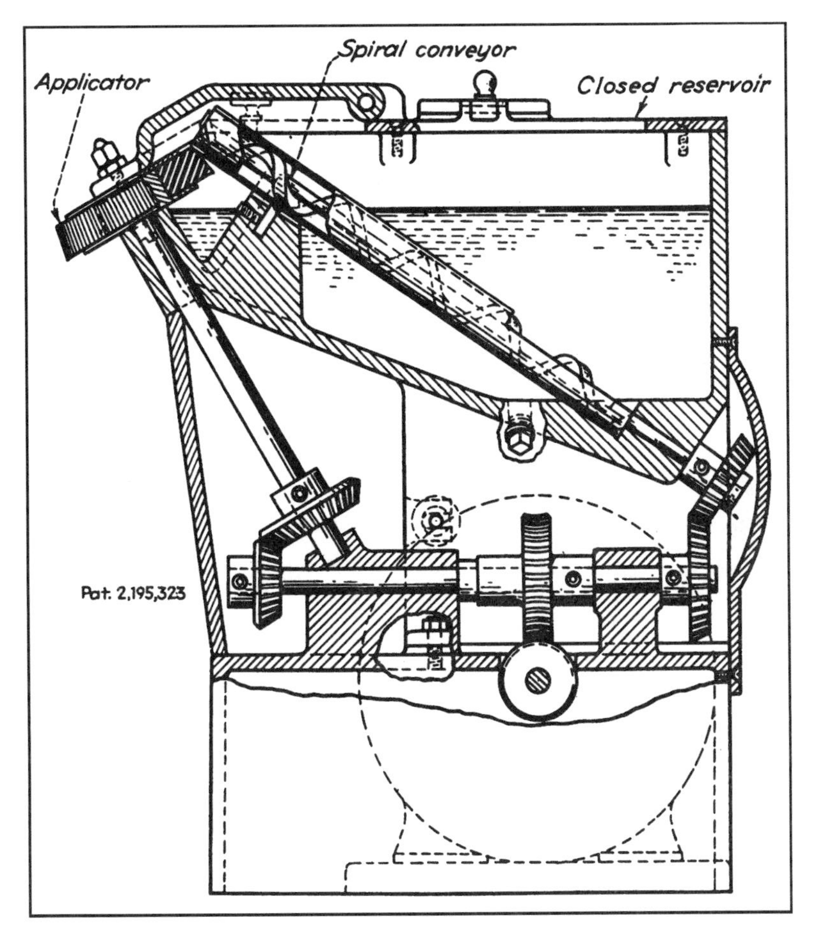

Fig. 3 An applicator wheel is fed by a spiral conveyor.

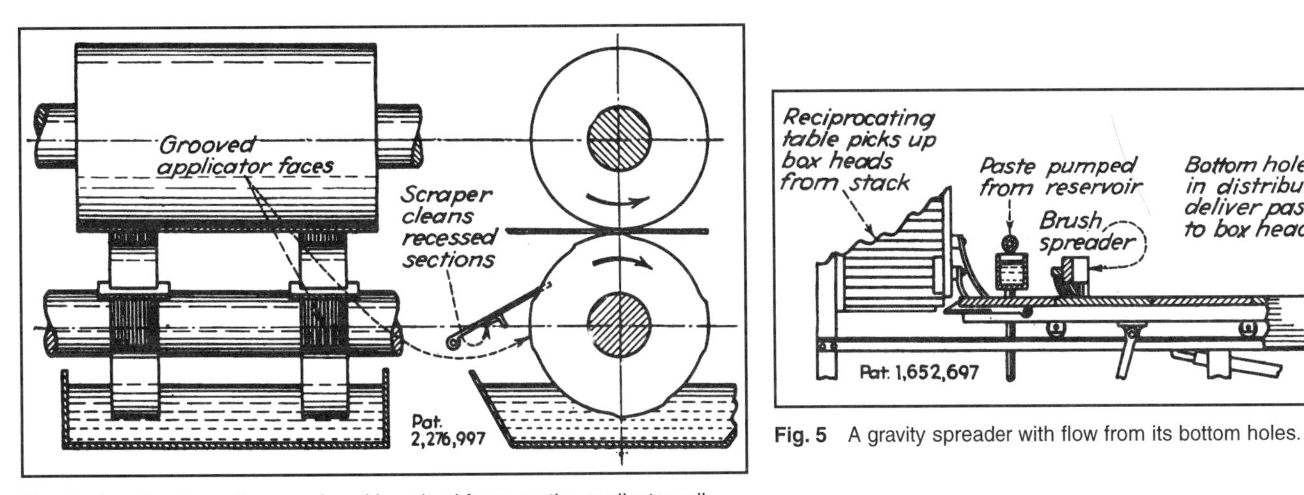

Fig. 4 An adhesive pattern produced by raised faces on the applicator roll.

Fig. 5 A gravity spreader with flow from its bottom holes.

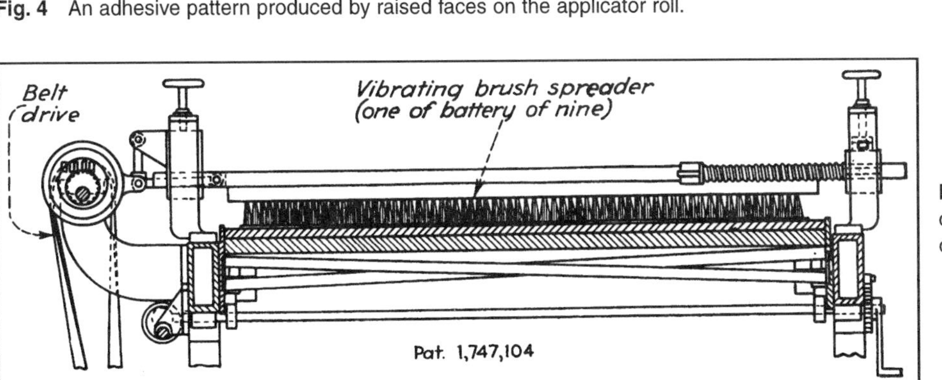

Fig. 6 Vibrating brushes spread the coating after the application by a cylindrical brush.

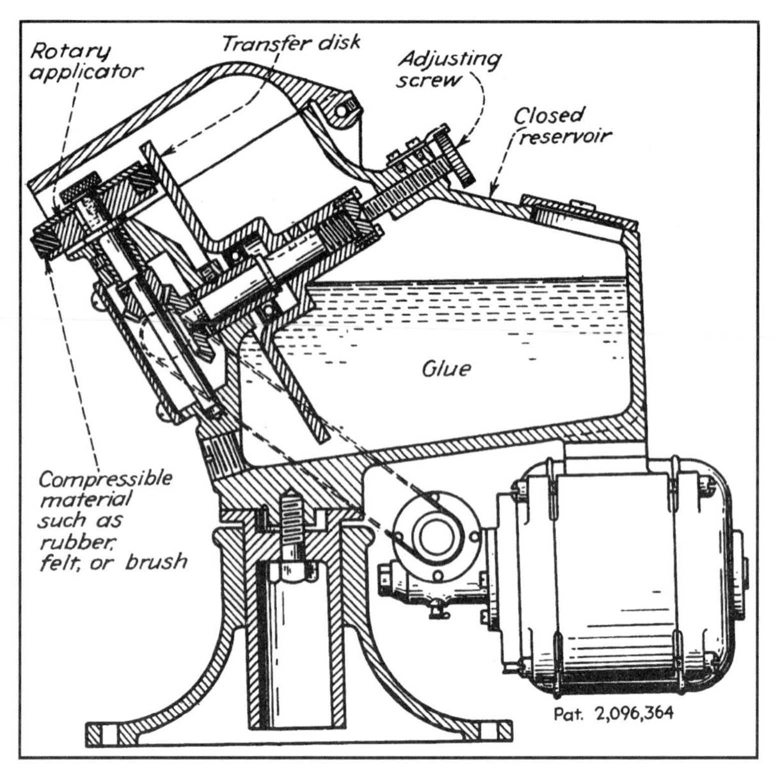

Fig. 7 This applicator wheel is fed by a transfer disk.

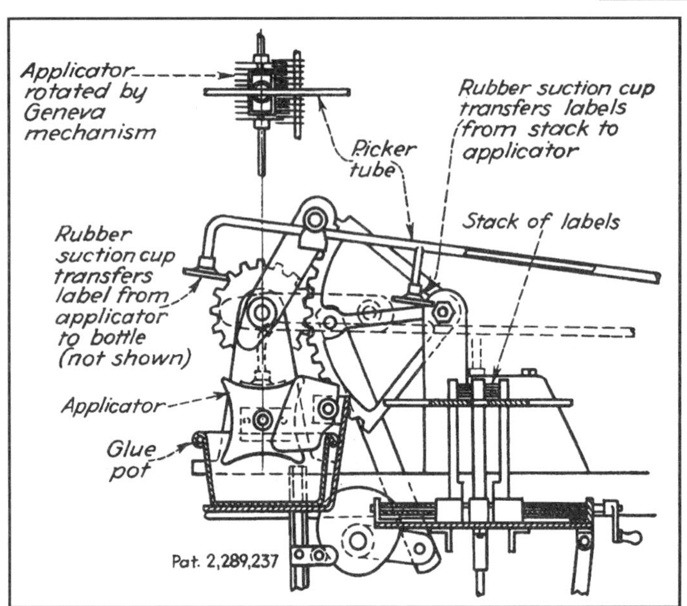

Fig. 8 This applicator surface, consisting of a series of plate edges, is rotated by a Geneva mechanism in the glue pot.

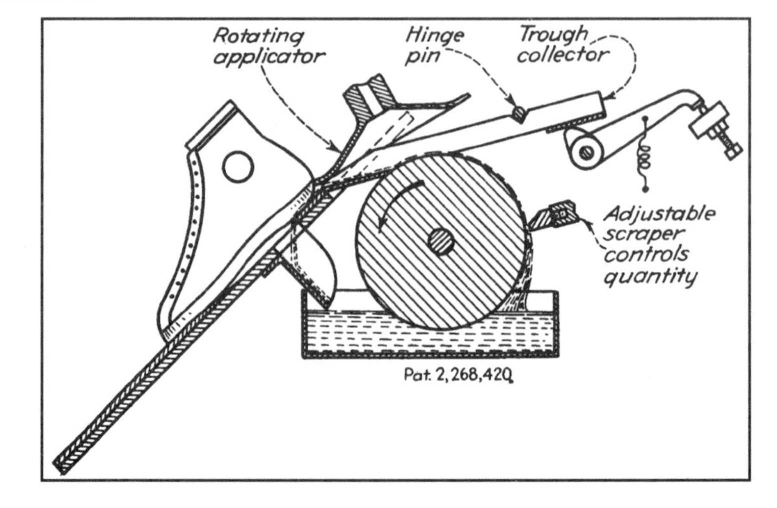

Fig. 9 A rotating applicator disk fed by a trough collector on a transfer drum.

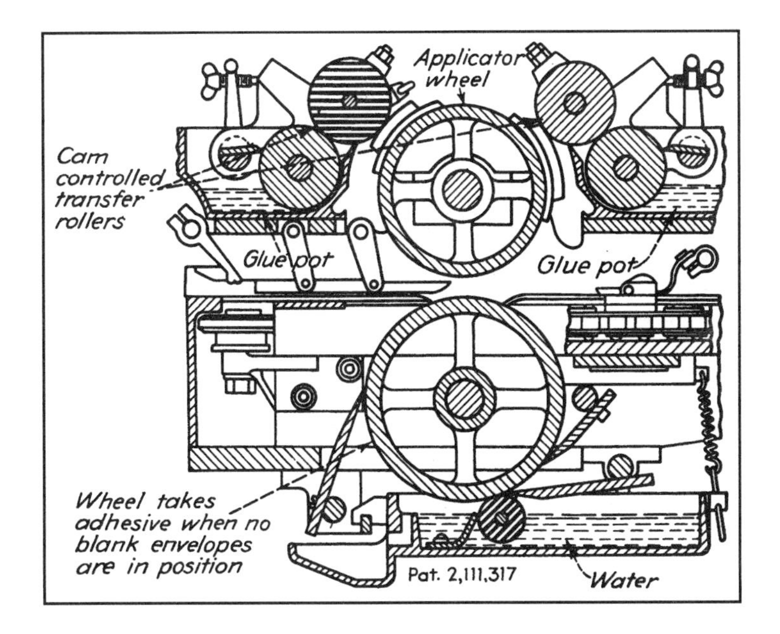

Fig. 10 Cam controlled transfer rollers supply applicator wheel pads with two kinds of adhesive.

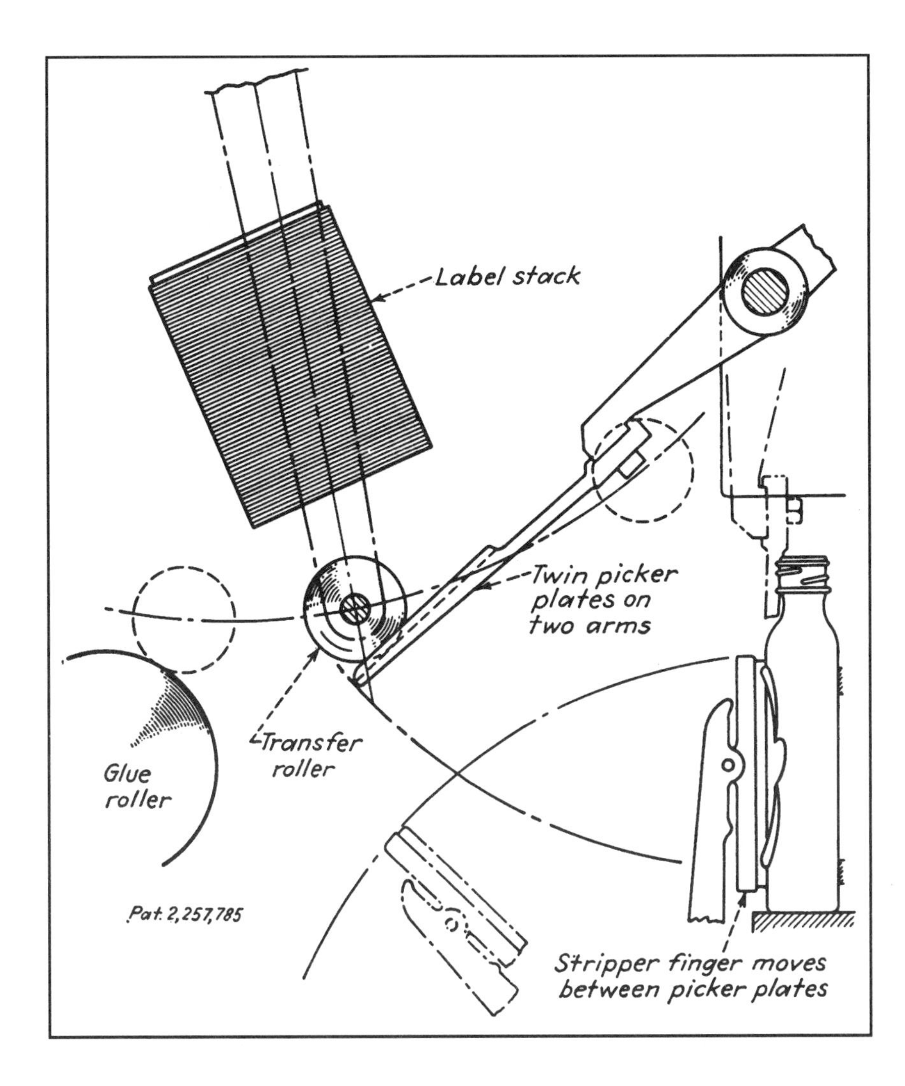

Fig. 11 The bottom label is spread with glue by two abutting glue-coated picker plates, which separate during contact with label stack, then carry the label to the bottle.

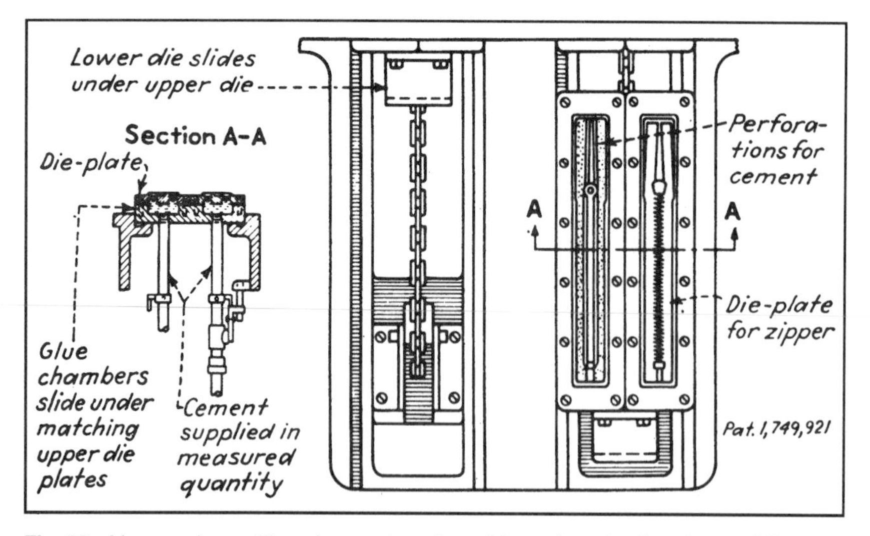

Fig. 12 Measured quantities of cement are forced through perforations in specially designed upper and lower die plates, which are closed hydraulically over zippers. Only the lower die is shown.

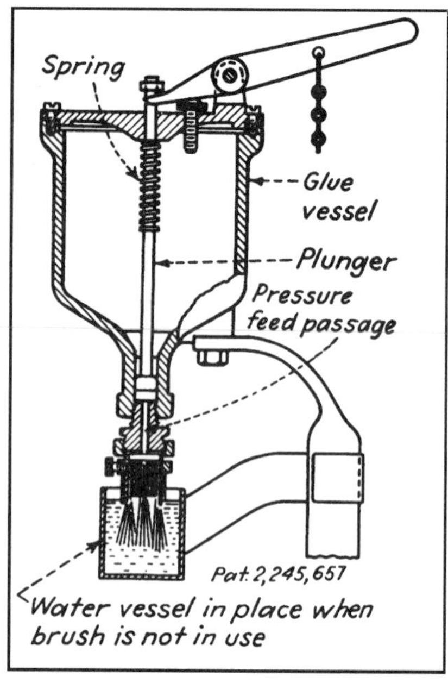

Fig. 13 A brush applicator is fed through passages between bristle tufts by a spring-operated plunger.

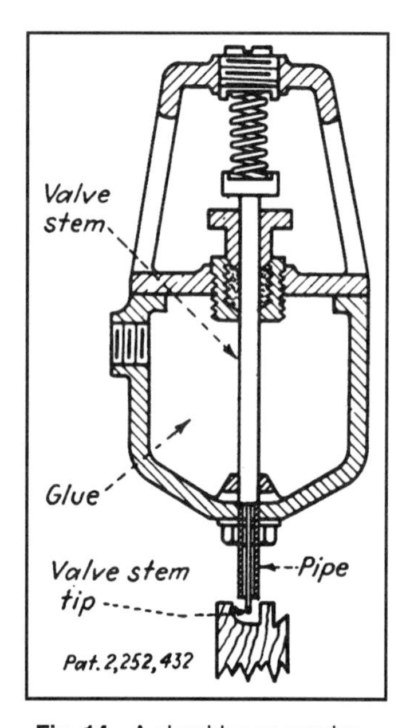

Fig. 14 A shoulder on a valve stem in a glue chamber retains glue until pressure on the tip opens the bottom valve.

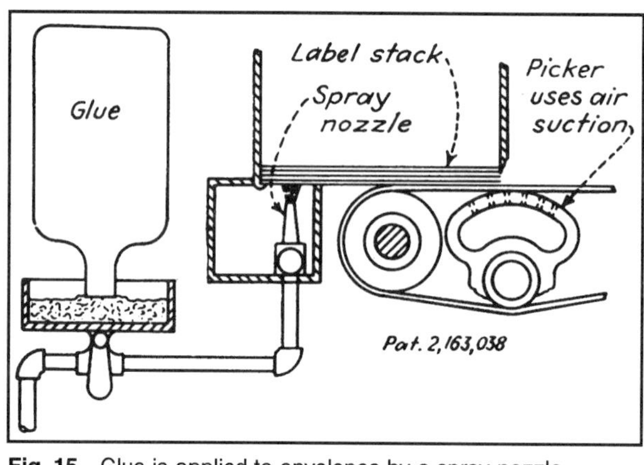

Fig. 15 Glue is applied to envelopes by a spray nozzle.

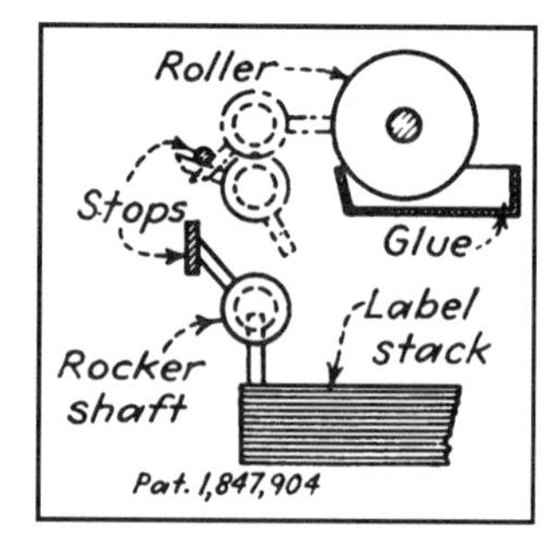

Fig. 16 A rocker shaft on a rack, which is moved vertically by a sector gear, carries glue on a contact bar from the roll to the label stack.

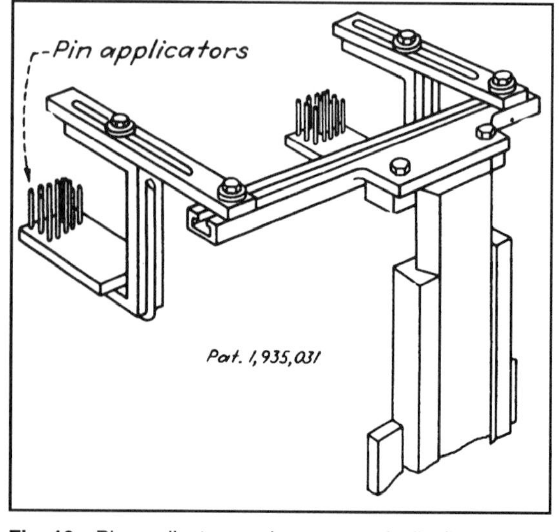

Fig. 18 Pin applicators reciprocate vertically, first immersing themselves in glue, then contacting the undersides of carton flaps in a desired pattern.

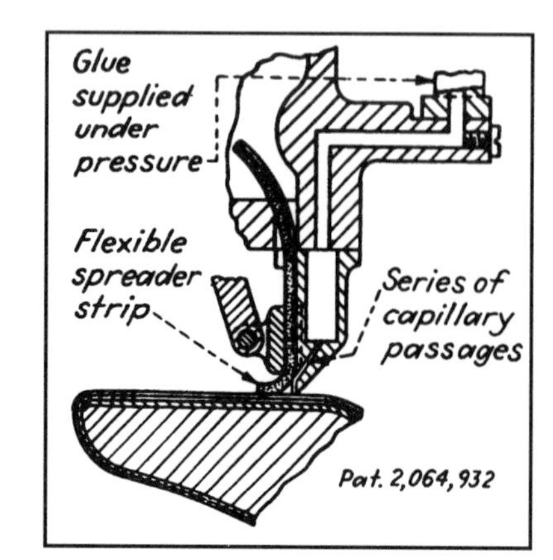

Fig. 17 Glue is extruded through nozzle on work.

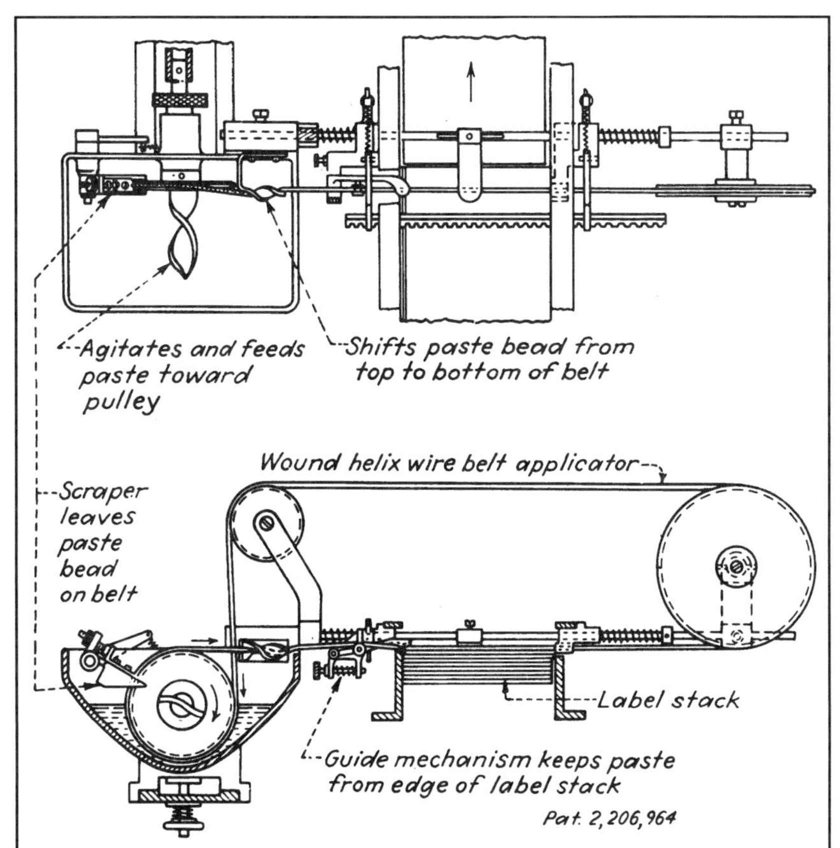

Fig. 19 A paste belt applicator passes around the pulley in a pastepot and slides over the label stack.

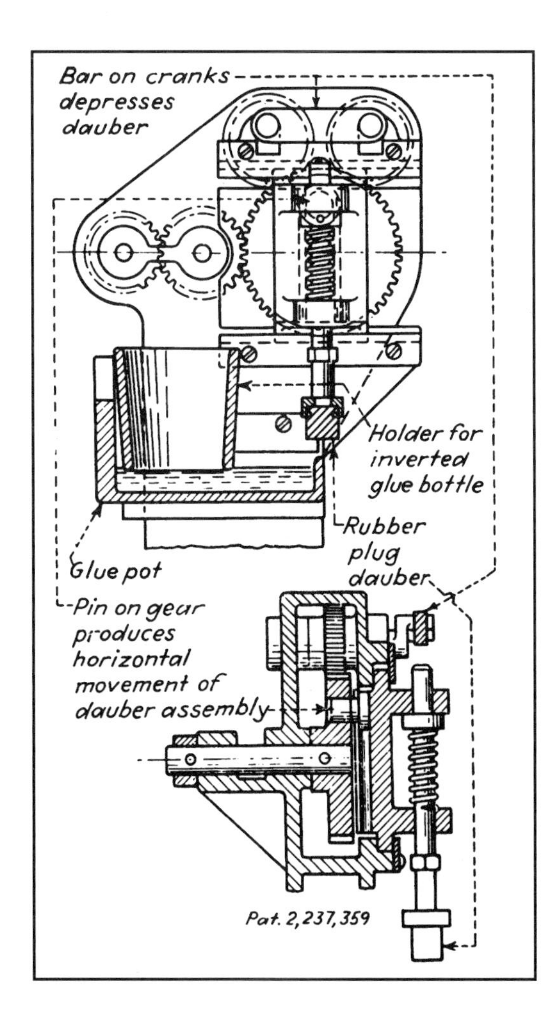

Fig. 20 A dauber assembly is moved horizontally between a glue pot and work by an eccentric pin on a gear. Vertical movements are produced by a crank-operated bar over the dauber shaft.

TWENTY-FOUR AUTOMATIC MECHANISMS FOR STOPPING UNSAFE MACHINES

Automatic stopping mechanisms that prevent machines from damaging themselves or destroying work in process are based on the principles of mechanics, electricity, hydraulics, and pneumatics.

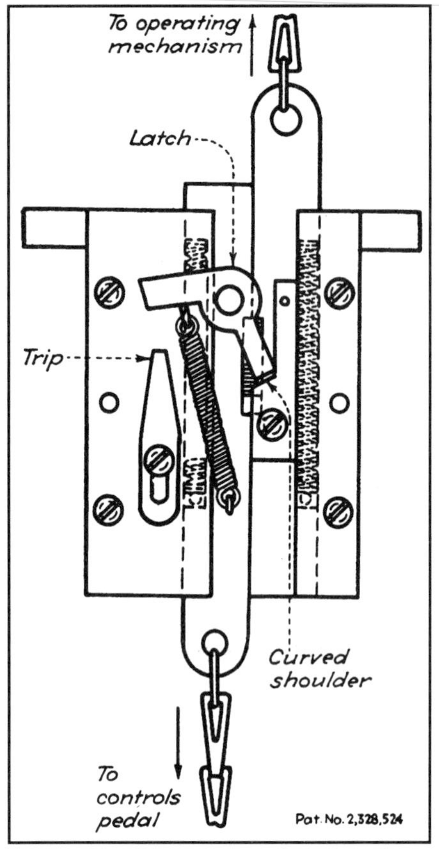

Fig. 1 A repetition of the machine cycle is prevented if a pedal remains depressed. The latch carried by the left slide pushes the right slide downward with a curved shoulder until the latch is disengaged by trip member.

Fig. 2 The gumming of the suction picker and label carrier when the label is not picked up by the suction is prevented by insufficient suction on a latch-operating cylinder, caused by open suction holes on the picker. When a latch-operating cylinder does not operate, the gum box holding latch returns to its holding position after cyclic removal by the cam and roller, thus preventing the gum box and rolls from rocking to make contact with the picker face.

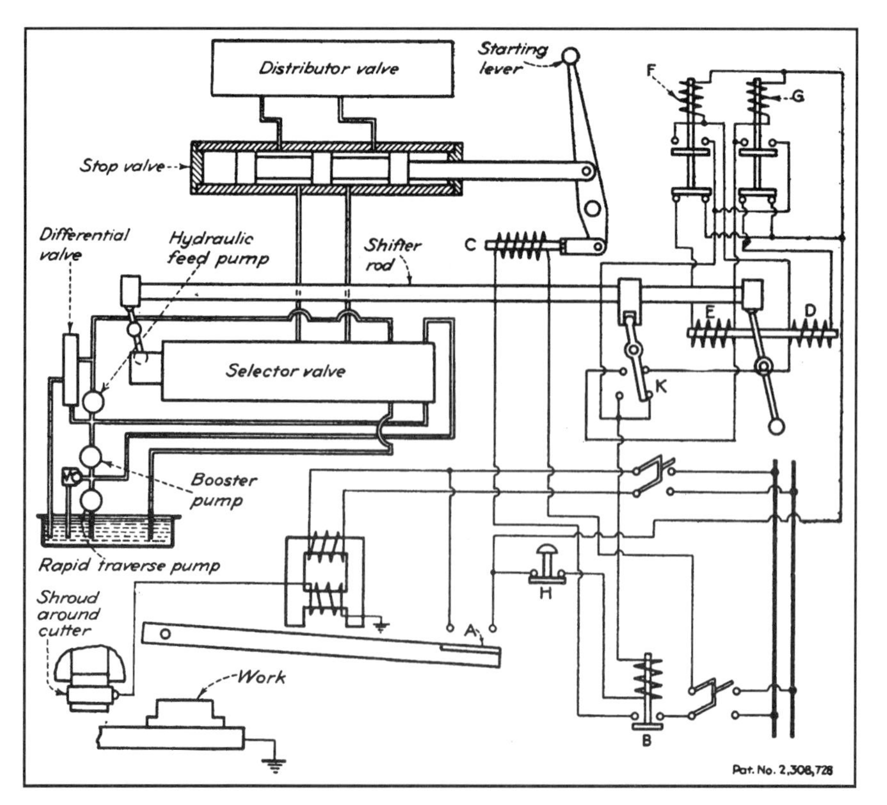

Fig. 3 Damage to the milling cutter, work, or fixtures is prevented by the shroud around the cutter. Upon contact with the work, the shroud closes the electric circuit through the relay, thus closing contact A. This causes contact B to close, thus energizing relay C to operate a stop valve. It also closes a circuit through relay D, thus reversing the selector valve by means of a shifter rod so that bed travel will reverse on starting. Simultaneously, relay F opens the circuit of relay E and closes a holding circuit that was broken by the shifter lever at K. Relay G also closes a holding circuit and opens a circuit through relay D. The starting lever, released by pushbutton H, releases contact A and returns the circuit to normal. If contact is made with the shroud when the bed travel is reversed, interchange the positions of D and E, with F and G in the sequence of operations.

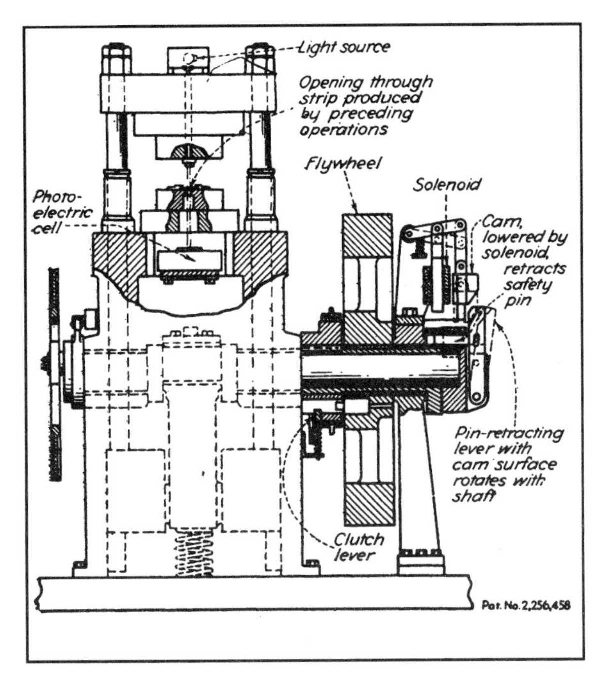

Fig. 4 A high-speed press is stopped when a metal strip advances improperly. A hole punched in the strip fails to match while the opening in the die block to permit a light beam to pass. When the light beam to the photoelectric cell is blocked, the solenoid which withdraws the clutch pin is activated.

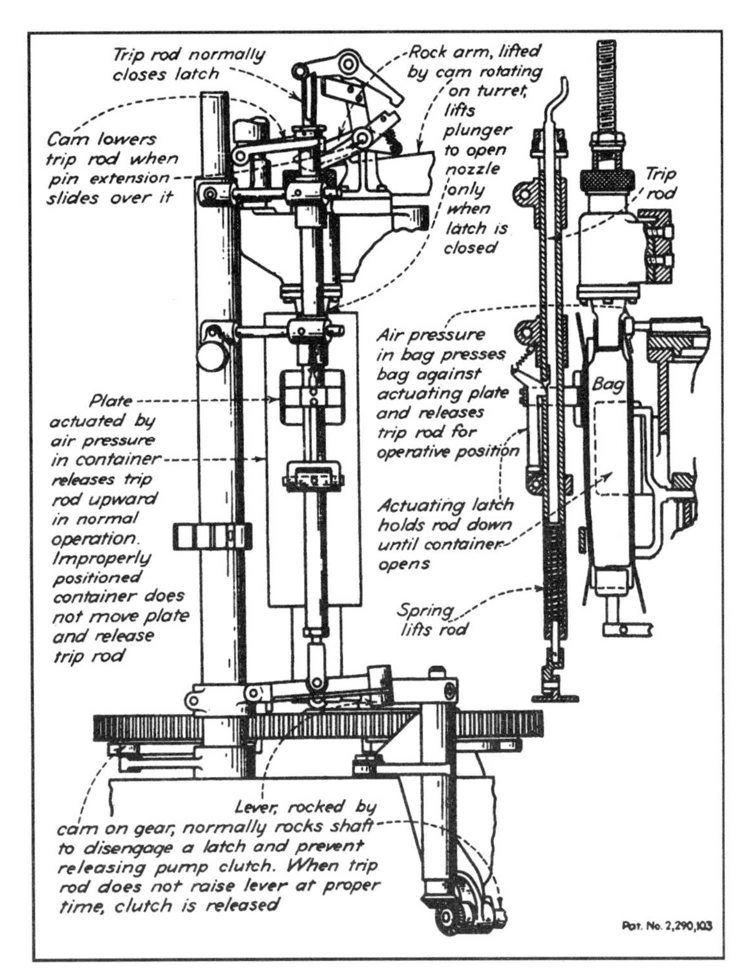

Fig. 6 A nozzle on the packaging machine does not open when the container is improperly positioned.

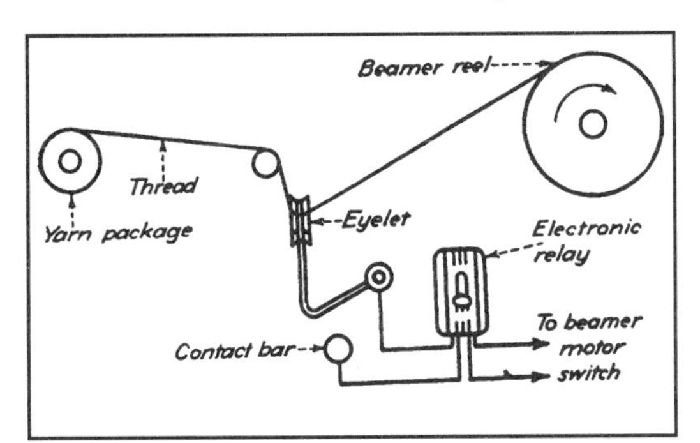

Fig. 5 A broken thread allows the contact bar to drop, thereby closing the electronic relay circuit; this stops the beamer reeling equipment.

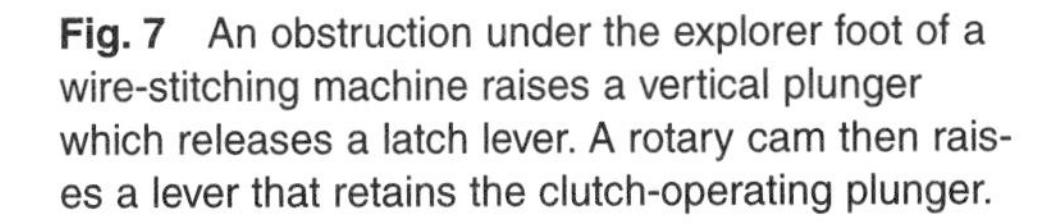

Fig. 7 An obstruction under the explorer foot of a wire-stitching machine raises a vertical plunger which releases a latch lever. A rotary cam then raises a lever that retains the clutch-operating plunger.

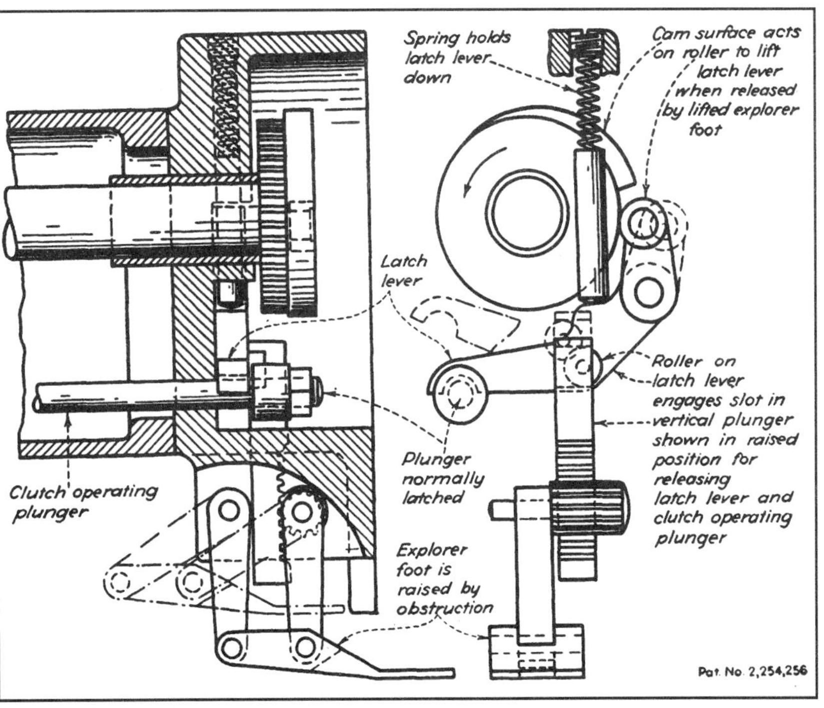

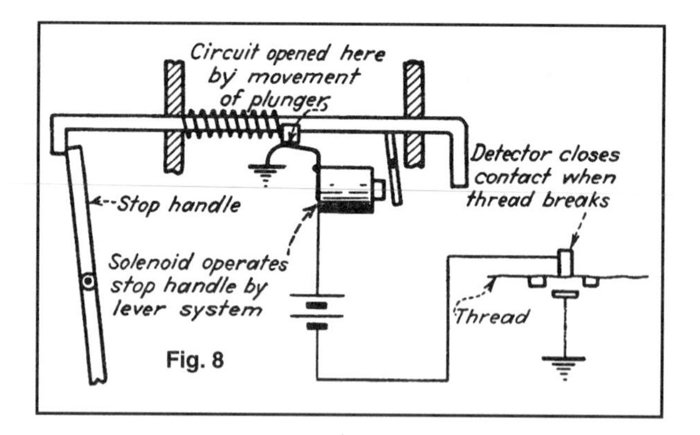

Fig. 8

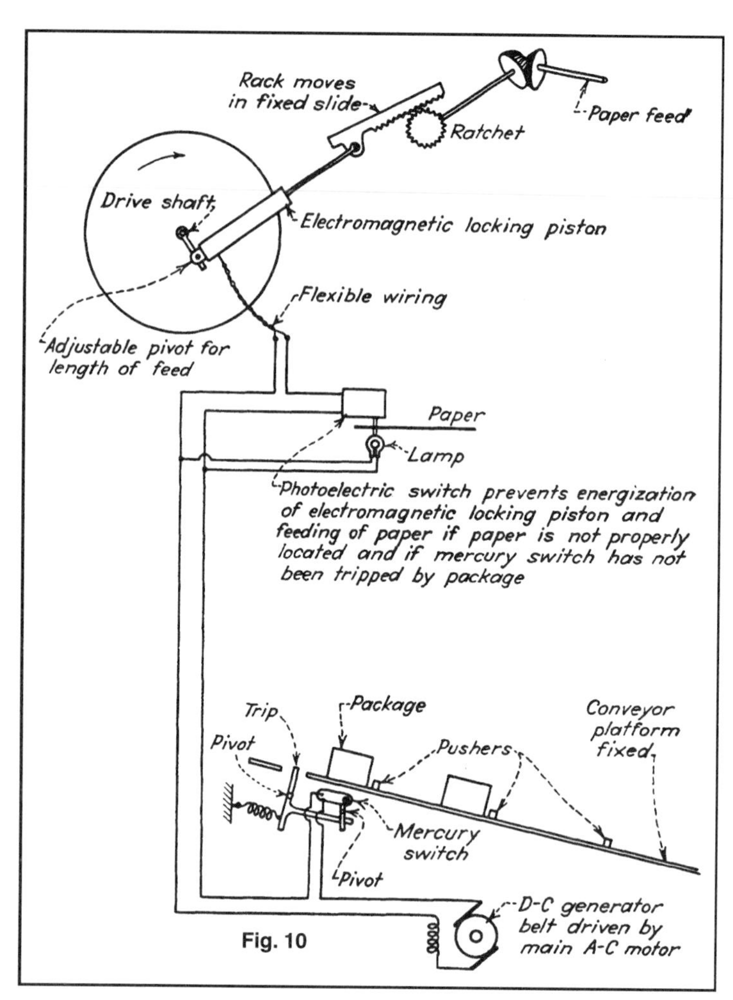

Fig. 10

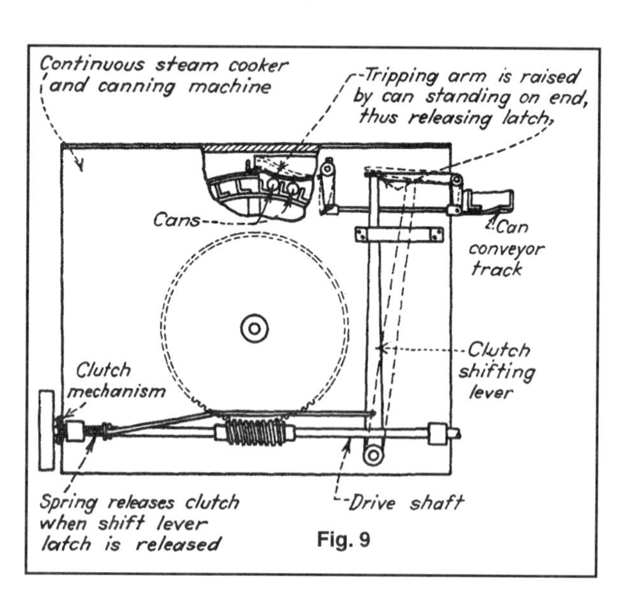

Fig. 9

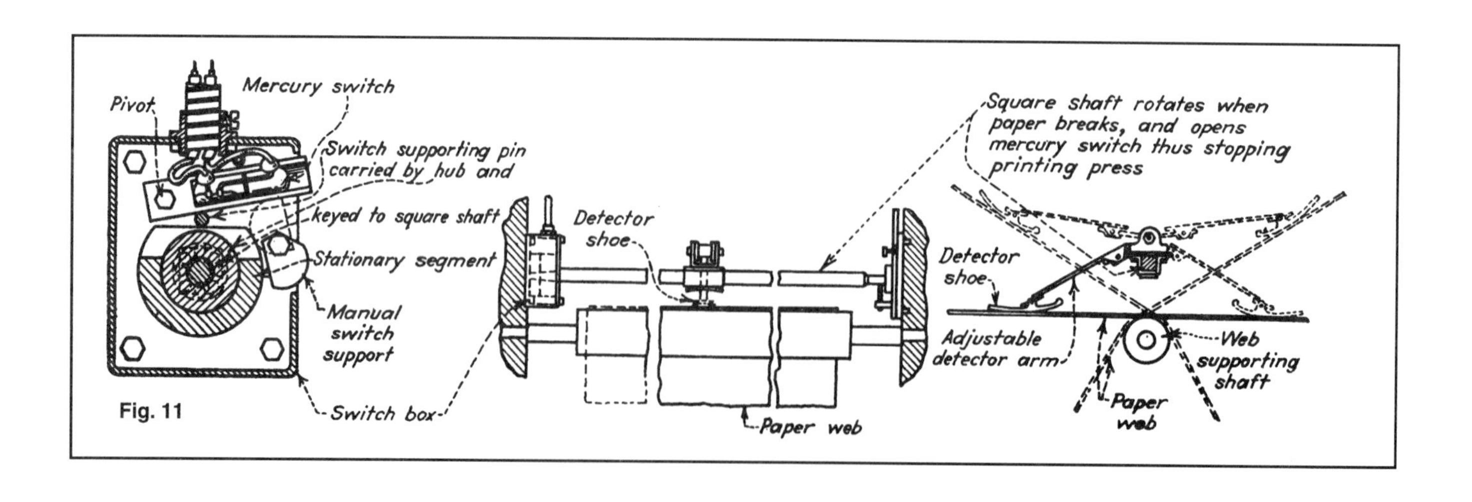

Fig. 11

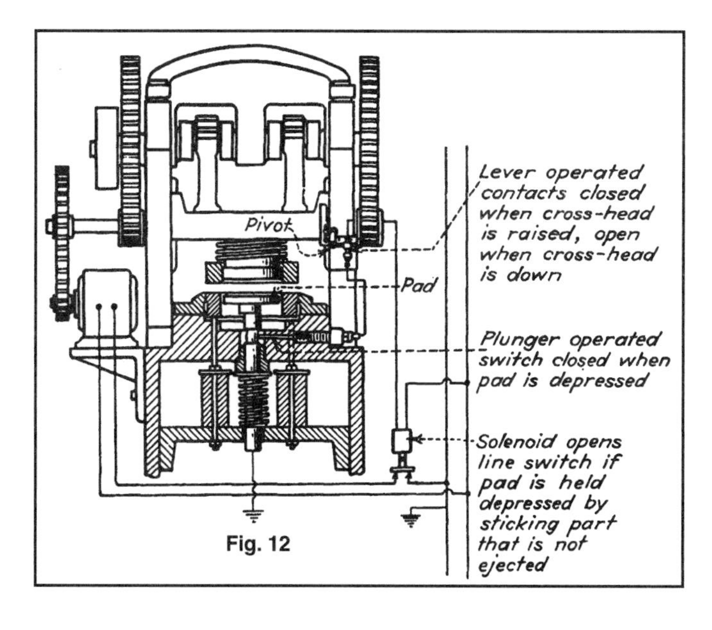

Fig. 12

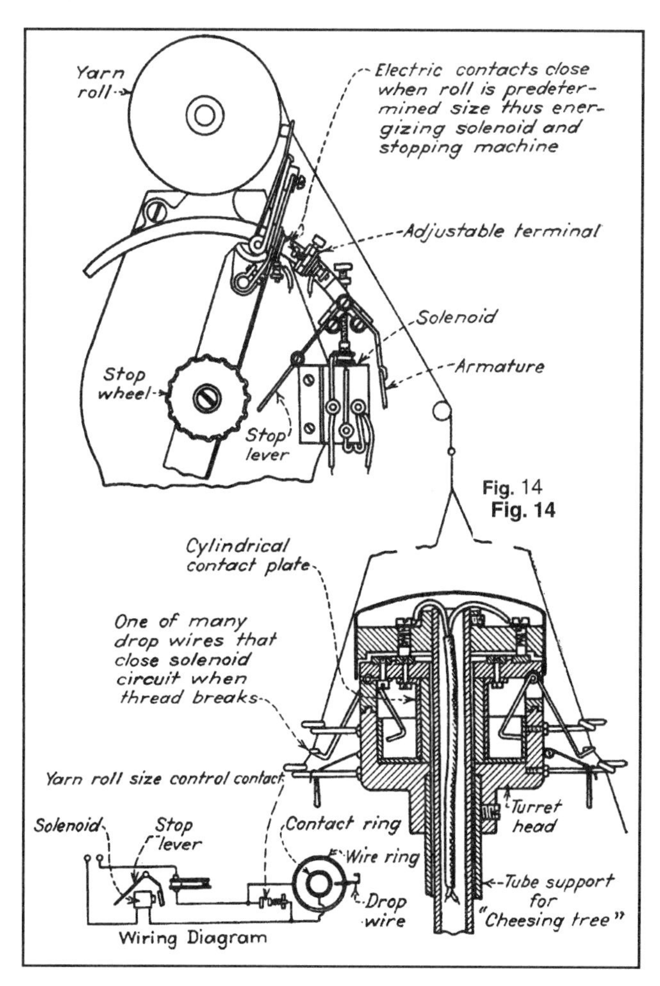

Fig. 14

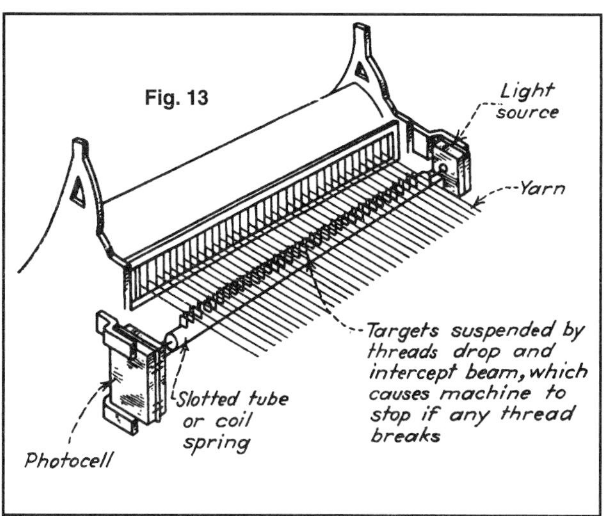

Fig. 13

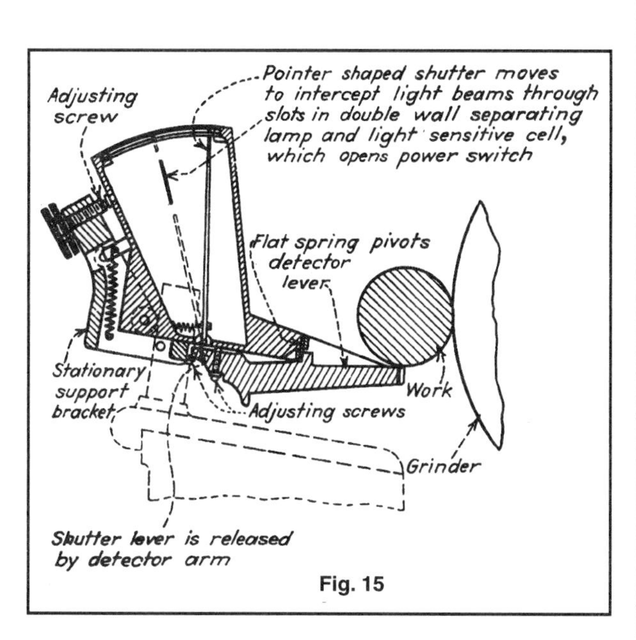

Fig. 15

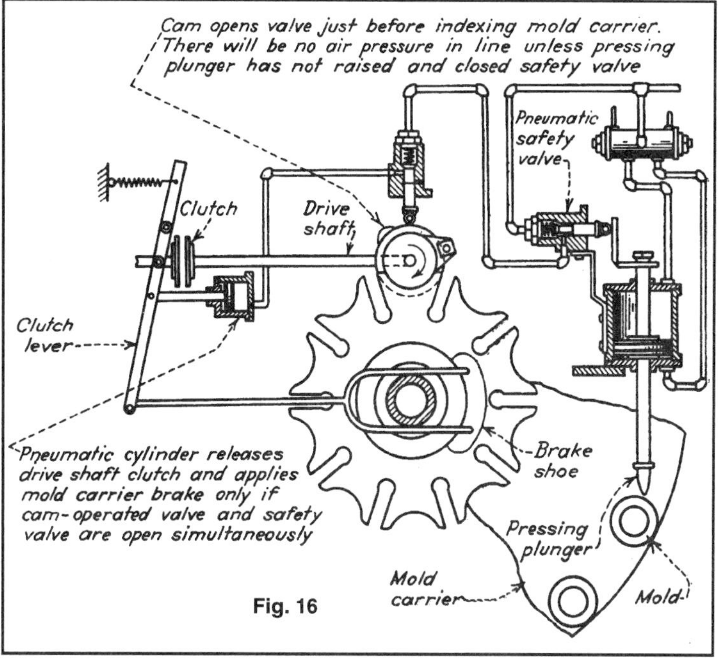

Fig. 16

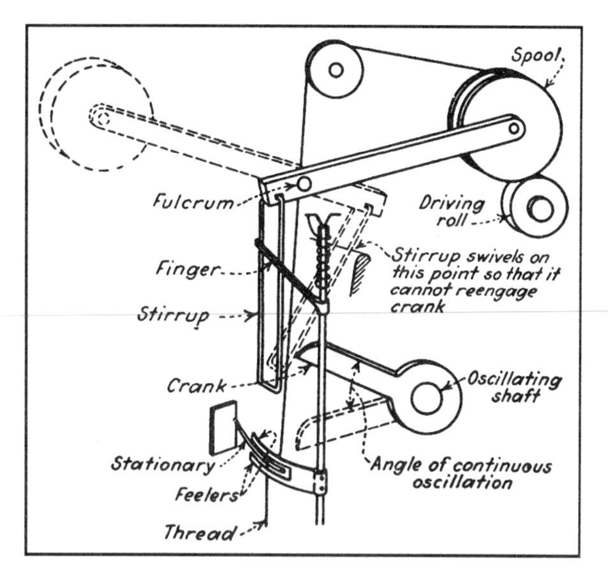

Fig. 17 A mechanism on a spooler. When a thread breaks, the feelers are released and the spiral spring causes the spindle with finger to rotate. The finger throws the stirrup into the path of the oscillating crank, which on its downward stroke throws the spool into the position shown dotted. The stirrup is then thrown out of the path of the oscillating crank.

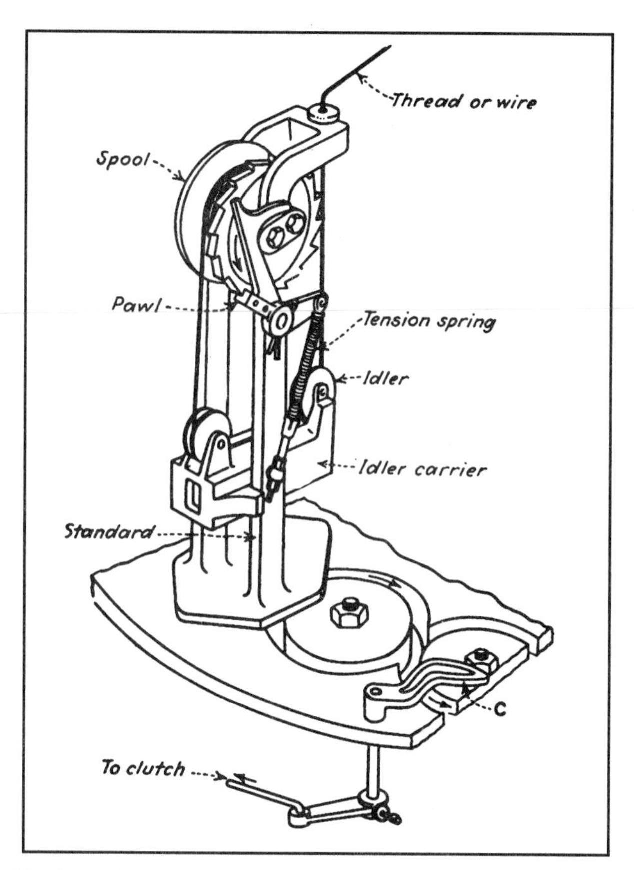

Fig. 18 A mechanism in use on tubular braiding machines. When the machine is braiding, tension on the wire or thread lifts the idler carrier which then releases the pawl from the ratchet on the spool flange and allows the spool to turn and unwind. When the machine stops, the tension on the wire is decreases, allowing the idler carrier to fall so that the pawl can engage the ratchet. If a wire breaks while the machine is running, the unsupported idler carrier falls to the base of the standard, and when the standard arrives at the station in the raceway adjacent to the cam *C*, the lug *L* on the idler carrier strikes the cam *C*, rotating it far enough to disengage a clutch on the driving shaft, thereby stopping the machine.

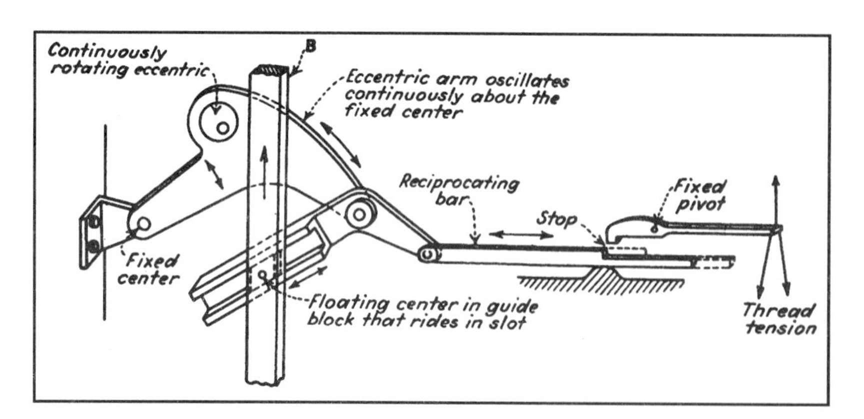

Fig. 19 When thread breaks, the stop drops and intercepts the reciprocating bar. On the next counter-clockwise oscillation of the eccentric arm, the bar B is raised. A feature of this design is that it permits the arm B to move up or down independently for a limited distance.

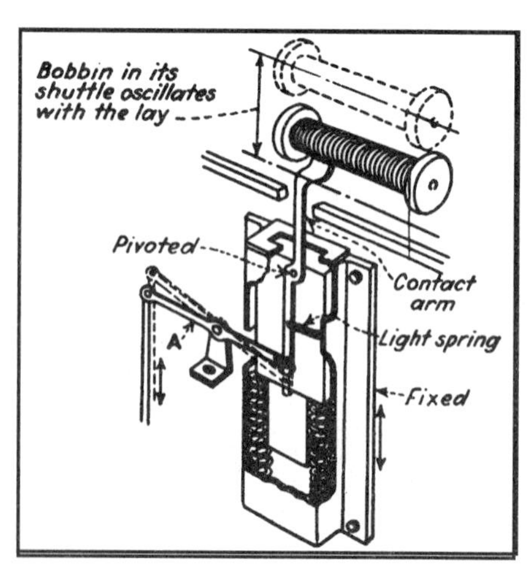

Fig. 20 A diagram of a mechanism that causes a bobbin changer to operate. If the contact arm does not slip on the bobbin, lever *A* will rotate to the position shown. But if contact with the bobbin center slips, when the bobbin is empty, lever A will not rotate to the position indicated by the dashed line. This will cause the bobbin changer to operate.

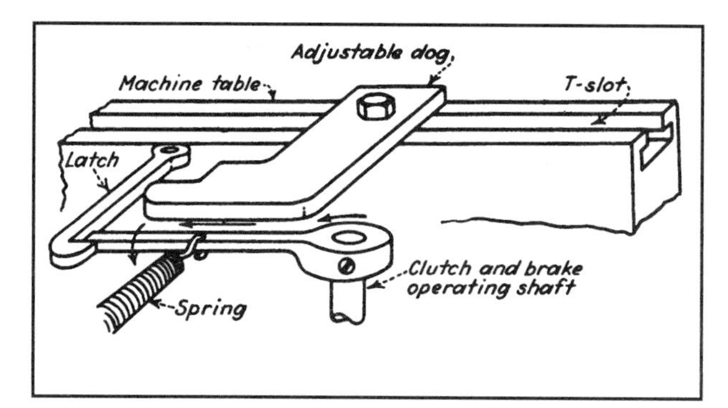

Fig. 21 A simple stop mechanism for limiting the stroke of a reciprocating machine member. Arrows indicate the direction of movements.

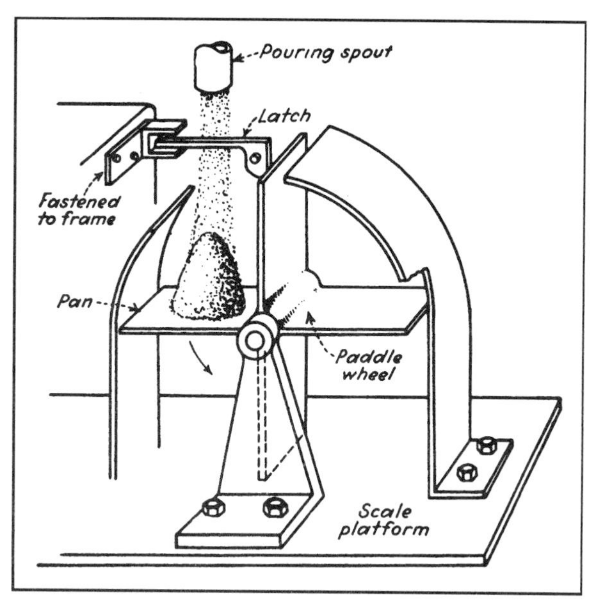

Fig. 22 When the predetermined weight of material has been poured on the pan, the movement of the scale beam pushes the latch out of engagement. This allows the paddle wheel to rotate and dump the load. The scale beam drops, thereby returning the latch to the holding position and stopping the wheel when the next vane hits the latch.

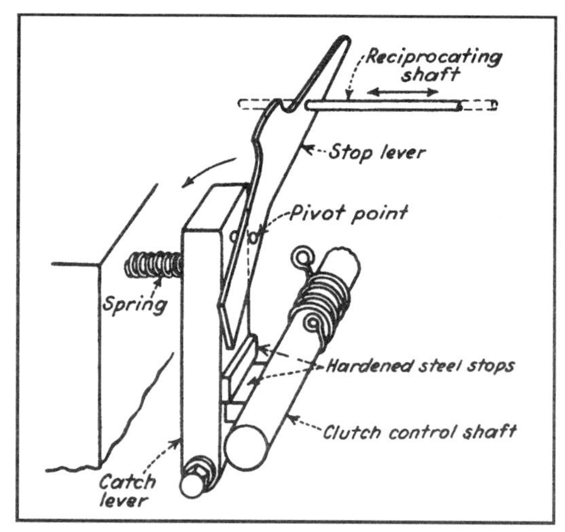

Fig. 23 In this textile machine, any movement that will rotate the stop lever counter-clockwise will move it into the path of the continuously reciprocating shaft. This will cause the catch lever to be pushed counter-clockwise, freeing the hardened steel stop on the clutch control shaft. A spiral spring then impels the clutch control shaft to rotate clockwise. That movement throws out the clutch and applies the brake. The initial movement of the stop lever can be caused by a breaking thread or a moving dog.

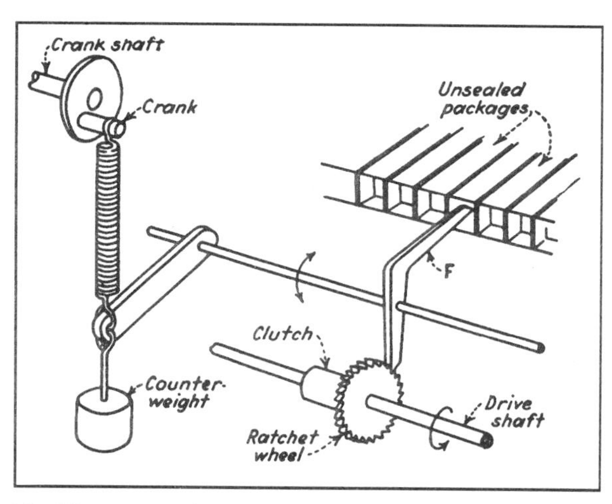

Fig. 24 some package-loading machines have provisions to stop the machine if a package passes the loading station without receiving an insert. Pawl finger *F* has a rocking motion imparted by the crankshaft, timed so that it enters the unsealed packages and is stopped against the contents. If the box is not filled, the finger enters a long distance. The pawl end at the bottom engages and holds a ratchet wheel on the driving clutch which disengages the machine-driving shaft.

SIX AUTOMATIC ELECTRICAL CIRCUITS FOR STOPPING TEXTILE MACHINES

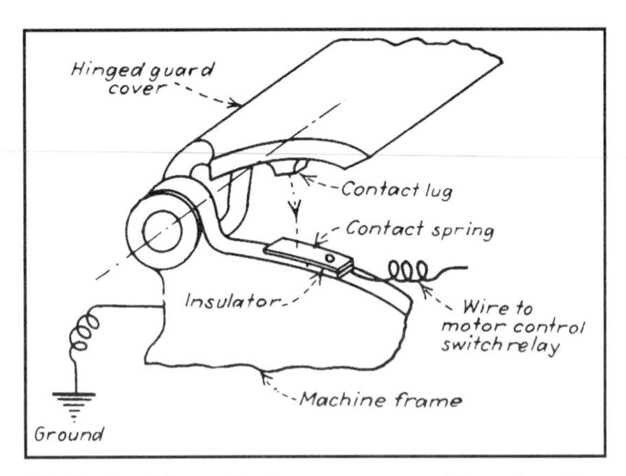

Fig. 1 A safety mechanism on some machines stops the motor when a guard cover is lifted. The circuit is complete only when the cover is down. In that position a contact lug establishes a metal-to-metal connection with the contact spring, completing the relay circuit.

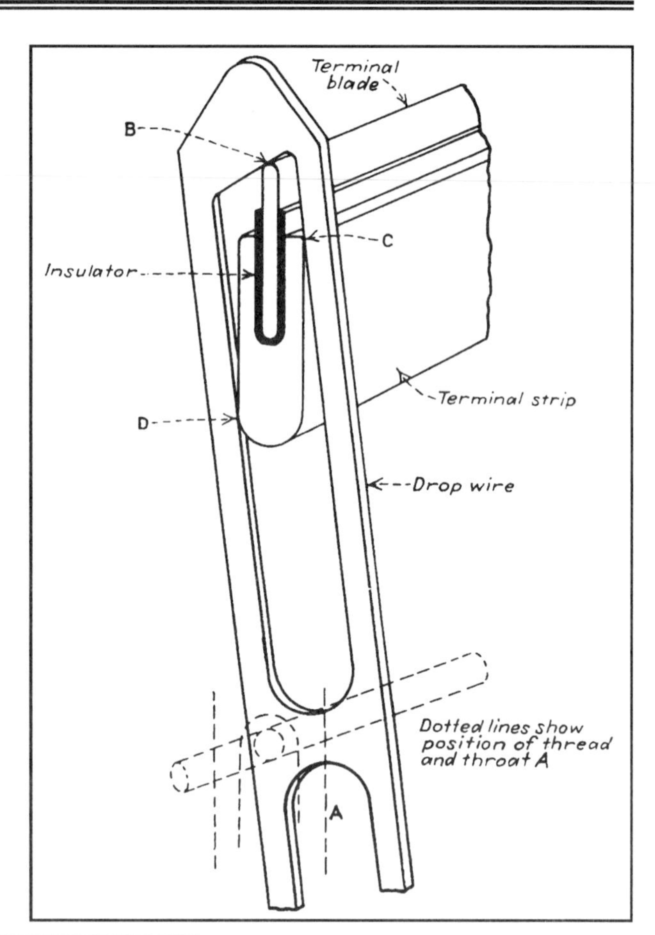

Fig. 2 An electrical three-point wedging "warp stop" is shown after a thread has broken and a drop wire has fallen and tilted to close the circuit. Dotted lines indicate the normal position of the drop wire when it is riding on a thread. When a thread breaks, the drop wire falls and strikes the top of the terminal blade B, the inclined top of the slot. This causes a wedging action that tilts the drop wire against the terminal strip at C and D, reinforcing the circuit-closing action.

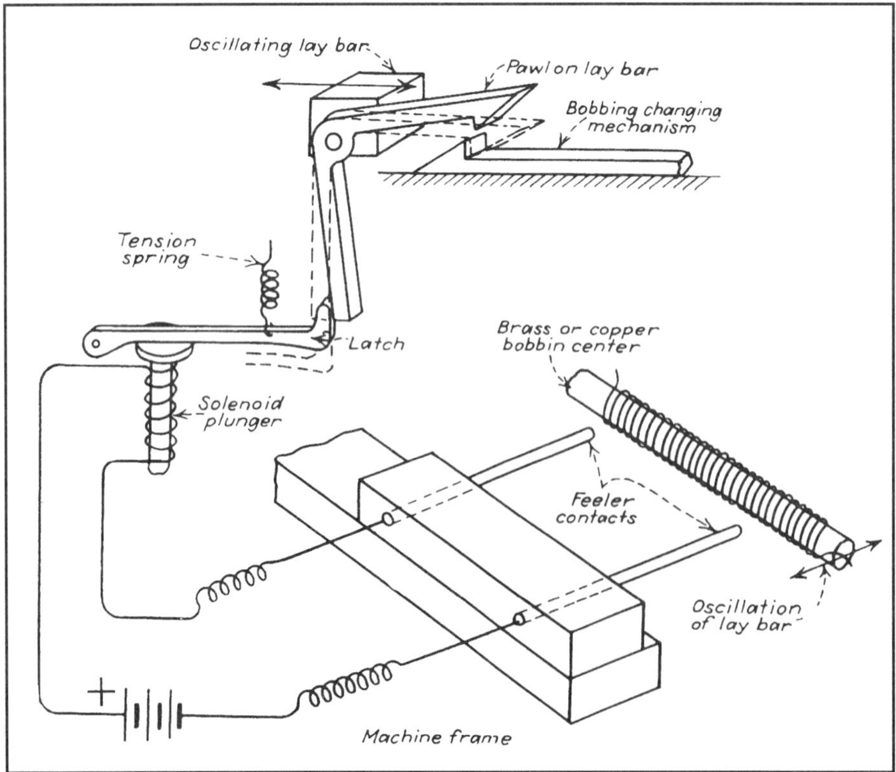

Fig. 3 Bobbin changer. When a bobbin is empty, the feeders contact the metal bobbin center, completing the circuit through a solenoid which pulls a latch. That causes the bobbin-changing mechanism to operator and put a new bobbin in the shuttle. As long as the solenoid remains deenergized, the pawl on the lay bar is raised clear of the hook on the bobbin-changing mechanism.

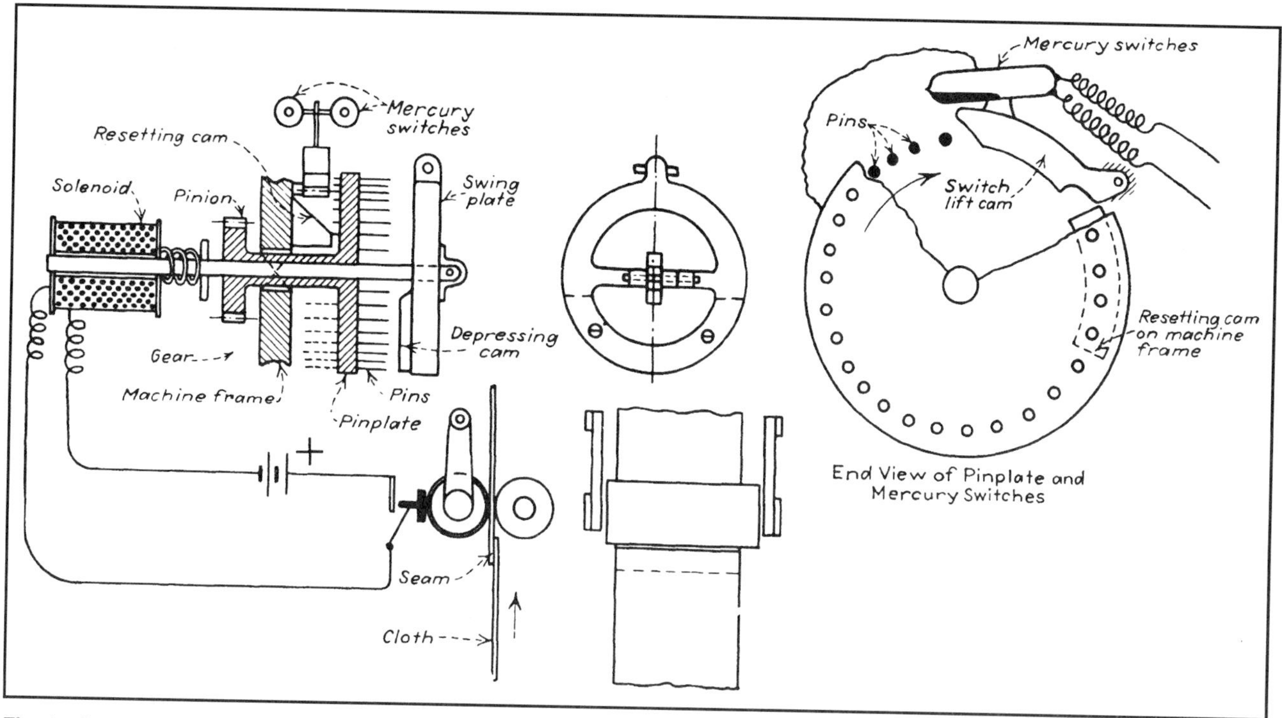

Fig. 4 Control for the automatic shear. When a seam of two thicknesses of cloth passes between the rolls, the swing roller is moved outward and closes a sensitive switch which energizes a solenoid. The solenoid pulls in an armature whose outer end is attached to the hinged ring where a cam plate is also fastened. The cam plate depresses the pins in a rotating plate. As the plate rotates, the depressed pins lift a hinged cam arm on which two mercury switches are mounted. When tilted, the switches complete circuits in two motor controls. A resetting cam for pushing the depressed pins back to their original position is fastened on the machine frame. The two motors are stopped and reversed until the seam has passed through rollers before they are stopped and reversed again.

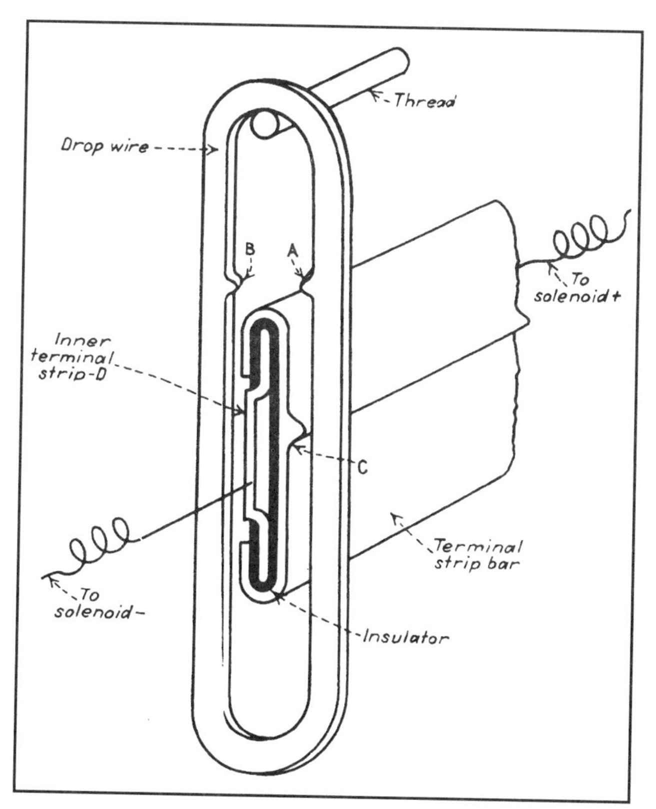

Fig. 5 Electric stop for a loom. When a thread breaks or slackens, the drop wire falls and contact A rides on contact C. The drop wire, supported off-center, swings so that contact B is pulled against the inner terminal strip D, completing the solenoid circuit.

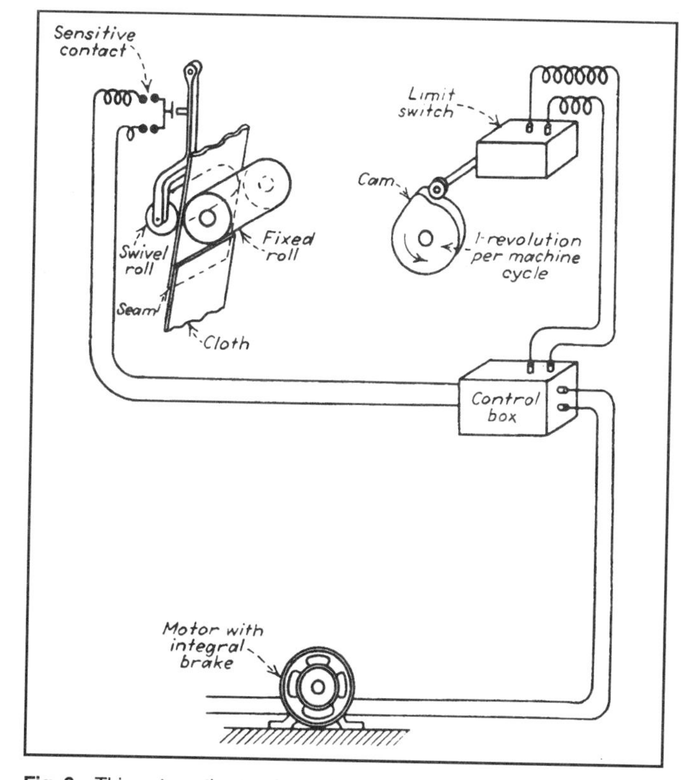

Fig. 6 This automatic stop for a folder or yarder machine always stops the machine in the same position when a seam in the cloth passes between the rolls. A seam passing between the rolls causes the swivel-mounted roll to lift slightly. This motion closes contacts in a sensitive switch that opens a relay in the control box. The next time the cam closes the limit switch, the power of the motor with the integral magnetic brake is shut off. The brake always stops the machine in the same place.

SIX AUTOMATIC MECHANISMS FOR ASSURING SAFE MACHINE OPERATION

The best automatic safety mechanisms are those that have been designed specifically for the machine on which they will be installed. When properly designed, automatic safety devices (1) do not reduce the operator's visibility, (2) do not interfere with the operator's performance, (3) do not make physical contact with the operator to pre-vent injury (e.g., by knocking his hand out of the way), (4) are fail-safe, (5) are sensitive enough to operate instantly, and (6) render the machine inoperative if anyone attempts to tamper with them or remove them.

Safety mechanisms can range from those that keep both of the operator's hands occupied on controls away from the work area to shields that completely enclose the work in progress on the machine and prevent machine operation unless the shield is securely in place. Many modern safety systems are triggered if any person or object breaks a light beam between a photoemitter and photoreceiver.

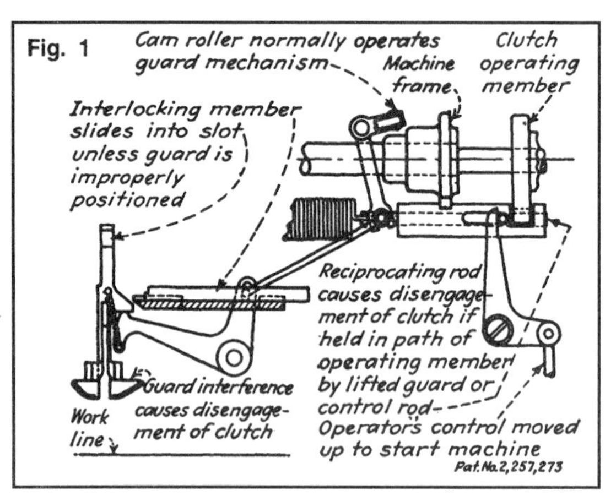

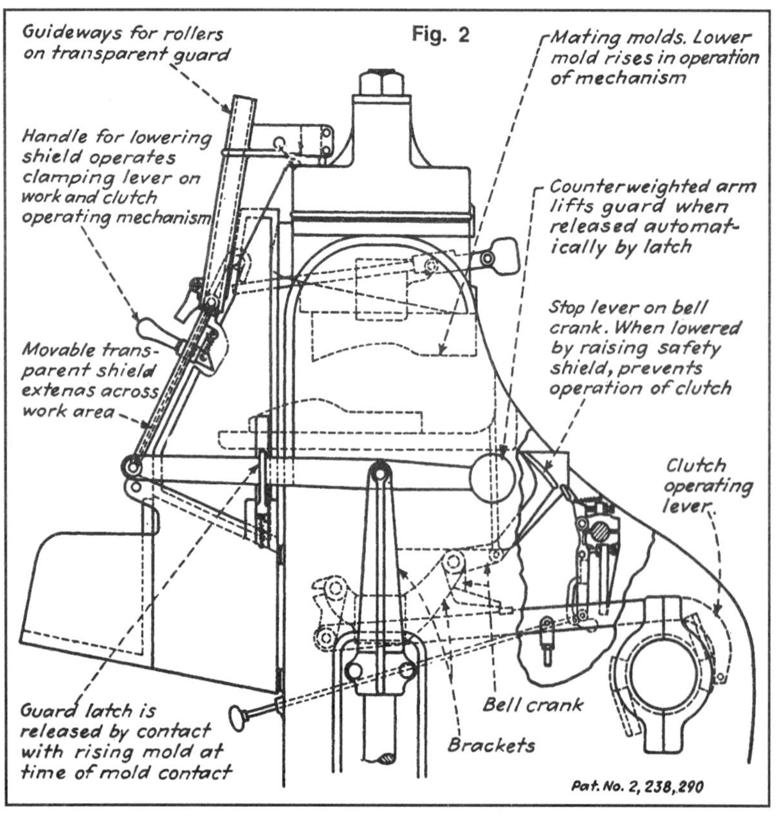

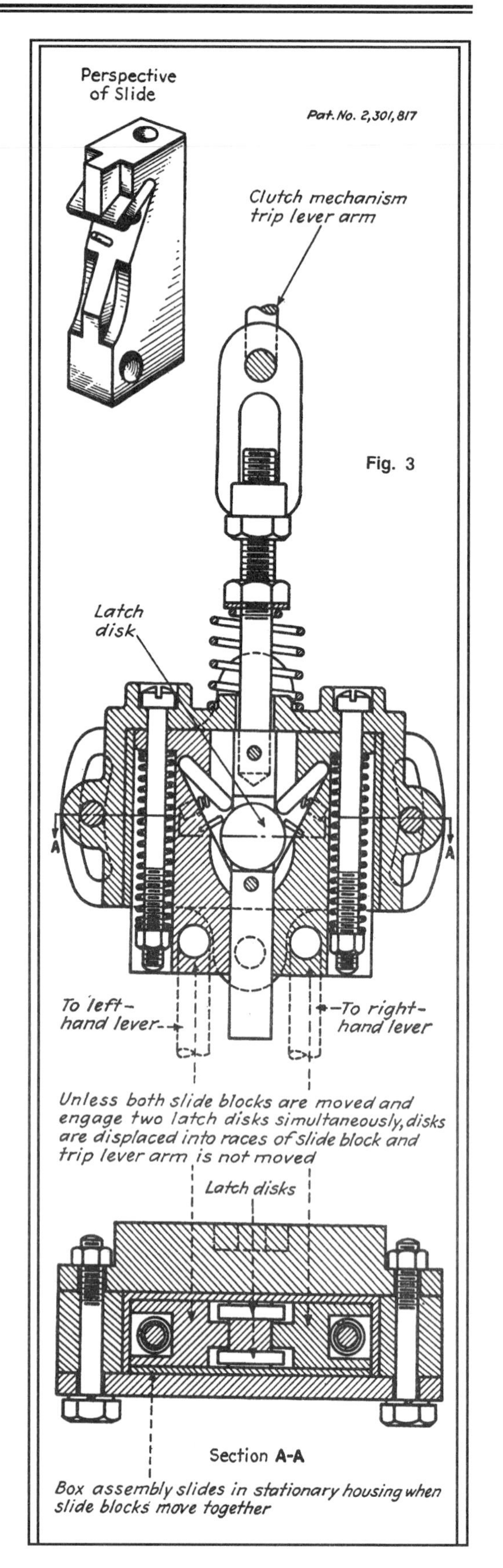

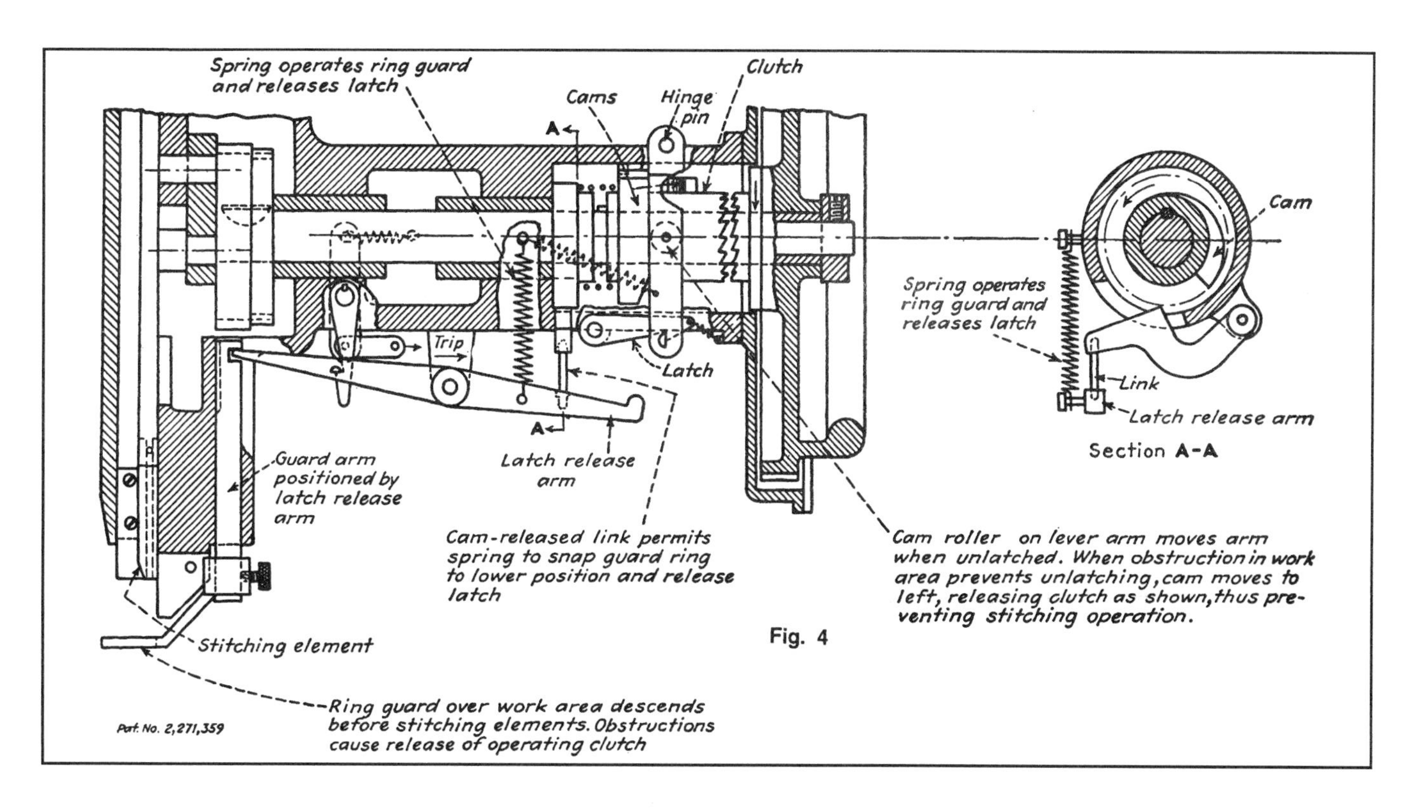

Fig. 4

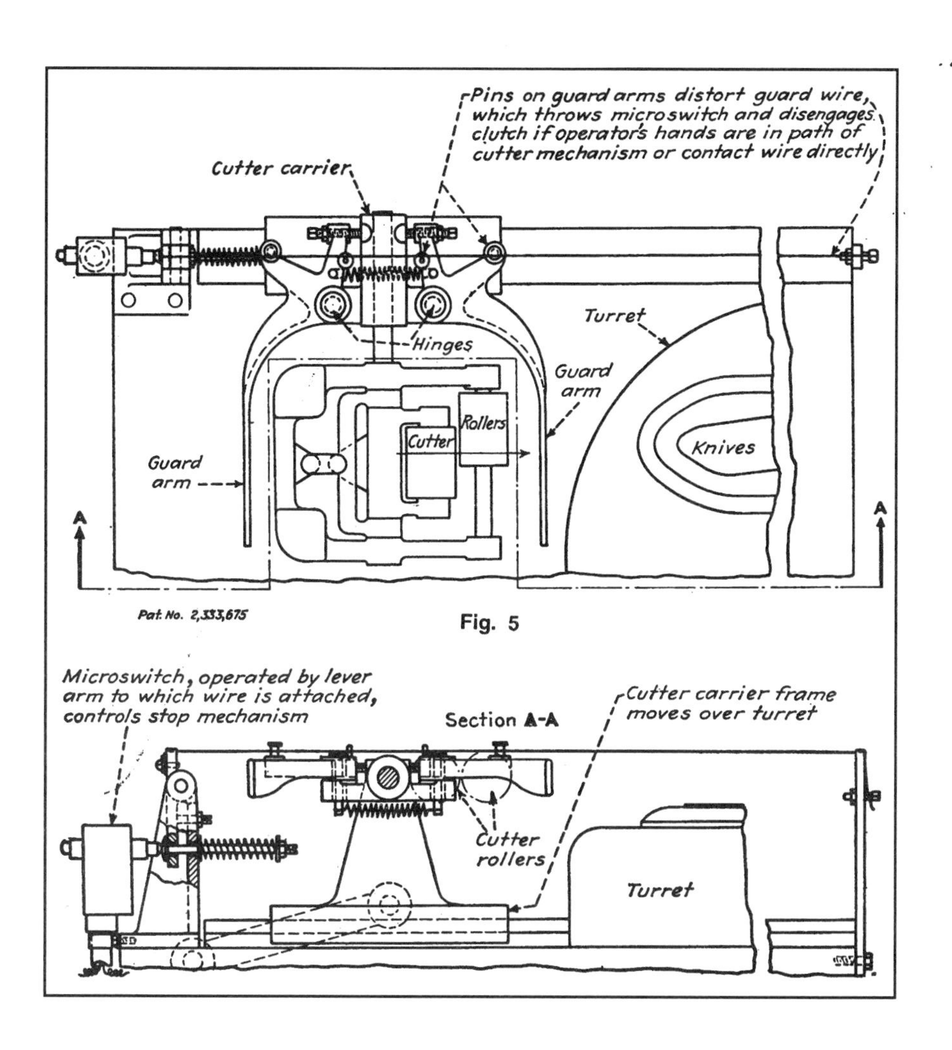

Fig. 5

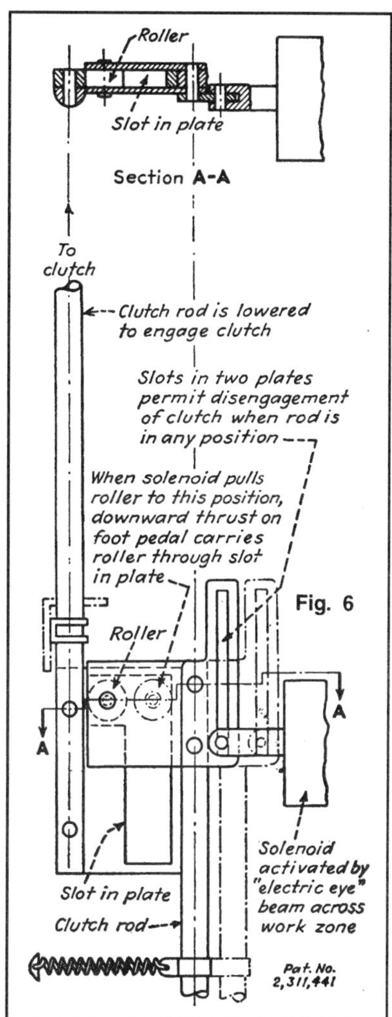

Fig. 6

CHAPTER 15

TORQUE, SPEED, TENSION, AND LIMIT CONTROL SYSTEMS

APPLICATIONS OF THE DIFFERENTIAL WINCH TO CONTROL SYSTEMS

Known for its mechanical advantage, the differential winch is a control mechanism that can supplement the gear and rack and four-bar linkage systems in changing rotary motion into linear. It can magnify displacement to meet the needs of delicate instruments or be varied almost at will to fulfill uncommon equations of motion.

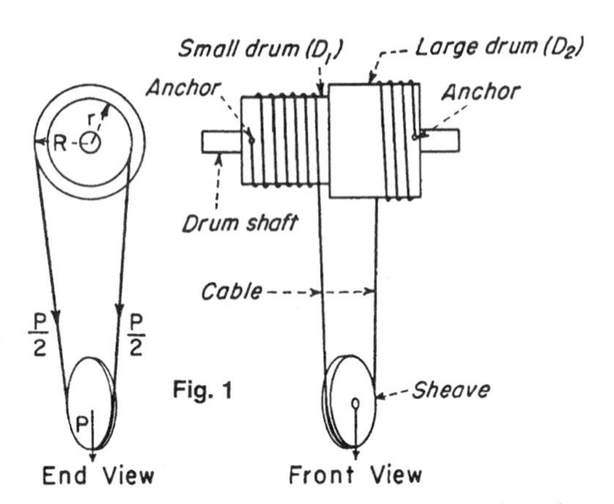

Fig. 1 A standard differential winch consists of two drums, D_1 and D_2, and a cable or chain which is anchored on both ends and wound clockwise around one drum and counterclockwise around the other. The cable supports a load-carrying sheave, and if the shaft is rotated clockwise, the cable, which unwinds from D_1 on to D_2, will raise the sheave a distance

$$\text{Sheave rise/rev} = \frac{2\pi R - 2\pi r}{2} = \pi(R - r)$$

The winch, which is not in equilibrium exerts a counterclockwise torque.

$$\text{Unbalanced torque} = \frac{P}{2}(R - r)$$

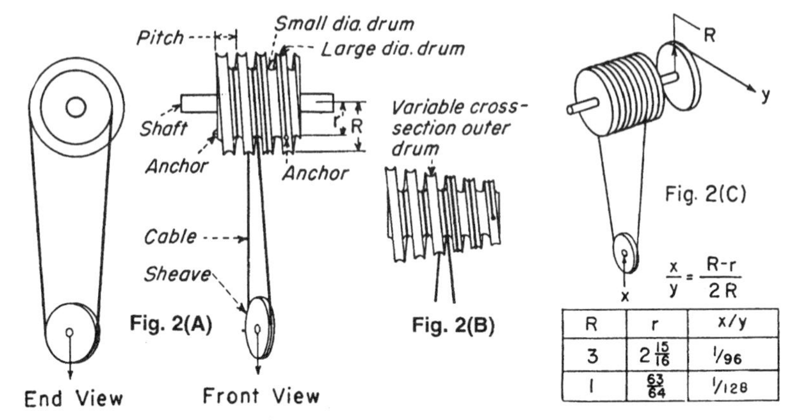

R	r	x/y
3	$2\frac{15}{16}$	1/96
1	$\frac{63}{64}$	1/128

Fig. 2(A) Hulse Differential Winch*. Two drums, which are in the form of worm threads contoured to guide the cables, concentrically occupy the same logitudinal space. This keeps the cables approximately at right angles to the shaft and eliminates cable shifting and rubbing, especially when used with variable cross sections as in Fig. 2(B). Any equation of motion can be satisfied by choosing suitable cross sections for the drums. Methods for resisting or supporting the axial thrust should be considered in some installations. Fig. 2(C) shows typical reductions in displacement. *Pat. No. 2,590,623

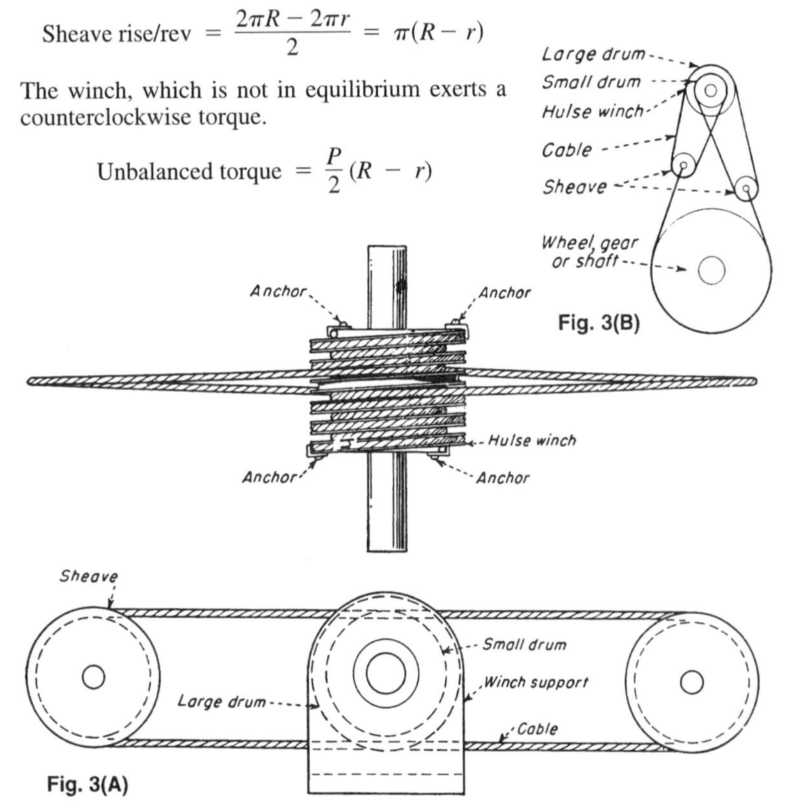

Fig. 3(A) A Hulse Winch with opposing sheaves. This arrangement, which uses two separate cables and four anchor points, can be considered as two winches back-to-back with one common set of drums. Variations in motion can be obtained by: (1) restraining in the sheaves so that when the system is rotated the drums will travel toward one of the sheaves; (2) restraining the drums and allowing the sheaves to travel. The distance between the sheaves will remain constant and is usually connected by a bar; (3) permitting the drums to move axially while restraining them transversely. When the system is rotated, drums will travel axially one pitch per revolution, and sheaves remain in the same plane perpendicular to the drum axis. This variation can be reversed by allowing sheaves to move axially; and (4) sheaves need not be opposite but can be arranged as in Fig. 3(B) to rotate a wheel.

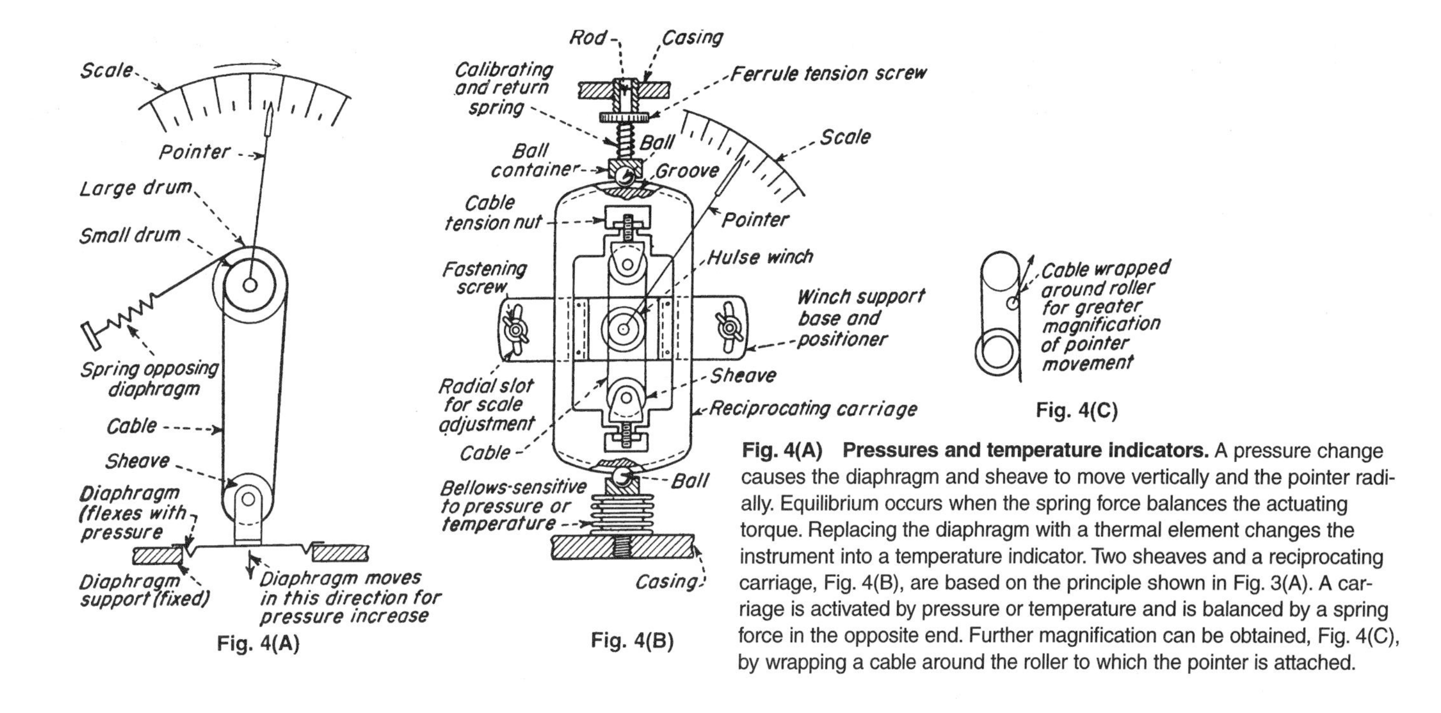

Fig. 4(A) Pressures and temperature indicators. A pressure change causes the diaphragm and sheave to move vertically and the pointer radially. Equilibrium occurs when the spring force balances the actuating torque. Replacing the diaphragm with a thermal element changes the instrument into a temperature indicator. Two sheaves and a reciprocating carriage, Fig. 4(B), are based on the principle shown in Fig. 3(A). A carriage is activated by pressure or temperature and is balanced by a spring force in the opposite end. Further magnification can be obtained, Fig. 4(C), by wrapping a cable around the roller to which the pointer is attached.

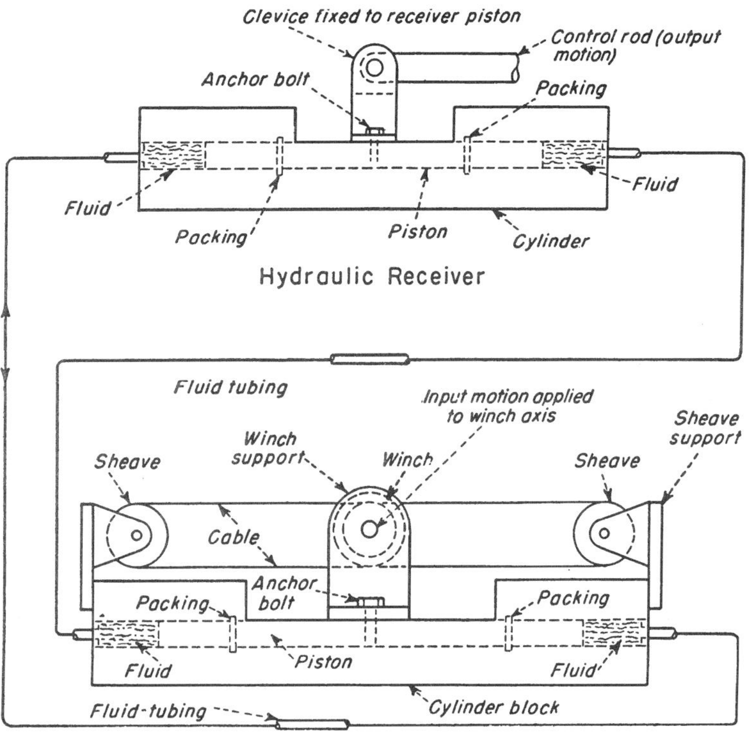

Fig. 5

Fig. 5 A hydraulic control system, actuated by a differential winch, performs remote precision positioning of a control rod with a minimum of applied torque. The sending piston, retained in a cylinder block, reciprocates back and forth from a torque applied to the winch shaft. Fluid is forced out from one end of the cylinder through the pipe lines to displace the receiving piston, which in turn activates a control rod. The receiver simultaneously displaces a similar amount of fluid from the opposite end back to the sender. By suitable valving, the sender can become a double-acting pump.

SIX WAYS TO PREVENT REVERSE ROTATION

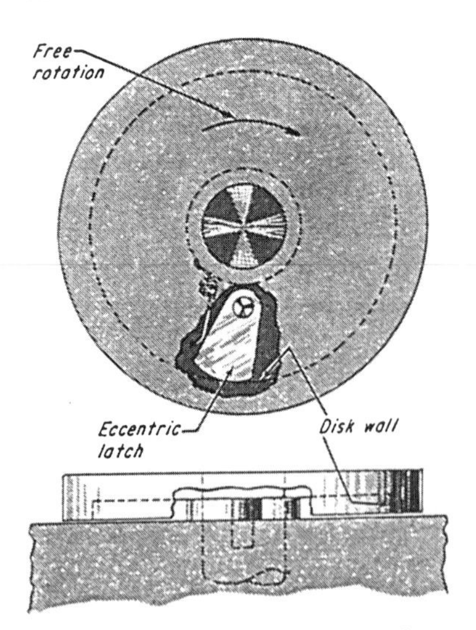

Fig. 1 An eccentric latch allows the shaft to rotate in one direction; any attempted reversal immediately causes the latch to wedge against the disk wall.

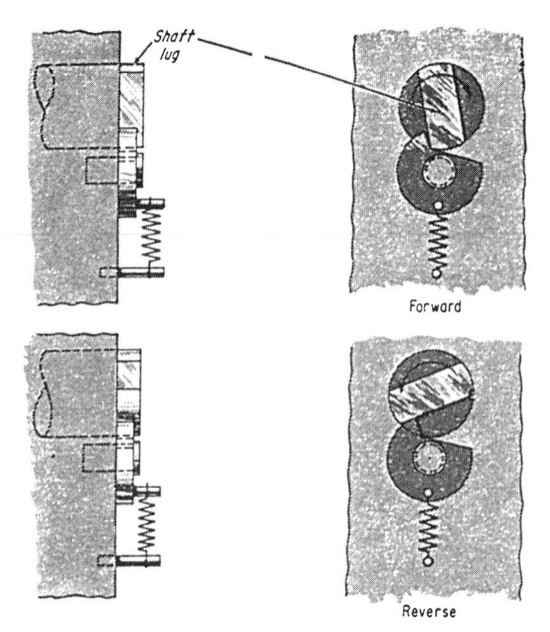

Fig. 2 A lug on a shaft pushes the notched disk free during normal rotation. The disk periphery stops the lug to prevent reverse rotation.

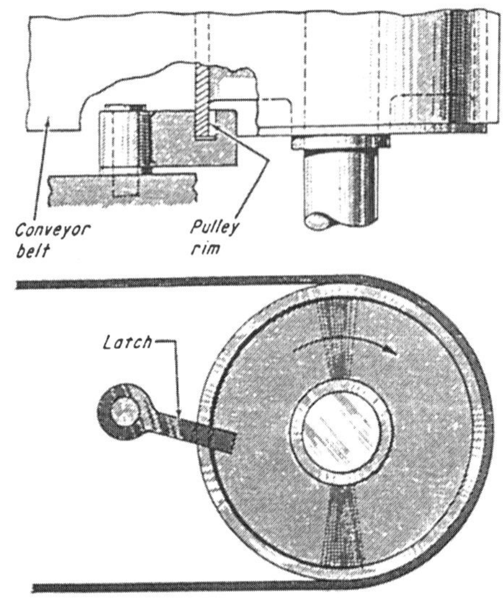

Fig. 3 A latch on the rim of the pulley is free only when the rotation is in the direction shown. This arrangement is ideal for conveyor-belt pulleys.

Fig. 4 Spring-loaded friction pads contact the right gear. The idler meshes and locks the gear set when the rotation is reversed.

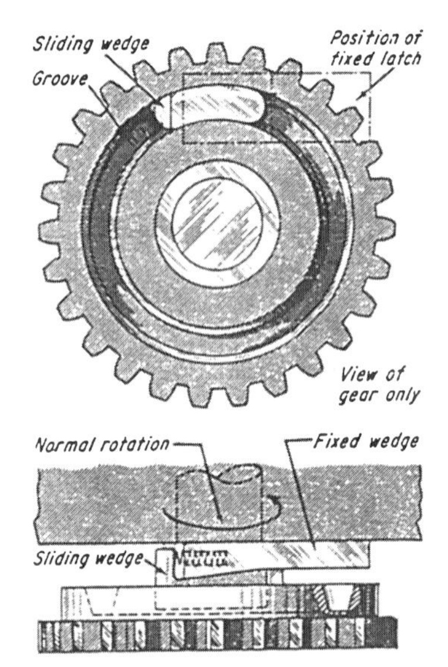

Fig. 5 A fixed wedge and sliding wedge tend to disengage when the gear is turning clockwise. The wedges jam in the reverse direction.

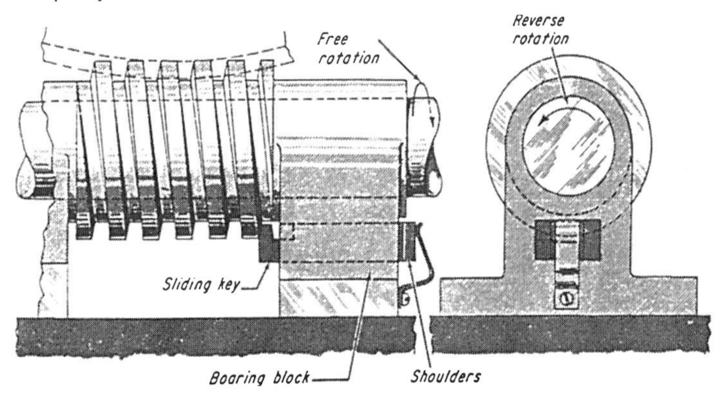

Fig. 6 A sliding key has a tooth which engages the worm threads. In reverse rotation, the key is pulled in until its shoulders contact the block.

CALIPER BRAKES KEEP PAPER TENSION IN WEB PRESSES

A simple cam-and-linkage arrangement (drawing) works in a team with two caliper disk brakes to provide automatic tension control for paper feeds on a web press.

In the feed system controlled tension must be maintained on the paper that's being drawn off at 1200 fpm from a roll up to 42 in. wide and 36 in. in diameter. Such rolls, when full, weigh 2000 lb. The press must also be able to make nearly instantaneous stops.

Friction-disk brakes are subject to lining wear, but they can make millions of stops before they need relining.

In the system, two pneumatic disk brakes made by Tol-O-Matic, Inc., Minneapolis, were mounted on each roll, gripping two separate 12-in. disks that provide maximum heat dissipation. To provide the desired constant-drag tension on the rolls, the brakes are always under air pressure. A dancer roll riding on the paper web can, however, override the brakes at any time. It operates a cam that adjusts a pressure regulator for controlling brake effort.

If the web should break or the paper run out on the roll, the dancer roll will allow maximum braking. The press can be stopped in less than one revolution.

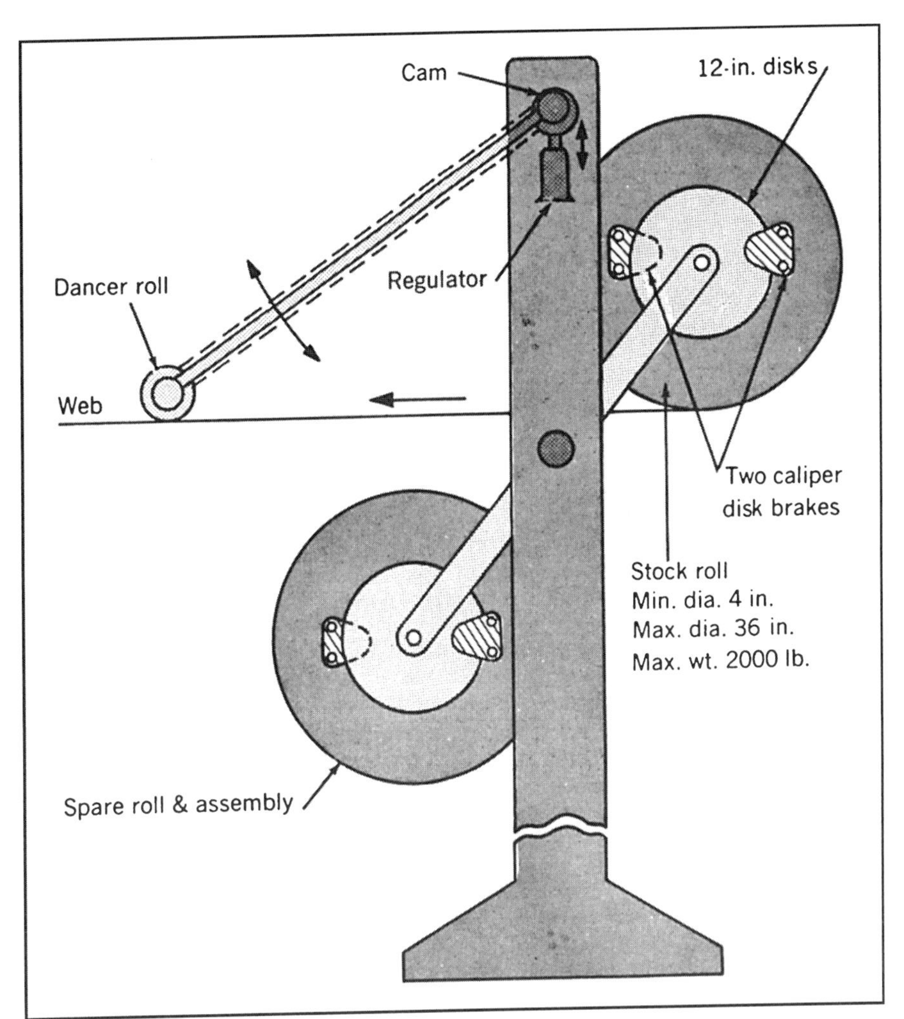

This linkage system works in combination with a regulator and caliper disk brakes to stop a press rapidly from a high speed, if the web should break.

CONTROL SYSTEM FOR PAPER CUTTING

Two clutch/brake systems, teamed with magnetic pickup sensors, cut paper sheets into exact lengths. One magnetic pickup senses the teeth on a rotating sprocket. The resulting pulses, which are related to the paper length, are counted, and a cutter wheel is actuated by the second clutch/brake system. The flywheel on the second system enhances the cutting force.

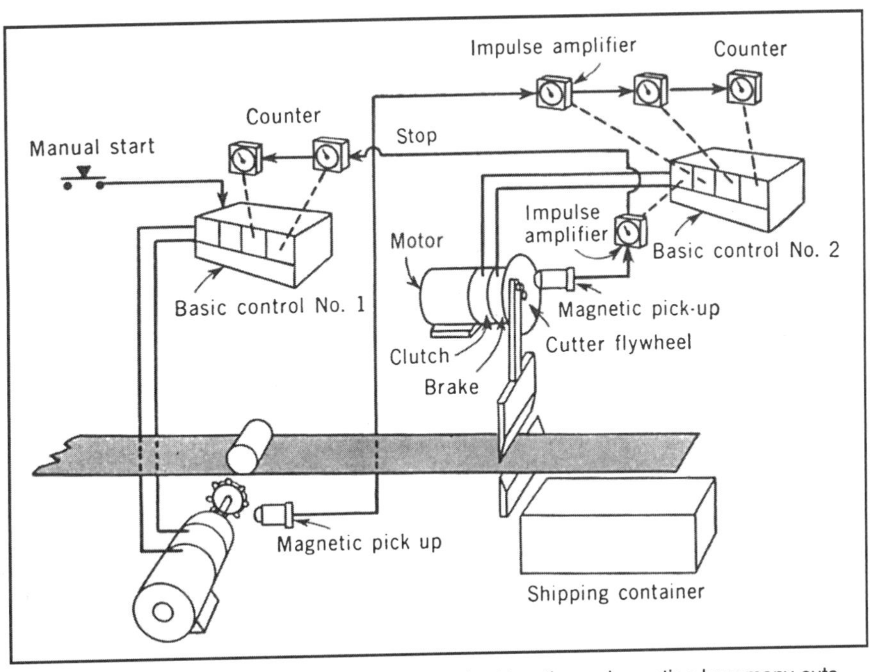

This control system makes cutting sheets to desired lengths and counting how many cuts are made simpler.

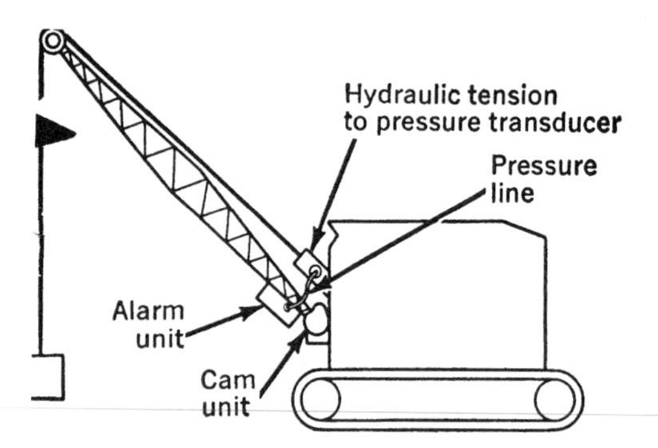

WARNING SYSTEM PREVENTS OVERLOADING OF BOOM

Cranes can now be protected against unsafe loading by a device whose movable electrical contacts are shifted by a combination of fluidic power and cam-and-gear arrangement (see drawing).

The device takes into consideration the two key factors in the safe loading of a crane boom: the boom angle (low angles create a greater overturning torque than high angles) and the compression load on the boom, which is greatest at high boom angles. Both factors are translated into inputs that are integrated to actuate the electrical warning system, which alerts the crane operator that a load is unsafe to lift.

How it works. In a prototype built for Thew-Lorain Inc. by US Gauge, Sellersville, Pennsylvania, a tension-to-pressure transducer (see drawing) senses the load on the cable and converts it into a hydraulic pressure that is proportional to the tension. This pressure is applied to a Bourdon-tube pressure gage with a rotating pointer that carries a small permanent magnet (see details in drawing). Two miniature magnetic reed switches are carried by another arm that moves on the same center as the pointer.

This arm is positioned by a gear and rack controlled by a cam, with a sinusoidal profile, that is attached to the cab. As the boom is raised or lowered, the cam shifts the position of the reed switches so they will come into close proximity with the magnet on the pointer and, sooner or later, make contact. The timing of this contact depends partly on the movement of the pointer that carries the magnet. On an independent path, the hydraulic pressure representing cable tension is shifting the pointer to the right or left on the dial.

When the magnet contacts the reed switches, the alarm circuit is closed, and it remains closed during a continuing pressure increase without retarding the movement of the point. In the unit built for Thew-Lorain, the switches were arranged in two stages: the first to trigger an amber warning light and second to light a red bulb and also sound an alarm bell.

Over-the-side or over-the-rear loading requires a different setting of the Bourdon pressure-gage unit than does over-the-front loading. A cam built into the cab pivot post actuated a selector switch.

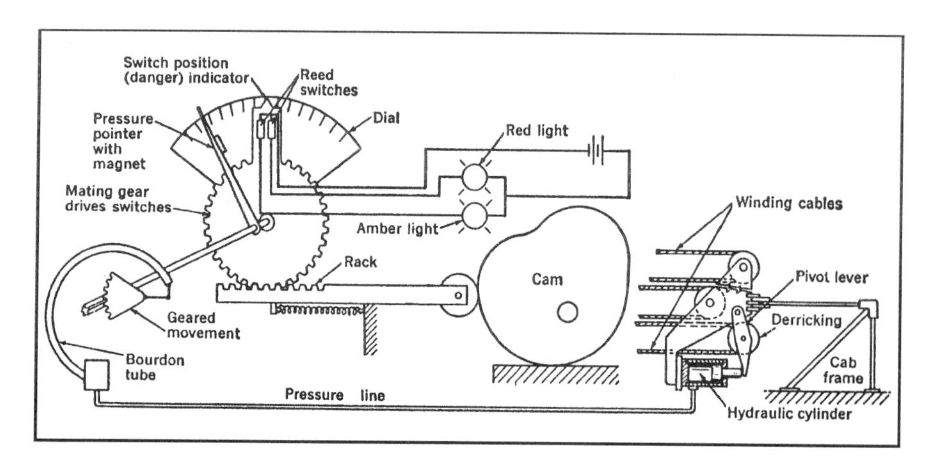

A cam on the cab positions an arm with reed switches according to boom angle; the pressure pointer reacts to cable tension.

LEVER SYSTEM MONITORS CABLE TENSION

A simple lever system solved the problem of how to keep track of varying tension loads on a cable as it is wound on its drum.

Thomas Grubbs of NASA's Manned Spacecraft Center in Houston devised the system, built around two pulleys mounted on a pivoted lever. The cable is passed between the pulleys (drawing) so an increase in cable tension causes the lever to pivot. This, in turn, pulls linearly on a flat metal tongue to which a strain gage has been cemented. Load on the lower pulley is proportional to tension on the cable. The stretching of the strain gage changes and electrical current that gives a continuous, direct reading of the cable tension.

The two pulleys on the pivoting lever are free to translate on the axes of rotation to allow proper positioning of the cable as it traverses the take-up drum.

A third pulley might be added to the two-pulley assembly to give some degree of adjustment to strain-gage sensitivity. Located in the plane of the other two pulleys, it would be positioned to reduce the strain on the tongue (for heavy loads) or increase the strain (for light loads).

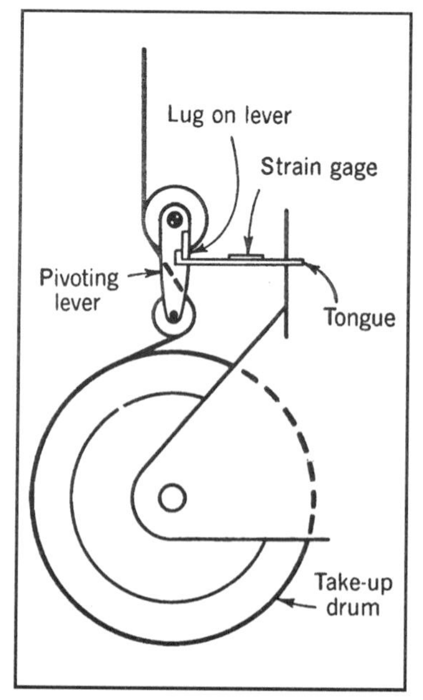

A load on the lower pulley varies with tension on the cable, and the pivoting of the lever gives a direct reading with a strain gage.

EIGHT TORQUE-LIMITERS PROTECT LIGHT-DUTY DRIVES

Light-duty drives break down when they are overloaded. These eight devices disconnect them from dangerous torque surges.

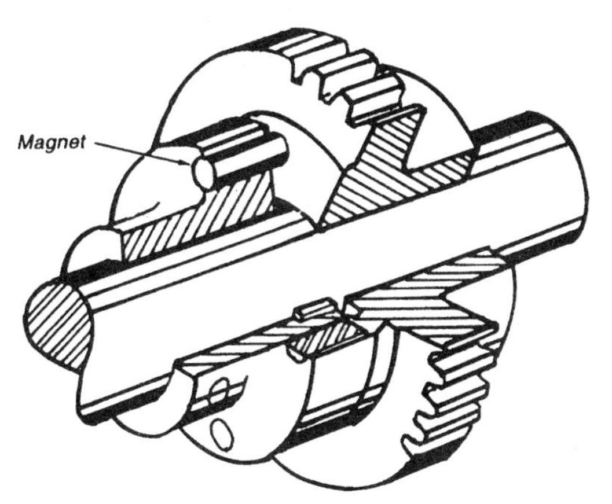

Fig. 1 Permanent magnets transmit torque in accordance with their numbers and size around the circumference of the clutch plate. Control of the drive in place is limited to removing magnets to reduce the drive's torque capacity.

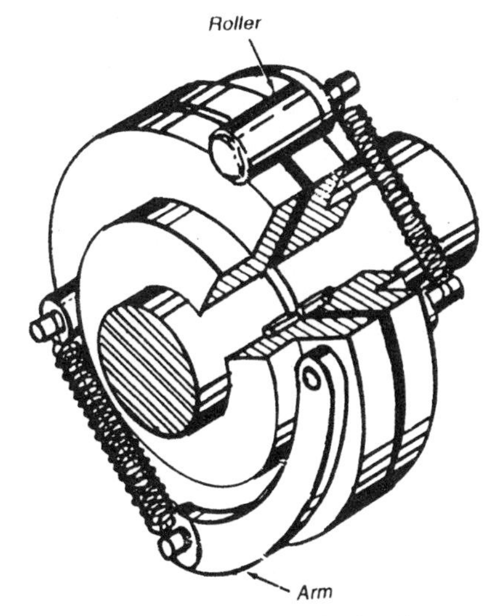

Fig. 2 Arms hold rollers in the slots that are cut across the disks mounted on the ends of butting shafts. Springs keep the roller in the slots, but excessive torque forces them out.

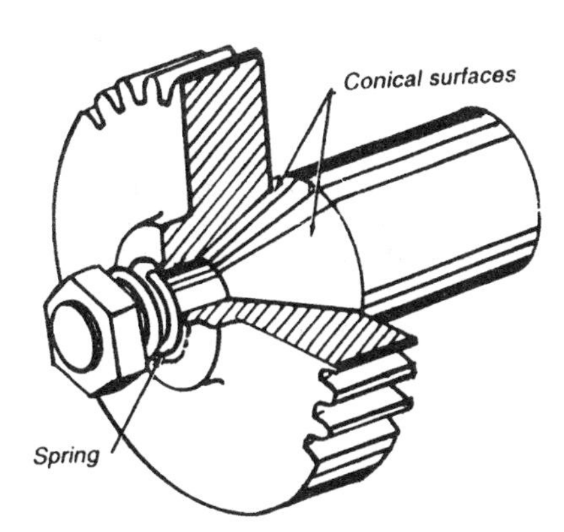

Fig. 3 A cone clutch is formed by mating a taper on the shaft to a beveled central hole in the gear. Increasing compression on the spring by tightening the nut increases the drive's torque capacity.

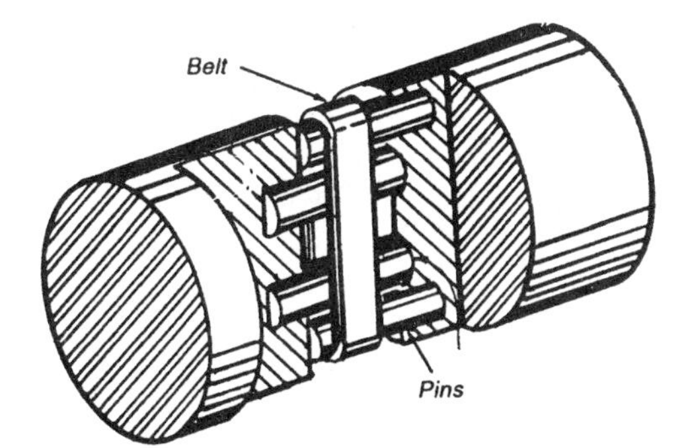

Fig. 4 A flexible belt wrapped around four pins transmits only the lightest loads. The outer pins are smaller than the inner pins to ensure contact.

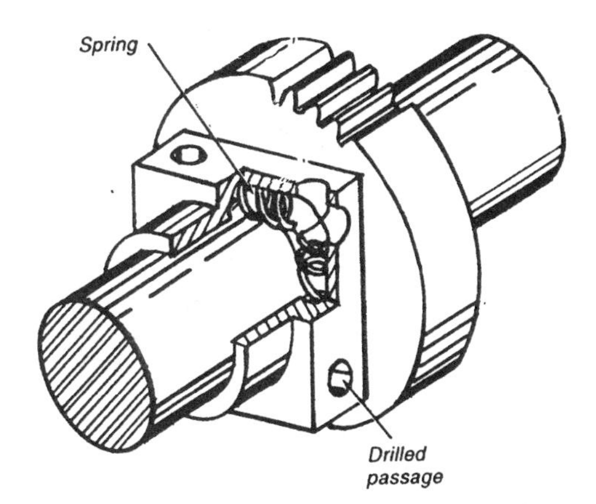

Fig. 5 Springs inside the block grip the shaft because they are distorted when the gear is mounted to the box on the shaft.

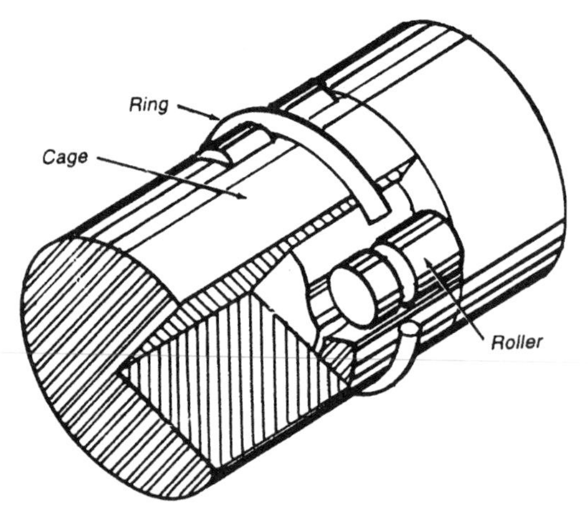

Fig. 6 The ring resists the natural tendency of the rollers to jump out of the grooves in the reduced end of one shaft. The slotted end of the hollow shaft acts as a cage.

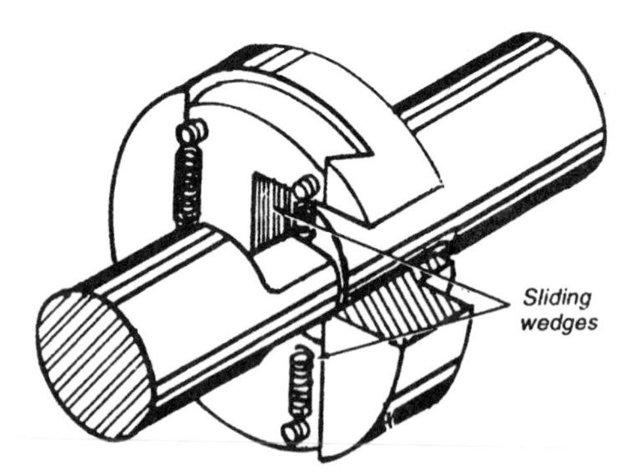

Fig. 7 Sliding wedges clamp down on the flattened end of the shaft. They spread apart when torque becomes excessive. The strength of the springs in tension that hold the wedges together sets the torque limit.

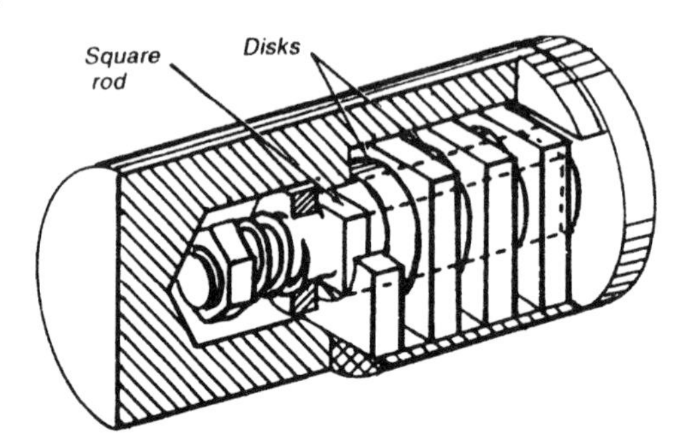

Fig. 8 Friction disks are compressed by an adjustable spring. Square disks lock into the square hole in the left shaft, and round disks lock onto the square rod on the right shaft.

THIRTEEN LIMITERS PREVENT OVERLOADING

These 13 "safety valves" give way if machinery jams, thus preventing serious damage.

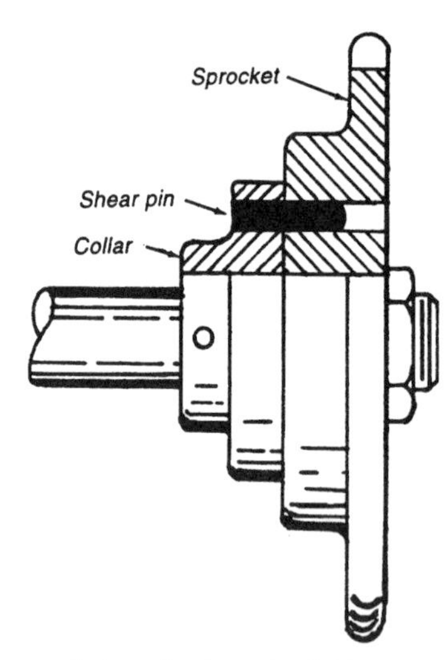

Fig. 1 A shear pin is a simple and reliable torque limiter. However, after an overload, removing the sheared pin stubs and replacing them with a new pin can be time consuming. Be sure that spare shear pins are available in a convenient location.

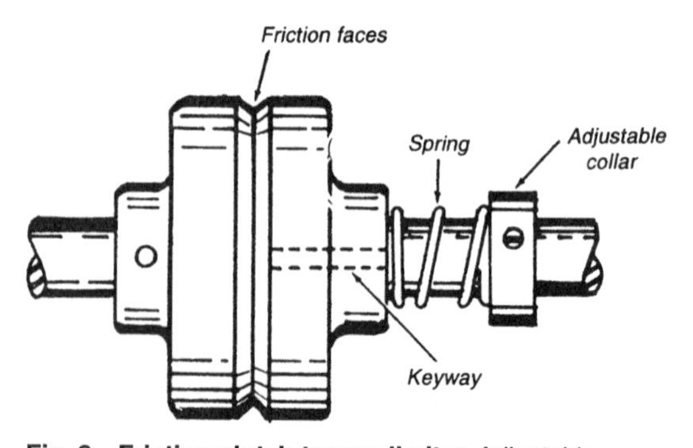

Fig. 2 Friction clutch torque limiter. Adjustable spring tension holds the two friction surfaces together to set the overload limit. As soon as an overload is removed, the clutch reengages. A drawback to this design is that a slipping clutch can destroy itself if it goes undetected.

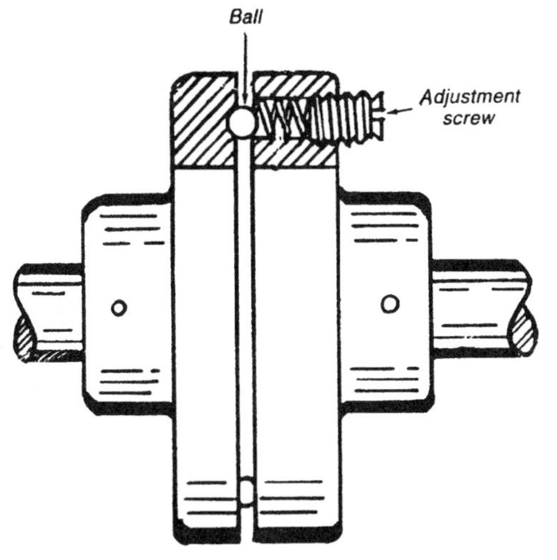

Fig. 3 Mechanical keys. A spring holds a ball in a dimple in the opposite face of this torque limiter until an overload forces it out. Once a slip begins, clutch face wear can be rapid. Thus, this limiter is not recommended for machines where overload is common.

Fig. 4 A cylinder cut at an angle forms a torque limiter. A spring clamps the opposing-angled cylinder faces together, and they separate from angular alignment under overload conditions. The spring tension sets the load limit.

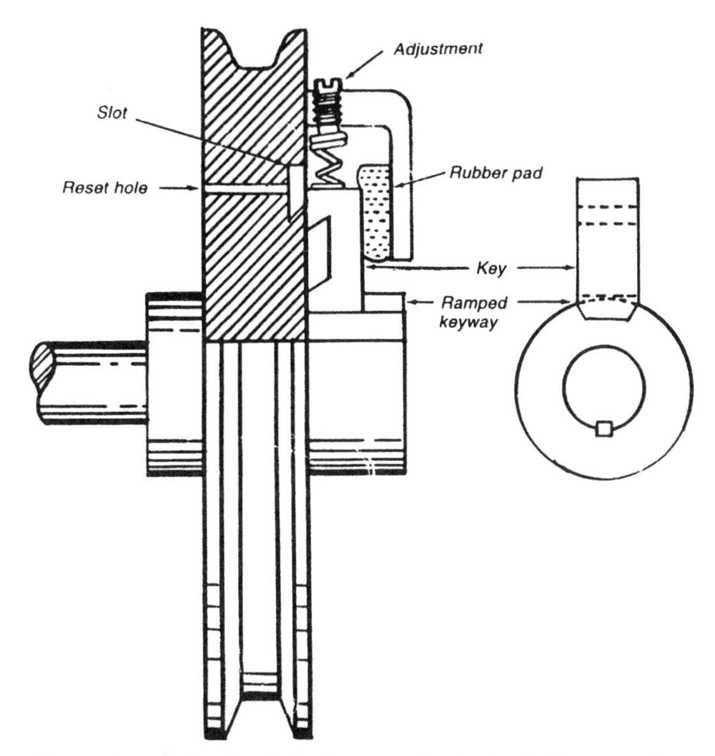

Fig. 5 A retracting key limits the torque in this clutch. The ramped sides of the keyway force the key outward against an adjustable spring. As the key moves outward, a rubber pad or another spring forces the key into a slot in the sheave. This holds the key out of engagement and prevents wear. To reset the mechanism, the key is pushed out of the slot with a tool in the reset hole of the sheave.

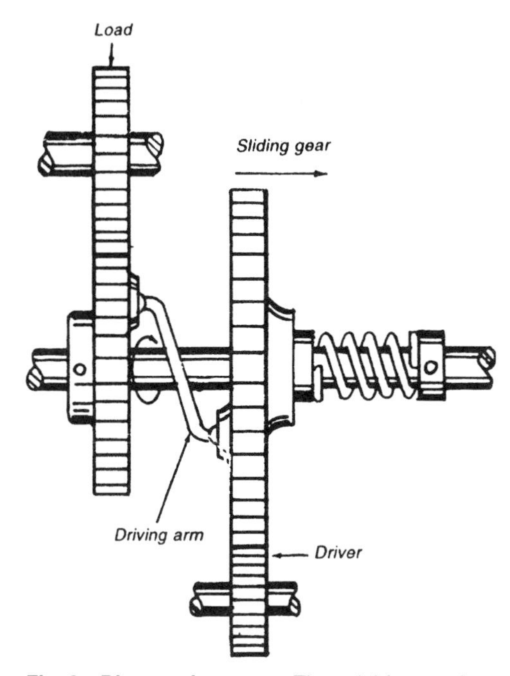

Fig. 6 Disengaging gears. The axial forces of a spring and driving arm are in balance in this torque limiter. An overload condition overcomes the force of the spring to slide the gears out of engagement. After the overload condition is removed, the gears must be held apart to prevent them from being stripped. With the driver off, the gears can safely be reset.

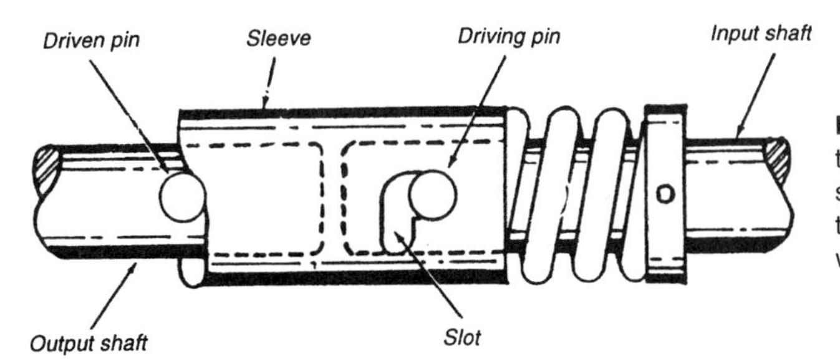

Fig. 7 A cammed sleeve connects the input and output shafts of this torque limiter. A driven pin pushes the sleeve to the right against the spring. When an overload occurs, the driving pin drops into the slot to keep the shaft disengaged. The limiter is reset by turning the output shaft backwards.

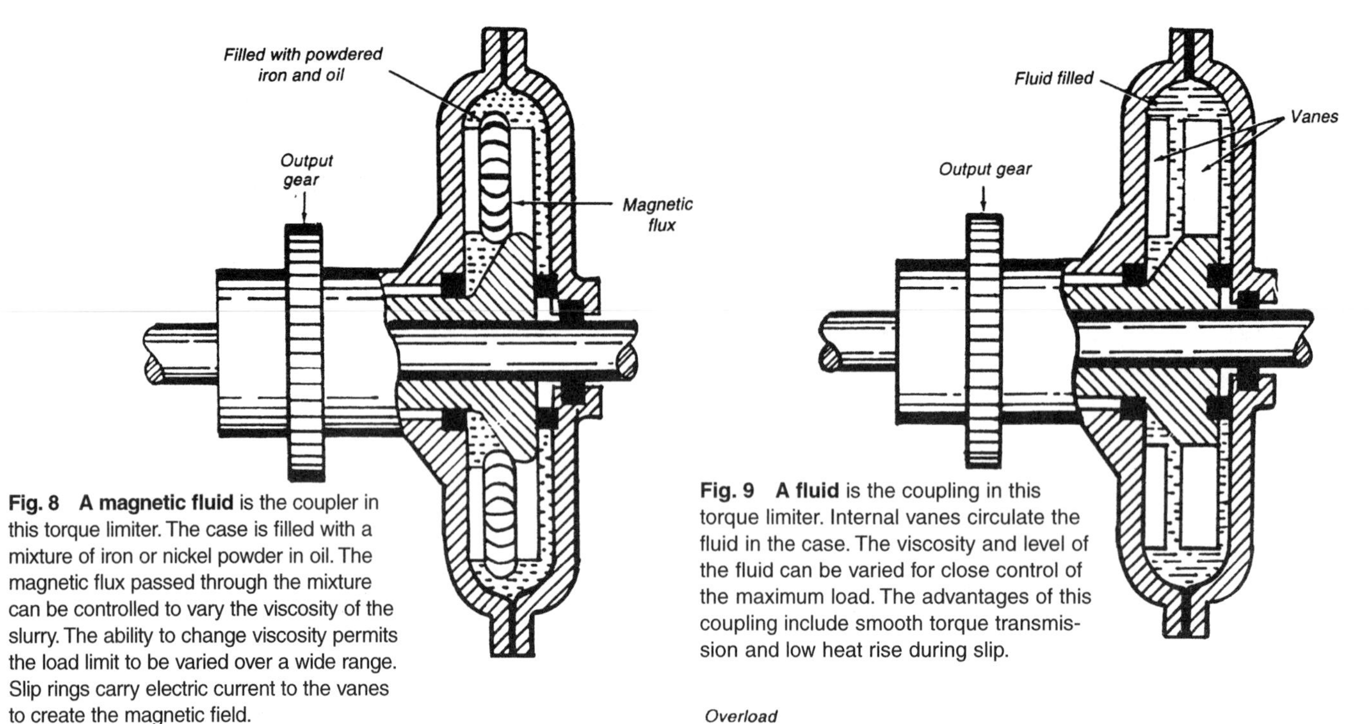

Fig. 8 A magnetic fluid is the coupler in this torque limiter. The case is filled with a mixture of iron or nickel powder in oil. The magnetic flux passed through the mixture can be controlled to vary the viscosity of the slurry. The ability to change viscosity permits the load limit to be varied over a wide range. Slip rings carry electric current to the vanes to create the magnetic field.

Fig. 9 A fluid is the coupling in this torque limiter. Internal vanes circulate the fluid in the case. The viscosity and level of the fluid can be varied for close control of the maximum load. The advantages of this coupling include smooth torque transmission and low heat rise during slip.

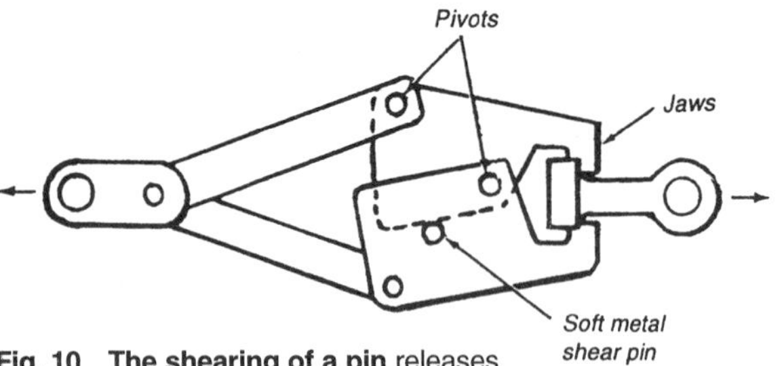

Fig. 10 The shearing of a pin releases tension in this coupling. A toggle-operated blade shears a soft pin so that the jaws open and release an excessive load. In an alternative design, a spring that keeps the jaws from spreading replaces the shear pin.

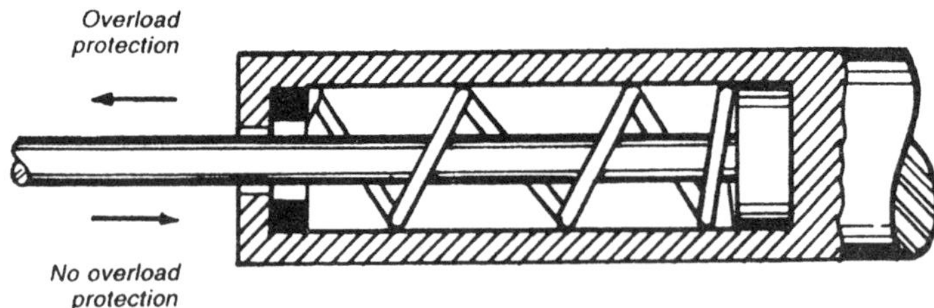

Fig. 11 A spring plunger provides reciprocating motion in this coupling. Overload can occur only when the rod is moving to the left. The spring is compressed under an overload condition.

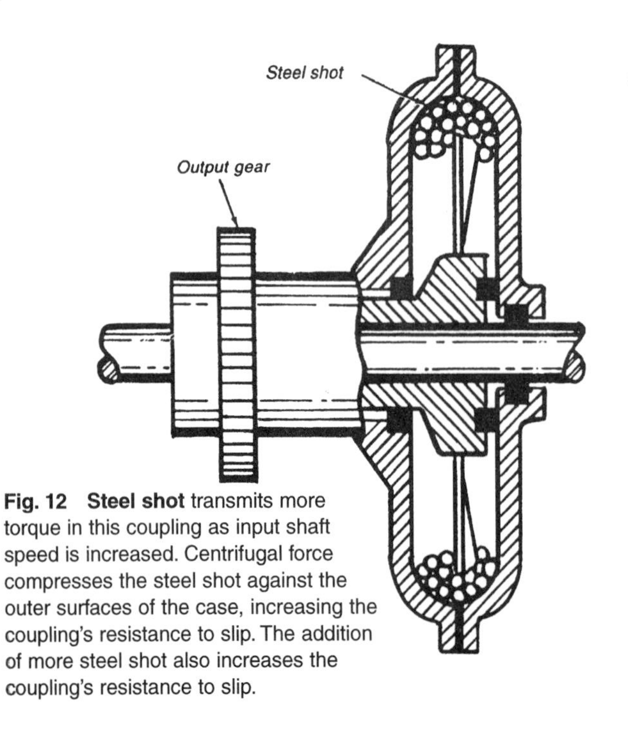

Fig. 12 Steel shot transmits more torque in this coupling as input shaft speed is increased. Centrifugal force compresses the steel shot against the outer surfaces of the case, increasing the coupling's resistance to slip. The addition of more steel shot also increases the coupling's resistance to slip.

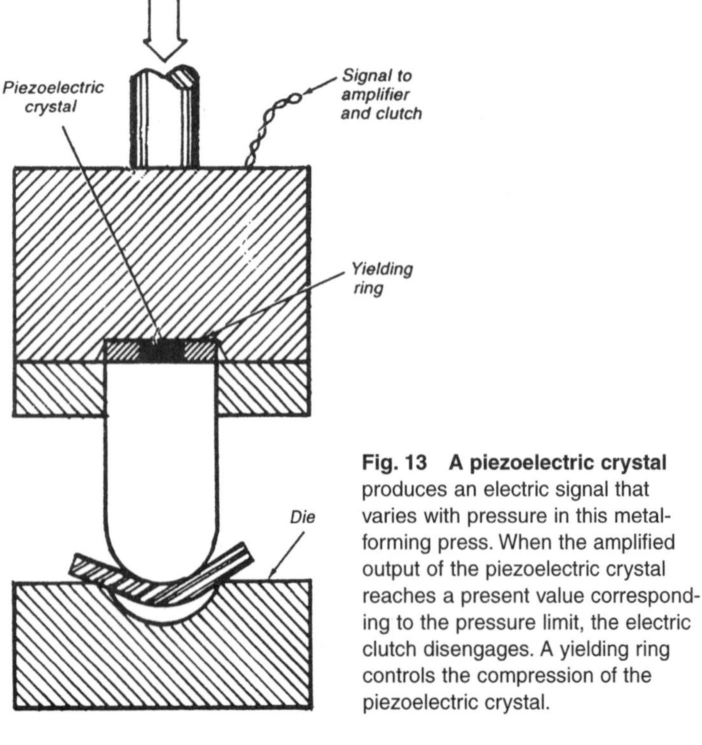

Fig. 13 A piezoelectric crystal produces an electric signal that varies with pressure in this metal-forming press. When the amplified output of the piezoelectric crystal reaches a present value corresponding to the pressure limit, the electric clutch disengages. A yielding ring controls the compression of the piezoelectric crystal.

SEVEN WAYS TO LIMIT SHAFT ROTATION

Traveling nuts, clutch plates, gear fingers, and pinned members form the basis of these ingenious mechanisms.

Mechanical stops are often required in automatic machinery and servomechanisms to limit shaft rotation to a given number of turns. Protection must be provided against excessive forces caused by abrupt stops and large torque requirements when machine rotation is reversed after being stopped.

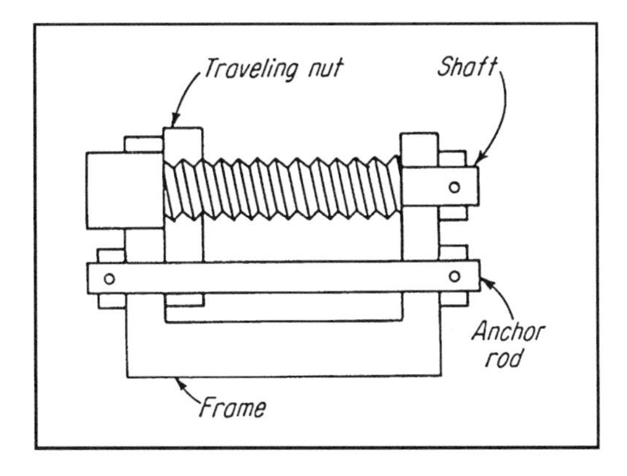

Fig. 1 A traveling nut moves along the threaded shaft until the frame prevents further rotation. This is a simple device, but the traveling nut can jam so tightly that a large torque is required to move the shaft from its stopped position. This fault is overcome at the expense of increased device length by providing a stop pin in the traveling nut.

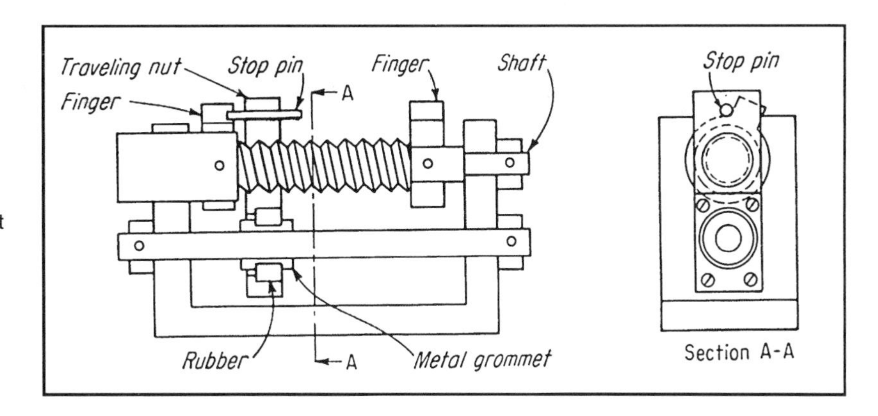

Fig. 2 The engagement between the pin and the rotating finger must be shorter than the thread pitch so the pin can clear the finger on the first reverse-turn. The rubber ring and grommet lessen the impact and provide a sliding surface. The grommet can be oil-impregnated metal.

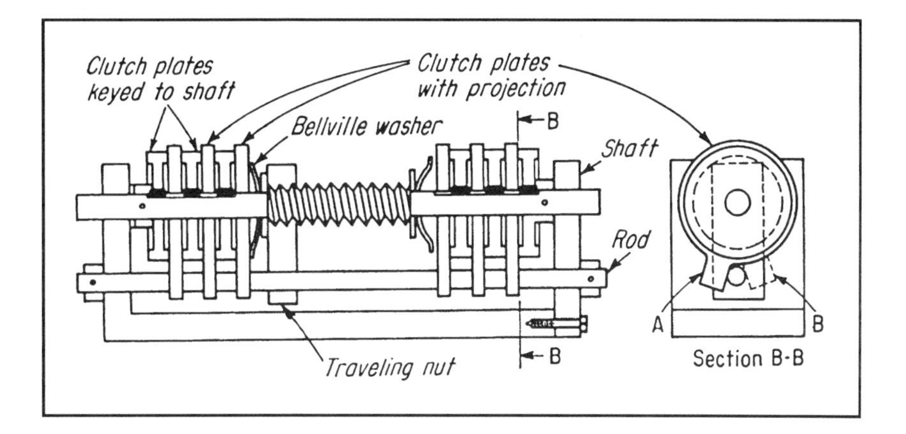

Fig. 3 Clutch plates tighten and stop their rotation as the rotating shaft moves the nut against the washer. When rotation is reversed, the clutch plates can turn with the shaft from A to B. During this movement, comparatively low torque is required to free the nut from the clutch plates. Thereafter, subsequent movement is free of clutch friction until the action is repeated at the other end of the shaft. The device is recommended for large torques because the clutch plates absorb energy well.

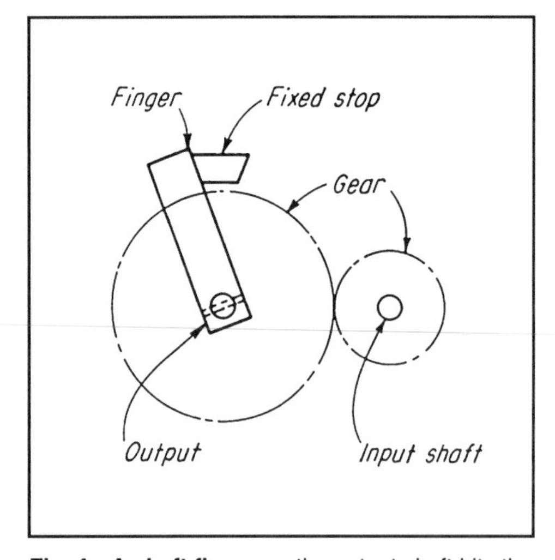

Fig. 4 A shaft finger on the output shaft hits the resilient stop after making less than one revolution. The force on the stop depends upon the gear ratio. The device is, therefore, limited to low ratios and few turns, unless a worm-gear setup is used.

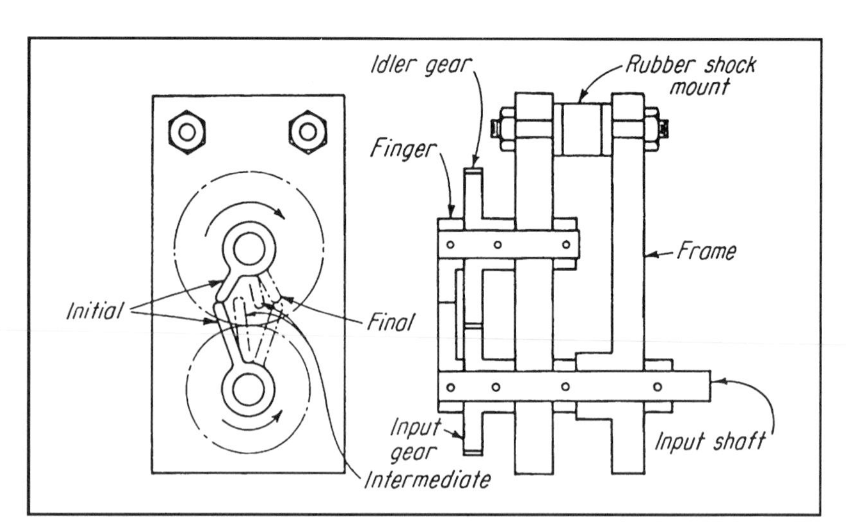

Fig. 5 Two fingers butt together at the initial and final positions to prevent rotation beyond these limits. A rubber shock-mount absorbs the impact load. A gear ratio of almost 1:1 ensures that the fingers will be out-of-phase with one another until they meet on the final turn. Example: Gears with 30 to 32 teeth limit shaft rotation to 25 turns. Space is saved here, but these gears are expensive.

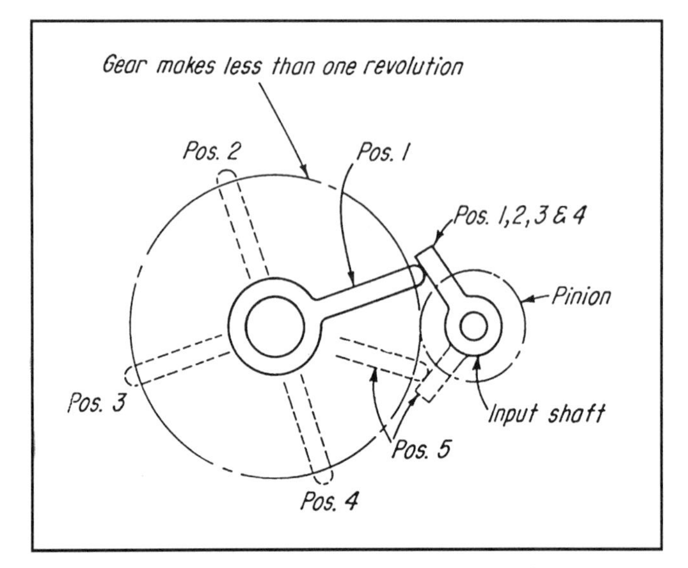

Fig. 6 A large gear ratio limits the idler gear to less than one turn. Stop fingers can be added to the existing gears in a train, making this design the simplest of all. The input gear, however, is limited to maximum of about five turns.

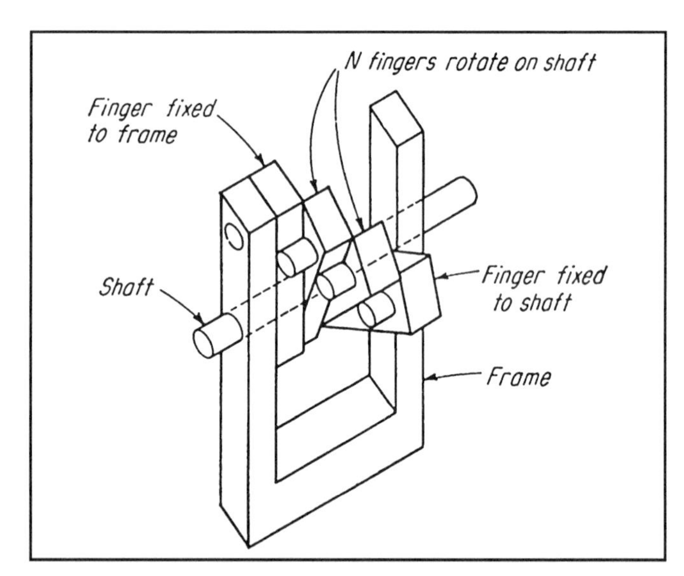

Fig. 7 Pinned fingers limit shaft turns to approximately N + 1 revolutions in either direction. Resilient pin-bushings would help reduce the impact force.

MECHANICAL SYSTEMS FOR CONTROLLING TENSION AND SPEED

The key to the successful operation of any continuous-processing system that is linked together by the material being processed is positive speed synchronization of the individual driving mechanisms. Typical examples of such a system are steel mill strip lines, textile processing equipment, paper machines, rubber and plastic processers, and printing presses. In each of these examples, the material will become wrinkled, marred, stretched or otherwise damaged if precise control is not maintained.

FIG. 1—PRIMARY INDICATORS

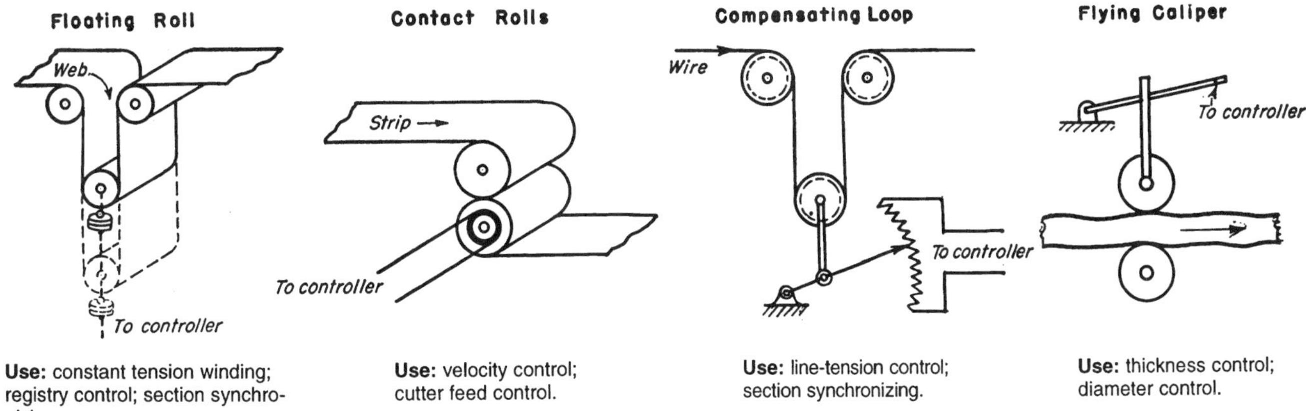

Use: constant tension winding; registry control; section synchronizing.

Use: velocity control; cutter feed control.

Use: line-tension control; section synchronizing.

Use: thickness control; diameter control.

FIG. 2—SECONDARY INDICATORS

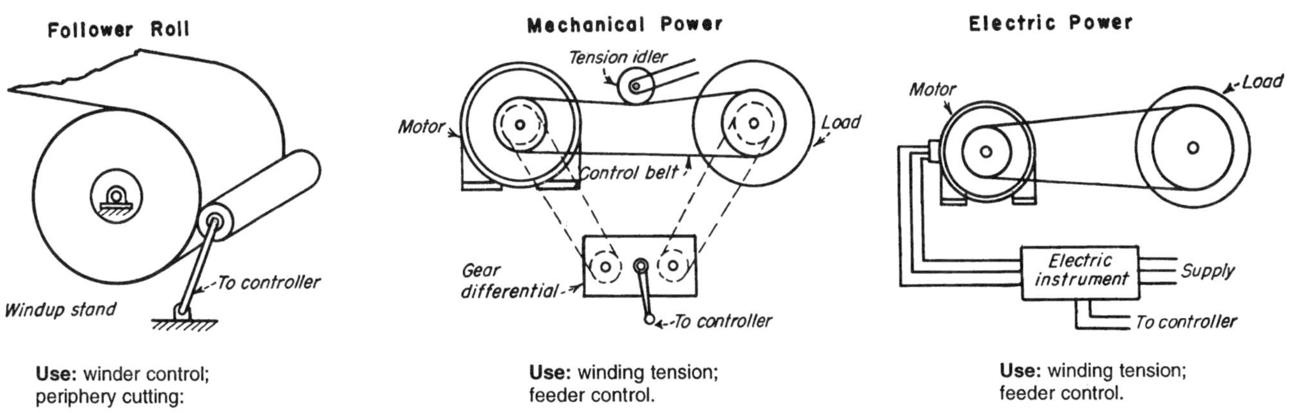

Use: winder control; periphery cutting:

Use: winding tension; feeder control.

Use: winding tension; feeder control.

FIG. 3—CONTROLLERS AND ACTUATORS

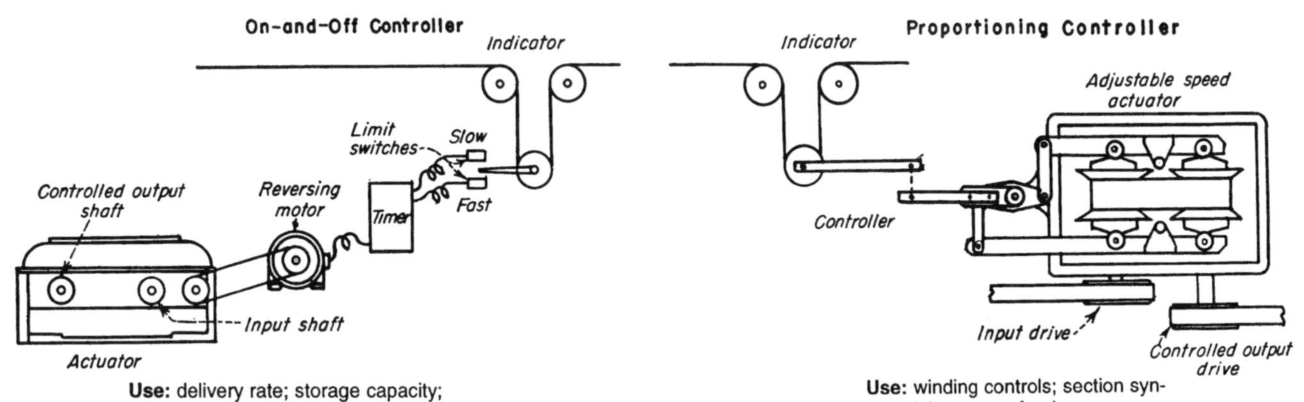

Use: delivery rate; storage capacity; rough section synchronizing.

Use: winding controls; section synchronizing; cutter feeds.

The automatic control for such a system contains three basic elements: The *signal device* or *indicator*, which senses the error to be corrected; the *controller*, which interprets the indicator signal and amplifies it, if necessary, to initiate control action; and the *transmission*, which operates from the controller to change the speed of the driving mechanism to correct the error.

Signal indicators for continuous systems are divided in two general classifications: *Primary indicators* that measure the change in speed or tension of the material by direct contact with the material; and *secondary indicators* that measure a change in the material from some reaction in the system that is proportional to the change.

The primary type is inherently more accurate because of its direct contact with the material. These indicators take the form of contact rolls, floating or compensating rolls, resistance bridges and flying calipers, as illustrated in Fig. 1. In each case, any change in the tension, velocity, or pressure of the material is indicated directly and immediately by a displacement or change in position of the indicator element. The primary indicator, therefore, shows deviation from an established norm, regardless of the factors that have caused the change.

Secondary indicators, shown in Fig. 2, are used in systems where the material cannot be in direct contact with the indicator or when the space limitations of a particular application make their use undesirable. This type of indicator introduces a basic inaccuracy into the control system which is the result of measuring an error in the material from a reaction that is not exactly proportional to the error. The control follows the summation of the errors in the material and the indicator itself.

The controlling devices, which are operated by the indicators, determine the degree of speed change required to correct the error, the rate at which the correction must be made, and the stopping point of the control action after the error has been corrected. The manner in which the corrective action of the controller is stopped determines both the accuracy of the control system and the kind of control equipment required.

Three general types of control action are illustrated in Fig. 3. Their selection for any individual application is based on the degree of control action required, the amount of power available for initiating the control, that is, the torque amplification required, and the space limitations of the equipment.

The on-and-off control with timing action is the simplest of the three types. It functions in this way: when the indicator is displaced, the timer contact energizes the control in the proper direction for correcting the error. The control action continues until the timer stops the action. After a short interval, the timer again energizes the control system and, if the error still exists, control action is continued in the same direction. Thus, the control process is a step-by-step response to make the correction and to stop the operation of the controller.

The proportioning controller corrects an error in the system, as shown by the indicator, by continuously adjusting the actuator to a speed that is in exact proportion to the displacement of the indicator. The diagram in Fig. 3 shows the proportioning controller in its simplest form as a direct link connection between the indicator and the actuating drive. However, the force amplification between the indicator and the drive is relatively low; thus it limits

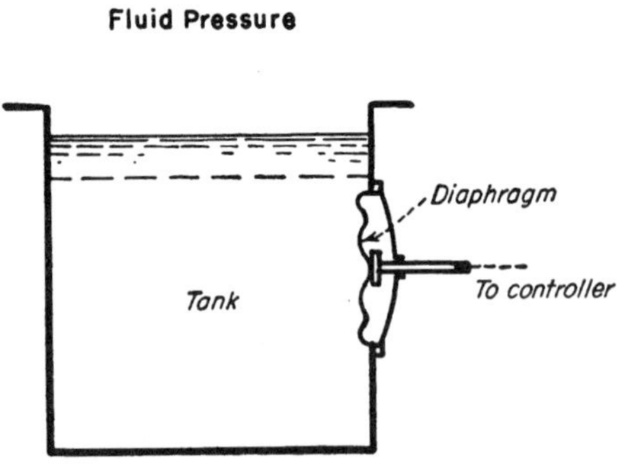

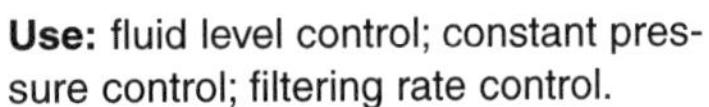
Use: fluid level control; constant pressure control; filtering rate control.

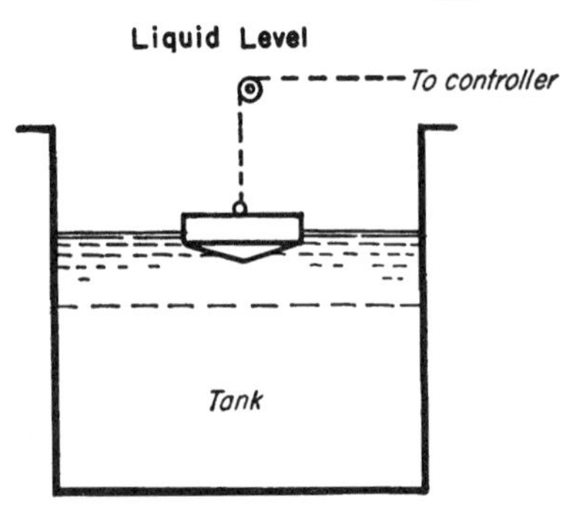

Use: Pumping rate control; system pressure control.

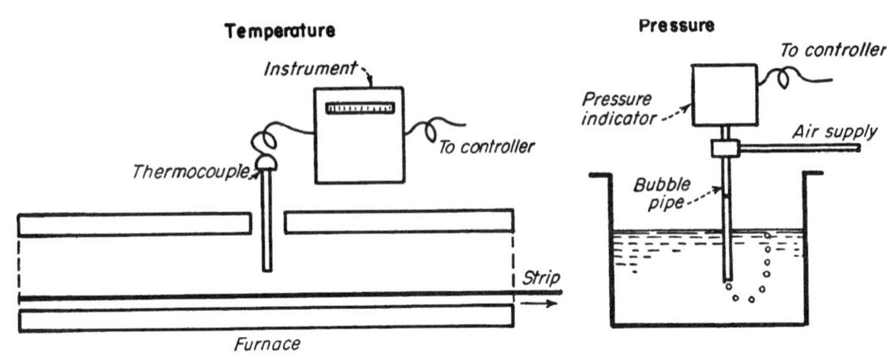

Use: annealing; drum dryers; kilns.

Use: fluid density; feeding rate; flow rate.

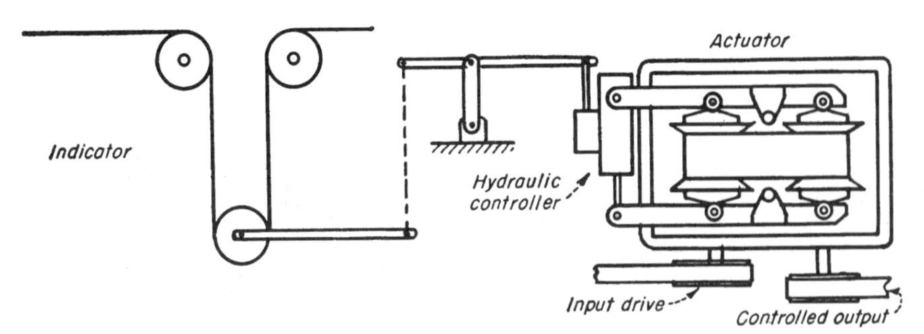

Use: constant tension winding; registry control; exact section synchronizing.

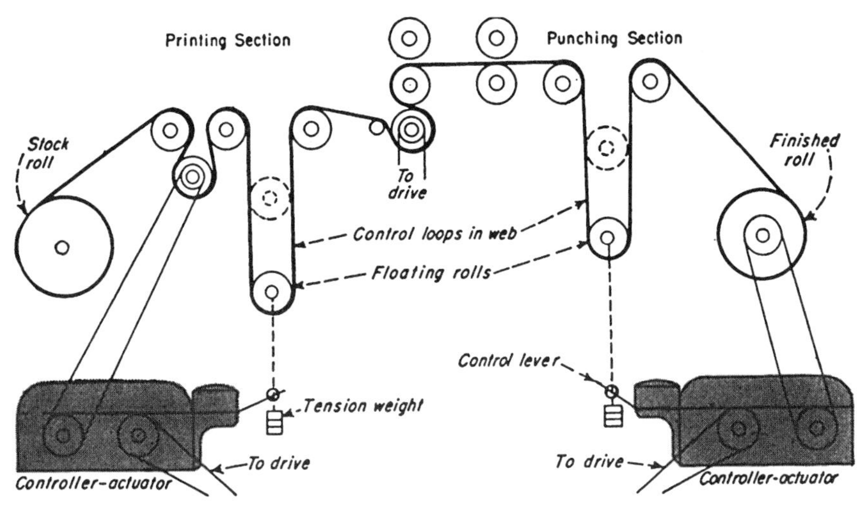

Fig 4 Floating rolls are direct indicators of speed and tension in the paper web. Controller-actuators adjust feed and windup rolls to maintain registry during printing.

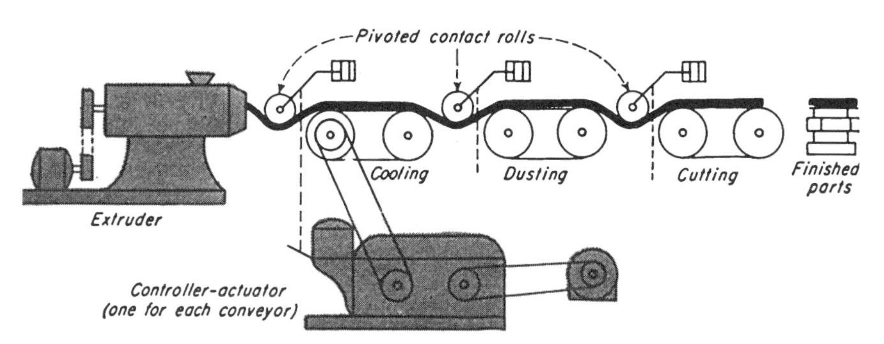

Fig. 5 Dimension control of extruded materials calls for primary indicators like the contact rolls shown. Their movements actuate conveyor control mechanisms.

this controller to applications where the indicator has sufficient operating force to adjust the speed of the variable-speed transmission directly.

The most accurate controller is the proportioning type with throttling action. Here, operation is in response to the rate or error indication. This controller, as shown in Fig. 3, is connected to a throttling valve, which operates a hydraulic servomechanism for adjusting the variable-speed transmission.

The throttling action of the valve provides a slow control action for small error correction or for continuous correction at a slow rate. For following large error, as shown by the indicator, the valve opens to the full position and makes the correction as rapidly as the variable-speed transmission will allow.

Many continuous processing systems can be automatically controlled with a packaged unit consisting of a simple, mechanical, variable-speed transmission and an accurate hydraulic controller.

This controller-transmission package can change the speed relationship at the driving points in the continuous system from any indicator that signals for correction by a displacement. It has anti-hunting characteristics because of the throttling action on the control valve, and is self-neutralizing because the control valve is part of the transmission adjustment system.

The rotary printing press is an example of a continuous processing system that requires automatic control. When making billing forms on a press, the printing plates are rubber, and the forms are printed on a continuous web or paper. The paper varies in texture, moisture content, flatness, elasticity, and finish. In addition, the length of the paper changes as the ink is applied.

In a typical application of this kind, the accuracy required for proper registry of the printing and hole punching must be held to a differential of $^1/_{32}$ in. in 15 ft of web. For this degree of accuracy, a floating or compensating roll, as shown in Fig. 4, serves as the indicator because it is the most accurate way to indicate changes in the length of the web by displacement. In this case, two floating rolls are combined with two separate controllers and actuators. The first controls the in-feed speed and tension of the paper stock, and the second controls the wind-up.

The in-feed is controlled by maintaining the turning speed of a set of feeding rolls that pull the paper off the stock roll. The second floating roll controls the speed of the wind-up mandrel. The web of paper is held to an exact value of tension between the feed rolls and the punching cylinder of the press by the in-feed control. It is also held between the punching cylinder and the wind-up roll. Hence, it is possible to control the tension in the web of different grades of paper and also adjust the relative length at these two points, thereby maintaining proper registry.

The secondary function of maintaining exact control of the tension in the paper as it is rewound after printing is to condition the paper and obtain a uniformly wound roll. This makes the web ready for subsequent operations.

The control of dimension or weight by tension and velocity regulation can be illustrated by applying the same general type of controller actuator to the take-odd conveyors in a extruder line such as those used in rubber and plastics processing. Two problems must be solved: First, to set the speed of the take-away conveyor at the extruder to match the variation in extrusion rate; and, second, to set the speeds of the subsequent conveyor sections to match the movement of the stock as it cools and tends to change dimension.

One way to solve these problems is to use the pivoted idlers or contact rolls as indicators, as shown in Fig. 5. The rolls contact the extruded material between each of the conveyor sections and control the speed of the driving mechanism of the following section. The material forms a slight catenary between the stations, and the change in the catenary length indicates errors in driving speeds.

The plasticity of the material prevents the use of a complete control loop. Thus, the contract roll must operate with very little resistance or force through a small operating angle.

The difficulties in winding or coiling a strip of thin steel that has been plated or pre-coated for painting on a continuous basis is typical of processing systems whose primary indicators cannot be used. While it is important that no contact be made with the prepared surface of the steel, it also desirable to rewind the strip after preparation in a coil that is sound and slip-free. An automatic, constant-tension winding control and a secondary indicator initiate the control action.

The control system shown in Fig. 6 is used in winding coils from 16 in. core diameter to 48 in. maximum diameter. The power to wind the coil is the controlling medium because, by maintaining constant winding power as the coil builds up, a constant value of strip tension can be held within the limits required. Actually, this method is so inaccurate that the losses in the driving equipment (which are a factor in the power being measured) are not constant; hence the strip tension changes slightly. This same factor enters into any control system that uses winding power as an index of control.

A torque-measuring belt that operates a differential controller measures the power of the winder. Then, in turn, the controller adjusts the variable-speed transmission. The change in speed between the source of power and the transmission is measured by the three-shaft gear differential, which is driven in tandem with the control belt. Any change in load across the control belt produces a change in speed between the driving and driven ends of the belt. The differential acts as the controller, because any change in speed between the two outside shafts of the differential results in a rotation or displacement of the center or control shaft. By connecting the control shaft of the differential directly to a screw-controlled variable-speed transmission, it is possible to adjust the transmission to correct any change in speed and power delivered by the belt.

This system is made completely automatic by establishing a neutralizing speed between the two input shaft of the differential (within the creep value of the belt). When there is no tension in the strip (e.g., when it is cut), the input speed to the actuator side of the differential is higher on the driven side than it is on the driving side of the differential. This unbalance reverses the rotation of the control shaft of the differential, which in turn resets the transmission to high speed required for starting the next coil on the rewinding mandrel.

In operation, any element in the system that tends to change strip tension causes a change in winding power. This change, in turn, is immediately compensated by the rotation (or tendency to rotate) of the controlling shaft in the differential. Hence, the winding mandrel speed is continuously and automatically corrected to maintain constant tension in the strip.

When the correct speed relationships are established in the controller, the system operates automatically for all conditions of operation. In addition, tension in the strip can be adjusted to any value by moving the tension idler on the control belt to increase or decrease the load capacity of the belt to match a desired strip tension.

There are many continuous processing systems that require constant velocity of the material during processing, yet do not require accurate control of the tension in the material. An example of this process is the annealing of wire that is pulled off stock reels through an annealing furnace and then rewound on a windup block.

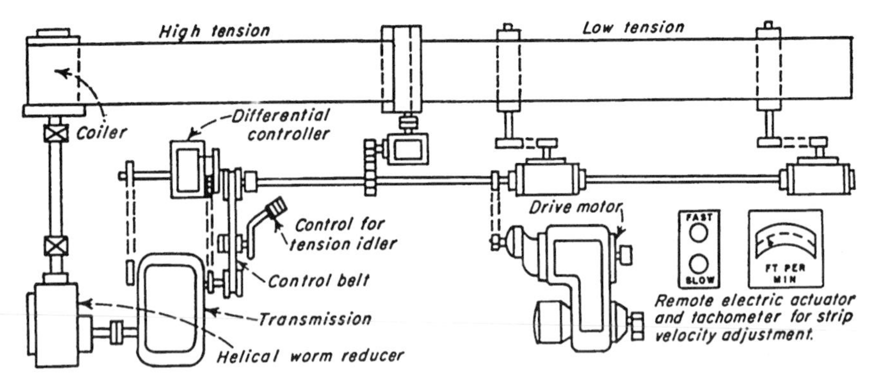

Fig. 6 The differential controller has a third shaft that signals the remote actuator when tension in sheet material changes. Coiler power is a secondary-control index.

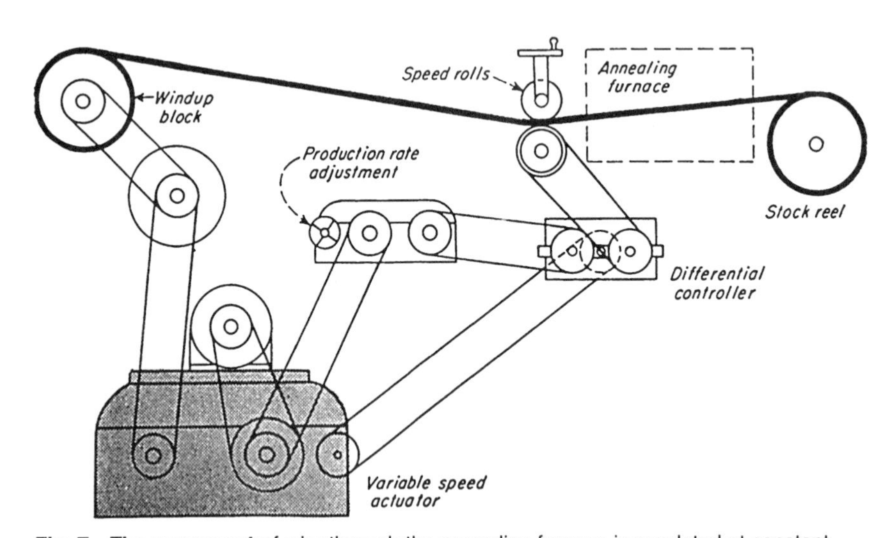

Fig. 7 The movement of wire through the annealing furnace is regulated at constant velocity by continuously retarding the speed of the windup reels to allow for wire build-up.

The wire must be passed through the furnace at a constant rate so that the annealing time is maintained at a fixed value. Because the wire is pulled through the furnace by the wind-up blocks, shown in Fig. 7, its rate of movement through the furnace would increase as the wire builds up on the reels unless a control slows down the reels.

A constant-velocity control that makes use of the wire as a direct indicator measures the speed of the wire to initiate a control action for adjusting the speed of the wind-up reel. In this case, the wire can be contacted directly, and a primary indicator in the form of a contact roll can register any change in speed. The contact roll drives one input shaft of the differential controller. The second input shaft is connected to the driving shaft of the variable-speed transmission to provide a reference speed. The third, or control, shaft will then rotate when any difference in speed exists between the two input shafts. Thus, if the control shaft is connected to a screw-regulated actuator, an adjustment is obtained for slowing down the wind-up blocks as the coils build up and the wire progresses through the furnace at a constant speed.

NINE DRIVES FOR CONTROLLING TENSION

Mechanical, electrical, and hydraulic methods for obtaining controlled tension on winding reels and similar drives, or for driving independent parts of a machine in synchronism.

MECHANICAL DRIVES

A band brake is used on coil winders, insulation winders, and similar machines where maintaining the tension within close limits is not required.

It is simple and economical, but tension will vary considerably. Friction drag at start-up might be several times that which occurs during running because of the difference between the coefficient of friction at the start and the coefficient of sliding friction. Sliding friction will be affected by moisture, foreign matter, and wear of the surfaces.

Capacity is limited by the heat radiating capacity of the brake at the maximum permissible running temperature.

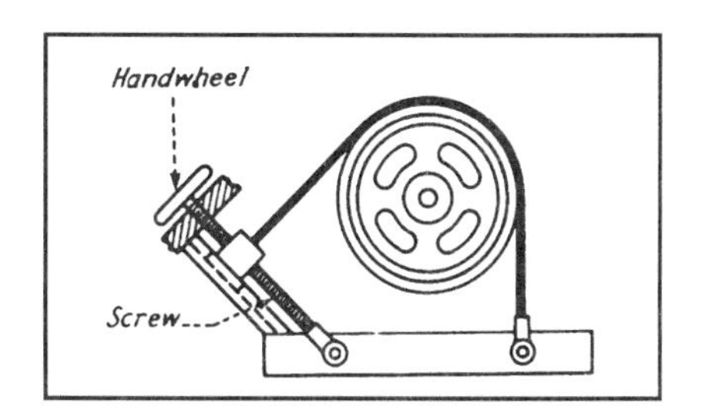

Differential drives can take many different forms, e.g., epicycle spur gears, bevel gear differentials, or worm gear differentials.

The braking device on the ring gear or spider could be a band brake, a fan, an impeller, an electric generator, or an electric drag element such as a copper disk rotating in a powerful magnetic field. A brake will give a drag or tension that is reasonably constant over a wide speed range. The other braking devices mentioned here will exert a torque that will vary widely with speed, but will be definite for any given speed of the ring gear or spider.

A definite advantage of any differential drive is that maximum driving torque can never exceed the torque developed by the braking device.

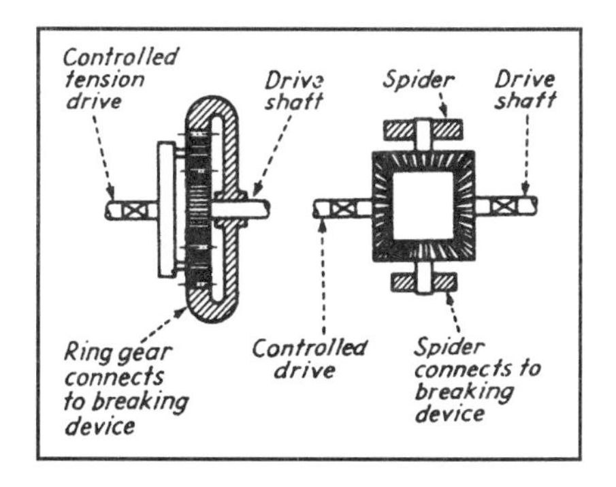

Differential gearing can be used to control a variable-speed transmission. If the ring gear and sun gear are to be driven in opposite directions from their respective shafts which are to be held in synchronism, the gear train can be designed so that the spider on which the planetary gears are mounted will not rotate when the shafts are running at the desired relative speeds. If one or the other of the shafts speeds ahead, the spider rotates correspondingly. The spider rotation changes the ratio of the variable-speed transmission unit.

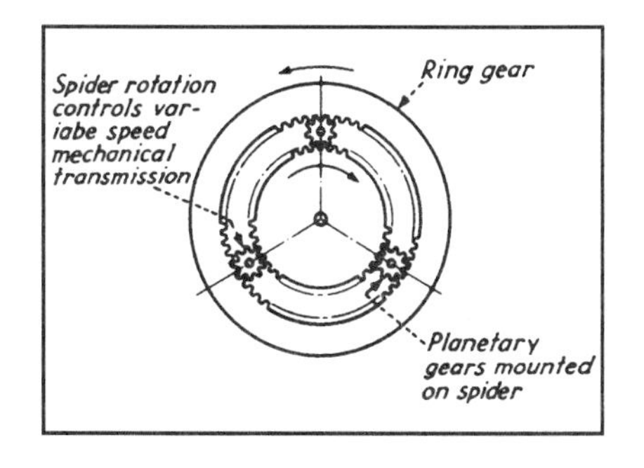

ELECTRICAL DRIVES

The shunt-field rheostat in a DC motor drive can be used to synchronize drives. When connected to a machine for processing paper, cloth, or other sheet material that is passing around a take-up roll, the movement of the take-up roll moves a control arm which is connected to the rheostat. This kind of drive is not suitable for wide changes of speed that exceed about a 2.5 to 1 ratio.

For wide ranges of speed, the rheostat is put in the shunt field of a DC generator that is driven by another motor. The voltage developed by the generator is controlled from zero to full voltage. The generator furnishes the current to the driving motor armature, and the fields of the driving motor are separately excited. Thus, the motor speed is controlled from zero to maximum.

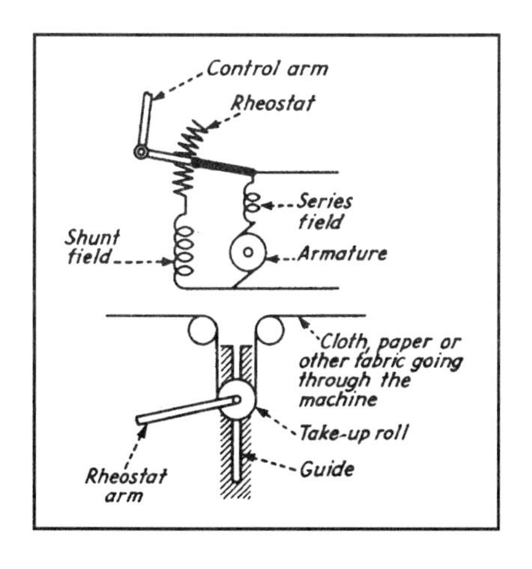

Selsyn motors can directly drive independent units in exact synchronism, provided their inertias are not too great. Regardless of loads and speeds, selsyn motors can be the controlling units. As an example, variable-speed mechanical transmission units with built-in selsyn motors are available for powering constant-tension drives or the synchronous driving of independent units.

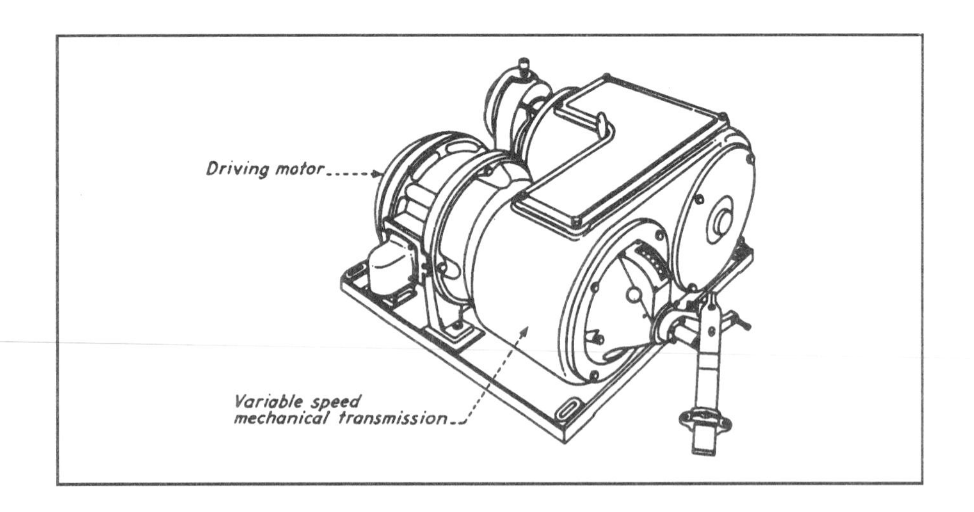

HYDRAULIC DRIVES

Hydraulic Control—Tension between successive pairs of rolls, or synchronism between successive units of a machine can be controlled automatically by hydraulic drives. Driving the variable delivery pump from one of the pairs of rolls automatically maintains an approximately constant relative speed between the two units, at all speeds and loads. The variations caused by oil leakage and similar factors are compensated automatically by the idler roll and linkage. They adjust the pilot valve that controls the displacement of the variable delivery pump.

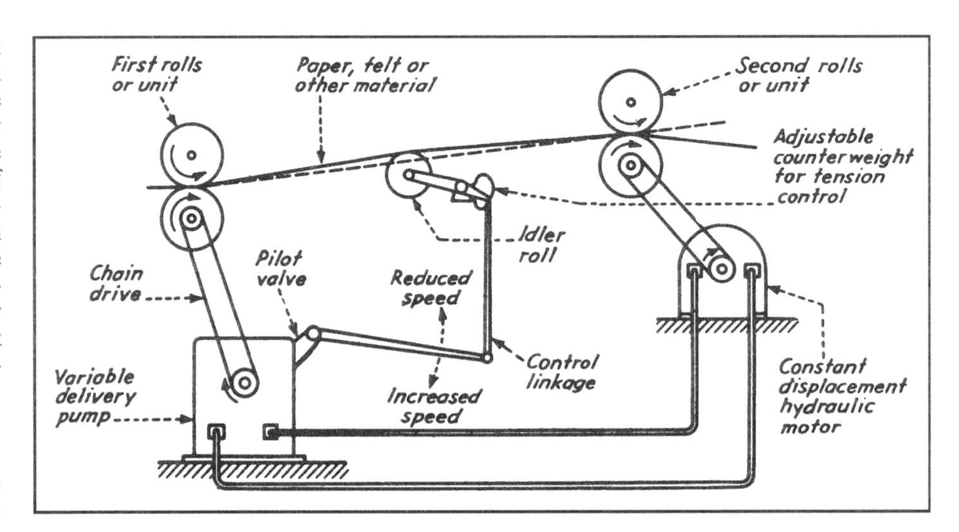

The counterweight on the idler roll is set for the desired tension in the felt, paper, or other material. Increased tension as a result of the high speed of the second pair of rolls depresses the idler roll. The control linkage then moves the pilot valve to decrease pump delivery, which slows the speed of the second pair of rolls. The reverse operations occur when the tension in the paper decreases, allowing the idler roll to move upwards.

If the material passing through the machine is too weak to operate a mechanical linkage, the desired control can be obtained by photoelectric devices. The hydraulic operation is exactly the same as that described for the hydraulic drives.

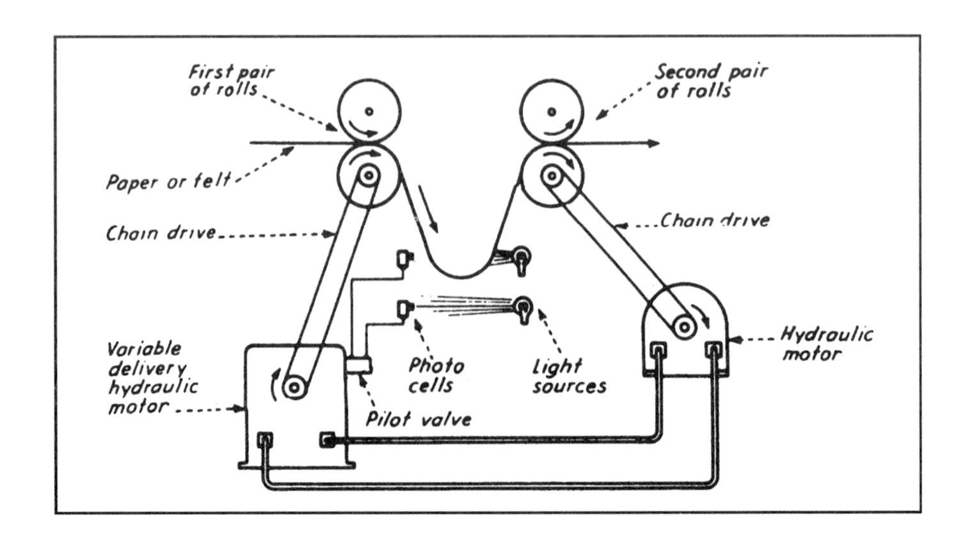

A band brake intended to obtain a friction drag will give variable tension. In this hydraulic drive, the winding tension is determined by the difference in torque exerted on the rewinder feed roll and the winding roll. The brake plays no part in establishing the tension.

The constant displacement hydraulic motor and the variable displacement hydraulic motor are connected in series with the variable delivery pump. Thus, the relative speeds of the two hydraulic motors will always remain substantially the same. The displacement of the variable speed motor is then adjusted to an amount slightly greater than the displacement of the constant-speed motor. This tends to give the winding roll a speed slightly greater than the feed roll speed. This determines the tension, because the winding roll cannot go faster than the feed roll. Both are in contact with the paper roll being wound. The pressure in the hydraulic line between the constant and variable displacement pumps will increase in proportion to the winding tension. For any setting of the winding speed controller on the variable delivery hydraulic pump, the motor speeds are generally constant. Thus, the surface speed of winding will remain substantially constant, regardless of the diameter of the roll being wound.

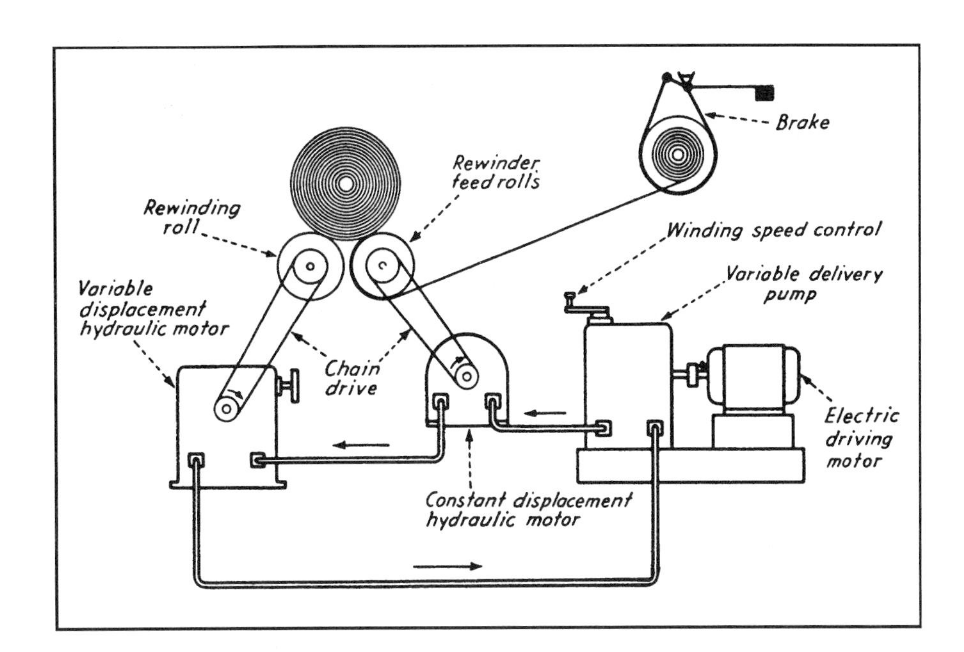

This is a hydraulic drive for fairly constant tension. The variable-delivery, constant-speed pumping unit supplies the oil to two constant displacement motors. One drives the apparatus that carries the fabric through the bath at a constant speed, and the other drives the winder. The two motors are in series, Motor A drives the winding reel, whose diameter increases from about 5 in. when the reel is empty to about 33 in. when the reel is full. Motor A is geared to the reel so that even when the reel is empty, the surface speed of paper travel will be somewhat faster than the mean rate of paper travel established by motor B, driving the apparatus. Only a small amount of oil will be bypassed through the choke located between the pressure and the return line.

When the roll is full, the revolutions per minute of the reel and its driving motor are only about one-seventh of the revolutions per minute when the reel is empty. More oil is forced through the choke when the reel is full because of the increased pressure in the line between the two motors. The pressure in this line increases as the reel diameter increases because the torque resistance encountered by the reel motor will be directly proportional to the reel diameter and because tension is constant. The larger the diameter of the fabric on the reel, the greater will be the torque exerted by the tension in the fabric. The installation is designed so that the torque developed by the motor driving the reel will be inversely proportional to the revolutions per minute of the reel. Hence, the tension on the fabric will remain fairly constant, regardless of the diameter of the reel. This drive is limited to about 3 hp, and it is relatively inefficient.

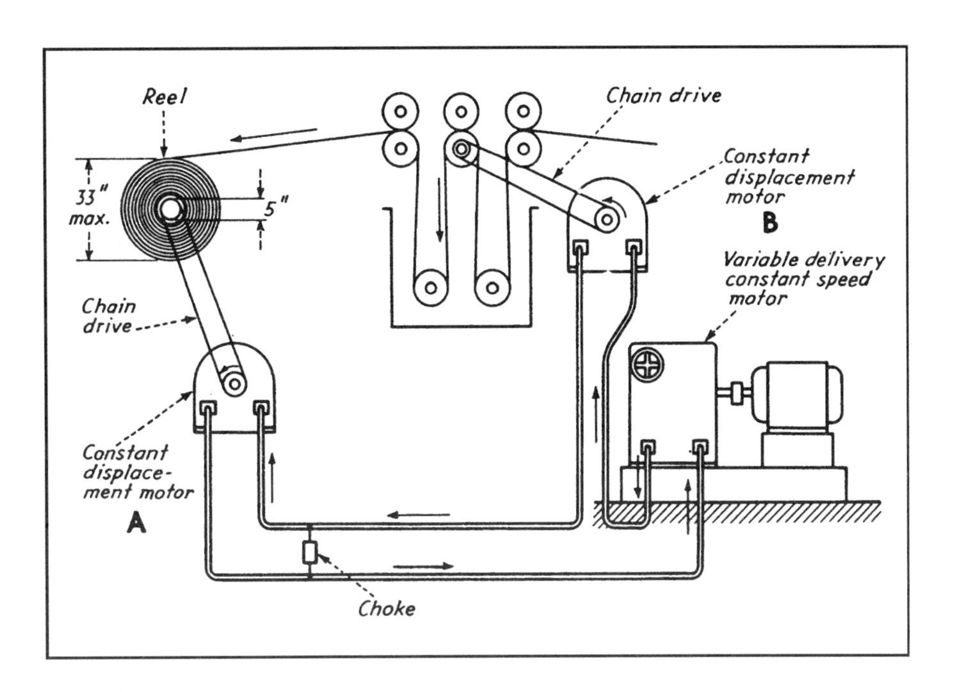

LIMIT SWITCHES IN MACHINERY

Limit switches, which confine or restrain the travel or rotation of moving parts within certain predetermined points, are actuated by varying methods. Some of these, such as cams, rollers, push-rods, and traveling nuts, are described and illustrated.

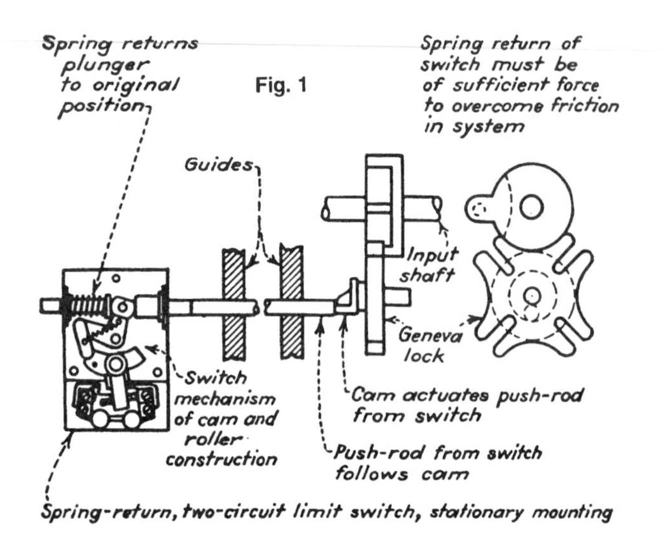

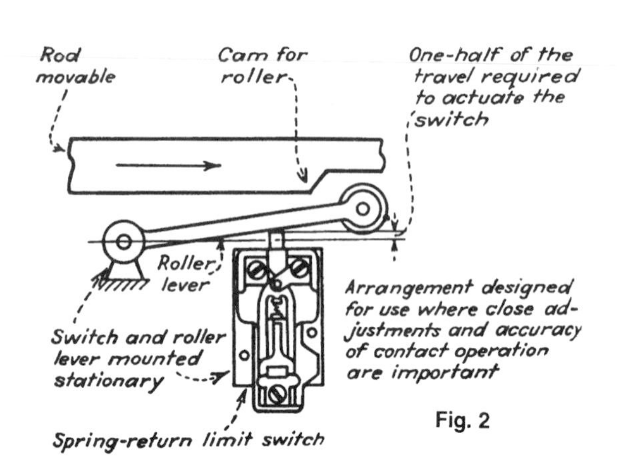

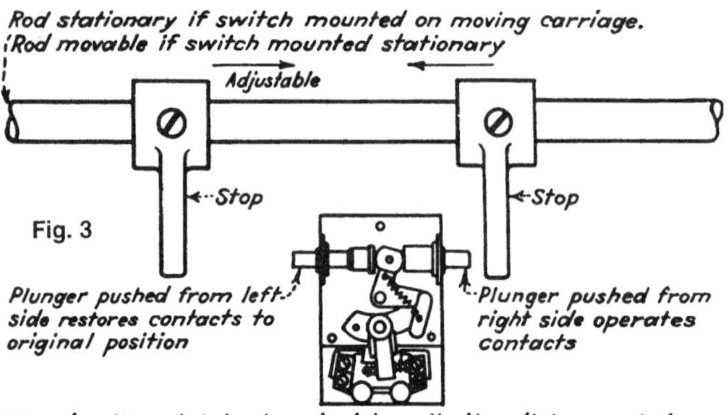

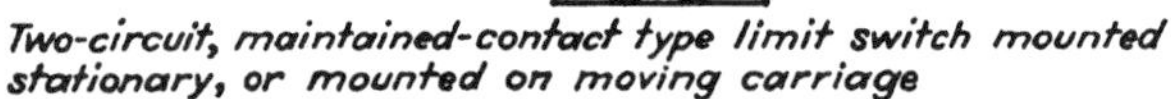

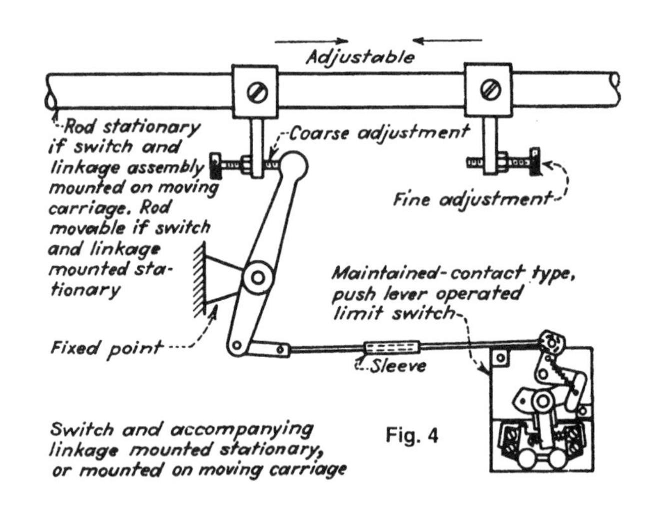

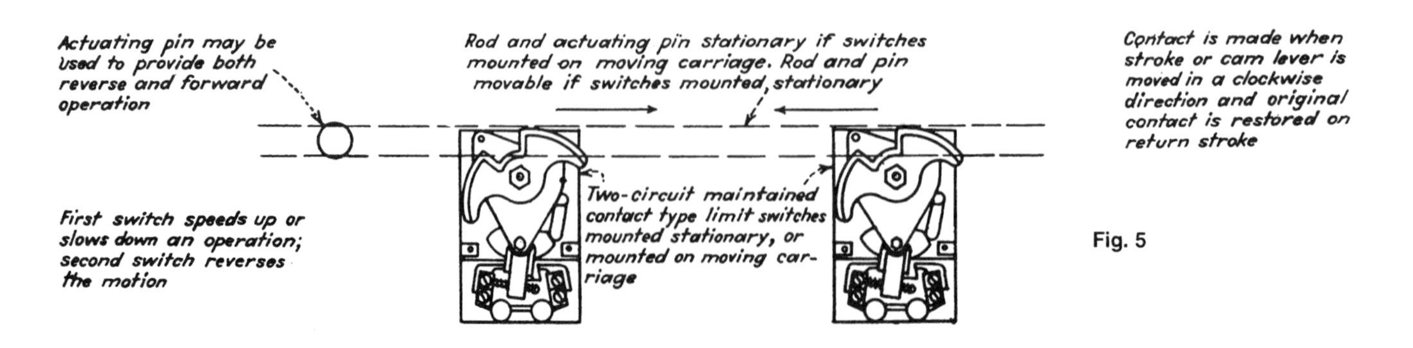

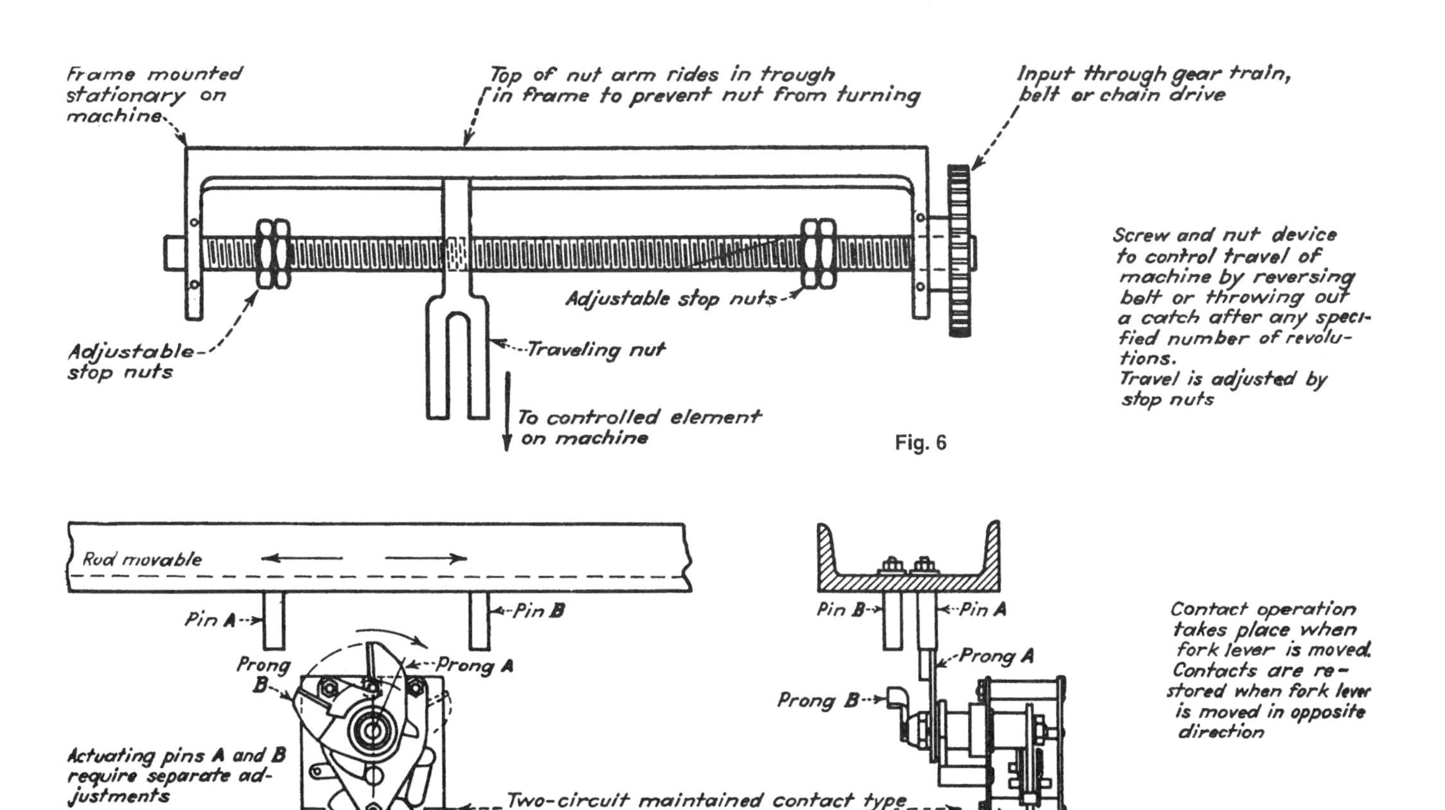

Fig. 6

Fig. 7

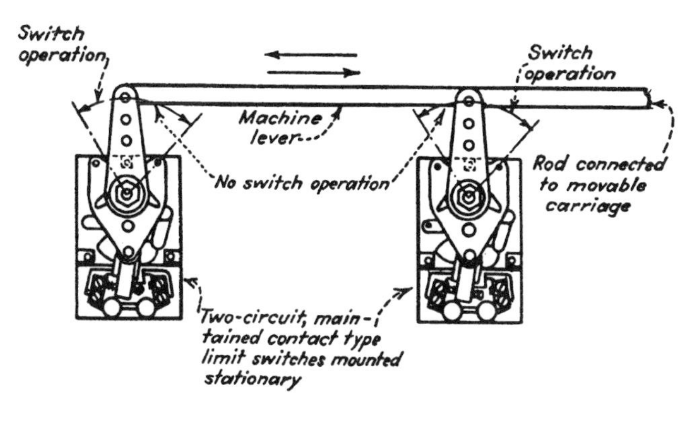

Fig. 8

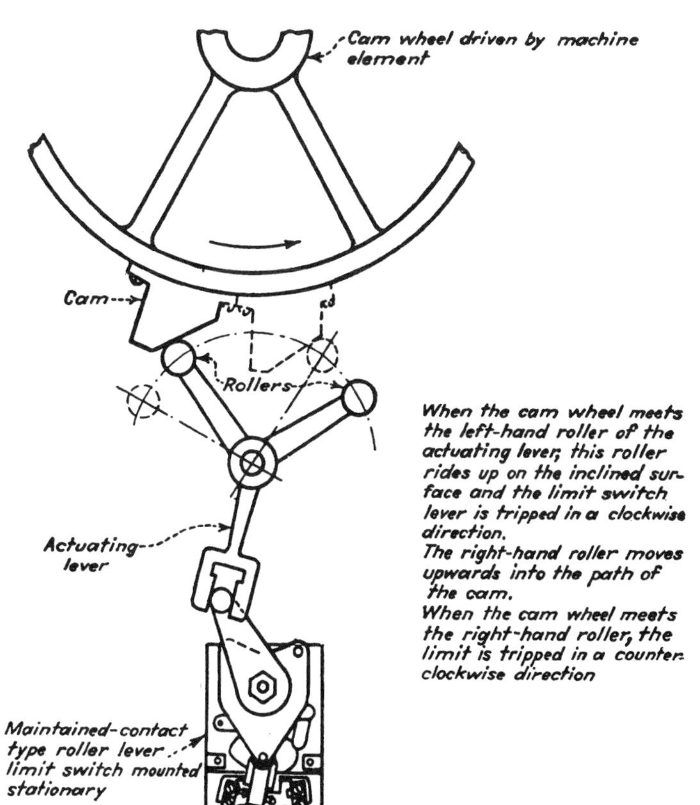

Fig. 9

Electrical contact arrangements

All contracts in normal position with limit switch unactuated

SINGLE POLE

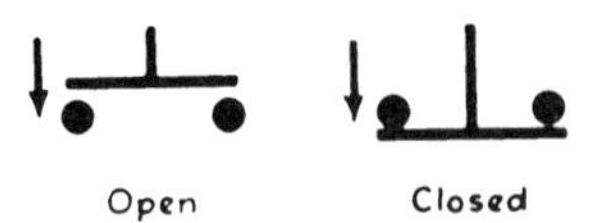

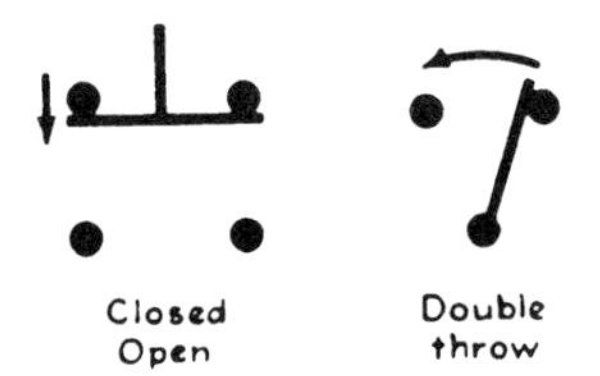

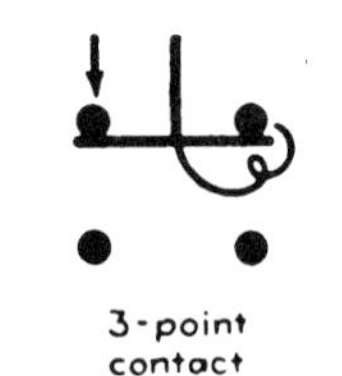

TWO POLE

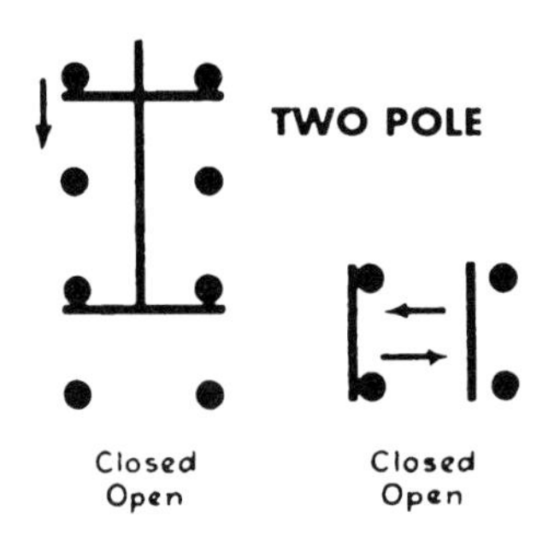

MULTI-CONTACT

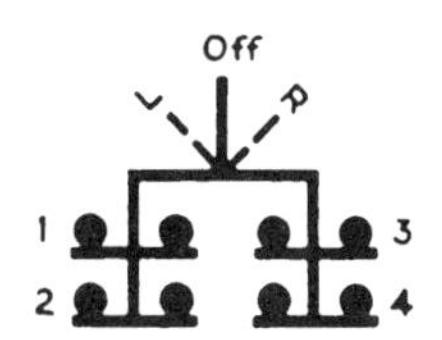

POS.	1	2	3	4
R	C	C	O	O
Off	C	C	C	C
L	O	O	C	C

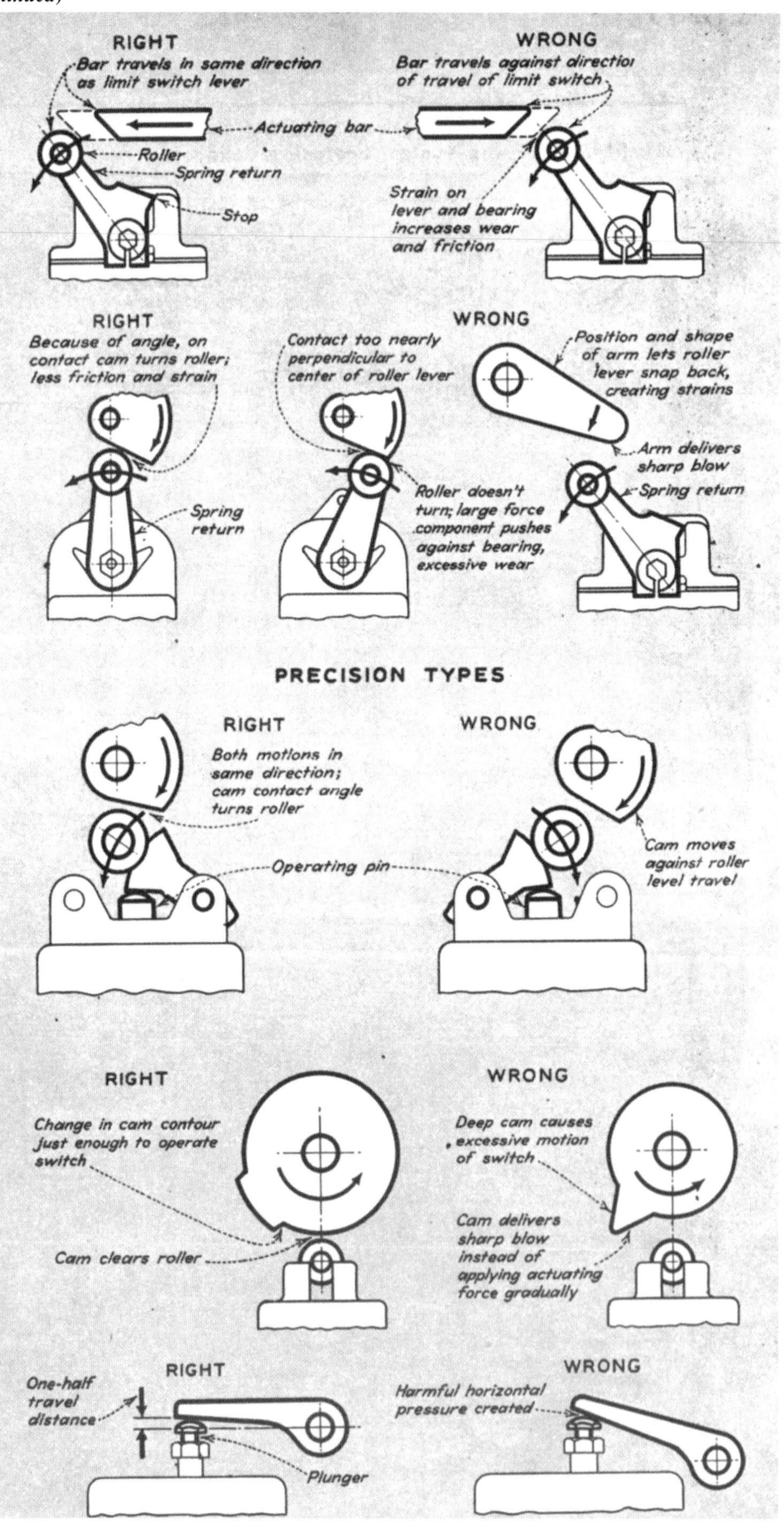

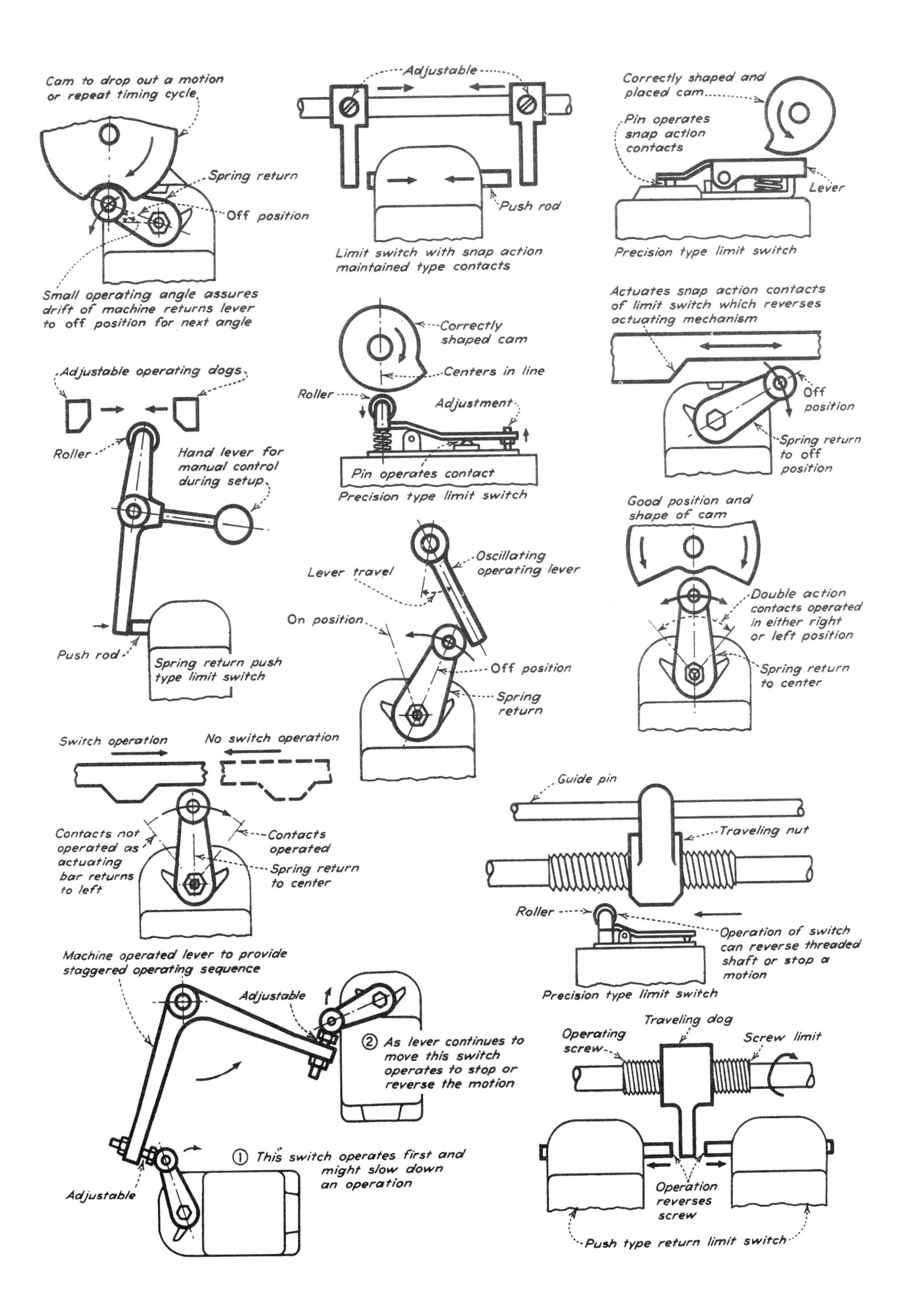

Cam to drop out a motion or repeat timing cycle
Spring return
Off position
Small operating angle assures drift of machine returns lever to off position for next angle
Adjustable
Push rod
Limit switch with snap action maintained type contacts
Correctly shaped and placed cam
Pin operates snap action contacts
Lever
Precision type limit switch
Actuates snap action contacts of limit switch which reverses actuating mechanism
Off position
Spring return to off position
Correctly shaped cam
Centers in line
Roller
Adjustment
Pin operates contact
Precision type limit switch
Adjustable operating dogs
Roller
Hand lever for manual control during setup
Push rod
Spring return push type limit switch
Good position and shape of cam
Double action contacts operated in either right or left position
Spring return to center
Oscillating operating lever
Lever travel
On position
Off position
Spring return
Switch operation
No switch operation
Contacts not operated as actuating bar returns to left
Contacts operated
Spring return to center
Guide pin
Traveling nut
Roller
Operation of switch can reverse threaded shaft or stop a motion
Precision type limit switch
Machine operated lever to provide staggered operating sequence
Adjustable
② As lever continues to move this switch operates to stop or reverse the motion
① This switch operates first and might slow down an operation
Adjustable
Traveling dog
Operating screw
Screw limit
Operation reverses screw
Push type return limit switch

NINE AUTOMATIC SPEED GOVERNORS

Speed governors, designed to maintain the speeds of machines within reasonably constant limits, regardless of loads, depend for their action upon centrifugal force or cam linkages. Other governors depend on pressure differentials and fluid velocities for their actuation.

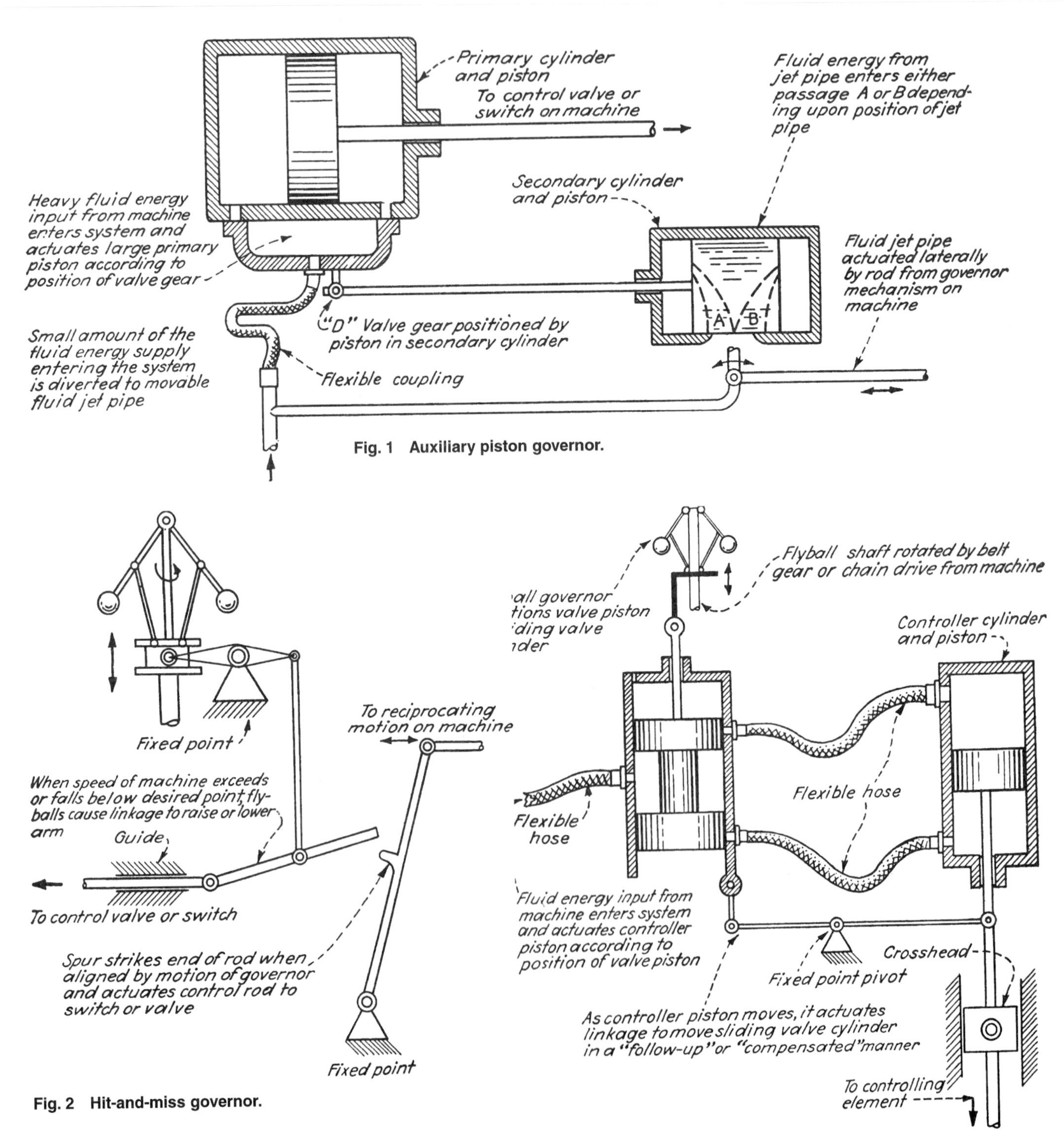

Fig. 1 Auxiliary piston governor.

Fig. 2 Hit-and-miss governor.

Fig. 3 Force-compensated regulator.

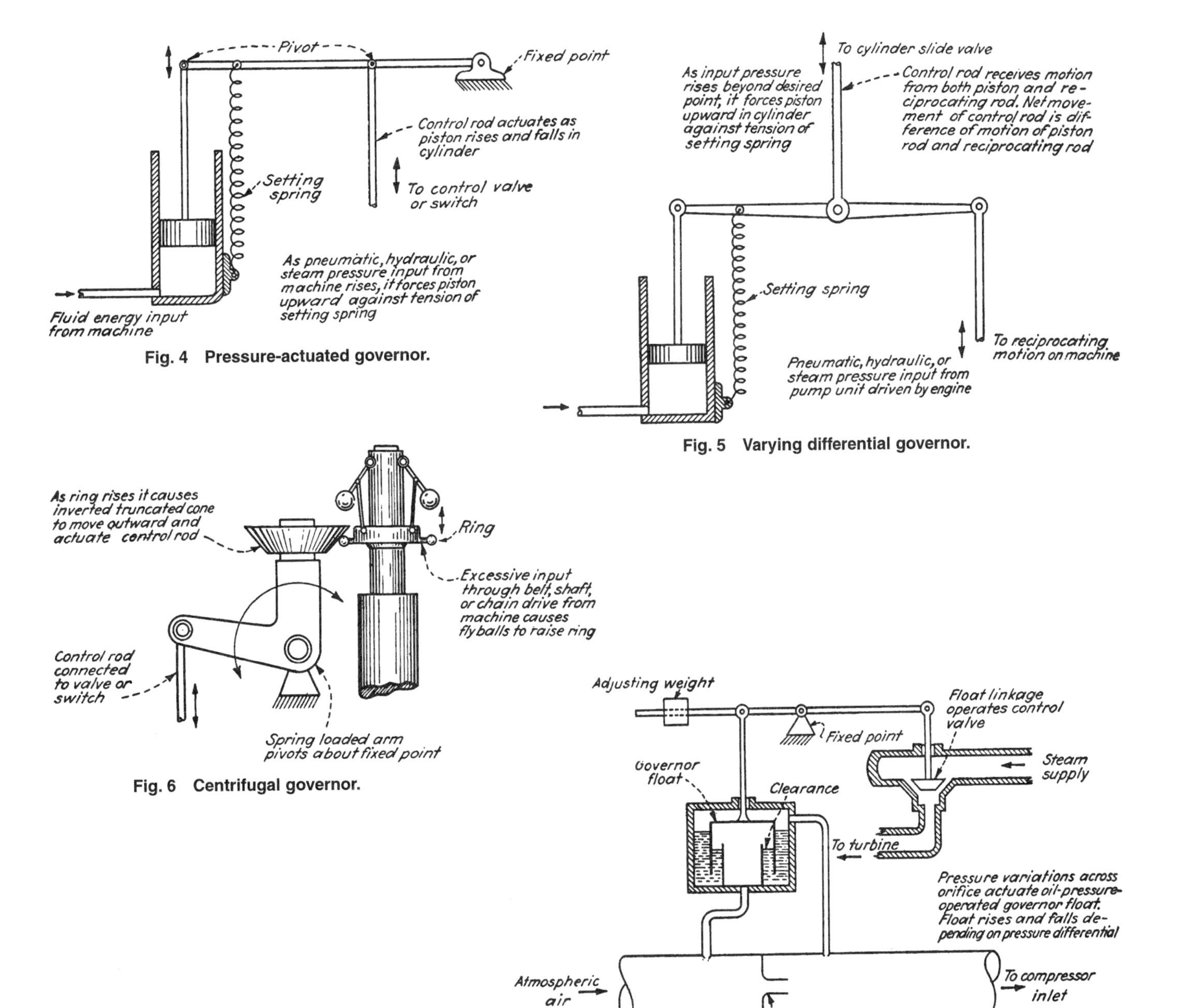

Fig. 4 Pressure-actuated governor.

Fig. 5 Varying differential governor.

Fig. 6 Centrifugal governor.

Fig. 7 Constant-volume governor.

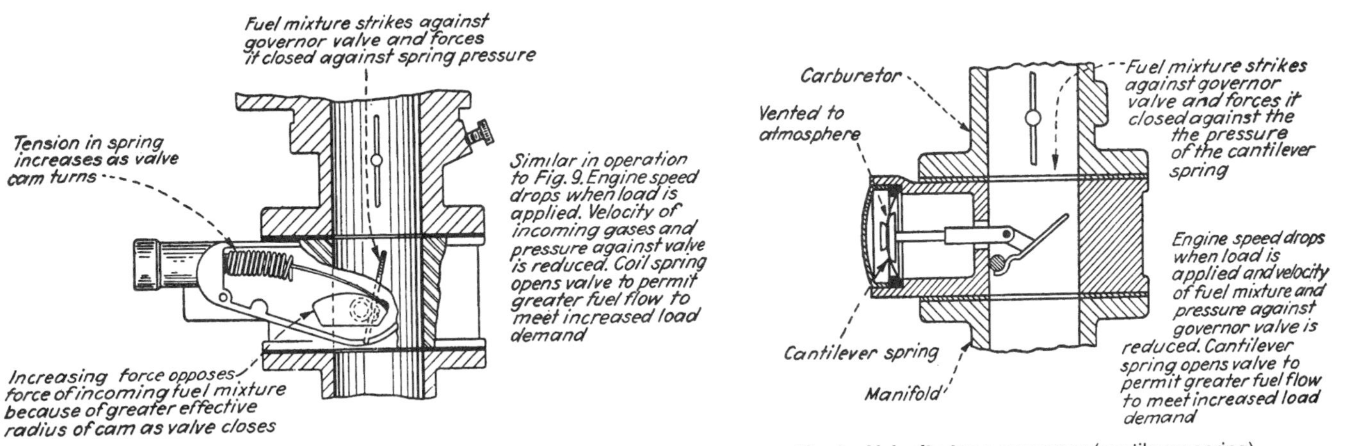

Fig. 8 Velocity-type governor (coil spring).

Fig. 9 Velocity-type governor (cantilever spring).

EIGHT SPEED CONTROL DEVICES FOR MECHANISMS

Friction devices, actuated by centrifugal force, automatically keep speed constant regardless of variations of load or driving force.

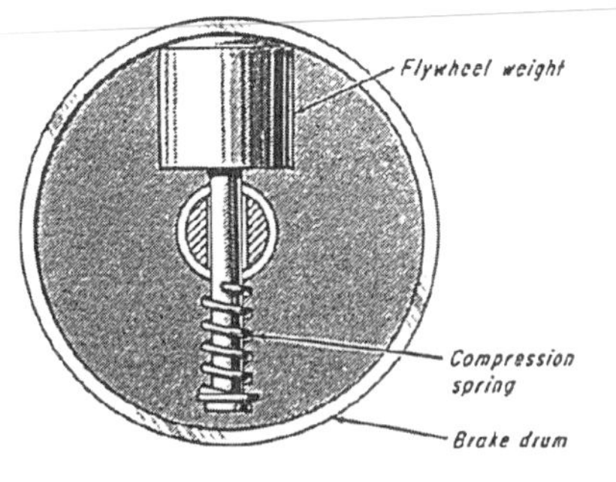

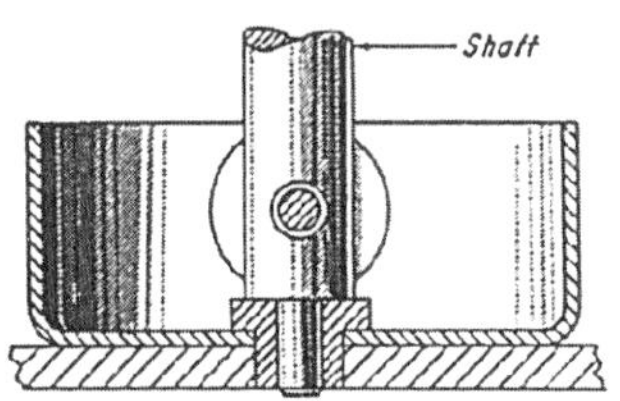

The weight is counterbalanced by a spring that brakes the shaft when the rotation speed becomes too fast. The braking surface is small.

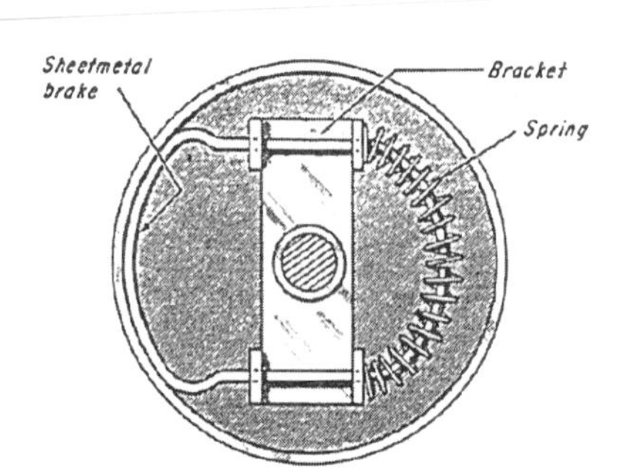

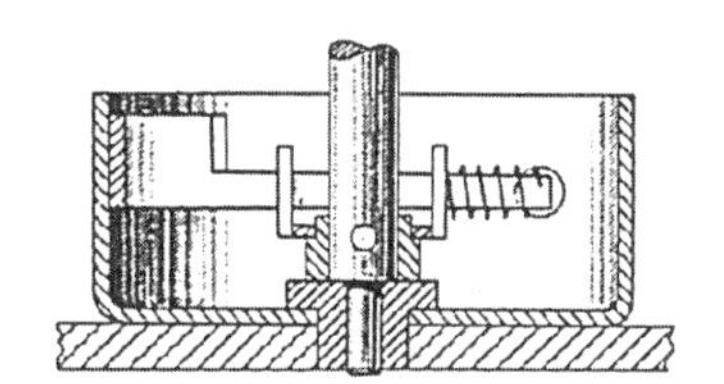

A sheetmetal brake provides a larger braking surface than in the previous brake. Braking is more uniform, and it generates less heat.

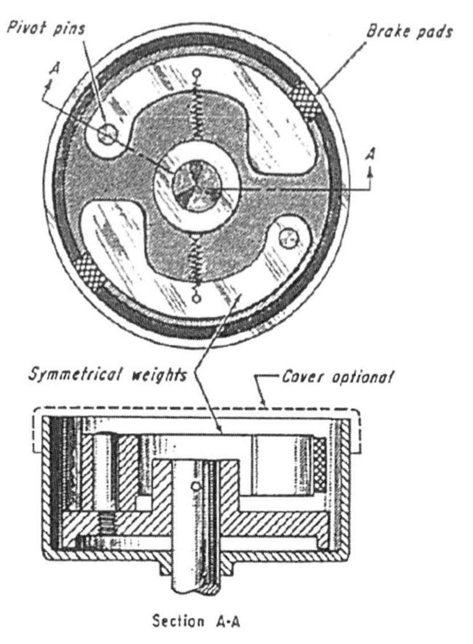

Symmetrical weights give an even braking action when they pivot outward. The entire action can be enclosed in a case.

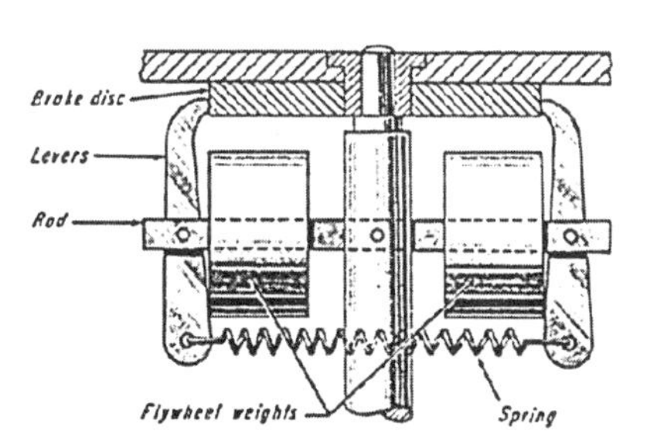

Weight-actuated levers make this arrangement suitable where high braking moments are required.

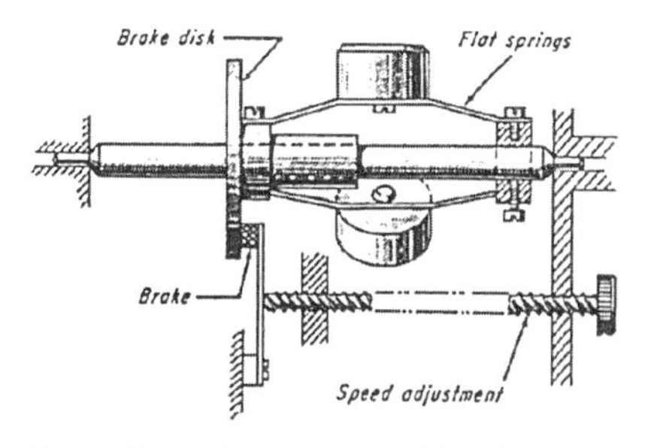

Three flat springs carry weights that provide a brake force upon rotation. A speed adjustment can be included.

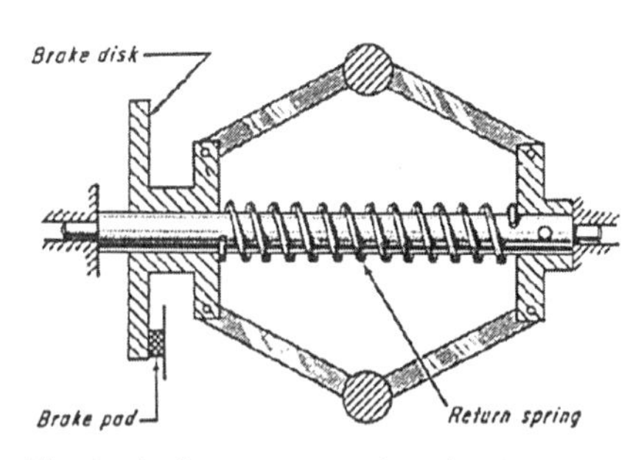

The typical governor action of swinging weights is applied here. As in the previous brake, adjustment is optional.

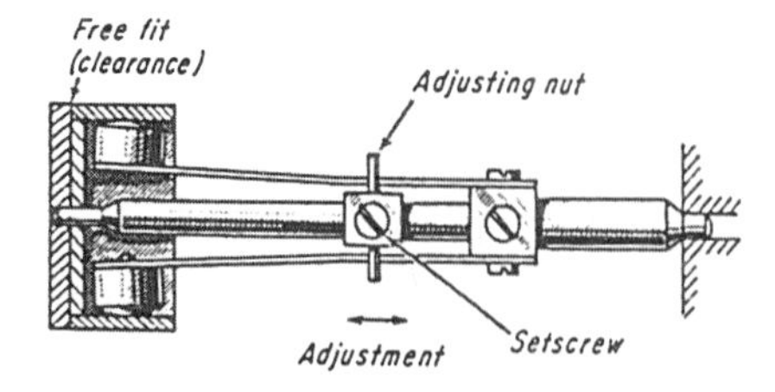

The adjustment of the speed at which this device starts to brake is quick and easy. The adjusting nut is locked in place with a setscrew.

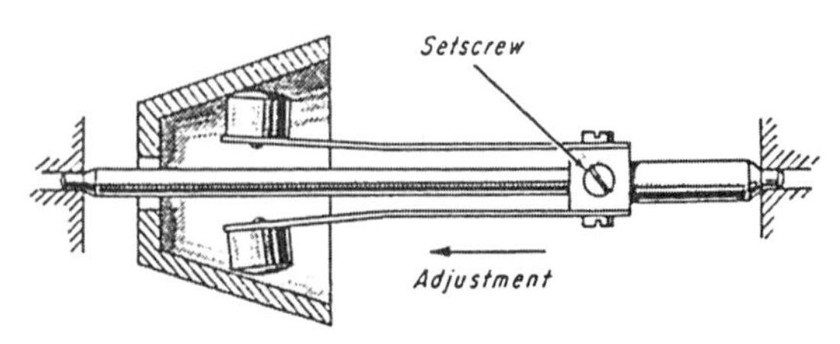

A tapered brake drum is another way to provide for varying speed-control. The adjustment is again locked.

CHAPTER 16

INSTRUMENTS AND CONTROLS: PNEUMATIC, HYDRAULIC, ELECTRIC, AND ELECTRONIC

TWENTY-FOUR MECHANISMS ACTUATED BY PNEUMATIC OR HYDRAULIC CYLINDERS

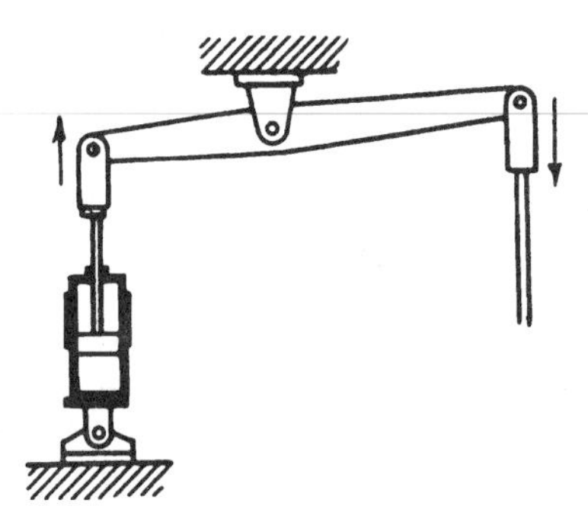

Fig. 1 A cylinder can be used with a first-class lever.

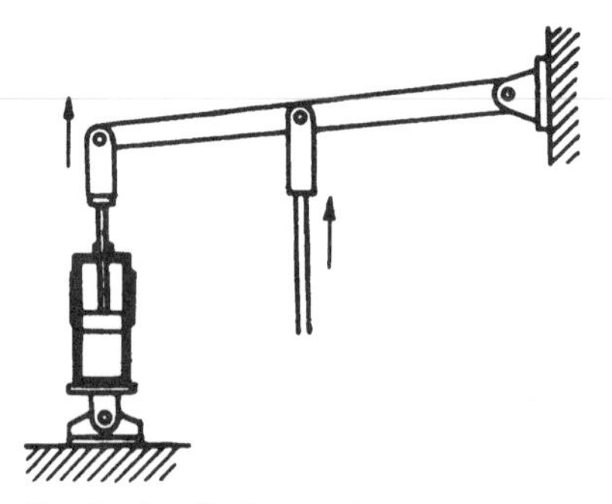

Fig. 2 A cylinder can be used with a second-class lever.

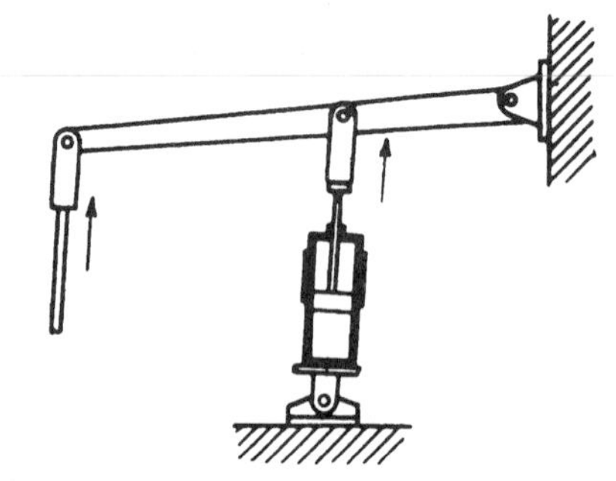

Fig. 3 A cylinder can be used with a third-class lever.

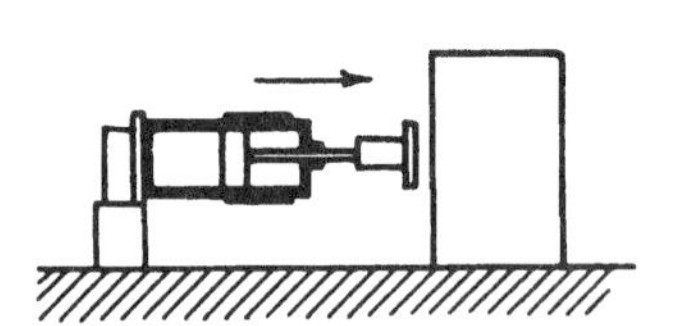

Fig. 4 A cylinder can be linked up directly to the load.

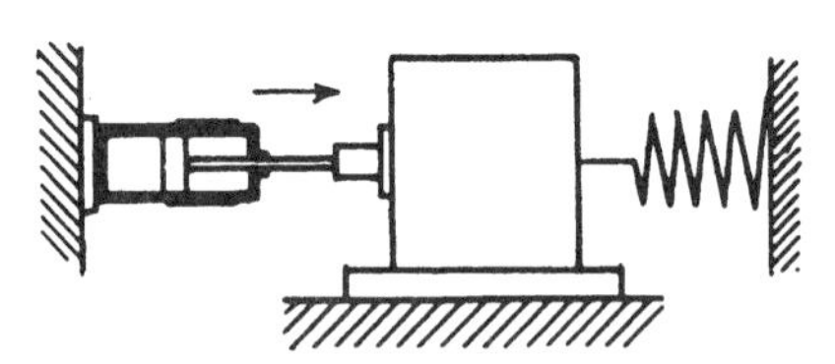

Fig. 5 A spring reduces the thrust at the end of the stroke.

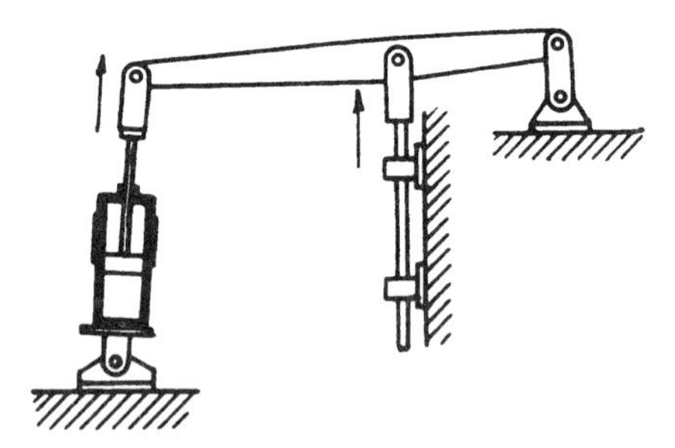

Fig. 6 The point of application of force follows the direction of thrust.

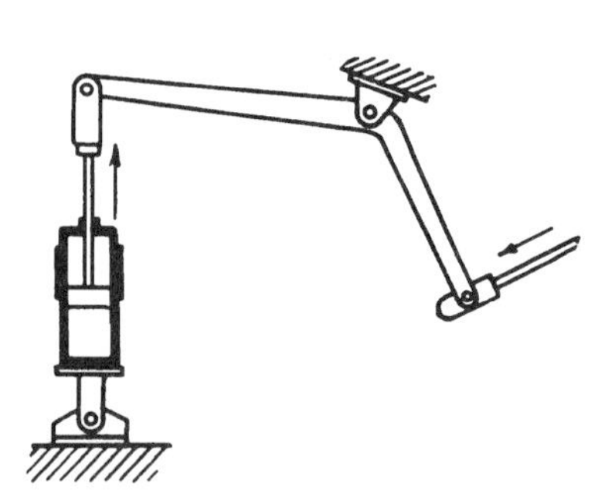

Fig. 7 A cylinder can be used with a bent lever.

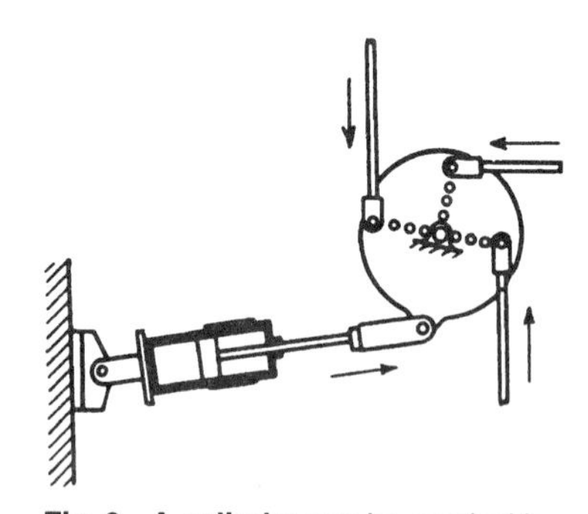

Fig. 8 A cylinder can be used with a trammel plate.

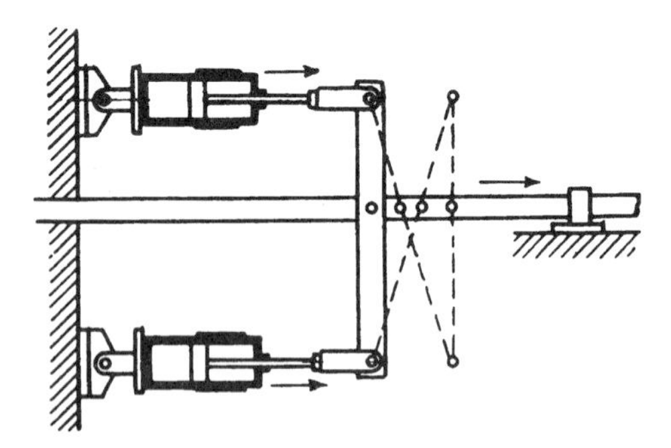

Fig. 9 Two pistons with fixed strokes position the load in any of four stations.

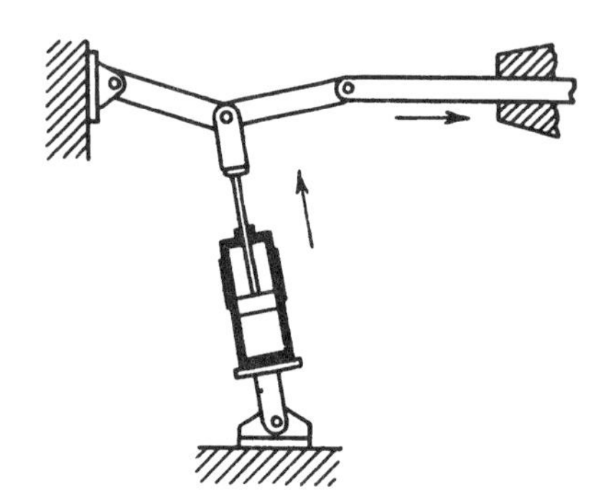

Fig. 10 A toggle can be actuated by the cylinder.

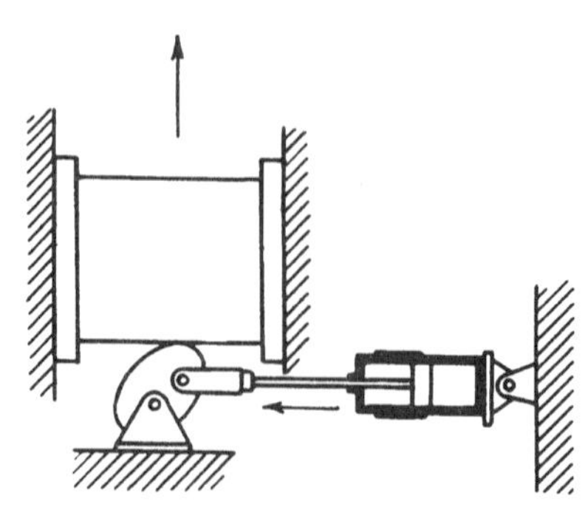

Fig. 11 The cam supports the load after the completion of the stroke.

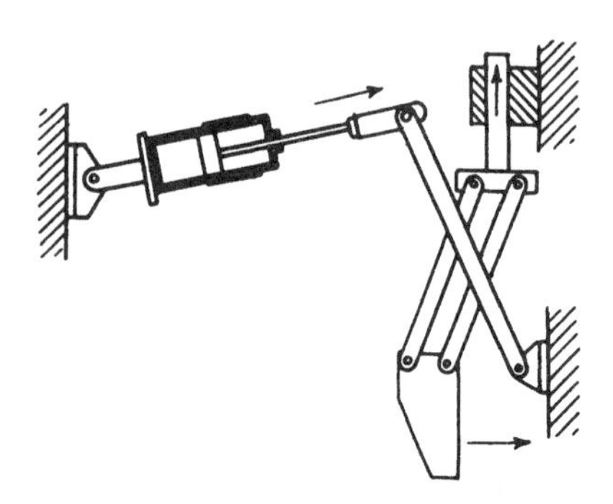

Fig. 12 Simultaneous thrusts in two different directions are obtained.

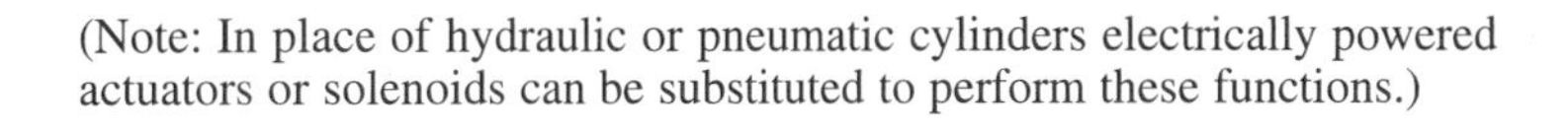
(Note: In place of hydraulic or pneumatic cylinders electrically powered actuators or solenoids can be substituted to perform these functions.)

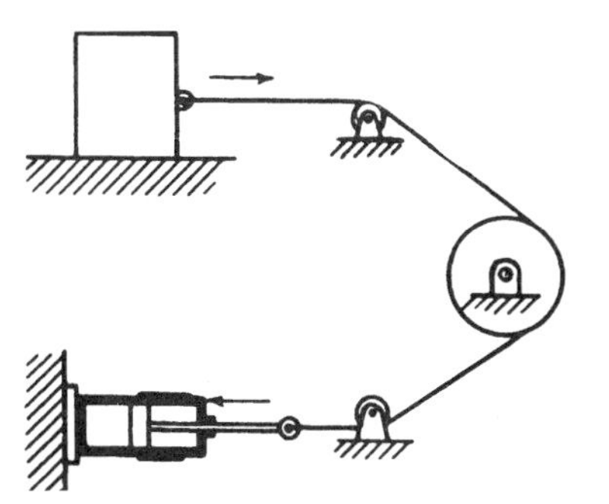

Fig. 13 **A force** is transmitted by a cable.

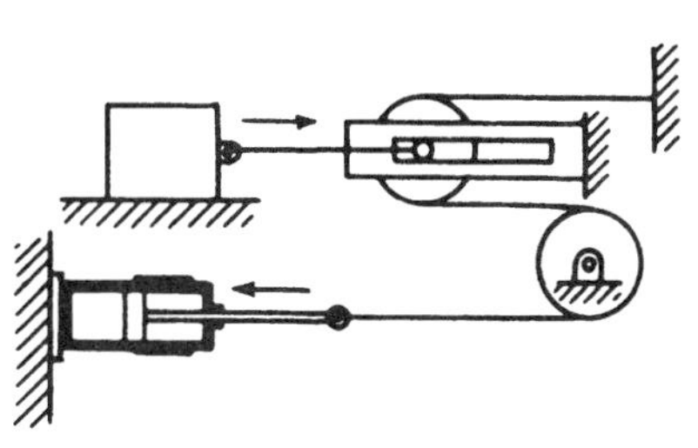

Fig. 14 **A force** can be modified by a system of pulleys.

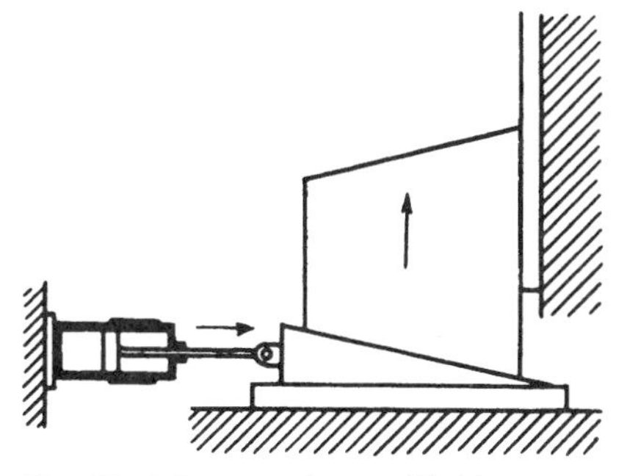

Fig. 15 A force can be modified by wedges.

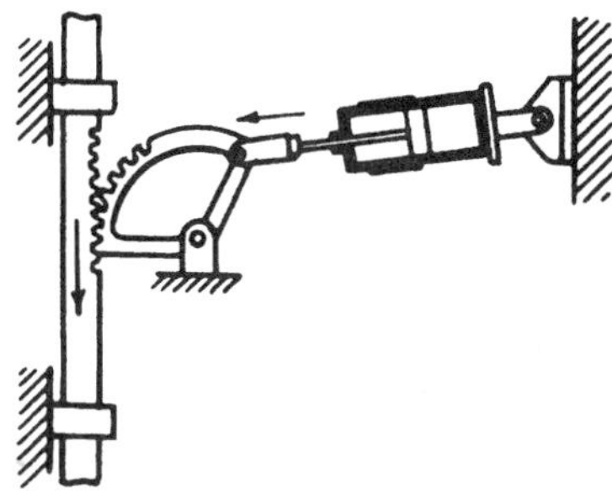

Fig. 16 **A gear sector** moves the rack perpendicular to the piston stroke.

Fig. 17 **A rack** turns the gear sector.

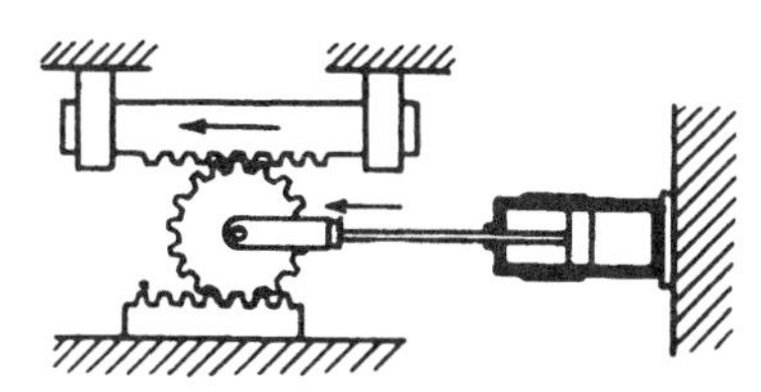

Fig. 18 **The motion** of a movable rack is twice that of the piston.

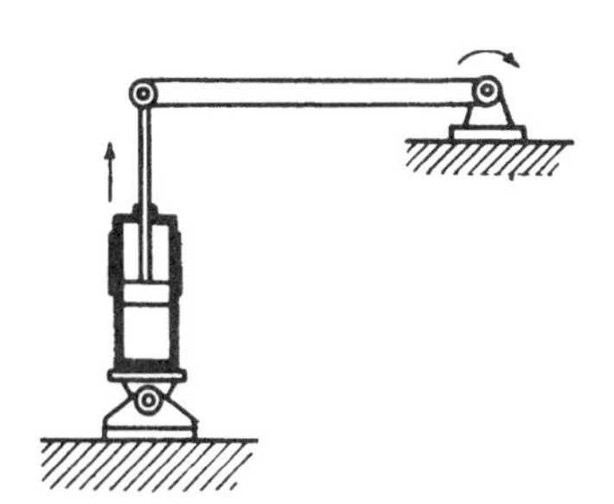

Fig. 19 **A torque** applied to the shaft can be transmitted to a distant point.

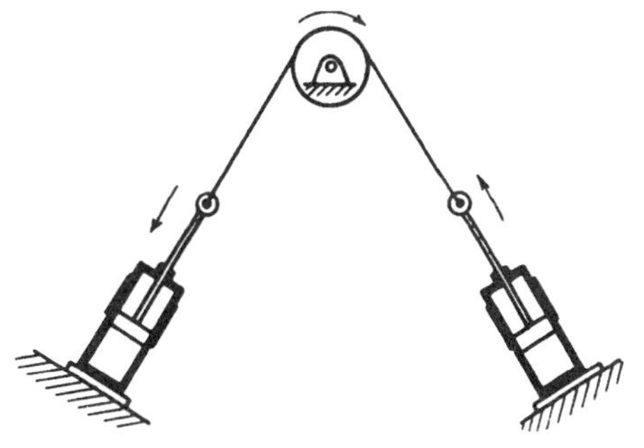

Fig. 20 **A torque** can also be applied to a shaft by a belt and pulley.

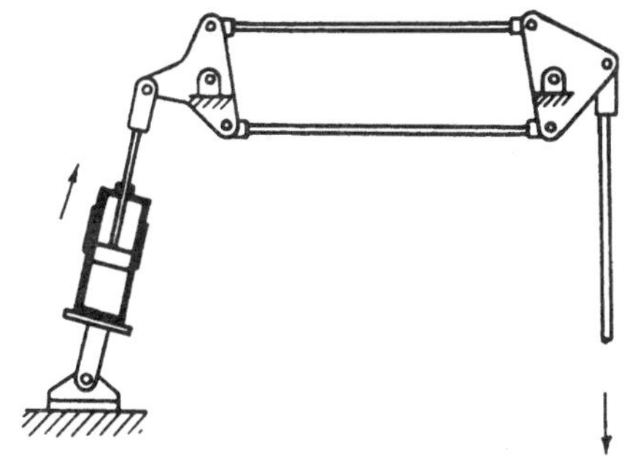

Fig. 21 **A motion** is transmitted to a distant point in the plane of motion.

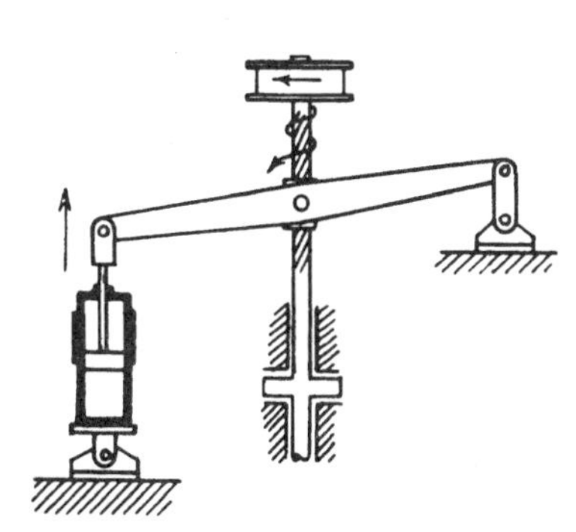

Fig. 22 **A steep screw nut** produces a rotation of the shaft.

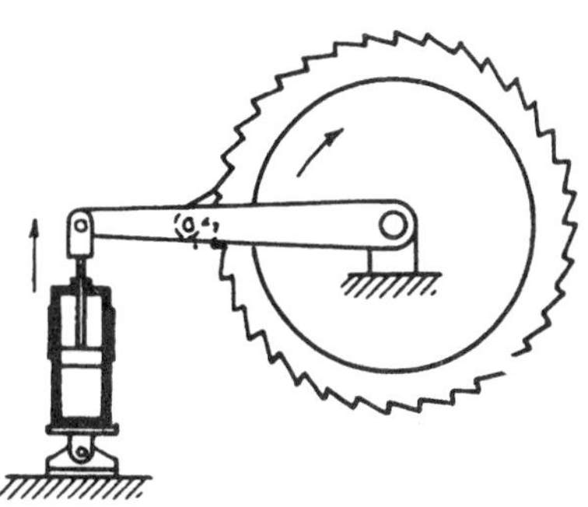

Fig. 23 **A single-sprocket wheel** produces rotation in the plane of motion.

Fig. 24 **A double-sprocket wheel** makes the rotation more nearly continuous.

FOOT-CONTROLLED BRAKING SYSTEM

This crane braking system (see figure) operates when the main line switch closes. The full depression of the master-cylinder foot-pedal compresses the brake-setting spring mounted on the hydraulic releasing cylinder. After the setting spring is fully compressed, the hydraulic pressure switch closes, completing the electric circuit and energizing the magnetic check valve. The setting spring remains compressed as long as the magnetic check valve is energized because the check valve traps the fluid in the hydraulic-releasing cylinder. Upon release of the foot peal, the brake lever arm is pulled down by the brake releasing spring, thus releasing the brake shoes.

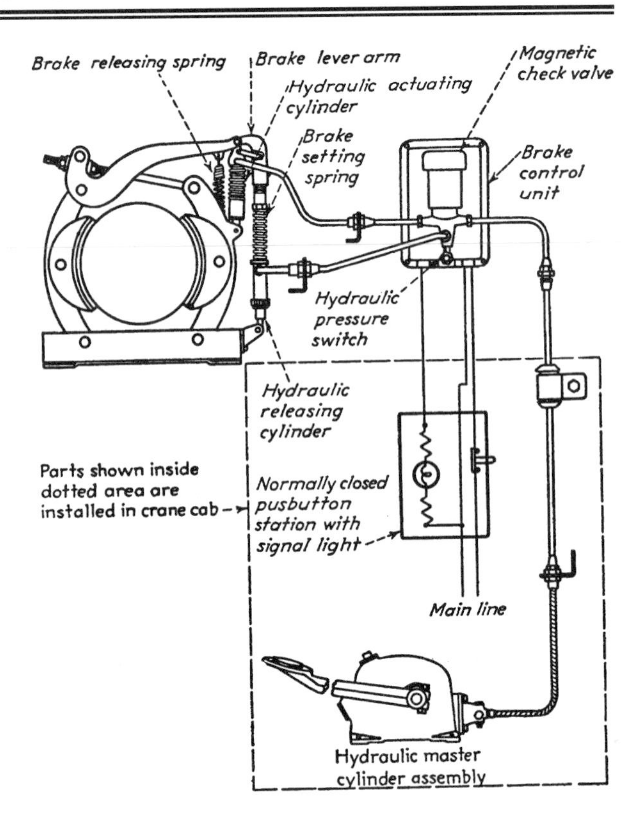

FIFTEEN TASKS FOR PNEUMATIC POWER

Suction can feed, hold, position, and lift parts, form plastic sheets, sample gases, test for leaks, convey solids, and deaerate liquids. Compressed air can convey materials, atomize and agitate liquids, speed heat transfer, support combustion, and protect cable.

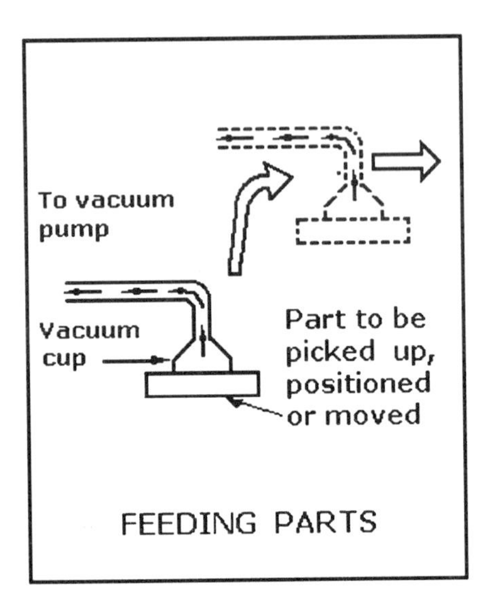

FEEDING PARTS

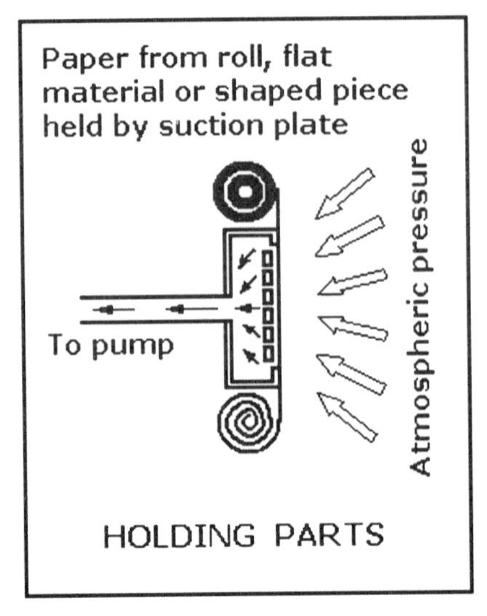

HOLDING PARTS

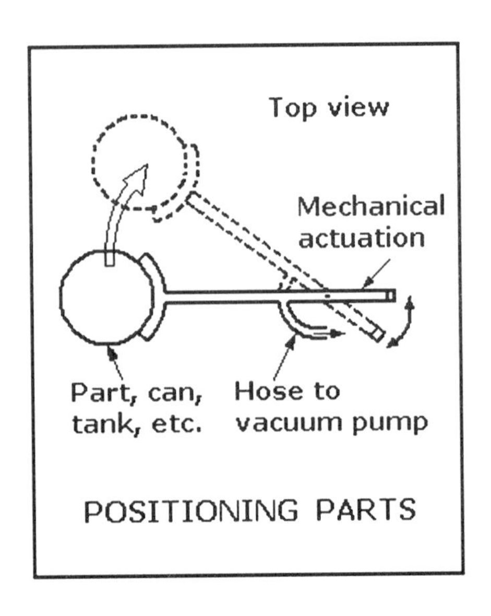

POSITIONING PARTS

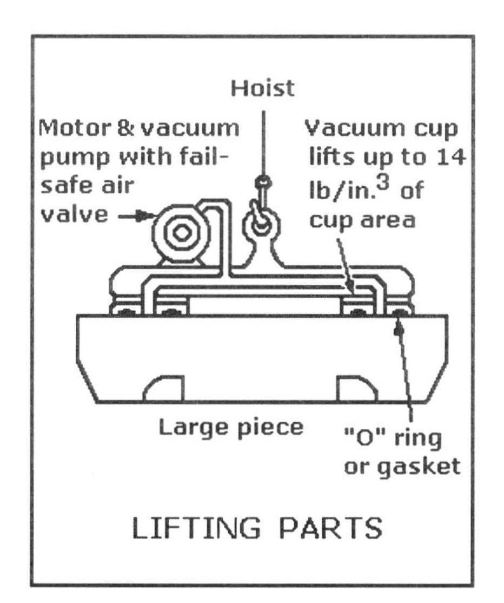
Hoist
Motor & vacuum pump with fail-safe air valve
Vacuum cup lifts up to 14 lb/in.3 of cup area
Large piece
"O" ring or gasket
LIFTING PARTS

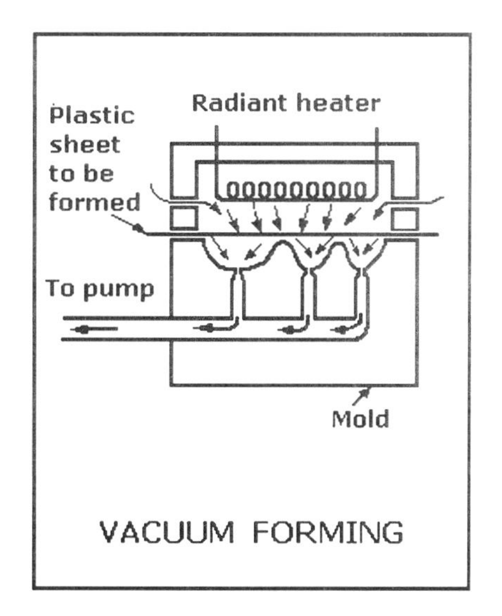
Plastic sheet to be formed
Radiant heater
To pump
Mold
VACUUM FORMING

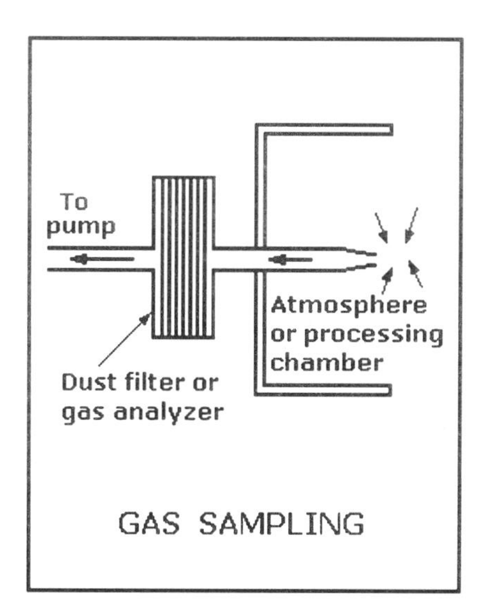
To pump
Atmosphere or processing chamber
Dust filter or gas analyzer
GAS SAMPLING

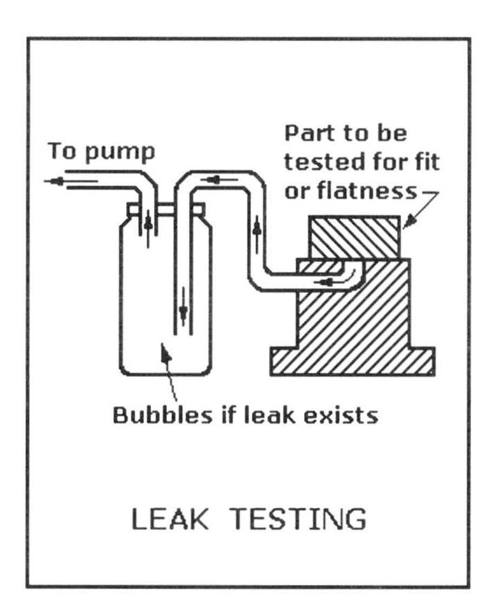
To pump
Part to be tested for fit or flatness
Bubbles if leak exists
LEAK TESTING

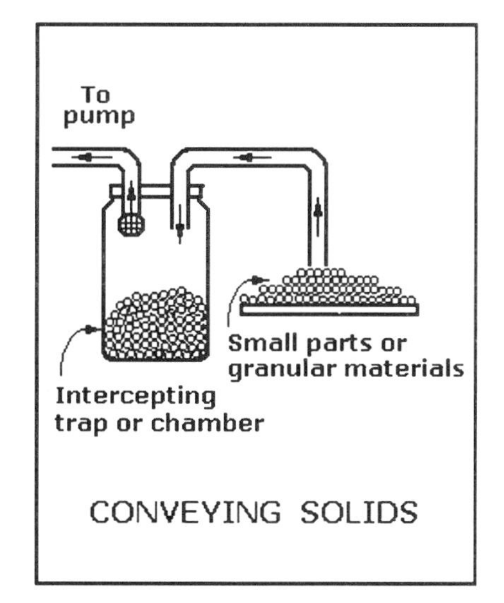
To pump
Small parts or granular materials
Intercepting trap or chamber
CONVEYING SOLIDS

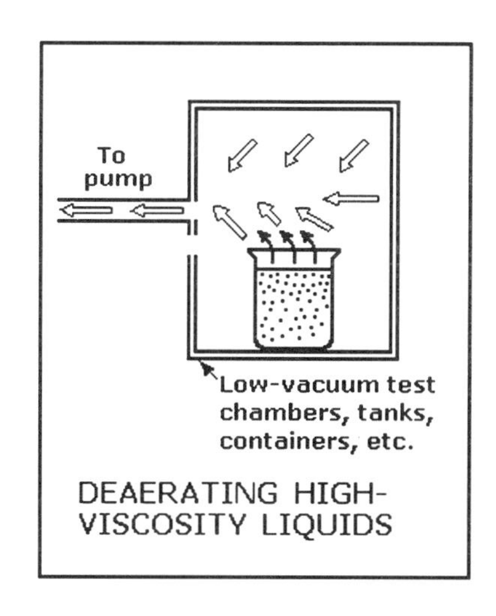
To pump
Low-vacuum test chambers, tanks, containers, etc.
DEAERATING HIGH-VISCOSITY LIQUIDS

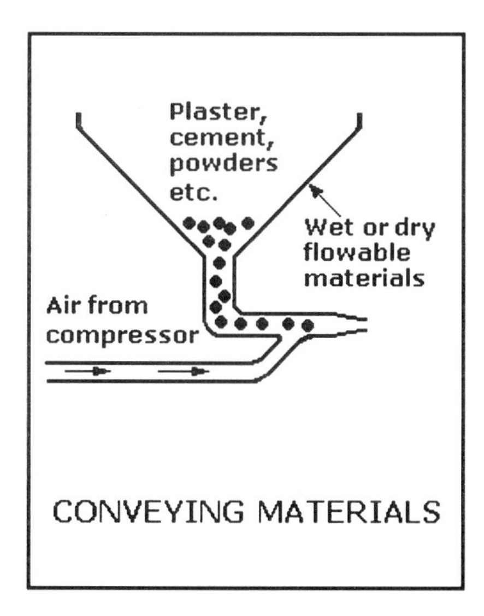
Plaster, cement, powders etc.
Wet or dry flowable materials
Air from compressor
CONVEYING MATERIALS

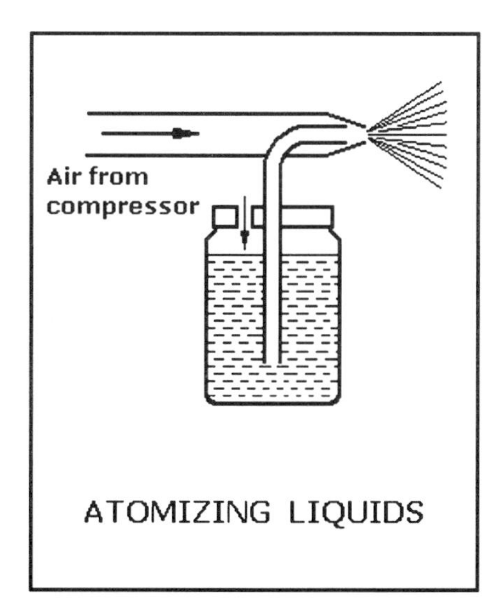
Air from compressor
ATOMIZING LIQUIDS

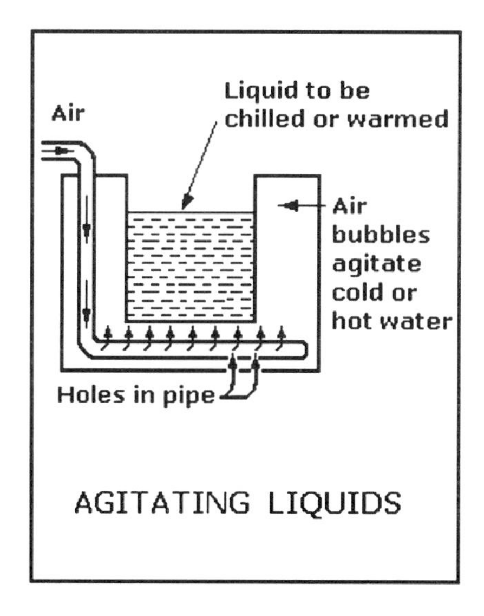
Air
Liquid to be chilled or warmed
Air bubbles agitate cold or hot water
Holes in pipe
AGITATING LIQUIDS

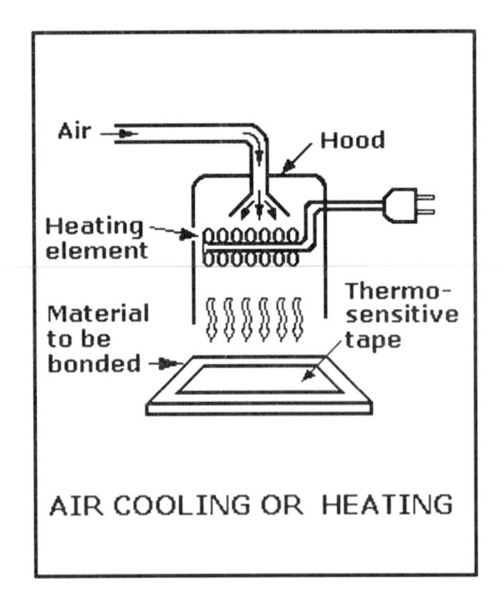

AIR COOLING OR HEATING

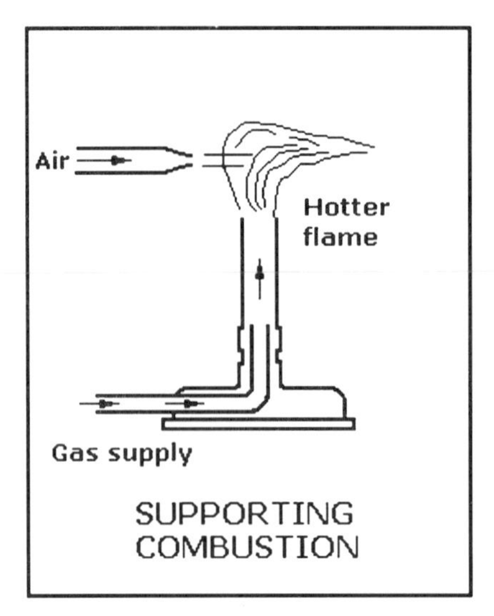

SUPPORTING COMBUSTION

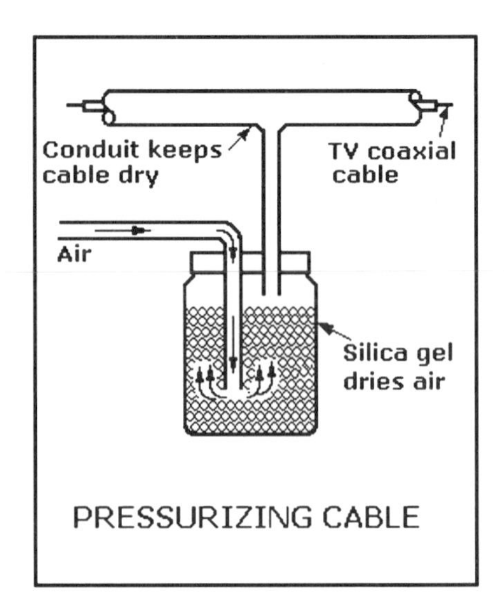

PRESSURIZING CABLE

TEN APPLICATIONS FOR METAL DIAPHRAGMS AND CAPSULES

Fig. 1

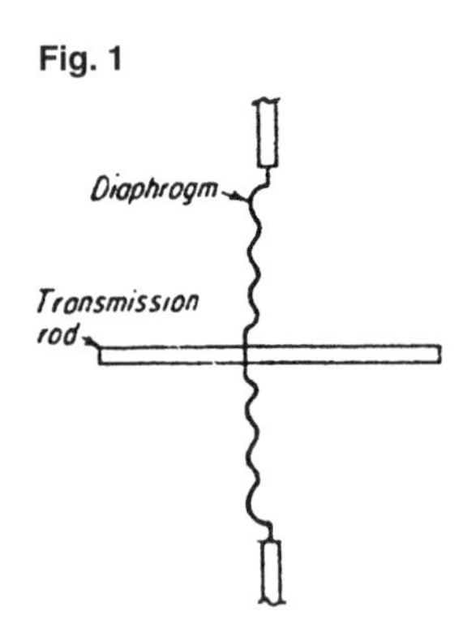

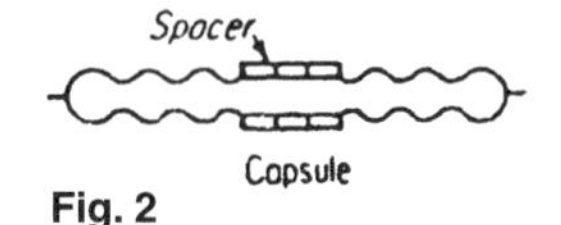

Fig. 2

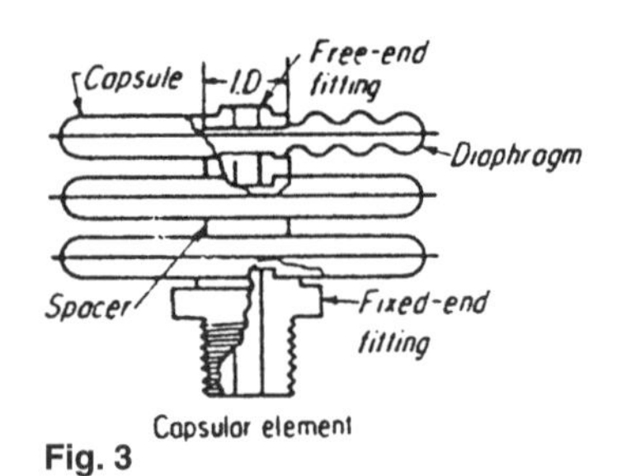

Fig. 3

A metal diaphragm is usually corrugated (Fig. 1) or formed to some irregular profile. It can be used as a flexible seal for an actuating rod. The capsule (Fig. 2) is an assembly of two diaphragms sealed together at their outer edges, usually by soldering, brazing, or welding. Two or more capsules assembled together are known as a capsular element (Fig. 3). End fittings for the capsules vary according to their function; the "fixed end" is fixed to the equipment. The "free end" moves the related components and linkages. The nested capsule (Fig. 4) requires less space and can be designed to withstand large external overpressures without damage.

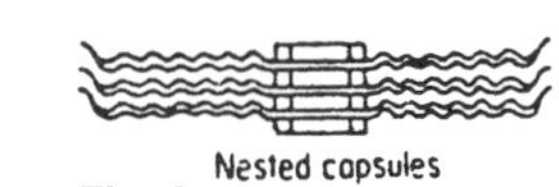

Fig. 4

Fig. 5

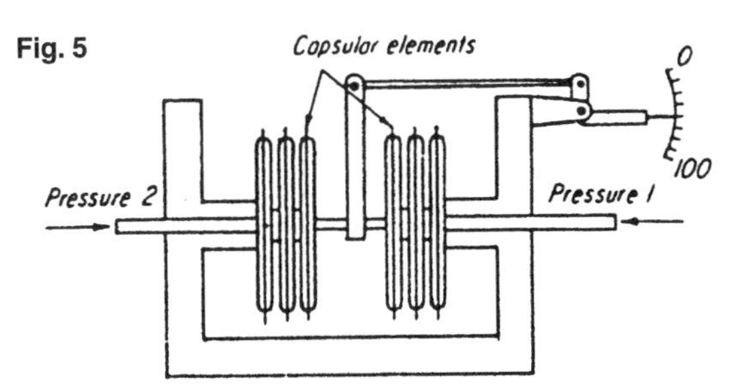

A differential pressure gage (Fig. 5) with opposing capsules can have either single or multicapsular elements. The multicapsular type gives greater movement to the indicator. Capsules give improved linearity over bellows for such applications as pressure-measuring devices. The force exerted by any capsule is equal to the total effective area of the diaphragms (about 40% actual area) multiplied by the pressure exerted on it. Safe pressure is the maximum pressure that can be applied to a diaphragm before hysteresis or set become objectionable.

Fig. 6

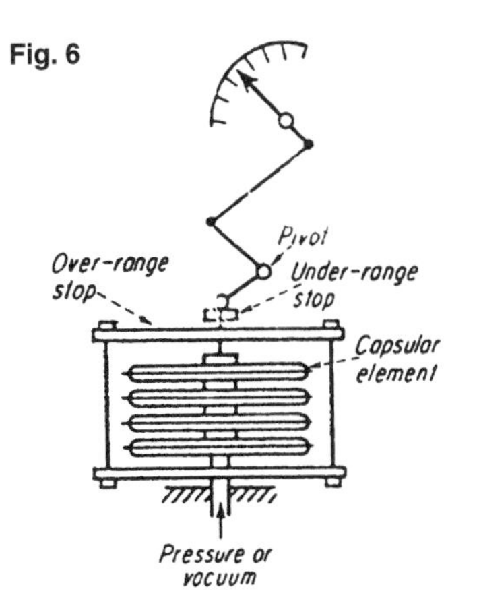

A pressure gage (Fig. 6) has a capsular element linked to a dial indicator by a three-bar linkage. Such a gage measures pressure or vacuum relative to prevailing atmospheric pressure. If greater angular motion of the indicator is required than can be obtained from the three-bar linkage, a quadrant and gear can be substituted.

Fig. 7

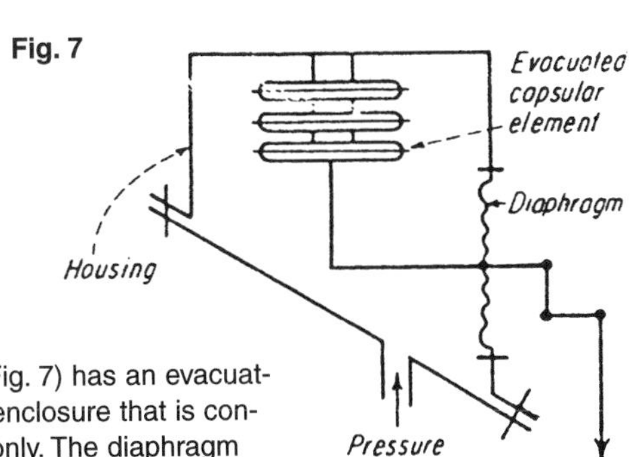

An absolute pressure gage (Fig. 7) has an evacuated capsular element inside an enclosure that is connected to the pressure source only. The diaphragm allows the linkage movement from the capsule to pass through a sealed chamber. This arrangement can also be used as a differential pressure gage by making a second pressure connection to the interior of the element.

Fig. 8

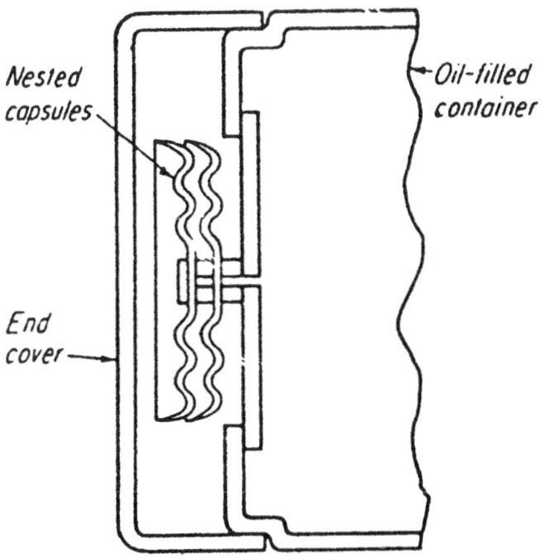

An expansion compensator (Fig. 8) for oil-filled systems takes up less space when the capsules are nested. In this application, one end of the capsule is open and connected to oil in the system; the other end is sealed. Capsule expansion prevents the internal oil pressure from increasing dangerously from thermal expansion. The capsule is protected by its end cover.

Fig. 9

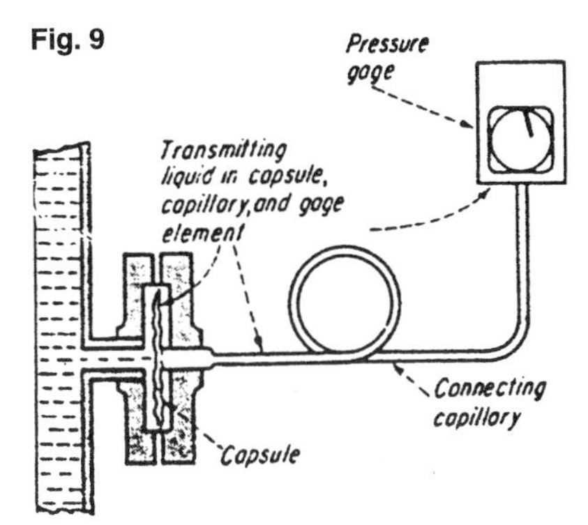

A capsule pressure-seal (Fig. 9) works like a thermometer system except that the bulb is replaced by a pressure-sensitive capsule. The capsule system is filled with a liquid such as silicone oil and is self-compensating for ambient and operating temperatures. When subjected to external pressure changes, the capsule expands or contracts until the internal system pressure just balances the external pressure surrounding the capsule.

Fig. 10

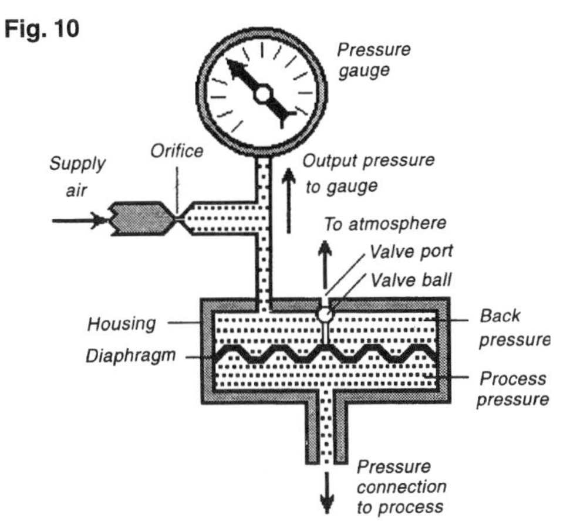

A force-balanced seal (Fig. 10) solves the problem, as in the seal of Fig. 9, for keeping corrosive, viscous or solids-bearing fluids out of the pressure gage. The air pressure on one side of a diaphragm is controlled so as to balance the other side of the diaphragm exactly. The pressure gage is connected to measure this balancing air pressure. The gage, therefore, reads an air pressure that is always exactly equal to the process pressure.

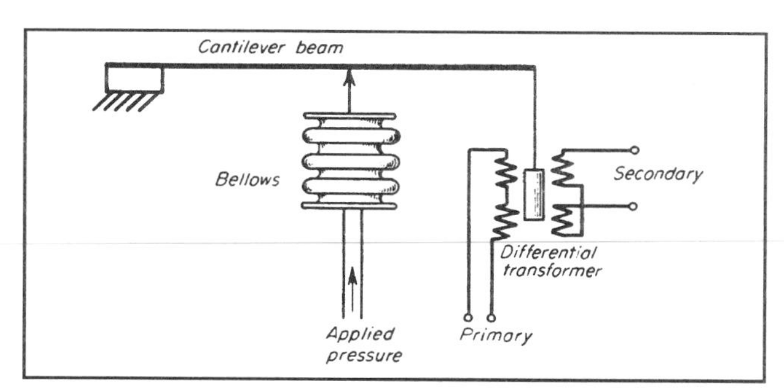

Gage pressure bellows transmitter. The bellows is connected to a cantilever beam with a needle bearing. The beam adopts a different position for every pressure; the transformer output varies with beam position. The bellows are available for ranges from 0–10 in. to 0–200 in. of water for pressure indication or control.

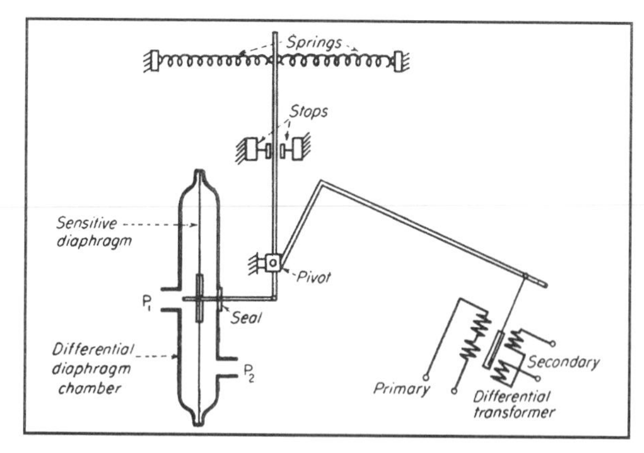

Differential diaphragm pressure transmitter. Differential pressures P_1 and P_2 act on the opposite sides of a sensitive diaphragm and move the diaphragm against the spring load. The diaphragm displacement, spring extension, and transformer core movement are proportional to the difference in pressure. The device can measure differentials as low as 0.005 in. of water. It can be installed as the primary element in a differential pressure flowmeter, or in a boiler windbox for a furnace-draft regulator.

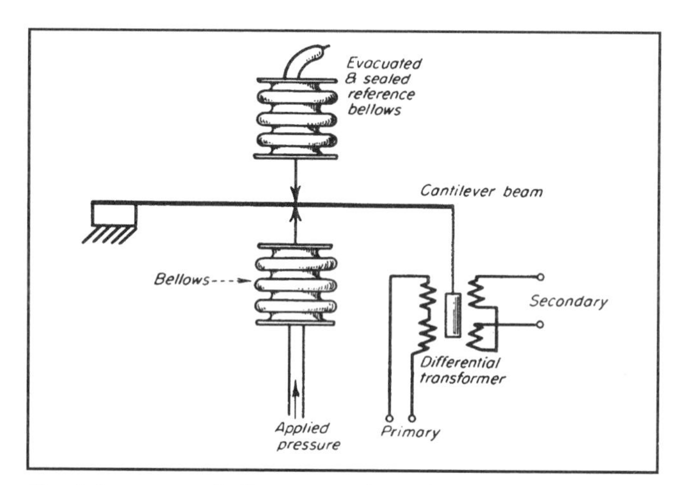

Absolute pressure bellows transmitter. This transmitter is similar to the differential diaphragm transmitter except for addition of a reference bellows which is evacuated and sealed. It can measure negative gage pressures with ranges from 0–50 mm to 0–30 in. of mercury. The reference bellows compensates for variations in atmospheric pressure.

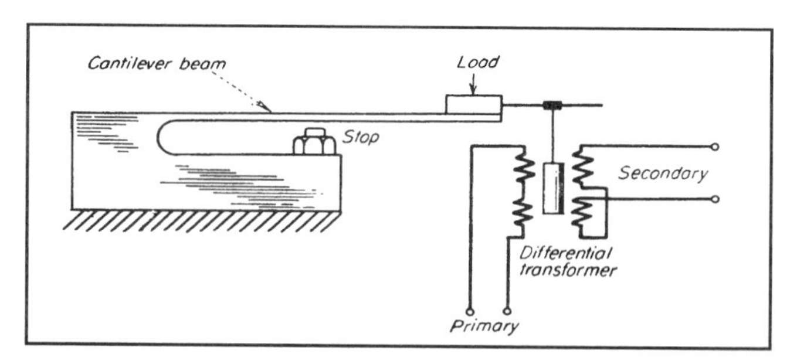

Cantilever load cell. The deflection of a cantilever beam and the displacement of a differential transformer core are proportional to the applied load. The stop prevents damage to the beam in the event of overload. Beams are available for ranges from 0–5 to 0–500 lb. And they can provide precise measurement of either tension or compression forces.

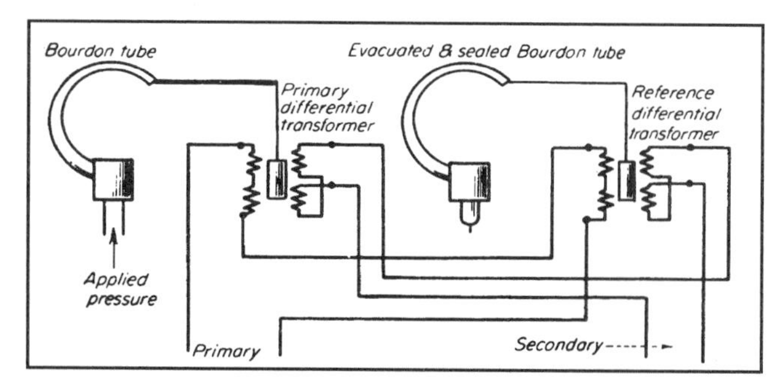

Absolute pressure Bourdon-tube transmitter. This device can indicate or control absolute pressures from 15 to 10,000 psi, depending on tube rating. The reference tube is evacuated and sealed, and compensates for variations in atmospheric pressure by changing the output of the reference differential transformer. The signal output consists of the algebraic sum of the outputs of both the primary and reference differential transformers.

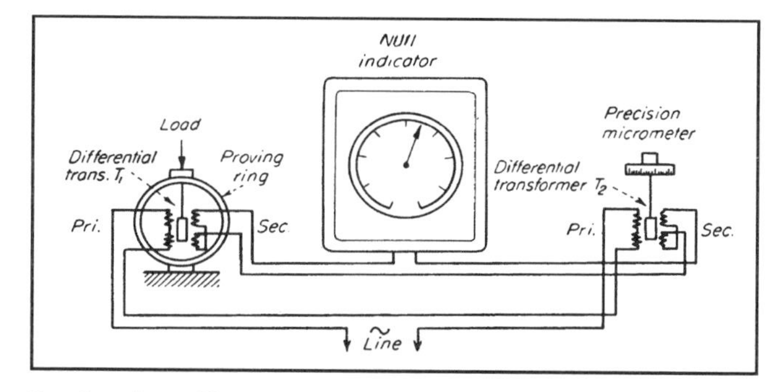

Proving ring. The core of the transmitting transformer, T_1, is fastened to the top of the proving ring, while the windings are stationary. The proving ring and transformer core deflect in proportion to the applied load. The signal output of the balancing transformer, T_2, opposes the output of T_1, so that at the balance point, the null point indicator reads zero. The core of the balancing transformer is actuated by a calibrated micrometer that indicates the proving ring deflection when the differential transformer outputs are equal and balanced.

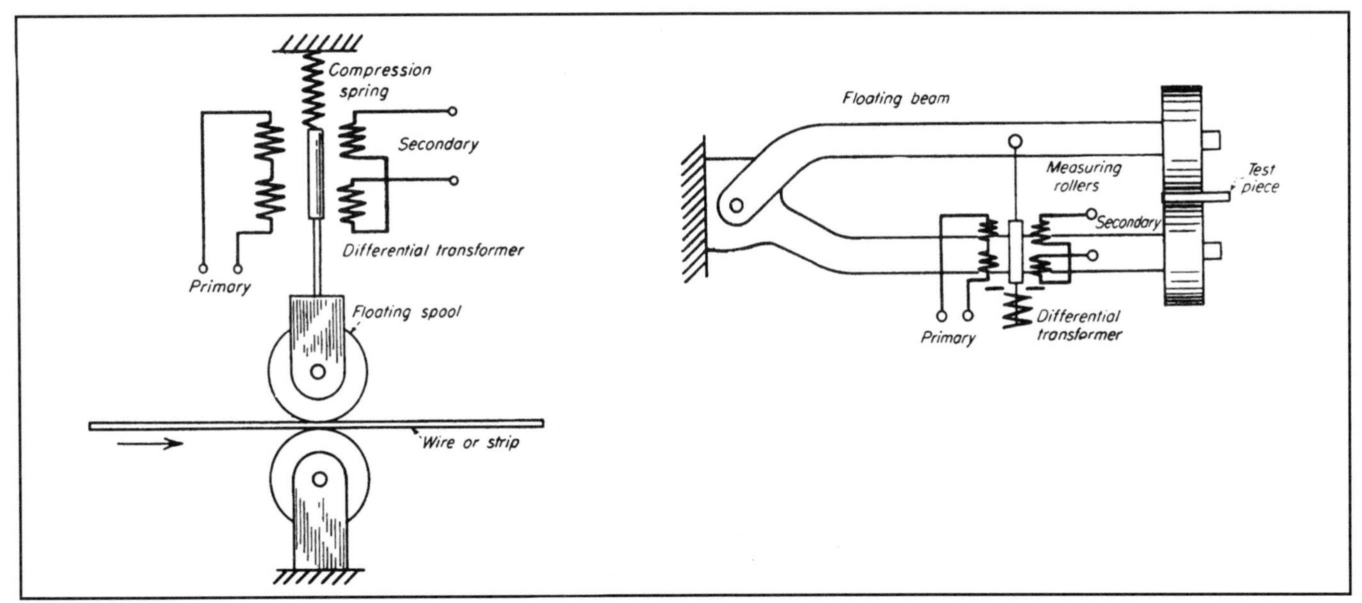

Gaging and calipering. The thickness of a moving wire or strip is gaged by the position of the floating spool and transformer core. If the core is at the null point for proper material thickness, the transformer output phase and magnitude indicate whether the material is too thick or thin and the amount of the error. The signal can be amplified to operate a controller, recorder, or indicator. The device at the right can function as a production caliper or as an accurate micrometer. If the transformer output is fed into a meter indicator with *go* and *no-go* bands, it becomes a convenient device for gaging items.

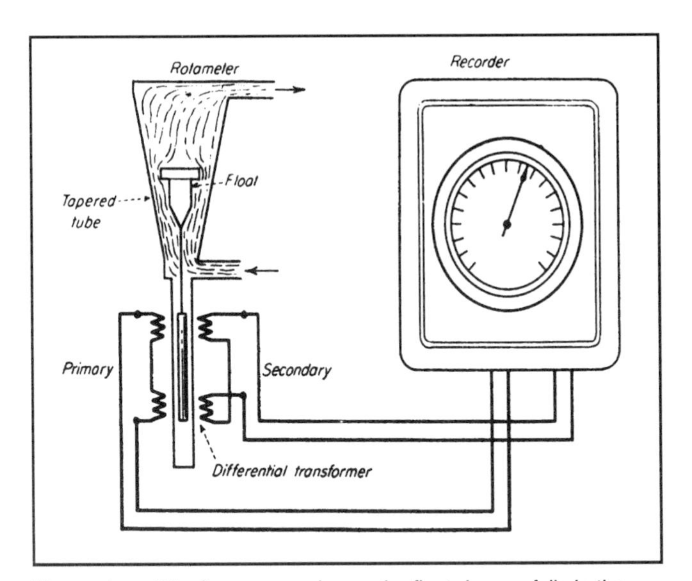

Flowmeter. The flow area varies as the float rises or falls in the tapered tube. High flows cause the float to rise, and low flows cause it to drop. The differential transformer core follows the float travel and generates an AC signal which is fed into a square-root recorder. A servo can be equipped with a square root cam to read on a linear chart. The transformer output can also be amplified and used to actuate a flow regulating valve so that the flowmeter becomes the primary element in a flow controller. Normally meter accuracy is better than 2%, but its flow range is limited.

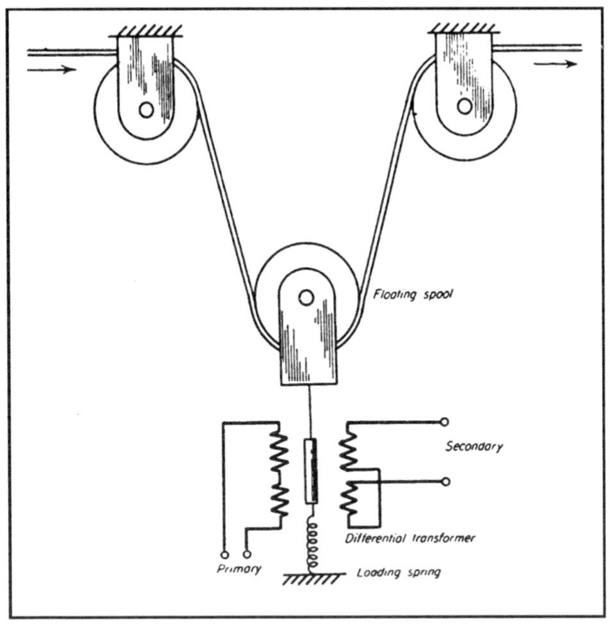

Tension control. The loading spring can be adjusted so that when the transformer core is at its null point, the proper tension is maintained in the wire. The amplified output of the transformer is transmitted to some kind of tension-controlling device which increases or reduces the tension in the wire, depending on the phase and magnitude of the applied differential transformer signal.

HIGH-SPEED ELECTRONIC COUNTERS

The electronic counter counts electrical pulses and gives a running display of accumulated pulses at any instant. Because the input is an electrical signal, a transducer is generally required to transform the nonelectrical signal into a usable input for the counter.

With a preset function on the counter, any number can be selected within the count capacity of the device. Once the counter reaches the preset number, it can open or close the relay to control some operation. The counter will either reset automatically or stop. A dual unit permits continuous control over two different count sequence operations. Two sets of predetermining switches are usually mounted on the front panel of the counter, but they can be mounted at a remote location. If two different numbers are programmed into the counter, it will alternately count the two selected numbers. Multiple presets are also available, but at higher cost.

In addition to performing two separate operations, a dual preset can control speeds, as shown in Fig. 1. In the metal shearing operation run at high speed, one preset switch can be used to slow the material down at a given distance before the second preset actuates the shearing. Then both switches automatically reset and start to measure again. The same presets could also be used. The same presets could also be used to alternately shear the material into two different lengths.

One form of measurement well adapted to high-speed counters is the measuring of continuous materials such as wire, rope, paper, textiles, or steel. Fig. 2 shows a coil-winding operation in which a counter stops the machine at a predetermined number of turns of wire.

A second application is shown in Fig. 3. Magazines are counted as they run off a press. A photoelectric pickup senses the alternate light and dark lines formed by the shadows of the folded edges of the magazines. At the predetermined number, a knife edge, actuated by the counter, separates the magazines into equal batches.

A third application is in machine-tool control. A preset counter can be paired with a transducer or pulse generator mounted on the feed mechanism. It could, for example, convert revolutions of screw feed, hence displacement, into pulses to be fed into the counter. A feed of 0.129 in. might represent a count of 129 to the counter.

When preset at that number, the counter could stop, advance, or reverse the feed mechanism.

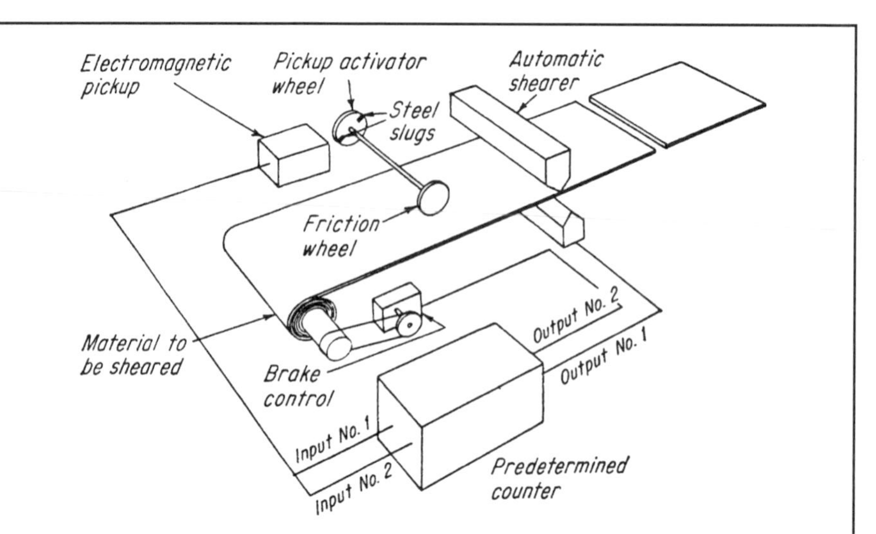

Fig. 1 A dual preset function on a high-speed counter controls the high-speed shearing operation. If the material is to be cut in 10-ft lengths and each pulse of the electromagnetic pickup represents 0.1 ft, the operator presets 100 into the first input channel. The second input is set to 90. When 90 pulses are counted, the second channel slows the material. Then when the counter reaches 100, the first channel actuates the shear. Both channels reset instantaneously and start the next cycle.

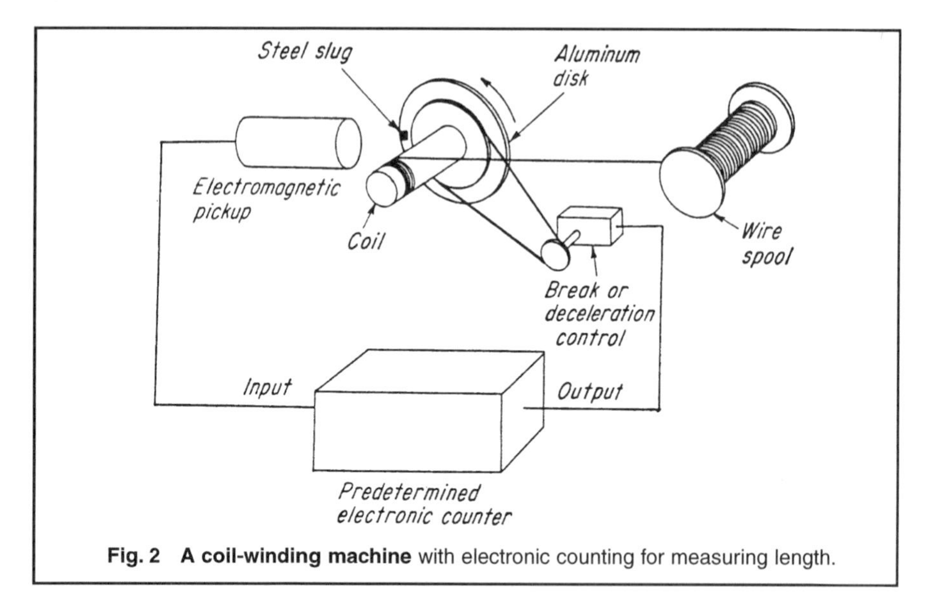

Fig. 2 A coil-winding machine with electronic counting for measuring length.

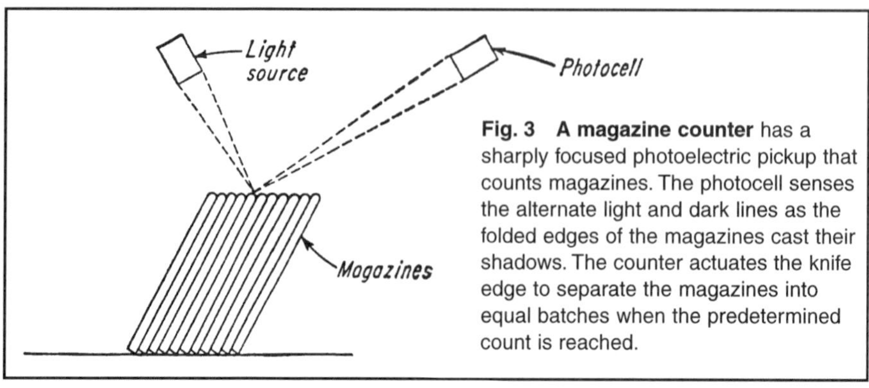

Fig. 3 A magazine counter has a sharply focused photoelectric pickup that counts magazines. The photocell senses the alternate light and dark lines as the folded edges of the magazines cast their shadows. The counter actuates the knife edge to separate the magazines into equal batches when the predetermined count is reached.

APPLICATIONS FOR PERMANENT MAGNETS

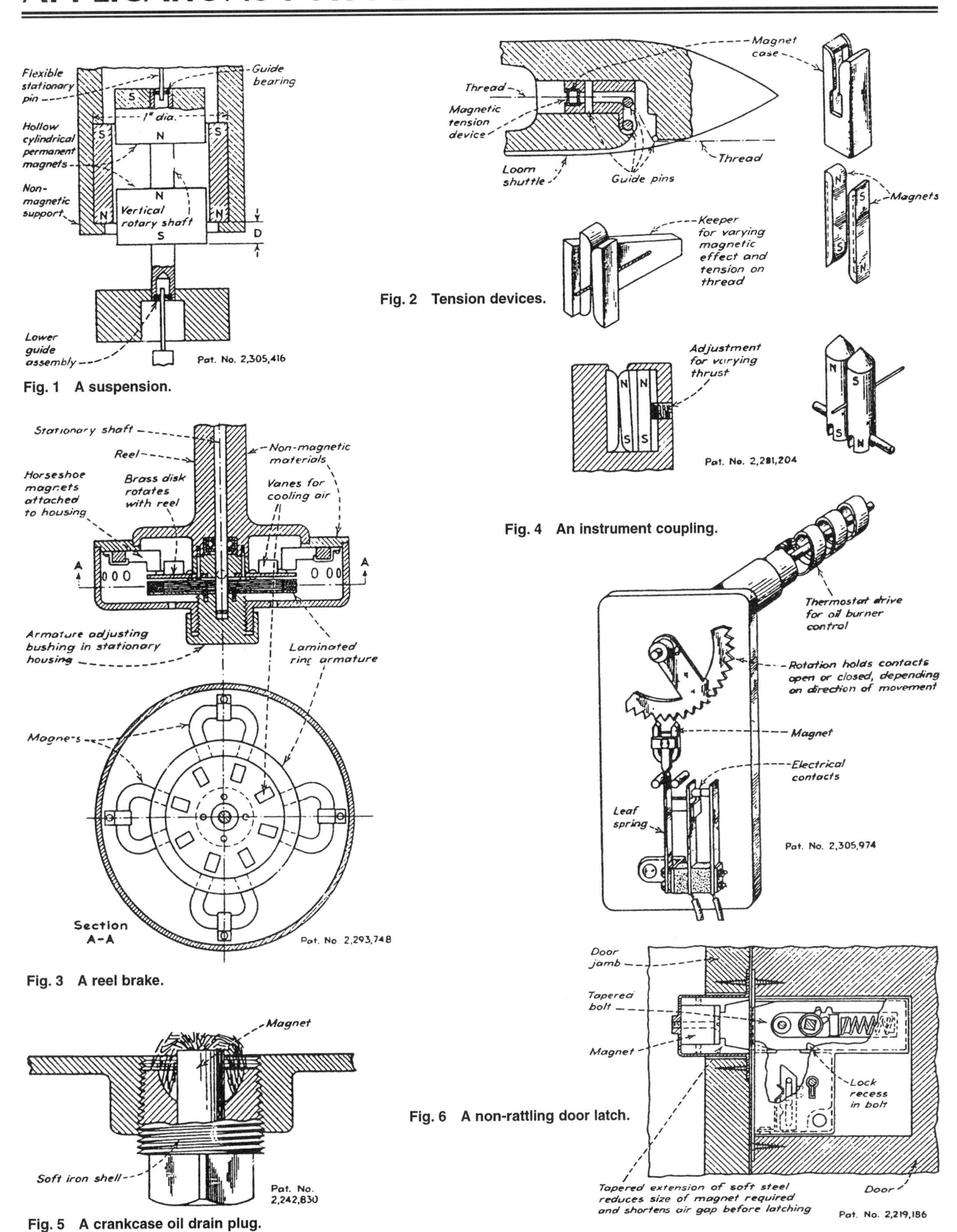

Fig. 1 A suspension.

Fig. 2 Tension devices.

Fig. 3 A reel brake.

Fig. 4 An instrument coupling.

Fig. 5 A crankcase oil drain plug.

Fig. 6 A non-rattling door latch.

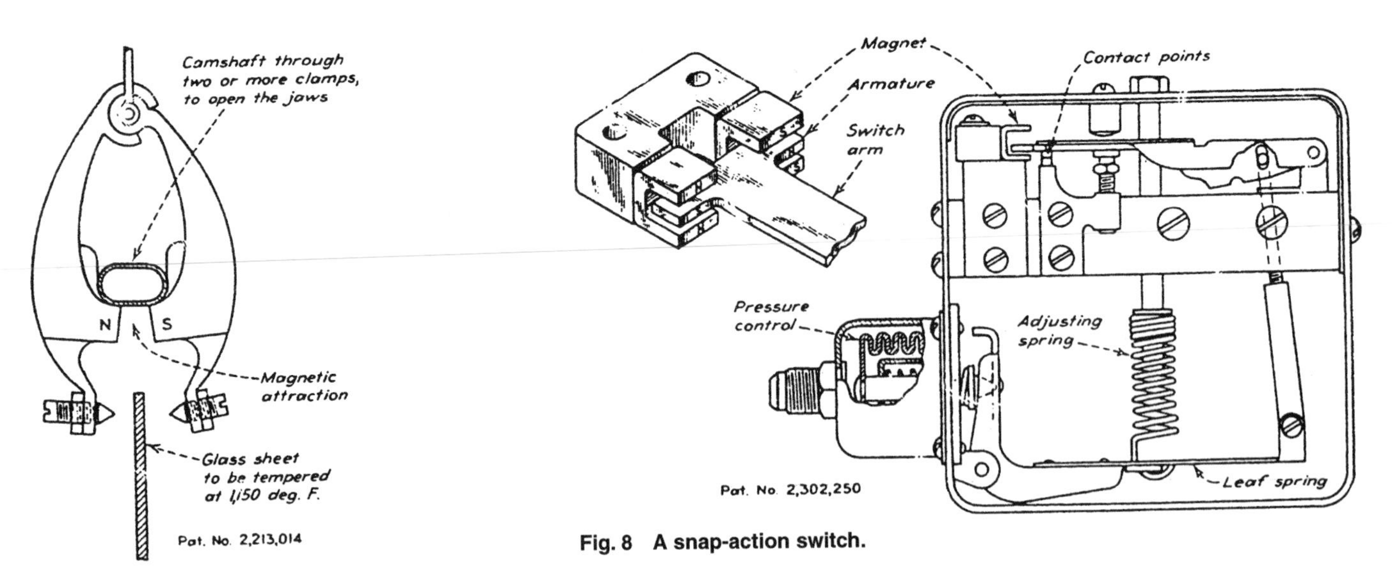

Fig. 7 A clamp.

Fig. 8 A snap-action switch.

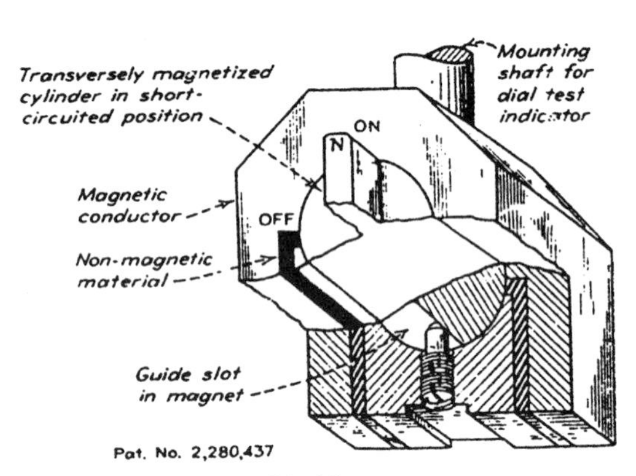

Fig. 9 An instrument holder.

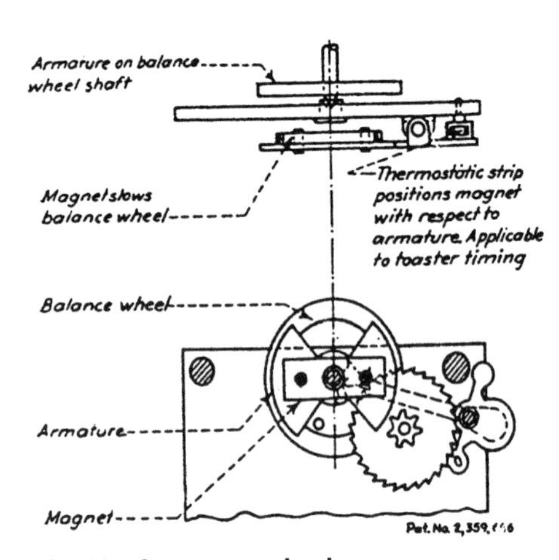

Fig. 10 An escape wheel.

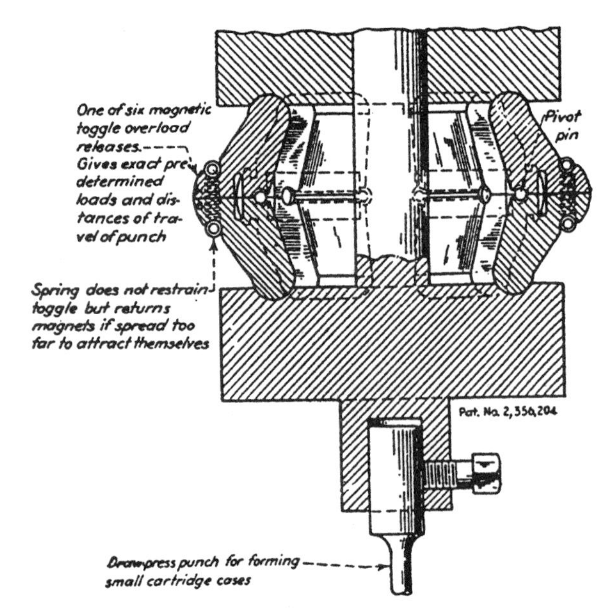

Fig. 11 A pressure release.

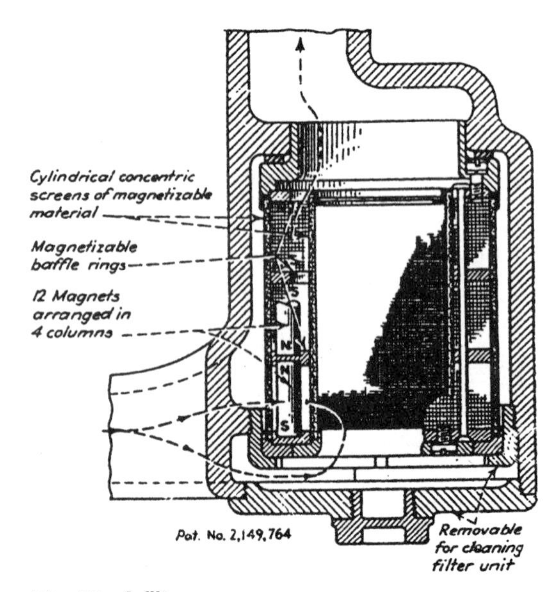

Fig. 12 A filter.

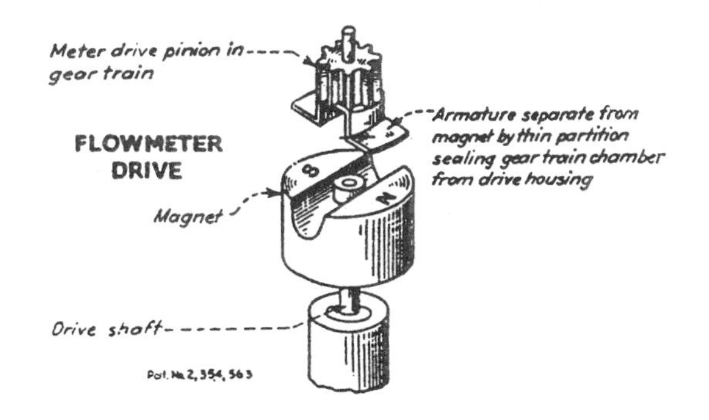

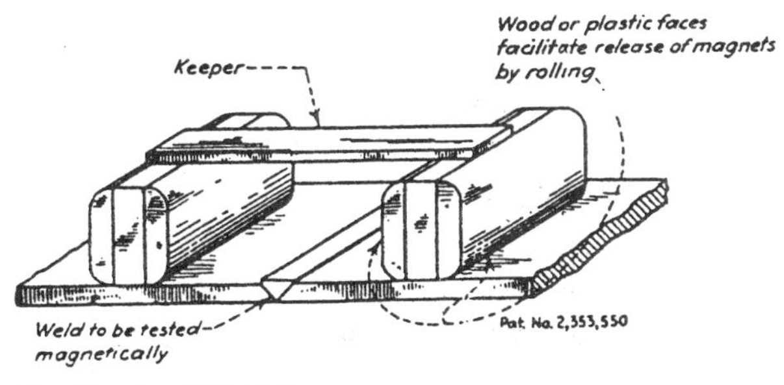

Fig. 13 A weld tester.

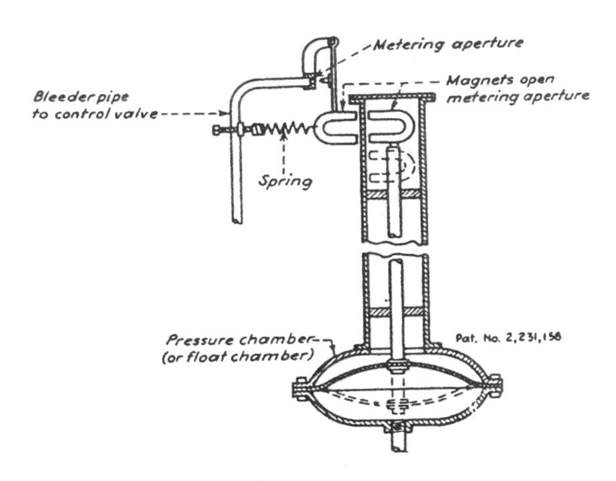

Fig. 14 A control device.

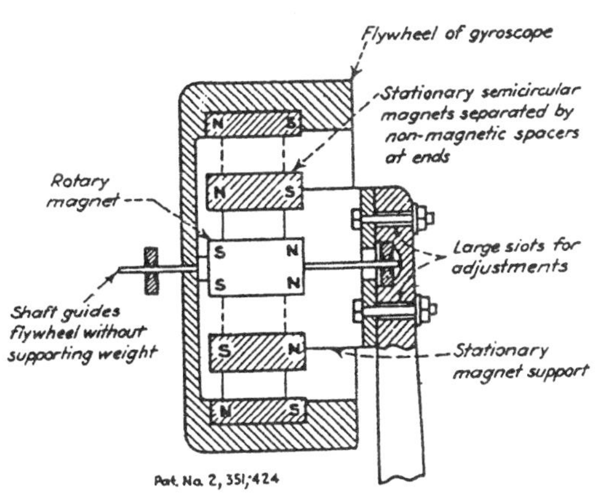

Fig. 15 A horizontal-shaft suspension.

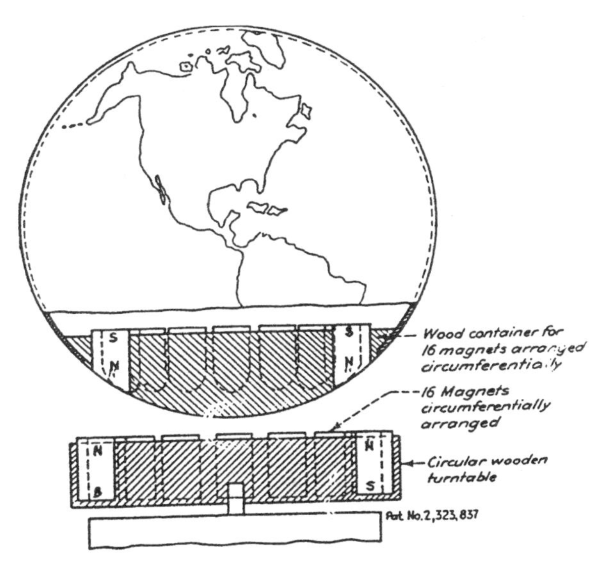

Fig. 16 A floating advertising display.

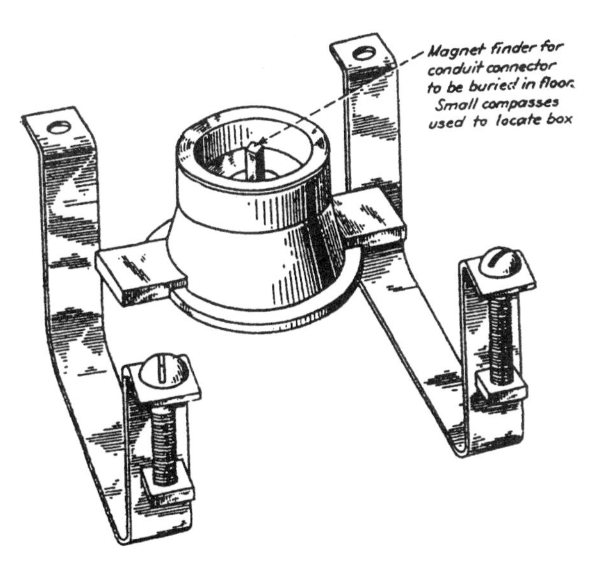

Fig. 17 A finder.

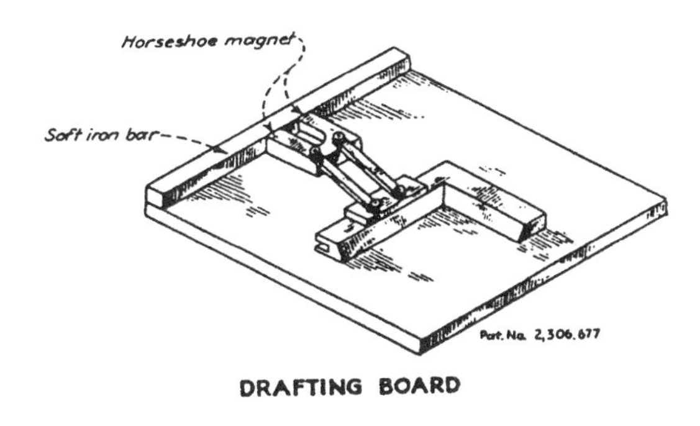

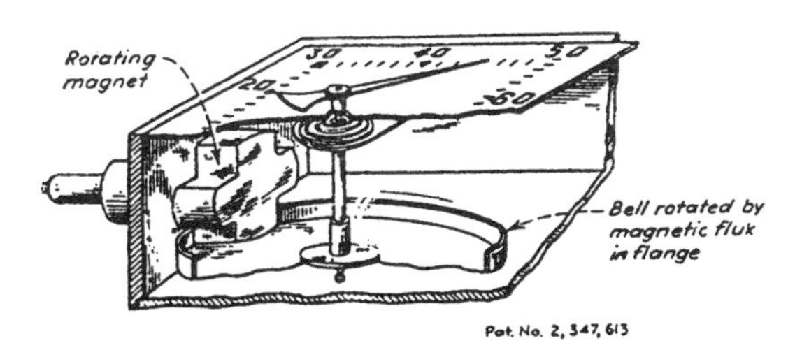

Fig. 18 A tachometer.

NINE ELECTRICALLY DRIVEN HAMMERS

The application of controlled impact forces can be as practical in specialized stationary machinery as in the portable electric hammers shown here. These mechanisms have been employed in vibrating concrete forms, nailing machines, and other special machinery. In portable hammers they are efficient in drilling, chiseling, digging, chipping, tamping, riveting, and similar

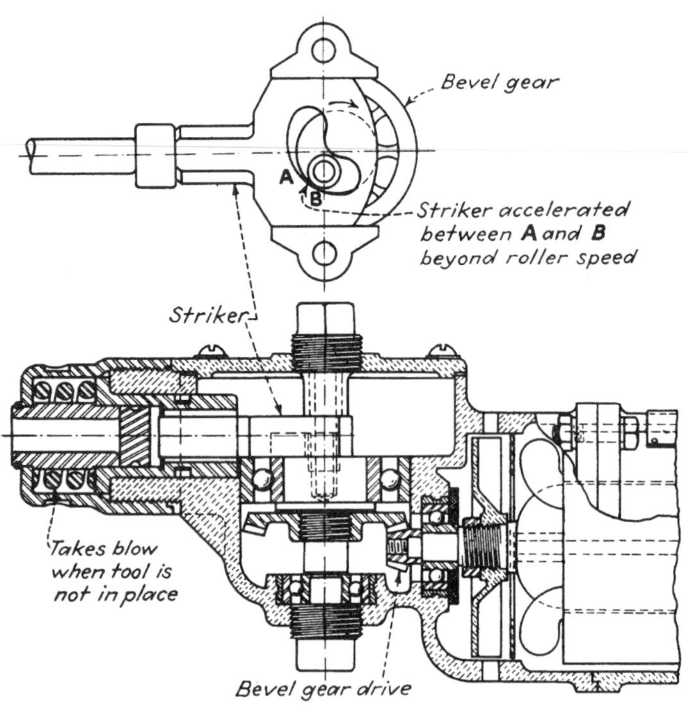

Fig. 1 A free-driving throw of the cam-slotted striker is produced by the eccentric stud roller during contact between points A and B of the slot. This accelerates the striker beyond the tangential speed of the roller for an instant before the striker is picked up for the return stroke.

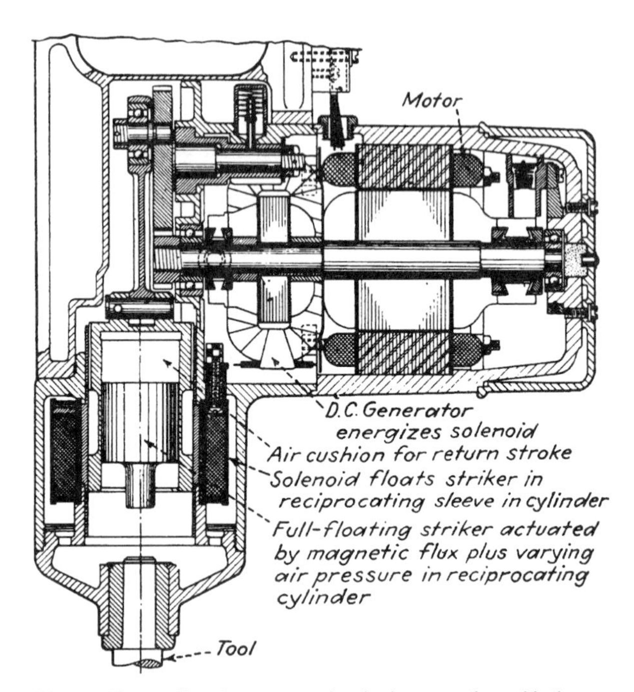

Fig. 3 The striker has no mechanical connection with the reciprocating drive in this hammer.

Fig. 2 The centrifugal force of two oppositely rotating weights throws the striker assembly of this hammer. The power connection is maintained by a sliding-splined shaft. The guide, not shown, prevents the rotation of the striker assembly.

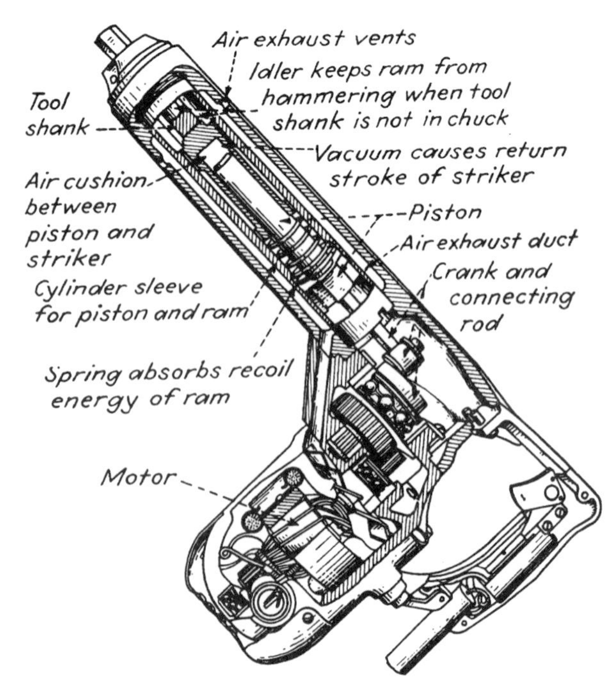

Fig. 4 This combination of mechanical, pneumatic, and spring action is included in this hammer.

operations where quick, concentrated blows are required. The striker mechanisms illustrated are operated by springs, cams, magnetic force, air and vacuum chambers, and centrifugal force. The drawings show only the striking mechanisms.

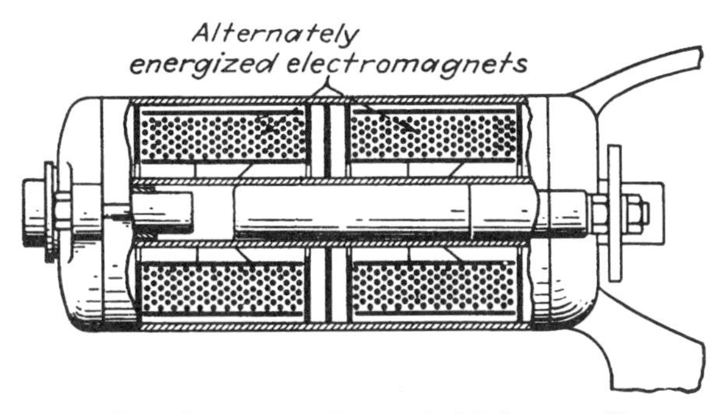

Fig. 5 Two electromagnets operate this hammer. The weight of the blows can be controlled by varying the electric current in the coils or timing the current reversals by an air-gap adjustment of the contacts.

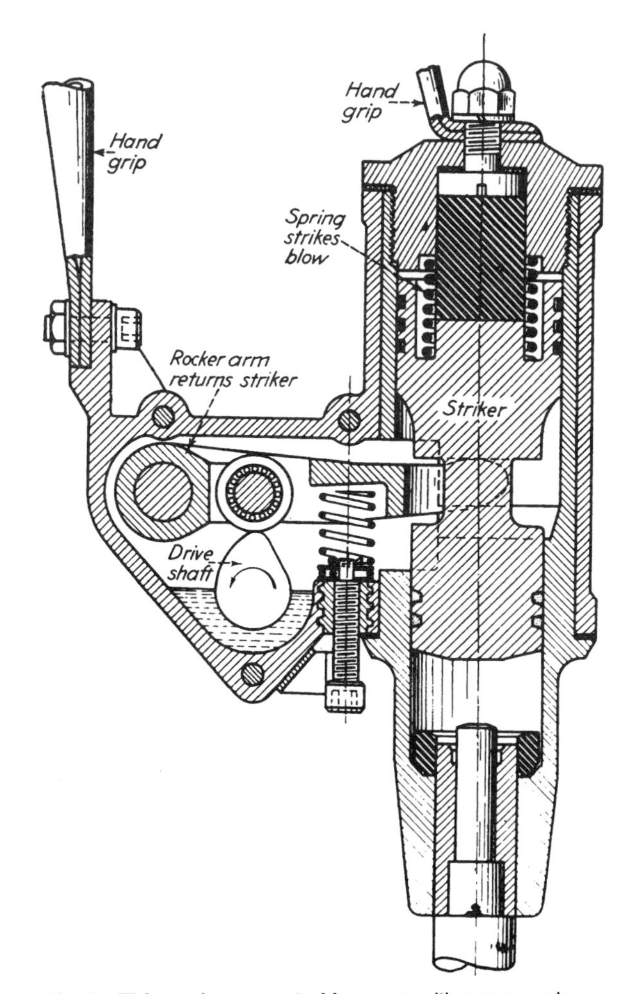

Fig. 6 This spring-operated hammer with a cam and rocker for the return stroke has a screw for adjusting the blow to be imparted.

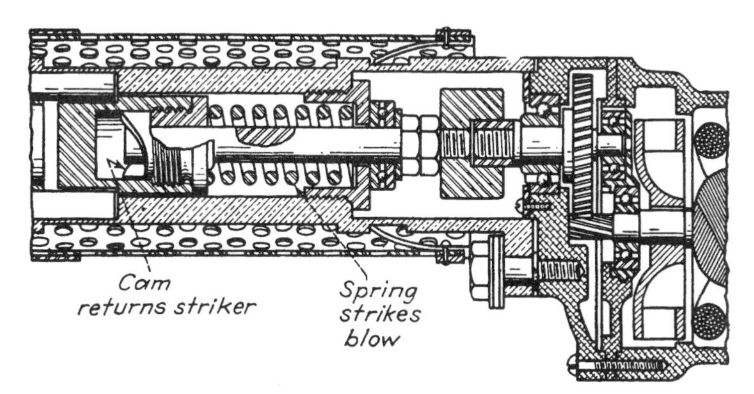

Fig. 7 This spring-operated hammer includes a shaft rotating in a female cam to return the striker.

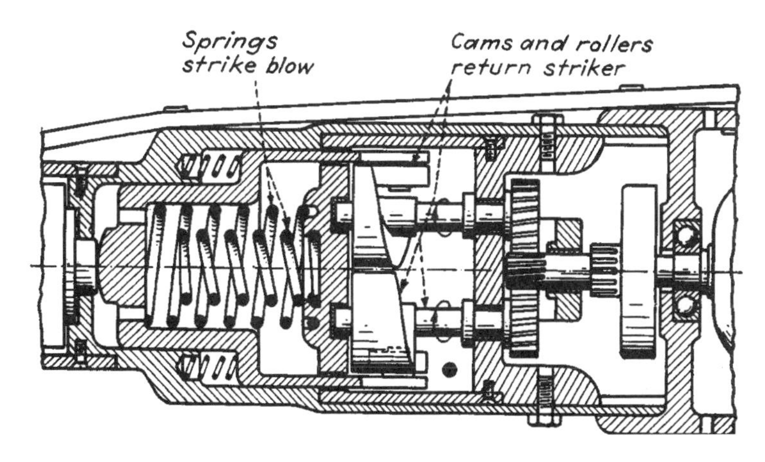

Fig. 8 This spring-operated hammer has two fixed rotating-barrel cams. They return the striker by means of two rollers on the opposite sides of the striker sleeve. Auxiliary springs prevent the striker from hitting the retaining cylinder. A means of rotating the tool, not shown, is also included in this hammer.

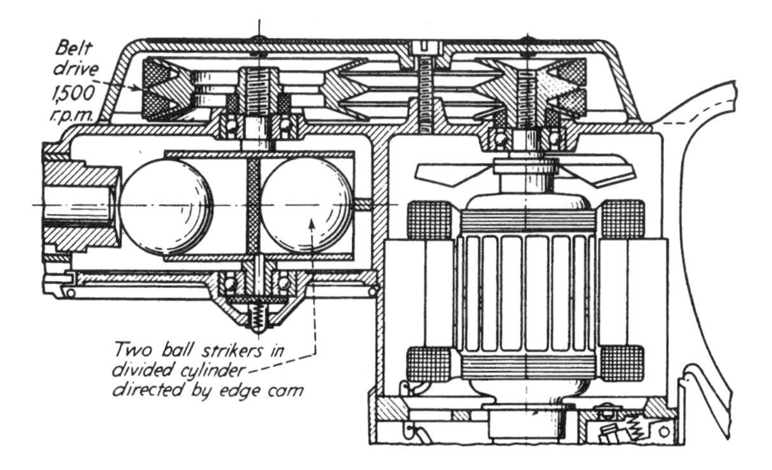

Fig. 9 Two steel balls rotated in a divided cylinder and steered by an edge cam develop centrifugal force to strike blows against the tool holder. The collar is held clear of the hammer by a compression spring when no tool is in the holder. A second spring cushions the blows when the motor is running, but the tool is not held against the work.

Sensitivity or change in deflection for a given temperature change depends upon the combination of meals selected and the dimensions of the bimetal element. Sensitivity increases with the square of the length and inversely with the thickness. The force developed for a given temperature change also depends on the type of bimetal. However, the allowable working load for the thermostatic strip increases with the width and the square of the thickness. Thus, the design of bimetal elements depends upon the relative importance of sensitivity and working load.

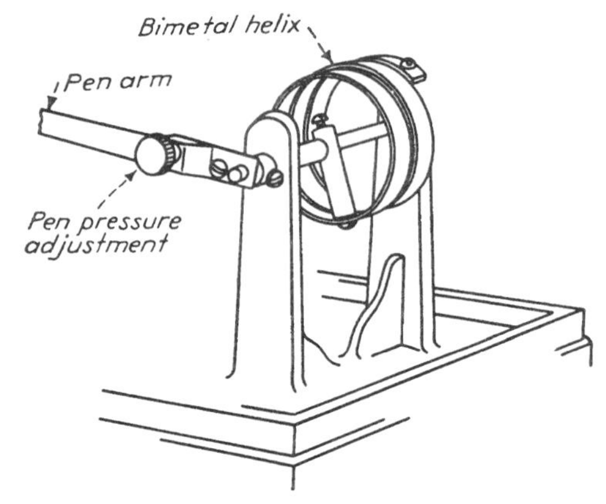

Fig. 1 This recording thermometer has a pen that is moved vertically across a revolving chart by a brass-invar bimetal element. To obtain sensitivity, the long movement of the pen requires a long strip of bimetal, which is coiled into a helix to save space. For accuracy, a relatively large cross section gives stiffness, although the large thickness requires increased length to obtain the desired sensitivity.

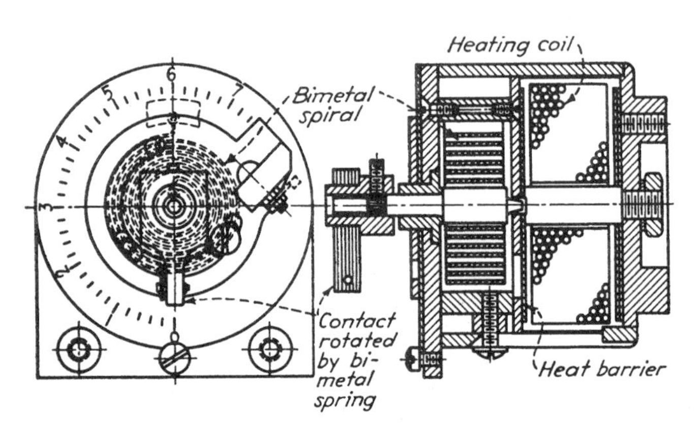

Fig. 3 This overload relay for large motors passes some of the motor current through a heating coil within the relay. Heat from the coil raises the temperature of a bimetal spiral which rotates a shaft carrying an electrical contact. To withstand the operating temperatures, it includes a heat-resistant bimetal spiral. It is coiled into the spiral form for compactness. Because of the large deflection needed, the spiral is long and thin, whereas the width is made large to provide the required contact pressure.

Heat barriers between the bimetal spiral and the heating coil make the temperature rise of the bimetal spiral closely follow the increase in temperature within the motor. Thus, momentary overloads do not cause sufficient heating to close the contacts. However, a continued overload will, in time, cause the bimetal spiral to rotate the contact arm around to the adjustable stationary contact, causing a relay to shut down the motor.

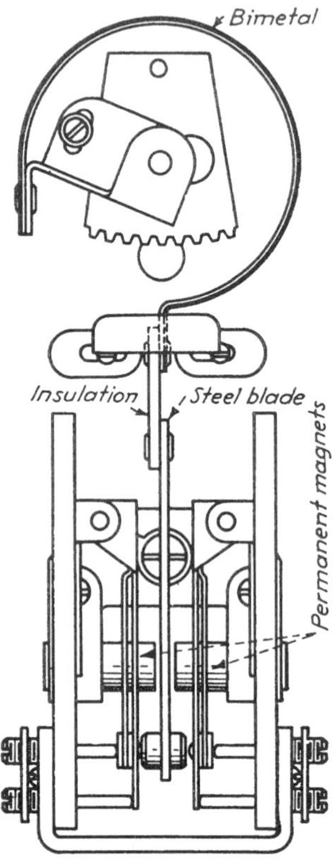

Fig. 2 Room temperatures in summer and winter are controlled over a wide range by a single, large-diameter coil of brass-invar in this thermometer. To prevent chattering, a small permanent magnet is mounted on each side of the steel contact blade. The magnetic attraction on the blade, which increases inversely with the square of the distance from the magnet, gives a snap action to the contacts.

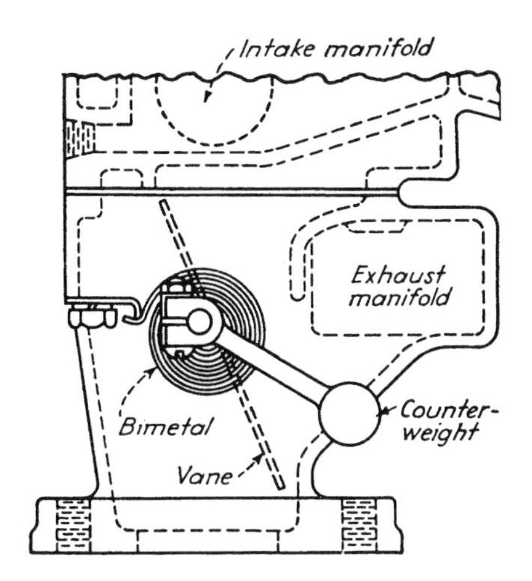

Fig. 4 Carburetor control. When the engine is cold, a vane in the exhaust passage to the "hot spot" is held open by a bimetal spring against the force of a small counterweight. When the thermostatic spiral is heated by the outside air or by the warm air stream from the radiator, the spring coils up and allows the weight to close the vane. Because high accuracy is not needed, a thin, flexible cross section with a long length provides the desired sensitivity.

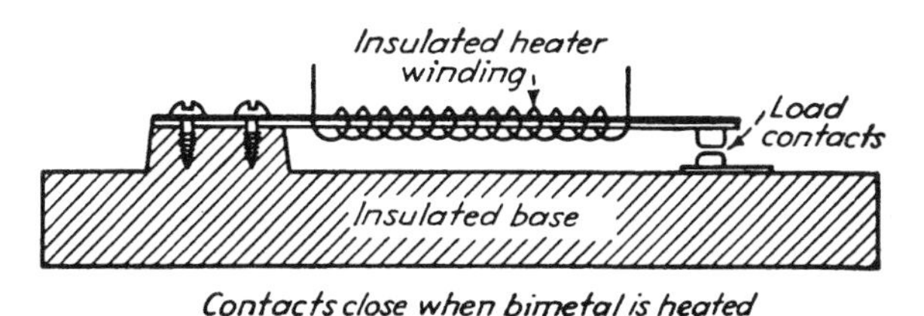

Fig. 5 Thermostatic relay. A constant current through an electrical heating coil around a straight bimetal strip gives a time-delay action. Because the temperature range is relatively large, high sensitivity is not necessary. Thus, a short, straight strip of bimetal is suitable. Because of its relatively heavy thickness, the strip is sufficiently stiff to close the contact firmly without chattering.

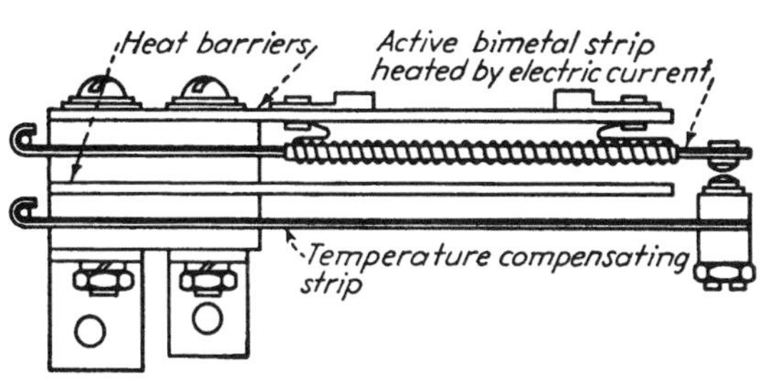

Fig. 6 The bimetal element in this time-delay relay protects mercury-vapor rectifiers. This relay closes the voltage circuit to the mercury tube only after the filament has had time to reach its normal operating temperature. To eliminate the effect of changes in room temperature on the length of the contact gap (and therefore the time interval) the stationary contact is carried by a second bimetal strip, similar to the heated element. Barriers of laminated plastic on both sides of the active bimetal strip shield the compensating strip and prevent air currents from affecting the heating rate. The relatively high temperature range allows the use of a straight, thick strip, but the addition of the compensating strip makes accurate timing possible with a short travel.

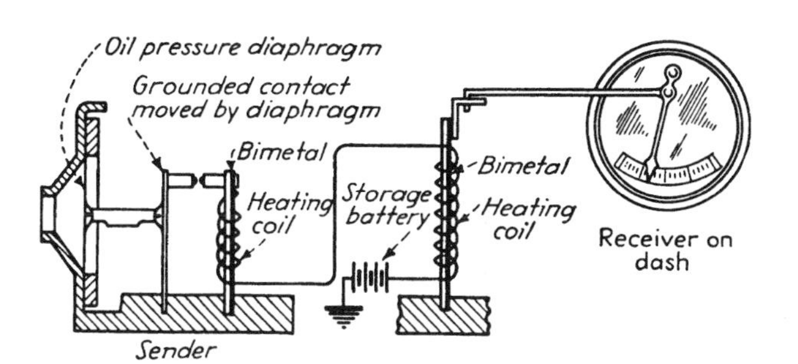

Fig. 7 Oil pressure, engine temperature, and gasoline level are indicated electrically on automobile dashboard instruments whose bimetal element is both the sender and receiver. A grounded contact at the sender completes an electric circuit through heaters around two similar bimetal strips. Because the same current flows around the two bimetal elements, their deflections are the same. But the sender element, when heated, will bend away from the grounded contact until the circuit is broken Upon cooling, the bimetal element again makes contact and the cycle continues. This allows the bimetal element to follow the movement of the grounded contact. For the oil-pressure gage, the grounded contact is attached to a diaphragm; for the temperature indicator, the contact is carried by another thermostatic bimetal strip; in the gasoline-level device, the contact is shifted by a cam on a shaft rotated by a float. Deflections on the receiving bimetal element are amplified through a linkage that operates a pointer over the scale of the receiving instrument. Because only small deflections are needed, the bimetal element is in the form of a short, stiff strip.

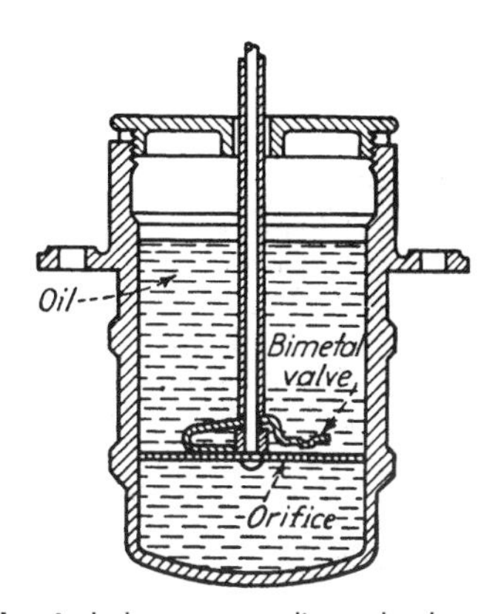

Fig. 8 Oil dashpots in heavy-capacity scales have a thermostatic control to compensate for changes in oil viscosity with temperature. A rectangular orifice in the plunger is covered by a swaged projection on the bimetal element. With a decrease in oil temperature, the oil viscosity increases, tending to increase the damping effect. But the bimetal element deflects upward, enlarging the orifice enough to keep the damping force constant. A wide bimetal strip provides sufficient stiffness so that the orifice will not be altered by the force of the flowing oil.

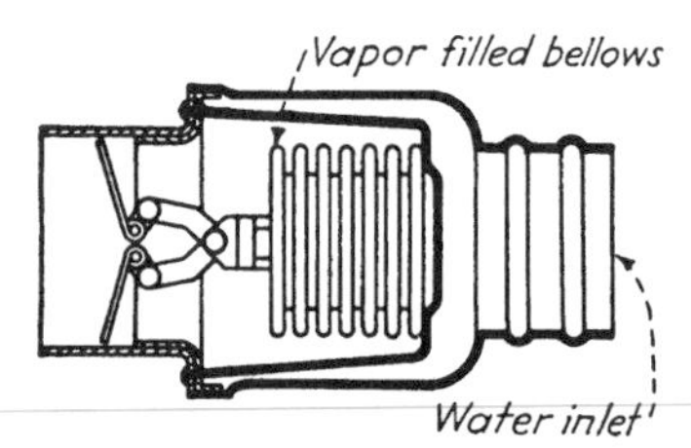

Fig. 9 Automobile cooling-water temperature is controlled by a self-contained bellows in the thermostat. As in the radiator air valve, the bellows itself is subjected to the temperature to be controlled. As the temperature of the water increases to about 140°F., the valve starts to open; at approximately 180°F., free flow is permitted. At intermediate temperatures, the valve opening is in proportion to the temperature.

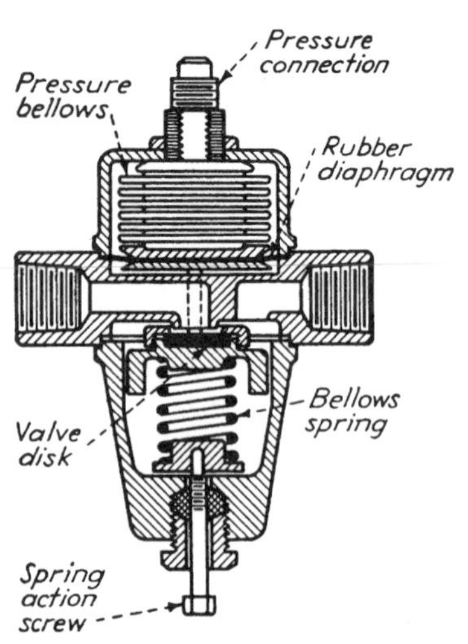

Fig. 10 A throttling circulating-water control valve for refrigeration plants has its valve opening vary with the pressure on the bellows. This valve controls the rate of flow of the cooling water through the condenser. A greater amount of water is required when the temperature, and therefore the pressure, increases. The pressure in the condenser is transmitted through a pipe to the valve bellows, thereby adjusting the flow of cooling water. The bronze bellows is protected from contact with the water by a rubber diaphragm.

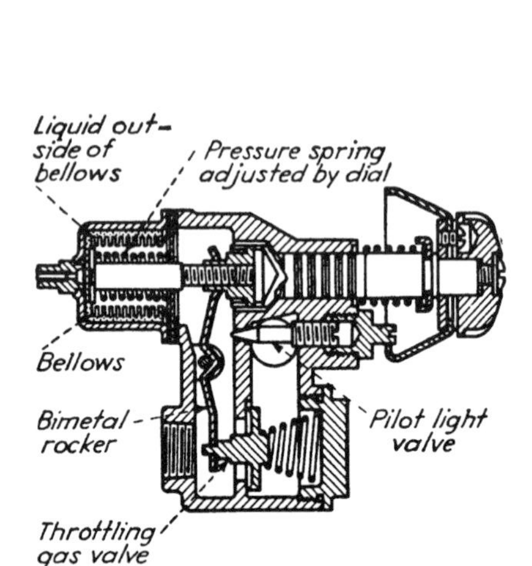

Fig. 11 An automatic gas-range control has a sealed thermostatic element consisting of a bulb, capillary tube, and bellows. As food is often placed near the bulb, a nontoxic liquid, chlorinated diphenyl, is in the liquid expansion system. The liquid is also non-flammable and has no corrosive effect on the phosphor-bronze bellows. By placing the liquid outside instead of inside the bellows, the working stresses are maximum at normal temperatures when the bellows bottoms on the cup. At elevated working temperatures, the expansion of the liquid compresses the bellows against the action of the extended spring. This, in turn, is adjusted by the knob. Changes in calibration caused by variations in ambient temperature are compensated by making the rocker arm of a bimetal suitable for high-temperature service.

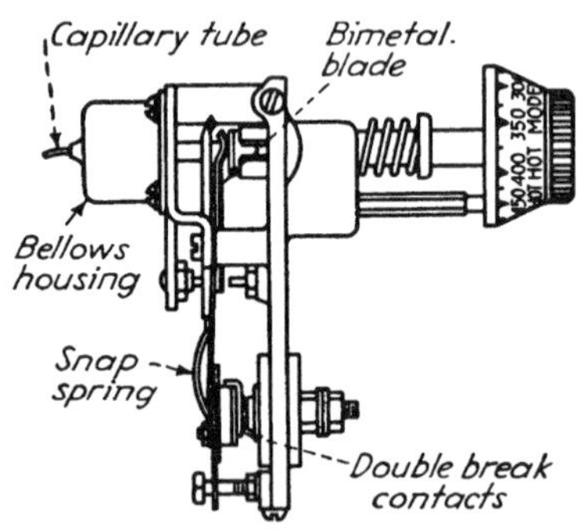

Fig. 12 For electric ranges, this thermostat has the same bellows unit as the gas-type control. But, instead of a throttling action, the thermostat opens and closes the electrical contacts with a snap action. To obtain sufficient force for the snap action, the control requires a temperature difference between *on* and *off* positions. For a control range from room temperature to 550°F., the differential in this instrument is ±10°F. With a smaller control range, the differential is proportionately less. The snap-action switch is made of beryllium copper, giving it high strength, better snap action, and longer life than is obtainable with phosphor bronze. Because of its corrosion resistance, the beryllium-copper blade requires no protective finish.

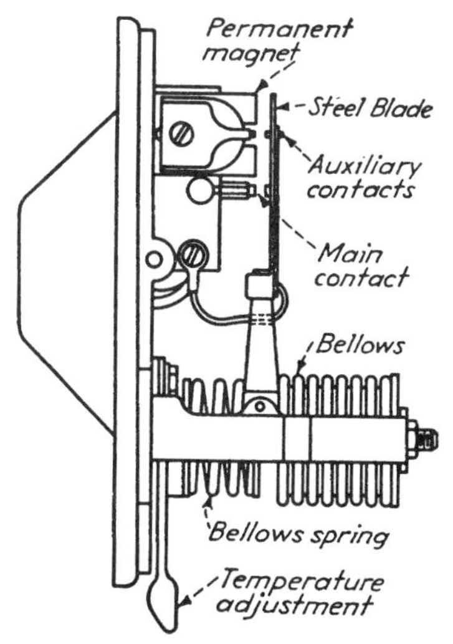

Fig. 13 For heavy-duty room-temperature controls, this thermostat has a bellows mechanism that develops a high force with small changes in temperature. The bellows is partly filled with liquid butane. At room temperatures this gas exhibits a large change in vapor pressure for small temperature differentials. Snap action of the electrical contact is obtained from a small permanent magnet that pulls the steel contact blade into firm contact when the bellows cools. Because of the firm contact, the device is rated at 20 A for noninductive loads. To avoid chattering or bounce under the impact delivered by the rapid magnetic closing action, small auxiliary contacts are carried on light spring blades. With the large force developed by the bellows, a temperature differential of only 2°F. is obtained.

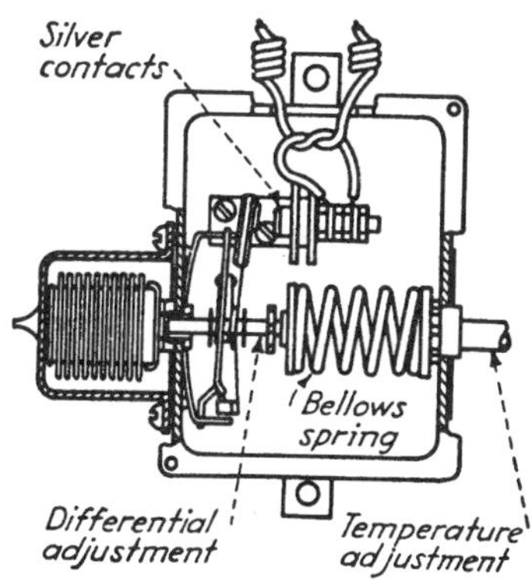

Fig. 14 Snap action in this refrigerator control is obtained from a bowed flat spring. The silver contacts carried on an extended end of the spring open or close rapidly when movement of the bellows actuates the spring. With this snap action, the contacts can control an alternating-current motor as large as $1\frac{1}{2}$ hp without auxiliary relays. Temperature differential is adjusted by changing the spacing between two collars on the bellows shaft passing through the contact spring. For the temperatures needed to freeze ice, the bellows system is partly filled with butane.

Fig. 15 In this refrigerator control, the necessary snap action is obtained from a toggle spring supported from a long arm moved by the bellows. With this form of toggle action, the contact pressure is at a maximum at the instant the contacts start to open. Thermostatic action is obtained from a vapor-filled system. Sulfur dioxide is the fill for typical refrigerating service or methyl chloride where lower temperatures are required. To reduce friction, the bellows makes point-contact with the bellows cup. Operating temperature is adjusted by changing the initial compression in the bellows spring. For resistance to corrosion, levers and blades are made from stainless steel with bronze pin bearings.

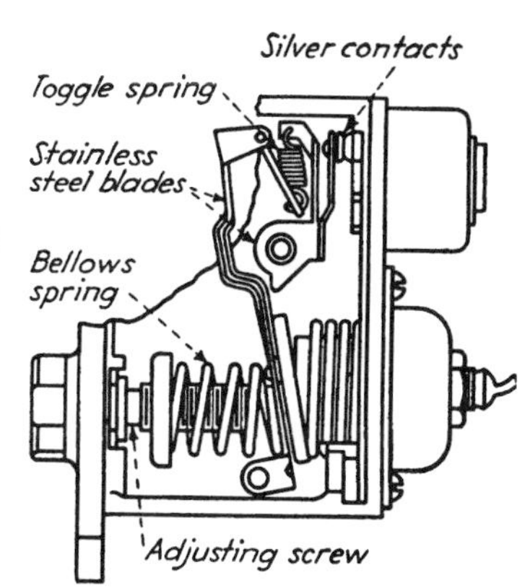

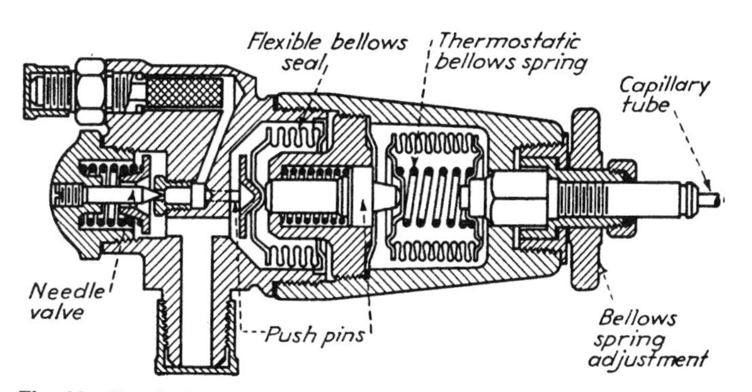

Fig. 16 Two bellows units in this thermostatic expansion valve control large refrigeration systems. A removable power bellows unit is operated by vapor pressure in a bulb attached to the evaporator output line. The second bellows serves as a flexible, gastight seal for the gas valve. A stainless-steel spring holds the valve closed until opened by pressure transmitted from the thermostatic bellows through a molded push pin.

EIGHT TEMPERATURE-REGULATING CONTROLS

Temperature regulators are either on-off or throttling. The characteristics of the process determine which should be used. Within each group, selection of a regulator is governed by the accuracy required, space limitations, simplicity, and cost.

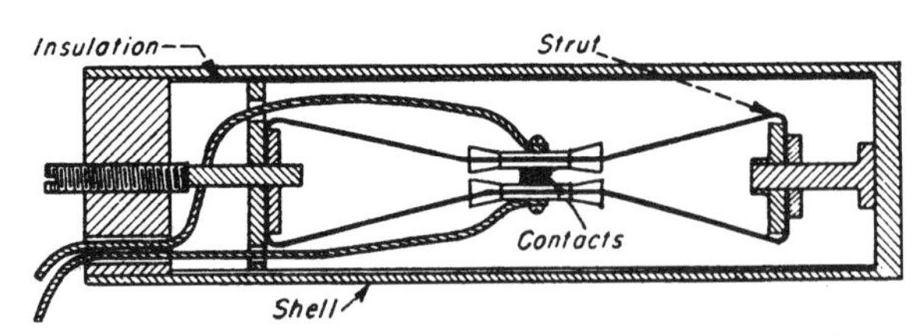

Fig. 1 A bimetallic sensor is simple, compact, and precise. Contacts mounted on low-expansion struts determine slow make-and-break action. A shell contracts or expands with temperature changes, opening or closing the electrical circuit that controls a heating or cooling unit. It is adjustable and resistant to shock and vibration. Its range is 100 to 1500°F, and it responds to a temperature changes of less than 0.5°F.

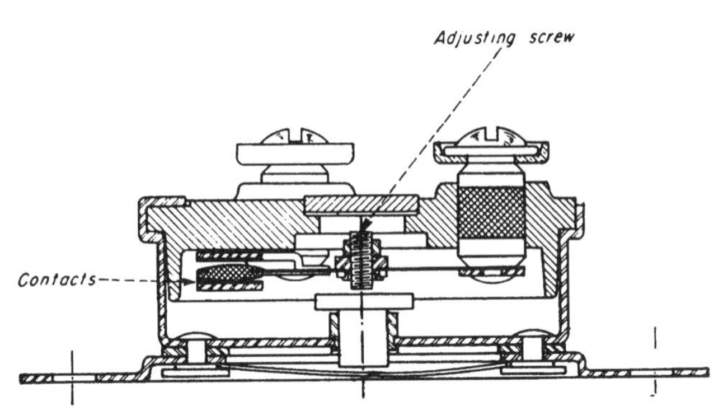

Fig. 2 This enclosed, disk-type, snap-action control has a fixed operating temperature. It is suitable for unit and space heaters, small hot water heaters, clothes dryers, and other applications requiring non-adjustable temperature control. It is useful where dirt, dust, oil, or corrosive atmospheres are present. It is available with various temperature differentials and with a manual reset. Depending on the model, its temperature setting range is from –10° to 550°F and its minimum differential can be 10, 20, 30, 40 or 50°F.

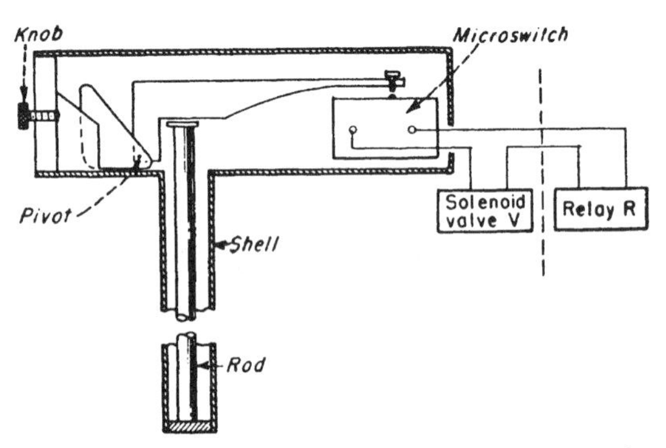

Fig. 3 This bimetallic unit has a rod with a low coefficient of expansion and a shell with a high coefficient of expansion. A microswitch gives snap action to the electrical control circuit. The current can be large enough to operate a solenoid valve or relay directly. The set point is adjusted by a knob which moves the pivot point of the lever. Its range is –20° to 175°F, and its accuracy is 0.25 to 0.50°.

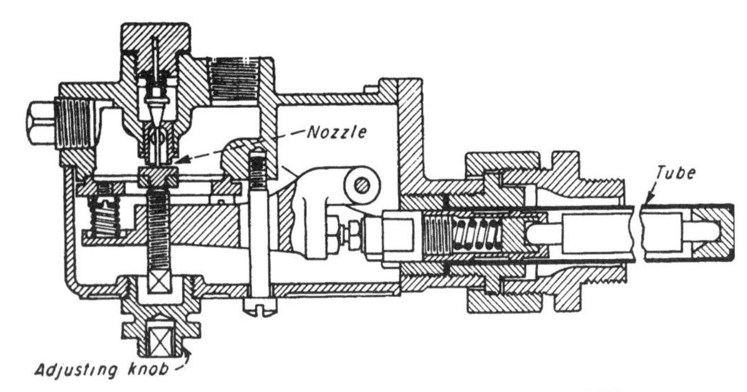

Fig. 4 This is a bimetallic-actuated, air-piloted control. The expansion of the rod causes an air signal (3 to 15 psi) to be transmitted to a heating or cooling pneumatic valve. The position of the pneumatic valve depends on the amount of air bled through the pilot valve of the control. This produces a throttling type of temperature control as contrasted to the on-off characteristic that is obtained with the three units described previously. Its range is 32 to 600°F, and its accuracy is ±1 to ±3°F, depending on the range.

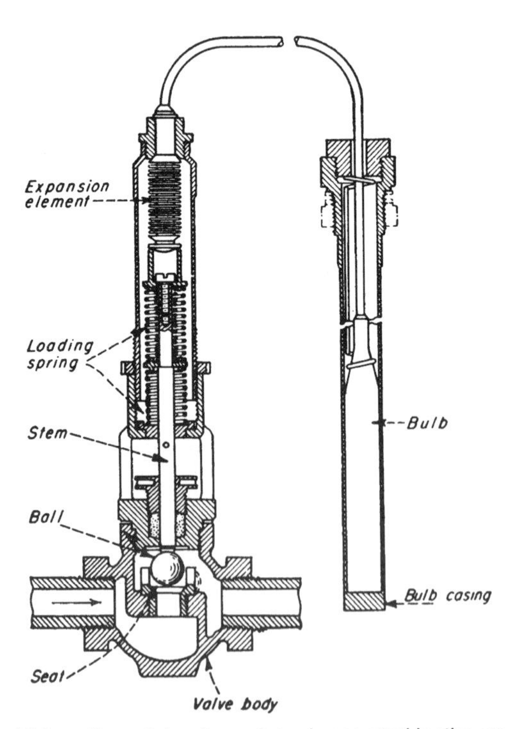

Fig. 5 This self-contained regulator is actuated by the expansion or contraction of liquid or gas in a temperature-sensitive bulb that is immersed in the medium being controlled. The signal is transmitted from the bulb to a sealed expansion element which opens or closes the ball valve. Its range is 20 to 270°F, and its accuracy is ±1°F. The maximum pressure rating is 100 psi for dead-end service and 200 psi for continuous flow.

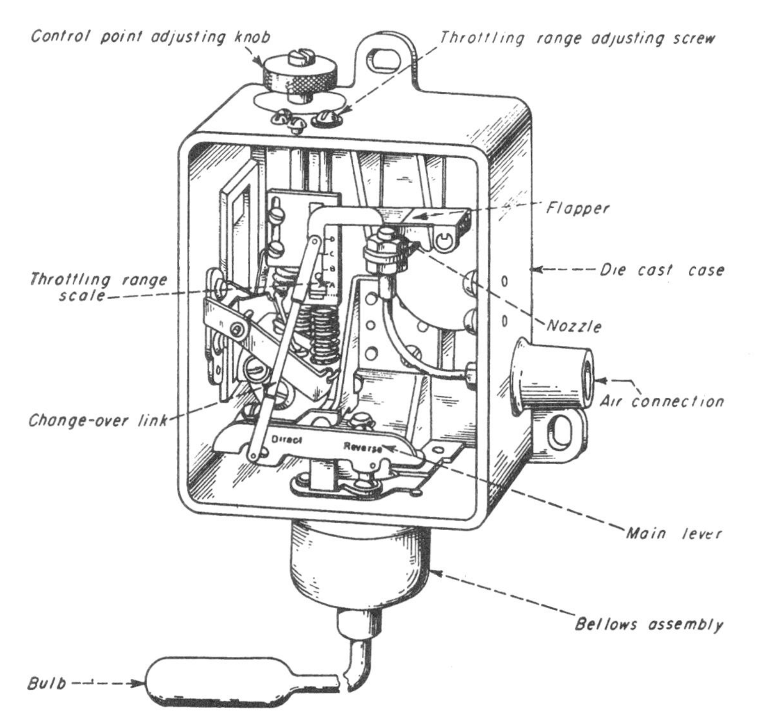

Fig. 6 This remote bulb, nonindicating regulator has a bellows assembly that operates a flapper. This allows air pressure in the control system to build up or bleed, depending on the position of the changeover link. The unit can be direct-or reverse-acting. A control knob adjusts the setting, and the throttling range adjustment determines the percentages of the control range in which full output pressure (3–15 psi) is obtained. Its range is 0 to 700°F, and its accuracy is about ±0.5% of full scale, depending on the way it is installed.

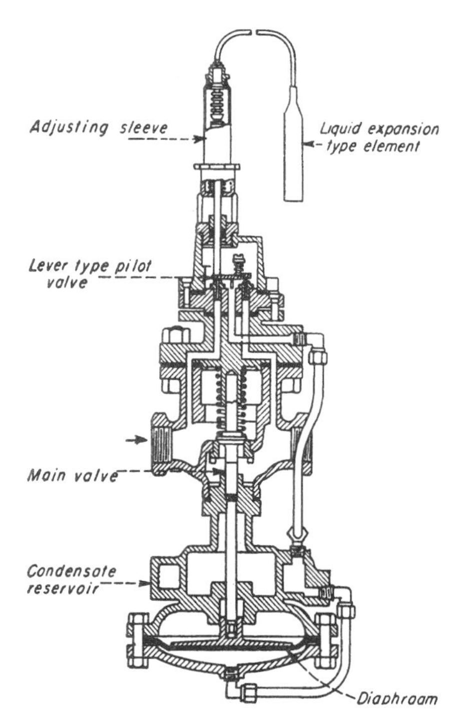

Fig. 7 This lever-type pilot valve is actuated by a temperature-sensitive bulb. The motion of the lever causes the water or steam being controlled to exert pressure on a diaphragm which opens or closes the main valve. Its temperature range is 20 to 270°F, and its accuracy is ±1 to 4°F. It is rated for 5 to 125 psi of steam and 5 to 175 psi of water.

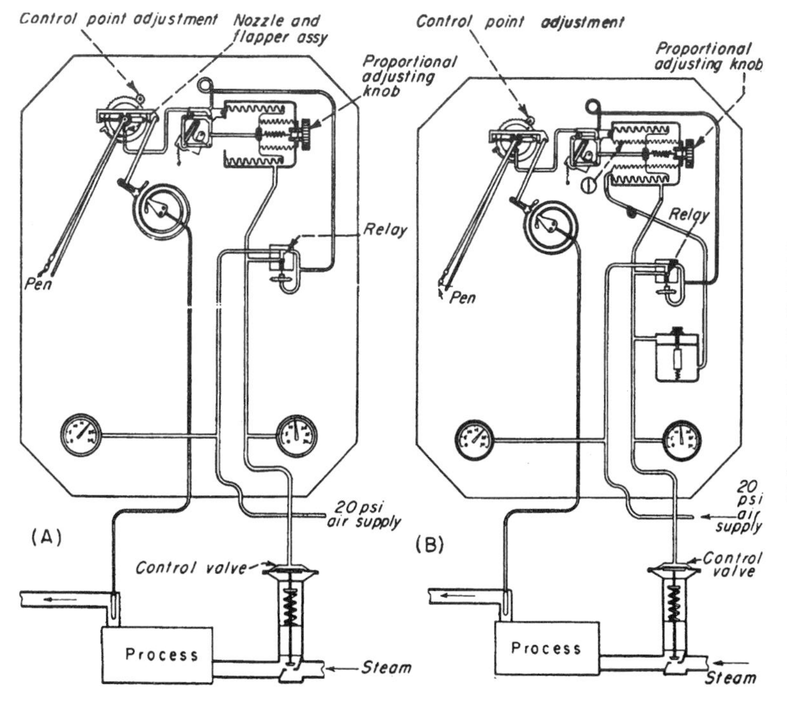

Fig. 8 These two recording and controlling instruments have adjustable proportional ranges. In both, air supply is divided by a relay valve. A small part goes through a nozzle and flapper assembly. The main part goes to the control valve. Unit B has an extra bellows for automatic resetting. It was designed for systems with continuously changing control points, and it can be used where both heating and cooling are required in one process. Both A and B are easily changed from direct to reverse acting. Its accuracy is 1% of its temperature range of –40 to 800°F.

SEVEN PHOTOELECTRIC CONTROLS

Typical applications are presented for reducing production costs and increasing operator safeguards by precisely and automatically controlling the feed, transfer, or inspection of products from one process stage to another.

Fig. 1 Automatic weighing and filling. The task is to fill each box with an exact quantity of products, such as screws. An electric feeder vibrates parts through a chute and into a box on a small balance. The photoelectric control is mounted at the rear of the scale. The light beam is restricted to very small dimensions by an optical slit. The control is positioned so that the light is interrupted by a balanced cantilever arm attached to the scale when the proper box weight is reached. The photoelectric control then stops the flow of parts by deenergizing the feeder. Simultaneously, an indexing mechanism is activated to remove the filled box and replace it with an empty one. The completion of indexing reenergizes the feeder, which starts the flow of screws.

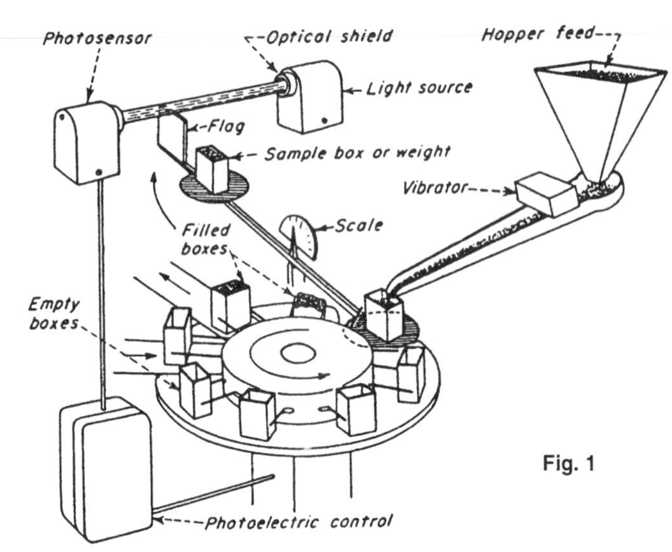

Fig. 1

Fig. 2 Operator safeguard. Most pressures are operated by a foot pedal that leaves the operator's hands free for loading and unloading. This creates a safety hazard. The use of mechanical gate systems reduces the speed of production. With photoelectric controls, a curtain of light is set up by a multiple series of photoelectric scanners and light sources. When a light beam is broken at any point by the operator's hand, the control energizes a locking mechanism that prevents the punch-press drive from being energized. A circuit or power failure causes the control to function as if the light beam were broken. In addition, the light beam frequently becomes the actuating control because the clutch is released as soon as the operator removes his hand from the die on the press table.

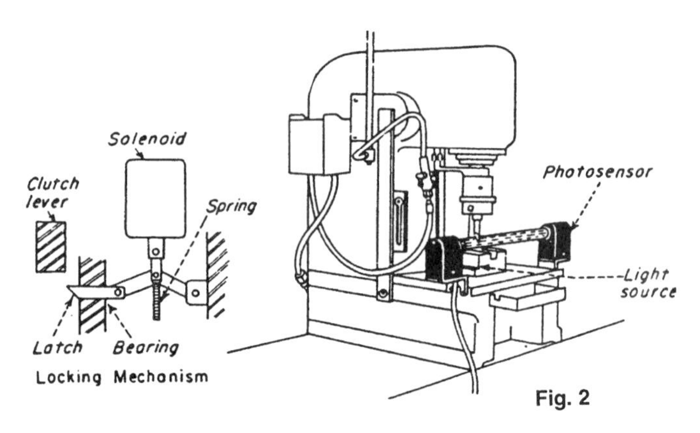

Fig. 2

Fig. 3 This apparatus sorts cartons of three different kinds of objects. Because the cartons containing objects differ widely in size, it is not feasible to sort by carton size and shape. A small strip of reflecting tape is put on the cartons by a packer during assembly. On one type of object, the strip is located along one edge of the bottom, and it extends almost to the middle. For the second type, the strip is located along the same edge, but from the middle to the opposite side. No tape is placed on the third type. Cartons are placed on the conveyor so that the tape is at right angles to the direction of travel. Photoelectric controls shown in A "see" the reflecting tape and operate a pusher-bar mechanism shown in B. This pushes the carton onto the proper distribution conveyor. Cartons without tape pass.

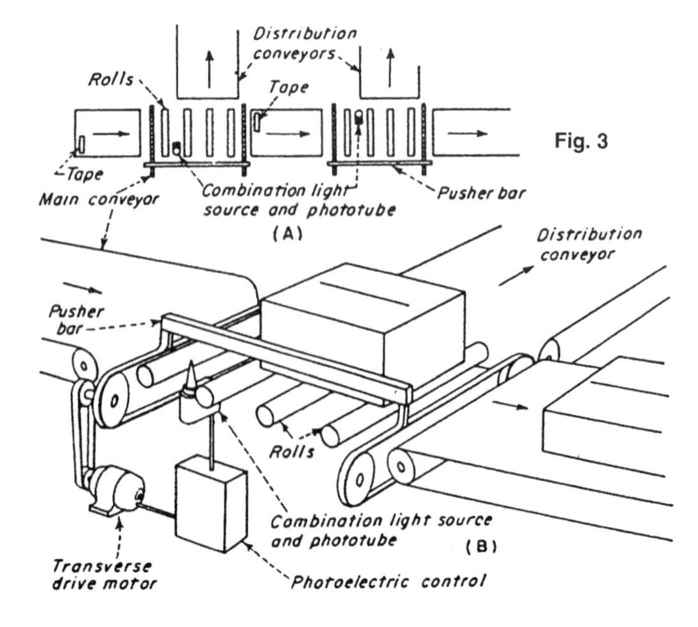

Fig. 3

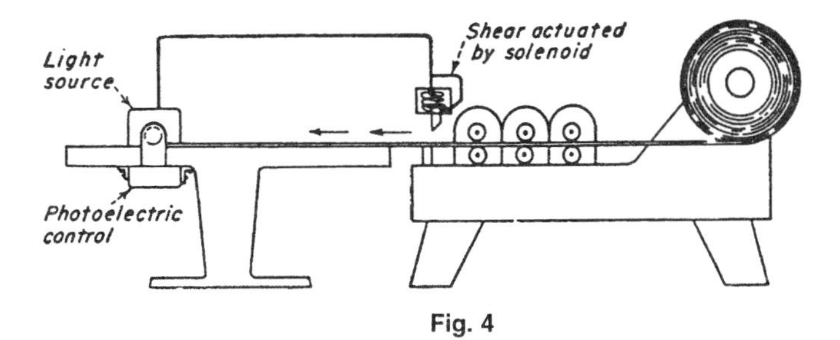

Fig. 4

Fig. 4 This cut-off machine has a photoelectric control for strip materials that lack sufficient mass to operate a mechanical limit switch satisfactorily. The forward end of the strip breaks the light beam, thus actuating the cut-off operation. The light source and the control are mounted on an adjustable stand at the end of the machine to vary the length of the finished stock.

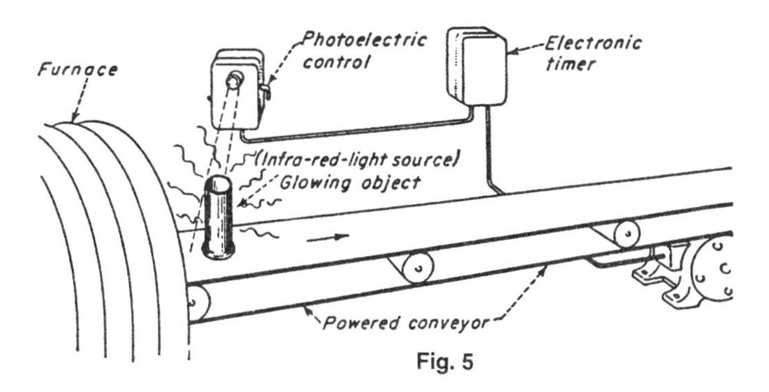

Fig. 5

Fig. 5 This heat-treating conveyor has an electronic timer paired with a photoelectric control to carry parts emerging from a furnace at 2300°F. The conveyor must operate only when a part is placed on it and only for the distance required to reach the next process stage. Parts are ejected onto the conveyor at varying rates. High temperatures caused failures when the mechanical switches were used. Glowing white-hot parts radiate infrared rays that actuate the photoelectric control as soon as a part comes in view. The control operates the conveyor that carries the parts away from the furnace and simultaneously starts the timer. The conveyor is kept running by a timer for the predetermined length of time required to position the part for the next operation.

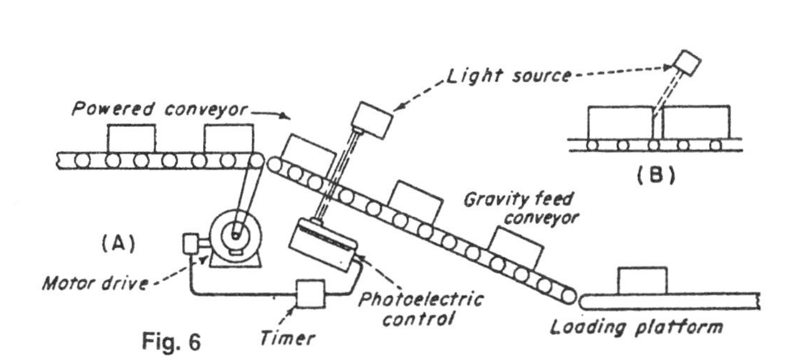

Fig. 6

Fig. 6 Jam detector. Cartons jamming on the conveyor cause losses in production and damage to cartons, products, and conveyors. Detection is accomplished with a photoelectric control that has a timer, as shown in (A). Each time a carton passes the light source, the control beam is broken. That starts the timing interval in the timer. The timing circuit is reset without relay action each time the beam is restored before the preset timing interval has elapsed. If a jam occurs, causing cartons to butt against each other, the light beam cannot reach the control. The timing circuit will then time-out, opening the load circuit. This stops the conveyor motor. By locating the light source at an angle with respect to the conveyor, as shown in (B), the power conveyor can be delayed if cartons are too close to each other but not butting each other.

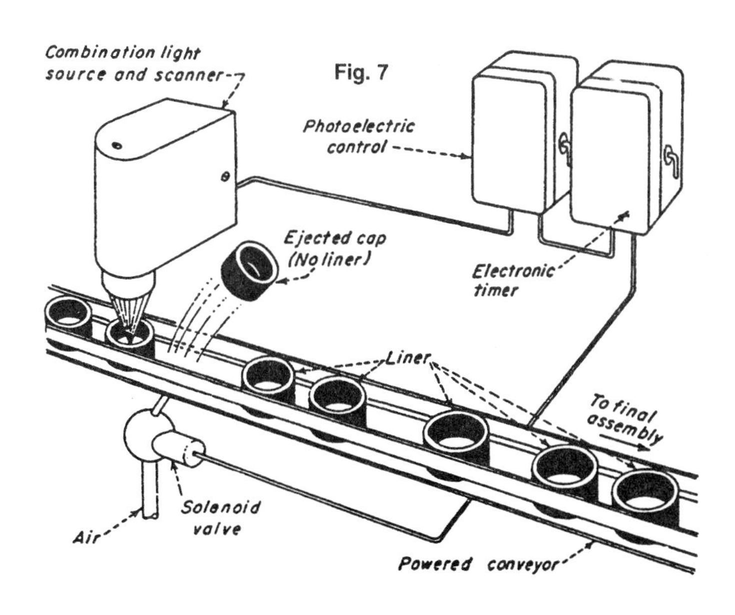

Fig. 7

Fig. 7 Automatic inspection. As steel caps are conveyed to final assembly, they pass an intermediate stage where an assembler inserts an insulation liner into a cap. The inspection point for the missing liners has a reflection-type photoelectric scanner which incorporates both a light source and photosensor with a common lens system to recognize the difference in reflection between the dark liner and the light steel cap instantly. When it detects a cap without a liner, a relay operates an airjet ejector that is controlled by a solenoid valve. The start and duration of the air blast is accurately controlled by a timer so that no other caps are displaced.

LIQUID LEVEL INDICATORS AND CONTROLLERS

Thirteen different systems of operation are shown. Each one represents at least one commercial instrument. Some of them are available in several modified forms.

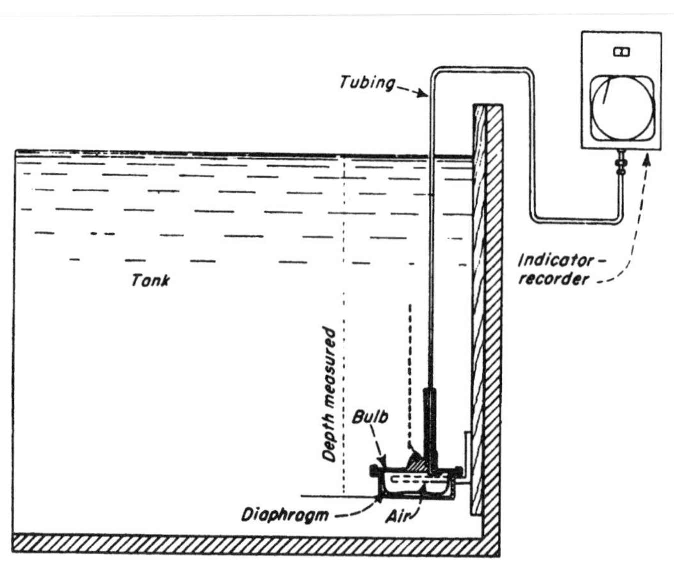

A diaphragm actuated indicator will work with any kind of liquid, whether it is flowing, turbulent, or carrying solid matter. A recorder can be mounted above or below the level of the tank or reservoir.

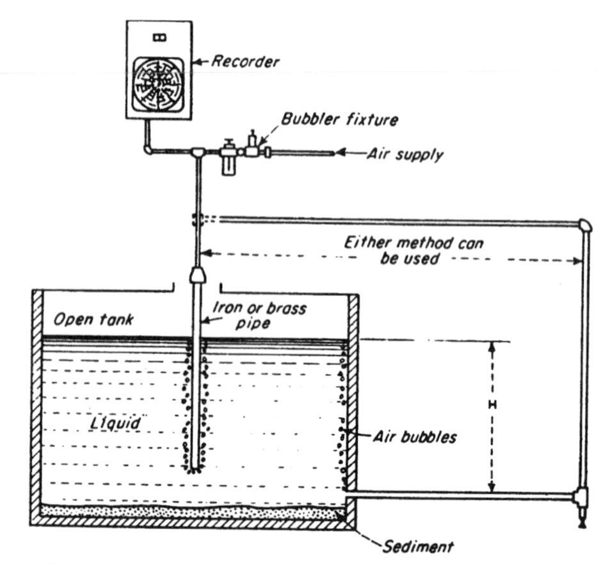

A bubbler-type recorder measures height H. It can be used with all kinds of liquids, including those carrying solids. A small amount of air is bled into a submerged pipe. A gage measures the air pressure that displaces the fluid.

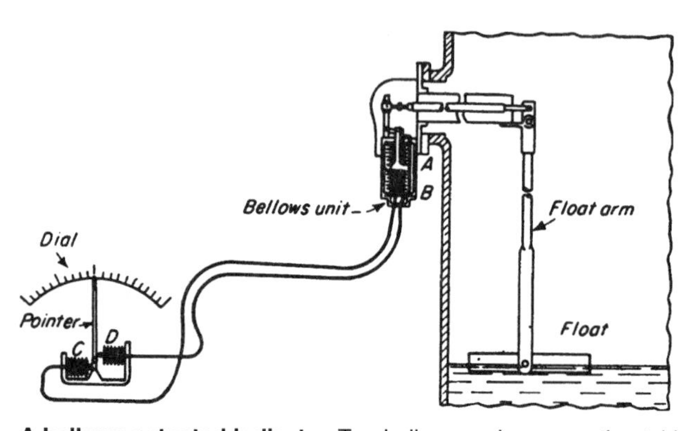

A bellows-actuated indicator. Two bellows and a connecting tubing are filled with incompressible fluid. A change in liquid level displaces the transmitting bellows and pointer.

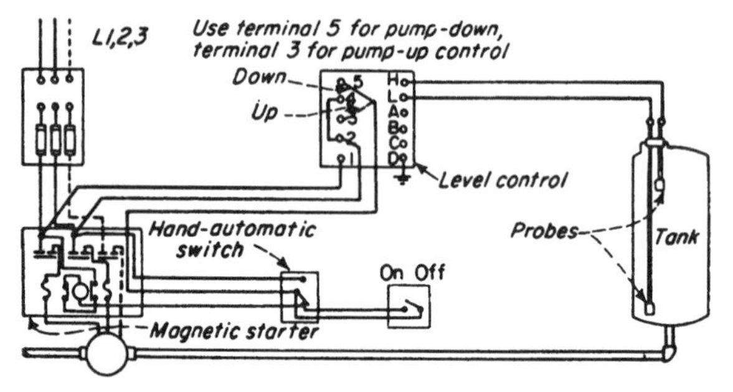

An electrical level controller. The positions of the probes determine the duration of pump operation. When a liquid touches the upper probe, a relay operates and the pump stops. Auxiliary contacts on the lower probe provide a relay-holding current until the liquid level drops below it.

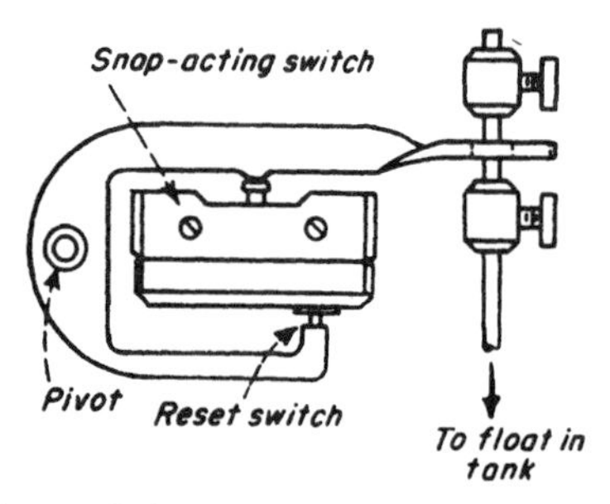

A float-switch controller. When liquid reaches a predetermined level, a float actuates a switch through a horseshoe-shaped arm. A switch can operate the valve or pump.

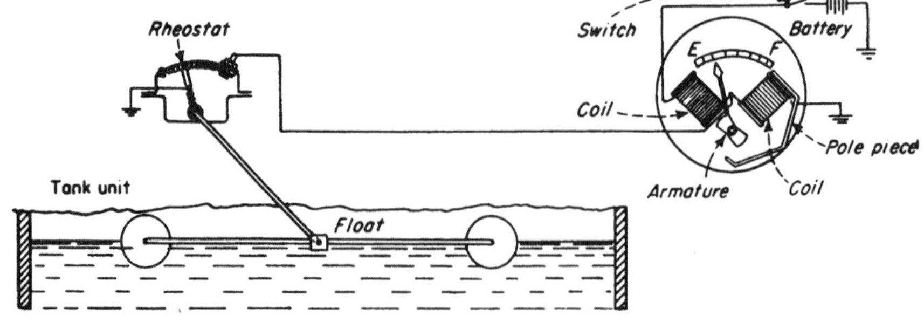

An automotive liquid-level indicator. The indicator and tank unit are connected by a single wire. As the liquid level in the tank increases, brush contact on the tank rheostat moves to the right, introducing an increasing amount of resistance into the circuit that grounds the F coil. The displacement of a needle from its empty mark is proportional to the amount of resistance introduced into this circuit.

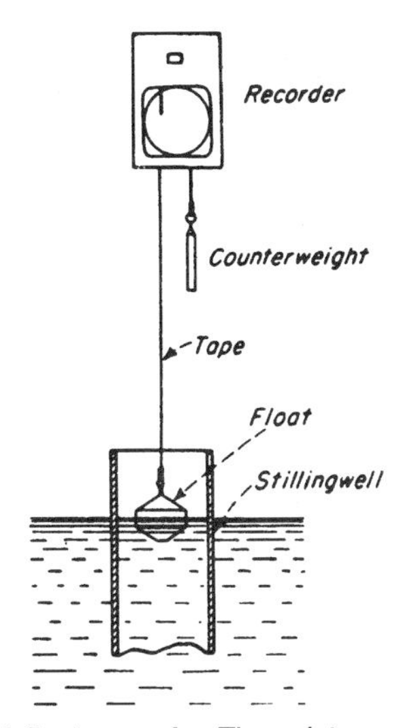

A float recorder. The pointer can be attached to a calibrated float tape to give an approximate instantaneous indication of fluid level.

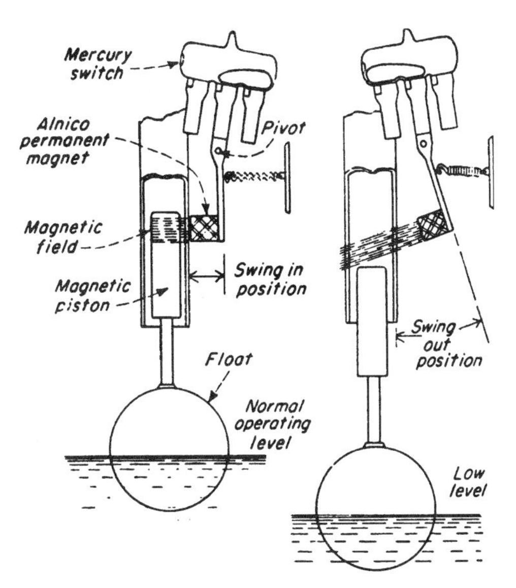

A magnetic liquid-level controller. When the liquid level is normal, the common-to-right leg circuit of the mercury switch is closed. When the liquid drops to a predetermined level, the magnetic piston is drawn below the magnetic field.

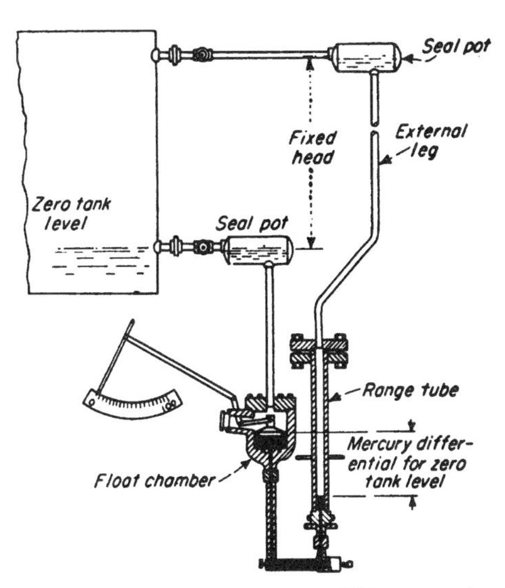

A differential pressure system. This system is applicable to liquids under pressure. The measuring element is a mercury manometer. A mechanical or electric meter body can be used. The seal pots protest the meter body.

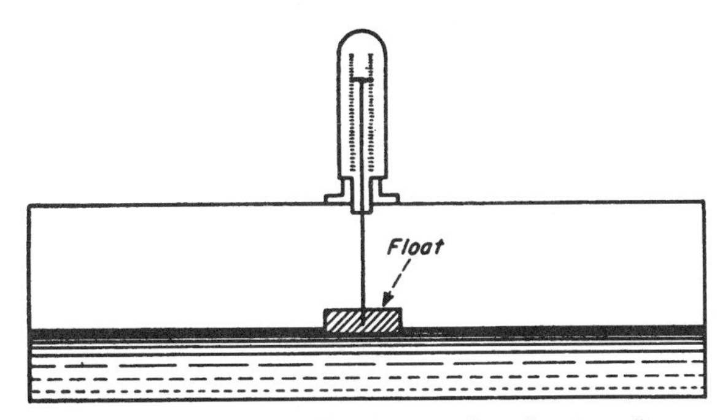

A direct-reading float gage. This inexpensive, direct-reading gage has a dial calibrated to the tank volume. A comparable gage, in terms of simplicity, has a needle connected through a right-angle arm to the float. As the liquid level drops, the float rotates the arm and the needle.

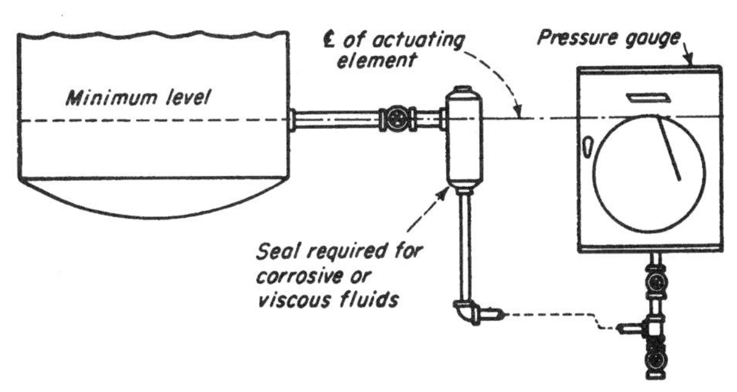

A pressure gage indicator for open vessels. The pressure of the liquid head is imposed directly upon the actuating element of the pressure gage. A center-line of the actuating element must coincide with the minimum level line if the gage is to read zero when the liquid reaches the minimum level.

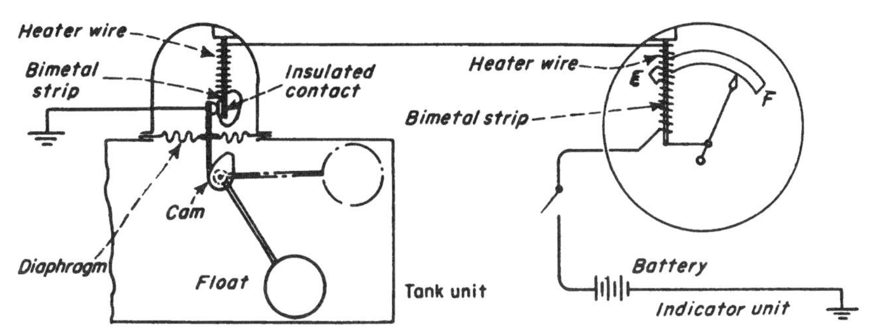

A bimetallic indicator. When the tank is empty, contacts in the tank unit just touch. With the switch closed, heaters cause both bimetallic strips to bend. This opens the contacts in the tank, and the bimetals cool, closing the circuit again. The cycle repeats about once per second. As the liquid level increases, the float forces the cam to bend the tank bimetal. This action is similar to that of the float gage, but the current and the needle displacement are increased.

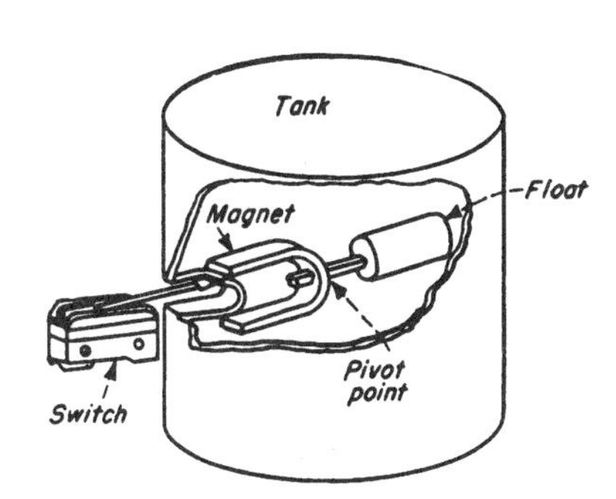

A switch-actuated level controller. This pump is actuated by the switch. The float pivots the magnet so that the upper pole attracts the switch contact. The tank wall serves as the other contact.

APPLICATIONS FOR EXPLOSIVE-CARTRIDGE DEVICES

Cartridge-actuated devices generate a punch that cuts cable and pipe, shears bolts for fast release, and provides emergency thrust.

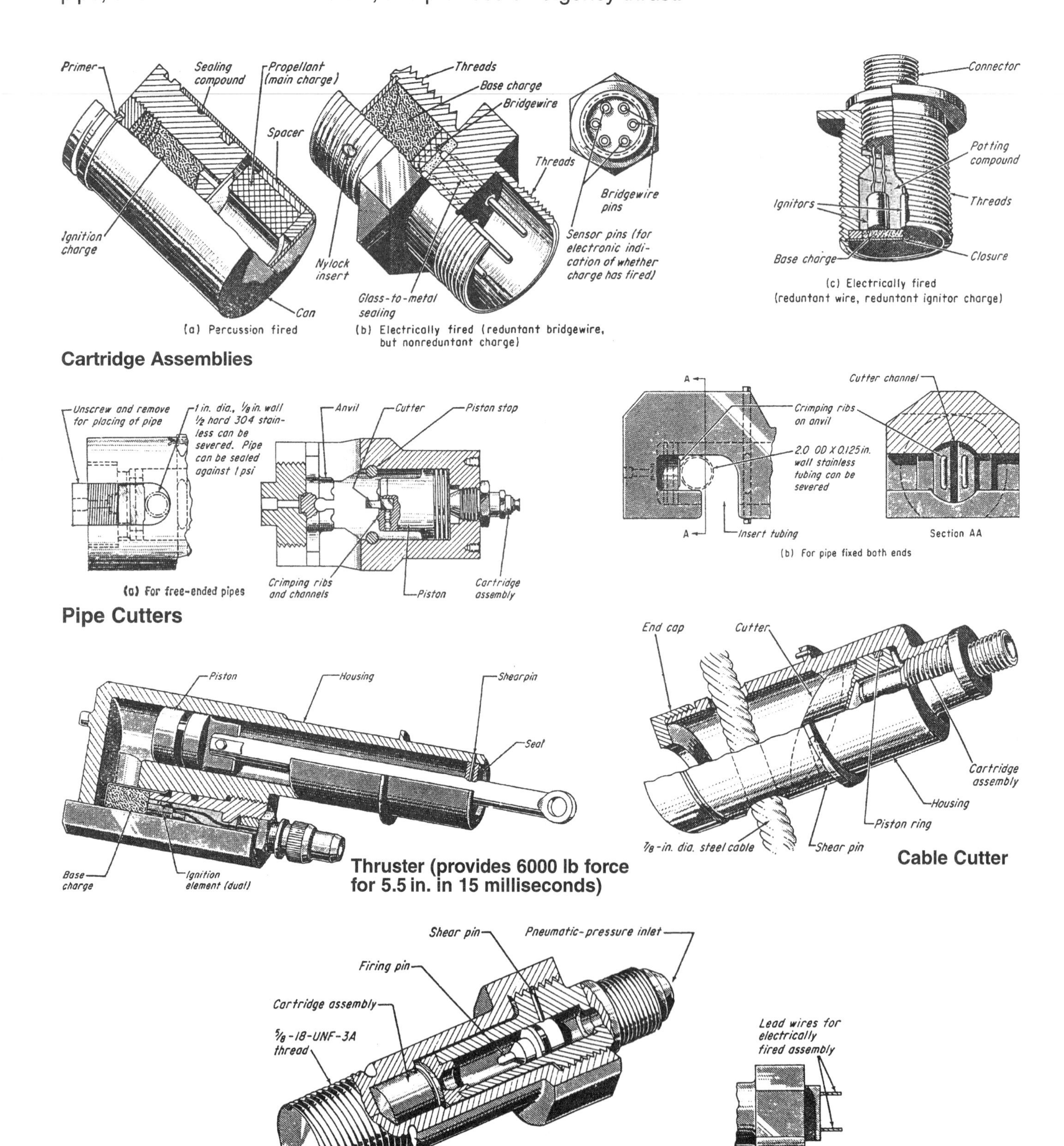

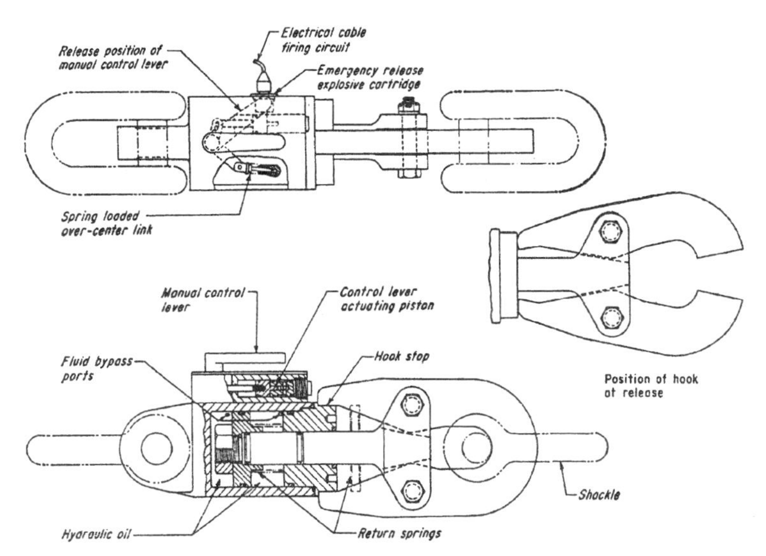

An emergency hook release lets the loads be jettisoned at any time. The hook is designed to release automatically if it is overloaded.

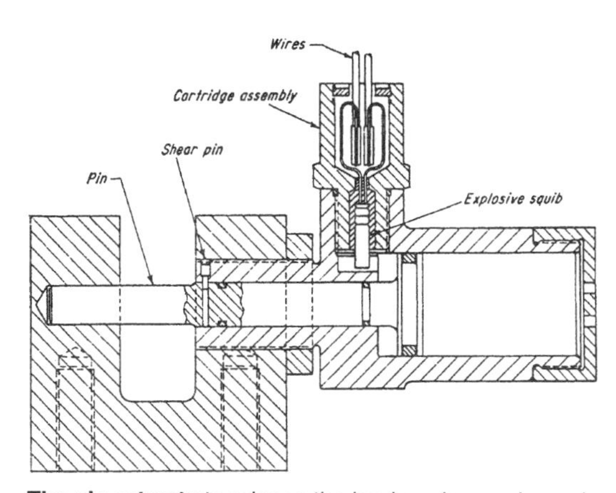

The pin retracts to release the load or clear a channel for free movement.

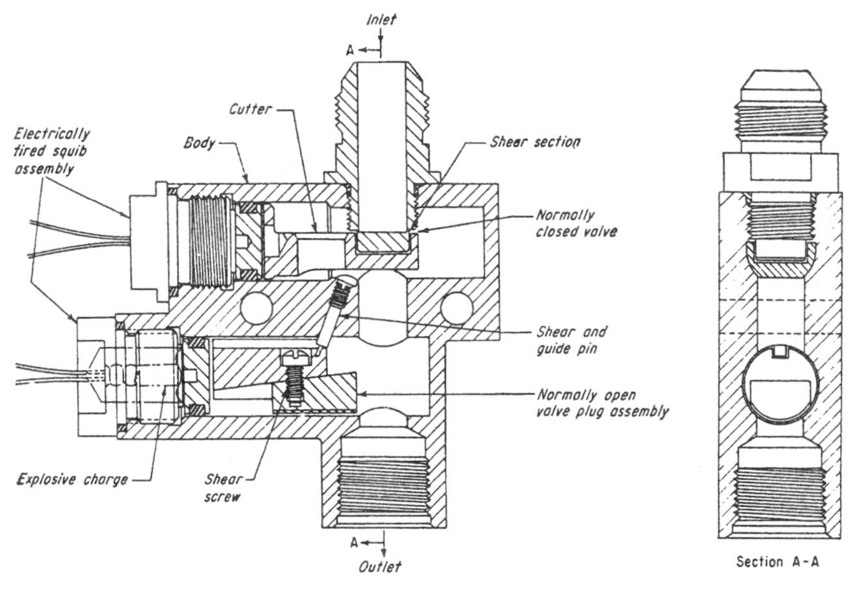

This dual valve is designed so that the flow will be started and stopped by the same unit. Firing one squib starts the flow; firing the other squib stops the flow.

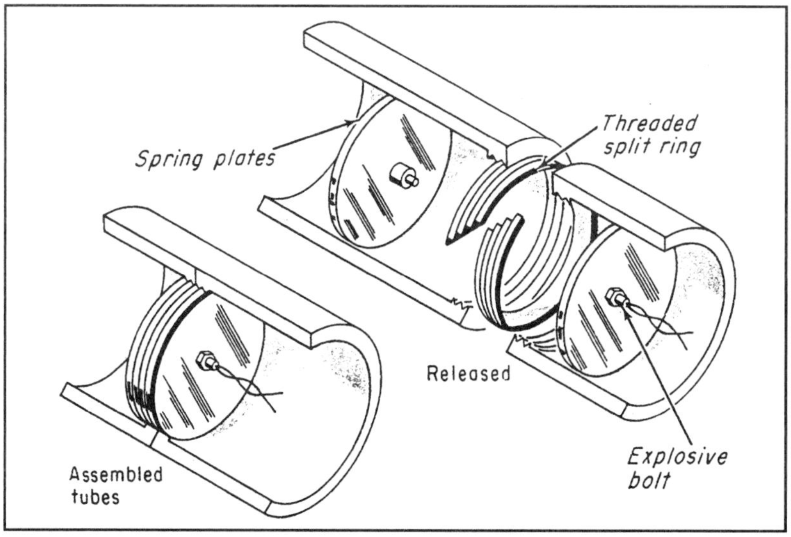

Quick Disconnector

A tube joint can be separated almost instantaneously by remote control with an explosive bolt and a split threaded ring, in a design developed by James Mayo of NASA's Langley Research Center, Hampton, Virginia.

External threads of the ring mesh with the internal threads of the members that are joined—and they must be separated quickly. The ring has a built-in spring characteristic that will assume a helically wound shape and reduce to a smaller diameter when not laterally constrained. During assembly, it is held to its expanded size by two spring plates whose rims fit into internal grooves machined in the split ring. The plates are fastened together by an explosive bolt and nut.

Upon ignition of the explosive bolt, the plates fly apart form the axial spring tension of the ring. The ring then contracts to its normally smaller diameter, releasing the two structural members.

The tube joint can be made in any size and configuration. The retaining media need not be limited to V-type screw threads.

A threaded split ring with a helical-spring response holds the ends of the tubes at a joint until the explosive bolt is fired; then it releases instantly.

CENTRIFUGAL, PNEUMATIC, HYDRAULIC, AND ELECTRIC GOVERNORS

Centrifugal governors are the most common—they are simple and sensitive and have high output force. There is more published information on centrifugal governors than on all other types combined.

In operation, centrifugal flyweights develop a force proportional to the square of the speed, modified by linkages, as required. In small engines the flyweight movement can actuate the fuel throttle directly. Larger engines require amplifiers or relays. This has lead to innumerable combinations of pilot pistons, linear actuators, dashpots, compensators, and gear boxes.

Centrifugal Governors

Control rod
Pivoted to hub
Flyweights
Eccentric motion of control rod
Spring

ACCELERATION GOVERNOR
(steam engine)

Adjustment nut
Spring
Airbleed valve

CENTRIFUGAL VALVE

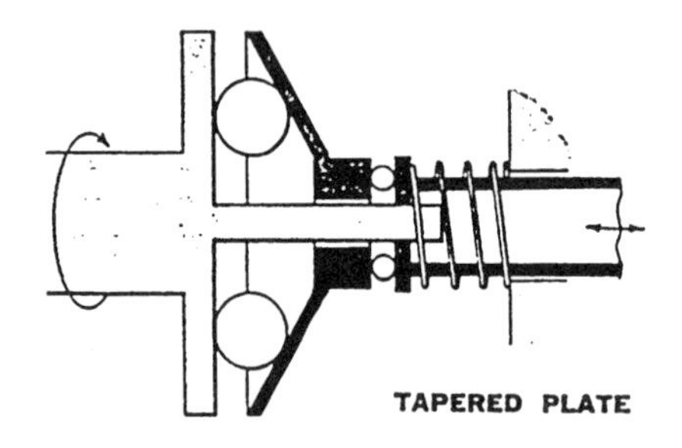

TAPERED PLATE

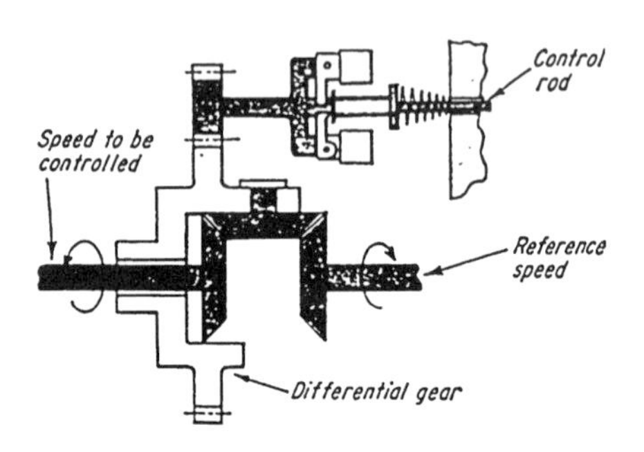

DIFFERENTIAL CENTRIFUGAL

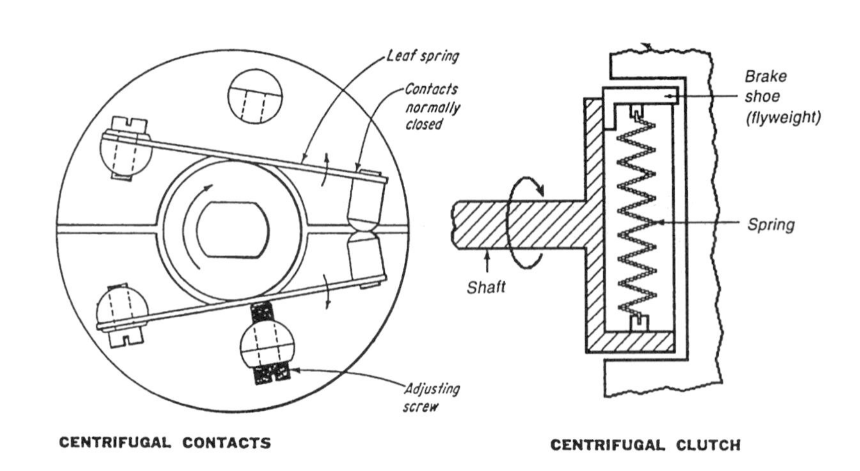

CENTRIFUGAL CONTACTS

CENTRIFUGAL CLUTCH

Pneumatic sensors are the most inexpensive and also the most inaccurate of all speed-measuring and governing components. Nevertheless, they are entirely adequate for many applications. The pressure or velocity of cooling or combustion air is used to measure and govern the speed of the engine.

Pneumatic Governors

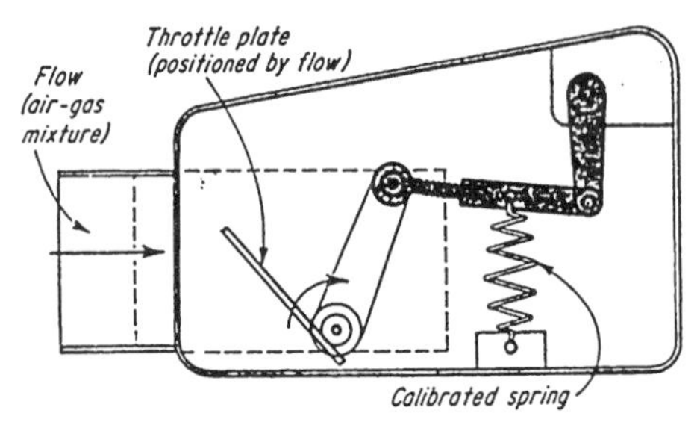

CARBURETOR-FLOW VELOCITY
(linkage)

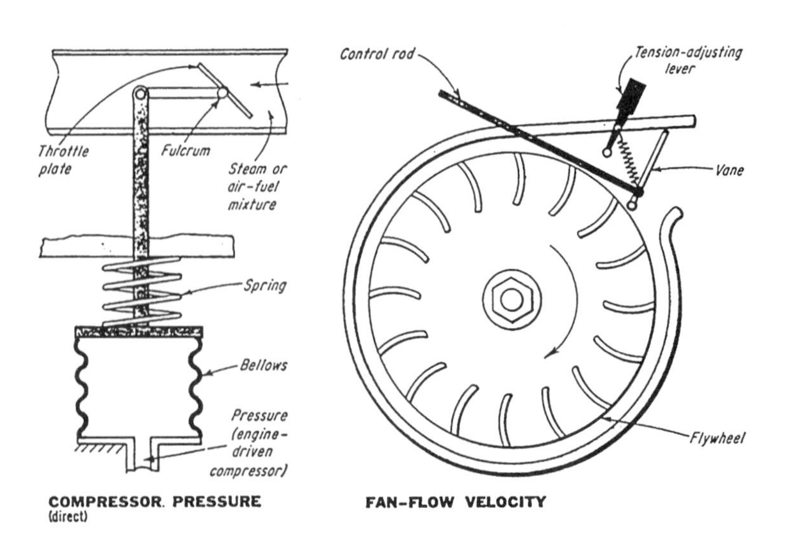

COMPRESSOR PRESSURE
(direct)

FAN-FLOW VELOCITY

More pneumatic governors

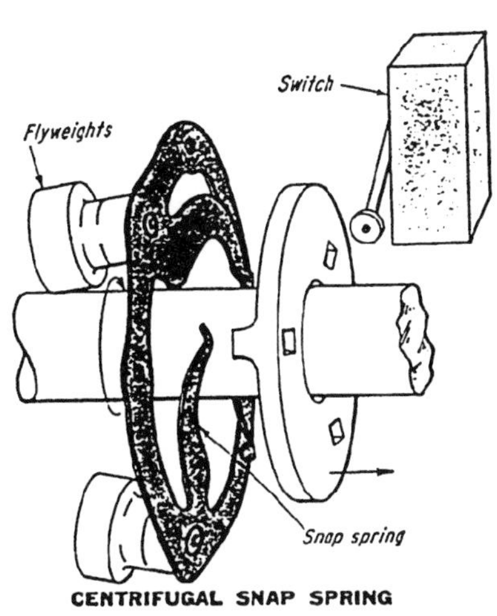

CENTRIFUGAL SNAP SPRING

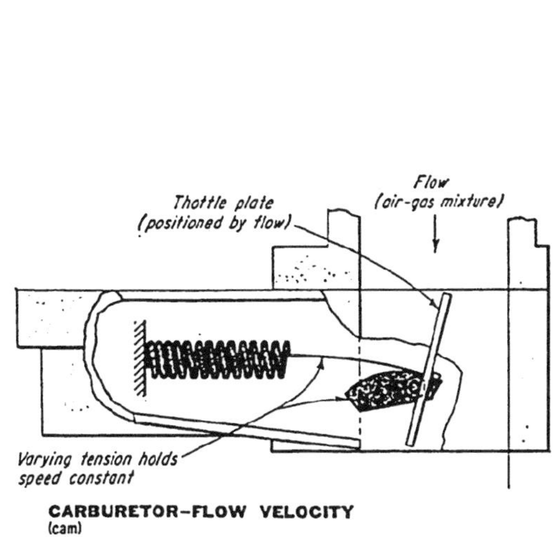

CARBURETOR-FLOW VELOCITY
(cam)

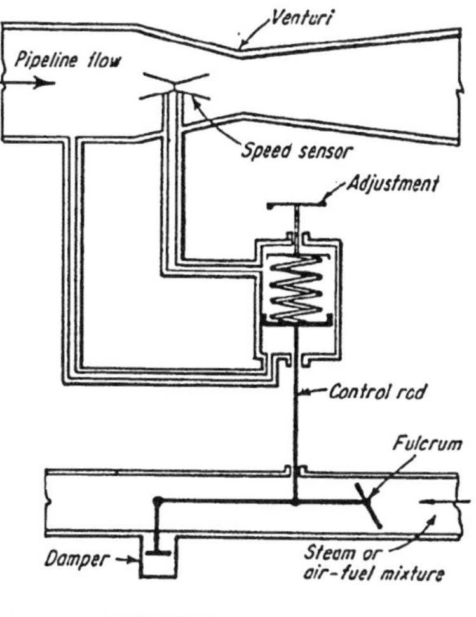

COMPRESSOR PRESSURE
(differential)

Hydraulic Governors

Hydraulic sensors measure the discharge pressure of engine-driven pumps. Pressure is proportional to the square of the speed of most pumps, although some have special impellers with linear pressure-speed characteristics.

Straight vanes are better than curved vanes because the pressure is less affected by the volume flow. Low pressures are preferred over high pressures because fluid friction is less.

Typical applications for these governors include farm tractors with diesel or gasoline engines, larger diesel engines, and small steam turbines.

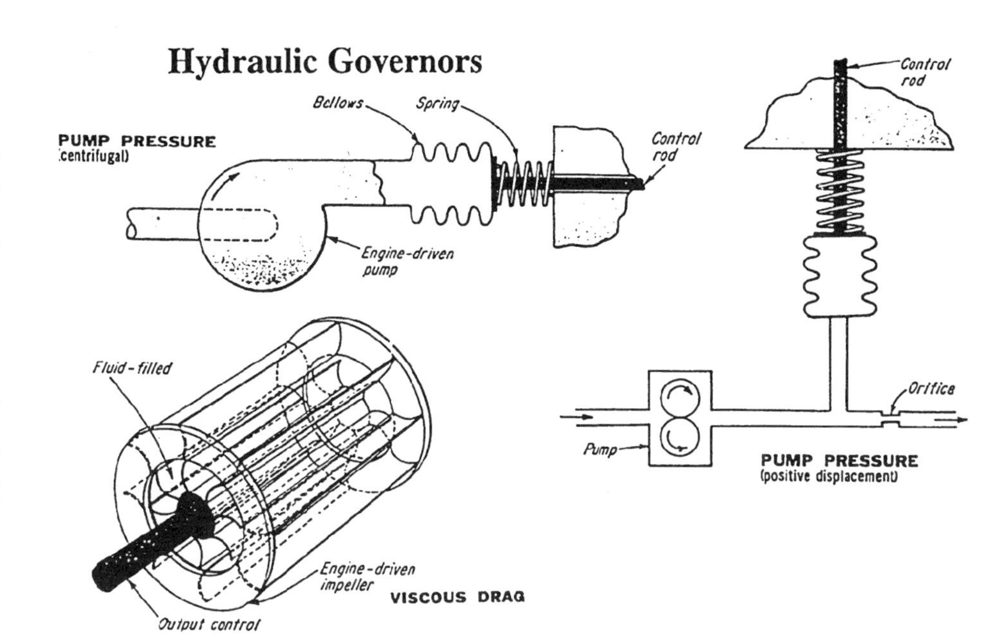

Electric Governors

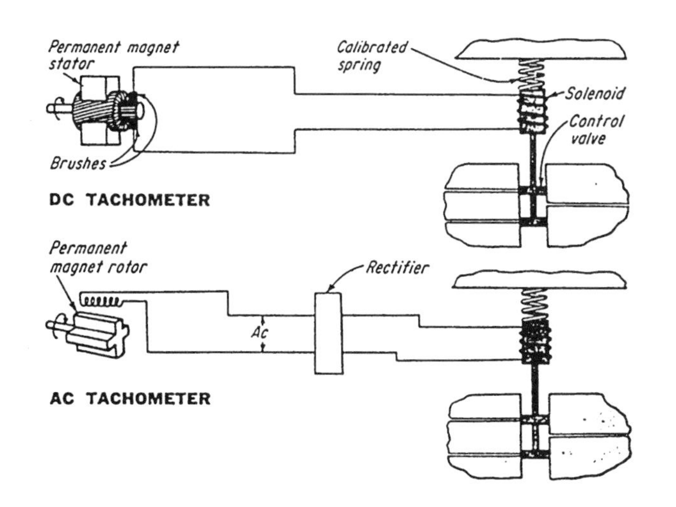

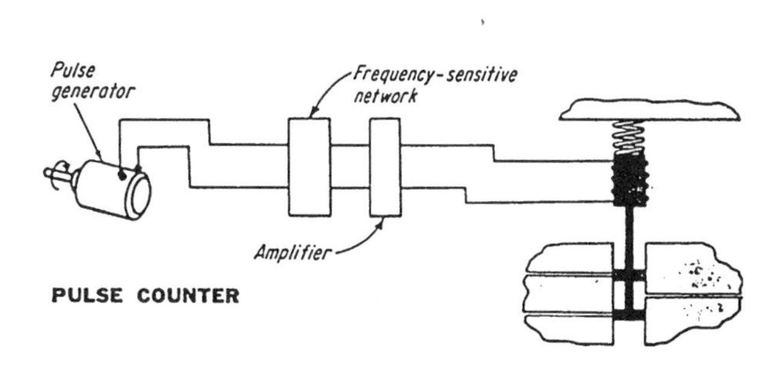

CHAPTER 17

COMPUTER-AIDED DESIGN CONCEPTS

INTRODUCTION TO COMPUTER-AIDED DESIGN

Computer Aided Design (CAD) is a computer-based technology that allows a designer to draw and label the engineering details of a product or project electronically on a computer screen while relegating drawing reproduction to a printer or X-Y plotter. It also permits designers in different locations to collaborate in the design process via a computer network and permits the drawing to be stored digitally in computer memory for ready reference. CAD has done for engineering graphics what the word processor did for writing. The introduction of CAD in the late 1960s changed the traditional method of drafting forever by relieving the designer of the tedious and time-consuming tasks of manual drawing from scratch, inking, and dimensioning on a conventional drawing board.

While CAD offers many benefits to designers or engineers never before possible, it does not relieve them of the requirement for extensive technical training and wide background knowledge of drawing standards and practice if professional work is to be accomplished. Moreover, in making the transition from the drawing board to the CAD workstation, the designer must spend the time and make the effort to master the complexities of the specific CAD software systems in use, particularly how to make the most effective use of the icons that appear on the screen.

The discovery of the principles of 3D isometric and perspective drawing in the Middle Ages resulted in a more realistic and accurate portrayal of objects than 2D drawings, and they conveyed at a glance more information about that object, but making a 3D drawing manually was then and is still more difficult and time-consuming, calling for a higher level of drawing skill. Another transition is required for the designer moving up from 2D to 3D drawing, contouring, and shading.

The D in CAD stands for design, but CAD in its present state is still essentially "computer-aided drawing" because the user, not the computer, must do the designing. Most commercial CAD programs permit lettering, callouts, and the entry of notes and parts lists, and some even offer the capability for calculating such physical properties as volume, weight, and center of gravity if the drawing meets certain baseline criteria. Meanwhile, CAD software developers are busy adding more automated features to their systems to move them closer to being true design programs and more user-friendly. For example, CAD techniques now available can perform analysis and simulation of the design as well as generate manufacturing instructions. These features are being integrated with the code for modeling the form and structure of the design.

In its early days, CAD required at least the computing power of a minicomputer and the available CAD software was largely application specific and limited in capability. CAD systems were neither practical nor affordable for most design offices and independent consultants. As custom software became more sophisticated and costly, even more powerful workstations were required to support them, raising the cost of entry into CAD even higher. Fortunately, with the rapid increases in the speed and power of microprocessors and memories, desktop personal computers rapidly began to close the gap with workstations even as their prices fell. Before long, high-end PCs become acceptable low-cost CAD platforms. When commercial CAD software producers addressed that market sector with lower-cost but highly effective software packages, their sales surged.

PCs that include high-speed microprocessors, Windows operating systems, and sufficient RAM and hard-drive capacity can now run software that rivals the most advanced custom Unix-based products of a few years ago. Now both 2D and 3D CAD software packages provide professional results when run on off-the-shelf personal computers. The many options available in commercial CAD software include.

- 2D drafting
- 3D wireframe and surface modeling
- 3D solid modeling
- 3D feature-based solid modeling
- 3D hybrid surface and solid modeling

Two-Dimensional Drafting

Two-dimensional drafting software for mechanical design is focused on drawing and dimensioning traditional engineering drawings. This CAD software was readily accepted by engineers, designers, and draftspersons with many years of experience. They felt comfortable with it because it automated their customary design changes, provided a way to make design changes quickly, and also permitted them to reuse their CAD data for new layouts.

A typical 2D CAD software package includes a complete library of geometric entities. It can also support curves, splines, and polylines as well as define hatching patterns and place hatching within complex boundaries. Other features include the ability to perform associative hatching and provide complete dimensioning. Some 2D packages can also generate bills of materials. 2D drawing and detailing software packages are based on ANSI, ISO, DIN, and JIS drafting standards.

In a 2D CAD drawing, an object must be described by multiple 2D views, generally three or more, to reveal profile and internal geometry from specific viewpoints. Each view of the object is created independently from other views. However, 2D views

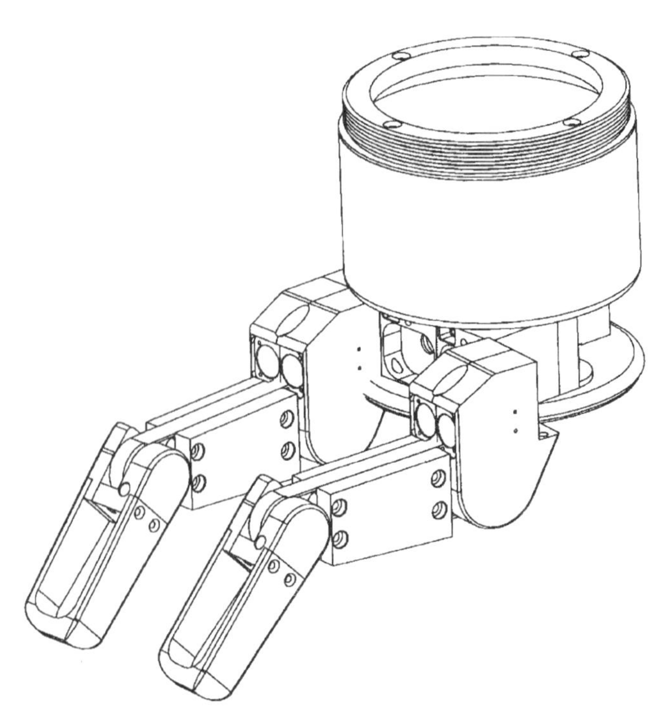

Robotic hand assembly with fingers 3D wireframe drawing.
Courtesy of SolidWorks Corporation

Robotic hand assembly with fingers 3D solid drawing. *Courtesy of SolidWorks Corporation*

typically contain many visible and hidden lines, dimensions, and other detailing features. Unless careful checks of the finished drawing are made, mistakes in drawing or dimensioning intricate details can be overlooked. These can lead to costly problems downstream in the product design cycle. Also, when a change is made, each view must be individually updated. One way to avoid this problem (or lessen the probability that errors will go undetected) is to migrate upward to a 3D CAD system.

Three-Dimensional Wireframe and Surface Modeling

A 3D drawing provides more visual impact than a 2D drawing because it portrays the subject more realistically and its value does not depend on the viewer's ability to read and interpret the multiple drawings in a 2D layout. Of more importance to the designer or engineer, the 3D presentation consolidates important information about a design, making it easier and faster to detect design flaws. Typically a 3D CAD model can be created with fewer steps than are required to produce a 2D CAD layout. Moreover, the data generated in producing a 3D model can be used to generate a 2D CAD layout, and this information can be preserved throughout the product design cycle. In addition, 3D models can be created in the orthographic or perspective modes and rotated to any position in 3D space.

The wireframe model, the simplest of the 3D presentations, is useful for most mechanical design work and might be all that is needed for many applications where 3D solid modeling is not required. It is the easiest 3D system to migrate to when making the transition from 2D to 3D drawing. A wireframe model is adequate for illustrating new concepts, and it can also be used to build on existing wireframe designs to create models of working assemblies.

Wireframe models can be quickly edited during the concept phase of the design without having to maintain complex solid-face relationships or parametric constraints. In wireframe modeling only edge information is stored, so data files can be significantly smaller than for other 3D modeling techniques. This can increase productivity and conserve available computer memory.

The unification of multiple 2D views into a single 3D view for modeling a complex machine design with many components permits the data for the entire machine to be stored and managed in a single wireframe file rather than many separate files. Also, model properties such as color, line style, and line width can be controlled independently to make component parts more visually distinctive.

The construction of a wireframe structure is the first step in the preparation of a 3D surface model. Many commercial CAD software packages include surface modeling with wireframe capability. The designer can then use available surface-modeling tools to apply a "skin" over the wireframework to convert it to a surface model whose exterior shape depends on the geometry of the wireframe.

One major advantage of surface modeling is its ability to provide the user with visual feedback. A wireframe model does not readily show any gaps, protrusions, and other defects. By making use of dynamic rotation features as well as shading, the designer is better able to evaluate the model. Accurate 2D views can also be generated from the surface model data for detailing purposes. Surface models can also be used to generate tool paths for numerically controlled (NC) machining. Computer-aided manufacturing (CAM) applications require accurate surface geometry for the manufacture of mechanical products.

Yet another application for surface modeling is its use in the preparation of photorealistic graphics of the end product. This capability is especially valued in consumer product design, where graphics stress the aesthetics of the model rather than its precision.

Some wireframe software also includes data translators, libraries of machine design elements and icons, and 2D drafting and detailing capability, which support design collaboration and compatibility among CAD, CAM, and computer-aided engineering (CAE) applications. Designers and engineers can store and use data accumulated during the design process. This data permits product manufacturers with compatible software to receive 2D and 3D wireframe data from other CAD systems.

Among the features being offered in commercial wireframe software are:

- Basic dimensioning, dual dimensioning, balloon notes, datums, and section lines.
- Automated geometric dimensioning and tolerancing (GD&T).
- Symbol creation, including those for weld and surface finish, with real-time edit or move capability and leaders.
- A library of symbols for sheet metal, welding, electrical piping, fluid power, and flow chart applications.

Data translators provide an effective and efficient means for transferring information from the source CAD design station to outside contract design offices, manufacturing plants, or engineering analysis consultants, job shops, and product development services. These include IGES, DXF, DWG, STL, CADL, and VRML.

Three-Dimensional Solid Modeling

CAD solid-modeling programs can perform many more functions than simple 3D wireframe modelers. These programs are used to form models that are solid objects rather than simple 3D line drawings. Because these models are represented as solids, they are the source of data that permits the physical properties of the parts to be calculated.

Some solid-modeling software packages provide fundamental analysis features. With the assignment of density values for a variety of materials to the solid model, such vital statistics as strength and weight can be determined. Mass properties such as area, volume, moment of inertia, and center of gravity can be calculated for regularly and irregularly shaped parts. Finite element analysis software permits the designer to investigate stress, kinematics, and other factors useful in optimizing a part or component in an assembly. Also, solid models can provide the basic data needed for rapid prototyping using stereolithography, and can be useful in CAM software programs.

Most CAD solid-model software includes a library of primitive 3D shapes such as rectangular prisms, spheres, cylinders, and cones. Using Boolean operations for forming unions, subtractions, and intersections, these components can be added, subtracted, intersected, and sectioned to form complex 3D assemblies. Shading can be used to make the solid model easier for the viewers to comprehend. Precise 2D standard, isometric, and auxiliary views as well as cross sections can be extracted from the solid modeling data, and the cross sections can be cross-hatched.

Three-Dimensional Feature-Based Solid Modeling

3D feature-based solid modeling starts with one or more wireframe profiles. It creates a solid model by extruding, sweeping, revolving, or skinning these profiles. Boolean operations can also be used on the profiles as well as the solids generated from these profiles. Solids can also be created by combining surfaces, including those with complex shapes. For example, this technique can be used to model streamlined shapes such as those of a ship's hull, racing-car's body, or aircraft.

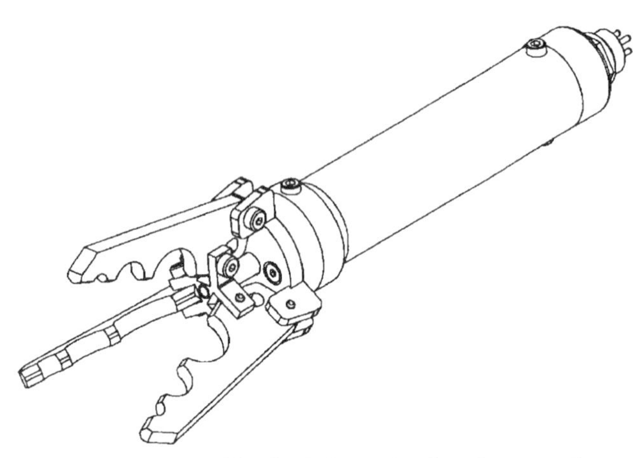

Robotic gripper assembly 3D wireframe drawing. *Courtesy of SolidWorks Corporation*

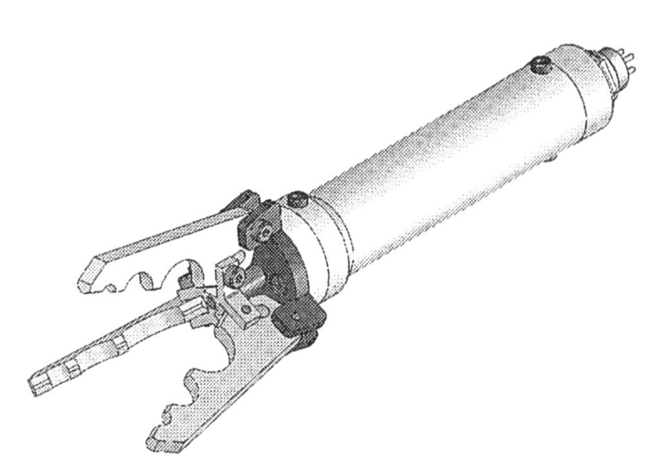

Robotic gripper assembly 3D solid drawing. *Courtesy of SolidWorks Corporation*

3D feature-based solid modeling allows the designer to create such features as holes, fillets, chamfers, bosses, and pockets, and combine them with specific edges and faces of the model. If a design change causes the edges or faces to move, the features can be regenerated so that they move with the changes to keep their original relationships.

However, to use this system effectively, the designer must make the right dimensioning choices when developing these models, because if the features are not correctly referenced, they could end up the wrong location when the model is regenerated. For example, a feature that is positioned from the edge of an

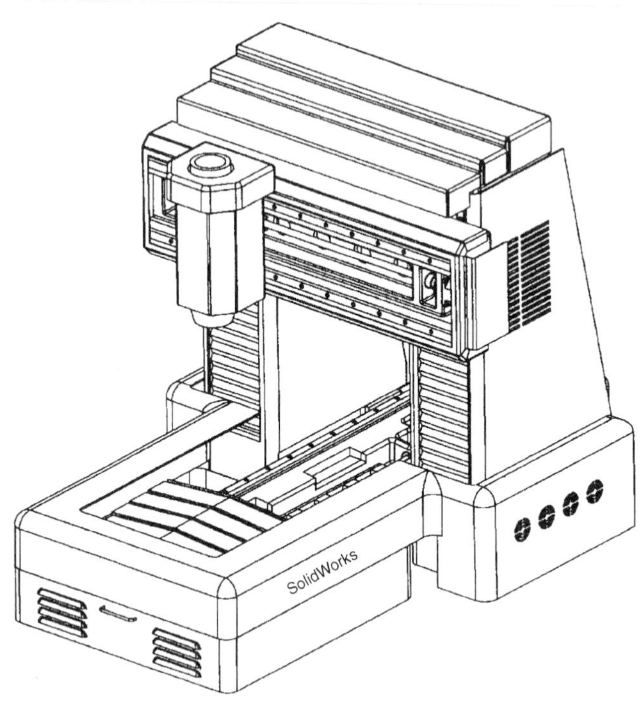

Milling machine 3D wireframe drawing. *Courtesy of SolidWorks Corporation*

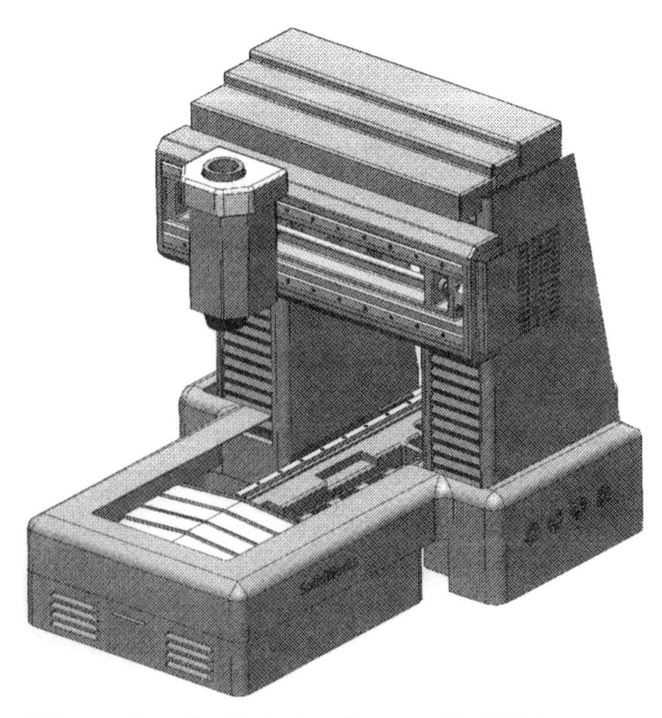

Milling machine 3D solid drawing. *Courtesy of SolidWorks Corporation*

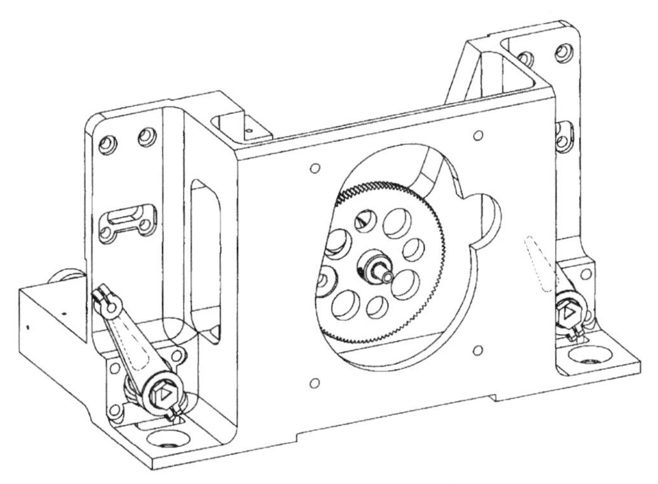

Machine subassembly 3D wireframe drawing. *Courtesy of SolidWorks Corporation*

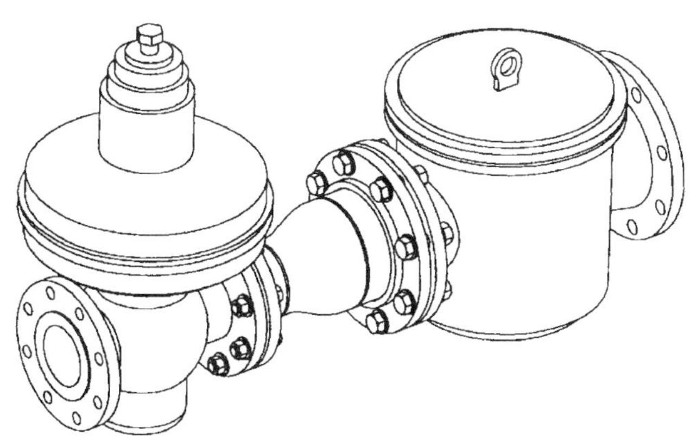

Piping adapter assembly 3D wireframe drawing. *Courtesy of SolidWorks Corporation*

object rather than from its center might no longer be centered when the model is regenerated. The way to avoid this is to add constraints to the model that will keep the feature at the center of the face.

The key benefit of the parametric feature of solid modeling is that it provides a method for facilitating change. It imposes dimensional constraints on the model that permit the design to meet specific requirements for size and shape. This software permits the use of constraint equations that govern relationships between parameters. If some parameters remain constant or a specific parameter depends on the values of others, these relationships will be maintained throughout the design process. This form of modeling is useful if the design is restricted by space allowed for the end product or if its parts such as pipes or wiring must mate precisely with existing pipes or conduits.

Thus, in a parametric model, each entity, such as a line or arc in a wireframe, or fillet, is constrained by dimensional parameters. For example, in the model of a rectangular object, these parameters can control its geometric properties such as the length, width, and height. The parametric feature allows the designer to make changes as required to create the desired model. This software uses stored historical records that have recorded the steps in producing the model so that if the parameters of the model are changed, the software refers to the stored history and repeats the sequence of operations to create a new model for regeneration. Parametric modeling can also be used in trial-and-error operations to determine the optimum size of a component best suited for an application, either from an engineering or aesthetic viewpoint, simply by adjusting the parameters and regenerating a new model.

Parametric modeling features will also allow other methods of relating entities. Design features can, for example, be located at the origin of curves, at the end of lines or arcs, at vertices, or at the midpoints of lines and faces, and they can also be located at a specified distance or at the end of a vector from these points. When the model is regenerated, these relationships will be maintained. Some software systems also allow geometric constraints between features. These can mandate that the features be parallel, tangent, or perpendicular.

Some parametric modeling features of software combine freeform solid modeling, parametric solid modeling, surface modeling, and wireframe modeling to produce true hybrid models. Its features typically include hidden line removal, associative layouts, photorealistic rendering, attribute masking, and level management.

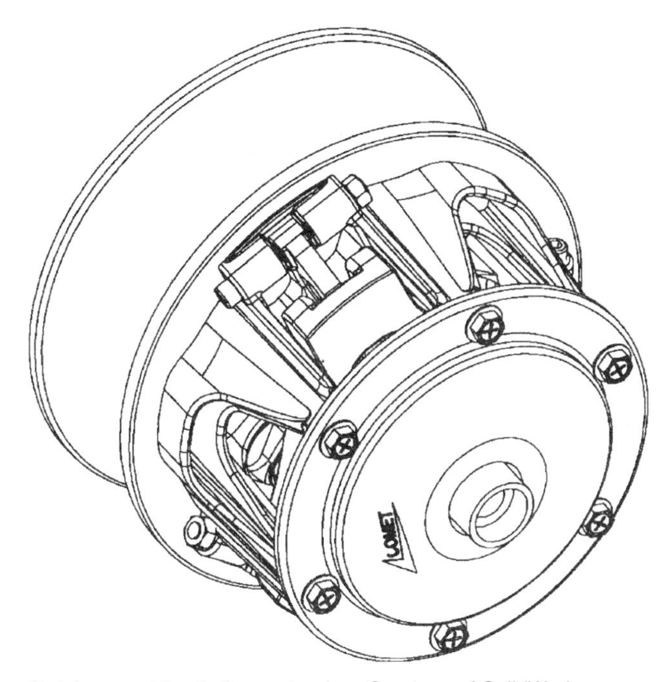

Clutch assembly wireframe drawing. *Courtesy of SolidWorks Corporation*

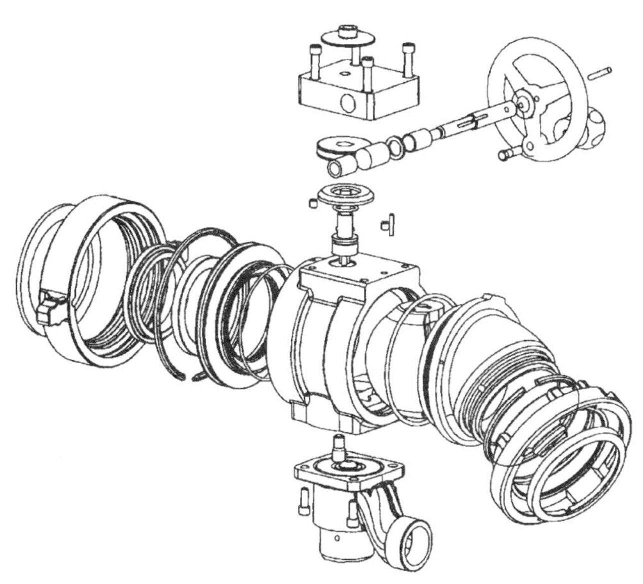

Firehose adapter assembly 3D wireframe drawing. *Courtesy of SolidWorks Corporation*

Three-Dimensional Hybrid Surface and Solid Modeling

Some modeling techniques are more efficient that others. For example, some are better for surfacing the more complex shapes as well as organic and freeform shapes. Consequently, commercial software producers offer 3D hybrid surface and solid-modeling suites that integrate 2D drafting and 3D wireframe with 3D surface and 3D solid modeling into a single CAD package. Included in these packages might also be software for photorealistic rendering and data translators to transport all types of data from the component parts of the package to other CAD or CAM software.

Glossary of Commonly Used CAD Terms

absolute coordinates: Distances measured from a fixed reference point, such as the origin, on the computer screen.

ANSI: An abbreviation for the American National Standards Institute.

associative dimensions: A method of dimensioning in CAD software that automatically updates dimension values when dimension size is changed.

Boolean modeling: A CAD 3D modeling technique that permits the user to add or subtract 3D shapes from one model to another.

Cartesian coordinates: A rectangular system for locating points in a drawing area in which the origin point is the 0,0 location and *X* represents length, *Y* width, and *Z* height. The surfaces between them can be designated as the X–Z, X–Y, and Y–*Z* planes.

composite drawing: A drawing containing multiple drawings in the form of CAD layers.

DXF: An abbreviation for Data Exchange Format, a standard format or translator for transferring data describing CAD drawings between different CAD programs.

FEM: An acronym for Finite Element Method for CAD structural design.

FTD: An abbreviation for File Transfer Protocol for upload and download of files to the Internet.

function: A task in a CAD program that can be completed by issuing a set of commands.

GD&T: An automated geometric, dimensioning, and tolerancing feature of CAD software.

GIS: An abbreviation for Geographic Information System.

IGES: An abbreviation for International Graphics Exchange Specification, a standard format or translator for transferring CAD data between different programs.

ISO: An abbreviation for International Standards Organization.

linear extrusion: A 3D technique that projects 2D into 3D shapes along a linear path.

MCAD: An abbreviation for mechanical CAD.

menu: A set of modeling functions or commands that are displayed on the computer screen. Options can be selected from the menu by a pointing device such as a mouse.

object snaps: A method for indicating point locations on existing drawings as references.

origin point: The 0,0 location in the coordinate system.

parametric modeling: CAD software that links the 3D drawing on the computer screen with data that sets dimensional and positional constraints.

polar coordinates: A coordinate system that locates points with an angle and radial distance from the origin, considered to be the center of a sphere.

polyline: A string of lines that can contain many connected line segments.

primitives: The basic elements of a graphics display such as points, lines, curves, polygons, and alphanumeric characters.

prototype drawing: A master drawing or template that includes preset computer defaults so that it can be reused in other applications.

radial extrusion: A 3D technique for projecting 2D into 3D shapes along a circular path.

spline: A flexible curve that can be drawn to connect a series of points in a smooth shape.

STL: An abbreviation for Solid Transfer Language, files created by a CAD system for use in rapid prototyping (RP).

tangent: A line in contact with the circumference of a circle that is at right angles to a line drawn between the contact point and the center of the circle.

CHAPTER 18
RAPID PROTOTYPING

RAPID PROTOTYPING FOCUSES ON BUILDING FUNCTIONAL PARTS

A three-dimensional (3-D) model makes it a lot easier to visualize the size and shape of a prospective new product than any two-dimensional (2-D) rendering or image. In the past, designers and engineers who wanted a 3-D model of a planned product to hold, hand around, and evaluate, had to order one custom-made from wood or metal; skilled model makers took a long time to build them, and they were expensive. If, as a result of evaluation, design changes were recommended, more time and money had to be spent in either correcting the model or ordering a new one. Fortunately, with the introduction of computer-aided rapid prototyping some 20 years ago, model or prototype-making was accelerated, and the cost of each model was drastically reduced. Since that time the technology has evolved: older processes have been improved and new ones have been introduced. Prototypes can now be made full size or scaled down from a variety of materials in a wide range of colors, and the technology has achieved global status.

Rapid prototyping (RP) is a class of computer-aided technologies for building 3-D prototypes from a range of materials based on data obtained from computer-aided design (CAD) drawings. The dimensional data in digital form taken from CAD drawings is converted into build directions for the 3-D model on various RP machines or systems. Typically these call for the construction of the model one layer at a time. The objective of all RP systems is the fabrication of prototypes, molds, and even functional parts or tools faster and more economically than they could be made by skilled persons using hand tools or conventional machine tools.

A 3-D prototype gives engineers, designers, and others concerned with the design and manufacture of a product a more convenient way to evaluate a proposed design and elicit comments on it. Persons participating in the evaluation process typically include manufacturing, marketing, and sales managers, suppliers, dealers, and even prospective customers. The model or models can be passed around a conference table for a hands-on inspection and review, providing an opportunity for all present to detect flaws, omissions, or objectionable features that could create manufacturing problems and lead to its failure in the marketplace. It is important that these problems be discovered early on in the design phase, before tooling is ordered. Early identification of design problems saves time and can eliminate the higher cost of making design corrections during the production phase. Any corrections that are identified can be made to the CAD drawings so they are reflected in a revised prototype. All RP prototypes can be helpful in reducing the time-to-market of the product, especially important for highly competitive consumer goods.

In addition to model making, some RP technologies have been adapted for the short-run manufacture of functional parts or tools. These RP processes are classified under the general heading of *rapid manufacturing* (RM), *solid free-form fabrication, computer-automated manufacturing,* and *layered manufacturing.* The same or comparable materials—plastic, metal, or ceramic—that would be used to mass-produce a product can be used for these short runs. Some RP technologies are now being used to manufacture replacement or spare parts for maintaining or repairing older existing or obsolete machines. These RP technologies eliminate the high cost and delay incurred in reproducing the original production tooling. Yet another offshoot of RP, *rapid tooling* (RT), focuses on the economical design and fabrication of certain specialized functional tools which can also be made from the same or equivalent materials as the mass-produced tools.

The term "rapid" as it applies to RP technologies is, of course, relative; even the fastest RP fabrication process takes from 3 to 72 hours, depending on the size and complexity of the prototype. Nevertheless, all of these methods are faster than the weeks or even months required to fashion a prototype by traditional handcrafted methods.

RAPID PROTOTYPE PROCESSES

Most commercial RP technologies are *layered* or *additive* processes and many call for post-process hand or machine finishing to obtain the desired finish or dimensional precision. Layers of plastic, paper, wax, and powdered metal or ceramic are combined to create solid or hollow 3-D objects. Some specialized RP processes focus on economical and rapid preparation of sand or wax molds for traditional metal casting. Magnesium, aluminum, iron, and steel parts have been cast from these molds. Other RP processes are *subtractive* meaning that the prototype or functional object is machined from solid blocks of material under computer control. The materials might be plastic resin or easily machined metals such as aluminum, magnesium, or their alloys. This machining is typically done automatically by computer numerically controlled (CNC) milling machines directed by software derived from CAD data.

RAPID PROTOTYPING STEPS

There are five steps in the production of a rapid prototype:

1. *Prepare a CAD Drawing of the Prototype*: It is necessary to produce a 3-D drawing file with the necessary dimensional data to prepare the software that will direct the RP build process.
2. *Convert the CAD Data to STL Format*: After the solid model is drawn, its dimensional data must be converted by specialized software to files that can be used by the RP equipment. Of the available software to do this, the STL format is considered the industry standard. It provides an approximation of the surface and solid characteristics of the model as a network of triangles called *facets*. The STL file contains only approximate information about the shape of the object because curved surfaces cannot be represented precisely by triangles. While increasing the number of facets improves the accuracy of the representation, it involves an accuracy-speed trade-off. The larger the file, the more time it will take to convert the file to build instructions; this means that it will take more time for the RP equipment to build the prototype.
3. *Convert the STL File into Cross-Sectional Layers*: In the third step, preprocessing software converts the STL file into software that will actually direct the RP system how to build the model in layers. Typically this software permits the size, location, and orientation of the model to be selected. The *X, Y, Z* coordinates of an RP build envelope are illustrated in Fig. 1. The size of this envelope for most RP processes is typically less than 1 ft^3. The shortest dimension is usually oriented along the *Z* direction because this direction directly relates to build time. The preprocessing software "slices" the STL model into many layers; the thickness of these layers depends on the RP process selected and the accuracy requirements. Many RP systems offer a range of "slice" thickness that can vary from 0.0025 to 0.5 in. (0.06 to 13 mm). The software can also generate the temporary structures needed to support the model while it is being built. Features such as overhangs, internal cavities, and thin-walled sections require support; many RP system manufacturers include their own proprietary preprocessing software which includes provisions for building these supports.
4. *Construct the Prototype in Layers*: The fourth step is the actual construction of the model. Using techniques described later, the RP system builds the prototype one layer at a time from materials that are principally polymers, paper, powdered metal, or powdered ceramic. Most RP systems are essentially autonomous, so they need very little human intervention.
5. *Clean and Finish the Model*: The fifth and final step is post-processing. This typically calls for removing the prototype from the system, removing any temporary structural supports and, in some cases, recycling excess materials. Prototypes made from photosensitive plastic resins usually must undergo a final ultraviolet (UV) curing step to complete the hardening of the prototype. Other finishing steps might include cleaning and sanding, sealing, painting, or polishing the prototype to improve its appearance and durability; however it might also need additional machining to improve its dimensional accuracy.

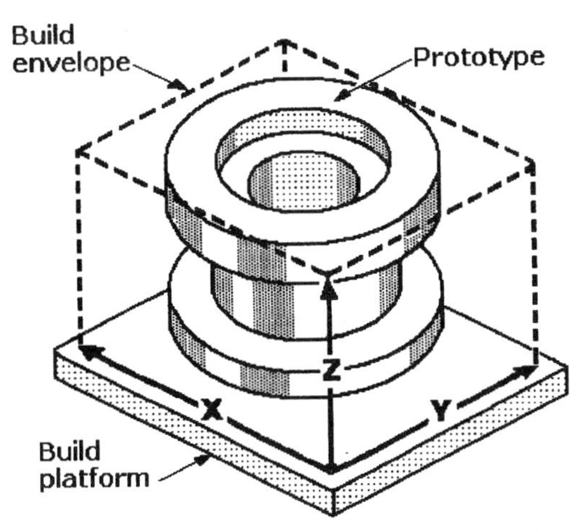

Fig. 1 Build Envelope: The rapid prototyping industry's term for the maximum size of any object or model that can be built with one of the RP technologies. It is given as length *X*, width *Y*, and height *Z* dimensions in inches or millimeters. The size of the object that can be built depends on the materials used in the process or the capacity of the RP equipment or system. A short Z dimension minimizes the layers required.

Each of the RP processes is focused on a specific market segment, taking into account its requirements for model size, durability, fabrication method, process speed, and prototype finish. Some RP processes are not suitable for building large models, and each process yields a model with a different finish. The choice of the most appropriate RP method for any given application depends on the urgency of the project and the cost and time saving of building an RP prototype versus producing the prototype by conventional model-making procedures.

COMMERCIAL RAPID PROTOTYPING CHOICES

In 2006 there were eight recognized additive RP systems and at least one subtractive RP system available commercially and supported by their manufacturers; each has its own unique strengths and drawbacks. A commercial system is one that is sold as a complete turnkey package that is either off-the-shelf or built to custom order using standardized components. Typical customers are industries or laboratories that require enough prototypes to justify purchasing the systems. For those organizations that need only a few per year, there are third-party contract service providers who will produce them in their own facilities. Some RP system manufacturers offer these services in their own shops using their own equipment. Of the many commercially available systems, six have become the most popular worldwide.

Most RP systems were developed in the United States, but some have been developed in Germany, the Netherlands, Japan, and Israel. RP systems have been sold around the world. In addition to direct sales by the RP system original equipment manufacturers (OEMs), there are at least 15 resellers of RP equipment; 11 of these are in the United States and the others are in Europe and Australia. In addition, there are contractors willing to provide various RP services in most industrialized countries.

In instances in which contracts are let to third-party service providers, the customer usually provides the requisite CAD data to the service provider. While organizations typically buy RP systems if internal demand for prototypes is sufficient to justify the cost of purchasing a system and training operators, some buy the equipment mainly because they want all information about their prototype designs to be kept confidential.

A measure of the popularity or acceptance of specific commercial RP technologies can be gained from the advertising of contract service providers. Allowing for the possibility of new entries into the ranks as well as dropouts, more than 200 service providers were in business in 2006. Of the 219 organizations that advertise the technologies they offered in 2005, 131 or 60% offered Stereolithography (SL), 45 or 20% offered Fused Deposition Modeling (FDM), 42 or 19% offered Selective Laser Sintering (SLS), and 27 or 12% offered Laminated Object Manufacturing (LOM).

Many service providers offer two or more of these technologies, while a smaller number offered technologies other than the four mentioned. These numbers include OEMs who offer RP services using their own proprietary systems. The fact that 47 or more than 20% of the service providers are located outside of the United States, predominately in Europe, Canada, Central and South America, and Asia, attests to the global acceptance of RP.

Some RP systems called 3-D office printers are self-contained autonomous manufacturing units housed in small desktop cabinets suitable for operation within an office environment. They include provisions for either containing or venting any smoke or fumes resulting from their processes. Although model size and choice of materials are limited, these small systems are relatively inexpensive compared with those capable of producing larger prototypes.

All commercial RP methods depend on computers, but four of them require lasers either to cut or fuse each lamination or provide sufficient heat to sinter or melt metal powders or plastic resins. The four processes that depend on lasers are Stereolithography (SL), Selective Laser Sintering (SLS), Laminated Object Manufacturing (LOM), and Directed Light Fabrication (DLF). The four processes that do not require lasers are Fused Deposition Modeling (FDM), Three Dimensional Printing (3DP), Direct Shell Production Casting (DSPC), and Solid Ground Curing (SGC).

All of the RP system OEMs mentioned in this chapter were included because of the availability of technical information about their systems; however, this chapter is not intended as a comprehensive overview of all RP systems that exist. Each system mentioned has been identified by the OEM's proprietary name, but their proprietary registered and trademarked names for their computer software, materials, and processes have not been included.

Stereolithography (SL)

Stereolithography (SL), also known as *3-D layering* or *3-D printing*, is a process that translates CAD design into solid objects using a combination of laser, photochemistry, and software technology. The digital data from a computer CAD drawing is first processed by software which slices the vertical dimensions of the product design into very thin cross sections. The basic build process is illustrated schematically in Fig. 2. The system is located in a sealed chamber to prevent the escape of fumes during the SL process.

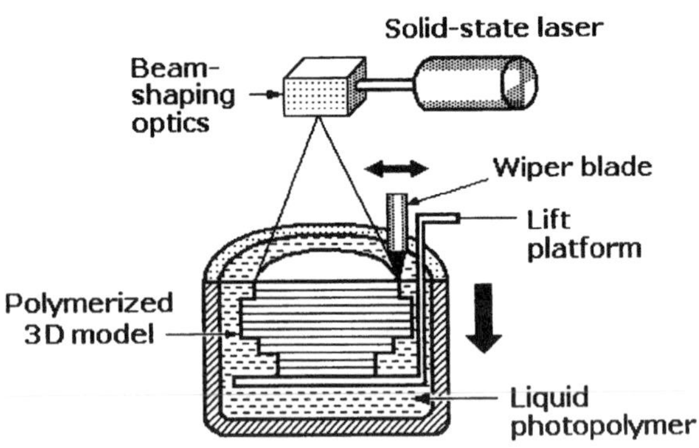

Fig. 2 Stereolithography (SL): A platform is immersed in a vat of liquid photopolymer to a depth equal to the thickness of the first layer of the 3-D prototype. A computer-controlled solid-state ultraviolet (UV)-energy-emitting laser outlines the first layer or "slice" of the model in the film of UV-curable photopolymer. The laser scans the outlined area to cure or solidify the first layer. The platform is then lowered into vat to a depth equal to the thickness of the first layer, and the outlining and surface curing process are repeated until the model is completed. The model is then removed from the vat of liquid polymer not exposed to UV radiation, cleaned, and finished by further UV curing to complete the hardening process.

A platform capable of vertical movement is positioned in a vat filled with 5 to 10 gallons of a clear liquid photopolymers such as urethane acrylate resins. The platform is first moved down below the table surface under computer control to a depth equal to the specified thickness of a prototype "slice" or layer; this permits the platform to be covered with a layer of liquid resin to that depth. A low-power UV laser beam, focused on the *X-Y* mirror of the beam-shaping and scanning system, traces the outline of the lowest cross-section slice of the prototype to be built. The laser is then directed to scan the area of resin traced to cure or harden it, forming the first layer. The UV radiation links the molecules of liquid polymer in chain-like formations. The layer can be hardened to a depth of 0.0025 to 0.0300 in. (0.06 to 0.8 mm).

The platform containing the first layer is then lowered into the resin vat to the same depth as the thickness of the first layer, and it is recoated with liquid photopolymer. The UV laser then repeats its previous steps of tracing the outline of the next layer and curing its traced area on top of the first layer. This SL process is repeated layer by layer until the build is complete. A typical SL prototype can be built within a few hours because the laser beam scans at speeds of up to 350 in./s (890 cm/s). The photopolymer that has not been scanned by the laser remains a liquid. The thinner the resin film (slice thickness), the higher the resolution; this means that the prototype will have a refined finish that requires little or no sanding or polishing. When the prototype surface finish is important, layer thicknesses are set for 0.0050 in. (0.13 mm) or less.

The photopolymer used in the SL process tends to curl or sag as it cures, so prototypes with overhangs or unsupported horizontal sections must be reinforced with supporting structures which can be walls, gussets, or columns. Without support, parts of the model would sag until they could break off before the polymer is fully cured. Instructions for forming these supports are included in the digitized fabrication data entered into the RP system. Each scan of the laser forms support layers where they are necessary while simultaneously forming the layers of the prototype.

When the build process is complete, the SL prototype is raised from the polymer vat, and the uncured or liquid resin is allowed to drain off; any excess can be removed manually from the prototype's surfaces. Postcuring is required because the process leaves the prototype only partially polymerized with about half of its fully cured strength. This is done by exposing the whole prototype to intense UV radiation in the enclosed chamber of the postcuring apparatus (PCA).

Finally, all supports are removed, and the model can be milled, drilled, bored, or tapped before being sanded or polished as needed to accept paint or sprayed-on metal. The liquid photopolymers used in SL are similar to the photosensitive UV-curable polymers used to form the photoresist masks on semiconductor wafers for the later wet or dry chemical etching and plating of circuit features. Resins can be formulated to solidify under either UV radiation or visible light.

The SL process was the first RP technology to gain commercial acceptance, and it still accounts for the largest base of installed RP systems. 3D Systems, Inc. of Valencia, California, manufactures stereolithographic equipment for its proprietary SLA stereolithography process. According to the company, SLA systems are installed at some of the largest manufacturing organizations in the world, and they are used in prototyping for a range of products from automobile parts to consumer appliances.

3D Systems is offering its new Viper si2 SLA system which it says is its first solid imaging system to combine standard and high-resolution prototype building in the same system. It uses a solid-state laser and an elevator with 0.0001 in. (0.0025 mm) vertical resolution and position repeatability of 0.0003 in. (0.0076 mm). The system can lift a weight of 20 lb (9.1 kg). In its standard build mode, the prototype can have a build volume of 10 cu in. (250 cu mm), and in its high-resolution build mode the prototype can have a build volume of 5 × 5 × 10 in. (125 × 125 × 250 mm).

The *Viper HA SLA* system has the same solid-state laser and elevator specifications as the Viper si2 SLA system, but in a single-vat configuration it has a standard mode build envelope of 10 × 10 × 2 in. (250 × 250 × 50.8 mm) and a high-resolution mode build envelope of 5 × 5 × 2 in. (125 × 125 × 50.8 mm). 3D Systems also offers the higher definition *SLA 7000* and *SLA 5000* systems which can form objects with a build volume of 20 × 20 × 23.6 in. (508 × 508 × 584 mm). The company says this is the largest build envelope of any SL system. The SLA 7000 system has a minimum build layer thickness of 0.001 in. (0.025 mm), but it more typically forms layers that are 0.004 in. (0.10 mm) thick. This compares with the SLA 5000 system's layer thickness of 0.002 in. (0.05 mm). Both systems cure photopolymers with solid-state lasers: 800 mW for the SLA 7000 and 216 mW for the SLA 5000.

3D Systems also offers the *InVision 3-D* and the *ThermoJet solid-object printers*. The ThermoJet printer can build models, patterns for investment casting, and molds for other casting applications. Both of these systems can build wax models suitable for conference room design demonstrations or for a manufacturer who wants to call attention to certain critical surface details. The *InVision 3-D* was developed by Solidvision Ltd. of Israel.

Selective Laser Sintering (SLS)

Selective Laser Sintering (SLS), developed at the University of Texas at Austin, is an RP process similar to SL stereolithography. It creates 3-D models from plastic, metal, or ceramic powders with heat generated by a carbon dioxide (CO_2) infrared (IR)-emitting laser, as shown schematically in Fig. 3. The prototype is fabricated in a build cylinder with a piston which acts as an elevator. This cylinder is positioned next to a powder-delivery cylinder filled with preheated powder. A piston within the delivery system rises to eject powder which is spread by a roller over the top of the build cylinder. Just before it is applied, the powder is heated further until its temperature is just below its melting point.

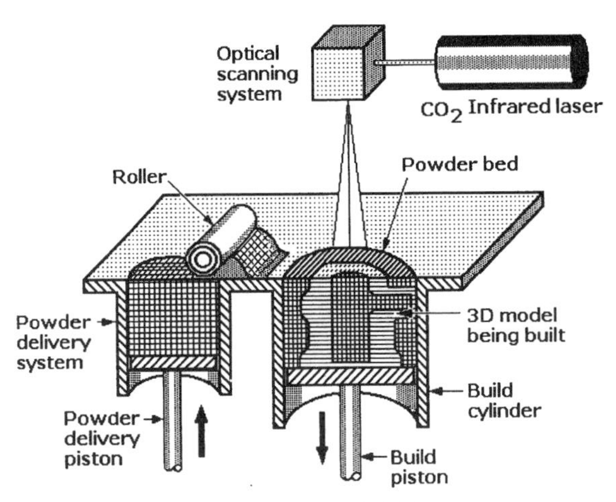

Fig. 3 Selective Laser Sintering (SLS): Plastic powder from a delivery system is spread by roller over a build piston positioned below the table at a depth equal to the thickness of a single layer of the 3-D model. The powder is then scanned by a computer-controlled carbon dioxide infrared (IR) laser that defines the first layer or "slice" and melts the powder so that it flows and hardens. The piston is then lowered to a depth equal to the first layer thickness, more powder is added, and the steps are repeated so that the second layer bonds to the first layer. This process is repeated until the model is completed. The model is then removed and finished. All unbonded plastic powder is recovered and mixed with new powder for use in the next process.

When the laser beam scans the thin layer of powder under the control of the computerized optical scanner system, it raises the temperature of the powder even further until it melts or sinters and flows together; it then forms a solid layer in a pattern derived from the STL-formatted CAD data. As in other RP processes, the piston or supporting platform is lowered to the depth of each slice or layer thickness after each layer is completed. The roller then spreads the next layer of powder over the previous layer. This procedure is repeated with each layer fused to the underlying layer until the 3-D prototype is completed.

The unsintered powder is brushed away and the part is removed. No final curing is required in the SLS process, but because the prototypes are sintered they are porous. A wax coating, for example, can be applied to the inner and outer porous surfaces, and they can be smoothed by manual or machine sanding or melting processes. No supports are required in SLS because overhangs and undercuts are supported by the compressed unsintered powder within the build cylinder.

Many different powdered materials have been used in the SLS process: these include polyamide, glass-filled polyamide, and aluminum-filled polyamide; polymer-coated metal powder is an alternative. One advantage of SLS is that the materials used in forming the prototypes are strong and stable enough to permit the prototype to be used in low-stress functional and environmental testing. The prototypes can also serve as molds or patterns for casting parts.

An SLS system can have as many as five major components: (1) the sinter station, (2) a build module on wheels for transferring the prototype between stations, (3) a thermal station for preheating the powder delivery station and build cylinder, (4) a breakout station for removing the prototype from the build module, and (5) a recycling station for mixing recycled and new powder. The systems also include a nitrogen generator and a new powder storage tank. Customer-installed piping transports new and recycled powder between stations. The SLS sinter station is enclosed in a nitrogen-filled chamber that is sealed and maintained at a temperature just below the melting point of the powder. The nitrogen prevents an explosion that could be caused by the rapid oxidation of the build powder.

3D Systems Corporation offers two SLS systems: the Sinterstation Pro 140 and Pro 230. Both systems include 70-W CO_2 UV lasers, and both can deposit layers in thicknesses from 0.004 in. (0.1 mm) to 0.006 in. (0.15 mm). The company also offers the HiQ and HiQ+ Hs systems. Both have maximum build envelopes of 14 × 12 × 17 in. (381 × 330 × 432 mm) with a limited build height if some accepted materials are used. The HiQ system includes a 30-W CO_2 laser with a maximum scan speed of 5 m/s, and the HiQ+ HS system has a 100-W CO_2 laser with a maximum scan speed of 10 m/s.

EOS GmbH recently introduced the EOSINT P 380i plastic laser-sintering system capable of 13.5 × 13.5 × 26.5 in. (340 × 340 × 600 mm) build envelopes. It has also upgraded its EOSINT P 700 system to have a build volume of 9400 cu in. and it has twin 50-W CO_2 lasers. The company also offers the EOSINT M 270 for Direct Metal Laser Sintering (DMSL). Its EOSINT S 750 with dual 100-W CO_2 lasers sinters foundry sand for making cores and molds for casting metals: these include magnesium, aluminum, steel, and iron.

Laminated Object Manufacturing (LOM)

The Laminated Object Manufacturing (LOM) process, shown schematically in Fig. 4, forms 3-D models by cutting, stacking, and bonding successive layers or laminations of paper coated with heat-activated adhesive. The CO_2 laser beam, directed by an optical system under STL-formatted CAD data control, cuts cross-sectional outlines of each layer of the prototype in the paper. The paper layers are successively bonded to previous layers to form a stack that is the prototype.

The paper that forms the bottom layer is unwound from a supply roll and pulled across the movable platform. The laser beam cuts the outline of each layer and crosshatches the waste material within and around the layer for easy removal after the prototype is completed. The outer waste paper, left over after the first layer has been cut and detached, is removed by a take-up roll. A new layer of paper that will be used to form the second layer is then pulled from a roll and positioned over the first layer; the laser cutting, cross-hatching, and web removal steps are repeated, and a heated roller applies pressure to bond the adhesive coating on the second layer to the first layer. This procedure is repeated until the final layer completes the prototype. The excess cross-hatched material, now in the form of stacked segments, is removed, revealing the finished 3-D model. The LOM process yields models that have wood-like finishes which can be sanded or polished before being sealed and painted.

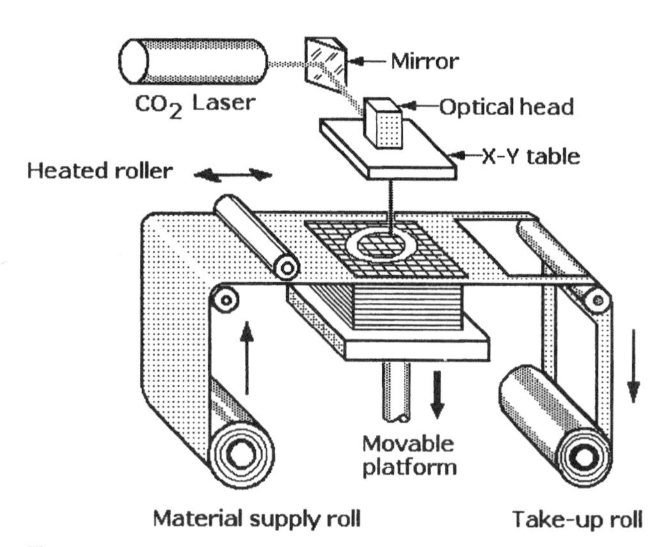

Fig. 4 Laminated Object Manufacturing (LOM): Adhesive-backed paper is fed across an elevator platform and a computer-controlled carbon dioxide infrared (CO_2-IR) laser cuts the outline of the first layer of the 3-D model and crosshatches the unused paper. The platform is then lowered and more paper is fed across the first layer. The laser cuts the next outline and a heated roller bonds the adhesive coating between layers. This process continues until all of the layers have been cut and bonded, and the crosshatched paper is removed, exposing the finished model which is then sealed and painted.

According to the developers of the LOM process, the use of inexpensive thick sheets of paper glued together to form monolithic blocks produces models that are more resistant to deformity and cost less to make than those formed by other RP processes. These models can be used directly as patterns for investment and sand casting or as forms for making silicone molds. LOM models can be larger than those made by most other RP processes—up to 30 × 20 × 20 in. (75 × 50 × 50 cm).

However, the LOM process has three controversial characteristics, any one of which could negate its merits and cancel a prospective purchase decision: the limitation on the laser's ability to cut through the thicker paper that would permit certain models to be made with thicker layers, resulting in faster building at lower cost; the requirement for sealing the edges of the layers against moisture penetration; and the requirement that the process be enclosed with continuous ventilation to dissipate smoke caused by the laser burning through the paper layers.

Kira America Corp., Frankville, Wisconsin, offers three LOM systems that perform what it calls Paper Lamination Technology (PLT): Katana, PLT-A3, and PLT-A4.

- The Katana, the smallest system, uses STL-formatted CAM data to produce models measuring 11.0 × 7.1 × 5.9 in. (280 × 180 × 150 mm) with paper thickness of 0.004 in. (0.10 mm) or 0.006 in. (0.15 mm).
- The PLT-A3 system can use CAD data formatted as STL, RPF, RPS, or JAMA-IGES to build with build volumes of up to 15.7 × 11.0 × 11.8 in. (400 × 280 × 300 mm) from paper thickness of 0.003 in. (0.08 mm) or 0.006 in. (0.15 mm).
- The PLT-A4, a smaller version of the PLT-A3 system, can build from the same formatted data with the same paper thicknesses but with build volumes of only 11.0 × 7.4 × 7.8 in. (280 × 190 × 200 mm).

CAM-LEM, Inc., Cleveland, Ohio, offers the CL-100 system that can perform LOM processing by laminating as many as five different materials: these include paper, "green" ceramic tape, and strips of metal. Layer thickness can range from 30 to 1300 μm (0.001 to 0.050 in.), although the most common materials used in the process have thickness between only 150 and 600 μm (0.006 to 0.024 in.). Layers of differing thickness or composition can be mixed in a single automated build cycle. The company adds that the CL-100 system fabricates full-strength ceramic or metal parts with internal hollow cavities and channels, and that "fugitive" materials can be interleaved to support voids or overhangs, as required. The CL-100 can build prototypes or sets of prototypes within its 6 × 6 × 6 in. (150 × 150 × 150 mm) build envelope.

CAM-LEM's process starts with CAD data converted into an outline on thin layers or slices. Individual layers of sheet material such as "green" ceramic tape are laser-cut to the specified contours. The resulting contoured layers are removed from the sheet stock and stacked to assemble a 3-D prototype conforming to the original CAD description. The assembly operation includes a tacking operation to fix the position of each layer with the correct orientation set by the CAD model.

After assembly, the layers undergo a furnace sintering step to bond the prototype layers into a monolithic structure. CAM-LEM reports that during this step, the boundaries between layers are erased so that the 3-D part appears the same as those produced by conventional ceramic manufacturing methods; the final prototype is said to have the correct geometric form as well as functional structure.

Fused Deposition Modeling (FDM)

The Fused Deposition Modeling (FDM) process, shown in Fig. 5, builds prototypes from melted thermoplastic filament. The monofilament, with a diameter of 0.070 in. (1.78 mm), is fed into a temperature-controlled FDM extrusion head where it is heated to a semiliquid state. It is then extruded and deposited in ultrathin, precise layers on a fixtureless platform under *X-Y* computer control. Successive laminations ranging in thickness from 0.002 to 0.030 in. (0.05 to 0.76 mm) with wall thickness of 0.010 to 0.125 in. (0.25 to 3.2 mm) adhere to each other by thermal fusion to form the 3-D model.

Structures needed to support overhanging or fragile structures in FDM modeling must be designed into the STL-formatted CAD data file and fabricated as part of the model. These supports can easily be removed in a later secondary operation. All functional components of an FDM system are contained within a temperature-controlled enclosure. Among the different kinds of inert, nontoxic filament materials being used in FDM are acrylonitrile butadiene styrene (ABS) polymer, high-impact strength ABS (ABSi), UV-curing plastics, and polyphenylsulfone. These materials all melt at temperatures between 180 and 220°F (82 to 104°C).

Stratasys, Inc. Eden Prairie, Minnesota, developed and patented FDM. The company's RPS build system can build accurate, complex models from materials such as ABS, polycarbonate, and polyphenylsulfone. Stratasys' proprietary software accepts STL-formatted CAD files for a selected prototype and automatically "slices" the 3-D model and orients the layers to fit within the build size limits of any of its RPS systems. The software also generates precise extrusion paths for building the models simultaneously with the commands needed to form any support structures.

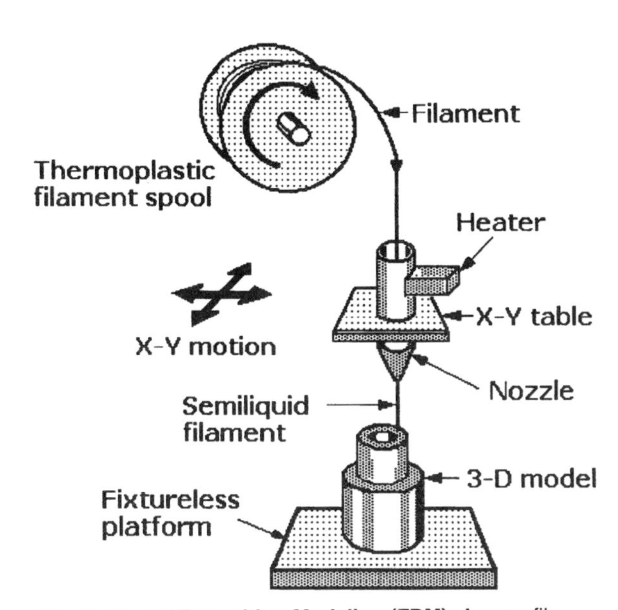

Fig. 5 Fused Deposition Modeling (FDM): A monofilament of a thermoplastic resin is unwound from a spool and passed through a heated extrusion nozzle mounted on a computer-controlled *X-Y* table free to move over a fixtureless platform. The 3-D model is formed as the nozzle extrudes hot filament in a pattern as it moves over the platform. The hot plastic filament binds to the layer below it and hardens to form a finished model. The process continues until the model is complete. This laserless process can form thin-walled, contoured concept models or molds for investment casting. The completed object is removed and sanded to improve its finish.

Stratasys' Prodigy Plus system can be used in an environmentally controlled office environment because it does not need special ventilation or use any toxic solvents. It builds models from ABS plastic resin with water-soluble support structures; this eliminates manual or solvent support removal. Models as large as 8 × 8 × 12 in. (200 × 200 × 300 mm) can be built with the system.

The company's compact, portable Eden260 can operate in any air-conditioned room and, like other Stratasys FDM machines, it requires no special venting. The Eden260 can build larger models than the Prodigy Plus. Prototypes are built from proprietary UV-curing plastic resins and any needed support structures are formed from a gel-like photopolymer plastic.

Stratasys' Eden333 can also build larger models than its Prodigy Plus. It uses Stratasys' proprietary technology to form layers as thin as 16 μm (0.0006 in.). The software selects the orientation of the model to be built based on its size or required build speed; it then automatically processes the STL file which determines the necessary support structures in real time. The software then creates a precise path for the eight jet heads which simultaneously deposit the proprietary UV-curing plastic resin for model building. Additional instructions are provided for a separate extrusion head that deposits the gel-like photopolymer for any needed support structures.

Stratasys also offers five other FDM systems: the Vantage i, S, and SE, the Titan, and Maxum.

- The Vantage systems build models from ABS and polycarbonate with either water-soluble or breakaway support structures; sizes range from 14 × 10 × 10 in. (350 × 250 × 250 mm) to 16 × 14 × 16 in. (400 × 350 × 400 mm).
- The Titan system builds models from polyphenylsulfone, ABS, and polycarbonate resins. As in other systems support structures can either be water-soluble or breakaway.
- The Maxum system can build its largest models to sizes up to 24 × 20 × 24.6 in. (600 × 500 × 600 mm). They can be made from ABS or ABSi, all with water-soluble supports.

Three Dimensional Printing (3DP)

The Three Dimensional Printing (3DP) or the ink-jet printing process, shown in Fig. 6, is similar to SLS except that a multichannel ink-jet head and liquid adhesive supply replace the laser. The powder supply cylinder is filled with starch and cellulose powder which is delivered to the work platform by elevating a piston. A roller then distributes a single layer of powder from the powder supply cylinder to the upper surface of a piston within the build cylinder. Finally, a multichannel ink-jet head sprays a water-based liquid adhesive onto the surface of the powder, bonding it in the shape of a horizontal layer or lamination of the model.

In successive steps, the build piston is lowered a distance equal to the thickness of each layer while the powder delivery piston pushes up fresh powder which the roller spreads over the previous layer on the build piston. This process is repeated until the 3-D model is complete. Any loose excess powder is brushed away, and wax is coated on the inner and outer surfaces of the model to improve its strength. The 3DP process was developed at the Three Dimensional Printing Laboratory at the MIT; it has subsequently been licensed to several companies.

Z Corporation, Somerville, Massachusetts, is one of the firms using the original MIT process to form 3-D models. It offers three 3-D printers: Zprinter 310 for affordable printing; Z810 System for large format printing; and Spectrum Z510 for high-definition color 3-D printing.

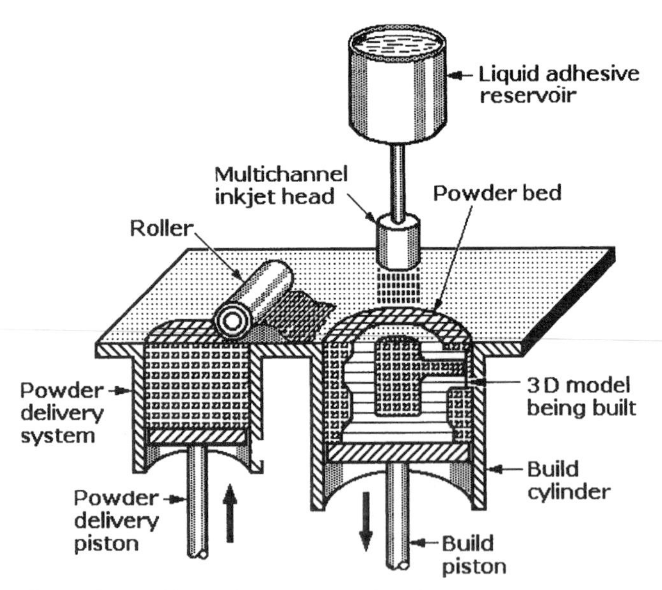

Fig. 6 Three Dimensional Printing (3DP): Plastic powder from a reservoir is spread across a platform by a roller onto an elevator piston set below the table surface at a depth equal to the thickness of one layer. Liquid adhesive is then sprayed on the powder to form the first layer of a 3-D model. The piston is lowered again to a depth equal to the thickness of a layer, and another layer of powder is applied to the piston. More adhesive is sprayed on to bond the second layer to the first layer. This procedure is repeated until the 3-D model is completed. As the final step, the model is removed and finished.

The *Z* Cast Direct Metal Casting process produces cast-metal parts from STL-formatted CAD file data. The process is said to be significantly faster and less expensive than other prototype casting methods. It prints molds and cores on a 3-D printer directly from digital data, eliminating the need for a pattern and core box production step, as in traditional sand casting. Metal is then poured into the 3-D printed molds.

Solidscape Inc., (formerly Sanders Prototype Inc.), Merrimack, New Hampshire, also supports 3DP. Solidscape offers two of these systems: R612 and T66. Both permit the direct casting of patterns made from materials with negligible coefficients of thermal expansion, a requirement for the prevention of ruptures in casting shells. Both systems can build patterns from silicone, RTV, epoxy, and other elastomeric materials, which are nonhazardous.

- The R612 benchtop model-making system uses a proprietary hot thermoplastic ink-jet spray that emits 0.003-in. diameter microdroplets. The CAD file data input can be formatted STL or SLC, and proprietary front-end software permits variations in layer thickness within a single-build structure. The maximum size of the models it can produce is 12 × 6 × 6 in. (30 × 15 × 15 cm), *X*, *Y*, and *Z* dimensions, respectively. It can form layers as thin as 0.0005 in. (0.013 mm) or as thick as 0.003 in. (0.08 mm). Achievable accuracy is said to be ±0.001 in. (0.025 mm) per inch in *X*, *Y*, and *Z* dimensions. Surface finish can range from 32 to 63 μin. (RMS), and minimum feature size is 0.010 in. (0.25 mm).
- The T66 benchtop model-making system has specifications similar to the R612, but the maximum size is 6 × 6 × 6 in. (15 × 15 × 15 cm); layer thickness can be from 0.0005 in. (0.13 mm) to 0.003 in. (0.08 mm). Accuracy, surface finish, and minimum feature size are said to equal those of the R612.

Directed Light Fabrication (DLF)

The Directed Light Fabrication (DLF) process, diagrammed in Fig. 7, uses a neodymium YAG (Nd: YAG) laser to fuse powdered metals to build 3-D models; they are more durable than models made from paper or plastics. The metal powders can be finely milled 300 and 400 series stainless steel, tungsten, nickel aluminides, molybdenum disilicide, copper, or aluminum. The technique is also called *Direct Metal Fusing*, *Laser Sintering*, or *Laser-Engineered Net Shaping (LENS)*.

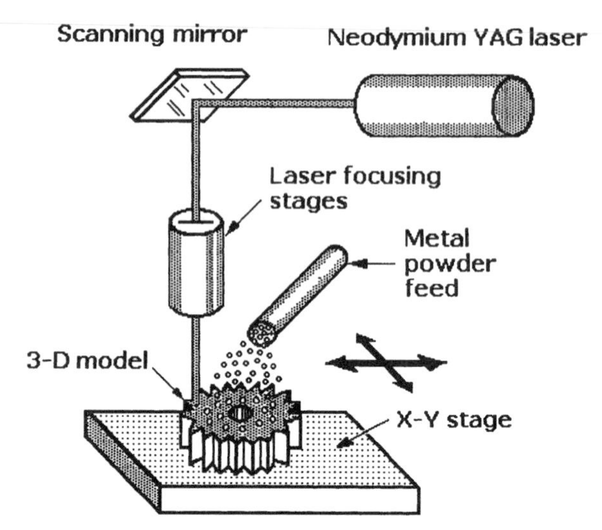

Fig. 7 Directed Light Fabrication (DLF): Fine metal powder is distributed on an *X-Y* work platform that is rotated under computer control beneath the beam of a neodymium YAG laser. The heat from the laser beam sinters the metal powder to form a thin layer of a 3-D object. The process is repeated until enough layers are built up and bonded to complete the object. More durable objects can be made from metal than plastic or paper. The sintered metal object is then heat-treated to increase its bond strength. Powdered aluminum, copper, stainless steel, tantalum, and other metals have been sintered to form functional tools or working parts.

The laser beam, under *X-Y* computer control, fuses the metal powder fed from a nozzle to form dense 3-D objects whose dimensions are said to be within a few thousandths of an inch of design tolerance. DLF is an outgrowth of nuclear weapons research done at the Los Alamos National Laboratory (LANL), Los Alamos, New Mexico. The laboratory has also been experimenting with the laser fusing of ceramic powders to fabricate ceramic parts as an alternative to metal parts.

Optomec Design Company, Albuquerque, New Mexico, offers its LENS process based on DLF technology. The work is performed by a turnkey manufacturing unit that includes software, a high-power laser, motion control, and other components. The LENS process builds metal parts directly from CAD designs by injecting metal powder into the focus spot of the laser beam; it then "prints" layers of metal to fabricate a part from the bottom up. Optomec reports that the LENS process can be used to make short production runs of new metal products or fabricate replacement parts for repairing older machines or equipment.

AeroMet Corporation, Eden Prairie, Minnesota, offers a proprietary process similar to DLF that it calls Laser Additive Manufacturing (LAM). The process concentrates on building titanium alloy parts for the aircraft industry. As in other additive RP processes, 3-D CAD renderings are converted to STL software from which application-specific instructions are derived. Titanium alloy powder is deposited on a target plate, and a high-power

18-kW CO_2 laser sinters the deposited titanium powder in stages until a nearly complete part is formed; no expensive molds or dies are used. The sintered part is then heat-treated before material from 0.020 to 0.050 in. (0.5 to 1.3 mm) thick is machined away to achieve the final dimensions.

Direct Shell Production Casting (DSPC)

The Direct Shell Production Casting (DSPC) process, based on technology developed at MIT and diagrammed in Fig. 8, is similar to the 3DP process. Primarily intended for forming molds or shells rather than 3-D models, DSPC also begins with a CAD file of the desired shell. Although DSPC is considered to be an RP technology, all 3-D models or prototypes must be produced by later casting processes.

Two specialized kinds of equipment are used in DSPC: a dedicated computer called a Shell-Design Unit (SDU) and a shell-processing unit (SPU). The CAD file is loaded into the SDU to generate the data needed to define the mold or shell; the resulting SDU software also modifies the original shell dimensions in the CAD file to compensate for ceramic shrinkage. This software can also direct the formation of filets and delete certain features in the shell such as holes or keyways which must be machined in the prototype after it is cast.

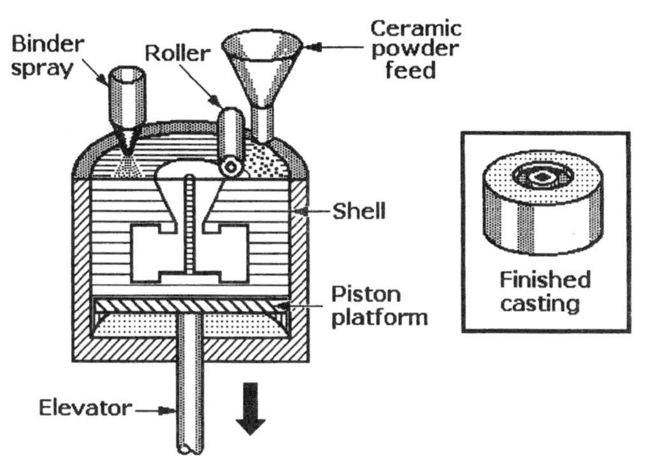

Fig. 8 Direct Shell Production Casting (DSPC): Ceramic powder is spread by roller over the surface of an elevator build piston that is recessed to the depth of a single 3-D mold layer. A binder is then sprayed on the powder under computer control, and the platform and first layer are lowered to the depth of the second layer. More power is spread over the build piston and more binder is sprayed on to bond the second layer to the first layer. This process is continued until the mold is completed. The bonded "green" ceramic shell is removed and furnace fired to make a durable functional mold. This is an RP process because it makes molds faster and cheaper than by conventional methods, thus permitting metal castings to be made more rapidly.

The movable platform in DSPC is the piston within the build cylinder. As in other RP technologies, it is lowered to a depth below the rim of the build cylinder equal to the thickness of each layer. Then, a thin layer of fine aluminum oxide (alumina) powder is spread by roller over the platform. Next, a fine jet of colloidal silica is sprayed onto the powder to bond it in the shape of a single mold or shell layer. The piston is then lowered for the next layer, and the process is repeated until all layers have been formed, completing the entire 3-D shell. The excess powder is removed and the object is furnace fired, converting the bonded powder to monolithic ceramic. After the shell has cooled, it is strong enough to withstand molten metal, and can function like a conventional investment casting mold. After the molten metal has cooled, the ceramic shell and any cores or gating are broken away from the prototype. It can then be finished by any of the methods normally used on metal castings.

Soligen Technologies, Northridge, California, offers its proprietary DSPC system for generating ceramic casting molds for metal parts and tools faster than conventional methods. Soligen Technologies Parts Now Division provides metal-castings services using its DSPC technology.

Solid Ground Curing (SGC)

Solid Ground Curing (SGC) (or the "solider process") is the multistep in-line process, as diagrammed in Fig. 9. It begins when a photomask for the first layer of the 3-D model is generated by the equipment shown at the far left of the figure. An electron gun writes a charge pattern of the photomask on a clear glass plate and opaque toner is transferred electrostatically to the plate, forming a photolithographic mask by a xerographic process. The photomask is then moved to the exposure station where it is aligned over a work platform and under a collimated UV lamp. The SGC process calls for the work platform to be moved sequentially right and left to complete the process.

Model building begins when the work platform is moved right to the resin application station; in that position a thin layer of photopolymer resin is applied to the top surface of the work platform and wiped to the desired thickness. The platform is then moved back to the left to the exposure station again; the UV lamp is turned on, and a shutter is opened for a few seconds to expose the resin layer to the mask pattern. Because the UV light is so intense, the layer is fully cured and no secondary curing is needed.

Following the curing step, the platform is moved again to the right to the wiper station where all of the unexposed resin is removed and discarded. The platform is moved right again to the wax application station where melted wax is applied and spread into the cavities left by the removal of the uncured resin. The platform continues its moves to the right to a station where the

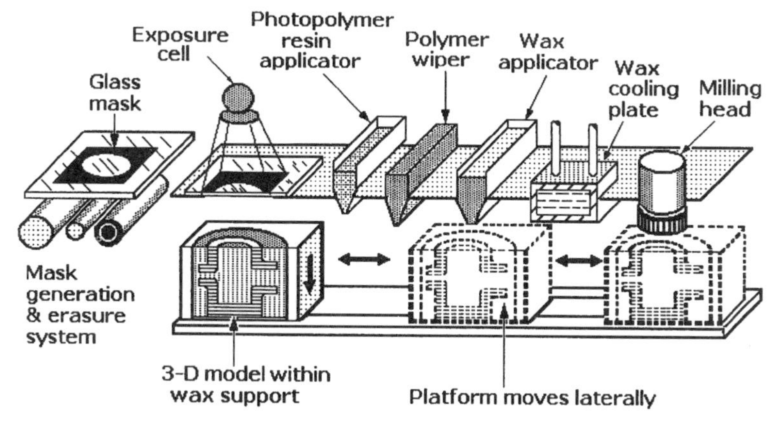

Fig. 9 Solid Ground Curing (SGC): First, a photomask is generated on a glass plate by a xerographic process. Liquid photopolymer is applied to the work platform to form the first layer of a 3-D model. The platform is moved under the photomask and a UV energy source defines and cures the layer. The platform then moves to a station and hot wax is applied over the layer to fill in margins and spaces. After the wax has hardened, excess polymer and wax is milled off to complete the first "slice." The first photomask is erased and replaced by a second mask on the same glass plate. The entire process is repeated as the platform moves back and forth under the work stations until the model is complete. Finally, the wax is removed by heating or hot water immersion to release the model.

wax is hardened by pressing it against a wax cooling plate. The platform continues to be moved to the right to the milling station where both the resin and wax layers are milled to a precise thickness. Then the platform is moved left to the resin application station where it is lowered to a depth equal to the thickness of the next layer and more resin is applied to continue the process.

Meanwhile, the opaque toner has been removed from the glass mask, and a new mask for the next layer is generated on the same plate. The complete platform movement cycle is repeated, and it will continue until the 3-D model encased in the wax matrix is completed. This wax matrix supports any overhangs or undercuts so extra support structures are not needed. Finally, the prototype is removed from the process equipment and the wax is either melted away or dissolved in washing chamber similar to a dishwasher. The surface of the 3-D model is then sanded or polished as necessary by hand or machine methods.

The SGC process is similar to *drop on demand ink-jet plotting:* This method depends on a dual ink-jet subsystem that travels on a precision *X-Y* drive carriage and deposits both thermoplastic and wax materials onto the build platform under CAD program control. The drive carriage also energizes a flatbed milling subsystem for obtaining the precise vertical height of each layer and the overall height of the object by milling off the excess material.

Cubital America Inc., Dearborn, Michigan, a subsidiary of Cubital Ltd. of Raanana, Israel, offers the Solider 4600/5600 equipment for building prototypes with the SGC process.

Desktop Prototyping

Desktop prototyping is a commercial subtractive method that is an alternative to the more common additive RP technologies. It begins with the preparation of a 3-D CAD drawing which is translated into applications-specific computer software. The software directs the operation of a small computer numerically controlled (CNC) milling machine within an enclosed cabinet. This machine is capable of carving or milling relatively small models or prototypes from solid blocks of plastic, wax, or soft metal.

Delft Spline Systems of Utrecht, the Netherlands, offers a self-contained CNC milling machine within a cabinet for desktop prototyping within an office environment. Proprietary Delft Spline Systems software converts CAD data into instructions for the machine that mills prototypes from blocks of wax, plastic, or soft metals. The company reports that this system is useful for making presentation and concept models, wax models for jewelry production, molds for tools, and products for both dental and orthopedic applications.

Research and Development in Rapid Prototyping

Many RP technologies are still experimental and have not yet achieved commercial status. Information about this research has been announced to the public by the laboratories performing direct manufacturing of prototypes or products from metal or ceramics, and some of the research is described in patents. Two of these experimental technologies are described here: Shape Deposition Manufacturing (SDM), Mold Shape Deposition Manufacturing (MSDM), and Robocasting.

While showing some commercial promise, these systems have not been organized by commercial OEMs as have the RP technologies described earlier. Nevertheless, where the resources and space are available, the equipment can be purchased and the facilities to perform these processes can be established. However, because of equipment cost, they are more likely to be performed in industrial, academic, and government laboratories.

Shape Deposition Manufacturing (SDM)

The Shape Deposition Manufacturing (SDM) process was developed at the SDM Laboratory of Carnegie Mellon University's Robotics Institute in Pittsburgh, Pennsylvania. The process, as shown in Fig. 10, is a variation of *Solid Free-Form* (SFF) fabrication. It can produce functional metal prototypes directly from CAD data. Successive layers of hard metal are deposited on a platform without masking for the direct manufacture of rugged functional parts; it is an alternative to conventional manufacturing without the added cost of specialized tooling needed for full-scale production.

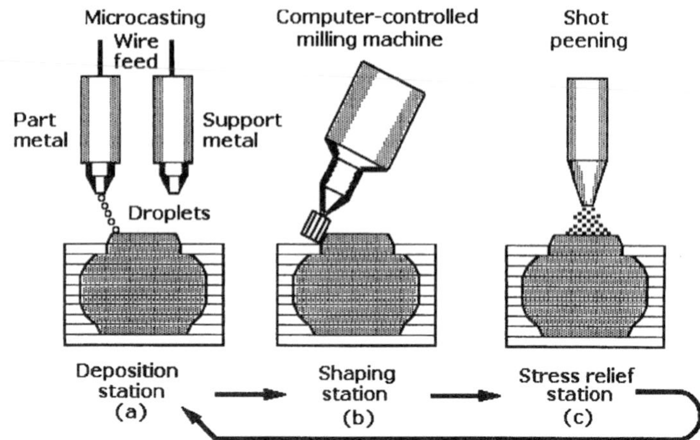

Fig. 10 Shape Deposition Manufacturing (SDM): Droplets of hot metal are sprayed on to form both the prototype and sacrificial support layers of the nearly complete structure shown at deposition station **(a).** After the first prototype layer is deposited, the work is then moved to shaping station **(b)** where the edges of that layer are machined, under computer control, to a precise shape and dimensions. The work is next moved to stress relief station **(c)** where the metal layer is shot-peened to relieve stress. Then the work is moved back to deposition station **(a)** where the first support layer and the next prototype layer are deposited. The sprayed metal droplets retain their heat long enough to remelt the previous layers on impact to form a metal bond with them. This three-step process continues until the prototype is complete. Finally, the sacrificial layers, which support all undercut features, are removed by acid etch to release the prototype.

The nearly complete structure of primary and sacrificial layers is shown at the deposition station (a). Software derived from CAD drawings of the prototype determines the number and thickness of layers and how they are to be deposited. The first primary metal layer is sprayed as hot metal droplets by an additive process called *microcasting* at the deposition station. This first layer and all of those that follow are deposited slightly oversize so that the outer edges of each layer can be machined to the specified shape and dimensions. After each layer is deposited it is moved to shaping station (b) where a computer controlled milling machine or grinder removes excess metal. Next, the work is moved to a stress-relief station (c) where it is shot-peened to relieve stresses that have built up in the layer. The work is then transferred back to deposition station (a) for the deposition of the next layer of primary metal and the sacrificial metal that will support and protect any overhanging layers. This three-step SDM build cycle is repeated until the prototype is complete.

The metal droplets that form each layer retain their heat long enough to remelt the previous layer on impact and form a strong metal bond with it between them; shot peening prevents warping. When the build process is complete, the sacrificial metal is etched away with acid, and any final grinding and polishing is performed. A successful combination of metals in SDM has been stainless steel for the prototype and copper for the sacrificial support layers.

The SDM Laboratory at CMU has investigated many techniques including thermal spraying and plasma or laser welding for depositing high-quality metals before it decided on microcasting.

It is a compromise between those two techniques that provides better results than either of the other techniques. It was found that the larger diameter metal droplets (1 to 3 mm) formed by microcasting retain their heat longer than the 50 μm droplets of conventional thermal spraying. SDM can form complex shaped parts rapidly while also permitting both the fabrication of multimaterial structures and the embedding of prefabricated components within the parts as they are shaped.

The CMU SDM laboratory has produced custom-made functional mechanical parts, and it has embedded prefabricated mechanical parts, electronic components, electronic circuits, and sensors in the metal layers during the SDM process. It has also made custom tools such as injection molds with internal cooling pipes and metal heat sinks with embedded copper pipes for heat redistribution.

Mold Shape Deposition Manufacturing (MSDM)

Both the Rapid Prototyping Laboratory at Stanford University, Palo Alto, California and the Robotics Institute at CMU have SDM laboratories, but the Stanford Laboratory developed its own version of SDM, called Mold SDM; it is also intended for building layered molds for casting ceramics and polymers.

The Mold SDM (MSDM) process, diagrammed in Fig. 11, uses wax to form the molds. The wax has the same function in MSDM as the sacrificial support metal in SDM; it occupies and supports the mold cavity as it is formed. Water-soluble photopolymer sacrificial support material in MSDM corresponds to the primary metal in SDM deposited to form the finished prototype. It is worth noting that no machining is performed in the mold SDM process.

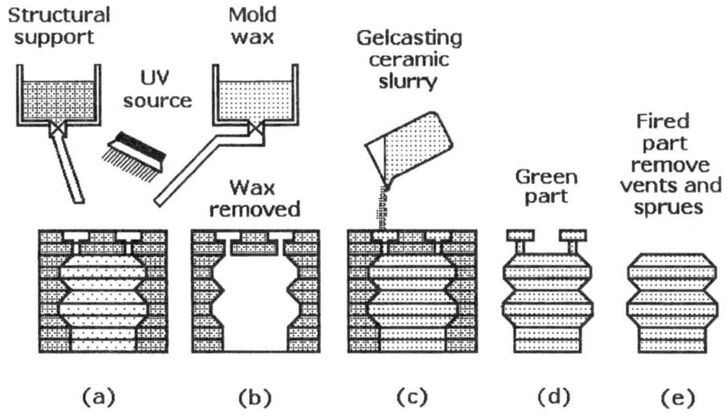

Fig. 11 Mold Shape Deposition Manufacturing (MSDM): Wax for a casting mold and water-soluble photopolymer to support the mold cavity are deposited under computer control in successive layers to build the structure shown at station **(a)**. UV radiation cures each successive polymer layer at the station. After the mold structure is complete, the polymer support material is washed from the wax mold cavity at station **(b)** leaving it empty. Gelcasting ceramic slurry is then poured in the wax cavity at station **(c)** to form a "green" ceramic prototype. The wax mold is then melted away, releasing the "green" prototype at **(d).** Finally, furnace firing hardens the ceramic part, and its vents and sprues are removed at station **(e).**

As in other RP technologies, the MSDM process starts with the conversion of data derived from a CAD drawing of the mold into software that determines both the optimum number of layers of wax and sacrificial support material and how they are to be deposited. Layer thickness depends on the complexity and contours of the mold to be built. Figure 11(a) shows the complete structure of mold wax and sacrificial photopolymer layers needed to form the mold. The structure is built one layer at a time under computer control. With the exception of the first layer of wax, both mold wax and support polymer are deposited in the same layer. The polymer is then cured by UV radiation. The built-up structure is then moved to a station (b) where the sacrificial photopolymer is removed by dissolving it in water, leaving the wax mold cavity empty. The gelcasting ceramic slurry is then poured into the wax mold at station (c) to form the ceramic part. The slurry is allowed to cure into a "green" semirigid, clay-like state. Next, the wax mold is melted away, releasing the "green" ceramic part, as shown in step (d), for furnace firing. After firing, the last step is the removal of the vents and sprues, as shown at (e). The kinds of materials used in MSDM have been expanded by making parts from a variety of polymer materials, and it has also been used to make preassembled mechanisms from both polymer and ceramic materials.

Robocasting

A rapid prototyping method called *robocasting* permits the fabrication of ceramic parts under computer control without molds or machining. Developed at Sandia National Laboratories in Albuquerque, New Mexico, the process can also mix metal with the ceramic slurry to form graded hybrid parts that resist cracking due to different rates of thermal expansion. Joe Cesarano, the Sandia engineer who developed the process, reports that robocast ceramics are denser than ceramic parts made by other RP processes. He added that the ability to make reliable metal-ceramic parts able to withstand very high temperatures is particularly useful in the manufacture of engine components.

A dense ceramic part can be free-formed by robocasting, dried, and baked in less than 24 hours. As in other types of RP, the designer can correct mistakes or improve on the part design and quickly verify the improvement. Cesarano said that traditional ceramic fabrication processes can take weeks to go from a design stage to completion. If a complicated ceramic part is to be made by the standard dry pressing method, the ceramic powder must first be compacted into a solid form or billet, and then the billet must be sculpted into its final shape by costly machining. Intricate ceramic parts can also be made by other techniques: These include slipcasting, gelcasting, and injection molding, but they require the design and manufacture of molds prior to fabrication.

Although still in the laboratory development stage, robocasting holds promise for producing ceramic parts in high quantities. Cesarano acknowledges that the success of robocasting depends on the development of ceramic slurries that actually contain more solid than liquid but are, nevertheless, able to flow. The ceramic slurry is dispensed by a computer-controlled syringe that remains in a fixed position while the platform on which slurry is deposited moves. The slurry must dry rapidly into a semisolid state so that the next layer can be accepted. The high solid content and the tailored properties of the slurry minimize shrinkage and permit a layer to dry within 10 to 15 seconds of being deposited.

After the part is formed by layering and completely dried, it is sintered to bond the particles by firing at temperatures typically from 1000 to 1700°C for about 2 hours. Where a design requirement calls for ceramics and metal combined in the same device, joining them together can be difficult because their differences in rates of thermal expansion can cause cracking at the material's interface. Robocasting makes it possible to shift gradually from one material to another, thus spreading the stress evenly; this will yield a more stable joint. Robocasting also allows the discrete placement of certain materials between the layers that will evaporate or burn away during the sintering process; this permits the formation of functional internal structures such as cooling channels within the part.

Resource

The following Web site provided information about the topics featured in this article:

The Rapid Prototyping Home Page
www.cc.utah.edu/~asn8200/rapid.html

CHAPTER 19

NEW DIRECTIONS IN MECHANICAL ENGINEERING

THE ROLE OF MICROTECHNOLOGY IN MECHANICAL ENGINEERING

The role of mechanical engineering has expanded dramatically over the past half century as a result of the introduction of electronics and computer science into this engineering discipline. While the basic physical principals governing mechanics and mechanical design have not changed, engineering practice has. Solid-state digital electronics and microprocessors have obsoleted many traditional instruments and devices: examples include the slide rule, mechanical timers, mechanical numerical displays, and mechanical computing mechanisms. Computers have been responsible for introducing radical changes in mechanical engineering: these include computer-aided design (CAD), computer-aided manufacturing (CAM), computer-based simulation, rapid prototyping (RP), and, of course, the computer's ability to perform complex calculations faster and more accurately than previous calculators.

Together these technologies have increased the kinds of projects that engage mechanical engineers (MEs), and they have introduced new and different educational requirements for entering the profession. While MEs continue to design mechanisms, machines, and mechanical devices, they have recently become participants in such new activities as electronic packaging, the mechanical design of computer hard, CD, and DVD drives, robotics, mechatronics, and microtechnology. They are also making contributions to nanotechnology, a subject which is merging the engineering disciplines with the physical and biological sciences.

MEs now work on megastructures and megamachines the size of aircraft carriers, cruise ships, and deep-ocean oil rigs. By contrast, they are also working on microelectromechanical systems (MEMS), built to micrometer scale (millionths of a meter) that are so small they must be viewed under a microscope. MEMS now include tiny motors, chain drives, torque converters, transmissions, accelerometers, pressure sensors, gyroscopes, and gear reduction units.

Microtechnology Today

MEMS fabrication technology was acquired from the manufacturing processes for large-scale integrated circuits such as microprocessors and memories. This technology calls for sequential masking and chemical etching steps that have made possible the sculpting of dynamic components from multiple layers of silicon. Because MEMS are capable of performing work, they qualify as machines. However, for many reasons both technical and economic, MEMS have yet to live up to the early predictions that they would all soon become mass-produced. However, some MEMS have jumped economic and technical hurdles to achieve success in the marketplace because they created a demand that led to high production levels and, as a result, significant reductions in unit price.

Among the more popular MEMS are acceleration sensors that can detect forces unleashed in vehicular collisions and trigger the deployment of airbags. More than 100 million of these sensors have been installed in motor vehicles worldwide. Analog Devices Inc., a major supplier of these devices, uses its standard, high-volume integrated circuit (IC) manufacturing technology called iMEMS to surface micromachine the precisely patterned sensor structures on a silicon wafer. The company's smallest acceleration sensor, in an aspirin-size 0.2 × 0.2 × 0.1 in. (5 × 5 × 2 mm) package, performs two-axis motion sensing. A single monolithic chip provides a digital output with low power drain and self-test features.

Analog Devices is offering programmable low-power gyroscopes made with its iMEMS technology. ADIS16250 gyros are complete integrated systems in single compact packages for measuring the angular rate of rotation. Applications include their use for platform stabilization, motion control, navigation, and robotics. Each gyro IC includes two identical MEMS polysilicon sensors called *resonators*. They are operated antiphase to give signals in opposite directions so that a differential output can be obtained. Each resonator contains a mass suspended by springs within an inner dither frame. Other springs suspend the dither frame inside a substrate frame. The inner springs tethering the mass allow it to oscillate only in one direction while the outer springs supporting the dither frame restrict its movement to one direction.

When electrostatically driven at their resonant frequencies, the masses produce Coriolis forces proportional to system movements around a single axis. Capacitors formed by movable sensing fingers on the dither frames interdigitated with fixed fingers on the substrate frame respond to these movements with instantaneous changes in capacitance values directly related to the angular rate of rotation of the gyro. The differential between the instantaneous values of these resonator capacitors is used to measure angular rate. This technique cancels the effects of environmental shock and vibration. The differential signal is fed to a series of on-chip electronic gain and demodulation stages to produce the electrical rate-signal output. ADIS16250 gyros are housed in standard IC packages measuring 0.4 × 0.4 × 0.2 in. (11 × 11 × 5.5 mm), and they are powered from 4.75 to 5.25 V DC supplies.

Another successful MEMS is the more complex digital micromirror device (DMD) shown in Fig. 1. Invented at Texas Instruments, it is an electromechanical transducer now known as

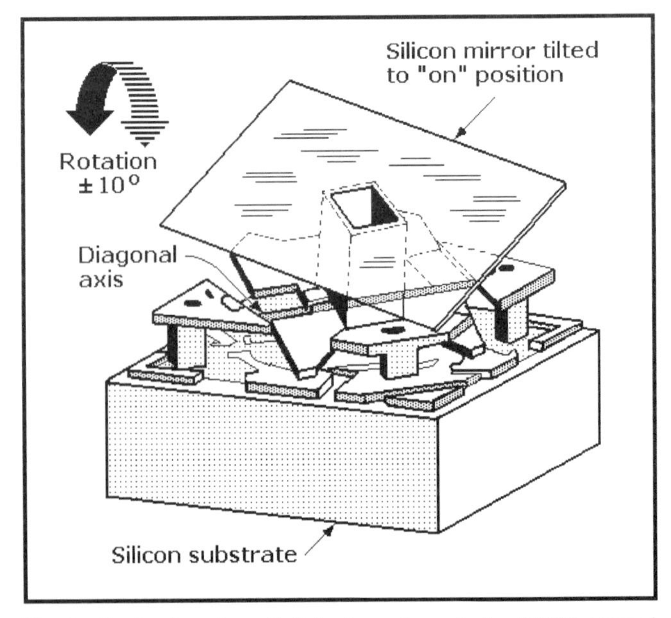

Fig. 1 Each mirror in a digital micromirror device (DMD) array of mirrors is mounted on a diagonal axis. When switched, it tilts one way to reflect light on a screen to form an "on" pixel or the other way to form a dark "off" pixel. A DMD with more than 2 million mirrors can form high-definition colored images with a resolution of 1920 × 1080.

a DLP chip (for Digital Light Processing); more than five million of these chips have been shipped. With the ability to translate video signals into multicolored video images, DLP chips are now integral components in large-screen digital TVs, projectors for home and business video presentations, and projection systems for displaying movies on theater-size screens.

Each DLP chip contains an array of two million hinge-mounted microscopic mirrors that have been formed on a silicon chip no larger than a finger nail. The micromirrors are sculpted by MEMS photolithographic masking and chemical etching processes. Each mirror measures less than one-fifth the width of a human hair (0.0004 in. or approximately 10 μm). The mirrors are hinged so that they can be deflected or switched from reflecting to nonreflecting positions thousands of times per second. This mechanical DLP chip performs all of the electronic and optical functions necessary to translate the input digital video signals into a sequence of mirror deflections, making it possible for them to form dynamic video images. Each mirror corresponds to a single pixel in a projected or displayed image. When synchronized with a light source and a multicolor projection lens system, the DLP's mirrors reflect full color images onto a screen or other suitable flat surface.

The use of MEMS for direct tire-pressure sensing is another growing market. Millions of indirect automotive tire-pressure monitoring systems have already been installed in high-end vehicles as optional accessories, but the demand for these systems could skyrocket if they are to become requirements for all motor vehicles as a safety feature. However, indirect systems are costly and their mandatory installation on lower-priced vehicles would raise their prices significantly. It is expected that direct monitoring systems, which include MEMS as pressure sensors, will result in lower priced and higher performing tire-pressure monitoring modules.

MEMS sensors would directly measure air pressure in each tire. They would be based on either a piezoresistive effect that would provide a voltage output when stressed by tire pressure changes or capacitive versions whose capacitance values would change with tire pressure. These MEMS would be installed in electronic modules positioned on each wheel rim of a vehicle. The modules might also include temperature and voltage sensors, an accelerometer, a microcontroller chip, a radio transmitter, an antenna, and a battery. The conditioned pressure data signals would first be transmitted wirelessly to a central receiver in each vehicle. The data from the receiver could then be displayed either as a warning light if the tire pressure is too low or continuously as a digital readout of actual pressure. The complete module is expected to weigh between 1.1 and 1.4 oz (30 and 40 grams).

Microtechnology Tomorrow

It has been estimated that about two-thirds of the components in cell phones and radios are filters, and they are mechanical devices. This, according to researchers at Stanford University, suggests that sensors, computers, and communications gear can all be highly miniaturized. The scientists have demonstrated that small, unhardened microscopic radios can transmit and receive messages while in Earth orbit. Conventional radios on satellites must be hardened against radiation, adding to their high cost. However, they could be replaced by large networks of low-cost expendable micron-scale radios. The scientists believe that clusters of expendable radios could be mounted on a spacecraft and, although some might be burned out by exposure to damaging radiation, the survivors could continue to transmit and receive.

Dr. Albert Pisano, chairmen of the Department of Mechanical Engineering at the University of California at Berkeley, believes that a market for tiny transceivers could emerge within five years. He also pointed out that most of the components of a typical radio transceiver are passive discrete devices, noting that MEMS devices can replace such discrete components as filters. "Since the MEMS components can be integrated with the transistor electronics, they can reduce the total parts count, the overall size, the cost of the radio," he asserted.

Professor Pisano said MEMS have a role to play in macrotechnology: he noted that with the use of MEMS sensors and actuators, jet engines, for example, could be made more reliable. He explained that heat conductivity can be adjusted and strain can be resolved to very low levels. MEMS sensors are located in jet engines at strategic sites to measure such factors as pressure, strain, vibration, temperature, and acoustic output. According to Dr. Pisano, microstrain gages have over 10,000 times the sensitivity of conventional metal-foil strain gages. MEMS extensometers, he noted, were capable of reading absolute strain with no drift. "With a gage length of 1 mm, this device can resolve strains to help with condition-based maintenance and structural health and monitoring," he reported. The professor went on to say that telemetry systems with sensors and transceivers containing MEMS components could fill a need for improved monitoring of power lines, freeways, and bridges by acting as early warning systems by reporting extreme stresses or strains in structures before they fail.

MEMS devices, he said, could perform the functions of microphones, optical communicators, actuators, uncooled infrared sensors, and inertial measurement units; they could also function as mass data storage devices. Microminiature communicators could also be powered by MEMS-based power generators and energy converters.

The California professor reported that subminiature batteries could be developed for converting thermal energy into electricity. He explained that the use of thermal energy at a small scale is practical, adding that existing tiny rotary engines measuring only microns in diameter would have higher efficiency than existing lithium-ion batteries which must be electrically charged, thus consuming energy. The microminiature thermal motors, powered by methane, could convert the gas into electrical energy. Dr. Pisano agrees that MEMS have not yet reached their early expectations because of the difficulties encountered in manufacturing and packaging them in quantity, but he also believes that progress in nanotechnology will enhance the long-term prospects for MEMS.

MICROMACHINES OPEN A NEW FRONTIER FOR MACHINE DESIGN

The technology for fabricating microelectromechanical systems (MEMS) micron-scale motors, valves, transducers, accelerometers, and other devices, was derived from the proven photolithographic and chemical etching processes used to fabricate silicon integrated circuits. This technology has opened a new field for the mechanical engineer that depends on the application of design rules and manufacturing techniques which differ radically from traditional mechanical engineering practice; they call for the use of materials and chemicals that, until recently, would have been unfamiliar to most mechanical engineers. In MEMS manufacture, silicon replaces steel, brass, aluminum, and other more familiar materials, and chemical etching removes excess material rather than milling, turning, or boring.

MEMS are so small that they can be clearly seen only when viewed under an electron microscope, and the normal laws of physics do not necessarily apply in powering devices this small. For example, MEMS motors are typically driven by electrostatic attraction rather than electromagnetism because when mechanisms are scaled down to this size, electromagnetism is too weak to be effective. Mechanical engineers working on MEMS have ventured into a dimensional realm that was formerly the province of microbiologists, atomic physicists, and microcircuit designers.

Among the more remarkable examples of MEMS made in laboratories are a microminiature electric-powered vehicle that can be parked on a pinhead, electric motors so small that they can easily fit inside the eye of a needle, and pumps and gear trains the size of grains of salt. Far from novelties that serve only to demonstrate the feasibility of the technology, many MEMS are now being mass-produced for automotive, electronic, optical, and biomedical applications. Biomedical researchers, for example, have been developing microminiature medicine dispensers that will reach the site of disease within the human body by traveling through arteries or veins. These natural conduits are also being considered for transporting microscale instruments to precise internal sites where they can perform remote-controlled microsurgery.

The practical MEMS devices now being manufactured in quantity include accelerometers, digital micromirror devices (DMDs), tire-pressure sensors, and modulators. The accelerometers trigger vehicular air bags; the DMDs project television and other video images on large screens; the tire-pressure monitors are included in vehicular tire-pressure monitoring systems; and the modulators convert electronic signals into optical signals in fiberoptic communication systems.

The Microactuators

The rotary micromotor, shown sectioned in Fig. 1, is an example of a microactuator driven by static electricity rather than electric current. Some of the experimental motors made to date have diameters of 0.1 to 0.2 mm and heights of 4 to 6 μm. The rotor, formed with spokes in a "rising sun" pattern, rests on a knife-edge bushing that minimizes frictional contact with motor's base substrate; this leaves it free to rotate around a central nail-shaped hub which prevents it from being displaced vertically. Insulating slots separate the stator into 20 electrically isolated commutator segments. The conductive inner surfaces of the commutator segments and outer surfaces of the rotor spokes form a rotating capacitor that responds to electrostatic forces produced by voltages applied sequentially to the stator segments; these forces cause the rotor to spin at high speeds.

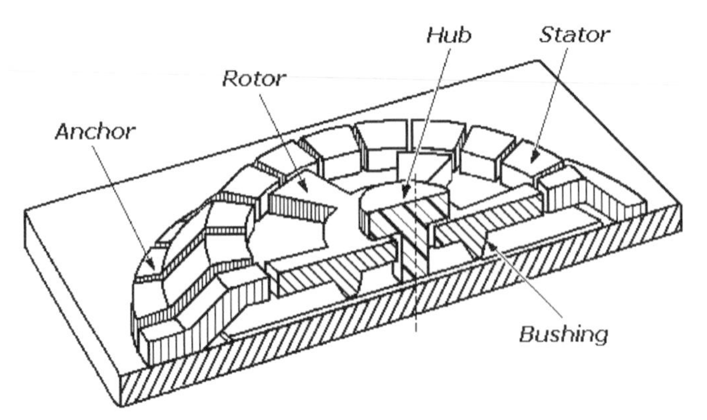

Fig. 1 A cross-section view of a typical micromotor that is driven electrostatically rather than electromagnetically.

Rotors of statically powered MEMS motors can reach speeds in excess of 10,000 rpm when driven by excitation voltages of 30 to 40 V. Some of these miniscule motors have been in continuous operation for 150 hours. Researchers in laboratories at the University of California at Berkeley and at Massachusetts Institute of Technology (MIT) have built these tiny machines.

Other successful microactuators that have been built include microvalves and micropumps; a section view of a typical microvalve is shown in Fig. 2. The diaphragm or control element of this microvalve flexes in a direction that is perpendicular to the valve seat in the base substrate. It can be moved by an embedded

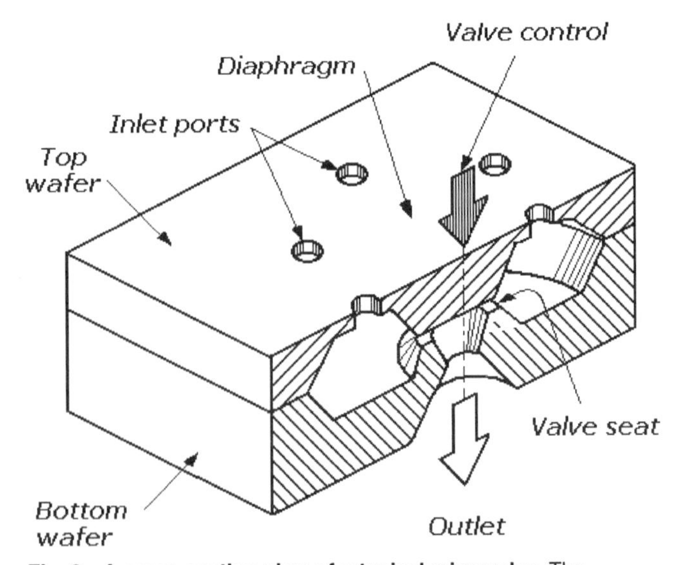

Fig. 2 A cross-section view of a typical microvalve. The diaphragm moves perpendicular to its base substrate. Diaphragms can be moved by an embedded piezoelectric film, by electrostatic forces, or by thermal expansion.

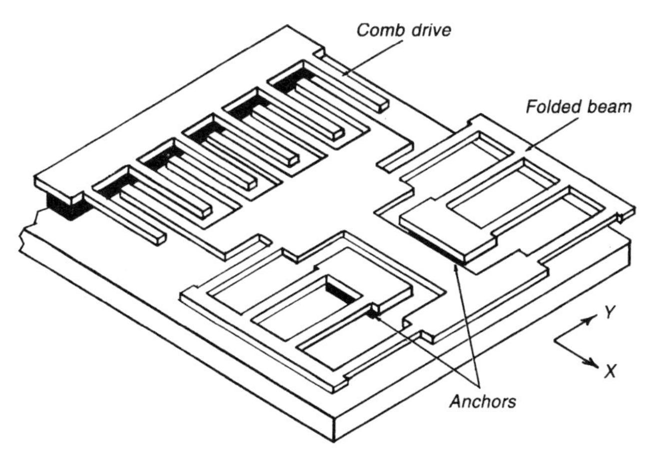

Fig. 3 This linear resonator consists of a pair of folded beams that are set in vibrational motion in the X direction by an electrostatically driven comb structure. Lateral or Y-direction motion is restrained by the geometry of the folded beams.

piezoelectric film, electrostatic force, or thermal expansion. There are applications for these microminiature valves and similar pumps in biomedical research because they are orders of magnitude smaller than conventional biomedical pumps and valves, and they require very little energy to drive them.

The linear resonator shown in Fig. 3 is another form of microactuator that is driven by electrostatic forces, but its operation depends on different principles than those that apply to the motor. The resonator consists of two major components: a comb drive and a folded-beam structure. The folded-beam structure includes a set of fingers or digits on its left end that is interleaved with a similar set of digits rigidly mounted on a pedestal at the left end of the resonator; together both sets of digits form the comb drive. However, the digits on the folded-beam structure are free to oscillate in the *X* direction because the structure includes twin-folded beams that are flexible, although the structure itself is rigidly mounted on two anchors. (The pedestal and anchors, grown on the same substrate, are identified by the black lines in the figure.)

The comb drive is set in oscillation electrostatically in the *X* direction. The flexible folded beams are dimensioned to resonate at a specific frequency when driven by electrostatic charges placed on the comb drive's digits, which act as capacitors. Both folded beams resonate simultaneously, but only in the *X* direction because motion in the lateral or *Y* direction is constrained by the geometries of the folded beams.

Microaccelerometers are microactuators that respond to external forces rather than an embedded piezoelectric film, electrostatic forces, or thermal expansion. Three different microminiature capacitive acceleration sensors are shown in Fig. 4. They are identified as (a) cantilever, (b) torsion-bar suspension, and (c) seismic mass suspended from a central pillar. The simple cantilever structure offers the highest sensitivity for a given size of suspension arm. Consequently, it can be made smaller for a desired sensitivity than any of the other configurations.

Capacitive sensing techniques require an AC voltage across the capacitor being measured. This, in turn, produces an electrostatic field which generates an attractive force between the capacitor plates. As the seismic mass deflects, the change in electrostatic field can be measured; with proper signal conditioning, the output will be proportional to acceleration. These accelerometers can operate as switches in either open-loop or closed-loop systems. The selection of the optimum structure for a specific application takes into account such key factors as sensitivity, stability, material fatigue, shock resistance, mass damping, output linearity, and temperature range, but other properties might also

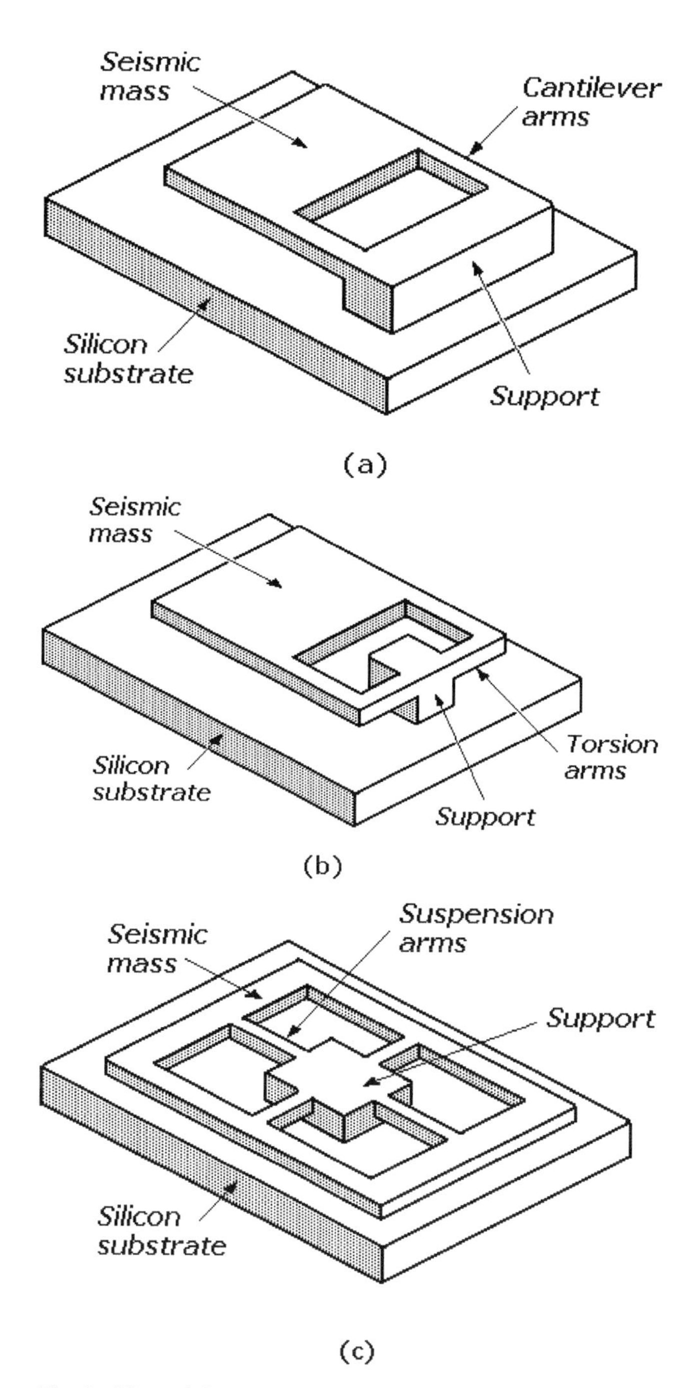

Fig. 4 Microminiature capacitive acceleration sensors: (a) cantilever, (b) torsion-bar suspension, and (c) seismic mass suspended from a central pillar.

be considered. MEMS capacitive accelerometers are now widely used in motor vehicles as sensors for detecting collisions and triggering airbag deployment.

Materials

At present, silicon remains the most popular material for fabricating MEMS because it has the most favorable properties of any of the materials tried for fabricating complex microminiature devices. The use of silicon also made it possible to fabricate integrated circuitry for signal processing on the same silicon substrate. However, some MEMS have also included parts made of aluminum and diamonds. The successful design and manufacture of billions of integrated circuits over the past 30 years has resulted in an extensive body of knowledge about the properties of silicon—how it can be grown, how its structure can be altered, how it can be milled chemically, and how slices of silicon can be permanently bonded.

Silicon is a very strong material with a modulus of elasticity that closely matches steel. It exceeds stainless steel in yield strength and aluminum in strength-to-weight ratio. In addition, it exhibits high thermal conductivity and a low thermal expansion coefficient. Because it lacks mechanical hysteresis, it is a near-perfect material for fabricating sensors and transducers. Moreover, silicon's sensitivity to stress, strain, and temperature lends itself to the fabrication of sensors that can easily communicate with electronic circuitry on the same substrate or chip for the transmission of electrical signals. However, silicon must be protected by encapsulation or hermetic sealing because it is susceptible to deterioration if exposed to air or moisture.

In building MEMS, silicon is chemically micromachined or etched into a wide variety of shapes rather than being machined by traditional cutting tools. Silicon, whether in the form of polysilicon or silicon nitride, and aluminum can be etched into many different shapes and contours in batch processes. In the micromachining process, mechanical structures are sculpted from a silicon wafer by selectively etching away sacrificial supporting layers or structures.

Masks produced by the photolithographic methods are used at various stages in the manufacturing process for selective etching of the silicon. They permit the formation of precise contours such as gear teeth, combs, beams, cantilever arms, and seismic masses. The furnace diffusion of chemical "dopants" in gaseous form into the silicon can alter its chemical makeup and change its electrical characteristics. Epitaxy is the process for growing multiple surface layers of material on the basic substrate, and deposition is the process for plating silicon surfaces with materials such as gold, silver, aluminum, or copper.

Power to the Micromachines

In theory, MEMS could be driven by any of four different forces: electromagnetic, thermal expansion, electrostatic, and piezoelectric effect. The choice of actuation method is generally determined by the end use of the device and its performance requirements. As stated earlier, electromagnetic forces are too weak to power MEMS actuators when scaled down to MEMS size. Moreover, thermal expansion is usually ruled out as a driving force for MEMS because of the excessive power required to concentrate enough heat in a small area to drive microminiature parts made of dissimilar materials. This leaves electrostatic force and piezoelectric effect as the only practical forces for powering MEMS.

Electrostatic Forces

Electrostatic force is attractive for actuating MEMS because, unlike magnetic force, it can be scaled down effectively to micro size. The basic requirement for the use of electrostatic forces for driving MEMS is the presence of two electrically conductive surfaces within the MEMS that will act as opposing capacitor plates. The electrostatic force applied is then directly proportional to the product of the square of the voltage across the two plates and inversely proportional to the square of the distance between the plates.

In the MEMS motor shown in Fig. 1, capacitors are formed by the end surfaces of the rotor spokes and the inner walls of the insulated stator segments. To drive this motor, voltage is applied sequentially to the stator segments, switching them on and off. Each rotor spoke is attracted to the nearest powered stator segment as voltage is switched to them in a rotating sequence, causing the whole rotor to spin as it follows the polarity changes. In this way, the rotor completes a revolution for multiple polarity changes in the stator elements.

The gaps separating the outer ends of the rotor spokes and the inner surfaces of the stator segments act as capacitor plates. However, the gaps are not likely to be uniform because of the microminiature size of this motor. The result is a variation in the electrostatic force with respect to time as the rotor spins, making the electrostatic force a nonlinear function of applied voltage. A MEMS motor can tolerate some variation in electrostatic force, but it can fail or stall for two reasons: if the gap between the opposing surfaces of stator and rotor is not concentric or smooth or if the rotor bearing surfaces are not smooth enough to prevent friction buildup.

The drawback of most surface-micromachined MEMS motors is that both rotor and stator are so thin that their opposing surface areas are too small to provide enough change in capacitance to sustain rotor rotation. One solution to this problem is the use the LIGA process to fabricate the motor because it can form thicker stators and rotors than can be achieved with surface micromachining. (The LIGA process is explained later in this chapter.)

Actuators have also been made in the form of vibrating microstructures with flexible suspensions. An example known as a linear resonator is shown in Fig. 3. It uses a different design to maximize the capacitance change so it can produce motion with larger amplitude by exploiting the advantage of the classical parallel-plate capacitor formula because only attractive forces are generated. The equation for energy stored in an electrostatic comb is:

$$E = \frac{CV^2}{2}$$

where E is the energy stored, C is the capacitance, and V is the voltage across the capacitor.

Surface-micromachined linear resonators with comb drives, as shown in Fig. 3, have multiple fingers or digits. When a voltage is applied, an attractive force is developed between the interleaved digits which then converge. The increase in capacitance is proportional to the number of fingers in the drive: many digits are required to generate enough force to maximize performance. Because the direction of motion of the digits, acting as capacitor plates in the electrostatic comb drive, is parallel along their length, the effective plate area with respect to spacing between the digits' plates remains constant. Consequently, capacitance with respect to the direction of motion is linear, and the induced force in the X direction is directly proportional to the square of the voltage applied across the plates. Comb-drive structures have been driven to deflect by as much as one-quarter of the comb finger length with DC voltages of 20 to 40 V.

The performance of a linear resonator with a comb drive will be degraded if the lateral gaps between the digits are not equal on both sides or if the digits are not straight and parallel. Any of those conditions would cause the digits to diverge at right angles to their intended direction of motion and, as a result, wedge together until the voltage is turned off. However, if they collided with sufficient force they might remain stuck permanently, destroying the linear resonator.

Piezoelectric Films

Microminiature transducers made as rigid beams and diaphragms with cores of polycrystalline zinc-oxide (ZnO) piezoelectric film can change their shape with the application of a voltage. A beam with a central piezoelectric layer of insulated polycrystalline ZnO several micrometers thick is illustrated in Fig. 5. This layer is then insulated on both sides and sandwiched between two conductive electrodes to form a rigid structure. When voltage is applied between the two external electrodes, the piezoelectrically induced stress in the ZnO film causes the structure to deflect.

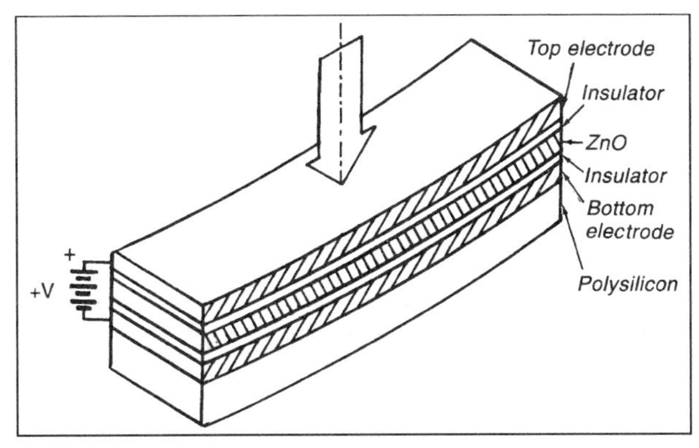

Fig. 5 A microminiature piezoelectric transducer is made as an insulated layer of polycrystalline zinc oxide (ZnO) sandwiched between two conductive electrodes to form a rigid bimetallic structure.

The converse of the piezoelectric effect can be obtained when the beam functions as a strain sensor; it converts strain applied to the beam into electrical signals that are proportional to the strain.

Bulk Micromachining

Bulk micromachining is a method for fabricating MEMS from a single wafer of silicon by chemically etching away silicon to sculpt the desired microstructure. The bulk silicon is obtained by slicing large, cylindrical, single crystals of purified silicon into thin wafers. Micromachining is done by removing the excess silicon from the wafers with a dry or wet chemical at a removal rate that depends on the orientation of the silicon crystal.

The etching process is controlled by etch masks that define the areas for material removal. The mask, produced by photolithography, is placed over a silicon wafer that has been coated with photoresist, a substance which becomes soluble after being exposed to ultraviolet (UV) energy. The photoresist is then chemically removed only from those areas of the wafer that have been exposed to UV through transparent windows in the mask. Only those areas of the wafer from which the photoresist has been removed to expose bare silicon can be chemically etched. The depth of silicon removal is a function of the time the etching process is allowed to continue.

Bulk micromachining is a process that includes methods for fusion bonding silicon substrates to form precise three-dimensional (3-D) structures such as micropumps and microvalves. Two or more etched wafers can be bonded by pressing them together and annealing them to form a permanent 3-D microstructure. This is how an internal or reentrant cavity such as the one shown in the cross-section view of a microvalve (Fig. 2) is formed.

Surface Micromachining

Surface micromachining is a manufacturing process that permits multilayer MEMS to be built. Initially developed for fabricating large-scale silicon integrated circuits, it calls for the deposition of multiple layers of various permanent and sacrificial materials on the surface of a silicon wafer in a sequence required to form complex structures. Free-standing movable structures such as microminiature motor rotors, gears, or chains can be sculpted chemically by etching away sacrificial layers. As in bulk micromachining, the process starts with a thin wafer of crystalline silicon which becomes the substrate. Structural material, usually polysilicon, is deposited on underlying layers of sacrificial materials such as silicon dioxide, silicon nitride, or phosphosilicate glass.

The etching of layers of material in surface micromachining is controlled by the same kinds of masks used in bulk micromachining, and the chemical etching techniques are the same as those used in that process. Each structural layer is sculpted to conform to the contours of the many different masks required. Structures that are intended to move or rotate are released from sacrificial layers by highly selective chemical etchants such as hydrofluoric acid. Etching can also remove sacrificial material deposited around the shafts or hubs patterned with the structure. This process provides enough clearance around the hub and between adjacent parts to permit gears or rotors to rotate freely; the hubs are also capped to restrict vertical movement of the gears or rotors as they rotate.

A MEMS motor, such as the one shown in Fig. 1, is surface micromachined in a series of deposition and masking steps in which alternate layers of permanent silicon and sacrificial material are deposited. After the sacrificial material is chemically removed, the motor structure—stator, rotor, and central capped shaft—is complete. The rotor is then free to rotate around the hub. During surface micromachining, electrically conductive pads and conductive paths are deposited by selective plating of gold or other suitable metal on the substrate to connect the power source to the commutator segments of the stator.

MULTILEVEL FABRICATION PERMITS MORE COMPLEX AND FUNCTIONAL MEMS

Researchers at Sandia National Laboratories, Albuquerque, New Mexico, have developed two surface micromachining processes for fabricating multilevel MEMS (microelectromechanical systems) from polysilicon that are more complex and functional than those made from two- and three-level processes. The processes are SUMMiT Technology, a four-level process in which one ground or electrical interconnect plane and three mechanical layers can be micromachined, and SUMMiT V Technology, a similar five-level process except that four mechanical layers can be micromachined. Sandia offers this technology under license agreement to qualified commercial IC producers.

According to Sandia researchers, polycrystalline silicon (also called polysilicon or poly) is an ideal material for making the microscopic mechanical systems. It is stronger than steel, with a strength of 2 to 3 GPa (assuming no surface flaws), whereas steel has a strength of 200 MPa to 1 GPa (depending on how it is processed). Also, polysilicon is extremely flexible, with a maximum strain before fracture of approximately 0.5%, and it does not readily fatigue.

Years of experience in working with polysilicon have been gained by commercial manufacturers of large-scale CMOS integrated circuits chips because it is used to form the gate structures of most CMOS transistors. Consequently, MEMS can be produced in large volumes at low cost in IC manufacturing facilities with standard production equipment and tools. The Sandia researchers report that because of these advantages, polysilicon surface micromachining is being pursued by many MEMS fabrication facilities.

The complexity of MEMS devices made from polysilicon is limited by the number of mechanical layers that can be deposited. For example, the simplest actuating comb drives can be made with one ground or electrical plane and one mechanical layer in a two-level process, but a three-level process with two mechanical layers permits micromachining mechanisms such as gears that rotate on hubs or movable optical mirror arrays. A four-level process such as SUMMiT permits mechanical linkages to be formed that connect actuator drives to gear trains. As a result, it is expected that entirely new kinds of complex and sophisticated micromachines will be fabricated with the five-level process.

According to the Sandia scientists, the primary difficulties encountered in forming the extra polysilicon layers for surface micromachining the more complex devices are residual film stress and device topography. The film stress can cause the mechanical layers to bow from the required flatness. This can cause the mechanism to function poorly or even prevent it from working. The scientists report that this has even been a problem in the fabrication of MEMS with only two mechanical layers.

To surmount the bowing problem, Sandia has developed a proprietary process for holding stress levels to values typically less than 5 MPa, thus permitting the successful fabrication and operation of two meshing gears, whose diameters are as large as 2000 μm.

The intricacies of device topography that make it difficult to pattern and etch successive polysilicon layers restrict the complexities of the devices that can be built successfully. Sandia has minimized that problem by developing a proprietary chemical-mechanical polishing (CMP) process called "planarizing" for forming truly flat top layers on the polysilicon. Because CMP is now so widely used in integrated circuit chip manufacture, it will allow MEMS to be batch fabricated by the SUMMiT processes using standard commercial IC fabrication equipment.

GALLERY OF MEMS ELECTRON-MICROSCOPE IMAGES

The Sandia National Laboratories, Albuquerque, New Mexico, have developed a wide range of microelectromechanical systems (MEMS). The scanning electron microscope (SEM) micrographs presented here show the range of these devices, and the captions describe their applications.

Fig. 1 Wedge Stepping Motor: This indexing motor can precisely index other MEMS components such as microgear trains. It can also position gears and index one gear tooth at a time at speeds of more than 200 teeth/s or less than 5 ms/step. An input of two simple input pulse signals will operate it. This motor can index gears in MEMS such as locking devices, counters, and odometers. It was built with Sandia's four-layer SUMMiT technology. Torque and indexing precision increase as the device is scaled up in size.

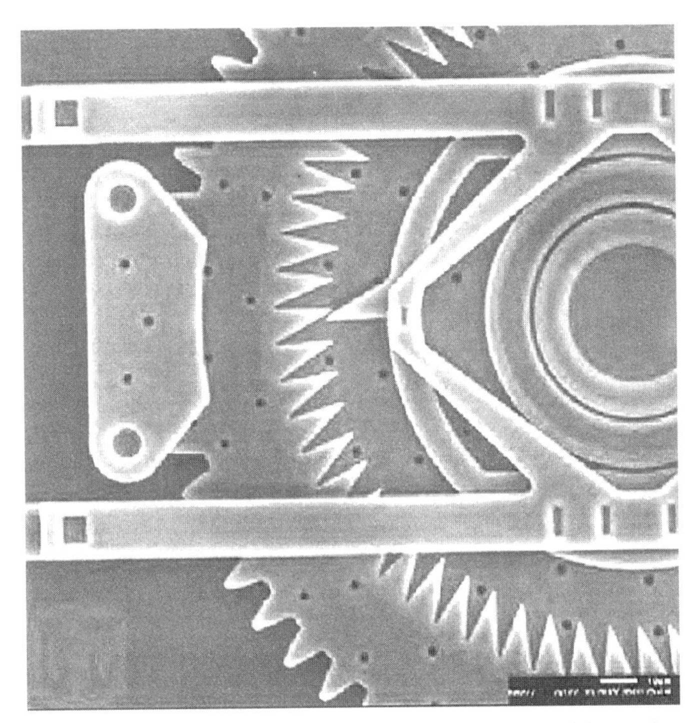

Fig. 2 **Wedge Stepping Motor:** A close-up view of one of the teeth of the indexing motor shown in Fig. 1.

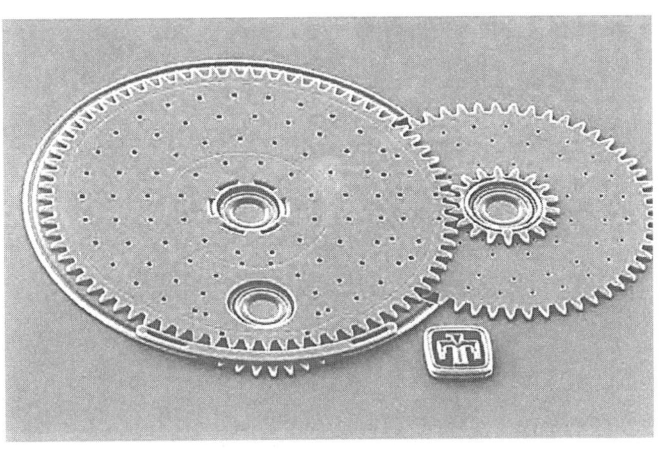

Fig. 3 **Torque Converter:** This modular transmission unit has an overall gear reduction ratio of 12 to 1. It consists of two multilevel gears, one with a gear reduction ratio of 3 to 1 and the other with a ratio of 4 to 1. A coupling gear within the unit permits cascading.

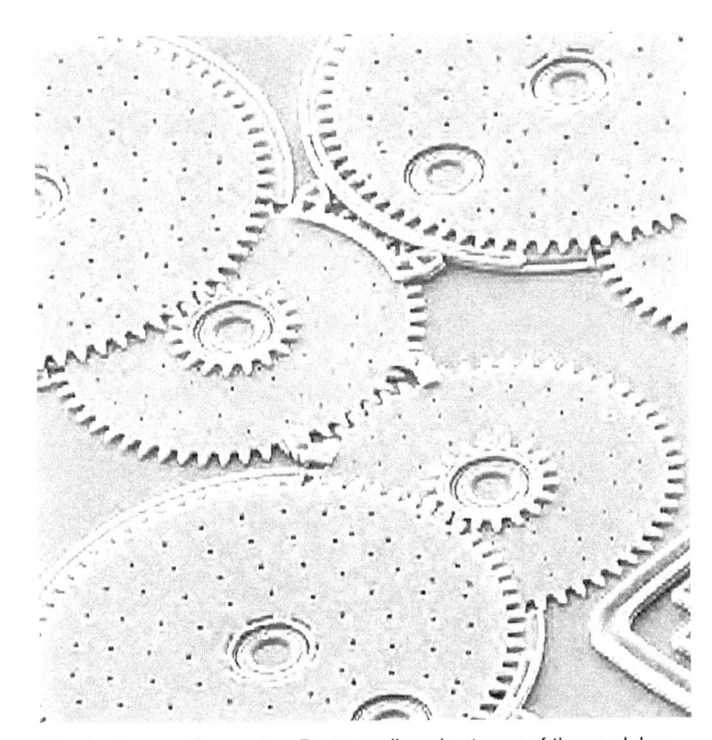

Fig. 4 **Torque Converter:** By cascading six stages of the modular 12-to-1 transmission units shown in Fig. 3, a 2,985,894-to-1 gear reduction ratio is obtained in a die area of less than 1 mm^2. The converter can step up or step down.

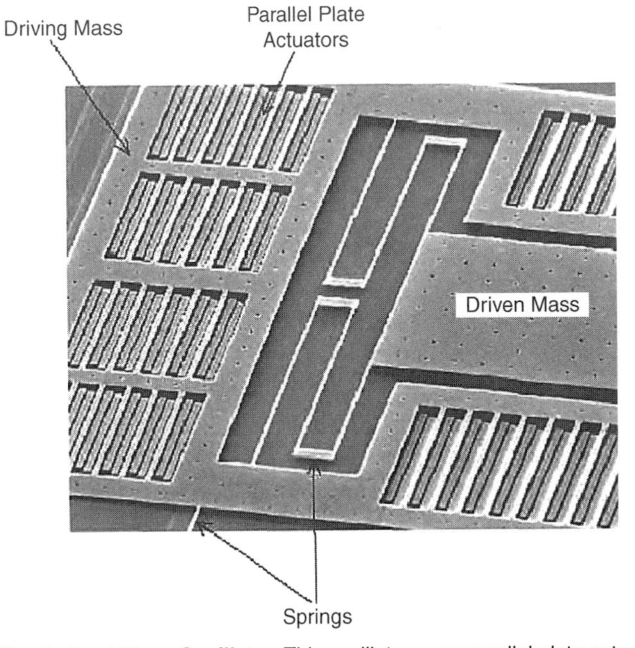

Fig. 5 **Dual-Mass Oscillator:** This oscillator uses parallel plate actuation and system dynamics to amplify motion. The 10-mm-long parallel plate actuators on the driving mass produce an amplified motion on the second mass when it is driven by a signal. The actuated mass remains nearly motionless, while the moving mass has an amplitude of approximately 4 μm when driven by a 4-V signal. It was designed to be part of a vibrating gyroscope.

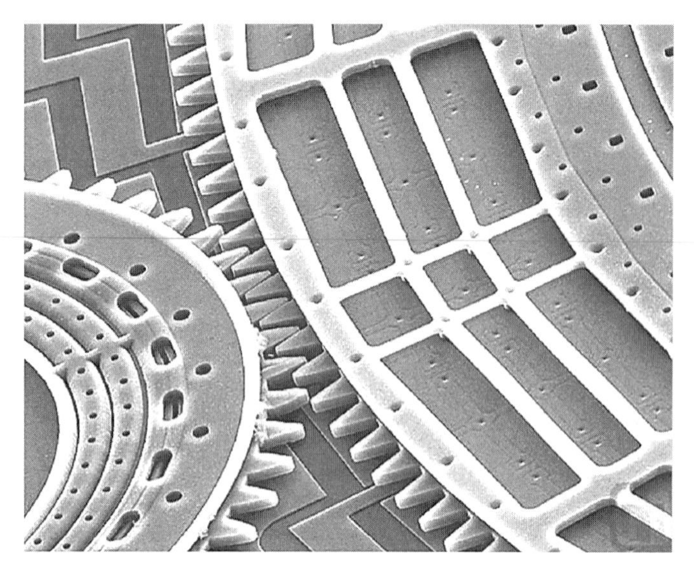

Fig. 6 Rotary Motor: This close-up shows part of a rotary motor that offers advantages over other MEMS actuators. Its operates on linearcomb drive principles, but the combs are bent in a circle to permit unlimited travel. The combs are embedded inside the rotor so that other micromachines can be powered directly from the rotor's perimeter. Built by Sandia's four-level SUMMiT technology, the motor is powered by a lower voltage and produces higher output torque than other MEMS actuators, but it still occupies a very small footprint. It can also operate as a stepper motor for precise positioning applications.

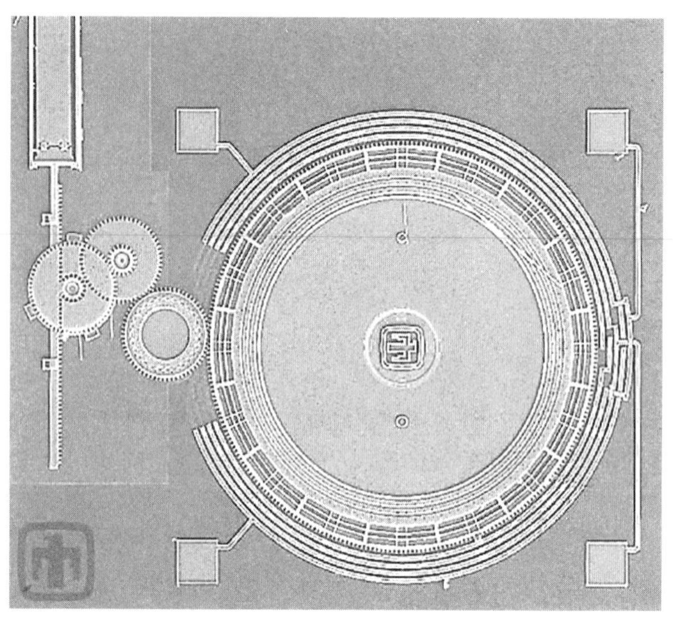

Fig. 7 Comb Drive Actuation: Two sets of comb-drive actuators (not shown) drive a set of linkages (upper right) to a set of rotary gears. The comb-drive actuators drive the linkages 90° out of phase with each other to rotate the small 19-tooth gear at rotational speeds in excess of 300,000 rpm. The operational lifetime of these small devices can exceed 8×10^9 revolutions. The smaller gear drives a larger 57-tooth (1.6-mm-diameter) gear that has been driven as fast as 4800 rpm.

Fig. 8 Micro Transmission: This transmission has sets of small and large gears mounted on the same shaft so that they interlock with other sets of gears to transfer power while providing torque multiplication and speed reduction. Its output gear is coupled to a double-level gear train.

Fig. 9 Microtransmission and Gear Reduction Unit: This mechanism is the same as that in Fig. 8 except that it performs a gear-reduction function. The microengine pinion gear, labeled A in the figure, meshes directly with the large 57-tooth gear, labeled B. A smaller 19-tooth gear, C, is positioned on top of gear B and is linked to B's hub. Because the gears are joined, both make the same number of turns per minute. The small gear essentially transmits the power of the larger gear over a shorter distance to turn the larger 61-tooth gear D. Two of the gear pairs (B and C, D and E) provide 12 times the torque of the engine. A linear rack F, capable of driving an external load, has been added to the final 17-tooth output gear E to provide a speed reduction/torque multiplication ratio of 9.6 to 1.

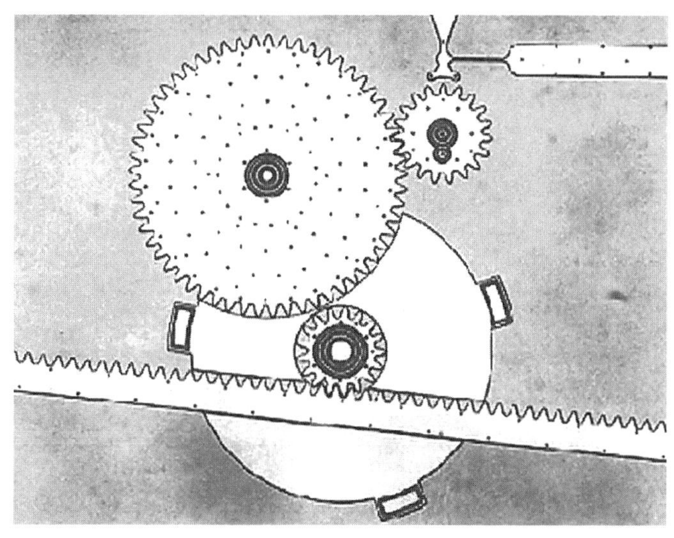

Fig. 10 Gear-Reduction Units: This micrograph shows the three lower-level gears (A, B, and E) as well as the rack (F) of the system shown in Fig. 9. The large flat area on the lower gear provides a planar surface for the fabrication of the large, upper-level 61-tooth gear (D).

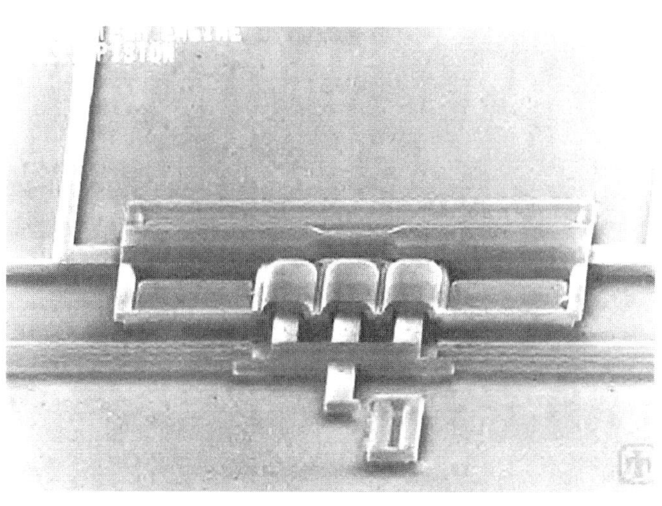

Fig. 11 Microsteam Engine: This is the world's smallest multipiston microsteam engine. Water inside the three compression cylinders is heated by electric current, and when it vaporizes, it pushes the pistons out. Capillary forces then retract the piston once current is removed.

Fig. 12 Microchain Drive: This 50-link microchain drive resembles a bicycle chain and sprocket assembly except that each link of the microchain could rest with atop a cross section of human hair with space to spare. The distance on centers between chain links is 50 µm. (The diameter of a human hair is 70 µm.) Because this microchain can rotate many sprockets, a single tiny MEMS motor coupled to the drive shaft of a single sprocket can rotate many other sprockets linked to the microchain. Typically, a single MEMS motor powers each MEMS device. Chain systems, unlike stroke systems, perform oscillatory motions permitting both continuous and intermittent drive translation.

Fig. 13 Detail of Microchain Link and Sprocket: A silicon microchain rather than a silicon microbelt is used in this MEMS because a silicon belt would have spring-like characteristics which would apply too much torque on any sprockets not aligned in the same plane. By contrast, each chain link is capable of ± 52° rotation with respect to the preceding link. This minimizes stress on the support structure. The longest span unsupported by sprockets or bracing is 500 microns, but a microchain tensioner permits longer spans. This multilevel surface-micromachined device was constructed with Sandia's Summit IV and Summit V technology.

MEMS CHIPS BECOME INTEGRATED MICROCONTROL SYSTEMS

The successful integration of MEMS (microelectromechanical systems) on CMOS integrated circuit chips has made it possible to produce "smart" control systems whose size, weight, and power requirements are significantly lower than those for other control systems. MEMS development has previously produced microminiature motors, sensors, gear trains, valves, and other devices that easily fit on a silicon microchip, but difficulties in powering these devices has inhibited their practical applications.

MEMS surface micromachining technology is a spin-off of conventional silicon IC fabrication technology, but fundamental differences in processing steps prevented their successful integration. The objective was to put both the control circuitry and mechanical device on the same substrate. However, the results of recent development work showed that they could be successfully merged.

It has been possible for many years to integrate the transistors, resistors, capacitors, and other electronic components needed for drive, control, and signal processing circuits on a single CMOS silicon chip, and many different MEMS have been formed on separate silicon chips. However, the MEMS required external control and signal-processing circuitry. It was clear that the best way to upgrade MEMS from laboratory curiosities to practical mechanical devices was to integrate them with their control circuitry. The batch fabrication of the electrical and mechanical sections on the same chip would offer the same benefits as other large-scale ICs—increased reliability and performance. Component count could be reduced, wire-bonded connections between the sections could be eliminated, minimizing power-wasting parasitics, and standard IC packaging could replace multichip hybrid packages to reduce product cost.

MEMS sections are fabricated by multilevel polysilicon surface micromachining that permits the formation of such intricate mechanisms as linear comb-drive actuators coupled to gear trains. This technology has produced micromotors, microactuators, microlocks, microsensors, microtransmissions, and micromirrors.

Early attempts to integrate CMOS circuitry with MEMS by forming the electronic circuitry on the silicon wafer before the MEMS devices met with only limited success. The aluminum electrical interconnects required in the CMOS process could not withstand the long, high-temperature annealing cycles needed to relieve stresses built up in the polysilicon mechanical layers of the MEMS. Tungsten interconnects that could withstand those high temperatures were tried, but the performance of the CMOS circuitry was degraded when the heat altered the doping profiles in the transistor junctions.

When the MEMS were formed before the CMOS sections, the thermal problems were eliminated, but the annealing procedure tended to warp the previously flat silicon wafers. Irregularities in the flatness or planarity of the wafer distorted the many photolithographic images needed in the masking steps required in CMOS processing. Any errors in registration can lower attainable resolution and cause circuit malfunction or failure.

Experiments showed that interleaving CMOS and MEMS process steps in a compromise improved yield but limited both the complexity and performance of the resulting system. In other experiments materials such as stacked aluminum and silicon dioxide layers were substituted for polysilicon as the mechanical layers, but the results turned out to be disappointing.

Each of these approaches had some merit for specific applications, but they all resulted in low yields. The researchers persevered in their efforts until they developed a method for embedding the MEMS in a trench below the surface of the silicon wafer before fabricating the CMOS. This is the procedure that now permits the sections to be built reliably on a single silicon chip.

Sandia's IMEMS Technology

Sandia National Laboratories, Albuquerque, New Mexico, working with the University of California's Berkeley Sensor and Actuator Center (BSAC), developed the unique method for forming the micromechanical section first in a 12-μm-deep "trench" and backfilling that trench with sacrificial silicon dioxide before forming the electronic section. This technique, called Integrated MicroElectroMechanical Systems (IMEMS), overcame the wafer-warping problem. Figure 1 is cross-section view of both sections combined on a single chip.

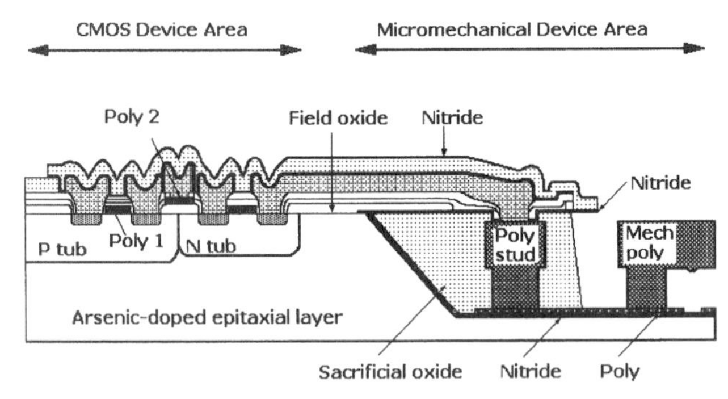

Fig. 1 A cross-section view of CMOS drive circuitry integrated on the same silicon chip with a microelectromechanical system.

The mechanical polysilicon devices are surface micromachined by methods similar to Sandia's SUMMiT process in the trench, using special photolithography methods. After the trench is filled with the silicon dioxide, the silicon wafer is annealed and that section is "planarized," or etched flat and flush with the rest of the wafer surface, by a process called chemical-mechanical polishing (CMP). After the CMOS section is complete, the sacrificial silicon dioxide in the trench is etched away, leaving the MEMS devices electrically interconnected with the adjacent CMOS circuitry.

Advantages of IMEMS

Sandia spokespersons say the IMEMS process is completely modular, meaning that the planarized wafers can be processed in any facility capable of processing CMOS, bipolar, and combinations of these processes. They add that modularity permits the mechanical devices and electronic circuitry to be optimized independently, making possible the development of high-performance microsystems.

Early Research and Development

Analog Devices Inc. (ADI) was one of the first companies to develop commercial surface-micromachined integrated-circuit accelerometers. ADI developed and marketed these accelerometer chips, demonstrating its capability and verifying commercial demand. Initially ADI built these devices by interleaving, combining, and customizing its internal manufacturing processes to produce the micromechanical devices with the same processes it used to produce monolithic electronic circuitry.

At the same time, researchers at BSAC developed the alternative process for replacing conventional aluminum interconnect layers with tungsten layers to enable the CMOS device to withstand the higher thermal stresses associated with subsequent micromechanical device processing. This process was later superseded by the joint BSAC–Sandia development of IMEMS.

Accelerometers

ADI offered the single-axis ADXL150 and dual-axis ADXL250, and Motorola Inc. offered the XMMAS40GWB. Both of ADI's integrated accelerometers are rated for $\pm 5\ g$ to $\pm 50\ g$. They have been in high-volume production since 1993. The company is now licensed to use Sandia's integrated MEMS/CMOS technology. Motorola is now offering the MMA1201P and MMA2200W single-axis IC accelerometers rated for $\pm 38\ g$.

These accelerometer chips differ in architecture and circuitry, but both work on the same principles. The surface micromachined sensor element is made by depositing polysilicon on a sacrificial oxide layer that is etched away, leaving the suspended sensor element. Figure 2 is a simplified view of the differential-capacitor sensor structure in an ADI accelerometer. It can be seen that two of the capacitor plates are fixed, and the center capacitor plate is on the polysilicon beam that deflects from its rest position in response to acceleration.

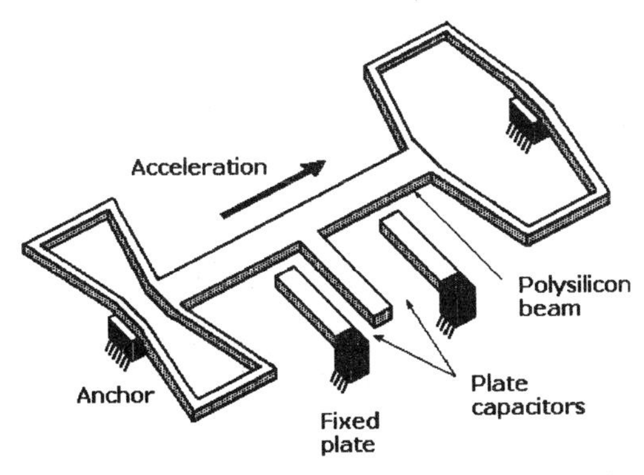

Fig. 2 A simplified view of the movement of a polysilicon beam in a surface-micromachined accelerometer moving in response to acceleration. The two fixed plates and one moving plate form a unit cell.

When the center plate deflects, its distance to one fixed plate increases while its distance to the other plate decreases. The change in distance is measured by the on-chip circuitry that converts it to a voltage proportional to acceleration. All of the circuitry, including a switched-capacitor filter needed to drive the sensor and convert the capacitance change to voltage, is on the chip. The only external component required is a decoupling capacitor.

Integrated-circuit accelerometers are now used primarily as airbag-deployment sensors in automobiles, but they are also finding many other applications. For example, they can be used to monitor and record vibration, control appliances, monitor the condition of mechanical bearings, and protect computer hard drives.

Three-Axis Inertial System

When the Defense Advanced Research Projects Agency (DARPA), an agency of the U.S. Department of Defense, initiated a program to develop a solid-state three-axis inertial measurement system, it found that the commercial IC accelerometers were not suitable components for the system it envisioned for two reasons: the accelerometers must be manually aligned and assembled, and this could result in unwanted variations in alignment, and the ICs lacked on-chip analog-to-digital converters (ADCs), so they could not meet DARPA's critical sensitivity specifications.

To overcome these limitations, BSAC designed a three-axis, force-balanced accelerometer system-on-a-chip for fabrication with Sandia's modular monolithic integration methods. It is said to exhibit an order of magnitude increase in sensitivity over the best commercially available single-axis integrated accelerometers. The Berkeley system also includes clock generation circuitry, a digital output, and photolithographic alignment of sense axes. Thus, the system provides full three-axis inertial measurement, and does not require the manual assembly and alignment of sense axes.

A combined *X*- and *Y*-axis rate gyro and a *Z*-axis rate gyro was also designed by researchers at BSAC. By using IMEMS technology, a full six-axis inertial measurement unit on a single chip was obtained. The 4- by 10-mm system is fabricated on the same silicon substrate as the three-axis accelerometer, and that chip will form the core of a future micro–navigation system. BSAC is teamed with ADI and Sandia Laboratories in this effort, with funds provided by DARPA's Microsystems Technology Office.

Micromechanical Actuators

Micromechanical actuators have not attained the popularity in commercial applications achieved by microminiature accelerometers, valves, and pressure sensors. The two principal drawbacks to their wider application have been their low torque characteristics and the difficulties encountered in coupling actuators to drive circuitry. Sandia has developed devices that can be made by its SUMMiT four-level polysilicon surface-micromachining process, such as the microengine pinion gear driving a 10 to 1 transmission shown in Fig. 3, to improve torque characteristics.

The SUMMiT process includes three mechanical layers of polysilicon in addition to a stationary level for grounding or electrical interconnection. These levels are separated by sacrificial silicon-dioxide layers. A total of eight mask levels are used in this process. An additional friction-reducing layer of silicon nitride is placed between the layers to form bearing surfaces.

If a drive comb, operating at a frequency of about 250,000 rpm, drives a 10-to-1 gear reduction unit, torque is traded off for speed. Torque is increased by a factor of 10 while speed is reduced to about 25,000 rpm. A second 10-to-1 gear reduction would increase torque by a factor of 100 while reducing speed to 2,500 rpm. That gear drives a rack and pinion slider that provides high-force linear motion. This gear train provides a speed-reduction/torque-multiplication ratio of 9.6 to 1.

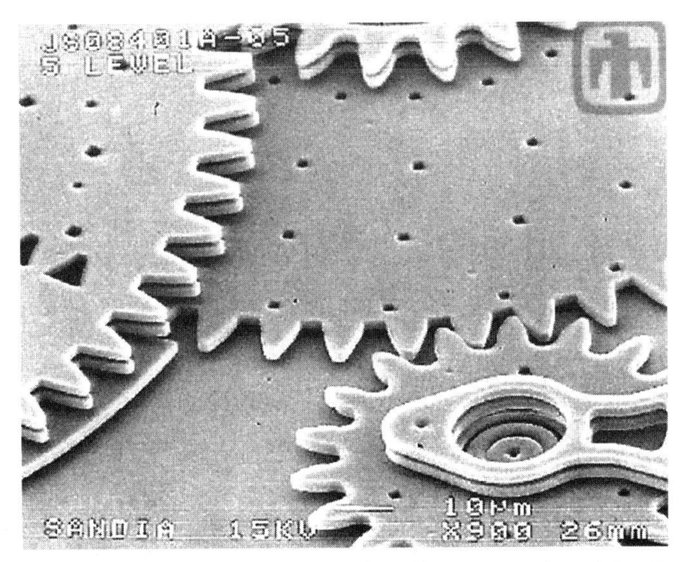

Fig. 3 This linear-rack gear reduction drive converts the rotational motion of a pinion gear to linear motion to drive a rack. *Courtesy of Sandia National Laboratories*

ALTERNATIVE MATERIALS FOR BUILDING MEMS

Researchers at Sandia National Laboratories, Albuquerque, New Mexico, have developed techniques for fabricating functional microelectromechanical systems (MEMS) components from ceramic and rare earth nanocomposites. Their methods include the formulation of nanocomposite mixtures, micromold fabrication, and micromolding. Sandia Laboratories reports that the fabrication of MEMS from ceramics will increase their robustness, high-temperature inertness, chemical and biological compatibility, magnetic properties, piezoelectric properties, and photochromism.

Silicon and its compounds are now the principal materials used in both the bulk and surface methods for micromachining MEMS. This technology has been derived from the photolithographic and acid etching technology for fabricating large-scale integrated circuits (ICs), permitting MEMS to be made at laboratories and commercial IC wafer fabrication facilities with standard IC manufacturing equipment. One advantage to the use of silicon is that the technology is compatible with complementary metal-oxide semiconductor (CMOS) IC production. As a result, MEMS can be integrated on the same chip with analog or digital ICs for driving, controlling, or communicating with the MEMS in a microminiature system.

Unfortunately, silicon MEMS made by surface micromachining are not intrinsically durable because active mechanical parts such as gears, rotors, or folded beams that have thickness measurable only in microns are subject to fracture. There is an inherent limitation on layer thickness imposed by the deposition and etching process of surface micromachining that has little or no effect on electronic circuitry. Consequently, researchers have been searching for alternative materials such as ceramics and composites to improve the versatility and durability of MEMS without sacrificing IC compatibility.

Sandia Laboratories researchers report that micromolded components can be made free-standing or assembled on substrates. By molding components from nanometer-sized particles, they were able to build them with lateral dimensions of a few microns. Their findings indicated that it is possible to mold micron-size mechanical parts from nanocomposites, and that the process is compatible with existing IC fabrication processes.

Ceramic nanoparticles less that 100 nm in diameter are now readily available. One example is aluminum oxide which offers excellent insulation and wear resistance properties as well as the ability to withstand intense heat. Other nanoparticles suitable for molding micron-sized mechanical components are samarium-cobalt, rare earth, and manganese-iron ferrite because of their magnetic properties.

One way to produce micromolded nanocomposites is to form them in high-aspect ratio micromolds made by the LIGA process. (See *LIGA: An Alternative Method for Making Microminiature Parts* in this chapter.) The molds are formed by x-ray lithography from polymethylmethacrylate (PMMA) on silicon substrates. The PMMA molds are filled with the nanoparticles embedded in a binder in which they are cured. The microparts are then polished flat or "planarized" and removed from the mold.

A variation on the LIGA process involves forming a dovetail-shaped cavity in the silicon substrate to anchor a coating of photoresist which is then patterned by conventional IC photolithographic processes and encased in PMMA. After the PMMA is polished back to reach the top of the photoresist, it is dissolved chemically, leaving a PMMA mold. This mold is filled with the nanoparticles embedded in a binder and cured before the microparts are planarized and removed from the mold.

In other research on the fabrication of more rugged microminiature parts, researchers at the University of California at Los Angeles (formerly working at Pennsylvania State University) developed a process called *microstereolithography*. They succeeded in forming 3-D ceramic structures with thicknesses of 50 μm to 1 mm by applying layers of alumina in thicknesses from 10 to 20 μm. Ceramic structures 5.7×10 μm have been produced. The researchers used the method to apply films of lead, zirconate, and titanate greater than 15 μm thick to silicon substrates. The process is a miniaturized version of the rapid prototyping stereolithography process. (See *Stereolithography* in Chap. 18 *Rapid Prototyping.*) An ultraviolet (UV) laser with its beam focused to a 1 to 2 μm width was used to perform the work. This beamwidth contrasts with widths of hundreds of microns used in conventional stereolithography.

Other material alternatives to silicon under consideration in laboratories for the manufacture of MEMS include stainless steel, aluminum, and titanium nanopowders in a binder, as well as bulk titanium. MEMS have also been formed by injection molding, and hot embossing has been used to stamp finely detailed microfeatures on them in thermoplastic. The LIGA process has produced highly detailed microscale tools from a list of metal alloys suitable for fabricating polymeric MEMS. Researchers have also been experimenting with the bulk micromachining of titanium as an alternative to silicon.

Plastic resins including polycarbonate and acrylic are seen as better materials for microfluidic structures because they offer the advantages of biocompatibility and resistance to solvents. A thermoplastic polymer microneedle with a diameter of about 100 μm was produced in the mechanical engineering laboratory at the University of California, Berkeley. It was formed in an aluminum mold on a 30-ton injection-molding machine.

Agilent Technologies, Palo Alto, California, produced a commercial polymeric microfluidic device for separating compounds such as proteins and peptides for analysis. Made from Kapton polyimide thermoplastic film, the microfluidic channels in the device were drilled by a laser, the same process used in making ink-jet printer heads. The device is mounted on a mass spectrometer for performing liquid chromatography.

LIGA: AN ALTERNATIVE METHOD FOR MAKING MICROMINIATURE PARTS

The Sandia National Laboratories, Livermore, California, is using a process called LIGA to form microminiature metal components as an alternative to producing them by the surface micromachining processes used to make microelectromechanical systems (MEMS). LIGA permits the fabrication of larger, thicker, and more durable components with greater height-to-width ratios. They can withstand high pressure and temperature excursions while providing more useful torques than polysilicon MEMS.

The acronym LIGA is derived from the German words for lithography, electroplating, and molding (Lithographie, Galvanoformung, and Abformung), a micromachining process originally developed at the Karlsruhe Nuclear Research Center in Karlsruhe, Germany, in the 1980s. Sandia Labs has produced a wide variety of LIGA microparts, including components for millimotors and miniature stepping motors. It has also made miniature accelerometers, robotic grippers, a heat exchanger, and a mass spectrometer. Sandia is carrying out an ongoing research and development program to improve the LIGA process and form practical microparts for various applications.

In the LIGA process, highly parallel X-rays from a synchrotron are focused through a mask containing thin 2D templates of the microparts to be formed. The X rays transfer the patterns to a substrate layered with PMMA (polymethylmethacrylate), a photoresist sensitive to X rays, on a metallized silicon or stainless-steel substrate. When the exposed layer of PMMA (better known as Plexiglas) is developed, the cavities left in the PMMA are the molds in which the microparts will be formed by electroplating. The thickness of the PMMA layer determines the large height-to-width ratio of the finished LIGA microparts. The resulting parts can be functional components or molds for replicating the parts in ceramic or plastic.

The highlights of the LIGA process as illustrated in Fig. 1 are

(a) An X-ray mask is prepared by a series of plating and lithographic steps. A metallized silicon wafer coated with photoresist is exposed to ultraviolet light through a preliminary mask containing the 2D patterns of the microparts to be produced. Development of the photoresist dissolves the resist from the plated surface of the wafer, forming the micropart pattern, which is plated in gold to a thickness of 8 to 30 μm. The remaining photoresist is then dissolved to finish the mask.

(b) Target substrate for forming microparts is prepared by solvent-bonding a layer of PMMA to a metallized-silicon or stainless-steel substrate.

(c) PMMA-coated substrate is then exposed to highly collimated parallel X rays from a synchrotron through the mask.

(d) PMMA is then chemically developed to dissolve the exposed areas down to the metallized substrate, etching deep cavities for forming microparts.

(e) Substrate is then electroplated to fill the cavities with metal, forming the microparts. The surface of the substrate is then lapped to finish the exposed surfaces of the microparts to the required height within ±5μm.

(f) Remaining PMMA is dissolved, exposing the 3D microparts, which can be separated from the metallized substrate or allowed to remain attached, depending on their application.

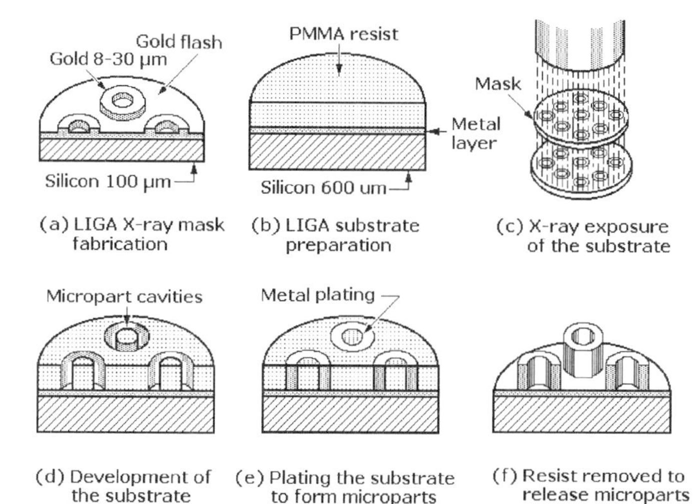

Fig. 1 Steps in fabricating microminiature parts by the LIGA process.

The penetrating power of the X rays from the synchrotron allows structures to be formed that have sharp, well-defined vertical surfaces or sidewalls. The minimum feature size is 20 μm, and microparts can be fabricated with thickness of 100 μm to 3 mm. The sidewall slope is about 1μm/mm. In addition to gold, microparts have been made from nickel, copper, nickel–iron, nickel–cobalt, and bronze.

An example of a miniature machine assembled from parts fabricated by LIGA is an electromagnetically actuated millimotor. With an 8-mm diameter and a height of 3 mm, it includes 20 LIGA parts as well as an EDM-machined permanent magnet and wound coils. The millimotor has run at speeds up to 1600 rpm, and it can provide torque in excess of 1 mN-m. Another example of a miniature machine built from LIGA parts is a size 5 stepper motor able to step in 1.8-deg increments. Both its rotor and stator were made from stacks of 50 laminations, each 1-mm thick.

According to Sandia researchers, the LIGA process is versatile enough to be an alternative to such precision machining methods as wire EDM for making miniature parts. The feature definition, radius, and sidewall texture produced by LIGA are said to be superior to those obtained by any precision metal cutting technique.

In an effort to form LIGA parts with higher aspect ratios, researchers at the University of Wisconsin in Madison teamed with the Brookhaven National Laboratory on Long Island to use the laboratory's 20,000-eV photon source to produce much higher levels of X-ray radiation than are used in other LIGA processes. The higher-energy X rays penetrate into the photoresist to depths of 1 cm or more, and they also pass more easily through the mask. This permitted the Wisconsin team to use thicker and stronger materials to make 4-in.-square masks rather than the standard 1- × 6-cm masks used in standard LIGA. Working with Honeywell, the team developed LIGA optical microswitches

The primary disadvantage to LIGA is its requirement for a synchrotron or other high-energy sources to image parallel X-rays on the PMMA covered substrate. In addition to their limited availability, these sources are expensive to build, install, and operate. Their use adds significantly to the cost of producing LIGA microstructures, especially for commercial applications.

MINIATURE MULTISPEED TRANSMISSIONS FOR SMALL MOTORS

Transmissions would be batch-fabricated using micromachining technologies. *NASA's Jet Propulsion Laboratory, Pasadena, California*

A design has been developed for manufacturing multispeed transmissions that are small enough to be used with minimotors—electromagnetic motors with power ratings of less than 1 W. Like similar, larger systems, such as those in automobiles, the proposed mechanism could be used to satisfy a wider dynamic range than could be achieved with fixed-ratio gearing. However, whereas typical transmission components are machined individually and then assembled, this device would be made using silicon batch-fabrication techniques, similar to those used to manufacture integrated circuits and sensors.

Until now, only fixed-ratio gear trains have been available for minimotors, affording no opportunity to change gears in operation to optimize for varying external conditions, or varying speed, torque, and power requirements. This is because conventional multispeed gear-train geometries and actuation techniques do not lend themselves to cost-effective miniaturization. In recent years, the advent of microelectromechanical systems (MEMS) and of micromachining techniques for making small actuators and gears has created the potential for economical mass production of multispeed transmissions for minimotors. In addition, it should be possible to integrate these mechanisms with sensors, such as tachometers and load cells, as well as circuits, to create integrated silicon systems, which could perform closed-loop speed or torque control under a variety of conditions. In comparison with a conventional motor/transmission assembly, such a package would be smaller and lighter, contain fewer parts, consume less power, and impose less of a computational burden on an external central processing unit (CPU).

Like conventional multispeed transmissions for larger motors, miniature multispeed transmissions would contain gears, clutches, and brakes. However, the designs would be more amenable to micromachining and batch fabrication. Gear stages would be nestled one inside the other (see figures 1, 2, and 3), rather than stacked one over the other, creating a more planar device. Actuators and the housing would be fabricated on separate layers. The complex mechanical linkages and bearings used to shift gears in conventional transmissions would not be practical at the small scales of interest here. Promising alternatives might include electrostatic-friction locks or piezoelectric actuators. For example, in the transmission depicted in the figure, piezoelectric clamps would serve as actuators in clutches and brakes.

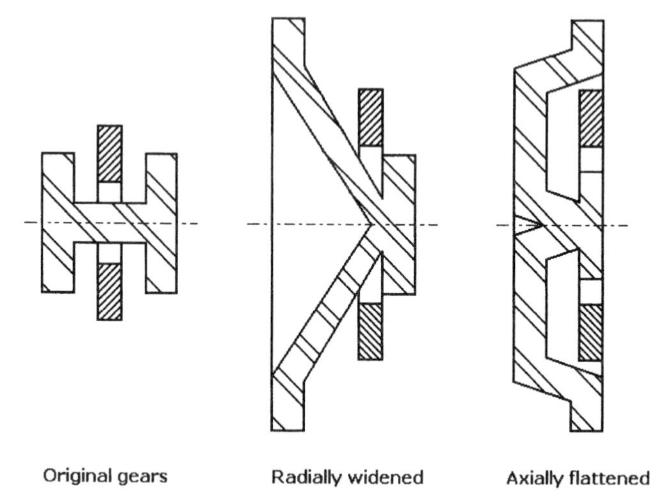

Fig. 2 Evolutionary stages in converting conventional gears to axially flattened gears.

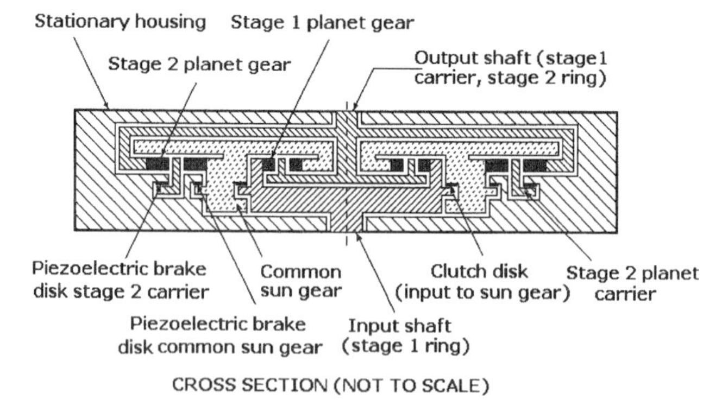

Fig. 3 This Miniature Transmission could be regarded as a flattened version of a conventional three-speed automatic transmission. The components would be fabricated by micromachining.

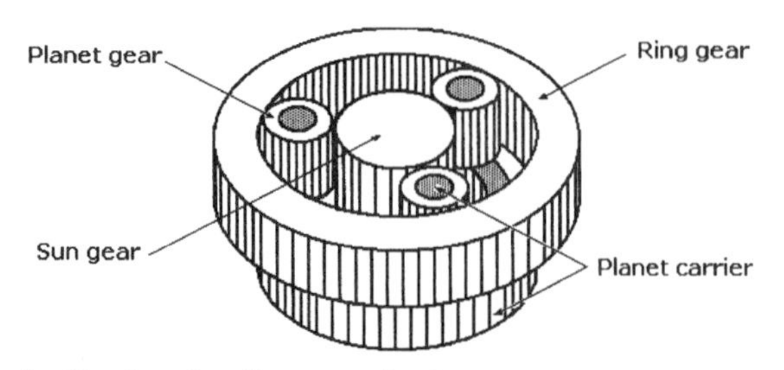

Fig. 1 Simple epicyclic gear train. Compound epicyclic gears in traditional automatic transmissions usually consist of simple epicyclics which are stacked one on top of the other along a radial axis.

The structures would be aligned and bonded, followed by a final etch to release the moving parts. The entire fabrication process can be automated, making it both precise and relatively inexpensive. The end product is a "gearbox on a chip," which can be "dropped" onto a compatible motor to make an integrated drive system.

This work was done by Indrani Chakraborty and Linda Miller of Caltech for **NASA's Jet Propulsion Laboratory**.

THE ROLE OF NANOTECHNOLOGY IN MECHANICAL ENGINEERING

Nanotechnology or *nanoscience* is defined as the science of designing, producing, and viewing structures at the molecular or nanometer scale (billionths of a meter). It has been receiving a lot of attention lately in both the media and scientific papers because of its potential for advancing not only the physical and biological sciences but mechanical and electronic engineering as well. The field promises a new era of stronger, lighter, and more versatile materials. Research in nanotechnology suggests the possibilities for creating such practical products as molecular transistors, artificial muscles, super strong aircraft and vehicles, capacitors that can replace batteries, and extremely strong thin films with applications in both electronics and structures. Nanoscale materials are also being used in low-cost, more efficient solar panels.

Nanotechnology is concerned with materials that range in size from 1 to 100 nanometers (nm). One nanometer equals 0.001 μm or, stated in another way, there are one-thousand nanometers in a micrometer (micron). Table 1 provides useful information about various objects in this size range. Although some very small silicon microchips have widths of only 130 μm, they are well beyond the size limits set for nanotechnology. Atoms have diameters of about 0.1 nm while molecules have diameters of about 1.0 nm. When objects approach atomic size, the behavior of individual atoms becomes more conspicuous, and the characteristics of common materials near this size change in unpredictable ways.

At the atomic or quantum level, the boundaries between biology, chemistry, physics, and electronics loose much of their meaning as those disciplines converge. Laboratories that develop promising nanosize inorganic materials are expected to explore the possibilities for creating hybrids of those materials and organic materials. For example, in research work, organic materials have been attached to inorganic carbon nanotubes to create microscopic transistors that can assemble themselves into larger arrays.

Table 1: Object Measurements in Nanometers

One nm = one billionth of a meter

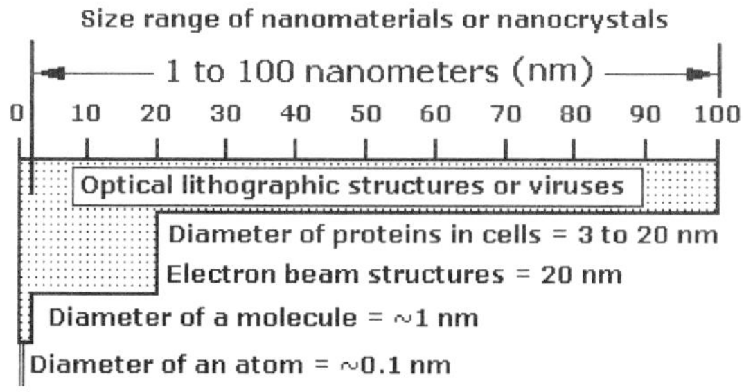

Size range of nanomaterials or nanocrystals

1 to 100 nanometers (nm)

0 10 20 30 40 50 60 70 80 90 100

Optical lithographic structures or viruses

Diameter of proteins in cells = 3 to 20 nm

Electron beam structures = 20 nm

Diameter of a molecule = ~1 nm

Diameter of an atom = ~0.1 nm

1 micrometer (μm) = 1000 nm

Shortest wavelength of visible light = 500 nm

Diameter of a human cell = 5,000 to 200,000 nm

Thickness of human hair = 50,000 to 70,000 nm

Smallest object that can be seen by the unaided human eye = 10,000 nm

Structures in the molecular size range are more likely to break than larger structures because their imperfections become more significant. Nanotechnology emphasizes the "bottom up" building of structures atom-by-atom as its most effective approach. This concept has caught the attention of researchers in the materials and biological sciences because it has significant implications for chemists and engineers searching for lighter and stronger materials. Nanotechnology could bring about a sea change in mechanical engineering because, unlike electronic systems, many of the tiny mechanisms or mechanical devices with dimensions measurable in nanometers that result will be able to interact directly with the physical world. They could, for example, include nanomotors and nanofluidic pumps. By contrast, electronic and computer engineers focus on devices that generate, transmit, or receive information in the form of electrical impulses.

Experts believe that nanotechnology will have its greatest impact in nanoelectronics, making it possible to produce molecular circuitry that is smaller, faster, and cheaper than silicon circuitry. One technique now being developed is the manufacture of molds from silicon that will permit the mass production of nanoscale circuitry. Photolithographic and chemical-etching processing of the silicon will produce molds which will contain the inverse of the structural features needed for "printing" vast quantities of nanoscale transistors. Elastomeric materials will be poured into the silicon molds, and after curing and removal, they will be "rubber stamps" for reproducing large arrays of transistorized circuitry. The stamps, with the necessary positive features for printing the circuitry, will then be coated with an "ink" made from nanomaterials in a suitable binder. By using a technique called "soft lithography," the stamps will imprint fine microscale and nanoscale transistor features on a suitable substrate. The goal of this effort is the production of billions if not trillions of molecular transistors on a chip.

Other approaches for producing extremely large-scale transistor arrays include the use of chains of molecules to make nanoprocessors and memory devices. However, it is generally believed that it will take many more years of R&D for molecular computing to become a practical technology, despite promising early results. There are now hundreds of academic research papers detailing present and future programs in nanotechnology and hundreds of business plans of companies that now or soon will be participating in commercial nanotechnology. However, the research papers and business plans reveal a preoccupation with nanoscale materials rather than devices. Examples of objectives mentioned in these papers and plans include the preparation of nanoparticles for coating fabric to make it stain-resistant, semiconductor nanoparticles or quantum dots for biological research, and nanocomposites for making plastics stronger and lighter.

Carbon nanotubes are now the hot topics for nanotechnology research; they exhibit remarkable strength and unusual electrical properties despite having dimensions measurable in atomic units. (See the sidebar *What Are Carbon Nanotubes?*) For example, nanotubes can act either as conductors or semiconductors. Scientists at IBM Watson Research Center believe that every electronic device on a nanochip could be made from nanotubes—sensors, transistors, light emitters, and interconnections. Sheets of self-supporting carbon nanotubes that are stronger than steel have been produced. These sheets are flexible and can be treated to emit light. Scientists at the University of Texas, Dallas (UTD), report that these sheets can act as solar cells to generate electricity from sunlight.

Recently MIT announced research leading to the possible replacement of chemical batteries with capacitors enhanced by nanomaterials. The report pointed out that while batteries produce voltage from a chemical reaction, capacitors store electrical energy between a pair of metal plates. The report noted the physical principle that the larger the area of capacitor plates and the smaller the space between them, the more energy a capacitor can hold. The MIT researchers covered the plates of its capacitor-type batteries with millions of carbon nanotubes, vastly expanding the surface areas of the capacitor plates.

The process is analogous to applying a layer of sponge to a surface to expand the amount of water that the surface can store. Aluminum capacitor manufacturers, for example, have long etched the surfaces of aluminum foil used as plate material to increase its porosity, thereby effectively increasing its surface area. The MIT researchers say that the porosity of the layer of conductive nanotubes permits it to store an amount of energy that is comparable to the storage capacity of a similar-sized battery. It is expected that these capacitors, when commercially available in perhaps five years, could be charged in minutes or seconds rather than hours. An additional benefit for these capacitor batteries is that they can be reused indefinitely, thus reducing or eliminating the toxic waste caused by the annual disposal of large numbers of conventional batteries in landfill.

Microfluidics is a technology that now combines MEMS with nanotechnology. As in macrofluidics, a practical system includes both pumps and valves. Microfluidics is expected to play an important role in both biomedical and chemical diagnostics as well as being the basis for microminiature devices capable of delivering medications to patients automatically according to a preset daily schedule. Microfluidic sensors are being developed for the manipulation of miniscule volumes of liquid to detect and measure very low levels of specific molecules. That application depends on the development of a microminiature device that can control the movement of liquids and molecules, making it possible to test and analyze microscopic biological samples and films.

A key component in both diagnostic and drug-delivery applications is an extremely fine microneedle. Typically made from silicon by a modified MEMS process, some of these needles have inside diameters or dimensions for the passage of fluids that are measurable in nanometers. For biomedical and chemical fluid analyzers, the needles include baffles to filter out unwanted microscopic organic matter or bacteria that would contaminate or degrade the test results; for drug delivery systems these hypodermic microneedles are fine enough to minimize or eliminate pain associated with needle insertion. Silicon microneedles with widths of 150 μm (150,000 nm) have been developed.

Researchers at the University of California, Berkeley, have developed a nanoscale silicon device that acts like a fluidic transistor. The researchers applied a voltage across the device that stopped and started the liquid flow and controlled the concentration of ions and molecules moving through the 35-nm high, 1-μm-wide channels of the device. This development is one response to the need in nanotechnology for diagnostics and sensors that manipulate very low volumes of liquid to detect and measure very low levels of specific molecules. The Berkeley team fabricated their device by the use of photolithography and chemical etching. The nanochannels were connected to three electrodes: one is located at each end of the channels, and the gate electrode spans and intersects the nanochannels. Negatively charged dye was put in the channels, and its movement was controlled by applying a positive voltage to the gate electrode.

WHAT ARE CARBON NANOTUBES?

Carbon nanotubes (CNTs) are the popular new materials in nanotechnology. They are expected to have many practical applications in materials science and mechanical, electronic, and optoelectronic devices: they are expected to be ingredients in conductive and high-strength composites and the basis for energy storage and conversion devices, sensors, and radiation sources. These nanotubes are also expected to form nanoscale transistors and circuits, frictionless nanobearings, nanoelectromechanical systems (NEMS), and nano-optoelectronic devices. CNTs are now in use as electron microscope probes, and they are seen as having a future as low-resistance electrical wires and connectors. Clearly the most promising of any nanomaterials, they were discovered in 1991 by Sumio Iijima, a scientist working at NEC in Japan.

Nanotubes can be formed from atoms other than carbon, but they do not offer the same versatility. Consequently, the term nanotube implies CNTs because they are well known as having remarkable tensile strength and useful electrical and thermal properties. Carbon is the only element that can bond itself to form extremely strong two-dimensional sheets which can be rolled into cylindrical nanotubes. Carbon can also be formed in other related geometries: soccer ball shapes called *buckminsterfullerenes*, *fullerenes*, or *buckyballs*, and cone shapes called *nanocones*. However, nanotubes can also be formed from the atoms of such materials as titanium dioxide, boron nitride, and silicon.

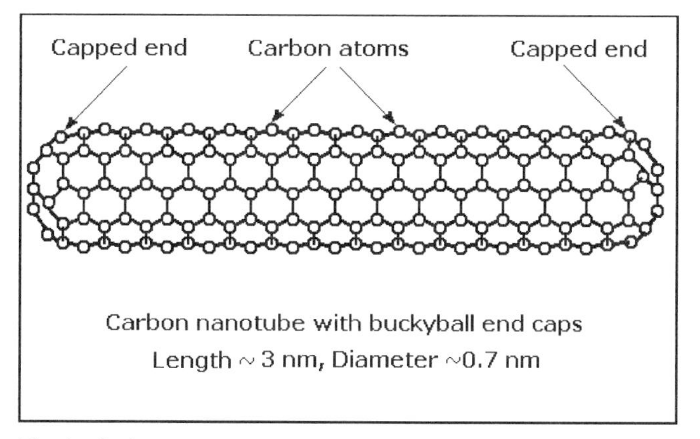

Fig. 1 Carbon nanotube with buckyball end caps (length ~3 nm, diameter ~0.7 nm).

Carbon nanotubes are miniscule convex cylindrical cages formed from carbon atoms; they usually arrange themselves in hexagonal patterns, as shown in Fig 1, but they can also arrange themselves in pentagonal patterns. These nanotubes are chemically grown, self-assembled structures which align themselves like trees in a forest. Their diameters typically range from 1 to 50 nm, and they depend on the diameters and thicknesses of the catalyst spots on which they are grown. By contrast, the lengths of nanotubes, measurable in microns, depend on how long they are allowed to grow and the conditions under which they are grown. There are both single-walled carbon nanotubes (SWCNTs) and multiwalled concentric carbon nanotubes (MWCNTs).

Single layers of graphite or graphene typically consist of hexagonal arrays of carbon atoms lying in the same plane like a flat section of wire screen. A single-wall carbon nanotube can be visualized as a single-layer of hexagonal-patterned chicken wire that has been rolled into a seamless cylinder. These nanotubes can have their ends capped by hemispherical carbon structures. The separation between the inner and outer nanotubes of MWCNTs is a radial distance about equal to that between natural layers of graphene.

A single perfect CNT is from 10 to 100 times stronger than steel per unit weight. In addition, CNTs can transmit 1000 times as much electrical current as copper, and their thermal conductivity is equal to that of diamonds. A nanotube's chiral angle—the angle between the axis of its hexagonal pattern and the axis of the tube—determines whether the tube is conductive (metallic) or semiconductive. Chiral angle is determined by the way the graphene tube has been rolled up. In addition to being conductive or semiconductive, nanotubes can also function as insulators. However, if there is a defect in the geometrical pattern of the atoms along its length, the carbon nanotube can change from a semiconductor to a conductor.

Research done at IBM Watson Research Center has demonstrated that the properties of CNTs depend on their diameters; CNTs with certain diameters can emit light at specific wavelengths. This property permits nanotubes to act as optical waveguides by transmitting data along electronic circuits more efficiently than copper wire.

Scientists at Rice University, Houston, Texas, are trying to create a prototype of a nanotube-based "quantum wire." They see the possibility that cables made from quantum wires could have larger diameters than existing power cables made from steel-reinforced aluminum and yet be lighter and stronger; the larger diameters would permit a tenfold increase in the capacity of existing electric power transmission towers. Quantum wires are seen as being able to perform at least as well as existing superconductors without the need for cooling equipment. This research on quantum wires is based on the discovery that, at nanometer scale, the unusual properties of quantum physics permit the wire to carry current without resistance. It was found that electrons could travel down a wire made of billions of overlapping carbon nanotubes with almost no energy loss.

One objective of the research at Rice University is the development of processes for forming fibers principally from well-aligned nanotubes in the so-called 5, 5-armchair configuration of carbon atoms; this arrangement is expected to offer the highest conductivity. The present fabrication techniques at Rice University produce mixtures of 150 different types of nanotubes, but differences in the properties of the nanotubes limit the wire's conductivity.

Carbon nanotubes are used as probe tips in two different kinds of electron microscopes: the atomic force microscope (AFM) and scanning tunneling microscope (STM). Researchers at IBM developed the STM, which provides images of atoms, and they coinvented AFM, which has become a standard tool for atomic-scale manipulation, making possible much of nanotechnology. Together they have the ability to magnify the images of nanomaterials and nanocrystals and manipulate individual atoms in their nanostructures. The STM uses carbon nanotubes because they are electrically conductive, and the AFM uses them because their nanoscale diameters buckle rather than break when deformed. Both probe tips can be modified with specific chemical or biological materials for high-resolution functional imaging.

Because carbon nanotubes are like minute bits of string, trillions of invisible nanotubes must be bound together to make useful macroscopic products capable of exploiting the extraordinary mechanical and electronic properties of the individual nanotubes.

Nanotechnologists at the University of Texas at Dallas (UTD), led by Dr. Ray H. Baughman and Australian scientist, Dr. Ken Atkinson from Commonwealth Scientific and Industrial Research Organization (CSIRO), have reported on a process for weaving carbon nanotubes into transparent sheets that are stronger than steel sheets of the same weight. The scientists have demonstrated that these sheets have practical applications in such products as organic light-emitting displays, low-noise electronic sensors, artificial muscles, and broadband polarized light sources which can be switched in one ten-thousandth of a second.

Carbon nanotube sheets are grown at speeds of up to 21 fpm by rotating trillions of ultra-long nanotubes in a dry-state process per minute for every 0.4 in. of sheet width. The Texas group created sheets of nanotubes that were so thin that an acre of the material would weigh only a quarter of a pound. In 2005, the group made a sheet 33-ft long, and the researchers are working diligently to expand the process. In addition to being good electrical conductors, the sheets can withstand more than 34,000 lb/in^2 of force without tearing, and they can endure temperatures as high as 840°F without losing their strength or conductivity. The Department of Defense has an interest in CNTs for reinforcing composite materials used to make helicopter blades, for manufacturing solar cells, and for use in robotics.

In more recent developments, scientists at MIT have developed a capacitor that has its plates coated with CNTs; this dramatically increases their surface area and permits them to hold more electric charge longer than capacitors of comparable size. The MIT scientists believe that these super capacitors will be competitive with conventional chemical batteries for powering electronic devices such as cell phones and radios.

NANOACTUATORS BASED ON ELECTROSTATIC FORCES ON DIELECTRICS

Large force-to-mass ratios could be achieved at the nanoscale.

NASA's Jet Propulsion Laboratory, Pasadena, California

The nanoactuators proposed here would exploit the forces exerted by electric fields on dielectric materials. The term "nanoactuators" as used here includes motors, manipulators, and other active mechanisms that have nanometer dimensions or are designed to manipulate objects that have nanometer dimensions.

The physical principle that explains nanoactuators can be demonstrated by a square parallel plate capacitor with a square dielectric plate inserted part way into the gap between the electrode plates as shown in Fig. 1. The equations for the properties of a parallel-plate capacitor show that the electrostatic field will pull the dielectric slab toward a central position in the gap with a force, F, given by

$$\text{Equation 1: } F = \frac{V^2(1 - 2)a}{2d}$$

where V is the potential applied between the electrode plates, 1 is the permittivity of the dielectric slab, 2 is the permittivity of air, a is the length of an electrode plate, and d is the dimension of the gap between the plates.

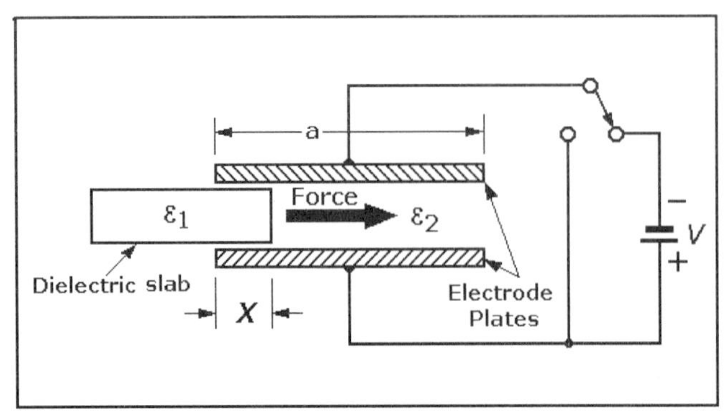

Fig. 1 Parallel plate capacitor: the electric field pulls a partially inserted dielectric slab further into a gap.

In the macroscopic domain, the force F is small but it becomes more significant in the microscopic domain. Equation 1 states that the force depends on the ratio between the capacitor dimensions but does not depend on the size. In other words, the force remains the same if the capacitor and the dielectric slab are reduced to nanometer dimensions. Nevertheless, the masses of all components are proportional to third power of their linear dimensions. Therefore, the force-to-mass ratio (and, consequently, the acceleration that can be imparted to the dielectric slab) is much larger at the nanoscale than at the macroscopic scale. The proposed actuators would exploit this effect.

A simple linear actuator based on a parallel-plate capacitor similar is shown in Fig. 2. It is similar to that shown in Fig. 1 except that the upper electrode plate is split into two parts (A and B) and the dielectric slab is slightly longer than either plate A or B. This actuator would be operated in a cycle. During the first half cycle, as shown in Fig. 2a, plate B would be grounded to the lower plate and plate A would be charged to a potential, V, with respect to the lower plate, causing the dielectric slab to be pulled under plate A. During the second half cycle, as shown in Fig. 2b, plate A would be grounded and plate B would be charged to potential V, causing the dielectric slab to be pulled under plate B. The reciprocal motion caused by alternation of the voltages on plates A and B could be used to drive a nanopump.

A rotary motor, shown in Fig. 3, would include a dielectric rotor sandwiched between a top and a bottom plate containing multiple electrodes arranged symmetrically in a circle. Voltages would be applied sequentially to electrode pairs 1 and 1a, then 2 and 2a, then 3 and 3a, attracting the dielectric rotor to sequential positions between the electrode pairs, thus causing it to rotate in a counterclockwise direction.

A micro- or nanomanipulator, shown in Fig. 4, would include lower and upper plates covered by rectangular grids of electrodes—in effect, a rectangular array of nanocapacitors. A dielectric or quasi-dielectric micro- or nanoparticle (a bacterium, virus, or molecule, for example) could be moved from an initial position

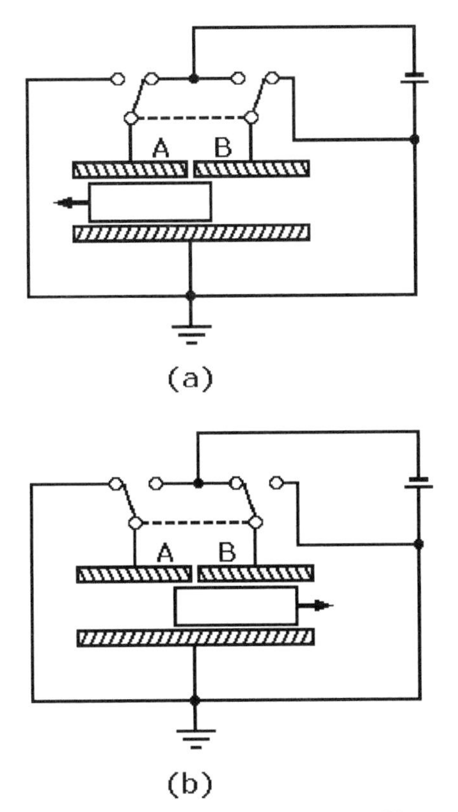

Fig. 2 Two stages of a reciprocating actuator. The actuator is driven by electrostatic forces.

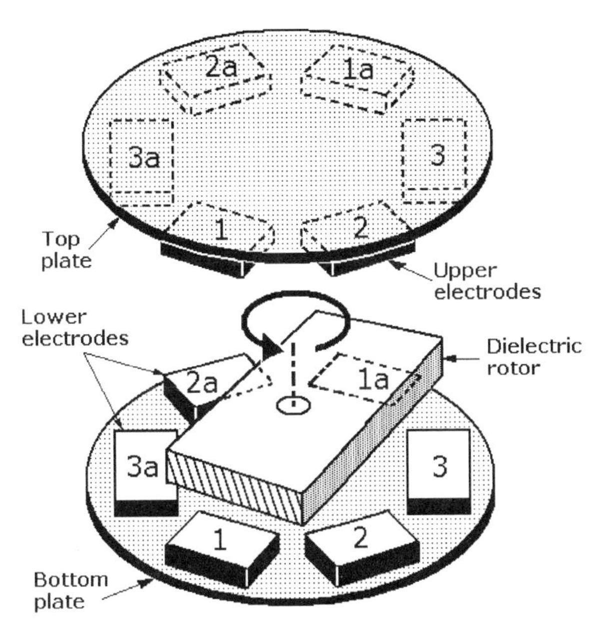

Fig. 3 Exploded view of a micromotor or a nanomotor.

on the grid to a final position on the grid by applying a potential sequentially to the pairs of electrodes along a path between the initial and final positions.

This work was done by Yu Wang of Caltech for **NASA's Jet Propulsion Laboratory**. *NPO-30747*

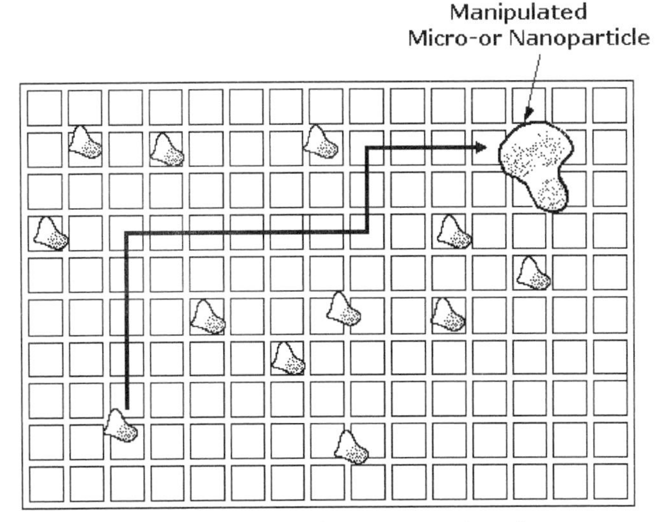

Fig. 4 A micro or nano capacitor array can pin and move a particle—bacteria, virus, or molecule to any position in the array if a voltage is alternated along the desired path of movement.

Resources

The following magazine articles and Web sites provided information included in this chapter:

Magazine articles for microtechnology and nanotechnology

Zach Zorich, "Carbon Nanotubes Burst Out of the Lab." *Discover Magazine*, January 2006, 29.

Erika Jonietz, "Quantum Wires, *Technology Review*, May 2005, 45.

Jia Chen, "Looking Past Silicon to Carbon Nanotubes," *Technology Review*, October 2005, 47.

Web sites for microtechnology and nanotechnology

American Elements
Nanotechnology Information Center
www.americanelements.com/nanotech.html

Mechanical Engineering magazine
Alan S. Brown, MEMS across the Valley of Death," April 2006
John De Gaspari, "Beyond Silicon," July 2005
www.memagazine.org/backissues/

Sandia National Laboratories
Microelectromechanical Systems (MEMS)
http://mems.sandia.gov/

Sandia National Laboratories
Liga Technology
http://www.ca.sandia.gov/liga/tech.html

Sandia National Laboratories
Micromachines Image Gallery
http://mems.sandia.gov/scripts/images.asp

University of Texas at Dallas
Carbon Nanotube Sheets
www.utdallas.edu/news/archive/2005/carbon-nanotube-sheets.html

INDEX